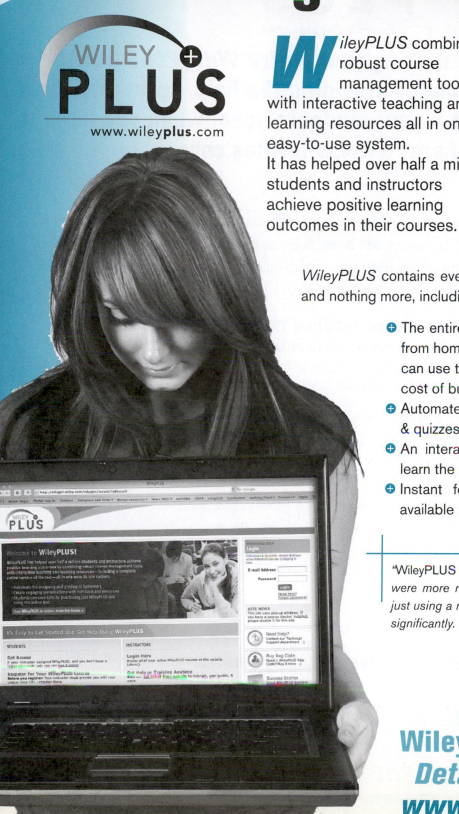

www.wileyplus.com

Wiley is committed to making your entire *WileyPLUS* experience productive & enjoyable by providing the help, resources, and personal support you & your students need, when you need it. It's all here: www.wileyplus.com –

TECHNICAL SUPPORT:

- A fully searchable knowledge base of FAQs and help documentation, available 24/7
- Live chat with a trained member of our support staff during business hours
- A form to fill out and submit online to ask any question and get a quick response
- **Instructor-only** phone line during business hours: 1.877.586.0192

FACULTY-LED TRAINING THROUGH THE WILEY FACULTY NETWORK:
Register online: www.wherefacultyconnect.com

Connect with your colleagues in a complimentary virtual seminar, with a personal mentor in your field, or at a live workshop to share best practices for teaching with technology.

1ST DAY OF CLASS...AND BEYOND!
Resources You & Your Students Need to Get Started
& Use *WileyPLUS* from the first day forward.

- 2-Minute Tutorials on how to set up & maintain your *WileyPLUS* course
- User guides, links to technical support & training options
- *WileyPLUS for Dummies*: Instructors' quick reference guide to using *WileyPLUS*
- Student tutorials & instruction on how to register, buy, and use *WileyPLUS*

YOUR *WileyPLUS* ACCOUNT MANAGER:

Your personal *WileyPLUS* connection for any assistance you need!

SET UP YOUR *WileyPLUS* COURSE IN MINUTES!

Selected *WileyPLUS* courses with QuickStart contain pre-loaded assignments & presentations created by subject matter experts who are also experienced *WileyPLUS* users.

Interested? See and try WileyPLUS *in action!*
Details and Demo: www.wileyplus.com

DIFFERENTIAL EQUATIONS
WITH BOUNDARY VALUE PROBLEMS

An Introduction to Modern Methods and Applications

DIFFERENTIAL EQUATIONS WITH BOUNDARY VALUE PROBLEMS

An Introduction to Modern Methods and Applications

James R. Brannan

Clemson University

William E. Boyce

Rensselaer Polytechnic Institute

WILEY

John Wiley & Sons, Inc.

PUBLISHER	Laurie Rosatone
ACQUISITIONS EDITOR	David Dietz
FREELANCE DEVELOPMENT EDITOR	Anne Scanlan-Rohrer
EDITORIAL ASSISTANT	Pamela Lashbrook
SENIOR PRODUCTION EDITOR	Ken Santor
MARKETING MANAGER	Sarah Davis
DESIGN DIRECTOR	Harry Nolan
COVER DESIGN	Hope Miller
COVER PHOTO	Dynamic Graphics Group / Creatas / Alamy
SENIOR ILLUSTRATION EDITOR	Sigmund Malinowski
SENIOR PHOTO EDITOR	Ellinor Wagner
MEDIA EDITOR	Melissa Edwards
MEDIA DEVELOPMENT PROJECT COORDINATOR	Elena Santa Maria

This book was set in LaTeX by Aptara®, Inc. and printed and bound by R.R. Donnelley – Willard. The cover was printed by R.R. Donnelley.

This book is printed on acid free paper. ∞

To order books or for customer service please, call 1-800-CALL WILEY (225-5945).

ISBN-13 978-0-470-41850-5

Printed in the United States of America

10 9 8 7 6 5 4 3 2 1

PREFACE

This is a textbook for a first course in differential equations. The book is intended for science and engineering majors who have completed the calculus sequence, but not necessarily a first course in linear algebra. It emphasizes a systems approach to the subject and integrates the use of modern computing technology in the context of contemporary applications from engineering and science.

Differential equations is an old and venerable subject that will always play a central role in describing phenomena that change over time. Indeed, at least one course in differential equations is part of the curriculum of most engineering and science students. Our goal in writing this text is to provide these students with both an introduction to, and a survey of, modern methods, applications, and theory of this beautiful and powerful mathematical apparatus that is likely to serve them well in their chosen field of study. The subject matter is presented in a manner consistent with the way practitioners use differential equations in their work; technology is used freely, with more emphasis on methods, modeling, graphical representation, qualitative concepts, and geometric intuition than on theoretical issues. For example, some important theoretical results, such as theorems guaranteeing existence and uniqueness of solutions to initial value problems, are not proved. Nevertheless, they are carefully stated, illustrated by examples, and used frequently.

Any student who studies major portions of this book, does a reasonable number of the section exercises, and completes some of the chapter projects will surely develop an appreciation for the power of differential equations to shed light on issues of societal importance. In addition, he or she will acquire skills in modeling, analysis, and computer simulation that will be useful in a wide variety of situations. It is for such students that we have written this book.

Major Features

▶ **Flexible Organization.** Chapters are arranged, and sections and projects are structured, to facilitate choosing from a variety of possible course configurations depending on desired course goals, topics, and depth of coverage.

▶ **Numerous and Varied Problems.** Throughout the text, section exercises of varying levels of difficulty give students hands-on experience in modeling, analysis, and computer experimentation.

▶ **Emphasis on Systems.** Systems of first order equations, a central and unifying theme of the text, are introduced early, in Chapter 3, and are used frequently thereafter.

▶ **Linear Algebra and Matrix Methods.** Two-dimensional linear algebra sufficient for the study of two first order equations, taken up in Chapter 3, is presented in Section 3.1. Linear algebra and matrix methods required for the study of linear systems of dimension n (Chapter 6) are treated in Appendix A.

▶ **ODE Architect.** The companion ODE Architect provides students with a user-friendly software tool for computing numerical approximations to solutions of systems of differential equations, and for constructing component plots, direction fields, and phase portraits.

▶ **Optional Computing Exercises.** In most cases, problems requesting computer generated solutions and graphics are optional.

▶ **Visual Elements.** In addition to a large number of illustrations and graphs within the text, physical representations of dynamical systems and animations available in ODE Architect enable students to visualize solutions routinely.

▶ **Contemporary Project Applications.** Optional projects at the ends of Chapters 2 through 10 integrate subject matter in the context of exciting, contemporary applications in science and engineering. Among these are controlling the attitude of a satellite, ray theory of wave propagation, uniformly distributing points on a sphere, and vibration analysis of tall buildings.

▶ **Laplace Transforms.** A detailed chapter on Laplace transforms discusses systems, discontinuous and impulsive input functions, transfer functions, feedback control systems, poles, and stability.

▶ **Control Theory.** Ideas and methods from the important application area of control theory are introduced in some examples, some projects, and in the last section on Laplace Transforms. All of this material is optional.

▶ **Recurring Themes and Applications.** Important themes, methods, and applications, such as dynamical system formulation, phase portraits, linearization, stability of equilibrium solutions, vibrating systems, and frequency response are revisited and reexamined in a variety of mathematical models under different mathematical settings.

▶ **Chapter Summaries.** A summary at the end of each chapter provides students and instructors with a birds-eye view of the most important ideas in the chapter.

▶ **Answers to Problems.** Answers to the problems are provided at the end of the book; many of them are accompanied by a figure.

Systems

The book emphasizes differential equations and applications within the systems framework. Two-dimensional linear systems appear early, in Chapter 3, following a treatment of first order equations in Chapters 1 and 2. Second order equations, taken up in Chapter 4, are not only dealt with directly, but are also presented in the context of first order systems. Higher order equations are subsumed within the logical extension from two dimensions to n dimensions in Chapter 6. We feel an early introduction to systems offers advantages over more traditional presentation formats for several reasons:

▶ The systems paradigm has not only permeated all fields of engineering and the natural sciences, but quantitative areas of the business sciences, such as economics and finance, as well. Most realistic problems in these disciplines consist of two or more components that interact in some manner. For such problems, the systems approach often facilitates modeling and analysis.

▶ Virtually all initial value problem solvers require that second and higher order scalar equations and systems be written as systems of first order equations. Since many of the problems and projects in this text are enhanced by numerical simulations and graphical presentations, it is important for the student to be able to cast problems as systems of first order equations early in the course.

▶ The student gets early exposure to the geometry of phase portraits along with the important relationship between eigenvalues and qualitative concepts such as long-term behavior of solutions and stability of equilibrium points.

▶ The systems approach helps promote a unified view of differential equations, in contrast to a perception shared by many students that the subject is a collection of somewhat distinct topics (first order equations, second order equations, nth order equations, and first order systems), each with its own method of solution.

Matrices and Linear Algebra

Students come to a differential equations course with widely varying knowledge of linear algebra and matrices. Some have had a full semester course, but many have had a much smaller exposure to the subject. In addition, many instructors prefer, or have time, to cover only two-dimensional systems, while others may wish to discuss systems in n dimensions. To accommodate these variations we present a two-tiered approach to systems of first order differential equations.

Chapter 3 deals primarily with systems of two linear homogeneous differential equations with constant coefficients. The algebraic skills that are needed to handle such systems are (1) solving two-dimensional linear algebraic systems, and (2) solving quadratic equations. These are skills that all differential equations students have, and Section 3.1 provides the opportunity to review them. What may be new to students in this section is the terminology and geometry associated with the eigenvalue problem for 2×2 matrices. Our experience is that students readily understand how to solve the eigenvalue problem in a two-dimensional setting, since they already know the underlying mathematics.

We want to emphasize that the treatment of two-dimensional linear systems in Chapter 3 is sufficient to make the rest of the book, except for Chapter 6, accessible to students. In other words, a quite limited knowledge of linear algebra and matrix methods is sufficient to handle most of the material in this book.

However, many important applications require consideration of systems of dimension greater than two. For those instructors who want to go beyond two-dimensional systems, we include a consideration of n-dimensional systems in Chapter 6. To deal with such systems effectively, we require additional machinery from linear algebra and matrix theory. Appendix A provides a summary of the necessary results from these areas required by Chapter 6. Readers with little or no previous exposure to this material will find it necessary to study some, or all, of Appendix A before proceeding on to Chapter 6. An alternative way to utilize Appendix A is to undertake the study of Chapter 6 directly, drawing upon necessary results from Appendix A as needed. The topics covered in Appendix A are treated in as independent a manner as possible, allowing the instructor to pick and choose on the fly. The depth of coverage will, of course, be determined by the background of the students and the course goals of the instructor. This approach may be preferred if the students already have an adequate background in matrix algebra, or if the calculation of eigenvalues and eigenvectors is to be performed primarily by using a computer or a calculator. Using this approach, coverage of Chapter 6 can be greatly streamlined. Appendix A provides all that is needed (and only what is needed) from matrix algebra to handle the systems of differential equations in Chapter 6. It is not an introduction to abstract linear algebra; rather, it focuses on the computational issues that are needed in Chapter 6.

Technology

Due to the availability of modern interactive computing environments such as MATLAB®, *Maple*®, and *Mathematica*®, it is now easy for students to approximate solutions of differential equations numerically, and to visualize them graphically. Indeed, most of the numerical calculations required by the problems in this book can be performed with the companion software, ODE Architect, a numerical differential equations solver package with a convenient user interface. In this text, the computer is used in two different ways:

▶ It is employed as a tool to help convey the proper subject matter of differential equations, especially through the use of graphics.

▶ In most realistic problems it is not possible to obtain closed-form analytic solutions of systems of equations. Even in a first course in differential equations, computer technology permits the treatment of serious, contemporary applications arising from problems in engineering and science. We include many such applications, some in the main text but primarily in the chapter projects.

Problems that require the use of a computer are marked with ODEA, indicating that ODE Architect may conveniently be used, or CAS, indicating that a computer algebra system should be used. While we feel that students will benefit from using the computer on those problems where numerical approximations or computer generated graphics are requested, in most problems it is clear that use of a computer, or even a graphing calculator, is optional. Furthermore, there are a large number of problems that do not require the use of a computer. Thus, the book can easily be used in a course without using any technology.

Projects

At the end of each chapter (except Chapter 1) we have included projects that deal with contemporary problems normally not included among traditional topics in differential equations.

▶ Many of the projects involve applications, drawn from a variety of disciplines, that illustrate either a physical principle or a prediction, design, control, or management strategy. To engineers and scientists in training, such problems effectively display the utility and importance of differential equations as a tool that can be used to effect change upon the world in which we live. Through these projects we are able to emphasize the active role that differential equations play in modern science and engineering in addition to their important, but more passive, role in simply describing phenomena that change over time.

▶ The projects integrate and/or extend the theory and methods presented in the core material in the context of specific problems difficult enough to require some critical and imaginative thinking. By demonstrating the usefulness of the mathematics, a project can be a valuable tool for enhancing students' comprehensive understanding of the subject matter.

▶ Many of the projects require modeling, analysis, numerical calculations, and interpretation of results. They are structured in the sense that the early exercises guide the reader through the modeling, analysis, and computations while later exercises are frequently more open-ended and demand more from the reader in the way of original and critical thinking. It is not necessary for a student to do all of the exercises to benefit from a project. Indeed, even the idea of the application can be an eye-opening experience for students.

▶ In our experience, a great deal of knowledge transfer occurs as a result of the student–student and teacher–student dialogue over the course of the project.

▶ For some, mathematical modeling may mean finding an accurate mathematical description of a given set of raw data. For example, following a discussion of a variety of different population models and their properties, the student may be asked to find one that best fits a set of historical population data. This is an interesting and worthwhile endeavor requiring some resourcefulness and knowledge of elementary methods on the part of the student. Lack of structure is, in fact, a desirable aspect and typifies problems that arise in biology, for example. For others, modeling may mean writing down an appropriate set of differential equations based on known physical laws. In such cases, relatively sophisticated methods may be required to analyze the problem. Few, if any, students are able to devise or derive such methods in any reasonable amount of time. Structured projects may introduce such methods and guide the student through the analysis. These types of projects provide important mathematical tools for the toolbox of the engineer or scientist in training. Most of the projects in this textbook are of this type.

▶ Projects vary in length and level of difficulty. Less demanding ones may be assigned on an individual basis. More challenging projects may be assigned to small groups or to individual students in an honors class. A few of the projects may require some programming guidance from the instructor. Assuming that all of the problems for each project are worked out, the table below indicates the approximate level of difficulty for the analysis and computational components: Beginning (B), Intermediate (I), or Advanced (A). As discussed above, ODEA and CAS indicate the suggested computational tool, although all computations can be performed using a computer algebra system. Level of difficulty can be adjusted by suitably restricting assigned problems. With respect to the analysis component, a B level of difficulty roughly corresponds to that of an end of the section problem, with I and A representing increasingly more challenging problems. With respect to the computational component, a B level of difficulty means that only direction fields, phase portraits, or component plots for standard differential equations or systems are requested. An I level of difficulty signifies that there may be discontinuities present, or that some data analysis, such as curve fitting, is required. Finally, an A level of difficulty generally means that some elementary programming, possibly involving functions and loop structures, is required.

Section	Project Title	Level of Difficulty	
		Analysis	Computation
2.P.1	Harvesting a Renewable Resource	B	
2.P.2	Designing a Drip Dispenser for a Hydrology Experiment	I	B-CAS
2.P.3	A Mathematical Model of a Groundwater Contaminant Source	I	I-CAS
2.P.4	Monte Carlo Option Pricing: Pricing Financial Options by Flipping a Coin	B	A-CAS
3.P.1	Eigenvalue-Placement Design of a Satellite Attitude Control System	I	B-ODEA
3.P.2	Estimating Rate Constants for an Open Two-Compartment Model	I	A-CAS
3.P.3	The Ray Theory of Wave Propagation	I	I-CAS
3.P.4	A Blood-Brain Pharmacokinetic Model	B	I-CAS
4.P.1	A Vibration Insulation Problem	I	I-CAS
4.P.2	Linearization of a Nonlinear Mechanical System	I	B-ODEA
4.P.3	A Spring-Mass Event Problem	I	B-ODEA
4.P.4	Uniformly Distributing Points on a Sphere	I	A-CAS
4.P.5	Euler-Lagrange Equations	A	
5.P.1	An Electric Circuit Problem	B	B-CAS
5.P.2	Effects of Pole Locations on Step Responses of Second Order Systems	I	B-CAS
5.P.3	The Watt Governor, Feedback Control, and Stability	A	I-CAS
6.P.1	A Compartment Model of Heat Flow in a Rod	B	I-CAS
6.P.2	Earthquakes and Tall Buildings	A	A-CAS
6.P.3	Controlling a Spring-Mass System to Equilibrium	A	A-CAS
7.P.1	Modeling of Epidemics	I	B-ODEA
7.P.2	Harvesting in a Competitive Environment	A	B-ODEA
7.P.3	The Rössler System	I	I-ODEA
8.P.1	Diffraction Through a Circular Aperture	I	
8.P.2	Hermite Polynomials and The Quantum Mechanical Harmonic Oscillator	A	B-CAS
8.P.3	Perturbation Methods	I	I-ODEA
9.P.1	Estimating the Diffusion Coefficient in the Heat Equation	I	I-CAS
9.P.2	The Transmission Line Problem	A	I-CAS
9.P.3	Solving Poisson's Equation by Finite Differences	B	A-CAS
10.P.1	Dynamic Behavior of a Hanging Cable	I	B-CAS
10.P.2	Advection-Dispersion: A Model for Solute Transport in Porous Media	A	A-CAS
10.P.3	Fisher's Equation for Population Growth and Dispersion	I	A-CAS

Computer Simulations

In most realistic problems it is not possible to obtain closed-form analytic solutions of systems of equations. Usually initial value problem (IVP) solvers contained in commercial software packages such as MATLAB, *Maple*, *Mathematica*, or in the companion ODE Architect Tool, are used to obtain numerical approximations to solutions. Our understanding of the problem then depends to a large extent on visualization and interpretation of plots of the graphs of these approximations, usually as one or more parameters in the problem are varied. Computer experiments of this type are called *computer simulations*. In combination with analytical theory, analytical methods, and qualitative techniques, computer simulations provide us with an additional powerful tool for studying the behavior of systems. Many modern engineering and scientific problems, for example, weather prediction, aircraft design, economic forecasting, epidemic modeling, and industrial processes are studied with the aid of computer simulations. Computer simulations of systems of differential equations may be viewed as part of a larger subject area frequently referred to as *computational science*. On a small scale, many of the section exercises and chapter projects expose the student to some of the following pragmatic issues that confront the computational scientist:

▶ There may be a considerable number of trial and error simulations required enroute to obtaining a successful set of results.

▶ Errors of one kind or another can occur in a number of different phases of the problem solving process: inaccuracies in the modeling, programming errors, failure of existence or uniqueness in the mathematical problem itself, and ill-behavior of the system relative to the numerical method employed to solve the IVP.

▶ For certain parameter values it is advantageous to have either exact solutions or analytical approximations that can be compared with the numerical approximation.

▶ Efficiently and effectively presenting and displaying output data generated from numerical simulations of complex problems can be a challenge, but it is also an aspect of the problem where imagination and creativity may be effectively used.

▶ Common sense and a healthy degree of skepticism when viewing output results, such as the numbers or the graphs, are desirable traits to develop.

▶ The better you understand your computational tool—its capabilities and limitations—the better off you are.

▶ The ability to conduct high quality simulation experiments is a craft that requires knowledge of several technical skills, and is developed through education and experience.

Numerical Algorithms

IVP solvers available in the companion ODE Architect Tool, MATLAB, *Maple*, or *Mathematica* should be used to perform most of the numerical calculations required in section exercises and chapter projects. Nevertheless, we introduce Euler's method for a scalar first order equation in Section 1.3. We return in Sections 2.7 and 2.8 to a discussion of errors and of methods more efficient than Euler's for approximating solutions numerically. In Section 3.7 we extend these methods to systems of first order equations. There are several reasons why we include introductory material on numerical algorithms in the text.

▶ Euler's method, in particular, gives great insight into the meaning of a dynamical system (system of first order differential equations) in that it provides a simple algorithm for advancing the state of the dynamical system in discrete time.

▶ The Runge-Kutta method, while less intuitive than the Euler method, shows how a cleverly designed algorithm can yield great rewards in terms of accuracy and efficiency.

▶ Numerical methods are an important component of the body of knowledge of differential equations. It is conceivable that many instructors may wish to provide their students with an introduction to some of the algorithms and their analysis.

▶ There are situations where a slight modification of an elementary numerical method, such as Euler's method, may provide a satisfactory scheme for approximating the solution of a problem. Illustrations of this are contained in the projects presented in Section 2.P.4, Monte Carlo Option Pricing, and Section 4.P.4, Uniform Distribution of Points on a Sphere.

Relation of This Text to Boyce and DiPrima

Brannan and Boyce is an offshoot, but not a new edition, of the well-known textbook by Boyce and DiPrima. Readers familiar with Boyce and DiPrima will doubtless recognize in the present book some of the hallmark features that distinguish that textbook.

To help avoid confusion among potential users of either text, the primary differences are described below:

▶ Boyce and DiPrima is more comprehensive and is laid out along fairly traditional lines. It includes all of the topics that are often included in a first course on differential equations. There are chapters on higher order linear equations, power series methods applied to both ordinary and regular singular points, two point boundary value problems, and partial differential equations using Fourier series methods. None of this material, except for brief references to higher order linear equations, appears in Brannan and Boyce.

▶ Brannan and Boyce is more sharply focused on the needs of students of engineering and science, whereas Boyce and DiPrima targets a somewhat more general audience, including engineers and scientists.

▶ Brannan and Boyce is intended to be more consistent with the way contemporary scientists and engineers actually use differential equations in the workplace.

▶ Brannan and Boyce emphasizes systems of first order equations, introducing them earlier, and also examining them in more detail than Boyce and DiPrima. Brannan and Boyce has an extensive appendix on matrix algebra to support the treatment of systems in n dimensions.

▶ Brannan and Boyce integrates the use of computers more thoroughly than Boyce and DiPrima. Brannan and Boyce introduces numerical approximation methods in Chapter 1, and assumes that most students will use computers to generate approximate solutions and graphs throughout the book.

▶ Brannan and Boyce emphasizes contemporary applications to a greater extent than Boyce and DiPrima, primarily through end of chapter projects.

▶ Brannan and Boyce makes somewhat more use of graphs, with more emphasis on phase plane displays, and uses engineering language (for example, state variables, transfer functions, gain functions, and poles) to a greater extent than Boyce and DiPrima.

Options for Course Structure

Chapter dependencies are shown in the following block diagram.

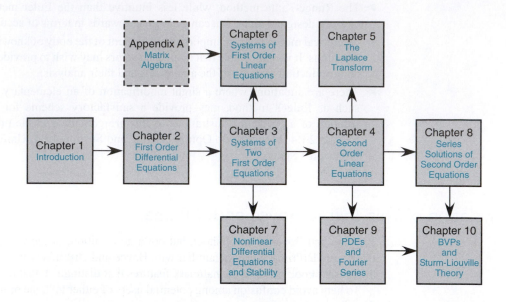

The book has much built-in flexibility and allows instructors to choose from many options. Depending on the course goals of the instructor and background of the students, selected sections may be covered lightly or even omitted.

▶ Chapters 5, 6, and 7 are independent of each other, and Chapters 6 and 7 are also independent of Chapter 4. It is possible to spend much class time on one of these chapters, or class time can be spread over two or more of them.

▶ The amount of time devoted to projects is entirely up to the instructor.

▶ For an honors class, a class consisting of students who have already had a course in linear algebra, or a course in which linear algebra is to be emphasized, Chapter 6 may be taken up immediately following Chapter 2. In this case, material from Appendix A, as well as sections, examples, and problems from Chapters 3 and 4 may be selected on an as needed, or desired, basis. This offers the possibility of spending more class time on Chapters 5, 7 and/or selected projects.

Comments on the use of projects. There are several ways of using projects in the course.

▶ For some of the projects, the work may be done outside of the classroom on an individual basis.

▶ Some of the lengthier, more challenging projects might be more appropriately assigned to small groups of students.

▶ Enough exposition is provided in a few of the projects so that students can derive some benefit from simply reading about the application. For example, students may be interested in James Watt's clever use of a centrifugal pendulum for controlling the speed of a steam engine as described in Section 5.P.3, or how systems of ordinary differential equations are used to describe wave propagation as described in Section 3.P.3. Of course, more benefit may be derived by doing some or all of the project exercises.

▶ Some instructors may wish to incorporate one or two projects into the course, with a lecture devoted to each one.

▶ An optional strategy, say for a modeling course, is to design the course to be "project driven". In this approach, selected projects are used to drive the discussion of the mathematics required to solve the problems posed in the project. The discussion of mathematical methods and theory are then intertwined with the project. This might be appropriate for some of the projects that are highly integrative in nature. For example, virtually every topic in Chapter 6, and then some, is utilized to solve the problem posed in Section 6.P.3, Controlling a Spring-Mass System to Equilibrium. Similarly, most of the machinery of Chapter 5 is required to solve the problem in Section 5.P.3, The Watt Governor, Feedback Control, and Stability. The projects then serve to motivate the mathematics.

▶ Projects can be given as "extra credit" assignments.

Acknowledgments

It is a pleasure to offer our grateful appreciation to the many people who have generously assisted in the preparation of this book.

To the individuals listed below who reviewed parts or all of the manuscript at various stages of its development:

G. Donald Allen, Texas A&M University
Frank Barnet, Frostburg State University
Daniel C. Biles, Western Kentucky University
Elizabeth Bonawitz, Virginia Polytechnic Institute and State University
Philip Boyland, University of Florida
Ronald Brent, University of Massachusetts–Lowell
Bernard Brooks, Rochester Institute of Technology
Luca Capogna, University of Arkansas
Zhixiong Chen, New Jersey City University
Branko Curgus, Western Washington University
Patrick De Leenheer, University of Florida
William Emerson, Metropolitan State College: Denver
Alejandro Engel, Rochester Institute of Technology
Mark Farris, Midwestern State University
Marcel B. Finan, Arkansas Tech University
Yuval Flicker, The Ohio State University
Jeffrey S. Fox, University of Colorado at Boulder
Moses Glasner, Pennsylvania State University
Yuri Godin, University of North Carolina at Charlotte
David Gurarie, Case Western Reserve University
Mansoor A. Haider, North Carolina State University
David Handron, Carnegie Mellon University
Donald Hartig, California Polytechnic University: San Luis Obispo
L. Thomas Hill, Lafayette College
Chung-wu Ho, Evergreen Valley College
Jack E. Hofer, California State Polytechnic University, Pomona
Michael Huff, Austin Community College
Michael G. Hurley, Case Western Reserve University
Michel Jabbour, University of Kentucky
Ronald Jorgensen, Milwaukee School of Engineering
Matthias Kawski, Arizona State University
Semen Koksal, Florida Institute of Technology

Julie Levandosky, Framingham State College
Gilbert N. Lewis, Michigan Technological University
Tiao Lu, University of North Carolina at Charlotte
Karen A. Marrongelle, Portland State University
Stephen McDowall, Western Washington University
Peter Mucha, The University of North Carolina at Chapel Hill
Guntram Mueller, University of Massachusetts–Lowell
Diego A. Murio, University of Cincinnati
Alexander Nabutovsky, Pennsylvania State University
Tejinder Neelon, California State University, San Marcos
Lisa Perrone, Tufts University
David E. Radford, University of Illinois at Chicago
Maurice Rahe, Texas A&M University
Joan Remski, The University of Michigan-Dearborn
Behzad Rouhani, The University of Texas at El Paso
Weihau Ruan, Purdue University Calumet
David Ryeburn, Simon Fraser University
Eileen Shugart, Virginia Polytechnic Institute and State University
Yurii Shylnov, Loyola University Chicago
Leonid Slavin, University of Connecticut
Avraham Soffer, Rutgers University
Nancy K. Stanton, University of Notre Dame
Eric de Sturler, Virginia Polytechnic Institute and State University
Tong Sun, Bowling Green State University
Marie Vanisko, California State University, Stanislaus
Tingxiu Wang, Oakton Community College
Arthur G. Wasserman, The University of Michigan
Arthur G. Wasserman, University of Michigan: Ann Arbor
Shangyou Zhang, University of Delaware

To Michael Brannan, for drawing several of the illustrations.

To Jing-en Pang (Clemson University), for providing valuable feedback on several of the end-of-chapter projects.

To Jim Peterson and Jim Reneke, colleagues at Clemson University, for reading parts of the manuscript and making useful suggestions on style and content.

To Dan Warner, also of Clemson University, for classroom testing an early version of the manuscript.

To David Ryeburn (Simon Fraser University), who examined the proof sheets with uncommon diligence, resulting in many corrections and clarifications.

To the editorial and production staff of John Wiley and Sons, Inc., identified on page vi, who were responsible for turning our manuscript into a finished book. In the process, they maintained the highest standards of professionalism.

We also wish to acknowledge the less tangible contributions of our friend and colleague, the late Richard DiPrima. Parts of this book draw extensively on the book on differential equations by Boyce and DiPrima. Since Dick DiPrima was an equal partner in creating the early editions of that book, his influence lives on more than twenty years after his untimely death.

Finally, and most important of all, we thank our wives, Cheryl and Elsa, for their understanding, encouragement, and patience throughout the writing and production of this book. Without their support it would have been much more difficult, if not impossible, for us to complete this project.

James R. Brannan
Clemson, South Carolina

William E. Boyce
Grafton, New York

March 1, 2009

Supplemental Resources for Instructors and Students

The *ODE Architect* tool is available in WileyPlus or via an access code provided with new copies of the text is included with every copy of the text. *ODE Architect* is a prize-winning, state-of-the-art NSF-sponsored learning software package, which is Windows-compatible. A solver tool allows you to build your own models with ODEs and study them in a truly interactive point-and-click environment. The *Architect* includes an interactive library of more than one hundred model differential equation systems with graphs of solutions. The *Architect* also has 14 interactive multimedia modules, which provide a range of models and phenomena, from a golf game to chaos.

An Instructor's Solutions Manual, ISBN 978-0-470-52654-5 includes solutions for all problems in the text.

A Student Solutions Manual, ISBN 978-0-470-41851-2 includes solutions for selected problems in the text.

A Companion Web site, www.wiley.com/college/brannan, provides a wealth of resources for students and instructors, including:

► PowerPoint slides of important ideas and graphics for study and note taking.

► Review and Study Outlines to help students prepare for quizzes and exams.

► Online Review Quizzes to enable students to test their knowledge of key concepts. For further review diagnostic feedback is provided that refers to pertinent sections in the text.

► Additional problems for use with Mathematica, Maple, and MATLAB, allowing opportunities for further exploration of important ideas in the course utilizing these computer algebra and numerical analysis packages.

WileyPLUS Expect More from Your Classroom Technology

This text is supported by *WileyPLUS*—a powerful and highly integrated suite of teaching and learning resources designed to bridge the gap between what happens in the classroom and what happens at home. *WileyPLUS* includes a complete online version of the text, algorithmically generated exercises, all of the text supplements, plus course and homework management tools, in one easy-to-use website.

Organized around the everyday activities you perform in class, *WileyPLUS* helps you:

► **Prepare and Present:** *WileyPLUS* lets you create class presentations quickly and easily using a wealth of Wiley-provided resources, including an online version of the textbook, PowerPoint slides, and more. You can adapt this content to meet the needs of your course.

► **Create Assignments:** *WileyPLUS* enables you to automate the process of assigning and grading homework or quizzes.

► **Track Student Progress:** An instructor's gradebook allows you to analyze individual and overall class results to determine students' progress and level of understanding.

► **Promote Strong Problem-Solving Skills:** *WileyPLUS* can link homework problems to the relevant section of the online text, providing students with context-sensitive help. *WileyPLUS* also features mastery problems that promote conceptual understanding of key topics and video walkthroughs of example problems.

► **Provide numerous practice opportunities:** Algorithmically generated problems provide unlimited self-practice opportunities for students, as well as problems for homework and testing.

▶ **Support Varied Learning Styles:** *WileyPLUS* includes the entire text in digital format, enhanced with varied problem types to support the array of different student learning styles in today's classrooms.

▶ **Administer Your Course:** You can easily integrate *WileyPLUS* with another course management system, gradebooks, or other resources you are using in your class, enabling you to build your course, your way.

WileyPLUS includes a wealth of instructor and student resources:

Student Solutions Manual: Includes worked-out solutions for all odd-numbered problems and study tips.

Instructor's Solutions Manual: Presents worked out solutions to all problems.

PowerPoint Lecture Notes: In each section of the book a corresponding set of lecture notes and worked out examples are presented as PowerPoint slides that are tied to the examples in the text.

View an online demo at www.wiley.com/college/wileyplus or contact your local Wiley representative for more details.

The Wiley Faculty Network—Where Faculty Connect

The Wiley Faculty Network is a faculty-to-faculty network promoting the effective use of technology to enrich the teaching experience. The Wiley Faculty Network facilitates the exchange of best practices, connects teachers with technology, and helps to enhance instructional efficiency and effectiveness. The network provides technology training and tutorials, including *WileyPLUS* training, online seminars, peer-to-peer exchanges of experiences and ideas, personalized consulting, and sharing of resources.

▶ **Connect with a Colleague.** Wiley Faculty Network mentors are faculty like you, from educational institutions around the country, who are passionate about enhancing instructional efficiency and effectiveness through best practices. You can engage a faculty mentor in an online conversation at www.wherefacultyconnect.com.

▶ **Participate in a Faculty-Led Online Seminar.** The Wiley Faculty Network provides you with virtual seminars led by faculty using the latest teaching technologies. In these seminars, faculty share their knowledge and experiences on discipline-specific teaching and learning issues. All you need to participate in a virtual seminar is high-speed Internet access and a phone line. To register for a seminar, go to www.wherefacultyconnect.com.

▶ **Connect with the Wiley Faculty Network**
Web: www.wherefacultyconnect.com
Phone: 1-866-4FACULTY

The Authors

James R. Brannan received his B.S. and M.S. degrees in Mathematics from Utah State University and his Ph.D. degree in Mathematics from Rensselaer Polytechnic Institute. He is a member of the Society of Industrial and Applied Mathematics and the Mathematical Association of America. He has taught applied mathematics at Clemson University for 27 years. He has published technical papers in the areas of mathematical biology, underwater acoustics, wave propagation, geophysical fluid dynamics, and groundwater modeling. Dr. Brannan was mentored by both Boyce and DiPrima, wrote the student solutions manual to the third edition of Boyce and DiPrima's *Elementary Differential Equations and Boundary Value Problems*, and contributed problems to the eighth edition of that text. During his tenure at Clemson he has developed a large number of modeling and computational project modules for differential equations and advanced engineering mathematics courses. He has also been exposed to a large number of problems and applications in differential equations, serving as a visiting scientist at U.S. Department of Energy and Navy research laboratories and as a consultant in applied mathematics to government and industry.

William E. Boyce received his B.A. degree in Mathematics from Rhodes College, and his M.S. and Ph.D. degrees in Mathematics from Carnegie-Mellon University. He is a member of the American Mathematical Society, the Mathematical Association of America, and the Society for Industrial and Applied Mathematics. He is currently the Edward P. Hamilton Distinguished Professor Emeritus of Science Education (Department of Mathematical Sciences) at Rensselaer. He is the author of numerous technical papers in boundary value problems and random differential equations and their applications. He is the author of several textbooks including two differential equations texts, and is the coauthor (with M.H. Holmes, J.G. Ecker and W.L. Siegmann) of a text on using Maple to explore calculus. He is also coauthor (with R.L. Borrelli and C.S. Coleman) of *Differential Equations Laboratory Workbook* (Wiley 1992), which received the EDUCOM Best Mathematics Curricular Innovation Award in 1993. Professor Boyce was a member of the NSF-sponsored CODEE (Consortium for Ordinary Differential Equations Experiments) that led to the widely-acclaimed *ODE Architect*. He has also been active in curriculum innovation and reform. Among other things, he was the initiator of the "Computers in Calculus" project at Rensselaer, partially supported by the NSF. In 1991 he received the William H. Wiley Distinguished Faculty Award given by Rensselaer.

CONTENTS

5 The Laplace Transform   306

6 Systems of First Order Linear Equations   390

7 Nonlinear Differential Equations and Stability   472

Matrices and Linear Algebra 777

 Complex Variables 823

CHAPTER ONE

Introduction

I n this chapter, we try to give perspective to your study of differential equations in several different ways. First, we formulate three problems to illustrate some of the basic ideas that we will return to and elaborate upon frequently throughout the remainder of the book. We introduce geometrical, analytical, and numerical methods for investigating the solutions of these problems. Later, we indicate several ways of classifying equations, in order to provide organizational structure for the book. The study of differential equations has attracted the attention of many of the world's greatest mathematicians during the past three centuries. Nevertheless, it remains a dynamic field of inquiry today, with many interesting open questions.

1.1 Some Basic Mathematical Models; Direction Fields

Before embarking on a serious study of differential equations (for example, by reading this book or major portions of it), you should have some idea of the possible benefits to be gained by doing so. For some students, the intrinsic interest of the subject itself is enough motivation, but for most it is the likelihood of important applications to other fields that makes the undertaking worthwhile.

Many of the principles, or laws, underlying the behavior of the natural world are statements or relations involving rates at which things happen. When expressed in mathematical terms, the relations are equations and the rates are derivatives. Equations containing derivatives are **differential equations**. Therefore, to understand and to investigate problems involving the motion of fluids, the flow of current in electric circuits, the dissipation of heat in solid objects, the propagation and detection of seismic waves, the increase or decrease of populations, and many others, it is necessary to know something about differential equations.

A differential equation that describes some physical process is often called a **mathematical model** of the process, and many such models are discussed throughout this book. In this section, we begin with three models leading to equations that are easy to solve. It is noteworthy that even the simplest differential equations provide useful models of important physical processes.

EXAMPLE 1

A Falling Object

Suppose that an object is falling in the atmosphere near sea level. Formulate a differential equation that describes the motion.

We begin by introducing letters to represent various quantities that may be of interest in this problem. The motion takes place during a certain time interval, so let us use t to denote time. Also, let us use v to represent the velocity of the falling object. The velocity will presumably change with time, so we think of v as a function of t; in other words, t is the independent variable and v is the dependent variable. The choice of units of measurement is somewhat arbitrary, and there is nothing in the statement of the problem to suggest appropriate units, so we are free to make any choice that seems reasonable. To be specific, let us measure time t in seconds and velocity v in meters/second. Further, we will assume that v is positive in the downward direction—that is, when the object is falling.

The physical law that describes the motion of everyday objects is Newton's second law, which states that the mass of the object times its acceleration is equal to the net force on the object. In mathematical terms this law is expressed by the equation

$$F = ma, \tag{1}$$

where m is the mass of the object, a is its acceleration, and F is the net force exerted on the object. To keep our units consistent, we will measure m in kilograms, a in meters/second^2, and F in newtons. Of course, a is related to v by $a = dv/dt$, so we can rewrite Eq. (1) in the form

$$F = m(dv/dt). \tag{2}$$

Next, consider the forces that act on the object as it falls. Gravity exerts a force equal to the weight of the object, or mg, where g is the acceleration due to gravity. In the units we have chosen, g has been determined experimentally to be approximately equal to 9.8 m/s^2 near the earth's surface. There is also a force due to air resistance, or drag, that is more difficult to model. This is not the place for an extended discussion of the drag force; suffice it to say that it is often assumed that the drag is proportional to the velocity, and we will make that assumption here. Thus the drag force has the magnitude γv, where γ is a constant called the drag coefficient. The numerical value of the drag coefficient varies widely from one object to another; smooth streamlined objects have much smaller drag coefficients than rough blunt ones. The physical units for γ are kg/s; if this seems peculiar, remember that γv must have the units of force, namely, kg·m/s^2.

In writing an expression for the net force F, we need to remember that gravity always acts in the downward (positive) direction, whereas drag acts in the upward (negative) direction, as shown in Figure 1.1.1. Thus

$$F = mg - \gamma v \tag{3}$$

and Eq. (2) then becomes

$$m\frac{dv}{dt} = mg - \gamma v. \tag{4}$$

Equation (4) is a mathematical model of an object falling in the atmosphere near sea level. Note that the model contains the three constants m, g, and γ. The constants m and γ depend very much on the particular object that is falling, and they are usually different for different objects. It is common to refer to them as parameters, since they may take on a range of values during the course of an experiment. On the other hand, the value of g is the same for all objects.

$$\gamma v$$

$$m$$

$$mg$$

FIGURE 1.1.1   Free-body diagram of the forces on a falling object.

To solve Eq. (4) we need to find a function $v = v(t)$ that satisfies the equation. It is not hard to do this, and we will show you how in the next section. For the present, however, let us see what we can learn about solutions without actually finding any of them. Our task is simplified slightly if we assign numerical values to m and γ, but the procedure is the same regardless of which values we choose. In the following example we use values of m and γ that are typical of hailstones.

EXAMPLE 2

A Falling Hailstone

Consider a hailstone whose mass is $m = 0.025$ kg and whose drag coefficient is $\gamma = 0.007$ kg/s. Write down the differential equation describing the motion of the hailstone as it falls. Then, without solving the equation, investigate the behavior of its solutions.

The differential equation for the falling hailstone is just Eq. (4) with the given values of m, g, and γ; that is,

$$0.025\frac{dv}{dt} = (0.025)(9.8) - 0.007v,$$

or

$$\frac{dv}{dt} = 9.8 - 0.28v. \tag{5}$$

To determine the qualitative behavior of solutions of Eq. (5), we can proceed from a geometrical viewpoint. Suppose that we choose a value for v. Then, by evaluating the right side of Eq. (5), we can find the corresponding value of dv/dt. For instance, if $v = 25$, then $dv/dt = 2.8$. This means that the slope of a solution $v = v(t)$ has the value 2.8 at any point where $v = 25$. We can display this information graphically in the tv-plane by drawing short line segments with slope 2.8 at several points on the line $v = 25$. Similarly, if $v = 40$, then $dv/dt = -1.4$, so we draw line segments with slope -1.4 at several points on the line $v = 40$. We obtain Figure 1.1.2 by proceeding in the same way with other values of v. Figure 1.1.2 is an example of what is called a **direction field** or sometimes a **slope field**.

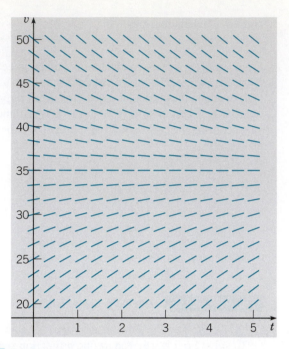

FIGURE 1.1.2 A direction field for Eq. (5).

The importance of Figure 1.1.2 is that each line segment is a tangent line to the graph of a solution of Eq. (5). Consequently, by looking at the direction field, we can visualize how solutions of Eq. (5) vary with time. On a printed copy of a direction field we can even sketch (approximately) graphs of solutions by drawing curves that are always tangent to line segments in the direction field.

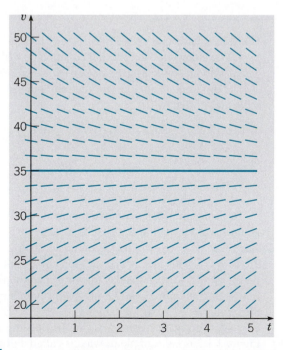

FIGURE 1.1.3 Direction field and equilibrium solution for Eq. (5).

Thus, even though we have not found any solutions of Eq. (5), we can nonetheless draw some qualitative conclusions about the behavior of its solutions. For instance, if v is less than a certain critical value, then all the line segments have positive slopes; hence $v(t)$ is increasing, and the speed of the hailstone increases as it falls. On the other hand, if v is greater than the critical value, then the line segments have negative slopes, $v(t)$ is decreasing, and the hailstone slows down as it falls. What is this critical value of v that separates hailstones whose speed is increasing from those whose speed is decreasing? Referring again to Eq. (5), we ask what value of v will cause dv/dt to be zero. The answer is $v = 9.8/0.28 = 35$ m/s.

In fact, the constant function $v(t) = 35$ is a solution of Eq. (5). To verify this statement, substitute $v(t) = 35$ into Eq. (5) and observe that each side of the equation is zero. Because it does not change with time, the solution $v(t) = 35$ is called an **equilibrium solution.** It is the solution that corresponds to a balance between gravity and drag. In Figure 1.1.3, we show the equilibrium solution $v(t) = 35$ superimposed on the direction field. From this figure, we can draw another conclusion, namely, that all other solutions seem to be converging to the equilibrium solution as t increases.

The approach illustrated in Example 2 can be applied equally well to the more general Eq. (4), where the parameters m and γ are unspecified positive numbers. The results are essentially identical to those of Example 2. The equilibrium solution of Eq. (4) is $v(t) = mg/\gamma$. Solutions below the equilibrium solution increase with time, those above it decrease with time, and all solutions approach the equilibrium solution as t becomes large.

▶ **Field Mice and Owls.** Now let us look at another, quite different, situation. Consider a population of field mice who inhabit a certain rural area. In the absence of predators we assume that the rate of change of the mouse population is proportional to the current population; for example, if the population doubles, then the number of births per unit time also doubles. This assumption is not a well-established physical law (as Newton's law of motion is in Example 1), but it is a common initial hypothesis[1] in a study of population growth. If we denote time by t and the mouse population by $p(t)$, then the assumption about population growth can be expressed by the equation

$$\frac{dp}{dt} = rp, \tag{6}$$

where the proportionality factor r is called the **rate constant** or **growth rate**.

Now let us suppose that the field mice are preyed upon by some owls, who also live in the same vicinity, and let us assume that the predation rate is a constant k. By modifying Eq. (6) to take this into account, we obtain the equation

$$\frac{dp}{dt} = rp - k, \tag{7}$$

where both r and k are positive. Thus the rate of change of the mouse population, dp/dt, is the net effect of the growth term rp and the predation term $-k$. Depending on the values of p, r, and k, the value of dp/dt may be of either sign.

---

[1] A somewhat better model of population growth is discussed in Section 2.5.

EXAMPLE 3

Suppose that the growth rate for the field mice is 0.5/month and that the owls kill 15 mice per day. Write down the differential equation for the mouse population and investigate its solutions graphically.

We naturally assume that p is the number of individuals in the mouse population at time t. We can choose our units for time to be whatever seems most convenient; the two obvious possibilities are days or months. If we choose to measure time in months, then the growth term is $0.5p$ and the predation term is -450, assuming an average month of 30 days. Thus Eq. (7) becomes

$$\frac{dp}{dt} = 0.5p - 450, \tag{8}$$

where each term has the units of mice/month.

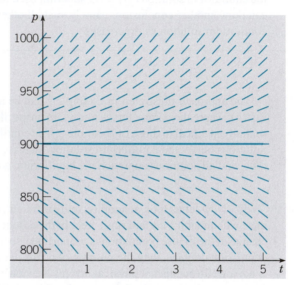

FIGURE 1.1.4   Direction field and equilibrium solution for Eq. (8).

A direction field for Eq. (8) is shown in Figure 1.1.4. For sufficiently large values of p it can be seen from the figure, or directly from Eq. (8) itself, that dp/dt is positive, so that solutions increase. On the other hand, for small values of p the opposite is the case. Again, the critical value of p that separates solutions that increase from those that decrease is the value of p for which dp/dt is zero. By setting dp/dt equal to zero in Eq. (8) and then solving for p, we find the equilibrium solution $p(t) = 900$ for which the growth term and the predation term in Eq. (8) are exactly balanced. The equilibrium solution is also shown in Figure 1.1.4.

Solutions of the more general equation (7), in which the growth rate and the predation rate are unspecified, behave very much like those of Eq. (8). The equilibrium solution of Eq. (7) is $p(t) = k/r$. Solutions above the equilibrium solution increase, while those below it decrease.

Comparing Examples 2 and 3, we note that in both cases the equilibrium solution separates increasing from decreasing solutions. In Example 2, other solutions converge to, or are attracted by, the equilibrium solution, so that after the object falls far enough, an observer will see it moving at very nearly the equilibrium velocity. On the other hand, in Example 3, other solutions diverge from, or are repelled by, the equilibrium solution. Solutions behave very differently depending on whether they start above or below the equilibrium solution. As time passes, an observer might see populations either much larger or much smaller than the equilibrium population, but the equilibrium solution itself will not, in practice, be observed. In both problems, however, the equilibrium solution is very important in understanding how solutions of the given differential equation behave.

You should keep in mind that both of the models discussed in this section have their limitations. The model (5) of the falling hailstone is valid only as long as the hailstone is falling freely, without encountering any obstacles. The population model (8) eventually predicts negative numbers of mice (if $p < 900$) or enormously large numbers (if $p > 900$). Both these predictions are unrealistic, so this model becomes unacceptable after a fairly short time interval.

▶ **Direction Fields.** The differential equations considered up to now, Eqs. (5) and (8), are somewhat special in that the right-hand sides of the equations do not depend explicitly on the independent variable t, but only on the dependent variable v or p. However, direction fields are perhaps even more valuable for visualizing the solutions of more general differential equations of the form

$$\frac{dy}{dt} = f(t, y). \tag{9}$$

Here f is a given function of the two variables t and y, sometimes referred to as the **rate function**. A useful direction field for equations of the form (9) can be constructed by evaluating f at each point of a rectangular grid consisting of at least a few hundred points. Then, at each point of the grid, a short line segment is drawn whose slope is the value of f at that point. Thus each line segment is tangent to the graph of the solution passing through that point. A direction field drawn on a fairly fine grid gives a good picture of the overall behavior of solutions of a differential equation. The construction of a direction field is often a worthwhile first step in the investigation of a differential equation.

Two observations are worth particular mention. First, in constructing a direction field, we do not have to solve Eq. (9), but merely to evaluate the given function $f(t, y)$ many times. Thus direction fields can be readily constructed even for equations that may be quite difficult to solve. Second, repeated evaluation of a given function is a task for which a computer is well suited, and you should usually use a computer to draw a direction field. All the direction fields shown in this book, such as the ones in this section, were computer-generated.

EXAMPLE 4

Heating and Cooling

Consider a building as a partly insulated box that is subject to external temperature fluctuations. Construct a model that describes the temperature fluctuations inside the building.

Let $u(t)$ and $T(t)$ be the internal and external temperatures, respectively, at time t. Our mathematical model is based on Newton's law of cooling, which states that the rate of change of $u(t)$, denoted by du/dt, is proportional to the temperature difference $u(t) - T(t)$. Thus we obtain the differential equation

$$\frac{du}{dt} = -k[u - T(t)], \tag{10}$$

where k is a positive constant. The minus sign immediately to the right of the equals sign in Eq. (10) is due to the fact that du/dt must be negative if $u(t) > T(t)$. Let us suppose that u

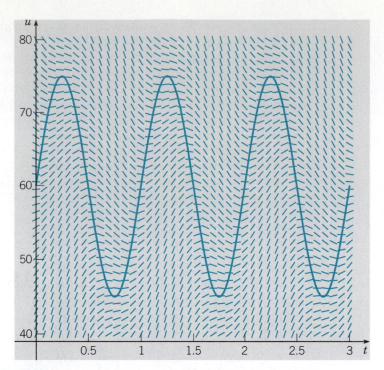

FIGURE 1.1.5 Direction field and graph of $T(t)$ for Eq. (11).

and T are measured in degrees Fahrenheit and that t is measured in days. Then k must have the units of 1/day.

Finally, let us suppose that $k = 1.5$ and that $T(t) = 60 + 15 \sin 2\pi t$. The expression for $T(t)$ corresponds to a daily variation of $15°$ above and below a mean temperature of $60°$. Then Eq. (10) becomes

$$\frac{du}{dt} = -1.5(u - 60 - 15 \sin 2\pi t). \tag{11}$$

Figure 1.1.5 shows a direction field for Eq. (11) and also a graph of the exterior temperature $T(t)$. The direction field is a bit more complicated than those shown earlier in this section because the slopes depend on the independent variable t as well as on the dependent variable u. Nevertheless, if you look carefully at Figure 1.1.5, you can visualize the behavior of the interior temperature u as a function of t. To help in the visualization you can, if you wish, sketch graphs of u versus t by drawing curves that are always tangent to line segments in the direction field. Note that there is a time lag in the response of the interior temperature (as indicated by the direction field) to the fluctuations of the outside temperature (as shown by the graph of $T(t)$).

▶ **Constructing Mathematical Models.** In applying differential equations to any of the numerous fields in which they are useful, it is necessary first to formulate the appropriate differential equation that describes, or models, the problem being investigated. In this section, we have looked at three examples of this modeling process, two drawn from physics and the other from ecology. In constructing future mathematical models yourself, you should recognize that each problem is different, and that successful modeling is not a skill that can be reduced to the observance of a set of prescribed rules. Indeed, constructing a satisfactory model is sometimes the most difficult part of the problem. Nevertheless, it may be helpful to list some steps that are often part of the process:

1. Identify the independent and dependent variables and assign letters to represent them. Often the independent variable is time.

2. Choose the units of measurement for each variable. In a sense, the choice of units is arbitrary, but some choices may be much more convenient than others. For example, we chose to measure time in seconds for the falling hailstone problem, in months for the population problem, and in days for the temperature problem.

3. Articulate the basic principle that underlies or governs the problem you are investigating. This may be a widely accepted physical law, such as Newton's law of motion, or it may be a more speculative assumption based on your own experience or observations. In any case, this step is likely not to be a purely mathematical one, but will require you to be familiar with the field in which the problem originates.

4. Express the principle or law in step 3 in terms of the variables you chose in step 1. This may be easier said than done. It may require the introduction of physical constants or parameters (such as the drag coefficient in Example 1) and the determination of appropriate values for them. Or it may involve the use of auxiliary or intermediate variables that must then be related to the primary variables.

5. Make sure that each term in your equation has the same physical units. If this is not the case, then your equation is wrong and you need to correct it. If the units agree, then your equation at least is dimensionally consistent, although it may have other shortcomings that this test does not reveal.

6. In the problems considered here, the result of step 4 is a single differential equation, which constitutes the desired mathematical model. Keep in mind, though, that in more complex problems the resulting mathematical model may be much more complicated, perhaps involving a system of several differential equations, for example.

PROBLEMS

ODEA In each of Problems 1 through 6, draw a direction field for the given differential equation. Based on the direction field, determine the behavior of y as $t \to \infty$. If this behavior depends on the initial value of y at $t = 0$, describe the dependency.

1. $y' = 3 - 2y$
2. $y' = 2y - 3$
3. $y' = 3 + 2y$
4. $y' = -1 - 2y$
5. $y' = 1 + 2y$
6. $y' = y + 2$

In each of Problems 7 through 10, write down a differential equation of the form $dy/dt = ay + b$ whose solutions have the required behavior as $t \to \infty$.

7. All solutions approach $y = 3$.
8. All solutions approach $y = 2/3$.
9. All other solutions diverge from $y = 2$.
10. All other solutions diverge from $y = 1/3$.

In each of Problems 11 through 14, draw a direction ODEA field for the given differential equation. Based on the direction field, determine the behavior of y as $t \to \infty$. If this behavior depends on the initial value of y at $t = 0$, describe this dependency. Note that in these problems the equations are not of the form $y' = ay + b$, and the behavior of their solutions is somewhat more complicated than in Examples 1, 2, and 3 in the text.

11. $y' = y(4 - y)$
12. $y' = -y(5 - y)$
13. $y' = y^2$
14. $y' = y(y - 2)^2$

Consider the following list of differential equations, some of which produced the direction fields shown in

Figures 1.1.6 through 1.1.11. In each of Problems 15 through 20, identify the differential equation that corresponds to the given direction field.

(a) $y' = 2y - 1$
(b) $y' = 2 + y$
(c) $y' = y - 2$
(d) $y' = y(y + 3)$
(e) $y' = y(y - 3)$
(f) $y' = 1 + 2y$
(g) $y' = -2 - y$
(h) $y' = y(3 - y)$
(i) $y' = 1 - 2y$
(j) $y' = 2 - y$

15. The direction field of Figure 1.1.6.

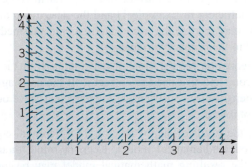

FIGURE 1.1.6 Direction field for Problem 15.

16. The direction field of Figure 1.1.7.

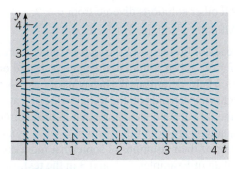

FIGURE 1.1.7 Direction field for Problem 16.

17. The direction field of Figure 1.1.8.
18. The direction field of Figure 1.1.9.
19. The direction field of Figure 1.1.10.
20. The direction field of Figure 1.1.11.

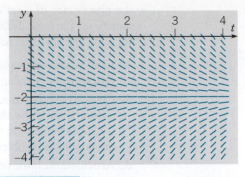

FIGURE 1.1.8 Direction field for Problem 17.

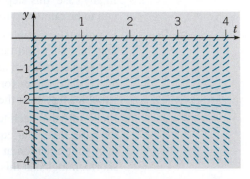

FIGURE 1.1.9 Direction field for Problem 18.

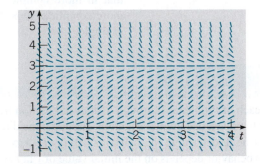

FIGURE 1.1.10 Direction field for Problem 19.

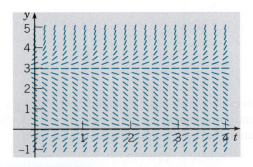

FIGURE 1.1.11 Direction field for Problem 20.

21. A pond initially contains 1,000,000 gal of water and an unknown amount of an undesirable chemical. Water containing 0.01 g of this chemical per gallon flows into the pond at a rate of 300 gal/h. The mixture flows out at the same rate, so the amount of water in the pond remains constant. Assume that the chemical is uniformly distributed throughout the pond.

    (a) Write a differential equation for the amount of chemical in the pond at any time.
    (b) How much of the chemical will be in the pond after a very long time? Does this limiting amount depend on the amount that was present initially?

22. A spherical raindrop evaporates at a rate proportional to its surface area. Write a differential equation for the volume of the raindrop as a function of time.

23. Newton's law of cooling states that the temperature of an object changes at a rate proportional to the difference between the temperature of the object itself and the temperature of its surroundings (the ambient air temperature in most cases). Suppose that the ambient temperature is 70°F and that the rate constant is 0.05 (min)$^{-1}$. Write a differential equation for the temperature of the object at any time.

24. A certain drug is being administered intravenously to a hospital patient. Fluid containing 5 mg/cm^3 of the drug enters the patient's bloodstream at a rate of 100 cm^3/h. The drug is absorbed by body tissues or otherwise leaves the bloodstream at a rate proportional to the amount present, with a rate constant of 0.4 h^{-1}.

    (a) Assuming that the drug is always uniformly distributed throughout the bloodstream, write a differential equation for the amount of the drug that is present in the bloodstream at any time.
    (b) How much of the drug is present in the bloodstream after a long time?

25. For small, slowly falling objects, the assumption made in the text that the drag force is proportional to the velocity is a good one. For larger, more rapidly falling objects, it is more accurate to assume that the drag force is proportional to the square of the velocity.[2]

    (a) Write a differential equation for the velocity of a falling object of mass m if the drag force is proportional to the square of the velocity.
    (b) Determine the limiting velocity after a long time.
    (c) If $m = 0.025$ kg, find the drag coefficient so that the limiting velocity is 35 m/s.
    (d) Using the data in part (c), draw a direction field and compare it with Figure 1.1.2.

In each of Problems 26 through 33, draw a direction field for the given differential equation. Based on the direction field, determine the behavior of y as $t \to \infty$. If this behavior depends on the initial value of y at $t = 0$, describe this dependency. Note that the right sides of these equations depend on t as well as y.

26. $y' = -2 + t - y$
27. $y' = te^{-2t} - 2y$
28. $y' = e^{-t} + y$
29. $y' = t + 2y$
30. $y' = 3 \sin t + 1 + y$
31. $y' = 2t - 1 - y^2$
32. $y' = -(2t + y)/2y$
33. $y' = y^3/6 - y - t^2/3$

1.2 Solutions of Some Differential Equations

In the preceding section we derived three differential equations:

$$m\frac{dv}{dt} = mg - \gamma v \tag{1}$$

[2]See Lyle N. Long and Howard Weiss, "The Velocity Dependence of Aerodynamic Drag: A Primer for Mathematicians," *American Mathematical Monthly 106* (1999), 2, pp. 127–135.

for the velocity of a falling object,

$$\frac{dp}{dt} = rp - k \tag{2}$$

for a population of field mice preyed on by owls, and

$$\frac{du}{dt} = -k[u - T(t)] \tag{3}$$

for the temperature in a building subject to a varying external temperature. We were able to draw some important qualitative conclusions about the behavior of solutions of these equations by considering the associated direction fields. To answer questions of a quantitative nature, however, we need to find the solutions themselves, and we now investigate how to do that.

EXAMPLE 1

Field Mice and Owls

(continued)

Consider the equation

$$\frac{dp}{dt} = 0.5p - 450, \tag{4}$$

which describes the interaction of certain populations of field mice and owls (see Eq. (8) of Section 1.1). Find solutions of this equation.

To solve Eq. (4), we need to find functions $p(t)$ that, when substituted into the equation, reduce it to an obvious identity. Here is one way to proceed. First, rewrite Eq. (4) in the form

$$\frac{dp}{dt} = \frac{p - 900}{2}, \tag{5}$$

or, if $p \neq 900$,

$$\frac{dp/dt}{p - 900} = \frac{1}{2}. \tag{6}$$

By the chain rule, the left side of Eq. (6) is the derivative of $\ln |p - 900|$ with respect to t, so we have

$$\frac{d}{dt} \ln |p - 900| = \frac{1}{2}. \tag{7}$$

Then, by integrating both sides of Eq. (7), we obtain

$$\ln |p - 900| = \frac{t}{2} + C, \tag{8}$$

where C is an arbitrary constant of integration. Therefore, by taking the exponential of both sides of Eq. (8), we find that

$$|p - 900| = e^{(t/2)+C} = e^C e^{t/2}, \tag{9}$$

or

$$p - 900 = \pm e^C e^{t/2}, \tag{10}$$

and finally

$$p = 900 + ce^{t/2}, \tag{11}$$

where $c = \pm e^C$ is also an arbitrary (nonzero) constant. Note that the constant function $p = 900$ is also a solution of Eq. (5) and that it is contained in the expression (11) if we

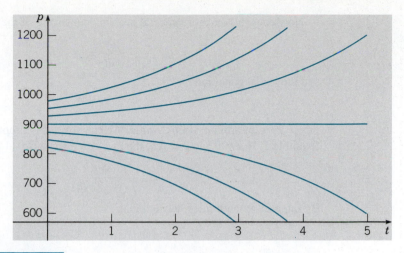

FIGURE 1.2.1   Graphs of Eq. (11) for several values of c.

allow c to take the value zero. Graphs of Eq. (11) for several values of c are shown in Figure 1.2.1.

Note that they have the character inferred from the direction field in Figure 1.1.4. For instance, solutions lying on either side of the equilibrium solution $p = 900$ tend to diverge from that solution.

In Example 1, we found infinitely many solutions of the differential equation (4), corresponding to the infinitely many values that the arbitrary constant c in Eq. (11) might have. This is typical of what happens when you solve a differential equation. The solution process involves an integration, which brings with it an arbitrary constant, whose possible values generate an infinite family of solutions.

Frequently, we want to focus our attention on a single member of the infinite family of solutions by specifying the value of the arbitrary constant. Most often, we do this indirectly by specifying instead a point that must lie on the graph of the solution. For example, to determine the constant c in Eq. (11), we could require that the population have a given value at a certain time, such as the value 850 at time $t = 0$. In other words, the graph of the solution must pass through the point $(0, 850)$. Symbolically, we can express this condition as

$$p(0) = 850. \tag{12}$$

Then, substituting $t = 0$ and $p = 850$ into Eq. (11), we obtain

$$850 = 900 + c.$$

Hence $c = -50$, and by inserting this value in Eq. (11), we obtain the desired solution, namely,

$$p = 900 - 50e^{t/2}. \tag{13}$$

The additional condition (12) that we used to determine c is an example of an **initial condition**. The differential equation (4) together with the initial condition (12) form an **initial value problem**.

Now consider the more general problem consisting of the differential equation

$$\frac{dy}{dt} = ay - b \tag{14}$$

and the initial condition

$$y(0) = y_0, \tag{15}$$

where y_0 is an arbitrary initial value. We can solve this problem by the same method as in Example 1. If $a \neq 0$ and $y \neq b/a$, then we can rewrite Eq. (14) as

$$\frac{dy/dt}{y - (b/a)} = a. \tag{16}$$

By integrating both sides, we find that

$$\ln|y - (b/a)| = at + C, \tag{17}$$

where C is arbitrary. Then, taking the exponential of both sides of Eq. (17) and solving for y, we obtain

$$y = (b/a) + ce^{at}, \tag{18}$$

where $c = \pm e^C$ is also arbitrary. Observe that $c = 0$ corresponds to the equilibrium solution $y = b/a$. For later reference, also note that the solution (18) consists of two terms. The exponential term $y = ce^{at}$ is a solution of the differential equation $y' = ay$, and the constant term $y = b/a$ is needed to accommodate the constant term $-b$ in Eq. (14).

The expression (18) contains all possible solutions of Eq. (14) and is called the **general solution**. The geometrical representation of the general solution (18) is an infinite family of curves called **integral curves**. Each integral curve is associated with a particular value of c and is the graph of the solution corresponding to that value of c. Satisfying an initial condition amounts to identifying the integral curve that passes through the given initial point. For example, the initial condition (15) requires that $c = y_0 - (b/a)$, so the solution of the initial value problem (14), (15) is

$$y = (b/a) + [y_0 - (b/a)]e^{at}. \tag{19}$$

To relate the solution (19) to Eq. (2), which models the field mouse population, we need only replace a by the growth rate r and b by the predation rate k. Then the solution (19) becomes

$$p = (k/r) + [p_0 - (k/r)]e^{rt}, \tag{20}$$

where p_0 is the initial population of field mice. The solution (20) confirms the conclusions reached on the basis of the direction field and Example 1. If $p_0 = k/r$, then from Eq. (20) it follows that $p = k/r$ for all t; this is the constant, or equilibrium, solution. If $p_0 \neq k/r$, then the behavior of the solution depends on the sign of the coefficient $p_0 - (k/r)$ of the exponential term in Eq. (20). If $p_0 > k/r$, then p grows exponentially with time t; if $p_0 < k/r$, then p decreases and eventually becomes zero, corresponding to extinction of the field mouse population. Negative values of p, while possible for the expression (20), make no sense in the context of this particular problem.

To put the falling-object equation (1) in the form (14), we must identify a with $-\gamma/m$ and b with $-g$. Making these substitutions in the solution (19), we obtain

$$v = (mg/\gamma) + [v_0 - (mg/\gamma)]e^{-\gamma t/m}, \tag{21}$$

where v_0 is the initial velocity. Again, this solution confirms the conclusions reached in Section 1.1 on the basis of a direction field. There is an equilibrium, or constant, solution

$v = mg/\gamma$, and all other solutions tend to approach this equilibrium solution. The speed of convergence to the equilibrium solution is determined by the exponent $-\gamma/m$. Thus, for a given mass m, the velocity approaches the equilibrium value more rapidly as the drag coefficient γ increases.

**EXAMPLE
2**

**A Falling
Hailstone**

(continued)

Suppose that, as in Example 2 of Section 1.1, we consider a falling hailstone of mass $m = 0.025$ kg and drag coefficient $\gamma = 0.007$ kg/s. Then the equation of motion (1) becomes

$$\frac{dv}{dt} = 9.8 - 0.28v. \tag{22}$$

Suppose the hailstone falls from a cloud that is 300 m above the ground. Find the velocity of the hailstone at any time t. How long will it take to fall to the ground, and how fast will it be moving at the time of impact?

The first step is to state an appropriate initial condition for Eq. (22). We assume that the hailstone starts to fall with an initial velocity zero, so we will use the initial condition

$$v(0) = 0. \tag{23}$$

The solution of Eq. (22) can be found by substituting the values of the coefficients into the solution (21), but we will proceed instead to solve Eq. (22) directly. First, rewrite the equation as

$$\frac{dv/dt}{v - 35} = -0.28. \tag{24}$$

By integrating both sides we obtain

$$\ln|v - 35| = -0.28t + C, \tag{25}$$

and then the general solution of Eq. (22) is

$$v = 35 + ce^{-0.28t}, \tag{26}$$

where c is arbitrary. To determine c, we substitute $t = 0$ and $v = 0$ from the initial condition (23) into Eq. (26), with the result that $c = -35$. Then the solution of the initial value problem (22), (23) is

$$v = 35(1 - e^{-0.28t}). \tag{27}$$

Equation (27) gives the velocity of the falling hailstone at any positive time (before it hits the ground, of course).

Graphs of the solution (26) for several values of c are shown in Figure 1.2.2, with the solution (27) shown by the heavy curve. It is evident that all solutions tend to approach the equilibrium solution $v = 35$. This confirms the conclusions we reached in Section 1.1 on the basis of the direction fields in Figures 1.1.2 and 1.1.3.

To find the velocity of the hailstone when it hits the ground, we need to know the time at which impact occurs. In other words, we need to determine how long it takes the hailstone to fall 300 m. To do this, we note that the distance x the hailstone has fallen is related to its velocity v by the equation $v = dx/dt$, or

$$\frac{dx}{dt} = 35(1 - e^{-0.28t}). \tag{28}$$

Consequently, by integrating both sides of Eq. (28), we have

$$x = 35t + 125e^{-0.28t} + c, \tag{29}$$

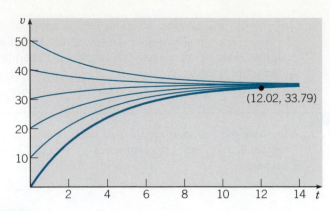

FIGURE 1.2.2 Graphs of the solution (26) for several values of c.

where c is an arbitrary constant of integration. The hailstone starts to fall when $t = 0$, so we know that $x = 0$ when $t = 0$. From Eq. (29), it follows that $c = -125$, so the distance the hailstone has fallen at time t is given by

$$x = 35t + 125e^{-0.28t} - 125. \tag{30}$$

Let T be the time at which the hailstone hits the ground; then $x = 300$ when $t = T$. By substituting these values in Eq. (30), we obtain the equation

$$35T + 125e^{-0.28T} - 425 = 0. \tag{31}$$

The value of T satisfying Eq. (31) can be readily approximated by a numerical process using a scientific calculator or computer, with the result that $T \cong 12.02$ s. At this time, the corresponding velocity v_T is found from Eq. (27) to be $v_T \cong 33.79$ m/s. The point (12.02,33.79) is also shown in Figure 1.2.2.

Let us now consider the equation

$$\frac{dy}{dt} + ky = b \sin t, \tag{32}$$

which might arise from the temperature equation (3) if the external temperature has a sinusoidal variation. Referring to the remark following Eq. (18), we infer that the solution of Eq. (32) might contain an exponential term satisfying the equation $y' + ky = 0$, as well as one or more terms accommodating the $b \sin t$ term in Eq. (32). Thus we are led to consider the expression

$$y = ce^{-kt} + A \sin t + B \cos t \tag{33}$$

as a possible solution[3] of Eq. (32). To determine whether this expression is an actual solution, we substitute it into the differential equation (32). We obtain

$$-kce^{-kt} + A \cos t - B \sin t + k\left[ce^{-kt} + A \sin t + B \cos t\right] = b \sin t. \tag{34}$$

---

[3]Here we are making an assumption as to the general form of the solution, and you may well find this assumption to be less than convincing. In solving a differential equation, we are looking for a mathematical function; the totality of such functions is enormous. If we can narrow our search to a relatively small subset of functions, such as those of the form (33), then it may be possible to accelerate the solution process very greatly. Of course, success depends on choosing a convenient set of functions that includes the solution that we seek. Experience helps a lot in making this choice.

Observe that the exponential terms cancel out, so the constant c remains undetermined; in other words, it is an arbitrary constant. The constants A and B are determined by collecting the coefficients of the sin t and cos t terms, respectively. From the sin t terms we obtain

$$kA - B = b \tag{35}$$

and from the cos t terms we have

$$A + kB = 0. \tag{36}$$

By solving Eqs. (35) and (36), we find that

$$A = \frac{kb}{k^2 + 1}, \quad B = -\frac{b}{k^2 + 1}. \tag{37}$$

Thus the general solution of Eq. (32) is

$$y = ce^{-kt} + \frac{b}{k^2 + 1}(k \sin t - \cos t). \tag{38}$$

The arbitrary constant c in Eq. (38) can be determined if an initial condition is prescribed.

The solution procedure used here is called the method of **undetermined coefficients** and is very useful for a rather restricted class of equations. It is explored, to some extent, in Problems 5 through 10, and is explained in detail in Section 4.6.

EXAMPLE 3

Heating and Cooling
(continued)

Let us return now to the situation described in Example 4 in Section 1.1. The internal temperature $u(t)$ in a building satisfies the differential equation (3),

$$\frac{du}{dt} = -k\big[u - T(t)\big],$$

where $T(t)$ is the external temperature and k is a positive constant. We assume that u and T are measured in degrees Fahrenheit and t in days. Note that there are two limiting, or extreme, cases: $k = 0$ and $k \to \infty$. If $k = 0$, then $du/dt = 0$ and the internal temperature does not change regardless of the external temperature fluctuations; in other words, the building is perfectly insulated. On the other hand, if $k \to \infty$, then $u = T(t)$; otherwise, the derivative du/dt becomes unbounded, which is contrary to experimental evidence. Thus, in this case, the internal temperature is always the same as the external temperature; that is, the building provides no barrier to the flow of heat.

We now assume, as in Example 4 of Section 1.1, that $k = 1.5$ and that

$$T(t) = 60 + 15 \sin 2\pi t.$$

Then Eq. (3) becomes

$$\frac{du}{dt} = -1.5\big[u - 60 - 15 \sin 2\pi t\big]. \tag{39}$$

Let us also suppose that the initial internal temperature is

$$u(0) = 70. \tag{40}$$

Solve the initial value problem (39), (40) and plot the solution for several days. Compare the variations of the internal and external temperatures.

We can find the general solution of Eq. (39) by using the method of undetermined coefficients that we just described. First we rewrite Eq. (39) in the form

$$\frac{du}{dt} + 1.5u = 90 + 22.5 \sin 2\pi t. \tag{41}$$

Then we assume that the solution consists of an exponential term $ce^{-1.5t}$ that satisfies the equation $u' + 1.5u = 0$, together with a constant term and $\sin 2\pi t$ and $\cos 2\pi t$ terms to accommodate the constant term and $\sin 2\pi t$ terms on the right side of Eq. (41). In other words, we assume that

$$u = ce^{-1.5t} + A + B \sin 2\pi t + D \cos 2\pi t, \tag{42}$$

where A, B, and D are constants to be determined by substituting u from Eq. (42) into Eq. (41). On carrying out this substitution, we find that the terms involving c drop out, so c remains undetermined (that is, arbitrary). Further, we find that $A = 60$, and that B and D satisfy the equations

$$2\pi B + 1.5D = 0, \quad 1.5B - 2\pi D = 22.5. \tag{43}$$

Solving these equations for B and D, we find that

$$B = \frac{(1.5)(22.5)}{(2\pi)^2 + (1.5)^2} \cong 0.808801, \quad D = -\frac{(2\pi)(22.5)}{(2\pi)^2 + (1.5)^2} \cong -3.387899,$$

correct to six decimal places. Thus the general solution of Eq. (41) is

$$u = ce^{-1.5t} + 60 + 0.808801 \sin 2\pi t - 3.387899 \cos 2\pi t. \tag{44}$$

The initial condition (40) requires that $c = 13.387899$ to six decimal places.

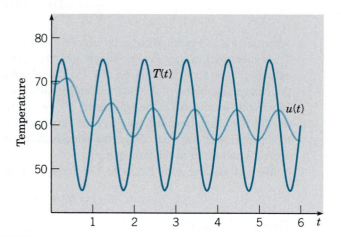

FIGURE 1.2.3   Graphs of the internal temperature $u(t)$ and the external temperature $T(t)$ for the problem (39), (40).

Figure 1.2.3 contains a plot of the solution of the initial value problem (39), (40) and a plot of the external temperature $T(t)$. Note that after two or three days the internal temperature $u(t)$ experiences an essentially steady oscillation about the mean value of 60; in other words, the effect of the exponential term in the solution (44) is no longer noticeable. Once this steady state has been reached, the total variation in the internal temperature $u(t)$ is approximately 7°F, much less than the variation of 30°F in the external temperature $T(t)$. Further, the maximum and minimum values of the $u(t)$ occur more than five hours later than the corresponding maximum and minimum values of $T(t)$. Finally, the local extreme values of the graph of $u(t)$ in Figure 1.2.3 seem to occur when the graphs of $u(t)$ and $T(t)$ cross each other. Do you think this is coincidental, or is this what you would expect?

Think about the results of this example and whether they are consistent with your own experience in your home or other places where you have spent time. In your experience, are the internal fluctuations substantially less than the external fluctuations? Is there a time lag of several hours between the extreme external temperatures and the corresponding extreme internal temperatures? What would be the effects of changing the value of k?

▶ **Further Remarks on Mathematical Modeling.** Up to this point we have related our discussion of differential equations to mathematical models of a falling hailstone, of a hypothetical relation between field mice and owls, and of the heating or cooling of a building by external temperature variations. The derivation of these models may have been plausible, and possibly even convincing, but you should remember that the ultimate test of any mathematical model is whether its predictions agree with observations or experimental results. We have no actual observations or experimental results to use for comparison purposes here, but there are several sources of possible discrepancies.

In the case of the falling object, the underlying physical principle (Newton's law of motion) is well established and widely applicable. However, the assumption that the drag force is proportional to the velocity is less certain. Even if this assumption is correct, the determination of the drag coefficient γ by direct measurement presents difficulties. Indeed, sometimes one finds the drag coefficient indirectly—for example, by measuring the time of fall from a given height and then calculating the value of γ that predicts this observed time.

The model of the field mouse population is subject to various uncertainties. The determination of the growth rate r and the predation rate k depends on observations of actual populations, which may be subject to considerable variation. The assumption that r and k are constants may also be questionable. For example, a constant predation rate becomes harder to sustain as the field mouse population becomes smaller. Further, the model predicts that a population above the equilibrium value will grow exponentially larger and larger. This seems at variance with the behavior of actual populations; see the further discussion of population dynamics in Section 2.5.

For the temperature variation model there may well be factors that affect the relation between inside and outside temperatures other than the simple temperature difference. For example, the direction and velocity of the wind may have some influence. A single value of the parameter k may also be inappropriate, since some parts of the building may be more permeable to the passage of heat than others. Even assuming that a single value of k is reasonable, the determination of an accurate value for it may not be easy.

If the differences between actual observations and a mathematical model's predictions are too great, then you need to consider refining the model, making more careful observations, or perhaps both. There is almost always a tradeoff between accuracy and simplicity. Both are desirable, but a gain in one usually involves a loss in the other. However, even if a mathematical model is incomplete or somewhat inaccurate, it may nevertheless be useful in explaining qualitative features of the problem under investigation. It may also give satisfactory results under some circumstances but not others. Thus you should always use good judgment and common sense in constructing mathematical models and in using their predictions.

PROBLEMS

1. Solve each of the following initial value problems and plot the solutions for several values of y_0. Then describe in a few words how the solutions resemble, and differ from, each other.

   (a) $dy/dt = -y + 5$, $y(0) = y_0$
   (b) $dy/dt = -2y + 5$, $y(0) = y_0$
   (c) $dy/dt = -2y + 10$, $y(0) = y_0$

2. Follow the instructions for Problem 1 for the following initial value problems:

   (a) $dy/dt = y - 5$, $y(0) = y_0$
   (b) $dy/dt = 2y - 5$, $y(0) = y_0$
   (c) $dy/dt = 2y - 10$, $y(0) = y_0$

3. Consider the differential equation

$$dy/dt = -ay + b,$$

   where both a and b are positive numbers.

   (a) Solve the differential equation.
   (b) Sketch the solution for several different initial conditions.
   (c) Describe how the solutions change under each of the following conditions:

      i. a increases.

      ii. b increases.

      iii. Both a and b increase, but the ratio b/a remains the same.

4. Consider the differential equation $dy/dt = ay - b$.

   (a) Find the equilibrium solution y_e.
   (b) Let $Y(t) = y - y_e$; thus $Y(t)$ is the deviation from the equilibrium solution. Find the differential equation satisfied by $Y(t)$.

   Undetermined Coefficients. The method of undetermined coefficients is a way to solve some equations of the form

$$dy/dt + ay = f(t), \qquad \text{(i)}$$

   where a is a constant and $f(t)$ is either a polynomial, an exponential function, a sine, or a cosine, or perhaps a sum of such functions. As the discussion in the text suggests, we seek a solution of Eq. (i) in the form

$$y = ce^{-at} + Y(t), \qquad \text{(ii)}$$

   where ce^{-at} is the general solution of $y' + ay = 0$ and $Y(t)$ is chosen to match the term(s) in $f(t)$.

More specifically, if $f(t)$ is a polynomial, then choose $Y(t)$ to be a polynomial of the same degree, with undetermined coefficients A, B, ...; if $f(t)$ is a multiple of e^{kt}, and $k \neq -a$, then choose $Y(t)$ to be Ae^{kt}; and if $f(t)$ is a multiple of $\sin bt$ or $\cos bt$, then choose $Y(t)$ to be $A \sin bt + B \cos bt$. The coefficients A, B, and so forth are found by substituting the expression for $Y(t)$ into the differential equation (i).

In each of Problems 5 through 10, use the method of undetermined coefficients to find the general solution of the given equation.

5. $y' + 2y = t - 3$

6. $y' - 3y = e^{-t}$

7. $y' + y = 3 \cos 2t$

8. $y' - 2y = 2 \sin t$

9. $y' + 2y = 2t + 3 \sin t$

10. $y' - 2y = 3e^t + t^2 + 1$

11. The field mouse population in Example 1 satisfies the differential equation

$$dp/dt = 0.5p - 450.$$

   (a) Find the time at which the population becomes extinct if $p(0) = 850$.
   (b) Find the time of extinction if $p(0) = p_0$, where $0 < p_0 < 900$.
   (c) Find the initial population p_0 if the population is to become extinct in 1 year.

12. Consider a population p of field mice that grows at a rate proportional to the current population, so that $dp/dt = rp$.

   (a) Find the rate constant r if the population doubles in 30 days.
   (b) Find r if the population doubles in N days.

13. The falling hailstone in Example 2 satisfies the initial value problem

$$dv/dt = 9.8 - 0.28v, \qquad v(0) = 0.$$

   (a) Find the time that must elapse for the hailstone to reach 98% of its limiting velocity.
   (b) How far does the hailstone fall in the time found in part (a)?

14. Modify Example 2 so that the falling hailstone experiences no air resistance.

   (a) Write down the modified initial value problem.

(b) Determine how long it takes the hailstone to reach the ground.

(c) Determine its velocity at the time of impact.

15. Consider the falling hailstone of mass 0.025 kg in Example 2, but assume now that the drag force is proportional to the square of the velocity.

    (a) If the limiting velocity is 35 m/s (the same as in Example 2), show that the equation of motion can be written as

    $$dv/dt = [(35)^2 - v^2]/125.$$

    Also see Problem 25 of Section 1.1.

    (b) If $v(0) = 0$, find an expression for $v(t)$ at any time.

    (c) Plot your solution from part (b) and the solution (27) from Example 2 on the same axes.

    (d) Based on your plots in part (c), compare the effect of a quadratic drag force with that of a linear drag force.

    (e) Find the distance $x(t)$ that the hailstone falls in time t.

    (f) Find the time T it takes the hailstone to fall 300 meters.

16. A radioactive material, such as the isotope thorium-234, disintegrates at a rate proportional to the amount currently present. If $Q(t)$ is the amount present at time t, then $dQ/dt = -rQ$, where $r > 0$ is the decay rate.

    (a) If 100 mg of thorium-234 decays to 82.04 mg in 1 week, determine the decay rate r.

    (b) Find an expression for the amount of thorium-234 present at any time t.

    (c) Find the time required for the thorium-234 to decay to one-half its original amount.

17. The **half-life** of a radioactive material is the time required for an amount of this material to decay to one-half its original value. Show that for any radioactive material that decays according to the equation $Q' = -rQ$, the half-life τ and the decay rate r satisfy the equation $r\tau = \ln 2$.

18. Radium-226 has a half-life of 1620 years. Find the time period during which a given amount of this material is reduced by one-quarter.

19. According to Newton's law of cooling (see Example 3), the temperature $u(t)$ of an object satisfies the differential equation

    $$du/dt = -k(u - T),$$

    where the ambient temperature T is assumed to be constant and k is a positive constant. Suppose that the initial temperature of the object is $u(0) = u_0$.

    (a) Find the temperature of the object at any time.

    (b) Let τ be the time at which the initial temperature difference $u_0 - T$ has been reduced by half. Find the relation between k and τ.

20. Suppose that a building loses heat in accordance with Newton's law of cooling (see Example 3 and Problem 19) and that the rate constant k has the value 0.15 h^{-1}. Assume that the interior temperature is 70°F when the heating system fails. If the external temperature is 10°F, how long will it take for the interior temperature to fall to 32°F?

21. Consider an electric circuit containing a capacitor, resistor, and battery; see Figure 1.2.4. The charge $q(t)$ on the capacitor satisfies the equation[4]

    $$R\frac{dq}{dt} + \frac{q}{C} = V,$$

    where R is the resistance, C is the capacitance, and v is the constant voltage supplied by the battery.

    (a) If $q(0) = 0$, find $q(t)$ at any time t, and sketch the graph of q versus t.

    (b) Find the limiting value q_L that $q(t)$ approaches after a long time.

    (c) Suppose that $q(t_1) = q_L$ and that at time $t = t_1$ the battery is removed and the circuit closed again. Find $q(t)$ for $t > t_1$ and sketch its graph.

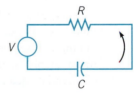

FIGURE 1.2.4   The electric circuit of Problem 21.

22. A pond containing 1,000,000 gal of water is initially free of a certain undesirable chemical (see Problem 21 of Section 1.1). Water containing

---

[4]This equation results from Kirchhoff's laws, which are stated in the discussion preceding Problem 21 in Section 3.2.

ODEA

0.01 g/gal of the chemical flows into the pond at a rate of 300 gal/h, and water also flows out of the pond at the same rate. Assume that the chemical is uniformly distributed throughout the pond.

(a) Let $Q(t)$ be the amount of the chemical in the pond at time t. Write down an initial value problem for $Q(t)$.

(b) Solve the problem in part (a) for $Q(t)$. How much chemical is in the pond after 1 year?

(c) At the end of 1 year, the source of the chemical in the pond is removed; thereafter pure water flows into the pond, and the mixture flows out at the same rate as before. Write down the initial value problem that describes this new situation.

(d) Solve the initial value problem in part (c). How much chemical remains in the pond after 1 additional year (2 years from the beginning of the problem)?

(e) How long does it take for $Q(t)$ to be reduced to 10 g?

(f) Plot $Q(t)$ versus t for 3 years.

23. Your swimming pool, containing 60,000 gal of water, has been contaminated by 5 kg of a nontoxic dye that leaves a swimmer's skin an unattractive green. The pool's filtering system can take water from the pool, remove the dye, and return the water to the pool at a flow rate of 200 gal/min.

(a) Write down the initial value problem for the filtering process; let $q(t)$ be the amount of dye (in grams) in the pool at any time t (in minutes).

(b) Solve the problem in part (a).

(c) You have invited several dozen friends to a pool party that is scheduled to begin in 4 h. You have also determined that the effect of the dye is imperceptible if its concentration is less than 0.02 g/gal. Is your filtering system capable of reducing the dye concentration to this level within 4 h?

(d) Find the time T at which the concentration of dye first reaches the value 0.02 g/gal.

(e) Find the flow rate that is sufficient to achieve the concentration 0.02 g/gal within 4 h.

1.3 Numerical Approximations: Euler's Method

We have seen in Section 1.1 that by drawing a direction field we can visualize qualitatively the behavior of solutions of a differential equation. In fact, if we use a fairly fine grid, then we obtain a direction field such as the one in Figure 1.3.1, which corresponds to the differential equation

$$\frac{dy}{dt} + \tfrac{1}{2}y = \tfrac{3}{2} - t. \tag{1}$$

Many tangent line segments at successive values of t almost touch each other in this figure. It takes only a bit of imagination to consider starting at a point on the y-axis and linking line segments for consecutive t-values in the grid, thereby producing a piecewise linear graph. Such a graph would apparently be an approximation to a solution of the differential equation. Of course, this raises some questions, including the following:

1. Can we carry out this linking of tangent lines in a simple and systematic manner?
2. If so, does the resulting piecewise linear function provide an approximation to an actual solution of the differential equation?
3. If so, can we say anything about the accuracy of the approximation?

It turns out that the answer to each question is affirmative. We will take up the first question here, and return to the other two in Section 2.7.

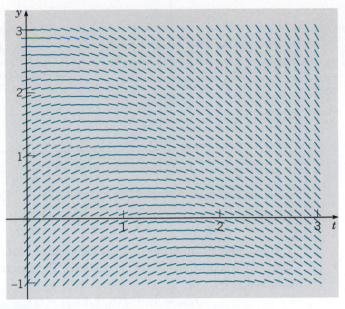

FIGURE 1.3.1 A direction field for $y' + \frac{1}{2} y = \frac{3}{2} - t$.

Suppose that we have a differential equation

$$\frac{dy}{dt} = f(t, y) \tag{2}$$

and a starting point, given by the initial condition

$$y(t_0) = y_0. \tag{3}$$

Suppose also that we have chosen a sequence of points $t_0, t_1, t_2, \ldots, t_n, \ldots$. Let the solution of the initial value problem (2), (3) be denoted by $y = \phi(t)$. Then the line tangent to the graph of $\phi(t)$ at the initial point (t_0, y_0) has the slope $f(t_0, y_0)$, and the equation of this tangent line is

$$y = y_0 + f(t_0, y_0)(t - t_0). \tag{4}$$

We can use the tangent line (4) to approximate the solution $\phi(t)$ in the interval $t_0 \leq t \leq t_1$. In particular, if we evaluate Eq. (4) at $t = t_1$, we obtain the value

$$y_1 = y_0 + f(t_0, y_0)(t_1 - t_0), \tag{5}$$

which is an approximation to the solution value $\phi(t_1)$; see Figure 1.3.2.

To proceed further, we can try to repeat the process. Unfortunately, we do not know the value $\phi(t_1)$ of the solution at t_1. The best we can do is to use the approximate value y_1 instead. Thus we construct the line through (t_1, y_1) with the slope $f(t_1, y_1)$,

$$y = y_1 + f(t_1, y_1)(t - t_1). \tag{6}$$

To approximate the value of $\phi(t)$ at t_2, we use Eq. (6), obtaining

$$y_2 = y_1 + f(t_1, y_1)(t_2 - t_1). \tag{7}$$

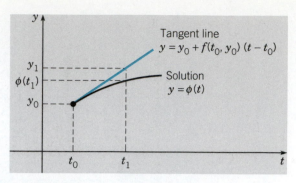

FIGURE 1.3.2   A tangent line approximation.

Continuing in this manner, we use the value of y calculated at each step to determine the slope of the approximation for the next step. The general expression for y_{n+1} in terms of t_n, t_{n+1}, and y_n is

$$y_{n+1} = y_n + f(t_n, y_n)(t_{n+1} - t_n), \qquad n = 0, 1, 2, \ldots. \tag{8}$$

Equation (8) is known as Euler's formula. If there is a uniform step size h between the points $t_0, t_1, t_2, \ldots$, then Euler's method is summarized by the equations

$$t_{n+1} = t_n + h, \qquad y_{n+1} = y_n + hf(t_n, y_n), \qquad n = 0, 1, 2, \ldots. \tag{9}$$

By introducing the notation $f_n = f(t_n, y_n)$, we can rewrite Eqs. (9) as

$$t_{n+1} = t_n + h, \qquad y_{n+1} = y_n + hf_n, \qquad n = 0, 1, 2, \ldots. \tag{10}$$

To use Euler's method you simply evaluate Eqs. (9) repeatedly (or use Eq. (8) if the step size is not constant). The result at each step is used to execute the next step. In this way you generate a sequence of values $y_1, y_2, y_3, \ldots$ that approximate the values of the solution $\phi(t)$ at the points $t_1, t_2, t_3, \ldots$. If, instead of a sequence of points, you need a function to approximate the solution $\phi(t)$, then you can use the piecewise linear function constructed from the collection of tangent line segments. That is, let y be given by Eq. (4) in $[t_0, t_1]$, by Eq. (6) in $[t_1, t_2]$, and in general by

$$y = y_n + f(t_n, y_n)(t - t_n) \tag{11}$$

in $[t_n, t_{n+1}]$.

EXAMPLE 1

Use Euler's method with a step size $h = 0.2$ to approximate the solution of the initial value problem

$$\frac{dy}{dt} + \tfrac{1}{2}y = \tfrac{3}{2} - t, \qquad y(0) = 1 \tag{12}$$

on the interval $0 \le t \le 1$.

For later comparison we note that the exact solution of this initial value problem is

$$y = 7 - 2t - 6e^{-t/2}. \tag{13}$$

This solution is readily found by the method of undetermined coefficients described in Section 1.2. Or, if you wish simply to verify that this solution is correct, you can substitute it into each of Eqs. (12).

To use Euler's method, note that $f(t, y) = -(1/2)y + 3/2 - t$, so using the initial values $t_0 = 0, y_0 = 1$, we have

$$f_0 = f(t_0, y_0) = f(0, 1) = -0.5 + 1.5 - 0 = 1.0.$$

Thus, from Eq. (4), the tangent line approximation for t in $[0, 0.2]$ is

$$y = 1 + 1.0(t - 0) = 1 + t. \tag{14}$$

Setting $t = 0.2$ in Eq. (14), we find the approximate value y_1 of the solution at $t = 0.2$, namely, $y_1 = 1.2$.

At the next step we have

$$f_1 = f(t_1, y_1) = f(0.2, 1.2) = -0.6 + 1.5 - 0.2 = 0.7,$$

and from Eq. (6),

$$y = 1.2 + 0.7(t - 0.2) = 1.06 + 0.7t \tag{15}$$

for t in $[0.2, 0.4]$. Evaluating the expression in Eq. (15) at $t = 0.4$, we obtain $y_2 = 1.34$.

TABLE 1.3.1 Results of Euler's Method with $h = 0.2$ for $y' + (1/2)y = 3/2 - t$, $\quad y(0) = 1$.

t	Exact	Euler with $h = 0.2$	Tangent line
0.0	1.00000	1.00000	$y = 1 + t$
0.2	1.17098	1.20000	$y = 1.06 + 0.7t$
0.4	1.28762	1.34000	$y = 1.168 + 0.43t$
0.6	1.35509	1.42600	$y = 1.3138 + 0.187t$
0.8	1.37808	1.46340	$y = 1.48876 - 0.0317t$
1.0	1.36082	1.45706	

Continuing in this manner, we obtain the results shown in Table 1.3.1. The first column contains the t-values separated by the step size $h = 0.2$. The third column shows the corresponding y-values computed from Euler's formula (10). The fourth column displays the tangent line approximations found from Eq. (11). The second column contains values of the solution (13) of the initial value problem (12), correct to five decimal places. The solution (13) and the tangent line approximation are also plotted in Figure 1.3.3.

As you can see from Table 1.3.1 and Figure 1.3.3, the approximations given by Euler's method for this problem are greater than the corresponding values of the exact solution. This is because the graph of the solution is concave down and therefore the tangent line approximations lie above the graph. It is also clear that the step size $h = 0.2$ is too large to produce a good approximation to the solution (13) on the interval $[0,1]$. Better results can

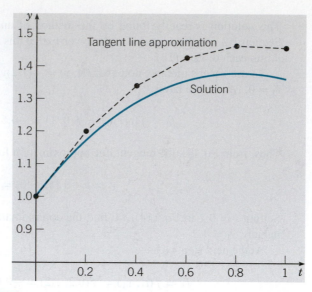

FIGURE 1.3.3 Plots of the solution and a tangent line approximation for the initial value problem (12).

be obtained by using a smaller step size, with a corresponding increase in the number of computational steps, as we will see in Example 2 below.

Euler's method dates from about 1768 and is the oldest numerical method for approximating the solution of a differential equation. It is simple in concept and easy to execute. The simplicity of Euler's method makes it a good way to begin to explore the numerical approximation of solutions of relatively simple differential equations.

The purpose of Example 1 is to show you the details of implementing a few steps of Euler's method so that it will be clear exactly what computations are being executed. Of course, computations such as these are usually done on a computer. Some software packages include code for the Euler method, while others do not. In any case, it is easy to write a computer program to carry out the calculations required to produce results such as those in Table 1.3.1. The outline of such a program is given below; the specific instructions can be written in any high-level programming language.

The Euler Method

Step 1. define $f(t, y)$

Step 2. input initial values $t0$ and $y0$

Step 3. input step size h and number of steps n

Step 4. output $t0$ and $y0$

Step 5. for j from 1 to n **do**

Step 6. $k1 = f(t, y)$
       $y = y + h * k1$
       $t = t + h$

Step 7. output t and y

Step 8. end

The output of this algorithm can be numbers listed on the screen or printed on a printer, as in the third column of Table 1.3.1. Alternatively, the calculated results can be displayed in graphical form, as in Figure 1.3.3.

EXAMPLE 2

Consider again the initial value problem (12),

$$\frac{dy}{dt} + \frac{1}{2}y = \frac{3}{2} - t, \qquad y(0) = 1.$$

Use Euler's method with various step sizes to calculate approximate values of the solution for $0 \leq t \leq 5$. Compare the calculated results with the corresponding values of the exact solution (13),

$$y = \phi(t) = 7 - 2t - 6e^{-t/2}.$$

We used step sizes $h = 0.1, 0.05, 0.025$, and 0.01, corresponding respectively to 50, 100, 200, and 500 steps, to go from $t = 0$ to $t = 5$. Some of the results of these calculations, along with the values of the exact solution, are presented in Table 1.3.2. All computed entries are rounded to five decimal places, although more digits were retained in the intermediate calculations.

TABLE 1.3.2

A Comparison of Exact Solution with Euler's Method for Several Step Sizes h for $y' + (1/2)y = 3/2 - t$, $\quad y(0) = 1$.

t	Exact	$h = 0.1$	$h = 0.05$	$h = 0.025$	$h = 0.01$
0.0	1.00000	1.00000	1.00000	1.00000	1.00000
1.0	1.36082	1.40758	1.38387	1.37227	1.36538
2.0	0.79272	0.84908	0.82061	0.80659	0.79825
3.0	−0.33878	−0.28783	−0.31349	−0.32618	−0.33375
4.0	−1.81201	−1.77107	−1.79163	−1.80184	−1.80795
5.0	−3.49251	−3.46167	−3.47710	−3.48481	−3.48943

In Figure 1.3.4, we have plotted the absolute value of the error (that is, the difference between the exact solution and its approximations) for each value of h and for each value of t as recorded in Table 1.3.2. The lines in this graph do not necessarily represent the error accurately in between the data points, but are included to make the plot more visually understandable.

What conclusions can we draw from the data in Table 1.3.2 and from Figure 1.3.4? In the first place, for a fixed value of t, the computed approximate values become more accurate as the step size h decreases. This is what we would expect, of course, but it is encouraging that the data confirm our expectations. For example, for $t = 1$ the approximate value with $h = 0.1$ is too large by about 3.43%, whereas the value with $h = 0.01$ is too large by only 0.34%. In this case, reducing the step size by a factor of 10 (and performing 10 times as many computations) also reduces the error by a factor of about 10. A second observation is that, for a fixed step size h, the approximations become more accurate as t increases. For instance, for $h = 0.1$ the error for $t = 5$ is only about 0.88%, compared with 3.43% for $t = 1$. For the data we have recorded the maximum error occurs at $t = 2$ in each case. An examination of data at intermediate points not recorded in Table 1.3.2 would reveal where the maximum error occurs for a given step size and how large it is.

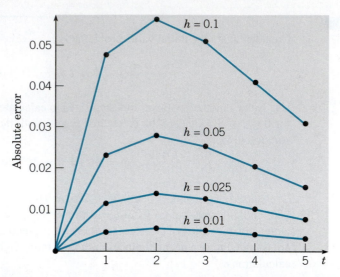

FIGURE 1.3.4 Plot of the absolute value of the error in using Euler's method for the initial value problem (12) for $h = 0.1, 0.05, 0.025$, and 0.01.

A plot of the maximum recorded error (that is, the error at $t = 2$) versus the step size h is shown in Figure 1.3.5. Each data point lies very close to a straight line through the origin, which means that the maximum error is very nearly proportional to h. From Figure 1.3.5 or from the data in Table 1.3.2 you can conclude that the value of the proportionality constant is about 0.56.

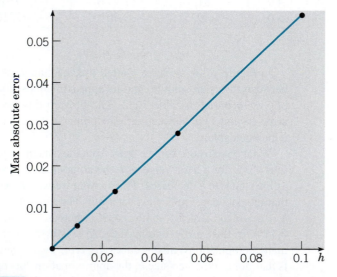

FIGURE 1.3.5 Plot of the maximum error versus stepsize h in using Euler's method for the initial value problem (12).

All in all, Euler's method seems to work rather well for this problem. Reasonably good results are obtained even for a moderately large step size $h = 0.1$, and the approximation can be improved by decreasing h.

Let us now look at another example.

EXAMPLE 3

Consider the initial value problem

$$\frac{dy}{dt} - 2y = 4 - t, \qquad y(0) = 1. \tag{16}$$

Use Euler's method with several step sizes to find approximate values of the solution on the interval $0 \le t \le 5$. Compare the results with the corresponding values of the exact solution.

The solution of the initial value problem (16) is

$$y = -\tfrac{7}{4} + \tfrac{1}{2}t + \tfrac{11}{4}e^{2t}. \tag{17}$$

It can be easily found by using the method of undetermined coefficients, or you can simply verify its validity by substitution into Eqs. (16).

Using the same range of step sizes as in Example 2, we obtain the results presented in Table 1.3.3.

TABLE 1.3.3 A Comparison of Exact Solution with Euler's Method for Several Step Sizes h for $y' - 2y = 4 - t$, $y(0) = 1$.

t	Exact	$h = 0.1$	$h = 0.05$	$h = 0.025$	$h = 0.01$
0.0	1.000000	1.000000	1.000000	1.000000	1.000000
1.0	19.06990	15.77728	17.25062	18.10997	18.67278
2.0	149.3949	104.6784	123.7130	135.5440	143.5835
3.0	1109.179	652.5349	837.0745	959.2580	1045.395
4.0	8197.884	4042.122	5633.351	6755.175	7575.577
5.0	60573.53	25026.95	37897.43	47555.35	54881.32

The data in Table 1.3.3 again confirm our expectation that, for a given value of t, accuracy improves as the step size h is reduced. For example, for $t = 1$ the percentage error diminishes from 17.3% when $h = 0.1$ to 2.1% when $h = 0.01$. However, the error increases fairly rapidly as t increases for a fixed h. Even for $h = 0.01$, the error at $t = 5$ is 9.4%, and it is much greater for larger step sizes. This is shown in Figure 1.3.6, which shows the absolute error versus t for each value of h. The maximum error always occurs at $t = 5$ and is plotted against h in Figure 1.3.7. Again the data points lie approximately on a straight line through the origin, so the maximum error is nearly proportional to the step size, as in Example 2. Now, however, the proportionality constant is greater than 50,000, or about 100,000 times greater than in Example 2.

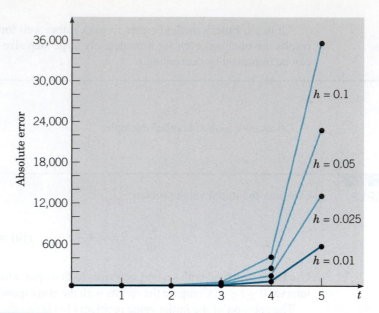

FIGURE 1.3.6 Plot of the absolute value of the error in using Euler's method for the initial value problem (16) for $h = 0.1$, 0.05, 0.025, and 0.01.

Of course, the accuracy that is needed depends on the purpose for which the results are intended, but the errors in Table 1.3.3 are too large for most scientific or engineering applications. To improve the situation, one might either try even smaller step sizes or else restrict the computations to a rather short interval away from the initial point. Nevertheless, it is clear that Euler's method is much less effective in this example than in Example 2.

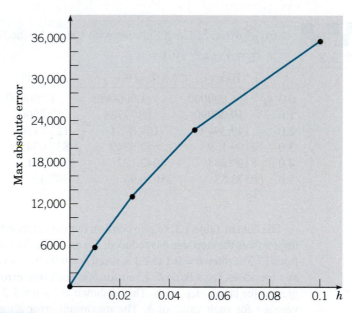

FIGURE 1.3.7 Plot of the maximum error versus stepsize h in using Euler's method for the initial value problem (16).

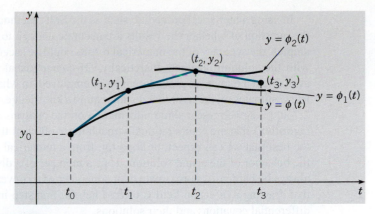

FIGURE 1.3.8   The Euler method.

To understand better what is happening in these examples, let us look again at Euler's method for the general initial value problem

$$\frac{dy}{dt} = f(t, y), \qquad y(t_0) = y_0, \tag{18}$$

whose solution we denote by $\phi(t)$. Recall that a first order differential equation has an infinite family of solutions, indexed by an arbitrary constant c, and that the initial condition picks out one member of this infinite family by determining the value of c. Thus $\phi(t)$ is the member of the infinite family of solutions that satisfies the initial condition $\phi(t_0) = y_0$.

At the first step, Euler's method uses the tangent line approximation to the graph of $y = \phi(t)$ passing through the initial point (t_0, y_0), and this produces the approximate value y_1 at t_1. Usually $y_1 \neq \phi(t_1)$, so at the second step, Euler's method uses the tangent line approximation not to $y = \phi(t)$, but to a nearby solution $y = \phi_1(t)$ that passes through the point (t_1, y_1). So it is at each following step. Euler's method uses a succession of tangent line approximations to a sequence of different solutions $\phi(t), \phi_1(t), \phi_2(t), \ldots$ of the differential equation. At each step the tangent line is constructed to the solution passing through the point determined by the result of the preceding step, as shown in Figure 1.3.8. The quality of the approximation after many steps depends strongly on the behavior of the set of solutions that pass through the points (t_n, y_n) for $n = 1,2,3,\ldots$.

In Examples 1 and 2, the general solution of the differential equation is

$$y = 7 - 2t + ce^{-t/2} \tag{19}$$

and the solution of the initial value problem (12) corresponds to $c = -6$. This family of solutions is a converging family since the term involving the arbitrary constant c approaches zero as $t \to \infty$. It does not matter very much which solutions we are approximating by tangent lines in the implementation of Euler's method, since all the solutions are getting closer and closer to each other as t increases.

On the other hand, in Example 3 the general solution of the differential equation is

$$y = -\tfrac{7}{4} + \tfrac{1}{2}t + ce^{2t}, \tag{20}$$

and this is a diverging family. Note that solutions corresponding to two nearby values of c separate arbitrarily far as t increases. In Example 3, we are trying to follow the solution for $c = 11/4$, but, in the use of Euler's method, we are actually at each step following another solution that separates from the desired one faster and faster as t increases. This explains why the errors in Example 3 are so much larger than those in Example 2.

In using a numerical procedure such as the Euler method, one must always keep in mind the question of whether the results are accurate enough to be useful. In the preceding examples, the accuracy of the numerical results could be ascertained directly by a comparison with the solution obtained analytically. Of course, usually the analytical solution is not available if a numerical procedure is to be employed, so what is needed are bounds for, or at least estimates of, the error that do not require a knowledge of the exact solution. In Sections 2.7 and 2.8, we present some information on the analysis of errors and also discuss other algorithms that are computationally much more efficient than the Euler method. However, the best that we can expect, or hope for, from a numerical approximation is that it reflects the behavior of the actual solution. Thus a member of a diverging family of solutions will always be harder to approximate than a member of a converging family. Finally, remember that drawing a direction field is often a helpful first step in understanding the behavior of differential equations and their solutions.

PROBLEMS

Many of the problems in this section call for fairly extensive numerical computations. The amount of computing that is reasonable for you to do depends strongly on the type of computing equipment that you have. A few steps of the requested calculations can be carried out on almost any pocket calculator—or even by hand if necessary. To do more, you will need a computer or at least a programmable calculator.

Remember also that numerical results may vary somewhat depending on how your program is constructed and on how your computer executes arithmetic steps, rounds off, and so forth. Minor variations in the last decimal place may be due to such causes and do not necessarily indicate that something is amiss. Answers in the back of the book are recorded to six digits in most cases, although more digits were retained in the intermediate calculations.

CAS In each of Problems 1 through 4:

(a) Find approximate values of the solution of the given initial value problem at $t = 0.1, 0.2, 0.3,$ and 0.4 using the Euler method with $h = 0.1$.

(b) Repeat part (a) with $h = 0.05$. Compare the results with those found in (a).

(c) Repeat part (a) with $h = 0.025$. Compare the results with those found in (a) and (b).

(d) Find the solution $y = \phi(t)$ of the given problem and evaluate $\phi(t)$ at $t = 0.1, 0.2, 0.3,$ and 0.4. Compare these values with the results of (a), (b), and (c).

1. $y' = 3 + t - y$, $\quad y(0) = 1$
2. $y' = 2y - 1$, $\quad y(0) = 1$
3. $y' = 0.5 - t + 2y$, $\quad y(0) = 1$
4. $y' = 3 \cos t - 2y$, $\quad y(0) = 0$

In each of Problems 5 through 10, draw a direction field for the given differential equation and state whether you think that the solutions are converging or diverging. ODEA

5. $y' = 5 - 3\sqrt{y}$
6. $y' = y(3 - ty)$
7. $y' = (4 - ty)/(1 + y^2)$
8. $y' = -ty + 0.1y^3$
9. $y' = t^2 + y^2$
10. $y' = (y^2 + 2ty)/(3 + t^2)$

In each of Problems 11 through 14, use Euler's method CAS to find approximate values of the solution of the given initial value problem at $t = 0.5, 1, 1.5, 2, 2.5,$ and 3:
(a) With $h = 0.1$.
(b) With $h = 0.05$.
(c) With $h = 0.025$.
(d) With $h = 0.01$.

11. $y' = 5 - 3\sqrt{y}$, $\qquad y(0) = 2$
12. $y' = y(3 - ty)$, $\qquad y(0) = 0.5$
13. $y' = (4 - ty)/(1 + y^2)$, $\qquad y(0) = -2$
14. $y' = -ty + 0.1y^3$, $\qquad y(0) = 1$

15. Consider the initial value problem CAS

$$y' = 3t^2/(3y^2 - 4), \qquad y(1) = 0.$$

(a) Use Euler's method with $h = 0.1$ to obtain approximate values of the solution at $t = 1.2, 1.4, 1.6,$ and 1.8.

(b) Repeat part (a) with $h = 0.05$.

(c) Compare the results of parts (a) and (b). Note that they are reasonably close for $t = 1.2, 1.4,$ and 1.6 but are quite different for $t = 1.8$. Also note (from the differential equation) that the line tangent to the solution is parallel to the y-axis when $y = \pm 2/\sqrt{3} \cong \pm 1.155$. Explain how this might cause such a difference in the calculated values.

CAS **16.** Consider the initial value problem

$$y' = t^2 + y^2, \qquad y(0) = 1.$$

Use Euler's method with $h = 0.1, 0.05, 0.025,$ and 0.01 to explore the solution of this problem for $0 \le t \le 1$. What is your best estimate of the value of the solution at $t = 0.8$? At $t = 1$? Are your results consistent with the direction field in Problem 9?

CAS **17.** Consider the initial value problem

$$y' = (y^2 + 2ty)/(3 + t^2), \qquad y(1) = 2.$$

Use Euler's method with $h = 0.1, 0.05, 0.025,$ and 0.01 to explore the solution of this problem for $1 \le t \le 3$. What is your best estimate of the value of the solution at $t = 2.5$? At $t = 3$? Are your results consistent with the direction field in Problem 10?

18. Consider the initial value problem   CAS

$$y' = -ty + 0.1y^3, \qquad y(0) = \alpha,$$

where α is a given number.

(a) Draw a direction field for the differential equation (or reexamine the one from Problem 8). Observe that there is a critical value of α in the interval $2 \le \alpha \le 3$ that separates converging solutions from diverging ones. Call this critical value α_0.

(b) Use Euler's method with $h = 0.01$ to estimate α_0. Do this by restricting α_0 to an interval $[a, b]$, where $b - a = 0.01$.

19. Consider the initial value problem   CAS

$$y' = y^2 - t^2, \qquad y(0) = \alpha,$$

where α is a given number.

(a) Draw a direction field for the differential equation. Observe that there is a critical value of α in the interval $0 \le \alpha \le 1$ that separates converging solutions from diverging ones. Call this critical value α_0.

(b) Use Euler's method with $h = 0.01$ to estimate α_0. Do this by restricting α_0 to an interval $[a, b]$, where $b - a = 0.01$.

1.4 Classification of Differential Equations

The main purpose of this book is to discuss some of the properties of solutions of differential equations, and to present some of the methods that have proved effective in finding solutions or, in many cases, approximating them. A **differential equation** is simply an equation containing one or more derivatives of the unknown function or functions. Since differential equations can have many different forms, in this section we describe several useful ways of classifying them. This will help to provide an organizational framework for our presentation.

▶ **Ordinary and Partial Differential Equations.** One of the more obvious classifications is based on whether the unknown function depends on a single independent variable or on several independent variables. In the first case, only ordinary derivatives appear in the differential equation, and it is said to be an **ordinary differential equation**. In the second case, the derivatives are partial derivatives, and the equation is called a **partial differential equation**. In this book, we consider only ordinary differential equations, which are more basic and somewhat simpler than partial differential equations.

▶ **Systems of Differential Equations.** Another classification of differential equations depends on the number of unknown functions that are involved. If there is a single function to be determined, then one equation is sufficient. However, if there are two or more unknown functions, then a system of equations is required. For example, the Lotka–Volterra, or predator–prey, equations are important in ecological modeling. They have the form

$$dx/dt = ax - \alpha xy$$
$$dy/dt = -cy + \gamma xy, \tag{1}$$

where $x(t)$ and $y(t)$ are the respective populations of the prey and predator species. The constants $a, \alpha, c,$ and γ are based on empirical observations and depend on the particular species being studied. Systems of equations are discussed in Chapters 3, 6, and 7; in particular, the Lotka–Volterra equations are examined in Section 7.4. In some areas of application, it is not unusual to encounter very large systems containing hundreds, or even many thousands, of equations.

▶ **Order.**

The **order** of a differential equation is the order of the highest derivative that appears in the equation. The equations in the preceding sections are all first order equations. The equation

$$ay'' + by' + cy = f(t), \tag{2}$$

where a, b, and c are given constants, and f is a given function, is a second order equation. Equation (2) is a useful model of several physical systems, and we will consider it in detail in Chapter 4. More generally, the equation

$$F[t, u(t), u'(t), \ldots, u^{(n)}(t)] = 0 \tag{3}$$

is an ordinary differential equation of the nth order. Equation (3) expresses a relation between the independent variable t and the values of the function u and its first n derivatives $u', u'',$ $\ldots, u^{(n)}$. It is convenient and customary in differential equations to write y for $u(t)$, with y', $y'', \ldots, y^{(n)}$ standing for $u'(t), u''(t), \ldots, u^{(n)}(t)$. Thus Eq. (3) is written as

$$F(t, y, y', \ldots, y^{(n)}) = 0. \tag{4}$$

For example,

$$y''' + 2e^t y'' + yy' = t^4 \tag{5}$$

is a third order differential equation for $y = u(t)$. Occasionally, other letters will be used instead of t and y for the independent and dependent variables; the meaning should be clear from the context.

We assume that it is always possible to solve a given ordinary differential equation for the highest derivative, obtaining

$$y^{(n)} = f(t, y, y', y'', \ldots, y^{(n-1)}). \tag{6}$$

We study only equations of the form (6), although in the process of solving them we often find it convenient to rewrite them in other forms.

▶ **Linear and Nonlinear Equations.** A crucial classification of differential equations is whether they are linear or nonlinear. The ordinary differential equation

$$F(t, y, y', \ldots, y^{(n)}) = 0$$

is said to be **linear** if F is a linear function of the variables $y, y', \ldots, y^{(n)}$. Thus the general linear ordinary differential equation of order n is

$$a_0(t)y^{(n)} + a_1(t)y^{(n-1)} + \cdots + a_n(t)y = g(t). \tag{7}$$

Most of the equations you have seen thus far in this book are linear; examples are the equations in Sections 1.1 and 1.2 describing the falling hailstone, the field mouse population,

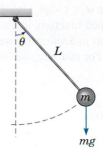

FIGURE 1.4.1   An oscillating pendulum.

and the heating and cooling of a building. Similarly, in this section, Eq. (2) is a linear differential equation. An equation that is not of the form (7) is a **nonlinear** equation. Each equation in the system (1) is nonlinear because of the terms that involve the product xy, and Eq. (5) is also nonlinear because of the term yy'.

A simple physical problem that leads to a nonlinear differential equation is the oscillating pendulum. The angle θ that an oscillating pendulum of length L makes with the vertical direction (see Figure 1.4.1) satisfies the equation

$$\frac{d^2\theta}{dt^2} + \frac{g}{L}\sin\theta = 0, \tag{8}$$

whose derivation is outlined in Problems 21 through 23. The presence of the term involving $\sin\theta$ makes Eq. (8) nonlinear.

The mathematical theory and methods for solving linear equations are highly developed. In contrast, for nonlinear equations the theory is more complicated, and analytical methods of solution are less satisfactory. In view of this, it is fortunate that many significant problems lead to linear ordinary differential equations or can be approximated by linear equations. For example, for the pendulum, if the angle θ is small, then $\sin\theta \cong \theta$ and Eq. (8) can be approximated by the linear equation

$$\frac{d^2\theta}{dt^2} + \frac{g}{L}\theta = 0. \tag{9}$$

This process of approximating a nonlinear equation by a linear one is called **linearization**; it is an extremely valuable way to deal with nonlinear equations. Nevertheless, there are many physical phenomena that simply cannot be represented adequately by linear equations. To study these phenomena, it is essential to deal with nonlinear equations. Consequently, numerical approximation methods are of much greater importance for nonlinear equations than for linear ones.

In an elementary text, it is natural to emphasize the simpler and more straightforward parts of the subject. Therefore the greater part of this book is devoted to linear equations and various methods for solving them. However, Chapter 7, as well as parts of Chapters 2 and 3, are concerned with nonlinear equations. Whenever it is appropriate, we point out why nonlinear equations are, in general, more difficult and why many of the techniques that are useful in solving linear equations cannot be applied to nonlinear equations.

▶ Solutions.   A **solution** of the ordinary differential equation (6) on the interval $\alpha < t < \beta$ is a function ϕ such that $\phi', \phi'', \ldots, \phi^{(n)}$ exist and satisfy

$$\phi^{(n)}(t) = f[t, \phi(t), \phi'(t), \ldots, \phi^{(n-1)}(t)] \tag{10}$$

for every t in $\alpha < t < \beta$. Unless stated otherwise, we assume that the function f of Eq. (6) is a real-valued function, and we are interested in obtaining real-valued solutions $y = \phi(t)$.

Recall that in Section 1.2 we found solutions of certain equations by a process of direct integration. For instance, we found that the equation

$$\frac{dp}{dt} = 0.5p - 450 \tag{11}$$

has the solution

$$p = 900 + ce^{t/2}, \tag{12}$$

where c is an arbitrary constant. It is often not so easy to find solutions of differential equations. However, if you find a function that you think may be a solution of a given equation, it is usually relatively easy to determine whether the function is actually a solution simply by substituting the function into the equation. For example, in this way it is easy to show that the function $y_1(t) = \cos t$ is a solution of

$$y'' + y = 0 \tag{13}$$

for all t. To confirm this, observe that $y_1'(t) = -\sin t$ and $y_1''(t) = -\cos t$; then it follows that $y_1''(t) + y_1(t) = 0$. In the same way, you can easily show that $y_2(t) = \sin t$ is also a solution of Eq. (13). Of course, this does not constitute a satisfactory way to solve most differential equations, because there are far too many possible functions for you to have a good chance of finding the correct one by a random choice. Nevertheless, you should realize that you can verify whether any proposed solution is correct by substituting it into the differential equation. For any problem that is important to you, this can be a very useful check. It is one that you should make a habit of considering.

▶ Some Important Questions. Although for Eqs. (11) and (13) we are able to verify that certain simple functions are solutions, in general we do not have such solutions readily available. Thus a fundamental question is the following: Does an equation of the form (6) always have a solution? The answer is "No." Merely writing down an equation of the form (6) does not necessarily mean that there is a function $y = \phi(t)$ that satisfies it. So, how can we tell whether some particular equation has a solution? This is the question of *existence* of a solution, and it is answered by theorems stating that under certain restrictions on the function f in Eq. (6), the equation always has solutions. However, this is not a purely mathematical concern, for at least two reasons. If a problem has no solution, we would prefer to know that fact before investing time and effort in a vain attempt to solve the problem. Further, if a sensible physical problem is modeled mathematically as a differential equation, then the equation should have a solution. If it does not, then presumably there is something wrong with the formulation. In this sense, an engineer or scientist has some check on the validity of the mathematical model.

If we assume that a given differential equation has at least one solution, the question arises as to how many solutions it has, and what additional conditions must be specified to single out a particular solution. This is the question of *uniqueness*. In general, solutions of differential equations contain one or more arbitrary constants of integration, as does the solution (12) of Eq. (11). Equation (12) represents an infinity of functions corresponding to the infinity of possible choices of the constant c. As we saw in Section 1.2, if p is specified at some time t, this condition will determine a value for c; even so, we have not yet ruled out the possibility that there may be other solutions of Eq. (11) that also have the prescribed value of p at the prescribed time t. The issue of uniqueness also has practical implications. If we are fortunate enough to find a solution of a given problem, and if we know that the problem has a unique solution, then we can be sure that we have completely solved the problem. If there may be other solutions, then perhaps we should continue to search for them.

A third important question is: Given a differential equation of the form (6), can we actually determine a solution, and if so, how? Note that if we find a solution of the given equation, we have at the same time answered the question of the existence of a solution. However, without knowledge of existence theory we might, for example, use a computer to find a numerical approximation to a "solution" that does not exist. On the other hand, even though we may know that a solution exists, it may be that the solution is not expressible in terms of the usual elementary functions—polynomial, trigonometric, exponential, logarithmic, and hyperbolic functions. Unfortunately, this is the situation for most differential equations. Thus, we discuss both elementary methods that can be used to obtain exact solutions of certain relatively simple problems, and also methods of a more general nature that can be used to find approximations to solutions of more difficult problems.

▶ **Computer Use in Differential Equations.** A computer can be an extremely valuable tool in the study of differential equations. For many years, computers have been used to execute numerical algorithms, such as those described in Sections 1.3, 2.8, and 3.7, to construct numerical approximations to solutions of differential equations. These algorithms have been refined to an extremely high level of generality and efficiency. A few lines of computer code, written in a high-level programming language and executed (often within a few seconds) on a relatively inexpensive computer, suffice to approximate the solutions of a wide range of differential equations to a high degree of accuracy. More sophisticated routines are also readily available. These routines combine the ability to handle very large and complicated systems with numerous diagnostic features that alert the user to possible problems as they are encountered.

The usual output from a numerical algorithm is a table of numbers, listing selected values of the independent variable and the corresponding values of the dependent variable. It is also easy to display the solution of a differential equation graphically, whether the solution has been approximated numerically or obtained as the result of an analytical procedure of some kind. Such a graphical display is often much more illuminating and helpful in understanding and interpreting the solution of a differential equation than a table of numbers or a complicated analytical formula. There are several well-crafted and relatively inexpensive special-purpose software packages for the graphical investigation of differential equations on the market. The widespread availability of personal computers has brought powerful computational and graphical capability within the reach of individual students. You should consider, in the light of your own circumstances, how best to take advantage of the available computing resources. You will surely find it enlightening to do so.

Another aspect of computer use that is very relevant to the study of differential equations is the availability of extremely powerful and general software packages that can perform a wide variety of mathematical operations. Among these are *Maple*®, *Mathematica*®, and MATLAB®, each of which can be used on various kinds of personal computers or workstations. All three of these packages can execute extensive numerical computations and have versatile graphical facilities. *Maple* and *Mathematica* also have very powerful symbolic capabilities. For example, they can perform the analytical steps involved in solving many differential equations, often in response to a single command. Anyone who expects to deal with differential equations in more than a superficial way should become familiar with at least one of these products and explore the ways in which it can be used.

For you, the student, these computing resources have an effect on how you should study differential equations. To become confident in using differential equations, it is essential to understand how the solution methods work, and this understanding is achieved, in part, by working out a sufficient number of examples in detail. However, eventually you should plan to delegate as many as possible of the routine (often repetitive) details to a computer, while you focus on the proper formulation of the problem and on the interpretation of the solution. Our viewpoint is that you should always try to use the best methods and tools available for each task. In particular, you should strive to combine numerical, graphical, and analytical methods so as to attain maximum understanding of the behavior of the solution and of the underlying process that the problem models. You should also remember that some tasks can best be done with pencil and paper, while others require a calculator or computer. Good judgment is often needed in selecting a judicious combination.

PROBLEMS

In each of Problems 1 through 6, determine the order of the given differential equation; also state whether the equation is linear or nonlinear.

1. $t^2 \dfrac{d^2 y}{dt^2} + t \dfrac{dy}{dt} + 2y = \sin t$

2. $(1 + y^2) \dfrac{d^2 y}{dt^2} + t \dfrac{dy}{dt} + y = e^t$

3. $\dfrac{d^4 y}{dt^4} + \dfrac{d^3 y}{dt^3} + \dfrac{d^2 y}{dt^2} + \dfrac{dy}{dt} + y = 1$

4. $\dfrac{dy}{dt} + ty^2 = 0$

5. $\dfrac{d^2 y}{dt^2} + \sin(t + y) = \sin t$

6. $\dfrac{d^3 y}{dt^3} + t \dfrac{dy}{dt} + (\cos^2 t)y = t^3$

In each of Problems 7 through 14, verify that each given function is a solution of the differential equation.

7. $y'' - y = 0;$ $y_1(t) = e^t,$ $y_2(t) = \cosh t$

8. $y'' + 2y' - 3y = 0;$ $y_1(t) = e^{-3t},$ $y_2(t) = e^t$

9. $ty' - y = t^2;$ $y = 3t + t^2$

10. $y'''' + 4y''' + 3y = t;$ $y_1(t) = t/3,$
    $y_2(t) = e^{-t} + t/3$

11. $2t^2 y'' + 3ty' - y = 0,$ $t > 0;$ $y_1(t) = t^{1/2},$
    $y_2(t) = t^{-1}$

12. $t^2 y'' + 5ty' + 4y = 0,$ $t > 0;$ $y_1(t) = t^{-2},$
    $y_2(t) = t^{-2} \ln t$

13. $y'' + y = \sec t,$ $0 < t < \pi/2;$
    $y = (\cos t) \ln \cos t + t \sin t$

14. $y' - 2ty = 1;$ $y = e^{t^2} \displaystyle\int_0^t e^{-s^2}\, ds + e^{t^2}$

In each of Problems 15 through 18, determine the values of r for which the given differential equation has solutions of the form $y = e^{rt}$.

15. $y' + 2y = 0$

16. $y'' - y = 0$

17. $y'' + y' - 6y = 0$

18. $y''' - 3y'' + 2y' = 0$

In each of Problems 19 and 20, determine the values of r for which the given differential equation has solutions of the form $y = t^r$ for $t > 0$.

19. $t^2 y'' + 4ty' + 2y = 0$

20. $t^2 y'' - 4ty' + 4y = 0$

21. Follow the steps indicated here to derive the equation of motion of a pendulum, Eq. (8) in the text. Assume that the rod is rigid and weightless, that the mass is a point mass, and that there is no friction or drag anywhere in the system.

    (a) Assume that the mass is in an arbitrary displaced position, indicated by the angle θ. Draw a free-body diagram showing the forces acting on the mass.

    (b) Apply Newton's law of motion in the direction tangential to the circular arc on which the mass moves. Then the tensile force in the rod does not enter the equation. Observe that you need to find the component of the gravitational force in the tangential direction. Observe also that the linear acceleration, as opposed to the angular acceleration, is $L d^2\theta/dt^2$, where L is the length of the rod.

    (c) Simplify the result from part (b) to obtain Eq. (8) in the text.

22. Another way to derive the pendulum equation (8) is based on the principle of conservation of energy.

    (a) Show that the kinetic energy T of the pendulum in motion is

$$T = \frac{1}{2} m L^2 \left(\frac{d\theta}{dt} \right)^2 .$$

(b) Show that the potential energy v of the pendulum, relative to its rest position, is

$$V = mgL(1 - \cos\theta).$$

(c) By the principle of conservation of energy, the total energy $E = T + V$ is constant. Calculate dE/dt, set it equal to zero, and show that the resulting equation reduces to Eq. (8).

23. A third derivation of the pendulum equation depends on the principle of angular momentum: the rate of change of angular momentum about any point is equal to the net external moment about the same point.

(a) Show that the angular momentum M, or moment of momentum, about the point of support is given by $M = mL^2 d\theta/dt$.

(b) Set dM/dt equal to the moment of the gravitational force, and show that the resulting equation reduces to Eq. (8). Note that positive moments are counterclockwise.

CHAPTER SUMMARY

Differential equations are used to model systems that change continuously in time. The mathematical investigation of a differential equation usually involves one or more of three general approaches: **geometrical, analytical,** and **numerical**.

Section 1.1   Geometrical   The right hand side of a differential equation $dy/dt = f(t, y)$ specifies the slope of a solution passing through the point y at time t. This information can be displayed geometrically by drawing a line segment with slope $f(t, y)$ at the point (t, y). A **direction field** consists of a large number of such line segments, usually drawn on a rectangular grid. A direction field provides a geometric representation of the *flow of solutions* of the differential equation.

Section 1.2   Analytical   Using elementary integration techniques, **general solutions** and solutions to **initial value problems** are found for the simple mathematical models introduced in Section 1.1.

Section 1.3   Numerical   For the initial value problem $y' = f(t, y)$, $y(t_0) = y_0$, the Euler method is the numerical approximation algorithm

$$t_{n+1} = t_n + h, \qquad y_{n+1} = y_n + hf(t_n, y_n), \qquad n = 0, 1, 2, \ldots$$

that geometrically consists of a finite number of connected line segments, each of whose slopes is determined by the slope at the initial point of each segment.

Section 1.4   Classification   It is necessary to classify differential equations according to various criteria for a sensible treatment of theory, solution methods, and solution behavior:

▶   ordinary differential equations and partial differential equations

▶   single (or scalar) equations and systems of equations

▶   linear equations and nonlinear equations

▶   first order equations and higher order equations

CHAPTER TWO

First Order Differential Equations

This chapter deals with differential equations of first order,

$$\frac{dy}{dt} = f(t, y), \tag{1}$$

where f is a given function of two variables. Any differentiable function $y = \phi(t)$ that satisfies this equation for all t in some interval is called a solution. Our object is to develop methods for finding solutions or, if that is not possible, approximating them. Unfortunately, for an arbitrary function f, there is no general method for solving Eq. (1) in terms of elementary functions. Instead, we will describe several methods, each of which is applicable to a certain subclass of first order equations. The most important of these are linear equations (Section 2.1), separable equations (Section 2.2), and exact equations (Section 2.6). In Section 2.5, we discuss another subclass of first order equations, autonomous equations, for which geometrical methods yield valuable information about solutions. Finally, in Sections 2.7 and 2.8, we revisit the question of constructing numerical approximations to solutions, and introduce algorithms more efficient than the Euler method of Section 1.3. Along the way, especially in Sections 2.3 and 2.5, we point out some of the many areas of application in which first order differential equations provide useful mathematical models.

2.1 Linear Equations; Method of Integrating Factors

If the function f in Eq. (1) depends linearly on the dependent variable y, then Eq. (1) is called a first order linear equation. We will usually write the general **first order linear equation** in the form

$$\frac{dy}{dt} + p(t)y = g(t), \tag{2}$$

where p and g are given functions of the independent variable t.

Recall that in Section 1.2 we were able to solve equations of the form (2) by a direct integration process when the coefficients p and g are constants. Further, if p is a constant and g is a sufficiently simple function (that is, a polynomial, exponential, sine, or cosine), then we were able to find solutions by using the method of undetermined coefficients. However, these methods fail if p is not a constant, or if g is a more complicated function. In this section, we will describe a method that works even in these more difficult cases. We owe this method to Leibniz; it involves multiplying the differential equation (2) by a certain function $\mu(t)$, chosen so that the resulting equation is readily integrable. The function $\mu(t)$ is called an **integrating factor**, and the main difficulty is to determine how to find it. We will introduce this method in a simple example, later showing how to extend it to other first order linear equations, including the general equation (2).

EXAMPLE 1

Solve the differential equation

$$\frac{dy}{dt} - 2y = 4 - t. \tag{3}$$

Plot the graphs of several solutions, and find the particular solution whose graph contains the point $(0, -2)$. Discuss the behavior of solutions as $t \to \infty$.

The first step is to multiply Eq. (3) by a function $\mu(t)$, as yet undetermined; thus

$$\mu(t)\frac{dy}{dt} - 2\mu(t)y = \mu(t)(4 - t). \tag{4}$$

The question now is whether we can choose $\mu(t)$ so that the left side of Eq. (4) is recognizable as the derivative of some particular expression. If so, then we can integrate Eq. (4), even though we do not know the function y. To guide our choice of the integrating factor $\mu(t)$, observe that the left side of Eq. (4) contains two terms and that the first term is part of the result of differentiating the product $\mu(t)y$. Thus, let us try to determine $\mu(t)$ so that the left side of Eq. (4) becomes the derivative of the expression $\mu(t)y$. If we compare the left side of Eq. (4) with the differentiation formula

$$\frac{d}{dt}\big[\mu(t)y\big] = \mu(t)\frac{dy}{dt} + \frac{d\mu(t)}{dt}y, \tag{5}$$

we note that the first terms are identical for any $\mu(t)$, and that the second terms also agree, provided that we choose $\mu(t)$ to satisfy

$$\frac{d\mu(t)}{dt} = -2\mu(t). \tag{6}$$

Therefore, our search for an integrating factor will be successful if we can find a solution of Eq. (6). Perhaps you can readily identify a function that satisfies Eq. (6). What well-known function from calculus has a derivative that is equal to -2 times the original function? More systematically, rewrite Eq. (6) as

$$\frac{d\mu(t)/dt}{\mu(t)} = -2, \tag{7}$$

which is equivalent to

$$\frac{d}{dt}\ln|\mu(t)| = -2. \tag{8}$$

Then it follows that

$$\ln|\mu(t)| = -2t + C, \tag{9}$$

or

$$\mu(t) = ce^{-2t}. \tag{10}$$

The function $\mu(t)$ given by Eq. (10) is an integrating factor for Eq. (3). Since we do not need the most general integrating factor, we will choose c to be one in Eq. (10) and use $\mu(t) = e^{-2t}$.

Now we return to Eq. (3), multiply it by the integrating factor e^{-2t}, and obtain

$$e^{-2t}\frac{dy}{dt} - 2e^{-2t}y = 4e^{-2t} - te^{-2t}. \tag{11}$$

By the choice we have made of the integrating factor, the left side of Eq. (11) is the derivative of $e^{-2t}y$, so that Eq. (11) becomes

$$\frac{d}{dt}(e^{-2t}y) = 4e^{-2t} - te^{-2t}. \tag{12}$$

By integrating both sides of Eq. (12) we obtain

$$e^{-2t}y = 4\int e^{-2t}\,dt - \int te^{-2t}\,dt + c, \tag{13}$$

or, using integration by parts on the second integral,

$$e^{-2t}y = -2e^{-2t} - \left[-\tfrac{1}{2}te^{-2t} - \tfrac{1}{4}e^{-2t}\right] + c, \tag{14}$$

where c is an arbitrary constant. Finally, on solving Eq. (14) for y, we have the general solution of Eq. (3), namely,

$$y = -\tfrac{7}{4} + \tfrac{1}{2}t + ce^{2t}. \tag{15}$$

To find the solution passing through the point $(0, -2)$, we set $t = 0$ and $y = -2$ in Eq. (15), obtaining $-2 = -(7/4) + c$. Thus $c = -1/4$, and the desired solution is

$$y = -\tfrac{7}{4} + \tfrac{1}{2}t - \tfrac{1}{4}e^{2t}. \tag{16}$$

Figure 2.1.1 includes the graphs of Eq. (15) for several values of c with a direction field in the background. The solution passing through $(0, -2)$ is shown by the heavy curve. The behavior of the family of solutions (15) for large values of t is determined by the term ce^{2t}. If $c \neq 0$, then the solution grows exponentially large in magnitude, with the same sign as c itself. Thus the solutions diverge as t becomes large. The boundary between solutions that ultimately grow positively from those that ultimately grow negatively occurs when $c = 0$. If we substitute $c = 0$ into Eq. (15) and then set $t = 0$, we find that $y = -7/4$ is the separation point on the y-axis. Note that, for this initial value, the solution is $y = -(7/4) + (1/2)t$; it grows positively, but linearly rather than exponentially.

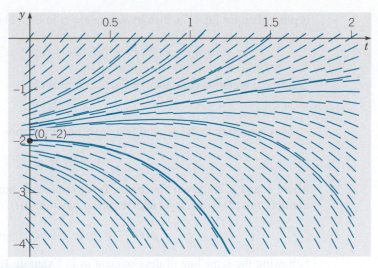

FIGURE 2.1.1 Integral curves of $y' - 2y = 4 - t$.

Proceeding exactly as in Example 1, we can apply the method of integrating factors to equations of the form

$$\frac{dy}{dt} + ay = g(t), \tag{17}$$

where a is any given constant, and $g(t)$ is a given function. We find that the integrating factor $\mu(t)$ must satisfy

$$\frac{d\mu}{dt} = a\mu, \tag{18}$$

rather than Eq. (6). Thus the integrating factor is $\mu(t) = e^{at}$. Multiplying Eq. (17) by $\mu(t)$, we obtain

$$e^{at}\frac{dy}{dt} + ae^{at}y = e^{at}g(t),$$

or

$$\frac{d}{dt}(e^{at}y) = e^{at}g(t). \tag{19}$$

By integrating both sides of Eq. (19) we find that

$$e^{at}y = \int e^{at}g(t)\,dt + c, \tag{20}$$

where c is an arbitrary constant. For many simple functions $g(t)$, we can evaluate the integral in Eq. (20) and express the solution y in terms of elementary functions, as in Example 1. However, for more complicated functions $g(t)$, it may be necessary to leave the solution in integral form. In this case,

$$y = e^{-at}\int_{t_0}^{t} e^{as}g(s)\,ds + ce^{-at}. \tag{21}$$

Note that in Eq. (21) we have used s to denote the integration variable to distinguish it from the independent variable t, and we have chosen some convenient value t_0 as the lower limit of integration. If the initial condition

$$y(t_0) = y_0$$

is prescribed for Eq. (17), then by setting $t = t_0$ in Eq. (21) we obtain

$$y_0 = e^{-at_0} \int_{t_0}^{t_0} e^{as} g(s)\, ds + ce^{-at_0} = 0 + ce^{-at_0}.$$

Thus the integration constant c in Eq. (21) is given by

$$c = e^{at_0} y_0.$$

Now we return to the general first order linear equation (2),

$$\frac{dy}{dt} + p(t)y = g(t),$$

where p and g are given functions. To determine an appropriate integrating factor, we multiply Eq. (2) by an as yet undetermined function $\mu(t)$, obtaining

$$\mu(t)\frac{dy}{dt} + p(t)\mu(t)y = \mu(t)g(t). \tag{22}$$

Following the same line of development as in Example 1, we see that the left side of Eq. (22) is the derivative of the product $\mu(t)y$, provided that $\mu(t)$ satisfies the equation

$$\frac{d\mu(t)}{dt} = p(t)\mu(t). \tag{23}$$

If we assume temporarily that $\mu(t)$ is positive, then we have

$$\frac{d\mu(t)/dt}{\mu(t)} = p(t),$$

and consequently

$$\ln \mu(t) = \int p(t)\, dt + k.$$

By choosing the arbitrary constant k to be zero, we obtain the simplest possible function for μ, namely,

$$\mu(t) = \exp \int p(t)\, dt. \tag{24}$$

Note that $\mu(t)$ is positive for all t, as we assumed. Returning to Eq. (22), we have

$$\frac{d}{dt}\left[\mu(t)y\right] = \mu(t)g(t). \tag{25}$$

Hence

$$\mu(t)y = \int \mu(t)g(t)\, dt + c, \tag{26}$$

where c is an arbitrary constant. Sometimes the integral in Eq. (26) can be evaluated in terms of elementary functions. However, in general, this is not possible, so the general solution of Eq. (2) is

$$y = \frac{1}{\mu(t)}\left[\int_{t_0}^{t} \mu(s)g(s)\, ds + c\right], \tag{27}$$

where again t_0 is some convenient lower limit of integration. Observe that Eq. (27) involves two integrations, one to obtain $\mu(t)$ from Eq. (24) and the other to determine y from Eq. (27).

EXAMPLE 2

Solve the initial value problem

$$ty' + 2y = 4t^2, \tag{28}$$
$$y(1) = 2. \tag{29}$$

In order to determine $p(t)$ and $g(t)$ correctly, we must first rewrite Eq. (28) in the standard form (2). Thus we have

$$y' + (2/t)y = 4t, \tag{30}$$

so $p(t) = 2/t$ and $g(t) = 4t$. To solve Eq. (30) we first compute the integrating factor $\mu(t)$:

$$\mu(t) = \exp \int \frac{2}{t}\, dt = e^{2 \ln |t|} = t^2.$$

On multiplying Eq. (30) by $\mu(t) = t^2$, we obtain

$$t^2 y' + 2ty = (t^2 y)' = 4t^3,$$

and therefore

$$t^2 y = t^4 + c,$$

where c is an arbitrary constant. It follows that

$$y = t^2 + \frac{c}{t^2} \tag{31}$$

is the general solution of Eq. (28). Integral curves of Eq. (28) for several values of c are shown in Figure 2.1.2. To satisfy the initial condition (29), it is necessary to choose $c = 1$; thus

$$y = t^2 + \frac{1}{t^2}, \qquad t > 0 \tag{32}$$

is the solution of the initial value problem (28), (29). This solution is shown by the heavy curve in Figure 2.1.2. Note that it becomes unbounded and is asymptotic to the positive y-axis as $t \to 0$ from the right. This is the effect of the infinite discontinuity in the coefficient $p(t)$ at the origin. The function $y = t^2 + (1/t^2)$ for $t < 0$ is not part of the solution of this initial value problem.

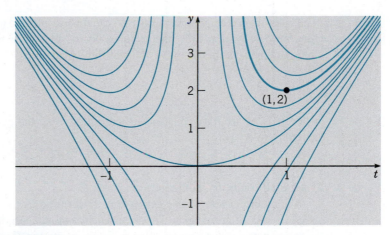

FIGURE 2.1.2 Integral curves of $ty' + 2y = 4t^2$.

This is the first example in which the solution fails to exist for some values of t. Again, this is due to the infinite discontinuity in $p(t)$ at $t = 0$, which restricts the solution to the interval $0 < t < \infty$.

Looking again at Figure 2.1.2, we see that some solutions (those for which $c > 0$) are asymptotic to the positive y-axis as $t \to 0$ from the right, while other solutions (for which

$c < 0$) are asymptotic to the negative y-axis. The solution for which $c = 0$, namely, $y = t^2$, remains bounded and differentiable even at $t = 0$. If we generalize the initial condition (29) to

$$y(1) = y_0, \tag{33}$$

then $c = y_0 - 1$ and the solution (32) becomes

$$y = t^2 + \frac{y_0 - 1}{t^2}, \qquad t > 0. \tag{34}$$

As in Example 1, this is another instance where there is a critical initial value, namely, $y_0 = 1$, that separates solutions that behave in two quite different ways.

EXAMPLE 3

Solve the initial value problem

$$2y' + ty = 2, \tag{35}$$

$$y(0) = 1. \tag{36}$$

First divide the differential equation (35) by two, obtaining

$$y' + (t/2)y = 1. \tag{37}$$

Thus $p(t) = t/2$, and the integrating factor is $\mu(t) = \exp(t^2/4)$. Then multiply Eq. (37) by $\mu(t)$, so that

$$e^{t^2/4}y' + \frac{t}{2}e^{t^2/4}y = e^{t^2/4}. \tag{38}$$

The left side of Eq. (38) is the derivative of $e^{t^2/4}y$, so by integrating both sides of Eq. (38) we obtain

$$e^{t^2/4}y = \int e^{t^2/4}\, dt + c. \tag{39}$$

The integral on the right side of Eq. (39) cannot be evaluated in terms of the usual elementary functions, so we leave the integral unevaluated. However, by choosing the lower limit of integration as the initial point $t = 0$, we can replace Eq. (39) by

$$e^{t^2/4}y = \int_0^t e^{s^2/4}\, ds + c, \tag{40}$$

where c is an arbitrary constant. It then follows that the general solution y of Eq. (35) is given by

$$y = e^{-t^2/4}\int_0^t e^{s^2/4}\, ds + ce^{-t^2/4}. \tag{41}$$

The initial condition (36) requires that $c = 1$.

The main purpose of this example is to illustrate that sometimes the solution must be left in terms of an integral. This is usually at most a slight inconvenience, rather than a serious obstacle. For a given value of t, the integral in Eq. (41) is a definite integral and can be approximated to any desired degree of accuracy by using readily available numerical integrators. By repeating this process for many values of t and plotting the results, you can obtain a graph of a solution. Alternatively, you can use a numerical approximation method, such as Euler's method or those discussed in Sections 2.7 and 2.8. Such methods proceed directly from the differential equation and need no expression for the solution. Software packages such as *Maple*, *Mathematica*, and MATLAB, among others, readily execute such procedures and produce graphs of solutions of differential equations.

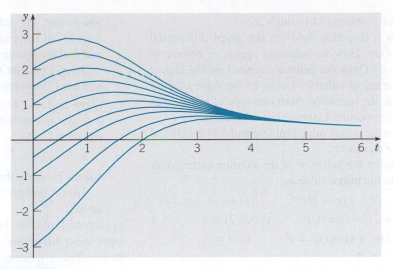

FIGURE 2.1.3 Integral curves of $2y' + ty = 2$.

Figure 2.1.3 displays graphs of the solution (41) for several values of c. From the figure, it may be plausible to conjecture that all solutions approach a limit as $t \to \infty$. The limit can be found analytically (see Problem 32).

PROBLEMS

ODEA In each of Problems 1 through 12:
(a) Draw a direction field for the given differential equation.
(b) Based on an inspection of the direction field, describe how solutions behave for large t.
(c) Find the general solution of the given differential equation, and use it to determine how solutions behave as $t \to \infty$.

1. $y' + 3y = t + e^{-2t}$
2. $y' - 2y = t^2 e^{2t}$
3. $y' + y = te^{-t} + 1$
4. $y' + (1/t)y = 3\cos 2t$, $t > 0$
5. $y' - 2y = 3e^t$
6. $ty' + 2y = \sin t$, $t > 0$
7. $y' + 2ty = 2te^{-t^2}$
8. $(1 + t^2)y' + 4ty = (1 + t^2)^{-2}$
9. $2y' + y = 3t$
10. $ty' - y = t^2 e^{-t}$, $t > 0$
11. $y' + y = 5\sin 2t$
12. $2y' + y = 3t^2$

In each of Problems 13 through 20, find the solution of the given initial value problem.

13. $y' - y = 2te^{2t}$, $y(0) = 1$
14. $y' + 2y = te^{-2t}$, $y(1) = 0$
15. $ty' + 2y = t^2 - t + 1$, $y(1) = \frac{1}{2}$, $t > 0$
16. $y' + (2/t)y = (\cos t)/t^2$, $y(\pi) = 0$, $t > 0$
17. $y' - 2y = e^{2t}$, $y(0) = 2$
18. $ty' + 2y = \sin t$, $y(\pi/2) = 1$, $t > 0$
19. $t^3 y' + 4t^2 y = e^{-t}$, $y(-1) = 0$, $t < 0$
20. $ty' + (t+1)y = t$, $y(\ln 2) = 1$, $t > 0$

In each of Problems 21 through 23:   ODEA
(a) Draw a direction field for the given differential equation. How do solutions appear to behave as t becomes large? Does the behavior depend on the choice of the initial value a? Let a_0 be the value of a for which the transition from one type of behavior to another occurs. Estimate the value of a_0.
(b) Solve the initial value problem and find the critical value a_0 exactly.
(c) Describe the behavior of the solution corresponding to the initial value a_0.

21. $y' - \frac{1}{2}y = 2\cos t$, $y(0) = a$
22. $2y' - y = e^{t/3}$, $y(0) = a$
23. $3y' - 2y = e^{-\pi t/2}$, $y(0) = a$

In each of Problems 24 through 26:

ODEA (a) Draw a direction field for the given differential equation. How do solutions appear to behave as $t \to 0$? Does the behavior depend on the choice of the initial value a? Let a_0 be the value of a for which the transition from one type of behavior to another occurs. Estimate the value of a_0.

(b) Solve the initial value problem and find the critical value a_0 exactly.

(c) Describe the behavior of the solution corresponding to the initial value a_0.

24. $ty' + (t + 1)y = 2te^{-t}$, $\quad y(1) = a$, $\quad t > 0$

25. $ty' + 2y = (\sin t)/t$, $\quad y(-\pi/2) = a$, $\quad t < 0$

26. $(\sin t)y' + (\cos t)y = e^t$, $\quad y(1) = a$, $\quad 0 < t < \pi$

27. Consider the initial value problem

$$y' + \tfrac{1}{2}y = 2\cos t, \qquad y(0) = -1.$$

Find the coordinates of the first local maximum point of the solution for $t > 0$.

28. Consider the initial value problem

$$y' + \tfrac{2}{3}y = 1 - \tfrac{1}{2}t, \qquad y(0) = y_0.$$

Find the value of y_0 for which the solution touches, but does not cross, the t-axis.

29. Consider the initial value problem

$$y' + \tfrac{1}{4}y = 3 + 2\cos 2t, \qquad y(0) = 0.$$

(a) Find the solution of this initial value problem and describe its behavior for large t.

(b) Determine the value of t for which the solution first intersects the line $y = 12$.

30. Find the value of y_0 for which the solution of the initial value problem

$$y' - y = 1 + 3\sin t, \qquad y(0) = y_0$$

remains finite as $t \to \infty$.

31. Consider the initial value problem

$$y' - \tfrac{3}{2}y = 3t + 2e^t, \qquad y(0) = y_0.$$

Find the value of y_0 that separates solutions that grow positively as $t \to \infty$ from those that grow negatively. How does the solution that corresponds to this critical value of y_0 behave as $t \to \infty$?

32. Show that all solutions of $2y' + ty = 2$ (see Eq. (35) of the text) approach a limit as $t \to \infty$, and find the limiting value.

Hint: Consider the general solution, Eq. (41), and use L'Hospital's rule on the first term.

33. Show that if a and λ are positive constants, and b is any real number, then every solution of the equation

$$y' + ay = be^{-\lambda t}$$

has the property that $y \to 0$ as $t \to \infty$.

Hint: Consider the cases $a = \lambda$ and $a \neq \lambda$ separately.

In each of Problems 34 through 37, construct a first order linear differential equation whose solutions have the required behavior as $t \to \infty$. Then solve your equation and confirm that the solutions do indeed have the specified property.

34. All solutions have the limit 3 as $t \to \infty$.

35. All solutions are asymptotic to the line $y = 3 - t$ as $t \to \infty$.

36. All solutions are asymptotic to the line $y = 2t - 5$ as $t \to \infty$.

37. All solutions approach the curve $y = 4 - t^2$ as $t \to \infty$.

38. Consider the initial value problem

$$y' + ay = g(t), \qquad y(t_0) = y_0.$$

Assume that a is a positive constant and that $g(t) \to g_0$ as $t \to \infty$. Show that $y(t) \to g_0/a$ as $t \to \infty$. Construct an example with a nonconstant $g(t)$ that illustrates this result.

39. **Variation of Parameters.** Consider the following method of solving the general linear equation of first order:

$$y' + p(t)y = g(t). \tag{i}$$

(a) If $g(t) = 0$ for all t, show that the solution is

$$y = A\exp\left[-\int p(t)\,dt\right], \tag{ii}$$

where A is a constant.

(b) If $g(t)$ is not everywhere zero, assume that the solution of Eq. (i) is of the form

$$y = A(t)\exp\left[-\int p(t)\,dt\right], \tag{iii}$$

where A is now a function of t. By substituting for y in the given differential equation, show that $A(t)$ must satisfy the condition

$$A'(t) = g(t) \exp\left[\int p(t)\,dt\right]. \quad \text{(iv)}$$

(c) Find $A(t)$ from Eq. (iv). Then substitute for $A(t)$ in Eq. (iii) and determine y. Verify that the solution obtained in this manner agrees with that of Eq. (27) in the text. This technique is known as the method of **variation of parameters**; it is discussed in detail in Section 4.8 in connection with second order linear equations.

In each of Problems 40 through 43, use the method of Problem 39 to solve the given differential equation.

40. $y' - 2y = t^2 e^{2t}$

41. $y' + (1/t)y = 3\cos 2t, \qquad t > 0$

42. $ty' + 2y = \sin t, \qquad t > 0$

43. $2y' + y = 3t^2$

2.2 Separable Equations

In Section 1.2, we used a process of direct integration to solve first order linear equations of the form

$$\frac{dy}{dt} = ay + b, \quad \text{(1)}$$

where a and b are constants. We will now show that this process is actually applicable to a much larger class of equations.

We will use x, rather than t, to denote the independent variable in this section for two reasons. In the first place, different letters are frequently used for the variables in a differential equation, and you should not become too accustomed to using a single pair. In particular, x often occurs as the independent variable. Further, we want to reserve t for another purpose later in the section.

The general first order equation is

$$\frac{dy}{dx} = f(x, y). \quad \text{(2)}$$

Linear equations were considered in the preceding section, but if Eq. (2) is nonlinear, then the method developed in Section 2.1 no longer applies. Here, we consider a subclass of first order equations that can be solved by direct integration.

To identify this class of equations, we first rewrite Eq. (2) in the form

$$M(x, y) + N(x, y)\frac{dy}{dx} = 0. \quad \text{(3)}$$

It is always possible to do this by setting $M(x, y) = -f(x, y)$ and $N(x, y) = 1$, but there may be other ways as well. If it happens that M is a function of x only and N is a function of y only, then Eq. (3) becomes

$$M(x) + N(y)\frac{dy}{dx} = 0. \quad \text{(4)}$$

Such an equation is said to be **separable**, because if it is written in the differential form

$$M(x)\,dx + N(y)\,dy = 0, \quad \text{(5)}$$

then, if you wish, terms involving each variable may be separated by the equals sign. The differential form (5) is also more symmetric and tends to diminish the distinction between independent and dependent variables.

A separable equation can be solved by integrating the functions M and N. We illustrate the process by an example and then discuss it in general for Eq. (4).

EXAMPLE 1

Show that the equation

$$\frac{dy}{dx} = \frac{x^2}{1-y^2} \tag{6}$$

is separable, and then find an equation for its integral curves.

If we write Eq. (6) as

$$-x^2 + (1-y^2)\frac{dy}{dx} = 0, \tag{7}$$

then it has the form (4) and is therefore separable. Next, observe that the first term in Eq. (7) is the derivative of $-x^3/3$ with respect to x. Further, if we think of y as a function of x, then by the chain rule

$$\frac{d}{dx}\left(y - \frac{y^3}{3}\right) = \frac{d}{dy}\left(y - \frac{y^3}{3}\right)\frac{dy}{dx} = (1-y^2)\frac{dy}{dx}.$$

Thus Eq. (7) can be written as

$$\frac{d}{dx}\left(-\frac{x^3}{3}\right) + \frac{d}{dx}\left(y - \frac{y^3}{3}\right) = 0,$$

or

$$\frac{d}{dx}\left(-\frac{x^3}{3} + y - \frac{y^3}{3}\right) = 0.$$

Therefore by integrating we obtain

$$-x^3 + 3y - y^3 = c, \tag{8}$$

where c is an arbitrary constant. Equation (8) is an equation for the integral curves of Eq. (6). A direction field and several integral curves are shown in Figure 2.2.1. Any differentiable function $y = \phi(x)$ that satisfies Eq. (8) is a solution of Eq. (6). An equation of the integral curve passing through a particular point (x_0, y_0) can be found by substituting x_0 and y_0 for x and y, respectively, in Eq. (8) and determining the corresponding value of c.

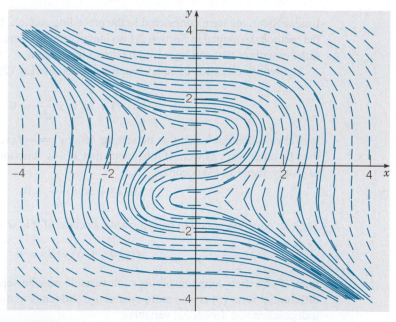

FIGURE 2.2.1 Direction field and integral curves of $y' = x^2/(1-y^2)$.

Essentially the same procedure can be followed for any separable equation. Returning to Eq. (4), let H_1 and H_2 be any antiderivatives of M and N, respectively. Thus

$$H_1'(x) = M(x), \qquad H_2'(y) = N(y), \tag{9}$$

and Eq. (4) becomes

$$H_1'(x) + H_2'(y)\frac{dy}{dx} = 0. \tag{10}$$

According to the chain rule, if y is a function of x, then

$$H_2'(y)\frac{dy}{dx} = \frac{dH_2(y)}{dy}\frac{dy}{dx} = \frac{d}{dx}H_2(y). \tag{11}$$

Consequently, we can write Eq. (10) as

$$\frac{d}{dx}[H_1(x) + H_2(y)] = 0. \tag{12}$$

By integrating Eq. (12), we obtain

$$H_1(x) + H_2(y) = c, \tag{13}$$

where c is an arbitrary constant. Any differentiable function $y = \phi(x)$ that satisfies Eq. (13) is a solution of Eq. (4); in other words, Eq. (13) defines the solution implicitly rather than explicitly. In practice, Eq. (13) is usually obtained from Eq. (5) by integrating the first term with respect to x and the second term with respect to y. This more direct procedure is illustrated in Examples 2 and 3 below.

If, in addition to the differential equation, an initial condition

$$y(x_0) = y_0 \tag{14}$$

is prescribed, then the solution of Eq. (4) satisfying this condition is obtained by setting $x = x_0$ and $y = y_0$ in Eq. (13). This gives

$$c = H_1(x_0) + H_2(y_0). \tag{15}$$

Substituting this value of c in Eq. (13) and noting that

$$H_1(x) - H_1(x_0) = \int_{x_0}^{x} M(s)\,ds, \qquad H_2(y) - H_2(y_0) = \int_{y_0}^{y} N(s)\,ds,$$

we obtain

$$\int_{x_0}^{x} M(s)\,ds + \int_{y_0}^{y} N(s)\,ds = 0. \tag{16}$$

Equation (16) is an implicit representation of the solution of the differential equation (4) that also satisfies the initial condition (14). You should bear in mind that the determination of an explicit formula for the solution requires that Eq. (16) be solved for y as a function of x. Unfortunately, it is often impossible to do this analytically; in such cases, you can resort to numerical methods to find approximate values of y for given values of x. Alternatively, if it is possible to solve for x in terms of y, then this can often be very helpful.

EXAMPLE 2

Solve the initial value problem

$$\frac{dy}{dx} = \frac{3x^2 + 4x + 2}{2(y - 1)}, \qquad y(0) = -1, \tag{17}$$

and determine the interval in which the solution exists.

The differential equation can be written as

$$2(y - 1) \, dy = (3x^2 + 4x + 2) \, dx.$$

Integrating the left side with respect to y and the right side with respect to x gives

$$y^2 - 2y = x^3 + 2x^2 + 2x + c, \tag{18}$$

where c is an arbitrary constant. To determine the solution satisfying the prescribed initial condition, we substitute $x = 0$ and $y = -1$ in Eq. (18), obtaining $c = 3$. Hence the solution of the initial value problem is given implicitly by

$$y^2 - 2y = x^3 + 2x^2 + 2x + 3. \tag{19}$$

To obtain the solution explicitly, we must solve Eq. (19) for y in terms of x. That is a simple matter in this case, since Eq. (19) is quadratic in y, and we obtain

$$y = 1 \pm \sqrt{x^3 + 2x^2 + 2x + 4}. \tag{20}$$

Equation (20) gives two solutions of the differential equation, only one of which, however, satisfies the given initial condition. This is the solution corresponding to the minus sign in Eq. (20), so we finally obtain

$$y = \phi(x) = 1 - \sqrt{x^3 + 2x^2 + 2x + 4} \tag{21}$$

as the solution of the initial value problem (17). Note that if the plus sign is chosen by mistake in Eq. (20), then we obtain the solution of the same differential equation that satisfies the initial condition $y(0) = 3$. Finally, to determine the interval in which the solution (21) is valid, we must find the interval (containing the initial point $x = 0$) in which the quantity under the radical is positive. The only real zero of this expression is $x = -2$, so the desired interval is $x > -2$. The solution of the initial value problem and some other integral curves of the differential equation are shown in Figure 2.2.2. Observe that the boundary of the interval of validity of the solution (21) is determined by the point $(-2, 1)$ at which the tangent line is vertical.

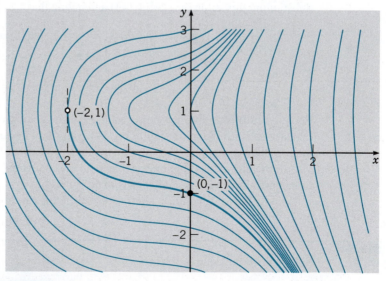

FIGURE 2.2.2 Integral curves of $y' = (3x^2 + 4x + 2)/2(y - 1)$.

EXAMPLE 3

Solve the equation

$$\frac{dy}{dx} = \frac{4x - x^3}{4 + y^3} \qquad (22)$$

and draw graphs of several integral curves. Also find the solution passing through the point (0,1) and determine its interval of validity.

Rewriting Eq. (22) as

$$(4 + y^3)\, dy = (4x - x^3)\, dx,$$

integrating each side, multiplying by 4, and rearranging the terms, we obtain

$$y^4 + 16y + x^4 - 8x^2 = c, \qquad (23)$$

where c is an arbitrary constant. Any differentiable function $y = \phi(x)$ that satisfies Eq. (23) is a solution of the differential equation (22). Graphs of Eq. (23) for several values of c are shown in Figure 2.2.3.

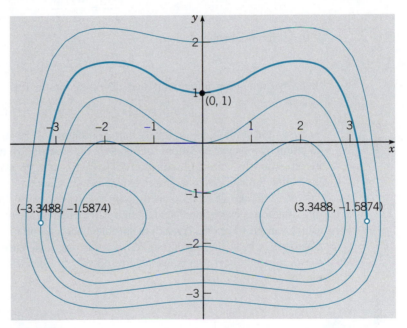

FIGURE 2.2.3 Integral curves of $y' = (4x - x^3)/(4 + y^3)$. The solution passing through (0,1) is shown by the heavy curve.

To find the particular solution passing through (0,1), we set $x = 0$ and $y = 1$ in Eq. (23) with the result that $c = 17$. Thus the solution in question is given implicitly by

$$y^4 + 16y + x^4 - 8x^2 = 17. \qquad (24)$$

It is shown by the heavy curve in Figure 2.2.3. The interval of validity of this solution extends on either side of the initial point as long as the function remains differentiable. From the figure, we see that the interval ends when we reach points where the tangent line is vertical. It follows, from the differential equation (22), that these are points where $4 + y^3 = 0$, or $y = (-4)^{1/3} \cong -1.5874$. From Eq. (24), the corresponding values of x are $x \cong \pm 3.3488$. These points are marked on the graph in Figure 2.2.3.

Sometimes an equation of the form (2),

$$\frac{dy}{dx} = f(x, y),$$

has a constant solution $y = y_0$. Such a solution is usually easy to find because if $f(x, y_0) = 0$ for some value y_0 and for all x, then the constant function $y = y_0$ is a solution of the differential equation (2). For example, the equation

$$\frac{dy}{dx} = \frac{(y - 3)\cos x}{1 + 2y^2} \tag{25}$$

has the constant solution $y = 3$. Other solutions of this equation can be found by separating the variables and integrating.

The investigation of a first order nonlinear equation can sometimes be facilitated by regarding both x and y as functions of a third variable t. Then

$$\frac{dy}{dx} = \frac{dy/dt}{dx/dt}. \tag{26}$$

If the differential equation is

$$\frac{dy}{dx} = \frac{F(x, y)}{G(x, y)}, \tag{27}$$

then, by comparing numerators and denominators in Eqs. (26) and (27), we obtain the system

$$dx/dt = G(x, y), \qquad dy/dt = F(x, y). \tag{28}$$

At first sight it may seem unlikely that a problem will be simplified by replacing a single equation by a pair of equations, but, in fact, the system (28) may well be more amenable to investigation than the single equation (27). Nonlinear systems of the form (28) are introduced in Section 3.6 and discussed more extensively in Chapter 7.

Note: In Example 2, it was not difficult to solve explicitly for y as a function of x. However, this situation is exceptional, and often it will be better to leave the solution in implicit form, as in Examples 1 and 3. Thus, in the problems below and in other sections where nonlinear equations appear, the words "solve the following differential equation" mean to find the solution explicitly if it is convenient to do so, but otherwise, to find an equation defining the solution implicitly.

PROBLEMS

In each of Problems 1 through 8, solve the given differential equation.

1. $y' = x^2/y$
2. $y' = x^2/y(1 + x^3)$
3. $y' + y^2 \sin x = 0$
4. $y' = (3x^2 - 1)/(3 + 2y)$
5. $y' = (\cos^2 x)(\cos^2 2y)$
6. $xy' = (1 - y^2)^{1/2}$
7. $\frac{dy}{dx} = \frac{x - e^{-x}}{y + e^y}$
8. $\frac{dy}{dx} = \frac{x^2}{1 + y^2}$

In each of Problems 9 through 20: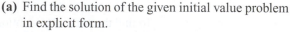
(a) Find the solution of the given initial value problem in explicit form.
(b) Plot the graph of the solution.
(c) Determine (at least approximately) the interval in which the solution is defined.

9. $y' = (1 - 2x)y^2$, $\quad y(0) = -1/6$
10. $y' = (1 - 2x)/y$, $\quad y(1) = -2$
11. $x\,dx + ye^{-x}dy = 0$, $\quad y(0) = 1$
12. $dr/d\theta = r^2/\theta$, $\quad r(1) = 2$
13. $y' = 2x/(y + x^2 y)$, $\quad y(0) = -2$
14. $y' = xy^3(1 + x^2)^{-1/2}$, $\quad y(0) = 1$

15. $y' = 2x/(1 + 2y)$, $\qquad y(2) = 0$

16. $y' = x(x^2 + 1)/4y^3$, $\qquad y(0) = -1/\sqrt{2}$

17. $y' = (3x^2 - e^x)/(2y - 5)$, $\qquad y(0) = 1$

18. $y' = (e^{-x} - e^x)/(3 + 4y)$, $\qquad y(0) = 1$

19. $\sin 2x \, dx + \cos 3y \, dy = 0$, $\qquad y(\pi/2) = \pi/3$

20. $y^2(1 - x^2)^{1/2} dy = \arcsin x \, dx$, $\qquad y(0) = 1$

Some of the results requested in Problems 21 through 28 can be obtained either by solving the given equations analytically or by plotting numerically generated approximations to the solutions. Try to form an opinion as to the advantages and disadvantages of each approach.

21. Solve the initial value problem

$$y' = (1 + 3x^2)/(3y^2 - 6y), \qquad y(0) = 1$$

and determine the interval in which the solution is valid.

Hint: To find the interval of definition, look for points where the integral curve has a vertical tangent.

22. Solve the initial value problem

$$y' = 3x^2/(3y^2 - 4), \qquad y(1) = 0$$

and determine the interval in which the solution is valid.

Hint: To find the interval of definition, look for points where the integral curve has a vertical tangent.

23. Solve the initial value problem

$$y' = 2y^2 + xy^2, \qquad y(0) = 1$$

and determine where the solution attains its minimum value.

24. Solve the initial value problem

$$y' = (2 - e^x)/(3 + 2y), \qquad y(0) = 0$$

and determine where the solution attains its maximum value.

25. Solve the initial value problem

$$y' = 2\cos 2x/(3 + 2y), \qquad y(0) = -1$$

and determine where the solution attains its maximum value.

26. Solve the initial value problem

$$y' = 2(1 + x)(1 + y^2), \qquad y(0) = 0$$

and determine where the solution attains its minimum value.

27. Consider the initial value problem   ◀ODEA

$$y' = ty(4 - y)/3, \qquad y(0) = y_0.$$

(a) Determine how the behavior of the solution as t increases depends on the initial value y_0.

(b) Suppose that $y_0 = 0.5$. Find the time T at which the solution first reaches the value 3.98.

28. Consider the initial value problem   ◀ODEA

$$y' = ty(4 - y)/(1 + t), \qquad y(0) = y_0 > 0.$$

(a) Determine how the solution behaves as $t \to \infty$.

(b) If $y_0 = 2$, find the time T at which the solution first reaches the value 3.99.

(c) Find the range of initial values for which the solution lies in the interval $3.99 < y < 4.01$ by the time $t = 2$.

29. Solve the equation

$$\frac{dy}{dx} = \frac{ay + b}{cy + d},$$

where a, b, c, and d are constants.

Homogeneous Equations. If the right side of the equation $dy/dx = f(x, y)$ can be expressed as a function of the ratio y/x only, then the equation is said to be homogeneous.[1] Such equations can always be transformed into separable equations by a change of the dependent variable. Problem 30 illustrates how to solve first order homogeneous equations.

30. Consider the equation

$$\frac{dy}{dx} = \frac{y - 4x}{x - y}. \tag{i}$$

(a) Show that Eq. (i) can be rewritten as

$$\frac{dy}{dx} = \frac{(y/x) - 4}{1 - (y/x)}; \tag{ii}$$

thus Eq. (i) is homogeneous.

(b) Introduce a new dependent variable v so that $v = y/x$, or $y = xv(x)$. Express dy/dx in terms of x, v, and dv/dx.

---

[1] The word "homogeneous" has different meanings in different mathematical contexts. The homogeneous equations considered here have nothing to do with the homogeneous equations that will occur in Chapter 3 and elsewhere.

(c) Replace y and dy/dx in Eq. (ii) by the expressions from part (b) that involve v and dv/dx. Show that the resulting differential equation is

$$v + x\frac{dv}{dx} = \frac{v-4}{1-v},$$

or

$$x\frac{dv}{dx} = \frac{v^2-4}{1-v}. \qquad \text{(iii)}$$

Observe that Eq. (iii) is separable.

(d) Solve Eq. (iii), obtaining v implicitly in terms of x.

(e) Find the solution of Eq. (i) by replacing v by y/x in the solution in part (d).

(f) Draw a direction field and some integral curves for Eq. (i). Recall that the right side of Eq. (i) actually depends only on the ratio y/x. This means that integral curves have the same slope at all points on any given straight line through the origin, although the slope changes from one line to another. Therefore the direction field and the integral curves are symmetric with respect to the origin. Is this symmetry property evident from your plot?

The method outlined in Problem 30 can be used for any homogeneous equation. That is, the substitution $y = xv(x)$ transforms a homogeneous equation into a separable equation. The latter equation can be solved by direct integration, and then replacing v by y/x gives the solution to the original equation.

In each of Problems 31 through 38:

(a) Show that the given equation is homogeneous.

(b) Solve the differential equation.

(c) Draw a direction field and some integral curves. **ODEA** Are they symmetric with respect to the origin?

31. $\dfrac{dy}{dx} = \dfrac{x^2 + xy + y^2}{x^2}$

32. $\dfrac{dy}{dx} = \dfrac{x^2 + 3y^2}{2xy}$

33. $\dfrac{dy}{dx} = \dfrac{4y - 3x}{2x - y}$

34. $\dfrac{dy}{dx} = -\dfrac{4x + 3y}{2x + y}$

35. $\dfrac{dy}{dx} = \dfrac{x + 3y}{x - y}$

36. $(x^2 + 3xy + y^2)\,dx - x^2\,dy = 0$

37. $\dfrac{dy}{dx} = \dfrac{x^2 - 3y^2}{2xy}$

38. $\dfrac{dy}{dx} = \dfrac{3y^2 - x^2}{2xy}$

2.3 Modeling with First Order Equations

Differential equations are of interest to nonmathematicians primarily because of the possibility of using them to investigate a wide variety of problems in engineering and in the physical, biological, and social sciences. One reason for this is that mathematical models and their solutions lead to equations relating the variables and parameters in the problem. These equations often enable you to make predictions about how the natural process will behave in various circumstances. It is often easy to vary parameters in the mathematical model over wide ranges, whereas this may be very time-consuming or expensive, if not impossible, in an experimental setting. Nevertheless, mathematical modeling and experiment or observation are both critically important and have somewhat complementary roles in scientific investigations. Mathematical models are validated by comparison of their predictions with experimental results. On the other hand, mathematical analyses may suggest the most promising directions for experimental exploration, and they may indicate fairly precisely what experimental data will be most helpful.

In Sections 1.1 and 1.2, we formulated and investigated a few simple mathematical models. We begin by recapitulating and expanding on some of the conclusions reached in those sections. Regardless of the specific field of application, there are three identifiable stages that are always present in the process of mathematical modeling.

▶ **Construction of the Model.** In this stage, you translate the physical situation into mathematical terms, often using the steps listed at the end of Section 1.1. Perhaps most critical at this stage is to state clearly the physical principle(s) that are believed to govern the process. For example, it has been observed that in some circumstances heat passes from a warmer to a cooler body at a rate proportional to the temperature difference, that objects move about in accordance with Newton's laws of motion, and that isolated insect populations grow at a rate proportional to the current population. Each of these statements involves a rate of change (derivative) and consequently, when expressed mathematically, leads to a differential equation. The differential equation is a mathematical model of the process.

It is important to realize that the mathematical equations are almost always only an approximate description of the actual process. For example, bodies moving at speeds comparable to the speed of light are not governed by Newton's laws, insect populations do not grow indefinitely as stated because of eventual limitations on their food supply, and heat transfer is affected by factors other than the temperature difference. Alternatively, one can adopt the point of view that the mathematical equations exactly describe the operation of a simplified or ideal physical model, which has been constructed (or imagined) so as to embody the most important features of the actual process. Sometimes, the process of mathematical modeling involves the conceptual replacement of a discrete process by a continuous one. For instance, the number of members in an insect population is an integer; however, if the population is large, it may seem reasonable to consider it to be a continuous variable and even to speak of its derivative.

▶ **Analysis of the Model.** Once the problem has been formulated mathematically, you are often faced with the problem of solving one or more differential equations or, failing that, of finding out as much as possible about the properties of the solution. It may happen that this mathematical problem is quite difficult, and if so, further approximations may be required at this stage to make the problem more susceptible to mathematical investigation. For example, a nonlinear equation may be approximated by a linear one, or a slowly varying coefficient may be replaced by a constant. Naturally, any such approximations must also be examined from the physical point of view to make sure that the simplified mathematical problem still reflects the essential features of the physical process under investigation. At the same time, an intimate knowledge of the physics of the problem may suggest reasonable mathematical approximations that will make the mathematical problem more amenable to analysis. This interplay of understanding of physical phenomena and knowledge of mathematical techniques and their limitations is characteristic of applied mathematics at its best, and it is indispensable in successfully constructing useful mathematical models of intricate physical processes.

▶ **Comparison with Experiment or Observation.** Finally, having obtained the solution (or at least some information about it), you must interpret this information in the context in which the problem arose. In particular, you should always check that the mathematical solution appears physically reasonable. If possible, calculate the values of the solution at selected points and compare them with experimentally observed values. Or ask whether the behavior of the solution after a long time is consistent with observations. Or examine the solutions corresponding to certain special values of parameters in the problem. Of course, the fact that the mathematical solution appears to be reasonable does not guarantee that it is correct. However, if the predictions of the mathematical model are seriously inconsistent with observations of the physical system it purports to describe, this suggests that errors have been made in solving the mathematical problem, that the mathematical model itself needs refinement, or that observations must be made with greater care.

The examples in this section are typical of applications in which first order differential equations arise.

EXAMPLE 1

Mixing

At time $t = 0$ a tank contains Q_0 lb of salt dissolved in 100 gal of water; see Figure 2.3.1. Assume that water containing 1/4 lb of salt/gal is entering the tank at a rate of r gal/min and that the well-stirred mixture is draining from the tank at the same rate. Set up the initial value problem that describes this flow process. Find the quantity of salt $Q(t)$ in the tank at any time, and also find the limiting quantity Q_L that is present after a very long time. If $r = 3$ and $Q_0 = 2Q_L$, find the time T after which the salt level is within 2% of Q_L. Also find the flow rate that is required if the value of T is not to exceed 45 min.

r gal/min, $\frac{1}{4}$ lb/gal

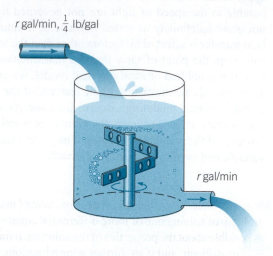

r gal/min

FIGURE 2.3.1   The water tank in Example 1.

We assume that salt is neither created nor destroyed in the tank. Therefore variations in the amount of salt are due solely to the flows in and out of the tank. More precisely, the rate of change of salt in the tank, dQ/dt, is equal to the rate at which salt is flowing in minus the rate at which it is flowing out. In symbols,

$$\frac{dQ}{dt} = \text{rate in} - \text{rate out}. \tag{1}$$

The rate at which salt enters the tank is the concentration 1/4 lb/gal times the flow rate r gal/min, or $(r/4)$ lb/min. To find the rate at which salt leaves the tank, we need to multiply the concentration of salt in the tank by the rate of outflow, r gal/min. Since the rates of flow in and out are equal, the volume of water in the tank remains constant at 100 gal, and since the mixture is "well-stirred," the concentration throughout the tank is the same, namely, $[Q(t)/100]$ lb/gal. Therefore the rate at which salt leaves the tank is $[rQ(t)/100]$ lb/min. Thus the differential equation governing this process is

$$\frac{dQ}{dt} = \frac{r}{4} - \frac{rQ}{100}. \tag{2}$$

The initial condition is

$$Q(0) = Q_0. \tag{3}$$

Upon thinking about the problem physically, we might anticipate that eventually the mixture originally in the tank will be essentially replaced by the mixture flowing in, whose concentration is 1/4 lb/gal. Consequently, we might expect that ultimately the amount of salt in the tank would be very close to 25 lb. We can also find the limiting amount $Q_L = 25$ by setting dQ/dt equal to zero in Eq. (2) and solving the resulting algebraic equation for Q.

To find $Q(t)$ at any time t, note that Eq. (2) is both linear and separable. Rewriting it in the usual form for a linear equation, we have

$$\frac{dQ}{dt} + \frac{rQ}{100} = \frac{r}{4}. \tag{4}$$

Thus the integrating factor is $e^{rt/100}$. Mutiplying by this factor and integrating, we obtain

$$e^{rt/100}Q(t) = \frac{r}{4}\frac{100}{r}e^{rt/100} + c,$$

so the general solution is

$$Q(t) = 25 + ce^{-rt/100}, \tag{5}$$

where c is an arbitrary constant. To satisfy the initial condition (3) we must choose $c = Q_0 - 25$. Therefore the solution of the initial value problem (2), (3) is

$$Q(t) = 25 + (Q_0 - 25)e^{-rt/100}, \tag{6}$$

or

$$Q(t) = 25(1 - e^{-rt/100}) + Q_0 e^{-rt/100}. \tag{7}$$

From Eq. (6) or (7), you can see that $Q(t) \to 25$ lb as $t \to \infty$, so the limiting value Q_L is 25, confirming our physical intuition. Further, $Q(t)$ approaches the limit more rapidly as r increases. In interpreting the solution (7), note that the second term on the right side is the portion of the original salt that remains at time t, while the first term gives the amount of salt in the tank due to the action of the flow processes. Plots of the solution for $r = 3$ and for several values of Q_0 are shown in Figure 2.3.2.

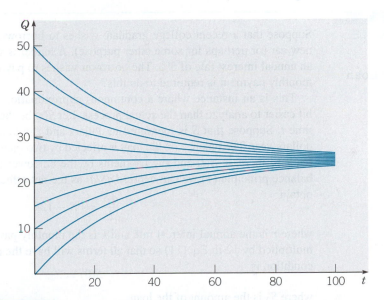

FIGURE 2.3.2 Solutions of the initial value problem (2), (3) for $r = 3$ and several values of Q_0.

Now suppose that $r = 3$ and $Q_0 = 2Q_L = 50$; then Eq. (6) becomes

$$Q(t) = 25 + 25e^{-0.03t}. \tag{8}$$

Since 2% of 25 is 0.5, we wish to find the time T at which $Q(t)$ has the value 25.5. Substituting $t = T$ and $Q = 25.5$ in Eq. (8) and solving for T, we obtain

$$T = (\ln 50)/0.03 \cong 130.4 \text{ (min)}. \tag{9}$$

To determine r so that $T = 45$, return to Eq. (6), set $t = 45$, $Q_0 = 50$, $Q(t) = 25.5$, and solve for r. The result is

$$r = (100/45) \ln 50 \cong 8.69 \text{ gal/min}. \tag{10}$$

Since this example is hypothetical, there are no experimental data for comparison. The flow rates can presumably be measured with considerable accuracy. However, the assumption of a uniform concentration of salt in the tank may be more questionable, since this may depend on how the liquid in the tank is stirred, whether the incoming flow is distributed or concentated in one location, and perhaps on the shape of the tank.

While the details in Example 1 have no special significance, the fundamental balance principle expressed by Eq. (1) can be applied in many different circumstances. Models similar to Example 1 are often used in problems involving a pollutant in a lake, or a drug in an organ of the body, for example, rather than a tank of salt water. In such cases, the flow rates may not be easy to determine or may vary with time. Similarly, the concentration may be far from uniform in some cases. Finally, the rates of inflow and outflow may be different, which means that the variation of the amount of liquid in the problem must also be taken into account. The next example illustrates the use of the balance principle (1) in a financial setting.

EXAMPLE 2

A Car Loan

Suppose that a recent college graduate wishes to borrow $20,000 in order to purchase a new car (or perhaps for some other purpose). A lender is willing to provide the loan with an annual interest rate of 8%. The borrower wishes to pay off the loan in four years. What monthly payment is required to do this?

This is an instance where a continuous approximation to a discrete process may be a bit easier to analyze than the actual process. Let $S(t)$ be the balance due on the loan at any time t. Suppose that S is measured in dollars and t in years. Then dS/dt has the units of dollars/year. The balance on the loan is affected by two factors: the accumulation of interest tends to increase $S(t)$ and the payments by the borrower tend to reduce it. Based on the balance principle (1) we can express dS/dt as the net effect of these two factors. Thus we obtain

$$\frac{dS}{dt} = rS - 12k, \tag{11}$$

where r is the annual interest rate and k is the monthly payment rate. Note that k must be multiplied by 12 in Eq. (11) so that all terms will have the units of dollars/year. The initial condition is

$$S(0) = S_0, \tag{12}$$

where S_0 is the amount of the loan.

For the situation stated in this example $r = 0.08$ and $S_0 = 20,000$, so we have the initial value problem

$$\frac{dS}{dt} = 0.08S - 12k, \qquad S(0) = 20,000. \tag{13}$$

If we rewrite the differential equation as

$$S' - 0.08S = -12k,$$

then the integrating factor is $e^{-0.08t}$, and after an integration we obtain

$$e^{-0.08t}S = \frac{12}{0.08}ke^{-0.08t} + c,$$

or

$$S = 150k + ce^{0.08t}. \tag{14}$$

From the initial condition it follows that $c = 20{,}000 - 150k$, so the solution of the initial value problem (13) is

$$S = 20{,}000e^{0.08t} - 150k(e^{0.08t} - 1). \tag{15}$$

To find the monthly payment needed to pay off the loan in four years, we set $t = 4$, $S = 0$, and solve Eq. (15) for k. The result is

$$k = \frac{20{,}000}{150}\frac{e^{0.32}}{e^{0.32} - 1} = \$486.88. \tag{16}$$

The total amount paid over the life of the loan is 48 times $486.88, or $23,370.24; thus the total interest payment is $3,370.24.

The solution (15) can also be used to answer other possible questions. For example, suppose that the borrower wants to limit the monthly payment to $450. One way to do this is to extend the period of the loan beyond four years, thereby increasing the number of payments. To find the required time period, set $k = 450$, $S = 0$, and solve for t, with the result that

$$t = \frac{\ln(27/19)}{0.08} \cong 4.39 \text{ yr}, \tag{17}$$

or about 53 months.

To assess the accuracy of this continuous model we can solve the problem more precisely (see Problem 11). The comparison shows that the continuous model understates the monthly payment by $1.38, or about 0.28 %.

The approach used in Example 2 can also be applied to the more general initial value problem (11), (12), whose solution is

$$S = S_0e^{rt} - 12\frac{k}{r}(e^{rt} - 1). \tag{18}$$

The result (18) can be used in a large number of financial circumstances, including various kinds of investment programs, as well as loans and mortgages. For an investment situation, r is the estimated rate of return (interest, dividends, capital gains) and k is the monthly rate of deposits or withdrawals. The first term in expression (18) is the part of $S(t)$ that is due to the return accumulated on the initial amount S_0, and the second term is the part that is due to the deposit or withdrawal rate k.

The advantage of stating the problem in this general way without specific values for S_0, r, or k lies in the generality of the resulting formula (18) for $S(t)$. With this formula we can readily compare the results of different investment programs or different rates of return.

EXAMPLE 3

Chemicals in a Pond

Consider a pond that initially contains 10 million gal of fresh water. Stream water containing an undesirable chemical flows into the pond at the rate of 5 million gal/yr, and the mixture in the pond flows out through an overflow culvert at the same rate. The concentration $\gamma(t)$ of chemical in the incoming water varies periodically with time t, measured in years, according to the expression $\gamma(t) = 2 + \sin 2t$ g/gal. Construct a mathematical model of this flow process and determine the amount of chemical in the pond at any time. Plot the solution and describe in words the effect of the variation in the incoming concentration.

Since the incoming and outgoing flows of water are the same, the amount of water in the pond remains constant at 10^7 gal. Let us denote the mass of the chemical by $Q(t)$, measured in grams. This example is similar to Example 1, and the same inflow/outflow principle applies. Thus

$$\frac{dQ}{dt} = \text{rate in} - \text{rate out},$$

where "rate in" and "rate out" refer to the rates at which the chemical flows into and out of the pond, respectively. The rate at which the chemical flows in is given by

$$\text{rate in} = (5 \times 10^6) \text{ gal/yr } (2 + \sin 2t) \text{ g/gal}. \tag{19}$$

The concentration of chemical in the pond is $Q(t)/10^7$ g/gal, so the rate of flow out is

$$\text{rate out} = (5 \times 10^6) \text{ gal/yr } [Q(t)/10^7] \text{ g/gal} = Q(t)/2 \text{ g/yr}. \tag{20}$$

Thus we obtain the differential equation

$$\frac{dQ}{dt} = (5 \times 10^6)(2 + \sin 2t) - \frac{Q(t)}{2}, \tag{21}$$

where each term has the units of g/yr.

To make the coefficients more manageable, it is convenient to introduce a new dependent variable defined by $q(t) = Q(t)/10^6$ or $Q(t) = 10^6 q(t)$. This means that $q(t)$ is measured in millions of grams, or megagrams. If we make this substitution in Eq. (21), then each term contains the factor 10^6, which can be canceled. If we also transpose the term involving $q(t)$ to the left side of the equation, we finally have

$$\frac{dq}{dt} + \tfrac{1}{2}q = 10 + 5 \sin 2t. \tag{22}$$

Originally, there is no chemical in the pond, so the initial condition is

$$q(0) = 0. \tag{23}$$

Equation (22) is linear, and although the right side is a function of time, the coefficient of q is a constant. Thus the integrating factor is $e^{t/2}$. Multiplying Eq. (22) by this factor and integrating the resulting equation, we obtain the general solution

$$q(t) = 20 - \frac{40}{17}\cos 2t + \frac{10}{17}\sin 2t + ce^{-t/2}. \tag{24}$$

The initial condition (23) requires that $c = -300/17$, so the solution of the initial value problem (22), (23) is

$$q(t) = 20 - \frac{40}{17}\cos 2t + \frac{10}{17}\sin 2t - \frac{300}{17}e^{-t/2}. \tag{25}$$

A plot of the solution (25) is shown in Figure 2.3.3, along with the line $q = 20$. The exponential term in the solution is important for small t, but it diminishes rapidly as t increases. Later, the solution consists of an oscillation, due to the $\sin 2t$ and $\cos 2t$ terms, about the constant level $q = 20$. Note that if the $\sin 2t$ term were not present in Eq. (22), then $q = 20$ would be the equilibrium solution of that equation.

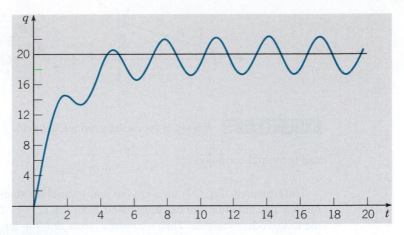

FIGURE 2.3.3 Solution of the initial value problem (22), (23).

Let us now consider the adequacy of the mathematical model itself for this problem. The model rests on several assumptions that have not yet been stated explicitly. In the first place, the amount of water in the pond is controlled entirely by the rates of flow in and out—none is lost by evaporation or by seepage into the ground, and none is gained by rainfall. The same is true of the chemical; it flows into and out of the pond, but none is absorbed by fish or other organisms living in the pond. In addition, we assume that the concentration of chemical in the pond is uniform throughout the pond. Whether the results obtained from the model are accurate depends strongly on the validity of these simplifying assumptions.

EXAMPLE 4

Escape Velocity

A body of constant mass m is projected away from the earth in a direction perpendicular to the earth's surface with an initial velocity v_0. Assuming that there is no air resistance, but taking into account the variation of the earth's gravitational field with distance, find an expression for the velocity during the ensuing motion. Also find the initial velocity that is required to lift the body to a given maximum altitude ξ above the surface of the earth, and find the least initial velocity for which the body will not return to the earth; the latter is the **escape velocity**.

Let the positive x-axis point away from the center of the earth along the line of motion, with $x = 0$ lying on the earth's surface; see Figure 2.3.4. The figure is drawn horizontally to remind you that gravity is directed toward the center of the earth, which is not necessarily downward from a perspective away from the earth's surface. The gravitational force acting on the body (that is, its weight) is inversely proportional to the square of the distance from the center of the earth and is given by $w(x) = -k/(x + R)^2$, where k is a constant, R is the radius of the earth, and the minus sign signifies that $w(x)$ is directed in the negative x direction. We know that on the earth's surface $w(0)$ is given by $-mg$, where g is the acceleration due to gravity at sea level. Therefore $k = mgR^2$ and

$$w(x) = -\frac{mgR^2}{(R + x)^2}. \tag{26}$$

Since there are no other forces acting on the body, the equation of motion is

$$m\frac{dv}{dt} = -\frac{mgR^2}{(R + x)^2}, \tag{27}$$

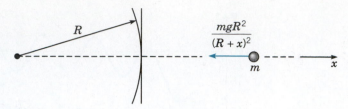

FIGURE 2.3.4   A body in the earth's gravitational field.

and the initial condition is

$$v(0) = v_0. \tag{28}$$

Unfortunately, Eq. (27) involves too many variables since it depends on t, x, and v. To remedy this situation, we can eliminate t from Eq. (27) by thinking of x, rather than t, as the independent variable. Then we must express dv/dt in terms of dv/dx by the chain rule; hence

$$\frac{dv}{dt} = \frac{dv}{dx}\frac{dx}{dt} = v\frac{dv}{dx},$$

and Eq. (27) is replaced by

$$v\frac{dv}{dx} = -\frac{gR^2}{(R+x)^2}. \tag{29}$$

Equation (29) is separable but not linear, so by separating the variables and integrating, we obtain

$$\frac{v^2}{2} = \frac{gR^2}{R+x} + c. \tag{30}$$

Since $x = 0$ when $t = 0$, the initial condition (28) at $t = 0$ can be replaced by the condition that $v = v_0$ when $x = 0$. Hence $c = (v_0^2/2) - gR$ and

$$v = \pm\sqrt{v_0^2 - 2gR + \frac{2gR^2}{R+x}}. \tag{31}$$

Note that Eq. (31) gives the velocity as a function of altitude rather than as a function of time. The plus sign must be chosen if the body is rising, and the minus sign if it is falling back to earth.

To determine the maximum altitude ξ that the body reaches, we set $v = 0$ and $x = \xi$ in Eq. (31) and then solve for ξ, obtaining

$$\xi = \frac{v_0^2 R}{2gR - v_0^2}. \tag{32}$$

Solving Eq. (32) for v_0, we find the initial velocity required to lift the body to the altitude ξ, namely,

$$v_0 = \sqrt{2gR\frac{\xi}{R+\xi}}. \tag{33}$$

The escape velocity v_e is then found by letting $\xi \to \infty$. Consequently,

$$v_e = \sqrt{2gR}. \tag{34}$$

The numerical value of v_e is approximately 6.9 mi/s, or 11.1 km/s.

The preceding calculation of the escape velocity neglects the effect of air resistance, so the actual escape velocity (including the effect of air resistance) is somewhat higher. On the other hand, the effective escape velocity can be significantly reduced if the body is transported a considerable distance above sea level before being launched. Both gravitational and frictional forces are thereby reduced; air resistance, in particular, diminishes quite rapidly with increasing altitude. You should keep in mind also that it may well be impractical to impart too large an initial velocity instantaneously; space vehicles, for instance, receive their initial acceleration during a period of a few minutes.

PROBLEMS

1. Consider a tank used in certain hydrodynamic experiments. After one experiment the tank contains 200 liters of a dye solution with a concentration of 1 g/liter. To prepare for the next experiment, the tank is to be rinsed with fresh water flowing in at a rate of 2 liters/min, the well-stirred solution flowing out at the same rate. Find the time that will elapse before the concentration of dye in the tank reaches 1% of its original value.

2. A tank initially contains 120 liters of pure water. A mixture containing a concentration of γ g/liter of salt enters the tank at a rate of 2 liters/min, and the well-stirred mixture leaves the tank at the same rate. Find an expression in terms of γ for the amount of salt in the tank at any time t. Also find the limiting amount of salt in the tank as $t \rightarrow \infty$.

3. A tank originally contains 100 gal of fresh water. Then water containing 1/2 lb of salt per gallon is poured into the tank at a rate of 2 gal/min, and the mixture is allowed to leave at the same rate. After 10 min the process is stopped, and fresh water is poured into the tank at a rate of 2 gal/min, with the mixture again leaving at the same rate. Find the amount of salt in the tank at the end of an additional 10 min.

4. A tank with a capacity of 500 gal originally contains 200 gal of water with 100 lb of salt in solution. Water containing 1 lb of salt per gallon is entering at a rate of 3 gal/min, and the mixture is allowed to flow out of the tank at a rate of 2 gal/min. Find the amount of salt in the tank at any time prior to the instant when the solution begins to overflow. Find the concentration (in pounds per gallon) of salt in the tank when it is on the point of overflowing. Compare this concentration with the theoretical limiting concentration if the tank had infinite capacity.

5. A tank contains 100 gal of water and 50 oz of salt. Water containing a salt concentration of $(1/4)[1 + (1/2) \sin t]$ oz/gal flows into the tank at a rate of 2 gal/min, and the mixture in the tank flows out at the same rate.

(a) Find the amount of salt in the tank at any time.

(b) Plot the solution for a time period long enough so that you see the ultimate behavior of the graph.

(c) The long-time behavior of the solution is an oscillation about a certain constant level. What is this level? What is the amplitude of the oscillation?

6. Suppose that a tank containing a certain liquid has an outlet near the bottom. Let $h(t)$ be the height of the liquid surface above the outlet at time t. Torricelli's principle states that the outflow velocity v at the outlet is equal to the velocity of a particle falling freely (with no drag) from the height h.

(a) Show that $v = \sqrt{2gh}$, where g is the acceleration due to gravity.

(b) By equating the rate of outflow to the rate of change of liquid in the tank, show that $h(t)$ satisfies the equation

$$A(h)\frac{dh}{dt} = -\alpha a\sqrt{2gh}, \qquad \text{(i)}$$

where $A(h)$ is the area of the cross section of the tank at height h and a is the area of the outlet. The constant α is a contraction coefficient that accounts for the observed fact that the cross section of the (smooth) outflow stream is smaller than a. The value of α for water is about 0.6.

(c) Consider a water tank in the form of a right circular cylinder that is 3 m high above the outlet. The radius of the tank is 1 m and the radius of the circular outlet is 0.1 m. If the tank is initially full of water, determine how long it takes to drain the tank down to the level of the outlet.

7. Suppose that a sum S_0 is invested at an annual rate of return r compounded continuously.

(a) Find the time T required for the original sum to double in value as a function of r.

(b) Determine T if $r = 7\%$.

(c) Find the return rate that must be achieved if the initial investment is to double in 8 years.

8. A young person with no initial capital invests k dollars per year at an annual rate of return r. Assume that investments are made continuously and that the return is compounded continuously.

(a) Determine the sum $S(t)$ accumulated at any time t.

(b) If $r = 7.5\%$, determine k so that $1 million will be available for retirement in 40 years.

(c) If $k = \$2000/\text{year}$, determine the return rate r that must be obtained to have $1 million available in 40 years.

9. A home buyer can afford to spend no more than $800/month on mortgage payments. Suppose that the interest rate is 9% and that the term of the mortgage is 20 years. Assume that interest is compounded continuously and that payments are also made continuously.

(a) Determine the maximum amount that this buyer can afford to borrow.

(b) Determine the total interest paid during the term of the mortgage.

10. A recent college graduate borrows $100,000 at an interest rate of 9% to purchase a condominium. Anticipating steady salary increases, the buyer expects to make payments at a monthly rate of $800(1+t/120)$, where t is the number of months since the loan was made.

(a) Assuming that this payment schedule can be maintained, when will the loan be fully paid?

(b) Assuming the same payment schedule, how large a loan could be paid off in exactly 20 years?

11. **A Difference Equation.** In this problem, we approach the loan problem in Example 2 from a discrete viewpoint. This leads to a difference equation rather than a differential equation.

(a) Let S_0 be the initial balance of the loan, and let S_n be the balance after n months. Show that

$$S_n = (1 + r)S_{n-1} - k, \quad n = 1, 2, 3, \ldots \quad (i)$$

where r is the monthly interest rate and k is the monthly payment. In Example 2, the annual interest rate is 8%, so here we take $r = 0.08/12$.

(b) Let $R = 1 + r$, so that Eq. (i) becomes

$$S_n = R S_{n-1} - k, \quad n = 1, 2, 3, \ldots \quad (ii)$$

Find S_1, S_2, and S_3.

(c) Use an induction argument to show that

$$S_n = R^n S_0 - \frac{R^n - 1}{R - 1} k \quad (iii)$$

for each positive integer n.

(d) Let $S_0 = 20{,}000$ and suppose that (as in Example 2) the loan is to be paid off in 48 months. Find the value of k and compare it with the result of Example 2.

12. An important tool in archeological research is radiocarbon dating, developed by the American chemist Willard F. Libby. This is a means of determining the age of certain wood and plant remains, hence of animal or human bones or artifacts found buried at the same levels. Radiocarbon dating is based on the fact that some wood or plant remains contain residual amounts of carbon-14, a radioactive isotope of carbon. This isotope is accumulated during the lifetime of the plant and begins to decay at its death. Since the half-life of carbon-14 is long (approximately 5730 years[2]), measurable amounts of carbon-14 remain after many thousands of years. If even a tiny fraction of the original amount of carbon-14 is still present, then by appropriate laboratory measurements the *proportion* of the original amount of carbon-14 that remains can be accurately determined. In other words, if $Q(t)$ is the amount of carbon-14 at time t and Q_0 is the original amount, then the ratio $Q(t)/Q_0$ can be determined, at least if this quantity is not too small. Present measurement techniques permit the use of this method for time periods of 50,000 years or more.

(a) Assuming that Q satisfies the differential equation $Q' = -rQ$, determine the decay constant r for carbon-14.

(b) Find an expression for $Q(t)$ at any time t, if $Q(0) = Q_0$.

(c) Suppose that certain remains are discovered in which the current residual amount of carbon-14 is 20% of the original amount. Determine the age of these remains.

---

[2] *McGraw-Hill Encyclopedia of Science and Technology* (8th ed.) (New York: McGraw-Hill, 1997), Vol. 5, p. 48.

13. The population of mosquitoes in a certain area increases at a rate proportional to the current population, and in the absence of other factors, the population doubles each week. There are 200,000 mosquitoes in the area initially, and predators (birds, bats, and so forth) eat 20,000 mosquitoes/day. Determine the population of mosquitoes in the area at any time.

ODEA 14. Suppose that a certain population has a growth rate that varies with time and that this population satisfies the differential equation

$$dy/dt = (0.5 + \sin t)y/5.$$

(a) If $y(0) = 1$, find (or estimate) the time τ at which the population has doubled. Choose other initial conditions and determine whether the doubling time τ depends on the initial population.

(b) Suppose that the growth rate is replaced by its average value $1/10$. Determine the doubling time τ in this case.

(c) Suppose that the term $\sin t$ in the differential equation is replaced by $\sin 2\pi t$; that is, the variation in the growth rate has a substantially higher frequency. What effect does this have on the doubling time τ?

(d) Plot the solutions obtained in parts (a), (b), and (c) on a single set of axes.

ODEA 15. Suppose that a certain population satisfies the initial value problem

$$dy/dt = r(t)y - k, \qquad y(0) = y_0,$$

where the growth rate $r(t)$ is given by $r(t) = (1 + \sin t)/5$, and k represents the rate of predation.

(a) Suppose that $k = 1/5$. Plot y versus t for several values of y_0 between $1/2$ and 1.

(b) Estimate the critical initial population y_c below which the population will become extinct.

(c) Choose other values of k and find the corresponding y_c for each one.

(d) Use the data you have found in parts (b) and (c) to plot y_c versus k.

16. Newton's law of cooling states that the temperature of an object changes at a rate proportional to the difference between its temperature and that of its surroundings. Suppose that the temperature of a cup of coffee obeys Newton's law of cooling. If the coffee has a temperature of $200°$F when freshly poured, and 1 min later has cooled to $190°$F in a room at $70°$F, determine when the coffee reaches a temperature of $150°$F.

17. Heat transfer from a body to its surroundings by ODEA radiation, based on the Stefan-Boltzmann law, is described by the differential equation

$$\frac{du}{dt} = -\alpha(u^4 - T^4), \qquad \text{(i)}$$

where $u(t)$ is the absolute temperature of the body at time t, T is the absolute temperature of the surroundings, and α is a constant depending on the physical parameters of the body. However, if u is much larger than T, then solutions of Eq. (i) are well approximated by solutions of the simpler equation

$$\frac{du}{dt} = -\alpha u^4. \qquad \text{(ii)}$$

Suppose that a body with initial temperature $2000°$K is surrounded by a medium with temperature $300°$K and that $\alpha = 2.0 \times 10^{-12} \, °\text{K}^{-3}/\text{s}$.

(a) Determine the temperature of the body at any time by solving Eq. (ii).

(b) Plot the graph of u versus t.

(c) Find the time τ at which $u(\tau) = 600$, that is, twice the ambient temperature. Up to this time, the error in using Eq. (ii) to approximate the solutions of Eq. (i) is no more than 1%.

18. Consider an insulated box (a building, perhaps) ODEA with internal temperature $u(t)$. According to Newton's law of cooling, u satisfies the differential equation

$$\frac{du}{dt} = -k[u - T(t)]. \qquad \text{(i)}$$

where $T(t)$ is the ambient (external) temperature. Suppose that $T(t)$ varies sinusoidally; for example, assume that $T(t) = T_0 + T_1 \cos \omega t$.

(a) Solve Eq. (i) and express $u(t)$ in terms of t, k, T_0, T_1, and ω. Observe that part of your solution approaches zero as t becomes large; this is called the transient part. The remainder of the solution is called the steady state; denote it by $S(t)$.

(b) Suppose that t is measured in hours and that $\omega = \pi/12$, corresponding to a period of 24 hours for $T(t)$. Further, let $T_0 = 60°$F, $T_1 = 15°$F, and $k = 0.2$/h. Draw graphs of $S(t)$ and $T(t)$ versus t on the same axes. From your graph, estimate the amplitude R of the oscillatory part of $S(t)$. Also estimate the time lag τ between corresponding maxima of $T(t)$ and $S(t)$.

(c) Let k, T_0, T_1, and ω now be unspecified. Write the oscillatory part of $S(t)$ in the form $R\cos[\omega(t - \tau)]$. Use trigonometric identities to find expressions for R and τ. Let T_1 and ω have the values given in part (b), and plot graphs of R and τ versus k.

19. Consider a lake of constant volume V containing at time t an amount $Q(t)$ of pollutant, evenly distributed throughout the lake with a concentration $c(t)$, where $c(t) = Q(t)/V$. Assume that water containing a concentration k of pollutant enters the lake at a rate r, and that water leaves the lake at the same rate. Suppose that pollutants are also added directly to the lake at a constant rate P. Note that the given assumptions neglect a number of factors that may, in some cases, be important—for example, the water added or lost by precipitation, absorption, and evaporation; the stratifying effect of temperature differences in a deep lake; the tendency of irregularities in the coastline to produce sheltered bays; and the fact that pollutants are not deposited evenly throughout the lake but (usually) at isolated points around its periphery. The results below must be interpreted in the light of the neglect of such factors as these.

   (a) If at time $t = 0$ the concentration of pollutant is c_0, find an expression for the concentration $c(t)$ at any time. What is the limiting concentration as $t \to \infty$?

   (b) If the addition of pollutants to the lake is terminated ($k = 0$ and $P = 0$ for $t > 0$), determine the time interval T that must elapse before the concentration of pollutants is reduced to 50% of its original value; to 10% of its original value.

   (c) Table 2.3.1 contains data[3] for several of the Great Lakes. Using these data, determine from part (b) the time T necessary to reduce the contamination of each of these lakes to 10% of the original value.

TABLE 2.3.1   Volume and Flow Data for the Great Lakes.

Lake	V (km^3 $\times 10^3$)	r (km^3/yr)
Superior	12.2	65.2
Michigan	4.9	158
Erie	0.46	175
Ontario	1.6	209

20. A ball with mass 0.15 kg is thrown upward with initial velocity 20 m/s from the roof of a building 30 m high. Neglect air resistance.

   (a) Find the maximum height above the ground that the ball reaches.

   (b) Assuming that the ball misses the building on the way down, find the time that it hits the ground.

   (c) Plot the graphs of velocity and position versus time.

21. Assume that conditions are as in Problem 20 except that there is a force due to air resistance of $|v|/30$, where the velocity v is measured in m/s.

   (a) Find the maximum height above the ground that the ball reaches.

   (b) Find the time that the ball hits the ground.

   (c) Plot the graphs of velocity and position versus time. Compare these graphs with the corresponding ones in Problem 20.

22. Assume that conditions are as in Problem 20 except that there is a force due to air resistance of $v^2/1325$, where the velocity v is measured in m/s.

   (a) Find the maximum height above the ground that the ball reaches.

   (b) Find the time that the ball hits the ground.

   (c) Plot the graphs of velocity and position versus time. Compare these graphs with the corresponding ones in Problems 20 and 21.

23. A sky diver weighing 180 lb (including equipment) falls vertically downward from an altitude of 5000 ft and opens the parachute after 10 s of free fall. Assume that the force of air resistance is $0.75|v|$ when the parachute is closed and $12|v|$ when the parachute is open, where the velocity v is measured in ft/s.

   (a) Find the speed of the sky diver when the parachute opens.

   (b) Find the distance fallen before the parachute opens.

   (c) What is the limiting velocity v_L after the parachute opens?

   (d) Determine how long the sky diver is in the air after the parachute opens.

   (e) Plot the graph of velocity versus time from the beginning of the fall until the skydiver reaches the ground.

[3]This problem is based on R. H. Rainey, "Natural Displacement of Pollution from the Great Lakes," *Science 155* (1967), pp. 1242–1243; the information in the table was taken from that source.

24. A rocket sled having an initial speed of 150 mi/h is slowed by a channel of water. Assume that, during the braking process, the acceleration a is given by $a(v) = -\mu v^2$, where v is the velocity and μ is a constant.

    (a) As in Example 4 in the text, use the relation $dv/dt = v(dv/dx)$ to write the equation of motion in terms of v and x.
    (b) If it requires a distance of 2000 ft to slow the sled to 15 mi/h, determine the value of μ.
    (c) Find the time τ required to slow the sled to 15 mi/h.

25. A body of constant mass m is projected vertically upward with an initial velocity v_0 in a medium offering a resistance $k|v|$, where k is a constant. Neglect changes in the gravitational force.

    (a) Find the maximum height x_m attained by the body and the time t_m at which this maximum height is reached.
    (b) Show that if $kv_0/mg < 1$, then t_m and x_m can be expressed as

    $$t_m = \frac{v_0}{g}\left[1 - \frac{1}{2}\frac{kv_0}{mg} + \frac{1}{3}\left(\frac{kv_0}{mg}\right)^2 - \cdots\right],$$

    $$x_m = \frac{v_0^2}{2g}\left[1 - \frac{2}{3}\frac{kv_0}{mg} + \frac{1}{2}\left(\frac{kv_0}{mg}\right)^2 - \cdots\right].$$

    (c) Show that the quantity kv_0/mg is dimensionless.

26. A body of mass m is projected vertically upward with an initial velocity v_0 in a medium offering a resistance $k|v|$, where k is a constant. Assume that the gravitational attraction of the earth is constant.

    (a) Find the velocity $v(t)$ of the body at any time.
    (b) Use the result of part (a) to calculate the limit of $v(t)$ as $k \to 0$, that is, as the resistance approaches zero. Does this result agree with the velocity of a mass m projected upward with an initial velocity v_0 in a vacuum?
    (c) Use the result of part (a) to calculate the limit of $v(t)$ as $m \to 0$, that is, as the mass approaches zero.

27. A body falling in a relatively dense fluid, oil for example, is acted on by three forces (see Figure 2.3.5): a resistive force R, a buoyant force B, and its weight w due to gravity. The buoyant force is equal to the weight of the fluid displaced by the object. For a slowly moving spherical body of radius a, the resistive force is given by Stokes' law,

$R = 6\pi\,\mu a|v|$, where v is the velocity of the body, and μ is the coefficient of viscosity of the surrounding fluid.

FIGURE 2.3.5 A body falling in a dense fluid.

(a) Find the limiting velocity of a solid sphere of radius a and density ρ falling freely in a medium of density ρ' and coefficient of viscosity μ.
(b) In 1910, R. A. Millikan studied the motion of tiny droplets of oil falling in an electric field. A field of strength E exerts a force Ee on a droplet with charge e. Assume that E has been adjusted so the droplet is held stationary ($v = 0$) and that w and B are as given above. Find an expression for e. Millikan repeated this experiment many times, and from the data that he gathered he was able to deduce the charge on an electron.

28. A mass of 0.25 kg is dropped from rest in a medium offering a resistance of $0.2|v|$, where v is measured in m/s.

    (a) If the mass is dropped from a height of 30 m, find its velocity when it hits the ground.
    (b) If the mass is to attain a velocity of no more than 10 m/s, find the maximum height from which it can be dropped.
    (c) Suppose that the resistive force is $k|v|$, where v is measured in m/s and k is a constant. If the mass is dropped from a height of 30 m and must hit the ground with a velocity of no more than 10 m/s, determine the coefficient of resistance k that is required.

29. Suppose that a rocket is launched straight up from the surface of the earth with initial velocity $v_0 = \sqrt{2gR}$, where R is the radius of the earth. Neglect air resistance.

(a) Find an expression for the velocity v in terms of the distance x from the surface of the earth.

(b) Find the time required for the rocket to go 240,000 miles (the approximate distance from the earth to the moon). Assume that $R = 4000$ miles.

ODEA 30. Let $v(t)$ and $w(t)$, respectively, be the horizontal and vertical components of the velocity of a batted (or thrown) baseball. In the absence of air resistance, v and w satisfy the equations

$$dv/dt = 0, \qquad dw/dt = -g.$$

(a) Show that

$$v = u \cos A, \qquad w = -gt + u \sin A,$$

where u is the initial speed of the ball and A is its initial angle of elevation.

(b) Let $x(t)$ and $y(t)$, respectively, be the horizontal and vertical coordinates of the ball at time t. If $x(0) = 0$ and $y(0) = h$, find $x(t)$ and $y(t)$ at any time t.

(c) Let $g = 32$ ft/s^2, $u = 125$ ft/s, and $h = 3$ ft. Plot the trajectory of the ball for several values of the angle A; that is, plot $x(t)$ and $y(t)$ parametrically.

(d) Suppose the outfield wall is at a distance L and has height H. Find a relation between u and A that must be satisfied if the ball is to clear the wall.

(e) Suppose that $L = 350$ ft and $H = 10$ ft. Using the relation in part (d), find (or estimate from a plot) the range of values of A that correspond to an initial velocity of $u = 110$ ft/s.

(f) For $L = 350$ and $H = 10$, find the minimum initial velocity u and the corresponding optimal angle A for which the ball will clear the wall.

ODEA 31. A more realistic model (than that in Problem 30) of a baseball in flight includes the effect of air resistance. In this case, the equations of motion are

$$dv/dt = -rv, \qquad dw/dt = -g - rw,$$

where r is the coefficient of resistance.

(a) Determine $v(t)$ and $w(t)$ in terms of initial speed u and initial angle of elevation A.

(b) Find $x(t)$ and $y(t)$ if $x(0) = 0$ and $y(0) = h$.

(c) Plot the trajectory of the ball for $r = 1/5$, $u = 125$, $h = 3$, and for several values of A. How do the trajectories differ from those in Problem 31 with $r = 0$?

(d) Assuming that $r = 1/5$ and $h = 3$, find the minimum initial velocity u and the optimal angle A for which the ball will clear a wall that is 350 ft distant and 10 ft high. Compare this result with that in Problem 30(f).

32. **Brachistochrone Problem.** One of the famous problems in the history of mathematics is the brachistochrone[4] problem: to find the curve along which a particle will slide without friction in the minimum time from one given point P to another Q, the second point being lower than the first but not directly beneath it (see Figure 2.3.6). This problem was posed by Johann Bernoulli in 1696 as a challenge problem to the mathematicians of his day. Correct solutions were found by Johann Bernoulli and his brother Jakob Bernoulli and by Isaac Newton, Gottfried Leibniz, and the Marquis de L'Hospital. The brachistochrone problem is important in the development of mathematics as one of the forerunners of the calculus of variations.

In solving this problem, it is convenient to take the origin as the upper point P and to orient the axes as shown in Figure 2.3.6. The lower point Q has coordinates (x_0, y_0). It is then possible to show that the curve of minimum time is given by a function $y = \phi(x)$ that satisfies the differential equation

$$(1 + y'^2)y = k^2, \tag{i}$$

where k^2 is a certain positive constant to be determined later.

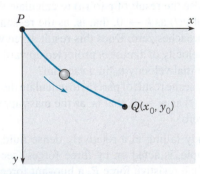

FIGURE 2.3.6 The brachistochrone.

[4]The word "brachistochrone" comes from the Greek words *brachistos*, meaning shortest, and *chronos*, meaning time.

(a) Solve Eq. (i) for y'. Why is it necessary to choose the positive square root?

(b) Introduce the new variable t by the relation

$$y = k^2 \sin^2 t. \qquad \text{(ii)}$$

Show that the equation found in part (a) then takes the form

$$2k^2 \sin^2 t \, dt = dx. \qquad \text{(iii)}$$

(c) Letting $\theta = 2t$, show that the solution of Eq. (iii) for which $x = 0$ when $y = 0$ is given by

$$x = k^2(\theta - \sin\theta)/2, \quad y = k^2(1 - \cos\theta)/2. \qquad \text{(iv)}$$

Equations (iv) are parametric equations of the solution of Eq. (i) that passes through $(0,0)$. The graph of Eqs. (iv) is called a **cycloid**.

(d) If we make a proper choice of the constant k, then the cycloid also passes through the point (x_0, y_0) and is the solution of the brachistochrone problem. Find k if $x_0 = 1$ and $y_0 = 2$.

2.4 Differences Between Linear and Nonlinear Equations

Up to now, we have been primarily concerned with showing that first order differential equations can be used to investigate many different kinds of problems in the natural sciences, and with presenting methods of solving such equations if they are either linear or separable. Now it is time to turn our attention to some more general questions about differential equations and to explore, in more detail, some important ways in which nonlinear equations differ from linear ones.

▶ **Existence and Uniqueness of Solutions.** So far, we have discussed a number of initial value problems, each of which had a solution and apparently only one solution. This raises the question of whether this is true of all initial value problems for first order equations. In other words, does every initial value problem have exactly one solution? This may be an important question even for nonmathematicians. If you encounter an initial value problem in the course of investigating some physical problem, you might want to know that it has a solution before spending very much time and effort in trying to find it. Further, if you are successful in finding one solution, you might be interested in knowing whether you should continue a search for other possible solutions or whether you can be sure that there are no other solutions. For linear equations, the answers to these questions are given by the following fundamental theorem.

THEOREM 2.4.1	If the functions p and g are continuous on an open interval $I: \alpha < t < \beta$ containing the point $t = t_0$, then there exists a unique function $y = \phi(t)$ that satisfies the differential equation $$y' + p(t)y = g(t) \qquad (1)$$ for each t in I, and that also satisfies the initial condition $$y(t_0) = y_0, \qquad (2)$$ where y_0 is an arbitrary prescribed initial value.

Observe that Theorem 2.4.1 states that the given initial value problem *has* a solution and also that the problem has *only one* solution. In other words, the theorem asserts both the *existence* and *uniqueness* of the solution of the initial value problem (1), (2). In addition, it states that the solution exists throughout any interval I containing the initial point t_0 in which the coefficients p and g are continuous. That is, the solution can be discontinuous or fail to exist only at points where at least one of p and g is discontinuous. Such points can often be identified at a glance.

The proof of this theorem is partly contained in the discussion in Section 2.1 leading to the formula (Eq. (26) in Section 2.1)

$$\mu(t)y = \int \mu(t)g(t)\,dt + c, \tag{3}$$

where (Eq. (24) in Section 2.1)

$$\mu(t) = \exp \int p(t)\,dt. \tag{4}$$

The derivation in Section 2.1 shows that if Eq. (1) has a solution, then it must be given by Eq. (3). By looking a little more closely at that derivation, we can also conclude that the differential equation (1) must indeed have a solution. Since p is continuous for $\alpha < t < \beta$, it follows that μ is defined in this interval and is a nonzero differentiable function. Upon multiplying Eq. (1) by $\mu(t)$ we obtain

$$[\mu(t)y]' = \mu(t)g(t). \tag{5}$$

Since both μ and g are continuous, the function μg is integrable, and Eq. (3) follows from Eq. (5). Further, the integral of μg is differentiable, so y as given by Eq. (3) exists and is differentiable throughout the interval $\alpha < t < \beta$. By substituting the expression for y from Eq. (3) into either Eq. (1) or Eq. (5), you can easily verify that this expression satisfies the differential equation throughout the interval $\alpha < t < \beta$. Finally, the initial condition (2) determines the constant c uniquely, so there is only one solution of the initial value problem, thus completing the proof.

Equation (4) determines the integrating factor $\mu(t)$ only up to a multiplicative factor that depends on the lower limit of integration. If we choose this lower limit to be t_0, then

$$\mu(t) = \exp \int_{t_0}^{t} p(s)\,ds, \tag{6}$$

and it follows that $\mu(t_0) = 1$. Using the integrating factor given by Eq. (6), and choosing the lower limit of integration in Eq. (3) also to be t_0, we obtain the general solution of Eq. (1) in the form

$$y = \frac{1}{\mu(t)}\left[\int_{t_0}^{t} \mu(s)g(s)\,ds + c\right]. \tag{7}$$

To satisfy the initial condition (2), we must choose $c = y_0$. Thus the solution of the initial value problem (1), (2) is

$$y = \frac{1}{\mu(t)}\left[\int_{t_0}^{t} \mu(s)g(s)\,ds + y_0\right], \tag{8}$$

where $\mu(t)$ is given by Eq. (6).

Turning now to nonlinear differential equations, we must replace Theorem 2.4.1 by a more general theorem, such as the following.

THEOREM 2.4.2	Let the functions f and $\partial f/\partial y$ be continuous in some rectangle $\alpha < t < \beta,\ \gamma < y < \delta$ containing the point (t_0, y_0). Then, in some interval $t_0 - h < t < t_0 + h$ contained in

$\alpha < t < \beta$, there is a unique solution $y = \phi(t)$ of the initial value problem

$$y' = f(t, y), \qquad y(t_0) = y_0. \tag{9}$$

Observe that the hypotheses in Theorem 2.4.2 reduce to those in Theorem 2.4.1 if the differential equation is linear. For then $f(t, y) = -p(t)y + g(t)$ and $\partial f(t, y)/\partial y = -p(t)$, so the continuity of f and $\partial f/\partial y$ is equivalent to the continuity of p and g in this case. The proof of Theorem 2.4.1 was comparatively simple because it could be based on the expression (3) that gives the solution of an arbitrary linear equation. There is no corresponding expression for the solution of the differential equation (9), so the proof of Theorem 2.4.2 is much more difficult. It is discussed in more advanced books on differential equations.

Here we note that the conditions stated in Theorem 2.4.2 are sufficient to guarantee the existence of a unique solution of the initial value problem (9) in some interval $t_0 - h < t < t_0 + h$, but they are not necessary. That is, the conclusion remains true under slightly weaker hypotheses about the function f. In fact, the existence of a solution (but not its uniqueness) can be established on the basis of the continuity of f alone.

An important geometrical consequence of the uniqueness parts of Theorems 2.4.1 and 2.4.2 is that the graphs of two solutions cannot intersect each other. Otherwise, there would be two solutions that satisfy the initial condition corresponding to the point of intersection, in violation of Theorem 2.4.1 or 2.4.2.

We now consider some examples.

EXAMPLE 1

Use Theorem 2.4.1 to find an interval in which the initial value problem

$$ty' + 2y = 4t^2, \tag{10}$$

$$y(1) = 2 \tag{11}$$

has a unique solution.

Rewriting Eq. (10) in the standard form (1), we have

$$y' + (2/t)y = 4t,$$

so $p(t) = 2/t$ and $g(t) = 4t$. Thus, for this equation, g is continuous for all t, while p is continuous only for $t < 0$ or for $t > 0$. The interval $t > 0$ contains the initial point; consequently, Theorem 2.4.1 guarantees that the problem (10), (11) has a unique solution on the interval $0 < t < \infty$. In Example 2 of Section 2.1, we found the solution of this initial value problem to be

$$y = t^2 + \frac{1}{t^2}, \qquad t > 0. \tag{12}$$

Now suppose that the initial condition (11) is changed to $y(-1) = 2$. Then Theorem 2.4.1 asserts the existence of a unique solution for $t < 0$. As you can readily verify, the solution is again given by Eq. (12), but now on the interval $-\infty < t < 0$.

EXAMPLE 2

Apply Theorem 2.4.2 to the initial value problem

$$\frac{dy}{dx} = \frac{3x^2 + 4x + 2}{2(y - 1)}, \qquad y(0) = -1. \tag{13}$$

Note that Theorem 2.4.1 is not applicable to this problem since the differential equation is nonlinear. To apply Theorem 2.4.2, observe that

$$f(x, y) = \frac{3x^2 + 4x + 2}{2(y - 1)}, \qquad \frac{\partial f}{\partial y}(x, y) = -\frac{3x^2 + 4x + 2}{2(y - 1)^2}.$$

Thus each of these functions is continuous everywhere except on the line $y = 1$. Consequently, a rectangle can be drawn about the initial point $(0, -1)$ in which both f and $\partial f/\partial y$ are continuous. Therefore Theorem 2.4.2 guarantees that the initial value problem has a unique solution in some interval about $x = 0$. However, even though the rectangle can be stretched infinitely far in both the positive and negative x directions, this does not necessarily mean that the solution exists for all x. Indeed, the initial value problem (13) was solved in Example 2 of Section 2.2 and the solution exists only for $x > -2$.

Now suppose we change the initial condition to $y(0) = 1$. The initial point now lies on the line $y = 1$ so no rectangle can be drawn about it within which f and $\partial f/\partial y$ are continuous. Consequently, Theorem 2.4.2 says nothing about possible solutions of this modified problem. However, if we separate the variables and integrate, as in Section 2.2, we find that

$$y^2 - 2y = x^3 + 2x^2 + 2x + c.$$

Further, if $x = 0$ and $y = 1$, then $c = -1$. Finally, by solving for y, we obtain

$$y = 1 \pm \sqrt{x^3 + 2x^2 + 2x}. \tag{14}$$

Equation (14) provides two functions that satisfy the given differential equation for $x > 0$ and also satisfy the initial condition $y(0) = 1$. Thus the initial value problem consisting of the differential equation (13) with the initial condition $y(0) = 1$ does not have a unique solution. The two solutions are shown in Figure 2.4.1.

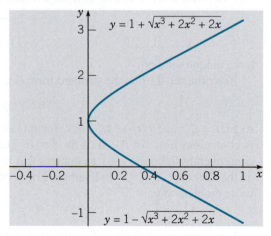

FIGURE 2.4.1   Nonunique solutions of the differential equation (13) with the initial condition $y(0) = 1$.

EXAMPLE 3

Consider the initial value problem

$$y' = y^{1/3}, \qquad y(0) = 0 \tag{15}$$

for $t \geq 0$. Apply Theorem 2.4.2 to this initial value problem and then solve the problem.

The function $f(t, y) = y^{1/3}$ is continuous everywhere, but $\partial f/\partial y$ does not exist when $y = 0$, and hence is not continuous there. Thus Theorem 2.4.2 does not apply to this problem and no conclusion can be drawn from it. However, by the remark following Theorem 2.4.2 the continuity of f does guarantee the existence of solutions, but not their uniqueness.

To understand the situation more clearly, we must actually solve the problem, which is easy to do since the differential equation is separable. Thus we have

$$y^{-1/3}dy = dt,$$

so

$$\tfrac{3}{2}y^{2/3} = t + c$$

and

$$y = \left[\tfrac{2}{3}(t+c)\right]^{3/2}.$$

The initial condition is satisfied if $c = 0$, so

$$y = \phi_1(t) = \left(\tfrac{2}{3}t\right)^{3/2}, \qquad t \geq 0 \tag{16}$$

satisfies both of Eqs. (15). On the other hand, the function

$$y = \phi_2(t) = -\left(\tfrac{2}{3}t\right)^{3/2}, \qquad t \geq 0 \tag{17}$$

is also a solution of the initial value problem. Moreover, the function

$$y = \psi(t) = 0, \qquad t \geq 0 \tag{18}$$

is yet another solution. Indeed, it is not hard to show that, for an arbitrary positive t_0, the functions

$$y = \chi(t) = \begin{cases} 0, & \text{if } 0 \leq t < t_0, \\ \pm\left[\tfrac{2}{3}(t - t_0)\right]^{3/2}, & \text{if } t \geq t_0 \end{cases} \tag{19}$$

are continuous, differentiable (in particular at $t = t_0$), and are solutions of the initial value problem (15). Hence this problem has an infinite family of solutions; see Figure 2.4.2, where a few of these solutions are shown.

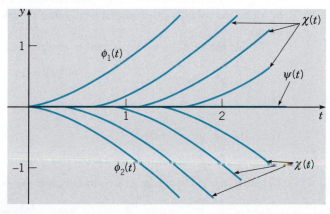

FIGURE 2.4.2   Several solutions of the initial value problem $y' = y^{1/3}$, $y(0) = 0$.

As already noted, the nonuniqueness of the solutions of the problem (15) does not contradict the existence and uniqueness theorem, since the theorem is not applicable if the initial point lies on the t-axis. If (t_0, y_0) is any point not on the t-axis, however, then the theorem guarantees that there is a unique solution of the differential equation $y' = y^{1/3}$ passing through (t_0, y_0).

▶ **Interval of Definition.** According to Theorem 2.4.1, the solution of a linear equation (1),

$$y' + p(t)y = g(t),$$

subject to the initial condition $y(t_0) = y_0$, exists throughout any interval about $t = t_0$ in which the functions p and g are continuous. Thus, vertical asymptotes or other discontinuities in the solution can occur only at points of discontinuity of p or g. For instance, the solutions in Example 1 (with one exception) are asymptotic to the y-axis, corresponding to the discontinuity at $t = 0$ in the coefficient $p(t) = 2/t$, but none of the solutions has any other point where it fails to exist and to be differentiable. The one exceptional solution shows that solutions may sometimes remain continuous even at points of discontinuity of the coefficients.

On the other hand, for a nonlinear initial value problem satisfying the hypotheses of Theorem 2.4.2, the interval in which a solution exists may be difficult to determine. The solution $y = \phi(t)$ is certain to exist as long as the point $[t, \phi(t)]$ remains within a region in which the hypotheses of Theorem 2.4.2 are satisfied. This is what determines the value of h in that theorem. However, since $\phi(t)$ is usually not known, it may be impossible to locate the point $[t, \phi(t)]$ with respect to this region. In any case, the interval in which a solution exists may have no simple relationship to the function f in the differential equation $y' = f(t, y)$. This is illustrated by the following example.

EXAMPLE 4

Solve the initial value problem

$$y' = y^2, \qquad y(0) = 1, \tag{20}$$

and determine the interval in which the solution exists.

Theorem 2.4.2 guarantees that this problem has a unique solution since $f(t, y) = y^2$ and $\partial f/\partial y = 2y$ are continuous everywhere. However, Theorem 2.4.2 does not give an interval in which the solution exists, and it would be a mistake to conclude that the solution exists for all t.

To find the solution, we separate the variables and integrate, with the result that

$$y^{-2}\, dy = dt \tag{21}$$

and

$$-y^{-1} = t + c.$$

Then, solving for y, we have

$$y = -\frac{1}{t + c}. \tag{22}$$

To satisfy the initial condition we must choose $c = -1$, so

$$y = \frac{1}{1 - t} \tag{23}$$

is the solution of the given initial value problem. Clearly, the solution becomes unbounded as $t \to 1$; therefore, the solution exists only in the interval $-\infty < t < 1$. There is no indication from the differential equation itself, however, that the point $t = 1$ is in any way remarkable. Moreover, if the initial condition is replaced by

$$y(0) = y_0, \tag{24}$$

then the constant c in Eq. (22) must be chosen to be $c = -1/y_0$, and it follows that

$$y = \frac{y_0}{1 - y_0 t} \tag{25}$$

is the solution of the initial value problem with the initial condition (24). Observe that the solution (25) becomes unbounded as $t \to 1/y_0$, so the interval of existence of the solution is $-\infty < t < 1/y_0$ if $y_0 > 0$, and is $1/y_0 < t < \infty$ if $y_0 < 0$. Figure 2.4.3 shows the solution for $y_0 > 0$. This example illustrates another feature of initial value problems for nonlinear equations; namely, the singularities of the solution may depend in an essential way on the initial conditions as well as on the differential equation.

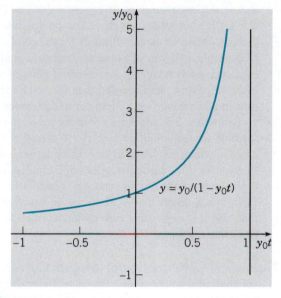

FIGURE 2.4.3 The solution (25) of the initial value problem $y' = y^2$, $y(0) = y_0 > 0$.

▶ **General Solution.** Another way in which linear and nonlinear equations differ concerns the concept of a general solution. For a first order linear equation it is possible to obtain a solution containing one arbitrary constant, from which all possible solutions follow by specifying values for this constant. For nonlinear equations this may not be the case; even though a solution containing an arbitrary constant may be found, there may be other solutions that cannot be obtained by giving values to this constant. For instance, for the differential equation $y' = y^2$ in Example 4, the expression in Eq. (22) contains an arbitrary constant, but does not include all solutions of the differential equation. To show this, observe that the function $y = 0$ for all t is certainly a solution of the differential equation, but it cannot be obtained from Eq. (22) by assigning a value to c. In this example, we might anticipate that something of this sort might happen because, to rewrite the original differential equation in the form (21), we must require that y is not zero. However, the existence of "additional" solutions is not uncommon for nonlinear equations; a less obvious example is given in Problem 22. Thus we will use the term "general solution" only when discussing linear equations.

▶ **Implicit Solutions.** Recall again that, for an initial value problem for a first order linear equation, Eq. (8) provides an explicit formula for the solution $y = \phi(t)$. As long as the necessary antiderivatives can be found, the value of the solution at any point can be determined merely by substituting the appropriate value of t into the equation. The situation for nonlinear equations is much less satisfactory. Usually, the best that we can hope for is to find an equation

$$F(t, y) = 0 \tag{26}$$

involving t and y that is satisfied by the solution $y = \phi(t)$. Even this can be done only for differential equations of certain particular types, of which separable equations are the most important. The equation (26) is called an integral, or first integral, of the differential equation, and (as we have already noted) its graph is an integral curve, or perhaps a family of integral curves. Equation (26), assuming it can be found, defines the solution implicitly; that is, for each value of t we must solve Eq. (26) to find the corresponding value of y. If Eq. (26) is simple enough, it may be possible to solve it for y by analytical means and thereby obtain an explicit formula for the solution. However, more frequently this will not be possible, and you will have to resort to a numerical calculation to determine the value of y for a given value of t. Once several pairs of values of t and y have been calculated, it is often helpful to plot them and then to sketch the integral curve that passes through them. You should arrange for a computer to do this for you, if possible.

Examples 2, 3, and 4 are nonlinear problems in which it is easy to solve for an explicit formula for the solution $y = \phi(t)$. On the other hand, Examples 1 and 3 in Section 2.2 are cases in which it is better to leave the solution in implicit form, and to use numerical means to evaluate it for particular values of the independent variable. The latter situation is more typical; unless the implicit relation is quadratic in y, or has some other particularly simple form, it is unlikely that it can be solved exactly by analytical methods. Indeed, more often than not, it is impossible even to find an implicit expression for the solution of a first order nonlinear equation.

▶ **Graphical or Numerical Construction of Integral Curves.** Because of the difficulty in obtaining exact analytical solutions of nonlinear differential equations, methods that yield approximate solutions or other qualitative information about solutions are of correspondingly greater importance. We have already described, in Section 1.1, how the direction field of a differential equation can be constructed. The direction field can often show the qualitative form of solutions and can also be helpful in identifying regions of the ty-plane where solutions exhibit interesting features that merit more detailed analytical or numerical investigation. Graphical methods for first order equations are discussed further in Section 2.5. An introduction to numerical approximation methods for first order equations was given in Section 1.3, and a more systematic discussion of numerical methods appears in Sections 2.7 and 2.8. However, it is not necessary to study the numerical algorithms themselves in order to use effectively one of the many software packages that generate and plot numerical approximations to solutions of initial value problems.

▶ **Summary.** The linear equation $y' + p(t)y = g(t)$ has several nice properties that can be summarized in the following statements:

  1. Assuming that the coefficients are continuous, there is a general solution, containing an arbitrary constant, that includes all solutions of the differential equation.

A particular solution that satisfies a given initial condition can be picked out by choosing the proper value for the arbitrary constant.

2. There is an expression for the solution, namely, Eq. (7) or Eq. (8). Moreover, although it involves two integrations, the expression is an explicit one for the solution $y = \phi(t)$ rather than an equation that defines ϕ implicitly.

3. The possible points of discontinuity, or singularities, of the solution can be identified (without solving the problem) merely by finding the points of discontinuity of the coefficients. Thus, if the coefficients are continuous for all t, then the solution also exists and is continuous for all t.

None of these statements is true, in general, of nonlinear equations. Although a nonlinear equation may well have a solution involving an arbitrary constant, there may also be other solutions. There is no general formula for solutions of nonlinear equations. If you are able to integrate a nonlinear equation, you are likely to obtain an equation defining solutions implicitly rather than explicitly. Finally, the singularities of solutions of nonlinear equations can usually be found only by solving the equation and examining the solution. It is likely that the singularities will depend on the initial condition as well as the differential equation.

PROBLEMS

In each of Problems 1 through 6, determine (without solving the problem) an interval in which the solution of the given initial value problem is certain to exist.

1. $(t - 3)y' + (\ln t)y = 2t$, $y(1) = 2$
2. $t(t - 4)y' + y = 0$, $y(2) = 1$
3. $y' + (\tan t)y = \sin t$, $y(\pi) = 0$
4. $(4 - t^2)y' + 2ty = 3t^2$, $y(-3) = 1$
5. $(4 - t^2)y' + 2ty = 3t^2$, $y(1) = -3$
6. $(\ln t)y' + y = \cot t$, $y(2) = 3$

In each of Problems 7 through 12, state where in the ty-plane the hypotheses of Theorem 2.4.2 are satisfied.

7. $y' = \dfrac{t - y}{2t + 5y}$

8. $y' = (1 - t^2 - y^2)^{1/2}$

9. $y' = \dfrac{\ln |ty|}{1 - t^2 + y^2}$

10. $y' = (t^2 + y^2)^{3/2}$

11. $\dfrac{dy}{dt} = \dfrac{1 + t^2}{3y - y^2}$

12. $\dfrac{dy}{dt} = \dfrac{(\cot t)y}{1 + y}$

In each of Problems 13 through 16, solve the given initial value problem and determine how the interval in which the solution exists depends on the initial value y_0.

13. $y' = -4t/y$, $y(0) = y_0$

14. $y' = 2ty^2$, $y(0) = y_0$
15. $y' + y^3 = 0$, $y(0) = y_0$
16. $y' = t^2/y(1 + t^3)$, $y(0) = y_0$

In each of Problems 17 through 20, draw a direction field and plot (or sketch) several solutions of the given differential equation. Describe how solutions appear to behave as t increases and how their behavior depends on the initial value y_0 when $t = 0$.

17. $y' = ty(3 - y)$
18. $y' = y(3 - ty)$
19. $y' = -y(3 - ty)$
20. $y' = t - 1 - y^2$

21. Consider the initial value problem $y' = y^{1/3}$, $y(0) = 0$ from Example 3 in the text.

    (a) Is there a solution that passes through the point $(1, 1)$? If so, find it.

    (b) Is there a solution that passes through the point $(2, 1)$? If so, find it.

    (c) Consider all possible solutions of the given initial value problem. Determine the set of values that these solutions have at $t = 2$.

22. (a) Verify that both $y_1(t) = 1 - t$ and $y_2(t) = -t^2/4$ are solutions of the initial value problem

$$y' = \frac{-t + (t^2 + 4y)^{1/2}}{2}, \qquad y(2) = -1.$$

Where are these solutions valid?

(b) Explain why the existence of two solutions of the given problem does not contradict the uniqueness part of Theorem 2.4.2.

(c) Show that $y = ct + c^2$, where c is an arbitrary constant, satisfies the differential equation in part (a) for $t \geq -2c$. If $c = -1$, the initial condition is also satisfied, and the solution $y = y_1(t)$ is obtained. Show that there is no choice of c that gives the second solution $y = y_2(t)$.

23. (a) Show that $\phi(t) = e^{2t}$ is a solution of $y' - 2y = 0$ and that $y = c\phi(t)$ is also a solution of this equation for any value of the constant c.

(b) Show that $\phi(t) = 1/t$ is a solution of $y' + y^2 = 0$ for $t > 0$ but that $y = c\phi(t)$ is not a solution of this equation unless $c = 0$ or $c = 1$. Note that the equation of part (b) is nonlinear, while that of part (a) is linear.

24. Show that if $y = \phi(t)$ is a solution of $y' + p(t)y = 0$, then $y = c\phi(t)$ is also a solution for any value of the constant c.

25. Let $y = y_1(t)$ be a solution of

$$y' + p(t)y = 0, \qquad \text{(i)}$$

and let $y = y_2(t)$ be a solution of

$$y' + p(t)y = g(t). \qquad \text{(ii)}$$

Show that $y = y_1(t) + y_2(t)$ is also a solution of Eq. (ii).

26. (a) Show that the solution (7) of the general linear equation (1) can be written in the form

$$y = cy_1(t) + y_2(t), \qquad \text{(i)}$$

where c is an arbitrary constant. Identify the functions y_1 and y_2.

(b) Show that y_1 is a solution of the differential equation

$$y' + p(t)y = 0, \qquad \text{(ii)}$$

corresponding to $g(t) = 0$.

(c) Show that y_2 is a solution of the full linear equation (1). We see later (for example, in Section 4.6) that solutions of higher order linear equations have a pattern similar to Eq. (i).

Bernoulli Equations. Sometimes it is possible to solve a nonlinear equation by making a change of the dependent variable that converts it into a linear equation. The most important such equation has the form

$$y' + p(t)y = q(t)y^n,$$

and is called a Bernoulli equation after Jakob Bernoulli. Problems 27 through 31 deal with equations of this type.

27. (a) Solve Bernoulli's equation when $n = 0$; when $n = 1$.

(b) Show that if $n \neq 0$, 1, then the substitution $v = y^{1-n}$ reduces Bernoulli's equation to a linear equation. This method of solution was found by Leibniz in 1696.

In each of Problems 28 through 31, the given equation is a Bernoulli equation. In each case, solve it by using the substitution mentioned in Problem 27(b).

28. $t^2 y' + 2ty - y^3 = 0, \qquad t > 0$

29. $y' = ry - ky^2$, $r > 0$ and $k > 0$. This equation is important in population dynamics and is discussed in detail in Section 2.5.

30. $y' = \epsilon y - \sigma y^3$, $\epsilon > 0$ and $\sigma > 0$. This equation occurs in the study of the stability of fluid flow.

31. $dy/dt = (\Gamma \cos t + T)y - y^3$, where Γ and T are constants. This equation also occurs in the study of the stability of fluid flow.

Discontinuous Coefficients. Linear differential equations sometimes occur in which one or both of the functions p and g have jump discontinuities. If t_0 is such a point of discontinuity, then it is necessary to solve the equation separately for $t < t_0$ and $t > t_0$. Afterward, the two solutions are matched so that y is continuous at t_0. This is accomplished by a proper choice of the arbitrary constants. The following two problems illustrate this situation. Note in each case that it is impossible also to make y' continuous at t_0.

32. Solve the initial value problem

$$y' + 2y = g(t), \qquad y(0) = 0,$$

where

$$g(t) = \begin{cases} 1, & 0 \leq t \leq 1, \\ 0, & t > 1. \end{cases}$$

33. Solve the initial value problem

$$y' + p(t)y = 0, \qquad y(0) = 1,$$

where

$$p(t) = \begin{cases} 2, & 0 \leq t \leq 1, \\ 1, & t > 1. \end{cases}$$

34. Consider the initial value problem

$$y' + p(t)y = g(t), \qquad y(t_0) = y_0. \qquad \text{(i)}$$

(a) Show that the solution of the initial value problem (i) can be written in the form

$$y = y_0 \exp\left(-\int_{t_0}^t p(s)\,ds\right)$$
$$+ \int_{t_0}^t \exp\left(-\int_s^t p(r)\,dr\right)g(s)\,ds. \tag{27}$$

(b) Assume that $p(t) \geq p_0 > 0$ for all $t \geq t_0$ and that $g(t)$ is bounded for $t \geq t_0$ (that is, there is a constant M such that $|g(t)| \leq M$ for all $t \geq t_0$). Show that the solution of the initial value problem (i) is bounded for $t \geq t_0$.

(c) Construct an example with nonconstant $p(t)$ and $g(t)$ that illustrates this result.

2.5 Autonomous Equations and Population Dynamics

An important class of first order equations are those in which the independent variable does not appear explicitly. Such equations are called **autonomous** and have the form

$$dy/dt = f(y). \tag{1}$$

We will discuss these equations in the context of the growth or decline of the population of a given species, an important issue in fields ranging from medicine to ecology to global economics. A number of other applications are mentioned in some of the problems. Recall that in Sections 1.1 and 1.2 we considered the special case of Eq. (1) in which $f(y) = ay + b$.

Equation (1) is separable, so the discussion in Section 2.2 is applicable to it, but the main purpose of this section is to show how geometrical methods can be used to obtain important qualitative information about solutions directly from the differential equation, without solving the equation. Of fundamental importance in this effort are the concepts of stability and instability of solutions of differential equations. These ideas were introduced informally in Chapter 1, but without using this terminology. They are discussed further here and will be examined in greater depth and in a more general setting in Chapters 3 and 7.

▶ **Exponential Growth.** Let $y = \phi(t)$ be the population of the given species at time t. The simplest hypothesis concerning the variation of population is that the rate of change of y is proportional to the current value of y. For example, if the population doubles, then the number of births in a given time period should also double. Thus we have

$$dy/dt = ry, \tag{2}$$

where the constant of proportionality r is called the **rate of growth or decline**, depending on whether it is positive or negative. Here, we assume that $r > 0$, so the population is growing.

Solving Eq. (2) subject to the initial condition

$$y(0) = y_0, \tag{3}$$

we obtain

$$y = y_0 e^{rt}. \tag{4}$$

Thus the mathematical model consisting of the initial value problem (2), (3) with $r > 0$ predicts that the population will grow exponentially for all time, as shown in Figure 2.5.1 for several values of y_0. Under ideal conditions, Eq. (4) has been observed to be reasonably accurate for many populations, at least for limited periods of time. However, it is clear that such ideal conditions cannot continue indefinitely; eventually, limitations on space, food supply, or other resources will reduce the growth rate and bring an end to uninhibited exponential growth.

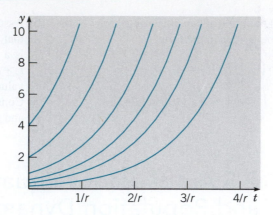

FIGURE 2.5.1 Exponential growth: y versus t for $dy/dt = ry$.

▶ **Logistic Growth.** To take account of the fact that the growth rate actually depends on the population, we replace the constant r in Eq. (2) by a function $h(y)$ and thereby obtain the modified equation

$$dy/dt = h(y)y. \tag{5}$$

We now want to choose $h(y)$ so that $h(y) \cong r > 0$ when y is small, $h(y)$ decreases as y grows larger, and $h(y) < 0$ when y is sufficiently large. The simplest function that has these properties is $h(y) = r - ay$, where a is also a positive constant. Using this function in Eq. (5), we obtain

$$dy/dt = (r - ay)y. \tag{6}$$

Equation (6) is known as the Verhulst equation or the **logistic equation**. It is often convenient to write the logistic equation in the equivalent form

$$\frac{dy}{dt} = r\left(1 - \frac{y}{K}\right)y, \tag{7}$$

where $K = r/a$. The constant r is called the **intrinsic growth rate**, that is, the growth rate in the absence of any limiting factors. The interpretation of K will become clear shortly.

Before proceeding to investigate the solutions of Eq. (7), let us look at a specific example.

EXAMPLE 1

Consider the differential equation

$$\frac{dy}{dt} = \left(1 - \frac{y}{3}\right)y. \tag{8}$$

Without solving the equation, determine the qualitative behavior of its solutions and sketch the graphs of a representative sample of them.

We will follow, more or less, the lines of our approach in Section 1.1. We could begin by drawing a direction field, but we will not do this because we want to show how to sketch solution curves without computer assistance. Recall from Section 1.1 that constant solutions are of particular importance. For such a solution, $dy/dt = 0$ for all t, so any constant solution of Eq. (8) must satisfy the algebraic equation

$$(1 - y/3)y = 0.$$

Thus the constant solutions are $y = \phi_1(t) = 0$ and $y = \phi_2(t) = 3$. These solutions are called **equilibrium solutions** of Eq. (8) because they correspond to no change or variation in the value of y as t increases. Equilibrium solutions are also often referred to as **critical points**.

To visualize other solutions of Eq. (8) and to sketch their graphs quickly, we can proceed in the following way. Let $f(y) = (1 - y/3)y$ and draw the graph of $f(y)$ versus y. Since f is a quadratic function, its graph is a parabola, opening downward, and with intercepts on the y-axis at $y = 0$ and $y = 3$. The graph is shown in Figure 2.5.2. Remember that $f(y_0)$ represents the slope of a line tangent to the graph of the solution passing through a point (t, y_0) in the ty-plane.

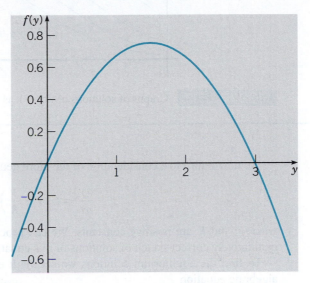

FIGURE 2.5.2   Graph of $f(y) = (1 - y/3)y$ versus y.

To draw a sketch of solutions of Eq. (8), the first step is to draw the equilibrium solutions $y = \phi_1(t) = 0$ and $y = \phi_2(t) = 3$; see Figure 2.5.3. Then from Figure 2.5.2 note that $f(y) > 0$ for $0 < y < 3$. Thus in the ty-plane solutions are increasing (have a positive slope) for $0 < y < 3$. A few of these solutions are shown in Figure 2.5.3. These solution curves flatten out near $y = 0$ and $y = 3$ because, from Figure 2.5.2, their slopes, given by $f(y)$, are near zero there. The slopes reach a maximum at $y = 3/2$, the vertex of the parabola. Observe also that $f(y)$ or dy/dt is increasing for $y < 3/2$ and decreasing for $y > 3/2$. This means that the graphs of y versus t are concave up for $y < 3/2$ and concave down for $y > 3/2$. In other words, solution curves have an inflection point as they cross the line $y = 3/2$.

For $y > 3$ you can see from Figure 2.5.2 that $f(y)$, or dy/dt, is negative and decreasing. Therefore the graphs of y versus t for this range of y are decreasing and concave up. They also become flatter as they approach the equilibrium solution $y = 3$. Some of these graphs are also shown in Figure 2.5.3.

None of the other solutions can intersect the equilibrium solutions $y = 0$ and $y = 3$ at a finite time. If they did, they would violate the uniqueness part of Theorem 2.4.2, which states that only one solution can pass through any given point in the ty-plane.

Finally, although we have drawn Figures 2.5.2 and 2.5.3 with a computer, it is important to understand that very similar qualitatively correct sketches can be drawn by hand, without any computer assistance, by following the steps described in this example.

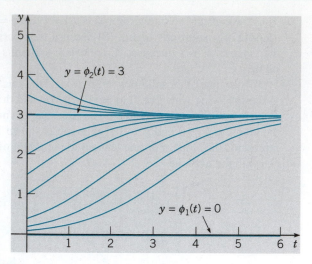

FIGURE 2.5.3   Graphs of solutions of Eq. (8): $y' = (1 - y/3)y$.

We now return to a consideration of the more general Eq. (7),

$$\frac{dy}{dt} = r\left(1 - \frac{y}{K}\right)y,$$

where r and K are positive constants. We can proceed, just as in Example 1, to draw a qualitatively correct sketch of solutions of this equation.

To find the equilibrium solutions we set dy/dt equal to zero and solve the resulting algebraic equation

$$r(1 - y/K)y = 0.$$

Thus, for Eq. (7), the equilibrium solutions are $y = \phi_1(t) = 0$ and $y = \phi_2(t) = K$.

Next we draw the graph of $f(y)$ versus y. In the case of Eq. (7), $f(y) = r(1 - y/K)y$, so the graph is the parabola shown in Figure 2.5.4. The intercepts are $(0, 0)$ and $(K, 0)$, corresponding to the critical points of Eq. (7), and the vertex of the parabola is $(K/2, rK/4)$. Observe that $dy/dt > 0$ for $0 < y < K$ (since $dy/dt = f(y)$); therefore, y is an increasing function of t when y is in this interval. This is indicated by the rightward-pointing arrows near the y-axis in Figure 2.5.4. Similarly, if $y > K$, then $dy/dt < 0$; hence, y is decreasing, as indicated by the leftward-pointing arrow in Figure 2.5.4.

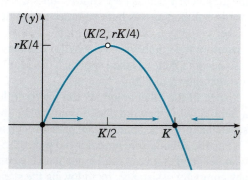

FIGURE 2.5.4 $f(y)$ versus y for $dy/dt = r(1 - y/K)y$.

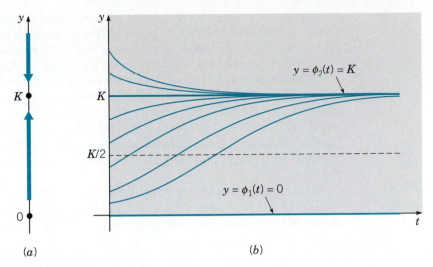

FIGURE 2.5.5   Logistic growth: $dy/dt = r(1 - y/K)y$. (a) The phase line. (b) Plots of y versus t.

In this context, the y-axis is often called the **phase line**, and it is reproduced in its more customary vertical orientation in Figure 2.5.5a. The dots at $y = 0$ and $y = K$ are the critical points, or equilibrium solutions. The arrows again indicate that y is increasing whenever $0 < y < K$ and that y is decreasing whenever $y > K$.

Further, from Figure 2.5.4, note that if y is near zero or K, then the slope $f(y)$ is near zero, so the solution curves are relatively flat. They become steeper as the value of y leaves the neighborhood of zero or K.

To sketch the graphs of solutions of Eq. (7) in the ty-plane, we start with the equilibrium solutions $y = 0$ and $y = K$; then we draw other curves that are increasing when $0 < y < K$, decreasing when $y > K$, and flatten out as y approaches either of the values 0 or K. Thus the graphs of solutions of Eq. (7) must have the general shape shown in Figure 2.5.5b, regardless of the values of r and K.

As in Example 1, the concavity of a solution curve changes as it passes through the value $y = K/2$ corresponding to the vertex of the parabola in Figure 2.5.4. Thus each solution curve has an inflection point when $y = K/2$. Further, Figure 2.5.5b may seem to show that other solutions intersect the equilibrium solution $y = K$, but this is impossible by the uniqueness part of Theorem 2.4.2. Thus, although other solutions may be asymptotic to the equilibrium solution as $t \to \infty$, they cannot intersect it at any finite time.

Finally, observe that K is the upper bound that is approached, but not exceeded, by growing populations starting below this value. Thus it is natural to refer to K as the **saturation level**, or the **environmental carrying capacity**, for the given species.

A comparison of Figures 2.5.1 and 2.5.5b reveals that solutions of the nonlinear equation (7) are strikingly different from those of the linear equation (1), at least for large values of t. Regardless of the value of K, that is, no matter how small the nonlinear term in Eq. (7), solutions of that equation approach a finite value as $t \to \infty$, whereas solutions of Eq. (1) grow (exponentially) without bound as $t \to \infty$. Thus even a tiny nonlinear term in the differential equation has a decisive effect on the solution for large t.

The same methods can be applied to the general autonomous equation (1),

$$dy/dt = f(y).$$

The **equilibrium solutions** of this equation can be found by locating the roots of $f(y) = 0$. The zeros of $f(y)$ are also called **critical points**.

Suppose that $y = y_1$ is an equilibrium solution, or critical point, of this equation; that is, $f(y_1) = 0$. Suppose further that $f'(y_1) > 0$. Then $f(y) < 0$ for values of y just below y_1 and $f(y) > 0$ for values of y just above y_1. In other words, if y is close to y_1, then $dy/dt < 0$ for $y < y_1$ and $dy/dt > 0$ for $y > y_1$. This means that solutions below the equilibrium solution $y = y_1$ are decreasing and those that are above it are increasing. Thus all solutions in the neighborhood of the equilibrium solution are moving away from it. For example, this is the situation at $y = 0$ for the logistic equation (7). On the other hand, if $y = y_1$ is a critical point, and if $f'(y_1) < 0$, then the situation is reversed. In this case, solutions above the equilibrium solution are decreasing, and those that are below it are increasing. Thus all nearby solutions are approaching the equilibrium solution. This is illustrated by the critical point $y = K$ for Eq. (7).

To carry the investigation one step further, we can determine the concavity of the solution curves and the location of inflection points by finding d^2y/dt^2. From the differential equation (1), we obtain (using the chain rule)

$$\frac{d^2y}{dt^2} = \frac{d}{dt}\frac{dy}{dt} = \frac{d}{dt}f(y) = f'(y)\frac{dy}{dt} = f'(y)f(y). \tag{9}$$

The graph of y versus t is concave up when $y'' > 0$, that is, when f and f' have the same sign. Similarly, it is concave down when $y'' < 0$, which occurs when f and f' have opposite signs. The signs of f and f' can be easily identified from the graph of $f(y)$ versus y. Inflection points may occur when $f'(y) = 0$.

In many situations, it is sufficient to have the qualitative information about a solution $y = \phi(t)$ of Eq. (7) that is shown in Figure 2.5.5b. This information was obtained entirely from the graph of $f(y)$ versus y, and without solving the differential equation (7). However, if we wish to have a more detailed description of logistic growth—for example, if we wish to know the value of the population at some particular time—then we must solve Eq. (7) subject to the initial condition (3). Provided that $y \neq 0$ and $y \neq K$, we can write Eq. (7) in the form

$$\frac{dy}{(1 - y/K)y} = r\,dt.$$

Using a partial fraction expansion on the left side, we have

$$\left(\frac{1}{y} + \frac{1/K}{1 - y/K}\right)dy = r\,dt.$$

Then, by integrating both sides, we obtain

$$\ln|y| - \ln\left|1 - \frac{y}{K}\right| = rt + c, \tag{10}$$

where c is an arbitrary constant of integration to be determined from the initial condition $y(0) = y_0$. We have already noted that if $0 < y_0 < K$, then y remains in this interval for all time. Thus in this case we can remove the absolute value bars in Eq. (10), and by taking the exponential of both sides, we find that

$$\frac{y}{1 - (y/K)} = Ce^{rt}, \tag{11}$$

where $C = e^c$. In order to satisfy the initial condition $y(0) = y_0$, we must choose $C = y_0/[1 - (y_0/K)]$. Using this value for C in Eq. (11) and solving for y, we obtain

$$y = \frac{y_0 K}{y_0 + (K - y_0)e^{-rt}}. \tag{12}$$

We have derived the solution (12) under the assumption that $0 < y_0 < K$. If $y_0 > K$, then the details of dealing with Eq. (10) are only slightly different, and we leave it to you to show that Eq. (12) is also valid in this case. Finally, note that Eq. (12) also contains the equilibrium solutions $y = \phi_1(t) = 0$ and $y = \phi_2(t) = K$ corresponding to the initial conditions $y_0 = 0$ and $y_0 = K$, respectively.

All the qualitative conclusions that we reached earlier by geometrical reasoning can be confirmed by examining the solution (12). In particular, if $y_0 = 0$, then Eq. (12) requires that $y(t) = 0$ for all t. If $y_0 > 0$, and if we let $t \to \infty$ in Eq. (12), then we obtain

$$\lim_{t \to \infty} y(t) = y_0 K / y_0 = K.$$

Thus, for each $y_0 > 0$, the solution approaches the equilibrium solution $y = \phi_2(t) = K$ asymptotically as $t \to \infty$. Therefore the constant solution $\phi_2(t) = K$ is said to be an **asymptotically stable solution** of Eq. (7) and the point $y = K$ is said to be an asymptotically stable equilibrium or critical point. After a long time, the population is close to the saturation level K regardless of the initial population size, as long as it is positive. Other solutions approach the equilibrium solution more rapidly as r increases.

On the other hand, the situation for the equilibrium solution $y = \phi_1(t) = 0$ is quite different. Even solutions that start very near zero grow as t increases and, as we have seen, approach K as $t \to \infty$. The solution $\phi_1(t) = 0$ is said to be an **unstable equilibrium solution** and $y = 0$ is an unstable equilibrium or critical point. This means that the only way to guarantee that the solution remains near zero is to make sure its initial value is *exactly* equal to zero.

More generally, if $y = y_1$ is an equilibrium solution of Eq. (1), then it is asymptotically stable if $f'(y_1) < 0$ and unstable if $f'(y_1) > 0$.

EXAMPLE 2

The logistic model has been applied to the natural growth of the halibut population in certain areas of the Pacific Ocean.[5] Let y, measured in kilograms, be the total mass, or biomass, of the halibut population at time t. The parameters in the logistic equation are estimated to have the values $r = 0.71/\text{yr}$ and $K = 80.5 \times 10^6$ kg. If the initial biomass is $y_0 = 0.25K$, find the biomass 2 years later. Also find the time τ for which $y(\tau) = 0.75K$.

It is convenient to scale the solution (12) to the carrying capacity K; thus we write Eq. (12) in the form

$$\frac{y}{K} = \frac{y_0/K}{(y_0/K) + [1 - (y_0/K)]e^{-rt}}. \tag{13}$$

Using the data given in the problem, we find that

$$\frac{y(2)}{K} = \frac{0.25}{0.25 + 0.75e^{-1.42}} \cong 0.5797.$$

Consequently, $y(2) \cong 46.7 \times 10^6$ kg.

To find τ we can first solve Eq. (13) for t. We obtain

$$e^{-rt} = \frac{(y_0/K)[1 - (y/K)]}{(y/K)[1 - (y_0/K)]};$$

[5] A good source of information on the population dynamics and economics involved in making efficient use of a renewable resource, with particular emphasis on fisheries, is the book by Clark listed in the references at the end of the book. The parameter values used here are given on page 53 of this book and were obtained from a study by H. S. Mohring.

hence,

$$t = -\frac{1}{r} \ln \frac{(y_0/K)[1 - (y/K)]}{(y/K)[1 - (y_0/K)]}. \tag{14}$$

Using the given values of r and y_0/K and setting $y/K = 0.75$, we find that

$$\tau = -\frac{1}{0.71} \ln \frac{(0.25)(0.25)}{(0.75)(0.75)} = \frac{1}{0.71} \ln 9 \cong 3.095 \text{ years.}$$

The graphs of y/K versus t for the given parameter values and for several initial conditions are shown in Figure 2.5.6.

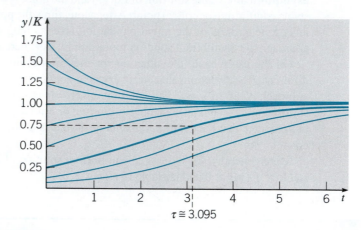

FIGURE 2.5.6 y/K versus t for population model of halibut in the Pacific Ocean.

▶ A Critical Threshold. We now turn to a consideration of the equation

$$\frac{dy}{dt} = -r\left(1 - \frac{y}{T}\right)y, \tag{15}$$

where r and T are given positive constants. Observe that (except for replacing the parameter K by T) this equation differs from the logistic equation (7) only in the presence of the minus sign on the right side. However, as we will see, the solutions of Eq. (15) behave very differently from those of Eq. (7).

For Eq. (15), the graph of $f(y)$ versus y is the parabola shown in Figure 2.5.7. The intercepts on the y-axis are the critical points $y = 0$ and $y = T$, corresponding to the equilibrium solutions $\phi_1(t) = 0$ and $\phi_2(t) = T$. If $0 < y < T$, then $dy/dt < 0$, and y decreases as t increases. On the other hand, if $y > T$, then $dy/dt > 0$, and y grows as t increases. Thus $\phi_1(t) = 0$ is an asymptotically stable equilibrium solution and $\phi_2(t) = T$ is an unstable one. Further, $f'(y)$ is negative for $0 < y < T/2$ and positive for $T/2 < y < T$, so the graph of y versus t is concave up and concave down, respectively, in these intervals. Also, $f'(y)$ is positive for $y > T$, so the graph of y versus t is also concave up there.

Figure 2.5.8a shows the phase line (the y-axis) for Eq. (15). The dots at $y = 0$ and $y = T$ are the critical points, or equilibrium solutions, and the arrows indicate where solutions are either increasing or decreasing.

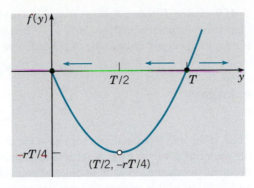

FIGURE 2.5.7 $f(y)$ versus y for $dy/dt = -r(1 - y/T)y$.

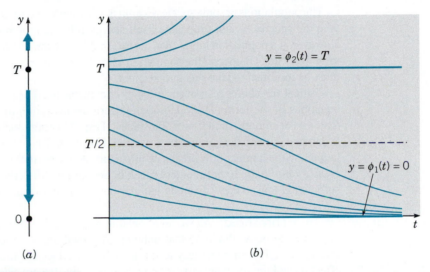

(a)   (b)

FIGURE 2.5.8   Growth with a threshold: $dy/dt = -r(1 - y/T)y$. (a) The phase line.
(b) Plots of y versus t.

Solution curves of Eq. (15) can now be sketched quickly. First draw the equilibrium solutions $y = 0$ and $y = T$. Then sketch curves in the strip $0 < y < T$ that are decreasing as t increases and change concavity as they cross the line $y = T/2$. Next draw some curves above $y = T$ that increase more and more steeply as t and y increase. Make sure that all curves become flatter as t approaches either zero or T. The result is Figure 2.5.8b, which is a qualitatively accurate sketch of solutions of Eq. (15) for any r and T. From this figure, it appears that as time increases, y either approaches zero or grows without bound, depending on whether the initial value y_0 is less than or greater than T. Thus T is a **threshold level**, below which growth does not occur.

We can confirm the conclusions that we have reached through geometrical reasoning by solving the differential equation (15). This can be done by separating the variables and integrating, just as we did for Eq. (7). However, if we note that Eq. (15) can be obtained from Eq. (7) by replacing K by T and r by $-r$, then we can make the same substitutions in the solution (12) and thereby obtain

$$y = \frac{y_0 T}{y_0 + (T - y_0)e^{rt}}, \tag{16}$$

which is the solution of Eq. (15) subject to the initial condition $y(0) = y_0$.

If $0 < y_0 < T$, then it follows from Eq. (16) that $y \to 0$ as $t \to \infty$. This agrees with our qualitative geometric analysis. If $y_0 > T$, then the denominator on the right side of Eq. (16) is zero for a certain finite value of t. We denote this value by t^* and calculate it from

$$y_0 - (y_0 - T)e^{rt^*} = 0,$$

which gives

$$t^* = \frac{1}{r} \ln \frac{y_0}{y_0 - T}. \tag{17}$$

Thus, if the initial population y_0 is above the threshold T, the threshold model predicts that the graph of y versus t has a vertical asymptote at $t = t^*$. In other words, the population becomes unbounded in a finite time, whose value depends on y_0, T, and r. The existence and location of this asymptote were not apparent from the geometric analysis, so, in this case, the explicit solution yields additional important qualitative, as well as quantitative, information.

The populations of some species exhibit the threshold phenomenon. If too few are present, then the species cannot propagate itself successfully and the population becomes extinct. However, if a population larger than the threshold level can be brought together, then further growth occurs. Of course, the population cannot become unbounded, so eventually Eq. (15) must be modified to take this into account.

Critical thresholds also occur in other circumstances. For example, in fluid mechanics, equations of the form (7) or (15) often govern the evolution of a small disturbance y in a *laminar* (or smooth) fluid flow. For instance, if Eq. (15) holds and $y < T$, then the disturbance is damped out and the laminar flow persists. However, if $y > T$, then the disturbance grows larger and the laminar flow breaks up into a turbulent one. In this case, T is referred to as the *critical amplitude*. Experimenters speak of keeping the disturbance level in a wind tunnel sufficiently low so they can study laminar flow over an airfoil, for example.

▶ **Logistic Growth with a Threshold.** As we mentioned in the last subsection, the threshold model (15) may need to be modified so that unbounded growth does not occur when y is above the threshold T. The simplest way to do this is to introduce another factor that will have the effect of making dy/dt negative when y is large. Thus we consider

$$\frac{dy}{dt} = -r \left(1 - \frac{y}{T}\right) \left(1 - \frac{y}{K}\right) y, \tag{18}$$

where $r > 0$ and $0 < T < K$.

The graph of $f(y)$ versus y is shown in Figure 2.5.9. In this problem, there are three critical points, $y = 0$, $y = T$, and $y = K$, corresponding to the equilibrium solutions $\phi_1(t) = 0$, $\phi_2(t) = T$, and $\phi_3(t) = K$, respectively. From Figure 2.5.9, it is clear that $dy/dt > 0$ for $T < y < K$, and consequently y is increasing there. The reverse is true for $y < T$ and for $y > K$. Consequently, the equilibrium solutions $\phi_1(t)$ and $\phi_3(t)$ are asymptotically stable, and the solution $\phi_2(t)$ is unstable.

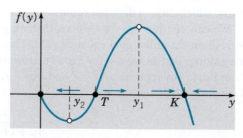

FIGURE 2.5.9 $f(y)$ versus y for $dy/dt = -r(1 - y/T)(1 - y/K)y$.

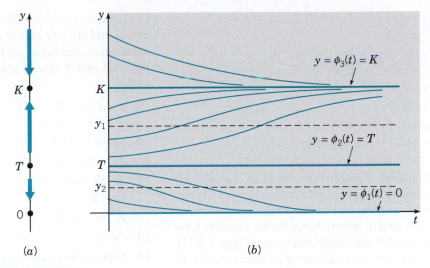

FIGURE 2.5.10 Logistic growth with a threshold: $dy/dt = -r(1 - y/T)(1 - y/K)y$. (a) The phase line. (b) Plots of y versus t.

The phase line for Eq. (18) is shown in Figure 2.5.10a, and the graphs of some solutions are sketched in Figure 2.5.10b. You should make sure that you understand the relation between these two figures, as well as the relation between Figures 2.5.9 and 2.5.10a. From Figure 2.5.10b we see that if y starts below the threshold T, then y declines to ultimate extinction. On the other hand, if y starts above T, then y eventually approaches the carrying capacity K. The inflection points on the graphs of y versus t in Figure 2.5.10b correspond to the maximum and minimum points, y_1 and y_2, respectively, on the graph of $f(y)$ versus y in Figure 2.5.9. These values can be obtained by differentiating the right side of Eq. (18) with respect to y, setting the result equal to zero, and solving for y. We obtain

$$y_{1,2} = (K + T \pm \sqrt{K^2 - KT + T^2})/3, \tag{19}$$

where the plus sign yields y_1 and the minus sign y_2.

A model of this general sort apparently describes the population of the passenger pigeon,[6] which was present in the United States in vast numbers until late in the nineteenth century. It was heavily hunted for food and for sport, and consequently its numbers were drastically reduced by the 1880s. Unfortunately, the passenger pigeon could apparently breed successfully only when present in a large concentration, corresponding to a relatively high threshold T. Although a reasonably large number of individual birds remained alive in the late 1880s, there were not enough in any one place to permit successful breeding, and the population rapidly declined to extinction. The last survivor died in 1914. The precipitous decline in the passenger pigeon population from huge numbers to extinction in a few decades was one of the early factors contributing to a concern for conservation in this country.

PROBLEMS

Problems 1 through 6 involve equations of the form $dy/dt = f(y)$. In each problem, sketch the graph of $f(y)$ versus y, determine the critical (equilibrium) points, and classify each one as asymptotically stable or unstable. Draw the phase line, and sketch several graphs of solutions in the ty-plane.

[6]See, for example, Oliver L. Austin, Jr., *Birds of the World* (New York: Golden Press, 1983), pp. 143–145.

1. $dy/dt = ay + by^2$, $a > 0$, $b > 0$,
   $y_0 \geq 0$
2. $dy/dt = ay + by^2$, $a > 0$, $b > 0$,
   $-\infty < y_0 < \infty$
3. $dy/dt = y(y - 1)(y - 2)$, $y_0 \geq 0$
4. $dy/dt = e^y - 1$, $-\infty < y_0 < \infty$
5. $dy/dt = e^{-y} - 1$, $-\infty < y_0 < \infty$
6. $dy/dt = -2(\arctan y)/(1 + y^2)$,
   $-\infty < y_0 < \infty$

7. **Semistable Equilibrium Solutions.** Sometimes a constant equilibrium solution has the property that solutions lying on one side of the equilibrium solution tend to approach it, whereas solutions lying on the other side depart from it (see Figure 2.5.11). In this case, the equilibrium solution is said to be **semistable**.

(a) Consider the equation

$$dy/dt = k(1 - y)^2,   \text{(i)}$$

where k is a positive constant. Show that $y = 1$ is the only critical point, with the corresponding equilibrium solution $\phi(t) = 1$.
(b) Sketch $f(y)$ versus y. Show that y is increasing as a function of t for $y < 1$ and also for $y > 1$. The phase line has upward-pointing arrows both below and above $y = 1$. Thus solutions below the equilibrium solution approach it, and those above it grow farther away. Therefore $\phi(t) = 1$ is semistable.
(c) Solve Eq. (i) subject to the initial condition $y(0) = y_0$ and confirm the conclusions reached in part (b).

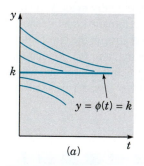
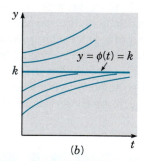

(a)

(b)

FIGURE 2.5.11   In both cases the equilibrium solution $\phi(t) = k$ is semistable. (a) $dy/dt \leq 0$; (b) $dy/dt \geq 0$.

Problems 8 through 13 involve equations of the form $dy/dt = f(y)$. In each problem, sketch the graph of $f(y)$ versus y, determine the critical (equilibrium) points, and classify each one as asymptotically stable, unstable, or semistable (see Problem 7). Draw the phase line, and sketch several graphs of solutions in the ty-plane.

8. $dy/dt = -k(y - 1)^2$, $k > 0$,
   $-\infty < y_0 < \infty$
9. $dy/dt = y^2(y^2 - 1)$, $-\infty < y_0 < \infty$
10. $dy/dt = y(1 - y^2)$, $-\infty < y_0 < \infty$
11. $dy/dt = ay - b\sqrt{y}$, $a > 0$, $b > 0$,
    $y_0 \geq 0$
12. $dy/dt = y^2(4 - y^2)$, $-\infty < y_0 < \infty$
13. $dy/dt = y^2(1 - y)^2$, $-\infty < y_0 < \infty$

14. Suppose that a certain population obeys the logistic equation $dy/dt = ry[1 - (y/K)]$.

(a) If $y_0 = K/3$, find the time τ at which the initial population has doubled. Find the value of τ corresponding to $r = 0.025$ per year.
(b) If $y_0/K = \alpha$, find the time T at which $y(T)/K = \beta$, where $0 < \alpha, \beta < 1$. Observe that $T \to \infty$ as $\alpha \to 0$ or as $\beta \to 1$. Find the value of T for $r = 0.025$ per year, $\alpha = 0.1$, and $\beta = 0.9$.

15. Another equation that has been used to model population growth is the Gompertz equation:

$$dy/dt = ry \ln(K/y),$$

where r and K are positive constants.

(a) Sketch the graph of $f(y)$ versus y, find the critical points, and determine whether each is asymptotically stable or unstable.
(b) For $0 \leq y \leq K$, determine where the graph of y versus t is concave up and where it is concave down.
(c) For each y in $0 < y \leq K$, show that dy/dt, as given by the Gompertz equation, is never less than dy/dt, as given by the logistic equation.

16. (a) Solve the Gompertz equation

$$dy/dt = ry \ln(K/y),$$

subject to the initial condition $y(0) = y_0$.
   Hint: You may wish to let $u = \ln(y/K)$.
(b) For the data given in Example 2 in the text ($r = 0.71$ per year, $K = 80.5 \times 10^6$ kg, $y_0/K = 0.25$), use the Gompertz model to find the predicted value of $y(2)$.

(c) For the same data as in part (b), use the Gompertz model to find the time τ at which $y(\tau) = 0.75K$.

ODEA **17.** Consider the equation

$$dy/dt = -0.25(1 - y)[1 - (y/4)]y.$$

(a) Draw a direction field and plot several solutions. Confirm that the solutions behave as shown in Figure 2.5.10b.
(b) For what values of y do inflection points occur?
(c) Suppose that $y(0) = 2$. Find the value of this solution when $t = 5$.
(d) Consider all solutions starting at $t = 0$ in the interval $2 \leq y \leq 6$. Find (to two decimal places) the smallest value of t for which all of these solutions lie in the interval $3.95 \leq y \leq 4.05$.

18. A pond forms as water collects in a conical depression of radius a and depth h. Suppose that water flows in at a constant rate k and is lost through evaporation at a rate proportional to the surface area.

(a) Show that the volume $V(t)$ of water in the pond at time t satisfies the differential equation

$$dV/dt = k - \alpha\pi(3a/\pi h)^{2/3}V^{2/3},$$

where α is the coefficient of evaporation.
(b) Find the equilibrium depth of water in the pond. Is the equilibrium asymptotically stable?
(c) Find a condition that must be satisfied if the pond is not to overflow.

19. Consider a cylindrical water tank of constant cross section A. Water is pumped into the tank at a constant rate k and leaks out through a small hole of area a in the bottom of the tank. From Torricelli's principle in hydrodynamics (see Problem 6 in Section 2.3), it follows that the rate at which water flows through the hole is $\alpha a\sqrt{2gh}$, where h is the current depth of water in the tank, g is the acceleration due to gravity, and α is a contraction coefficient that satisfies $0.5 \leq \alpha \leq 1.0$.

(a) Show that the depth of water in the tank at any time satisfies the equation

$$dh/dt = (k - \alpha a\sqrt{2gh})/A.$$

(b) Determine the equilibrium depth h_e of water, and show that it is asymptotically stable. Observe that h_e does not depend on A.

Epidemics. The use of mathematical methods to study the spread of contagious diseases goes back at least to some work by Daniel Bernoulli in 1760 on smallpox. In more recent years, many mathematical models have been proposed and studied for many different diseases.[7] Problems 20 through 22 deal with a few of the simpler models and the conclusions that can be drawn from them. Similar models have also been used to describe the spread of rumors and of consumer products.

20. Suppose that a given population can be divided into two parts: those who have a given disease and can infect others, and those who do not have it but are susceptible. Let x be the proportion of susceptible individuals and y the proportion of infectious individuals; then $x + y = 1$. Assume that the disease spreads by contact between sick and well members of the population and that the rate of spread dy/dt is proportional to the number of such contacts. Further, assume that members of both groups move about freely among each other, so the number of contacts is proportional to the product of x and y. Since $x = 1 - y$, we obtain the initial value problem

$$dy/dt = \alpha y(1 - y), \qquad y(0) = y_0, \qquad \text{(i)}$$

where α is a positive proportionality factor, and y_0 is the initial proportion of infectious individuals.

(a) Find the equilibrium points for the differential equation (i) and determine whether each is asymptotically stable, semistable, or unstable.
(b) Solve the initial value problem (i) and verify that the conclusions you reached in part (a) are correct. Show that $y(t) \to 1$ as $t \to \infty$, which means that ultimately the disease spreads through the entire population.

21. Some diseases (such as typhoid fever) are spread largely by *carriers*, individuals who can transmit the disease but who exhibit no overt symptoms. Let x and y, respectively, denote the proportion of susceptibles and carriers in the population. Suppose that carriers are identified and removed from the population at a rate β, so

$$dy/dt = -\beta y. \qquad \text{(i)}$$

---

[7] A standard source is the book by Bailey listed in the references. The models in Problems 20 through 22 are discussed by Bailey in Chapters 5, 10, and 20, respectively.

Suppose also that the disease spreads at a rate proportional to the product of x and y; thus

$$dx/dt = -\alpha xy. \qquad (ii)$$

(a) Determine y at any time t by solving Eq. (i) subject to the initial condition $y(0) = y_0$.
(b) Use the result of part (a) to find x at any time t by solving Eq. (ii) subject to the initial condition $x(0) = x_0$.
(c) Find the proportion of the population that escapes the epidemic by finding the limiting value of x as $t \to \infty$.

22. Daniel Bernoulli's work in 1760 had the goal of appraising the effectiveness of a controversial inoculation program against smallpox, which at that time was a major threat to public health. His model applies equally well to any other disease that, once contracted and survived, confers a lifetime immunity.

Consider the cohort of individuals born in a given year ($t = 0$), and let $n(t)$ be the number of these individuals surviving t years later. Let $x(t)$ be the number of members of this cohort who have not had smallpox by year t and who are therefore still susceptible. Let β be the rate at which susceptibles contract smallpox, and let ν be the rate at which people who contract smallpox die from the disease. Finally, let $\mu(t)$ be the death rate from all causes other than smallpox. Then dx/dt, the rate at which the number of susceptibles changes, is given by

$$dx/dt = -[\beta + \mu(t)]x. \qquad (i)$$

The first term on the right side of Eq. (i) is the rate at which susceptibles contract smallpox, and the second term is the rate at which they die from all other causes. Also

$$dn/dt = -\nu\beta x - \mu(t)n, \qquad (ii)$$

where dn/dt is the death rate of the entire cohort, and the two terms on the right side are the death rates due to smallpox and to all other causes, respectively.

(a) Let $z = x/n$ and show that z satisfies the initial value problem

$$dz/dt = -\beta z(1 - \nu z), \qquad z(0) = 1. \quad (iii)$$

Observe that the initial value problem (iii) does not depend on $\mu(t)$.
(b) Find $z(t)$ by solving Eq. (iii).

(c) Bernoulli estimated that $\nu = \beta = 1/8$. Using these values, determine the proportion of 20-year-olds who have not had smallpox.

Note: On the basis of the model just described and the best mortality data available at the time, Bernoulli calculated that if deaths due to smallpox could be eliminated ($\nu = 0$), then approximately 3 years could be added to the average life expectancy (in 1760) of 26 years 7 months. He therefore supported the inoculation program.

Bifurcation Points. For an equation of the form

$$dy/dt = f(a, y), \qquad (i)$$

where a is a real parameter, the critical points (equilibrium solutions) usually depend on the value of a. As a steadily increases or decreases, it often happens that at a certain value of a, called a **bifurcation point**, critical points come together, or separate, and equilibrium solutions may either be lost or gained. Bifurcation points are of great interest in many applications, because near them the nature of the solution of the underlying differential equation is undergoing an abrupt change. For example, in fluid mechanics a smooth (laminar) flow may break up and become turbulent. Or an axially loaded column may suddenly buckle and exhibit a large lateral displacement. Or, as the amount of one of the chemicals in a certain mixture is increased, spiral wave patterns of varying color may suddenly emerge in an originally quiescent fluid. Problems 23 through 25 describe three types of bifurcations that can occur in simple equations of the form (i).

23. Consider the equation

$$dy/dt = a - y^2. \qquad (ii)$$

(a) Find all of the critical points for Eq. (ii). Observe that there are no critical points if $a < 0$, one critical point if $a = 0$, and two critical points if $a > 0$.
(b) Draw the phase line in each case and determine whether each critical point is asymptotically stable, semistable, or unstable.
(c) In each case, sketch several solutions of Eq. (ii) in the ty-plane.
(d) If we plot the location of the critical points as a function of a in the ay-plane, we obtain Figure 2.5.12. This is called the **bifurcation diagram** for Eq. (ii). The bifurcation at $a = 0$ is called a **saddle–node** bifurcation. This name is more natural in the context of second order systems, which are discussed in Chapter 7.

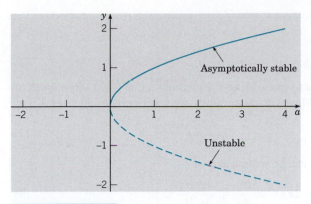

FIGURE 2.5.12 Bifurcation diagram for $y' = a - y^2$.

24. Consider the equation

$$dy/dt = ay - y^3 = y(a - y^2). \qquad \text{(iii)}$$

(a) Again consider the cases $a < 0$, $a = 0$, and $a > 0$. In each case, find the critical points, draw the phase line, and determine whether each critical point is asymptotically stable, semistable, or unstable.

(b) In each case, sketch several solutions of Eq. (iii) in the ty-plane.

(c) Draw the bifurcation diagram for Eq. (iii), that is, plot the location of the critical points versus a. For Eq. (iii), the bifurcation point at $a = 0$ is called a **pitchfork bifurcation**; your diagram may suggest why this name is appropriate.

25. Consider the equation

$$dy/dt = ay - y^2 = y(a - y). \qquad \text{(iv)}$$

(a) Again consider the cases $a < 0$, $a = 0$, and $a > 0$. In each case, find the critical points, draw the phase line, and determine whether each critical point is asymptotically stable, semistable, or unstable.

(b) In each case, sketch several solutions of Eq. (iv) in the ty-plane.

(c) Draw the bifurcation diagram for Eq. (iv). Observe that for Eq. (iv) there are the same number of critical points for $a < 0$ and $a > 0$ but that their stability has changed. For $a < 0$, the equilibrium solution $y = 0$ is asymptotically stable and $y = a$ is unstable, while for $a > 0$ the situation is reversed. Thus there has been an **exchange of stability** as a passes through the bifurcation point $a = 0$. This type of bifurcation is called a **transcritical bifurcation**.

26. **Chemical Reactions.** A second order chemical reaction involves the interaction (collision) of one molecule of a substance P with one molecule of a substance Q to produce one molecule of a new substance X; this is denoted by $P + Q \rightarrow X$. Suppose that p and q, where $p \neq q$, are the initial concentrations of P and Q, respectively, and let $x(t)$ be the concentration of X at time t. Then $p - x(t)$ and $q - x(t)$ are the concentrations of P and Q at time t, and the rate at which the reaction occurs is given by the equation

$$dx/dt = \alpha(p - x)(q - x), \qquad \text{(i)}$$

where α is a positive constant.

(a) If $x(0) = 0$, determine the limiting value of $x(t)$ as $t \rightarrow \infty$ without solving the differential equation. Then solve the initial value problem and find $x(t)$ for any t.

(b) If the substances P and Q are the same, then $p = q$ and Eq. (i) is replaced by

$$dx/dt = \alpha(p - x)^2. \qquad \text{(ii)}$$

If $x(0) = 0$, determine the limiting value of $x(t)$ as $t \rightarrow \infty$ without solving the differential equation. Then solve the initial value problem and determine $x(t)$ for any t.

2.6 Exact Equations and Integrating Factors

For first order equations, there are a number of integration methods that are applicable to various classes of problems. We have already discussed linear equations and separable equations. Here, we consider a class of equations known as exact equations for which there is also a well-defined method of solution. Keep in mind, however, that those first order equations that can be solved by elementary integration methods are rather special; most first order equations cannot be solved in this way.

EXAMPLE
1

Solve the differential equation

$$2x + y^2 + 2xyy' = 0. \qquad (1)$$

The equation is neither linear nor separable, so the methods suitable for those types of equations are not applicable here. However, observe that the function $\psi(x, y) = x^2 + xy^2$ has the property that

$$2x + y^2 = \frac{\partial \psi}{\partial x}, \qquad 2xy = \frac{\partial \psi}{\partial y}. \qquad (2)$$

Therefore the differential equation can be written as

$$2x + y^2 + 2xyy' = \frac{\partial \psi}{\partial x} + \frac{\partial \psi}{\partial y}\frac{dy}{dx} = 0. \qquad (3)$$

Assuming that y is a function of x and using the chain rule, it follows that

$$\frac{\partial \psi}{\partial x} + \frac{\partial \psi}{\partial y}\frac{dy}{dx} = \frac{d\psi}{dx} = \frac{d}{dx}(x^2 + xy^2) = 0. \qquad (4)$$

Therefore, by integrating with respect to x we obtain

$$\psi(x, y) = x^2 + xy^2 = c, \qquad (5)$$

where c is an arbitrary constant. Equation (5) defines solutions of Eq. (1) implicitly.

The integral curves of Eq. (1) are the level curves, or contour lines, of the function $\psi(x, y)$ given by Eq. (5). Contour plotting routines in modern software packages are a convenient way to plot a representative sample of integral curves for a differential equation, once $\psi(x, y)$ has been determined. This is an alternative to using a numerical approximation method, such as Euler's method, to approximate solutions of the differential equation. Some integral curves for Eq. (1) are shown in Figure 2.6.1.

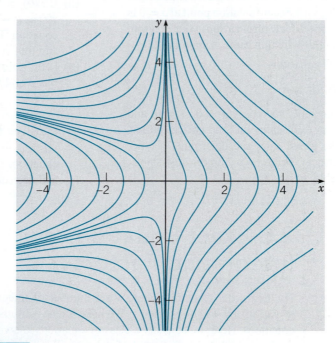

FIGURE 2.6.1 Integral curves of Eq. (1).

In solving Eq. (1), the key step was the recognition that there is a function ψ that satisfies Eqs. (2). More generally, let the differential equation

$$M(x, y) + N(x, y)y' = 0 \qquad (6)$$

be given. Suppose that we can identify a function ψ such that

$$\frac{\partial \psi}{\partial x}(x, y) = M(x, y), \qquad \frac{\partial \psi}{\partial y}(x, y) = N(x, y), \qquad (7)$$

and such that $\psi(x, y) = c$ defines $y = \phi(x)$ implicitly as a differentiable function of x. Then

$$M(x, y) + N(x, y)y' = \frac{\partial \psi}{\partial x} + \frac{\partial \psi}{\partial y}\frac{dy}{dx} = \frac{d}{dx}\psi[x, \phi(x)]$$

and the differential equation (6) becomes

$$\frac{d}{dx}\psi[x, \phi(x)] = 0. \qquad (8)$$

In this case, Eq. (6) is said to be an **exact** differential equation. Solutions of Eq. (6), or the equivalent Eq. (8), are given implicitly by

$$\psi(x, y) = c, \qquad (9)$$

where c is an arbitrary constant.

In Example 1, it was relatively easy to see that the differential equation was exact and, in fact, easy to find its solution, by recognizing the required function ψ. For more complicated equations, it may not be possible to do this so easily. A systematic way of determining whether a given differential equation is exact is provided by the following theorem.

THEOREM 2.6.1

Let the functions M, N, M_y, and N_x, where subscripts denote partial derivatives, be continuous in the rectangular[8] region R: $\alpha < x < \beta, \gamma < y < \delta$. Then Eq. (6),

$$M(x, y) + N(x, y)y' = 0,$$

is an exact differential equation in R if and only if

$$M_y(x, y) = N_x(x, y) \qquad (10)$$

at each point of R. That is, there exists a function ψ satisfying Eqs. (7),

$$\psi_x(x, y) = M(x, y), \qquad \psi_y(x, y) = N(x, y),$$

if and only if M and N satisfy Eq. (10).

The proof of this theorem has two parts. First, we show that if there is a function ψ such that Eqs. (7) are true, then it follows that Eq. (10) is satisfied. Computing M_y and N_x from Eqs. (7), we obtain

$$M_y(x, y) = \psi_{xy}(x, y), \qquad N_x(x, y) = \psi_{yx}(x, y). \qquad (11)$$

---

[8]It is not essential that the region be rectangular, only that it be simply connected. In two dimensions, this means that the region has no holes in its interior. Thus, for example, rectangular or circular regions are simply connected, but an annular region is not. More details can be found in most books on advanced calculus.

Since M_y and N_x are continuous, it follows that ψ_{xy} and ψ_{yx} are also continuous. This guarantees their equality, and Eq. (10) follows.

We now show that if M and N satisfy Eq. (10), then Eq. (6) is exact. The proof involves the construction of a function ψ satisfying Eqs. (7),

$$\psi_x(x, y) = M(x, y), \qquad \psi_y(x, y) = N(x, y).$$

We begin by integrating the first of Eqs. (7) with respect to x, holding y constant. We obtain

$$\psi(x, y) = Q(x, y) + h(y), \tag{12}$$

where $Q(x, y)$ is any differentiable function such that $\partial Q(x, y)/\partial x = M(x, y)$. For example, we might choose

$$Q(x, y) = \int_{x_0}^{x} M(s, y)\, ds, \tag{13}$$

where x_0 is some specified constant in $\alpha < x_0 < \beta$. The function h in Eq. (12) is an arbitrary differentiable function of y, playing the role of the arbitrary constant. Now we must show that it is always possible to choose $h(y)$ so that the second of Eqs. (7) is satisfied, that is, $\psi_y = N$. By differentiating Eq. (12) with respect to y and setting the result equal to $N(x, y)$, we obtain

$$\psi_y(x, y) = \frac{\partial Q}{\partial y}(x, y) + h'(y) = N(x, y).$$

Then, solving for $h'(y)$, we have

$$h'(y) = N(x, y) - \frac{\partial Q}{\partial y}(x, y). \tag{14}$$

In order for us to determine $h(y)$ from Eq. (14), the right side of Eq. (14), despite its appearance, must be a function of y only. To establish that this is true, we can differentiate the quantity in question with respect to x, obtaining

$$\frac{\partial N}{\partial x}(x, y) - \frac{\partial}{\partial x}\frac{\partial Q}{\partial y}(x, y). \tag{15}$$

By interchanging the order of differentiation in the second term of Eq. (15), we have

$$\frac{\partial N}{\partial x}(x, y) - \frac{\partial}{\partial y}\frac{\partial Q}{\partial x}(x, y),$$

or, since $\partial Q/\partial x = M$,

$$\frac{\partial N}{\partial x}(x, y) - \frac{\partial M}{\partial y}(x, y),$$

which is zero on account of Eq. (10). Hence, despite its apparent form, the right side of Eq. (14) does not, in fact, depend on x. Then we find $h(y)$ by integrating Eq. (14), and upon substituting this function in Eq. (12), we obtain the required function $\psi(x, y)$. This completes the proof of Theorem 2.6.1.

It is possible to obtain an explicit expression for $\psi(x, y)$ in terms of integrals (see Problem 17), but in solving specific exact equations, it is usually simpler and easier just to repeat the procedure used in the preceding proof. That is, integrate $\psi_x = M$ with respect to x, including an arbitrary function of $h(y)$ instead of an arbitrary constant, and then differentiate the result with respect to y and set it equal to N. Finally, use this last equation to solve for $h(y)$. The next example illustrates this procedure.

EXAMPLE 2

Solve the differential equation

$$(y\cos x + 2xe^y) + (\sin x + x^2 e^y - 1)y' = 0. \tag{16}$$

It is easy to see that

$$M_y(x, y) = \cos x + 2xe^y = N_x(x, y),$$

so the given equation is exact. Thus there is a $\psi(x, y)$ such that

$$\psi_x(x, y) = y \cos x + 2xe^y = M(x, y),$$
$$\psi_y(x, y) = \sin x + x^2 e^y - 1 = N(x, y).$$

Integrating the first of these equations with respect to x, we obtain

$$\psi(x, y) = y \sin x + x^2 e^y + h(y). \tag{17}$$

Then, finding ψ_y from Eq. (17) and setting the result equal to N gives

$$\psi_y(x, y) = \sin x + x^2 e^y + h'(y) = \sin x + x^2 e^y - 1.$$

Thus $h'(y) = -1$ and $h(y) = -y$. The constant of integration can be omitted since any solution of the preceding differential equation is satisfactory; we do not require the most general one. Substituting for $h(y)$ in Eq. (17) gives

$$\psi(x, y) = y \sin x + x^2 e^y - y.$$

Hence solutions of Eq. (16) are given implicitly by

$$y \sin x + x^2 e^y - y = c. \tag{18}$$

If an initial condition is prescribed, then it determines the value of c corresponding to the integral curve passing through the given initial point. For example, if the initial condition is $y(3) = 0$, then $c = 9$. Some integral curves of Eq. (16) are shown in Figure 2.6.2; the one passing through (3.0) is heavier than the others.

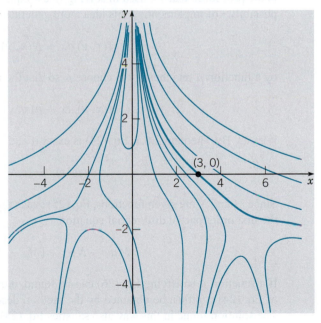

FIGURE 2.6.2   Integral curves of Eq. (16); the heavy curve is the integral curve through the initial point (3.0).

EXAMPLE 3

Solve the differential equation

$$(3xy + y^2) + (x^2 + xy)y' = 0. \tag{19}$$

Here,

$$M_y(x, y) = 3x + 2y, \qquad N_x(x, y) = 2x + y;$$

since $M_y \neq N_x$, the given equation is not exact. To see that it cannot be solved by the procedure described above, let us seek a function ψ such that

$$\psi_x(x, y) = 3xy + y^2 = M(x, y), \qquad \psi_y(x, y) = x^2 + xy = N(x, y). \tag{20}$$

Integrating the first of Eqs. (20) gives

$$\psi(x, y) = \tfrac{3}{2}x^2y + xy^2 + h(y), \tag{21}$$

where h is an arbitrary function of y only. To try to satisfy the second of Eqs. (20), we compute ψ_y from Eq. (21) and set it equal to N, obtaining

$$\tfrac{3}{2}x^2 + 2xy + h'(y) = x^2 + xy$$

or

$$h'(y) = -\tfrac{1}{2}x^2 - xy. \tag{22}$$

Since the right side of Eq. (22) depends on x as well as y, it is impossible to solve Eq. (22) for $h(y)$. Thus there is no $\psi(x, y)$ satisfying both of Eqs. (20).

▶ **Integrating Factors.** It is sometimes possible to convert a differential equation that is not exact into an exact equation by multiplying the equation by a suitable integrating factor. Recall that this is the procedure that we used in solving linear equations in Section 2.1. To investigate the possibility of implementing this idea more generally, let us multiply the equation

$$M(x, y)\, dx + N(x, y)\, dy = 0 \tag{23}$$

by a function μ and then try to choose μ so that the resulting equation

$$\mu(x, y)M(x, y)\, dx + \mu(x, y)N(x, y)\, dy = 0 \tag{24}$$

is exact. By Theorem 2.6.1, Eq. (24) is exact if and only if

$$(\mu M)_y = (\mu N)_x. \tag{25}$$

Since M and N are given functions, Eq. (25) states that the integrating factor μ must satisfy the first order partial differential equation

$$M\mu_y - N\mu_x + (M_y - N_x)\mu = 0. \tag{26}$$

If a function μ satisfying Eq. (26) can be found, then Eq. (24) will be exact. The solution of Eq. (24) can then be obtained by the method described in the first part of this section. The solution found in this way also satisfies Eq. (23), since the integrating factor μ can be canceled out of Eq. (24).

A partial differential equation of the form (26) may have more than one solution. If this is the case, any such solution may be used as an integrating factor of Eq. (23). This possible nonuniqueness of the integrating factor is illustrated in Example 4.

Unfortunately, Eq. (26), which determines the integrating factor μ, is ordinarily at least as hard to solve as the original equation (23). Therefore, although in principle, integrating factors are powerful tools for solving differential equations, in practice, they can be found only in special cases. The most important situations in which simple integrating factors can be found occur when μ is a function of only one of the variables x or y, instead of both. Let us determine a condition on M and N so that Eq. (23) has an integrating factor μ that depends on x only. Assuming that μ is a function of x only, we have

$$(\mu M)_y = \mu M_y, \qquad (\mu N)_x = \mu N_x + N \frac{d\mu}{dx}.$$

Thus, if $(\mu M)_y$ is to equal $(\mu N)_x$, it is necessary that

$$\frac{d\mu}{dx} = \frac{M_y - N_x}{N}\mu. \tag{27}$$

If $(M_y - N_x)/N$ is a function of x only, then there is an integrating factor μ that also depends only on x. Further, $\mu(x)$ can be found by solving Eq. (27), which is both linear and separable.

A similar procedure can be used to determine a condition under which Eq. (23) has an integrating factor depending only on y; see Problem 23.

EXAMPLE 4

Find an integrating factor for the equation

$$(3xy + y^2) + (x^2 + xy)y' = 0 \tag{19}$$

and then solve the equation.

In Example 3, we showed that this equation is not exact. Let us determine whether it has an integrating factor that depends on x only. On computing the quantity $(M_y - N_x)/N$, we find that

$$\frac{M_y(x, y) - N_x(x, y)}{N(x, y)} = \frac{3x + 2y - (2x + y)}{x^2 + xy} = \frac{x + y}{x(x + y)} = \frac{1}{x}. \tag{28}$$

Thus there is an integrating factor μ that is a function of x only, and it satisfies the differential equation

$$\frac{d\mu}{dx} = \frac{\mu}{x}. \tag{29}$$

Hence

$$\mu(x) = x. \tag{30}$$

Multiplying Eq. (19) by this integrating factor, we obtain

$$(3x^2y + xy^2) + (x^3 + x^2y)y' = 0. \tag{31}$$

The latter equation is exact, and it is easy to show that its solutions are given implicitly by

$$x^3y + \tfrac{1}{2}x^2y^2 = c. \tag{32}$$

Solutions may also be readily found in explicit form since Eq. (32) is quadratic in y. Some integral curves of Eq. (19) are shown in Figure 2.6.3.

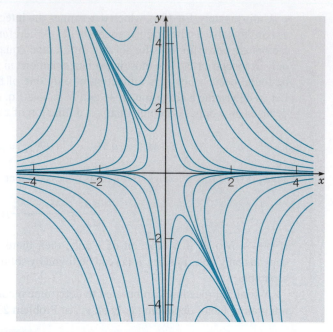

FIGURE 2.6.3 Integral curves of Eq. (19).

You may also verify that a second integrating factor of Eq. (19) is

$$\mu(x, y) = \frac{1}{xy(2x + y)},$$

and that the same solution is obtained, though with much greater difficulty, if this integrating factor is used (see Problem 32).

PROBLEMS

In each of Problems 1 through 12:

(a) Determine whether the equation is exact. If it is exact, then:

(b) Solve the equation.

ODEA (c) Use a computer to draw several integral curves.

1. $(2x + 3) + (2y - 2)y' = 0$

2. $(2x + 4y) + (2x - 2y)y' = 0$

3. $(3x^2 - 2xy + 2) \, dx + (6y^2 - x^2 + 3) \, dy = 0$

4. $(2xy^2 + 2y) + (2x^2y + 2x)y' = 0$

5. $\dfrac{dy}{dx} = -\dfrac{4x + 2y}{2x + 3y}$

6. $\dfrac{dy}{dx} = -\dfrac{4x - 2y}{2x - 3y}$

7. $(e^x \sin y - 2y \sin x) \, dx$
   $\qquad + (e^x \cos y + 2 \cos x) \, dy = 0$

8. $(e^x \sin y + 3y) \, dx - (3x - e^x \sin y) \, dy = 0$

9. $(ye^{xy} \cos 2x - 2e^{xy} \sin 2x + 2x) \, dx$
   $\qquad + (xe^{xy} \cos 2x - 3) \, dy = 0$

10. $(y/x + 6x) \, dx + (\ln x - 2) \, dy = 0, \qquad x > 0$

11. $(x \ln y + xy) \, dx + (y \ln x + xy) \, dy = 0;$
    $\qquad x > 0, \quad y > 0$

12. $\dfrac{x \, dx}{(x^2 + y^2)^{3/2}} + \dfrac{y \, dy}{(x^2 + y^2)^{3/2}} = 0$

In each of Problems 13 and 14, solve the given initial value problem and determine, at least approximately, where the solution is valid.

13. $(2x - y) \, dx + (2y - x) \, dy = 0, \qquad y(1) = 3$

14. $(9x^2 + y - 1) \, dx - (4y - x) \, dy = 0, \quad y(1) = 0$

In each of Problems 15 and 16, find the value of b for which the given equation is exact, and then solve it using that value of b.

15. $(xy^2 + bx^2y) \, dx + (x + y)x^2 \, dy = 0$

16. $(ye^{2xy} + x)\,dx + bxe^{2xy}\,dy = 0$

17. Assume that Eq. (6) meets the requirements of Theorem 2.6.1 in a rectangle R and is therefore exact. Show that a possible function $\psi(x, y)$ is

$$\psi(x, y) = \int_{x_0}^{x} M(s, y_0)\,ds + \int_{y_0}^{y} N(x, t)\,dt,$$

where (x_0, y_0) is a point in R.

18. Show that any separable equation

$$M(x) + N(y)y' = 0$$

is also exact.

In each of Problems 19 through 22:

(a) Show that the given equation is not exact but becomes exact when multiplied by the given integrating factor.

(b) Solve the equation.

ODEA **(c)** Use a computer to draw several integral curves.

19. $x^2y^3 + x(1 + y^2)y' = 0, \qquad \mu(x, y) = 1/xy^3$

20.

$$\left(\frac{\sin y}{y} - 2e^{-x}\sin x\right)\,dx$$

$$+ \left(\frac{\cos y + 2e^{-x}\cos x}{y}\right)\,dy = 0, \quad \mu(x, y) = ye^x$$

21. $y\,dx + (2x - ye^y)\,dy = 0, \qquad \mu(x, y) = y$

22. $(x + 2)\sin y\,dx + x\cos y\,dy = 0,$
$\mu(x, y) = xe^x$

23. Show that if $(N_x - M_y)/M = Q$, where Q is a function of y only, then the differential equation

$$M + Ny' = 0$$

has an integrating factor of the form

$$\mu(y) = \exp\int Q(y)\,dy.$$

24. Show that if $(N_x - M_y)/(xM - yN) = R$, where R depends on the quantity xy only, then the differential equation

$$M + Ny' = 0$$

has an integrating factor of the form $\mu(xy)$. Find a general formula for this integrating factor.

In each of Problems 25 through 31:

(a) Find an integrating factor and solve the given equation.

(b) Use a computer to draw several integral curves. ODEA

25. $(3x^2y + 2xy + y^3)\,dx + (x^2 + y^2)\,dy = 0$

26. $y' = e^{2x} + y - 1$

27. $dx + (x/y - \sin y)\,dy = 0$

28. $y\,dx + (2xy - e^{-2y})\,dy = 0$

29. $e^x\,dx + (e^x\cot y + 2y\csc y)\,dy = 0$

30. $[4(x^3/y^2) + (3/y)]\,dx + [3(x/y^2) + 4y]\,dy = 0$

31. $\left(3x + \dfrac{6}{y}\right) + \left(\dfrac{x^2}{y} + 3\dfrac{y}{x}\right)\dfrac{dy}{dx} = 0$
Hint: See Problem 24.

32. Use the integrating factor $\mu(x, y) = [xy(2x+y)]^{-1}$ to solve the differential equation

$$(3xy + y^2) + (x^2 + xy)y' = 0.$$

Verify that the solution is the same as that obtained in Example 4 with a different integrating factor.

2.7 Accuracy of Numerical Methods

In Section 1.3, we introduced the Euler, or tangent line, method for approximating the solution of an initial value problem

$$y' = f(t, y), \tag{1}$$

$$y(t_0) = y_0. \tag{2}$$

This method involves the repeated evaluation of the expressions

$$t_{n+1} = t_n + h, \tag{3}$$

$$y_{n+1} = y_n + hf(t_n, y_n) \tag{4}$$

for $n = 0, 1, 2, \ldots$. The result is a set of approximate values $y_1, y_2, \ldots$ at the mesh points $t_1, t_2, \ldots$. We assume, for simplicity, that the step size h is constant, although this is not necessary. In this section, we will begin to investigate the errors that may occur in this numerical approximation process.

Some examples of Euler's method appear in Section 1.3. As another example, consider the initial value problem

$$y' = 1 - t + 4y,$$ (5)

$$y(0) = 1.$$ (6)

Equation (5) is a first order linear equation, and it is easily verified that the solution satisfying the initial condition (6) is

$$y = \phi(t) = \tfrac{1}{4}t - \tfrac{3}{16} + \tfrac{19}{16}e^{4t}.$$ (7)

Since the exact solution is known, we do not need numerical methods to approximate the solution of the initial value problem (5), (6). On the other hand, the availability of the exact solution makes it easy to determine the accuracy of any numerical procedure that we use on this problem. We will use this problem in this section and the next to illustrate and compare different numerical methods. The solutions of Eq. (5) diverge rather rapidly from each other, so we should expect that it will be fairly difficult to approximate the solution (7) well over any considerable interval. Indeed, this is the reason for choosing this particular problem; it will be relatively easy to observe the benefits of using more accurate methods.

EXAMPLE 1

Using the Euler formula (4) and step sizes $h = 0.05$, 0.025, 0.01, and 0.001, determine approximate values of the solution $y = \phi(t)$ of the problem (5), (6) on the interval $0 \le t \le 2$.

The indicated calculations were carried out on a computer, and some of the results are shown in Table 2.7.1. Their accuracy is not particularly impressive. For $h = 0.01$ the percentage error is 3.85% at $t = 0.5$, 7.49% at $t = 1.0$, and 14.4% at $t = 2.0$. The corresponding percentage errors for $h = 0.001$ are 0.40%, 0.79%, and 1.58%, respectively. Observe that if $h = 0.001$, then it requires 2000 steps to traverse the interval from $t = 0$ to $t = 2$. Thus considerable computation is needed to obtain even reasonably good accuracy for this problem using the Euler method. When we discuss other numerical approximation methods in Section 2.8, we will find that it is possible to obtain comparable or better accuracy with much larger step sizes and many fewer computational steps.

TABLE 2.7.1 A Comparison of Results for the Numerical Solution of $y' = 1 - t + 4y$, $y(0) = 1$ Using the Euler Method for Different Step Sizes h.

t	$h = 0.05$	$h = 0.025$	$h = 0.01$	$h = 0.001$	Exact
0.0	1.0000000	1.0000000	1.0000000	1.0000000	1.0000000
0.1	1.5475000	1.5761188	1.5952901	1.6076289	1.6090418
0.2	2.3249000	2.4080117	2.4644587	2.5011159	2.5053299
0.3	3.4333560	3.6143837	3.7390345	3.8207130	3.8301388
0.4	5.0185326	5.3690304	5.6137120	5.7754845	5.7942260
0.5	7.2901870	7.9264062	8.3766865	8.6770692	8.7120041
1.0	45.588400	53.807866	60.037126	64.382558	64.897803
1.5	282.07187	361.75945	426.40818	473.55979	479.25919
2.0	1745.6662	2432.7878	3029.3279	3484.1608	3540.2001

To begin to investigate the errors in using numerical approximations, and also to suggest ways to construct more accurate algorithms, it is helpful to mention some alternative ways to look at the Euler method.

First, let us write the differential equation (1) at the point $t = t_n$ in the form

$$\frac{d\phi}{dt}(t_n) = f[t_n, \phi(t_n)]. \tag{8}$$

Then we approximate the derivative in Eq. (8) by the corresponding (forward) difference quotient, obtaining

$$\frac{\phi(t_{n+1}) - \phi(t_n)}{t_{n+1} - t_n} \cong f[t_n, \phi(t_n)]. \tag{9}$$

Finally, if we replace $\phi(t_{n+1})$ and $\phi(t_n)$ by their approximate values y_{n+1} and y_n, respectively, and solve for y_{n+1}, we obtain the Euler formula (4).

Another way to proceed is to write the problem as an integral equation. Since $y = \phi(t)$ is a solution of the initial value problem (1), (2), by integrating from t_n to t_{n+1} we obtain

$$\int_{t_n}^{t_{n+1}} \phi'(t)\, dt = \int_{t_n}^{t_{n+1}} f[t, \phi(t)]\, dt,$$

or

$$\phi(t_{n+1}) = \phi(t_n) + \int_{t_n}^{t_{n+1}} f[t, \phi(t)]\, dt. \tag{10}$$

The integral in Eq. (10) is represented geometrically as the area under the curve in Figure 2.7.1 between $t = t_n$ and $t = t_{n+1}$. If we approximate the integral by replacing $f[t, \phi(t)]$ by its value $f[t_n, \phi(t_n)]$ at $t = t_n$, then we are approximating the actual area by the area of the shaded rectangle. In this way we obtain

$$\phi(t_{n+1}) \cong \phi(t_n) + f[t_n, \phi(t_n)](t_{n+1} - t_n)$$
$$= \phi(t_n) + hf[t_n, \phi(t_n)]. \tag{11}$$

Finally, to obtain an approximation y_{n+1} for $\phi(t_{n+1})$, we make a second approximation by replacing $\phi(t_n)$ by its approximate value y_n in Eq. (11). This gives the Euler formula $y_{n+1} = y_n + hf(t_n, y_n)$. A more accurate algorithm can be obtained by approximating the integral more accurately. This is discussed in Section 2.8.

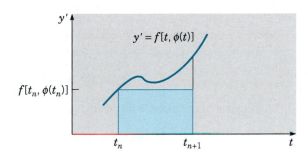

FIGURE 2.7.1   Integral derivation of the Euler method.

A third approach is to assume that the solution $y = \phi(t)$ has a Taylor series about the point t_n. Then

$$\phi(t_n + h) = \phi(t_n) + \phi'(t_n)h + \phi''(t_n)\frac{h^2}{2!} + \cdots,$$

or

$$\phi(t_{n+1}) = \phi(t_n) + f[t_n, \phi(t_n)]h + \phi''(t_n)\frac{h^2}{2!} + \cdots. \tag{12}$$

If the series is terminated after the first two terms, and $\phi(t_{n+1})$ and $\phi(t_n)$ are replaced by their approximate values y_{n+1} and y_n, we again obtain the Euler formula (4). If more terms in the series are retained, a more accurate formula is obtained. Further, by using a Taylor series with a remainder, it is possible to estimate the magnitude of the error in the formula.

▶ **Errors in Numerical Approximations.** The use of a numerical procedure, such as the Euler method, to approximate the solution of an initial value problem raises a number of questions that must be answered before the numerical approximation can be accepted as satisfactory. One of these is the question of **convergence**. That is, as the step size h tends to zero, do the values of the numerical approximation $y_1, y_2, \ldots, y_n, \ldots$ approach the corresponding values of the actual solution? If we assume that the answer is affirmative, there remains the important practical question of how rapidly the numerical approximation converges to the solution. In other words, how small a step size is needed in order to guarantee a given level of accuracy? We want to use a step size that is small enough to ensure the required accuracy, but not too small. An unnecessarily small step size slows down the calculations, makes them more expensive, and may even cause a loss of accuracy in some cases.

There are two fundamental sources of error in approximating the solution of an initial value problem numerically. Let us first assume that our computer is such that we can carry out all computations with complete accuracy; that is, we can retain infinitely many decimal places. The difference E_n between the solution $y = \phi(t)$ of the initial value problem (1), (2) and its numerical approximation y_n at t_n is given by

$$E_n = \phi(t_n) - y_n \tag{13}$$

and is known as the **global truncation error**. It arises from two causes: First, at each step we use an approximate formula to determine y_{n+1}; second, the input data at each step are only approximately correct. For example, in calculating y_{n+1}, we use y_n rather than (the unknown) $\phi(t_n)$, and, in general, $\phi(t_n)$ is not equal to y_n. If we assume that $y_n = \phi(t_n)$, then the only error in going one further step is due to the use of an approximate formula. This error is known as the **local truncation error** e_n.

The second fundamental source of error is that we carry out the computations in arithmetic with only a finite number of digits. This leads to a **round-off error** R_n defined by

$$R_n = y_n - Y_n, \tag{14}$$

where Y_n is the value *actually computed* from the given numerical method.

The absolute value of the total error in computing $\phi(t_n)$ is given by

$$|\phi(t_n) - Y_n| = |\phi(t_n) - y_n + y_n - Y_n|. \tag{15}$$

Making use of the triangle inequality, $|a + b| \le |a| + |b|$, we obtain, from Eq. (15),

$$|\phi(t_n) - Y_n| \le |\phi(t_n) - y_n| + |y_n - Y_n|$$
$$\le |E_n| + |R_n|. \tag{16}$$

Thus the total error is bounded by the sum of the absolute values of the global truncation and round-off errors. For the numerical procedures discussed in this book, it is possible to obtain useful estimates of the global truncation error. However, we limit our discussion primarily to the local truncation error, which is somewhat simpler. The round-off error is clearly more random in nature. It depends on the type of computer used, the sequence in which the computations are carried out, the method of rounding off, and so forth. An analysis of round-off error is beyond the scope of this book, but it is possible to say more about it than one might at first expect.[9] Some of the dangers from round-off error are discussed in Problems 23 through 25.

▶ **Local Truncation Error for the Euler Method.** Let us assume that the solution $y = \phi(t)$ of the initial value problem (1), (2) has a continuous second derivative in the interval of interest. To ensure this, we can assume that f, f_t, and f_y are continuous. Observe that if f has these properties and if ϕ is a solution of the initial value problem (1), (2), then

$$\phi'(t) = f[t, \phi(t)],$$

and, by the chain rule,

$$\phi''(t) = f_t[t, \phi(t)] + f_y[t, \phi(t)]\phi'(t)$$
$$= f_t[t, \phi(t)] + f_y[t, \phi(t)]f[t, \phi(t)]. \tag{17}$$

Since the right side of this equation is continuous, ϕ'' is also continuous.

Then, making use of a Taylor polynomial with a remainder to expand ϕ about t_n, we obtain

$$\phi(t_n + h) = \phi(t_n) + \phi'(t_n)h + \tfrac{1}{2}\phi''(\bar{t}_n)h^2, \tag{18}$$

where $\bar{t}_n$ is some point in the interval $t_n < \bar{t}_n < t_n + h$. Subtracting Eq. (4) from Eq. (18), and noting that $\phi(t_n + h) = \phi(t_{n+1})$ and $\phi'(t_n) = f[t_n, \phi(t_n)]$, we find that

$$\phi(t_{n+1}) - y_{n+1} = [\phi(t_n) - y_n] + h\{f[t_n, \phi(t_n)] - f(t_n, y_n)\} + \tfrac{1}{2}\phi''(\bar{t}_n)h^2. \tag{19}$$

To compute the local truncation error, we apply Eq. (19) to the true solution $\phi(t)$; that is, we take y_n to be $\phi(t_n)$. Then we immediately see from Eq. (19) that the local truncation error e_{n+1} is

$$e_{n+1} = \phi(t_{n+1}) - y_{n+1} = \tfrac{1}{2}\phi''(\bar{t}_n)h^2. \tag{20}$$

Thus the local truncation error for the Euler method is proportional to the square of the step size h, and the proportionality factor depends on the second derivative of the solution ϕ. The expression given by Eq. (20) depends on n and, in general, is different for each step. A uniform bound, valid on an interval $[a, b]$, is given by

$$|e_n| \leq Mh^2/2, \tag{21}$$

where M is the maximum of $|\phi''(t)|$ on the interval $[a, b]$. Since Eq. (21) is based on a consideration of the worst possible case—that is, the largest possible value of $|\phi''(t)|$—it may well be a considerable overestimate of the actual local truncation error in some parts of the interval $[a, b]$. The primary difficulty in using Eq. (20) or (21) lies in estimating $|\phi''(t)|$ or M. However, the central fact expressed by these equations is that the local truncation error is proportional to h^2. Thus, if h is multiplied by 1/2, then the error is multiplied by 1/4, and so on. Further, the proportionality factor depends on the second derivative of the solution, so Euler's method works best on problems whose solutions have relatively small second derivatives.

One use of Eq. (21) is to choose a step size that will result in a local truncation error no greater than some given tolerance level. For example, if the local truncation error must be no greater than ϵ, then from Eq. (21) we have

$$h \leq \sqrt{2\epsilon/M}. \tag{22}$$

---

[9]See, for example, the book by Henrici listed in the references.

More important than the local truncation error is the global truncation error E_n. The analysis for estimating E_n is more difficult than that for e_n. However, knowing the local truncation error, we can make an *intuitive* estimate of the global truncation error at a fixed $T > t_0$ as follows. Suppose that we take n steps in going from t_0 to $T = t_0 + nh$. In each step, the error is at most $Mh^2/2$; thus the error in n steps is at most $nMh^2/2$. Noting that $n = (T - t_0)/h$, we find that the global truncation error for the Euler method in going from t_0 to T is bounded by

$$n\frac{Mh^2}{2} = (T - t_0)\frac{Mh}{2}. \tag{23}$$

This argument is not complete since it does not take into account the effect that an error at one step will have in succeeding steps. Nevertheless, it can be shown that the global truncation error in using the Euler method on a finite interval is no greater than a constant times h. The Euler method is called a first order method because its global truncation error is proportional to the first power of the step size.

Because it is more accessible, we will hereafter use the local truncation error as our principal measure of the accuracy of a numerical method and for comparing different methods. If we have *a priori* information about the solution of the given initial value problem, we can use the result (20) to obtain more precise information about how the local truncation error varies with t. As an example, consider the illustrative problem

$$y' = 1 - t + 4y, \qquad y(0) = 1 \tag{24}$$

on the interval $0 \le t \le 2$. Let $y = \phi(t)$ be the solution of the initial value problem (24). Then, as noted previously,

$$\phi(t) = (4t - 3 + 19e^{4t})/16$$

and therefore

$$\phi''(t) = 19e^{4t}.$$

Equation (20) then states that

$$e_{n+1} = \frac{19e^{4\bar{t}_n}h^2}{2}, \qquad t_n < \bar{t}_n < t_n + h. \tag{25}$$

The appearance of the factor 19 and the rapid growth of e^{4t} explain why the results in Table 2.7.1 are not very accurate.

For instance, using $h = 0.05$, the error in the first step is

$$e_1 = \phi(t_1) - y_1 = \frac{19e^{4\bar{t}_0}(0.0025)}{2}, \qquad 0 < \bar{t}_0 < 0.05.$$

It is clear that e_1 is positive, and since $e^{4\bar{t}_0} < e^{0.2}$, we have

$$e_1 \le \frac{19e^{0.2}(0.0025)}{2} \cong 0.02901. \tag{26}$$

Note also that $e^{4\bar{t}_0} > 1$; hence, $e_1 > 19(0.0025)/2 = 0.02375$. The actual error is 0.02542. It follows from Eq. (25) that the error becomes progressively worse with increasing t. This is also clearly shown by the results in Table 2.7.1. Similar computations for bounds for the local truncation error give

$$1.0617 \cong \frac{19e^{3.8}(0.0025)}{2} \le e_{20} \le \frac{19e^4(0.0025)}{2} \cong 1.2967 \tag{27}$$

in going from 0.95 to 1.0 and

$$57.96 \cong \frac{19e^{7.8}(0.0025)}{2} \le e_{40} \le \frac{19e^8(0.0025)}{2} \cong 70.80 \tag{28}$$

in going from 1.95 to 2.0.

These results indicate that, for this problem, the local truncation error is about 2500 times larger near $t = 2$ than near $t = 0$. Thus, to reduce the local truncation error to an acceptable level throughout $0 \le t \le 2$, one must choose a step size h based on an analysis near $t = 2$. Of course, this step size will be much smaller than necessary near $t = 0$. For example, to achieve a local truncation error of 0.01 for this problem, we need a step size of about 0.00059 near $t = 2$ and a step size of about 0.032 near $t = 0$. The use of a uniform step size that is smaller than necessary over much of the interval results in more calculations than necessary, more time consumed, and possibly more danger of unacceptable round-off errors.

Another approach is to keep the local truncation error approximately constant throughout the interval by gradually reducing the step size as t increases. In the example problem, we would need to reduce h by a factor of about 50 in going from $t = 0$ to $t = 2$. A method that provides for variations in the step size is called **adaptive**. All modern computer codes for solving differential equations have the capability of adjusting the step size as needed. We will return to this question in the next section.

PROBLEMS

CAS In each of Problems 1 through 6, find approximate values of the solution of the given initial value problem at $t = 0.1, 0.2, 0.3,$ and 0.4.

(a) Use the Euler method with $h = 0.05$.
(b) Use the Euler method with $h = 0.025$.

1. $y' = 3 + t - y,$ $y(0) = 1$
2. $y' = 5t - 3\sqrt{y},$ $y(0) = 2$
3. $y' = 2y - 3t,$ $y(0) = 1$
4. $y' = 2t + e^{-ty},$ $y(0) = 1$
5. $y' = \dfrac{y^2 + 2ty}{3 + t^2},$ $y(0) = 0.5$
6. $y' = (t^2 - y^2)\sin y,$ $y(0) = -1$

CAS In each of Problems 7 through 12, find approximate values of the solution of the given initial value problem at $t = 0.5, 1.0, 1.5,$ and 2.0.

(a) Use the Euler method with $h = 0.025$.
(b) Use the Euler method with $h = 0.0125$.

7. $y' = 0.5 - t + 2y,$ $y(0) = 1$
8. $y' = 5t - 3\sqrt{y},$ $y(0) = 2$
9. $y' = \sqrt{t + y},$ $y(0) = 3$
10. $y' = 2t + e^{-ty},$ $y(0) = 1$
11. $y' = (4 - ty)/(1 + y^2),$ $y(0) = -2$
12. $y' = (y^2 + 2ty)/(3 + t^2),$ $y(0) = 0.5$

13. Complete the calculations leading to the entries in CAS columns three and four of Table 2.7.1.

14. Using three terms in the Taylor series given in CAS Eq. (12) and taking $h = 0.1$, determine approximate values of the solution of the illustrative example $y' = 1 - t + 4y,$ $y(0) = 1$ at $t = 0.1$ and 0.2. Compare the results with those using the Euler method and with the exact values.
Hint: If $y' = f(t, y)$, what is y''?

In each of Problems 15 and 16, estimate the local truncation error for the Euler method in terms of the solution $y = \phi(t)$. Obtain a bound for e_{n+1} in terms of t and $\phi(t)$ that is valid on the interval $0 \le t \le 1$. By using a formula for the solution, obtain a more accurate error bound for e_{n+1}. For $h = 0.1$, compute a bound for e_1 and compare it with the actual error at $t = 0.1$. Also compute a bound for the error e_4 in the fourth step.

15. $y' = 2y - 1,$ $y(0) = 1$
16. $y' = 0.5 - t + 2y,$ $y(0) = 1$

In each of Problems 17 through 20, obtain a formula for the local truncation error for the Euler method in terms of t and the solution ϕ.

17. $y' = t^2 + y^2,$ $y(0) = 1$

18. $y' = 5t - 3\sqrt{y}, \qquad y(0) = 2$

19. $y' = \sqrt{t + y}, \qquad y(1) = 3$

20. $y' = 2t + e^{-ty}, \qquad y(0) = 1$

CAS 21. Consider the initial value problem

$$y' = \cos 5\pi t, \qquad y(0) = 1.$$

(a) Determine the solution $y = \phi(t)$ and draw a graph of $y = \phi(t)$ for $0 \le t \le 1$.

(b) Determine approximate values of $\phi(t)$ at $t = 0.2, 0.4$, and 0.6 using the Euler method with $h = 0.2$. Draw a broken-line graph for the approximate solution and compare it with the graph of the exact solution.

(c) Repeat the computation of part (b) for $0 \le t \le 0.4$, but take $h = 0.1$.

(d) Show by computing the local truncation error that neither of these step sizes is sufficiently small. Determine a value of h to ensure that the local truncation error is less than 0.05 throughout the interval $0 \le t \le 1$. That such a small value of h is required results from the fact that max $|\phi''(t)|$ is large.

CAS 22. Using a step size $h = 0.05$ and the Euler method, but retaining only three digits throughout the computations, determine approximate values of the solution at $t = 0.1, 0.2, 0.3$, and 0.4 for each of the following initial value problems.

(a) $y' = 1 - t + 4y, \qquad y(0) = 1$

(b) $y' = 3 + t - y, \qquad y(0) = 1$

(c) $y' = 2y - 3t, \qquad y(0) = 1$

Compare the results with those obtained in Example 1 and in Problems 1 and 3. The small differences between some of those results rounded to three digits and the present results are due to round-off error. The round-off error would become important if the computation required many steps.

23. The following problem illustrates a danger that occurs because of round-off error when nearly equal numbers are subtracted and the difference is then multiplied by a large number. Evaluate the quantity

$$1000 \cdot \begin{vmatrix} 6.010 & 18.04 \\ 2.004 & 6.000 \end{vmatrix}$$

in the following ways:

(a) First round each entry in the determinant to two digits.

(b) First round each entry in the determinant to three digits.

(c) Retain all four digits. Compare this value with the results in parts (a) and (b).

24. The distributive law $a(b - c) = ab - ac$ does not hold, in general, if the products are rounded off to a smaller number of digits. To show this in a specific case, take $a = 0.22, b = 3.19$, and $c = 2.17$. After each multiplication, round off the last digit.

25. In this section, we stated that the global trunca- CAS tion error for the Euler method applied to an initial value problem over a fixed interval is no more than a constant times the step size h. In this problem, we show you how to obtain some experimental evidence in support of this statement. Consider the initial value problem in Example 1 for which some numerical approximations are given in Table 2.7.1. Observe that, for each step size, the maximum error E occurs at the end point $t = 2$. Now let us assume that $E = Ch^p$, where the constants C and p are to be determined. By taking the logarithm of each side of this equation, we obtain

$$\ln E = \ln C + p \ln h,$$

which is the equation of a straight line in the $(\ln h)$ $(\ln E)$-plane. The slope of this line is the value of the exponent p and the intercept on the ln E-axis determines the value of C.

(a) Using the data in Table 2.7.1, calculate the maximum error E for each of the given values of h.

(b) Plot ln E versus ln h for the four data points that you obtained in part (a).

(c) Do the points in part (b) lie approximately on a single straight line? If so, then this is evidence that the assumed expression for E is correct.

(d) Estimate the slope of the line in part (c). If the statement in the text about the magnitude of the global truncation error is correct, then the slope should be no greater than one.

Note: Your estimate of the slope p depends on how you choose the straight line. If you have a curve fitting routine in your software, you can use it to determine the straight line that best fits the data. Otherwise, you may wish to resort to less precise methods. For example, you could calculate the slopes of the line segments joining (one or more) pairs of data points, and then average your results.

2.8 Improved Euler and Runge-Kutta Methods

Since for many problems the Euler method requires a very small step size to produce sufficiently accurate results, much effort has been devoted to the development of more efficient methods. In this section, we will discuss two of these methods. Consider the initial value problem

$$y' = f(t, y), \qquad y(t_0) = y_0 \tag{1}$$

and let $y = \phi(t)$ denote its solution. Recall from Eq. (10) of Section 2.7 that, by integrating the given differential equation from t_n to t_{n+1}, we obtain

$$\phi(t_{n+1}) = \phi(t_n) + \int_{t_n}^{t_{n+1}} f[t, \phi(t)]\, dt. \tag{2}$$

The Euler formula

$$y_{n+1} = y_n + hf(t_n, y_n) \tag{3}$$

is obtained by replacing $f[t, \phi(t)]$ in Eq. (2) by its approximate value $f(t_n, y_n)$ at the left endpoint of the interval of integration.

▶ **Improved Euler Formula.** A better approximate formula can be obtained if the integrand in Eq. (2) is approximated more accurately. One way to do this is to replace the integrand by the average of its values at the two endpoints, namely, $\{f[t_n, \phi(t_n)] + f[t_{n+1}, \phi(t_{n+1})]\}/2$. This is equivalent to approximating the area under the curve in Figure 2.8.1 between $t = t_n$ and $t = t_{n+1}$ by the area of the shaded trapezoid. Further, we replace $\phi(t_n)$ and $\phi(t_{n+1})$ by their respective approximate values y_n and y_{n+1}. In this way we obtain, from Eq. (2),

$$y_{n+1} = y_n + \frac{f(t_n, y_n) + f(t_{n+1}, y_{n+1})}{2} h. \tag{4}$$

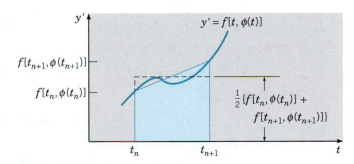

FIGURE 2.8.1   Derivation of the improved Euler method.

Since the unknown y_{n+1} appears as one of the arguments of f on the right side of Eq. (4), this equation defines y_{n+1} implicitly rather than explicitly. Depending on the nature of the function f, it may be fairly difficult to solve Eq. (4) for y_{n+1}. This difficulty can be overcome by replacing y_{n+1} on the right side of Eq. (4) by the value obtained using the Euler formula (3). Thus

$$y_{n+1} = y_n + \frac{f(t_n, y_n) + f[t_n + h, y_n + hf(t_n, y_n)]}{2} h$$

$$= y_n + \frac{f_n + f(t_n + h, y_n + hf_n)}{2} h, \tag{5}$$

where t_{n+1} has been replaced by $t_n + h$.

112 | Chapter 2   First Order Differential Equations

Equation (5) gives an explicit formula for computing y_{n+1}, the approximate value of $\phi(t_{n+1})$, in terms of the data at t_n. This formula is known as the **improved Euler formula** or the **Heun formula**. The improved Euler formula is an example of a two-stage method; that is, we first calculate $y_n + hf_n$ from the Euler formula and then use this result to calculate y_{n+1} from Eq. (5). The improved Euler formula (5) does represent an improvement over the Euler formula (3) because the local truncation error in using Eq. (5) is proportional to h^3, while for the Euler method it is proportional to h^2. This error estimate for the improved Euler formula is established in Problem 14. It can also be shown that for a finite interval the global truncation error for the improved Euler formula is bounded by a constant times h^2, so this method is a second order method. Note that this greater accuracy is achieved at the expense of more computational work, since it is now necessary to evaluate $f(t, y)$ twice in order to go from t_n to t_{n+1}.

If $f(t, y)$ depends only on t and not on y, then solving the differential equation $y' = f(t, y)$ reduces to integrating $f(t)$. In this case, the improved Euler formula (5) becomes

$$y_{n+1} - y_n = \frac{h}{2}[f(t_n) + f(t_n + h)], \tag{6}$$

which is just the trapezoid rule for numerical integration.

EXAMPLE 1

Use the improved Euler formula (5) to calculate approximate values of the solution of the initial value problem

$$y' = 1 - t + 4y, \qquad y(0) = 1. \tag{7}$$

To make clear exactly what computations are required, we show a couple of steps in detail. For this problem $f(t, y) = 1 - t + 4y$; hence,

$$f_n = 1 - t_n + 4y_n$$

and

$$f(t_n + h, y_n + hf_n) = 1 - (t_n + h) + 4(y_n + hf_n).$$

Further, $t_0 = 0$, $y_0 = 1$, and $f_0 = 1 - t_0 + 4y_0 = 5$. If $h = 0.025$, then

$$f(t_0 + h, y_0 + hf_0) = 1 - 0.025 + 4[1 + (0.025)(5)] = 5.475.$$

Then, from Eq. (5),

$$y_1 = 1 + (0.5)(5 + 5.475)(0.025) = 1.1309375. \tag{8}$$

At the second step we must calculate

$$f_1 = 1 - 0.025 + 4(1.1309375) = 5.49875,$$
$$y_1 + hf_1 = 1.1309375 + (0.025)(5.49875) = 1.26840625,$$

and

$$f(t_2, y_1 + hf_1) = 1 - 0.05 + 4(1.26840625) = 6.023625.$$

Then, from Eq. (5),

$$y_2 = 1.1309375 + (0.5)(5.49875 + 6.023625)(0.025) = 1.2749671875. \tag{9}$$

Further results for $0 \leq t \leq 2$ obtained by using the improved Euler method with $h = 0.025$ and $h = 0.01$ are given in Table 2.8.1. To compare the results of the improved Euler method with those of the Euler method, note that the improved Euler method requires two evaluations of f at each step, while the Euler method requires only one. This is significant because typically most of the computing time in each step is spent in evaluating f, so counting these evaluations is a reasonable way to estimate the total computing effort. Thus, for a given step size h, the improved Euler method requires twice as many evaluations of f as the Euler method. Alternatively, the improved Euler method for step size h requires the same number of evaluations of f as the Euler method with step size $h/2$.

TABLE 2.8.1

A Comparison of Results Using the Euler and Improved Euler Methods for the Initial Value Problem $y' = 1 - t + 4y$, $y(0) = 1$

| | Euler | | Improved Euler | | |
| | $h = 0.01$ | $h = 0.001$ | $h = 0.025$ | $h = 0.01$ | |
t					Exact
0	1.0000000	1.0000000	1.0000000	1.0000000	1.0000000
0.1	1.5952901	1.6076289	1.6079462	1.6088585	1.6090418
0.2	2.4644587	2.5011159	2.5020618	2.5047827	2.5053299
0.3	3.7390345	3.8207130	3.8228282	3.8289146	3.8301388
0.4	5.6137120	5.7754845	5.7796888	5.7917911	5.7942260
0.5	8.3766865	8.6770692	8.6849039	8.7074637	8.7120041
1.0	60.037126	64.382558	64.497931	64.830722	64.897803
1.5	426.40818	473.55979	474.83402	478.51588	479.25919
2.0	3029.3279	3484.1608	3496.6702	3532.8789	3540.2001

By referring to Table 2.8.1, you can see that the improved Euler method with $h = 0.025$ gives much better results than the Euler method with $h = 0.01$. Note that to reach $t = 2$ with these step sizes, the improved Euler method requires 160 evaluations of f, while the Euler method requires 200. More noteworthy is that the improved Euler method with $h = 0.025$ is also slightly more accurate than the Euler method with $h = 0.001$ (2000 evaluations of f). In other words, with something like one-twelfth of the computing effort, the improved Euler method yields results for this problem that are comparable to, or a bit better than, those generated by the Euler method. This illustrates that, compared to the Euler method, the improved Euler method is clearly more efficient, yielding substantially better results or requiring much less total computing effort, or both.

The percentage errors at $t = 2$ for the improved Euler method are 1.23% for $h = 0.025$ and 0.21% for $h = 0.01$.

▶ **Variation of Step Size.** In Section 2.7, we mentioned the possibility of adjusting the step size as a calculation proceeds so as to maintain the local truncation error at a more or less constant level. The goal is to use no more steps than necessary and, at the same time, to keep some control over the accuracy of the approximation. Here we will describe how this can be done. Suppose that after n steps we have reached the point (t_n, y_n). We choose a step size h and calculate y_{n+1}. Next we need to estimate the error we have made in calculating y_{n+1}. Not knowing the actual solution, the best that we can do is to use a more accurate method and repeat the calculation starting from (t_n, y_n). For example, if we used the Euler method for the original calculation, we might repeat it with the improved Euler method. Then the difference between the two calculated values is an estimate e_{n+1}^{est} of the error in using the original method. If the estimated error is different from the error tolerance ϵ, then we adjust the step size and repeat the calculation. To make this adjustment efficiently it is crucial to know how the local truncation error e_{n+1} depends on the step size h. For the Euler method, the local truncation error is proportional to h^2, so to bring the estimated error down (or up) to the tolerance level ϵ, we must multiply the original step size by the factor $\sqrt{\epsilon/e_{n+1}^{\text{est}}}$.

To illustrate this procedure, consider the example problem (7),

$$y' = 1 - t + 4y, \qquad y(0) = 1.$$

You can verify that after one step with $h = 0.1$ we obtain the values 1.5 and 1.595 from the Euler method and the improved Euler method, respectively. Thus the estimated error in using the Euler method is 0.095. If we have chosen an error tolerance of 0.05, for instance, then we need to adjust the step size downward by the factor $\sqrt{0.05/0.095} \cong 0.73$. Rounding downward to be conservative, let us choose the adjusted step size $h = 0.07$. Then, from the Euler formula, we obtain

$$y_1 = 1 + (0.07)f(0, 1) = 1.35 \cong \phi(0.07).$$

Using the improved Euler method, we obtain $y_1 = 1.39655$, so the estimated error in using the Euler formula is 0.04655, which is slightly less than the specified tolerance. The actual error, based on a comparison with the solution itself, is somewhat greater, namely, 0.05122.

We can follow the same procedure at each step of the calculation, thereby keeping the local truncation error approximately constant throughout the entire numerical process. Modern adaptive codes for solving differential equations adjust the step size as they proceed in very much this way, although they use more accurate formulas than the Euler and improved Euler formulas. Consequently, they are able to achieve both efficiency and accuracy by using very small steps only where they are really needed.

▶ Runge–Kutta Method.   The Euler and improved Euler methods belong to what is now called the Runge–Kutta class of numerical approximation methods. Here we discuss the method originally developed by Runge and Kutta. This method is now called the classic fourth order, four-stage Runge–Kutta method, but it is often referred to simply as *the* Runge–Kutta method, and we will follow this practice for brevity. This method has a local truncation error that is proportional to h^5. Thus it is two orders of magnitude more accurate than the improved Euler method and three orders of magnitude better than the Euler method. It is relatively simple to use and is sufficiently accurate to handle many problems efficiently. This is especially true of adaptive Runge–Kutta methods, in which provision is made to vary the step size as needed. We return to this issue at the end of the section.

The Runge–Kutta method involves a weighted average of values of $f(t, y)$ at different points in the interval $t_n \leq t \leq t_{n+1}$. It is given by

$$t_{n+1} = t_n + h, \qquad y_{n+1} = y_n + h \left(\frac{k_{n1} + 2k_{n2} + 2k_{n3} + k_{n4}}{6} \right), \qquad (10)$$

where

$$\begin{aligned} k_{n1} &= f(t_n, y_n) \\ k_{n2} &= f\left(t_n + \tfrac{1}{2}h, y_n + \tfrac{1}{2}hk_{n1}\right), \\ k_{n3} &= f\left(t_n + \tfrac{1}{2}h, y_n + \tfrac{1}{2}hk_{n2}\right), \\ k_{n4} &= f\left(t_n + h, y_n + hk_{n3}\right). \end{aligned} \qquad (11)$$

The sum $(k_{n1} + 2k_{n2} + 2k_{n3} + k_{n4})/6$ can be interpreted as an average slope. Note that k_{n1} is the slope at the left end of the interval, k_{n2} is the slope at the midpoint using the Euler formula to go from t_n to $t_n + h/2$, k_{n3} is a second approximation to the slope at the midpoint, and k_{n4} is the slope at $t_n + h$ using the Euler formula and the slope k_{n3} to go from t_n to $t_n + h$.

Although in principle it is not difficult to show that Eq. (10) differs from the Taylor expansion of the solution ϕ by terms that are proportional to h^5, the algebra is rather lengthy.[10] Thus we will simply accept the fact that the local truncation error in using Eq. (10) is proportional to h^5 and that for a finite interval the global truncation error is at most a constant times h^4. The earlier description of this method as a fourth order, four-stage method reflects the facts that the global truncation error is of fourth order in the step size h and that there are four intermediate stages in the calculation (the calculation of $k_{n1}, \ldots, k_{n4}$).

Clearly the Runge–Kutta formula, Eqs. (10) and (11), is more complicated than the formulas discussed previously. This is of relatively little significance, however, since it is not hard to write a computer program to implement this method. Such a program has the same structure as the algorithm for the Euler method outlined in Section 1.3.

Note that if f does not depend on y, then

$$k_{n1} = f(t_n), \qquad k_{n2} = k_{n3} = f(t_n + h/2), \qquad k_{n4} = f(t_n + h), \qquad (12)$$

and Eq. (10) reduces to

$$y_{n+1} - y_n = \frac{h}{6}[f(t_n) + 4f(t_n + h/2) + f(t_n + h)]. \qquad (13)$$

Equation (13) can be identified as Simpson's rule for the approximate evaluation of the integral of $y' = f(t)$. The fact that Simpson's rule has an error proportional to h^5 is consistent with the local truncation error in the Runge–Kutta formula.

EXAMPLE 2

Use the Runge–Kutta method to calculate approximate values of the solution $y = \phi(t)$ of the initial value problem

$$y' = 1 - t + 4y, \qquad y(0) = 1. \qquad (14)$$

Taking $h = 0.2$, we have

$$k_{01} = f(0, 1) = 5; \qquad hk_{01} = 1.0,$$
$$k_{02} = f(0 + 0.1, 1 + 0.5) = 6.9; \qquad hk_{02} = 1.38,$$
$$k_{03} = f(0 + 0.1, 1 + 0.69) = 7.66; \qquad hk_{03} = 1.532,$$
$$k_{04} = f(0 + 0.2, 1 + 1.532) = 10.928.$$

Thus

$$y_1 = 1 + \frac{0.2}{6}[5 + 2(6.9) + 2(7.66) + 10.928]$$
$$= 1 + 1.5016 = 2.5016.$$

Further results using the Runge–Kutta method with $h = 0.2$, $h = 0.1$, and $h = 0.05$ are given in Table 2.8.2. Note that the Runge–Kutta method yields a value at $t = 2$ that differs from the exact solution by only 0.122% if the step size is $h = 0.1$, and by only 0.00903% if $h = 0.05$. In the latter case, the error is less than one part in ten thousand, and the calculated value at $t = 2$ is correct to four digits.

For comparison, note that both the Runge–Kutta method with $h = 0.05$ and the improved Euler method with $h = 0.025$ require 160 evaluations of f to reach $t = 2$. The improved Euler method yields a result at $t = 2$ that is in error by 1.23%. Although this error may be

---

[10]See, for example, Chapter 3 of the book by Henrici listed in the references.

acceptable for some purposes, it is more than 135 times the error yielded by the Runge–Kutta method with comparable computing effort. Note also that the Runge–Kutta method with $h = 0.2$, or 40 evaluations of f, produces a value at $t = 2$ with an error of 1.40%, which is only slightly greater than the error in the improved Euler method with $h = 0.025$, or 160 evaluations of f. Thus we see again that a more accurate algorithm is more efficient; it produces better results with similar effort, or similar results with less effort.

TABLE 2.8.2   A Comparison of Results Using the Improved Euler and Runge–Kutta Methods for the Initial Value Problem $y' = 1 - t + 4y$, $y(0) = 1$.

	Improved Euler	Runge–Kutta			Exact
t	$h = 0.025$	$h = 0.2$	$h = 0.1$	$h = 0.05$	
0	1.0000000	1.0000000	1.0000000	1.0000000	1.0000000
0.1	1.6079462		1.6089333	1.6090338	1.6090418
0.2	2.5020618	2.5016000	2.5050062	2.5053060	2.5053299
0.3	3.8228282		3.8294145	3.8300854	3.8301388
0.4	5.7796888	5.7776358	5.7927853	5.7941197	5.7942260
0.5	8.6849039		8.7093175	8.7118060	8.7120041
1.0	64.497931	64.441579	64.858107	64.894875	64.897803
1.5	474.83402		478.81928	479.22674	479.25919
2.0	3496.6702	3490.5574	3535.8667	3539.8804	3540.2001

The classic Runge–Kutta method suffers from the same shortcoming as other methods with a fixed step size for problems in which the local truncation error varies widely over the interval of interest. That is, a step size that is small enough to achieve satisfactory accuracy in some parts of the interval may be much smaller than necessary in other parts of the interval. This has stimulated the development of adaptive Runge–Kutta methods that provide for modifying the step size automatically as the computation proceeds, so as to maintain the local truncation error near or below a specified tolerance level. As explained earlier in this section, this requires the estimation of the local truncation error at each step. One way to do this is to repeat the computation with a fifth order method—which has a local truncation error proportional to h^6—and then to use the difference between the two results as an estimate of the error. If this is done in a straightforward (unsophisticated) manner, then the use of the fifth order method requires at least five more evaluations of f at each step, in addition to those required originally by the fourth order method. However, if we make an appropriate choice of the intermediate points and the weighting coefficients in the expressions for k_{n1}, ... in a certain fourth order Runge–Kutta method, then these expressions can be used again, together with one additional stage, in a corresponding fifth order method. This results in a substantial gain in efficiency. It turns out that this can be done in more than one way.[11] The resulting adaptive Runge–Kutta methods are very powerful and efficient means of numerically approximating the solutions of an enormous class of initial

[11]The first widely used fourth and fifth order Runge–Kutta pair was developed by Erwin Fehlberg in the late 1960's. Its popularity was considerably enhanced by the appearance in 1977 of its Fortran implementation RKF45 by Lawrence F. Shampine and H. A. Watts.

value problems. Specific implementations of one or more of them are widely available in commercial software packages.

PROBLEMS

CAS In each of Problems 1 through 6, find approximate values of the solution of the given initial value problem at $t = 0.1, 0.2, 0.3$, and 0.4. Compare the results with those obtained by the Euler method in Section 2.7 and with the exact solution (if available).

(a) Use the improved Euler method with $h = 0.05$.
(b) Use the improved Euler method with $h = 0.025$.
(c) Use the improved Euler method with $h = 0.0125$.
(d) Use the Runge-Kutta method with $h = 0.1$.
(e) Use the Runge-Kutta method with $h = 0.05$.

1. $y' = 3 + t - y$, $y(0) = 1$
2. $y' = 5t - 3\sqrt{y}$, $y(0) = 2$
3. $y' = 2y - 3t$, $y(0) = 1$
4. $y' = 2t + e^{-ty}$, $y(0) = 1$
5. $y' = \dfrac{y^2 + 2ty}{3 + t^2}$, $y(0) = 0.5$
6. $y' = (t^2 - y^2)\sin y$, $y(0) = -1$

CAS In each of Problems 7 through 12, find approximate values of the solution of the given initial value problem at $t = 0.5, 1.0, 1.5$, and 2.0.

(a) Use the improved Euler method with $h = 0.025$.
(b) Use the improved Euler method with $h = 0.0125$.
(c) Use the Runge-Kutta method with $h = 0.1$.
(d) Use the Runge-Kutta method with $h = 0.05$.

7. $y' = 0.5 - t + 2y$, $y(0) = 1$
8. $y' = 5t - 3\sqrt{y}$, $y(0) = 2$
9. $y' = \sqrt{t + y}$, $y(0) = 3$
10. $y' = 2t + e^{-ty}$, $y(0) = 1$
11. $y' = (4 - ty)/(1 + y^2)$, $y(0) = -2$
12. $y' = (y^2 + 2ty)/(3 + t^2)$, $y(0) = 0.5$

CAS 13. Complete the calculations leading to the entries in columns four and five of Table 2.8.1.

CAS 14. Confirm the results in Table 2.8.2 by executing the indicated computations.

CAS 15. Consider the initial value problem

$$y' = t^2 + y^2, \qquad y(0) - 1.$$

(a) Draw a direction field for this equation.
(b) Use the Runge-Kutta method to find approximate values of the solution at $t = 0.8, 0.9$, and 0.95. Choose a small enough step size so that

you believe your results are accurate to at least four digits.

(c) Try to extend the calculations in part (b) to obtain an accurate approximation to the solution at $t = 1$. If you encounter difficulties in doing this, explain why you think this happens. The direction field in part (a) may be helpful.

16. Consider the initial value problem   CAS

$$y' = 3t^2/(3y^2 - 4), \qquad y(0) = 0.$$

(a) Draw a direction field for this equation.
(b) Estimate how far the solution can be extended to the right. Let t_M be the right endpoint of the interval of existence of this solution. What happens at t_M to prevent the solution from continuing farther?
(c) Use the Runge-Kutta method with various step sizes to determine an approximate value of t_M.
(d) If you continue the computation beyond t_M, you can continue to generate values of y. What significance, if any, do these values have?
(e) Suppose that the initial condition is changed to $y(0) = 1$. Repeat parts (b) and (c) for this problem.

17. In this problem, we establish that the local truncation error for the improved Euler formula is proportional to h^3. If we assume that the solution ϕ of the initial value problem $y' = f(t, y), y(t_0) = y_0$ has derivatives that are continuous through the third order (f has continuous second partial derivatives), then it follows that

$$\phi(t_n + h) = \phi(t_n) + \phi'(t_n)h + \frac{\phi''(t_n)}{2!}h^2$$
$$+ \frac{\phi'''(\bar{t}_n)}{3!}h^3,$$

where $t_n < \bar{t}_n < t_n + h$. Assume that $y_n = \phi(t_n)$.

(a) Show that, for y_{n+1} as given by Eq. (5),

$$e_{n+1} = \phi(t_{n+1}) - y_{n+1}$$
$$= \frac{\phi''(t_n)h - \{f[t_n + h, y_n + hf(t_n, y_n)] \quad f(t_n, y_n)\}}{2!}h$$
$$+ \frac{\phi'''(\bar{t}_n)h^3}{3!}. \tag{i}$$

(b) Making use of the facts that $\phi''(t) = f_t[t, \phi(t)] + f_y[t, \phi(t)]\phi'(t)$ and that the Taylor approximation with a remainder for a function $F(t, y)$ of two variables is

$$F(a + h, b + k) = F(a, b) + F_t(a, b)h$$
$$+ F_y(a, b)k$$
$$+ \frac{1}{2!}(h^2 F_{tt} + 2hk F_{ty} + k^2 F_{yy})\Big|_{x=\xi, y=\eta}$$

where ξ lies between a and $a + h$ and η lies between b and $b + k$, show that the first term on the right side of Eq. (i) is proportional to h^3 plus higher order terms. This is the desired result.

(c) Show that if $f(t, y)$ is linear in t and y, then $e_{n+1} = \phi'''(\bar{t}_n)h^3/6$, where $t_n < \bar{t}_n < t_{n+1}$.

Hint: What are f_{tt}, f_{ty}, and f_{yy}?

18. Consider the improved Euler method for solving the illustrative initial value problem $y' = 1 - t + 4y, y(0) = 1$. Using the result of Problem 17(c) and the exact solution of the initial value problem, determine e_{n+1} and a bound for the error at any step on $0 \le t \le 2$. Compare this error with the one obtained in Eq. (25) of Section 2.7 using the Euler method. Also obtain a bound for e_1 for $h = 0.05$, and compare it with Eq. (26) of Section 2.7.

In each of Problems 19 and 20, use the actual solution $\phi(t)$ to determine e_{n+1} and a bound for e_{n+1} at any step on $0 \le t \le 1$ for the improved Euler method for the given initial value problem. Also obtain a bound for e_1 for $h = 0.1$, and compare it with the similar estimate for the Euler method and with the actual error using the improved Euler method.

19. $y' = 2y - 1$, $\quad y(0) = 1$
20. $y' = 0.5 - t + 2y$, $\quad y(0) = 1$

In each of Problems 21 through 24, carry out one step CAS of the Euler method and of the improved Euler method, using the step size $h = 0.1$. Suppose that a local truncation error no greater than 0.0025 is required. Estimate the step size that is needed for the Euler method to satisfy this requirement at the first step.

21. $y' = 0.5 - t + 2y$, $\quad y(0) = 1$
22. $y' = 5t - 3\sqrt{y}$, $\quad y(0) = 2$
23. $y' = \sqrt{t + y}$, $\quad y(0) = 3$
24. $y' = (y^2 + 2ty)/(3 + t^2)$, $\quad y(0) = 0.5$

CHAPTER SUMMARY

In this chapter we discuss a number of special solution methods for first order equations $dy/dt = f(t, y)$. The most important types of equations that can be solved analytically are **linear, separable**, and **exact** equations. For equations that cannot be solved by symbolic analytic methods, it is necessary to resort to geometrical and numerical methods.

Some aspects of the qualitative theory of differential equations are also introduced in this chapter: existence and uniqueness of solutions; stability properties of equilibrium solutions to autonomous equations.

Section 2.1 Linear Equations $dy/dt + p(t)y = g(t)$. Multiplying by the integrating factor $\mu(t) = \exp(\int p(t)dt)$ leads to $d(\mu y)/dt = \mu g$. Integrating then yields $y = \mu^{-1}[\int \mu g \, dt + c]$.

Section 2.2 Separable Equations $M(x) + N(y)dy/dx = 0$ can be solved by direct integration.

Section 2.3 We discuss mathematical models for several types of problems that lead to either linear or separable equations: mixing tanks, compound interest, projectile motion, heating and cooling, and radioactive decay.

Section 2.4 Qualitative Theory existence and uniqueness of solutions to initial value problems.

▶ Conditions guaranteeing existence and uniqueness of solutions are given in Theorems 2.4.1 and 2.4.2 for linear and nonlinear equations, respectively.

▶ We show examples of initial value problems where solutions are not unique or become unbounded in finite time.

Section 2.5 Qualitative Theory
autonomous equations, equilibrium solutions, and their stability characteristics.

▶ Autonomous equations are of the form $dy/dt = f(y)$.

▶ Critical points (equilibrium solutions) are solutions of $f(y) = 0$.

▶ Whether an equilibrium solution is **asymptotically stable** or **unstable** determines to a great extent the long-time (asymptotic) behavior of solutions.

Section 2.6 Exact Equations
$M(x, y)dx + N(x, y)dy = 0$ is exact if and only if $\partial M/\partial y = \partial N/\partial x$.

▶ Direct integration of an exact equation leads to implicitly defined solutions $F(x, y) = c$ where $\partial F/\partial x = M$ and $\partial F/\partial y = N$.

▶ Some differential equations can be made exact if a special integrating factor can be found.

Section 2.7

▶ Numerical approximations to solutions of initial value problems involve two types of error: (i) **truncation error** (local and global), and (ii) **round-off error**.

▶ For the Euler method, the global truncation error is proportional to the step size h and the local truncation error is proportional to h^2.

Section 2.8
Two numerical approximation methods more sophisticated and efficient than the Euler method, the **improved Euler** method and the **Range-Kutta** method, are described and illustrated by examples.

PROJECTS

Project 1 Harvesting a Renewable Resource

Suppose that the population y of a certain species of fish (for example, tuna or halibut) in a given area of the ocean is described by the logistic equation

$$dy/dt = r(1 - y/K)y.$$

If the population is subjected to harvesting at a rate $H(y, t)$ members per unit time, then the harvested population is modeled by the differential equation

$$dy/dt = r(1 - y/K)y - H(y, t). \tag{1}$$

Although it is desirable to utilize the fish as a food source, it is intuitively clear that if too many fish are caught, then the fish population may be reduced below a useful level and possibly even driven to extinction. The following problems explore some of the questions involved in formulating a rational strategy for managing the fishery.

Project 1 PROBLEMS

1. **Constant Effort Harvesting**. At a given level of effort, it is reasonable to assume that the rate at which fish are caught depends on the population y: the more fish there are, the easier it is to catch them. Thus we assume that the rate at which fish are caught is given by $H(y, t) = Ey$, where E is a positive constant, with units of 1/time, that measures the total effort made to harvest the given species of fish. With this choice for $H(y, t)$, Eq. (1) becomes

$$dy/dt = r(1 - y/K)y - Ey. \qquad (i)$$

This equation is known as the **Schaefer model** after the biologist M. B. Schaefer, who applied it to fish populations.

(a) Show that if $E < r$, then there are two equilibrium points, $y_1 = 0$ and $y_2 = K(1 - E/r) > 0$.

(b) Show that $y = y_1$ is unstable and $y = y_2$ is asymptotically stable.

(c) A sustainable yield Y of the fishery is a rate at which fish can be caught indefinitely. It is the product of the effort E and the asymptotically stable population y_2. Find Y as a function of the effort E. The graph of this function is known as the yield–effort curve.

(d) Determine E so as to maximize Y and thereby find the **maximum sustainable yield** Y_m.

2. **Constant Yield Harvesting**. In this problem, we assume that fish are caught at a constant rate h independent of the size of the fish population, that is,

the harvesting rate $H(y, t) = h$. Then y satisfies

$$dy/dt = r(1 - y/K)y - h = f(y). \qquad (ii)$$

The assumption of a constant catch rate h may be reasonable when y is large but becomes less so when y is small.

(a) If $h < rK/4$, show that Eq. (ii) has two equilibrium points y_1 and y_2 with $y_1 < y_2$; determine these points.

(b) Show that y_1 is unstable and y_2 is asymptotically stable.

(c) From a plot of $f(y)$ versus y, show that if the initial population $y_0 > y_1$, then $y \to y_2$ as $t \to \infty$, but if $y_0 < y_1$, then y decreases as t increases. Note that $y = 0$ is not an equilibrium point, so if $y_0 < y_1$, then extinction will be reached in a finite time.

(d) If $h > rK/4$, show that y decreases to zero as t increases regardless of the value of y_0.

(e) If $h = rK/4$, show that there is a single equilibrium point $y = K/2$ and that this point is semistable (see Problem 7, Section 2.5). Thus the maximum sustainable yield is $h_m = rK/4$, corresponding to the equilibrium value $y = K/2$. Observe that h_m has the same value as Y_m in Problem 1(d). The fishery is considered to be overexploited if y is reduced to a level below $K/2$.

Project 2 Designing a Drip Dispenser for a Hydrology Experiment

In order to make laboratory measurements of water filtration and saturation rates in various types of soils under the condition of steady rainfall, a hydrologist wishes to design drip dispensing containers in such a way that the water drips out at a nearly constant rate. The containers are supported above glass cylinders that contain the soil samples (Figure 2.P.1). The hydrologist elects to use the following differential equation, based on Torricelli's principle (see Problem 6, Section 2.3), to help solve the design problem,

$$A(h)\frac{dh}{dt} = -\alpha a\sqrt{2gh}. \qquad (1)$$

In Eq. (1), $h(t)$ is the height of the liquid surface above the dispenser outlet at time t, $A(h)$ is the cross-sectional area of the dispenser at height h, a is the area of the outlet, and α is a measured contraction coefficient that accounts for the observed fact that the cross section of the (smooth) outflow stream is smaller than a. Note that the hydrologist is using a laminar flow model as a guide in designing the shape of the container. Forces due to surface tension at the tiny outlet are ignored in the design problem. Once the shape

FIGURE 2.P.1 Water dripping into a soil sample.

of the container has been determined, the outlet aperture is adjusted to a desired drip rate that will remain nearly constant for an extended period of time. Of course, since surface tension forces are not accounted for in Eq. (1), the equation is not a valid model for the output flow rate when the aperture is so small that the water drips out. Nevertheless, once the hydrologist sees and interprets the results based on her design strategy, she feels justified in using Eq. (1).

Project 2 PROBLEMS

1. Assume that the shape of the dispensers are surfaces of revolution so that $A(h) = \pi [r(h)]^2$ where $r(h)$ is the radius of the container at height h. For each of the h dependent cross-sectional radii prescribed below in (i)–(v),

   (a) create a surface plot of the surface of revolution, and

   (b) find numerical approximations of solutions of Eq. (1) for $0 \le t \le 60$:

   i. $r(h) = r_1, \qquad 0 \le h \le H$

   ii. $r(h) = r_0 + (r_1 - r_0)h/H, \qquad 0 \le h \le H$

   iii. $r(h) = r_0 + (r_1 - r_0)\sqrt{h/H},$
   $0 \le h \le H$

   iv. $r(h) = r_0 \exp\left[(h/H)\ln(r_1/r_0)\right]$
   $0 \le h \le H$

   v. $r(h) = \dfrac{r_0 r_1 H}{r_1 H - (r_1 - r_0)h}, \qquad 0 \le h \le H.$

   Use the parameter values specified in the following table:

   $$r_0 = 0.1 \text{ ft}$$
   $$r_1 = 1 \text{ ft}$$
   $$\alpha = 0.6$$
   $$a = 0.001 \text{ ft}^2$$

   In addition, use the initial condition $h(0) = H$, where the initial height H of water in each of the containers is determined by requiring that the initial volume of water satisfies

   $$V(0) = \int_0^H \pi r^2(h) \, dh = 1 \text{ ft}^3.$$

   Determine the qualitative shape of the container such that the output flow rate given by the right-hand side of Eq. (1), $F_R(t) = \alpha a \sqrt{2gh(t)}$, varies slowly during the early stages of the experiment. As a design criterion, consider plotting the ratio $R = F_R(t)/F_R(0)$ for $0 \le t \le 60$, where values of R near 1 are most desirable. Based on the results of your computer experiments, sketch the shape of what a suitable container should look like.

2. After viewing the results of her computer experiments, it slowly dawns on the hydrologist that the "optimal shape" of the container is consistent with what would be expected based on the conceptualization that the water in the ideal container would consist of a collection of small parcels of water of mass m, all possessing the same amount of potential energy. If laboratory spatial constraints were not

an issue, what would be the ideal "shape" of each container? Perform a computer experiment that sup-

ports your conclusions based on potential energy considerations.

Project 3 A Mathematical Model of a Groundwater Contaminant Source

Chlorinated solvents such as trichloroethylene (TCE) are a common cause of environmental contamination[12] at thousands of government and private industry facilities. TCE and other chlorinated organics, collectively referred to as dense nonaqueous phase liquids (DNAPLs), are denser than water and only slightly soluble in water. DNAPLs tend to accumulate as a separate phase below the water table and provide a long-term source of groundwater contamination. A downstream contaminant plume is formed by the process of dissolution of DNAPL into water flowing through the source region as shown in Figure 2.P.2.

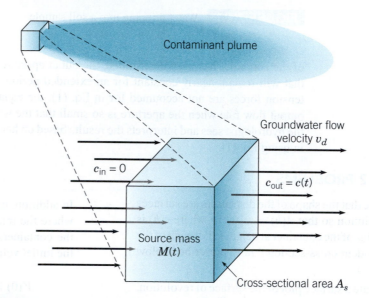

FIGURE 2.P.2 Conceptual model of DNAPL source.

In this project, we study a first order differential equation that describes the time dependent rate of dissolved contaminant discharge leaving the source zone and entering the plume.[13]

Parameters and variables relevant to formulating a mathematical model of contaminant discharge from the source region are defined in the following table:

A_s = cross-sectional area of the source region

v_d = Darcy groundwater flow velocity.[14]

$m(t)$ = total DNAPL mass in source region

---

[12]Falta, R.W., P.S. Rao, and N. Basu, "Assessing the Impacts of Partial Mass Depletion in DNAPL Source Zones: I. Analytical Modeling of Source Strength Functions and Plume Response," *Journal of Contaminant Hydrology* 78, (2005)(4) pp 259–280.

[13]The output of this model can then be used as input into another mathematical model that in turn describes the processes of advection, adsorption, dispersion, and degradation of contaminant within the plume.

[14]In porous media flow, the Darcy flow velocity v_d is defined by $v_d = Q/A$ where A is a cross-sectional area available for flow and Q is the volumetric flow rate (volume/time) through A.

$c_s(t)$ = concentration (flow averaged) of dissolved contaminant leaving the source zone

m_0 = initial DNAPL mass in source region

c_0 = source zone concentration (flow averaged) corresponding to an initial source zone mass of m_0

The equation describing the rate of DNAPL mass discharge from the source region is

$$\frac{dm}{dt} = -A_s v_d c_s(t), \tag{1}$$

while an algebraic relationship between $c_s(t)$ and $m(t)$ is postulated in the form of a power law,

$$\frac{c_s(t)}{c_0} = \left[\frac{m(t)}{m_0}\right]^\gamma, \tag{2}$$

in which $\gamma > 0$ is empirically determined. Combining Eqs. (1) and (2) (Problem 1) yields a first order differential equation

$$\frac{dm}{dt} = -\alpha m^\gamma \tag{3}$$

that models the dissolution of DNAPL into the groundwater flowing through the source region.

Project 3 PROBLEMS

1. Derive Eq. (3) from Eqs. (1) and (2) and show that $\alpha = v_d A_s c_0 / m_0^\gamma$.

2. Additional processes due to biotic and abiotic degradation contributing to source decay can be accounted for by adding a decay term to (3) that is proportional to $m(t)$,

$$m'(t) = -\alpha m^\gamma - \lambda m, \tag{i}$$

where λ is the associated decay rate constant. Find solutions of Eq. (i) using the initial condition $m(0) = m_0$ for the following cases: (i) $\gamma = 1$, (ii) $\gamma \neq 1$ and $\lambda = 0$, (iii) $\gamma \neq 1$ and $\lambda \neq 0$. Then find expressions for $c_s(t)$ using Eq. (2).
   Hint: Eq. (i) is a type of nonlinear equation known as a Bernoulli equation. A method for solving Bernoulli equations is discussed in Problem 27 of Section 2.4.

3. Show that when $\gamma \geq 1$ the source has an infinite lifetime but if $0 < \gamma < 1$ the source has a finite lifetime. In the latter case, find the time that the DNAPL source mass attains the value zero.

ODEA 4. Assume the following values for the parameters: $m_0 = 1620$ kg, $c_0 = 100$ mg/l, $A_s = 30$ m^2, $v_d = 20$ m/yr, $\lambda = 0$. Use the solutions obtained in Problem 2 to plot graphs of $c_s(t)$ for each of the following cases: (i) $\gamma = 0.5$ for $0 \leq t \leq t_f$ where $c_s(t_f) = 0$, (ii) $\gamma = 2$ for $0 \leq t \leq 100$ years.

5. **Effects of Partial Source Remediation.** ODEA

   (a) Assume that a source remediation process results in a 90% reduction in the initial amount of DNAPL mass in the source region. Repeat Problem 4 with m_0 and c_0 in Eq. (2) replaced by $m_1 = (0.1)\, m_0$ and $c_1 = (0.1)^\gamma c_0$, respectively. Compare the graphs of $c_s(t)$ in this case with the graphs obtained in Problem 4.

   (b) Assume that the 90% efficient source remediation process is not applied until $t_1 = 10$ years have elapsed following the initial deposition of the contaminant. Under this scenario, plot the graphs of $c_s(t)$ using the parameters and initial conditions of Problem 4. In this case, use Eq. (2) to compute concentration for $0 \leq t < t_1$. Following remediation, use the initial condition $m(t_1) = m_1 = 0.1 m(t_1 - 0) = 0.1 \lim_{t \uparrow t_1} m(t)$ for Eq. (i) and use the following modification of Eq. (2),

   $$\frac{c_s(t)}{c_1} = \left[\frac{m(t)}{m_1}\right]^\gamma, \qquad t > t_1, \tag{5}$$

   where $c_1 = (0.1)^\gamma c(t_1 - 0) = (0.1)^\gamma \lim_{t \uparrow t_1} c(t)$ to compute concentrations for times $t > t_1$. Compare the graphs of $c_s(t)$ in this case with the graphs obtained in Problems 4 and 5(a). Can you draw any conclusions about the possible effectiveness of source remediation? If so, what are they?

Project 4 Monte Carlo Option Pricing: Pricing Financial Options by Flipping a Coin

A discrete model for change in price of a stock over a time interval $[0, T]$ is

$$S_{n+1} = S_n + \mu\, S_n\, \Delta t + \sigma\, S_n\, \varepsilon_{n+1}\, \sqrt{\Delta t}, \qquad S_0 = s \tag{1}$$

where $S_n = S(t_n)$ is the stock price at time $t_n = n\Delta t$, $n = 0, \ldots, N - 1$, $\Delta t = T/N$, μ is the annual growth rate of the stock, and σ is a measure of the stock's annual price volatility or tendency to fluctuate. Highly volatile stocks have large values for σ, for example, values ranging from 0.2 to 0.4. Each term in the sequence $\varepsilon_1, \varepsilon_2, \ldots$ takes on the value 1 or -1 depending on whether the outcome of a coin tossing experiment is heads or tails, respectively. Thus, for each $n = 1, 2, \ldots$

$$\varepsilon_n = \begin{cases} 1 & \text{with probability} = 1/2 \\ -1 & \text{with probability} = 1/2. \end{cases} \tag{2}$$

A sequence of such numbers can easily be created by using one of the random number generators available in most mathematical computer software applications. Given such a sequence, the difference equation (1) can then be used to simulate a **sample path** or **trajectory** of stock prices, $\{s, S_1, S_2, \ldots S_N\}$. The "random" terms $\sigma\, S_n \varepsilon_{n+1} \sqrt{\Delta t}$ on the right-hand side of (1) can be thought of as "shocks" or "disturbances" that model fluctuations in the stock price. By repeatedly simulating stock price trajectories and computing appropriate averages, it is possible to obtain estimates of the price of a **European call option**, a type of financial derivative. A statistical simulation algorithm of this type is called a **Monte Carlo method**.

A European call option is a contract between two parties, a holder and a writer, whereby, for a premium paid to the writer, the holder acquires the right (but not the obligation) to purchase the stock at a future date T (the **expiration date**) at a price K (the **strike price**) agreed upon in the contract. If the buyer elects to exercise the option on the expiration date, the writer is obligated to sell the underlying stock to the buyer at the price K. Thus, the option has, associated with it, a **payoff function**

$$f(S) = \max(S - K, 0) \tag{3}$$

where $S = S(T)$ is the price of the underlying stock at the time T when the option expires (see Figure 2.P.3).

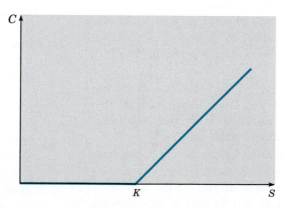

FIGURE 2.P.3 The value of a call option at expiration is $C = \max(S - K, 0)$ where K is the strike price of the option and $S = S(T)$ is the stock price at expiration.

Eq. (3) is the value of the option at time T since, if $S(T) > K$, the holder can purchase, at price K, stock with market value $S(T)$ and thereby make a profit equal to $S(T) - K$ not counting the option premium. If $S(T) < K$, the holder will simply let the option expire since it would be irrational to purchase stock at a price that exceeds the market value. The option valuation problem is to determine the correct and fair price of the option at the time that the holder and writer enter into the contract.[15]

To estimate the price of a call option using a Monte Carlo method, an ensemble

$$\left\{ S_N^{(k)} = S^{(k)}(T), \ k = 1, \ldots, M \right\}$$

of M stock prices at expiration is generated using the difference equation

$$S_{n+1}^{(k)} = S_n^{(k)} + r\, S_n^{(k)}\, \Delta t + \sigma\, S_n^{(k)}\, \varepsilon_{n+1}^{(k)}\, \sqrt{\Delta t}, \qquad S_0^{(k)} = s. \tag{4}$$

For each $k = 1, \ldots, M$, the difference equation (4) is identical to (1) except that the growth rate μ is replaced by the annual rate of interest r that it costs the writer to borrow money. Option pricing theory requires that the average value of the payoffs $\left\{ f(S_N^{(k)}), k = 1, \ldots, M \right\}$ be equal to the compounded total return obtained by investing the option premium, $\hat{C}(s)$, at rate r over the life of the option,

$$\frac{1}{M} \sum_{k=1}^{M} f(S_N^{(k)}) = (1 + r\Delta t)^N\, \hat{C}(s). \tag{5}$$

Solving (5) for $\hat{C}(s)$ yields the Monte Carlo estimate

$$\hat{C}(s) = (1 + r\Delta t)^{-N} \left\{ \frac{1}{M} \sum_{k=1}^{M} f(S_N^{(k)}) \right\} \tag{6}$$

for the option price. Thus, the Monte Carlo estimate $\hat{C}(s)$ is the present value of the average of the payoffs computed using the rules of compound interest.

Project 4 PROBLEMS

1. Show that Euler's method applied to the differential equation

$$\frac{dS}{dt} = \mu S \tag{i}$$

yields Eq. (1) in the absence of random disturbances, that is, when $\sigma = 0$.

CAS 2. Simulate five sample trajectories of (1) for the following parameter values and plot the trajectories on the same set of coordinate axes: $\mu = 0.12$, $\sigma = 0.1$, $T = 1$, $s = \$40$, $N = 254$. Then repeat the experiment using the value $\sigma = 0.25$ for the volatility. Do the sample trajectories generated in the latter case appear to exhibit a greater degree of variability in their behavior?

Hint: For the ε_n's it is permissible to use a random number generator that creates normally distributed random numbers with mean zero and variance one.

3. Use the difference equation (4) to generate CAS an ensemble of stock prices $S_N^{(k)} = S^{(k)}(N\Delta t)$, $k = 1, \ldots, M$ (where $T = N\Delta t$) and then use formula (6) to compute a Monte Carlo estimate of the value of a five month call option ($T = 5/12$ years) for the following parameter values: $r = 0.06$, $\sigma = 0.2$, and $K = \$50$. Find estimates corresponding to current stock prices of $S(0) = s = \$45$, $\$50$, and $\$55$. Use $N = 200$ time steps for each trajectory and $M \cong 10{,}000$ sample trajectories for each Monte Carlo estimate.[16] Check the accuracy of your results by comparing the Monte Carlo approximation with

[15]The 1997 Nobel Prize in Economics was awarded to Robert C. Merton and Myron S. Scholes for their work, along with Fischer Black, in developing the Black-Scholes options pricing model.

[16]As a rule of thumb, you may assume that the sampling error in these Monte Carlo estimates is proportional to $1/\sqrt{M}$. Using software packages such as MATLAB that allow vector operations where all M trajectories can be simulated simultaneously greatly speeds up the calculations.

the value computed from the exact Black-Scholes formula

$$C(s) = \frac{s}{2}\,\text{erfc}(-d_1/\sqrt{2}) - \frac{K}{2}\,e^{-rT}\text{erfc}(-d_2/\sqrt{2}) \tag{ii}$$

where

$$d_1 = \frac{1}{\sigma\sqrt{T}}\left[\ln(s/k) + (r + \sigma^2/2)T\right],$$

$$d_2 = d_1 - \sigma\sqrt{T}$$

and erfc(x) is the complementary error function,

$$\text{erfc}(x) = \frac{2}{\sqrt{\pi}}\int_x^\infty e^{-t^2}\,dt.$$

CAS **4. Variance Reduction by Antithetic Variates.** A simple and widely used technique for increasing the efficiency and accuracy of Monte Carlo simulations in certain situations with little additional increase in computational complexity is the method of antithetic variates. For each $k = 1, \ldots, M$ use the sequence $\left\{\varepsilon_1^{(k)}, \ldots, \varepsilon_{N-1}^{(k)}\right\}$ in (4) to simulate

a payoff $f(S_N^{(k+)})$ and also use the sequence $\left\{-\varepsilon_1^{(k)}, \ldots, -\varepsilon_{N-1}^{(k)}\right\}$ in (4) to simulate an associated payoff $f(S_N^{(k-)})$. Thus, the payoffs are simulated in pairs $\left\{(f(S_N^{(k+)}), f(S_N^{(k-)}))\right\}$. A modified Monte Carlo estimate is then computed by replacing each payoff $f(S_N^{(k)})$ in (6) by the average $[f(S_N^{(k+)}) + f(S_N^{(k-)})]/2$,

$$\hat{C}_{AV}(s) = (1 + r\Delta t)^{-N}\left\{\frac{1}{M}\sum_{k=1}^M \frac{f(S_N^{(k+)}) + f(S_N^{(k-)})}{2}\right\}. \tag{iii}$$

Use the parameters specified in Problem 3 to compute several (say 20 or so) option price estimates using (6) and an equivalent number of option price estimates using (iii). For each of the two methods, plot a histogram of the estimates and compute the mean and standard deviation of the estimates. Comment on the accuracies of the two methods.

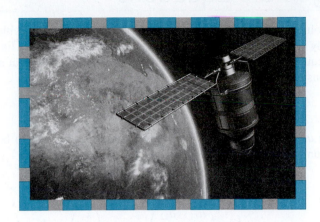

CHAPTER THREE

Systems of Two First Order Equations

The first two chapters of this book contain the material most essential for dealing with single first order differential equations. To proceed further, there are two natural paths that we might follow. The first is to take up the study of second order equations and the other is to consider systems of two first order equations. There are many important problems in various areas of application that lead to each of these types of problems, so both are important and we will eventually discuss both. They are also closely related to each other, as we will show. Our approach is to introduce systems of two first order equations in this chapter and to take up second order equations in Chapter 4.

There are many problem areas that involve several components linked together in some way. For example, electrical networks have this character, as do some problems in mechanics and in other fields. In these and similar cases, there are two (or more) dependent variables and the corresponding mathematical problem consists of a system of two (or more) differential equations, which can always be written as a system of first order equations. In this chapter, we consider only systems of two first order equations and we focus most of our attention on systems of the simplest kind: two first order linear equations with constant coefficients. Our goals are to show what kinds of solutions such a system may have and how the solutions can be determined and displayed graphically, so that they can be easily visualized.

3.1 Systems of Two Linear Algebraic Equations

The solution of a system of two linear differential equations with constant coefficients is directly related to the solution of an associated system of two linear algebraic equations. Consequently, we start by reviewing the properties of such linear algebraic systems.[1]

Consider the system

$$a_{11}x_1 + a_{12}x_2 = b_1, \tag{1}$$

$$a_{21}x_1 + a_{22}x_2 = b_2,$$

where $a_{11}, \ldots, a_{22}, b_1$, and b_2 are given and x_1 and x_2 are to be determined. In matrix notation, we can write the system (1) as

$$\mathbf{Ax} = \mathbf{b}, \tag{2}$$

where

$$\mathbf{A} = \begin{pmatrix} a_{11} & a_{12} \\ a_{21} & a_{22} \end{pmatrix}, \quad \mathbf{x} = \begin{pmatrix} x_1 \\ x_2 \end{pmatrix}, \quad \mathbf{b} = \begin{pmatrix} b_1 \\ b_2 \end{pmatrix}. \tag{3}$$

Here $\mathbf{A}$ is a given 2×2 matrix, $\mathbf{b}$ is a given 2×1 column vector, and $\mathbf{x}$ is a 2×1 column vector to be determined.

To see what kinds of solutions the system (1) or (2) may have, it is helpful to visualize the situation geometrically. If a_{11} and a_{12} are not both zero, then the first equation in the system (1) corresponds to a straight line in the x_1x_2-plane, and similarly for the second equation. There are three distinct possibilities for two straight lines in a plane: they may intersect at a single point, they may be parallel and nonintersecting, or they may be coincident. In the first case, the system (1) or (2) is satisfied by a single pair of values of x_1 and x_2. In the second case, the system has no solutions; that is, there is no point that lies on both lines. In the third case, the system has infinitely many solutions, since every point on one line also lies on the other. The following three examples illustrate these possibilities.

EXAMPLE 1

Solve the system

$$3x_1 - x_2 = 8, \tag{4}$$

$$x_1 + 2x_2 = 5.$$

We can solve this system in a number of ways. For instance, from the first equation we have

$$x_2 = 3x_1 - 8. \tag{5}$$

Then, substituting this expression for x_2 in the second equation, we obtain

$$x_1 + 2(3x_1 - 8) = 5,$$

or $7x_1 = 21$, from which $x_1 = 3$. Then, from Eq. (5), $x_2 = 1$. Thus the solution of the system (4) is $x_1 = 3$, $x_2 = 1$. In other words, the point $(3, 1)$ is the unique point of intersection of the two straight lines corresponding to the equations in the system (4). See Figure 3.1.1.

---

[1] We believe that much of the material in this section will be familiar to you. A more extensive discussion of linear algebra and matrices appears in Appendix A.

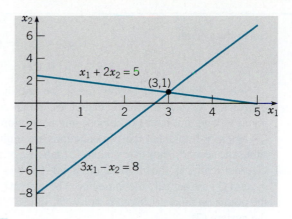

FIGURE 3.1.1 Geometrical interpretation of the system (4).

EXAMPLE 2

Solve the system

$$x_1 + 2x_2 = 1, \tag{6}$$
$$x_1 + 2x_2 = 5.$$

We can see at a glance that these two equations have no solution, since $x_1 + 2x_2$ cannot simultaneously take on the values 1 and 5. Proceeding more formally, as in Example 1, we can solve the first equation for x_1, with the result that $x_1 = 1 - 2x_2$. On substituting this expression for x_1 in the second equation, we obtain the false statement that $1 = 5$. Of course, you should not regard this as a demonstration that the numbers 1 and 5 are equal. Rather, you should conclude that the two equations in the system (6) are incompatible or inconsistent, and have no solution. The geometrical interpretation of the system (6) is shown in Figure 3.1.2. The two lines are parallel and therefore have no points in common.

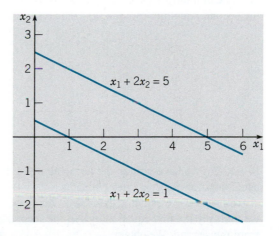

FIGURE 3.1.2 Geometrical interpretation of the system (6).

EXAMPLE 3

Solve the system

$$2x_1 + 4x_2 = 10,$$
$$x_1 + 2x_2 = 5. \tag{7}$$

Solving the second equation for x_1, we find that $x_1 = 5 - 2x_2$. Then, substituting this expression for x_1 in the first equation, we obtain $2(5 - 2x_2) + 4x_2 = 10$, or $10 = 10$. This result is true, but does not impose any restriction on x_1 or x_2. On looking at the system (7) again, we note that the two equations are multiples of each other; the first is just two times the second. Thus every point that satisfies one of the equations also satisfies the other. Geometrically, as shown in Figure 3.1.3, the two lines described by the equations in the system (7) are actually the same line. The system (7) has an infinite set of solutions—all of the points on this line. In other words, all values of x_1 and x_2 such that $x_1 + 2x_2 = 5$ satisfy the system (7).

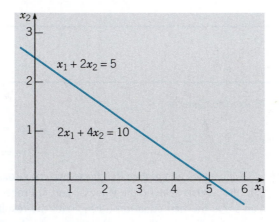

FIGURE 3.1.3   Geometrical interpretation of the system (7).

Let us now return to the system (1) and find its solution. To eliminate x_2 from the system (1) we can multiply the first equation by a_{22}, the second equation by a_{12}, and then subtract the second equation from the first. The result is

$$(a_{11}a_{22} - a_{12}a_{21})x_1 = a_{22}b_1 - a_{12}b_2, \tag{8}$$

so

$$x_1 = \frac{a_{22}b_1 - a_{12}b_2}{a_{11}a_{22} - a_{12}a_{21}} = \frac{\begin{vmatrix} b_1 & a_{12} \\ b_2 & a_{22} \end{vmatrix}}{\begin{vmatrix} a_{11} & a_{12} \\ a_{21} & a_{22} \end{vmatrix}}. \tag{9}$$

In Eq. (9), we have written the numerator and denominator of the expression for x_1 in the usual notation for 2×2 determinants.

To find a corresponding expression for x_2 we can eliminate x_1 from Eqs. (1). To do this, we multiply the first equation by a_{21}, the second by a_{11}, and then subtract the first equation from the second. We obtain

$$(a_{11}a_{22} - a_{12}a_{21})x_2 = a_{11}b_2 - a_{21}b_1, \tag{10}$$

so

$$x_2 = \frac{a_{11}b_2 - a_{21}b_1}{a_{11}a_{22} - a_{12}a_{21}} = \frac{\begin{vmatrix} a_{11} & b_1 \\ a_{21} & b_2 \end{vmatrix}}{\begin{vmatrix} a_{11} & a_{12} \\ a_{21} & a_{22} \end{vmatrix}}. \tag{11}$$

The denominator in Eqs. (9) and (11) is the **determinant of coefficients** of the system (1), or the determinant of the matrix **A**. We will denote it by det(**A**), or sometimes by Δ. Thus

$$\Delta = \det(\mathbf{A}) = \begin{vmatrix} a_{11} & a_{12} \\ a_{21} & a_{22} \end{vmatrix} = a_{11}a_{22} - a_{12}a_{21}. \tag{12}$$

As long as $\det(\mathbf{A}) \neq 0$, the expressions in Eqs. (9) and (11) give the unique values of x_1 and x_2 that satisfy the system (1). The solution of the system (1) in terms of determinants, given by Eqs. (9) and (11), is known as **Cramer's rule**.

The condition that $\det(\mathbf{A}) \neq 0$ has a simple geometric interpretation. Observe that the slope of the line given by the first equation in the system (1) is $-a_{11}/a_{12}$, as long as $a_{12} \neq 0$. Similarly, the slope of the line given by the second equation is $-a_{21}/a_{22}$, provided that $a_{22} \neq 0$. If the slopes are different, then

$$-\frac{a_{11}}{a_{12}} \neq -\frac{a_{21}}{a_{22}},$$

which is equivalent to

$$a_{11}a_{22} - a_{12}a_{21} = \det(\mathbf{A}) \neq 0.$$

Of course, if the slopes are different, then the lines intersect at a single point, whose coordinates are given by Eqs. (9) and (11). We leave it to you to consider what happens if either a_{12} or a_{22} or both are zero. Thus we have the following important result.

| THEOREM 3.1.1 | The system (1),

$$a_{11}x_1 + a_{12}x_2 = b_1,$$

$$a_{21}x_1 + a_{22}x_2 = b_2,$$

has a unique solution if and only if the determinant

$$\Delta = a_{11}a_{22} - a_{12}a_{21} \neq 0.$$

The solution is given by Eqs. (9) and (11). If $\Delta = 0$, then the system (1) has either no solution or infinitely many.

We now introduce two matrices of special importance, as well as some associated terminology. The 2×2 identity matrix is denoted by **I** and is defined to be

$$\mathbf{I} = \begin{pmatrix} 1 & 0 \\ 0 & 1 \end{pmatrix}. \tag{13}$$

Note that the product of **I** with any 2×2 matrix or with any 2×1 vector is just the matrix or vector itself.

For a given 2×2 matrix **A** there may be another 2×2 matrix **B** such that $\mathbf{AB} = \mathbf{BA} = \mathbf{I}$. There may be no such matrix **B**, but if there is, then it can be shown that there is only one. The matrix **B** is called the **inverse** of **A** and is denoted by $\mathbf{B} = \mathbf{A}^{-1}$.

If $\mathbf{A}^{-1}$ exists, then $\mathbf{A}$ is called **nonsingular** or **invertible**. On the other hand, if $\mathbf{A}^{-1}$ does not exist, then $\mathbf{A}$ is said to be **singular** or **noninvertible**. In Problem 37, we ask you to show that if $\mathbf{A}$ is given by Eq. (3), then $\mathbf{A}^{-1}$, when it exists, is given by

$$\mathbf{A}^{-1} = \frac{1}{\det(\mathbf{A})} \begin{pmatrix} a_{22} & -a_{12} \\ -a_{21} & a_{11} \end{pmatrix}. \tag{14}$$

It is easy to verify the correctness of Eq. (14) simply by multiplying $\mathbf{A}$ and $\mathbf{A}^{-1}$ together. Equation (14) strongly suggests that $\mathbf{A}$ is nonsingular if and only if $\det(\mathbf{A}) \neq 0$, and this is in fact true. If $\det(\mathbf{A}) = 0$, then $\mathbf{A}$ is singular, and conversely.

We now return to the system (2). If $\mathbf{A}$ is nonsingular, multiply each side of Eq. (2) on the left by $\mathbf{A}^{-1}$. This gives

$$\mathbf{A}^{-1} \mathbf{A} \mathbf{x} = \mathbf{A}^{-1} \mathbf{b},$$

or

$$\mathbf{I} \mathbf{x} = \mathbf{A}^{-1} \mathbf{b},$$

or

$$\mathbf{x} = \mathbf{A}^{-1} \mathbf{b}. \tag{15}$$

It is straightforward to show that the result (15) agrees with Eqs. (9) and (11).

▶ **Homogeneous Systems.** If $b_1 = b_2 = 0$ in the system (1), then the system is said to be **homogeneous**; otherwise, it is called **nonhomogeneous**. Thus the general system of two linear homogeneous algebraic equations has the form

$$a_{11}x_1 + a_{12}x_2 = 0, \tag{16}$$

$$a_{21}x_1 + a_{22}x_2 = 0,$$

or, in matrix notation,

$$\mathbf{A}\mathbf{x} = \mathbf{0}. \tag{17}$$

For the homogeneous system (16) the corresponding straight lines must pass through the origin. Thus the lines always have at least one point in common, namely, the origin. If the two lines coincide, then every point on each line also lies on the other and the system (16) has infinitely many solutions. The two lines cannot be parallel and nonintersecting. In most applications, the solution $x_1 = 0$, $x_2 = 0$ is of little interest and it is often called the **trivial solution**. According to Eqs. (9) and (11), or Eq. (15), this is the only solution when $\det(\mathbf{A}) \neq 0$, that is, when $\mathbf{A}$ is nonsingular. Nonzero solutions occur if and only if $\det(\mathbf{A}) = 0$, that is, when $\mathbf{A}$ is singular. We summarize these results in the following theorem.

THEOREM 3.1.2	The homogeneous system (16) always has the trivial solution $x_1 = 0$, $x_2 = 0$, and this is the only solution when $\det(\mathbf{A}) \neq 0$. Nontrivial solutions exist if and only if $\det(\mathbf{A}) = 0$. In this case, unless $\mathbf{A} = \mathbf{0}$, all solutions are proportional to any nontrivial solution; in other words, they lie on a line through the origin. If $\mathbf{A} = \mathbf{0}$, then every point in the x_1x_2-plane is a solution of Eqs. (16).

The following examples illustrate the two possible cases.

EXAMPLE 4

Solve the system

$$3x_1 - x_2 = 0, \tag{18}$$

$$x_1 + 2x_2 = 0.$$

From the first equation, we have $x_2 = 3x_1$. Then, substituting into the second equation, we obtain $7x_1 = 0$, or $x_1 = 0$. Then $x_2 = 0$ also. Note that the determinant of coefficients has the value 7 (which is not zero) so this confirms the first part of Theorem 3.1.2 in this case. Figure 3.1.4 shows the two lines corresponding to the equations in the system (18).

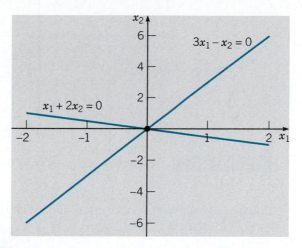

FIGURE 3.1.4 Geometrical interpretation of the system (18).

EXAMPLE 5

Solve the system

$$2x_1 + 4x_2 = 0, \tag{19}$$
$$x_1 + 2x_2 = 0.$$

From the second equation, we have $x_1 = -2x_2$. Then, from the first equation, we obtain $-4x_2 + 4x_2 = 0$, or $0 = 0$. Thus x_2 is not determined, but remains arbitrary. If $x_2 = c$, where c is an arbitrary constant, then $x_1 = -2c$. Thus solutions of the system (19) are of the form $(-2c, c)$, or $c(-2, 1)$, where c is any number. The system (19) has an infinite set of solutions, all of which are proportional to $(-2, 1)$, or to any other nontrivial solution. In the system (19) the two equations are multiples of each other and the determinant of coefficients has the value zero. See Figure 3.1.5.

▶ **Eigenvalues and Eigenvectors..** The equation $\mathbf{y} = \mathbf{Ax}$, where $\mathbf{A}$ is a given 2×2 matrix, can be viewed as a transformation, or mapping, of a two-dimensional vector $\mathbf{x}$ to a new two-dimensional vector $\mathbf{y}$. For example, suppose that

$$\mathbf{A} = \begin{pmatrix} 1 & 1 \\ 4 & 1 \end{pmatrix}, \qquad \mathbf{x} = \begin{pmatrix} 1 \\ 1 \end{pmatrix}, \tag{20}$$

Then

$$\mathbf{y} = \mathbf{Ax} = \begin{pmatrix} 1 & 1 \\ 4 & 1 \end{pmatrix} \begin{pmatrix} 1 \\ 1 \end{pmatrix} = \begin{pmatrix} 2 \\ 5 \end{pmatrix}; \tag{21}$$

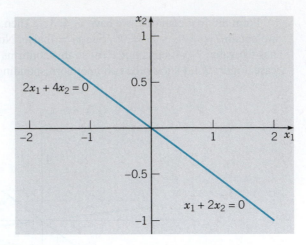

$$x_2$$

$2x_1 + 4x_2 = 0$

$x_1 + 2x_2 = 0$

FIGURE 3.1.5 Geometrical interpretation of the system (19).

thus the original vector $\mathbf{x}$ has been transformed into the new vector $\mathbf{y}$. Similarly, if $\mathbf{A}$ is given by Eq. (20) and $\mathbf{x} = (2, -1)$, then

$$\mathbf{y} = \mathbf{A}\mathbf{x} = \begin{pmatrix} 1 & 1 \\ 4 & 1 \end{pmatrix} \begin{pmatrix} 2 \\ -1 \end{pmatrix} = \begin{pmatrix} 1 \\ 7 \end{pmatrix}, \tag{22}$$

and so on.

In many applications it is of particular importance to find those vectors that a given matrix transforms into vectors that are multiples of the original vectors. In other words, we want $\mathbf{y}$ to be a multiple of $\mathbf{x}$; that is, $\mathbf{y} = \lambda\mathbf{x}$, where λ is some (scalar) constant. In this case, the equation $\mathbf{y} = \mathbf{A}\mathbf{x}$ becomes

$$\mathbf{A}\mathbf{x} = \lambda\mathbf{x}. \tag{23}$$

If $\mathbf{x} = \mathbf{0}$, then Eq. (23) is true for any $\mathbf{A}$ and for any λ, so we require $\mathbf{x}$ to be a *nonzero* vector. Then, since $\mathbf{I}\mathbf{x} = \mathbf{x}$, we can rewrite Eq. (23) in the form

$$\mathbf{A}\mathbf{x} = \lambda\mathbf{I}\mathbf{x}, \tag{24}$$

or

$$(\mathbf{A} - \lambda\mathbf{I})\mathbf{x} = \mathbf{0}. \tag{25}$$

To see the elements of $\mathbf{A} - \lambda\mathbf{I}$, we write

$$\mathbf{A} - \lambda\mathbf{I} = \begin{pmatrix} a_{11} & a_{12} \\ a_{21} & a_{22} \end{pmatrix} - \begin{pmatrix} \lambda & 0 \\ 0 & \lambda \end{pmatrix} = \begin{pmatrix} a_{11} - \lambda & a_{12} \\ a_{21} & a_{22} - \lambda \end{pmatrix}. \tag{26}$$

Recall that we are looking for nonzero vectors $\mathbf{x}$ that satisfy the homogeneous system (25). By Theorem 3.1.2, nonzero solutions of this system occur if and only if the determinant of coefficients is zero. Thus we require that

$$\det(\mathbf{A} - \lambda\mathbf{I}) = \begin{vmatrix} a_{11} - \lambda & a_{12} \\ a_{21} & a_{22} - \lambda \end{vmatrix} = 0. \tag{27}$$

Writing Eq. (27) in expanded form, we obtain

$$(a_{11} - \lambda)(a_{22} - \lambda) - a_{12}a_{21} = \lambda^2 - (a_{11} + a_{22})\lambda + a_{11}a_{22} - a_{12}a_{21} = 0. \tag{28}$$

Equation (28) is a quadratic equation in λ, so it has two roots λ_1 and λ_2. The values λ_1 and λ_2 are called **eigenvalues** of the given matrix **A**. By replacing λ by λ_1 in Eq. (25) and solving the resulting equation for **x**, we obtain the **eigenvector** $\mathbf{x}_1$ corresponding to the eigenvalue λ_1. In a similar way, we find the eigenvector $\mathbf{x}_2$ that corresponds to the second eigenvalue λ_2. The eigenvectors are not determined uniquely, but only up to an arbitrary constant multiplier.

Equation (28), which determines the eigenvalues, is called the **characteristic equation** of the matrix **A**. The constant term in this equation is just the determinant of **A**. The coefficient of λ in Eq. (28) involves the quantity $a_{11} + a_{22}$, the sum of the diagonal elements of **A**. This expression is called the trace of **A**, or tr(**A**). Thus the characteristic equation is sometimes written as

$$\lambda^2 - \text{tr}(\mathbf{A})\lambda + \det(\mathbf{A}) = 0. \tag{29}$$

We are assuming that the elements of **A** are real numbers. Consequently, the coefficients in the characteristic equation (28) are also real. As a result, the eigenvalues λ_1 and λ_2 may be real and different, real and equal, or complex conjugates. The following examples illustrate the calculation of eigenvalues and eigenvectors in each of these cases.

EXAMPLE 6

Find the eigenvalues and eigenvectors of the matrix

$$\mathbf{A} = \begin{pmatrix} 1 & 1 \\ 4 & 1 \end{pmatrix}. \tag{30}$$

In this case, Eq. (25) becomes

$$\begin{pmatrix} 1-\lambda & 1 \\ 4 & 1-\lambda \end{pmatrix} \begin{pmatrix} x_1 \\ x_2 \end{pmatrix} = \begin{pmatrix} 0 \\ 0 \end{pmatrix}. \tag{31}$$

The characteristic equation is

$$(1-\lambda)^2 - 4 = \lambda^2 - 2\lambda - 3 = (\lambda - 3)(\lambda + 1) = 0, \tag{32}$$

so the eigenvalues are $\lambda_1 = 3$ and $\lambda_2 = -1$.

To find the eigenvector $\mathbf{x}_1$ associated with the eigenvalue λ_1, we substitute $\lambda = 3$ in Eq. (31). Thus we obtain

$$\begin{pmatrix} -2 & 1 \\ 4 & -2 \end{pmatrix} \begin{pmatrix} x_1 \\ x_2 \end{pmatrix} = \begin{pmatrix} 0 \\ 0 \end{pmatrix}. \tag{33}$$

Observe that the rows in Eq. (33) are proportional to each other (as required by the vanishing of the determinant of coefficients), so we need only consider one row of this equation. Consequently, $-2x_1 + x_2 = 0$, or $x_2 = 2x_1$, while x_1 remains arbitrary. Thus

$$\mathbf{x}_1 = \begin{pmatrix} c \\ 2c \end{pmatrix} = c \begin{pmatrix} 1 \\ 2 \end{pmatrix}, \tag{34}$$

where c is an arbitrary constant. From Eq. (34), we see that there is an infinite set of eigenvectors associated with the eigenvalue λ_1. It is usually convenient to choose one member of this set to represent the entire set. For example, in this case, we might choose

$$\mathbf{x}_1 = \begin{pmatrix} 1 \\ 2 \end{pmatrix}, \tag{35}$$

and even refer to it as *the* eigenvector corresponding to λ_1. However, you should never forget that there are actually infinitely many other eigenvectors, each of which is proportional to the chosen representative.

In the same way, we can find the eigenvector $\mathbf{x}_2$ corresponding to the eigenvalue λ_2. By substituting $\lambda = -1$ in Eq. (31), we obtain

$$\begin{pmatrix} 2 & 1 \\ 4 & 2 \end{pmatrix} \begin{pmatrix} x_1 \\ x_2 \end{pmatrix} = \begin{pmatrix} 0 \\ 0 \end{pmatrix}. \tag{36}$$

Thus $x_2 = -2x_1$, so the eigenvector $\mathbf{x}_2$ is

$$\mathbf{x}_2 = \begin{pmatrix} 1 \\ -2 \end{pmatrix}, \tag{37}$$

or any vector proportional to this one.

EXAMPLE 7

Find the eigenvalues and eigenvectors of the matrix

$$\mathbf{A} = \begin{pmatrix} -\frac{1}{2} & 1 \\ -1 & -\frac{1}{2} \end{pmatrix}. \tag{38}$$

In this case, we obtain, from Eq. (25),

$$\begin{pmatrix} -\frac{1}{2} - \lambda & 1 \\ -1 & -\frac{1}{2} - \lambda \end{pmatrix} \begin{pmatrix} x_1 \\ x_2 \end{pmatrix} = \begin{pmatrix} 0 \\ 0 \end{pmatrix}. \tag{39}$$

The characteristic equation is

$$\left(-\tfrac{1}{2} - \lambda\right)^2 + 1 = \lambda^2 + \lambda + \tfrac{5}{4} = 0, \tag{40}$$

so the eigenvalues are

$$\lambda_1 = -\tfrac{1}{2} + i, \quad \lambda_2 = -\tfrac{1}{2} - i. \tag{41}$$

For $\lambda = \lambda_1$ Eq. (39) reduces to

$$\begin{pmatrix} -i & 1 \\ -1 & -i \end{pmatrix} \begin{pmatrix} x_1 \\ x_2 \end{pmatrix} = \begin{pmatrix} 0 \\ 0 \end{pmatrix}. \tag{42}$$

Thus $x_2 = ix_1$ and the eigenvector $\mathbf{x}_1$ corresponding to the eigenvalue λ_1 is

$$\mathbf{x}_1 = \begin{pmatrix} 1 \\ i \end{pmatrix}, \tag{43}$$

or any vector proportional to this one. In a similar way, we find the eigenvector $\mathbf{x}_2$ corresponding to λ_2, namely,

$$\mathbf{x}_2 = \begin{pmatrix} 1 \\ -i \end{pmatrix}. \tag{44}$$

Observe that $\mathbf{x}_1$ and $\mathbf{x}_2$ are also complex conjugates. This will always be the case when the matrix $\mathbf{A}$ has real elements and a pair of complex conjugate eigenvalues.

**EXAMPLE
8**

Find the eigenvalues and eigenvectors of the matrix

$$\mathbf{A} = \begin{pmatrix} 1 & -1 \\ 1 & 3 \end{pmatrix}. \tag{45}$$

From Eq. (25), we obtain

$$\begin{pmatrix} 1-\lambda & -1 \\ 1 & 3-\lambda \end{pmatrix} \begin{pmatrix} x_1 \\ x_2 \end{pmatrix} = \begin{pmatrix} 0 \\ 0 \end{pmatrix}. \tag{46}$$

Consequently, the characteristic equation is

$$(1-\lambda)(3-\lambda) + 1 = \lambda^2 - 4\lambda + 4 = (\lambda-2)^2 = 0, \tag{47}$$

and the eigenvalues are $\lambda_1 = \lambda_2 = 2$.

Returning to Eq. (46) and setting $\lambda = 2$, we find that

$$\begin{pmatrix} -1 & -1 \\ 1 & 1 \end{pmatrix} \begin{pmatrix} x_1 \\ x_2 \end{pmatrix} = \begin{pmatrix} 0 \\ 0 \end{pmatrix}. \tag{48}$$

Hence $x_2 = -x_1$, so there is an eigenvector

$$\mathbf{x}_1 = \begin{pmatrix} 1 \\ -1 \end{pmatrix}. \tag{49}$$

As usual, any other (nonzero) vector proportional to $\mathbf{x}_1$ is also an eigenvector.

However, in contrast to the two preceding examples, in this case there is only one distinct family of eigenvectors, which is typified by the vector $\mathbf{x}_1$ in Eq. (49). This situation is common when a matrix $\mathbf{A}$ has a repeated eigenvalue.

The following example shows that it is also possible for a repeated eigenvalue to be accompanied by two distinct eigenvectors.

**EXAMPLE
9**

Find the eigenvalues and eigenvectors of the matrix

$$\mathbf{A} = \begin{pmatrix} 2 & 0 \\ 0 & 2 \end{pmatrix}. \tag{50}$$

In this case, Eq. (25) becomes

$$\begin{pmatrix} 2-\lambda & 0 \\ 0 & 2-\lambda \end{pmatrix} \begin{pmatrix} x_1 \\ x_2 \end{pmatrix} = \begin{pmatrix} 0 \\ 0 \end{pmatrix}. \tag{51}$$

Thus the characteristic equation is

$$(2-\lambda)^2 = 0, \tag{52}$$

and the eigenvalues are $\lambda_1 = \lambda_2 = 2$. Returning to Eq. (51) and setting $\lambda = 2$, we obtain

$$\begin{pmatrix} 0 & 0 \\ 0 & 0 \end{pmatrix} \begin{pmatrix} x_1 \\ x_2 \end{pmatrix} = \begin{pmatrix} 0 \\ 0 \end{pmatrix}. \tag{53}$$

Thus no restriction is placed on x_1 and x_2; in other words, every nonzero vector in the plane is an eigenvector of this matrix $\mathbf{A}$. For example, we can choose as eigenvectors

$$\mathbf{x}_1 = \begin{pmatrix} 1 \\ 0 \end{pmatrix}, \qquad \mathbf{x}_2 = \begin{pmatrix} 0 \\ 1 \end{pmatrix}, \tag{54}$$

or any other pair of nonzero vectors that are not proportional to each other.

Sometimes a matrix depends on a parameter and, in this case, its eigenvalues also depend on the parameter.

EXAMPLE 10

Consider the matrix

$$\mathbf{A} = \begin{pmatrix} 2 & \alpha \\ -1 & 0 \end{pmatrix}, \tag{55}$$

where α is a parameter. Find the eigenvalues of $\mathbf{A}$ and describe their dependence on α.

The characteristic equation is

$$(2 - \lambda)(-\lambda) + \alpha = \lambda^2 - 2\lambda + \alpha = 0, \tag{56}$$

so the eigenvalues are

$$\lambda = \frac{2 \pm \sqrt{4 - 4\alpha}}{2} = 1 \pm \sqrt{1 - \alpha}. \tag{57}$$

Observe that, from Eq. (57), the eigenvalues are real and different when $\alpha < 1$, real and equal when $\alpha = 1$, and complex conjugates when $\alpha > 1$. As α varies, the case of equal eigenvalues occurs as a transition between the other two cases.

PROBLEMS

In each of Problems 1 through 12:
(a) Find all solutions of the given system of equations.
(b) Sketch the graph of each equation in the system. Are the lines intersecting, parallel, or coincident?

1. $2x_1 + 3x_2 = 7, \quad -3x_1 + x_2 = -5$
2. $x_1 - 2x_2 = 10, \quad 2x_1 + 3x_2 = 6$
3. $x_1 + 3x_2 = 0, \quad 2x_1 - x_2 = 0$
4. $-x_1 + 2x_2 = 4, \quad 2x_1 - 4x_2 = -6$
5. $2x_1 - 3x_2 = 4, \quad x_1 + 2x_2 = -5$
6. $3x_1 - 2x_2 = 0, \quad -6x_1 + 4x_2 = 0$
7. $2x_1 - 3x_2 = 6, \quad -4x_1 + 6x_2 = -12$
8. $4x_1 + x_2 = 0, \quad 4x_1 - 3x_2 = -12$
9. $x_1 + 4x_2 = 10, \quad 4x_1 + x_2 = 10$
10. $x_1 + x_2 = -1, \quad -x_1 + 2x_2 = 4$
11. $4x_1 - 3x_2 = 0, \quad -2x_1 + 5x_2 = 0$
12. $2x_1 + 5x_2 = 0, \quad 4x_1 + 10x_2 = 0$

In each of Problems 13 through 32, find the eigenvalues and eigenvectors of the given matrix.

13. $\mathbf{A} = \begin{pmatrix} 3 & -2 \\ 2 & -2 \end{pmatrix}$

14. $\mathbf{A} = \begin{pmatrix} 3 & -2 \\ 4 & -1 \end{pmatrix}$

15. $\mathbf{A} = \begin{pmatrix} 3 & -4 \\ 1 & -1 \end{pmatrix}$

16. $\mathbf{A} = \begin{pmatrix} 1 & -2 \\ 3 & -4 \end{pmatrix}$

17. $\mathbf{A} = \begin{pmatrix} -1 & -4 \\ 1 & -1 \end{pmatrix}$

18. $\mathbf{A} = \begin{pmatrix} \frac{5}{4} & \frac{3}{4} \\ -\frac{3}{4} & -\frac{1}{4} \end{pmatrix}$

19. $\mathbf{A} = \begin{pmatrix} -\frac{3}{2} & 1 \\ -\frac{1}{4} & -\frac{1}{2} \end{pmatrix}$

20. $\mathbf{A} = \begin{pmatrix} 2 & -1 \\ 3 & -2 \end{pmatrix}$

21. $\mathbf{A} = \begin{pmatrix} 2 & -5 \\ 1 & -2 \end{pmatrix}$ 22. $\mathbf{A} = \begin{pmatrix} 6 & 3 \\ 2 & 1 \end{pmatrix}$

23. $\mathbf{A} = \begin{pmatrix} 1 & 1 \\ 4 & -2 \end{pmatrix}$ 24. $\mathbf{A} = \begin{pmatrix} 2 & -\frac{5}{2} \\ \frac{9}{5} & -1 \end{pmatrix}$

25. $\mathbf{A} = \begin{pmatrix} -3 & \frac{5}{2} \\ -\frac{5}{2} & 2 \end{pmatrix}$ 26. $\mathbf{A} = \begin{pmatrix} 1 & -1 \\ 5 & -3 \end{pmatrix}$

27. $\mathbf{A} = \begin{pmatrix} 1 & \frac{4}{3} \\ -\frac{9}{4} & -3 \end{pmatrix}$ 28. $\mathbf{A} = \begin{pmatrix} -2 & 1 \\ 1 & -2 \end{pmatrix}$

29. $\mathbf{A} = \begin{pmatrix} 1 & 2 \\ -5 & -1 \end{pmatrix}$ 30. $\mathbf{A} = \begin{pmatrix} -1 & -\frac{1}{2} \\ 2 & -3 \end{pmatrix}$

31. $\mathbf{A} = \begin{pmatrix} \frac{5}{4} & \frac{3}{4} \\ \frac{3}{4} & \frac{5}{4} \end{pmatrix}$ 32. $\mathbf{A} = \begin{pmatrix} 2 & \frac{1}{2} \\ -\frac{1}{2} & 1 \end{pmatrix}$

In each of Problems 33 through 36:
(a) Find the eigenvalues of the given matrix.
(b) Describe how the nature of the eigenvalues depends on the parameter α in the matrix $\mathbf{A}$.

33. $\mathbf{A} = \begin{pmatrix} 2 & \alpha \\ 1 & -3 \end{pmatrix}$ 34. $\mathbf{A} = \begin{pmatrix} 3 & 4 \\ -\alpha & 2 \end{pmatrix}$

35. $\mathbf{A} = \begin{pmatrix} 1 & 2 \\ 3 & \alpha \end{pmatrix}$ 36. $\mathbf{A} = \begin{pmatrix} 1 & -\alpha \\ 2\alpha & 3 \end{pmatrix}$

37. If $\det(\mathbf{A}) \neq 0$, derive the result in Eq. (14) for $\mathbf{A}^{-1}$.
38. Show that $\lambda = 0$ is an eigenvalue of the matrix $\mathbf{A}$ if and only if $\det(\mathbf{A}) = 0$.

3.2 Systems of Two First Order Linear Differential Equations

We begin our discussion of systems of differential equations with an example that is basically an extension of the mixing problem in Example 1 in Section 2.3.

EXAMPLE 1

Two Interconnected Tanks

Consider the two interconnected tanks shown in Figure 3.2.1. Tank 1 initially contains 30 gal of water and 55 oz of salt and Tank 2 initially contains 20 gal of water and 26 oz of salt. Water containing 1 oz/gal of salt flows into Tank 1 at a rate of 1.5 gal/min. The mixture flows from Tank 1 to Tank 2 at a rate of 3 gal/min. Water containing 3 oz/gal of salt also flows into Tank 2 at a rate of 1 gal/min (from the outside). The mixture drains from Tank 2 at a rate of 4 gal/min, of which some flows back into Tank 1 at a rate of 1.5 gal/min, while the remainder leaves the system. Formulate a system of differential equations and initial conditions that describes this flow process.

Observe first that the volume of water in each tank remains constant since the total rates of flow in and out of each tank are the same: 3 gal/min in Tank 1 and 4 gal/min in Tank 2. However, the amount of salt in each tank can be expected to change as time goes on, so let us denote by $Q_1(t)$ and $Q_2(t)$ the amount of salt in each tank at time t. We measure t in minutes, Q_1 and Q_2 in ounces. The starting values of Q_1 and Q_2 are given in the statement of the example, namely,

$$Q_1(0) = 55, \qquad Q_2(0) = 26. \tag{1}$$

The variation of salt in each tank is due entirely to the flows in and out of the tank, so we can apply the fundamental balance principle to each tank:

rate of change of salt in tank = rate of flow of salt in − rate of flow of salt out.   (2)

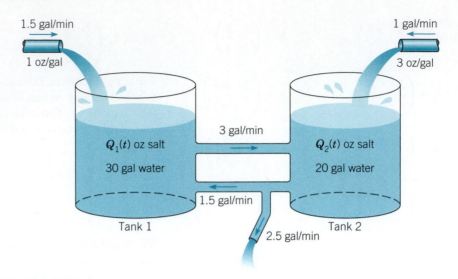

FIGURE 3.2.1 The interconnected tanks in Example 1.

Recall that this is the same principle that we used in the single tank problem in Example 1 of Section 2.3. For Tank 1, the rate of change of salt is dQ_1/dt oz/min. Each flow rate of salt (in oz/min) is given by the product of a concentration of salt (in oz/gal) and a flow rate of water (in gal/min). The rate of flow of salt into Tank 1 consists of two terms:

$$\text{rate of flow from outside the system} = 1 \text{ oz/gal} \times 1.5 \text{ gal/min} = 1.5 \text{ oz/min};$$

$$\text{rate of flow from Tank 2} = \frac{Q_2(t)}{20} \text{ oz/gal} \times 1.5 \text{ gal/min} = 0.075\, Q_2(t) \text{ oz/min}.$$

The rate of flow out of Tank 1 is given by

$$\text{rate of flow out of Tank 1} = \frac{Q_1(t)}{30} \text{ oz/gal} \times 3 \text{ gal/min} = 0.1\, Q_1(t) \text{ oz/min}.$$

Substituting these expressions into Eq. (2) and rearranging the terms, we obtain the differential equation

$$\frac{dQ_1}{dt} = -0.1\, Q_1 + 0.075\, Q_2 + 1.5. \tag{3}$$

In a similar way, we can also apply the balance principle (2) to Tank 2 and thereby obtain the differential equation

$$\frac{dQ_2}{dt} = 0.1\, Q_1 - 0.2\, Q_2 + 3. \tag{4}$$

The pair of differential equations (3) and (4), together with the initial conditions (1), constitutes a mathematical model for the flow in the two tanks. If we are able to solve these equations for $Q_1(t)$ and $Q_2(t)$, we will be able to predict how much salt will be in each tank at any future time.

Equations (3) and (4) are an example of a **system** of differential equations. Each of these equations contains both of the unknown functions Q_1 and Q_2, so the equations cannot be solved separately, but must be investigated together. A system, such as Eqs. (3) and (4), is a generalization of the differential equation that occurred in Example 1 in Section 2.3 for the mixing problem in a single tank. We will proceed here much as we did when we first encountered differential equations in Sections 1.1 and 1.2. First, we will visualize graphically how solutions of this system behave and then, in the next section, we will discuss a way of finding the solutions.

▶ **Matrix Notation and Systems Terminology.** In dealing with systems of equations, it is helpful to introduce the notation of vectors and matrices. This not only saves a great deal of space and facilitates calculations but also emphasizes the similarity between systems of equations and single (scalar) equations, which we discussed in the preceding chapters. We begin by rewriting Eqs. (3) and (4) in the form

$$\begin{pmatrix} dQ_1/dt \\ dQ_2/dt \end{pmatrix} = \begin{pmatrix} -0.1 & 0.075 \\ 0.1 & -0.2 \end{pmatrix} \begin{pmatrix} Q_1 \\ Q_2 \end{pmatrix} + \begin{pmatrix} 1.5 \\ 3 \end{pmatrix}. \tag{5}$$

Next we define the vectors $\mathbf{Q}$ and $\mathbf{b}$ and the matrix $\mathbf{K}$ to be

$$\mathbf{Q} = \begin{pmatrix} Q_1 \\ Q_2 \end{pmatrix}, \qquad \mathbf{b} = \begin{pmatrix} 1.5 \\ 3 \end{pmatrix}, \qquad \mathbf{K} = \begin{pmatrix} -0.1 & 0.075 \\ 0.1 & -0.2 \end{pmatrix}. \tag{6}$$

Then Eq. (5) takes the form

$$\frac{d\mathbf{Q}}{dt} = \mathbf{K}\mathbf{Q} + \mathbf{b}. \tag{7}$$

The variables Q_1 and Q_2 are called **state variables**, since their values at any time t describe the state of the system, that is, the amount of salt in each tank. Similarly, the vector $\mathbf{Q} = Q_1\mathbf{i} + Q_2\mathbf{j}$ is called the **state vector** of the system. The Q_1Q_2-plane itself is called the **state plane**, or more commonly the **phase plane**. As time advances, the tip of the vector $\mathbf{Q}$ traces a curve in the phase plane. This curve, often called a **trajectory** or **orbit**, displays graphically the changing state or evolution of the system.

If $Q_1(t)$ and $Q_2(t)$ are expressions that give Q_1 and Q_2 as functions of t, then for any given value of t we can calculate the corresponding point (Q_1, Q_2) in the phase plane. We construct the trajectory corresponding to $Q_1(t)$ and $Q_2(t)$ by repeating the calculation for several values of t, and then drawing a curve through the resulting points. The trajectory is then a curve in the two-dimensional Q_1Q_2 state space; in fact, it is just the parametric plot of Q_2 versus Q_1 with t as the parameter. The trajectory can also be constructed by approximating solutions of the system (7) numerically and then plotting a curve through the points generated in this way. This is how the trajectories in Figure 3.2.3 were produced.

▶ **Graphical Representations.** There are three types of plots that can be very helpful in visualizing the behavior of solutions of systems such as (7). One is a **direction field**, similar to those that we saw in Chapters 1 and 2. To draw a direction field for the system (7) in the $Q_1 Q_2$-plane, we begin by choosing a point $\mathbf{Q} = (Q_1, Q_2)$ and evaluating the right side of Eq. (7), namely $\mathbf{K}\mathbf{Q} + \mathbf{b}$, at that point. The result is the vector $d\mathbf{Q}/dt$ at the given point. This vector is tangent to the trajectory of the system (7) passing through the point (Q_1, Q_2). By drawing this vector as an arrow starting at the given point, we are able to display the direction in which solutions pass through the point. Repeating this calculation at many points, we obtain the plot shown in Figure 3.2.2.

Just as for the direction fields that we saw in Chapters 1 and 2 for single first order equations, we are able to infer the general behavior of trajectories, or solutions, of the system (7) by looking at the direction field in Figure 3.2.2. For example, in this case it appears that all trajectories are approaching a certain point in the first quadrant. Of course, you can also plot a representative sample of the trajectories. Such a plot is called a **phase portrait**. A phase portrait for the system (7) is shown in Figure 3.2.3.

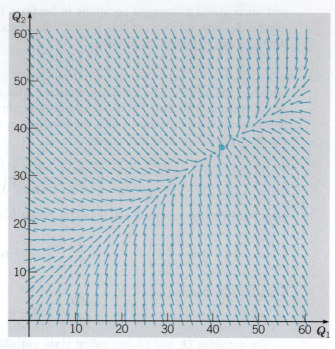

FIGURE 3.2.2   A direction field for the system (7).

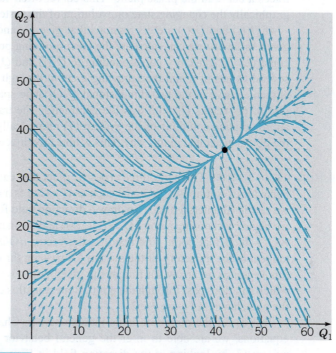

FIGURE 3.2.3   A phase portrait for the system (7).

Phase portraits are especially useful in showing the overall behavior of all solutions of a system such as (7). However, they do not show the detailed dependence of any particular solution on time t. For this purpose, you can also plot Q_1 and Q_2 versus t. Such plots are called **component plots**. In Figure 3.2.4, we show the component plots of Q_1 and Q_2 that satisfy the system (7) and the initial conditions (1).

The system (7) is relatively easy to solve and we will show you how in the next section. In the meantime, we assume that you have software that can produce plots similar to those in Figures 3.2.2 through 3.2.4. Such software packages plot direction fields and compute numerical approximations to solutions, which can be displayed as phase plane trajectories and component plots. They are very useful, not only for plotting solutions of systems that can easily be solved, but especially for investigating systems that are more difficult, if not impossible, to solve analytically.

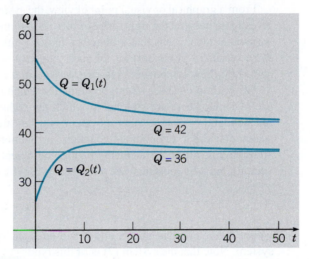

FIGURE 3.2.4 Component plots for solutions of the system (7) and the initial conditions (1).

▶ **General Systems of Two First Order Linear Equations.** The system (5) or (7) belongs to a much more general class of problems, which we obtain by replacing the constant coefficients on the right side by functions of the independent variable t. Thus we obtain the general system of two first order linear differential equations,

$$\begin{pmatrix} dx/dt \\ dy/dt \end{pmatrix} = \begin{pmatrix} p_{11}(t)x + p_{12}(t)y + g_1(t) \\ p_{21}(t)x + p_{22}(t)y + g_2(t) \end{pmatrix}. \tag{8}$$

Using vector notation, we can write the linear system (8) as

$$\frac{d\mathbf{x}}{dt} = \mathbf{P}(t)\mathbf{x} + \mathbf{g}(t), \tag{9}$$

where

$$\mathbf{x} = \begin{pmatrix} x \\ y \end{pmatrix}, \qquad \mathbf{P}(t) = \begin{pmatrix} p_{11}(t) & p_{12}(t) \\ p_{21}(t) & p_{22}(t) \end{pmatrix}, \quad \text{and} \quad \mathbf{g}(t) = \begin{pmatrix} g_1(t) \\ g_2(t) \end{pmatrix}.$$

Again, we refer to x and y as state variables, to $\mathbf{x}$ as the state vector, and to the xy-plane as the state plane, or (usually) the phase plane.

The system (8) is called a **first order linear system of dimension two** because it consists of first order linear equations and because its state space (the xy-plane) is two dimensional. Further, if $\mathbf{g}(t) = \mathbf{0}$ for all t, that is, if $g_1(t) = g_2(t) = 0$ for all t, then the system (9) or (8) is said to be **homogeneous**. Otherwise, it is **nonhomogeneous**. For example, the system (5), or (7), is nonhomogeneous.

Frequently, there will also be given initial conditions

$$x(t_0) = x_0, \quad y(t_0) = y_0, \tag{10}$$

or

$$\mathbf{x}(t_0) = \mathbf{x}_0, \tag{11}$$

where $\mathbf{x}_0 = x_0\mathbf{i} + y_0\mathbf{j}$. Equations (8) and (10), or in vector form Eqs. (9) and (11), form an **initial value problem**.

A **solution** of the system (8) consists of two differentiable functions $x = \phi(t)$ and $y = \psi(t)$ that satisfy Eqs. (8) for all values of t in some interval I. In vector terminology a solution is a vector $\mathbf{x} = \boldsymbol{\phi}(t) = \phi(t)\mathbf{i} + \psi(t)\mathbf{j}$ that satisfies Eq. (9) for all t in I.

From a graphical or visual point of view, the systems (8) and (9) are relatively difficult to investigate because the right sides of these equations depend explicitly on the independent variable t. This means that a direction field for these systems changes with time. For the same reason, a phase portrait for such a system is also not useful. It is still possible, of course, to draw component plots of x versus t and y versus t for such systems.

We wish to consider mainly systems for which direction fields and phase portraits *are* useful tools. These are systems in which the right sides do not depend explicitly on the independent variable t. Such a system is called **autonomous**. Recall that we discussed scalar first order autonomous equations in Section 2.5. For the linear system (9) to be autonomous, all of the elements of the coefficient matrix $\mathbf{P}$ and the components of the vector $\mathbf{g}$ must be constants. We will usually use the notation

$$\frac{d\mathbf{x}}{dt} = \mathbf{A}\mathbf{x} + \mathbf{b}, \tag{12}$$

where $\mathbf{A}$ is a constant matrix and $\mathbf{b}$ is a constant vector, to denote an autonomous linear system. Note that the system (7) is of this type.

▶ **Existence and Uniqueness of Solutions.** It may be important to know that the initial value problem (8), (10), or (9), (11) has a unique solution, and there is a theorem (stated below) that asserts that this is the case. This theorem is analogous to Theorem 2.4.1 that deals with initial value problems for first order linear scalar equations. The importance of such a theorem to nonmathematicians is first, that it ensures that a problem you are trying to solve actually has a solution; second, that if you are successful in finding a solution, you can be sure that it is the only one; and third, that it promotes confidence in using numerical approximation methods when you are sure that there is actually something to approximate.

THEOREM 3.2.1	Let each of the functions $p_{11}, \ldots, p_{22}, g_1$, and g_2 be continuous on an open interval $I: \alpha < t < \beta$, let t_0 be any point in I, and let x_0 and y_0 be any given numbers. Then there exists a unique solution of the system (8)

$$\begin{pmatrix} dx/dt \\ dy/dt \end{pmatrix} = \begin{pmatrix} p_{11}(t)x + p_{12}(t)y + g_1(t) \\ p_{21}(t)x + p_{22}(t)y + g_2(t) \end{pmatrix}$$

that also satisfies the initial conditions (10)

$$x(t_0) = x_0, \quad y(t_0) = y_0.$$

Further, the solution exists throughout the interval I.

In some cases, the existence of solutions can be demonstrated by actually finding them, and much of this book is devoted to that goal. However, a proof of Theorem 3.2.1, in general, is too difficult to give here; it may be found in many more advanced books on differential equations.

Observe that the interval of existence of the solution is the entire interval I in which the hypotheses are satisfied. Further, the initial values x_0 and y_0 are completely arbitrary. Finally, for the linear autonomous system (12), the coefficients are continuous for all t; consequently, the solution exists and is unique on the entire t-axis.

▶ **Autonomous Systems.** Let us now consider autonomous systems, for which direction fields and phase portraits are effective tools. Recall that in Section 2.5 we found that equilibrium, or constant, solutions were of particular importance in the study of single first order autonomous equations. We will see that the same is true for autonomous systems of equations.

For the linear autonomous system (12), we find the **equilibrium solutions**, or **critical points**, by setting $d\mathbf{x}/dt$ equal to zero. In this way, we obtain the linear algebraic system

$$\mathbf{Ax} = -\mathbf{b}. \tag{13}$$

If the coefficient matrix $\mathbf{A}$ has an inverse, as we usually assume, then the system (13) has a single solution, namely, $\mathbf{x} = -\mathbf{A}^{-1}\mathbf{b}$. This is then the only critical point of the system (12). However, if $\mathbf{A}$ is singular, then Eq. (13) has either no solution or infinitely many.

It is important to understand that critical points are found by solving algebraic, rather then differential, equations. As we will see later, the behavior of trajectories in the vicinity of critical points can also be determined by algebraic methods. Thus a good deal of information about solutions of autonomous systems can be found without actually solving the system.

EXAMPLE 2

Consider again the two tank problem in Example 1. Find the critical point and describe the behavior of the trajectories in its neighborhood.

The system of differential equations for this problem is Eq. (5) or (7). To find the critical point, we set dQ_1/dt and dQ_2/dt equal to 0 in Eq. (5) and solve the resulting system of linear algebraic equations for Q_1 and Q_2. We have

$$\begin{pmatrix} -0.1 & 0.075 \\ 0.1 & -0.2 \end{pmatrix} \begin{pmatrix} Q_1 \\ Q_2 \end{pmatrix} = -\begin{pmatrix} 1.5 \\ 3 \end{pmatrix}, \tag{14}$$

whence $Q_1 = 42$ and $Q_2 = 36$. Keep in mind that this point is itself a trajectory. If the initial conditions

$$Q_1(0) = 42, \quad Q_2(0) = 36 \tag{15}$$

are prescribed, then the corresponding solution of the system (5) consists of the two constant functions $Q_1(t) = 42$ and $Q_2(t) = 36$. From the phase portrait in Figure 3.2.3, it appears that all other trajectories are approaching the critical point as $t \to \infty$. However, no other trajectory can actually reach the critical point at a finite time, since this would violate the uniqueness part of Theorem 3.2.1. This behavior is borne out by the component plots in Figure 3.2.4, which indicate that $Q_1 = 42$ and $Q_2 = 36$ are horizontal asymptotes for the solutions shown there.

▶ **Transformation of a Second Order Equation to a System of First Order Equations.** One or more higher order differential equations can always be transformed into a system of first order equations. Thus such systems can be considered as the most fundamental problem area in differential equations. For example, almost all numerical algorithms for approximating solutions of differential equations are written for systems of first order equations.

To illustrate how this transformation can be done let us consider the second order equation

$$y'' + p(t)y' + q(t)y = g(t), \tag{16}$$

where p, q, and g are given functions. We introduce new variables x_1 and x_2 defined as

$$x_1 = y, \quad x_2 = y'. \tag{17}$$

From Eqs. (17), we note that

$$x_1' = x_2. \tag{18}$$

Further, from the second of Eqs. (17), we have $y'' = x_2'$, and then, by writing Eq. (16) in terms of x_1 and x_2, we obtain

$$x_2' + p(t)x_2 + q(t)x_1 = g(t),$$

or

$$x_2' = -q(t)x_1 - p(t)x_2 + g(t). \tag{19}$$

Equations (18) and (19) form a system of two first order equations that is equivalent to the original Eq. (16). Using matrix notation we can write this system as

$$\mathbf{x}' = \begin{pmatrix} 0 & 1 \\ -q(t) & -p(t) \end{pmatrix} \mathbf{x} + \begin{pmatrix} 0 \\ g(t) \end{pmatrix}. \tag{20}$$

EXAMPLE 3

Consider the differential equation

$$u'' + 0.25u' + 2u = 3\sin t. \tag{21}$$

Suppose that initial conditions

$$u(0) = 2, \quad u'(0) = -1 \tag{22}$$

are also given. As we will show in the next chapter, this initial value problem can serve as a model for a vibrating spring-mass system. Transform this problem into an equivalent one for a system of first order equations.

We start by letting $x_1 = u$ and $x_2 = u'$. Then one equation is just $x_1' = x_2$. A second equation is obtained by writing Eq. (21) in terms of x_1 and x_2. The result is the system

$$x_1' = x_2, \quad x_2' = -2x_1 - 0.25x_2 + 3\sin t. \tag{23}$$

The initial conditions (22) lead directly to

$$x_1(0) = 2, \quad x_2(0) = -1. \tag{24}$$

The initial value problem (23), (24) is equivalent to Eqs. (21), (22). In matrix notation, we write this initial value problem as

$$\mathbf{x}' = \begin{pmatrix} 0 & 1 \\ -2 & -0.25 \end{pmatrix} \mathbf{x} + \begin{pmatrix} 0 \\ 3\sin t \end{pmatrix}, \quad \mathbf{x}(0) = \begin{pmatrix} 2 \\ -1 \end{pmatrix}. \tag{25}$$

PROBLEMS

In each of Problems 1 through 8, state whether the given system is autonomous or nonautonomous and also whether it is homogeneous or nonhomogeneous.

1. $x' = y, \quad y' = x + 4$

2. $x' = x + 2y + \sin t, \quad y' = -x + y - \cos t$

3. $x' = -2tx + y, \quad y' = 3x - y$

4. $x' = x + 2y + 4, \quad y' = -2x + y - 3$

5. $x' = 3x - y, \quad y' = x + 2y$

6. $x' = -x + ty, \quad y' = tx - y$

7. $x' = x + y + 4, \quad y' = -2x + (\sin t)y$

8. $x' = 3x - 4y, \quad y' = x + 3y$

ODEA In each of Problems 9 through 14:

(a) Find the equilibrium solution, or critical point, for the given system.

(b) Draw a direction field centered at the critical point.

(c) Describe how solutions of the system behave in the vicinity of the critical point.

9. $x' = -x + y + 1, \quad y' = x + y - 3$

10. $x' = -x - 4y - 4, \quad y' = x - y - 6$

11. $x' = -0.25x - 0.75y + 8,$
    $y' = 0.5x + y - 11.5$

12. $x' = -2x + y - 11, \quad y' = -5x + 4y - 35$

13. $x' = x + y - 3, \quad y' = -x + y + 1$

14. $x' = -5x + 4y - 35, \quad y' = -2x + y - 11$

In each of Problems 15 through 18, transform the given equation into a system of first order equations.

15. $u'' + 0.5u' + 2u = 0$

16. $2u'' + 0.5u' + 8u = 6 \sin 2t$

17. $t^2 u'' + tu' + (t^2 - 0.25)u = 0$

18. $t^2 u'' + 3tu' + 5u = t^2 + 4$

In each of Problems 19 and 20, transform the given initial value problem into an initial value problem for two first order equations.

19. $u'' + 0.25u' + 4u = 2 \cos 3t,$
    $u(0) = 1, \quad u'(0) = -2$

20. $tu'' + u' + tu = 0, \quad u(1) = 1, \quad u'(1) = 0$

Electric Circuits. The theory of electric circuits, such as that shown in Figure 3.2.5, consisting of inductors, resistors, and capacitors, is based on Kirchhoff's laws: (1) The algebraic sum of currents at each node (or junction) is zero, and (2) the algebraic sum of voltages across the elements in each closed loop is zero. In addition to Kirchhoff's laws, we also have the relation between the current $i(t)$ in amperes through each circuit element and the voltage $v(t)$ in volts across the element:

$v = Ri, \qquad R = $ resistance in ohms;

$C\dfrac{dv}{dt} = i, \qquad C = $ capacitance in farads;[2]

$L\dfrac{di}{dt} = v, \qquad L = $ inductance in henrys.

Kirchhoff's laws and the current–voltage relation for each circuit element provide a system of algebraic and differential equations from which the voltage and current throughout the circuit can be determined. Problems 21 through 23 illustrate the procedure just described.

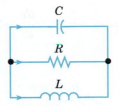

FIGURE 3.2.5 A parallel LRC circuit; see Problem 21.

21. Consider the circuit shown in Figure 3.2.5. Let i_1, i_2, and i_3 be the currents through the capacitor, resistor, and inductor, respectively. Likewise, let v_1, v_2, and v_3 be the corresponding voltages. The arrows denote the arbitrarily chosen directions in which currents and voltages will be taken to be positive.

(a) Applying Kirchhoff's second law to the upper loop in the circuit, show that

$$v_1 - v_2 = 0. \qquad \text{(i)}$$

In a similar way, show that

$$v_2 - v_3 = 0. \qquad \text{(ii)}$$

(b) Applying Kirchhoff's first law to either node in the circuit, show that

$$i_1 + i_2 + i_3 = 0. \qquad \text{(iii)}$$

(c) Use the current–voltage relation through each element in the circuit to obtain the equations

$$Cv_1' = i_1, \qquad v_2 = Ri_2, \qquad Li_3' = v_3. \qquad \text{(iv)}$$

(d) Eliminate v_2, v_3, i_1, and i_2 among Eqs. (i) through (iv) to obtain

$$Cv_1' = -i_3 - \frac{v_1}{R}, \qquad Li_3' = v_1. \qquad \text{(v)}$$

These equations form a system of two equations for the variables v_1 and i_3.

---

[2]Actual capacitors typically have capacitances measured in microfarads. We use farad as the unit for numerical convenience.

22. Consider the circuit shown in Figure 3.2.6. Use the method outlined in Problem 21 to show that the current i through the inductor and the voltage v across the capacitor satisfy the system of differential equations

$$\frac{di}{dt} = -i - v, \qquad \frac{dv}{dt} = 2i - v.$$

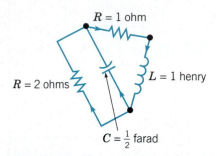

FIGURE 3.2.6 The circuit in Problem 22.

$R = 1$ ohm

$R = 2$ ohms

$L = 1$ henry

$C = \frac{1}{2}$ farad

23. Consider the circuit shown in Figure 3.2.7. Use the method outlined in Problem 21 to show that the current i through the inductor and the voltage v across the capacitor satisfy the system of differential equations

$$L\frac{di}{dt} = -R_1 i - v, \qquad C\frac{dv}{dt} = i - \frac{v}{R_2}.$$

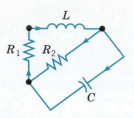

FIGURE 3.2.7 The circuit in Problem 23.

24. Consider two interconnected tanks similar to those in Figure 3.2.1. Tank 1 initially contains 60 gal of water and Q_1^0 oz of salt, and Tank 2 initially contains 100 gal of water and Q_2^0 oz of salt. Water containing q_1 oz/gal of salt flows into Tank 1 at a rate of 3 gal/min. The mixture in Tank 1 flows out at a rate of 4 gal/min, of which half flows into Tank 2 while the remainder leaves the system. Water containing q_2 oz/gal of salt also flows into Tank 2 from the outside at the rate of 1 gal/min. The mixture in Tank 2 leaves the tank at a rate of 3 gal/min, of which some flows back into Tank 1 at a rate of 1 gal/min, while the rest leaves the system.

   (a) Draw a diagram that depicts the flow process described above. Let $Q_1(t)$ and $Q_2(t)$, respectively, be the amount of salt in each tank at time t. Write down differential equations and initial conditions for Q_1 and Q_2 that model the flow process.

   (b) Find the equilibrium values Q_1^E and Q_2^E in terms of the concentrations q_1 and q_2.

   (c) Is it possible (by adjusting q_1 and q_2) to obtain $Q_1^E = 60$ and $Q_2^E = 50$ as an equilibrium state?

   (d) Describe which equilibrium states are possible for this system for various values of q_1 and q_2.

3.3 Homogeneous Linear Systems with Constant Coefficients

In the preceding section, we noted that a linear autonomous system of first order differential equations has the form

$$\frac{d\mathbf{x}}{dt} = \mathbf{A}\mathbf{x} + \mathbf{b}, \tag{1}$$

where $\mathbf{A}$ and $\mathbf{b}$ are a constant matrix and a constant vector, respectively. In Example 1 of that section, we showed that a certain two-tank flow problem can be modeled by the equations

$$\begin{pmatrix} dQ_1/dt \\ dQ_2/dt \end{pmatrix} = \begin{pmatrix} -0.1 & 0.075 \\ 0.1 & -0.2 \end{pmatrix} \begin{pmatrix} Q_1 \\ Q_2 \end{pmatrix} + \begin{pmatrix} 1.5 \\ 3 \end{pmatrix}, \tag{2}$$

which are of the form (1). The initial conditions

$$Q_1(0) = 55, \quad Q_2(0) = 26 \tag{3}$$

were also given. Further, in Example 2, we found that the system (2) has a single critical point, or equilibrium solution, given by $Q_1^E = 42$ and $Q_2^E = 36$.

EXAMPLE 1

Two Interconnected Tanks Revisited

Consider the two-tank problem modeled by the system (2). Let $x_1(t)$ and $x_2(t)$ be the deviations of $Q_1(t)$ and $Q_2(t)$ from their respective equilibrium values, that is,

$$x_1(t) = Q_1(t) - 42, \quad x_2(t) = Q_2(t) - 36. \tag{4}$$

Find the system of differential equations and the initial conditions satisfied by $x_1(t)$ and $x_2(t)$.

Rewrite Eqs. (4) in the form

$$Q_1(t) = 42 + x_1(t), \quad Q_2(t) = 36 + x_2(t). \tag{5}$$

Then, by substituting for Q_1 and Q_2 in the first of Eqs. (2) from Eqs. (5), we obtain

$$(42 + x_1)' = -0.1(42 + x_1) + 0.075(36 + x_2) + 1.5$$

or

$$\begin{aligned} x_1' &= -4.2 - 0.1\,x_1 + 2.7 + 0.075\,x_2 + 1.5 \\ &= -0.1\,x_1 + 0.075\,x_2, \end{aligned} \tag{6}$$

since the constant terms on the right side add to zero. Proceeding in the same way with the second of Eqs. (2), we also find that

$$\begin{aligned} x_2' &= 0.1(42 + x_1) - 0.2(36 + x_2) + 3 \\ &= 0.1\,x_1 - 0.2\,x_2. \end{aligned} \tag{7}$$

If we write Eqs. (6) and (7) in vector form, we have

$$\begin{pmatrix} dx_1/dt \\ dx_2/dt \end{pmatrix} = \begin{pmatrix} -0.1 & 0.075 \\ 0.1 & -0.2 \end{pmatrix} \begin{pmatrix} x_1 \\ x_2 \end{pmatrix}, \tag{8}$$

or

$$\frac{d\mathbf{x}}{dt} = \begin{pmatrix} -0.1 & 0.075 \\ 0.1 & -0.2 \end{pmatrix} \mathbf{x}, \tag{9}$$

where $\mathbf{x}(t) = x_1(t)\mathbf{i} + x_2(t)\mathbf{j}$. In terms of x_1 and x_2, the initial conditions (3) become

$$x_1(0) = 13, \quad x_2(0) = -10. \tag{10}$$

Observe that, by introducing the variables x_1 and x_2 defined by Eq. (4), we have transformed the nonhomogeneous system (2) into the homogeneous system (9) with the same coefficient matrix. The variables x_1 and x_2 specify the amount of salt in each tank relative to their respective equilibrium values, while Q_1 and Q_2 are the actual amounts of salt in each tank. The result of using x_1 and x_2 instead of Q_1 and Q_2 is to simplify the system (2) by eliminating the nonhomogeneous term. In geometrical language, we have shifted coordinates so that the critical point (42,36) in the Q_1Q_2-plane is now located at the origin in the x_1x_2-plane. This type of coordinate transformation is often useful in simplifying a problem.

If we assume that the matrix $\mathbf{A}$ is nonsingular, then the system (1) has a single critical point. We can then proceed, just as in Example 1, to transform the nonhomogeneous system (1) into the homogeneous system

$$\frac{d\mathbf{x}}{dt} = \mathbf{A}\mathbf{x} \tag{11}$$

with the same coefficient matrix (see Problem 29). If we are able to solve the homogeneous system (11), then we can use the reverse transformation to obtain the solution of the original system (1).

On the other hand, if $\mathbf{A}$ is singular, there may be no critical points, and then it will not be possible to reduce the nonhomogeneoous system to a homogeneous one in this manner. Similarly, if the nonhomogeneous term $\mathbf{b}$ depends on t, then it is no longer possible to eliminate it by a simple change of variables similar to Eqs. (4) or (5). However, even in these cases, if the homogeneous system (11) can be solved, then the nonhomogeneous system (1) can also be solved by well-established methods that we will discuss in Section 4.8. Thus the homogeneous system (11) is the more fundamental problem and we will focus most of our attention on it. To begin to understand how we might solve a linear homogeneous system with constant coefficients, let us look first at a particularly simple example.

EXAMPLE 2

Consider the system

$$\frac{d\mathbf{x}}{dt} = \begin{pmatrix} -1 & 0 \\ 0 & -4 \end{pmatrix} \mathbf{x}. \tag{12}$$

Find solutions of the system (12) and then find the particular solution that satisfies the initial condition

$$\mathbf{x}(0) = \begin{pmatrix} 2 \\ 3 \end{pmatrix}. \tag{13}$$

The most important feature of this system is apparent if we write it in scalar form, that is,

$$x_1' = -x_1, \qquad x_2' = -4x_2. \tag{14}$$

Each equation involves only one of the unknown variables; as a result, the two equations can be solved independently. By solving Eqs. (14) we obtain

$$x_1 = c_1 e^{-t}, \qquad x_2 = c_2 e^{-4t}, \tag{15}$$

where c_1 and c_2 are arbitrary constants. Then, by writing the solution (15) in vector form, we have

$$\mathbf{x} = \begin{pmatrix} c_1 e^{-t} \\ c_2 e^{-4t} \end{pmatrix} = c_1 \begin{pmatrix} e^{-t} \\ 0 \end{pmatrix} + c_2 \begin{pmatrix} 0 \\ e^{-4t} \end{pmatrix} = c_1 e^{-t} \begin{pmatrix} 1 \\ 0 \end{pmatrix} + c_2 e^{-4t} \begin{pmatrix} 0 \\ 1 \end{pmatrix}. \tag{16}$$

Note that this solution consists of two terms, each of which involves a vector multiplied by a certain exponential function.

To satisfy the initial conditions (13), we can set $t=0$ in Eq. (16); then

$$\mathbf{x}(0) = c_1 \begin{pmatrix} 1 \\ 0 \end{pmatrix} + c_2 \begin{pmatrix} 0 \\ 1 \end{pmatrix} = \begin{pmatrix} 2 \\ 3 \end{pmatrix}. \tag{17}$$

Consequently, we must choose $c_1 = 2$ and $c_2 = 3$. The solution of the system (12) that satisfies the initial conditions (13) is

$$\mathbf{x} = 2e^{-t} \begin{pmatrix} 1 \\ 0 \end{pmatrix} + 3e^{-4t} \begin{pmatrix} 0 \\ 1 \end{pmatrix}. \tag{18}$$

▶ **Extension to a General System.** We now turn to a consideration of a general system of two first order linear homogeneous differential equations with constant coefficients. We will usually write such a system in the form

$$\frac{d\mathbf{x}}{dt} = \mathbf{Ax}, \tag{19}$$

where

$$\mathbf{x} = \begin{pmatrix} x_1 \\ x_2 \end{pmatrix}, \qquad \mathbf{A} = \begin{pmatrix} a_{11} & a_{12} \\ a_{21} & a_{22} \end{pmatrix}. \tag{20}$$

The elements of the matrix $\mathbf{A}$ are given real constants and the components of the vector $\mathbf{x}$ are to be determined.

To solve Eq. (19), we are guided by the form of the solution in Example 2 and assume that

$$\mathbf{x} = e^{\lambda t}\mathbf{v}, \tag{21}$$

where $\mathbf{v}$ and λ are a constant vector and a scalar, respectively, to be determined. By substituting from Eq. (21) into Eq. (19) and noting that $\mathbf{v}$ and λ do not depend on t, we obtain

$$\lambda e^{\lambda t}\mathbf{v} = \mathbf{A}e^{\lambda t}\mathbf{v}.$$

Further, $e^{\lambda t}$ is never zero, so we have

$$\mathbf{Av} = \lambda\mathbf{v}, \tag{22}$$

or

$$(\mathbf{A} - \lambda\mathbf{I})\mathbf{v} = \mathbf{0}, \tag{23}$$

where $\mathbf{I}$ is the 2×2 identity matrix.

We showed in Section 3.1 that Eq. (23) is precisely the equation that determines the eigenvalues and eigenvectors of the matrix $\mathbf{A}$. Thus

$$\mathbf{x} = e^{\lambda t}\mathbf{v}$$

is a solution of

$$\frac{d\mathbf{x}}{dt} = \mathbf{Ax}$$

provided that λ is an eigenvalue and $\mathbf{v}$ is a corresponding eigenvector of the coefficient matrix $\mathbf{A}$. The eigenvalues λ_1 and λ_2 are the roots of the characteristic equation

$$\det(\mathbf{A} - \lambda\mathbf{I}) = \begin{vmatrix} a_{11} - \lambda & a_{12} \\ a_{21} & a_{22} - \lambda \end{vmatrix} = (a_{11} - \lambda)(a_{22} - \lambda) - a_{12}a_{21}$$

$$= \lambda^2 - (a_{11} + a_{22})\lambda + a_{11}a_{22} - a_{12}a_{21} = 0. \tag{24}$$

For each eigenvalue, we can solve the system (23) and thereby obtain the corresponding eigenvector $\mathbf{v}_1$ or $\mathbf{v}_2$. Recall that the eigenvectors are determined only up to an arbitrary constant multiplier.

Since the elements of $\mathbf{A}$ are real-valued, the eigenvalues may be real and unequal, real and equal, or complex conjugates. We will restrict our discussion in this section to the first case, and will defer consideration of the latter two possibilities until the following two sections.

▶ **Real and Unequal Eigenvalues.** We assume now that λ_1 and λ_2 are real and unequal. Then, using the eigenvalues and the corresponding eigenvectors, we can write down two solutions of Eq. (19), namely,

$$\mathbf{x}_1(t) = e^{\lambda_1 t}\mathbf{v}_1, \qquad \mathbf{x}_2(t) = e^{\lambda_2 t}\mathbf{v}_2. \tag{25}$$

Next, we form the linear combination

$$\mathbf{x} = c_1\mathbf{x}_1(t) + c_2\mathbf{x}_2(t), \tag{26}$$

where c_1 and c_2 are arbitrary constants. This expression is also a solution of Eq. (19). To show this, simply substitute from Eq. (26) into Eq. (19). This gives

$$\frac{d\mathbf{x}}{dt} - \mathbf{A}\mathbf{x} = \frac{d\left[c_1\mathbf{x}_1(t) + c_2\mathbf{x}_2(t)\right]}{dt} - \mathbf{A}\left[c_1\mathbf{x}_1(t) + c_2\mathbf{x}_2(t)\right]$$

$$= c_1\frac{d\mathbf{x}_1(t)}{dt} + c_2\frac{d\mathbf{x}_2(t)}{dt} - c_1\mathbf{A}\mathbf{x}_1(t) - c_2\mathbf{A}\mathbf{x}_2(t)$$

$$= c_1\left[\frac{d\mathbf{x}_1(t)}{dt} - \mathbf{A}\mathbf{x}_1(t)\right] + c_2\left[\frac{d\mathbf{x}_2(t)}{dt} - \mathbf{A}\mathbf{x}_2(t)\right]$$

$$= \mathbf{0} + \mathbf{0} = \mathbf{0}. \tag{27}$$

Observe that in the last line of Eq. (27) we have used the fact that both $\mathbf{x}_1(t)$ and $\mathbf{x}_2(t)$ are solutions of Eq. (19). However, we made no use of the particular form of $\mathbf{x}_1(t)$ and $\mathbf{x}_2(t)$ given by Eq. (25). Thus, in Eq. (27), $\mathbf{x}_1(t)$ and $\mathbf{x}_2(t)$ can be any solutions of Eq. (19). We formalize this result as a theorem.

THEOREM 3.3.1	**Principle of Superposition.** Suppose that $\mathbf{x}_1(t)$ and $\mathbf{x}_2(t)$ are solutions of Eq. (19), $$\frac{d\mathbf{x}}{dt} = \mathbf{A}\mathbf{x}.$$ Then the expression (26) $$\mathbf{x} = c_1\mathbf{x}_1(t) + c_2\mathbf{x}_2(t),$$ where c_1 and c_2 are arbitrary constants, is also a solution.

This theorem expresses one of the fundamental properties of linear homogeneous systems. Starting with two specific solutions, you can immediately generate a much larger (doubly infinite, in fact) family of solutions. The linear combination of $\mathbf{x}_1$ and $\mathbf{x}_2$ given by Eq. (26) with arbitrary coefficients c_1 and c_2 is called the **general solution** of Eq. (19). Suppose now that there is a prescribed initial condition

$$\mathbf{x}(t_0) = \mathbf{x}_0, \tag{28}$$

where t_0 is any given value of t, and $\mathbf{x}_0$ is any given constant vector. Is it possible to choose the constants c_1 and c_2 in Eq. (26) so as to satisfy the initial condition (28)? By substituting from Eq. (26) into Eq. (28) we obtain

$$c_1\mathbf{x}_1(t_0) + c_2\mathbf{x}_2(t_0) = \mathbf{x}_0, \tag{29}$$

or, in more detail,

$$\begin{pmatrix} x_{11}(t_0) & x_{12}(t_0) \\ x_{21}(t_0) & x_{22}(t_0) \end{pmatrix} \begin{pmatrix} c_1 \\ c_2 \end{pmatrix} = \begin{pmatrix} x_{10} \\ x_{20} \end{pmatrix}. \tag{30}$$

The notation in Eq. (30) is such that, for example, x_{12} is the first component of the vector $\mathbf{x}_2$, and so on. Thus the first subscript identifies the component of a vector and the second subscript identifies the vector itself. Equations (30) can be solved uniquely for c_1 and c_2 for any values of x_{10} and x_{20} if and only if the determinant of the coefficient matrix is nonzero. It is possible to write down expressions for c_1 and c_2 that satisfy Eqs. (30), but it is usually preferable just to solve this system of equations whenever it is necessary.

The determinant

$$W[\mathbf{x}_1, \mathbf{x}_2](t) = \begin{vmatrix} x_{11}(t) & x_{12}(t) \\ x_{21}(t) & x_{22}(t) \end{vmatrix} \tag{31}$$

is called the **Wronskian determinant**, or more simply the **Wronskian**, of the two vectors $\mathbf{x}_1$ and $\mathbf{x}_2$. If $\mathbf{x}_1$ and $\mathbf{x}_2$ are given by Eqs. (25), then their Wronskian is

$$W[\mathbf{x}_1, \mathbf{x}_2](t) = \begin{vmatrix} v_{11}e^{\lambda_1 t} & v_{12}e^{\lambda_2 t} \\ v_{21}e^{\lambda_1 t} & v_{22}e^{\lambda_2 t} \end{vmatrix} = \begin{vmatrix} v_{11} & v_{12} \\ v_{21} & v_{22} \end{vmatrix} e^{(\lambda_1 + \lambda_2)t}. \tag{32}$$

The exponential function is never zero, so whether $W[\mathbf{x}_1, \mathbf{x}_2](t)$ is zero depends entirely on the determinant whose columns are the eigenvectors of the coefficient matrix $\mathbf{A}$. It is possible to show (see Appendix A) that this determinant is nonzero whenever the eigenvectors $\mathbf{v}_1$ and $\mathbf{v}_2$ correspond to eigenvalues λ_1 and λ_2 that are different, as we are assuming here. Two solutions $\mathbf{x}_1(t)$ and $\mathbf{x}_2(t)$ of Eq. (19) whose Wronskian is not zero are referred to as a **fundamental set of solutions**. Thus the solutions (25) form such a set.

The results that we have obtained here are a special case of a more general result, which we state as a theorem.

THEOREM 3.3.2

Suppose that $\mathbf{x}_1(t)$ and $\mathbf{x}_2(t)$ are two solutions of Eq. (19),

$$\frac{d\mathbf{x}}{dt} = \mathbf{A}\mathbf{x},$$

and that their Wronskian is not zero. Then $\mathbf{x}_1(t)$ and $\mathbf{x}_2(t)$ form a fundamental set of solutions, and the general solution of Eq. (19) is given by Eq. (26),

$$\mathbf{x} = c_1\mathbf{x}_1(t) + c_2\mathbf{x}_2(t),$$

where c_1 and c_2 are arbitrary constants. If there is a given initial condition $\mathbf{x}(t_0) = \mathbf{x}_0$, where $\mathbf{x}_0$ is any constant vector, then this condition determines the constants c_1 and c_2 uniquely.

This theorem expresses one of the basic properties of systems of linear differential equations. So far, we have discussed it only for the case in which the coefficient matrix $\mathbf{A}$ has eigenvalues that are real and different. In the following two sections, we will see that Theorem 3.3.2 is also valid when the eigenvalues are complex or repeated. This theorem will also reappear in a more general setting in Chapter 6.

EXAMPLE 3

Two Interconnected Tanks (revisited)

Consider again the two-tank flow problem from Example 1, which is modeled by Eq. (9). We rewrite this equation as

$$\frac{d\mathbf{x}}{dt} = \begin{pmatrix} 0.1 & 0.075 \\ 0.1 & -0.2 \end{pmatrix} \mathbf{x} = \mathbf{A}\mathbf{x}. \tag{33}$$

Draw a direction field for this system and find its general solution. Then plot a phase portrait and several component plots.

The direction field for Eq. (33) in Figure 3.3.1 clearly shows that all solutions approach the origin. Observe that this direction field is the same as the one in Figure 3.2.2, except that the critical point is now at the origin.

To solve the system (33), we let $\mathbf{x} = e^{\lambda t}\mathbf{v}$; then we obtain the algebraic system

$$\begin{pmatrix} -0.1 - \lambda & 0.075 \\ 0.1 & -0.2 - \lambda \end{pmatrix} \begin{pmatrix} v_1 \\ v_2 \end{pmatrix} = \begin{pmatrix} 0 \\ 0 \end{pmatrix}, \tag{34}$$

which is Eq. (23) for the system (33). Thus the characteristic equation is

$$(-0.1 - \lambda)(-0.2 - \lambda) - (0.075)(0.1) = \lambda^2 + 0.3\lambda + 0.0125$$
$$= (\lambda + 0.25)(\lambda + 0.05) = 0, \tag{35}$$

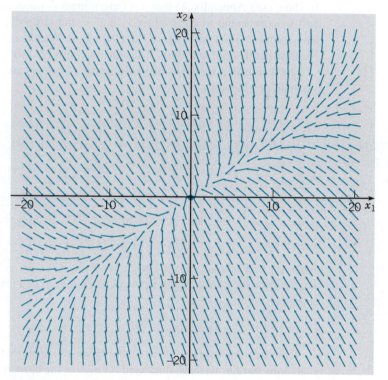

FIGURE 3.3.1 A direction field for the system (33).

so the eigenvalues are $\lambda_1 = -0.25$ and $\lambda_2 = -0.05$. For $\lambda = -0.25$, Eq. (34) becomes

$$\begin{pmatrix} 0.15 & 0.075 \\ 0.1 & 0.05 \end{pmatrix} \begin{pmatrix} v_1 \\ v_2 \end{pmatrix} = \begin{pmatrix} 0 \\ 0 \end{pmatrix}. \tag{36}$$

Thus $2v_1 + v_2 = 0$, so the eigenvector $\mathbf{v}_1$ corresponding to the eigenvalue $\lambda_1 = -0.25$ can be taken to be

$$\mathbf{v}_1 = \begin{pmatrix} 1 \\ -2 \end{pmatrix}. \tag{37}$$

Similarly, corresponding to $\lambda = -0.05$, Eq. (34) yields

$$\begin{pmatrix} -0.05 & 0.075 \\ 0.1 & -0.15 \end{pmatrix} \begin{pmatrix} v_1 \\ v_2 \end{pmatrix} = \begin{pmatrix} 0 \\ 0 \end{pmatrix}. \tag{38}$$

Hence $2v_1 - 3v_2 = 0$, so the eigenvector $\mathbf{v}_2$ can be chosen as

$$\mathbf{v}_2 = \begin{pmatrix} 3 \\ 2 \end{pmatrix}. \tag{39}$$

Observe that Eqs. (36) and (38) determine the eigenvectors only up to an arbitrary multiplicative constant. We have chosen the constants so that $\mathbf{v}_1$ and $\mathbf{v}_2$ have small integer components, but any vectors proportional to those given by Eqs. (37) and (39) could also be used.

The corresponding solutions of the differential equation are

$$\mathbf{x}_1(t) = e^{-0.25t} \begin{pmatrix} 1 \\ -2 \end{pmatrix}, \qquad \mathbf{x}_2(t) = e^{-0.05t} \begin{pmatrix} 3 \\ 2 \end{pmatrix}. \tag{40}$$

The Wronskian of these solutions is

$$W[\mathbf{x}_1, \mathbf{x}_2](t) = \begin{vmatrix} e^{-0.25t} & 3e^{-0.05t} \\ -2e^{-0.25t} & 2e^{-0.05t} \end{vmatrix} = 8e^{-0.3t}, \tag{41}$$

which is never zero. Hence the solutions $\mathbf{x}_1$ and $\mathbf{x}_2$ form a fundamental set, and the general solution of the system (33) is

$$\mathbf{x} = c_1 \mathbf{x}_1(t) + c_2 \mathbf{x}_2(t)$$

$$= c_1 e^{-0.25t} \begin{pmatrix} 1 \\ -2 \end{pmatrix} + c_2 e^{-0.05t} \begin{pmatrix} 3 \\ 2 \end{pmatrix}, \tag{42}$$

where c_1 and c_2 are arbitrary constants. The solution satisfying the initial conditions (10) is found by solving the system

$$\begin{pmatrix} 1 & 3 \\ -2 & 2 \end{pmatrix} \begin{pmatrix} c_1 \\ c_2 \end{pmatrix} = \begin{pmatrix} 13 \\ -10 \end{pmatrix} \tag{43}$$

for c_1 and c_2, with the result that $c_1 = 7$ and $c_2 = 2$.

To visualize the two-parameter family of solutions (42), it is helpful to consider its trajectories in the x_1x_2-plane for various values of the constants c_1 and c_2. We start with $\mathbf{x} = c_1 \mathbf{x}_1(t)$ or, in scalar form,

$$x_1 = c_1 e^{-0.25t}, \qquad x_2 = -2c_1 e^{-0.25t}.$$

By eliminating t between these two equations, we see that this solution lies on the straight line $x_2 = -2x_1$; see Figure 3.3.2. This is the line through the origin in the direction of the eigenvector $\mathbf{v}_1$. If we look on the solution as the trajectory of a moving particle, then the particle is in the fourth quadrant when $c_1 > 0$ and in the second quadrant when $c_1 < 0$. In either case, the particle moves toward the origin as t increases. Next consider $\mathbf{x} = c_2 \mathbf{x}_2(t)$, or

$$x_1 = 3c_2 e^{-0.05t}, \qquad x_2 = 2c_2 e^{-0.05t}.$$

This solution lies on the line $x_2 = (2/3)x_1$, whose direction is determined by the eigenvector $\mathbf{v}_2$. The solution is in the first quadrant when $c_2 > 0$ and in the third quadrant when $c_2 < 0$, as shown in Figure 3.3.2. In both cases, the particle moves toward the origin as t increases. The solution (42) is a combination of $\mathbf{x}_1(t)$ and $\mathbf{x}_2(t)$, so all solutions approach the origin at $t \to \infty$. For large t, the term $c_2\mathbf{x}_2(t)$ is dominant and the term $c_1\mathbf{x}_1(t)$ is negligible in comparison. Thus all solutions for which $c_2 \neq 0$ approach the origin tangent to the line $x_2 = (2/3)x_1$ as $t \to \infty$. If you look backward in time and let $t \to -\infty$, then the term $c_1 e^{-0.25t}\mathbf{v}_1$ is the dominant one (unless $c_1 = 0$). Consequently, all trajectories for which $c_1 \neq 0$ are asymptotic to lines parallel to $x_2 = -2x_1$ as $t \to -\infty$.

Figure 3.3.2 shows the trajectories corresponding to $\mathbf{x}_1(t)$ and $\mathbf{x}_2(t)$, as well as several other trajectories of the system (33). In other words, it is a phase portrait. Observe that Figure 3.3.2 is essentially the same as Figure 3.2.3, except that now the critical point is at the origin. The pattern of trajectories in Figure 3.3.2 is typical of all two-dimensional systems $\mathbf{x}' = \mathbf{Ax}$ whose eigenvalues are real, different, and of the same sign. The origin is called a **node** for such a system. If the eigenvalues were positive rather than negative, then the trajectories would be similar but traversed in the outward direction. Nodes are asymptotically stable if the eigenvalues are negative and unstable if the eigenvalues are positive. Asymptotically stable nodes and unstable nodes are also referred to as **nodal sinks** and **nodal sources**, respectively. Although Figure 3.3.2 was computer generated, it is important to realize that you can sketch a qualitatively correct phase portrait as soon as you have found the eigenvalues and eigenvectors of the coefficient matrix $\mathbf{A}$.

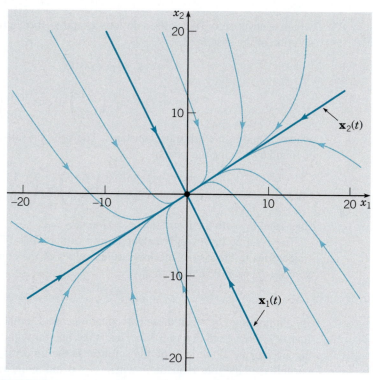

FIGURE 3.3.2 A phase portrait for the system (33).

Some typical component plots of x_1 versus t for the system (33) are shown in Figure 3.3.3. Observe that each graph approaches the t-axis asymptotically as t increases, corresponding to a trajectory that approaches the origin in Figure 3.3.2. The behavior of x_2 as a function of t is similar.

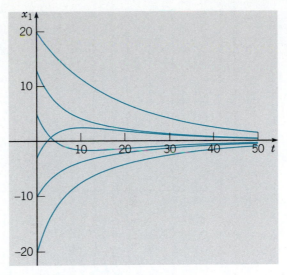

FIGURE 3.3.3 Typical component plots of x_1 versus t for the system (33).

EXAMPLE 4

Consider the system

$$\mathbf{x}' = \begin{pmatrix} 1 & 1 \\ 4 & 1 \end{pmatrix} \mathbf{x}. \tag{44}$$

Plot a direction field and determine the qualitative behavior of solutions. Then find the general solution and draw several trajectories.

A direction field for this system is shown in Figure 3.3.4. From this figure it is easy to see that a typical solution departs from the neighborhood of the origin and ultimately has a slope of approximately 2 in either the first or the third quadrant.

To find solutions explicitly, we assume that $\mathbf{x} = e^{\lambda t}\mathbf{v}$ and substitute for $\mathbf{x}$ in Eq. (44). This results in the system of algebraic equations

$$\begin{pmatrix} 1 - \lambda & 1 \\ 4 & 1 - \lambda \end{pmatrix} \begin{pmatrix} v_1 \\ v_2 \end{pmatrix} = \begin{pmatrix} 0 \\ 0 \end{pmatrix}. \tag{45}$$

The characteristic equation is

$$\begin{vmatrix} 1 - \lambda & 1 \\ 4 & 1 - \lambda \end{vmatrix} = (1 - \lambda)^2 - 4 = \lambda^2 - 2\lambda - 3$$

$$= (\lambda - 3)(\lambda + 1) = 0. \tag{46}$$

Thus the eigenvalues are $\lambda_1 = 3$ and $\lambda_2 = -1$. If $\lambda = 3$, then the system (45) reduces to the single equation

$$-2v_1 + v_2 = 0. \tag{47}$$

Thus $v_2 = 2v_1$, and the eigenvector corresponding to $\lambda_1 = 3$ can be taken as

$$\mathbf{v}_1 = \begin{pmatrix} 1 \\ 2 \end{pmatrix}. \tag{48}$$

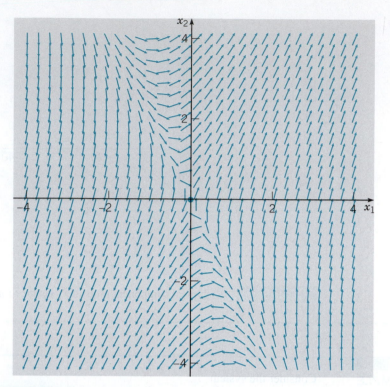

FIGURE 3.3.4 Direction field for the system (44).

Similarly, corresponding to $\lambda_2 = -1$, we find that $v_2 = -2v_1$, so the eigenvector is

$$\mathbf{v}_2 = \begin{pmatrix} 1 \\ -2 \end{pmatrix}. \tag{49}$$

The corresponding solutions of the differential equation are

$$\mathbf{x}_1(t) = e^{3t} \begin{pmatrix} 1 \\ 2 \end{pmatrix}, \qquad \mathbf{x}_2(t) = e^{-t} \begin{pmatrix} 1 \\ -2 \end{pmatrix}. \tag{50}$$

The Wronskian of these solutions is

$$W[\mathbf{x}_1, \mathbf{x}_2](t) = \begin{vmatrix} e^{3t} & e^{-t} \\ 2e^{3t} & -2e^{-t} \end{vmatrix} = -4e^{2t}, \tag{51}$$

which is never zero. Hence the solutions $\mathbf{x}_1$ and $\mathbf{x}_2$ form a fundamental set, and the general solution of the system (44) is

$$\mathbf{x} = c_1 \mathbf{x}_1(t) + c_2 \mathbf{x}_2(t)$$

$$= c_1 e^{3t} \begin{pmatrix} 1 \\ 2 \end{pmatrix} + c_2 e^{-t} \begin{pmatrix} 1 \\ -2 \end{pmatrix}, \tag{52}$$

where c_1 and c_2 are arbitrary constants.

To visualize the solution (52) in the x_1x_2-plane we start with $\mathbf{x} = c_1\mathbf{x}_1(t)$ or, in scalar form,

$$x_1 = c_1 e^{3t}, \qquad x_2 = 2c_1 e^{3t}.$$

By eliminating t between these two equations, we see that this solution lies on the straight line $x_2 = 2x_1$; see Figure 3.3.5. This is the line through the origin in the direction of the eigenvector $\mathbf{v}_1$. If we look at the solution as the trajectory of a moving particle, then the particle is in the first quadrant when $c_1 > 0$ and in the third quadrant when $c_1 < 0$. In either case, the particle departs from the origin as t increases. Next consider $\mathbf{x} = c_2 \mathbf{x}_2(t)$, or

$$x_1 = c_2 e^{-t}, \qquad x_2 = -2c_2 e^{-t}.$$

This solution lies on the line $x_2 = -2x_1$, whose direction is determined by the eigenvector $\mathbf{v}_2$. The solution is in the fourth quadrant when $c_2 > 0$ and in the second quadrant when $c_2 < 0$, as shown in Figure 3.3.5. In both cases, the particle moves toward the origin as t increases. The solution (52) is a combination of $\mathbf{x}_1(t)$ and $\mathbf{x}_2(t)$. For large t the term $c_1\mathbf{x}_1(t)$ is dominant and the term $c_2\mathbf{x}_2(t)$ is negligible. Thus all solutions for which $c_1 \neq 0$ are asymptotic to the line $x_2 = 2x_1$ as $t \to \infty$. Similarly, all solutions for which $c_2 \neq 0$ are asymptotic to the line $x_2 = -2x_1$ as $t \to -\infty$. The graphs of several trajectories comprise the phase portrait in Figure 3.3.5. The pattern of trajectories in this figure is typical of all two-dimensional systems $\mathbf{x}' = \mathbf{A}\mathbf{x}$ for which the eigenvalues are real and of opposite signs. The origin is called a **saddle point** in this case. Saddle points are always unstable because almost all trajectories depart from them as t increases.

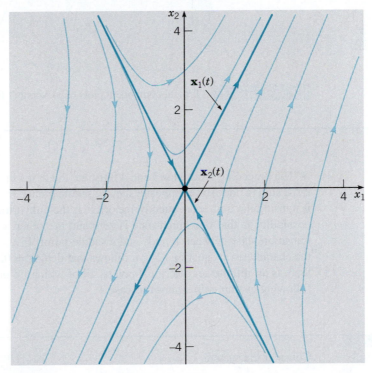

FIGURE 3.3.5 A phase portrait for the system (44); the origin is a saddle point.

A computer was used to plot the phase portrait in Figure 3.3.5, but you can draw a qualitatively accurate sketch of the trajectories as soon as you know the eigenvalues and eigenvectors.

You can also draw component plots of x_1 or x_2 versus t. Some typical plots of x_1 versus t are shown in Figure 3.3.6, and those of x_2 versus t are similar. For certain initial conditions, it follows that $c_1 = 0$ in Eq. (52) so that $x_1 = c_2e^{-t}$ and $x_1 \to 0$ as $t \to \infty$. One such graph is shown in Figure 3.3.6, corresponding to a trajectory that approaches the origin in Figure 3.3.5. For most initial conditions, however, $c_1 \neq 0$ and x_1 is given by $x_1 = c_1 e^{3t} + c_2e^{-t}$. Then the presence of the positive exponential term causes x_1 to grow exponentially in magnitude as t increases. Several graphs of this type are shown in Figure 3.3.6, corresponding to trajectories that depart from the neighborhood of the origin in Figure 3.3.5.

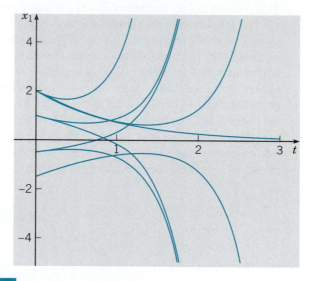

FIGURE 3.3.6 Typical component plots of x_1 versus t for the system (44).

The preceding Examples 3 and 4 illustrate the cases in which the eigenvalues are of the same sign (a node) or of opposite signs (a saddle point). In both of these cases, the matrix **A** is nonsingular and, consequently, the origin is the only critical point of the system. Another possibility is that one eigenvalue is zero and the other is not. This situation occurs as a transition state between a node and a saddle point. If $\lambda = 0$ is an eigenvalue, then from the characteristic equation (24), it follows that $\det(\mathbf{A}) = 0$, so **A** is singular. Consequently, there is an infinite set of critical points, all of which lie on a line through the origin. The following example illustrates this case.

EXAMPLE 5

Consider the system

$$\mathbf{x}' = \mathbf{Ax} = \begin{pmatrix} -1 & 4 \\ \frac{1}{2} & -2 \end{pmatrix} \mathbf{x}. \tag{53}$$

Find the critical points. Then solve the system, draw a phase portrait, and describe how the solutions behave.

Observe that $\det(\mathbf{A}) = 0$, so $\mathbf{A}$ is singular. Solutions of $\mathbf{Ax} = \mathbf{0}$ satisfy $x_1 = 4x_2$, so each point on this line is a critical point of the system (53). To solve the system, we assume that $\mathbf{x} = e^{\lambda t}\mathbf{v}$ and substitute for $\mathbf{x}$ in Eq. (53). This results in the system of algebraic equations

$$\begin{pmatrix} -1 - \lambda & 4 \\ \frac{1}{2} & -2 - \lambda \end{pmatrix} \begin{pmatrix} v_1 \\ v_2 \end{pmatrix} = \begin{pmatrix} 0 \\ 0 \end{pmatrix}. \tag{54}$$

The characteristic equation is

$$(-1 - \lambda)(-2 - \lambda) - 2 = \lambda^2 + 3\lambda = 0, \tag{55}$$

so the eigenvalues are $\lambda_1 = 0$ and $\lambda_2 = -3$. For $\lambda = 0$, it follows from Eq. (54) that $v_1 = 4v_2$ so the eigenvector corresponding to λ_1 is

$$\mathbf{v}_1 = \begin{pmatrix} 4 \\ 1 \end{pmatrix}. \tag{56}$$

In a similar way, we obtain the eigenvector corresponding to $\lambda_2 = -3$, namely,

$$\mathbf{v}_2 = \begin{pmatrix} -2 \\ 1 \end{pmatrix}. \tag{57}$$

Thus two solutions of the system (53) are

$$\mathbf{x}_1(t) = \begin{pmatrix} 4 \\ 1 \end{pmatrix}, \quad \mathbf{x}_2(t) = e^{-3t} \begin{pmatrix} -2 \\ 1 \end{pmatrix}. \tag{58}$$

The Wronskian of these two solutions is $6e^{-3t}$, which is not zero, so the general solution of the system (53) is

$$\mathbf{x} = c_1 \mathbf{x}_1(t) + c_2 \mathbf{x}_2(t) = c_1 \begin{pmatrix} 4 \\ 1 \end{pmatrix} + c_2 e^{-3t} \begin{pmatrix} -2 \\ 1 \end{pmatrix}. \tag{59}$$

If initial conditions are given, they will determine appropriate values for c_1 and c_2.

The solution $\mathbf{x}_1(t)$ is independent of t, that is, it is a constant solution. If $c_2 = 0$, then $\mathbf{x}(t)$ is proportional to $\mathbf{x}_1(t)$ and thus corresponds to a point on the line determined by the eigenvector $\mathbf{v}_1$. Such a solution remains stationary for all time and its trajectory is a single point. This behavior is consistent with the fact that every point on this line is a critical point, as we noted earlier. On the other hand, if $c_1 = 0$, then $\mathbf{x}(t)$ is proportional to $\mathbf{x}_2(t)$. In this case, the trajectory approaches the origin along the line determined by the eigenvector $\mathbf{v}_2$.

A phase portrait for the system (53) is shown in Figure 3.3.7. Any solution starting at a point on the line of critical points remains fixed for all time at its starting point. A solution starting at any other point in the plane moves on a line parallel to $\mathbf{v}_2$ toward the point of intersection of this line with the line of critical points. The phase portrait in Figure 3.3.7 is typical of that for any two-dimensional system $\mathbf{x}' = \mathbf{Ax}$ with one zero eigenvalue and one negative eigenvalue. If the nonzero eigenvalue is positive rather than negative, then the pattern of trajectories is also similar to Figure 3.3.7, but the direction of motion is away from the line of critical points rather than toward it.

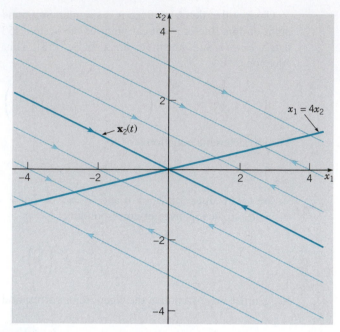

FIGURE 3.3.7 A phase portrait for the system (53).

PROBLEMS

ODEA In each of Problems 1 through 12, find the general solution of the given system of equations. Also draw a direction field and a phase portrait. Describe the behavior of the solutions as $t \to \infty$.

1. $\mathbf{x}' = \begin{pmatrix} 3 & -2 \\ 2 & -2 \end{pmatrix} \mathbf{x}$

2. $\mathbf{x}' = \begin{pmatrix} 1 & -2 \\ 3 & -4 \end{pmatrix} \mathbf{x}$

3. $\mathbf{x}' = \begin{pmatrix} 2 & -1 \\ 3 & -2 \end{pmatrix} \mathbf{x}$

4. $\mathbf{x}' = \begin{pmatrix} 1 & 1 \\ 4 & -2 \end{pmatrix} \mathbf{x}$

5. $\mathbf{x}' = \begin{pmatrix} 4 & -3 \\ 8 & -6 \end{pmatrix} \mathbf{x}$

6. $\mathbf{x}' = \begin{pmatrix} -2 & 1 \\ 1 & -2 \end{pmatrix} \mathbf{x}$

7. $\mathbf{x}' = \begin{pmatrix} \frac{5}{4} & \frac{3}{4} \\ \frac{3}{4} & \frac{5}{4} \end{pmatrix} \mathbf{x}$

8. $\mathbf{x}' = \begin{pmatrix} -\frac{3}{4} & -\frac{7}{4} \\ \frac{1}{4} & \frac{5}{4} \end{pmatrix} \mathbf{x}$

9. $\mathbf{x}' = \begin{pmatrix} -\frac{1}{4} & -\frac{3}{4} \\ \frac{1}{2} & 1 \end{pmatrix} \mathbf{x}$

10. $\mathbf{x}' = \begin{pmatrix} 5 & -1 \\ 3 & 1 \end{pmatrix} \mathbf{x}$

11. $\mathbf{x}' = \begin{pmatrix} -2 & 1 \\ -5 & 4 \end{pmatrix} \mathbf{x}$

12. $\mathbf{x}' = \begin{pmatrix} 3 & 6 \\ -1 & -2 \end{pmatrix} \mathbf{x}$

ODEA In each of Problems 13 through 16, solve the given initial value problem. Draw component plots of x_1 and x_2 versus t. Describe the behavior of the solution as $t \to \infty$.

13. $\mathbf{x}' = \begin{pmatrix} 1 & -2 \\ 3 & -4 \end{pmatrix} \mathbf{x}, \qquad \mathbf{x}(0) = \begin{pmatrix} 3 \\ 1 \end{pmatrix};$
    see Exercise 2

14. $\mathbf{x}' = \begin{pmatrix} 2 & -1 \\ 3 & -2 \end{pmatrix} \mathbf{x}, \qquad \mathbf{x}(0) = \begin{pmatrix} 2 \\ 5 \end{pmatrix};$
    see Exercise 3

15. $\mathbf{x}' = \begin{pmatrix} 5 & -1 \\ 3 & 1 \end{pmatrix} \mathbf{x}, \qquad \mathbf{x}(0) = \begin{pmatrix} 2 \\ -1 \end{pmatrix};$
    see Exercise 10

16. $\mathbf{x}' = \begin{pmatrix} -2 & 1 \\ -5 & 4 \end{pmatrix} \mathbf{x}, \qquad \mathbf{x}(0) = \begin{pmatrix} 1 \\ 3 \end{pmatrix};$
    see Exercise 11

In each of Problems 17 through 24, the eigenvalues and eigenvectors of a matrix $\mathbf{A}$ are given. Consider the corresponding system $\mathbf{x}' = \mathbf{A}\mathbf{x}$. Without using a computer, draw each of the following graphs.
(a) Sketch a phase portrait of the system.
(b) Sketch the trajectory passing through the initial point (2, 3).

(c) For the trajectory in part (b), sketch the component plots of x_1 versus t and of x_2 versus t on the same set of axes.

17. $\lambda_1 = -1, \quad \mathbf{v}_1 = \begin{pmatrix} -1 \\ 2 \end{pmatrix};$

$\lambda_2 = -2, \quad \mathbf{v}_2 = \begin{pmatrix} 1 \\ 2 \end{pmatrix}$

18. $\lambda_1 = 1, \quad \mathbf{v}_1 = \begin{pmatrix} -1 \\ 2 \end{pmatrix};$

$\lambda_2 = -2, \quad \mathbf{v}_2 = \begin{pmatrix} 1 \\ 2 \end{pmatrix}$

19. $\lambda_1 = -1, \quad \mathbf{v}_1 = \begin{pmatrix} -1 \\ 2 \end{pmatrix};$

$\lambda_2 = 2, \quad \mathbf{v}_2 = \begin{pmatrix} 1 \\ 2 \end{pmatrix}$

20. $\lambda_1 = 1, \quad \mathbf{v}_1 = \begin{pmatrix} 1 \\ 2 \end{pmatrix};$

$\lambda_2 = 2, \quad \mathbf{v}_2 = \begin{pmatrix} 1 \\ -2 \end{pmatrix}$

21. $\lambda_1 = 0.5, \quad \mathbf{v}_1 = \begin{pmatrix} 1 \\ 4 \end{pmatrix};$

$\lambda_2 = -0.5, \quad \mathbf{v}_2 = \begin{pmatrix} 4 \\ 1 \end{pmatrix}$

22. $\lambda_1 = -0.5, \quad \mathbf{v}_1 = \begin{pmatrix} 2 \\ 1 \end{pmatrix};$

$\lambda_2 = -0.8, \quad \mathbf{v}_2 = \begin{pmatrix} -1 \\ 2 \end{pmatrix}$

23. $\lambda_1 = 0.3, \quad \mathbf{v}_1 = \begin{pmatrix} 1 \\ -2 \end{pmatrix};$

$\lambda_2 = 0.6, \quad \mathbf{v}_2 = \begin{pmatrix} 1 \\ 3 \end{pmatrix}$

24. $\lambda_1 = 1.5, \quad \mathbf{v}_1 = \begin{pmatrix} -1 \\ 2 \end{pmatrix};$

$\lambda_2 = -1, \quad \mathbf{v}_2 = \begin{pmatrix} 3 \\ 1 \end{pmatrix}$

25. Consider the two-tank system in Example 3 of this section. Find the time T such that $|x_1(t)| \le 0.5$ and $|x_2(t)| \le 0.5$ for all $t \ge T$.

26. Consider the system

$$\mathbf{x}' = \begin{pmatrix} -1 & -1 \\ -\alpha & -1 \end{pmatrix} \mathbf{x}.$$

(a) Solve the system for $\alpha = 0.5$. What are the eigenvalues of the coefficient matrix? Classify the equilibrium point at the origin as to type.

(b) Solve the system for $\alpha = 2$. What are the eigenvalues of the coefficient matrix? Classify the equilibrium point at the origin as to type.

(c) In parts (a) and (b), solutions of the system exhibit two quite different types of behavior. Find the eigenvalues of the coefficient matrix in terms of α and determine the value of α between 0.5 and 2 where the transition from one type of behavior to the other occurs.

Electric Circuits. Problems 27 and 28 are concerned with the electric circuit described by the system of differential equations in Problem 23 of Section 3.2:

$$\frac{d}{dt} \begin{pmatrix} i \\ v \end{pmatrix} = \begin{pmatrix} -\dfrac{R_1}{L} & -\dfrac{1}{L} \\ \dfrac{1}{C} & -\dfrac{1}{CR_2} \end{pmatrix} \begin{pmatrix} i \\ v \end{pmatrix}. \tag{i}$$

27. (a) Find the general solution of Eq. (i) if $R_1 = 1$ ohm, $R_2 = 3/5$ ohm, $L=2$ henrys, and $C = 2/3$ farad.

(b) Show that $i(t) \to 0$ and $v(t) \to 0$ as $t \to \infty$ regardless of the initial values $i(0)$ and $v(0)$.

28. Consider the preceding system of differential equations (i).

(a) Find a condition on R_1, R_2, C, and L that must be satisfied if the eigenvalues of the coefficient matrix are to be real and different.

(b) If the condition found in part (a) is satisfied, show that both eigenvalues are negative. Then show that $i(t) \to 0$ and $v(t) \to 0$ as $t \to \infty$ regardless of the initial conditions.

29. Consider the system $\mathbf{x}' = \mathbf{A}\mathbf{x} + \mathbf{b}$. Assume that $\mathbf{A}$ has an inverse so that the only critical point of the system is $\mathbf{x}_C = -\mathbf{A}^{-1}\mathbf{b}$. Let $\mathbf{u} = \mathbf{x} - \mathbf{x}_C$. Show that $\mathbf{u}$ satisfies the corresponding homogeneous system $\mathbf{u}' = \mathbf{A}\mathbf{u}$.

3.4 Complex Eigenvalues

In this section, we again consider a system of two linear homogeneous equations with constant coefficients

$$\mathbf{x}' = \mathbf{A}\mathbf{x}, \tag{1}$$

where the coefficient matrix $\mathbf{A}$ is real-valued. If we seek solutions of the form $\mathbf{x} = e^{\lambda t}\mathbf{v}$, then it follows, as in Section 3.3, that λ must be an eigenvalue and $\mathbf{v}$ a corresponding eigenvector of the coefficient matrix $\mathbf{A}$. Recall that the eigenvalues λ_1 and λ_2 of $\mathbf{A}$ are the roots of the quadratic equation

$$\det(\mathbf{A} - \lambda\mathbf{I}) = 0, \tag{2}$$

and that the corresponding eigenvectors satisfy

$$(\mathbf{A} - \lambda\mathbf{I})\mathbf{v} = \mathbf{0}. \tag{3}$$

Since $\mathbf{A}$ is real, the coefficients in Eq. (2) for λ are real, and any complex eigenvalues must occur in conjugate pairs. For example, if $\lambda_1 = \mu + i\nu$, where μ and ν are real, is an eigenvalue of $\mathbf{A}$, then so is $\lambda_2 = \mu - i\nu$. Before looking further at the general system (1), we consider an example.

**EXAMPLE
1**

Consider the system

$$\mathbf{x}' = \begin{pmatrix} -\frac{1}{2} & 1 \\ -1 & -\frac{1}{2} \end{pmatrix} \mathbf{x}. \tag{4}$$

Draw a direction field for this system. Find a fundamental set of solutions and display them graphically in a phase portrait and component plots.

A direction field for the system (11) is shown in Figure 3.4.1. This plot suggests that the trajectories in the phase plane spiral clockwise toward the origin.

To find a fundamental set of solutions we assume that

$$\mathbf{x} = e^{\lambda t}\mathbf{v} \tag{5}$$

and obtain the set of linear algebraic equations

$$\begin{pmatrix} -\frac{1}{2} - \lambda & 1 \\ -1 & -\frac{1}{2} - \lambda \end{pmatrix} \begin{pmatrix} v_1 \\ v_2 \end{pmatrix} = \begin{pmatrix} 0 \\ 0 \end{pmatrix} \tag{6}$$

for the eigenvalues and eigenvectors of $\mathbf{A}$. The characteristic equation is

$$\begin{vmatrix} -\frac{1}{2} - \lambda & 1 \\ -1 & -\frac{1}{2} - \lambda \end{vmatrix} = \lambda^2 + \lambda + \frac{5}{4} = 0; \tag{7}$$

therefore the eigenvalues are $\lambda_1 = -\frac{1}{2} + i$ and $\lambda_2 = -\frac{1}{2} - i$. For $\lambda_1 = -\frac{1}{2} + i$ we obtain from Eq. (6)

$$\begin{pmatrix} -i & 1 \\ -1 & -i \end{pmatrix} \begin{pmatrix} v_1 \\ v_2 \end{pmatrix} = \begin{pmatrix} 0 \\ 0 \end{pmatrix}. \tag{8}$$

In scalar form, the first of these equations is $-iv_1 + v_2 = 0$ and the second equation is $-i$ times the first. Thus we have $v_2 = iv_1$, so the eigenvector corresponding to the eigenvalue

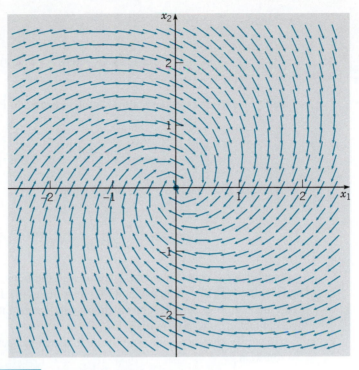

FIGURE 3.4.1 A direction field for the system (4).

λ_1 is

$$\mathbf{v}_1 = \begin{pmatrix} 1 \\ i \end{pmatrix} \tag{9}$$

or any multiple of this vector. A similar calculation for the second eigenvalue $\lambda_2 = -\frac{1}{2} - i$ leads to

$$\mathbf{v}_2 = \begin{pmatrix} 1 \\ -i \end{pmatrix}. \tag{10}$$

Thus, the eigenvectors as well as the eigenvalues are complex conjugates. The corresponding solutions of the system (4) are

$$\mathbf{x}_1(t) = e^{(-1/2+i)t} \begin{pmatrix} 1 \\ i \end{pmatrix}, \qquad \mathbf{x}_2(t) = e^{(-1/2-i)t} \begin{pmatrix} 1 \\ -i \end{pmatrix}. \tag{11}$$

The Wronskian of $\mathbf{x}_1(t)$ and $\mathbf{x}_2(t)$ is readily calculated to be $-2ie^{-t}$, which is never zero, so these solutions form a fundamental set. However, $\mathbf{x}_1$ and $\mathbf{x}_2$ are complex valued, and for many purposes it is desirable to find a fundamental set of real-valued solutions. To do this we start by finding the real and imaginary parts[3] of $\mathbf{x}_1(t)$. First, we use the properties of exponents to write

$$\mathbf{x}_1(t) = e^{-t/2}e^{it} \begin{pmatrix} 1 \\ i \end{pmatrix}. \tag{12}$$

Then, using Euler's formula for e^{it}, namely

$$e^{it} = \cos t + i \sin t, \tag{13}$$

[3] You will find a summary of the necessary elementary results from complex variables in Appendix B.

we obtain

$$\mathbf{x}_1(t) = e^{-t/2}(\cos t + i \sin t)\begin{pmatrix} 1 \\ i \end{pmatrix}. \tag{14}$$

Finally, by carrying out the multiplication indicated in Eq. (14), we find that

$$\mathbf{x}_1(t) = \begin{pmatrix} e^{-t/2}\cos t \\ -e^{-t/2}\sin t \end{pmatrix} + i\begin{pmatrix} e^{-t/2}\sin t \\ e^{-t/2}\cos t \end{pmatrix} = \mathbf{u}(t) + i\mathbf{w}(t). \tag{15}$$

The real and imaginary parts of $\mathbf{x}_1(t)$, that is,

$$\mathbf{u}(t) = \begin{pmatrix} e^{-t/2}\cos t \\ -e^{-t/2}\sin t \end{pmatrix}, \quad \mathbf{w}(t) = \begin{pmatrix} e^{-t/2}\sin t \\ e^{-t/2}\cos t \end{pmatrix} \tag{16}$$

are real-valued vector functions. They are also solutions of the system (4). One way to show this is simply to substitute $\mathbf{u}(t)$ and $\mathbf{w}(t)$ for $\mathbf{x}$ in Eq. (4). However, this is a very special case of a much more general result, which we will demonstrate shortly, so for the moment let us accept that $\mathbf{u}(t)$ and $\mathbf{w}((t)$ satisfy Eq. (4).

To verify that $\mathbf{u}(t)$ and $\mathbf{w}(t)$ are a fundamental set of solutions, we compute their Wronskian:

$$W[\mathbf{u}, \mathbf{w}](t) = \begin{vmatrix} e^{-t/2}\cos t & e^{-t/2}\sin t \\ -e^{-t/2}\sin t & e^{-t/2}\cos t \end{vmatrix}$$

$$= e^{-t}(\cos^2 t + \sin^2 t) = e^{-t}. \tag{17}$$

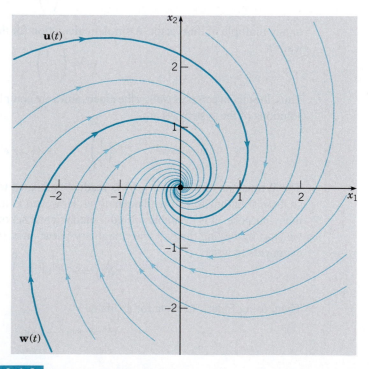

FIGURE 3.4.2   A phase portrait for the system (4).

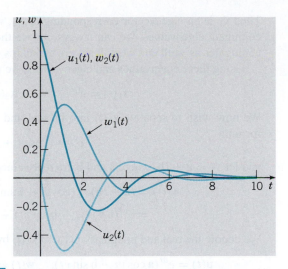

FIGURE 3.4.3 Component plots for the solutions $\mathbf{u}(t)$ and $\mathbf{w}(t)$ of the system (4).

Since the Wronskian is never zero, it follows that $\mathbf{u}(t)$ and $\mathbf{w}(t)$ constitute a fundamental set of (real-valued) solutions of the system (4). Consequently, every solution of Eq. (4) is a linear combination of $\mathbf{u}(t)$ and $\mathbf{w}(t)$, and the general solution of Eq. (4) is

$$\mathbf{x} = c_1\mathbf{u}(t) + c_2\mathbf{w}(t) = c_1\begin{pmatrix} e^{-t/2}\cos t \\ -e^{-t/2}\sin t \end{pmatrix} + c_2\begin{pmatrix} e^{-t/2}\sin t \\ e^{-t/2}\cos t \end{pmatrix}, \tag{18}$$

where c_1 and c_2 are arbitrary constants.

The trajectories of the solutions $\mathbf{u}(t)$ and $\mathbf{w}(t)$ are shown in Figure 3.4.2. Since

$$\mathbf{u}(0) = \begin{pmatrix} 1 \\ 0 \end{pmatrix}, \qquad \mathbf{w}(0) = \begin{pmatrix} 0 \\ 1 \end{pmatrix},$$

the trajectories of $\mathbf{u}(t)$ and $\mathbf{w}(t)$ pass through the points $(1, 0)$ and $(0, 1)$, respectively. Other solutions of the system (4) are linear combinations of $\mathbf{u}(t)$ and $\mathbf{w}(t)$, and trajectories of a few of these solutions are also shown in Figure 3.4.2. Each trajectory spirals toward the origin in the clockwise direction as $t \to \infty$, making infinitely many circuits about the origin. This is due to the fact that the solutions (16) are products of decaying exponential and sine or cosine factors.

Figure 3.4.3 shows the component plots of $\mathbf{u}(t)$ and $\mathbf{w}(t)$. Note that $u_1(t)$ and $w_2(t)$ are identical, so their plots coincide, while $u_2(t)$ and $w_1(t)$ are negatives of each other. Each plot represents a decaying oscillation in time. Since other solutions are linear combinations of $\mathbf{u}(t)$ and $\mathbf{w}(t)$, their component plots are also decaying oscillations.

▶ **Extension to a General System.** We can proceed just as in Example 1 in the case of a general two-dimensional system $\mathbf{x}' = \mathbf{Ax}$ with complex conjugate eigenvalues. Suppose that the eigenvalues are $\lambda_1 = \mu + iv$ and $\lambda_2 = \bar{\lambda}_1 = \mu - iv$. Suppose also that $\mathbf{v}_1$ is an eigenvector corresponding to λ_1. Then λ_1 and $\mathbf{v}_1$ satisfy

$$(\mathbf{A} - \lambda_1\mathbf{I})\mathbf{v}_1 = \mathbf{0}. \tag{19}$$

By taking the conjugate of Eq. (19) and remembering that $\mathbf{A}$, $\mathbf{I}$, and $\mathbf{0}$ are real-valued, we obtain

$$(\mathbf{A} - \bar{\lambda}_1\mathbf{I})\bar{\mathbf{v}}_1 = \mathbf{0}. \tag{20}$$

Thus $\overline{\mathbf{v}}_1$ is an eigenvector corresponding to $\overline{\lambda}_1$, which is λ_2. Thus, for a pair of complex conjugate eigenvalues, we can always choose the eigenvectors so that they are complex conjugates as well, and we will always make this choice.

Using these eigenvalues and eigenvectors, we obtain two solutions of the system (1):

$$\mathbf{x}_1(t) = e^{(\mu+i\nu)t}\mathbf{v}_1, \quad \mathbf{x}_2(t) = e^{(\mu-i\nu)t}\overline{\mathbf{v}}_1. \tag{21}$$

We now wish to separate $\mathbf{x}_1(t)$ into its real and imaginary parts. Recall that, by Euler's formula,

$$e^{i\nu t} = \cos \nu t + i \sin \nu t \tag{22}$$

and let $\mathbf{v}_1 = \mathbf{a} + i\mathbf{b}$, where $\mathbf{a}$ and $\mathbf{b}$ are real-valued. Then from Eq. (21) we have

$$\mathbf{x}_1(t) = (\mathbf{a} + i\mathbf{b})e^{\mu t}(\cos \nu t + i \sin \nu t)$$
$$= e^{\mu t}(\mathbf{a} \cos \nu t - \mathbf{b} \sin \nu t) + ie^{\mu t}(\mathbf{a} \sin \nu t + \mathbf{b} \cos \nu t). \tag{23}$$

We denote the real and imaginary parts of $\mathbf{x}_1(t)$ by

$$\mathbf{u}(t) = e^{\mu t}(\mathbf{a} \cos \nu t - \mathbf{b} \sin \nu t), \quad \mathbf{w}(t) = e^{\mu t}(\mathbf{a} \sin \nu t + \mathbf{b} \cos \nu t), \tag{24}$$

respectively. A similar calculation, starting from $\mathbf{x}_2(t)$ in Eq. (21), leads to

$$\mathbf{x}_2(t) = \mathbf{u}(t) - i\mathbf{w}(t). \tag{25}$$

Thus the solutions $\mathbf{x}_1(t)$ and $\mathbf{x}_2(t)$ are also complex conjugates.

Next, we want to show that $\mathbf{u}(t)$ and $\mathbf{w}(t)$ are solutions of Eq. (1). Since $\mathbf{x}_1(t)$ is a solution, we can write

$$0 = \mathbf{x}_1' - A\mathbf{x}_1 = \mathbf{u}' + i\mathbf{w}' - A(\mathbf{u} + i\mathbf{w})$$
$$= \mathbf{u}' - A\mathbf{u} + i(\mathbf{w}' - A\mathbf{w}). \tag{26}$$

A complex number (vector) is zero if and only if both its real and imaginary parts are zero, so we conclude that

$$\mathbf{u}' - A\mathbf{u} = 0, \quad \mathbf{w}' - A\mathbf{w} = 0; \tag{27}$$

therefore $\mathbf{u}(t)$ and $\mathbf{w}(t)$ are (real-valued) solutions of Eq. (1).

Finally, we want to calculate the Wronskian of $\mathbf{u}(t)$ and $\mathbf{w}(t)$ so as to determine whether they form a fundamental set of solutions. Let $\mathbf{a} = a_1\mathbf{i} + a_2\mathbf{j}$ and $\mathbf{b} = b_1\mathbf{i} + b_2\mathbf{j}$. Then

$$W[\mathbf{u}, \mathbf{w}](t) = \begin{vmatrix} e^{\mu t}(a_1 \cos \nu t - b_1 \sin \nu t) & e^{\mu t}(a_1 \sin \nu t + b_1 \cos \nu t) \\ e^{\mu t}(a_2 \cos \nu t - b_2 \sin \nu t) & e^{\mu t}(a_2 \sin \nu t + b_2 \cos \nu t) \end{vmatrix}. \tag{28}$$

A straightforward calculation shows that

$$W[\mathbf{u}, \mathbf{w}](t) = (a_1 b_2 - a_2 b_1)e^{2\mu t}. \tag{29}$$

Therefore the Wronskian is not zero if and only if $a_1 b_2 - a_2 b_1 \neq 0$. It is possible to show that if $\nu \neq 0$, then $a_1 b_2 - a_2 b_1 \neq 0$, and the solutions $\mathbf{u}(t)$ and $\mathbf{w}(t)$ given by Eq. (24) form a fundamental set of solutions of the system (1). The general solution can be written as

$$\mathbf{x} = c_1\mathbf{u}(t) + c_2\mathbf{w}(t), \tag{30}$$

where c_1 and c_2 are arbitrary constants.

To summarize, to solve the system (1) when it has complex eigenvalues, proceed just as in Example 1. That is, find the eigenvalues and eigenvectors, observing that they are complex conjugates. Then write down $\mathbf{x}_1(t)$ and separate it into its real and imaginary parts $\mathbf{u}(t)$ and $\mathbf{w}(t)$, respectively. Finally, form a linear combination of $\mathbf{u}(t)$ and $\mathbf{w}(t)$, as shown in Eq. (30). Of course, if complex-valued solutions are acceptable, you can simply use the solutions $\mathbf{x}_1(t)$ and $\mathbf{x}_2(t)$. Thus Theorem 3.3.2 is also valid when the eigenvalues are complex.

The phase portrait in Figure 3.4.2 is typical of all two-dimensional systems $\mathbf{x}' = \mathbf{Ax}$ whose eigenvalues are complex with negative real part. The origin is called a **spiral point** and is asymptotically stable because all trajectories approach it as t increases. Such a spiral point is often called a **spiral sink**. For a system whose eigenvalues have a positive real part, the trajectories are similar to those in Figure 3.4.2, but the direction of motion is away from the origin and the trajectories become unbounded. In this case, the origin is unstable and is often called a **spiral source**.

If the real part of the eigenvalues is zero, then there is no exponential factor in the solution and the trajectories neither approach the origin nor become unbounded. Instead they repeatedly traverse a closed curve about the origin. An example of this behavior can be seen in Figure 3.4.5. In this case, the origin is called a **center** and is said to be stable, but not asymptotically stable. In all three cases, the direction of motion may be either clockwise, as in Example 1, or counterclockwise, depending on the elements of the coefficient matrix $\mathbf{A}$.

EXAMPLE 2

Consider the system

$$\mathbf{x}' = \begin{pmatrix} \frac{1}{2} & -\frac{5}{4} \\ 2 & -\frac{1}{2} \end{pmatrix} \mathbf{x}. \tag{31}$$

Draw a direction field for this system. Then find its general solution and also the solution that satisfies the initial conditions

$$\mathbf{x}(0) = \begin{pmatrix} -1 \\ -2 \end{pmatrix}. \tag{32}$$

Draw a phase portrait for the system and component plots for the solution satisfying the initial conditions (32).

A direction field for this system is shown in Figure 3.4.4. From this plot, it appears that the trajectories move around the origin in the counterclockwise direction. Whether they slowly approach the origin, or slowly depart from it, or do neither cannot be conclusively determined from a direction field.

To solve the system, we assume that $\mathbf{x} = e^{\lambda t}\mathbf{v}$, substitute in Eq. (31), and thereby obtain the algebraic system $\mathbf{Av} = \lambda \mathbf{v}$, or

$$\begin{pmatrix} \frac{1}{2} - \lambda & -\frac{5}{4} \\ 2 & -\frac{1}{2} - \lambda \end{pmatrix} \begin{pmatrix} v_1 \\ v_2 \end{pmatrix} = \begin{pmatrix} 0 \\ 0 \end{pmatrix}. \tag{33}$$

The eigenvalues are found from the characteristic equation

$$\left(\tfrac{1}{2} - \lambda\right)\left(-\tfrac{1}{2} - \lambda\right) - \left(-\tfrac{5}{4}\right)(2) = \lambda^2 + \tfrac{9}{4} = 0, \tag{34}$$

so the eigenvalues are $\lambda_1 = 3i/2$ and $\lambda_2 = -3i/2$. By substituting λ_1 for λ in Eq. (33) and then solving this system, we find the corresponding eigenvector $\mathbf{v}_1$. We can find the other eigenvector $\mathbf{v}_2$ in a similar way, or we can simply take $\mathbf{v}_2$ to be the complex conjugate of $\mathbf{v}_1$. Either way we obtain

$$\mathbf{v}_1 = \begin{pmatrix} 5 \\ 2 - 6i \end{pmatrix}, \qquad \mathbf{v}_2 = \begin{pmatrix} 5 \\ 2 + 6i \end{pmatrix}. \tag{35}$$

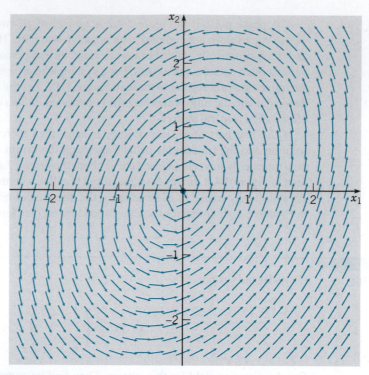

FIGURE 3.4.4 A direction field for the system (31).

Hence two (complex-valued) solutions of the system (31) are

$$\mathbf{x}_1(t) = e^{3it/2} \begin{pmatrix} 5 \\ 2 - 6i \end{pmatrix}, \quad \mathbf{x}_2(t) = e^{-3it/2} \begin{pmatrix} 5 \\ 2 + 6i \end{pmatrix}, \tag{36}$$

and the general solution of Eq. (31) can be expressed as a linear combination of $\mathbf{x}_1(t)$ and $\mathbf{x}_2(t)$ with arbitrary coefficients.

To find real-valued solutions, we can separate $\mathbf{x}_1(t)$ into its real and imaginary parts. Using Euler's formula, we have

$$\mathbf{x}_1(t) = \left[\cos(3t/2) + i \sin(3t/2) \right] \begin{pmatrix} 5 \\ 2 - 6i \end{pmatrix}$$

$$= \begin{pmatrix} 5 \cos(3t/2) \\ 2 \cos(3t/2) + 6 \sin(3t/2) \end{pmatrix} + i \begin{pmatrix} 5 \sin(3t/2) \\ 2 \sin(3t/2) - 6 \cos(3t/2) \end{pmatrix}. \tag{37}$$

Thus we can also write the general solution of Eq. (31) in the form

$$\mathbf{x} = c_1 \begin{pmatrix} 5 \cos(3t/2) \\ 2 \cos(3t/2) + 6 \sin(3t/2) \end{pmatrix} + c_2 \begin{pmatrix} 5 \sin(3t/2) \\ 2 \sin(3t/2) - 6 \cos(3t/2) \end{pmatrix}, \tag{38}$$

where c_1 and c_2 are arbitrary constants.

To satisfy the initial conditions (32) we set $t = 0$ in Eq. (38) and obtain

$$c_1 \begin{pmatrix} 5 \\ 2 \end{pmatrix} + c_2 \begin{pmatrix} 0 \\ -6 \end{pmatrix} = \begin{pmatrix} -1 \\ -2 \end{pmatrix}. \tag{39}$$

Therefore $c_1 = -1/5$ and $c_2 = 4/15$. Using these values in Eq. (38), we obtain the solution that satifies the initial conditions (32). The scalar components of this solution are easily found to be

$$x_1(t) = -\cos(3t/2) + \tfrac{4}{3}\sin(3t/2), \quad x_2(t) = -2\cos(3t/2) - \tfrac{2}{3}\sin(3t/2). \tag{40}$$

A phase portrait for the system (31) appears in Figure 3.4.5 with the heavy curve showing the trajectory passing through the point $(-1, -2)$. All of the trajectories (except the origin itself) are closed curves surrounding the origin, and each one is traversed in the counterclockwise direction, as the direction field in Figure 3.4.4 indicated. Closed trajectories correspond to eigenvalues whose real part is zero. In this case, the solution contains no exponential factor, but consists only of sine and cosine terms, and therefore is periodic. The closed trajectory is traversed repeatedly, with one period corresponding to a full circuit. The periodic nature of the solution is shown clearly by the component plots in Figure 3.4.6, which show the graphs of $x_1(t)$ and $x_2(t)$ given in Eq. (40). The period of each solution is $4\pi/3$.

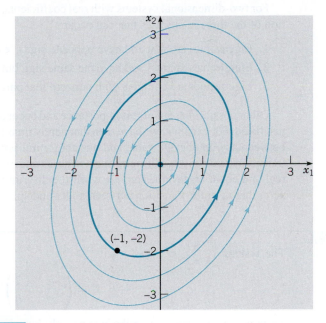

FIGURE 3.4.5   A phase portrait for the system (31); the heavy curve passes through $(-1, -2)$.

As mentioned previously, when the eigenvalues are purely imaginary the critical point at the origin is called a center. Because nearby trajectories remain near the critical point but do not approach it, the critical point is said to be stable, but not asymptotically stable. A center is often a transition state between spiral points that are asymptotically stable or unstable, depending on the sign of the real part of the eigenvalues. The trajectories in Figure 3.4.5 appear to be elliptical, and it can be shown (see Problem 23) that this is always the case for centers of two-dimensional linear systems with constant coefficients.

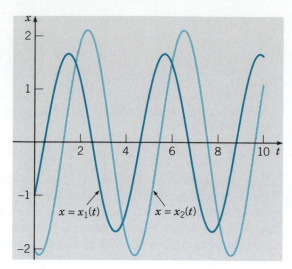

FIGURE 3.4.6   Plots of the components of the solution passing through $(-1, -2)$; they are given by Eq. (40).

For two-dimensional systems with real coefficients, we have now completed our description of the three main cases that can occur.

1. Eigenvalues are real and have opposite signs; $\mathbf{x} = \mathbf{0}$ is a saddle point.
2. Eigenvalues are real and have the same sign but are unequal; $\mathbf{x} = \mathbf{0}$ is a node.
3. Eigenvalues are complex with nonzero real part; $\mathbf{x} = \mathbf{0}$ is a spiral point.

Other possibilities are of less importance and occur as transitions between two of the cases just listed. For example, we have seen that purely imaginary eigenvalues occur at a transition between asymptotically stable and unstable spiral points. A zero eigenvalue occurs at the transition between a saddle point and a node, as one of the eigenvalues changes sign. Finally, real and equal eigenvalues appear when the discriminant of the characteristic equation is zero, that is, at the transition between nodes and spiral points.

EXAMPLE 3

The system

$$\mathbf{x}' = \begin{pmatrix} \alpha & 2 \\ -2 & 0 \end{pmatrix} \mathbf{x} \tag{41}$$

contains a parameter α. Describe how the solutions depend qualitatively on α; in particular, find the critical values of α at which the qualitative behavior of the trajectories in the phase plane changes markedly.

The behavior of the trajectories is controlled by the eigenvalues of the coefficient matrix. The characteristic equation is

$$\lambda^2 - \alpha\lambda + 4 = 0, \tag{42}$$

so the eigenvalues are

$$\lambda = \frac{\alpha \pm \sqrt{\alpha^2 - 16}}{2}. \tag{43}$$

From Eq. (43), it follows that the eigenvalues are complex conjugates for $-4 < \alpha < 4$ and are real otherwise. Thus two critical values are $\alpha = -4$ and $\alpha = 4$, where the eigenvalues change from real to complex, or vice versa. For $\alpha < -4$, both eigenvalues are negative, so all trajectories approach the origin, which is an asymptotically stable node. For $\alpha > 4$ both eigenvalues are positive, so the origin is again a node, this time unstable; all trajectories (except $\mathbf{x} = \mathbf{0}$) become unbounded. In the intermediate range, $-4 < \alpha < 4$, the eigenvalues are complex and the trajectories are spirals. However, for $-4 < \alpha < 0$, the real part of the eigenvalues is negative, the spirals are directed inward, and the origin is asymptotically stable, whereas for $0 < \alpha < 4$, the real part of the eigenvalues is positive and the origin is unstable. Thus $\alpha = 0$ is also a critical value where the direction of the spirals changes from inward to outward. For this value of α, the origin is a center and the trajectories are closed curves about the origin, corresponding to solutions that are periodic in time. The other critical values, $\alpha = \pm 4$, yield eigenvalues that are real and equal. In this case, the origin is again a node, but the phase portrait differs somewhat from those in Section 3.3. We take up this case in Section 3.5.

PROBLEMS

ODEA In each of Problems 1 through 6, express the general solution of the given system of equations in terms of real-valued functions. Also, draw a direction field and a phase portrait. Describe the behavior of the solutions as $t \to \infty$.

1. $\mathbf{x}' = \begin{pmatrix} 3 & -2 \\ 4 & -1 \end{pmatrix} \mathbf{x}$
2. $\mathbf{x}' = \begin{pmatrix} -1 & -4 \\ 1 & -1 \end{pmatrix} \mathbf{x}$

3. $\mathbf{x}' = \begin{pmatrix} 2 & -5 \\ 1 & -2 \end{pmatrix} \mathbf{x}$
4. $\mathbf{x}' = \begin{pmatrix} 2 & -\frac{5}{2} \\ \frac{9}{5} & -1 \end{pmatrix} \mathbf{x}$

5. $\mathbf{x}' = \begin{pmatrix} 1 & -1 \\ 5 & -3 \end{pmatrix} \mathbf{x}$
6. $\mathbf{x}' = \begin{pmatrix} 1 & 2 \\ -5 & -1 \end{pmatrix} \mathbf{x}$

ODEA In each of Problems 7 through 10, find the solution of the given initial value problem. Draw component plots of the solution and describe the behavior of the solution as $t \to \infty$.

7. $\mathbf{x}' = \begin{pmatrix} -1 & -4 \\ 1 & -1 \end{pmatrix} \mathbf{x}, \qquad \mathbf{x}(0) = \begin{pmatrix} 4 \\ -3 \end{pmatrix};$

   See Problem 2.

8. $\mathbf{x}' = \begin{pmatrix} 2 & -5 \\ 1 & 2 \end{pmatrix} \mathbf{x}, \qquad \mathbf{x}(0) = \begin{pmatrix} 3 \\ 2 \end{pmatrix};$

   See Problem 3.

9. $\mathbf{x}' = \begin{pmatrix} 1 & -5 \\ 1 & -3 \end{pmatrix} \mathbf{x}, \qquad \mathbf{x}(0) = \begin{pmatrix} 1 \\ 1 \end{pmatrix}$

10. $\mathbf{x}' = \begin{pmatrix} -3 & 2 \\ -1 & -1 \end{pmatrix} \mathbf{x}, \qquad \mathbf{x}(0) = \begin{pmatrix} 1 \\ -2 \end{pmatrix}$

In each of Problems 11 and 12:
(a) Find the eigenvalues of the given system.
(b) Choose an initial point (other than the origin) and sketch the corresponding trajectory in the x_1x_2-plane.
(c) For your trajectory in part (b), sketch the graphs of x_1 versus t and of x_2 versus t.

11. $\mathbf{x}' = \begin{pmatrix} \frac{3}{4} & -2 \\ 1 & -\frac{5}{4} \end{pmatrix} \mathbf{x}$
12. $\mathbf{x}' = \begin{pmatrix} -\frac{4}{5} & 2 \\ -1 & \frac{6}{5} \end{pmatrix} \mathbf{x}$

In each of Problems 13 through 20, the coefficient matrix contains a parameter α. In each of these problems: **ODEA**
(a) Determine the eigenvalues in terms of α.
(b) Find the critical value or values of α where the qualitative nature of the phase portrait for the system changes.
(c) Draw a phase portrait for a value of α slightly below, and for another value slightly above, each critical value.

13. $\mathbf{x}' = \begin{pmatrix} \alpha & 1 \\ -1 & \alpha \end{pmatrix} \mathbf{x}$
14. $\mathbf{x}' = \begin{pmatrix} 0 & -5 \\ 1 & \alpha \end{pmatrix} \mathbf{x}$

15. $\mathbf{x}' = \begin{pmatrix} 2 & -5 \\ \alpha & -2 \end{pmatrix} \mathbf{x}$
16. $\mathbf{x}' = \begin{pmatrix} \frac{5}{4} & \frac{3}{4} \\ \alpha & \frac{5}{4} \end{pmatrix} \mathbf{x}$

17. $\mathbf{x}' = \begin{pmatrix} -1 & \alpha \\ -1 & -1 \end{pmatrix} \mathbf{x}$ 18. $\mathbf{x}' = \begin{pmatrix} 3 & \alpha \\ -6 & -4 \end{pmatrix} \mathbf{x}$

19. $\mathbf{x}' = \begin{pmatrix} \alpha & 10 \\ -1 & -4 \end{pmatrix} \mathbf{x}$ 20. $\mathbf{x}' = \begin{pmatrix} 4 & \alpha \\ 8 & -6 \end{pmatrix} \mathbf{x}$

21. Consider the electric circuit shown in Figure 3.4.7. Suppose that $R_1 = R_2 = 4$ ohms, $C = 1/2$ farad, and $L = 8$ henrys.

(a) Show that this circuit is described by the system of differential equations

$$\frac{d}{dt}\begin{pmatrix} i \\ v \end{pmatrix} = \begin{pmatrix} -\frac{1}{2} & -\frac{1}{8} \\ 2 & -\frac{1}{2} \end{pmatrix}\begin{pmatrix} i \\ v \end{pmatrix}, \qquad (i)$$

where i is the current through the inductor and v is the voltage across the capacitor.

Hint: See Problem 21 of Section 3.2.

(b) Find the general solution of Eqs. (i) in terms of real-valued functions.

(c) Find $i(t)$ and $v(t)$ if $i(0) = 2$ amperes and $v(0) = 3$ volts.

(d) Determine the limiting values of $i(t)$ and $v(t)$ as $t \to \infty$. Do these limiting values depend on the initial conditions?

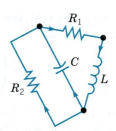

FIGURE 3.4.7 The circuit in Problem 21.

22. The electric circuit shown in Figure 3.4.8 is described by the system of differential equations

$$\frac{d}{dt}\begin{pmatrix} i \\ v \end{pmatrix} = \begin{pmatrix} 0 & \frac{1}{L} \\ -\frac{1}{C} & -\frac{1}{RC} \end{pmatrix}\begin{pmatrix} i \\ v \end{pmatrix}, \qquad (i)$$

where i is the current through the inductor and v is the voltage across the capacitor. These dif-

ferential equations were derived in Problem 21 of Section 3.2.

(a) Show that the eigenvalues of the coefficient matrix are real and different if $L > 4R^2C$. Show that they are complex conjugates if $L < 4R^2C$.

(b) Suppose that $R = 1$ ohm, $C = 1/2$ farad, and $L = 1$ henry. Find the general solution of the system (i) in this case.

(c) Find $i(t)$ and $v(t)$ if $i(0) = 2$ amperes and $v(0) = 1$ volt.

(d) For the circuit of part (b) determine the limiting values of $i(t)$ and $v(t)$ as $t \to \infty$. Do these limiting values depend on the initial conditions?

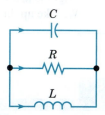

FIGURE 3.4.8 The circuit in Problem 22.

23. In this problem, we indicate how to show that the trajectories are ellipses when the eigenvalues are purely imaginary. Consider the system

$$\begin{pmatrix} x \\ y \end{pmatrix}' = \begin{pmatrix} a_{11} & a_{12} \\ a_{21} & a_{22} \end{pmatrix}\begin{pmatrix} x \\ y \end{pmatrix}. \qquad (i)$$

(a) Show that the eigenvalues of the coefficient matrix are purely imaginary if and only if

$$a_{11} + a_{22} = 0, \qquad a_{11}a_{22} - a_{12}a_{21} > 0. \qquad (ii)$$

(b) The trajectories of the system (i) can be found by converting Eqs. (i) into the single equation

$$\frac{dy}{dx} = \frac{dy/dt}{dx/dt} = \frac{a_{21}x + a_{22}y}{a_{11}x + a_{12}y}. \qquad (iii)$$

Use the first of Eqs. (ii) to show that Eq. (iii) is exact.

(c) By integrating Eq. (iii) show that

$$a_{21}x^2 + 2a_{22}xy - a_{12}y^2 = k, \qquad (iv)$$

where k is a constant. Use Eqs. (ii) to conclude that the graph of Eq. (iv) is always an ellipse.

Hint: What is the discriminant of the quadratic form in Eq. (iv)?

3.5 Repeated Eigenvalues

We continue our consideration of two-dimensional linear homogeneous systems with constant coefficients

$$\mathbf{x}' = \mathbf{A}\mathbf{x} \tag{1}$$

with a discussion of the case in which the matrix $\mathbf{A}$ has a repeated eigenvalue. This case occurs when the discriminant of the characteristic equation is zero, and it is a transition state between a node and a spiral point. There are two essentially different phenomena that can occur: the repeated eigenvalue may have two independent eigenvectors, or it may have only one. The first possibility is the simpler one to deal with, so we will start with it.

EXAMPLE 1

Solve the system $\mathbf{x}' = \mathbf{A}\mathbf{x}$, where

$$\mathbf{A} = \begin{pmatrix} -1 & 0 \\ 0 & -1 \end{pmatrix}. \tag{2}$$

Draw a direction field, a phase portrait, and typical component plots.

To solve the system we assume, as usual, that $\mathbf{x} = e^{\lambda t}\mathbf{v}$. Then, from Eq. (2), we obtain

$$\begin{pmatrix} -1-\lambda & 0 \\ 0 & -1-\lambda \end{pmatrix} \begin{pmatrix} v_1 \\ v_2 \end{pmatrix} = \begin{pmatrix} 0 \\ 0 \end{pmatrix}, \tag{3}$$

The characteristic equation is $(1+\lambda)^2 = 0$, so the eigenvalues are $\lambda_1 = \lambda_2 = -1$. To determine the eigenvectors, we set λ first equal to λ_1 and then to λ_2 in Eq. (3). In either case, we have

$$\begin{pmatrix} 0 & 0 \\ 0 & 0 \end{pmatrix} \begin{pmatrix} v_1 \\ v_2 \end{pmatrix} = \begin{pmatrix} 0 \\ 0 \end{pmatrix}. \tag{4}$$

Thus there are no restrictions on v_1 and v_2; in other words, we can choose them arbitrarily. It is convenient to choose $v_1 = 1$ and $v_2 = 0$ for λ_1 and to choose $v_1 = 0$ and $v_2 = 1$ for λ_2. Thus we obtain two solutions of the given system:

$$\mathbf{x}_1(t) = e^{-t} \begin{pmatrix} 1 \\ 0 \end{pmatrix}, \quad \mathbf{x}_2(t) = e^{-t} \begin{pmatrix} 0 \\ 1 \end{pmatrix}. \tag{5}$$

The general solution is

$$\mathbf{x} = c_1 \mathbf{x}_1(t) + c_2 \mathbf{x}_2(t). \tag{6}$$

Alternatively, you can also solve this system by starting from the scalar equations

$$x_1' = -x_1, \quad x_2' = -x_2, \tag{7}$$

which follow directly from Eq. (2). These equations can be integrated immediately, with the result that

$$x_1(t) = c_1 e^{-t}, \quad x_2(t) = c_2 e^{-t}. \tag{8}$$

Equations (8) are just the scalar form of Eq. (6).

A direction field and a phase portrait for the system (2) are shown in Figures 3.5.1 and 3.5.2, respectively. The trajectories lie on straight lines through the origin and (because

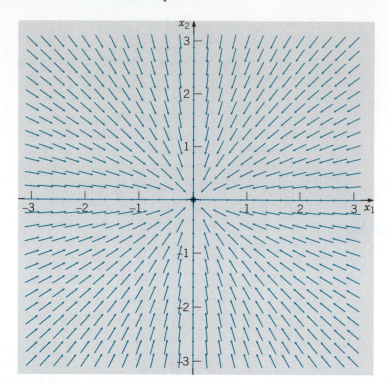

FIGURE 3.5.1 A direction field for the system (2).

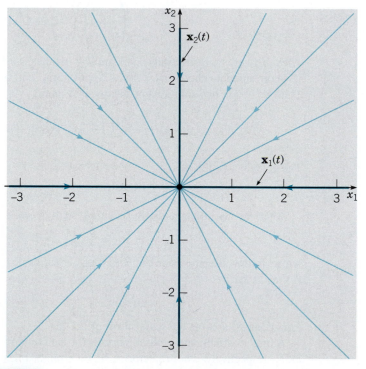

FIGURE 3.5.2 A phase portrait for the system (2).

the eigenvalues are negative) they approach the origin as $t \to \infty$. That this must be so is apparent from Eqs. (8). By eliminating t between these two equations, we find that

$$\frac{x_2(t)}{x_1(t)} = \frac{c_2}{c_1}. \tag{9}$$

Thus the ratio $x_2(t)/x_1(t)$ is a constant, so the trajectory lies on a line through the origin. The value of the constant in a particular case is determined by the initial conditions.

Typical component plots are shown in Figure 3.5.3. Each graph is proportional to the graph of e^{-t}, with the proportionality constant determined by the initial condition.

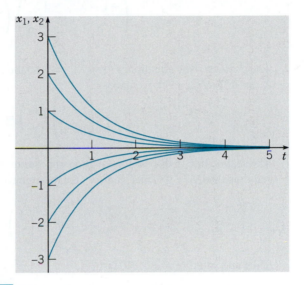

FIGURE 3.5.3   Typical component plots for the system (2).

It is possible to show that the only 2×2 matrices with a repeated eigenvalue and two independent eigenvectors are the diagonal matrices with the eigenvalues along the diagonal. Such matrices form a rather special class, since each of them is proportional to the identity matrix. The system in Example 1 is entirely typical of this class of systems. The origin is called a **proper node** or, sometimes, a **star point**. It is either asymptotically stable or unstable, according to whether the repeated eigenvalue is negative or positive. The trajectories lie along straight lines through the origin. They are traversed in the inward direction if the eigenvalues are negative and in the outward direction if they are positive. The component plots are simply the graphs of the appropriate exponential function.

We now take up the (more common) case in which repeated eigenvalues occur in a nondiagonal matrix. In this case, the repeated eigenvalue is accompanied by only a single independent eigenvector. This has implications for the solution of the corresponding system of differential equations because there is only one solution of our assumed form $\mathbf{x} = e^{\lambda t}\mathbf{v}$. To find a fundamental set of solutions, we must seek an additional solution of a different form. Before discussing the general problem of this type, we look at a relatively simple example.

EXAMPLE 2

Consider the system

$$\mathbf{x}' = \mathbf{Ax} = \begin{pmatrix} -\frac{1}{2} & 1 \\ 0 & -\frac{1}{2} \end{pmatrix} \mathbf{x}. \tag{10}$$

Find the eigenvalues and eigenvectors of the coefficient matrix, and then find the general solution of the system (10). Draw a direction field, phase portrait, and component plots.

Assuming that $\mathbf{x} = e^{\lambda t}\mathbf{v}$, we immediately obtain the algebraic system

$$\begin{pmatrix} -\frac{1}{2} - \lambda & 1 \\ 0 & -\frac{1}{2} - \lambda \end{pmatrix} \begin{pmatrix} v_1 \\ v_2 \end{pmatrix} = \begin{pmatrix} 0 \\ 0 \end{pmatrix}. \tag{11}$$

The characteristic equation is $\left(\frac{1}{2} + \lambda\right)^2 = 0$, so the eigenvalues are $\lambda_1 = \lambda_2 = -1/2$. Setting $\lambda = -1/2$ in Eq. (11), we have

$$\begin{pmatrix} 0 & 1 \\ 0 & 0 \end{pmatrix} \begin{pmatrix} v_1 \\ v_2 \end{pmatrix} = \begin{pmatrix} 0 \\ 0 \end{pmatrix}, \tag{12}$$

The second line in this vector equation imposes no restriction on v_1 and v_2, but the first line requires that $v_2 = 0$. Thus we may take the eigenvector $\mathbf{v}_1$ to be

$$\mathbf{v}_1 = \begin{pmatrix} 1 \\ 0 \end{pmatrix}. \tag{13}$$

Thus one solution of Eq. (10) is

$$\mathbf{x}_1(t) = e^{-t/2} \begin{pmatrix} 1 \\ 0 \end{pmatrix}, \tag{14}$$

but there is no second solution of the assumed form. Where should we look to find a second solution, and hence the general solution, of this system?

Let us consider the scalar equations corresponding to the vector equation (10). They are

$$x_1' = -\tfrac{1}{2}x_1 + x_2, \quad x_2' = -\tfrac{1}{2}x_2. \tag{15}$$

Because the second of these equations does not involve x_1, we can solve this equation for x_2. The equation is linear and separable and has the solution

$$x_2 = c_2 e^{-t/2}, \tag{16}$$

where c_2 is an arbitrary constant. Then, by substituting from Eq. (16) for x_2 in the first of Eqs. (15), we obtain an equation for x_1:

$$x_1' + \tfrac{1}{2}x_1 = c_2 e^{-t/2}. \tag{17}$$

Equation (17) is a first order linear equation with the integrating factor $\mu(t) = e^{t/2}$. On multiplying Eq. (17) by $\mu(t)$ and integrating, we find that

$$x_1 = c_2 t e^{-t/2} + c_1 e^{-t/2}, \tag{18}$$

where c_1 is another arbitrary constant. Now writing the scalar solutions (16) and (18) in vector form, we obtain

$$\mathbf{x} = \begin{pmatrix} x_1 \\ x_2 \end{pmatrix} = e^{-t/2} \begin{pmatrix} c_1 + c_2 t \\ c_2 \end{pmatrix}$$

$$= c_1 e^{-t/2} \begin{pmatrix} 1 \\ 0 \end{pmatrix} + c_2 e^{-t/2} \left[\begin{pmatrix} 1 \\ 0 \end{pmatrix} t + \begin{pmatrix} 0 \\ 1 \end{pmatrix} \right]. \tag{19}$$

The first term on the right side of Eq. (19) is the solution we found earlier in Eq. (14). However the second term on the right side of Eq. (19) is a new solution that we did not obtain before. This solution has the form

$$\mathbf{x} = t e^{-t/2} \mathbf{v} + e^{-t/2} \mathbf{w}. \tag{20}$$

Now that we know the form of the second solution, we can obtain it more directly by assuming an expression for $\mathbf{x}$ of the form (20), with unknown vector coefficients $\mathbf{v}$ and $\mathbf{w}$, and then substituting in Eq. (10). This leads to the equation

$$e^{-t/2} \mathbf{v} - \tfrac{1}{2} e^{-t/2} (\mathbf{v}t + \mathbf{w}) = e^{-t/2} \mathbf{A} (\mathbf{v}t + \mathbf{w}). \tag{21}$$

Equating coefficients of $te^{-t/2}$ and $e^{-t/2}$ on each side of Eq. (21) gives the conditions

$$\left(\mathbf{A} + \tfrac{1}{2} \mathbf{I} \right) \mathbf{v} = \mathbf{0} \tag{22}$$

and

$$\left(\mathbf{A} + \tfrac{1}{2} \mathbf{I} \right) \mathbf{w} = \mathbf{v} \tag{23}$$

for the determination of $\mathbf{v}$ and $\mathbf{w}$. Equation (22) is satisfied if $\mathbf{v}$ is the eigenvector $\mathbf{v}_1$, given by Eq. (13), associated with the eigenvalue $\lambda = -1/2$. Since $\det(\mathbf{A} + \tfrac{1}{2} \mathbf{I})$ is zero, we might expect that Eq. (23) has no solution. However, if we write out this equation in full, we have

$$\begin{pmatrix} 0 & 1 \\ 0 & 0 \end{pmatrix} \begin{pmatrix} w_1 \\ w_2 \end{pmatrix} = \begin{pmatrix} 1 \\ 0 \end{pmatrix}. \tag{24}$$

The first row of Eq. (24) requires that $w_2 = 1$, and the second row of Eq. (24) puts no restriction on either w_1 or w_2. Thus we can choose $w_1 = k$, where k is any constant. By substituting the results we have obtained in Eq. (20), we find a second solution of the system (10):

$$\mathbf{x}_2(t) = t e^{-t/2} \begin{pmatrix} 1 \\ 0 \end{pmatrix} + e^{-t/2} \begin{pmatrix} 0 \\ 1 \end{pmatrix} + k e^{-t/2} \begin{pmatrix} 1 \\ 0 \end{pmatrix}. \tag{25}$$

Since the third term on the right side of Eq. (25) is proportional to $\mathbf{x}_1(t)$, we need not include it in $\mathbf{x}_2(t)$; in other words, we choose $k = 0$. You can easily verify that the Wronskian of $\mathbf{x}_1(t)$ and $\mathbf{x}_2(t)$ is

$$W[\mathbf{x}_1, \mathbf{x}_2](t) = e^{-t}, \tag{26}$$

which is never zero. Thus the general solution of the system (10) is given by a linear combination of $\mathbf{x}_1(t)$ and $\mathbf{x}_2(t)$, that is, by Eq. (19).

A direction field and phase portrait for the system (10) are shown in Figures 3.5.4 and 3.5.5, respectively. In Figure 3.5.5, the trajectories of $\mathbf{x}_1(t)$ and $\mathbf{x}_2(t)$ are indicated by the heavy curves. As $t \to \infty$ every trajectory approaches the origin tangent to the x_1-axis. This is because the dominant term in the solution (19) for large t is the $te^{-t/2}$ term in $\mathbf{x}_2(t)$. If the

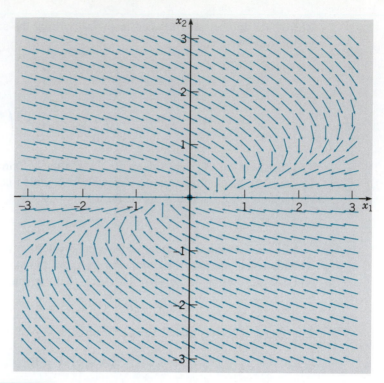

FIGURE 3.5.4 A direction field for the system (10).

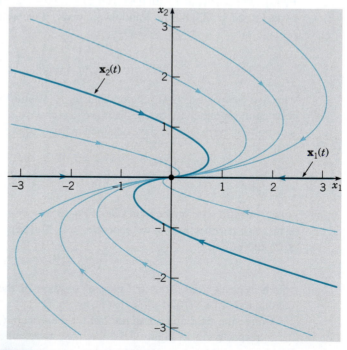

FIGURE 3.5.5 A phase portrait for the system (10).

initial conditions are such that $c_2 = 0$, then the solution is proportional to $\mathbf{x}_1(t)$, which lies along the x_1-axis. Figure 3.5.6 shows some representative plots of x_1 versus t. The component $x_2(t)$ is a purely exponential function (without any t factor) so its graphs resemble those in Figure 3.5.3.

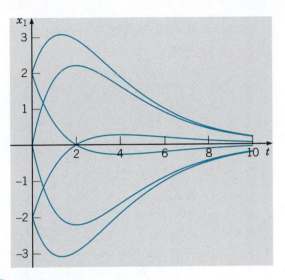

FIGURE 3.5.6 Typical plots of x_1 versus t for the system (10).

We can proceed in exactly the same way in the general case of Eq. (1). Suppose that $\lambda = \lambda_1$ is a repeated eigenvalue of the matrix $\mathbf{A}$ and that there is only one independent eigenvector $\mathbf{v}_1$. Then one solution (similar to Eq. (14)) is

$$\mathbf{x}_1(t) = e^{\lambda_1 t} \mathbf{v}_1, \tag{27}$$

where $\mathbf{v}_1$ satisfies

$$(\mathbf{A} - \lambda_1 \mathbf{I})\mathbf{v} = \mathbf{0}. \tag{28}$$

A second solution (similar to Eq. (25)) is

$$\mathbf{x}_2(t) = te^{\lambda_1 t} \mathbf{v}_1 + e^{\lambda_1 t} \mathbf{w}, \tag{29}$$

where $\mathbf{w}$ satisfies

$$(\mathbf{A} - \lambda_1 \mathbf{I})\mathbf{w} = \mathbf{v}_1. \tag{30}$$

Even though $\det(\mathbf{A} - \lambda_1 \mathbf{I}) = 0$, it can be shown that it is always possible to solve Eq. (30) for $\mathbf{w}$. The vector $\mathbf{w}$ is called a **generalized eigenvector** corresponding to the eigenvalue λ_1. It is possible to show that the Wronskian of $\mathbf{x}_1(t)$ and $\mathbf{x}_2(t)$ is not zero, so these solutions form a fundamental set. Thus the general solution of $\mathbf{x}' = \mathbf{A}\mathbf{x}$ in this case is a linear combination of $\mathbf{x}_1(t)$ and $\mathbf{x}_2(t)$ with arbitrary coefficients, as stated in Theorem 3.3.2.

In the case where the 2×2 matrix $\mathbf{A}$ has a repeated eigenvalue and only one eigenvector, the origin is called an **improper** or **degenerate node**. It is asymptotically stable when the eigenvalues are negative and all trajectories approach the origin as $t \to \infty$. If a positive eigenvalue is repeated, then the trajectories (except for $\mathbf{x} = \mathbf{0}$ itself) become unbounded and the improper node is unstable. In either case, the phase portrait resembles Figures 3.5.5 or 3.5.8. The trajectories are directed inward or outward according to whether the eigenvalues are negative or positive.

EXAMPLE 3

Find a fundamental set of solutions of

$$\mathbf{x}' = \mathbf{A}\mathbf{x} = \begin{pmatrix} 1 & -1 \\ 1 & 3 \end{pmatrix} \mathbf{x} \tag{31}$$

and draw a phase portrait for this system.

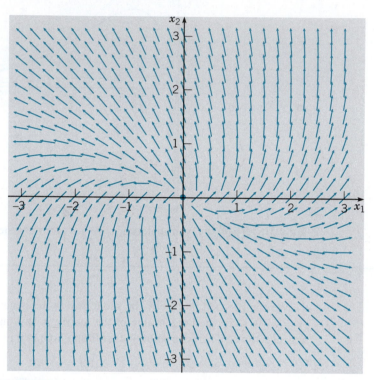

FIGURE 3.5.7 A direction field for the system (31).

A direction field for the system (31) is shown in Figure 3.5.7. From this figure it appears that all nonzero solutions depart from the origin. To solve Eq. (31), we start by finding the eigenvalues and eigenvectors of the coefficient matrix $\mathbf{A}$. They satisfy the algebraic equation $(\mathbf{A} - \lambda\,\mathbf{I})\,\mathbf{v} = \mathbf{0}$, or

$$\begin{pmatrix} 1 - \lambda & -1 \\ 1 & 3 - \lambda \end{pmatrix} \begin{pmatrix} v_1 \\ v_2 \end{pmatrix} = \begin{pmatrix} 0 \\ 0 \end{pmatrix}. \tag{32}$$

The eigenvalues are the roots of the characteristic equation

$$\det(\mathbf{A} - \lambda\mathbf{I}) = \begin{vmatrix} 1 - \lambda & -1 \\ 1 & 3 - \lambda \end{vmatrix} = \lambda^2 - 4\lambda + 4 = 0. \tag{33}$$

Thus the two eigenvalues are $\lambda_1 = \lambda_2 = 2$.

To determine the eigenvectors, we must return to Eq. (32) and use the value 2 for λ. This gives

$$\begin{pmatrix} -1 & -1 \\ 1 & 1 \end{pmatrix} \begin{pmatrix} v_1 \\ v_2 \end{pmatrix} = \begin{pmatrix} 0 \\ 0 \end{pmatrix}. \tag{34}$$

Hence we obtain the single condition $v_1 + v_2 = 0$, which determines v_2 in terms of v_1, or vice versa. Thus the eigenvector corresponding to the eigenvalue $\lambda = 2$ is

$$\mathbf{v}_1 = \begin{pmatrix} 1 \\ -1 \end{pmatrix}, \tag{35}$$

or any nonzero multiple of this vector. Consequently, one solution of the system (31) is

$$\mathbf{x}_1(t) = e^{2t} \begin{pmatrix} 1 \\ -1 \end{pmatrix}, \tag{36}$$

but there is no second solution of the form $\mathbf{x} = e^{\lambda t}\mathbf{v}$.

To find a second solution, we need to assume that

$$\mathbf{x} = te^{2t}\mathbf{v} + e^{2t}\mathbf{w}, \tag{37}$$

where $\mathbf{v}$ and $\mathbf{w}$ are constant vectors. Upon substituting this expression for $\mathbf{x}$ in Eq. (31), we obtain

$$2te^{2t}\mathbf{v} + e^{2t}(\mathbf{v} + 2\mathbf{w}) = \mathbf{A}(te^{2t}\mathbf{v} + e^{2t}\mathbf{w}). \tag{38}$$

Equating coefficients of te^{2t} and e^{2t} on each side of Eq. (38) gives the conditions

$$(\mathbf{A} - 2\mathbf{I})\mathbf{v} = \mathbf{0} \tag{39}$$

and

$$(\mathbf{A} - 2\mathbf{I})\mathbf{w} = \mathbf{v} \tag{40}$$

for the determination of $\mathbf{v}$ and $\mathbf{w}$. Equation (39) is satisfied if $\mathbf{v}$ is an eigenvector of $\mathbf{A}$ corresponding to the eigenvalue $\lambda = 2$, that is, $\mathbf{v}^T = (1, -1)$. Then Eq. (40) becomes

$$\begin{pmatrix} -1 & -1 \\ 1 & 1 \end{pmatrix} \begin{pmatrix} w_1 \\ w_2 \end{pmatrix} = \begin{pmatrix} 1 \\ -1 \end{pmatrix}. \tag{41}$$

Thus we have

$$-w_1 - w_2 = 1,$$

so if $w_1 = k$, where k is arbitrary, then $w_2 = -k - 1$. If we write

$$\mathbf{w} = \begin{pmatrix} k \\ -1 - k \end{pmatrix} = \begin{pmatrix} 0 \\ -1 \end{pmatrix} + k \begin{pmatrix} 1 \\ -1 \end{pmatrix}, \tag{42}$$

then by substituting for $\mathbf{v}$ and $\mathbf{w}$ in Eq. (37), we obtain

$$\mathbf{x} = te^{2t} \begin{pmatrix} 1 \\ -1 \end{pmatrix} + e^{2t} \begin{pmatrix} 0 \\ -1 \end{pmatrix} + ke^{2t} \begin{pmatrix} 1 \\ -1 \end{pmatrix}. \tag{43}$$

The last term in Eq. (43) is merely a multiple of the first solution $\mathbf{x}_1(t)$ and may be ignored, but the first two terms constitute a new solution:

$$\mathbf{x}_2(t) = te^{2t} \begin{pmatrix} 1 \\ -1 \end{pmatrix} + e^{2t} \begin{pmatrix} 0 \\ -1 \end{pmatrix}. \tag{44}$$

An elementary calculation shows that $W[\mathbf{x}_1, \mathbf{x}_2](t) = -e^{4t}$, and therefore $\mathbf{x}_1$ and $\mathbf{x}_2$ form a fundamental set of solutions of the system (31). The general solution is

$$\mathbf{x} = c_1\mathbf{x}_1(t) + c_2\mathbf{x}_2(t)$$

$$= c_1 e^{2t} \begin{pmatrix} 1 \\ -1 \end{pmatrix} + c_2 \left[te^{2t} \begin{pmatrix} 1 \\ -1 \end{pmatrix} + e^{2t} \begin{pmatrix} 0 \\ -1 \end{pmatrix} \right]. \tag{45}$$

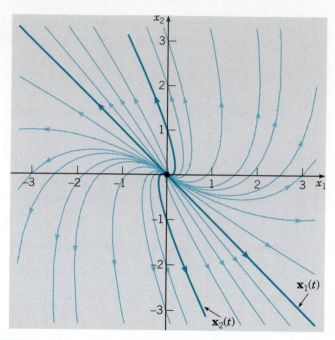

FIGURE 3.5.8   A phase portrait for the system (31).

A phase portrait for the system (31) is shown in Figure 3.5.8. The trajectories of $\mathbf{x}_1(t)$ and $\mathbf{x}_2(t)$ are shown by the heavy curves. It is clear that all solutions (except the equilibrium solution $\mathbf{x} = \mathbf{0}$) become unbounded at $t \to \infty$. Thus the improper node at the origin is unstable. It is possible to show that as $t \to -\infty$, all solutions approach the origin tangent to the line $x_2 = -x_1$ determined by the eigenvector. Similarly, as $t \to \infty$, each trajectory is asymptotic to a line of slope -1. Some typical plots of x_1 versus t are shown in Figure 3.5.9.

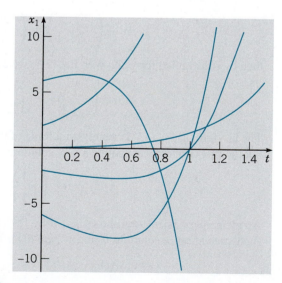

FIGURE 3.5.9   Some plots of x_1 versus t for the system (31).

▶ **Summary of Results.** This completes our investigation of the possible behavior of solutions of a two-dimensional linear homogeneous system with constant coefficients. When the coefficient matrix **A** has a nonzero determinant there is a single equilibrium solution, or critical point, which is located at the origin. By reflecting on the possibilities explored in this section and in the two preceding ones, and by examining the corresponding figures, we can make several observations:

1. After a long time, each individual trajectory exhibits one of only three types of behavior. As $t \to \infty$, each trajectory becomes unbounded, approaches the critical point $\mathbf{x} = \mathbf{0}$, or repeatedly traverses a closed curve, corresponding to a periodic solution, that surrounds the critical point.

2. Viewed as a whole, the pattern of trajectories in each case is relatively simple. To be more specific, through each point (x_0, y_0) in the phase plane there is only one trajectory; thus the trajectories do not cross each other. Do not be misled by the figures, in which it sometimes appears that many trajectories pass through the critical point $\mathbf{x} = \mathbf{0}$. In fact, the only solution passing through the origin is the equilibrium solution $\mathbf{x} = \mathbf{0}$. The other solutions that appear to pass through the origin actually only approach this point as $t \to \infty$ or $t \to -\infty$.

3. In each case, the set of all trajectories is such that one of three situations occurs.

    a. All trajectories approach the critical point $\mathbf{x} = \mathbf{0}$ as $t \to \infty$. This is the case if the eigenvalues are real and negative or complex with negative real part. The origin is either a nodal or a spiral sink.

    b. All trajectories remain bounded but do not approach the origin as $t \to \infty$. This is the case if the eigenvalues are purely imaginary. The origin is a center.

    c. Some trajectories, and possibly all trajectories except $\mathbf{x} = \mathbf{0}$, become unbounded as $t \to \infty$. This is the case if at least one of the eigenvalues is positive or if the eigenvalues have a positive real part. The origin is a nodal source, a spiral source, or a saddle point.

The situations described in 3(a), (b), and (c) above illustrate the concepts of asymptotic stability, stability, and instability, respectively, of the equilibrium solution $\mathbf{x} = \mathbf{0}$ of the system (1),

$$\mathbf{x}' = \mathbf{A}\mathbf{x}.$$

The precise definitions of these terms are given in Section 7.1, but their basic meaning should be clear from the geometrical discussion in this section. The information that we

TABLE 3.5.1

Stability Properties of Linear Systems $\mathbf{x}' = \mathbf{A}\mathbf{x}$ with $\det(\mathbf{A} - \lambda\mathbf{I}) = 0$ and $\det \mathbf{A} \neq 0$.

Eigenvalues	Type of Critical Point	Stability
$\lambda_1 > \lambda_2 > 0$	Node	Unstable
$\lambda_1 < \lambda_2 < 0$	Node	Asymptotically stable
$\lambda_2 < 0 < \lambda_1$	Saddle point	Unstable
$\lambda_1 = \lambda_2 > 0$	Proper or improper node	Unstable
$\lambda_1 = \lambda_2 < 0$	Proper or improper node	Asymptotically stable
$\lambda_1, \lambda_2 = \mu \pm i\nu$	Spiral point	
$\quad \mu > 0$		Unstable
$\quad \mu < 0$		Asymptotically stable
$\lambda_1 = i\nu, \lambda_2 = -i\nu$	Center	Stable

have obtained about the system (1) is summarized in Table 3.5.1. Also see Problems 15 and 16.

PROBLEMS

ODEA In each of Problems 1 through 6, find the general solution of the given system of equations. Also draw a direction field and a phase portrait. Describe how the solutions behave as $t \to \infty$.

1. $\mathbf{x}' = \begin{pmatrix} 3 & -4 \\ 1 & -1 \end{pmatrix} \mathbf{x}$ 2. $\mathbf{x}' = \begin{pmatrix} \frac{5}{4} & \frac{3}{4} \\ -\frac{3}{4} & -\frac{1}{4} \end{pmatrix} \mathbf{x}$

3. $\mathbf{x}' = \begin{pmatrix} -\frac{3}{2} & 1 \\ -\frac{1}{4} & -\frac{1}{2} \end{pmatrix} \mathbf{x}$ 4. $\mathbf{x}' = \begin{pmatrix} -3 & \frac{5}{2} \\ -\frac{5}{2} & 2 \end{pmatrix} \mathbf{x}$

5. $\mathbf{x}' = \begin{pmatrix} -1 & -\frac{1}{2} \\ 2 & -3 \end{pmatrix} \mathbf{x}$ 6. $\mathbf{x}' = \begin{pmatrix} 2 & \frac{1}{2} \\ -\frac{1}{2} & 1 \end{pmatrix} \mathbf{x}$

ODEA In each of Problems 7 through 12, find the solution of the given initial value problem. Draw the trajectory of the solution in the $x_1 x_2$-plane and also draw the component plots of x_1 versus t and of x_2 versus t.

7. $\mathbf{x}' = \begin{pmatrix} 1 & -4 \\ 4 & -7 \end{pmatrix} \mathbf{x}, \qquad \mathbf{x}(0) = \begin{pmatrix} 3 \\ 2 \end{pmatrix}$

8. $\mathbf{x}' = \begin{pmatrix} -\frac{5}{2} & \frac{3}{2} \\ -\frac{3}{2} & \frac{1}{2} \end{pmatrix} \mathbf{x}, \qquad \mathbf{x}(0) = \begin{pmatrix} 3 \\ -1 \end{pmatrix}$

9. $\mathbf{x}' = \begin{pmatrix} 2 & \frac{3}{2} \\ -\frac{3}{2} & -1 \end{pmatrix} \mathbf{x}, \qquad \mathbf{x}(0) = \begin{pmatrix} 3 \\ -2 \end{pmatrix}$

10. $\mathbf{x}' = \begin{pmatrix} \frac{5}{4} & \frac{3}{4} \\ -\frac{3}{4} & -\frac{1}{4} \end{pmatrix} \mathbf{x}, \qquad \mathbf{x}(0) = \begin{pmatrix} 2 \\ 3 \end{pmatrix}$;

See Problem 2.

11. $\mathbf{x}' = \begin{pmatrix} -3 & \frac{5}{2} \\ -\frac{5}{2} & 2 \end{pmatrix} \mathbf{x}, \qquad \mathbf{x}(0) = \begin{pmatrix} 3 \\ 1 \end{pmatrix}$;

See Problem 4.

12. $\mathbf{x}' = \begin{pmatrix} 2 & \frac{1}{2} \\ -\frac{1}{2} & 1 \end{pmatrix} \mathbf{x}, \qquad \mathbf{x}(0) = \begin{pmatrix} 1 \\ 3 \end{pmatrix}$;

See Problem 6.

13. Show that all solutions of the system

$$\mathbf{x}' = \begin{pmatrix} a & b \\ c & d \end{pmatrix} \mathbf{x}$$

approach zero as $t \to \infty$ if and only if $a + d < 0$ and $ad - bc > 0$.

14. Consider again the electric circuit in Problem 22 of Section 3.4. This circuit is described by the system of differential equations

$$\frac{d}{dt}\begin{pmatrix} i \\ v \end{pmatrix} = \begin{pmatrix} 0 & \frac{1}{L} \\ -\frac{1}{C} & -\frac{1}{RC} \end{pmatrix} \begin{pmatrix} i \\ v \end{pmatrix}.$$

(a) Show that the eigenvalues are real and equal if $L = 4R^2 C$.
(b) Suppose that $R = 1$ ohm, $C = 1$ farad, and $L = 4$ henrys. Suppose also that $i(0) = 1$ ampere and $v(0) = 2$ volts. Find $i(t)$ and $v(t)$.

15. Consider the linear system

$$dx/dt = a_{11}x + a_{12}y, \qquad dy/dt = a_{21}x + a_{22}y,$$

where $a_{11}, \ldots, a_{22}$ are real constants. Let $p = a_{11} + a_{22}$, $q = a_{11}a_{22} - a_{12}a_{21}$, and $\Delta = p^2 - 4q$. Observe that p and q are the trace and determinant, respectively, of the coefficient matrix of the given system. Show that the critical point $(0, 0)$ is a

(a) Node if $q > 0$ and $\Delta \geq 0$;
(b) Saddle point if $q < 0$;
(c) Spiral point if $p \neq 0$ and $\Delta < 0$;
(d) Center if $p = 0$ and $q > 0$.

Hint: These conclusions can be obtained by studying the eigenvalues λ_1 and λ_2. It may also be helpful to establish, and then to use, the relations $\lambda_1 \lambda_2 = q$ and $\lambda_1 + \lambda_2 = p$.

16. Continuing Problem 15, show that the critical point $(0, 0)$ is

(a) Asymptotically stable if $q > 0$ and $p < 0$;
(b) Stable if $q > 0$ and $p = 0$;
(c) Unstable if $q < 0$ or $p > 0$.

The results of Problems 15 and 16 are summarized visually in Figure 3.5.10.

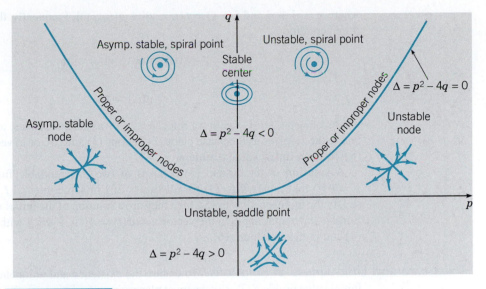

FIGURE 3.5.10 Stability diagram.

3.6 A Brief Introduction to Nonlinear Systems

In Section 3.2, we introduced the general two-dimensional first order linear system

$$\begin{pmatrix} dx/dt \\ dy/dt \end{pmatrix} = \begin{pmatrix} p_{11}(t)x + p_{12}(t)y + g_1(t) \\ p_{21}(t)x + p_{22}(t)y + g_2(t) \end{pmatrix} \tag{1}$$

or

$$\frac{d\mathbf{x}}{dt} = \mathbf{P}(t)\mathbf{x} + \mathbf{g}(t), \tag{2}$$

where

$$\mathbf{x} = \begin{pmatrix} x \\ y \end{pmatrix}, \qquad \mathbf{P}(t) = \begin{pmatrix} p_{11}(t) & p_{12}(t) \\ p_{21}(t) & p_{22}(t) \end{pmatrix}, \quad \text{and} \quad \mathbf{g}(t) = \begin{pmatrix} g_1(t) \\ g_2(t) \end{pmatrix}.$$

Of course, two-dimensional systems that are not of the form (1) or (2) may also occur. Such systems are said to be **nonlinear**.

A general two-dimensional first order system has the form

$$\begin{pmatrix} dx/dt \\ dy/dt \end{pmatrix} = \begin{pmatrix} f(t, x, y) \\ g(t, x, y) \end{pmatrix}, \tag{3}$$

where f and g are essentially arbitrary continuous functions. Using vector notation, we have

$$\frac{d\mathbf{x}}{dt} = \mathbf{f}(t, \mathbf{x}), \tag{4}$$

where $\mathbf{x} = x\mathbf{i} + y\mathbf{j}$ and $\mathbf{f}(t, \mathbf{x}) = f(t, x, y)\mathbf{i} + g(t, x, y)\mathbf{j}$. Frequently, there will also be given initial conditions

$$x(t_0) = x_0, \quad y(t_0) = y_0, \tag{5}$$

or

$$\mathbf{x}(t_0) = \mathbf{x}_0, \tag{6}$$

where $\mathbf{x}_0 = x_0\mathbf{i} + y_0\mathbf{j}$. Again we refer to x and y as state variables, to $\mathbf{x}$ as the state vector, and to the xy-plane as the phase plane. Equations (3) and (5), or in vector form Eqs. (4) and (6), form an **initial value problem**.

A **solution** of the system (3) consists of two differentiable functions $x = \phi(t)$ and $y = \psi(t)$ that satisfy Eqs. (3) for all values of t in some interval I. If $\phi(t)$ and $\psi(t)$ also satisfy the initial conditions (5), then they are a solution of the initial value problem (3), (5). Similar statements apply to the vector $\mathbf{x} = \boldsymbol{\phi}(t) = \phi(t)\mathbf{i} + \psi(t)\mathbf{j}$ with respect to the system (4) and initial conditions (6).

▶ **Existence and Uniqueness of Solutions.** To ensure the existence and uniqueness of solutions of the initial value problem (3), (5), we must place some restrictions on the functions f and g. The following theorem is analogous to Theorem 2.4.2 for first order nonlinear scalar equations.

THEOREM 3.6.1	Let each of the functions f and g and the partial derivatives $\partial f/\partial x$, $\partial f/\partial y$, $\partial g/\partial x$, and $\partial g/\partial y$ be continuous in a region R of txy-space defined by $\alpha < t < \beta$, $\alpha_1 < x < \beta_1$, $\alpha_2 < y < \beta_2$, and let the point (t_0, x_0, y_0) be in R. Then there is an interval $|t - t_0| < h$ in which there exists a unique solution of the system of differential equations (3)

$$\begin{pmatrix} dx/dt \\ dy/dt \end{pmatrix} = \begin{pmatrix} f(t, x, y) \\ g(t, x, y) \end{pmatrix}$$

that also satisfies the initial conditions (5)

$$x(t_0) = x_0, \quad y(t_0) = y_0.$$

Note that in the hypotheses of Theorem 3.6.1, nothing is said about the partial derivatives of f and g with respect to the independent variable t. Note also that, in the conclusion, the length $2h$ of the interval in which the solution exists is not specified exactly, and in some cases it may be very short. As for first order nonlinear scalar equations, the interval of existence of solutions here may bear no obvious relation to the functions f and g, and often depends also on the initial conditions.

▶ **Autonomous Systems.** It is usually impossible to solve nonlinear systems exactly by analytical methods. Therefore for such systems graphical methods and numerical approximations become even more important. In the next section, we will extend our discussion of approximate numerical methods to two-dimensional systems. Here we will consider systems for which direction fields and phase portraits are of particular importance. These are systems that do not depend explicitly on the independent variable t. In other words, the functions f and g in Eqs. (3) depend only on x and y and not on t. Or, in Eq. (4), the vector $\mathbf{f}$ depends only on $\mathbf{x}$ and not t. Such a system is called **autonomous**, and can be written in the form

$$\frac{dx}{dt} = f(x, y), \quad \frac{dy}{dt} = g(x, y). \tag{7}$$

In vector notation we have

$$\frac{d\mathbf{x}}{dt} = \mathbf{f}(\mathbf{x}). \tag{8}$$

In earlier sections of this chapter, we have examined linear autonomous systems. We now want to extend the discussion to nonlinear autonomous systems.

We have seen that equilibrium, or constant, solutions are of particular importance in the study of single first order autonomous equations and of two-dimensional linear autonomous systems. We will see that the same is true for nonlinear autonomous systems of equations. To find equilibrium, or constant, solutions of the system (7), we set dx/dt and dy/dt equal to zero, and solve the resulting equations

$$f(x, y) = 0, \quad g(x, y) = 0 \tag{9}$$

for x and y. Any solution of Eqs. (9) is a point in the phase plane that is a trajectory of an equilibrium solution. Such points are called **equilibrium points** or **critical points**. Depending on the particular forms of f and g, the nonlinear system (7) can have any number of critical points, ranging from none to infinitely many.

Sometimes the trajectories of a two-dimensional autonomous system can be found by solving a related first order differential equation. From Eqs. (7) we have

$$\frac{dy}{dx} = \frac{dy/dt}{dx/dt} = \frac{g(x, y)}{f(x, y)}, \tag{10}$$

which is a first order equation in the variables x and y. If Eq. (10) can be solved by any of the methods in Chapter 2, and if we write solutions (implicitly) as

$$H(x, y) = c, \tag{11}$$

then Eq. (11) is an equation for the trajectories of the system (7). In other words, the trajectories lie on the level curves of $H(x, y)$. Recall that there is no general way of solving Eq. (10) to obtain the function H, so this approach is applicable only in special cases.

Now let us look at some examples.

EXAMPLE 1

Consider the system

$$\frac{dx}{dt} = x - y, \quad \frac{dy}{dt} = 2x - y - x^2. \tag{12}$$

Find a function $H(x, y)$ such that the trajectories of the system (12) lie on the level curves of H. Find the critical points and draw a phase portrait for the given system. Describe the behavior of its trajectories.

To find the critical points we must solve the equations

$$x - y = 0, \quad 2x - y - x^2 = 0. \tag{13}$$

From the first equation we have $y = x$; then the second equation yields $x - x^2 = 0$. Thus $x = 0$ or $x = 1$, and it follows that the critical points are $(0, 0)$ and $(1, 1)$. To determine the trajectories, note that for this system, Eq. (10) becomes

$$\frac{dy}{dx} = \frac{2x - y - x^2}{x - y}. \tag{14}$$

Equation (14) is an exact equation, as discussed in Section 2.6. Proceeding as described in that section, we find that solutions satisfy

$$H(x, y) = x^2 - xy + \tfrac{1}{2}y^2 - \tfrac{1}{3}x^3 = c, \tag{15}$$

where c is an arbitrary constant. To construct a phase portrait, you can either draw some of the level curves of $H(x, y)$, or you can plot some solutions of the system (12). In either case, you need some computer assistance to produce a plot such as that in Figure 3.6.1. The direction of motion on the trajectories can be determined by drawing a direction field, or by evaluating dx/dt and dy/dt at one or two selected points. From Figure 3.6.1, it is clear that trajectories behave quite differently near the two critical points. Observe that there is one trajectory that departs from $(1, 1)$ as $t \to -\infty$, loops around the other critical point (the origin), and returns to $(1, 1)$ as $t \to \infty$. Inside this loop there are trajectories that lie on closed curves surrounding $(0, 0)$. These trajectories correspond to periodic solutions that repeatedly pass through the same points in the phase plane. Trajectories that lie outside the loop ultimately appear to leave the plot window in a southeasterly direction (as $t \to \infty$) or in a northeasterly direction (as $t \to -\infty$).

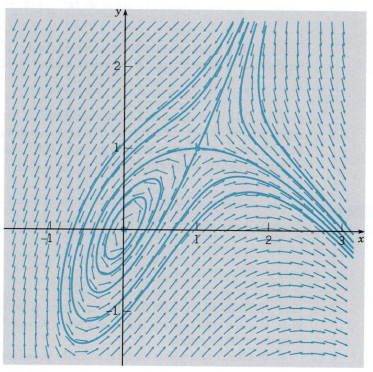

FIGURE 3.6.1 A phase portrait for the system (12).

It is also possible to construct component plots of particular solutions, and two are shown in Figures 3.6.2 and 3.6.3. In Figure 3.6.2, there is a plot of x versus t for the solution that satisfies the initial conditions

$$x(0) = \tfrac{1}{2}, \quad y(0) = \tfrac{1}{2}.$$

This graph confirms that the motion is periodic and enables you to estimate the period, which was not possible from the phase portrait. A plot of y versus t for this solution is similar.

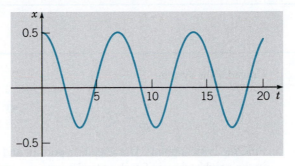

FIGURE 3.6.2 A plot of x versus t for the solution of the system (12) that passes through the point $(\frac{1}{2}, \frac{1}{2})$ at $t = 0$.

A plot of a typical unbounded solution appears in Figure 3.6.3, which shows the graph of x versus t for the solution that satisfies the initial conditions

$$x(0) = 3, \quad y(0) = 0.$$

It appears from this figure that there may be a vertical asymptote for a value of t between 1.5 and 1.6, although you should be cautious about drawing such a conclusion from a single small plot. In Problem 22, we outline how you can show conclusively that there is an asymptote in this particular case. The existence of a vertical asymptote means that x becomes unbounded in a finite time, rather than as $t \to \infty$. A plot of y versus t for this solution is similar, except that y becomes unbounded in the negative direction. The relative magnitudes of x and y when they are both large can be determined by keeping only the most

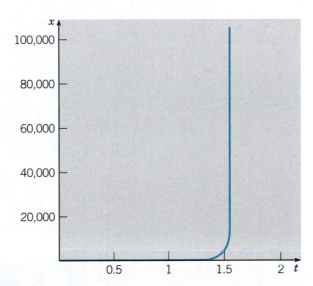

FIGURE 3.6.3 A plot of x versus t for the solution of the system (12) that passes through the point $(3, 0)$ at $t = 0$.

dominant terms in Eq. (15). In this way, we find that

$$y^2 \cong \tfrac{2}{3}x^3$$

when both $|x|$ and $|y|$ are very large.

**EXAMPLE
2**

Consider the system

$$\frac{dx}{dt} = (2+x)(y-x), \quad \frac{dy}{dt} = (2-x)(y+x). \tag{16}$$

Find the critical points. Draw a phase portrait and describe the behavior of the trajectories in the neighborhood of each critical point.

To find the critical points we must solve the equations

$$(2+x)(y-x) = 0, \quad (2-x)(y+x) = 0. \tag{17}$$

One way to satisfy the first equation is by choosing $x = -2$; then to satisfy the second equation we must choose $y = 2$. Similarly, the second equation can be satisfied by choosing $x = 2$; then the first equation requires that $y = 2$. If $x \neq 2$ and $x \neq -2$, then Eqs. (17) can only be satisfied if $y - x = 0$ and $y + x = 0$. The only solution of this pair of equations is the origin. Thus the system (16) has three critical points: $(-2, 2)$, $(2, 2)$, and $(0, 0)$.

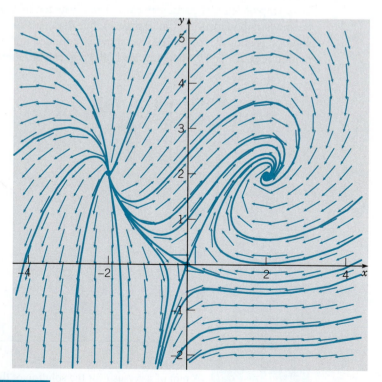

FIGURE 3.6.4 A phase portrait for the system (16).

A phase portrait for Eqs. (16) is shown in Figure 3.6.4. The critical point at (2, 2) attracts other trajectories in the upper right-hand part of the phase plane. These trajectories spiral around the critical point as they approach it. All trajectories near the point (−2, 2), except the critical point itself, depart from this neighborhood. Some approach the point (2, 2), as we have seen, while others appear to go infinitely far away. Near the origin, there are two trajectories that approach the origin in the second and fourth quadrants. Trajectories that lie above these two trajectories approach (2, 2), while those below them seem to go infinitely far away in the third quadrant.

If you look more closely at Figures 3.6.1 and 3.6.4, you will notice that, in the neighborhood of each critical point, the pattern of trajectories resembles one of the patterns found in Sections 3.3 through 3.5 for linear systems with constant coefficients. In particular, in Figure 3.6.1 it appears that the point (1, 1) is a saddle point and the point (0, 0) is a center. Similarly, in Figure 3.6.4 it appears that (2, 2) is a spiral point, (−2, 2) is a node, and (0, 0) is a saddle point. This apparent relation between nonlinear autonomous systems and linear systems with constant coefficients is not accidental and we will explore it more fully in Chapter 7. In the meantime, many of the problems following this section give you a chance to draw phase portraits similar to Figures 3.6.1 and 3.6.4 and to draw conclusions about the critical points from them.

PROBLEMS

For each of the systems in Problems 1 through 6:
(a) Find an equation of the form $H(x, y) = c$ satisfied by the solutions of the given system.
(b) Without using a computer, sketch some level curves of the function $H(x, y)$.
(c) For $t > 0$, sketch the trajectory corresponding to the given initial condition and indicate the direction of motion for increasing t.

1. $dx/dt = -x$, $\quad dy/dt = -2y$; $\quad x(0) = 4$, $y(0) = 2$

2. $dx/dt = -x$, $\quad dy/dt = 2y$; $\quad x(0) = 4$, $y(0) = 2$

3. $dx/dt = -x$, $\quad dy/dt = 2y$; $\quad x(0) = 4$, $y(0) = 0$

4. $dx/dt = 2y$, $\quad dy/dt = 8x$; $\quad x(0) = 2$, $y(0) = -1$

5. $dx/dt = 2y$, $\quad dy/dt = 8x$; $\quad x(0) = 1$, $y(0) = -3$

6. $dx/dt = 2y$, $\quad dy/dt = -8x$; $\quad x(0) = 1$, $y(0) = 2$

CAS For each of the systems in Problems 7 through 12:
(a) Find all of the critical points.

(b) Find an equation of the form $H(x, y) = c$ satisfied by solutions of the given system.
(c) Using a computer, plot several level curves of the function H. These are trajectories of the given system. Indicate the direction of motion on each trajectory.
(d) Describe the behavior of the trajectories near each critical point.

7. $dx/dt = 2x - y$, $\quad dy/dt = x - 2y$

8. $dx/dt = -x + y$, $\quad dy/dt = x + y$

9. $dx/dt = 2x - 4y$, $\quad dy/dt = 2x - 2y$

10. $dx/dt = -x + y + x^2$, $\quad dy/dt = y - 2xy$

11. $dx/dt = 2x^2y - 3x^2 - 4y$, $dy/dt = -2xy^2 + 6xy$

12. $dx/dt = 3x - x^2$, $\quad dy/dt = 2xy - 3y + 2$

For each of the systems in Problems 13 through 20:   ODEA
(a) Find all the critical points.
(b) Use a computer to draw a direction field and phase portrait for the system.
(c) From the plots in part (b), describe how the trajectories behave in the vicinity of each critical point.

13. $dx/dt = x - xy$, $\quad dy/dt = y + 2xy$

14. $dx/dt = 2 - y$, $\quad dy/dt = y - x^2$

15. $dx/dt = x - x^2 - xy,$

$dy/dt = \frac{1}{2}y - \frac{1}{4}y^2 - \frac{3}{4}xy$

16. $dx/dt = -(x - y)(1 - x - y),$

$dy/dt = x(2 + y)$

17. $dx/dt = y(2 - x - y),$

$dy/dt = -x - y - 2xy$

18. $dx/dt = (2 + x)(y - x),$

$dy/dt = y(2 + x - x^2)$

19. $dx/dt = -x + 2xy,$ $dy/dt = y - x^2 - y^2$

20. $dx/dt = y,$ $dy/dt = x - \frac{1}{6}x^3 - \frac{1}{5}y$

21. **(a)** Consider the system in Example 1. Draw a component plot of x versus t for several of the periodic solutions in the vicinity of the origin.

(b) From the plots in part **(a)** estimate the period and amplitude of each of the solutions. Is the period the same regardless of the amplitude? If not, how is the period related to the amplitude?

22. In this problem, we indicate how to find the asymptote suggested by Figure 3.6.3.

(a) Show that if x and y satisfy the initial conditions $x(0) = 3$ and $y(0) = 0$, then the constant

c in Eq. (15) is zero. Then show that Eq. (15) can be rewritten in the form

$$(x - y)^2 + x^2 - \frac{2}{3}x^3 = 0. \tag{i}$$

(b) From the first of Eqs. (12) recall that $x - y = x'$. Use this fact with Eq. (i) to obtain the differential equation

$$x'^2 = \frac{2}{3}x^3 - x^2, \tag{ii}$$

or

$$x' = \frac{1}{3}x\sqrt{6x - 9}. \tag{iii}$$

Why must the positive square root be chosen?

(c) Show that the solution of Eq. (iii) that satisfies the initial condition $x(0) = 3$ is

$$x = \frac{3}{2} + \frac{3}{2}\tan^2\left(\frac{1}{2}t + \frac{1}{4}\pi\right). \tag{iv}$$

Hint: In solving Eq. (iii), you may find it helpful to use the substitution $s^2 = 6x - 9$.

(d) Use the result of part **(c)** to show that the solution has a vertical asymptote at $t = \pi/2$. Compare this result with the graph in Figure 3.6.3.

(e) From Eq. (iv) there is another vertical asymptote at $t = -3\pi/2$. What is the significance of this asymptote?

3.7 Numerical Methods for Systems of First Order Equations

In Sections 1.3, 2.7, and 2.8, we have discussed numerical methods for approximating the solutions of initial value problems for a single first order differential equation. These methods can also be applied to a system of first order equations. The algorithms are the same for nonlinear and for linear equations, so we will not restrict ourselves to linear equations in this section. We consider a system of two first order equations

$$x' = f(t, x, y), \qquad y' = g(t, x, y), \tag{1}$$

with the initial conditions

$$x(t_0) = x_0, \qquad y(t_0) = y_0. \tag{2}$$

The functions f and g are assumed to satisfy the conditions of Theorem 3.6.1 so that the initial value problem (1), (2) has a unique solution in some interval of the t-axis containing the point t_0. We wish to determine approximate values $x_1, x_2, \ldots, x_n, \ldots$ and $y_1, y_2, \ldots, y_n, \ldots$ of the solution $x = \phi(t), y = \psi(t)$ at the points $t_n = t_0 + nh$ with $n = 1, 2, \ldots$.

In vector notation, the initial value problem (1), (2) can be written as

$$\mathbf{x}' = \mathbf{f}(t, \mathbf{x}), \qquad \mathbf{x}(t_0) = \mathbf{x}_0, \tag{3}$$

where $\mathbf{x}$ is the vector with components x and y, $\mathbf{f}$ is the vector function with components f and g, and $\mathbf{x}_0$ is the vector with components x_0 and y_0. The Euler, improved Euler, and Runge-Kutta methods can be readily generalized to handle systems of two (or more) equations. All that is needed (formally) is to replace the scalar variable x by the vector $\mathbf{x}$ and the scalar function f by the vector function $\mathbf{f}$ in the appropriate equations. For example, the scalar Euler formula

$$t_{n+1} = t_n + h, \quad x_{n+1} = x_n + hf_n \tag{4}$$

is replaced by

$$t_{n+1} = t_n + h, \quad \mathbf{x}_{n+1} = \mathbf{x}_n + h\mathbf{f}_n, \tag{5}$$

where $\mathbf{f}_n = \mathbf{f}(t_n, \mathbf{x}_n)$. In component form we have

$$\begin{pmatrix} x_{n+1} \\ y_{n+1} \end{pmatrix} = \begin{pmatrix} x_n \\ y_n \end{pmatrix} + h \begin{pmatrix} f(t_n, x_n, y_n) \\ g(t_n, x_n, y_n) \end{pmatrix}. \tag{6}$$

The initial conditions are used to determine $\mathbf{f}_0$, which is the vector tangent to the trajectory of the solution $\mathbf{x} = \boldsymbol{\phi}(t)$ at the initial point in the xy-plane. We move in the direction of this tangent vector for a time step h in order to find the next point $\mathbf{x}_1$. Then we calculate a new tangent vector $\mathbf{f}_1$, move along it for a time step h to find $\mathbf{x}_2$, and so forth.

In a similar way, the Runge–Kutta method (Section 2.8) can be extended to a system. For the step from t_n to t_{n+1} we have

$$t_{n+1} = t_n + h, \quad \mathbf{x}_{n+1} = \mathbf{x}_n + (h/6)(\mathbf{k}_{n1} + 2\mathbf{k}_{n2} + 2\mathbf{k}_{n3} + \mathbf{k}_{n4}), \tag{7}$$

where

$$\begin{aligned} \mathbf{k}_{n1} &= \mathbf{f}(t_n, \mathbf{x}_n), \\ \mathbf{k}_{n2} &= \mathbf{f}[t_n + (h/2), \mathbf{x}_n + (h/2)\mathbf{k}_{n1}], \\ \mathbf{k}_{n3} &= \mathbf{f}[t_n + (h/2), \mathbf{x}_n + (h/2)\mathbf{k}_{n2}], \\ \mathbf{k}_{n4} &= \mathbf{f}(t_n + h, \mathbf{x}_n + h\mathbf{k}_{n3}). \end{aligned} \tag{8}$$

The vector equations (3), (5), (7), and (8) are, in fact, valid in any number of dimensions. All that is needed is to interpret the vectors as having n components rather than two.

EXAMPLE 1

Determine approximate values of the solution $x = \phi(t)$, $y = \psi(t)$ of the initial value problem

$$x' = -x + 4y, \quad y' = x - y, \tag{9}$$

$$x(0) = 2, \quad y(0) = -0.5, \tag{10}$$

at the point $t = 0.2$. Use the Euler method with $h = 0.1$ and the Runge–Kutta method with $h = 0.2$. Compare the results with the values of the exact solution:

$$\phi(t) = \frac{e^t + 3e^{-3t}}{2}, \quad \psi(t) = \frac{e^t - 3e^{-3t}}{4}. \tag{11}$$

Note that the differential equations (9) form a linear homogeneous system with constant coefficients. Consequently, the solution is easily found by using the methods described in this chapter. In particular, the eigenvalues of the coefficient matrix are 1 and -3, so the origin is a saddle point for this system.

To approximate the solution numerically, let us first use the Euler method. For this problem $f_n = -x_n + 4y_n$ and $g_n = x_n - y_n$; hence

$$f_0 = -2 + (4)(-0.5) = -4, \qquad g_0 = 2 - (-0.5) = 2.5.$$

Then, from the Euler formulas (5) and (6), we obtain

$$x_1 = 2 + (0.1)(-4) = 1.6, \qquad y_1 = -0.5 + (0.1)(2.5) = -0.25.$$

At the next step

$$f_1 = -1.6 + (4)(-0.25) = -2.6, \qquad g_1 = 1.6 - (-0.25) = 1.85.$$

Consequently,

$$x_2 = 1.6 + (0.1)(-2.6) = 1.34, \qquad y_2 = -0.25 + (0.1)(1.85) = -0.065.$$

The values of the exact solution, correct to six decimal places, are $\phi(0.2) = 1.433919$ and $\psi(0.2) = -0.106258$. Thus the values calculated from the Euler method are in error by about 0.0939 and 0.0413, respectively, corresponding to percentage errors of about 6.5% and 38.8%.

Now let us use the Runge–Kutta method to approximate $\phi(0.2)$ and $\psi(0.2)$. With $h = 0.2$ we obtain the following values from Eqs. (8):

$$\mathbf{k}_{01} = \begin{pmatrix} f(2, -0.5) \\ g(2, -0.5) \end{pmatrix} = \begin{pmatrix} -4 \\ 2.5 \end{pmatrix};$$

$$\mathbf{k}_{02} = \begin{pmatrix} f(1.6, -0.25) \\ g(1.6, -0.25) \end{pmatrix} = \begin{pmatrix} -2.6 \\ 1.85 \end{pmatrix};$$

$$\mathbf{k}_{03} = \begin{pmatrix} f(1.74, -0.315) \\ g(1.74, -0.315) \end{pmatrix} = \begin{pmatrix} -3.00 \\ 2.055 \end{pmatrix};$$

$$\mathbf{k}_{04} = \begin{pmatrix} f(1.4, -0.089) \\ g(1.4, -0.089) \end{pmatrix} = \begin{pmatrix} -1.756 \\ 1.489 \end{pmatrix}.$$

Then, substituting these values in Eq. (7), we obtain

$$\mathbf{x}_1 = \begin{pmatrix} 2 \\ -0.5 \end{pmatrix} + \frac{0.2}{6} \begin{pmatrix} -16.956 \\ 11.799 \end{pmatrix} = \begin{pmatrix} 1.4348 \\ -0.1067 \end{pmatrix}.$$

These values of x_1 and y_1 are in error by about 0.000881 and 0.000442, respectively, corresponding to percentage errors of about 0.0614% and 0.416%.

This example again illustrates the great gains in accuracy that are obtainable by using a more accurate approximation method, such as the Runge–Kutta method. In the calculations we have just outlined, the Runge–Kutta method requires only twice as many function evaluations as the Euler method, but the error in the Runge–Kutta method is about 100 times less than in the Euler method.

Of course, these computations can be continued and Figure 3.6.1 shows the results of extending them as far as $t = 1$. The Euler method with $h = 0.1$ gives values that are qualitatively fairly accurate, but are quantitatively in error by several per cent. The Runge–Kutta approximations with $h = 0.2$ differ from the true values of the solution only in the third or fourth decimal place, and are indistinguishable from them in the plot.

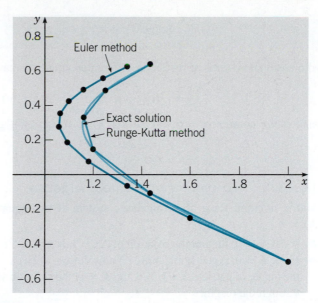

FIGURE 3.7.1 Approximations to the solution of the initial value problem (9), (10) using the Euler method ($h = 0.1$) and the Runge-Kutta method ($h = 0.2$).

PROBLEMS

CAS In each of Problems 1 through 6, determine approximate values of the solution $x = \phi(t)$, $y = \psi(t)$ of the given initial value problem at $t = 0.2, 0.4, 0.6, 0.8$, and 1.0. Compare the results obtained by different methods and different step sizes.

(a) Use the Euler method with $h = 0.1$.

(b) Use the Runge–Kutta method with $h = 0.2$.

(c) Use the Runge–Kutta method with $h = 0.1$.

1. $x' = x + y + t,\quad y' = 4x - 2y;$

   $x(0) = 1,\quad y(0) = 0$

2. $x' = 2x + ty,\quad y' = xy;$

   $x(0) = 1,\quad y(0) = 1$

3. $x' = -tx - y - 1,\quad y' = x;$

   $x(0) = 1,\quad y(0) = 1$

4. $x' = x - y + xy,\quad y' = 3x - 2y - xy;$

   $x(0) = 0,\quad y(0) = 1$

5. $x' = x(1 - 0.5x - 0.5y),$

   $y' = y(-0.25 + 0.5x);\quad x(0) = 4,\quad y(0) = 1$

6. $x' = \exp(-x + y) - \cos x,$

   $y' = \sin(x - 3y);\quad x(0) = 1,\quad y(0) = 2$

7. Consider the example problem $x' = -x + 4y$, CAS $y' = x - y$ with the initial conditions $x(0) = 2$ and $y(0) = -0.5$. Use the Runge–Kutta method to find approximate values of the solution of this problem on the interval $0 \le t \le 1$. Start with $h = 0.2$ and then repeat the calculation with step sizes $h = 0.1, 0.05, \ldots$, each half as long as in the preceding case. Continue the process until the first six digits of the solution at $t = 1$ are unchanged for successive step sizes. Determine whether these digits are accurate by comparing them with the exact solution given in Eqs. (11) in the text.

8. Consider the initial value problem CAS

   $$x'' + t^2 x' + 3x = t,\quad x(0) = 1,\quad x'(0) = 2.$$

   Convert this problem to a system of two first order equations, and determine approximate values of the solution at $t = 0.5$ and $t = 1.0$ using the Runge–Kutta method with $h = 0.1$.

CHAPTER SUMMARY

Two-dimensional systems of first order differential equations have the form

$$\frac{dx}{dt} = f(t, x, y), \qquad \frac{dy}{dt} = g(t, x, y)$$

or, using vector notation $\mathbf{x} = x\mathbf{i} + y\mathbf{j}$ and $\mathbf{f}(t, \mathbf{x}) = f(t, x, y)\mathbf{i} + g(t, x, y)\mathbf{j}$,

$$\frac{dx}{dt} = \mathbf{f}(t, \mathbf{x})$$

Section 3.1 Two-dimensional Linear Algebra

▶ Matrix notation for a linear algebraic system of two equations in two unknowns is $\mathbf{Ax} = \mathbf{b}$.

  ▶ If $\det \mathbf{A} \neq 0$, the unique solution of $\mathbf{Ax} = \mathbf{b}$ is $\mathbf{x} = \mathbf{A}^{-1}\mathbf{b}$.
  ▶ If $\det \mathbf{A} = 0$, $\mathbf{Ax} = \mathbf{b}$ may have (i) no solution, or (ii) a straight line of solutions in the plane; in particular, if $\mathbf{b} = \mathbf{0}$ and $\mathbf{A} \neq \mathbf{0}$, the solution set is a straight line passing through the origin.

▶ **The eigenvalue problem:** $(\mathbf{A} - \lambda\mathbf{I})\mathbf{x} = \mathbf{0}$. The **eigenvalues** of $\mathbf{A}$ are solutions of the **characteristic equation** $\det(\mathbf{A} - \lambda\mathbf{I}) = 0$. An **eigenvector** for the eigenvalue λ is a nonzero solution of $(\mathbf{A} - \lambda\mathbf{I})\mathbf{x} = \mathbf{0}$. Eigenvalues may be real and different, real and equal, or complex conjugates.

Section 3.2 Systems of Two First Order Linear Equations

$$\text{variable coefficient: } \mathbf{x}' = \mathbf{P}(t)\mathbf{x} + \mathbf{g}(t), \qquad \text{autonomous: } \mathbf{x}' = \mathbf{Ax} + \mathbf{b}$$

▶ **Existence and uniqueness of solutions.** If the entries of $\mathbf{P}(t)$ and $\mathbf{g}(t)$ are continuous on I, then a unique solution to the initial value problem $\mathbf{x}' = \mathbf{P}(t)\mathbf{x} + \mathbf{g}(t)$, $\mathbf{x}(t_0) = \mathbf{x}_0$, $t_0 \in I$ exists for all $t \in I$.

▶ **Graphical techniques:** (i) component plots, and for autonomous systems (ii) direction fields, and (iii) phase portraits.

▶ **Critical points (equilibrium solutions)** of linear autonomous systems are solutions of $\mathbf{Ax} + \mathbf{b} = \mathbf{0}$.

▶ **Second order linear equations** $y'' + p(t)y' + q(t)y = g(t)$ can be transformed into systems of two first order linear equations

$$\mathbf{x}' = \begin{pmatrix} 0 & 1 \\ -q(t) & -p(t) \end{pmatrix} \mathbf{x} + \begin{pmatrix} 0 \\ g(t) \end{pmatrix}$$

where $\mathbf{x} = y\mathbf{i} + y'\mathbf{j}$.

Section 3.3 Homogeneous Systems With Constant Coefficients: $\mathbf{x}' = \mathbf{Ax}$

▶ Two solutions $\mathbf{x}_1(t)$ and $\mathbf{x}_2(t)$ to $\mathbf{x}' = \mathbf{Ax}$ form a **fundamental set of solutions** if their **Wronskian**

$$W[\mathbf{x}_1, \mathbf{x}_2](t) = \begin{vmatrix} x_{11}(t) & x_{12}(t) \\ x_{21}(t) & x_{22}(t) \end{vmatrix} = x_{11}(t)x_{22}(t) - x_{12}(t)x_{21}(t) \neq 0.$$

If $\mathbf{x}_1(t)$ and $\mathbf{x}_2(t)$ are a fundamental set, then the **general solution** to $\mathbf{x}' = \mathbf{Ax}$ is $\mathbf{x} = c_1\mathbf{x}_1(t) + c_2\mathbf{x}_2(t)$ where c_1 and c_2 are arbitrary constants.

▶ When $\mathbf{A}$ has real eigenvalues $\lambda_1 \neq \lambda_2$ with corresponding eigenvectors $\mathbf{v}_1$ and $\mathbf{v}_2$,

  ▶ a general solution of $\mathbf{x}' = \mathbf{A}\mathbf{x}$ is $\mathbf{x} = c_1 e^{\lambda_1 t} \mathbf{v}_1 + c_2 e^{\lambda_2 t} \mathbf{v}_2$,

  ▶ if $\det(\mathbf{A}) \neq 0$, the only critical point (the origin) is (i) a **node** if the eigenvalues have the same algebraic sign, or (ii) a **saddle point** if the eigenvalues are of opposite sign.

Section 3.4 Complex Eigenvalues

▶ If the eigenvalues of $\mathbf{A}$ are $\mu \pm i\nu$, $\nu \neq 0$, with corresponding eigenvectors $\mathbf{a} \pm i\mathbf{b}$, a fundamental set of real vector solutions of $\mathbf{x}' = \mathbf{A}\mathbf{x}$ consists of
$\mathrm{Re}\{\exp[(\mu + i\nu)t][\mathbf{a} + i\mathbf{b}]\} = \exp(\mu t)(\cos \nu t \mathbf{a} - \sin \nu t \mathbf{b})$ and
$\mathrm{Im}\{\exp[(\mu + i\nu)t][\mathbf{a} + i\mathbf{b}]\} = \exp(\mu t)(\sin \nu t \mathbf{a} + \cos \nu t \mathbf{b})$.

▶ If $\mu \neq 0$, then the critical point (the origin) is a **spiral point**. If $\mu = 0$, then the critical points is a **center**.

Section 3.5 Repeated Eigenvalues

▶ If $\mathbf{A}$ has a single repeated eigenvalue λ, then a general solution of $\mathbf{x}' = \mathbf{A}\mathbf{x}$ is

  (i) $\mathbf{x} = c_1 e^{\lambda t} \mathbf{v}_1 + c_2 e^{\lambda t} \mathbf{v}_2$ if $\mathbf{v}_1$ and $\mathbf{v}_2$ are independent eigenvectors, or

  (ii) $\mathbf{x} = c_1 e^{\lambda t} \mathbf{v} + c_2 e^{\lambda t}(\mathbf{w} + t\mathbf{v})$ where $(\mathbf{A} - \lambda\mathbf{I})\mathbf{w} = \mathbf{v}$ if $\mathbf{v}$ is the only eigenvector of $\mathbf{A}$.

▶ The critical point at the origin is a **proper node** if there are two independent eigenvectors, and an **improper** or **degenerate node** if there is only one eigenvector.

Section 3.6 Nonlinear Systems

$$\text{nonautonomous: } \mathbf{x}' = \mathbf{f}(t, \mathbf{x}), \qquad \text{autonomous: } \mathbf{x}' = \mathbf{f}(\mathbf{x})$$

▶ Theorem 3.6.1 provides conditions that guarantee, locally in time, existence and uniqueness of solutions to the initial value problem $\mathbf{x}' = \mathbf{f}(t, \mathbf{x})$, $\mathbf{x}(t_0) = \mathbf{x}_0$.

▶ Examples of two-dimensional nonlinear autonomous systems suggest that locally their solutions behave much like solutions of linear systems.

Section 3.7 Numerical Approximation Methods for Systems

The Euler and Runge-Kutta methods described in Chapters 1 and 2 are extended to systems of first order equations, and are illustrated for a typical two-dimensional system.

PROJECTS

Project 1 Eigenvalue-Placement Design of a Satellite Attitude Control System

In a simplified description of the attitude of a satellite, we restrict its motion so that it is free to rotate only about a single fixed axis as shown in Figure 3.P.1.

The proper attitude angle, $\theta = 0$, is maintained by a **feedback control system** consisting of an attitude error detector, a set of thrusters, and a modulator that converts an electrical error signal into thrustor torque that corrects the orientation. The differential equation describing the orientation of the satellite is

$$I\theta'' = u(t) \tag{1}$$

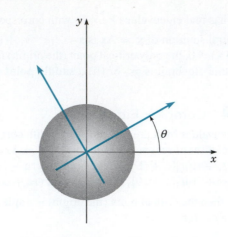

FIGURE 3.P.1 The attitude angle θ of a satellite confined to rotate about the z-axis (coming out of the page) and measured relative to a set of reference axes determined by sensor datum.

where I is the satellite's moment of inertia and $u(t)$ is the torque applied by the thrustors. For simplicity, we assume that $I = 1$. Equation (1) is converted to a system of two first order equations by introducing the state variables $x_1 = \theta$ and $x_2 = \theta'$. The resulting state equations are

$$\begin{aligned} x_1' &= x_2 \\ x_2' &= u(t). \end{aligned} \tag{2}$$

The set of equations in Eq. (2) expressed in matrix notation is

$$\mathbf{x}' = \begin{pmatrix} 0 & 1 \\ 0 & 0 \end{pmatrix} \mathbf{x} + \begin{pmatrix} 0 \\ 1 \end{pmatrix} u(t) \tag{3}$$

where

$$\mathbf{x} = \begin{pmatrix} x_1 \\ x_2 \end{pmatrix}. \tag{4}$$

The design of a suitable feedback control system is achieved by expressing the task in terms of an eigenvalue-placement problem by assuming that the input representing thrustor torque, $u(t)$, is specified as

$$u(t) = -K_1 x_1(t) - K_2 x_2(t) = \begin{pmatrix} -K_1 & -K_2 \end{pmatrix} \mathbf{x}(t), \tag{5}$$

that is, the torque that is ultimately "fed back" to the satellite is a weighted sum of its state components. The constants K_1 and K_2 are referred to as **gain constants**. Assuming that the gain constants are nonnegative, the minus signs in Eq. (5) imply that the torque generated by the thrustors is in a rotational direction opposite to that of the angular error and angular velocity. This is an example of **negative feedback control**. The control system is represented schematically in Figure 3.P.2. For a satellite subject to external disturbance inputs $f(t)$ (we have set $f(t) = 0$ in Figure 3.P.2) that tend to perturb the state of the system away from the zero state, $\mathbf{x} = 0\mathbf{i} + 0\mathbf{j}$, negative feedback will tend to return the state to the desired zero state.

A control system where a transformation of the output is fed back to the system as input is called a **closed-loop** control system. Negative feedback is indicated by the minus signs at the circular junction point at the left side of Figure 3.P.2.

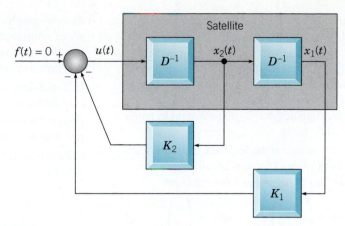

FIGURE 3.P.2 Schematic diagram for the satellite control system. The symbol D^{-1} represents an integration operation, $D^{-1}u = \int u(t)dt$. Thus, if $f(t)$ and $u(t)$ are known functions of t and $\mathbf{x}(0) = 0\mathbf{i} + 0\mathbf{j}$, then $x_2(t) = D^{-1}[u(t) + f(t)]$ and $x_1(t) = D^{-1}x_2(t)$.

The state equations for the closed-loop system are

$$\mathbf{x}'(t) = \begin{pmatrix} 0 & 1 \\ 0 & 0 \end{pmatrix}\mathbf{x} - \begin{pmatrix} 0 \\ 1 \end{pmatrix}\begin{pmatrix} K_1 & K_2 \end{pmatrix}\mathbf{x}$$

$$= \begin{pmatrix} 0 & 1 \\ 0 & 0 \end{pmatrix}\mathbf{x} - \begin{pmatrix} 0 & 0 \\ K_1 & K_2 \end{pmatrix}\mathbf{x}$$

$$= \begin{pmatrix} 0 & 1 \\ -K_1 & -K_2 \end{pmatrix}\mathbf{x}. \tag{6}$$

The eigenvalue-placement design problem then consists of choosing eigenvalues of

$$\mathbf{A_F} = \begin{pmatrix} 0 & 1 \\ -K_1 & -K_2 \end{pmatrix} \tag{7}$$

to achieve specified performance objectives. The design problem is explored in the following exercises.

Project 1 PROBLEMS

1. If the eigenvalues of $\mathbf{A_F}$ in Eq. (7) are prescribed to be λ_1 and λ_2, show that the corresponding gain constants are related to the eigenvalues by $K_1 = \lambda_1 \lambda_2$ and $K_2 = -(\lambda_1 + \lambda_2)$.

2. Find general solutions of $\mathbf{x}' = \mathbf{A_F}\mathbf{x}$ for each of the following cases: (a) λ_1, λ_2 real and distinct, (b) $\lambda_1 = \lambda_2$, and (c) $\lambda_1 = \alpha + i\beta$, $\lambda_2 = \alpha - i\beta$.

3. Suppose that design specifications are stated as follows:

   ▶ Engineering constraints on the gain constants are $0 \le K_1 \le 32$ and $0 \le K_2 \le 8$,

▶ The eigenvalues selected for $\mathbf{A_F}$ must be such that the first component, $x_1(t)$, of the initial value problem

$$\mathbf{x}' = \mathbf{A_F}\mathbf{x}, \qquad \mathbf{x}(0) = \begin{pmatrix} 0 \\ 1 \end{pmatrix} \tag{i}$$

decays at a faster rate than for any other choice of eigenvalues subject to the above constraints on K_1 and K_2.

Under the specified constraints, sketch the admissible regions in (a) the $\lambda_1\lambda_2$-plane in the case that λ_1 and λ_2 are real, and in **(b)** the complex plane if λ_1 and λ_2 are complex conjugates. Then select the "optimal" eigenvalues by comparing plots of $x_1(t)$ obtained from Eq. (i) for several different choices of eigenvalues in the admissible regions. Draw a phase portrait of the resulting feedback control system and classify the origin of the x_1x_2-plane as a node, spiral point, or saddle point.

4. Test the performance of the feedback control system designed in Problem 3 by numerically approximating the solution of the initial value problem

$$\mathbf{x}' = \mathbf{A}_F\mathbf{x} + \begin{pmatrix} 0 \\ f(t) \end{pmatrix}, \qquad \mathbf{x}(0) = \begin{pmatrix} 0 \\ 0 \end{pmatrix} \qquad \text{(ii)}$$

using the "noisy" input disturbance

$$f(t) = .1 \sin t - .25 \cos 3t - .3 \sin 7t + .5 \cos 13t.$$

Plot the graphs of $x_1(t)$, $x_2(t)$, and $f(t)$ on the same set of coordinate axes for $0 \le t \le 10$. What is the maximum value attained by $f(t)$ and what is the maximum value attained by $x_1(t)$?

5. Is it possible to control the satellite's attitude by choosing either $K_1 = 0$ or $K_2 = 0$? Explain.

6. Repeat Problem 4 in the absence of feedback control, that is, with $K_1 = 0$ and $K_2 = 0$.

7. If there is no feedback control, the unforced satellite system is

$$\mathbf{x}'(t) = \begin{pmatrix} 0 & 1 \\ 0 & 0 \end{pmatrix} \mathbf{x}. \qquad \text{(iii)}$$

What are the critical points of the system (iii) and what are their stability properties?

Project 2 Estimating Rate Constants for an Open Two-Compartment Model

Physiological systems are often modeled by dividing them into distinct functional units or compartments. A simple two-compartment model used to describe the evolution in time of a single intravenous drug dose (or a chemical tracer) is shown in Figure 3.P.3. The central compartment, consisting of blood and extracellular water, is rapidly diffused with the drug.

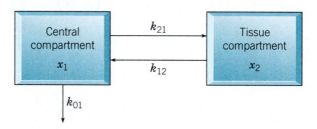

FIGURE 3.P.3 A two-compartment open model of a physiological system.

The second compartment, known as the tissue compartment, contains tissues that equilibrate more slowly with the drug. If x_1 is the concentration of drug in the blood and x_2 is its concentration in the tissue, the compartment model is described by the following system:

$$x_1' = -(k_{01} + k_{21})x_1 + k_{12}x_2$$
$$x_2' = k_{21}x_1 - k_{12}x_2 \qquad \text{(1)}$$

or $\mathbf{x}' = \mathbf{K}\mathbf{x}$, where

$$\mathbf{K} = \begin{pmatrix} K_{11} & K_{12} \\ K_{21} & K_{22} \end{pmatrix} = \begin{pmatrix} -k_{01} - k_{21} & k_{12} \\ k_{21} & -k_{12} \end{pmatrix}. \qquad \text{(2)}$$

Here the rate constant k_{21} is the fraction per unit time of drug in the blood compartment transferred to the tissue compartment; k_{12} is the fraction per unit time of drug in the tissue compartment transferred to the blood; and k_{01} is the fraction per unit time of drug eliminated from the system.

In this project, we illustrate a method for estimating the rate constants by using time dependent measurements of concentrations to estimate the eigenvalues and eigenvectors of the rate matrix $\mathbf{K}$ in Eq. (2) from which estimates of all rate constants can be computed.

Project 2 PROBLEMS

1. Assume that all the rate constants in Eq. (1) are positive.

   (a) Show that the eigenvalues of the matrix $\mathbf{K}$ are real, distinct, and negative.
   Hint: Show that the discriminant of the characteristic polynomial of $\mathbf{K}$ is positive.

   (b) If λ_1 and λ_2 are the eigenvalues of $\mathbf{K}$, show that $\lambda_1 + \lambda_2 = -(k_{01} + k_{12} + k_{21})$ and $\lambda_1 \lambda_2 = k_{12}k_{01}$.

2. **Estimating Eigenvalues and Eigenvectors of K from Transient Concentration Data.** Denote by $\mathbf{x}^*(t_k) = x_1^*(t_k)\mathbf{i} + x_2^*(t_k)\mathbf{j}$, $k = 1, 2, 3, \ldots$ measurements of the concentrations in each of the compartments. We assume that the eigenvalues of $\mathbf{K}$ satisfy $\lambda_2 < \lambda_1 < 0$. Denote the eigenvectors of λ_1 and λ_2 by

$$\mathbf{v_1} = \begin{pmatrix} v_{11} \\ v_{21} \end{pmatrix} \quad \text{and} \quad \mathbf{v_2} = \begin{pmatrix} v_{12} \\ v_{22} \end{pmatrix},$$

respectively. The solution of Eq. (1) can be expressed as

$$\mathbf{x}(t) = \alpha e^{\lambda_1 t}\mathbf{v_1} + \beta e^{\lambda_2 t}\mathbf{v_2} \qquad \text{(i)}$$

where α and β, assumed to be nonzero, depend on initial conditions. From Eq. (i) we note that

$$\mathbf{x}(t) = e^{\lambda_1 t}\left[\alpha\mathbf{v_1} + \beta e^{(\lambda_2 - \lambda_1)t}\mathbf{v_2}\right] \sim \alpha e^{\lambda_1 t}\mathbf{v_1}$$

$$\text{if } e^{(\lambda_2 - \lambda_1)t} \sim 0. \quad \text{(ii)}$$

   (a) For values of t such that $e^{(\lambda_2 - \lambda_1)t} \sim 0$, explain why the graphs of $\ln x_1(t)$ and $\ln x_2(t)$ should be approximately straight lines with slopes equal to λ_1 and intercepts equal to $\ln \alpha v_{11}$ and $\ln \alpha v_{21}$, respectively. Thus estimates of λ_1, αv_{11} and αv_{21} may be obtained by fitting straight lines to the data $\ln x_1^*(t_n)$ and $\ln x_2^*(t_n)$ corresponding to values of t_n, where graphs of the logarithms of the data are approximately linear as shown in Figure 3.P.4.

   (b) Given that both components of the data $\mathbf{x}^*(t_n)$ are accurately represented by a sum of expo-

nential functions of the form (i), explain how to find estimates of λ_2, βv_{12} and βv_{22} using the residual data $\mathbf{x}_r^*(t_n) = \mathbf{x}^*(t_n) - \hat{\mathbf{v}}_1^{(\alpha)}e^{\hat{\lambda}_1 t_n}$, where estimates of λ_1 and $\alpha\mathbf{v_1}$ are denoted by $\hat{\lambda}_1$ and $\hat{\mathbf{v}}_1^{(\alpha)}$, respectively.[4]

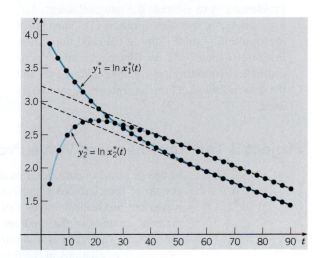

FIGURE 3.P.4 Graphs of the data $y_{1n}^* = \ln x_1^*(t_n)$ and $y_{2n}^* = \ln x_2^*(t_n)$ are approximately straight lines for values of t_n such that $e^{(\lambda_2 - \lambda_1)t_n} \sim 0$.

3. **Computing the entries of K from its eigenvalues and eigenvectors.** Assume that the eigenvalues λ_1 and λ_2 and corresponding eigenvectors $\mathbf{v_1}$ and $\mathbf{v_2}$ of $\mathbf{K}$ are known. Show that the entries of the matrix $\mathbf{K}$ must satisfy the following systems of equations:

$$\begin{pmatrix} v_{11} & v_{21} \\ v_{12} & v_{22} \end{pmatrix}\begin{pmatrix} K_{11} \\ K_{12} \end{pmatrix} = \begin{pmatrix} \lambda_1 v_{11} \\ \lambda_2 v_{12} \end{pmatrix} \qquad \text{(iii)}$$

and

$$\begin{pmatrix} v_{11} & v_{21} \\ v_{12} & v_{22} \end{pmatrix}\begin{pmatrix} K_{21} \\ K_{22} \end{pmatrix} = \begin{pmatrix} \lambda_1 v_{21} \\ \lambda_2 v_{22} \end{pmatrix} \qquad \text{(iv)}$$

[4]The procedure outlined here is called the **method of exponential peeling**. The method can be extended to cases where more than two exponential functions are required to represent the component concentrations. There must be one compartment for each exponential decay term. See, for example, D. Van Liew (1967), *J. Theoret. Biol.*, **16**, 43.

or, using matrix notation, $\mathbf{KV} = \mathbf{V\Lambda}$, where

$$\mathbf{V} = \begin{pmatrix} v_{11} & v_{12} \\ v_{21} & v_{22} \end{pmatrix} \text{ and } \mathbf{\Lambda} = \begin{pmatrix} \lambda_1 & 0 \\ 0 & \lambda_2 \end{pmatrix}.$$

4. Given estimates $\hat{K}_{ij}$ of the entries of $\mathbf{K}$ and estimates $\hat{\lambda}_1$ and $\hat{\lambda}_2$ of the eigenvalues of $\mathbf{K}$, show how to obtain an estimate $\hat{k}_{01}$ of k_{01} using the relations in Problem 1(**b**).

5. Table 3.P.1 lists drug concentration measurements made in blood and tissue compartments over a period of 100 minutes. Use the method described in Problems 2–4 to estimate the rate coefficients k_{01}, k_{12}, and k_{21} in the system model (1). Then solve the resulting system using initial conditions from line one of Table 3.P.1. Verify the accuracy of your estimates by plotting the solution components and the data in Table 3.P.1 on the same set of coordinate axes.

TABLE 3.P.1 Compartment Concentration Measurements

time(min)	x_1 (mg/ml)	x_2 (mg/ml)
0.000	0.623	0.000
7.143	0.374	0.113
14.286	0.249	0.151
21.429	0.183	0.157
28.571	0.145	0.150
35.714	0.120	0.137
42.857	0.103	0.124
50.000	0.089	0.110
57.143	0.078	0.098
64.286	0.068	0.087
71.429	0.060	0.077
78.571	0.053	0.068
85.714	0.047	0.060
92.857	0.041	0.053
100.000	0.037	0.047

Project 3 The Ray Theory of Wave Propagation

In this project, we consider a model of sound wave propagation that leads to a nonlinear system of four first order equations that, in general, must be solved numerically. Sound waves are mechanical waves that can be propagated in solids, liquids, and gases. The material particles of the medium transmitting the wave oscillate in the direction of the wave itself. The result is a continuous train of compressions and rarefactions observed as pressure oscillations in the direction of propagation. Surfaces of constant pressure are called **wavefronts**. Space curves that are orthogonal to the wavefronts are called **ray paths** or simply **rays**.

The atmosphere and the ocean are stratified in the sense that ambient properties vary primarily with height or depth. As a consequence, even over large horizontal distances, we may assume that sound speed varies only in the vertical direction. Thus in two dimensions we consider a point sound source located at $(x, y) = (x_0, y_0)$ in a fluid medium where the sound speed $c(y)$ depends only on the vertical coordinate y (see Figure 3.P.5).

The dependency of sound speed on depth causes a gradual bending of the ray paths, a phenomenon known as **refraction**. The initial value problem that describes the path of a

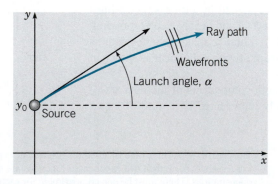

FIGURE 3.P.5 Ray emitted by a point source located at $(0, y_0)$.

ray launched from the point (x_0, y_0) with a launch angle α, measured from the horizontal, is

$$\frac{d}{ds}\left(\frac{1}{c}\frac{dx}{ds}\right) = 0, \qquad x(0) = x_0, \quad x'(0) = \cos\alpha$$
$$\frac{d}{ds}\left(\frac{1}{c}\frac{dy}{ds}\right) = -\frac{c'(y)}{c^2}, \qquad y(0) = y_0, \quad y'(0) = \sin\alpha. \tag{1}$$

Thus the ray model is a system of two second order equations. The independent variable s in (1) is the arc length of the ray measured from the source. Equations (1) can be derived from a version of Fermat's Principle known as **Snell's Law** (see Problem 3). To convert the equations in (1) to a first order system, we choose four state variables as follows:

$$x_1 = x,$$
$$x_2 = y,$$
$$x_3 = x'/c, \tag{2}$$
$$x_4 = y'/c.$$

Differentiating each equation in (2) yields

$$x_1' = c\,x_3$$
$$x_2' = c\,x_4$$
$$x_3' = 0 \tag{3}$$
$$x_4' = -\frac{1}{c^2}\frac{dc}{dx_2}(x_2).$$

where (1) is used to obtain the last two equations and we refer back to the table (2) each time one of the original dependent variables or its derivative appears so that it can be replaced by an appropriate state variable. Note that $c(x_2) = c(y)$. Initial conditions for (3) are obtained by using the table (2) to transfer the initial conditions in (1) to initial conditions for the state variables,

$$x_1(0) = x_0, \quad x_2(0) = y_0, \quad x_3(0) = (\cos\alpha)/c_0, \quad x_4(0) = (\sin\alpha)/c_0, \tag{4}$$

where $c_0 = c(y_0)$. Since $x_3' = 0$, x_3 is a constant determined by the third initial condition in (4). This results in a reduction in dimension of (3) from 4 to 3,

$$x_1' = (c\,\cos\alpha)/c_0$$
$$x_2' = c\,x_4 \tag{5}$$
$$x_4' = -\frac{1}{c^2}\frac{dc}{dx_2}(x_2).$$

If we wished, we could relabel the state variables in (5), for example, $y_1 = x_1$, $y_2 = x_2$, $y_3 = x_4$. However, with the power of modern computing resources, most scientists and engineers would simply compute the solution to the dimension four initial value problem (3), (4) since the additional time required is negligible compared to the time required to compute the solution of the three-dimensional system.

The graph on the left in Figure 3.P.6 exhibits acoustical velocity as a function of depth. The profile displays a prominent $c(y)$ minimum at a depth around 1300 meters and is characteristic of ocean water in a variety of equatorial and moderate latitudes. This profile is due to the opposing effects of temperature and pressure. At first, temperature dominates, the decreasing temperature resulting in decreasing c. After a certain depth, the increase in hydrostatic pressure gives an increasing c. The right-hand graph in Figure 3.P.6 illustrates a number of different ray paths launched at various angles from a source located at the depth where the sound speed profile has a minimum. The ray paths illustrate how sound may be trapped in zones of minimum velocity. These zones are referred to as **acoustic waveguides**

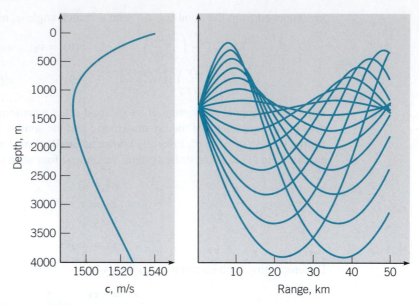

FIGURE 3.P.6 Model underwater sound speed profile and corresponding ray paths launched from a source at a depth of 1300 m.

or **sound channels**. Such channels allow the propagation of sound over very great distances with relatively little loss of intensity. A physical understanding of why sound rays bend away from zones of relatively higher c toward zones of relatively lower c in stratified fluids is provided by **Huygens' Principle**: *All points on a wavefront can be considered as a point source for the production of spherical secondary wavelets. After a time t, the new position of the wavefront will be the surface of tangency to these secondary wavelets.* The diagram in Figure 3.P.7 illustrates the application of Huygens' principle to a vertical wavefront propagating in a medium where the sound speed increases in the positive y direction.

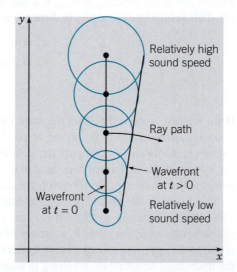

FIGURE 3.P.7 Spherical secondary wavelets are used to advance a wavefront. The sound rays bend away from a zone of relatively higher sound speed toward a zone of relatively lower sound speed.

Project 3 PROBLEMS

1. Solve the initial value problem (3), (4) in the case that the sound speed $c(y) = c_0$ is a constant. Explain why the solution shows that the ray paths are straight lines.

ODEA 2. Give a numerical demonstration that for linear sound speed profiles, $c(y) = c_0 + c_1 y$, the ray paths are arcs of circles by (i) computing the curvature
$$\kappa = \sqrt{(x''(s))^2 + (y''(s))^2}$$
at each point along a ray path, and (ii) plotting the ray path. For convenience, use parameter values $c_0 \sim 1$, $c_1 \sim 2$, and an arc length interval $0 \le s \le 1$.

3. For a sound speed that varies with respect to y, Snell's Law states that
$$\frac{\cos \theta}{c(y)} = \text{constant}$$
at each point along the ray path. Assuming that a ray path $x(s)\mathbf{i} + y(s)\mathbf{j}$ is parameterized with respect to arc length s, use the fact that the tangent vector $x'(s)\mathbf{i} + y'(s)\mathbf{j}$ is of unit length to derive the ray equations (1) (see Figure 3.P.8).

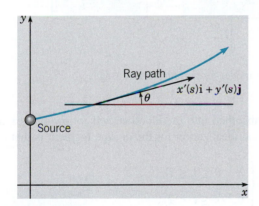

FIGURE 3.P.8

ODEA 4. Find numerical approximations of the solutions of (3), (4) over the interval $0 \le s \le 4000$ using the sound speed profile $c(y) = 1500 + .01y^2$. Place the point source at a depth of $y_0 = 0$ and plot a representative sample of ray paths, y as a function of x, and put all of them on the same set of coordinate axes (as in Figure 3.P.6). Then repeat the simulation with the sound source placed at $y_0 = -200$ and again at $y_0 = 100$. You should observe acoustical **shadow zones** and several acoustical **convergence zones** that depend on the vertical location of the sound source. Give an interpretation of what you think the physical significance of these zones are.

5. In this problem, we wish to conduct ray path simulations based on sound speed versus depth measurements, (y_i, c_i), $i = 1, \ldots, 7$, made in a certain region of the mid-Atlantic Ocean during the summer. The data are listed in Table 3.P.2. Conventionally, depth is assumed to be positive in the downward direction. Assume also that the depth of the ocean floor is approximately a constant 4000 meters over a range of several hundred kilometers. [CAS]

TABLE 3.P.2 Sound Speed Versus Depth Measurements

depth (m)	0	500	1000	1500	2000	3000	4000
sound speed (m/s)	1498	1486	1482	1484	1488	1502	1516

(a) Use the method of least squares to fit a 3rd degree polynomial
$$\hat{c}(\hat{y}) = \hat{c}_0 + \hat{c}_1 \hat{y} + \hat{c}_2 \hat{y}^2 + \hat{c}_3 \hat{y}^3 \qquad \text{(i)}$$
to the scaled data $(\hat{y}_i, \hat{c}_i) = (y_i/\sigma, c_i/\sigma)$, $i = 1, \ldots, 7$ in Table 3.P.2 where $\sigma = 1000$. Thus choose the coefficients $\hat{c}_0$, $\hat{c}_1$, $\hat{c}_2$, and $\hat{c}_3$ so that the function $G(\hat{c}_1, \hat{c}_2, \hat{c}_3, \hat{c}_4) = \sum_{i=1}^{4} [\hat{c}(\hat{y}_i) - \hat{c}_i]^2$ is minimized. Plot the original data and the fitted model, $c(y) = \sigma \hat{c}(y/\sigma)$, on the same set of coordinate axes. Why scale the data?

(b) Use the sound speed model $c(y) = \sigma \hat{c}(y/\sigma)$ found in Problem 5(a) in the system (3) to compute a number of ray paths launched from a source depth of $y_0 = 1000$ meters out to a range of 150 kilometers. Since much of the energy carried by rays that repeatedly reflect off the surface or bottom is lost, choose a cone of launch angles α, $-\alpha_s \le \alpha \le \alpha_s$, where α_s is the launch angle of a ray that just grazes the surface. Plot the ray paths on the same set of coordinate axes. At which ranges and depths, if any, do there appear to be convergence zones, that is, zones where sound energy is focused?

(c) Repeat Problem 5(b) for (i) a source at a depth $y_0 = 750$ meters, and (ii) a source at a depth $y_0 = 1500$ meters.

Project 4 A Blood-Brain Pharmacokinetic Model

Pharmacokinetics is the study of the time variation of drug and metabolite levels in the various fluids and tissues of the body. The discipline frequently makes use of compartment models to interpret data. In this problem, we consider a simple blood-brain compartment model (Figure 3.P.9),

$$\text{Compartment } 1 \equiv \text{Blood},$$
$$\text{Compartment } 2 \equiv \text{Brain},$$

that could be used to help estimate dosage strengths of an orally administered antidepressant drug. The rate at which the drug moves from compartment i to compartment j is denoted by the rate constant k_{ji} while the rate at which the drug is removed from the blood is represented by the rate constant K.

A pharmaceutical company must weigh many factors into determining drug dosage parameters; of particular importance are dosage strengths that will provide flexibility to a physician in determining individual dosage regimens to conveniently maintain concentration levels at effective therapeutic values while minimizing local irritation and other adverse side effects.

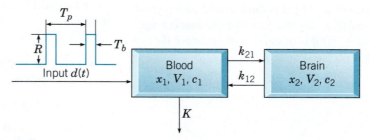

FIGURE 3.P.9 A two-compartment model for periodic drug dosages.

Assuming that the drug is rapidly absorbed into the blood stream following its introduction into the stomach, a mathematical idealization for the dosage regimen is that of a periodic square wave

$$d(t) = \begin{cases} R, & 0 \le t \le T_b \\ 0, & T_b \le t < T_p, \end{cases}$$

where R is the rate of uptake (milligrams/hour) into the bloodstream, T_b is the time period during which the drug is absorbed into the bloodstream following oral administration, and T_p is the length of time between doses.

Project 4 PROBLEMS

1. If $x_j(t)$ represents the amount of drug (milligrams) in compartment j, $j = 1, 2$, use Figure 3.P.9 and the mass balance law

$$\frac{dx_j}{dt} = \text{compartment } j \text{ input rate}$$
$$-\text{compartment } j \text{ output rate}, \quad \text{(i)}$$

to show that x_1 and x_2 satisfy the system

$$\frac{dx_1}{dt} = -(K + k_{21})x_1 + k_{12}x_2 + d(t)$$
$$\frac{dx_2}{dt} = k_{21}x_1 - k_{12}x_2. \quad \text{(ii)}$$

2. If $c_i(t)$ denotes the concentration of the drug and V_i denotes the apparent volume of distribution in compartment i, use the relation $c_i = x_i/V_i$ to show that the system (ii) is transformed into

$$\frac{dc_1}{dt} = -(K + k_{21})c_1 + \frac{k_{12}V_2}{V_1}c_2 + \frac{1}{V_1}d(t)$$

$$\frac{dc_2}{dt} = \frac{V_1 k_{21}}{V_2}c_1 - k_{12}c_2. \tag{iii}$$

ODEA 3. Assuming that $x_1(0) = 0$ and $x_2(0) = 0$, use the parameter values listed in the table below to perform numerical simulations of the system (iii) with the goal of recommending two different encapsulated dosage strengths $A = RT_b$ for distribution.

k_{21}	k_{12}	K	V_1	V_2	T_b
.29/h	.31/h	.16/h	6 L	.25 L	1 h

Use the following guidelines to arrive at your recommendations:

▶ It is desirable to keep the target concentration levels in the brain as close as possible to constant levels between 10 mg/L and 30 mg/L, depending on the individual patient. The therapeutic range must be above the minimum effective concentration and below the minimum toxic concentration. For the purpose of this project, we will specify that concentration fluctuations should not exceed 25% of the average of the steady-state response.

▶ As a matter of convenience, a lower frequency of administration is better than a higher frequency of administration; once every 24 hours or once every 12 hours is best. Once every 9.5 hours is unacceptable and more than 4 times per day is unacceptable. Multiple doses are acceptable, that is, "take two capsules every 12 hours."

4. If a dosage is missed, explain through the simulations why it is best to skip the dose rather than to try to "catch up" by doubling the next dose, given that it is dangerous and possibly fatal to overdose on the drug. Or, does it not really matter in the case of the given parameter values? **ODEA**

5. Suppose the drug can be packaged in a timed-release form so that $T_b = 8$ hours and R is adjusted accordingly. Does this change your recommendations? **ODEA**

Second Order Linear Equations

I n Chapter 3 we discussed systems of two first order equations, with primary emphasis on homogeneous linear equations with constant coefficients. In this chapter we will begin to consider second order linear equations, both homogeneous and nonhomogeneous. Since second order equations can always be transformed into a system of two first order equations, this may seem redundant. However, second order equations naturally arise in many areas of application, and it is important to be able to deal with them directly. One cannot go very far in the development of fluid mechanics, heat conduction, wave motion, or electromagnetic phenomena without encountering second order linear differential equations.

4.1 Definitions and Examples

A **second order differential equation** is an equation involving the independent variable t, and an unknown function or dependent variable $y = y(t)$ along with its first and second derivatives. We will assume that it is always possible to solve for the second derivative so that the equation has the form

$$y'' = f(t, y, y') \tag{1}$$

where f is some prescribed function. Usually, we will denote the independent variable by t since time is often the independent variable in physical problems, but sometimes we will use x instead. We will use y, or occasionally some other letter, to designate the dependent variable.

A **solution** of Eq. (1) on an interval I is a function $y = \phi(t)$, twice continuously differentiable on I, such that

$$\phi''(t) = f(t, \phi(t), \phi'(t)) \tag{2}$$

for all values of $t \in I$.

An **initial value problem** for a second order equation on an interval I consists of Eq. (1) together with two initial conditions

$$y(t_0) = y_0, \qquad y'(t_0) = y_1, \tag{3}$$

prescribed at a point $t_0 \in I$, where y_0 and y_1 are any given numbers. Thus $y = \phi(t)$ is a **solution of the initial value problem** (1), (3) on I if, in addition to satisfying Eq. (2) on I, $\phi(t_0) = y_0$ and $\phi'(t_0) = y_1$.

Remark. Observe that the initial conditions for a second order equation prescribe not only a particular point (t_0, y_0) through which the graph of the solution must pass, but also the slope $y'(t_0) = y_1$ of the graph at that point (Figure 4.1.1).

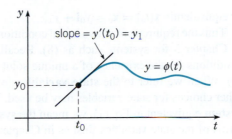

FIGURE 4.1.1   Initial conditions for a second order equation require that $y(t_0)$ and $y'(t_0)$ be prescribed.

It is reasonable to expect that two initial conditions are needed for a second order equation because, roughly speaking, two integrations are required to find a solution and each integration introduces an arbitrary constant. Presumably, two initial conditions will suffice to determine values for these two constants.

Linear Equations

The differential equation (1) is said to be **linear** if it can be written in the **standard form**

$$y'' + p(t)y' + q(t)y = g(t) \tag{4}$$

where the coefficient of y'' is equal to 1. The coefficients p, q, and g can be arbitrary functions of the independent variable t, but y, y', and y'' can appear in no other way except as designated by the form of Eq. (4).

Equation (4) is said to be **homogeneous** if the term $g(t)$ is zero for all t. Otherwise, the equation is **nonhomogeneous**, and the term $g(t)$ is referred to as the nonhomogeneous term.

A slightly more general form of a linear second order equation is

$$P(t)y'' + Q(t)y' + R(t)y = G(t) \tag{5}$$

Of course, if $P(t) \neq 0$, we can divide Eq. (5) by $P(t)$ and thereby obtain Eq. (4) with

$$p(t) = \frac{Q(t)}{P(t)}, \qquad q(t) = \frac{R(t)}{P(t)}, \qquad \text{and} \qquad g(t) = \frac{G(t)}{P(t)}. \tag{6}$$

Equation (5) is said to be a **constant coefficient** equation if P, Q, and R are constants. In this case, Eq. (5) reduces to

$$ay'' + by' + cy = g(t) \tag{7}$$

where $a \neq 0$, b, and c are given constants and we have replaced $G(t)$ by $g(t)$. Otherwise, Eq. (5) has **variable coefficients**.

Dynamical System Formulation

Recall from Section 3.2 that Eq. (1) can be converted to a system of first order equations of dimension two by introducing the state variables $x_1 = y$ and $x_2 = y'$. Then

$$\begin{aligned} x_1' &= x_2, \\ x_2' &= f(t, x_1, x_2), \end{aligned} \tag{8}$$

or, using vector notation, $\mathbf{x}' = \mathbf{f}(t, \mathbf{x}) = x_2\mathbf{i} + f(t, x_1, x_2)\mathbf{j}$, where $\mathbf{x} = x_1\mathbf{i} + x_2\mathbf{j}$. Initial conditions for the system (8), obtained from (3), are

$$x_1(t_0) = y_0, \qquad x_2(t_0) = y_1, \tag{9}$$

or equivalently $\mathbf{x}(t_0) = \mathbf{x}_0 = y_0\mathbf{i} + y_1\mathbf{j}$.

Thus the requirement of two initial conditions for Eq. (1) is consistent with our experience in Chapter 3 for systems such as (8). Recall, in particular, Theorem 3.6.1, which gives conditions for the existence of a unique solution of the initial value problem Eqs. (8) and (9). When we refer to the **state variables** for Eq. (1), we mean both y and y', although other choices for state variables may be used. In addition, when we refer to the **dynamical system** equivalent to Eq. (1), we mean the system of first order equations (8) expressed in terms of the state variables. Just as in Chapter 3, the evolution of the system state in time is graphically represented as a continuous **trajectory**, or **orbit**, through the phase plane or state space. It is sometimes helpful to think of an orbit as the path of a particle moving in accordance with the system of differential equations (8). The initial conditions (9) determine the starting point of the moving particle.[1] Note that if $\phi(t)$ is a solution of Eq. (1) on I, then $\boldsymbol{\phi}(t) = \phi(t)\mathbf{i} + \phi'(t)\mathbf{j}$ is a solution of the system (8) on I (Figure 4.1.2).

Special cases of (8) corresponding to Eqs. (4) and (7) are the linear systems

$$\begin{aligned} x_1' &= x_2, \\ x_2' &= -q(t)x_1 - p(t)x_2 + g(t), \end{aligned} \tag{10}$$

and

$$\begin{aligned} x_1' &= x_2, \\ x_2' &= -\frac{c}{a}x_1 - \frac{b}{a}x_2 + \frac{1}{a}g(t), \end{aligned} \tag{11}$$

respectively.

If the function f on the right side of Eq. (1) is independent of t so that $f(t, y, y') = f(y, y')$, then the system (8) is **autonomous**. In this case, critical points of the system (8) are solutions of the pair of equations $x_2 = 0$ and $f(x_1, x_2) = 0$. All critical points lie on the x_1 axis with coordinates $(\bar{x}_1, 0)$, where $\bar{x}_1$ is any solution of $f(\bar{x}_1, 0) = 0$. If $f(\bar{x}_1, 0) = 0$, $\partial f/\partial x_1(\bar{x}_1, 0) \neq 0$, and $\partial f/\partial x_1$ and $\partial f/\partial x_2$ are continuous at $(\bar{x}_1, 0)$, then $(\bar{x}_1, 0)$ is an **isolated critical point**, that is, there is a neighborhood of $(\bar{x}_1, 0)$ that contains no other

---

[1] For example, when f is continuous, Euler's method, the numerical approximation method discussed in Section 3.7, can be used to advance the state in discrete time by computing the sequence $\mathbf{x}_1 = \mathbf{x}_0 + h\mathbf{f}(t_0, \mathbf{x}_0)$, $\mathbf{x}_2 = \mathbf{x}_1 + h\mathbf{f}(t_1, \mathbf{x}_1), \ldots$.

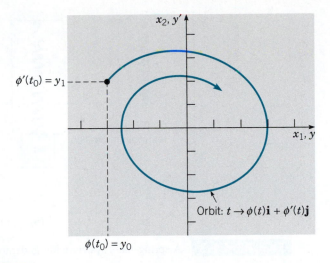

x_2, y'

$\phi'(t_0) = y_1$

x_1, y

Orbit: $t \to \phi(t)\mathbf{i} + \phi'(t)\mathbf{j}$

$\phi(t_0) = y_0$

FIGURE 4.1.2 Initial conditions for the system (8) correspond to the initial conditions $y(t_0) = y_0, y'(t_0) = y_1$ for the second order equation $y'' = f(t, y, y')$.

critical points. The linear, constant coefficient system (11) is autonomous if $g(t) = g_0$, a constant. In this case, Eq. (11) has a unique critical point at $(0, g_0/c)$ if the coefficient c is nonzero. If $c = 0$ and $g_0 \neq 0$, then there are no critical points. Finally, if $g_0 = 0$ and $c = 0$, then the set of critical points is the entire x_1-axis.

We now present three applications arising from problems in mechanics and circuit theory. The mathematical formulation in each case consists of a second order linear equation with constant coefficients.

The Spring-Mass System

Vibrational or oscillatory behavior is observed in many mechanical, electrical, and biological systems. Understanding the motion of a mass on a spring is the first step in the investigation of more complex vibrating systems. The principles involved are common to many problems. The differential equation that describes the motion of a spring-mass system is arguably the most important equation in an introductory course in differential equations for the following reasons:

▶ It involves translating a physical description of a simple mechanical system into a prototypical mathematical model, namely, a linear, second order differential equation with constant coefficients.

▶ Understanding the behavior of solutions as parameters vary, or as external input forces are added, is fundamental to understanding the qualitative behavior of solutions of both linear, second order equations with variable coefficients and second order nonlinear equations.

▶ Mathematical properties of solutions are easily interpreted in terms of the physical system.

We consider a vertical spring of natural length l attached to a horizontal support as shown in Figure 4.1.3(a).

Next we suppose that a mass of magnitude m is attached to the lower end of the spring and slowly lowered so as to achieve its equilibrium position, as shown in Figure 4.1.3(b). The mass causes an elongation L of the spring.

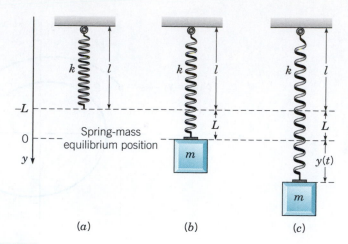

(a) (b) (c)

FIGURE 4.1.3 A spring-mass system with no damping or forcing.

Our goal is to investigate motions of the mass that might be caused by an external force acting upon it, or by an initial displacement of the mass away from its equilibrium position. We consider only motions along a vertical line. Let the y-axis be vertical, with the positive direction downward, and with the origin at the equilibrium position of the mass; see Figure 4.1.3(c). Then $y(t)$ denotes the displacement of the mass from its equilibrium position, and thus $y'(t)$ is its velocity. The equilibrium state of the spring-mass system corresponds to $y = 0$ and $y' = 0$, that is, both state variables are zero.

To derive an equation that describes possible motions of the mass, we need to examine the forces that may act upon it. We start with the force exerted on the mass by the spring. Denote by Δy the departure of the spring from its natural length so that if a displacement Δy from its natural length occurs, the length of the spring is $l + \Delta y$. We first assume that the force exerted by the spring on the mass is described by a function $F_s(\Delta y)$ satisfying the following properties:

▶ $F_s(0) = 0$; the spring exerts no force if $\Delta y = 0$.

▶ $F_s(\Delta y) < 0$ if $\Delta y > 0$; in an elongated state, the spring exerts a force in the upward direction.

▶ $F_s(\Delta y) > 0$ if $\Delta y < 0$; in a compressed state, the spring exerts a force in the downward direction.

Thus the direction of the force is always opposite to the displacement of the lower endpoint of the spring relative to its natural length. An example of such a function is $F_s(\Delta y) = -k\Delta y - \epsilon(\Delta y)^3$ where $k > 0$ and ϵ are constants (see Figure 4.1.4).

The spring is called a **hardening spring** if $\epsilon > 0$ and a **softening spring** if $\epsilon < 0$. If in a state of motion the maximum displacement of the spring from its natural length is small so that $\epsilon(\Delta y)^3$ is always negligible relative to $k\Delta y$, it is natural to discard the nonlinear term $\epsilon(\Delta y)^3$ and simply assume that the spring force is proportional to Δy,

$$F_s(\Delta y) = -k\Delta y. \tag{12}$$

Equation (12) is known as **Hooke's law**. It provides an excellent approximation to the behavior of real springs as long as they are not stretched or compressed too far. Unless stated otherwise, we will always assume that our springs are adequately described by Hooke's law. The **spring constant** k, sometimes referred to as the **stiffness** of the spring, is the magnitude of the spring force per unit of elongation. Thus, very stiff springs have large values of k.

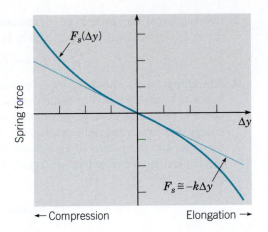

←Compression   Elongation →

FIGURE 4.1.4 Spring force $F_s(\Delta y) = -k\Delta y - \epsilon(\Delta y)^3$ of a hardening spring ($\epsilon > 0$) as a function of departure Δy from the natural length l of the spring. A compressed spring ($\Delta y < 0$) exerts a force in the positive direction while an elongated spring ($\Delta y > 0$) exerts a force in the negative direction. Hooke's Law assumes the linear approximation $F_s(\Delta y) \cong -k\Delta y$.

In the equilibrium state, there are two forces acting at the point where the mass is attached to the spring; see Figure 4.1.5.

The gravitational force, or weight of the mass, acts downward and has magnitude mg, where g is the acceleration due to gravity. The force F_s, due to the elongated spring, acts upward, that is, in the negative direction. Since the mass is in equilibrium, the spring and gravitational forces balance each other, which means that

$$mg - kL = 0. \tag{13}$$

For a given weight $w = mg$, one can measure L and then use Eq. (13) to determine k.

FIGURE 4.1.5 Force diagram for a spring-mass system.

In the corresponding dynamic problem, we are interested in studying the motion of the mass when it is initially displaced from equilibrium or acted on by an external force. Then $y(t)$ is related to the forces acting on the mass through Newton's law of motion,

$$my''(t) = F_{\text{net}}(t), \tag{14}$$

where y'' is the acceleration of the mass and F_{net} is the net force acting on the mass. Observe that both y and F_{net} are functions of time. In determining F_{net}, there are four separate forces that must be considered:

1. **(Gravitational force)** The weight $w = mg$ of the mass always acts downward.

2. (**Spring force**) The spring force F_s is assumed to be proportional to the total elongation $\Delta y = L + y$ following Hooke's law, Eq. (12) (see Figure 4.1.3(c)),

$$F_s = -k(L + y). \tag{15}$$

3. (**Damping force**) The damping or resistive force F_d always acts in the direction opposite to the direction of motion of the mass. This force may arise from several sources: resistance from the air or other medium in which the mass moves, internal energy dissipation due to the extension or compression of the spring, friction between the mass and the guides (if any) that constrain its motion to one dimension, or a mechanical device (dashpot; see Figure 4.1.6) that imparts a resistive force to the mass.

FIGURE 4.1.6   A damped spring-mass system.

In any case, we assume that the resistive force is proportional to the speed $|dy/dt|$ of the mass; this is usually referred to as viscous damping. If $dy/dt > 0$, then y is increasing, so the mass is moving downward. Then F_d is directed upward and is given by

$$F_d(t) = -\gamma y'(t), \tag{16}$$

where γ is a positive constant of proportionality known as the **damping constant**. On the other hand, if $dy/dt < 0$, then y is decreasing, the mass is moving upward, and F_d is directed downward. In this case, $F_d = \gamma |y'(t)|$; since $|y'(t)| = -y'(t)$, it follows that $F_d(t)$ is again given by Eq. (16). Thus, regardless of the direction of motion of the mass, the damping force is always expressed by Eq. (16).

Remark. The damping force may be rather complicated, and the assumption that it is modeled adequately by Eq. (16) may be open to question. Some dashpots do behave as Eq. (16) states, and if the other sources of dissipation are small, it may be possible to neglect them altogether or to adjust the damping constant γ to approximate them. An important benefit of the assumption (16) is that it leads to a linear (rather than a nonlinear) differential equation. In turn, this means that a thorough analysis of the system is straightforward, as we will show in the next section.

4. (**External forces or inputs**) An applied external force $F(t)$ is directed downward or upward as $F(t)$ is positive or negative. This could be a force due to the motion of the

mount to which the spring is attached, or it could be a force applied directly to the mass. Often the external force is periodic.

With $F_{net}(t) = mg + F_s(t) + F_d(t) + F(t)$ we can now rewrite Newton's law (14) as

$$my''(t) = mg + F_s(t) + F_d(t) + F(t)$$
$$= mg - k[L + y(t)] - \gamma y'(t) + F(t). \qquad (17)$$

Since $mg - kL = 0$ by Eq. (13), it follows that the equation of motion of the mass is a second order linear equation with constant coefficients,

$$my''(t) + \gamma y'(t) + ky(t) = F(t), \qquad (18)$$

where the constants m, γ, and k are positive.

Remark. It is important to understand that Eq. (18) is only an approximate equation for the displacement $y(t)$. In particular, both Eqs. (12) and (16) should be viewed as approximations for the spring force and the damping force, respectively. In our derivation, we have also neglected the mass of the spring in comparison with the mass of the attached body.

The complete formulation of the vibration problem requires that we specify two initial conditions, namely, the initial position y_0 and the initial velocity $y'(0) = v_0$ of the mass:

$$y(0) = y_0, \qquad y'(0) = v_0. \qquad (19)$$

Four cases of Eq. (18) that are of particular interest are listed in the table below.

Unforced, undamped oscillator:	$my''(t) + ky(t) = 0$
Unforced, damped oscillator:	$my''(t) + \gamma y'(t) + ky(t) = 0$
Forced, undamped oscillator:	$my''(t) + ky(t) = F(t)$
Forced, damped oscillator:	$my''(t) + \gamma y'(t) + ky(t) = F(t)$

Analytical solutions and properties of unforced oscillators and forced oscillators are studied in Sections 4.5 and 4.7, respectively.

EXAMPLE 1

A mass weighing 4 lb stretches a spring 2 in. Suppose that the mass is displaced an additional 6 inches in the positive direction and then released. The mass is in a medium that exerts a viscous resistance of 6 lb when the mass has a velocity of 3 ft/s. Under the assumptions discussed in this section, formulate the initial value problem that governs the motion of the mass.

The required initial value problem consists of the differential equation (18) and initial conditions (19), so our task is to determine the various constants that appear in these equations. The first step is to choose the units of measurement. Based on the statement of the problem, it is natural to use the English rather than the metric system of units. The only time unit mentioned is the second, so we will measure t in seconds. On the other hand, both the foot and the inch appear in the statement as units of length. It is immaterial which one we use, but having made a choice, we must be consistent. To be definite, let us measure the displacement y in feet.

Since nothing is said in the statement of the problem about an external force, we assume that $F(t) = 0$. To determine m note that

$$m = \frac{w}{g} = \frac{4 \text{ lb}}{32 \text{ ft/s}^2} = \frac{1}{8} \frac{\text{lb·s}^2}{\text{ft}}.$$

The damping coefficient γ is determined from the statement that $\gamma y'$ is equal to 6 lb when y' is 3 ft/s. Therefore

$$\gamma = \frac{6 \text{ lb}}{3 \text{ ft/s}} = 2 \frac{\text{lb·s}}{\text{ft}}.$$

The spring constant k is found from the statement that the mass stretches the spring by 2 in, or 1/6 ft. Thus

$$k = \frac{4 \text{ lb}}{1/6 \text{ ft}} = 24 \frac{\text{lb}}{\text{ft}}.$$

Consequently, Eq. (18) becomes

$$\tfrac{1}{8}y'' + 2y' + 24y = 0,$$

or

$$y'' + 16y' + 192y = 0. \tag{20}$$

The initial conditions are

$$y(0) = \tfrac{1}{2}, \qquad y'(0) = 0. \tag{21}$$

The second initial condition is implied by the word "released" in the statement of the problem, which we interpret to mean that the mass is set in motion with no initial velocity.

The Linearized Pendulum

Consider the configuration shown in Figure 4.1.7, in which a mass m is attached to one end of a rigid, but weightless, rod of length L.

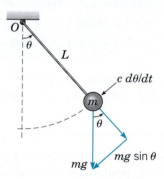

FIGURE 4.1.7   The gravitational force component $mg \sin \theta$ and damping force $cd\theta/dt$ are tangent to the path of the mass and act in the clockwise direction when $\theta > 0$ and $d\theta/dt > 0$.

The other end of the rod is supported at the origin O, and the rod is free to rotate in the plane of the paper. The position of the pendulum is described by the angle θ between the rod and the downward vertical direction, with the counterclockwise direction taken as positive. Since the mass is constrained to move along a circular path of radius L, the distance of the mass from its vertical downward position measured along the arc is $L\theta$. The instantaneous tangential velocity and instantaneous tangential acceleration are then given by $L\theta'$ and $L\theta''$. We consider two forces acting on the mass. The component of gravitational force tangent to the path of motion is $-mg \sin \theta$. Note that if θ is positive this force acts in the clockwise direction; if θ is negative, then the force is counterclockwise. We also assume the presence of a damping force proportional to the angular velocity, $-cd\theta/dt$, where c is positive, that is always opposite to the direction of motion. This force may arise from resistance from the air or other medium in which the mass moves, friction in the pivot mechanism, or some mechanical device that imparts a resistive force to the pendulum arm or bob. The differential equation that describes the motion of the pendulum is derived by applying Newton's second law $ma = F_{net}$ along the line tangent to the path of motion,

$$mL\frac{d^2\theta}{dt^2} = -c\frac{d\theta}{dt} - mg\sin\theta, \tag{22}$$

or

$$\frac{d^2\theta}{dt^2} + \gamma\frac{d\theta}{dt} + \omega^2\sin\theta = 0 \tag{23}$$

where $\gamma = c/mL$ and $\omega^2 = g/L$. Due to the presence of the term $\sin\theta$, Eq. (23) cannot be written in the form (4) of a linear equation. Thus Eq. (23) is **nonlinear**. However, if in its dynamic state the angle of deflection θ is always small, then $\sin\theta \cong \theta$ (see Figure 4.1.8.) and Eq. (23) can be approximated by the equation,

$$\frac{d^2\theta}{dt^2} + \gamma\frac{d\theta}{dt} + \omega^2\theta = 0, \tag{24}$$

a constant coefficient linear equation.

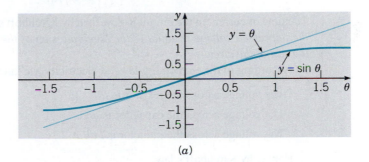

(a)

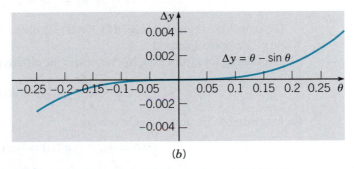

(b)

FIGURE 4.1.8 (a) Graphs of $y = \sin\theta$ and $y = \theta$. (b) The difference $|\Delta y| = |\theta - \sin\theta| < .0026$ for $|\theta| \le .25$ radians.

The process of approximating a nonlinear equation by a linear one is called **linearization**. It is an extremely useful method for studying nonlinear systems that operate near an equilibrium point, which, in terms of state variables for the simple pendulum, corresponds to $\theta = 0$ and $\theta' = 0$.

The Series RLC Circuit

A third example of a second order linear differential equation with constant coefficients is a model of flow of electric current in the simple series circuit shown in Figure 4.1.9.

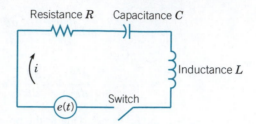

FIGURE 4.1.9 The series *RLC* circuit.

The current i, measured in amperes, is a function of time t. The resistance R (ohms), the capacitance C (farads), and the inductance L (henrys) are all positive and are assumed to be known constants. The impressed voltage e (volts) is a given function of time. Another physical quantity that enters the discussion is the total charge q (coulombs) on the capacitor at time t. The relation between charge q and current i is

$$i = dq/dt. \tag{25}$$

The flow of current in the circuit is governed by Kirchhoff's voltage law: *In a closed circuit, the impressed voltage is equal to the algebraic sum of the voltages across the elements in the rest of the circuit.*

According to the elementary laws of electricity, we know that

The voltage across the resistor is iR.

The voltage across the capacitor is q/C.

The voltage across the inductor is Ldi/dt.

Hence, by Kirchhoff's law,

$$L\frac{di}{dt} + Ri + \frac{1}{C}q = e(t). \tag{26}$$

The units have been chosen so that 1 volt = 1 ohm · 1 ampere = 1 coulomb/1 farad = 1 henry · 1 ampere/1 second.

Substituting for i from Eq. (25), we obtain the differential equation

$$Lq'' + Rq' + \frac{1}{C}q = e(t) \tag{27}$$

for the charge q. The initial conditions are

$$q(t_0) = q_0, \qquad q'(t_0) = i(t_0) = i_0. \tag{28}$$

Thus we must know the charge on the capacitor and the current in the circuit at some initial time t_0.

Alternatively, we can obtain a differential equation for the current i by differentiating Eq. (27) with respect to t, and then substituting for di/dt from Eq. (25). The result is

$$Li'' + Ri' + \frac{1}{C}i = e'(t), \tag{29}$$

with the initial conditions

$$i(t_0) = i_0, \qquad i'(t_0) = i_0'. \tag{30}$$

From Eq. (26) it follows that

$$i_0' = \frac{e(t_0) - Ri_0 - (1/C)q_0}{L}. \tag{31}$$

Hence i_0' is also determined by the initial charge and current, which are physically measurable quantities.

Remark. The most important conclusion from this discussion is that the flow of current in the circuit is described by an initial value problem of precisely the same form as the one that describes the motion of a spring–mass system or a pendulum exhibiting small amplitude oscillations. This is a good example of the unifying role of mathematics: Once you know how to solve second order linear equations with constant coefficients, you can interpret the results in terms of mechanical vibrations, electric circuits, or any other physical situation that leads to the same differential equation.

PROBLEMS

In Problems 1 through 5, determine whether the differential equation is linear or nonlinear.

1. $y'' + ty = 0$, (Airy's equation)
2. $y'' + y' + y + y^3 = 0$ (Duffing's equation)
3. $(1 - x^2)y'' - 2xy' + \alpha(\alpha + 1)y = 0$ (Legendre's equation)
4. $x^2 y'' + xy' + (x^2 - v^2)y = 0$ (Bessel's equation)
5. $y'' + \mu(1 - y^2)y' + y = 0$ (van der Pol's equation)

6. A mass weighing 8 lb stretches a spring 6 in. What is the spring constant for this spring?

7. A 10 kg mass attached to a vertical spring is slowly lowered to its equilibrium position. If the resulting change in the length of the spring from its rest length is 70 cm, what is the spring constant for this spring?

For each spring-mass system or electric circuit in Problems 8 through 15, write down the appropriate initial value problem based on the physical description.

8. A mass weighing 2 lb stretches a spring 6 in. The mass is pulled down an additional 3 in and then released. Assume there is no damping.

9. A mass of 100 g stretches a spring 5 cm. The mass is set in motion from its equilibrium position with a downward velocity of 10 cm/s. Assume there is no damping.

10. A mass weighing 3 lb stretches a spring 3 in. The mass is pushed upward, contracting the spring a distance of 1 in, and then set in motion with a downward velocity of 2 ft/s. Assume there is no damping.

11. A series circuit has a capacitor of 0.25×10^{-6} farad and an inductor of 1 henry. The initial charge

on the capacitor is 10^{-6} coulomb and there is no initial current.

12. A mass of 20 g stretches a spring 5 cm. Suppose that the mass is also attached to a viscous damper with a damping constant of 400 dyne·s/cm. The mass is pulled down an additional 2 cm and then released.

13. A mass weighing 16 lb stretches a spring 3 in. The mass is attached to a viscous damper with a damping constant of 2 lb·s/ft. The mass is set in motion from its equilibrium position with a downward velocity of 3 in/s.

14. A spring is stretched 10 cm by a force of 3 N (newtons). A mass of 2 kg is hung from the spring and is also attached to a viscous damper that exerts a force of 3 N when the velocity of the mass is 5 m/s. The mass is pulled down 5 cm below its equilibrium position and given an initial downward velocity of 10 cm/s.

15. A series circuit has a capacitor of 10^{-5} farad, a resistor of 3×10^2 ohms, and an inductor of 0.2 henry. The initial charge on the capacitor is 10^{-6} coulomb and there is no initial current.

16. Suppose that a mass m slides without friction on a horizontal surface. The mass is attached to a spring with spring constant k, as shown in Figure 4.1.10, and is also subject to viscous air resistance with coefficient γ and an applied external force $F(t)$. Assuming that the force exerted on the mass by the spring obeys Hooke's Law, show that the displacement $x(t)$ of the mass from its equilibrium position satisfies Eq. (18). How does the derivation of the equation of motion in this case differ from the derivation given in the text?

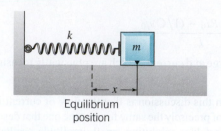

FIGURE 4.1.10 A mass attached to a spring and constrained to one dimensional motion on a horizontal, frictionless surface.

17. Duffing's Equation

(a) For the spring-mass system shown in Figure 4.1.10, derive the differential equation for the displacement $x(t)$ of the mass from its equilibrium position using the force law $F_s(\Delta x) = -k\Delta x - \epsilon(\Delta x)^3$ where Δx represents the change in length of the spring from its natural length. Assume that damping forces and external forces are not present.

(b) Find the linearized equation for the differential equation derived in part (a) under the assumption that, in its dynamic state, the maximum displacement of the mass from its equilibrium position is small.

18. A body of mass m is attached between two springs with spring constants k_1 and k_2 as shown in Figure 4.1.11. The springs are at their rest length when the system is in the equilibrium state. Assume that the mass slides without friction but the motion is subject to viscous air resistance with coefficient γ. Find the differential equation satisfied by the displacement $x(t)$ of the mass from its equilibrium position.

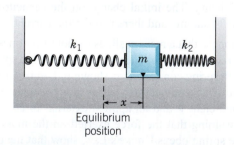

FIGURE 4.1.11 A mass attached between two springs and constrained to one dimensional motion on a horizontal, frictionless surface.

19. A cubic block of side l and mass density ρ per unit volume is floating in a fluid of mass density ρ_0 per unit volume, where $\rho_0 > \rho$. If the block is slightly depressed and then released, it oscillates in the vertical direction. Assuming that the viscous damping of the fluid and air can be neglected, derive the differential equation of motion for this system.

Hint: Use Archimedes' principle: An object that is completely or partially submerged in a fluid is acted on by an upward (buoyant) force equal to the weight of the displaced fluid.

In Problems 20 through 24, we specify the mass, damping constant, and spring constant of an unforced spring-mass system ► **ODEA**

$$my'' + \gamma y' + ky = 0. \qquad (i)$$

Convert (i) to a planar system for the state vector $\mathbf{y} = x_1\mathbf{i} + x_2\mathbf{j} \overset{\text{def}}{=} y\mathbf{i} + y'\mathbf{j}$ and use an IVP solver to:

(a) Draw component plots of the solution of the IVP.

(b) Draw a direction field and phase portrait for the system.

(c) From the plot(s) in part (b), determine whether each critical point is asymptotically stable, stable, or unstable, and classify it as to type.

20. $m = 1$ kg, $\gamma = 0$ kg/s, $k = 1$ kg/s^2, $y(0) = 1$ m, $y'(0) = 0$ m/s (weak spring)

21. $m = 1$ kg, $\gamma = 0$ kg/s, $k = 16$ kg/s^2, $y(0) = 1$ m, $y'(0) = 0$ m/s (stiff spring)

22. $m = 1$ kg, $\gamma = 1$ kg/s, $k = 4$ kg/s^2, $y(0) = 1$ m, $y'(0) = 0$ m/s (weak damping)

23. $m = 1$ kg, $\gamma = 3$ kg/s, $k = 4$ kg/s^2, $y(0) = 1$ m, $y'(0) = 0$ m/s (strong damping)

24. $m = 1$ kg, $\gamma = -1$ kg/s, $k = 4$ kg/s^2, $y(0) = 1$ m, $y'(0) = 0$ m/s (negative damping)

25. The Linear Versus the Nonlinear Pendulum. ► **ODEA** Convert Eq. (23) and Eq. (24) to planar systems using the state vector $\mathbf{x} = x_1\mathbf{i} + x_2\mathbf{j} = \theta\mathbf{i} + \theta'\mathbf{j}$ for each equation. For each of the following cases of parameter values and initial conditions use an IVP solver to draw, on the same set of coordinate axes, plots of the first component of both the nonlinear and linearized system.

(a) $\gamma = 0$, $\omega^2 = 1$, $\theta(0) = \pi/8$, $\theta'(0) = 0$

(b) $\gamma = 0$, $\omega^2 = 1$, $\theta(0) = \pi/4$, $\theta'(0) = 0$

Based on the graphical output, discuss the dependence of accuracy of the linear approximation on size of initial displacement and operating time. Are the periods of the nonlinear pendulum and the linear pendulum identical?

ODEA **26.** (a) Numerical simulations as well as intuition based on an understanding of the origin and significance of each term in the equation suggest that solutions of the undamped spring-mass equation

$$my'' + ky = 0 \qquad (i)$$

are oscillatory. Use substitution to determine a value of r such that $y_1 = \cos rt$ and $y_2 = \sin rt$ are both solutions of Eq. (i). Then show that $y = c_1 y_1 + c_2 y_2$ is also a solution of $my'' + ky = 0$ for any constants c_1 and c_2.

(b) Use the results of part (a) to find analytical solutions of the following initial value problems for $my'' + ky = 0$.

   i. $m = 1$ kg, $k = 1$ kg/s^2, $y(0) = 1$ m, $y'(0) = -1$ m/s (weak spring)

   ii. $m = 1$ kg, $k = 16$ kg/s^2, $y(0) = 1$ m, $y'(0) = -1$ m/s (stiff spring)

(c) Plot graphs of the two solutions obtained above on the same set of coordinate axes and discuss the amplitude and frequency of oscillations.

4.2 Theory of Second Order Linear Homogeneous Equations

Recall that in Sections 3.2 and 3.3 we stated some general properties of the solutions of systems of two first order linear equations. Some of these results were stated in the context of linear systems with constant coefficients, while others were stated for more general systems. Now we want to extend all of these results to systems with variable coefficients. Further, since a single second order equation can easily be reduced to a system of two first order equations, these results also hold, as special cases, for second order linear equations as well. However, our main emphasis in this chapter is on second order linear equations, so we will state and discuss these important theorems separately for such equations.

Existence and Uniqueness of Solutions

Consider the system

$$dx_1/dt = p_{11}(t)x_1 + p_{12}(t)x_2 + g_1(t),$$
$$dx_2/dt = p_{21}(t)x_1 + p_{22}(t)x_2 + g_2(t), \qquad (1)$$

where the coefficients $p_{11}(t), \dots, p_{22}(t)$ and the nonhomogeneous terms $g_1(t)$ and $g_2(t)$ are continuous on an open interval I. In matrix notation, this system is written as

$$\frac{d\mathbf{x}}{dt} = \mathbf{P}(t)\mathbf{x} + \mathbf{g}(t), \qquad (2)$$

where

$$\mathbf{x} = \begin{pmatrix} x_1 \\ x_2 \end{pmatrix}, \quad \mathbf{P}(t) = \begin{pmatrix} p_{11}(t) & p_{12}(t) \\ p_{21}(t) & p_{22}(t) \end{pmatrix} \quad \text{and} \quad \mathbf{g}(t) = \begin{pmatrix} g_1(t) \\ g_2(t) \end{pmatrix}. \qquad (3)$$

In addition to the system (1) or (2), suppose that initial conditions

$$x_1(t_0) = x_{10}, \quad x_2(t_0) = x_{20} \qquad (4)$$

are given. Here t_0 is any point in I and x_{10} and x_{20} are any given numbers. In vector notation the initial conditions (4) become

$$\mathbf{x}(t_0) = \mathbf{x}_0. \qquad (5)$$

Theorem 3.2.1 stated that the initial value problem (1) and (4), or (2) and (5), has a unique solution on the interval I.

Now consider the second order linear equation

$$y'' + p(t)y' + q(t)y = g(t), \qquad (6)$$

where p, q, and g are continuous functions on the interval I. Initial conditions for Eq. (6) are of the form

$$y(t_0) = y_0, \quad y'(t_0) = y_1. \tag{7}$$

Equation (6) is easily transformed into a system of two first order equations by letting $x_1 = y$ and $x_2 = y'$. Then

$$x_1' = x_2, \quad x_2' = -q(t)x_1 - p(t)x_2 + g(t). \tag{8}$$

In matrix form, this system can be written as

$$\frac{d\mathbf{x}}{dt} = \begin{pmatrix} 0 & 1 \\ -q(t) & -p(t) \end{pmatrix} \mathbf{x} + \begin{pmatrix} 0 \\ g(t) \end{pmatrix}, \tag{9}$$

which is of the form (2). Similarly, the initial conditions (7) can be written as

$$\mathbf{x}(t_0) = \begin{pmatrix} y_0 \\ y_1 \end{pmatrix} = \mathbf{x}_0, \tag{10}$$

which is of the form (5). Then Theorem 3.2.1 applies to the initial value problem (9), (10) and ensures the existence of a unique solution $\mathbf{x} = \boldsymbol{\phi}(t)$ on I. The first component of this solution is the unique solution $y = \phi(t)$ of the initial value problem (6), (7). We state this result in the following theorem.

THEOREM 4.2.1

Let $p(t)$, $q(t)$, and $g(t)$ be continuous on an open interval I, let t_0 be any point in I, and let y_0 and y_0' be any given numbers. Then there exists a unique solution $y = \phi(t)$ of the differential equation (6),

$$y'' + p(t)y' + q(t)y = g(t),$$

that also satisfies the initial conditions (7),

$$y(t_0) = y_0, \quad y'(t_0) = y_1.$$

Further, the solution exists throughout the interval I.

If the coefficients $p(t)$ and $q(t)$ are constants, and if the nonhomogeneous term $g(t)$ is not too complicated, then it is possible to solve the initial value problem (6), (7) by elementary methods. We will show you how to do this later in this chapter. Of course, the construction of a solution demonstrates the existence part of Theorem 4.2.1, although not the uniqueness part. A general proof of Theorem 4.2.1 is fairly difficult, and we do not discuss it here. We will, however, accept Theorem 4.2.1 as true and make use of it whenever necessary.

EXAMPLE 1

Find the longest interval in which the solution of the initial value problem

$$(t^2 - 3t)y'' + ty' - (t + 3)y = 0, \qquad y(1) = 2, \quad y'(1) = 1 \tag{11}$$

is certain to exist.

If the given differential equation is written in the form of Eq. (6), then $p(t) = 1/(t-3)$, $q(t) = -(t + 3)/t(t-3)$, and $g(t) = 0$. The only points of discontinuity of the coefficients are $t = 0$ and $t = 3$. Therefore, the longest open interval, containing the initial point $t_0 = 1$, in which all the coefficients are continuous is $0 < t < 3$. Thus this is the longest interval in which Theorem 4.2.1 guarantees that the solution exists (see Figure 4.2.1).

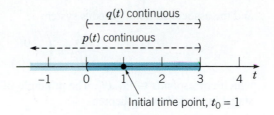

The maximum interval of existence and uniqueness for the initial value problem (11) guaranteed by Theorem 4.2.1 is $I = (0, 3)$.

In developing the theory of linear differential equations, it is helpful to introduce a differential operator notation. For any function y that is twice differentiable on I, we define the **differential operator** L by the equation

$$L[y] = y'' + py' + qy. \tag{12}$$

Note that $L[y]$ is a function on I. The value of $L[y]$ at a point t is

$$L[y](t) = y''(t) + p(t)y'(t) + q(t)y(t).$$

For example, if $p(t) = t^2$, $q(t) = 1+t$, and $y(t) = \sin 3t$, then

$$
\begin{aligned}
L[y](t) &= (\sin 3t)'' + t^2(\sin 3t)' + (1+t)\sin 3t \\
&= -9\sin 3t + 3t^2\cos 3t + (1+t)\sin 3t \\
&= 3t^2\cos 3t + (t-8)\sin 3t.
\end{aligned}
$$

Thus, L maps the function $\sin 3t$ to the function $3t^2\cos 3t + (t-8)\sin 3t$. In particular, at $t = \pi/6$,

$$L[y](\pi/6) = 3\left(\frac{\pi^2}{36}\right)\cos(\pi/2) + (\pi/6 - 8)\sin(\pi/2) = \pi/6 - 8.$$

The operator L is often written as $L = D^2 + pD + q$, where D is the derivative operator,

$$D = \frac{d}{dt}.$$

In terms of $L = y'' + py' + qy$, the nonhomogeneous differential equation $y'' + p(t)y' + q(t)y = g(t)$ is expressed as $L[y] = g$ and the corresponding homogeneous equation $y'' + p(t)y' + q(t)y = 0$ is expressed as $L[y] = 0$.

The Principle of Superposition for Linear Homogeneous Equations

In Theorem 3.3.1, we established a result known as the principle of superposition for a system of two first order linear homogeneous equations with constant coefficients. The result also holds for systems whose coefficients are not constants, and indeed it is valid for systems of arbitrary size as well, as we will see in Chapter 7. However, for the present, we consider the homogeneous system corresponding to Eq. (2), namely,

$$\mathbf{x}' = \mathbf{P}(t)\mathbf{x}, \tag{13}$$

where the elements of $\mathbf{P}$ are continuous functions of the independent variable t. We also consider the second order homogeneous equation

$$L[y] = y'' + p(t)y' + q(t)y = 0 \tag{14}$$

and the associated homogeneous system

$$\frac{d\mathbf{x}}{dt} = \begin{pmatrix} 0 & 1 \\ -q(t) & -p(t) \end{pmatrix} \mathbf{x}, \tag{15}$$

which corresponds to Eq. (9). The principle of superposition, as it applies to Eq. (13), is stated in the following theorem.

THEOREM 4.2.2

Suppose that $\mathbf{x}_1$ and $\mathbf{x}_2$ are solutions of the system (13)

$$\mathbf{x}' = \mathbf{P}(t)\mathbf{x},$$

where each element of $\mathbf{P}(t)$ is continuous. Then the linear combination

$$\mathbf{x} = c_1\mathbf{x}_1(t) + c_2\mathbf{x}_2(t) \tag{16}$$

is also a solution for any values of the constants c_1 and c_2.

Proof

The theorem is proved by substituting from Eq. (16) for $\mathbf{x}$ in Eq. (13) and using elementary properties of differentiation. We have

$$\frac{d\mathbf{x}}{dt} - \mathbf{P}(t)\mathbf{x} = \frac{d\left[c_1\mathbf{x}_1(t) + c_2\mathbf{x}_2(t)\right]}{dt} - \mathbf{P}(t)\left[c_1\mathbf{x}_1(t) + c_2\mathbf{x}_2(t)\right]$$

$$= c_1\frac{d\mathbf{x}_1(t)}{dt} + c_2\frac{d\mathbf{x}_2(t)}{dt} - c_1\mathbf{P}(t)\mathbf{x}_1(t) - c_2\mathbf{P}(t)\mathbf{x}_2(t)$$

$$= c_1\left[\frac{d\mathbf{x}_1(t)}{dt} - \mathbf{P}(t)\mathbf{x}_1(t)\right] + c_2\left[\frac{d\mathbf{x}_2(t)}{dt} - \mathbf{P}(t)\mathbf{x}_2(t)\right]$$

$$= \mathbf{0} + \mathbf{0} = \mathbf{0}. \tag{17}$$

The last line of Eq. (17) follows from the preceding one because $\mathbf{x}_1$ and $\mathbf{x}_2$ are solutions of Eq. (13). ∎

Theorem 4.2.2 applies, of course, to the system (15), provided that $p(t)$ and $q(t)$ are continuous. By looking at the first row of this result, we obtain the principle of superposition for the second order equation (14). We state it as a separate theorem.

THEOREM 4.2.3

Suppose that $p(t)$ and $q(t)$ are continuous and that y_1 and y_2 are two solutions of Eq. (14),

$$y'' + p(t)y' + q(t)y = 0.$$

Then the linear combination

$$y = c_1y_1(t) + c_2y_2(t) \tag{18}$$

is also a solution for any values of the constants c_1 and c_2.

Wronskians and Fundamental Sets of Solutions

The importance of Theorems 4.2.2 and 4.2.3 is that they enable you to greatly enlarge the set of solutions that you have, provided of course that you are able to find two solutions to start with. The next question is whether the enlarged solution set is sufficient for us to satisfy a given set of initial conditions by an appropriate choice of the constants c_1 and c_2. For the constant coefficient system $\mathbf{x}' = \mathbf{A}\mathbf{x}$, this question was answered in Theorem 3.3.2.

The same result also holds for the system (13) whose coefficients are functions of t, and it can be established by the same argument. We briefly repeat the development here.

Let us suppose that $\mathbf{x}_1$ and $\mathbf{x}_2$ are solutions of Eq. (13),

$$\mathbf{x}' = \mathbf{P}(t)\mathbf{x}$$

and consider the linear combination (16),

$$\mathbf{x} = c_1\mathbf{x}_1(t) + c_2\mathbf{x}_2(t).$$

By Theorem 4.2.2, this expression is also a solution for any values of c_1 and c_2. To satisfy the initial condition (5),

$$\mathbf{x}(t_0) = \mathbf{x}_0,$$

where $\mathbf{x}_0$ is any given vector, we must choose c_1 and c_2 to satisfy

$$c_1\mathbf{x}_1(t_0) + c_2\mathbf{x}_2(t_0) = \mathbf{x}_0. \tag{19}$$

Writing out Eq. (19) more fully, we have

$$\begin{pmatrix} x_{11}(t_0) & x_{12}(t_0) \\ x_{21}(t_0) & x_{22}(t_0) \end{pmatrix} \begin{pmatrix} c_1 \\ c_2 \end{pmatrix} = \begin{pmatrix} x_{10} \\ x_{20} \end{pmatrix}. \tag{20}$$

Recall that the first subscript refers to the component of a vector, while the second subscript identifies the vector itself. Thus, for example, x_{21} is the second component of the vector $\mathbf{x}_1$. Equation (20) has a unique solution for c_1 and c_2 regardless of the values of x_{10} and x_{20} if and only if the coefficient matrix in Eq. (20) is nonsingular, that is, if and only if the determinant of this matrix is nonzero. Since t_0 can be any point in I, the determinant must be nonzero throughout I.

As in Section 3.3, the determinant

$$W[\mathbf{x}_1, \mathbf{x}_2](t) = \begin{vmatrix} x_{11}(t) & x_{12}(t) \\ x_{21}(t) & x_{22}(t) \end{vmatrix} \tag{21}$$

is called the Wronskian of $\mathbf{x}_1$ and $\mathbf{x}_2$. Two solutions $\mathbf{x}_1$ and $\mathbf{x}_2$ whose Wronskian is not zero form a fundamental set of solutions. A linear combination of a fundamental set of solutions with arbitrary coefficients constitutes the general solution of the system (13). Since every possible set of initial conditions can be satisfied by a proper choice of c_1 and c_2, the general solution includes all solutions of the system (13). This last statement rests on the uniqueness part of Theorem 3.2.1, which asserts that only one solution can satisfy a given set of initial conditions. We have thus proved the following theorem.

THEOREM 4.2.4

Suppose that $\mathbf{x}_1$ and $\mathbf{x}_2$ are two solutions of Eq. (13),

$$\mathbf{x}' = \mathbf{P}(t)\mathbf{x}$$

and that their Wronskian is not zero on I. Then $\mathbf{x}_1$ and $\mathbf{x}_2$ form a fundamental set of solutions of Eq. (13), and the general solution is given by Eq. (16),

$$\mathbf{x} = c_1\mathbf{x}_1(t) + c_2\mathbf{x}_2(t),$$

where c_1 and c_2 are arbitrary constants. If there is a given initial condition $\mathbf{x}(t_0) = \mathbf{x}_0$, then this condition determines the constants c_1 and c_2 uniquely.

Let us now consider the second order equation (14),

$$y'' + p(t)y' + q(t)y = 0,$$

and the system (15)

$$\frac{d\mathbf{x}}{dt} = \begin{pmatrix} 0 & 1 \\ -q(t) & -p(t) \end{pmatrix}\mathbf{x}.$$

Theorem 4.2.4 applies to the system (15) so long as $p(t)$ and $q(t)$ are continuous. To interpret the result in terms of Eq. (14), we need to remember that $x_1 = y$ and $x_2 = y'$. Further, if y_1 and y_2 are two solutions of Eq. (14), then $x_{11} = y_1$ and $x_{21} = y_1'$. Similarly, $x_{12} = y_2$ and $x_{22} = y_2'$. Then it follows from Eq. (21) that the Wronskian of y_1 and y_2 is

$$W[y_1, y_2](t) = \begin{vmatrix} y_1(t) & y_2(t) \\ y_1'(t) & y_2'(t) \end{vmatrix} = y_1(t)y_2'(t) - y_1'(t)y_2(t). \tag{22}$$

Then, restating Theorem 4.2.4 for the second order linear equation (14), we have the following theorem.

THEOREM 4.2.5

Let y_1 and y_2 be two solutions of Eq. (14),

$$y'' + p(t)y' + q(t)y = 0,$$

and assume that their Wronskian is not zero on I. Then y_1 and y_2 form a fundamental set of solutions, and the general solution is given by Eq. (18),

$$y = c_1 y_1(t) + c_2 y_2(t),$$

where c_1 and c_2 are arbitrary constants. If there are given initial conditions $y(t_0) = y_0$ and $y'(t_0) = y_1$, then these conditions determine c_1 and c_2 uniquely.

Theorems 4.2.4 and 4.2.5 clarify the structure of the solutions of all systems of two linear homogeneous first order equations, and of all second order linear homogeneous equations, respectively. Assuming that the coefficients are continuous in each case, the general solution is just a linear combination of some pair of solutions whose Wronskian determinant is not zero. In Chapter 3, we saw that it is relatively easy to find such a pair of solutions of a system with constant coefficients. We will show in this chapter that it is also easy to do this for a second order linear equation with constant coefficients. For systems or equations with variable coefficients, the task of finding a suitable pair of solutions is usually much more challenging.

Abel's Equation for the Wronskian

The Wronskian, as defined by Eq. (21) for a system of first order equations or by Eq. (22) for a second order equation, can be calculated once we know two solutions. However, it turns out that the Wronskian can also be determined directly from the differential equation(s), without knowing any solutions. This result has some important consequences, so we state it in the following theorem.

THEOREM 4.2.6

(**Abel's Theorem**). The Wronskian of two solutions of the system (13),

$$\mathbf{x}' = \mathbf{P}(t)\mathbf{x},$$

is given by

$$W(t) = c\left[\exp\int \operatorname{tr}(\mathbf{P}(t))\, dt\right] = c\, \exp\int [p_{11}(t) + p_{22}(t)]\, dt, \tag{23}$$

where c is a constant that depends on the pair of solutions. The Wronskian of two solutions of the second order equation (14),

$$y'' + p(t)y' + q(t)y = 0,$$

is given by

$$W(t) = c\, \exp\left[-\int p(t)\, dt\right], \tag{24}$$

where again c is a constant that depends on the pair of solutions.

Proof

The proof of this theorem involves the construction and solution of a differential equation satisfied by the Wronskian. We start with the Wronskian for the system (13); from Eq. (21) we have

$$W(t) = \begin{vmatrix} x_{11}(t) & x_{12}(t) \\ x_{21}(t) & x_{22}(t) \end{vmatrix} = x_{11}(t)x_{22}(t) - x_{12}(t)x_{21}(t). \tag{25}$$

Hence the derivative of $W(t)$ is

$$W'(t) = x'_{11}(t)x_{22}(t) + x_{11}(t)x'_{22}(t) - x'_{12}(t)x_{21}(t) - x_{12}(t)x'_{21}(t). \tag{26}$$

The next step is to replace all of the differentiated factors on the right side of Eq. (26) by substituting for them from the system of differential equations. For example, we have

$$x'_{11}(t) = p_{11}(t)x_{11}(t) + p_{12}(t)x_{21}(t)$$

and similarly for $x'_{22}(t)$, $x'_{12}(t)$, and $x'_{21}(t)$. By making these substitutions we obtain

$$\begin{aligned} W' = \ &(p_{11}x_{11} + p_{12}x_{21})x_{22} + x_{11}(p_{21}x_{12} + p_{22}x_{22}) \\ &-(p_{11}x_{12} + p_{12}x_{22})x_{21} - x_{12}(p_{21}x_{11} + p_{22}x_{21}), \end{aligned} \tag{27}$$

where for brevity we have omitted the independent variable t in each term. On examining the terms in Eq. (27), we observe that those involving p_{12} and p_{21} cancel. By rearranging the remaining terms, we find that

$$W' = (p_{11} + p_{22})(x_{11}x_{22} - x_{12}x_{21}),$$

or

$$\frac{dW}{dt} = [p_{11}(t) + p_{22}(t)]\, W = \operatorname{tr}(\mathbf{P}(t))\, W. \tag{28}$$

Equation (28) is a first order separable or linear differential equation that is satisfied by the Wronskian. Its solution is readily found and is given by Eq. (23) in Theorem 4.2.6.

To establish Eq. (24), we can apply the result (23) to the system (15) that corresponds to the second order equation. For this system, $p_{11}(t) + p_{22}(t) = -p(t)$; making this substitution in Eq. (23) immediately yields Eq. (24). Equation (24) can also be derived directly from Eq. (22) by a process similar to the derivation of Eq. (23) that is given above. ∎

EXAMPLE 2

Find the Wronskian of any pair of solutions of the system

$$\mathbf{x}' = \begin{pmatrix} 1 & 1 \\ 4 & 1 \end{pmatrix} \mathbf{x}. \tag{29}$$

The trace of the coefficient matrix is 2; therefore by Eq. (23) the Wronskian of any pair of solutions is

$$W(t) = ce^{2t}. \tag{30}$$

In Example 4 of Section 3.3, we found two solutions of the system (29) to be

$$\mathbf{x}_1(t) = \begin{pmatrix} 1 \\ 2 \end{pmatrix} e^{3t}, \qquad \mathbf{x}_2(t) = \begin{pmatrix} 1 \\ -2 \end{pmatrix} e^{-t}.$$

The Wronskian of these two solutions is

$$W(t) = \begin{vmatrix} e^{3t} & e^{-t} \\ 2e^{3t} & -2e^{-t} \end{vmatrix} = -4e^{2t}.$$

Thus for this pair of solutions the multiplicative constant in Eq. (30) is -4. For other pairs of solutions the multiplicative constant in the Wronskian may change, but the exponential part will always be the same.

EXAMPLE 3

Find the Wronskian of any pair of solutions of

$$(1 - t)y'' + ty' - y = 0. \tag{31}$$

First we rewrite Eq. (31) in the standard form (14):

$$y'' + \frac{t}{1 - t} y' - \frac{1}{1 - t} y = 0. \tag{32}$$

Thus $p(t) = t/(1-t)$ and, omitting the constant of integration,

$$-\int p(t)\, dt = \int \frac{t}{t - 1}\, dt = \int \left(1 + \frac{1}{t - 1} \right) dt = t + \ln |t - 1|.$$

Consequently,

$$\exp\left[-\int p(t)\, dt \right] = |t - 1| e^t$$

and Theorem 4.2.6 gives

$$W(t) = c(t - 1)e^t. \tag{33}$$

Note that $|t - 1| = \pm (t - 1)$ and that the $\pm$ sign has been incorporated into the arbitrary constant c in Eq. (33).

You can verify that $y_1(t) = t$ and $y_2(t) = e^t$ are two solutions of Eq. (31) by substituting these two functions into the differential equation. The Wronskian of these two solutions is

$$W(t) = \begin{vmatrix} t & e^t \\ 1 & e^t \end{vmatrix} = (t - 1)e^t. \tag{34}$$

Thus, for these two solutions, the constant in Eq. (33) is one.

An important property of the Wronskian follows immediately from Theorem 4.2.6. If $p_{11}(t)$ and $p_{22}(t)$ in Eq. (23) or $p(t)$ in Eq. (24) are continuous on I, then the exponential functions in these equations are always positive. If the constant c is nonzero, then the Wronskian is never zero on I. On the other hand, if $c = 0$, then the Wronskian is zero for all t in I. Thus the Wronskian is either never zero or always zero. In a particular problem you can determine which alternative is the actual one by evaluating the Wronskian at some single convenient point.

It may seem at first that Eq. (31) in Example 3 is a counterexample to this assertion about the Wronskian because $W(t)$ given by Eq. (34) is obviously zero for $t = 1$ and nonzero otherwise. However, by writing Eq. (31) in the form (32), we see that the coefficients $p(t)$ and $q(t)$ become unbounded as $t \to 1$ and hence are discontinuous there. In any interval I not including $t = 1$, the coefficients are continuous and the Wronskian is nonzero, as claimed.

PROBLEMS

In each of Problems 1 through 6, determine the longest interval in which the given initial value problem is certain to have a unique twice differentiable solution. Do not attempt to find the solution.

1. $ty'' + 3y = t$, $\quad y(1) = 1$, $\quad y'(1) = 2$
2. $(t-1)y'' - 3ty' + 4y = \sin t$, $\quad y(-2) = 2$, $y'(-2) = 1$
3. $t(t-4)y'' + 3ty' + 4y = 2$, $\quad y(3) = 0$, $y'(3) = -1$
4. $y'' + (\cos t)y' + 3(\ln|t|)y = 0$, $\quad y(2) = 3$, $y'(2) = 1$
5. $(x-3)y'' + xy' + (\ln|x|)y = 0$, $\quad y(1) = 0$, $y'(1) = 1$
6. $(x-2)y'' + y' + (x-2)(\tan x)y = 0$, $y(3) = 1$, $\quad y'(3) = 2$

In each of Problems 7 through 12, find the Wronskian of the given pair of functions.

7. e^{2t}, $\quad e^{-3t/2}$ $\qquad$ 8. $\cos t$, $\quad \sin t$
9. e^{-2t}, $\quad te^{-2t}$ $\qquad$ 10. x, $\quad xe^x$
11. $e^t \sin t$, $\quad e^t \cos t$ $\qquad$ 12. $\cos^2 \theta$, $\quad 1 + \cos 2\theta$

13. Verify that $y_1(t) = t^2$ and $y_2(t) = t^{-1}$ are two solutions of the differential equation $t^2 y'' - 2y = 0$ for $t > 0$. Then show that $c_1 t^2 + c_2 t^{-1}$ is also a solution of this equation for any c_1 and c_2.

14. Consider the differential operator T defined by $T[y] = yy'' + (y')^2$. Show that $T[y] = 0$ is a nonlinear equation by three methods below.
    (a) Explain why $T[y] = 0$ cannot be put in the form of a linear equation $L[y] = y'' + py' + qy = 0$.
    (b) Show that T fails to satisfy the definition of a linear operator and therefore the equation $T[y] = 0$ is a nonlinear equation.

(c) Verify that $y_1(t) = 1$ and $y_2(t) = t^{1/2}$ are solutions of the differential equation $T[y] = 0$ for $t > 0$ but $c_1 + c_2 t^{1/2}$ is not, in general, a solution of this equation.

15. Can $y = \sin(t^2)$ be a solution on an interval containing $t = 0$ of an equation $y'' + p(t)y' + q(t)y = 0$ with continuous coefficients? Explain your answer.

16. If the Wronskian W of f and g is $3e^{4t}$, and if $f(t) = e^{2t}$, find $g(t)$.

17. If the Wronskian W of f and g is $t^2 e^t$, and if $f(t) = t$, find $g(t)$.

18. If $W[f, g]$ is the Wronskian of f and g, and if $u = 2f - g$, $v = f + 2g$, find the Wronskian $W[u, v]$ of u and v in terms of $W[f, g]$.

19. If the Wronskian of f and g is $t \cos t - \sin t$, and if $u = f + 3g$, $v = f - g$, find the Wronskian of u and v.

In each of Problems 20 through 23, verify that the functions y_1 and y_2 are solutions of the given differential equation. Do they constitute a fundamental set of solutions?

20. $y'' + 4y = 0$; $\quad y_1(t) = \cos 2t$, $\quad y_2(t) = \sin 2t$
21. $y'' - 2y' + y = 0$; $\quad y_1(t) = e^t$, $y_2(t) = te^t$
22. $x^2 y'' - x(x+2)y' + (x+2)y = 0$, $\quad x > 0$; $y_1(x) = x$, $\quad y_2(x) = xe^x$
23. $(1 - x \cot x)y'' - xy' + y = 0$, $\quad 0 < x < \pi$; $y_1(x) = x$, $\quad y_2(x) = \sin x$

24. Consider the equation $y'' - y' - 2y = 0$.
    (a) Show that $y_1(t) = e^{-t}$ and $y_2(t) = e^{2t}$ form a fundamental set of solutions.

(b) Let $y_3(t) = -2e^{2t}$, $y_4(t) = y_1(t) + 2y_2(t)$, and $y_5(t) = 2y_1(t) - 2y_3(t)$. Are $y_3(t)$, $y_4(t)$, and $y_5(t)$ also solutions of the given differential equation?

(c) Determine whether each of the following pairs form a fundamental set of solutions: $[y_1(t), y_3(t)]$; $[y_2(t), y_3(t)]$; $[y_1(t), y_4(t)]$; $[y_4(t), y_5(t)]$.

4.3 Linear Homogeneous Equations with Constant Coefficients

In this section we study the problem of finding fundamental sets of solutions of the linear homogeneous second order differential equation with constant coefficients

$$ay'' + by' + cy = 0, \tag{1}$$

where a, b, and c are given real numbers. We assume that $a \neq 0$, since otherwise Eq. (1) is only of first order. We encountered this type of equation in Section 4.1 as a model for unforced spring-mass systems, the linearized pendulum, and RLC circuits.

One way we might proceed is to transform Eq. (1) into a system of first order equations, by letting $x_1 = y$ and $x_2 = y'$. This leads to the system

$$\mathbf{x}' = \begin{pmatrix} 0 & 1 \\ -c/a & -b/a \end{pmatrix} \mathbf{x}. \tag{2}$$

We can solve Eq. (2) using the eigenvalue methods discussed in Sections 3.3 through 3.5. The first component of the vector solution then provides the solution of Eq. (1) (see Problems 17 and 18). However, it is usually easier to solve Eq. (1) directly, so this is what we will do.

The method described here does not apply to the linear equations $y'' + py' + qy = 0$ or $Py'' + Qy' + Ry = 0$ if the coefficients are not constants. In such cases, infinite series methods or numerical methods are required to find either exact solutions or approximations of solutions.

Characteristic Equations

Guided by our experience in solving systems such as (2), let us look for solutions of Eq. (1) that are of the form $y = e^{\lambda t}$, where λ is a parameter to be determined. If we substitute $y = e^{\lambda t}$ into Eq. (1), we obtain

$$a\lambda^2 e^{\lambda t} + b\lambda e^{\lambda t} + ce^{\lambda t} = (a\lambda^2 + b\lambda + c)e^{\lambda t} = 0.$$

Since $e^{\lambda t}$ can never take on the value zero, we conclude that $y = e^{\lambda t}$ is a solution of Eq. (1) if and only if λ satisfies the equation

$$Z(\lambda) = a\lambda^2 + b\lambda + c = 0. \tag{3}$$

The polynomial $Z(\lambda) = a\lambda^2 + b\lambda + c$ is called the **characteristic polynomial** and Eq. (3) is called the **characteristic equation** for the differential equation (1). It is easy to show that Eq. (3) is the same as the characteristic equation that determines the eigenvalues of the coefficient matrix of the system (2).

Since Eq. (3) is a quadratic equation with real coefficients, it has two roots

$$\lambda_1 = \frac{-b + \sqrt{b^2 - 4ac}}{2a} \quad \text{and} \quad \lambda_2 = \frac{-b - \sqrt{b^2 - 4ac}}{2a}. \tag{4}$$

There are three cases to consider depending on whether the discriminant $b^2 - 4ac$ is positive, zero, or negative:

> **1.** If $b^2 - 4ac > 0$, the roots are real and distinct, $\lambda_1 \neq \lambda_2$.
> **2.** If $b^2 - 4ac = 0$, the roots are real and equal, $\lambda_1 = \lambda_2$.
> **3.** If $b^2 - 4ac < 0$, the roots are complex conjugates, $\lambda_1 = \mu + i\nu$ and $\lambda_2 = \mu - i\nu$ where $\mu = -b/2a$ and $\nu = \sqrt{4ac - b^2}/2a$.

In each case, the nature of the roots determines the particular form of the solutions in the fundamental set for Eq. (1). The rest of this section is devoted to the study of the two cases corresponding to real roots while the case of complex conjugate roots is taken up in Section 4.4.

Characteristic Equations with Distinct Real Roots

Assume that the roots λ_1 and λ_2 of the characteristic equation (3) are real and distinct, that is, $\lambda_1 \neq \lambda_2$. This case occurs when the discriminant $b^2 - 4ac$ of Eq. (3) is positive and the factored form of the characteristic polynomial is

$$Z(\lambda) = a(\lambda - \lambda_1)(\lambda - \lambda_2), \qquad \lambda_1 \neq \lambda_2.$$

Then $y_1(t) = e^{\lambda_1 t}$ and $y_2(t) = e^{\lambda_2 t}$ are two solutions of Eq. (1) on $I = (-\infty, \infty)$. The Wronskian of y_1 and y_2 is

$$W[y_1, y_2](t) = \begin{vmatrix} e^{\lambda_1 t} & e^{\lambda_2 t} \\ \lambda_1 e^{\lambda_1 t} & \lambda_2 e^{\lambda_2 t} \end{vmatrix} = (\lambda_2 - \lambda_1)e^{(\lambda_1 + \lambda_2)t}.$$

Depending on the algebraic sign of $\lambda_2 - \lambda_1 \neq 0$, $W[y_1, y_2](t)$ is either strictly positive or strictly negative on I. By Theorem 4.2.5 the general solution of Eq. (1) is

$$y = c_1 e^{\lambda_1 t} + c_2 e^{\lambda_2 t}. \tag{5}$$

If we want to find the particular member of the family of solutions (5) that satisfies the initial conditions

$$y(t_0) = y_0, \qquad y'(t_0) = y_1,$$

we substitute $t = t_0$ and $y = y_0$ in Eq. (5) to obtain

$$c_1 e^{\lambda_1 t_0} + c_2 e^{\lambda_2 t_0} = y_0. \tag{6}$$

Next, we differentiate Eq. (5) and set $t = t_0$ and $y'(t_0) = y_1$ in the result to obtain

$$c_1 \lambda_1 e^{\lambda_1 t_0} + c_2 \lambda_2 e^{\lambda_2 t_0} = y_1. \tag{7}$$

On solving Eqs. (6) and (7) simultaneously for c_1 and c_2, we find that

$$c_1 = \frac{y_1 - y_0 \lambda_2}{\lambda_1 - \lambda_2} e^{-\lambda_1 t_0}, \qquad c_2 = \frac{y_0 \lambda_1 - y_1}{\lambda_1 - \lambda_2} e^{-\lambda_2 t_0}. \tag{8}$$

With the values of c_1 and c_2 given by Eq. (8), the expression (5) is the unique solution (by Theorem 4.2.1) of the initial value problem

$$ay'' + by' + cy = 0, \qquad y(t_0) = y_0, \quad y'(t_0) = y_1. \tag{9}$$

EXAMPLE 1

Find the general solution of

$$y'' + 5y' + 6y = 0. \tag{10}$$

We assume that $y = e^{\lambda t}$, and it then follows that λ must be a root of the characteristic equation

$$\lambda^2 + 5\lambda + 6 = (\lambda + 2)(\lambda + 3) = 0.$$

Thus the possible values of λ are $\lambda_1 = -2$ and $\lambda_2 = -3$. The general solution of Eq. (10) is therefore

$$y = c_1 e^{-2t} + c_2 e^{-3t}. \tag{11}$$

EXAMPLE 2

Find the solution of the initial value problem

$$y'' + 5y' + 6y = 0, \qquad y(0) = 2, \quad y'(0) = 3. \tag{12}$$

The general solution of the differential equation was found in Example 1 and is given by Eq. (11). To satisfy the first initial condition, we set $t = 0$ and $y = 2$ in Eq. (11); thus c_1 and c_2 must satisfy

$$c_1 + c_2 = 2. \tag{13}$$

To use the second initial condition, we must first differentiate Eq. (11). This gives $y' = -2c_1 e^{-2t} - 3c_2 e^{-3t}$. Then, setting $t = 0$ and $y' = 3$, we obtain

$$-2c_1 - 3c_2 = 3. \tag{14}$$

By solving Eqs. (13) and (14), we find that $c_1 = 9$ and $c_2 = -7$. Using these values in the expression (11), we obtain the solution

$$y = 9e^{-2t} - 7e^{-3t} \tag{15}$$

of the initial value problem (12). The graph of the solution is shown in Figure 4.3.1.

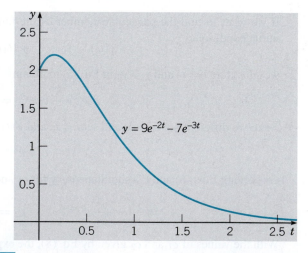

FIGURE 4.3.1   Solution of $y'' + 5y' + 6y = 0$, $y(0) = 2$, $y'(0) = 3$.

EXAMPLE 3

Formulate Eq. (10) as a dynamical system, discuss the corresponding phase portrait, and draw the trajectory associated with the solution (15).

The dynamical system corresponding to the state vector $\mathbf{x} = x_1\mathbf{i} + x_2\mathbf{j} = y\mathbf{i} + y'\mathbf{j}$ is

$$\mathbf{x}' = \begin{pmatrix} 0 & 1 \\ -6 & -5 \end{pmatrix}\mathbf{x}. \tag{16}$$

By using Eq. (11) together with its derivative, $x_2 = y' = -2c_1e^{-2t} - 3c_2e^{-3t}$, or by solving the system (16) as in Section 3.3, we obtain

$$\mathbf{x} = \begin{pmatrix} x_1 \\ x_2 \end{pmatrix} = c_1 \begin{pmatrix} 1 \\ -2 \end{pmatrix} e^{-2t} + c_2 \begin{pmatrix} 1 \\ -3 \end{pmatrix} e^{-3t} = c_1\mathbf{x}_1(t) + c_2\mathbf{x}_2(t). \tag{17}$$

For each choice of c_1 and c_2, a plot of the curve $t \to \mathbf{x}$ yields a corresponding phase plane trajectory. In particular, the choice $c_2 = 0$ yields $x_1 = c_1e^{-2t}, x_2 = -2c_1e^{-2t}$ corresponding to the straight line trajectory $x_2 = -2x_1$ and the choice $c_1 = 0$ yields $x_1 = c_2e^{-3t}, x_2 = -3c_2e^{-3t}$ corresponding to the straight line trajectory $x_2 = -3x_1$ (see Figure 4.3.2). The choice $c_1 = 9, c_2 = -7$ gives the trajectory associated with the solution (15) of Example 2 (the heavy trajectory in Figure 4.3.2). An alternative parametrization of the orbits in terms of initial conditions is possible by using (8) to express c_1 and c_2 in terms of y_0 and y_1. The phase portrait drawn in Figure 4.3.2 shows that the critical point $(0, 0)$ is an asymptotically stable node.

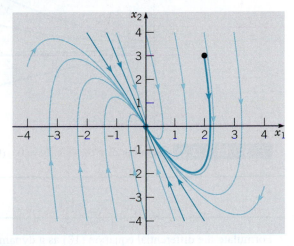

FIGURE 4.3.2 Phase portrait for the dynamical system (3) associated with Eq. (10).

EXAMPLE 4

Find solutions of the initial value problems

$$y'' + y' - 2y = 0, \qquad y(0) = 1, \quad y'(0) = y_1. \tag{18}$$

for the four initial slopes $y_1 = -4, -2, 0, 2$.

If $y = e^{\lambda t}$, then the characteristic equation is

$$\lambda^2 + \lambda - 2 = 0$$

and its roots are $\lambda = -2$ and $\lambda = 1$. Therefore the general solution of the differential equation is

$$y = c_1 e^{-2t} + c_2 e^t. \tag{19}$$

Applying the initial conditions, we obtain the following two equations for c_1 and c_2:

$$c_1 + c_2 = 1, \qquad -2c_1 + c_2 = y_1.$$

The solution of these equations is $c_1 = (1 - y_1)/3$, $c_2 = (2 + y_1)/3$, and the solution of the initial value problem (18), parameterized by initial slope y_1, is

$$y = \frac{1 - y_1}{3} e^{-2t} + \frac{2 + y_1}{3} e^t. \tag{20}$$

Graphs of the solutions for each of the four initial slopes $y'(0) = y_1 = -4, -2, 0, 2$ are shown in Figure 4.3.3.

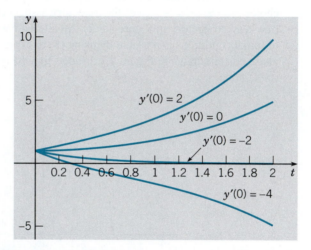

FIGURE 4.3.3 Solution of $y'' + y' - 2y = 0$, $y(0) = 1$, $y'(0) = -4, -2, 0, 2$.

EXAMPLE 5

Formulate the differential equation (18) as a dynamical system, discuss the phase portrait, and draw the trajectories corresponding to the four initial slopes specified in Example 4.

The dynamical system corresponding to the state vector $\mathbf{x} = x_1 \mathbf{i} + x_2 \mathbf{j} = y \mathbf{i} + y' \mathbf{j}$ is

$$x_1' = x_2,$$
$$x_2' = 2x_1 - x_2. \tag{21}$$

Equation (19) together with its derivative, $x_2 = y' = -2c_1 e^{-2t} + c_2 e^t$, gives us a two parameter family of trajectories,

$$\mathbf{x} = \begin{pmatrix} x_1 \\ x_2 \end{pmatrix} = \begin{pmatrix} c_1 e^{-2t} + c_2 e^t \\ -2c_1 e^{-2t} + c_2 e^t \end{pmatrix}. \tag{22}$$

For each choice of c_1 and c_2, a plot of the curve $t \to (x_1(t), x_2(t))$ yields a corresponding phase plane trajectory. The phase portrait for the dynamical system (21) is drawn in Figure 4.3.4.

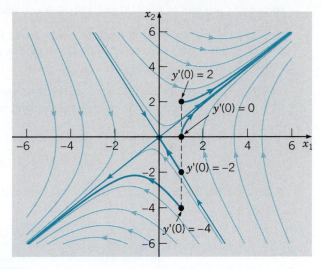

FIGURE 4.3.4   Phase portrait for the dynamical system (21) associated with $y'' + y' - 2y = 0$.

The choice $c_2 = 0$ yields $x_1 = c_1 e^{-2t}$, $x_2 = -2c_1 e^{-2t}$ corresponding to trajectories that lie on the line $x_2 = -2x_1$ and the choice $c_1 = 0$ yields $x_1 = c_2 e^t$, $x_2 = c_2 e^t$ corresponding to trajectories on the line $x_2 = x_1$. Note that all solutions corresponding to initial conditions on the line $x_2 = -2x_1$ approach the origin $(0, 0)$ as $t \to \infty$. The heavy trajectories in Figure 4.3.4 correspond to solutions of the initial value problem (18) using $y_1 = -4, -2, 0, 2$. From the phase portrait it is clear that $(0, 0)$ is a saddle point, and hence it is unstable. Note that the solution corresponding to the initial condition $y(0) = 1$, $y'(0) = y_1 = -2$ yields $c_1 = 1$ and $c_2 = 0$, that is, the solution is $x_1 = e^{-2t}$, $x_2 = -2e^{-2t}$ which means that the orbit lies on the line $x_2 = -2x_1$.

Characteristic Equations with Real Repeated Roots; Reduction of Order

We now consider the case when the two roots of the characteristic equation

$$a\lambda^2 + b\lambda + c = 0 \tag{23}$$

are equal, that is, $\lambda_1 = \lambda_2$. This is the case if the discriminant $b^2 - 4ac$ of Eq. (23) is zero. Then

$$\lambda_1 = \lambda_2 = -b/2a \tag{24}$$

and the factored form of the characteristic polynomial is

$$Z(\lambda) = a(\lambda - \lambda_1)^2.$$

Since λ_1 is a root of Eq. (23), the function $y_1 = e^{\lambda_1 t}$ is a solution of Eq. (1). In the case of repeated roots, we show directly that

$$y_2(t) = ty_1(t) = te^{\lambda_1 t} = te^{-bt/2a} \tag{25}$$

is also a solution of Eq. (1). Substituting y_2 into Eq. (1) we find that

$$(a\lambda_1^2 + b\lambda_1 + c)te^{\lambda_1 t} + (2a\lambda_1 + b)e^{\lambda_1 t} = 0$$

because λ_1 is a root of Eq. (23) and $2a\lambda_1 + b = 0$ since $\lambda_1 = -b/2a$.

Remark. The solution $y_2(t)$ in Eq. (25) can be found by applying the eigenvalue method to Eq. (2) (see Problem 18) or by applying the method of reduction of order, which is discussed below.

Thus $y_1(t) = e^{\lambda_1 t}$ and $y_2(t) = te^{\lambda_1 t}$ are two solutions of Eq. (1) on $I = (-\infty, \infty)$ whenever $b^2 - 4ac = 0$. To confirm that $\{y_1, y_2\}$ is a fundamental set, we compute the Wronskian,

$$W[y_1, y_2](t) = \begin{vmatrix} e^{\lambda_1 t} & te^{\lambda_1 t} \\ \lambda_1 e^{\lambda_1 t} & (1 + \lambda_1 t)e^{\lambda_1 t} \end{vmatrix} = e^{2\lambda_1 t} > 0.$$

It follows from Theorem 4.2.5 that if $\lambda_1 = \lambda_2$, then the general solution of Eq. (1) is

$$y = c_1 e^{\lambda_1 t} + c_2 te^{\lambda_1 t}, \qquad \lambda_1 = -\frac{b}{2a}. \tag{26}$$

EXAMPLE 6

Find the solution of the initial value problem

$$y'' - y' + 0.25y = 0, \qquad y(0) = 2, \quad y'(0) = \tfrac{1}{3}. \tag{27}$$

The characteristic equation is

$$\lambda^2 - \lambda + 0.25 = 0,$$

so the roots are $\lambda_1 = \lambda_2 = 1/2$. Thus the general solution of the differential equation is

$$y = c_1 e^{t/2} + c_2 te^{t/2}. \tag{28}$$

The first initial condition requires that

$$y(0) = c_1 = 2.$$

To satisfy the second initial condition, we first differentiate Eq. (28) and then set $t = 0$. This gives

$$y'(0) = \tfrac{1}{2}c_1 + c_2 = \tfrac{1}{3},$$

so $c_2 = -2/3$. Thus the solution of the initial value problem is

$$y = 2e^{t/2} - \tfrac{2}{3}te^{t/2}. \tag{29}$$

The graph of this solution is shown in Figure 4.3.5.

Let us now modify the initial value problem (27) by changing the initial slope; to be specific, let the second initial condition be $y'(0) = 2$. The solution of this modified problem is

$$y = 2e^{t/2} + te^{t/2},$$

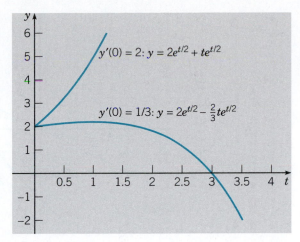

FIGURE 4.3.5 Solutions of $y'' - y' + 0.25y = 0$, $y(0) = 2$, with $y'(0) = 1/3$ and $y'(0) = 2$, respectively.

and its graph is also shown in Figure 4.3.5. The graphs shown in this figure suggest that there is a critical initial slope, with a value between 1/3 and 2, that separates solutions that grow positively from those that ultimately grow negatively. We discuss this question further in the next example.

EXAMPLE 7

Draw and discuss the phase portrait of $y'' - y' + 0.25y = 0$.

Letting $x_1 = y$ and $x_2 = y'$, as usual, we obtain the dynamical system

$$\mathbf{x}' = \begin{pmatrix} 0 & 1 \\ -0.25 & 1 \end{pmatrix} \mathbf{x}. \tag{30}$$

This system can be solved as in Section 3.5. Alternatively, you can use Eq. (28) and its derivative, $x_2 = y' = c_1 e^{t/2}/2 + c_2(1 + t/2)e^{t/2}$, to obtain

$$\mathbf{x} = c_1 \begin{pmatrix} 1 \\ 0.5 \end{pmatrix} e^{t/2} + c_2 \left[\begin{pmatrix} 0 \\ 1 \end{pmatrix} + \begin{pmatrix} 1 \\ 0.5 \end{pmatrix} t \right] e^{t/2}. \tag{31}$$

For each choice of c_1 and c_2, a plot of the curve $t \to \langle x_1(t), x_2(t) \rangle$ yields a corresponding phase plane trajectory. The phase portrait is drawn in Figure 4.3.6. The choice $c_2 = 0$ yields $x_1 = c_1 e^{t/2}$, $x_2 = c_1 e^{t/2}/2$ corresponding to trajectories that lie on the line $x_2 = x_1/2$.

The heavy trajectories in Figure 4.3.6 correspond to the solutions of $y'' - y' + 0.25y = 0$ using the initial conditions of Example 6. Since the slope of the straight line trajectories in Figure 4.3.6 is 1/2, it follows that $\lim_{t \to \infty} y(t) = \infty$ for any solution of $y'' - y' + 0.25y = 0$ such that $y'(0) > y(0)/2$ and that $\lim_{t \to \infty} y(t) = -\infty$ for any solution such that $y'(0) < y(0)/2$. Referring to the preceding example in which $y(0) = 2$, the critical initial slope that separates the two types of asymptotic behavior is $y'(0) = 1$. The critical point $(0, 0)$ is clearly an unstable node.

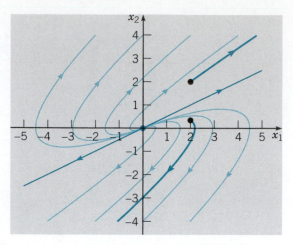

FIGURE 4.3.6   Phase portrait for the dynamical system associated with $y'' - y' + 0.25y = 0$.

As we saw in Section 3.5, the geometrical behavior of solutions in the case of repeated roots is similar to that when the roots are real and different. If the exponents are either positive or negative, then the magnitude of the solution grows or decays accordingly; the linear factor t has little influence. However, if the repeated root is zero, then the differential equation is $y'' = 0$ and the general solution is a first degree polynomial in t. In this case, all points on the x_1-axis are critical points of the associated dynamical system.

Reduction of Order

Given one solution y_1 of a second order linear homogeneous equation

$$y'' + p(t)y' + q(t)y = 0, \tag{32}$$

a systematic procedure for deriving a second solution y_2 such that $\{y_1, y_2\}$ is a fundamental set is known as the method of **reduction of order**. To find a second solution, let

$$y = v(t)y_1(t); \tag{33}$$

then

$$y' = v'(t)y_1(t) + v(t)y_1'(t)$$

and

$$y'' = v''(t)y_1(t) + 2v'(t)y_1'(t) + v(t)y_1''(t).$$

Substituting for y, y', and y'' in Eq. (32) and collecting terms, we find that

$$y_1 v'' + (2y_1' + py_1)v' + (y_1'' + py_1' + qy_1)v = 0. \tag{34}$$

Since y_1 is a solution of Eq. (32), the coefficient of v in Eq. (34) is zero, so that Eq. (34) becomes

$$y_1 v'' + (2y_1' + py_1)v' = 0. \tag{35}$$

Despite its appearance, Eq. (35) is actually a first order equation for the function v' and can be solved either as a first order linear equation or as a separable equation. Once v' has been found, then v is obtained by an integration. Finally, y is determined from Eq. (33). This procedure is called the method of reduction of order, because the crucial step is the solution of a first order differential equation for v' rather than the original second order equation for y. Although it is possible to write down a formula for $v(t)$, we will illustrate how this method works by examples instead.

EXAMPLE 8

Given that $b^2 - 4ac = 0$, and consequently that $y_1(t) = e^{\lambda_1 t}$ with $\lambda_1 = -b/2a$ is a solution of
$$ay'' + by' + cy = 0, \tag{36}$$
find a second solution y_2 such that $W[y_1, y_2](t) \neq 0$.

Dividing Eq. (36) by a, we identify the coefficients $p(t) = b/a$ and $q(t) = c/a$ in Eq. (32). Thus $2y_1' + py_1 = (-b/a + b/a)e^{-bt/2a} = 0$ so Eq. (35) reduces to $y_1 v'' = 0$ or $v'' = 0$ since y_1 is never zero. Therefore,
$$v(t) = c_1 t + c_2.$$

Hence, from Eq. (33), we have
$$y = c_1 t e^{-bt/2a} + c_2 e^{-bt/2a}. \tag{37}$$

Thus y is a linear combination of the two solutions
$$y_1(t) = e^{-bt/2a}, \qquad y_2(t) = t e^{-bt/2a}. \tag{38}$$

Since we have already shown that $W[y_1, y_2](t) = e^{2\lambda_1 t} > 0$, the solutions y_1 and y_2 given by Eq. (38) are a fundamental set of solutions. Further, Eq. (37) is the general solution of Eq. (36) when the roots of the characteristic equation are equal. In other words, in this case, there is one exponential solution corresponding to the repeated root, and a second solution is obtained by multiplying the exponential solution by t.

EXAMPLE 9

Given that $y_1(t) = t^{-1}$ is a solution of
$$2t^2 y'' + 3ty' - y = 0, \qquad t > 0, \tag{39}$$
find a second solution y_2 such that $W[y_1, y_2](t) \neq 0$.

We set $y = v(t)\, t^{-1}$; then
$$y' = v't^{-1} - vt^{-2}, \qquad y'' = v''t^{-1} - 2v't^{-2} + 2vt^{-3}.$$

Substituting for y, y', and y'' in Eq. (31) and collecting terms, we obtain
$$2t^2(v''t^{-1} - 2v't^{-2} + 2vt^{-3}) + 3t(v't^{-1} - vt^{-2}) - vt^{-1}$$
$$= 2tv'' + (-4 + 3)v' + (4t^{-1} - 3t^{-1} - t^{-1})v$$
$$= 2tv'' - v' = 0. \tag{40}$$

Note that the coefficient of v is zero, as it should be. This provides a useful check on our algebra.

Separating the variables in Eq. (40) and solving for $v'(t)$, we find that
$$v'(t) = ct^{1/2};$$

then
$$v(t) = \tfrac{2}{3}ct^{3/2} + k.$$

It follows that
$$y = \tfrac{2}{3}ct^{1/2} + kt^{-1}, \tag{41}$$

where c and k are arbitrary constants. The second term on the right side of Eq. (41) is a multiple of $y_1(t)$ and can be dropped. Neglecting the arbitrary multiplicative constant appearing in the first term on the right side of Eq. (41) and selecting $y_2(t) = t^{1/2}$, we find that $W[y_1, y_2](t) = 3t^{-3/2}/2 > 0$ for $t > 0$.

PROBLEMS

ODEA The equations in Problems 1 through 8 have distinct, real, characteristic roots.

(a) Find the general solution in each case.

(b) Use the general solution obtained in part (a) to find a two-parameter family of trajectories $\mathbf{x} = x_1\mathbf{i} + x_2\mathbf{j} = y\mathbf{i} + y'\mathbf{j}$ of the corresponding dynamical system. Then use a computer to draw a phase portrait, including any straight line orbits, from this family of trajectories.

(c) From the plot(s) in part (b), determine whether each critical point is asymptotically stable, stable, or unstable, and classify it as to type.

1. $y'' + 2y' - 3y = 0$ 2. $y'' + 3y' + 2y = 0$
3. $6y'' - y' - y = 0$ 4. $2y'' - 3y' + y = 0$
5. $y'' + 5y' = 0$ 6. $4y'' - 9y = 0$
7. $y'' - 9y' + 9y = 0$ 8. $y'' - 2y' - 2y = 0$

ODEA The equations in Problems 9 through 16 have repeated, real, characteristic roots.

(a) Find the general solution in each case.

(b) Use the general solution obtained in part (a) to find a two-parameter family of trajectories $\mathbf{x} = x_1\mathbf{i} + x_2\mathbf{j} = y\mathbf{i} + y'\mathbf{j}$ of the corresponding dynamical system. Then use a computer to draw a phase portrait, including any straight line orbits, from this family of trajectories.

(c) From the plot(s) in part (b), determine whether each critical point is asymptotically stable, stable, or unstable, and classify it as to type.

9. $y'' - 2y' + y = 0$
10. $9y'' + 6y' + y = 0$
11. $4y'' - 4y' + y = 0$
12. $4y'' + 12y' + 9y = 0$
13. $25y'' - 20y' + 4y = 0$
14. $y'' - 6y' + 9y = 0$
15. $y'' + 4y' + 4y = 0$
16. $9y'' - 24y' + 16y = 0$

17. Use the eigenvalue method to find the following fundamental set of solutions of Eq. (2) in the case that $b^2 - 4ac > 0$:

$$\mathbf{x}_1(t) = e^{\lambda_1 t}\begin{pmatrix} 1 \\ \lambda_1 \end{pmatrix}, \qquad \mathbf{x}_2(t) = e^{\lambda_2 t}\begin{pmatrix} 1 \\ \lambda_2 \end{pmatrix},$$

where λ_1 and λ_2 are the real and distinct characteristic roots shown in Eq. (4). Conclude that, in this case, the general solution (5) of Eq. (1) corresponds to the first component of $c_1\mathbf{x}_1 + c_2\mathbf{x}_2$.

18. Use the eigenvalue method to find the following fundamental set of solutions of Eq. (2) in the case that $b^2 - 4ac = 0$:

$$\mathbf{x}_1(t) = e^{\lambda_1 t}\begin{pmatrix} 1 \\ \lambda_1 \end{pmatrix}, \qquad \mathbf{x}_2(t) = e^{\lambda_1 t}\begin{pmatrix} t \\ 1 + \lambda_1 t \end{pmatrix},$$

where $\lambda_1 = -b/2a$. Conclude that, in this case, the general solution (26) of Eq. (1) corresponds to the first component of $c_1\mathbf{x}_1 + c_2\mathbf{x}_2$.

In each of Problems 19 through 29, solve the given initial value problem. Sketch the graph of its solution and describe its behavior for increasing t. **ODEA**

19. $y'' + y' - 2y = 0$, $y(0) = 1$, $y'(0) = 1$
20. $9y'' - 12y' + 4y = 0$, $y(0) = 2$, $y'(0) = -1$
21. $y'' + 4y' + 3y = 0$, $y(0) = 2$, $y'(0) = -1$
22. $6y'' - 5y' + y = 0$, $y(0) = 4$, $y'(0) = 0$
23. $y'' - 6y' + 9y = 0$, $y(0) = 0$, $y'(0) = 2$
24. $y'' + 3y' = 0$, $y(0) = -2$, $y'(0) = 3$
25. $y'' + 4y' + 4y = 0$, $y(-1) = 2$, $y'(-1) = 1$
26. $y'' + 5y' + 3y = 0$, $y(0) = 1$, $y'(0) = 0$
27. $2y'' + y' - 4y = 0$, $y(0) = 0$, $y'(0) = 1$
28. $y'' + 8y' - 9y = 0$, $y(1) = 1$, $y'(1) = 0$
29. $4y'' - y = 0$, $y(-2) = 1$, $y'(-2) = -1$

30. Find a differential equation whose general solution is $y = c_1 e^{2t} + c_2 e^{-3t}$.

31. Find a differential equation whose general solution is $y = c_1 e^{-2t} + c_2 t e^{-2t}$.

In each of Problems 32 and 33, determine the values of α, if any, for which all solutions tend to zero as $t \to \infty$. Also determine the values of α, if any, for which all (nonzero) solutions become unbounded as $t \to \infty$.

32. $y'' - (2\alpha - 1)y' + \alpha(\alpha - 1)y = 0$

33. $y'' + (3 - \alpha)y' - 2(\alpha - 1)y = 0$

34. If the roots of the characteristic equation are real, show that a solution of $ay'' + by' + cy = 0$ can take on the value zero at most once.

35. Consider the equation $ay'' + by' + cy = d$, where a, b, c, and d are constants.
   (a) Find all equilibrium or constant solutions of this differential equation.
   (b) Let y_e denote an equilibrium solution, and let $Y = y - y_e$. Thus Y is the deviation of a solution y from an equilibrium solution. Find the differential equation satisfied by Y.

36. Consider the equation $ay'' + by' + cy = 0$, where a, b, and c are constants with $a > 0$. Find conditions on a, b, and c such that the roots of the characteristic equation are:
   (a) real, different, and negative
   (b) real with opposite signs
   (c) real, different, and positive

In each of Problems 37 through 44, use the method of reduction of order to find a second solution y_2 of the given differential equation such that $\{y_1, y_2\}$ is a fundamental set of solutions on the given interval.

37. $t^2 y'' - 4ty' + 6y = 0$, $t > 0$; $y_1(t) = t^2$

38. $t^2 y'' + 2ty' - 2y = 0$, $t > 0$; $y_1(t) = t$

39. $t^2 y'' + 3ty' + y = 0$, $t > 0$; $y_1(t) = t^{-1}$

40. $t^2 y'' - t(t+2)y' + (t+2)y = 0$, $t > 0$; $y_1(t) = t$

41. $xy'' - y' + 4x^3 y = 0$, $x > 0$; $y_1(x) = \sin x^2$

42. $(x-1)y'' - xy' + y = 0$, $x > 1$; $y_1(x) = e^x$

43. $x^2 y'' - (x - 0.1875)y = 0$, $x > 0$; $y_1(x) = x^{1/4} e^{2\sqrt{x}}$

44. $x^2 y'' + xy' + (x^2 - 0.25)y = 0$, $x > 0$; $y_1(x) = x^{-1/2} \sin x$

45. The differential equation
$$xy'' - (x + N)y' + Ny = 0,$$
where N is a nonnegative integer, has been discussed by several authors.[2] One reason why it is interesting is that it has an exponential solution and a polynomial solution.
   (a) Verify that one solution is $y_1(x) = e^x$.
   (b) Show that a second solution has the form $y_2(x) = ce^x \int x^N e^{-x}\,dx$. Calculate $y_2(x)$ for $N = 1$ and $N = 2$. Convince yourself that, with $c = -1/N!$,
$$y_2(x) = 1 + \frac{x}{1!} + \frac{x^2}{2!} + \cdots + \frac{x^N}{N!}.$$
   Note that $y_2(x)$ is exactly the first $N + 1$ terms in the Taylor series about $x = 0$ for e^x, that is, for $y_1(x)$.

46. The differential equation
$$y'' + \delta(xy' + y) = 0$$
arises in the study of the turbulent flow of a uniform stream past a circular cylinder. Verify that $y_1(x) = \exp(-\delta x^2/2)$ is one solution and then find the general solution in the form of an integral.

4.4 Characteristic Equations with Complex Roots

We now consider the constant coefficient equation
$$ay'' + by' + cy = 0 \tag{1}$$
in the case that the discriminant $b^2 - 4ac$ of the characteristic equation
$$a\lambda^2 + b\lambda + c = 0 \tag{2}$$

[2]T. A. Newton, "On Using a Differential Equation to Generate Polynomials," *American Mathematical Monthly 81* (1974), pp. 592–601. Also see the references given there.

is negative. It follows that the roots of Eq. (2) are conjugate complex numbers

$$\lambda_1 = \mu + iv \quad \text{and} \quad \lambda_2 = \mu - iv, \tag{3}$$

where

$$\mu = -\frac{b}{2a} \tag{4}$$

and

$$v = \frac{\sqrt{4ac - b^2}}{2a} \tag{5}$$

are real. These roots result in the pair of complex-valued solutions

$$z_1(t) = \exp[(\mu + iv)t] \quad \text{and} \quad z_2(t) = \exp[(\mu - iv)t]. \tag{6}$$

Employing the method used in Section 3.4, we obtain real-valued solutions of Eq. (1) by taking the real and imaginary parts of z_1 (or z_2),

$$y_1(t) = \operatorname{Re} z_1(t) = \frac{1}{2}z_1(t) + \frac{1}{2}z_2(t) = e^{\mu t} \cos vt \tag{7}$$

and

$$y_2(t) = \operatorname{Im} z_1(t) = \frac{1}{2i}z_1(t) - \frac{1}{2i}z_2(t) = e^{\mu t} \sin vt. \tag{8}$$

By direct computation, you can show that the Wronskian of y_1 and y_2 is

$$W[y_1, y_2](t) = ve^{2\mu t}. \tag{9}$$

Thus, as long as $v \neq 0$, the Wronskian W is not zero, so y_1 and y_2 form a fundamental set of solutions. (Of course, if $v = 0$, then the roots are real and the discussion in this section is not applicable.) Consequently, if the roots of the characteristic equation are complex numbers $\mu \pm iv$, with $v \neq 0$, then the general solution of Eq. (1) is

$$y = c_1 e^{\mu t} \cos vt + c_2 e^{\mu t} \sin vt, \tag{10}$$

where c_1 and c_2 are arbitrary constants. Note that the solution (10) can be written down as soon as the values of μ and v are known. In Problem 28 you are asked to find (10) by applying the eigenvalue method to the dynamical system equivalent to Eq. (1).

EXAMPLE 1

Find the general solution of

$$y'' + y' + y = 0. \tag{11}$$

The characteristic equation is

$$\lambda^2 + \lambda + 1 = 0,$$

and its roots are

$$\lambda = \frac{-1 \pm (1 - 4)^{1/2}}{2} = -\frac{1}{2} \pm i\frac{\sqrt{3}}{2}.$$

Thus $\mu = -1/2$ and $v = \sqrt{3}/2$, so the general solution of Eq. (11) is

$$y = c_1 e^{-t/2} \cos(\sqrt{3}t/2) + c_2 e^{-t/2} \sin(\sqrt{3}t/2). \tag{12}$$

EXAMPLE 2

Find the general solution of

$$y'' + 9y = 0. \tag{13}$$

The characteristic equation is $\lambda^2 + 9 = 0$ with the roots $\lambda = \pm 3i$; thus $\mu = 0$ and $\nu = 3$. The general solution is

$$y = c_1 \cos 3t + c_2 \sin 3t. \tag{14}$$

Note that if the real part of the roots is zero, as in this example, then there is no exponential factor in the solution.

EXAMPLE 3

Find the solution of the initial value problem

$$16y'' - 8y' + 145y = 0, \qquad y(0) = -2, \quad y'(0) = 1. \tag{15}$$

The characteristic equation is $16\lambda^2 - 8\lambda + 145 = 0$ and its roots are $\lambda = 1/4 \pm 3i$. Thus the general solution of the differential equation is

$$y = c_1 e^{t/4} \cos 3t + c_2 e^{t/4} \sin 3t. \tag{16}$$

To apply the first initial condition, we set $t = 0$ in Eq. (16). This gives

$$y(0) = c_1 = -2.$$

For the second initial condition, we must differentiate Eq. (16) and then set $t = 0$. In this way, we find that

$$y'(0) = \tfrac{1}{4}c_1 + 3c_2 = 1,$$

from which $c_2 = 1/2$. Using these values of c_1 and c_2 in Eq. (16), we obtain

$$y = -2e^{t/4} \cos 3t + \tfrac{1}{2}e^{t/4} \sin 3t \tag{17}$$

as the solution of the initial value problem (15).

We will discuss the geometrical properties of solutions such as these more fully in Section 4.5, so we will be very brief here. Each of the solutions y_1 and y_2 in Eqs. (7) and (8) represents an oscillation, because of the trigonometric factors, and also either grows or decays exponentially, depending on the sign of μ (unless $\mu = 0$). In Example 1, we have $\mu = -1/2 < 0$, so solutions are decaying oscillations. The graph of a typical solution of Eq. (11) is shown in Figure 4.4.1(a) along with phase plane trajectories $t \to \mathbf{x} = x_1\mathbf{i} + x_2\mathbf{j} = y\mathbf{i} + y'\mathbf{j}$ of the equivalent dynamical system shown in Figure 4.4.1(b). Oscillations with decaying amplitude clearly show that the critical point (0,0) is an asymptotically stable spiral point. If $\mu = 0$ as in Example 2, the solutions are pure oscillations without growth or decay (Figure 4.4.2(a)). The origin is a center and therefore stable (Figure 4.4.2(b)). In Example 3, $\mu = 1/4 > 0$ so solutions of the differential equation (15) are growing oscillations. The graph of the solution (17) of the given initial value problem is shown in Figure 4.4.3(a). The corresponding phase portrait in Figure 4.4.3(b) shows that the origin is an unstable spiral point. You should compare the phase portraits shown in these figures with those that occurred in Section 3.4.

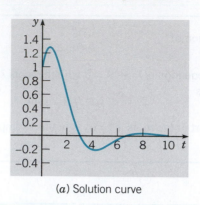

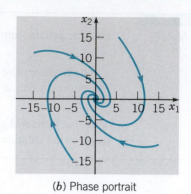

(a) Solution curve

(b) Phase portrait

FIGURE 4.4.1 (a) A typical solution of $y'' + y' + y = 0$. (b) Phase plane trajectories $t \rightarrow \mathbf{x} = x_1\mathbf{i} + x_2\mathbf{j} = y\mathbf{i} + y'\mathbf{j}$ of the equivalent dynamical system.

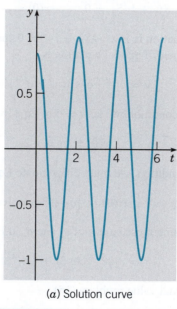

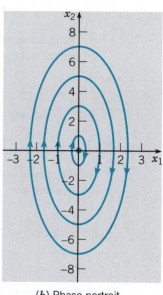

(a) Solution curve

(b) Phase portrait

FIGURE 4.4.2 (a) A typical solution of $y'' + 9y = 0$. (b) Phase plane trajectories $t \rightarrow \mathbf{x} = x_1\mathbf{i} + x_2\mathbf{j} = y\mathbf{i} + y'\mathbf{j}$ of the equivalent dynamical system.

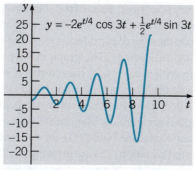

$y = -2e^{t/4} \cos 3t + \frac{1}{2}e^{t/4} \sin 3t$

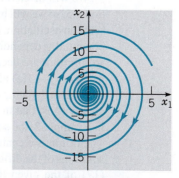

(a) Solution curve

(b) Phase portrait

FIGURE 4.4.3 (a) Solution of $16y'' - 8y' + 145y = 0$, $y(0) = -2$, $y'(0) = 1$. (b) Phase plane trajectories $t \rightarrow \mathbf{x} = x_1\mathbf{i} + x_2\mathbf{j} = y\mathbf{i} + y'\mathbf{j}$ of the dynamical system equivalent to $16y'' - 8y' + 145y = 0$.

Cauchy-Euler Equations

Now that we know how to solve linear constant coefficient equations it is appropriate to study a special class of variable coefficient equations that can be transformed into equations with constant coefficients. A (homogeneous) second order **Cauchy-Euler equation** is one that can be expressed in the form

$$ax^2 y'' + bxy' + cy = 0. \tag{18}$$

Note that the independent variable used in Eq. (18) is x instead of t. The distinguishing feature of Eq. (18) is that the power of the independent variable in each coefficient matches the order of the derivative. For example,

$$2xy'' + 3y' - \frac{1}{x}y = 0, \quad x > 0$$

is a Cauchy-Euler equation since multiplication by x transforms it into

$$2x^2 y'' + 3xy' - y = 0, \quad x > 0.$$

On the other hand, $\qquad 2y'' + 3xy' + 4y = 0, \quad x > 0$

is not a Cauchy-Euler equation since it cannot be placed in the form (18).

Theorem 4.2.1, concerning existence and uniqueness of solutions, can be applied to Eq. (18) if we divide through by the leading coefficient to put it in the form

$$y'' + \frac{b}{ax}y' + \frac{c}{ax^2}y = 0. \tag{19}$$

If initial conditions for Eqs. (18) or (19) are prescribed at $x_0 > 0$, say $y(x_0) = y_0$ and $y'(x_0) = y_1$, then Theorem 4.2.1 guarantees that a unique solution of the initial value problem exists on the interval $I = (0, \infty)$ since the only point at which the coefficients of Eq. (19) fail to be continuous is $x = 0$. If initial conditions are prescribed at $x_0 < 0$, then the maximum interval in which a unique solution is guaranteed is $I = (-\infty, 0)$.

The special form of a Cauchy-Euler equation allows us to find a fundamental set of solutions in terms of elementary functions by first transforming the equation into one with constant coefficients. Thus we introduce the new independent variable $z = \ln x$ and consider only the interval $x > 0$. The chain rule for derivatives applied to $y(x) = y(z(x))$ yields

$$\frac{dy}{dx} = \frac{dy}{dz}\frac{dz}{dx} = \frac{1}{x}\frac{dy}{dz}. \tag{20}$$

A second application of the chain rule to Eq. (20) gives

$$\frac{d^2 y}{dx^2} = \frac{1}{x^2}\frac{d^2 y}{dz^2} - \frac{1}{x^2}\frac{dy}{dz}. \tag{21}$$

Substituting the expressions on the right sides of Eqs. (20) and (21) for dy/dx and d^2y/dx^2, respectively, in Eq. (18), results in the constant coefficient equation

$$a\frac{d^2 y}{dz^2} + (b - a)\frac{dy}{dz} + cy = 0. \tag{22}$$

The general solution of Eq. (18) is obtained by replacing z by $\ln x$ in the general solution of Eq. (22). If λ_1 and λ_2 denote the roots of the characteristic equation

$$a\lambda^2 + (b - a)\lambda + c = 0 \tag{23}$$

for Eq. (22), the general solutions of Eq. (18) are expressed in Table 4.4.1.

TABLE 4.4.1 General solutions of the Cauchy-Euler equation $ax^2y'' + bxy' + cy = 0$ for $0 < x < \infty$.

Roots	General solution
λ_1 and λ_2 real, $\lambda_1 \neq \lambda_2$	$y = c_1 e^{\lambda_1 z} + c_2 e^{\lambda_2 z} = c_1 x^{\lambda_1} + c_2 x^{\lambda_2}, \quad x > 0$
λ_1 and λ_2 real, $\lambda_1 = \lambda_2$	$y = c_1 e^{\lambda_1 z} + c_2 z e^{\lambda_1 z} = c_1 x^{\lambda_1} + c_2 x^{\lambda_1} \ln x, \quad x > 0$
$\lambda_1 = \mu + iv, \lambda_2 = \mu - iv$	$y = e^{\mu z}[c_1 \cos(vz) + c_2 \sin(vz)]$ $= x^{\mu}[c_1 \cos(v \ln x) + c_2 \sin(v \ln x)], \quad x > 0$

The roots λ_1 and λ_2 are solutions of the characteristic equation $a\lambda^2 + (b - a)\lambda + c = 0$.
The general solutions of Eq. (18) in $I = (-\infty, 0)$ can be found by first transforming the equation to one on $I = (0, \infty)$ by introducing a new independent variable $\hat{x} = -x$. The transformed equation, in terms of the independent variable $\hat{x} \in (0, \infty)$, is (see Problem 46)

$$a\hat{x}^2 y'' + b\hat{x} y' + cy = 0, \qquad \hat{x} > 0. \tag{24}$$

Thus the general solutions of Eq. (24) are identical to the general solutions that appear in Table 4.4.1 with x replaced by $\hat{x}$. To get back to the general solutions of Eq. (18) in $I = (-\infty, 0)$, we then substitute $-x$ for $\hat{x}$. Thus general solutions of Eq. (18) in $I = (-\infty, 0)$ are identical to those in Table 4.4.1 except that x must be replaced by $-x$.

EXAMPLE 4

Solve

$$2x^2 y'' + 3xy' - y = 0, \qquad x > 0. \tag{25}$$

Changing the independent variable to $z = \ln x$ in Eq. (25) yields the constant coefficient equation

$$2\frac{d^2y}{dz^2} + \frac{dy}{dz} - y = 0. \tag{26}$$

The roots of the characteristic equation for (26), $2\lambda^2 + \lambda - 1 = 0$, are $\lambda_1 = 1/2$ and $\lambda_2 = -1$. Therefore the general solution of Eq. (25) is

$$y = c_1 x^{1/2} + c_2 x^{-1}, \qquad x > 0. \tag{27}$$

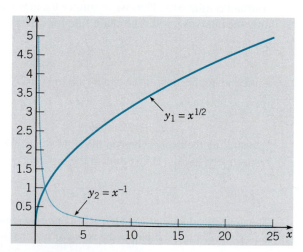

FIGURE 4.4.4 A fundamental set of solutions for $2x^2y'' + 3xy' - y = 0$ on $0 < x < \infty$: $y_1(x) = x^{1/2}$ and $y_2(x) = x^{-1}$.

It is worth noting that solutions of constant coefficient equations near $x = 0$ cannot exhibit the type of behavior displayed by either of the solutions comprising the fundamental set of Eq. (25), $y_1(x) = x^{1/2}$ and $y_2(x) = x^{-1}$ (see Figure 4.4.4). This is a consequence of the singularity at $x = 0$ in Eq. (25).

EXAMPLE 5

Solve

$$y'' + \frac{5}{x}y' + \frac{4}{x^2}y = 0, \qquad x < 0. \tag{28}$$

Multiplying Eq. (28) by x^2 followed by a change in the independent variable to $\hat{x} = -x$ yields

$$\hat{x}^2 y'' + 5\hat{x}y' + 4y = 0, \qquad \hat{x} > 0 \tag{29}$$

where the primes indicate derivatives with respect to $\hat{x}$. Setting $z = \ln \hat{x}$ in Eq. (29) yields

$$\frac{d^2 y}{dz^2} + 4\frac{dy}{dz} + 4y = 0. \tag{30}$$

The roots of the characteristic equation for Eq. (30), $\lambda^2 + 4\lambda + 4 = 0$, are $\lambda_1 = \lambda_2 = -2$. Thus the general solution of Eq. (28) is

$$y = c_1 e^{-2z} + c_2 z e^{-2z} = c_1 \hat{x}^{-2} + c_2 \hat{x}^{-2} \ln \hat{x} = c_1 x^{-2} + c_2 x^{-2} \ln(-x), \qquad x < 0. \tag{31}$$

Graphs of $y_1 = x^{-2}$ and $y_2 = x^{-2} \ln(-x)$ are shown in Figure 4.4.5.

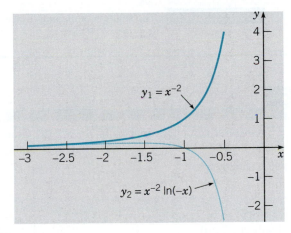

FIGURE 4.4.5 A fundamental set of solutions for $y'' + (5/x)y' + (4/x^2)y = 0$ on $-\infty < x < 0$: $y_1 = x^{-2}$ and $y_2 = x^{-2} \ln(-x)$.

EXAMPLE 6

Find the solution of the initial value problem

$$4x^2 y'' + 8xy' + 65y = 0, \qquad y(1) = 2, \quad y'(1) = 3. \tag{32}$$

Setting $z = \ln x$ in the differential equation yields

$$4\frac{d^2 y}{dz^2} + 4\frac{dy}{dz} + 65y = 0. \tag{33}$$

The roots of the associated characteristic equation, $4\lambda^2 + 4\lambda + 65 = 0$, are $\lambda = -1/2 \pm 4i$ so the general solution of Eq. (32) is

$$y = x^{-1/2}\left[c_1 \cos(4\ln x) + c_2 \sin(4\ln x)\right], \qquad x > 0. \tag{34}$$

Since $y'(x) = -(1/2)x^{-1}y(x) + 4x^{-3/2}\left[-c_1 \sin(4\ln x) + c_2 \cos(4\ln x)\right]$, the initial conditions in Eq. (32) imply that $c_1 = 2$ and $-1/2\,c_1 + 4c_2 = 3$. Thus $c_2 = 1$ so the solution of the initial value problem (32) is

$$y = x^{-1/2}\left[2\cos(4\ln x) + \sin(4\ln x)\right].$$

The graph of y is shown in Figure 4.4.6.

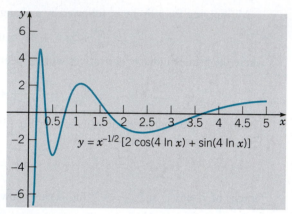

y = x^{-1/2} [2 cos(4 ln x) + sin(4 ln x)]

FIGURE 4.4.6 The solution of the initial value problem $4x^2 y'' + 8xy' + 65y = 0$, $y(1) = 2$, $y'(1) = 3$.

PROBLEMS

See Appendix B for a review of complex variables to help solve Problems 1 through 9.

1. Prove the following statements:
   (a) $\overline{z_1 + z_2} = \bar{z}_1 + \bar{z}_2$.
   (b) $\overline{z_1 z_2} = \bar{z}_1 \bar{z}_2$.
   (c) $\overline{z_1/z_2} = \bar{z}_1/\bar{z}_2$ for $z_2 \ne 0$.
   (d) $z = 0$ if and only if $|z| = 0$.
   (e) $|z_1 z_2| = |z_1||z_2|$.
   (f) $|z_1/z_2| = |z_1|/|z_2|$ if $z_2 \ne 0$.

2. **The Triangle Inequality.** Apply the law of cosines to the triangle figure on the right to show that $|z_1 + z_2| \le |z_1| + |z_2|$.
   Under what conditions is it true that $|z_1 + z_2| = |z_1| + |z_2|$?

3. Let $z_1 = x_1 + iy_1$ and $z_2 = x_2 + iy_2$ and let $|z_1| = r_1$, $|z_2| = r_2$, $\theta_1 = \arg(z_1)$ and $\theta_2 = \arg(z_2)$.
   (a) Use the polar coordinate representations $z_1 = r_1 \cos\theta_1 + ir_1 \sin\theta_1$ and $z_2 = r_2 \cos\theta_2 + ir_2 \sin\theta_2$ to show that $z_1 z_2 = |z_1||z_2|e^{i(\theta_1+\theta_2)}$.

(b) Graphically illustrate the result in part (a) using $z_1 = \sqrt{2} + i\sqrt{2}$ and $z_2 = -\sqrt{2} + i\sqrt{2}$.

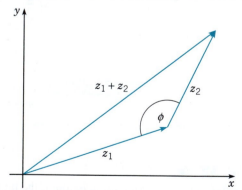

In each of Problems 4 through 9, use Euler's formula to write the given expression in the form $a + ib$.

4. $\exp(1 + 2i)$
5. $\exp(2 - 3i)$
6. $e^{i\pi}$
7. $e^{2-(\pi/2)i}$
8. 2^{1-i}
9. π^{-1+2i}

10. Suppose that $b^2 - 4ac < 0$ and consider the initial value problem $ay'' + by' + cy = 0$, $y(0) = y_0$, $y'(0) = y_1$ where y_0 and y_1 are real numbers.

(a) Assuming a general solution of the form

$$y(t) = c_1 e^{(\mu+i\nu)t} + c_2 e^{(\mu-i\nu)t}$$

where μ and ν are defined in Eqs. (4) and (5), respectively, show that the requirement that the initial conditions be real numbers imply that c_1 and c_2 are not arbitrary but in fact must be complex conjugates of one another, that is, $c_2 = \bar{c}_1$.

(b) Part (a) implies that if the initial conditions are real numbers, then the general solution is of the form

$$y(t) = ce^{(\mu+i\nu)t} + \bar{c}e^{(\mu-i\nu)t}.$$

Why does this imply that $y(t)$ is real?

(c) If we assume that $c = \alpha + i\beta$, how are α and β related to c_1 and c_2 in Eq. (10)?

ODEA In each of Problems 11 through 20,

(a) Find and express the general solution in terms of real functions.

(b) Use the general solution obtained in part (a) to find a two-parameter family of trajectories $\mathbf{x} = x_1\mathbf{i} + x_2\mathbf{j} = y\mathbf{i} + y'\mathbf{j}$ of the corresponding dynamical system. Then use a computer to draw a phase portrait, including any straight line orbits, from this family of trajectories.

(c) From the plot(s) in part (b), determine whether each critical point is asymptotically stable, stable, or unstable, and classify it as to type.

11. $y'' - 2y' + 2y = 0$ **12.** $y'' - 2y' + 6y = 0$

13. $y'' + 2y' - 8y = 0$ **14.** $y'' + 2y' + 2y = 0$

15. $y'' + 6y' + 13y = 0$ **16.** $y'' + 2y' + 1.25y = 0$

17. $4y'' + 9y = 0$ **18.** $9y'' + 9y' - 4y = 0$

19. $y'' + y' + 1.25y = 0$ **20.** $y'' + 4y' + 6.25y = 0$

ODEA In each of Problems 21 through 26, find the solution of the given initial value problem. Plot the graph of the solution and describe its behavior for increasing t.

21. $y'' + 4y = 0$, $y(0) = 0$, $y'(0) = 1$

22. $y'' + 4y' + 5y = 0$, $y(0) = 1$, $y'(0) = 0$

23. $y'' - 2y' + 5y = 0$, $y(\pi/2) = 0$, $y'(\pi/2) = 2$

24. $y'' + y = 0$, $y(\pi/3) = 2$, $y'(\pi/3) = -4$

25. $y'' + y' + 1.25y = 0$, $y(0) = 3$, $y'(0) = 1$

26. $y'' + 2y' + 2y = 0$, $y(\pi/4) = 2$, $y'(\pi/4) = -2$

27. Show that $W[e^{\mu t}\cos \nu t, e^{\mu t}\sin \nu t] = \nu e^{2\mu t}$.

28. Use the eigenvalue method to find the following fundamental set of complex-valued solutions of

$$\mathbf{x}' = \begin{pmatrix} 0 & 1 \\ -c/a & -b/a \end{pmatrix} \mathbf{x}$$

in the case that $b^2 - 4ac < 0$,

$$\mathbf{x}_1(t) = e^{\lambda_1 t}\begin{pmatrix} 1 \\ \lambda_1 \end{pmatrix}, \qquad \mathbf{x}_2(t) = e^{\lambda_2 t}\begin{pmatrix} 1 \\ \lambda_2 \end{pmatrix},$$

where λ_1 and λ_2 are the conjugate complex roots in Eq. (3). Conclude that, in this case, the fundamental set of real-valued solutions of Eq. (1) that appear in Eq. (6) can be obtained by taking the real and imaginary parts of the first component of $\mathbf{x}_1(t)$.

29. In this problem we outline a derivation of Euler's formula, $e^{it} = \cos t + i \sin t$, different from that presented in Appendix B.

(a) Show that $y_1(t) = \cos t$ and $y_2(t) = \sin t$ are a fundamental set of solutions of $y'' + y = 0$, that is, show that they are solutions and that their Wronskian is not zero.

(b) Show (formally) that $y = e^{it}$ is also a solution of $y'' + y = 0$. Therefore,

$$e^{it} = c_1\cos t + c_2 \sin t \qquad \text{(i)}$$

for some constants c_1 and c_2. Why is this so?

(c) Set $t = 0$ in Eq. (i) to show that $c_1 = 1$.

(d) Assuming that $d/dt\, e^{\lambda t} = \lambda e^{\lambda t}$ is true if λ is a complex number, differentiate Eq. (i) and then set $t = 0$ to conclude that $c_2 = i$. Use the values of c_1 and c_2 in Eq. (i) to arrive at Euler's formula.

30. Using Euler's formula, $e^{it} = \cos t + i \sin t$, show that

$$\cos t = (e^{it} + e^{-it})/2, \qquad \sin t = (e^{it} - e^{-it})/2i.$$

31. If $\lambda = \mu + i\nu$ and $e^{\lambda t}$ is defined by $e^{\lambda t} = e^{\mu t}(\cos \nu t + i \sin \nu t)$, show that $e^{(\lambda_1+\lambda_2)t} = e^{\lambda_1 t}e^{\lambda_2 t}$ for any complex numbers λ_1 and λ_2 (see Problem 3).

32. If $\lambda = \mu + i\nu$ and $e^{\lambda t}$ is defined by $e^{\lambda t} = e^{\mu t}(\cos \nu t + i \sin \nu t)$, show that

$$\frac{d}{dt}e^{\lambda t} = \lambda e^{\lambda t}.$$

33. Let the real-valued functions p and q be continuous on the open interval I, and let $y = \phi(t) = u(t) + iv(t)$ be a complex-valued solution of

$$y'' + p(t)y' + q(t)y = 0, \qquad \text{(ii)}$$

where u and v are real-valued functions. Show that u and v are also solutions of Eq. (ii).

Hint: Substitute $y = \phi(t)$ in Eq. (ii) and separate into real and imaginary parts.

In each of Problems 34 through 41, use the method discussed under "Cauchy-Euler Equations" to find a general solution to the given equation for $x > 0$.

34. $x^2 y'' + xy' + y = 0$

35. $x^2 y'' + 4xy' + 2y = 0$

36. $x^2 y'' + 3xy' + 1.25y = 0$

37. $x^2 y'' - 4xy' - 6y = 0$

38. $x^2 y'' - 2y = 0$

39. $x^2 y'' - 3xy' + 4y = 0$

40. $x^2 y'' + 2xy' + 4y = 0$

41. $2x^2 y'' - 4xy' + 6y = 0$

In each of Problems 42 through 45, find the solution of the given initial value problem. Plot the graph of the solution and describe how the solution behaves as $x \to 0$.

42. $2x^2 y'' + xy' - 3y = 0$,　　$y(1) = 1, y'(1) = 1$

43. $4x^2 y'' + 8xy' + 17y = 0$,　　$y(1) = 2$, $y'(1) = -3$

44. $x^2 y'' - 3xy' + 4y = 0$,　　$y(-1) = 2$, $y'(-1) = 3$

45. $x^2 y'' + 3xy' + 5y = 0$, $y(1) = 1, y'(1) = -1$

46. Show that the change in the independent variable, $x = -\hat{x}$, transforms the Cauchy-Euler equation $ax^2 y'' + bxy' + cy = 0$ for $x < 0$ into $a\hat{x}^2 y'' + b\hat{x} y' + cy = 0$ for $\hat{x} > 0$.

4.5 Mechanical and Electrical Vibrations

In Section 4.1 the mathematical models derived for the spring-mass system, the linearized pendulum, and the RLC circuit all turned out to be linear constant coefficient differential equations which, in the absence of a forcing function, are of the form

$$ay'' + by' + cy = 0. \tag{1}$$

To adapt Eq. (1) to a specific application merely requires interpretation of the coefficients in terms of the physical parameters that characterize the application. Using the theory and methods developed in Sections 4.2 through 4.4, we are able to solve Eq. (1) completely for all possible parameter values and initial conditions. Thus, Eq. (1) provides us with an important class of problems which illustrates the linear theory described in Section 4.2 and solution methods developed in Sections 4.3 and 4.4.

Undamped Free Vibrations

Recall that the equation of motion for the damped spring-mass system with external forcing is

$$my''(t) + \gamma y'(t) + ky(t) = F(t). \tag{2}$$

Equation (2) and the pair of conditions

$$y(0) = y_0, \qquad y'(0) = v_0, \tag{3}$$

that specify initial position y_0 and initial velocity v_0 provide a complete formulation of the vibration problem. If there is no external force, then $F(t) = 0$ in Eq. (2). Let us also suppose that there is no damping, so that $\gamma = 0$. This is an idealized configuration of the system, seldom (if ever) completely attainable in practice. However, if the actual damping is very small, then the assumption of no damping may yield satisfactory results over short to moderate time intervals. In this case, the equation of motion (2) reduces to

$$my'' + ky = 0. \tag{4}$$

If we divide Eq. (4) by m it becomes

$$y'' + \omega_0^2 y = 0, \tag{5}$$

where

$$\omega_0^2 = k/m. \tag{6}$$

The characteristic equation for Eq. (5) is

$$\lambda^2 + \omega_0^2 = 0, \tag{7}$$

and the corresponding characteristic roots are $\lambda = \pm i\,\omega_0$. It follows that the general solution of Eq. (5) is

$$y = A \cos \omega_0 t + B \sin \omega_0 t. \tag{8}$$

Substituting from Eq. (8) into the initial conditions (3) determines the integration constants A and B in terms of initial position and velocity, $A = y_0$ and $B = v_0/\omega_0$.

In discussing the solution of Eq. (5), it is convenient to rewrite Eq. (8) in the **phase-amplitude** form

$$y = R \cos(\omega_0 t - \delta). \tag{9}$$

To see the relationship between Eqs. (8) and (9), use the trigonometric identity for the cosine of the difference of the two angles, $\omega_0 t$ and δ, to rewrite Eq. (9) as

$$y = R \cos \delta \cos \omega_0 t + R \sin \delta \sin \omega_0 t. \tag{10}$$

By comparing Eq. (10) with Eq. (8), we find that A, B, R, and δ are related by the equations

$$A = R \cos \delta, \qquad B = R \sin \delta. \tag{11}$$

From these two equations, we see that (R, δ) is simply the polar coordinate representation of the point with cartesian coordinates (A, B) (Figure 4.5.1).

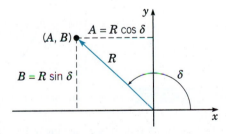

FIGURE 4.5.1 Relation between (R, δ) in Eq. (9) and (A, B) in Eq. (8).

Thus,

$$R = \sqrt{A^2 + R^2} \tag{12}$$

while δ satisfies

$$\cos \delta = \frac{A}{\sqrt{A^2 + B^2}}, \qquad \sin \delta = \frac{B}{\sqrt{A^2 + B^2}}. \tag{13}$$

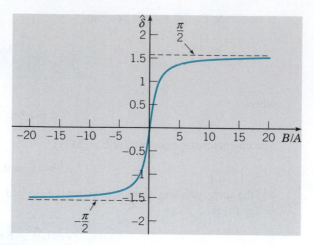

FIGURE 4.5.2 The principal branch of the arctangent function.

Let arctan (B/A) be the angle that lies in the principal branch of the inverse tangent function, that is, in the interval $-\pi/2 < \hat{\delta} < \pi/2$ (Figure 4.5.2). Then the values of δ given by

$$\delta = \begin{cases} \arctan(B/A), & \text{if } A > 0, B \geq 0 \text{ (1st quadrant)} \\ \pi + \arctan(B/A), & \text{if } A < 0 \text{ (2nd or 3rd quadrant)} \\ 2\pi + \arctan(B/A), & \text{if } A > 0, B < 0 \text{ (4th quadrant)} \\ \pi/2, & \text{if } A = 0, B > 0 \\ 3\pi/2, & \text{if } A = 0, B < 0 \end{cases}$$

will lie in the interval $[0, 2\pi)$.

The graph of Eq. (9), or the equivalent Eq. (8), for a typical set of initial conditions is shown in Figure 4.5.3.

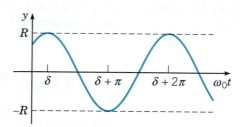

FIGURE 4.5.3 Simple harmonic motion $y = R \cos(\omega_0 t - \delta)$.

The graph is a displaced cosine wave that describes a periodic, or simple harmonic, motion of the mass. The **period** of the motion is

$$T = \frac{2\pi}{\omega_0} = 2\pi \left(\frac{m}{k}\right)^{1/2}. \tag{14}$$

The circular frequency $\omega_0 = \sqrt{k/m}$, measured in radians per unit time, is called the **natural frequency** of the vibration. The maximum displacement R of the mass from equilibrium is the **amplitude** of the motion. The dimensionless parameter δ is called the **phase**, or phase angle, and measures the displacement of the wave from its normal position corresponding to $\delta = 0$.

Note that the motion described by Eq. (9) has a constant amplitude that does not diminish with time. This reflects the fact that, in the absence of damping, there is no way for the system to dissipate the energy imparted to it by the initial displacement and velocity. Further, for a given mass m and spring constant k, the system always vibrates at the same frequency ω_0, regardless of the initial conditions. However, the initial conditions do help to determine the amplitude of the motion. Finally, observe from Eq. (14) that T increases as m increases, so larger masses vibrate more slowly. On the other hand, T decreases as k increases, which means that stiffer springs cause the system to vibrate more rapidly.

EXAMPLE 1

Suppose that a mass weighing 10 lb stretches a spring 2 in. If the mass is displaced an additional 2 in and is then set in motion with an initial upward velocity of 1 ft/s, determine the position of the mass at any later time. Also determine the period, amplitude, and phase of the motion.

The spring constant is $k = 10\,\text{lb}/2\,\text{in} = 60\,\text{lb/ft}$, and the mass is $m = w/g = 10/32\,\text{lb·s}^2/\text{ft}$. Hence the equation of motion reduces to

$$y'' + 192y = 0, \tag{15}$$

and the general solution is
$$y = A\cos(8\sqrt{3}\,t) + B\sin(8\sqrt{3}\,t).$$

The solution satisfying the initial conditions $y(0) = 1/6$ ft and $y'(0) = -1$ ft/s is

$$y = \frac{1}{6}\cos(8\sqrt{3}\,t) - \frac{1}{8\sqrt{3}}\sin(8\sqrt{3}\,t), \tag{16}$$

that is, $A = 1/6$ and $B = -1(8\sqrt{3})$. The natural frequency is $\omega_0 = \sqrt{192} \cong 13.856$ rad/s, so the period is $T = 2\pi/\omega_0 \cong 0.45345$ s. The amplitude R and phase δ are found from Eqs. (12) and (13). We have

$$R^2 = \frac{1}{36} + \frac{1}{192} = \frac{19}{576}, \quad \text{so} \quad R \cong 0.18162 \text{ ft.}$$

and since $A > 0$ and $B < 0$, the angle δ lies in the fourth quadrant,

$$\delta = 2\pi + \arctan(-\sqrt{3}/4) \cong 5.87455 \text{ rad.}$$

The graph of the solution (16) is shown in Figure 4.5.4.

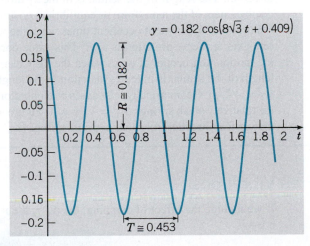

$$y = 0.182\cos\!\left(8\sqrt{3}\,t + 0.409\right)$$

FIGURE 4.5.4   An undamped free vibration: $y'' + 192y = 0$, $y(0) = 1/6$, $y'(0) = -1$.

EXAMPLE 2

Draw and discuss the phase portrait for the dynamical system formulation of Eq. (5). The dynamical system corresponding to Eq. (5) is

$$\mathbf{x}' = \begin{pmatrix} 0 & 1 \\ -\omega_0^2 & 0 \end{pmatrix} \mathbf{x}. \tag{17}$$

A phase portrait can be drawn by plotting trajectories of the system (17) for a typical value of ω_0 or by using the phase amplitude form of the solution given by Eq. (9). If we use the latter approach, then the state vector is $\mathbf{x} = x_1\mathbf{i} + x_2\mathbf{j} = y\mathbf{i} + y'\mathbf{j} = R\cos(\omega_0 t - \delta)\mathbf{i} - \omega_0 R\sin(\omega_0 t - \delta)\mathbf{j}$. It follows directly that the trajectories lie on the elliptical integral curves

$$\frac{x_1^2}{R^2} + \frac{x_2^2}{\omega_0^2 R^2} = 1,$$

as shown in Figure 4.5.5.

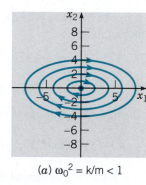
(a) $\omega_0^2 = k/m < 1$

(b) $\omega_0^2 = k/m > 1$

FIGURE 4.5.5 Phase portraits of the dynamical system corresponding to Eq. (5). Case (a) corresponds to relatively weak springs, $k < m$ while case (b) corresponds to relatively stiff springs, $k > m$.

The critical point at $(0, 0)$ is a stable center. To determine the direction of motion about the origin, we can use Eq. (17) to evaluate $\mathbf{x}'$ for a convenient point $\mathbf{x}$. For example, if we choose $\mathbf{x} = (0, 1)$, then $\mathbf{x}' = (1, 0)$. Thus, at the point $(0, 1)$, the vector $\mathbf{x}'$ points from the second quadrant into the first. Consequently, trajectories encircle the origin in the clockwise direction. The length of the semiaxis in the x_1 direction, R, corresponds to the maximum displacement of the mass from its equilibrium position, that is, the amplitude of the motion. Maximum displacement occurs at times when the velocity $x_2 = y' = 0$. The length of the semiaxis in the x_2 direction is $\omega_0 R$. This value corresponds to the maximum velocity of the motion and occurs at times when the displacement is zero. Note that the ratio of the length of the semiaxis in the x_2 direction to the length of the semiaxis in the x_1 direction is $\omega_0 = \sqrt{k/m}$. Geometrically this causes an elongation of the orbits in the horizontal direction for relatively weak springs ($\omega_0^2 = k/m < 1$, Figure 4.5.5(a)) and elongation of the orbits in the vertical direction for relatively stiff springs ($\omega_0^2 = k/m > 1$, Figure 4.5.5(b)).

Damped Free Vibrations

If we include the effect of damping, the differential equation governing the motion of the mass is

$$my'' + \gamma y' + ky = 0. \tag{18}$$

We are especially interested in examining the effect of variations in the damping coefficient γ for given values of the mass m and spring constant k. The roots of the corresponding characteristic equation are

$$\lambda_1, \lambda_2 = \frac{-\gamma \pm \sqrt{\gamma^2 - 4km}}{2m} = \frac{\gamma}{2m}\left(-1 \pm \sqrt{1 - \frac{4km}{\gamma^2}}\right). \tag{19}$$

Depending on the sign of $\gamma^2 - 4km$, the solution y has one of the following forms:

$$\text{i.} \qquad y = Ae^{\lambda_1 t} + Be^{\lambda_2 t} \qquad \text{if } \gamma^2 - 4km > 0, \tag{20}$$

$$\text{ii.} \qquad y = (A + Bt)e^{-\gamma t/2m} \qquad \text{if } \gamma^2 - 4km = 0, \tag{21}$$

$$\text{iii.} \qquad y = e^{-\gamma t/2m}(A \cos \nu t + B \sin \nu t) \qquad \text{if } \gamma^2 - 4km < 0, \tag{22}$$

$$\text{where } \nu = \frac{(4km - \gamma^2)^{1/2}}{2m} > 0.$$

Since m, γ, and k are positive, $\gamma^2 - 4km$ is always less than γ^2. Hence, if $\gamma^2 - 4km \geq 0$, then the values of λ_1 and λ_2 given by Eq. (19) are *negative*. If $\gamma^2 - 4km < 0$, then the values of λ_1 and λ_2 are complex, but with *negative* real parts. Thus, in all cases, the solution y tends to zero as $t \to \infty$. This occurs regardless of the values of the arbitrary constants A and B, that is, regardless of the initial conditions. This confirms our intuitive expectation, namely, that damping gradually dissipates the energy initially imparted to the system, and consequently the motion dies out with increasing time.

The most important case is the third one, which occurs when the damping is small. If we let $A = R \cos \delta$ and $B = R \sin \delta$ in Eq. (22), then we obtain

$$y = Re^{-\gamma t/2m} \cos(\nu t - \delta). \tag{23}$$

The displacement y lies between the curves $y = \pm Re^{-\gamma t/2m}$; hence it resembles a cosine wave whose amplitude decreases as t increases. A typical example is sketched in Figure 4.5.6. The motion is called a damped oscillation or a damped vibration. The amplitude factor R depends on m, γ, k, and the initial conditions.

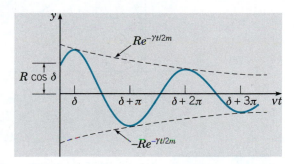

FIGURE 4.5.6 Damped vibration; $y = Re^{-\gamma t/2m} \cos(\nu t - \delta)$.

Although the motion is not periodic, the parameter ν determines the frequency with which the mass oscillates back and forth; consequently, ν is called the **quasi frequency**. By comparing ν with the frequency ω_0 of undamped motion, we find that

$$\frac{\nu}{\omega_0} = \frac{(4km - \gamma^2)^{1/2}/2m}{\sqrt{k/m}} = \left(1 - \frac{\gamma^2}{4km}\right)^{1/2} \cong 1 - \frac{\gamma^2}{8km}. \tag{24}$$

The last approximation is valid when $\gamma^2/4km$ is small. We refer to this situation as "small damping." Thus the effect of small damping is to reduce slightly the frequency of the oscillation. By analogy with Eq. (14), the quantity $T_d = 2\pi/\nu$ is called the **quasi period**. It is the time between successive maxima or successive minima of the position of the mass, or between successive passages of the mass through its equilibrium position while going in the same direction. The relation between T_d and T is given by

$$\frac{T_d}{T} = \frac{\omega_0}{\nu} = \left(1 - \frac{\gamma^2}{4km}\right)^{-1/2} \cong 1 + \frac{\gamma^2}{8km}, \tag{25}$$

where again the last approximation is valid when $\gamma^2/4km$ is small. Thus small damping increases the quasi period.

Equations (24) and (25) reinforce the significance of the dimensionless ratio $\gamma^2/4km$. It is not the magnitude of γ alone that determines whether damping is large or small, but the magnitude of γ^2 compared to $4km$. When $\gamma^2/4km$ is small, then damping has a small effect on the quasi frequency and quasi period of the motion. On the other hand, if we want to study the detailed motion of the mass for all time, then we can *never* neglect the damping force, no matter how small.

As $\gamma^2/4km$ increases, the quasi frequency ν decreases and the quasi period T_d increases. In fact, $\nu \to 0$ and $T_d \to \infty$ as $\gamma \to 2\sqrt{km}$. As indicated by Eqs. (20), (21), and (22), the nature of the solution changes as γ passes through the value $2\sqrt{km}$. This value is known as **critical damping**. The motion is said to be **underdamped** for values of $\gamma < 2\sqrt{km}$; while for values of $\gamma > 2\sqrt{km}$, the motion is said to be **overdamped**. In the critically damped and overdamped cases given by Eqs. (21) and (20), respectively, the mass creeps back to its equilibrium position but does not oscillate about it, as for small γ. Two typical examples of critically damped motion are shown in Figure 4.5.7, and the situation is discussed further in Problems 21 and 22.

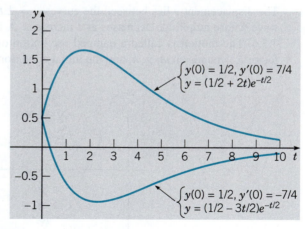

FIGURE 4.5.7 Critically damped motions: $y'' + y' + 0.25y = 0$; $y = (A + Bt)e^{-t/2}$.

EXAMPLE 3

The motion of a certain spring–mass system is governed by the differential equation

$$y'' + 0.125y' + y = 0, \tag{26}$$

where y is measured in feet and t in seconds. If $y(0) = 2$ and $y'(0) = 0$, determine the position of the mass at any time. Find the quasi frequency and the quasi period, as well as the time at which the mass first passes through its equilibrium position. Find the time τ such that $|y(t)| < 0.1$ for all $t > \tau$. Draw the orbit of the initial value problem in phase space.

The solution of Eq. (26) is

$$y = e^{-t/16}\left[A\cos\frac{\sqrt{255}}{16}t + B\sin\frac{\sqrt{255}}{16}t\right].$$

To satisfy the initial conditions we must choose $A = 2$ and $B = 2/\sqrt{255}$; hence, the solution of the initial value problem is

$$y = e^{-t/16}\left(2\cos\frac{\sqrt{255}}{16}t + \frac{2}{\sqrt{255}}\sin\frac{\sqrt{255}}{16}t\right)$$

$$= \frac{32}{\sqrt{255}}e^{-t/16}\cos\left(\frac{\sqrt{255}}{16}t - \delta\right), \tag{27}$$

where $\tan\delta = 1/\sqrt{255}$, so $\delta \cong 0.06254$. The displacement of the mass as a function of time is shown in Figure 4.5.8. For purposes of comparison, we also show the motion if the damping term is neglected.

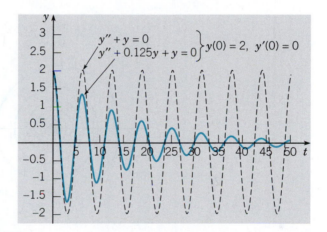

FIGURE 4.5.8 Vibration with small damping (solid curve) and with no damping (dashed curve).

The quasi frequency is $\nu = \sqrt{255}/16 \cong 0.998$ and the quasi period is $T_d = 2\pi/\nu \cong 6.295$ s. These values differ only slightly from the corresponding values (1 and 2π, respectively) for the undamped oscillation. This is also evident from the graphs in Figure 4.5.8, which rise and fall almost together. The damping coefficient is small in this example, only one-sixteenth of the critical value, in fact. Nevertheless, the amplitude of the oscillation is reduced rather rapidly. Figure 4.5.9 shows the graph of the solution for $40 \leq t \leq 60$, together with the graphs of $y = \pm 0.1$. From the graph, it appears that τ is about 47.5, and by a more precise calculation we find that $\tau \cong 47.5149$ s.

To find the time at which the mass first passes through its equilibrium position, we refer to Eq. (27) and set $\sqrt{255}t/16 - \delta$ equal to $\pi/2$, the smallest positive zero of the cosine function. Then, by solving for t, we obtain

$$t = \frac{16}{\sqrt{255}}\left(\frac{\pi}{2} + \delta\right) \cong 1.637 \text{ s}.$$

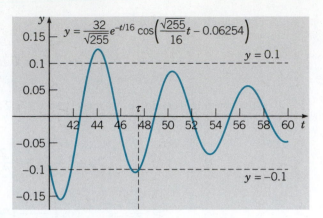

FIGURE 4.5.9 Solution of Example 3; determination of τ.

The phase plane orbit $t \to \mathbf{x}(t) = x_1(t)\mathbf{i} + x_2(t)\mathbf{j} = y(t)\mathbf{i} + y'(t)\mathbf{j}$ associated with Eq. (27) is shown in Figure 4.5.10.

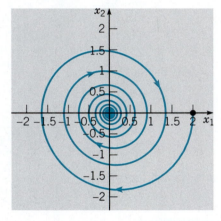

FIGURE 4.5.10 The phase plane orbit associated with the initial value problem $y'' + 0.125y' + y = 0$, $y(0) = 2$, $y'(0) = 0$.

The origin is an asymptotically stable spiral point, characteristic of damped oscillatory motion.

PROBLEMS

In each of Problems 1 through 4, determine ω_0, R, and δ so as to write the given expression in the form $y = R\cos(\omega_0 t - \delta)$.

1. $y = 3\cos 2t + 4\sin 2t$

2. $y = -\cos t + \sqrt{3}\sin t$

3. $y = 4\cos 3t - 2\sin 3t$

4. $y = -2\cos \pi t - 3\sin \pi t$

5. **(a)** A mass weighing 2 lb stretches a spring 6 in. If the mass is pulled down an additional 3 in and then released, and if there is no damping, determine the position y of the mass at any time t. Plot y versus t. Find the frequency, period, and amplitude of the motion.

(b) Draw a phase portrait of the equivalent dynamical system that includes the trajectory corresponding to the initial value problem in part (a).

ODEA 6. (a) A mass of 100 g stretches a spring 5 cm. If the mass is set in motion from its equilibrium position with a downward velocity of 10 cm/s, and if there is no damping, determine the position y of the mass at any time t. When does the mass first return to its equilibrium position?

(b) Draw a phase portrait of the equivalent dynamical system that includes the trajectory corresponding to the initial value problem in part (a).

7. A mass weighing 3 lb stretches a spring 3 in. If the mass is pushed upward, contracting the spring a distance of 1 in, and then set in motion with a downward velocity of 2 ft/s, and if there is no damping, find the position y of the mass at any time t. Determine the frequency, period, amplitude, and phase of the motion.

8. A series circuit has a capacitor of 0.25×10^{-6} farad and an inductor of 1 henry. If the initial charge on the capacitor is 10^{-6} coulomb and there is no initial current, find the charge q on the capacitor at any time t.

ODEA 9. (a) A mass of 20 g stretches a spring 5 cm. Suppose that the mass is also attached to a viscous damper with a damping constant of 400 dyne·s/cm. If the mass is pulled down an additional 2 cm and then released, find its position y at any time t. Plot y versus t. Determine the quasi frequency and the quasi period. Determine the ratio of the quasi period to the period of the corresponding undamped motion. Also find the time τ such that $|y(t)| < 0.05$ cm for all $t > \tau$.

(b) Draw a phase portrait of the equivalent dynamical system that includes the trajectory corresponding to the initial value problem in part (a).

ODEA 10. A mass weighing 16 lb stretches a spring 3 in. The mass is attached to a viscous damper with a damping constant of 2 lb·s/ft. If the mass is set in motion from its equilibrium position with a downward velocity of 3 in/s, find its position y at any time t. Plot y versus t. Determine when the mass first returns to its equilibrium position. Also find the time τ such that $|y(t)| < 0.01$ in for all $t > \tau$.

11. (a) A spring is stretched 10 cm by a force of 3 N ODEA (newtons). A mass of 2 kg is hung from the spring and is also attached to a viscous damper that exerts a force of 3 N when the velocity of the mass is 5 m/s. If the mass is pulled down 5 cm below its equilibrium position and given an initial downward velocity of 10 cm/s, determine its position y at any time t. Find the quasi frequency ν and the ratio of ν to the natural frequency of the corresponding undamped motion.

(b) Draw a phase portrait of the equivalent dynamical system that includes the trajectory corresponding to the initial value problem in part (a).

12. (a) A series circuit has a capacitor of 10^{-5} farad, ODEA a resistor of 3×10^2 ohms, and an inductor of 0.2 henry. The initial charge on the capacitor is 10^{-6} coulomb and there is no initial current. Find the charge q on the capacitor at any time t.

(b) Draw a phase portrait of the equivalent dynamical system that includes the trajectory corresponding to the initial value problem in part (a).

13. A certain vibrating system satisfies the equation $y'' + \gamma y' + y = 0$. Find the value of the damping coefficient γ for which the quasi period of the damped motion is 50% greater than the period of the corresponding undamped motion.

14. Show that the period of motion of an undamped vibration of a mass hanging from a vertical spring is $2\pi \sqrt{L/g}$, where L is the elongation of the spring due to the mass and g is the acceleration due to gravity.

15. Show that the solution of the initial value problem

$$my'' + \gamma y' + ky = 0, \quad y(t_0) = y_0, \quad y'(t_0) = y_1$$

can be expressed as the sum $y = v + w$, where v satisfies the initial conditions $v(t_0) = y_0, v'(t_0) = 0$, w satisfies the initial conditions $w(t_0) = 0$, $w'(t_0) = y_1$, and both v and w satisfy the same differential equation as u. This is another instance of superposing solutions of simpler problems to obtain the solution of a more general problem.

16. Show that $A \cos \omega_0 t + B \sin \omega_0 t$ can be written in the form $r \sin(\omega_0 t - \theta)$. Determine r and θ in terms of A and B. If $R \cos(\omega_0 t - \delta) = r \sin(\omega_0 t - \theta)$, determine the relationship among R, r, δ, and θ.

17. A mass weighing 8 lb stretches a spring 1.5 in. The mass is also attached to a damper with coefficient γ. Determine the value of γ for which the system is critically damped. Be sure to give the units for γ.

18. If a series circuit has a capacitor of $C = 0.8 \times 10^{-6}$ farad and an inductor of $L = 0.2$ henry, find the resistance R so that the circuit is critically damped.

19. Assume that the system described by the equation $my'' + \gamma y' + ky = 0$ is either critically damped or overdamped. Show that the mass can pass through the equilibrium position at most once, regardless of the initial conditions.

   Hint: Determine all possible values of t for which $y = 0$.

20. Assume that the system described by the equation $my'' + \gamma y' + ky = 0$ is critically damped and that the initial conditions are $y(0) = y_0$, $y'(0) = v_0$. If $v_0 = 0$, show that $y \to 0$ as $t \to \infty$ but that y is never zero. If y_0 is positive, determine a condition on v_0 that will ensure that the mass passes through its equilibrium position after it is released.

21. **Logarithmic Decrement.**
   (a) For the damped oscillation described by Eq. (23), show that the time between successive maxima is $T_d = 2\pi/\nu$.
   (b) Show that the ratio of the displacements at two successive maxima is given by $\exp(\gamma T_d/2m)$. Observe that this ratio does not depend on which pair of maxima is chosen. The natural logarithm of this ratio is called the logarithmic decrement and is denoted by Δ.
   (c) Show that $\Delta = \pi\gamma/m\nu$. Since m, ν, and Δ are quantities that can be measured easily for a mechanical system, this result provides a convenient and *practical* method for determining the damping constant of the system, which is more difficult to measure directly. In particular, for the motion of a vibrating mass in a viscous fluid, the damping constant depends on the viscosity of the fluid. For simple geometric shapes, the form of this dependence is known, and the preceding relation allows the experimental determination of the viscosity. This is one of the most accurate ways of determining the viscosity of a gas at high pressure.

22. Referring to Problem 21, find the logarithmic decrement of the system in Problem 10.

23. For the system in Problem 17, suppose that $\Delta = 3$ and $T_d = 0.3$ s. Referring to Problem 21, determine the value of the damping coefficient γ.

24. The position of a certain spring–mass system satisfies the initial value problem

$$\tfrac{3}{2}y'' + ky = 0, \qquad y(0) = 2, \quad y'(0) = v.$$

   If the period and amplitude of the resulting motion are observed to be π and 3, respectively, determine the values of k and v.

25. Consider the initial value problem

$$y'' + \gamma y' + y = 0, \qquad y(0) = 2, \quad y'(0) = 0.$$

   We wish to explore how long a time interval is required for the solution to become "negligible" and how this interval depends on the damping coefficient γ. To be more precise let us seek the time τ such that $|y(t)| < 0.01$ for all $t > \tau$. Note that critical damping for this problem occurs for $\gamma = 2$.
   (a) Let $\gamma = 0.25$ and determine τ, or at least estimate it fairly accurately from a plot of the solution.
   (b) Repeat part (a) for several other values of γ in the interval $0 < \gamma < 1.5$. Note that τ steadily decreases as γ increases for γ in this range.
   (c) Create a graph of τ versus γ by plotting the pairs of values found in parts (a) and (b). Is the graph a smooth curve?
   (d) Repeat part (b) for values of γ between 1.5 and 2. Show that τ continues to decrease until γ reaches a certain critical value γ_0, after which τ increases. Find γ_0 and the corresponding minimum value of τ to two decimal places.
   (e) Another way to proceed is to write the solution of the initial value problem in the form (23). Neglect the cosine factor and consider only the exponential factor and the amplitude R. Then find an expression for τ as a function of γ. Compare the approximate results obtained in this way with the values determined in parts (a), (b), and (d).

26. Consider the initial value problem

$$my'' + \gamma y' + ky = 0, \quad y(0) = y_0, \quad y'(0) = v_0.$$

   Assume that $\gamma^2 < 4km$.
   (a) Solve the initial value problem.

(b) Write the solution in the form $y(t) = R \exp(-\gamma t/2m) \cos(\nu t - \delta)$. Determine R in terms of m, γ, k, y_0, and v_0.

(c) Investigate the dependence of R on the damping coefficient γ for fixed values of the other parameters.

27. Use the differential equation derived in Problem 19 of Section 4.1 to determine the period of vertical oscillations of a cubic block floating in a fluid under the stated conditions.

ODEA 28. The position of a certain undamped spring–mass system satisfies the initial value problem

$$y'' + 2y = 0, \qquad y(0) = 0, \quad y'(0) = 2.$$

(a) Find the solution of this initial value problem.

(b) Plot y versus t and y' versus t on the same axes.

(c) Draw the phase portrait for the dynamical system equivalent to $y'' + 2y = 0$. Include the trajectory corresponding to the initial conditions $y(0) = 0, y'(0) = 2$.

ODEA 29. The position of a certain spring–mass system satisfies the initial value problem

$$y'' + \tfrac{1}{4}y' + 2y = 0, \qquad y(0) = 0, \quad y'(0) = 2.$$

(a) Find the solution of this initial value problem.

(b) Plot y versus t and y' versus t on the same axes.

(c) Draw the phase portrait for the dynamical system equivalent to $y'' + (1/4)y' + 2y = 0$. Include the trajectory corresponding to the initial conditions $y(0) = 0, y'(0) = 2$.

30. In the absence of damping, the motion of a spring–mass system satisfies the initial value problem

$$my'' + ky = 0, \qquad y(0) = a, \quad y'(0) = b.$$

(a) Show that the kinetic energy initially imparted to the mass is $mb^2/2$ and that the potential energy initially stored in the spring is $ka^2/2$, so that initially the total energy in the system is $(ka^2 + mb^2)/2$.

(b) Solve the given initial value problem.

(c) Using the solution in part (b), determine the total energy in the system at any time t. Your result should confirm the principle of conservation of energy for this system.

31. If the restoring force of a nonlinear spring satisfies ODEA the relation

$$F_s(\Delta x) = -k\Delta x - \epsilon(\Delta x)^3,$$

where $k > 0$, then the differential equation for the displacement $x(t)$ of the mass from its equilibrium position satisfies the differential equation (see Problem 17, Section 4.1)

$$mx'' + \gamma x' + kx + \epsilon x^3 = 0.$$

Assume that the initial conditions are

$$x(0) = 0, \quad x'(0) = 1.$$

(a) Find $x(t)$ when $\epsilon = 0$ and also determine the amplitude and period of the motion.

(b) Let $\epsilon = 0.1$. Plot a numerical approximation to the solution. Does the motion appear to be periodic? Estimate the amplitude and period.

(c) Repeat part (c) for $\epsilon = 0.2$ and $\epsilon = 0.3$.

(d) Plot your estimated values of the amplitude A and the period T versus ϵ. Describe the way in which A and T, respectively, depend on ϵ.

(e) Repeat parts (c), (d), and (e) for negative values of ϵ.

4.6 Nonhomogeneous Equations; Method of Undetermined Coefficients

We now return to the nonhomogeneous equation

$$L[y] = y'' + p(t)y' + q(t)y = g(t), \tag{1}$$

where p, q, and g are given (continuous) functions on the open interval I. The equation

$$L[y] = y'' + p(t)y' + q(t)y = 0, \tag{2}$$

in which $g(t) = 0$ and p and q are the same as in Eq. (1), is called the homogeneous equation corresponding to Eq. (1). The following two results describe the structure of solutions of the nonhomogeneous equation (1) and provide a basis for constructing its general solution.

THEOREM 4.6.1

If Y_1 and Y_2 are two solutions of the nonhomogeneous equation (1), then their difference $Y_1 - Y_2$ is a solution of the corresponding homogeneous equation (2). If, in addition, y_1 and y_2 are a fundamental set of solutions of Eq. (2), then

$$Y_1(t) - Y_2(t) = c_1 y_1(t) + c_2 y_2(t), \tag{3}$$

where c_1 and c_2 are certain constants.

Proof

To prove this result, note that Y_1 and Y_2 satisfy the equations

$$L[Y_1](t) = g(t), \qquad L[Y_2](t) = g(t). \tag{4}$$

Subtracting the second of these equations from the first, we have

$$L[Y_1](t) - L[Y_2](t) = g(t) - g(t) = 0. \tag{5}$$

However,

$$L[Y_1] - L[Y_2] = L[Y_1 - Y_2],$$

so Eq. (5) becomes

$$L[Y_1 - Y_2](t) = 0. \tag{6}$$

Equation (6) states that $Y_1 - Y_2$ is a solution of Eq. (2). Finally, since all solutions of Eq. (2) can be expressed as linear combinations of a fundamental set of solutions by Theorem 4.2.5, it follows that the solution $Y_1 - Y_2$ can be so written. Hence Eq. (3) holds and the proof is complete. ∎

THEOREM 4.6.2

The general solution of the nonhomogeneous equation (1) can be written in the form

$$\boxed{y = \phi(t) = c_1 y_1(t) + c_2 y_2(t) + Y(t),} \tag{7}$$

where y_1 and y_2 are a fundamental set of solutions of the corresponding homogeneous equation (2), c_1 and c_2 are arbitrary constants, and Y is some specific solution of the nonhomogeneous equation (1).

Proof

The proof of Theorem 4.6.2 follows quickly from Theorem 4.6.1. Note that Eq. (3) holds if we identify Y_1 with an arbitrary solution ϕ of Eq. (1) and Y_2 with the specific solution Y. From Eq. (3) we thereby obtain

$$\phi(t) - Y(t) = c_1 y_1(t) + c_2 y_2(t), \tag{8}$$

which is equivalent to Eq. (7). Since ϕ is an arbitrary solution of Eq. (1), the expression on the right side of Eq. (7) includes all solutions of Eq. (1); thus it is natural to call it the general solution of Eq. (1). ∎

General Solution Strategy

Theorem 4.6.2 states that to solve the nonhomogeneous equation (1), we must do three things:

1. Find the general solution $c_1 y_1(t) + c_2 y_2(t)$ of the corresponding homogeneous equation. This solution is frequently called the **complementary solution** and may be denoted by $y_c(t)$.

2. Find some single solution $Y(t)$ of the nonhomogeneous equation. Often this solution is referred to as a **particular solution**.

3. Add together the functions found in the two preceding steps.

We have already discussed how to find $y_c(t)$, at least when the homogeneous equation (2) has constant coefficients. In the remainder of this section, we will focus on a special method of finding a particular solution $Y(t)$ of the nonhomogeneous equation (1) known as the **method of undetermined coefficients**. In Section 4.8, we present a general method known as the **method of variation of parameters**. Each has some advantages and some possible shortcomings.

Method of Undetermined Coefficients

The method of undetermined coefficients requires that we make an initial assumption about the form of the particular solution $Y(t)$, but with the coefficients left unspecified. We then substitute the assumed expression into Eq. (1) and attempt to determine the coefficients so as to satisfy that equation. If we are successful, then we have found a solution of the differential equation (1) and can use it for the particular solution $Y(t)$. If we cannot determine the coefficients, then this means that there is no solution of the form that we assumed. In this case, we may modify the initial assumption and try again.

The main advantage of the method of undetermined coefficients is that it is straightforward to execute once the assumption is made as to the form of $Y(t)$. Its major limitation is that it is useful primarily for equations for which we can easily write down the correct form of the particular solution in advance. For this reason, this method is usually used only for problems in which the homogeneous equation has constant coefficients and the nonhomogeneous term is restricted to a relatively small class of functions. In particular, we consider only nonhomogeneous terms that consist of polynomials, exponential functions, sines, cosines, or sums or products of such functions. Despite this limitation, the method of undetermined coefficients is useful for solving many problems that have important applications. However, the algebraic details may become tedious, and a computer algebra system can be very helpful in practical applications. We will illustrate the method of undetermined coefficients by several simple examples and then summarize some rules for using it.

EXAMPLE 1

Find a particular solution of

$$y'' - 3y' - 4y = 3e^{2t}. \tag{9}$$

We seek a function Y such that the combination $Y''(t) - 3Y'(t) - 4Y(t)$ is equal to $3e^{2t}$. Since the exponential function reproduces itself through differentiation, the most plausible way to achieve the desired result is to assume that $Y(t)$ is some multiple of e^{2t}, that is,

$$Y(t) = Ae^{2t},$$

where the coefficient A is yet to be determined. To find A, we calculate

$$Y'(t) = 2Ae^{2t}, \qquad Y''(t) = 4Ae^{2t},$$

and substitute for y, y', and y'' in Eq. (9). We obtain

$$(4A - 6A - 4A)e^{2t} = 3e^{2t}.$$

Hence $-6Ae^{2t}$ must equal $3e^{2t}$, so $A = -1/2$. Thus a particular solution is

$$Y(t) = -\tfrac{1}{2}e^{2t}. \tag{10}$$

EXAMPLE 2

Find a particular solution of
$$y'' - 3y' - 4y = 2\sin t. \tag{11}$$

By analogy with Example 1, let us first assume that $Y(t) = A \sin t$, where A is a constant to be determined. On substituting in Eq. (11) and rearranging the terms, we obtain

$$-5A \sin t - 3A \cos t = 2\sin t,$$

or

$$(2 + 5A) \sin t + 3A \cos t = 0. \tag{12}$$

Equation (12) can hold on an interval only if the coefficients of $\sin t$ and $\cos t$ are both zero. Thus we must have $2 + 5A = 0$ and also $3A = 0$. These contradictory requirements mean that there is no choice of the constant A that makes Eq. (12) true for all t. Thus we conclude that our assumption concerning $Y(t)$ is inadequate. The appearance of the cosine term in Eq. (12) suggests that we modify our original assumption to include a cosine term in $Y(t)$, that is,

$$Y(t) = A \sin t + B \cos t,$$

where A and B are to be determined. Then

$$Y'(t) = A \cos t - B \sin t, \qquad Y''(t) = -A \sin t - B \cos t.$$

By substituting these expressions for y, y', and y'' in Eq. (11) and collecting terms, we obtain

$$(-A + 3B - 4A) \sin t + (-B - 3A - 4B) \cos t = 2\sin t. \tag{13}$$

To satisfy Eq. (12), we must match the coefficients of $\sin t$ and $\cos t$ on each side of the equation. Thus A and B must satisfy the equations

$$-5A + 3B = 2, \qquad -3A - 5B = 0.$$

Hence $A = -5/17$ and $B = 3/17$, so a particular solution of Eq. (11) is

$$Y(t) = -\tfrac{5}{17} \sin t + \tfrac{3}{17} \cos t.$$

The method illustrated in the preceding examples can also be used when the right side of the equation is a polynomial. Thus, to find a particular solution of

$$y'' - 3y' - 4y = 4t^2 - 1, \tag{14}$$

we initially assume that $Y(t)$ is a polynomial of the same degree as the nonhomogeneous term, that is, $Y(t) = At^2 + Bt + C$.

To summarize our conclusions up to this point:

1. If the nonhomogeneous term $g(t)$ in Eq. (1) is an exponential function $e^{\alpha t}$, then assume that $Y(t)$ is proportional to the same exponential function.

2. If $g(t)$ is $\sin \beta t$ or $\cos \beta t$, then assume that $Y(t)$ is a linear combination of $\sin \beta t$ and $\cos \beta t$.

3. If $g(t)$ is a polynomial, then assume that $Y(t)$ is a polynomial of like degree.

The same principle extends to the case where $g(t)$ is a product of any two, or all three, of these types of functions, as the next example illustrates.

EXAMPLE 3

Find a particular solution of

$$y'' - 3y' - 4y = -8e^t \cos 2t. \qquad (15)$$

In this case, we assume that $Y(t)$ is the product of e^t and a linear combination of $\cos 2t$ and $\sin 2t$, that is,

$$Y(t) = Ae^t \cos 2t + Be^t \sin 2t.$$

The algebra is more tedious in this example, but it follows that

$$Y'(t) = (A + 2B)e^t \cos 2t + (-2A + B)e^t \sin 2t$$

and

$$Y''(t) = (-3A + 4B)e^t \cos 2t + (-4A - 3B)e^t \sin 2t.$$

By substituting these expressions in Eq. (15), we find that A and B must satisfy

$$10A + 2B = 8, \qquad 2A - 10B = 0.$$

Hence $A = 10/13$ and $B = 2/13$. Therefore a particular solution of Eq. (15) is

$$Y(t) = \tfrac{10}{13}e^t \cos 2t + \tfrac{2}{13}e^t \sin 2t.$$

Superposition Principle for Nonhomogeneous Equations

Now suppose that $g(t)$ is the sum of two terms, $g(t) = g_1(t) + g_2(t)$, and suppose that Y_1 and Y_2 are solutions of the equations

$$ay'' + by' + cy = g_1(t) \qquad (16)$$

and

$$ay'' + by' + cy = g_2(t), \qquad (17)$$

respectively. Then $Y_1 + Y_2$ is a solution of the equation

$$ay'' + by' + cy = g(t). \qquad (18)$$

To prove this statement, substitute $Y_1(t) + Y_2(t)$ for y in Eq. (18) and make use of Eqs. (16) and (17). A similar conclusion holds if $g(t)$ is the sum of any finite number of terms. The practical significance of this result is that for an equation whose nonhomogeneous function $g(t)$ can be expressed as a sum, one can consider instead several simpler equations and then add the results together. The following example is an illustration of this procedure.

EXAMPLE 4

Find a particular solution of

$$y'' - 3y' - 4y = 3e^{2t} + 2\sin t - 8e^t \cos 2t. \qquad (19)$$

By splitting up the right side of Eq. (19), we obtain the three equations

$$y'' - 3y' - 4y = 3e^{2t},$$
$$y'' - 3y' - 4y = 2\sin t,$$

and

$$y'' - 3y' - 4y = -8e^t \cos 2t.$$

Solutions of these three equations have been found in Examples 1, 2, and 3, respectively. Therefore a particular solution of Eq. (19) is their sum, namely,

$$Y(t) = -\tfrac{1}{2}e^{2t} + \tfrac{3}{17}\cos t - \tfrac{5}{17}\sin t + \tfrac{10}{13}e^t\cos 2t + \tfrac{2}{13}e^t\sin 2t.$$

The procedure illustrated in these examples enables us to solve a fairly large class of problems in a reasonably efficient manner. However, there is one difficulty that sometimes occurs. The next example illustrates how it arises.

EXAMPLE 5

Find a particular solution of

$$y'' - 3y' - 4y = 2e^{-t}. \tag{20}$$

Proceeding as in Example 1, we assume that $Y(t) = Ae^{-t}$. By substituting in Eq. (20), we then obtain

$$(A + 3A - 4A)e^{-t} = 2e^{-t}. \tag{21}$$

Since the left side of Eq. (21) is zero, there is no choice of A that satisfies this equation. Therefore, there is no particular solution of Eq. (20) of the assumed form. The reason for this possibly unexpected result becomes clear if we solve the homogeneous equation

$$y'' - 3y' - 4y = 0 \tag{22}$$

that corresponds to Eq. (20). A fundamental set of solutions of Eq. (22) is $y_1(t) = e^{-t}$ and $y_2(t) = e^{4t}$. Thus our assumed particular solution of Eq. (20) is actually a solution of the homogeneous equation (22). Consequently, it cannot possibly be a solution of the nonhomogeneous equation (20). To find a solution of Eq. (20), we must therefore consider functions of a somewhat different form.

At this stage, we have several possible alternatives. One is simply to try to guess the proper form of the particular solution of Eq. (20). Another is to solve this equation in some different way and then to use the result to guide our assumptions if this situation arises again in the future (see Problem 27 for another solution method). Still another possibility is to seek a simpler equation where this difficulty occurs and to use its solution to suggest how we might proceed with Eq. (20). Adopting the latter approach, we look for a first order equation analogous to Eq. (20). One possibility is

$$y' + y = 2e^{-t}. \tag{23}$$

If we try to find a particular solution of Eq. (23) of the form Ae^{-t}, we will fail because e^{-t} is a solution of the homogeneous equation $y' + y = 0$. However, from Section 2.1 we already know how to solve Eq. (23). An integrating factor is $\mu(t) = e^t$, and by multiplying by $\mu(t)$ and then integrating both sides, we obtain the solution

$$y = 2te^{-t} + ce^{-t}. \tag{24}$$

The second term on the right side of Eq. (24) is the general solution of the homogeneous equation $y' + y = 0$, but the first term is a solution of the full nonhomogeneous equation (23). Observe that it involves the exponential factor e^{-t} multiplied by the factor t. This is the clue that we were looking for.

We now return to Eq. (20) and assume a particular solution of the form $Y(t) = Ate^{-t}$. Then

$$Y'(t) = Ae^{-t} - Ate^{-t}, \qquad Y''(t) = -2Ae^{-t} + Ate^{-t}. \tag{25}$$

Substituting these expressions for y, y', and y'' in Eq. (20), we obtain $-5A = 2$, so $A = -2/5$. Thus a particular solution of Eq. (20) is

$$Y(t) = -\tfrac{2}{5}te^{-t}. \tag{26}$$

The outcome of Example 5 suggests a modification of the principle stated previously: if the assumed form of the particular solution duplicates a solution of the corresponding homogeneous equation, then modify the assumed particular solution by multiplying it by t. Occasionally, this modification will be insufficient to remove all duplication with the solutions of the homogeneous equation, in which case it is necessary to multiply by t a second time. For a second order equation, it will never be necessary to carry the process further than this.

Summary: Method of Undetermined Coefficients

We now summarize the steps involved in finding the solution of an initial value problem consisting of a nonhomogeneous equation of the form

$$ay'' + by' + cy = g(t), \tag{27}$$

where the coefficients a, b, and c are constants, together with a given set of initial conditions:

1. Find the general solution of the corresponding homogeneous equation.
2. Make sure that the function $g(t)$ in Eq. (27) belongs to the class of functions discussed in this section; that is, be sure it involves nothing more than exponential functions, sines, cosines, polynomials, or sums or products of such functions. If this is not the case, use the method of variation of parameters (discussed in Section 4.8).
3. If $g(t) = g_1(t) + \cdots + g_n(t)$, that is, if $g(t)$ is a sum of n terms, then form n subproblems, each of which contains only one of the terms $g_1(t), \ldots, g_n(t)$. The ith subproblem consists of the equation

$$ay'' + by' + cy = g_i(t),$$

where i runs from 1 to n.
4. For the ith subproblem, assume a particular solution $Y_i(t)$ consisting of the appropriate exponential function, sine, cosine, polynomial, or combination thereof. If there is any duplication in the assumed form of $Y_i(t)$ with the solutions of the homogeneous equation (found in step 1), then multiply $Y_i(t)$ by t, or (if necessary) by t^2, so as to remove the duplication. See Table 4.6.1.
5. Find a particular solution $Y_i(t)$ for each of the subproblems. Then the sum $Y_1(t) + \cdots + Y_n(t)$ is a particular solution of the full nonhomogeneous equation (27).
6. Form the sum of the general solution of the homogeneous equation (step 1) and the particular solution of the nonhomogeneous equation (step 5). This is the general solution of the nonhomogeneous equation.
7. Use the initial conditions to determine the values of the arbitrary constants remaining in the general solution.

For some problems, this entire procedure is easy to carry out by hand, but in many cases it requires considerable algebra. Once you understand clearly how the method works, a computer algebra system can be of great assistance in executing the details.

TABLE 4.6.1	The Particular Solution of $ay'' + by' + cy = g_i(t)$.

$g_i(t)$	$Y_i(t)$
$P_n(t) = a_0 t^n + a_1 t^{n-1} + \cdots + a_n$	$t^s(A_0 t^n + A_1 t^{n-1} + \cdots + A_n)$
$P_n(t)e^{\alpha t}$	$t^s(A_0 t^n + A_1 t^{n-1} + \cdots + A_n)e^{\alpha t}$
$P_n(t)e^{\alpha t} \begin{cases} \sin \beta t \\ \cos \beta t \end{cases}$	$t^s[(A_0 t^n + A_1 t^{n-1} + \cdots + A_n)e^{\alpha t} \cos \beta t$
	$+ (B_0 t^n + B_1 t^{n-1} + \cdots + B_n)e^{\alpha t} \sin \beta t]$

Notes. Here s is the smallest nonnegative integer ($s = 0$, 1, or 2) that will ensure that no term in $Y_i(t)$ is a solution of the corresponding homogeneous equation. Equivalently, for the three cases, s is the number of times 0 is a root of the characteristic equation, α is a root of the characteristic equation, and $\alpha + i\beta$ is a root of the characteristic equation, respectively.

EXAMPLE 6

In each of the following problems, use Table 4.6.1 to determine a suitable form for the particular solution if the method of undetermined coefficients is to be used.

(a) $y'' = 3t^3 - t$ (b) $y'' - y' - 2y = -3te^{-t} + 2\cos 4t$

(c) $y'' + 2y' + 5y = t^2 e^{-t} \sin 2t$ (d) $y'' + y = \tan t$

Solution

▶

(a) The general solution of $y'' = 0$ is $y = c_1 1 + c_2 t$. Since $g(t) = 3t^3 - t$ is a third degree polynomial, we assume that $Y(t) = t^s [A_3 t^3 + A_2 t^2 + A_1 t + A_0]$. Here we must take $s = 2$ to ensure that none of the functions in the assumed form for $Y(t)$ appear in the fundamental set. Thus

$$Y(t) = A_3 t^5 + A_2 t^4 + A_1 t^3 + A_0 t^2.$$

(b) The general solution of $y'' - y' - 2y = 0$ is $y = c_1 e^{-t} + c_2 e^{2t}$. We identify two sub-problems corresponding to the nonhomogeneous terms $g_1(t) = -3te^{-t}$ and $g_2(t) = 2\cos 4t$. Since g_1 is the exponential function e^{-t} multiplied by a first degree polynomial, we assume that $Y_1(t) = t^s[(A_1 t + A_0)e^{-t}]$. Since e^{-t} is a solution of the homogeneous equation, we must take $s = 1$. Thus

$$Y_1(t) = (A_1 t^2 + A_0 t)e^{-t}.$$

The correct form for Y_2 is $Y_2(t) = t^s [B_0 \cos 4t + C_0 \sin 4t]$. Neither of the terms $\cos 4t$ or $\sin 4t$ are solutions of the homogeneous equation so we set $s = 0$ and obtain

$$Y_2(t) = B_0 \cos 4t + C_0 \sin 4t.$$

Substituting the expression for Y_1 into $y'' - y' - 2y = -3te^{-t}$ will determine A_0 and A_1 while substituting the expression for Y_2 into $y'' - y' - 2y = 2\cos 4t$ will determine B_0 and C_0. With these coefficients ascertained, the general solution of $y'' - y' - 2y = -3te^{-t} + 2\cos 4t$ is $y = c_1 e^{-t} + c_2 e^{2t} + Y_1(t) + Y_2(t)$.

(c) The general solution of $y'' + 2y' + 5y = 0$ is $y = c_1 e^{-t} \cos 2t + c_2 e^{-t} \sin 2t$. In this case, $g(t) = t^2 e^{-t} \sin 2t$, that is, a second degree polynomial times $e^{-t} \sin 2t$. Since it is necessary to include both sine and cosine functions even if only one or the other is present in the nonhomogeneous expression, the correct form for $Y(t)$ is

$$Y(t) = t^s \left[(A_0 t^2 + A_1 t + A_2)e^{-t} \cos 2t + (B_0 t^2 + B_1 t + B_2)e^{-t} \sin 2t \right].$$

We must then choose $s = 1$ to ensure that none of the terms in the assumed form for $Y(t)$ are solutions of the homogeneous equation. Thus

$$Y(t) = (A_0t^3 + A_1t^2 + A_2t)e^{-t}\cos 2t + (B_0t^3 + B_1t^2 + B_2t)e^{-t}\sin 2t.$$

(d) The method of undetermined coefficients is not applicable to this problem since the nonhomogeneous function $g(t) = \tan t$ does not lie in the class of functions consisting of linear combinations of products of polynomials, exponential, sine, and cosine functions. However, a particular solution can be obtained by the method of variation of parameters to be discussed in Section 4.8.

The method of undetermined coefficients is self-correcting in the sense that if one assumes too little for $Y(t)$, then a contradiction is soon reached that usually points the way to the modification that is needed in the assumed form. On the other hand, if one assumes too many terms, then some unnecessary work is done and some coefficients turn out to be zero, but at least the correct answer is obtained.

PROBLEMS

In each of Problems 1 through 12, find the general solution of the given differential equation.

1. $y'' - 2y' - 3y = 3e^{2t}$
2. $y'' + 2y' + 5y = 3\sin 2t$
3. $y'' - 2y' - 3y = -3te^{-t}$
4. $y'' + 2y' = 3 + 4\sin 2t$
5. $y'' + 9y = t^2e^{3t} + 6$
6. $y'' + 2y' + y = 2e^{-t}$
7. $2y'' + 3y' + y = t^2 + 3\sin t$
8. $y'' + y = 3\sin 2t + t\cos 2t$
9. $u'' + \omega_0^2 u = \cos\omega t, \qquad \omega^2 \neq \omega_0^2$
10. $u'' + \omega_0^2 u = \cos\omega_0 t$
11. $y'' + y' + 4y = 2\sinh t$
    Hint: $\sinh t = (e^t - e^{-t})/2$
12. $y'' - y' - 2y = \cosh 2t$
    Hint: $\cosh t = (e^t + e^{-t})/2$

In each of Problems 13 through 18, find the solution of the given initial value problem.

13. $y'' + y' - 2y = 2t, \qquad y(0) = 0, \quad y'(0) = 1$
14. $y'' + 4y = t^2 + 3e^t, \qquad y(0) = 0, \quad y'(0) = 2$
15. $y'' - 2y' + y = te^t + 4,$
    $y(0) = 1, \quad y'(0) = 1$
16. $y'' - 2y' - 3y = 3te^{2t},$
    $y(0) = 1, \quad y'(0) = 0$
17. $y'' + 4y = 3\sin 2t, \qquad y(0) = 2, \quad y'(0) = -1$
18. $y'' + 2y' + 5y = 4e^{-t}\cos 2t,$
    $y(0) = 1, \quad y'(0) = 0$

In each of Problems 19 through 26: CAS
(a) Determine a suitable form for $Y(t)$ if the method of undetermined coefficients is to be used.
(b) Use a computer algebra system to find a particular solution of the given equation.

19. $y'' + 3y' = 2t^4 + t^2e^{-3t} + \sin 3t$
20. $y'' + y = t(1 + \sin t)$
21. $y'' - 5y' + 6y = e^t\cos 2t + e^{2t}(3t + 4)\sin t$
22. $y'' + 2y' + 2y = 3e^{-t} + 2e^{-t}\cos t + 4e^{-t}t^2\sin t$
23. $y'' - 4y' + 4y = 2t^2 + 4te^{2t} + t\sin 2t$
24. $y'' + 4y = t^2\sin 2t + (6t + 7)\cos 2t$
25. $y'' + 3y' + 2y = e^t(t^2 + 1)\sin 2t + 3e^{-t}\cos t + 4e^t$
26. $y'' + 2y' + 5y = 3te^{-t}\cos 2t - 2te^{-2t}\cos t$
27. Consider the equation

$$y'' - 3y' - 4y = 2e^{-t} \qquad \text{(i)}$$

from Example 5. Recall that $y_1(t) = e^{-t}$ and $y_2(t) = e^{4t}$ are solutions of the corresponding homogeneous equation. Adapting the method of reduction of order (Section 4.3), seek a solution of the nonhomogeneous equation of the form $Y(t) = v(t)y_1(t) = v(t)e^{-t}$, where $v(t)$ is to be determined.
(a) Substitute $Y(t)$, $Y'(t)$, and $Y''(t)$ into Eq. (i) and show that $v(t)$ must satisfy $v'' - 5v' = 2$.
(b) Let $w(t) = v'(t)$ and show that $w(t)$ must satisfy $w' - 5w = 2$. Solve this equation for $w(t)$.

(c) Integrate $w(t)$ to find $v(t)$ and then show that

$$Y(t) = -\tfrac{2}{5}te^{-t} + \tfrac{1}{5}c_1e^{4t} + c_2e^{-t}.$$

The first term on the right side is the desired particular solution of the nonhomogeneous equation. Note that it is a product of t and e^{-t}.

Nonhomogeneous Cauchy-Euler Equations. In each of Problems 28 through 31, find the general solution by using the change of variable $z = \ln x$ to transform the equation into one with constant coefficients.

28. $x^2y'' - 3xy' + 4y = \ln x$
29. $x^2y'' + 7xy' + 5y = x$
30. $x^2y'' - 2xy' + 2y = 3x^2 + 2 \ln x$
31. $x^2y'' + xy' + 4y = \sin(\ln x)$

32. Determine the general solution of

$$y'' + \lambda^2 y = \sum_{m=1}^{N} a_m \sin m\pi t,$$

where $\lambda > 0$ and $\lambda \neq m\pi$ for $m = 1, \ldots, N$.

33. In many physical problems, the nonhomogeneous ▸ODEA term may be specified by different formulas in different time periods. As an example, determine the solution $y = \phi(t)$ of

$$y'' + y = \begin{cases} t, & 0 \le t \le \pi, \\ \pi e^{\pi - t}, & t > \pi, \end{cases}$$

satisfying the initial conditions $y(0) = 0$ and $y'(0) = 1$. Assume that y and y' are also continuous at $t = \pi$. Plot the nonhomogeneous term and the solution as functions of time.

Hint: First solve the initial value problem for $t \le \pi$; then solve for $t > \pi$, determining the constants in the latter solution from the continuity conditions at $t = \pi$.

34. Follow the instructions in Problem 33 to solve the ▸ODEA differential equation

$$y'' + 2y' + 5y = \begin{cases} 1, & 0 \le t \le \pi/2, \\ 0, & t > \pi/2 \end{cases}$$

with the initial conditions $y(0) = 0$ and $y'(0) = 0$.

4.7 Forced Vibrations, Frequency Response, and Resonance

We will now investigate the situation in which a periodic external force is applied to a spring–mass system. The behavior of this simple system models that of many oscillatory systems with an external force due, for example, to a motor attached to the system. We will first consider the case in which damping is present and will look later at the idealized special case in which there is assumed to be no damping.

Forced Vibrations with Damping

Recall that the equation of motion for a damped spring–mass system with external forcing, $F(t)$, is

$$my'' + \gamma y' + ky = F(t), \tag{1}$$

where m, γ, and k are the mass, damping coefficient, and spring constant, respectively. Dividing through Eq. (1) by m puts it in the form

$$y'' + 2\delta y' + \omega_0^2 y = f(t), \tag{2}$$

where $\delta = \gamma/(2m)$, $\omega_0^2 = k/m$ and $f(t) = F(t)/m$.

A useful approach to the analysis of a vibrating mechanical system is to study its response to harmonic inputs. While we could select $f(t)$ to be of the form $A \cos \omega t$ (or $A \sin \omega t$), the ensuing analysis is in fact less complicated if we select a complex valued exponential function as the input, $f(t) = Ae^{i\omega t} = A(\cos \omega t + i \sin \omega t)$. Thus, we wish to find the general solution of

$$y'' + 2\delta y' + \omega_0^2 y = Ae^{i\omega t}. \tag{3}$$

Note that the solutions $y_1(t)$ and $y_2(t)$ of the homogeneous equation depend on the roots λ_1 and λ_2 of the characteristic equation $\lambda^2 + 2\delta\lambda + \omega_0^2 = 0$. Since δ and ω_0^2 are positive, it follows that λ_1 and λ_2 are either real and negative or are complex conjugates with negative real part. Consequently, the forcing function on the right hand side of Eq. (3) cannot be a solution of the homogeneous equation. The correct form to assume for the particular solution using the method of undetermined coefficients is therefore $Y(t) = Ce^{i\omega t}$. Substituting $Y(t)$ into the left side of Eq. (3) and solving for C results in

$$Y(t) = G(i\omega)Ae^{i\omega t} \tag{4}$$

where

$$G(i\omega) = \frac{1}{\omega_0^2 - \omega^2 + 2i\delta\omega}, \tag{5}$$

or

$$G(i\omega) = \frac{\omega_0^2 - \omega^2 - 2i\delta\omega}{\left(\omega_0^2 - \omega^2\right)^2 + 4\delta^2\omega^2}, \tag{6}$$

where we have multiplied both numerator and denominator of the right side of Eq. (5) by the conjugate of the denominator. The general solution of Eq. (3) is of the form

$$y = y_c(t) + Y(t) \tag{7}$$

where $y_c(t) = c_1 y_1(t) + c_2 y_2(t)$ is the general solution of the homogeneous equation and the constants c_1 and c_2 depend on initial conditions. Since $y_c(t) \to 0$ as $t \to \infty$, it is called the **transient solution**. In many applications, it is of little importance and (depending on the value of δ) may well be undetectable after only a few seconds. The transient solution enables us to satisfy whatever initial conditions may be imposed. With increasing time, the energy put into the system by the initial displacement and velocity is dissipated through the damping force, and the motion then becomes the response of the system to the external force. Without damping, the effect of the initial conditions would persist for all time.

The remaining term in Eq. (7), $Y(t) = G(i\omega)Ae^{i\omega t}$, or in the real case corresponding to the input $f(t) = A\cos\omega t$,

$$Y_{\text{Re}}(t) = \text{Re } Y(t) = A\frac{(\omega_0^2 - \omega^2)\cos\omega t + 2\delta\omega\sin\omega t}{(\omega_0^2 - \omega^2)^2 + 4\delta^2\omega^2}, \tag{8}$$

does not die out as t increases but persists indefinitely, or as long as the external force is applied. It represents a steady oscillation with the same frequency as the external force and is called the **steady-state solution**, the **steady-state response**, the **steady-state output**, or the **forced response**.

The Frequency Response Function

It is convenient to represent $G(i\omega)$ in Eq. (6) in complex exponential form (see Figure 4.7.1),

$$G(i\omega) = |G(i\omega)|e^{-i\phi(\omega)} = |G(i\omega)|\{\cos\phi(\omega) - i\sin\phi(\omega)\} \tag{9}$$

where

$$|G(i\omega)| = \left[G(i\omega)\overline{G(i\omega)}\right]^{1/2} = \frac{1}{\sqrt{\left(\omega_0^2 - \omega^2\right)^2 + 4\delta^2\omega^2}} \tag{10}$$

and $\phi(\omega)$, $0 \leq \phi(\omega) < \pi$, is given by

$$\phi(\omega) = \arccos\left\{\frac{\omega_0^2 - \omega^2}{\sqrt{\left(\omega_0^2 - \omega^2\right)^2 + 4\delta^2\omega^2}}\right\}. \tag{11}$$

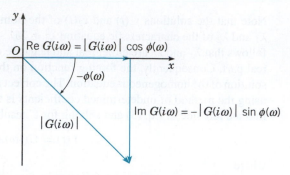

FIGURE 4.7.1 Polar coordinate representation of the frequency response function $G(i\omega)$.

Using Eq. (9) we can express the particular solution (4) as

$$Y(t) = G(i\omega)Ae^{i\omega t} = |G(i\omega)|e^{-i\phi(\omega)}Ae^{i\omega t} = A|G(i\omega)|e^{i(\omega t - \phi(\omega))}. \tag{12}$$

In the case of the real input, $f(t) = A\cos\omega t$, the real-valued particular solution (8), expressed in phase-amplitude form, arises by taking the real part of $Y(t)$ in (12),

$$Y_{\text{Re}}(t) = \text{Re } Y(t) = A|G(i\omega)|\cos(\omega t - \phi). \tag{13}$$

From Eq. (12), we see that the steady-state response is simply the product of $G(i\omega)$ with the input $Ae^{i\omega t}$. The function $G(i\omega)$ is called the **frequency response** of the system. The absolute value of the frequency response, $|G(i\omega)|$, is called the **gain** of the frequency response and the angle

$$\phi(\omega) = -\arg(G(i\omega)) = \arccos\left\{\frac{\omega_0^2 - \omega^2}{\sqrt{(\omega_0^2 - \omega^2)^2 + 4\delta^2\omega^2}}\right\}$$

is called the **phase** of the frequency response. As indicated by expressions (12) and (13), the harmonic input is modified in two ways as it passes through the spring-mass system (see Figure 4.7.2):

1. The amplitude of the output equals the amplitude of the harmonic input amplified or attenuated by the gain factor, $|G(i\omega)|$.
2. There is a phase shift in the steady-state output of magnitude $\phi(\omega)$ relative to the harmonic input.

We now investigate how the gain $|G(i\omega)|$ and the phase shift $\phi(\omega)$ depend on the frequency ω of the harmonic input. For low-frequency excitation, that is, as $\omega \to 0$, it follows from Eq. (10) that $|G(i\omega)| \to 1/\omega_0^2 = m/k$. At the other extreme, for very high-frequency excitation, Eq. (10) implies that $|G(i\omega)| \to 0$ as $\omega \to \infty$. At an intermediate value of ω, the amplitude may have a maximum. To find this maximum point, we can differentiate $|G(i\omega)|$ with respect to ω and set the result equal to zero. In this way, we find that the maximum amplitude occurs when $\omega = \omega_{\text{max}}$, where

$$\omega_{\text{max}}^2 = \omega_0^2 - 2\delta^2 = \omega_0^2 - \frac{\gamma^2}{2m^2} = \omega_0^2\left(1 - \frac{\gamma^2}{2mk}\right). \tag{14}$$

Note that $\omega_{\text{max}} < \omega_0$ and that ω_{max} is close to ω_0 when the damping coefficient γ is small. The maximum value of $|G(i\omega)|$ is

$$|G(i\omega_{\text{max}})| = \frac{m}{\gamma\omega_0\sqrt{1 - (\gamma^2/4mk)}} \cong \frac{m}{\gamma\omega_0}\left(1 + \frac{\gamma^2}{8mk}\right), \tag{15}$$

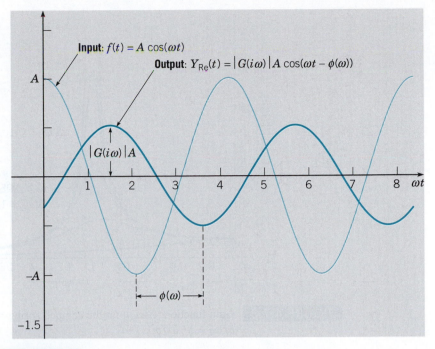

FIGURE 4.7.2 The steady-state response $Y_{Re} = |G(i\omega)|A\cos(\omega t - \phi(\omega))$ of a spring-mass system due to the harmonic input $f(t) = A\cos\omega t$.

where the last expression is an approximation for small γ. If $\gamma^2/mk > 2$, then ω_{max}, as given by Eq. (14), is imaginary. In this case, the maximum value of $|G(i\omega)|$ occurs for $\omega = 0$, and $|G(i\omega)|$ is a monotone decreasing function of ω. Recall that critical damping occurs when $\gamma^2/mk = 4$.

For small γ, it follows from Eq. (15) that $|G(i\omega_{max})| \cong m/\gamma\omega_0$. Thus, for lightly damped systems, the gain $|G(i\omega)|$ when ω is near ω_0 is quite large, and the smaller the value of γ, the more pronounced is this effect. This phenomenon is known as **resonance**, and it is often an important design consideration. Resonance can be either good or bad, depending on the circumstances. It must be taken very seriously in the design of structures, such as buildings and bridges, where it can produce instabilities that might lead to the catastrophic failure of the structure. On the other hand, resonance can be put to good use in the design of instruments, such as seismographs, that are intended to detect weak periodic incoming signals.

Figure 4.7.3 contains some representative graphs of $|G(i\omega)|/|G(0)| = \omega_0^2|G(i\omega)|$ versus ω/ω_0 for several values of $\delta = \gamma/2m$. Note particularly how the peaks in the curves sharpen near $\omega/\omega_0 = 1$ as δ gets smaller. The limiting case as $\delta \to 0$ is also shown. It follows from Eq. (10) that $|G(i\omega)| \to 1/|\omega_0^2 - \omega^2|$ as $\gamma \to 0$ and hence

$$|G(i\omega)|/|G(0)| = \omega_0^2|G(i\omega)| \to \left|1 - \frac{\omega^2}{\omega_0^2}\right|^{-1}, \qquad \gamma \to 0$$

is asymptotic to the vertical line at $\omega/\omega_0 = 1$, as shown in the figure. As the damping in the system increases, the peak response gradually diminishes.

Figure 4.7.3 also illustrates the usefulness of dimensionless variables. You can easily verify that each of the quantities $|G(i\omega)|/|G(0)| = \omega_0^2|G(i\omega)|$, ω/ω_0, and $\delta/\omega_0 = \gamma/2\sqrt{mk}$ is dimensionless. The importance of this observation is that the number of significant parameters in the problem has been reduced to three rather than the five that appear in

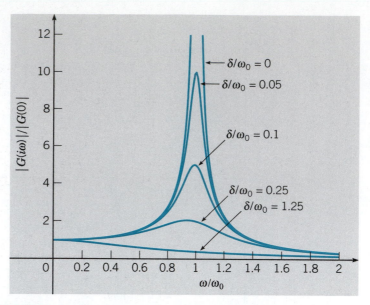

FIGURE 4.7.3 Gain function $|G(i\omega)|$ for the damped spring-mass system: $\delta/\omega_0 = \gamma/2\sqrt{mk}$.

Eq. (3). Thus only one family of curves, of which a few are shown in Figure 4.7.3, describes the response-versus-frequency behavior of all systems governed by Eq. (3).

The phase angle ϕ also depends in an interesting way on ω. For ω near zero, it follows from Eq. (11) that $\cos\phi \cong 1$. Thus $\phi \cong 0$, and the response is nearly in phase with the excitation, meaning that they rise and fall together and, in particular, assume their respective maxima nearly together and their respective minima nearly together. For $\omega = \omega_0$, we find that $\cos\phi = 0$, so $\phi = \pi/2$. In this case, the response lags behind the excitation by $\pi/2$, that is, the peaks of the response occur $\pi/2$ later than the peaks of the excitation, and similarly for the valleys. Finally, for ω very large, we have $\cos\phi \cong -1$. Thus $\phi \cong \pi$, so that the response

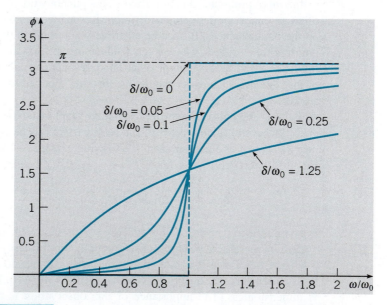

FIGURE 4.7.4 Phase function $\phi(\omega)$ for the damped spring-mass system: $\delta/\omega_0 = \gamma/2\sqrt{mk}$.

is nearly out of phase with the excitation. This means that the response is minimum when the excitation is maximum, and vice versa. Figure 4.7.4 shows the graphs of ϕ versus ω/ω_0 for several values of $\delta/\omega_0 = \gamma/2\sqrt{mk}$. For small damping, the phase transition from near $\phi = 0$ to near $\phi = \pi$ occurs rather abruptly, whereas for larger values of the damping parameter, the transition takes place more gradually.

EXAMPLE 1

Consider the initial value problem

$$y'' + 0.125y' + y = 3\cos\omega t, \qquad y(0) = 2, \quad y'(0) = 0. \tag{16}$$

Show plots of the solution for different values of the forcing frequency ω, and compare them with corresponding plots of the forcing function.

For this system, we have $\omega_0 = 1$ and $\delta/\omega_0 = .0625$. Its unforced motion was discussed in Example 3 of Section 4.5, and Figure 4.5.8 shows the graph of the solution of the unforced problem. Figures 4.7.5, 4.7.6, and 4.7.7 show the solution of the forced problem (16) for $\omega = 0.3$, $\omega = 1$, and $\omega = 2$, respectively. The graph of the corresponding forcing function is also shown in each figure. In this example the amplitude of the harmonic input, A, is equal to 3.

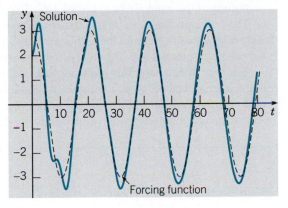

FIGURE 4.7.5 A forced vibration with damping; solution of
$y'' + 0.125y' + y = 3\cos(3t/10)$, $y(0) = 2$, $y'(0) = 0$.

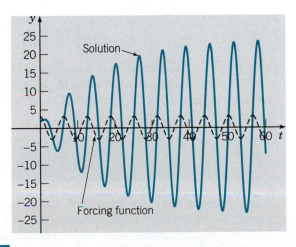

FIGURE 4.7.6 A forced vibration with damping; solution of
$y'' + 0.125y' + y = 3\cos t$, $y(0) = 2$, $y'(0) = 0$.

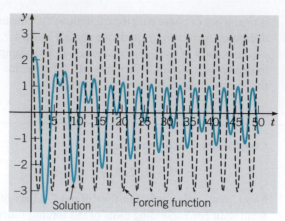

Solution   Forcing function

FIGURE 4.7.7   A forced vibration with damping; solution of
$y'' + 0.125y' + y = 3 \cos 2t, y(0) = 2, y'(0) = 0.$

Figure 4.7.5 shows the low-frequency case, $\omega/\omega_0 = 0.3$. After the initial transient response is substantially damped out, the remaining steady-state response is essentially in phase with the excitation, and the amplitude of the response is slightly larger than the amplitude of the input. To be specific, $A|G(.3i)| \cong 3.2939$ and $\phi \cong 0.041185$.

The resonant case, $\omega/\omega_0 = 1$, is shown in Figure 4.7.6. Here the amplitude of the steady-state response is eight times the amplitude of the harmonic input, and the figure also shows the predicted phase lag of $\pi/2$ relative to the external force.

The case of comparatively high-frequency excitation is shown in Figure 4.7.7. Observe that the amplitude of the steady forced response is approximately one-third the amplitude of the harmonic input and that the phase difference between the excitation and response is approximately π. More precisely, we find that $A|G(2i)| \cong 0.99655$ and that $\phi \cong 3.0585$.

Forced Vibrations without Damping

We now assume that $\delta = \gamma/2m = 0$ in Eq. (2), thereby obtaining the equation of motion of an undamped forced oscillator

$$y'' + \omega_0^2 y = A \cos \omega t \tag{17}$$

where we have assumed that $f(t) = A \cos \omega t$. The form of the general solution of Eq. (17) is different, depending on whether the forcing frequency ω is different from or equal to the natural frequency $\omega_0 = \sqrt{k/m}$ of the unforced system. First consider the case $\omega \neq \omega_0$; then the general solution of Eq. (17) is

$$y = c_1 \cos \omega_0 t + c_2 \sin \omega_0 t + \frac{A}{(\omega_0^2 - \omega^2)} \cos \omega t. \tag{18}$$

The constants c_1 and c_2 are determined by the initial conditions. The resulting motion is, in general, the sum of two periodic motions of different frequencies (ω_0 and ω) and amplitudes.

It is particularly interesting to suppose that the mass is initially at rest, so that the initial conditions are $y(0) = 0$ and $y'(0) = 0$. Then the energy driving the system comes entirely from the external force, with no contribution from the initial conditions. In this case, it turns out that the constants c_1 and c_2 in Eq. (18) are given by

$$c_1 = -\frac{A}{(\omega_0^2 - \omega^2)}, \qquad c_2 = 0, \tag{19}$$

and the solution of Eq. (17) is

$$y = \frac{A}{(\omega_0^2 - \omega^2)}(\cos \omega t - \cos \omega_0 t). \tag{20}$$

This is the sum of two periodic functions of different periods but the same amplitude. Making use of the trigonometric identities for $\cos(A \pm B)$ with $A = (\omega_0 + \omega)t/2$ and $B = (\omega_0 - \omega)t/2$, we can write Eq. (20) in the form

$$y = \left[\frac{2A}{(\omega_0^2 - \omega^2)} \sin \frac{(\omega_0 - \omega)t}{2}\right] \sin \frac{(\omega_0 + \omega)t}{2}. \tag{21}$$

If $|\omega_0 - \omega|$ is small, then $\omega_0 + \omega$ is much greater than $|\omega_0 - \omega|$. Consequently, $\sin(\omega_0 + \omega)t/2$ is a rapidly oscillating function compared to $\sin(\omega_0 - \omega)t/2$. Thus the motion is a rapid oscillation with frequency $(\omega_0 + \omega)/2$ but with a slowly varying sinusoidal amplitude

$$\frac{2A}{|\omega_0^2 - \omega^2|} \left| \sin \frac{(\omega_0 - \omega)t}{2} \right|.$$

This type of motion, possessing a periodic variation of amplitude, exhibits what is called a **beat**. For example, such a phenomenon occurs in acoustics when two tuning forks of nearly equal frequency are excited simultaneously. In this case, the periodic variation of amplitude is quite apparent to the unaided ear. In electronics, the variation of the amplitude with time is called **amplitude modulation**.

EXAMPLE 2

Solve the initial value problem

$$y'' + y = 0.5 \cos 0.8t, \qquad y(0) = 0, \quad y'(0) = 0, \tag{22}$$

and plot the solution.

In this case, $\omega_0 = 1$, $\omega = 0.8$, and $A = 0.5$, so from Eq. (21) the solution of the given problem is

$$y = 2.77778 \sin 0.1t \sin 0.9t. \tag{23}$$

A graph of this solution is shown in Figure 4.7.8.

The amplitude variation has a slow frequency of 0.1 and a corresponding slow period of 20π. Note that a half-period of 10π corresponds to a single cycle of increasing and then decreasing amplitude. The displacement of the spring–mass system oscillates with a relatively fast frequency of 0.9, which is only slightly less than the natural frequency ω_0.

Now imagine that the forcing frequency ω is further increased, say to $\omega = 0.9$. Then the slow frequency is halved to 0.05, and the corresponding slow half-period is doubled to 20π. The multiplier 2.7778 also increases substantially, to 5.2632. However, the fast frequency is only marginally increased, to 0.95. Can you visualize what happens as ω takes on values closer and closer to the natural frequency $\omega_0 = 1$?

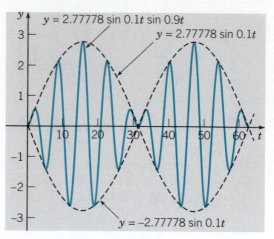

FIGURE 4.7.8 A beat; solution of $y'' + y = 0.5 \cos 0.8t$, $y(0) = 0$, $y'(0) = 0$; $y = 2.77778 \sin 0.1t \sin 0.9t$.

Now let us return to Eq. (17) and consider the case of resonance, where $\omega = \omega_0$, that is, the frequency of the forcing function is the same as the natural frequency of the system. Then the nonhomogeneous term $A \cos \omega t$ is a solution of the homogeneous equation. In this case, the solution of Eq. (17) is

$$y = c_1 \cos \omega_0 t + c_2 \sin \omega_0 t + \frac{A}{2\omega_0} t \sin \omega_0 t. \tag{24}$$

Because of the term $t \sin \omega_0 t$, the solution (24) predicts that the motion will become unbounded as $t \to \infty$ regardless of the values of c_1 and c_2; see Figure 4.7.9 for a typical example. Of course, in reality, unbounded oscillations do not occur. As soon as y becomes large, the mathematical model on which Eq. (17) is based is no longer valid, since the assumption that the spring force depends linearly on the displacement requires that y be

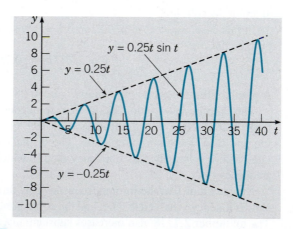

FIGURE 4.7.9 Resonance; solution of $y'' + y = 0.5 \cos t$, $y(0) = 0$, $y'(0) = 0$; $y = 0.25t \sin t$.

small. As we have seen, if damping is included in the model, the predicted motion remains bounded. However, the response to the input function $A \cos \omega t$ may be quite large if the damping is small and ω is close to ω_0.

PROBLEMS

In each of Problems 1 through 4, write the given expression as a product of two trigonometric functions of different frequencies.

1. $\cos 9t - \cos 7t$ 2. $\sin 7t - \sin 6t$
3. $\cos \pi t + \cos 2\pi t$ 4. $\sin 3t + \sin 4t$

5. A mass weighing 4 lb stretches a spring 1.5 in. The mass is displaced 2 inches in the positive direction from its equilibrium position and released with no initial velocity. Assuming that there is no damping and that the mass is acted on by an external force of $2 \cos 3t$ lb, formulate the initial value problem describing the motion of the mass.

6. A mass of 5 kg stretches a spring 10 cm. The mass is acted on by an external force of $10 \sin(t/2)$ N (newtons) and moves in a medium that imparts a viscous force of 2 N when the speed of the mass is 4 cm/s. If the mass is set in motion from its equilibrium position with an initial velocity of 3 cm/s, formulate the initial value problem describing the motion of the mass.

 7. (a) Find the solution of Problem 5.
  (b) Plot the graph of the solution.
  (c) If the given external force is replaced by a force $A \exp(i\omega t)$ of frequency ω, find the frequency response $G(i\omega)$, the gain $|G(i\omega)|$ and the phase $\phi(\omega) = -\arg(G(i\omega))$. Then find the value of ω for which resonance occurs.

CAS 8. (a) Find the solution of the initial value problem in Problem 6.
  (b) Identify the transient and steady-state parts of the solution.
  (c) Plot the graph of the steady-state solution.
  (d) If the given external force is replaced by a force $A \exp(i\omega t)$ of frequency ω, find the frequency response $G(i\omega)$, the gain $|G(i\omega)|$, and the phase $\phi(\omega) = -\arg(G(i\omega))$. Then find the value of ω for which the gain is maximum. Plot the graphs of $|G(i\omega)|$ and $\phi(\omega)$

ODEA 9. If an undamped spring–mass system with a mass that weighs 6 lb and a spring constant 1 lb/in is suddenly set in motion at $t = 0$ by an external force of $4 \cos 7t$ lb, determine the position of the mass

at any time and draw a graph of the displacement versus t.

10. A mass that weighs 8 lb stretches a spring 6 in. The system is acted on by an external force of $8 \sin 8t$ lb. If the mass is pulled down 3 in and then released, determine the position of the mass at any time. Determine the first four times at which the velocity of the mass is zero.

11. A spring is stretched 6 in by a mass that weighs 8 lb. The mass is attached to a dashpot mechanism that has a damping constant of 0.25 lb·s/ft and is acted on by an external force of $4 \cos 2t$ lb.
  (a) Determine the steady-state response of this system.
  (b) If the given mass is replaced by a mass m, determine the value of m for which the amplitude of the steady-state response is maximum.

12. A spring–mass system has a spring constant of 3 N/m. A mass of 2 kg is attached to the spring, and the motion takes place in a viscous fluid that offers a resistance numerically equal to the magnitude of the instantaneous velocity. If the system is driven by an external force of $(3 \cos 3t - 2 \sin 3t)$ N, determine the steady-state response.

13. Furnish the details in determining when the gain function given by Eq. (10) is maximum, that is, show that ω_{max}^2 and $|G(i\omega_{max})|$ are given by Eqs. (14) and (15), respectively.

14. Find the solution of the initial value problem

$$y'' + y = F(t), \qquad y(0) = 0, \quad y'(0) = 0,$$

where
$$F(t) = \begin{cases} At, & 0 \le t \le \pi, \\ A(2\pi - t), & \pi < t \le 2\pi, \\ 0, & 2\pi < t. \end{cases}$$

Hint: Treat each time interval separately, and match the solutions in the different intervals by requiring that y and y' be continuous functions of t.

15. A series circuit has a capacitor of 0.25×10^{-6} farad, a resistor of 5×10^3 ohms, and an inductor

of 1 henry. The initial charge on the capacitor is zero. If a 12-volt battery is connected to the circuit and the circuit is closed at $t = 0$, determine the charge on the capacitor at $t = 0.001$ s, at $t = 0.01$ s, and at any time t. Also determine the limiting charge as $t \to \infty$.

CAS 16. Consider a vibrating system described by the initial value problem

$$y'' + 0.25y' + 2y = 2 \cos \omega t,$$
$$y(0) = 0, \quad y'(0) = 2.$$

(a) Determine the steady-state part of the solution of this problem.
(b) Find the gain function $|G(i\omega)|$ of the system.
(c) Plot $|G(i\omega)|$ and $\phi(\omega) = -\arg(|G(i\omega)|)$ versus ω.
(d) Find the maximum value of $|G(i\omega)|$ and the frequency ω for which it occurs.

ODEA 17. Consider the forced but undamped system described by the initial value problem

$$y'' + y = 3 \cos \omega t, \quad y(0) = 0, \quad y'(0) = 0.$$

(a) Find the solution $y(t)$ for $\omega \neq 1$.
(b) Plot the solution $y(t)$ versus t for $\omega = 0.7$, $\omega = 0.8$, and $\omega = 0.9$. Describe how the response $y(t)$ changes as ω varies in this interval. What happens as ω takes on values closer and closer to 1? Note that the natural frequency of the unforced system is $\omega_0 = 1$.

ODEA 18. Consider the vibrating system described by the initial value problem

$$y'' + y = 3 \cos \omega t, \quad y(0) = 1, \quad y'(0) = 1.$$

(a) Find the solution for $\omega \neq 1$.
(b) Plot the solution $y(t)$ versus t for $\omega = 0.7$, $\omega = 0.8$, and $\omega = 0.9$. Compare the results with those of Problem 17, that is, describe the effect of the nonzero initial conditions.

ODEA 19. For the initial value problem in Problem 18, plot y' versus y for $\omega = 0.7$, $\omega = 0.8$, and $\omega = 0.9$, that is, draw the phase plot of the solution for these values of ω. Use a t interval that is long enough so the phase plot appears as a closed curve. Mark your curve with arrows to show the direction in which it is traversed as t increases.

Problems 20 through 22 deal with the initial value ODEA problem

$$y'' + 0.125y' + 4y = f(t), \quad y(0) = 2, \quad y'(0) = 0.$$

In each of these problems:
(a) Plot the given forcing function $f(t)$ versus t, and also plot the solution $y(t)$ versus t on the same set of axes. Use a t interval that is long enough so the initial transients are substantially eliminated. Observe the relation between the amplitude and phase of the forcing term and the amplitude and phase of the response. Note that $\omega_0 = \sqrt{k/m} = 2$.
(b) Draw the phase plot of the solution, that is, plot y' versus y.

20. $f(t) = 3 \cos(t/4)$
21. $f(t) = 3 \cos 2t$
22. $f(t) = 3 \cos 6t$

23. A spring–mass system with a hardening spring ODEA (Section 4.1) is acted on by a periodic external force. In the absence of damping, suppose that the displacement of the mass satisfies the initial value problem

$$y'' + y + 0.2y^3 = \cos \omega t,$$
$$y(0) = 0, \quad y'(0) = 0.$$

(a) Let $\omega = 1$ and plot a computer-generated solution of the given problem. Does the system exhibit a beat?
(b) Plot the solution for several values of ω between 1/2 and 2. Describe how the solution changes as ω increases.

24. Suppose that the system of Problem 23 is modified ODEA to include a damping term and that the resulting initial value problem is

$$y'' + 0.2y' + y + 0.2y^3 = \cos \omega t,$$
$$y(0) = 0, \quad y'(0) = 0.$$

(a) Plot a computer-generated solution of the given problem for several values of ω between 1/2 and 2, and estimate the amplitude, say $G_H(\omega)$, of the steady response in each case.
(b) Using the data from part (a), plot the graph of $G_H(\omega)$ versus ω. For what frequency ω is the amplitude greatest?
(c) Compare the results of parts (a) and (b) with the corresponding results for the linear spring.

4.8 Variation of Parameters

In this section, we describe another method for finding a particular solution of a nonhomogeneous equation. The method, known as **variation of parameters** or **variation of constants**, is due to Lagrange and complements the method of undetermined coefficients rather well. The main advantage of variation of parameters is that it is a *general method*; in principle at least, it can be applied to any linear nonhomogeneous equation or system. It requires no detailed assumptions about the form of the solution. In this section, we use this method to derive an integral representation for the particular solution of an arbitrary linear nonhomogeneous first order system of dimension 2. An analogous representation for the particular solution of an arbitrary second order linear nonhomogeneous equation then follows as a special case.

Variation of Parameters for Linear First Order Systems of Dimension 2

First consider the nonhomogeneous system

$$\mathbf{x}' = \mathbf{P}(t)\mathbf{x} + \mathbf{g}(t), \tag{1}$$

where each entry of

$$\mathbf{P}(t) = \begin{pmatrix} p_{11}(t) & p_{12}(t) \\ p_{21}(t) & p_{22}(t) \end{pmatrix} \quad \text{and} \quad \mathbf{g}(t) = \begin{pmatrix} g_1(t) \\ g_2(t) \end{pmatrix} \tag{2}$$

is continuous on an open interval I. Assume that

$$\mathbf{x}_1(t) = \begin{pmatrix} x_{11}(t) \\ x_{21}(t) \end{pmatrix} \quad \text{and} \quad \mathbf{x}_2(t) = \begin{pmatrix} x_{12}(t) \\ x_{22}(t) \end{pmatrix}$$

are a fundamental set of solutions for the homogeneous system $\mathbf{x}' = \mathbf{P}(t)\mathbf{x}$ obtained from Eq. (1) by setting $\mathbf{g}(t) = \mathbf{0}$. Thus

$$\mathbf{x}_1' = \mathbf{P}(t)\mathbf{x}_1, \quad \mathbf{x}_2' = \mathbf{P}(t)\mathbf{x}_2, \tag{3}$$

and the Wronskian of $\mathbf{x}_1$ and $\mathbf{x}_2$ is nonzero throughout I,

$$W[\mathbf{x}_1, \mathbf{x}_2](t) = \begin{vmatrix} x_{11}(t) & x_{12}(t) \\ x_{21}(t) & x_{22}(t) \end{vmatrix} \neq 0, \quad t \in I. \tag{4}$$

Finding a particular solution of Eq. (1) via the method of variation of parameters is facilitated by introducing the convenient matrix notation

$$\mathbf{X}(t) = [\mathbf{x}_1(t), \mathbf{x}_2(t)] = \begin{pmatrix} x_{11}(t) & x_{12}(t) \\ x_{21}(t) & x_{22}(t) \end{pmatrix} \tag{5}$$

by juxtaposing the column vectors $\mathbf{x}_1$ and $\mathbf{x}_2$. The matrix (5) is referred to as a **fundamental matrix** for $\mathbf{x}' = \mathbf{P}(t)\mathbf{x}$ since the columns of $\mathbf{X}(t)$ are a fundamental set of solutions. A column by column comparison shows that $\mathbf{X}(t)$ satisfies the matrix differential equation

$$\mathbf{X}'(t) = \left[\mathbf{x}_1'(t), \mathbf{x}_2'(t)\right] = [\mathbf{P}(t)\mathbf{x}_1(t), \mathbf{P}(t)\mathbf{x}_2(t)] = \mathbf{P}(t)\mathbf{X}(t). \tag{6}$$

Thus Eq. (6) is equivalent to the pair of equations (3).

Recall that the general solution of $\mathbf{x}' = \mathbf{P}(t)\mathbf{x}$ has the form

$$\mathbf{x} = c_1\mathbf{x}_1(t) + c_2\mathbf{x}_2(t). \tag{7}$$

The task of finding a particular solution of Eq. (1) is simplified by replacing the constants c_1 and c_2 in Eq. (7) by functions $u_1(t)$ and $u_2(t)$, respectively,

$$\mathbf{x} = u_1(t)\mathbf{x}_1(t) + u_2(t)\mathbf{x}_2(t) = \mathbf{X}(t)\mathbf{u}(t). \tag{8}$$

The component functions of $\mathbf{u}(t) = (u_1(t), u_2(t))^T$ must then be determined. The validity of the assumption (8) can subsequently be justified by direct verification of the resulting solution (see Problem 1(**b**)).

Substituting the right side of Eq. (8) into Eq. (1) yields the equation

$$\mathbf{X}'(t)\mathbf{u}(t) + \mathbf{X}(t)\mathbf{u}'(t) = \mathbf{P}(t)\mathbf{X}(t)\mathbf{u}(t) + \mathbf{g}(t)$$

which simplifies to

$$\mathbf{X}(t)\mathbf{u}'(t) = \mathbf{g}(t) \tag{9}$$

as a consequence of Eq. (6). Note that we have used the easily verified differentiation rule for the product $\mathbf{Xu}$, $(\mathbf{Xu})' = \mathbf{X}'\mathbf{u} + \mathbf{Xu}'$ (see Problem 1(**a**)). Thus the components of $\mathbf{u}'$ satisfy the linear algebraic system of equations

$$x_{11}(t)u_1'(t) + x_{12}(t)u_2'(t) = g_1(t),$$
$$x_{21}(t)u_1'(t) + x_{22}(t)u_2'(t) = g_2(t). \tag{10}$$

Since $\det \mathbf{X}(t) = W[\mathbf{x}_1, \mathbf{x}_2](t) \neq 0$, $\mathbf{X}^{-1}(t)$ exists and is given by

$$\mathbf{X}^{-1}(t) = \frac{1}{W[\mathbf{x}_1, \mathbf{x}_2](t)} \begin{pmatrix} x_{22}(t) & -x_{12}(t) \\ -x_{21}(t) & x_{11}(t) \end{pmatrix}. \tag{11}$$

In terms of $\mathbf{X}^{-1}(t)$ the solution of Eq. (9) can be represented as

$$\mathbf{u}'(t) = \mathbf{X}^{-1}(t)\mathbf{g}(t). \tag{12}$$

Thus, for $\mathbf{u}(t)$, we can select any vector from the class of vectors that satisfy Eq. (12). Since each component of $\mathbf{u}$ is determined up to an arbitrary additive constant, we denote $\mathbf{u}(t)$ by

$$\mathbf{u}(t) = \int \mathbf{X}^{-1}(t)\mathbf{g}(t)\, dt + \mathbf{c} = \begin{pmatrix} \int \dfrac{x_{22}(t)g_1(t) - x_{12}(t)g_2(t)}{W[\mathbf{x}_1, \mathbf{x}_2](t)}\, dt + c_1 \\[2mm] \int \dfrac{x_{11}(t)g_2(t) - x_{21}(t)g_1(t)}{W[\mathbf{x}_1, \mathbf{x}_2](t)}\, dt + c_2 \end{pmatrix} \tag{13}$$

where the constant vector $\mathbf{c} = (c_1, c_2)^T$ is arbitrary. Substituting for $\mathbf{u}(t)$ in Eq. (8) gives the solution $\mathbf{x}$ of the system (1),

$$\mathbf{x} = \mathbf{X}(t)\mathbf{c} + \mathbf{X}(t)\int \mathbf{X}^{-1}(t)\mathbf{g}(t)\, dt. \tag{14}$$

Since $\mathbf{c}$ is arbitrary, any initial condition at a point t_0 can be satisfied by an appropriate choice of $\mathbf{c}$. Thus, every solution of the system (1) is contained in the expression given by Eq. (14). Therefore, it is the general solution of Eq. (1). Note that the first term on the right side of Eq. (14) is the general solution of the corresponding homogeneous system $\mathbf{x}' = \mathbf{P}(t)\mathbf{x}$, and the second term is a particular solution of Eq. (1).

The above results are summarized in the following theorem.

THEOREM 4.8.1

Assume that the entries of the matrices $\mathbf{P}(t)$ and $\mathbf{g}(t)$ in Eq. (2) are continuous on an open interval I and that $\mathbf{x}_1$ and $\mathbf{x}_2$ are a fundamental set of solutions of the homogeneous equation $\mathbf{x}' = \mathbf{P}(t)\mathbf{x}$ corresponding to the nonhomogeneous equation (1)

$$\mathbf{x}' = \mathbf{P}(t)\mathbf{x} + \mathbf{g}(t).$$

Then a particular solution of Eq. (1) is

$$\mathbf{x}_p(t) = \mathbf{X}(t) \int \mathbf{X}^{-1}(t)\mathbf{g}(t)\, dt \tag{15}$$

where the fundamental matrix $\mathbf{X}(t)$ is defined by Eq. (5). Moreover, the general solution of Eq. (1) is

$$\mathbf{x}(t) = c_1\mathbf{x}_1(t) + c_2\mathbf{x}_2(t) + \mathbf{x}_p(t). \tag{16}$$

Remark. There may be two major difficulties in using the method of variation of parameters. One is the determination of $\mathbf{x}_1(t)$ and $\mathbf{x}_2(t)$, a fundamental set of solutions of the homogeneous equation $\mathbf{x}' = \mathbf{P}(t)\mathbf{x}$. If the coefficients in that equation are not constants, these solutions are generally not easy to obtain. The other possible difficulty is in the evaluation of the integrals appearing in Eq. (15) and this depends entirely on the nature of $\mathbf{X}^{-1}(t)$ and $\mathbf{g}(t)$.

EXAMPLE 1

Find the solution of the initial value problem

$$\mathbf{x}' = \begin{pmatrix} 1 & -4 \\ 2 & -5 \end{pmatrix}\mathbf{x} + \begin{pmatrix} 10\cos t \\ 2e^{-t} \end{pmatrix}, \quad \mathbf{x}(0) = \begin{pmatrix} 10 \\ 4 \end{pmatrix}. \tag{17}$$

Applying the eigenvalue method to the homogeneous equation

$$\mathbf{x}' = \begin{pmatrix} 1 & -4 \\ 2 & -5 \end{pmatrix}\mathbf{x}$$

yields the fundamental solution set

$$\mathbf{x}_1(t) = e^{-t}\begin{pmatrix} 2 \\ 1 \end{pmatrix} \quad \text{and} \quad \mathbf{x}_2(t) = e^{-3t}\begin{pmatrix} 1 \\ 1 \end{pmatrix}.$$

Thus the corresponding fundamental matrix and its inverse are given by

$$\mathbf{X}(t) = \begin{pmatrix} 2e^{-t} & e^{-3t} \\ e^{-t} & e^{-3t} \end{pmatrix} \quad \text{and} \quad \mathbf{X}^{-1}(t) = \begin{pmatrix} e^{t} & -e^{t} \\ -e^{3t} & 2e^{3t} \end{pmatrix}.$$

Since the nonhomogeneous term in the initial value problem (17) is $\mathbf{g}(t) = \begin{pmatrix} 10\cos t \\ 2e^{-t} \end{pmatrix}$, a particular solution is given by

$$\mathbf{x}_p(t) = \mathbf{X}(t) \int \mathbf{X}^{-1}(t)\mathbf{g}(t)\, dt$$

$$= \begin{pmatrix} 2e^{-t} & e^{-3t} \\ e^{-t} & e^{-3t} \end{pmatrix} \int \begin{pmatrix} e^t & -e^t \\ -e^{3t} & 2e^{3t} \end{pmatrix} \begin{pmatrix} 10\cos t \\ 2e^{-t} \end{pmatrix} dt$$

$$= \begin{pmatrix} 7\cos t + 9\sin t + 2(1 - 2t)e^{-t} \\ 2\cos t + 4\sin t + 2(1 - t)e^{-t} \end{pmatrix}.$$

It follows that the general solution of the nonhomogeneous equation in Eq. (17) is

$$\mathbf{x} = c_1 \begin{pmatrix} 2e^{-t} \\ e^{-t} \end{pmatrix} + c_2 \begin{pmatrix} e^{-3t} \\ e^{-3t} \end{pmatrix} + \begin{pmatrix} 7\cos t + 9\sin t + 2(1 - 2t)e^{-t} \\ 2\cos t + 4\sin t + 2(1 - t)e^{-t} \end{pmatrix}. \tag{18}$$

The initial condition prescribed in the initial value problem (17) requires that

$$c_1 \begin{pmatrix} 2 \\ 1 \end{pmatrix} + c_2 \begin{pmatrix} 1 \\ 1 \end{pmatrix} + \begin{pmatrix} 9 \\ 4 \end{pmatrix} = \begin{pmatrix} 10 \\ 4 \end{pmatrix}$$

or

$$\begin{pmatrix} 2 & 1 \\ 1 & 1 \end{pmatrix} \begin{pmatrix} c_1 \\ c_2 \end{pmatrix} = \begin{pmatrix} 1 \\ 0 \end{pmatrix}.$$

Thus $c_1 = 1$, $c_2 = -1$, and the solution of the initial value problem (17) is

$$\mathbf{x} = \begin{pmatrix} -e^{-3t} + 7\cos t + 9\sin t + 2(2 - 2t)e^{-t} \\ -e^{-3t} + 2\cos t + 4\sin t + (3 - 2t)e^{-t} \end{pmatrix}.$$

Variation of Parameters for Linear Second Order Equations

The method of variation of parameters used to find a particular solution of Eq. (1) will now be used to find a particular solution of

$$y'' + p(t)y' + q(t)y = g(t) \tag{19}$$

by applying the method to the dynamical system equivalent to Eq. (19),

$$\mathbf{x}' = \begin{pmatrix} 0 & 1 \\ -q(t) & -p(t) \end{pmatrix} \mathbf{x} + \begin{pmatrix} 0 \\ g(t) \end{pmatrix}, \tag{20}$$

where

$$\mathbf{x}(t) = \begin{pmatrix} x_1(t) \\ x_2(t) \end{pmatrix} = \begin{pmatrix} y(t) \\ y'(t) \end{pmatrix}.$$

If $\{y_1, y_2\}$ is a fundamental set of solutions of $y'' + p(t)y' + q(t)y = 0$, then

$$\mathbf{x}_1(t) = \begin{pmatrix} y_1(t) \\ y_1'(t) \end{pmatrix} \quad \text{and} \quad \mathbf{x}_2(t) = \begin{pmatrix} y_2(t) \\ y_2'(t) \end{pmatrix}$$

are a fundamental set of solutions for the homogeneous system obtained from Eq. (20) by setting $g(t) = 0$. A direct method for finding a particular solution of Eq. (19) can be found by using the substitutions

$$\mathbf{X}(t) = \begin{pmatrix} y_1(t) & y_2(t) \\ y_1'(t) & y_2'(t) \end{pmatrix} \quad \text{and} \quad \mathbf{g}(t) = \begin{pmatrix} 0 \\ g(t) \end{pmatrix} \tag{21}$$

in the steps leading from Eq. (8) to Eq. (14). Using these substitutions in Eq. (8), the assumed form for the particular solution of Eq. (20) is

$$\mathbf{x}_p(t) = u_1(t) \begin{pmatrix} y_1(t) \\ y_1'(t) \end{pmatrix} + u_2(t) \begin{pmatrix} y_2(t) \\ y_2'(t) \end{pmatrix}. \tag{22}$$

The first component of $\mathbf{x}_p(t)$ in (22) provides us with the form for the particular solution $Y(t)$ of Eq. (19),

$$Y(t) = u_1(t) y_1(t) + u_2(t) y_2(t). \tag{23}$$

Using the expressions for $\mathbf{X}(t)$ and $\mathbf{g}(t)$ in (21) in Eq. (10) then provides us with the following algebraic system for the components of $\mathbf{u}' = (u_1', u_2')$,

$$y_1(t) u_1'(t) + y_2(t) u_2'(t) = 0,$$
$$y_1'(t) u_1'(t) + y_2'(t) u_2'(t) = g(t). \tag{24}$$

The solution of Eq. (24) is

$$u_1'(t) = -\frac{y_2(t) g(t)}{W[y_1, y_2](t)}, \qquad u_2'(t) = \frac{y_1(t) g(t)}{W[y_1, y_2](t)}, \tag{25}$$

where $W[y_1, y_2](t)$ is the Wronskian of y_1 and y_2. Note that division by W is permissible since y_1 and y_2 are a fundamental set of solutions of $y'' + p(t)y' + q(t)y = 0$, and therefore their Wronskian is nonzero. By integrating Eqs. (25) we find the desired functions $u_1(t)$ and $u_2(t)$, namely,

$$u_1(t) = -\int \frac{y_2(t) g(t)}{W[y_1, y_2](t)} \, dt + c_1, \qquad u_2(t) = \int \frac{y_1(t) g(t)}{W[y_1, y_2](t)} \, dt + c_2. \tag{26}$$

If the integrals in Eqs. (26) can be evaluated in terms of elementary functions, then we substitute the results in Eq. (23), thereby obtaining the general solution of Eq. (19). More generally, the solution can always be expressed in terms of integrals, as stated in the following theorem.

THEOREM 4.8.2

If the functions p, q, and g are continuous on an open interval I, and if the functions y_1 and y_2 are a fundamental set of solutions of the homogeneous equation $y'' + p(t)y' + q(t)y = 0$ corresponding to the nonhomogeneous equation (19),

$$y'' + p(t)y' + q(t)y = g(t),$$

then a particular solution of Eq. (19) is

$$Y(t) = -y_1(t) \int \frac{y_2(t) g(t)}{W[y_1, y_2](t)} \, dt + y_2(t) \int \frac{y_1(t) g(t)}{W[y_1, y_2](t)} \, dt. \tag{27}$$

The general solution is

$$y = c_1 y_1(t) + c_2 y_2(t) + Y(t), \tag{28}$$

as prescribed by Theorem 4.6.2.

Remarks.

i. Note that the conclusions of Theorem 4.8.2 are predicated on the assumption that the leading coefficient in Eq. (19) is 1. If, for example, a second order equation is in the form $P(t)y'' + Q(t)y' + R(t)y = G(t)$, then dividing the equation by $P(t)$ will bring it into the form of Eq. (19) with $g(t) = G(t)/P(t)$.

ii. A major advantage of the method of variation of parameters, particularly when applied to the important case of linear second order equations, is that Eq. (27) provides an expression for the particular solution $Y(t)$ in terms of an arbitrary forcing function $g(t)$. This expression is a good starting point if you wish to investigate the effect of changes to the forcing function, or if you wish to analyze the response of a system to a number of different forcing functions.

EXAMPLE 2

Find a particular solution of

$$y'' + 4y = 3 \csc t. \tag{29}$$

Observe that this problem is not a good candidate for the method of undetermined coefficients, as described in Section 4.6, because the nonhomogeneous term $g(t) = 3 \csc t$ involves a quotient with $\sin t$ in the denominator. Some experimentation should convince you that a method for finding a particular solution based on naive assumptions about its form is unlikely to succeed. For example, if we assume a particular solution of the form $Y = A \csc t$, we find that $Y'' = A \csc t + 2A \csc t \cot^2 t$, suggesting that it may be difficult or impossible to guess a correct form for Y. Thus we seek a particular solution of Eq. (29) using the method of variation of parameters.

Since a fundamental set of solutions of $y'' + 4y = 0$ is $y_1 = \cos 2t$ and $y_2 = \sin 2t$, according to Eq. (23), the form of the particular solution that we seek is

$$Y(t) = u_1(t) \cos 2t + u_2(t) \sin 2t \tag{30}$$

where u_1 and u_2 are to be determined. In practice, a good starting point for determining u_1 and u_2 is the system of algebraic equations (24). Substituting $y_1 = \cos 2t$, $y_2 = \sin 2t$, and $g(t) = 3 \csc t$ into Eq. (24) gives

$$\cos 2t \, u_1'(t) + \sin 2t \, u_2'(t) = 0,$$
$$-2 \sin 2t \, u_1'(t) + 2 \cos 2t \, u_2'(t) = 3 \csc t. \tag{31}$$

The solution of the system (31) is

$$u_1'(t) = -\frac{3 \csc t \sin 2t}{2} = -3 \cos t, \tag{32}$$

and

$$u_2'(t) = \frac{3 \cos t \cos 2t}{\sin 2t} = \frac{3(1 - 2\sin^2 t)}{2 \sin t} = \frac{3}{2} \csc t - 3 \sin t, \tag{33}$$

where we have used the double-angle formula to simplify the expression for u_2'. Having obtained $u_1'(t)$ and $u_2'(t)$, we next integrate to find $u_1(t)$ and $u_2(t)$. The result is

$$u_1(t) = -3 \sin t + c_1 \tag{34}$$

and

$$u_2(t) = \frac{3}{2} \ln|\csc t - \cot t| + 3 \cos t + c_2. \tag{35}$$

On substituting these expressions in Eq. (30), we have

$$y = -3 \sin t \cos 2t + \frac{3}{2} \ln |\csc t - \cot t| \sin 2t + 3 \cos t \sin 2t$$
$$+ c_1 \cos 2t + c_2 \sin 2t.$$

Finally, by using the double-angle formulas once more, we obtain

$$y = 3 \sin t + \frac{3}{2} \ln |\csc t - \cot t| \sin 2t + c_1 \cos 2t + c_2 \sin 2t. \tag{36}$$

The terms in Eq. (36) involving the arbitrary constants c_1 and c_2 are the general solution of the corresponding homogeneous equation, while the remaining terms are a particular solution of the nonhomogeneous equation (29). Thus Eq. (36) is the general solution of Eq. (29).

PROBLEMS

1. (a) If $\mathbf{X}(t) = \begin{pmatrix} x_{11}(t) & x_{12}(t) \\ x_{21}(t) & x_{22}(t) \end{pmatrix}$ and $\mathbf{u}(t) = \begin{pmatrix} u_1(t) \\ u_2(t) \end{pmatrix}$, show that $(\mathbf{Xu})' = \mathbf{X'u} + \mathbf{Xu'}$.

   (b) Assuming that $\mathbf{X}(t)$ is a fundamental matrix for $\mathbf{x}' = \mathbf{P}(t)\mathbf{x}$ and that $\mathbf{u}(t) = \int \mathbf{X}^{-1}(t)\mathbf{g}(t)\,dt$, use the result of part (a) to verify that $\mathbf{x}_p(t)$ given by Eq. (15) satisfies Eq. (1), $\mathbf{x}' = \mathbf{P}(t)\mathbf{x} + \mathbf{g}(t)$.

In each of Problems 2 through 5, use the method of variation of parameters to find a particular solution using the given fundamental set of solutions $\{\mathbf{x}_1, \mathbf{x}_2\}$.

2. $\mathbf{x}' = \begin{pmatrix} -3 & 1 \\ 1 & -3 \end{pmatrix} \mathbf{x} + \begin{pmatrix} 1 \\ 4t \end{pmatrix}$,

   $\mathbf{x}_1 = e^{-2t} \begin{pmatrix} 1 \\ 1 \end{pmatrix}$, $\mathbf{x}_2 = e^{-4t} \begin{pmatrix} 1 \\ -1 \end{pmatrix}$

3. $\mathbf{x}' = \begin{pmatrix} -1 & -2 \\ 0 & 1 \end{pmatrix} \mathbf{x} + \begin{pmatrix} e^{-t} \\ t \end{pmatrix}$,

   $\mathbf{x}_1 = e^{-t} \begin{pmatrix} 1 \\ 0 \end{pmatrix}$, $\mathbf{x}_2 = e^{t} \begin{pmatrix} -1 \\ 1 \end{pmatrix}$

4. $\mathbf{x}' = \begin{pmatrix} -1 & 0 \\ -1 & -1 \end{pmatrix} \mathbf{x} + \begin{pmatrix} -1 \\ t \end{pmatrix}$,

   $\mathbf{x}_1 = e^{-t} \begin{pmatrix} 0 \\ 1 \end{pmatrix}$, $\mathbf{x}_2 = e^{-t} \begin{pmatrix} -1 \\ t \end{pmatrix}$

5. $\mathbf{x}' = \begin{pmatrix} 0 & 1 \\ -1 & 0 \end{pmatrix} \mathbf{x} + \begin{pmatrix} \cos t \\ -\sin t \end{pmatrix}$,

   $\mathbf{x}_1 = \begin{pmatrix} \cos t \\ -\sin t \end{pmatrix}$, $\mathbf{x}_2 = \begin{pmatrix} \sin t \\ \cos t \end{pmatrix}$

In each of Problems 6 through 9, find the solution of the specified initial value problem.

6. The equation in Problem 2 with initial condition $\mathbf{x}(0) = (2, -1)^T$.

7. The equation in Problem 3 with initial condition $\mathbf{x}(0) = (-1, 1)^T$.

8. The equation in Problem 4 with initial condition $\mathbf{x}(0) = (1, 0)^T$.

9. The equation in Problem 5 with initial condition $\mathbf{x}(0) = (1, 1)^T$.

In each of Problems 10 through 13, use the method of variation of parameters to find a particular solution of the given differential equation. Then check your answer by using the method of undetermined coefficients.

10. $y'' - 5y' + 6y = 2e^t$

11. $y'' - y' - 2y = 2e^{-t}$

12. $y'' + 2y' + y = 3e^{-t}$

13. $4y'' - 4y' + y = 16e^{t/2}$

In each of Problems 14 through 21, find the general solution of the given differential equation. In Problems 20 and 21, g is an arbitrary continuous function.

14. $y'' + y = \tan t$, $\quad 0 < t < \pi/2$

15. $y'' + 9y = 9 \sec^2 3t$, $\quad 0 < t < \pi/6$

16. $y'' + 4y' + 4y = t^{-2}e^{-2t}$, $\quad t > 0$

17. $y'' + 4y = 3\csc 2t$, $\quad 0 < t < \pi/2$

18. $4y'' + y = 2\sec(t/2)$, $\quad -\pi < t < \pi$

19. $y'' - 2y' + y = e^t/(1 + t^2)$

20. $y'' - 5y' + 6y = g(t)$

21. $y'' + 4y = g(t)$

In each of Problems 22 through 27, verify that the given functions y_1 and y_2 satisfy the corresponding homogeneous equation; then find a particular solution of the given nonhomogeneous equation. In Problems 26 and 27, g is an arbitrary continuous function.

22. $t^2 y'' - t(t+2)y' + (t+2)y = 2t^3$, $t > 0$;
    $y_1(t) = t$, $y_2(t) = te^t$

23. $ty'' - (1+t)y' + y = t^2 e^{2t}$, $\quad t > 0$;
    $y_1(t) = 1 + t$, $\quad y_2(t) = e^t$

24. $(1-t)y'' + ty' - y = 2(t-1)^2 e^{-t}$, $0 < t < 1$;
    $y_1(t) = e^t$, $\quad y_2(t) = t$

25. $x^2 y'' + xy' + (x^2 - 0.25)y = 3x^{3/2}\sin x$,
    $x > 0$; $y_1(x) = x^{-1/2}\sin x$, $y_2(x) = x^{-1/2}\cos x$

26. $(1-x)y'' + xy' - y = g(x)$, $\quad 0 < x < 1$;
    $y_1(x) = e^x$, $\quad y_2(x) = x$

27. $x^2 y'' + xy' + (x^2 - 0.25)y = g(x)$, $\quad x > 0$;
    $y_1(x) = x^{-1/2}\sin x$, $\quad y_2(x) = x^{-1/2}\cos x$

In each of Problems 28 through 31, find the general solution of the nonhomogeneous Cauchy-Euler equation.

28. $t^2 y'' - 2y = 3t^2 - 1$, $\quad t > 0$

29. $x^2 y'' - 3xy' + 4y = x^2 \ln x$, $\quad x > 0$

30. $t^2 y'' - 2ty' + 2y = 4t^2$, $\quad t > 0$

31. $t^2 y'' + 7ty' + 5y = t$, $\quad t > 0$

32. Show that the solution of the initial value problem

$$L[y] = y'' + p(t)y' + q(t)y = g(t),$$
$$y(t_0) = y_0, \quad y'(t_0) = y_1$$

can be written as $y = u(t) + v(t)$, where u and v are solutions of the two initial value problems

$$L[u] = 0, \quad u(t_0) = y_0, \quad u'(t_0) = y_1,$$

$$L[v] = g(t), \quad v(t_0) = 0, \quad v'(t_0) = 0,$$

respectively. In other words, the nonhomogeneities in the differential equation and in the initial conditions can be dealt with separately. Observe that u is easy to find if a fundamental set of solutions of $L[u] = 0$ is known.

33. By choosing the lower limit of integration in Eq. (27) in the text as the initial point t_0, show

that $Y(t)$ becomes

$$Y(t) = \int_{t_0}^t \frac{y_1(\tau)y_2(t) - y_1(t)y_2(\tau)}{y_1(\tau)y_2'(\tau) - y_1'(\tau)y_2(\tau)} g(\tau)\,d\tau.$$

Show that $Y(t)$ is a solution of the initial value problem

$$L[y] = g(t), \quad y(t_0) = 0, \quad y'(t_0) = 0.$$

Thus Y can be identified with v in Problem 32.

34. (a) Use the result of Problem 33 to show that the solution of the initial value problem

$$y'' + y = g(t), \quad y(t_0) = 0, \quad y'(t_0) = 0$$

is

$$y = \int_{t_0}^t \sin(t - s)g(s)\,ds.$$

(b) Use the result of Problem 32 to find the solution of the initial value problem

$$y'' + y = g(t), \quad y(0) = y_0, \quad y'(0) = y_1.$$

35. Use the result of Problem 33 to find the solution of the initial value problem

$$L[y] = (D - a)(D - b)y = g(t),$$
$$y(t_0) = 0, \quad y'(t_0) = 0,$$

where a and b are real numbers with $a \neq b$.

36. Use the result of Problem 33 to find the solution of the initial value problem

$$L[y] = [D^2 - 2\lambda D + (\lambda^2 + \mu^2)]y = g(t),$$
$$y(t_0) = 0, \quad y'(t_0) = 0.$$

Note that the roots of the characteristic equation are $\lambda \pm i\mu$.

37. Use the result of Problem 33 to find the solution of the initial value problem

$$L[y] = (D - a)^2 y = g(t),$$
$$y(t_0) = 0, \quad y'(t_0) = 0,$$

where a is any real number.

38. By combining the results of Problems 35 through 37, show that the solution of the initial value problem
$$L[y] = (D^2 + bD + c)y = g(t),$$
$$y(t_0) = 0, \quad y'(t_0) = 0,$$

where b and c are constants, has the form

$$y = \phi(t) = \int_{t_0}^t K(t - s)g(s)\,ds. \qquad \text{(i)}$$

The function K depends only on the solutions y_1 and y_2 of the corresponding homogeneous equation and is independent of the nonhomogeneous term. Once K is determined, all nonhomogeneous problems involving the same differential operator L are reduced to the evaluation of an integral. Note also that, although K depends on both t and s, only the combination $t - s$ appears, so K is actually a function of a single variable. When we think of $g(t)$ as the input to the problem and of $\phi(t)$ as the output, it follows from Eq. (i) that the output depends on the input over the entire interval from the initial point t_0 to the current value t. The integral in Eq. (i) is called the **convolution** of K and g, and K is referred to as the **kernel**.

39. The method of reduction of order (Section 4.3) can also be used for the nonhomogeneous equation

$$y'' + p(t)y' + q(t)y = g(t), \qquad (i)$$

provided one solution y_1 of the corresponding homogeneous equation is known. Let $y = v(t)y_1(t)$ and show that y satisfies Eq. (i) if v is a solution of

$$y_1(t)v'' + [2y_1'(t) + p(t)y_1(t)]v' = g(t). \qquad (ii)$$

Equation (ii) is a first order linear equation for v'. Solving this equation, integrating the result, and then multiplying by $y_1(t)$ lead to the general solution of Eq. (i).

In each of Problems 40 and 41, use the method outlined in Problem 39 to solve the given differential equation.

40. $ty'' - (1+t)y' + y = t^2 e^{2t}, \quad t > 0;$
    $y_1(t) = 1 + t \qquad$ (see Problem 23)

41. $(1-t)y'' + ty' - y = 2(t-1)^2 e^{-t},$
    $0 < t < 1; \qquad y_1(t) = e^t \qquad$ (see Problem 24)

CHAPTER SUMMARY

Section 4.1 Many simple vibrating systems are modeled by second order linear equations. Mathematical descriptions of spring-mass systems and series RLC circuits lead directly to such equations. Using a technique known as **linearization**, second order linear equations are often used as approximate models of nonlinear second order systems that operate near an equilibrium point. An example of this is the pendulum undergoing small oscillations about its downward hanging equilibrium state.

Section 4.2

▶ The theory of second order linear equations

$$y'' + p(t)y' + q(t)y = g(t), \qquad p, q \text{ and } g \text{ continuous on } I$$

follows from the theory of first order linear systems of dimension two by converting the equation into an equivalent dynamical system for $\mathbf{x} = (y, y')^T$,

$$\mathbf{x}' = \begin{pmatrix} 0 & 1 \\ -q(t) & -p(t) \end{pmatrix} \mathbf{x} + \begin{pmatrix} 0 \\ g(t) \end{pmatrix}.$$

▶ Two solutions y_1 and y_2 to the homogeneous equation

$$y'' + p(t)y' + q(t)y = 0, \qquad p \text{ and } q \text{ continuous on } I$$

form a **fundamental set** on I if their **Wronskian**

$$W[y_1, y_2](t) = \begin{vmatrix} y_1(t) & y_2(t) \\ y_1'(t) & y_2'(t) \end{vmatrix} = y_1(t)y_2'(t) - y_1'(t)y_2(t)$$

is nonzero for some (and hence all) $t \in I$. If y_1 and y_2 are a fundamental set of solutions to the homogeneous equation, then a **general solution** is

$$y = c_1 y_1(t) + c_2 y_2(t)$$

where c_1 and c_2 are arbitrary constants.

Section 4.3 Constant Coefficient Equations

The form of a general solution to $ay'' + by' + cy = 0$, $a \neq 0$, depends on the roots

$$\lambda_1 = \frac{-b + \sqrt{b^2 - 4ac}}{2a} \quad \text{and} \quad \lambda_2 = \frac{-b - \sqrt{b^2 - 4ac}}{2a}$$

of the **characteristic equation** $a\lambda^2 + b\lambda + c = 0$:

▸ **Real and distinct roots.** If $b^2 - 4ac > 0$, λ_1 and λ_2 are real and distinct and a general solution is

$$y = c_1 e^{\lambda_1 t} + c_2 e^{\lambda_2 t}.$$

▸ **Repeated roots.** If $b^2 - 4ac = 0$, $\lambda = \lambda_1 = \lambda_2 = -\dfrac{b}{2a}$ is a repeated root and a general solution is

$$y = c_1 e^{\lambda t} + c_2 t e^{\lambda t}.$$

Section 4.4

▸ **Complex roots.** If $b^2 - 4ac < 0$, the roots of $a\lambda^2 + b\lambda + c = 0$ are complex, $\lambda_1 = \mu + iv$ and $\lambda_2 = \mu - iv$, and a general solution of $ay'' + by' + cy = 0$ is

$$y = c_1 e^{\mu t} \cos vt + c_2 e^{\mu t} \sin vt.$$

▸ **Cauchy-Euler Equations** $ax^2 y'' + bxy' + cy = 0, x > 0$
Changing the independent variable to $z = \ln x$ transforms the equation into the following constant coefficient equation

$$a\frac{d^2 y}{dz^2} + (b - a)\frac{dy}{dz} + cy = 0.$$

Section 4.5

▸ The solution of the undamped spring-mass system $my'' + ky = 0$ can be expressed using phase-amplitude notation as $y = R \cos(\omega_0 t - \phi)$.

▸ For the damped spring-mass system $my'' + \gamma y' + ky = 0$, the motion is **overdamped** if $\gamma^2 - 4mk > 0$, **critically damped** if $\gamma^2 - 4mk = 0$, and **underdamped** if $\gamma^2 - 4mk < 0$.

Section 4.6

▸ If $\mathbf{x}_p$ is any particular solution to $\mathbf{x}' = \mathbf{P}(t)\mathbf{x} + \mathbf{g}(t)$ and $\{\mathbf{x}_1, \mathbf{x}_2\}$ is a fundamental set of solutions to the corresponding homogeneous equation, then a general solution to the nonhomogeneous equation is

$$\mathbf{x} = c_1 \mathbf{x}_1(t) + c_2 \mathbf{x}_2(t) + \mathbf{x}_p(t)$$

where c_1 and c_2 are arbitrary constants.

▸ If Y is any particular solution to $y'' + p(t)y' + q(t)y = g(t)$ and $\{y_1, y_2\}$ is a fundamental set of solutions to the corresponding homogeneous equation, then a general solution of the nonhomogeneous equation is

$$y = c_1 y_1(t) + c_2 y_2(t) + Y(t)$$

where c_1 and c_2 are arbitrary constants.

▸ **Undetermined Coefficients** $ay'' + by' + cy = g(t), \ a \neq 0$
This special method for finding a particular solution is primarily useful if the equation has constant coefficients and $g(t)$ is a polynomial, an exponential function, a sine function, a cosine function, or a linear combination of products of these functions. In this case the appropriate form for a particular solution, given in Table 4.6.1, can be determined.

Section 4.7

▶ The steady state response to $my'' + \gamma y' + ky = Ae^{i\omega t}$ can be expressed in the form

$$Y = A|G(i\omega)|e^{i(\omega t - \phi(\omega))}$$

where $G(i\omega)$ is called the **frequency response** of the system and $|G(i\omega)|$ and $\phi(\omega)$ are the **gain** and the **phase** of the frequency response, respectively.

▶ Information about the steady-state response as a function of the frequency ω of the input signal is contained in $G(i\omega)$. Frequencies at which $|G(i\omega)|$ are sharply peaked are called **resonant frequencies** of the system.

Section 4.8   Variation of Parameters Variation of parameters is a general method for finding particular solutions.

▶ If $\mathbf{x}_1$ and $\mathbf{x}_2$ are a fundamental set for $\mathbf{x}' = \mathbf{P}(t)\mathbf{x}$, then a particular solution to $\mathbf{x}' = \mathbf{P}(t)\mathbf{x} + \mathbf{g}(t)$ is

$$\mathbf{x}_p(t) = \mathbf{X}(t) \int \mathbf{X}^{-1}(t)\mathbf{g}(t)\,dt$$

where $\mathbf{X}(t) = [\mathbf{x}_1(t), \mathbf{x}_2(t)]$.

▶ Applying the last result to the nonhomogeneous second order equation, $y'' + p(t)y' + q(t)y = g(t)$, yields a particular solution

$$Y(t) = -y_1(t) \int \frac{y_2(t)g(t)}{W[y_1, y_2](t)}\,dt + y_2(t) \int \frac{y_1(t)g(t)}{W[y_1, y_2](t)}\,dt$$

where y_1 and y_2 are a fundamental set for the corresponding homogeneous equation.

PROJECTS

Project 1 A Vibration Insulation Problem

Passive isolation systems are sometimes used to insulate delicate equipment from unwanted vibrations. For example, in order to insulate electrical monitoring equipment from vibrations present in the floor of an industrial plant, the equipment may be placed on a platform supported by flexible mountings resting on the floor. A simple physical model for such a system is shown in Figure 4.P.1 where the mountings are modeled as an equivalent linear spring with spring constant k, the combined mass of the platform and equipment is m, and viscous damping with damping coefficient γ is assumed. Assume also that only vertical

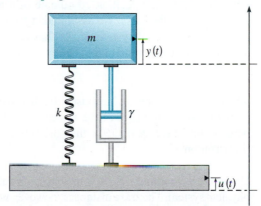

FIGURE 4.P.1   Equipment supported by flexible mountings.

motion occurs. In this project, we use this model to illustrate an important fundamental principle in the design of a passive isolation system.

Project 1 PROBLEMS

1. Denote by $y(t)$ the displacement of the platform from its equilibrium position relative to a fixed frame of reference. Relative to the same fixed reference frame, let $u(t)$ represent the displacement of the plant floor from its equilibrium position. Show that the differential equation for the motion of the platform is

$$y'' + 2\zeta\omega_0 y' + \omega_0^2 y = 2\zeta\omega_0 u' + \omega_0^2 u \qquad \text{(i)}$$

where $\omega_0 = \sqrt{k/m}$ and $\zeta = \gamma/2\sqrt{mk}$ is a dimensionless parameter known as the **viscous damping ratio**.

2. Denote by $G(i\omega)$ the frequency response of Eq. (i), that is, the ratio of the steady state response to the harmonic input $u(t) = Ae^{i\omega t}$. The **transmissibility**, T_R, of the platform mounting system is then defined to be the corresponding gain function, $T_R = |G(i\omega)|$. Show that T_R can be expressed as

$$T_R = \frac{\sqrt{1 + \left(2\zeta\dfrac{\omega}{\omega_0}\right)^2}}{\sqrt{\left[1 - \left(\dfrac{\omega}{\omega_0}\right)^2\right]^2 + \left[2\zeta\dfrac{\omega}{\omega_0}\right]^2}}.$$

3. Plot the graphs of T_R versus the dimensionless ratio ω/ω_0 for $\zeta = .1, .5, 1, 2$. For what values of ω/ω_0 is $T_R = 1$? Explain why the graphs imply that it is desirable that the mountings have a low natural frequency in order to isolate the vibration source from the equipment platform, and that using low stiffness isolators is one way to achieve this. CAS

4. The vibrations in the floor of an industrial plant lie in the range 16–75 Hz. The combined mass of the equipment and platform is 38 kg and the viscous damping ratio of the suspension is 0.2. Find the maximum value of the spring stiffness if the amplitude of the transmitted vibration is to be less than 10% of the amplitude of the floor vibration over the given frequency range.

5. Test the results of your design strategy for the situation described in Problem 4 by performing several numerical simulations of Eq. (i) in Problem 1 using various values of k as well as various input frequencies while holding the value of ζ fixed. For each simulation, plot the input $u(t)$ and the response $y(t)$ on the same set of coordinate axes. Do the numerical simulations support your theoretical results? ODEA

Project 2 Linearization of a Nonlinear Mechanical System

An undamped, one degree-of-freedom mechanical system with forces dependent on position is modeled by a second order differential equation

$$mx'' + f(x) = 0 \qquad \text{(1)}$$

where x denotes the position coordinate of a particle of mass m and $-f(x)$ denotes the force acting on the mass. We assume that $f(0) = 0$ so $\mathbf{x} = 0\mathbf{i} + 0\mathbf{j}$ is a critical point of the equivalent dynamical system. If $f(x)$ has two derivatives at $x = 0$, then its second degree Taylor formula with remainder at $x = 0$ is

$$f(x) = f'(0)x + \tfrac{1}{2}f''(z)x^2 \qquad \text{(2)}$$

where $0 < z < x$ and we have used the fact that $f(0) = 0$.

If the operating range of the system is such that $f''(z)x^2/2$ is always negligible compared to the linear term $f'(0)x$, then the nonlinear equation (1) may be approximated by its linearization,

$$mx'' + f'(0)x = 0. \qquad \text{(3)}$$

Under these conditions Eq. (3) may provide valuable information about the motion of the physical system. There are instances, however, where nonlinear behavior of $f(x)$, represented by the remainder term $f''(z)x^2/2$ in Eq. (2), is not negligible relative to $f'(0)x$ and Eq. (3) is a poor approximation to the actual system. In such cases, it is necessary to study the nonlinear system directly. In this project, we explore some of these questions in the context of a simple nonlinear system consisting of a mass attached to a pair of identical springs and confined to motion in the horizontal direction on a frictionless surface as shown in Figure 4.P.2. The springs are assumed to obey Hooke's law, $F_s(\Delta y) = -k\Delta y$ where Δy is the change in length of each spring from its equilibrium length L. When the mass is at its equilibrium position $x = 0$, both springs are assumed to have length $L + h$ with $h \geq 0$. Thus in the rest state both springs are either elongated and under tension ($h > 0$) or at their natural rest length and under zero tension ($h = 0$).

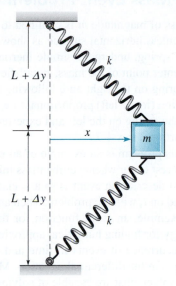

FIGURE 4.P.2 Horizontal motion of a mass attached to two identical springs.

Project 2 PROBLEMS

1. Show that the differential equation describing the motion of the mass in Figure 4.P.2 is

$$mx'' + 2kx \left[1 - \frac{L}{\sqrt{(L+h)^2 + x^2}} \right] = 0. \qquad (i)$$

CAS 2. **(a)** Find the linearization of Eq. (i) at $x = 0$.

   (b) In the case that $h > 0$, what is the natural frequency of the linearized system? Explain how the natural frequency of the linearized system depends on h and L.

   (c) On the same set of coordinate axes, plot the graphs of

$$y = f(x) = 2kx \left[1 - \frac{L}{\sqrt{(L+h)^2 + x^2}} \right],$$

   the 1st degree Taylor polynomial of f, $y_1 = f'(0)x$, and the third degree Taylor polynomial of f, $y_3 = f(0) + f'(0)x + f''(0)x^2/2 +$

$f'''(0)x^3/6$. Construct these plots for each of the following sets of parameter values:

   i. $L = 1, k = 1, h = 1$

   ii. $L = 1, k = 1, h = .5$

   iii. $L = 1, k = 1, h = .1$

   iv. $L = 1, k = 1, h = 0$.

   (d) Find the dependence on h of the approximate range of values of x such that $|y - y_1| \approx |y_3 - y_1| < .1$.

   (e) In the case that $h = 0$, explain why the solution of the linearized equation obtained in Problem 2 is a poor approximation to the solution of the nonlinear equation (i) for any initial conditions other than $x(0) = 0, x'(0) = 0$.

3. Subject to the initial conditions $x(0) = 0, x'(0) = 1$, ODEA draw the graph of the numerical approximation of the solution of Eq. (1) for $0 \leq t \leq 30$. On the same set of coordinate axes, draw the graph of the solution of the linearized equation using the same initial

conditions. Do this for each of the following sets of parameter values:

i. $m = 1$, $L = 1$, $k = 1$, and $h = 1$

ii. $m = 1$, $L = 1$, $k = 1$, and $h = .5$

iii. $m = 1$, $L = 1$, $k = 1$, and $h = .1$

Compare the accuracy of the period length and amplitude exhibited by the solution of the linearized equation to the period length and amplitude of the solution of the nonlinear equation as the value of h decreases.

Project 3 A Spring-Mass Event Problem

A mass of magnitude m is confined to one dimensional motion between two springs on a frictionless horizontal surface as shown in Figure 4.P.3. The mass, which is unattached to either spring, undergoes simple inertial motion whenever the distance from the origin of the center point of the mass, x, satisfies $|x| < L$. When $x \geq L$, the mass is in contact with the spring on the right and, following Hooke's Law, experiences a force in the negative x direction (to the left) proportional to $x - L$. Similarly, when $x \leq -L$, the mass is in contact with the spring on the left and experiences a force in the positive x direction (to the right) proportional to $x + L$.

This problem is an example of an **event problem** in differential equations. The events of interest occur whenever the mass initiates or terminates contact with a spring. A typical way to describe an event is to associate an event function $g(t, x)$, which may or may not depend on t, with the problem. Then an event is said to occur at time t^* if $g(t^*, x(t^*)) = 0$. For example, an event function for this problem could be $g(x) = (x - L)(x + L)$. One strategy for finding numerical approximations to solutions of an event problem is to locate the occurrence of events in time and restart the integration there so as to deal with the changes in the differential equation. Most modern commercial software packages contain ODE solvers that are capable of solving event problems.

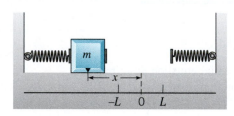

FIGURE 4.P.3 A mass bouncing back and forth between two springs.

Project 3 PROBLEMS

1. Assuming that both springs have spring constant k and that there is a damping force proportional to velocity x' with damping constant γ, write down the differential equation that describes the motion of the mass. Note that the expression for the force function depends on which of the three different parts of its domain x lies within.

2. The **Heaviside**, or **unit step function**, is defined by

$$u(x - c) = \begin{cases} 1, & x \geq c \\ 0, & x < c. \end{cases}$$

Use the unit step function to express the differential equation in Problem 1 in a single line.

3. Is the differential equation derived in Problems 1 and 2 linear or nonlinear? Explain why.

4. In the case that the damping constant $\gamma > 0$, find the critical points of the differential equation in Problem 2 and discuss their stability properties.

5. Consider the case of an undamped problem ($\gamma = 0$) using the parameter values $L = 1$, $m = 1$, $k = 1$ and initial conditions $x(0) = 2$, $x'(0) = 0$. Find the solution $x(t)$ of the initial value problem for $0 \leq t \leq t_1^*$

and $t_1^* \leq t < t_2^*$, where t_1^* and t_2^* are the times of the first and second events. Describe the physical situation corresponding to each of these two events.

ODEA 6. Consider the damped problem using the parameter values $L = 1$, $m = 1$, $\gamma = 0.1$, and $k = 1$.
   (a) Use a computer to draw a direction field of the corresponding dynamical system.
   (b) If you have access to computer software that is capable of solving event problems, solve for and plot the graphs of $x(t)$ and $x'(t)$ for the following sets of initial conditions:

   i. $x(0) = 2$, $x'(0) = 0$
   ii. $x(0) = 5$, $x'(0) = 0$.

   Give a physical explanation of why the limiting values of the trajectories as $t \to \infty$ depend on the initial conditions.

   (c) Draw a phase portrait for the equivalent dynamical system.

7. Describe some other physical problems that could be formulated as event problems in differential equations.

Project 4 Uniformly Distributing Points on a Sphere

In this project, we consider the problem of uniformly distributing n points on a sphere. The solution of this problem has applications ranging from certain computer graphics algorithms to optimally placing dimples on a golf ball. A class of geometric solutions can be obtained by considering the regular polyhedra. A **polyhedron** is a solid whose surface consists of polygonal faces. The polyhedron is **regular** if all of the polygonal faces are congruent and all the angles at vertices are equal. For example, the tetrahedron is a regular polyhedron with four triangular faces and four vertices. The geometric solution to the problem of uniformly distributing four points on the sphere corresponds to the positions of the vertices of an inscribed tetrahedron. The cube is a regular polyhedron that has six square faces and eight vertices. By a uniform distribution of eight points on a sphere, we mean the positions of the vertices of an inscribed cube. The number of solutions that can be found using these geometrical objects is rather limited since it can be proved that there are exactly five regular polyhedra as shown in Figure 4.P.4.

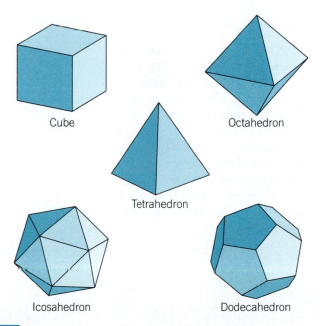

Cube

Octahedron

Tetrahedron

Icosahedron

Dodecahedron

FIGURE 4.P.4 The five regular polyhedra or platonic solids.

The numbers of faces, vertices, and edges are displayed in Table 4.P.1.

TABLE 4.P.1	Geometric characteristics of the regular polyhedra.			
Polyhedron	Face Type	Faces	Vertices	Edges
tetrahedron	triangle	4	4	6
octahedron	triangle	8	6	12
cube	square	6	8	12
icosahedron	triangle	20	12	30
dodecahedron	pentagon	12	20	30

The generalization of the geometric problem to numbers of vertices $n \notin \{4, 6, 8, 12, 20\}$ is that the solution corresponds to the locations of vertices of the polyhedron (obviously not regular) of maximum volume that can be inscribed within the sphere.

Another approach to finding solutions for arbitrary values of n is based on the following physical analog: *distribute n electrons on the surface of the sphere in some manner, let them move around under the influence of a mutually repulsive, inverse square law of force while imposing damping forces on the motion and constraining the electrons so that they remain on the surface of the sphere.* The electrons will try to stay as far away as possible from one another and the damping forces will cause energy to bleed out of the system. The end result is that the charges will eventually come to an equilibrium arrangement of minimum electrostatic potential. We will refer to this as the solution of the electrostatic problem. The purpose of this project is to solve the electrostatic problem by (i) formulating a first order system of differential equations describing the motion of the charges, (ii) developing a modified Euler algorithm for numerically approximating the solution, and (iii) comparing the approximations with the above geometric solutions.

Project 4 PROBLEMS

1. Consider a system of n electrons, each with mass $m_i = 1$ and position coordinates $\mathbf{q}_i = \langle q_{1i}, q_{2i}, q_{3i} \rangle$, $i = 1, \ldots, n$, that are constrained to move on the surface of a sphere. Assume a damping force on each charge proportional to velocity with damping constant α and that the force of repulsion between any pair of electrons is inversely proportional to the square of the distance between the pair with a constant of proportionality equal to β. Apply the vector form of Newton's second law, $m\mathbf{a} = \mathbf{f}$, to each of the charges to derive the following system of n second order equations,

$$\mathbf{q}_i'' = -\alpha \mathbf{q}_i' - \beta \sum_{\substack{j=1 \\ j \neq i}}^{n} \frac{\mathbf{q}_j - \mathbf{q}_i}{||\mathbf{q}_j - \mathbf{q}_i||^3}, \qquad \text{(i)}$$

where $||\mathbf{q}_j - \mathbf{q}_i|| =$
$\sqrt{(q_{j1} - q_{i1})^2 + (q_{j2} - q_{i2})^2 + (q_{j3} - q_{i3})^2}$
(see Figure 4.P.5).

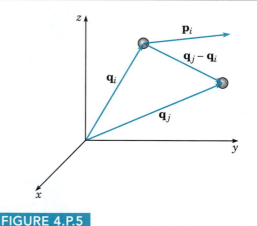

FIGURE 4.P.5

In addition to satisfying Eq. (i), the coordinates of each electron must satisfy $||\mathbf{q}_i|| = 1$.

2. Introduce velocity vectors $\mathbf{p}_i = \mathbf{q}_i'$ and show that Eq. (i) in Problem 1 can be transformed to the first order system

$$\mathbf{q}_i' = \mathbf{p}_i$$
$$\mathbf{p}_i' = -\beta \sum_{\substack{j=1 \\ j \neq i}}^{n} \frac{\mathbf{q}_j - \mathbf{q}_i}{||\mathbf{q}_j - \mathbf{q}_i||^3} - \alpha \mathbf{p}_i \qquad (i)$$

for each $i = 1, \ldots, n$. In terms of the $3n$-dimensional position and velocity vectors $\mathbf{q} = \langle \mathbf{q}_1, \ldots, \mathbf{q}_n \rangle$ and $\mathbf{p} = \langle \mathbf{p}_1, \ldots, \mathbf{p}_n \rangle$ and the $3n$-dimensional vector field

$$\mathbf{f}(\mathbf{q}, \mathbf{p}) = \left\langle -\beta \sum_{\substack{j=1 \\ j \neq 1}}^{n} \frac{\mathbf{q}_j - \mathbf{q}_1}{||\mathbf{q}_j - \mathbf{q}_1||^3} - \alpha \mathbf{p}_1, \ldots, \right.$$
$$\left. -\beta \sum_{\substack{j=1 \\ j \neq n}}^{n} \frac{\mathbf{q}_j - \mathbf{q}_n}{||\mathbf{q}_j - \mathbf{q}_n||^3} - \alpha \mathbf{p}_n \right\rangle,$$

show that Eq. (i) can be expressed as an autonomous system

$$\frac{d}{dt} \begin{bmatrix} \mathbf{q} \\ \mathbf{p} \end{bmatrix} = \begin{bmatrix} \mathbf{p} \\ \mathbf{f}(\mathbf{q}, \mathbf{p}) \end{bmatrix}. \qquad (ii)$$

CAS 3. Given the initial conditions, $\mathbf{q}(0) = \mathbf{q}_0$ and $\mathbf{p}(0) = \mathbf{0}$ use the following two-stage version of Euler's method to approximate the solution of Eq. (ii) in Problem 2 numerically,

$$\begin{bmatrix} \hat{\mathbf{q}}^{(m+1)} \\ \mathbf{p}^{(m+1)} \end{bmatrix} = \begin{bmatrix} \mathbf{q}^{(m)} \\ \mathbf{p}^{(m)} \end{bmatrix} + h \begin{bmatrix} \mathbf{p}^{(m)} \\ \mathbf{f}(\mathbf{q}^{(m)}, \mathbf{p}^{(m)}) \end{bmatrix}, \qquad (i)$$

$$\mathbf{q}^{(m+1)} = \left\langle \frac{\hat{\mathbf{q}}_1^{(m+1)}}{||\hat{\mathbf{q}}_1^{(m+1)}||}, \ldots, \frac{\hat{\mathbf{q}}_n^{(m+1)}}{||\hat{\mathbf{q}}_n^{(m+1)}||} \right\rangle, \qquad (ii)$$

where the second stage (ii) retracts each electron's position vector back to the surface of the unit sphere following the Euler step (i) which causes the electrons to separate themselves as well as *fly off* the sphere. While the position of the ith electron at the $(m+1)$st Euler step, $\hat{\mathbf{q}}_i^{(m+1)}$, will not lie on the surface of the unit sphere, the normalized position vector computed in the second stage (ii), $\mathbf{q}_i^{(m+1)}$, satisfies $||\mathbf{q}_i^{(m+1)}|| = 1$.
We suggest the following guidelines:

▶ Use the values $\alpha = 2$ and $\beta = 1$, but you may wish to experiment with other values.
▶ Use a time-step $h \approx 0.1$ or less, but you may wish to experiment with this to speed up convergence.

▶ Stop the Euler iteration when two successive states $\langle \mathbf{q}^{(m)}, \mathbf{p}^{(m)} \rangle$ and $\langle \mathbf{q}^{(m+1)}, \mathbf{p}^{(m+1)} \rangle$ are sufficiently close together, for example, when $||\mathbf{q}^{(m+1)} - \mathbf{q}^{(m)}|| + ||\mathbf{p}^{(m+1)} - \mathbf{p}^{(m)}|| < \epsilon$ for very small $\epsilon > 0$.
▶ Plot the potential energy

$$V(m) = V(\mathbf{q}^{(m)})$$
$$= \beta \sum_{i=1}^{n-1} \sum_{j=i+1}^{n} \frac{1}{||\mathbf{q}_j^{(m)} - \mathbf{q}_i^{(m)}||} \qquad (iii)$$

of the configuration of electrons as a function of $m = 1, 2, 3, \ldots$ to confirm that $V(m)$ is tending towards a local minimum.
▶ Choose initial conditions so that the initial configuration of the electrons does not possess a symmetry. For example, do not align the electrons on a great circle to begin with. One simple way to do this is to use spherical coordinates for the initial positions $q_{1i} = \sin\phi_i \cos\theta_i$, $q_{2i} = \sin\phi_i \sin\theta_i$, $q_{3i} = \cos\phi_i$ and use a random number generator to generate uniformly distributed random values of $\phi_i \in [0, \pi]$ and $\theta_i \in [0, 2\pi]$.

Compare the dynamically generated electrostatic solutions with the geometric solutions provided by the regular polyhedra for each of the cases $n = 4, 6, 8, 12, 20$. We illustrate this for the case $n = 6$. The two-stage Euler algorithm was used to approximate the solution of the system (ii) in Problem 2 using a stopping criterion of $\epsilon = 10^{-8}$. For each electron, the Euclidean distance to each of the other electrons was then computed and sorted from shortest to longest as shown in Table 4.P.2.
The fact that for all of the electrons there are only two identical distance classes, one consisting of the distance to four nearest neighbors and the second consisting of the distance to an antipodal vertex, is consistent with the geometry of an octahedron. For values of n corresponding to the other regular polyhedra a tabularized list of distances between electrons similar to Table 4.P.2, coupled with some geometric reasoning, permits a comparison between the geometric and electrostatic solutions. Note that distance values should be constant along each row of the table if the equilibrium positions actually correspond to vertices of a regular polyhedron.

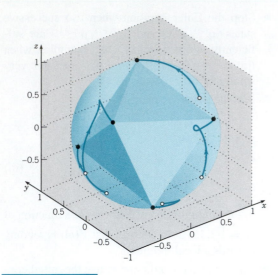

FIGURE 4.P.6 Trajectories of a system of six electrons confined to the surface of a sphere. The equilibrium configuration corresponds to the vertices of an inscribed octahedron.

In Figure 4.P.6 we show the trajectories of the electrons as they move from their initial locations to their equilibrium positions and then use the equilibrium positions as vertices of an inscribed octahedron.

 **4.** For the case $n = 8$, choose for an initial configuration the vertices of a cube, for example, $(q_{11}^{(0)}, q_{21}^{(0)}, q_{31}^{(0)}) = (1/\sqrt{3}, 1/\sqrt{3}, 1/\sqrt{3})$, etc. Then compare the limiting configuration with that obtained by using a random initial configuration. In particular, compare the limiting values of the potential energy $V(m)$ in Eq. (iii) of Problem 3 using the different initial configurations. Does this tell you anything about the behavior of the potential energy as a function of the configuration coordinates?

Now conduct computer simulations to find equilibrium distributions for values of $n \notin \{4, 6, 8, 12, 20\}$. To help make sense of your results, especially for large n, you may wish to plot on the same set of coordinate axes the graphs of $\bar{d}_j - \sigma_j$, $\bar{d}_j$, and $\bar{d}_j + \sigma_j$ as a function of $j = 1, \ldots, n-1$, where

$$\bar{d}_j = \frac{1}{n} \sum_{i=1}^{n} d_{ij}, \quad \sigma_j = \left[\frac{1}{n} \sum_{i=1}^{n} (d_{ij} - \bar{d}_j)^2 \right]^{1/2},$$

and

$$d_{ij} = \text{distance between electron}$$
$$i \text{ and its } j\text{th nearest neighbor.}$$

Thus, $\bar{d}_j$ and σ_j are the mean and standard deviation of the distance to the jth nearest neighbor for each $j = 1, \ldots, n-1$. Note that for any of the regular polyhedra, $\sigma_j = 0$ for each j.

5. In the table of distances to nearest neighbors, what is the geometric significance if the distance to the kth nearest neighbor is the same for all electrons?

6. In the table of distances to nearest neighbors, what is the geometric significance if two or more columns are identical?

7. If the simulation has achieved approximate equilibrium after M steps, why might you wish to check to see if $q_i^{(M)} p_i^{(M)} / \|p_i^{(M)}\| \cong 1$ for each $i = 1, \ldots, n$.

8. Why is a high degree of accuracy in approximating the solution of Eq. (ii) of Problem 2 during the transient phase of the calculations not an issue of major importance in this problem? Are we justified in using the Euler method?

9. Why do golf balls have dimples and how many dimples do they have? Why should the dimples be uniformly distributed over the surface of the golf ball?

TABLE 4.P.2 Table of distances from each electron to 1st nearest neighbor, 2nd nearest neighbor, etc.

	Electron 1	Electron 2	Electron 3	Electron 4	Electron 5	Electron 6
1st	1.4142	1.4142	1.4142	1.4142	1.4142	1.4142
2nd	1.4142	1.4142	1.4142	1.4142	1.4142	1.4142
3rd	1.4142	1.4142	1.4142	1.4142	1.4142	1.4142
4th	1.4142	1.4142	1.4142	1.4142	1.4142	1.4142
5th	2	2	2	2	2	2

Project 5 Euler-Lagrange Equations

Recall from calculus that a function $f(x)$ has a horizontal tangent at $x = a$ if $f'(a) = 0$. Such a point is called a stationary point; it may be, but does not have to be, a local maximum or minimum of $f(x)$. Similarly, for a function $f(x, y)$ of two variables, necessary conditions for (a, b) to be a stationary point of $f(x, y)$ are that $f_x(a, b) = 0$ and $f_y(a, b) = 0$. In this project, we explain how a generalization of this idea leads to several of the differential equations treated in this textbook. The generalization requires finding necessary conditions for a stationary point (a function) of a functional (a function of a function). For mechanical systems with one degree of freedom the functional is frequently of the form

$$J(y) = \int_a^b F(x, y, y')dx \qquad (1)$$

where $y = y(x)$. Consider, for example, the undamped pendulum of Figure 4.P.7. A fundamental principle of classical mechanics is the **Principle of Stationary Action:**[3] *The trajectory of an object in state space is the one that yields a stationary value for the integral with respect to time of the kinetic energy minus the potential energy.* If we denote kinetic energy by K and potential energy by V, the integral $\int_0^T (K - V)dt$ is called the **action integral** of the system. If $\theta(t)$ is the angular displacement of the pendulum, then the kinetic energy of the pendulum is $(1/2)mL^2(\theta')^2$ and its potential energy is $mgL(1 - \cos\theta)$. Therefore the Principle of Stationary Action requires that $\theta(t)$ must be a function for which the action integral,

$$J(\theta) = \int_0^T \left[\frac{1}{2}mL^2(\theta')^2 - mgL(1 - \cos\theta)\right] dt, \qquad (2)$$

is stationary. Note that $J(\theta)$ is of the form (1). Often a stationary point of the action integral is a minimum but can be a saddle-point or even a maximum. According to the Principle of Stationary Action, of all possible paths $t \to \langle\theta(t), \theta'(t)\rangle$ that the pendulum can follow in state space, the one observed in nature is the path for which the value of (2) is stationary. In this project, we want to deduce a characterization, which turns out to be a differential equation, of the stationary points (functions) of $J(y)$ in Eq. (1).

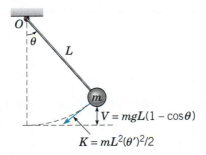

$$V = mgL(1 - \cos\theta)$$

$$K = mL^2(\theta')^2/2$$

FIGURE 4.P.7 An undamped pendulum in motion has kinetic energy $K = mL^2(\theta')^2/2$ and potential energy $V = mgL(1 - \cos\theta)$.

Thus we consider the problem of finding a function $y(x)$ that is a stationary point for the functional (1) where $y(x)$ is to be chosen from a class of suitably smooth functions (two or more continuous derivatives) that satisfy the boundary conditions

$$y(a) = A, \qquad y(b) = B.$$

[3]It can be shown that the Principle of Stationary Action is equivalent to Newton's laws of motion.

Suppose we give the name $\bar{y}(x)$ to such a stationary point. A very useful method for solving this problem is to examine the behavior of Eq. (1) for functions in a neighborhood of $\bar{y}(x)$. This is accomplished by considering small variations or perturbations to $\bar{y}(x)$ of the form $\epsilon\eta(x)$,

$$y(x) = \bar{y}(x) + \epsilon\eta(x), \tag{3}$$

where $0 < \epsilon \ll 1$ and $\eta(x)$ satisfies $\eta(a) = 0$ and $\eta(b) = 0$ so that $y(a) = A$ and $y(b) = B$ (Figure 4.P.8).

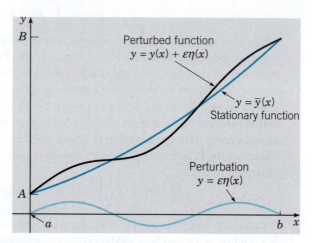

FIGURE 4.P.8 Perturbation of a function.

The parameter ϵ controls the size of the perturbation while $\eta(x)$ controls the "direction" of the perturbation. If we assume that $\bar{y}(x)$ is indeed a stationary point for (1), then it is necessary that, for any fixed perturbation direction $\eta(x)$, the function

$$\phi(\epsilon) = J(\bar{y} + \epsilon\eta) = \int_a^b F(x, \bar{y} + \epsilon\eta, \bar{y}' + \epsilon\eta')dx \tag{4}$$

must, as a function of the single variable ϵ, be stationary at $\epsilon = 0$ (Figure 4.P.9).

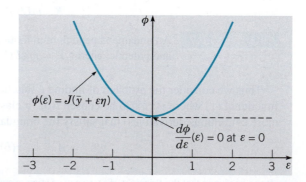

FIGURE 4.P.9 Graph of $J(\bar{y} + \epsilon\eta)$ near $\epsilon = 0$ in the case that $J(y)$ has a minimum at $\bar{y}$.

Therefore, we take the derivative of $\phi(\epsilon) = J(\bar{y} + \epsilon \eta)$ in Eq. (4) at $\epsilon = 0$ and set the result equal to zero,

$$
\begin{aligned}
\frac{d\phi}{d\epsilon}(0) &= \left. \frac{d}{d\epsilon} J(\bar{y} + \epsilon \eta) \right|_{\epsilon=0} = \int_a^b \left[\frac{\partial F}{\partial y}(x, \bar{y}, \bar{y}')\eta + \frac{\partial F}{\partial y'}(x, \bar{y}, \bar{y}')\eta' \right] dx \\
&= \left. \frac{\partial F}{\partial y'}(x, \bar{y}, \bar{y}')\eta \right|_{x=a}^b + \int_a^b \left[\frac{\partial F}{\partial y}(x, \bar{y}, \bar{y}') - \frac{\partial^2 F}{\partial x \partial y'}(x, \bar{y}, \bar{y}') \right] \eta \, dx \\
&= \int_a^b \left[\frac{\partial F}{\partial y}(x, \bar{y}, \bar{y}') - \frac{\partial^2 F}{\partial x \partial y'}(x, \bar{y}, \bar{y}') \right] \eta \, dx = 0,
\end{aligned}
\tag{5}
$$

where we have used integration by parts and the fact that $\eta(a) = \eta(b) = 0$ to set the boundary terms that arise to zero. Since the last integral is equal to zero for any function $\eta(x)$ satisfying $\eta(a) = \eta(b) = 0$, we conclude that the part of the integrand multiplying $\eta(x)$ must be the zero function, that is,

$$
\frac{\partial F}{\partial y}(x, \bar{y}, \bar{y}') - \frac{\partial^2 F}{\partial x \partial y'}(x, \bar{y}, \bar{y}') = 0,
$$

or, reverting to the notation y instead of $\bar{y}$,

$$
\frac{\partial F}{\partial y}(x, y, y') - \frac{\partial^2 F}{\partial x \partial y'}(x, y, y') = 0.
\tag{6}
$$

Equation (6) is known as the **Euler-Lagrange equation**[4] for the functional (1): if $J(y)$ in Eq. (1) *has a stationary function $y(x)$, then it is necessary that $y(x)$ satisfy the differential equation* (6).

EXAMPLE 1

Find the Euler-Lagrange equation for the action integral (2) of the simple undamped pendulum.

Note that the integrand in Eq. (2) is

$$
F(t, \theta, \theta') = \frac{1}{2} m L^2 (\theta')^2 - mgL(1 - \cos \theta).
$$

Then $F_\theta = -mgL \sin \theta$, $F_{\theta'} = m L^2 \theta'$, and $F_{\theta't} = m L^2 \theta''$. Thus the Euler-Lagrange equation $F_\theta - F_{\theta't} = 0$ is $-mgL \sin \theta - m L^2 \theta'' = 0$ or $\theta'' + (g/L) \sin \theta = 0$.

In the Problems that follow, several of the differential equations that appear in this textbook are derived by finding Euler-Lagrange equations of appropriate functionals.

[4]The argument that allows us to conclude that the bracketed function appearing in the last integral in Eq. (5) is zero is as follows. If this function is continuous and not zero (say, positive) at some point $x = \zeta$ in the interval, then it must be positive throughout a subinterval about $x = \zeta$. If we then choose an $\eta(x)$ that is positive inside this subinterval and zero outside, then the last integral in Eq. (5) will be positive—a contradiction.

Project 5 PROBLEMS

Problems 1 through 3 are concerned with one degree-of-freedom systems.

1. **Falling Bodies.** Let $y(t)$ be the height above the Earth's surface of a body of mass m subject only to the Earth's gravitational acceleration g. Find the action integral and Euler-Lagrange equation for $y(t)$.

2. **A Spring-Mass System.** A mass m supported by a frictionless horizontal surface is attached to a spring which is in turn attached to a fixed support. Assuming that motion is restricted to one direction, let $x(t)$ represent the displacement of the mass from its equilibrium position. This position corresponds to the spring being at its rest length. If the spring is compressed, then $x < 0$ and the mass is left of its equilibrium position. If the spring is elongated, then $x > 0$ and the mass is right of its equilibrium position. Assume that the potential energy stored in the spring is given by $kx^2/2$ where the parameter k is the **spring constant** of the spring. Find the action integral and Euler-Lagrange equation for $x(t)$.

3. **The Brachistochrone Problem.** Find the curve $y(x)$ connecting $y(a) = A$ to $y(b) = B$ (assume $a < b$ and $B < A$) along which a particle under the influence of gravity g will slide without friction in minimum time. This is the famous Brachistochrone problem discussed in Problem 32 of Section 2.3. Bernoulli showed that the appropriate functional for the sliding time is

$$T = \int_0^T dt = \int_0^L \frac{dt}{ds} ds = \int_0^L \frac{ds}{v} = \frac{1}{\sqrt{2g}} \int_0^L \frac{ds}{\sqrt{y}}$$

$$= \int_a^b \sqrt{\frac{1 + y'^2}{y}} dx. \tag{i}$$

(a) Justify each of the equalities in Eq. (i).

(b) From the functional on the right in Eq. (i), find the Euler-Lagrange equation for the curve $y(x)$.

Problems 4 and 5 are concerned with systems that have two degrees of freedom. The generalization of (1) to a functional with two degrees of freedom is

$$J(x, y) = \int_a^b F(t, x, y, x', y') dt \tag{7}$$

where t is a parameter, not necessarily time. Necessary conditions, stated as differential equations, for a vector-valued function $t \rightarrow \langle \bar{x}(t), \bar{y}(t) \rangle$ to be a stationary function of Eq. (7) are found by an analysis that is analogous to that leading up to Eq. (6). Consider the

perturbed functional

$$\phi(\epsilon) = J(\bar{x} + \epsilon\xi, \bar{y} + \epsilon\eta)$$

$$= \int_a^b F(t, \bar{x} + \epsilon\xi, \bar{y} + \epsilon\eta, \bar{x}' + \epsilon\xi', \bar{y}' + \epsilon\eta') dt \tag{8}$$

where the perturbation vector $\langle \xi(t), \eta(t) \rangle$ satisfies

$$\langle \xi(a), \eta(a) \rangle = \langle 0, 0 \rangle \quad \text{and} \quad \langle \xi(b), \eta(b) \rangle = \langle 0, 0 \rangle. \tag{9}$$

Thus, for $\phi(\epsilon)$ defined by Eq. (8), we compute $\phi'(0) = 0$, integrate by parts, and use the endpoint conditions (9) to arrive at

$$\int_a^b \left\{ [F_x - F_{x't}]\xi + [F_y - F_{y't}]\eta \right\} dt = 0. \tag{10}$$

Since Eq. (10) must hold for any ξ and η satisfying the conditions (9), we conclude that each factor multiplying ξ and η in the integrand must equal zero,

$$F_x - F_{x't} = 0$$

$$F_y - F_{y't} = 0. \tag{11}$$

Thus the Euler-Lagrange equations for the functional (7) are given by the system (11).

4. **A Two Mass, Three Spring System.** Consider the mechanical system consisting of two masses and three springs shown in Figure 4.P.10.

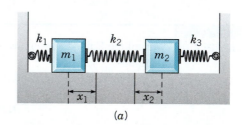

(a)

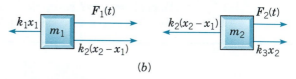

(b)

FIGURE 4.P.10 A two mass, three spring system.

Find the action integral for this system and the Euler-Lagrange equations for the positions $x_1(t)$ and $x_2(t)$ of the two masses.

5. **The Ray Equations.** In two dimensions, we consider a point sound source located at $(x, y) = (x_0, y_0)$ in a fluid medium where the sound speed

$c(x, y)$ can vary with location (see Figure 3.P.5 and Project 3 in Chapter 3 for a discussion of ray theory). The initial value problem that describes the path of a ray launched from the point (x_0, y_0) with a launch angle α, measured from the horizontal, is

$$\frac{d}{ds}\left(\frac{1}{c}\frac{dx}{ds}\right) = -\frac{c_x}{c^2}, \qquad x(0) = x_0, \quad x'(0) = \cos\alpha$$ (i)

$$\frac{d}{ds}\left(\frac{1}{c}\frac{dy}{ds}\right) = -\frac{c_y}{c^2}, \qquad y(0) = y_0, \quad y'(0) = \sin\alpha$$

where $c_x = \partial c/\partial x$ and $c_y = \partial c/\partial y$. The independent variable s in Eq. (vii) is the arc length of the ray measured from the source. Derive the system (i) from **Fermat's Principle**: *Rays, in physical space, are paths along which the transit time is minimum.* If $P : t \rightarrow \langle x(t), y(t) \rangle$, $a \leq t \leq b$ is a

parametric representation of a ray path from (x_0, y_0) to (x_1, y_1) in the xy-plane and sound speed is prescribed by $c(x, y)$, then the transit time functional is

$$T = \int_P \frac{ds}{c} = \int_a^b \frac{\sqrt{x'^2 + y'^2}}{c} dt$$ (ii)

where we have used the differential arc length relation

$$ds = \sqrt{x'^2 + y'^2}\, dt.$$ (iii)

Find the Euler-Lagrange equations from the functional representation on the right in Eq. (ii) and use Eq. (iii) to rewrite the resulting system using arc length s as the independent variable.

6. Carry out the calculations that lead from Eq. (8) to Eq. (11) in the discussion preceding Problem 4.

CHAPTER FIVE

The Laplace Transform

A n integral transform is a relation of the form

$$F(s) = \int_{\alpha}^{\beta} K(t, s)\, f(t)\, dt \tag{1}$$

which transforms or *maps* a given function $f(t)$ into another function $F(s)$. The function $K(t,s)$ in (1) is called the **kernel** of the transform and the function $F(s)$ is called the **transform** of $f(t)$. It is possible that $\alpha = -\infty$ or $\beta = \infty$, or both. If $\alpha = 0$, $\beta = \infty$, and $K(t, s) = e^{-st}$, Eq. (1) takes the form

$$F(s) = \int_{0}^{\infty} e^{-st} f(t)\, dt. \tag{2}$$

In this case, $F(s)$ is called the **Laplace transform** of $f(t)$.

In general, the parameter s can be a complex number but in most of this chapter we assume that s is real. It is customary to refer to $f(t)$ as a function, or signal, in the time or "t-domain" and $F(s)$ as its representation in the "s-domain." The Laplace transform is commonly used in engineering to study input-output relations of linear systems, to analyze feedback control systems, and to study electric circuits. One of its primary applications is to convert the problem of solving a constant coefficient linear differential equation in the t-domain into a problem involving algebraic operations in the s-domain. The general idea, illustrated by the block diagram in Figure 5.0.1, is as follows: Use the relation (2) to transform the problem for an unknown function f into a simpler problem for F, then solve this simpler problem to find F, and finally recover the desired function f from its transform F. This last step is known as "inverting the transform." The Laplace transform is particularly convenient for solving many practical engineering problems that involve mechanical or electrical systems acted on by discontinuous or impulsive forcing terms; for such problems the methods described in Chapter 4 are often rather awkward to use. In this chapter, we describe how the Laplace transform method works, emphasizing problems typical of those that arise in engineering applications.

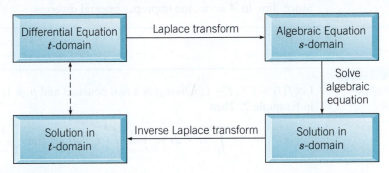

FIGURE 5.0.1 Laplace transform method for solving differential equations.

5.1 Definition of the Laplace Transform

▶ **Improper Integrals.** Since the Laplace transform involves an integral from zero to infinity, a knowledge of improper integrals of this type is necessary to appreciate the subsequent development of the properties of the transform. We provide a brief review of such improper integrals here. If you are already familiar with improper integrals, you may wish to skip over this review. On the other hand, if improper integrals are new to you, then you should probably consult a calculus book, where many more details and examples will be found.

An improper integral over an unbounded interval is defined as a limit of integrals over finite intervals; thus

$$\int_a^\infty f(t)\,dt = \lim_{A\to\infty} \int_a^A f(t)\,dt, \tag{1}$$

where A is a positive real number. If the integral from a to A exists for each $A > a$, and if the limit as $A \to \infty$ exists, then the improper integral is said to **converge** to that limiting value. Otherwise the integral is said to **diverge**, or to fail to exist. The following examples illustrate both possibilities.

EXAMPLE 1

Let $f(t) = e^{ct}$, $t \geq 0$, where c is a real nonzero constant. Then

$$\int_0^\infty e^{ct}\,dt = \lim_{A\to\infty} \int_0^A e^{ct}\,dt = \lim_{A\to\infty} \frac{e^{ct}}{c}\bigg|_0^A$$

$$= \lim_{A\to\infty} \frac{1}{c}(e^{cA} - 1).$$

It follows that the improper integral converges to the value $-1/c$ if $c < 0$ and diverges if $c > 0$. If $c = 0$, the integrand $f(t)$ is the constant function with value 1, and the integral again diverges.

EXAMPLE 2

Let $f(t) = 1/t$, $t \geq 1$. Then

$$\int_1^\infty \frac{dt}{t} = \lim_{A \to \infty} \int_1^A \frac{dt}{t} = \lim_{A \to \infty} \ln A.$$

Since $\lim_{A \to \infty} \ln A = \infty$, the improper integral diverges.

EXAMPLE 3

Let $f(t) = t^{-p}$, $t \geq 1$, where p is a real constant and $p \neq 1$; the case $p = 1$ was considered in Example 2. Then

$$\int_1^\infty t^{-p}\, dt = \lim_{A \to \infty} \int_1^A t^{-p}\, dt = \lim_{A \to \infty} \frac{1}{1-p}(A^{1-p} - 1).$$

As $A \to \infty$, $A^{1-p} \to 0$ if $p > 1$, but $A^{1-p} \to \infty$ if $p < 1$. Hence $\int_1^\infty t^{-p}\, dt$ converges to the value $1/(p - 1)$ for $p > 1$ but (incorporating the result of Example 2) diverges for $p \leq 1$. These results are analogous to those for the infinite series $\sum_{n=1}^\infty n^{-p}$.

▶ The Laplace Transform

DEFINITION 5.1.1

Let $f(t)$ be a function on $[0, \infty)$. The Laplace transform of f is the function F defined by the integral,

$$F(s) = \int_0^\infty e^{-st} f(t)\, dt. \tag{2}$$

The domain of $F(s)$ is the set of all values of s for which the integral in Eq. (2) converges. The Laplace transform of f is denoted by both F and $\mathcal{L}\{f\}$.

Remark. We will use t (representing time) as the independent variable of functions defined by lowercase letters such as x, y, or f. The corresponding Laplace transforms will be denoted by the same letter in uppercase:

$$\mathcal{L}\{f\} = F, \quad \mathcal{L}\{y\} = Y, \quad \mathcal{L}\{x\} = X.$$

We will use s to denote the independent variable of the transformed function,

$$\mathcal{L}\{f\}(s) = F(s), \quad \mathcal{L}\{y\}(s) = Y(s), \quad \mathcal{L}\{x\}(s) = X(s).$$

The Laplace transform of the function f is the function $\mathcal{L}\{f\}$, or F, and the Laplace transform evaluated at a number s is $\mathcal{L}\{f\}(s)$, or $F(s)$. For the purposes of clarity, simplicity, or efficiency, we may depart from standard function notational conventions. For example, we may use $\mathcal{L}\{f\}$ or $\mathcal{L}\{f(t)\}$ to represent $\mathcal{L}\{f\}(s)$. Thus $\mathcal{L}\{t\}$ could be used to represent the statement "$\mathcal{L}\{f\}(s)$ where f is defined by $f(t) = t$, $t > 0$."

Before we discuss the linearity property of the Laplace transform and sufficient conditions under which the transformation in Eq. (2) makes sense, we compute the Laplace transforms of some common functions.

EXAMPLE 4

Let $f(t) = 1$, $t \geq 0$. Then, as in Example 1,

$$\mathcal{L}\{1\} = \int_0^\infty e^{-st}\,dt = -\lim_{A\to\infty} \frac{e^{-st}}{s}\Big|_0^A = \frac{1}{s}, \qquad s > 0.$$

EXAMPLE 5

Let $f(t) = e^{at}$, $t \geq 0$ and a real. Then, again referring to Example 1,

$$\mathcal{L}\{e^{at}\} = \int_0^\infty e^{-st}e^{at}\,dt = \int_0^\infty e^{-(s-a)t}\,dt$$

$$= \frac{1}{s-a}, \qquad s > a.$$

EXAMPLE 6

Let $f(t) = e^{(a+bi)t}$, $t \geq 0$. As in the previous example

$$\mathcal{L}\{e^{(a+bi)t}\} = \int_0^\infty e^{-st}e^{(a+bi)t}\,dt = \int_0^\infty e^{-(s-a-bi)t}\,dt$$

$$= \frac{1}{s-a-bi}, \qquad s > a.$$

▶ **Linearity of the Laplace Transform.** The following theorem establishes the fact that the Laplace transform is a **linear operator**. This basic property is used frequently throughout the chapter.

THEOREM 5.1.2

Suppose that f_1 and f_2 are two functions whose Laplace transforms exist for $s > a_1$ and $s > a_2$, respectively. In addition, let c_1 and c_2 be real or complex numbers. Then, for s greater than the maximum of a_1 and a_2,

$$\mathcal{L}\{c_1 f_1(t) + c_2 f_2(t)\} = c_1 \mathcal{L}\{f_1(t)\} + c_2 \mathcal{L}\{f_2(t)\}. \tag{3}$$

Proof

$$\mathcal{L}\{c_1 f_1(t) + c_2 f_2(t)\} = \lim_{A\to\infty} \int_0^A e^{-st}[c_1 f_1(t) + c_2 f_2(t)]\,dt$$

$$= c_1 \lim_{A\to\infty} \int_0^A e^{-st} f_1(t)\,dt + c_2 \lim_{A\to\infty} \int_0^A e^{-st} f_2(t)\,dt$$

$$= c_1 \int_0^\infty e^{-st} f_1(t)\,dt + c_2 \int_0^\infty e^{-st} f_2(t)\,dt$$

where the last two integrals converge if $s > \max\{a_1, a_2\}$. ∎

Remark. The sum in Eq. (3) readily extends to an arbitrary number of terms,

$$\mathcal{L}\{c_1 f_1(t) + \cdots + c_n f_n(t)\} = c_1 \mathcal{L}\{f_1(t)\} + \cdots + c_n \mathcal{L}\{f_n(t)\}.$$

EXAMPLE 7

Find the Laplace transform of $f(t) = \sin at$, $t \geq 0$.

Recall that $\sin at$ can be represented as a linear combination of complex exponential functions,

$$\sin at = \frac{1}{2i}\left(e^{iat} - e^{-iat}\right).$$

Using the linearity of $\mathcal{L}$ as expressed in Eq. (3) and the result from Example 6 gives

$$\mathcal{L}\{\sin at\} = \frac{1}{2i}\left[\mathcal{L}\{e^{iat}\} - \mathcal{L}\{e^{-iat}\}\right] = \frac{1}{2i}\left(\frac{1}{s - ia} - \frac{1}{s + ia}\right) = \frac{a}{s^2 + a^2}, \qquad s > 0.$$

EXAMPLE 8

Find the Laplace transform of $f(t) = 2 + 5e^{-2t} - 3\sin 4t$, $t \geq 0$.

Extending Eq. (3) to a linear combination of three terms gives

$$\mathcal{L}\{f(t)\} = 2\mathcal{L}\{1\} + 5\mathcal{L}\{e^{-2t}\} - 3\mathcal{L}\{\sin 4t\}.$$

Then, using the results from Examples 4, 5, and 7, we find that

$$\mathcal{L}\{f(t)\} = \frac{2}{s} + \frac{5}{s + 2} - \frac{12}{s^2 + 16}, \qquad s > 0.$$

If $f(t)$ contains factors that are powers of t, it may be possible to compute the Laplace transform by using integration by parts.

EXAMPLE 9

Find the Laplace transform of $f(t) = t\cos at$, $t \geq 0$.

Expressing $\cos at$ in terms of complex exponential functions, $\cos at = \left(e^{iat} + e^{-iat}\right)/2$, we can write the Laplace transform of f in the form

$$\mathcal{L}\{t\cos at\} = \int_0^\infty e^{-st} t\cos at \, dt = \frac{1}{2}\int_0^\infty \left(te^{-(s-ia)t} + te^{-(s+ia)t}\right) dt.$$

Using integration by parts, we find that

$$\int_0^\infty te^{-(s-ia)t} \, dt = -\left[\frac{te^{-(s-ia)t}}{s - ia} + \frac{e^{-(s-ia)t}}{(s - ia)^2}\right]_0^\infty = \frac{1}{(s - ia)^2}, \qquad s > 0.$$

Similarly, we have

$$\int_0^\infty te^{-(s+ia)t} \, dt = \frac{1}{(s + ia)^2}, \qquad s > 0.$$

Thus

$$\mathcal{L}\{t\cos at\} = \frac{1}{2}\left[\frac{1}{(s - ia)^2} + \frac{1}{(s + ia)^2}\right] = \frac{s^2 - a^2}{(s^2 + a^2)^2}, \qquad s > 0.$$

The definitions that follow allow us to specify a large class of functions for which the Laplace transform is guaranteed to exist.

▶ Piecewise Continuous Functions

DEFINITION 5.1.3	A function f is said to be **piecewise continuous** on an interval $\alpha \leq t \leq \beta$ if the interval can be partitioned by a finite number of points $\alpha = t_0 < t_1 < \cdots < t_n = \beta$ so that:

1. f is continuous on each open subinterval $t_{i-1} < t < t_i$, and
2. f approaches a finite limit as the endpoints of each subinterval are approached from within the subinterval.

In other words, f is piecewise continuous on $\alpha \leq t \leq \beta$ if it is continuous at all but possibly finitely many points of $[\alpha, \beta]$, at each of which the function has a finite jump discontinuity. If f is piecewise continuous on $\alpha \leq t \leq \beta$ for every $\beta > \alpha$, then f is said to be piecewise continuous on $t \geq \alpha$. The graph of a piecewise continuous function is shown in Figure 5.1.1.

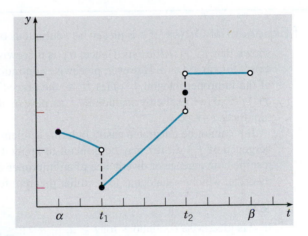

FIGURE 5.1.1   A piecewise continuous function.

We remark that continuous functions belong to the class of piecewise continuous functions and that for points of continuity the left- and right-hand limits are equal.

As shown in the next example, finding the Laplace transform of a piecewise continuous function may require expressing the right-hand side of Eq. (2) as a sum of integrals over subintervals for which the defining expressions for f change from subinterval to subinterval.

EXAMPLE 10

Find the Laplace transform of

$$f(t) = \begin{cases} e^{2t}, & 0 \leq t < 1, \\ 4, & 1 \leq t. \end{cases}$$

We observe that f is continuous on $(0,1)$ and $(1,\infty)$. Furthermore, $\lim_{\substack{t \to 1 \\ t < 1}} f(t) = e^2$ and $\lim_{\substack{t \to 1 \\ t > 1}} f(t) = 4$, so f has a single finite jump discontinuity at $t = 1$. Thus f is piecewise continuous on $t \geq 0$. Since the expressions defining $f(t)$ differ on each of the two subintervals $0 \leq t < 1$ and $1 \leq t$, we break the integral in Eq. (2) into two separate parts. Thus

$$F(s) = \int_0^\infty e^{-st} f(t)\, dt$$

$$= \int_0^1 e^{-st} e^{2t}\, dt + \int_1^\infty e^{-st} \cdot 4\, dt$$

$$= \int_0^1 e^{-(s-2)t}\, dt + 4 \lim_{A \to \infty} \int_1^A e^{-st}\, dt$$

$$= -\frac{e^{-(s-2)t}}{s-2}\bigg|_{t=0}^1 - 4 \lim_{A \to \infty} \frac{e^{-st}}{s}\bigg|_{t=1}^A$$

$$= \frac{1}{s-2} - \frac{e^{-(s-2)}}{s-2} - 4 \lim_{A \to \infty} \left[\frac{e^{-As}}{s} - \frac{e^{-s}}{s} \right]$$

$$= \frac{1}{s-2} - \frac{e^{-(s-2)}}{s-2} + 4\frac{e^{-s}}{s}, \qquad s > 0, s \neq 2.$$

▶ **Functions of Exponential Order.** If f is piecewise continuous on the interval $a \leq t \leq A$, then it can be shown that $\int_a^A f(t)\, dt$ exists. Hence, if f is piecewise continuous for $t \geq a$, then $\int_a^A f(t)\, dt$ exists for each $A > a$. However, piecewise continuity is not enough to ensure convergence of the improper integral $\int_a^\infty f(t)\, dt$, as the preceding examples show. Obviously, $f(t)$ (or $f(t)e^{-st}$ if we explicitly include e^{-st} as part of the integrand) must vanish sufficiently rapidly as $t \to \infty$.

If f cannot be integrated easily in terms of elementary functions, the definition of convergence of $\int_a^\infty f(t)\, dt$ may be difficult to apply. Frequently, the most convenient way to test the convergence or divergence of an improper integral is by the following comparison theorem, which is analogous to a similar theorem for infinite series.

THEOREM 5.1.4	If f is piecewise continuous for $t \geq a$, if $\lvert f(t) \rvert \leq g(t)$ when $t \geq M$ for some positive constant M, and if $\int_M^\infty g(t)\, dt$ converges, then $\int_a^\infty f(t)\, dt$ also converges. On the other hand, if $f(t) \geq g(t) \geq 0$ for $t \geq M$, and if $\int_M^\infty g(t)\, dt$ diverges, then $\int_a^\infty f(t)\, dt$ also diverges.

The proof of this result from calculus will not be given here. It is made plausible, however, by comparing the areas represented by $\int_M^\infty g(t)\, dt$ and $\int_M^\infty \lvert f(t) \rvert\, dt$. The functions most useful for comparison purposes are e^{ct} and t^{-p}, which we considered in Examples 1, 2, and 3. In particular, functions that increase no faster than an exponential function as $t \to \infty$ are useful for stating the the existence theorem for Laplace transforms.

DEFINITION 5.1.5	A function $f(t)$ is of **exponential order** (as $t \to +\infty$) if there exist real constants $M \geq 0, K > 0$, and a such that $$\lvert f(t) \rvert \leq K e^{at} \qquad (4)$$ when $t \geq M$.

Remark. To show that $f(t)$ is of exponential order, it suffices to show that $f(t)/e^{at}$ is bounded for all t sufficiently large.

EXAMPLE 11

Determine which of the following functions are of exponential order: (a) $f(t) = \cos at$, (b) $f(t) = t^2$, and (c) $f(t) = e^{t^2}$.

(a) Since $|\cos at| \leq 1$ for all t, inequality (4) holds if we choose $M = 0, K = 1$, and $a = 0$. The choice of constants is not unique. For example, $M = 10, K = 2$, and $a = 3$ would also suffice.

(b) By l'Hôpital's rule

$$\lim_{t \to \infty} \frac{t^2}{e^{at}} = \lim_{t \to \infty} \frac{2t}{ae^{at}} = \lim_{t \to \infty} \frac{2}{a^2 e^{at}} = 0$$

if $a > 0$. Suppose we choose $a = 1$. Then from the definition of limit, there exists an M such that $t^2/e^t \leq 0.1$, say, for all $t \geq M$. Then $|f(t)| = t^2 \leq 0.1e^t$ for all $t \geq M$. Thus $f(t)$ is of exponential order.

(c) Since

$$\lim_{t \to \infty} \frac{e^{t^2}}{e^{at}} = \lim_{t \to \infty} e^{t(t-a)} = \infty$$

no matter how large we choose a, $f(t) = e^{t^2}$ is not of exponential order.

Existence of the Laplace Transform

The following theorem guarantees that the Laplace transform F exists if f is a piecewise continuous function of exponential order. In this chapter, we deal almost exclusively with such functions.

THEOREM 5.1.6

Suppose

i. f is piecewise continuous on the interval $0 \leq t \leq A$ for any positive A, and

ii. f is of exponential order, that is, there exist real constants $M \geq 0, K > 0$, and a such that $|f(t)| \leq Ke^{at}$ when $t \geq M$.

Then the Laplace transform $\mathcal{L}\{f(t)\} = F(s)$, defined by Eq. (2), exists for $s > a$.

Proof

To establish this theorem, we must show that the integral in Eq. (2) converges for $s > a$. Splitting the improper integral into two parts, we have

$$F(s) = \int_0^\infty e^{-st} f(t) \, dt = \int_0^M e^{-st} f(t) \, dt + \int_M^\infty e^{-st} f(t) \, dt. \tag{5}$$

The first integral on the right side of Eq. (5) exists by hypothesis (i) of the theorem; hence, the existence of $F(s)$ depends on the convergence of the second integral. By hypothesis (ii) we have, for $t \geq M$,

$$|e^{-st} f(t)| \leq Ke^{-st}e^{at} = Ke^{(a-s)t},$$

and thus, by Theorem 5.1.4, $F(s)$ exists provided that $\int_M^\infty e^{(a-s)t} \, dt$ converges. Referring to Example 1 with c replaced by $a - s$, we see that this latter integral converges when $a - s < 0$, which establishes Theorem 5.1.6.

The Laplace transforms that appear in Examples 5 through 10 all have the property of vanishing as $s \to \infty$. This property is shared by all Laplace transforms of piecewise continuous functions of exponential order as stated in the next corollary.

COROLLARY 5.1.7

If $f(t)$ satisfies the hypotheses of Theorem 5.1.6, then

$$|F(s)| \leq L/s \tag{6}$$

for some constant L as $s \to \infty$. Thus,

$$\lim_{s \to \infty} F(s) = 0. \tag{7}$$

A proof of Corollary 5.1.7 is outlined in Problem 35.

Remarks

1. In general, the parameter s may be complex, and the full power of the Laplace transform becomes available only when we regard $F(s)$ as a function of a complex variable. For functions that satisfy the hypotheses of Theorem 5.1.6, the Laplace transform exists for all values of s in the set $\{s : \text{Re } s > a\}$. This follows from the proof of Theorem 5.1.6 since, if $s = \mu + i\nu$ is complex, $|e^{-st}| = |e^{-\mu t}(\cos \nu t - i \sin \nu t)| = e^{-\mu t} = e^{-\text{Re}(s)t}$.

2. While the hypotheses of Theorem 5.1.6 are sufficient for the existence of the Laplace transform of $f(t)$, they are not necessary. In Problem 37, you are asked to compute the Laplace transform of t^p where $p > -1$. In particular, you will show that the Laplace transform of $t^{-1/2}$ is $\sqrt{\pi/s}$.

3. Corollary 5.1.7 implies that if $\lim_{s \to \infty} F(s) \neq 0$, then F is not the Laplace transform of a piecewise continuous function of exponential order. Most of the Laplace transforms in this chapter are rational functions of s (quotients of polynomials in s). In order for such functions to be Laplace transforms of piecewise continuous functions of exponential order, Corollary 5.1.7 implies that it is necessary that the degree of the polynomial in the numerator be less than than the degree of the polynomial in the denominator.

4. In Section 5.7, we will encounter a "generalized function," denoted by $\delta(t)$, such that $\mathcal{L}\{\delta(t)\} = 1$. Thus, by Corollary 5.1.7, $\delta(t)$ is not a piecewise continuous function of exponential order.

PROBLEMS

In each of Problems 1 through 4, sketch the graph of the given function. In each case, determine whether f is continuous, piecewise continuous, or neither on the interval $0 \leq t \leq 3$.

1. $f(t) = \begin{cases} t^2, & 0 \leq t \leq 1 \\ 2 + t, & 1 < t \leq 2 \\ 6 - t, & 2 < t \leq 3 \end{cases}$

2. $f(t) = \begin{cases} t^2, & 0 \leq t \leq 1 \\ (t-1)^{-1}, & 1 < t \leq 2 \\ 1, & 2 < t \leq 3 \end{cases}$

3. $f(t) = \begin{cases} t^2, & 0 \leq t \leq 1 \\ 1, & 1 < t \leq 2 \\ 3 - t, & 2 < t \leq 3 \end{cases}$

4. $f(t) = \begin{cases} t, & 0 \leq t \leq 1 \\ 3 - t, & 1 < t \leq 2 \\ 1, & 2 < t \leq 3 \end{cases}$

In each of Problems 5 through 12, determine whether the given function is of exponential order. If it is, find suitable values for M, K, and a in inequality (4) of Definition 5.1.5.

5. $3e^{5t}$

6. $-e^{-3t}$

7. $e^{2t} \sin 3t$

8. t^{10}

9. $\cosh t^2$

10. $\sin t^4$

11. $1/(1+t)$

12. e^{t^3}

13. Find the Laplace transform of each of the following functions:
  (a) t
  (b) t^2
  (c) t^n, where n is a positive integer

In each of Problems 14 through 17, find the Laplace transform of the given function.

14. $f(t) = \begin{cases} t, & 0 \le t \le 1 \\ 1, & 1 < t \end{cases}$

15. $f(t) = \begin{cases} 0, & 0 \le t \le 1 \\ 1, & 1 < t \le 2 \\ 0, & 2 < t \end{cases}$

16. $f(t) = \begin{cases} 0, & 0 \le t \le 1 \\ e^{-t}, & 1 < t \end{cases}$

17. $f(t) = \begin{cases} t^2, & 0 \le t \le 1 \\ 3-t, & 1 < t \le 2 \\ 1, & 2 < t \end{cases}$

Recall that $\cosh bt = (e^{bt} + e^{-bt})/2$ and $\sinh bt = (e^{bt} - e^{-bt})/2$. In each of Problems 18 through 21, find the Laplace transform of the given function; a and b are real constants.

18. $\cosh bt$

19. $\sinh bt$

20. $e^{at} \cosh bt$

21. $e^{at} \sinh bt$

In each of Problems 22 through 24, use the facts that $\cos bt = (e^{ibt} + e^{-ibt})/2$, $\sin bt = (e^{ibt} - e^{-ibt})/2i$ to find the Laplace transform of the given function; a and b are real constants.

22. $\cos bt$

23. $e^{at} \sin bt$

24. $e^{at} \cos bt$

In each of Problems 25 through 30, using integration by parts, find the Laplace transform of the given function; n is a positive integer and a is a real constant.

25. te^{at}

26. $t \sin at$

27. $t \cosh at$

28. $t^n e^{at}$

29. $t^2 \sin at$

30. $t^2 \sinh at$

In each of Problems 31 through 34, determine whether the given integral converges or diverges.

31. $\int_0^\infty (t^2+1)^{-1}\,dt$

32. $\int_0^\infty te^{-t}\,dt$

33. $\int_1^\infty t^{-2}e^t\,dt$

34. $\int_0^\infty e^{-t}\cos t\,dt$

35. A Proof of Corollary 5.1.7
  (a) Starting from (5), use the fact that $|f(t)|$ is bounded on $[0, M]$ (since $f(t)$ is piecewise continuous) to show that if $s > \max(a, 0)$, then

$$|F(s)| \le \max_{0 \le t \le M} |f(t)| \int_0^M e^{-st}\,dt$$
$$+ K \int_M^\infty e^{-(s-a)t}\,dt$$
$$= \max_{0 \le t \le M} |f(t)| \frac{1 - e^{-sM}}{s}$$
$$+ \frac{K}{s-a} e^{-(s-a)M}.$$

  (b) Argue that there is a constant K_1 such that
$$Ke^{-(s-a)M}/(s-a) < K_1/s$$
for s sufficiently large.
  (c) Conclude that for s sufficiently large, $|F(s)| \le L/s$ where
$$L = \max_{0 \le t \le M} |f(t)| + K_1.$$

36. The Gamma Function. The gamma function is denoted by $\Gamma(p)$ and is defined by the integral
$$\Gamma(p+1) = \int_0^\infty e^{-x}x^p\,dx. \qquad \text{(i)}$$
The integral converges for all $p > -1$.
  (a) Show that, for $p > 0$,
$$\Gamma(p+1) = p\Gamma(p).$$
  (b) Show that $\Gamma(1) = 1$.
  (c) If p is a positive integer n, show that
$$\Gamma(n+1) = n!.$$
Since $\Gamma(p)$ is also defined when p is not an integer, this function provides an extension of the factorial function to nonintegral values of the independent variable. Note that it is also consistent to define $0! = 1$.
  (d) Show that, for $p > 0$,
$$p(p+1)(p+2)\cdots(p+n-1)$$
$$= \Gamma(p+n)/\Gamma(p).$$

Thus $\Gamma(p)$ can be determined for all positive values of p if $\Gamma(p)$ is known in a single interval of unit length, say, $0 < p \leq 1$. It is possible to show that $\Gamma(1/2) = \sqrt{\pi}$. Find $\Gamma(3/2)$ and $\Gamma(11/2)$.

37. Consider the Laplace transform of t^p, where $p > -1$.

(a) Referring to Problem 36, show that

$$\mathcal{L}\{t^p\} = \int_0^\infty e^{-st}t^p\,dt = \frac{1}{s^{p+1}}\int_0^\infty e^{-x}x^p\,dx$$

$$= \Gamma(p+1)/s^{p+1}, \qquad s > 0.$$

(b) Let p be a positive integer n in (a). Show that

$$\mathcal{L}\{t^n\} = n!/s^{n+1}, \qquad s > 0.$$

(c) Show that

$$\mathcal{L}\{t^{-1/2}\} = \frac{2}{\sqrt{s}}\int_0^\infty e^{-x^2}\,dx, \qquad s > 0.$$

It is possible to show that

$$\int_0^\infty e^{-x^2}\,dx = \frac{\sqrt{\pi}}{2};$$

hence,

$$\mathcal{L}\{t^{-1/2}\} = \sqrt{\pi/s}, \qquad s > 0.$$

(d) Show that

$$\mathcal{L}\{t^{1/2}\} = \sqrt{\pi}/(2s^{3/2}), \qquad s > 0.$$

5.2 Properties of the Laplace Transform

In the preceding section, we computed the Laplace transform of several functions $f(t)$ directly from the definition

$$\mathcal{L}\{f\}(s) = \int_0^\infty e^{-st}f(t)\,dt.$$

In this section, we present a number of operational properties of the Laplace transform that greatly simplify the task of obtaining explicit expressions for $\mathcal{L}\{f\}$. In addition, these properties enable us to use the transform to solve initial value problems for linear differential equations with constant coefficients.

▶ Laplace Transform of $e^{ct}f(t)$

THEOREM 5.2.1	If $F(s) = \mathcal{L}\{f(t)\}$ exists for $s > a$, and if c is a constant, then
	$$\mathcal{L}\{e^{ct}f(t)\} = F(s-c), \qquad s > a + c. \qquad (1)$$

Proof

The proof of this theorem merely requires the evaluation of $\mathcal{L}\{e^{ct}f(t)\}$. Thus

$$\mathcal{L}\{e^{ct}f(t)\} = \int_0^\infty e^{-st}e^{ct}f(t)\,dt = \int_0^\infty e^{-(s-c)t}f(t)\,dt$$

$$= F(s-c),$$

which is Eq. (1). Since $F(s)$ exists for $s > a$, $F(s-c)$ exists for $s - c > a$, or $s > a + c$. ∎

According to Theorem 5.2.1, multiplication of $f(t)$ by e^{ct} results in a translation of the transform $F(s)$ a distance c in the positive s direction if $c > 0$. (If $c < 0$ the translation is, of course, in the negative direction.)

EXAMPLE 1

Find the Laplace transform of $g(t) = e^{-2t} \sin 4t$.

The Laplace transform of $f(t) = \sin 4t$ is $F(s) = 4/(s^2 + 16)$ (see Example 7 in Section 5.1). Using Theorem 5.2.1 with $c = -2$, the Laplace transform of g is

$$G(s) = \mathcal{L}\{e^{-2t} f(t)\} = F(s+2) = \frac{4}{(s+2)^2 + 16}, \qquad s > -2. \tag{2}$$

▶ **Laplace Transform of Derivatives of f(t).** The usefulness of the Laplace transform, in connection with solving initial value problems, rests primarily on the fact that the transform of f' (and higher order derivatives) is related in a simple way to the transform of f. The relationship is expressed in the following theorem.

THEOREM 5.2.2

Suppose that f is continuous and f' is piecewise continuous on any interval $0 \le t \le A$. Suppose further that f and f' are of exponential order with a as specified in Theorem 5.1.6. Then $\mathcal{L}\{f'(t)\}$ exists for $s > a$, and moreover

$$\mathcal{L}\{f'(t)\} = s\mathcal{L}\{f(t)\} - f(0). \tag{3}$$

Proof

To prove this theorem, we consider the integral

$$\int_0^A e^{-st} f'(t)\, dt.$$

If f' has points of discontinuity in the interval $0 \le t \le A$, let them be denoted by $t_1, t_2, \ldots, t_n$. Then we can write this integral as

$$\int_0^A e^{-st} f'(t)\, dt = \int_0^{t_1} e^{-st} f'(t)\, dt + \int_{t_1}^{t_2} e^{-st} f'(t)\, dt + \cdots + \int_{t_n}^A e^{-st} f'(t)\, dt.$$

Integrating each term on the right by parts yields

$$\int_0^A e^{-st} f'(t)\, dt = e^{-st} f(t)\Big|_0^{t_1} + e^{-st} f(t)\Big|_{t_1}^{t_2} + \cdots + e^{-st} f(t)\Big|_{t_n}^A$$

$$+ s\left[\int_0^{t_1} e^{-st} f(t)\, dt + \int_{t_1}^{t_2} e^{-st} f(t)\, dt + \cdots + \int_{t_n}^A e^{-st} f(t)\, dt\right].$$

Since f is continuous, the contributions of the integrated terms at $t_1, t_2, \ldots, t_n$ cancel. Combining the integrals gives

$$\int_0^A e^{-st} f'(t)\, dt = e^{-sA} f(A) - f(0) + s\int_0^A e^{-st} f(t)\, dt.$$

As $A \to \infty$, $e^{-sA} f(A) \to 0$ whenever $s > a$. Hence, for $s > a$,

$$\mathcal{L}\{f'(t)\} = s\mathcal{L}\{f(t)\} - f(0),$$

which establishes the theorem.

EXAMPLE 2

Verify that Theorem 5.2.2 holds for $g(t) = e^{-2t} \sin 4t$ and its derivative $g'(t) = -2e^{-2t} \sin 4t + 4e^{-2t} \cos 4t$.

The Laplace transform of $g(t)$, computed in Example 1, is $G(s) = 4/[(s+2)^2 + 16]$. By Theorem 5.2.2

$$\mathcal{L}\{g'(t)\} = sG(s) - g(0) = \frac{4s}{(s+2)^2 + 16}$$

since $g(0) = 0$. On the other hand, from Example 7 and Problem 22 in Section 5.1, we know that $\mathcal{L}\{\sin 4t\} = 4/(s^2 + 16)$ and $\mathcal{L}\{\cos 4t\} = s/(s^2 + 16)$. Using the linearity property of the Laplace transform and Theorem 5.2.1 it follows that

$$\mathcal{L}\{g'(t)\} = -2\mathcal{L}\{e^{-2t} \sin 4t\} + 4\mathcal{L}\{e^{-2t} \cos 4t\}$$

$$= \frac{-8}{(s+2)^2 + 16} + \frac{4(s+2)}{(s+2)^2 + 16}$$

$$= \frac{4s}{(s+2)^2 + 16}.$$

If f' and f'' satisfy the same conditions that are imposed on f and f', respectively, in Theorem 5.2.2, then it follows that the Laplace transform of f'' also exists for $s > a$ and is given by

$$\mathcal{L}\{f''(t)\} = s^2 \mathcal{L}\{f(t)\} - sf(0) - f'(0). \tag{4}$$

Indeed, provided the function f and its derivatives satisfy suitable conditions, an expression for the transform of the nth derivative $f^{(n)}$ can be derived by successive applications of Theorem 5.2.2. The result is given in the following corollary.

COROLLARY 5.2.3

Suppose that

i. the functions $f, f', \ldots, f^{(n-1)}$ are continuous and that $f^{(n)}$ is piecewise continuous on any interval $0 \le t \le A$, and

ii. $f, f', \ldots, f^{(n-1)}, f^{(n)}$ are of exponential order with a as specified in Theorem 5.1.6.

Then $\mathcal{L}\{f^{(n)}(t)\}$ exists for $s > a$ and is given by

$$\mathcal{L}\{f^{(n)}(t)\} = s^n \mathcal{L}\{f(t)\} - s^{n-1} f(0) - \cdots - sf^{(n-2)}(0) - f^{(n-1)}(0). \tag{5}$$

EXAMPLE 3

Assume that the solution of the following initial value problem satisfies the hypotheses of Corollary 5.2.3. Find its Laplace transform.

$$y'' + 2y' + 5y = e^{-t}, \qquad y(0) = 1, \quad y'(0) = -3. \tag{6}$$

Taking the Laplace transform of both sides of the differential equation in the initial value problem (6) and using the linearity property of $\mathcal{L}$ yields

$$\mathcal{L}\{y''\} + 2\mathcal{L}\{y'\} + 5\mathcal{L}\{y\} = \mathcal{L}\{e^{-t}\}.$$

Letting $Y = \mathcal{L}\{y\}$ and applying formula (5) in Corollary 5.2.3 to each of the derivative terms and using the fact that $\mathcal{L}\{e^{-t}\} = 1/(s + 1)$ then yields

$$\underbrace{s^2 Y(s) - sy(0) - y'(0)}_{\mathcal{L}\{y''(t)\}(s)} + 2[\underbrace{sY(s) - y(0)}_{\mathcal{L}\{y'(t)\}(s)}] + 5Y(s) = \frac{1}{s + 1}$$

or

$$s^2 Y(s) - s \cdot 1 + 3 + 2[sY(s) - 1] + 5Y(s) = \frac{1}{s + 1}$$

where we have replaced $y(0)$ and $y'(0)$ using the initial conditions specified in Eqs. (6). Solving this last algebraic equation for $Y(s)$ gives

$$Y(s) = \frac{s - 1}{s^2 + 2s + 5} + \frac{1}{(s + 1)(s^2 + 2s + 5)} = \frac{s^2}{(s + 1)(s^2 + 2s + 5)}.$$

EXAMPLE 4

In the series RLC circuit shown in Figure 5.2.1, denote the current in the circuit by $i(t)$, the total charge on the capacitor by $q(t)$, and the impressed voltage by $e(t)$.

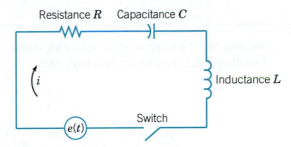

FIGURE 5.2.1 The series RLC circuit.

Assuming that $i(t)$ and $q(t)$ satisfy the hypotheses of Theorem 5.2.2, find the Laplace transforms of $q(t)$ and $i(t)$.

The relation between charge q and current i is

$$i = dq/dt \tag{7}$$

while Kirchhoff's second law (*in a closed circuit the impressed voltage is equal to the algebraic sum of the voltages across the elements in the rest of the circuit*) yields the equation

$$L\frac{di}{dt} + Ri + \frac{1}{C}q = e(t). \tag{8}$$

Taking the Laplace transform of Eqs. (7) and (8) and using Theorem 5.2.2 gives the system of algebraic equations

$$I(s) = sQ(s) \quad q(0) \tag{9}$$

and

$$L[sI(s) - i(0)] + RI(s) + \frac{1}{C}Q(s) = E(s) \tag{10}$$

where $I(s)$, $Q(s)$ and $E(s)$ are the Laplace transforms of $i(t)$, $q(t)$, and $e(t)$, respectively. Using matrix notation, Eqs. (9) and (10) may be written as

$$\begin{pmatrix} s & -1 \\ 1/C & Ls + R \end{pmatrix} \begin{pmatrix} Q(s) \\ I(s) \end{pmatrix} = \begin{pmatrix} q(0) \\ Li(0) + E(s) \end{pmatrix}. \tag{11}$$

The solution of Eq. (11) is

$$\begin{pmatrix} Q(s) \\ I(s) \end{pmatrix} = \frac{1}{Ls^2 + Rs + 1/C} \begin{pmatrix} Ls + R & 1 \\ -1/C & s \end{pmatrix} \begin{pmatrix} q(0) \\ Li(0) + E(s) \end{pmatrix}.$$

Thus

$$Q(s) = \frac{(Ls + R)q(0) + Li(0) + E(s)}{Ls^2 + Rs + 1/C} \tag{12}$$

and

$$I(s) = \frac{Lsi(0) + sE(s) - q(0)/C}{Ls^2 + Rs + 1/C}. \tag{13}$$

Note that $Q(s)$ and $I(s)$ depends on the initial current in the circuit, $i(0)$, and the initial charge on the capacitor, $q(0)$.

EXAMPLE 5

Assume that the solution of the following initial value problem satisfies the hypotheses of Corollary 5.2.3. Find its Laplace transform.

$$\frac{d^4 y}{dt^4} - y = 0, \tag{14}$$

$$y(0) = 0, \quad \frac{dy}{dt}(0) = 0, \quad \frac{d^2 y}{dt^2}(0) = 0, \quad \frac{d^3 y}{dt^3}(0) = 1. \tag{15}$$

Taking the Laplace transform of both sides of the differential equation (14) and using the linearity property of $\mathcal{L}$ yields

$$\mathcal{L}\left\{\frac{d^4 y}{dt^4}\right\} - \mathcal{L}\{y\} = 0. \tag{16}$$

The right-hand side is zero since the Laplace transform of the zero function in the t-domain is the zero function in the s-domain. Letting $Y = \mathcal{L}\{y\}$, applying formula (5) in Corollary 5.2.3 to the derivative term, and utilizing the initial conditions (15) gives

$$\mathcal{L}\left\{\frac{d^4 y}{dt^4}\right\} = s^4 Y(s) - s^3 \overbrace{y(0)}^{0} - s^2 \overbrace{\frac{dy}{dt}(0)}^{0} - s \overbrace{\frac{d^2 y}{dt^2}(0)}^{0} - \overbrace{\frac{d^3 y}{dt^3}(0)}^{1} = s^4 Y(s) - 1. \tag{17}$$

Using Eq. (17) in Eq. (16) yields the algebraic equation

$$s^4 Y(s) - 1 - Y(s) = 0,$$

the solution of which is

$$Y(s) = \frac{1}{s^4 - 1}. \tag{18}$$

▶ **Laplace Transform of $t^n f(t)$.**

THEOREM 5.2.4

Suppose that f is (i) piecewise continuous on any interval $0 \le t \le A$, and (ii) of exponential order with a as specified in Theorem 5.1.6. Then for any positive integer n

$$\mathcal{L}\{t^n f(t)\} = (-1)^n F^{(n)}(s), \qquad s > a. \tag{19}$$

Proof

$$F^{(n)}(s) = \frac{d^n}{ds^n} \int_0^\infty e^{-st} f(t)\, dt = \int_0^\infty \frac{d^n}{ds^n}(e^{-st}) f(t)\, dt$$

$$= \int_0^\infty (-t)^n e^{-st} f(t)\, dt = (-1)^n \int_0^\infty t^n e^{-st} f(t)\, dt$$

$$= (-1)^n \mathcal{L}\{t^n f(t)\}.$$

Thus

$$\mathcal{L}\{t^n f(t)\} = (-1)^n F^{(n)}(s).$$

The operation of interchanging the order of differentiation with respect to s and integration with respect to t is justified by a theorem from advanced calculus known as Leibniz's rule. ∎

COROLLARY 5.2.5

For any integer $n \ge 0$

$$\mathcal{L}\{t^n\} = \frac{n!}{s^{n+1}}, \qquad s > 0. \tag{20}$$

Proof

If $f(t) = 1$, $F(s) = \mathcal{L}\{f(t)\} = 1/s$ as shown in Example 4 of Section 5.1. Then

$$F^{(n)}(s) = \frac{d^n}{ds^n}\left(\frac{1}{s}\right) = \frac{(-1)^n n!}{s^{n+1}}.$$

The result (20) follows by setting $f(t) = 1$ and $F^{(n)}(s) = \dfrac{(-1)^n n!}{s^{n+1}}$ in Eq. (19). ∎

PROBLEMS

In each of Problems 1 through 10, find the Laplace transform of the given function. Assume that a and b are real numbers and n is a positive integer.

1. $f(t) = e^{-2t} \sin 4t$
2. $f(t) = e^{3t} \cos 2t$
3. $f(t) = t^3 - 4t^2 + 5$
4. $f(t) = t \cos t$
5. $f(t) = e^{-4t}(t^2 + 1)^2$
6. $f(t) = t^5 e^{2t}$
7. $f(t) = t^2 \sin bt$
8. $f(t) = t^n e^{at}$
9. $f(t) = t e^{at} \sin bt$
10. $f(t) = t e^{at} \cos bt$
11. (a) Let $F(s) = \mathcal{L}\{f(t)\}$ where $f(t)$ is piecewise continuous and of exponential order on $[0, \infty)$. Show that

$$\mathcal{L}\left\{ \int_0^t f(\tau)\, d\tau \right\} = \frac{1}{s} F(s). \tag{i}$$

Hint: Let $g_1(t) = \int_0^t f(t_1)\,dt_1$ and note that $g_1'(t) = f(t)$. Then use Theorem 5.2.2.

(b) Show that for $n \geq 2$,

$$\mathcal{L}\left\{\int_0^t \int_0^{t_n} \cdots \int_0^{t_2} f(t_1)\,dt_1 \cdots dt_n\right\} = \frac{1}{s^n}F(s). \quad \text{(ii)}$$

In each of Problems 12 through 21, transform the given initial value problem into an algebraic equation for $Y = \mathcal{L}\{y\}$ in the s-domain. Then find the Laplace transform of the solution of the initial value problem.

12. $y'' + 2y' - 2y = 0$,
$\quad y(0) = 1, \quad y'(0) = 1$

13. $9y'' + 12y' + 4y = 0$,
$\quad y(0) = 2, \quad y'(0) = -1$

14. $y'' + 4y' + 3y = 0$,
$\quad y(0) = 2, \quad y'(0) = -1$

15. $6y'' + 5y' + y = 0$,
$\quad y(0) = 4, \quad y'(0) = 0$

16. $y'' - 2y' + 2y = t^2 e^t + 4$,
$\quad y(0) = 1, \quad y'(0) = 1$

17. $y'' - 2y' - 3y = t^2 + 4$,
$\quad y(0) = 1, \quad y'(0) = 0$

18. $y'' + 4y = 3e^{-2t}\sin 2t$,
$\quad y(0) = 2, \quad y'(0) = -1$

19. $y'' + 2y' + 5y = t\cos 2t$,
$\quad y(0) = 1, \quad y'(0) = 0$

20. $y''' + y'' + y' + y = 0$,
$\quad y(0) = 1, \quad y'(0) = 0 \quad y''(0) = -2$

21. $y'''' - y = te^{-t}$, $\quad y(0) = 0, \quad y'(0) = 0$
$\quad y''(0) = 0 \quad y'''(0) = 1$

22. In Section 4.1, the differential equation for the charge on the capacitor of the RLC circuit shown in Figure 5.2.1 was shown to be

$$Lq'' + Rq' + \frac{1}{C}q = e(t). \quad \text{(iii)}$$

Take the Laplace transform of Eq. (iii) to derive Eq. (12) for $Q(s)$ and then use Eq. (9) to derive Eq. (13) for $I(s)$.

In each of Problems 23 through 27, find the Laplace transform $Y(s) = \mathcal{L}\{y\}$ of the solution of the given initial value problem. A method of determining the inverse transform is developed in Section 5.5.

23. $y'' + 4y = \begin{cases} 1, & 0 \leq t < \pi, \\ 0, & \pi \leq t < \infty; \end{cases}$
$\quad y(0) = 1, \quad y'(0) = 0$

24. $y'' + y = \begin{cases} t, & 0 \leq t < 1, \\ 0, & 1 \leq t < \infty; \end{cases}$
$\quad y(0) = 0, \quad y'(0) = 0$

25. $y'' + 4y = \begin{cases} t, & 0 \leq t < 1, \\ 1, & 1 \leq t < \infty; \end{cases}$
$\quad y(0) = 0, \quad y'(0) = 0$

26. A tank originally contains 100 gal of fresh water. Then water containing 1/2 lb of salt per gal is poured into the tank at a rate of 2 gal/min, and the mixture is allowed to leave at the same rate. After 10 minutes the salt water solution flowing into the tank suddenly switches to fresh water flowing in at a rate of 2 gal/min while the solution continues to leave the tank at the same rate. Find the Laplace transform of the amount of salt $y(t)$ in the tank.

27. A damped oscillator with mass m, damping constant γ, and spring constant k, is subjected to an external force $F(t) = F_0 t$ over the time interval $0 \leq t \leq T$. The external force is then removed at time T. Find the Laplace transform of the displacement $y(t)$ of the mass assuming that the oscillator is initially in the equilibrium state.

28. The Laplace transforms of certain functions can be found conveniently from their Taylor series expansions.

(a) Using the Taylor series for $\sin t$,

$$\sin t = \sum_{n=0}^{\infty} \frac{(-1)^n t^{2n+1}}{(2n+1)!},$$

and assuming that the Laplace transform of this series can be computed term by term, verify that

$$\mathcal{L}\{\sin t\} = \frac{1}{s^2 + 1}, \quad s > 1.$$

(b) Let

$$f(t) = \begin{cases} (\sin t)/t, & t \neq 0, \\ 1, & t = 0. \end{cases}$$

Find the Taylor series for f about $t = 0$. Assuming that the Laplace transform of this function can be computed term by term, verify that

$$\mathcal{L}\{f(t)\} = \arctan(1/s), \quad s > 1.$$

(c) The Bessel function of the first kind of order zero, J_0, has the Taylor series

$$J_0(t) = \sum_{n=0}^{\infty} \frac{(-1)^n t^{2n}}{2^{2n}(n!)^2}.$$

Assuming that the following Laplace transforms can be computed term by term, verify that

$$L\{J_0(t)\} = (s^2 + 1)^{-1/2}, \qquad s > 1,$$

and

$$L\{J_0(\sqrt{t})\} = s^{-1} e^{-1/4s}, \qquad s > 0.$$

29. For each of the following initial value problems, use Theorem 5.2.4 to find the differential equation satisfied by $Y(s) = L\{\phi(t)\}$, where $y = \phi(t)$ is the solution of the given initial value problem.

(a) $y'' - ty = 0; \qquad y(0) = 1, \quad y'(0) = 0$
(Airy's equation)

(b) $(1 - t^2)y'' - 2ty' + \alpha(\alpha + 1)y = 0;$
$y(0) = 0, y'(0) = 1$ (Legendre's equation)
Note that the differential equation for $Y(s)$ is of first order in part **(a)**, but of second order in part **(b)**. This is due to the fact that t appears at most to the first power in the equation of part **(a)**, whereas it appears to the second power in that of part **(b)**. This illustrates that the Laplace transform is not often useful in solving differential equations with variable coefficients, unless all the coefficients are at most linear functions of the independent variable.

5.3 The Inverse Laplace Transform

In the preceding section, we showed how the linearity of L and Corollary 5.2.3 provide the tools to convert linear differential equations with constant coefficients into algebraic equations in the s-domain. For example, applying these tools to the initial value problem in Example 3,

$$y'' + 2y' + 5y = e^{-t}, \qquad y(0) = 1, \quad y'(0) = -3, \tag{1}$$

led to the algebraic equation

$$(s^2 + 2s + 5)Y(s) - s + 1 = \frac{1}{s+1}. \tag{2}$$

The solution of Eq. (2),

$$Y(s) = \frac{s^2}{(s+1)(s^2 + 2s + 5)},$$

is presumably the Laplace transform of the solution of the initial value problem (1). Determining the function $y = \phi(t)$ corresponding to the transform $Y(s)$ is the the main difficulty in solving initial value problems by the transform technique. This problem is known as the inversion problem for the Laplace transform. The following theorem, which we state without proof, allows us to define the notion of inverse Laplace transform.

▶ Existence of the Inverse Transform

THEOREM 5.3.1

If $f(t)$ and $g(t)$ are piecewise continuous and of exponential order on $[0, \infty)$ and $F = G$ where $F = L\{f\}$ and $G = L\{g\}$, then $f(t) = g(t)$ at all points where both f and g are continuous. In particular, if f and g are continuous on $[0, \infty)$, then $f(t) = g(t)$ for all $t \in [0, \infty)$.

Remark. Thus if $F = G$ and f and g are piecewise continuous on $[0, \infty)$, we identify f with g even though they may differ at a finite number of points or even at an infinite number of points such as $t = 1, 2, 3, \ldots$ but they cannot be different over any interval of positive length on the

t-axis. For example, the following two functions, which have different values at $t = 1$,

$$f(t) = \begin{cases} 1, & 0 \le t < 1 \\ 0, & 1 \le t, \end{cases} \quad \text{and} \quad g(t) = \begin{cases} 1, & 0 \le t \le 1 \\ 0, & 1 < t, \end{cases}$$

both have the same Laplace transform, $(1 - e^{-s})/s$. An infinite number of functions can be constructed, each having $(1 - e^{-s})/s$ as its transform. However, in a practical sense these functions are essentially the same. This "lack of uniqueness" is relatively harmless and causes no problems in the application of Laplace transforms.

Theorem 5.3.1 justifies the following definition.

DEFINITION 5.3.2	If $f(t)$ is piecewise continuous and of exponential order on $[0, \infty)$ and $\mathcal{L}\{f(t)\} = F(s)$, then we call f the **inverse Laplace transform**[1] of F, and denote it by $$f = \mathcal{L}^{-1}\{F\}.$$

The fact that there is a one-to-one correspondence (essentially) between functions and their Laplace transforms suggests the compilation of a table, such as Table 5.3.1, giving the transforms of functions frequently encountered, and vice versa. The entries in the second column of Table 5.3.1 are the transforms of those in the first column. Perhaps more important, the functions in the first column are the inverse transforms of those in the second column. Thus, for example, if the transform of the solution of a differential equation is known, the solution itself can often be found merely by looking it up in the table. Some of the entries in Table 5.3.1 have been produced in examples and theorems, or appear as problems, in Sections 5.1 and 5.2. Others will be developed later in the chapter. The third column of the table indicates where the derivation of the given transforms may be found. Although Table 5.3.1 is sufficient for the examples and problems in this book, much larger tables are also available. Powerful computer algebra systems can also be used to find both Laplace transforms and inverse Laplace transforms. There also exist numerical algorithms for approximating inverse Laplace transforms that cannot be found using either a table or a computer algebra system.

EXAMPLE 1

Determine $\mathcal{L}^{-1}\{F\}$, where

a. $F(s) = \dfrac{4}{s^2 + 16}$,   b. $F(s) = \dfrac{6}{(s + 2)^4}$,   c. $F(s) = \dfrac{s + 1}{s^2 + 2s + 5}$.

(a) Using line 5 of Table 5.3.1 gives

$$\mathcal{L}^{-1}\left\{\frac{4}{s^2 + 16}\right\} = \mathcal{L}^{-1}\left\{\frac{4}{s^2 + 4^2}\right\} = \sin 4t.$$

---

[1] There is a general formula for the inverse Laplace transform that requires integration in the complex s-plane, but its use requires a knowledge of the theory of functions of a complex variable, and we do not consider it in this book. It is possible however, to develop many important properties of the Laplace transform, and to solve many interesting problems, without the use of complex variables.

TABLE 5.3.1	Elementary Laplace Transforms		
	$f(t) = \mathcal{L}^{-1}\{F(s)\}$	$F(s) = \mathcal{L}\{f(t)\}$	Notes
1.	1	$\dfrac{1}{s}, \quad s > 0$	Sec. 5.1; Ex. 4
2.	e^{at}	$\dfrac{1}{s-a}, \quad s > a$	Sec. 5.1; Ex. 5
3.	$t^n, \quad n = $ positive integer	$\dfrac{n!}{s^{n+1}}, \quad s > 0$	Sec. 5.2; Cor. 5.2.5
4.	$t^p, \quad p > -1$	$\dfrac{\Gamma(p+1)}{s^{p+1}}, \quad s > 0$	Sec. 5.1; Prob. 37
5.	$\sin at$	$\dfrac{a}{s^2 + a^2}, \quad s > 0$	Sec. 5.1; Ex. 7
6.	$\cos at$	$\dfrac{s}{s^2 + a^2}, \quad s > 0$	Sec. 5.1; Prob. 22
7.	$\sinh at$	$\dfrac{a}{s^2 - a^2}, \quad s > |a|$	Sec. 5.1; Prob. 19
8.	$\cosh at$	$\dfrac{s}{s^2 - a^2}, \quad s > |a|$	Sec. 5.1; Prob. 18
9.	$e^{at} \sin bt$	$\dfrac{b}{(s-a)^2 + b^2}, \quad s > a$	Sec. 5.1; Prob. 23
10.	$e^{at} \cos bt$	$\dfrac{s-a}{(s-a)^2 + b^2}, \quad s > a$	Sec. 5.1; Prob. 24
11.	$t^n e^{at}, n = $ positive integer	$\dfrac{n!}{(s-a)^{n+1}}, \quad s > a$	Sec. 5.2; Prob. 8
12.	$u_c(t)$	$\dfrac{e^{-cs}}{s}, \quad s > 0$	Sec. 5.5
13.	$u_c(t) f(t-c)$	$e^{-cs} F(s)$	Sec. 5.5
14.	$e^{ct} f(t)$	$F(s-c)$	Sec. 5.2
15.	$\displaystyle\int_0^t f(t-\tau)g(\tau)\, d\tau$	$F(s)G(s)$	Sec. 5.8
16.	$\delta(t-c)$	e^{-cs}	Sec. 5.7
17.	$f^{(n)}(t)$	$s^n F(s) - s^{n-1} f(0) -$ $\cdots - f^{(n-1)}(0)$	Sec. 5.2
18.	$t^n f(t)$	$(-1)^n F^{(n)}(s)$	Sec. 5.2; Thm. 5.2.4

(b) Using either line 11 or lines 3 and 14 of Table 5.3.1 we find that

$$\mathcal{L}^{-1}\left\{\frac{6}{(s+2)^4}\right\} = \mathcal{L}^{-1}\left\{\frac{3!}{s^4}\bigg|_{s \to s+2}\right\} = e^{-2t}\mathcal{L}^{-1}\left\{\frac{3!}{s^4}\right\} = e^{-2t}t^3.$$

(c) Completing the square in the denominator, $s^2 + 2s + 5 = (s+1)^2 + 4$, explicitly reveals the translation $s \to s + 1$. Then using either lines 6 and 14 or line 10 of Table 5.3.1 gives

$$\mathcal{L}^{-1}\left\{\frac{s+1}{s^2 + 2s + 5}\right\} = \mathcal{L}^{-1}\left\{\frac{s+1}{(s+1)^2 + 2^2}\right\} = \mathcal{L}^{-1}\left\{\frac{s}{s^2 + 2^2}\bigg|_{s \to s+1}\right\}$$

$$= e^{-t}\mathcal{L}^{-1}\left\{\frac{s}{s^2 + 2^2}\right\} = e^{-t}\cos 2t.$$

▶ **Linearity of $\mathcal{L}^{-1}$.** A table of Laplace transforms, such as Table 5.3.1, facilitates the task of finding the inverse Laplace transform of transforms that can easily be put into the form of those transforms of basic functions that appear in the table. However, it is not possible to create a table large enough to include all of the transforms that occur in applications. The next theorem states that $\mathcal{L}^{-1}$ is a linear operator. This property extends $\mathcal{L}^{-1}$ to all linear combinations of transforms that appear in the table.

THEOREM 5.3.3

Assume that $f_1 = \mathcal{L}^{-1}\{F_1\}$ and $f_2 = \mathcal{L}^{-1}\{F_2\}$ are piecewise continuous and of exponential order on $[0, \infty)$. Then for any constants c_1 and c_2,

$$\mathcal{L}^{-1}\{c_1 F_1 + c_2 F_2\} = c_1 \mathcal{L}^{-1}\{F_1\} + c_2 \mathcal{L}^{-1}\{F_2\} = c_1 f_1 + c_2 f_2.$$

Proof

Using the linearity of $\mathcal{L}$ we have

$$\mathcal{L}\{c_1 f_1 + c_2 f_2\} = c_1 F_1 + c_2 F_2.$$

Since $c_1 f_1 + c_2 f_2$ is piecewise continuous and of exponential order on $[0, \infty)$, the result follows from Definition 5.3.2. ∎

Remark. Equality in the conclusion of Theorem 5.3.3 is in the sense discussed in the remark immediately following Theorem 5.3.1.

Theorem 5.3.3 extends to a linear combination of n Laplace transforms

$$\mathcal{L}^{-1}\{c_1 F_1 + \cdots + c_n F_n\} = c_1 \mathcal{L}^{-1}\{F_1\} + \cdots + c_n \mathcal{L}^{-1}\{F_n\}.$$

EXAMPLE 2

Determine $\mathcal{L}^{-1}\left\{ \dfrac{2}{(s+2)^4} + \dfrac{3}{s^2+16} + \dfrac{5(s+1)}{s^2+2s+5} \right\}$.

Using the linearity of $\mathcal{L}^{-1}$ and the results of Example 1 gives

$$\mathcal{L}^{-1}\left\{ \frac{2}{(s+2)^4} + \frac{3}{s^2+16} + \frac{5(s+1)}{s^2+2s+5} \right\}$$

$$= \frac{1}{3}\mathcal{L}^{-1}\left\{ \frac{6}{(s+2)^4} \right\} + \frac{3}{4}\mathcal{L}^{-1}\left\{ \frac{4}{s^2+16} \right\} + 5\mathcal{L}^{-1}\left\{ \frac{s+1}{s^2+2s+5} \right\}$$

$$= \frac{1}{3}e^{-2t}t^3 + \frac{3}{4}\sin 4t + 5e^{-t}\cos 2t.$$

EXAMPLE 3

Find the inverse Laplace transform of $F(s) = \dfrac{3s+1}{s^2-4s+20}$.

By completing the square, the denominator can be written as $(s-2)^2 + 16$. This explicitly exposes the translation $s \to s-2$ in the s-domain. The same translation is introduced in the numerator by rewriting it as $3s + 1 = 3(s-2) + 7$. Thus

$$F(s) = \frac{3s+1}{s^2-4s+20} = \frac{3(s-2)+7}{(s-2)^2+16} = 3\frac{s-2}{(s-2)^2+4^2} + \frac{7}{4}\frac{4}{(s-2)^2+4^2}.$$

Using the linearity of $\mathcal{L}^{-1}$ and using lines 9 and 10 in Table 5.3.1, we find that

$$\mathcal{L}^{-1}\left\{\frac{3s+1}{s^2-4s+20}\right\}$$

$$= 3\mathcal{L}^{-1}\left\{\frac{s-2}{(s-2)^2+4^2}\right\} + \frac{7}{4}\mathcal{L}^{-1}\left\{\frac{4}{(s-2)^2+4^2}\right\}$$

$$= 3e^{2t}\cos 4t + \frac{7}{4}e^{2t}\sin 4t.$$

Partial Fractions

Most of the Laplace transforms that arise in the study of differential equations are rational functions of s, that is, functions of the form

$$F(s) = \frac{P(s)}{Q(s)} \qquad (3)$$

where $Q(s)$ is a polynomial of degree n and $P(s)$ is a polynomial of degree less than n. For example, the Laplace transform found in Example 3, Section 5.2,

$$Y(s) = \frac{s^2}{(s+1)(s^2+2s+5)}, \qquad (4)$$

is of the form (3) with $P(s) = s^2$ and $Q(s) = (s+1)(s^2+2s+5)$. Recall from calculus that the method of partial fractions involves the decomposition of rational functions into an equivalent sum of simpler rational functions which have numerators that are of degree one or zero and denominators of degree one or two, possibly raised to an integer power. The inverse Laplace transforms of these simpler rational functions can then often be found in a Laplace transform table.

EXAMPLE 4

The partial fraction expansion of the rational function (4),

$$Y(s) = \frac{s^2}{(s+1)(s^2+2s+5)} = \frac{1}{4}\frac{1}{s+1} + \frac{3}{4}\frac{s+1}{(s+1)^2+4} - \frac{2}{(s+1)^2+4},$$

can be directly verified by placing the terms on the right over the common denominator $(s+1)(s^2+2s+5) = (s+1)\left[(s+1)^2+4\right]$. Then, using the linearity of $\mathcal{L}^{-1}$ and lines 2, 9, and 10 in Table 5.3.1, we find that

$$\mathcal{L}^{-1}\{Y(s)\} = \mathcal{L}^{-1}\left\{\frac{s^2}{(s+1)(s^2+2s+5)}\right\}$$

$$= \frac{1}{4}\mathcal{L}^{-1}\left\{\frac{1}{s+1}\right\} + \frac{3}{4}\mathcal{L}^{-1}\left\{\frac{s+1}{(s+1)^2+4}\right\} - \mathcal{L}^{-1}\left\{\frac{2}{(s+1)^2+4}\right\}$$

$$= \frac{1}{4}e^{-t} + \frac{3}{4}e^{-t}\cos 2t - e^{-t}\sin 2t.$$

The last expression is the solution of the initial value problem posed in Example 3, Section 5.2, $y'' + 2y' + 5y = e^{-t}$, $y(0) = 1$, $y'(0) = -3$.

EXAMPLE 5

The partial fraction expansion of $Y(s)$ found in Example 5, Section 5.2,

$$Y(s) = \frac{1}{s^4 - 1} = \frac{1}{4}\frac{1}{s-1} - \frac{1}{4}\frac{1}{s+1} - \frac{1}{2}\frac{1}{s^2+1},$$

is verified by placing the terms on the right over a common denominator. Using the linearity of $\mathcal{L}^{-1}$ and lines 2 and 5 of Table 5.3.1, we then find that $\mathcal{L}^{-1}\{Y(s)\}$ is given by

$$\mathcal{L}^{-1}\left\{\frac{1}{s^4-1}\right\} = \frac{1}{4}\mathcal{L}^{-1}\left\{\frac{1}{s-1}\right\} - \frac{1}{4}\mathcal{L}^{-1}\left\{\frac{1}{s+1}\right\} - \frac{1}{2}\mathcal{L}^{-1}\left\{\frac{1}{s^2+1}\right\}$$

$$= \frac{1}{4}e^t - \frac{1}{4}e^{-t} - \frac{1}{2}\sin t.$$

Substitution shows that the last expression satisfies the initial value problem (14) and (15) in Example 5, Section 5.2.

We now review systematic methods for finding partial fraction decompositions of rational functions. Each expansion problem can be solved by considering the following three cases that arise.

1. **Nonrepeated Linear Factors**.
    If the denominator $Q(s)$ in Eq. (3) admits the factorization

$$Q(s) = (s - s_1)(s - s_2)\cdots(s - s_n)$$

where $s_1, s_2, \ldots, s_n$ are distinct, then $F(s)$ can be expanded as

$$F(s) = \frac{P(s)}{Q(s)} = \frac{a_1}{s - s_1} + \frac{a_2}{s - s_2} + \cdots + \frac{a_n}{s - s_n} \tag{5}$$

where the a_j are constants that need to be determined.

2. **Repeated Linear Factors**.
    If any root s_j of $Q(s)$ is of multiplicity k, that is, the factor $s - s_j$ appears exactly k times in the factorization of $Q(s)$, then the jth term in the right-hand side of Eq. (5) must be changed to

$$\frac{a_{j_1}}{s - s_j} + \frac{a_{j_2}}{(s - s_j)^2} + \cdots + \frac{a_{j_k}}{(s - s_j)^k} \tag{6}$$

where the constants $a_{j_1}, \ldots, a_{j_k}$ need to be determined.

3. **Quadratic Factors**.
    Complex conjugate roots of $Q(s)$, $\mu_j + i\nu_j$ and $\mu_j - i\nu_j$ where $\nu_j \neq 0$, give rise to quadratic factors of $Q(s)$ of the form

$$\left[s - (\mu_j + i\nu_j)\right]\left[s - (\mu_j - i\nu_j)\right] = (s - \mu_j)^2 + \nu_j^2.$$

If the roots $\mu_j + i\nu_j$ and $\mu_j - i\nu_j$ are of multiplicity k, that is, k is the highest power of $(s - \mu_j)^2 + \nu_j^2$ that divides $Q(s)$, then the partial fraction expansion of $F(s)$ must include the terms

$$\frac{a_{j_1}(s - \mu_j) + b_{j_1}\nu_j}{(s - \mu_j)^2 + \nu_j^2} + \frac{a_{j_2}(s - \mu_j) + b_{j_2}\nu_j}{\left[(s - \mu_j)^2 + \nu_j^2\right]^2} + \cdots + \frac{a_{j_k}(s - \mu_j) + b_{j_k}\nu_j}{\left[(s - \mu_j)^2 + \nu_j^2\right]^k} \tag{7}$$

where the constants $a_{j_1}, b_{j_1}, \ldots, a_{j_k}, b_{j_k}$ need to be determined.

In the following examples, we illustrate two methods commonly used to find linear systems of algebraic equations for the undetermined coefficients.

EXAMPLE 6

Find the partial fraction decomposition of $F(s) = (s-2)/(s^2 + 4s - 5)$ and compute $\mathcal{L}^{-1}\{F\}$.

Since $s^2 + 4s - 5 = (s-1)(s+5)$, we seek the partial fraction decomposition

$$\frac{s-2}{s^2+4s-5} = \frac{a_1}{s-1} + \frac{a_2}{s+5}. \tag{8}$$

The terms on the right-hand side of Eq. (8) can be combined by using the common denominator $(s-1)(s+5) = s^2 + 4s - 5$,

$$\frac{s-2}{s^2+4s-5} = \frac{a_1(s+5) + a_2(s-1)}{s^2+4s-5}. \tag{9}$$

Equality in (9) holds if and only if the numerators of the rational functions are equal,

$$s - 2 = a_1(s+5) + a_2(s-1). \tag{10}$$

Equation (10) is also easily obtained by multiplying both sides of Eq. (8) by $s^2 + 4s - 5 = (s+5)(s-1)$ and canceling those factors common to both numerator and denominator.

Matching Polynomial Coefficients. Since two polynomials are equal if and only if coefficients of corresponding powers of s are equal, we rewrite Eq. (10) in the form $s - 2 = (a_1 + a_2)s + 5a_1 - a_2$. Matching coefficients of $s^0 = 1$ and $s^1 = s$ yields the pair of equations

$$s^0 : -2 = 5a_1 - a_2$$
$$s^1 : 1 = a_1 + a_2.$$

The solution of this system is $a_1 = -1/6$ and $a_2 = 7/6$ so

$$\frac{s-2}{s^2+4s-5} = -\frac{1}{6}\frac{1}{s-1} + \frac{7}{6}\frac{1}{s+5}.$$

Matching Function Values. The functions of s that appear on both sides of Eq. (10) must be equal for all values of s. Since there are two unknowns to be determined, a_1 and a_2, we evaluate Eq. (10) at two different values of s to obtain a system of two independent equations for a_1 and a_2. Although there are many possibilities, choosing values of s corresponding to roots of $Q(s)$ often yields a particularly simple system,

$$s = 1 : -1 = 6a_1$$
$$s = -5 : -7 = -6a_2.$$

Thus, $a_1 = -1/6$ and $a_2 = 7/6$ as before. If we choose to evaluate Eq. (10) at $s = 0$ and $s = -1$ instead, we obtain the more complicated system of equations

$$s = 0 : -2 = 5a_1 - a_2$$
$$s = -1 : -3 = 4a_1 - 2a_2$$

with the same solution obtained above. It follows that

$$\mathcal{L}^{-1}\left\{\frac{s-2}{s^2+4s-5}\right\} = -\frac{1}{6}e^t + \frac{7}{6}e^{-5t}.$$

EXAMPLE 7

Determine $\mathcal{L}^{-1}\left\{\dfrac{s^2+20s+31}{(s+2)^2(s-3)}\right\}$.

Since $s+2$ is a linear factor of multiplicity two and $s-3$ is a linear factor of multiplicity one in the denominator, the appropriate form for the partial fraction decomposition is

$$\frac{s^2+20s+31}{(s+2)^2(s-3)} = \frac{a}{s+2}+\frac{b}{(s+2)^2}+\frac{c}{s-3}.$$

Multiplying both sides by $(s+2)^2(s-3)$ and canceling factors common to numerator and denominator yields the equation

$$s^2+20s+31 = a(s+2)(s-3)+b(s-3)+c(s+2)^2. \tag{11}$$

If we choose to match function values in Eq. (11) at $s=-2$, $s=3$ and $s=0$, we obtain the system

$$\begin{aligned}
s=-2: \quad -5 &= &-5b \\
s=3: \quad 100 &= & 25c \\
s=0: \quad 31 &= -6a &-3b &+4c,
\end{aligned}$$

which has the solution $b=1$, $c=4$, and $a=-3$. Thus

$$\frac{s^2+20s+31}{(s+2)^2(s-3)} = -\frac{3}{s+2}+\frac{1}{(s+2)^2}+\frac{4}{s-3}. \tag{12}$$

If Eq. (11) is written in the form

$$s^2+20s+31 = (a+c)s^2+(-a+b+4c)s-6a-3b+4c$$

and we match polynomial coefficients, we get the system

$$\begin{aligned}
s^0: \quad 31 &= -6a &-3b &+4c \\
s^1: \quad 20 &= -a &+b &+4c \\
s^2: \quad 1 &= a & &+c
\end{aligned}$$

which has the same solution, $a=-3$, $b=1$, $c=4$.

Applying $\mathcal{L}^{-1}$ to the partial fraction decomposition in Eq. (12) then yields

$$\mathcal{L}^{-1}\left\{\frac{s^2+20s+31}{(s+2)^2(s-3)}\right\} = -3\mathcal{L}^{-1}\left\{\frac{1}{s+2}\right\}+\mathcal{L}^{-1}\left\{\frac{1}{(s+2)^2}\right\}+4\mathcal{L}^{-1}\left\{\frac{1}{s-3}\right\}$$

$$= -3e^{-2t}+te^{-2t}+4e^{3t}.$$

EXAMPLE 8

Determine $\mathcal{L}^{-1}\left\{\dfrac{14s^2+70s+134}{(2s+1)(s^2+6s+13)}\right\}$.

Since the discriminant of the quadratic factor $s^2+6s+13$ is negative, its roots are complex so we complete the square, writing it as $(s+3)^2+2^2$. Thus we assume a partial fraction decomposition of the form

$$\frac{14s^2+70s+134}{(2s+1)(s^2+6s+13)} = \frac{a(s+3)+b\cdot 2}{(s+3)^2+2^2}+\frac{c}{2s+1}.$$

Using $a(s + 3) + b \cdot 2$ for the linear factor in the numerator on the right instead of $as + b$ anticipates table entries having the forms

$$\frac{(s - \beta)}{(s - \beta)^2 + \alpha^2} \quad \text{and} \quad \frac{\alpha}{(s - \beta)^2 + \alpha^2}$$

with $\alpha = 2$ and $\beta = -3$. Multiplying both sides by $(2s + 1)(s^2 + 6s + 13) = (2s + 1)((s + 3)^2 + 2^2)$ yields the equation

$$14s^2 + 70s + 134 = [a(s + 3) + b \cdot 2](2s + 1) + c\left[(s + 3)^2 + 2^2\right]. \tag{13}$$

Evaluating Eq. (13) at $s = 0$, $s = -3$, and $s = -1/2$ gives the system

$$
\begin{array}{rlll}
s = 0: & 134 & = 3a + & 2b + 13c \\
s = -3: & 50 & = & -10b + 4c \\
s = -1/2: & \frac{205}{2} & = & \frac{41}{4}c
\end{array}
$$

which has the solution $c = 10$, $b = -1$, $a = 2$. Thus

$$\frac{14s^2 + 70s + 134}{(2s + 1)(s^2 + 6s + 13)} = 2\frac{(s + 3)}{(s + 3)^2 + 2^2} - \frac{2}{(s + 3)^2 + 2^2} + \frac{10}{2s + 1}$$

so

$$\mathcal{L}^{-1}\left\{\frac{14s^2 + 70s + 134}{(2s + 1)(s^2 + 6s + 13)}\right\}$$

$$= 2\mathcal{L}^{-1}\left\{\frac{(s + 3)}{(s + 3)^2 + 2^2}\right\} - \mathcal{L}^{-1}\left\{\frac{2}{(s + 3)^2 + 2^2}\right\} + 5\mathcal{L}^{-1}\left\{\frac{1}{s + 1/2}\right\}$$

$$= 2e^{-3t}\cos 2t - e^{-3t}\sin 2t + 5e^{-t/2}.$$

Thus the appropriate form for the partial fraction decomposition of $P(s)/Q(s)$ is based on the linear and irreducible quadratic factors, counting multiplicities, of $Q(s)$ and is determined by the rules in the cases discussed above. In many textbook problems, where $Q(s)$ has rational roots for instance, a computer algebra system can be used to find inverse transforms directly without need of a partial fraction decomposition. In real world applications where the degree of $Q(s)$ is greater than or equal to three, computer algorithms are available that approximate partial fraction decompositions. A polynomial root finder can also be used to assist in finding a partial fraction decomposition of $P(s)/Q(s)$.

PROBLEMS

In each of Problems 1 through 8, find the unknown constants in the given partial fraction expansion.

1. $\dfrac{s - 18}{(s + 2)(s - 3)} = \dfrac{a}{s + 2} + \dfrac{b}{s - 3}$

2. $\dfrac{3s + 4}{(s + 2)^2} = \dfrac{a_1}{s + 2} + \dfrac{a_2}{(s + 2)^2}$

3. $\dfrac{-3s^2 + 32 - 14s}{(s + 4)(s^2 + 4)} = \dfrac{a}{s + 4} + \dfrac{bs + c \cdot 2}{s^2 + 4}$

4. $\dfrac{s^2 + 2s + 2}{s^3} = \dfrac{a_1}{s} + \dfrac{a_2}{s^2} + \dfrac{a_3}{s^3}$

5. $\dfrac{3s^2 - 8s + 5}{(s + 1)(s^2 - 2s + 5)} = \dfrac{a}{s + 1} + \dfrac{b(s - 1) + c \cdot 2}{(s - 1)^2 + 4}$

6. $\dfrac{-2s^3 - 8s^2 + 8s + 6}{(s + 3)(s + 1)s^2}$

$$= \dfrac{a_1}{s} + \dfrac{a_2}{s^2} + \dfrac{b}{s + 3} + \dfrac{c}{s + 1}$$

7. $\dfrac{8s^3 - 15s - s^5}{(s^2 + 1)^3} = \dfrac{a_1 s + b_1}{s^2 + 1} + \dfrac{a_2 s + b_2}{(s^2 + 1)^2}$
$\qquad\qquad\qquad\qquad + \dfrac{a_3 s + b_3}{(s^2 + 1)^3}$

8. $\dfrac{s^3 + 3s^2 + 3s + 1}{(s^2 + 2s + 5)^2} = \dfrac{a_1(s + 1) + b_1 \cdot 2}{(s + 1)^2 + 4}$
$\qquad\qquad\qquad + \dfrac{a_2(s + 1) + b_2 \cdot 2}{\left[(s + 1)^2 + 4\right]^2}$

In each of problems 9 through 24, use the linearity of $\mathcal{L}^{-1}$, partial fraction expansions, and Table 5.3.1 to find the inverse Laplace transform of the given function.

9. $\dfrac{3}{s^2 + 4}$

10. $\dfrac{4}{(s - 1)^3}$

11. $\dfrac{2}{s^2 + 3s - 4}$

12. $\dfrac{3s}{s^2 - s - 6}$

13. $\dfrac{2s + 2}{s^2 + 2s + 5}$

14. $\dfrac{2s - 3}{s^2 - 4}$

15. $\dfrac{2s + 1}{s^2 - 2s + 2}$

16. $\dfrac{8s^2 - 4s + 12}{s(s^2 + 4)}$

17. $\dfrac{1 - 2s}{s^2 + 4s + 5}$

18. $\dfrac{2s - 3}{s^2 + 2s + 10}$

19. $3 \dfrac{3s + 2}{(s - 2)(s + 2)(s + 1)}$

20. $2 \dfrac{s^3 - 2s^2 + 8}{s^2(s - 2)(s + 2)}$

21. $4 \dfrac{s^2 - 3s + 11}{(s^2 - 4s + 8)(s + 3)}$

22. $2 \dfrac{s^3 + 3s^2 + 4s + 3}{(s^2 + 1)(s^2 + 4)}$

23. $\dfrac{s^3 - 2s^2 - 6s - 6}{(s^2 + 2s + 2)s^2}$

24. $\dfrac{s^2 + 3}{(s^2 + 2s + 2)^2}$

In each of problems 25 through 28, use a computer algebra system to find the inverse Laplace transform of the given function. **CAS**

25. $\dfrac{s^3 - 2s^2 - 6s - 6}{s^4 + 4s^3 + 24s^2 + 40s + 100}$

26. $\dfrac{s^3 - 2s^2 - 6s - 6}{s^7 - 4s^6 + 5s^5}$

27. $\dfrac{s^3 - 2s^2 - 6s - 6}{s^8 - 2s^7 - 2s^6 + 16s^5 - 20s^4 - 8s^3 + 56s^2 - 64s + 32}$

28. $\dfrac{s^3 - 2s^2 - 6s - 6}{s^7 - 5s^6 + 5s^5 - 25s^4 + 115s^3 - 63s^2 + 135s - 675}$

5.4 Solving Differential Equations with Laplace Transforms

The block diagram in Figure 5.0.1 shows the main steps used to solve initial value problems by the method of Laplace transforms. The mathematical steps required to carry out each stage in the process, presented in Sections 5.1, 5.2, and 5.3, are summarized below.

1. Using the linearity of $\mathcal{L}$, its operational properties (for example, knowing how derivatives transform), and a table of Laplace transforms if necessary, the initial value problem for a linear constant coefficient differential equation is transformed into an algebraic equation in the s-domain.

2. Solving the algebraic equation gives the Laplace transform, say $Y(s)$, of the solution of the initial value problem. This step is illustrated by Examples 3, 4, and 5 in Section 5.2.

3. The solution of the initial value problem, $y(t) = \mathcal{L}^{-1}\{Y(s)\}$, is found by using partial fraction decompositions, the linearity of $\mathcal{L}^{-1}$, and a table of Laplace transforms. Partial fraction expansions and need for a table can be avoided by using a computer algebra system or other advanced computer software functions to evaluate $\mathcal{L}^{-1}\{Y(s)\}$.

In this section, we present examples ranging from first order equations to systems of equations that illustrate the entire process.

EXAMPLE 1

Find the solution of the initial value problem

$$y' + 2y = \sin 4t, \quad y(0) = 1. \tag{1}$$

Applying the Laplace transform to both sides of the differential equation gives

$$sY(s) - y(0) + 2Y(s) = \frac{4}{s^2 + 16}$$

where we have used line 17 of Table 5.3.1 to transform $y'(t)$. Substituting for $y(0)$ from the initial condition and solving for $Y(s)$, we obtain

$$Y(s) = \frac{1}{s+2} + \frac{4}{(s+2)(s^2+16)} = \frac{s^2 + 20}{(s+2)(s^2+16)}.$$

The partial fraction expansion of $Y(s)$ is of the form

$$\frac{s^2 + 20}{(s+2)(s^2+16)} = \frac{a}{s+2} + \frac{bs + c \cdot 4}{s^2 + 16}, \tag{2}$$

where the linear factor $bs + c \cdot 4$ in the numerator on the right anticipates a table entry of the form $\dfrac{\mu}{s^2 + \mu^2}$. Multiplying both sides of Eq. (2) by $(s+2)(s^2+16)$ gives the equation

$$s^2 + 20 = a(s^2 + 16) + (bs + c \cdot 4)(s + 2). \tag{3}$$

Evaluating Eq. (3) at $s = -2$, $s = 0$, and $s = 2$ yields the system

$$
\begin{aligned}
s = -2: \quad 24 &= 20a \\
s = 0: \quad 20 &= 16a &&+ 8c \\
s = 2: \quad 24 &= 20a + 8b + 16c.
\end{aligned}
\tag{4}
$$

The solution of the system (4) is $a = 6/5$, $b = -1/5$, and $c = 1/10$. Therefore the partial fraction expansion of Y is

$$Y(s) = \frac{6}{5}\frac{1}{s+2} - \frac{1}{5}\frac{s}{s^2+16} + \frac{1}{10}\frac{4}{s^2+16}.$$

It follows that the solution of the given initial value problem is

$$y(t) = \frac{6}{5}e^{-2t} - \frac{1}{5}\cos 4t + \frac{1}{10}\sin 4t.$$

EXAMPLE 2

Find the solution of the differential equation

$$y'' + y = e^{-t}\cos 2t, \tag{5}$$

satisfying the initial conditions

$$y(0) = 2, \qquad y'(0) = 1. \tag{6}$$

Taking the Laplace transform of the differential equation, we get

$$s^2 Y(s) - sy(0) - y'(0) + Y(s) = \frac{s+1}{(s+1)^2 + 4}.$$

Substituting for $y(0)$ and $y'(0)$ from the initial conditions and solving for $Y(s)$, we obtain

$$Y(s) = \frac{2s+1}{s^2+1} + \frac{s+1}{(s^2+1)\left[(s+1)^2 + 4\right]}. \tag{7}$$

Instead of using a common denominator to combine terms on the right side of (7) into a single rational function, we choose to compute their inverse Laplace transforms separately. The partial fraction expansion of $(2s + 1)/(s^2 + 1)$ can be written down by observation,

$$\frac{2s+1}{s^2+1} = 2\frac{s}{s^2+1} + \frac{1}{s^2+1}.$$

Thus

$$\mathcal{L}^{-1}\left\{\frac{2s+1}{s^2+1}\right\} = 2\mathcal{L}^{-1}\left\{\frac{s}{s^2+1}\right\} + \mathcal{L}^{-1}\left\{\frac{1}{s^2+1}\right\} = 2\cos t + \sin t. \tag{8}$$

The form of the partial fraction expansion for the second term on the right-hand side of (7) is

$$\frac{s+1}{(s^2+1)\left[(s+1)^2 + 4\right]} = \frac{as+b}{s^2+1} + \frac{c(s+1) + d\cdot 2}{(s+1)^2 + 4}. \tag{9}$$

Multiplying both sides of (9) by $(s^2 + 1)\left[(s+1)^2 + 4\right]$ and canceling factors common to denominator and numerator on the right yields

$$s + 1 = (as + b)\left[(s+1)^2 + 4\right] + [c(s+1) + d\cdot 2](s^2 + 1).$$

Expanding and collecting coefficients of like powers of s then gives the polynomial equation

$$s + 1 = (a + c)s^3 + (2a + b + c + 2d)s^2 + (5a + 2b + c)s + 5b + c + 2d.$$

Comparing coefficients of like powers of s, we have

$$
\begin{aligned}
s^0 : &\quad 5b + c + 2d = 1, \\
s^1 : &\quad 5a + 2b + c = 1, \\
s^2 : &\quad 2a + b + c + 2d = 0, \\
s^3 : &\quad a + c = 0.
\end{aligned}
$$

Consequently, $a = 1/10$, $b = 3/10$, $c = -1/10$, and $d = -1/5$, from which it follows that

$$
\frac{s+1}{(s^2+1)\left[(s+1)^2 + 4\right]}
$$

$$
= \frac{1}{10}\frac{s}{s^2+1} + \frac{3}{10}\frac{1}{s^2+1} - \frac{1}{10}\frac{s+1}{(s+1)^2 + 4} - \frac{1}{5}\frac{2}{(s+1)^2 + 4}.
$$

Thus

$$
\mathcal{L}^{-1}\left\{\frac{s+1}{(s^2+1)\left[(s+1)^2 + 4\right]}\right\}
$$

$$
= \frac{1}{10}\cos t + \frac{3}{10}\sin t - \frac{1}{10}e^{-t}\cos 2t - \frac{1}{5}e^{-t}\sin 2t. \tag{10}
$$

The solution of the given initial value problem follows by summing the results in Eqs. (8) and (10),

$$y(t) = \frac{21}{10}\cos t + \frac{13}{10}\sin t - \frac{1}{10}e^{-t}\cos 2t - \frac{1}{5}e^{-t}\sin 2t.$$

Characteristic Polynomials and Laplace Transforms of Differential Equations

Consider the general second order linear equation with constant coefficients

$$ay'' + by' + cy = f(t), \tag{11}$$

with initial conditions prescribed by

$$y(0) = y_0, \qquad y'(0) = y_1. \tag{12}$$

Taking the Laplace transform of Eq. (11), we obtain

$$a[s^2 Y(s) - sy(0) - y'(0)] + b[s Y(s) - y(0)] + cY(s) = F(s) \tag{13}$$

or

$$(as^2 + bs + c)Y(s) - (as + b)y(0) - ay'(0) = F(s), \tag{14}$$

where $F(s)$ is the transform of $f(t)$. By solving Eq. (14) for $Y(s)$, we find that

$$Y(s) = \frac{(as + b)y(0) + ay'(0)}{as^2 + bs + c} + \frac{F(s)}{as^2 + bs + c}. \tag{15}$$

Some important observations can be made about the steps leading from Eq. (11) to Eq. (15). First, observe that the polynomial $Z(s) = as^2 + bs + c$ that multiplies $Y(s)$ in Eq. (14) and subsequently appears in the denominator on the right side of Eq. (15) is precisely the characteristic polynomial associated with Eq. (11). Second, note that the coefficient $as + b$ of $y(0)$ in the numerator in the first term on the right-hand side of Eq. (15) can be obtained by canceling the constant c in $Z(s)$ and then dividing by s,

$$\frac{as^2 + bs + \cancel{c}}{s} = as + b.$$

Similarly, the coefficient a of $y'(0)$ in the numerator in the first term on the right side of Eq. (15) can be obtained by canceling the constant in $as + b$ and again dividing by s,

$$\frac{as + \cancel{b}}{s} = a.$$

Observance of this pattern allows us to pass directly from Eq. (11) to Eq. (15). Substituting for $y(0)$ and $y'(0)$ from the initial conditions then gives

$$Y(s) = \frac{(as + b)y_0 + ay_1}{as^2 + bs + c} + \frac{F(s)}{as^2 + bs + c}.$$

Since the use of a partial fraction expansion of $Y(s)$ to determine $y(t)$ requires us to factor the polynomial $Z(s) = as^2 + bs + c$, the use of Laplace transforms does not avoid the necessity of finding roots of the characteristic equation.

The pattern described above extends easily to nth order linear equations with constant coefficients,

$$a_n y^{(n)} + a_{n-1} y^{(n-1)} + \cdots + a_1 y' + a_0 y = f(t). \tag{16}$$

Since the characteristic polynomial associated with Eq. (16) is

$$Z(s) = a_n s^n + a_{n-1} s^{n-1} + \cdots + a_1 s + a_0,$$

the Laplace transform $Y(s) = \mathcal{L}\{y(t)\}$ is given by

$$Y(s) = \frac{[a_n s^{n-1} + \cdots + a_1] y(0) + \cdots + [a_n s + a_{n-1}] y^{(n-2)}(0) + a_n y^{(n-1)}(0)}{a_n s^n + a_{n-1} s^{n-1} + \cdots + a_1 s + a_0}$$

$$+ \frac{F(s)}{a_n s^n + a_{n-1} s^{n-1} + \cdots + a_1 s + a_0}.$$

EXAMPLE 3

Find the solution of the differential equation

$$y'''' + 2y'' + y = 0 \tag{17}$$

subject to the initial conditions

$$y(0) = 1, \quad y'(0) = -1, \quad y''(0) = 0, \quad y'''(0) = 2. \tag{18}$$

Since the characteristic polynomial associated with Eq. (17) is $s^4 + 2s^2 + 1$, $Y(s)$ is given by

$$Y(s) = \frac{(s^3 + 2s)y(0) + (s^2 + 2)y'(0) + sy''(0) + y'''(0)}{s^4 + 2s^2 + 1}. \tag{19}$$

Substituting the prescribed values for the initial conditions then gives

$$Y(s) = \frac{s^3 - s^2 + 2s}{s^4 + 2s^2 + 1} = \frac{s^3 - s^2 + 2s}{(s^2 + 1)^2}. \tag{20}$$

The form for the partial fraction expansion of $Y(s)$ is

$$\frac{s^3 - s^2 + 2s}{(s^2 + 1)^2} = \frac{as + b}{s^2 + 1} + \frac{cs + d}{(s^2 + 1)^2}. \tag{21}$$

Thus the undetermined coefficients must be selected to satisfy

$$s^3 - s^2 + 2s = as^3 + bs^2 + (a + c)s + b + d.$$

Equating coefficients of like powers of s gives $a = 1$, $b = -1$, $a + c = 2$, and $b + d = 0$. The latter two equations require that $c = 1$ and $d = 1$ so that

$$Y(s) = \frac{s - 1}{s^2 + 1} + \frac{s + 1}{(s^2 + 1)^2}. \tag{22}$$

It is clear that $\mathcal{L}^{-1}\{(s - 1)/(s^2 + 1)\} = \cos t - \sin t$ but the second term on the right requires some consideration. From line 18 in Table 5.3.1 we know that

$$\mathcal{L}\{t \sin t\} = -\frac{d}{ds}\frac{1}{s^2 + 1} = \frac{2s}{(s^2 + 1)^2}$$

so that

$$\mathcal{L}^{-1}\left\{\frac{s}{(s^2 + 1)^2}\right\} = \frac{1}{2}t \sin t.$$

Similarly,

$$\mathcal{L}\{t \cos t\} = -\frac{d}{ds}\frac{s}{s^2 + 1} = \frac{s^2 - 1}{(s^2 + 1)^2} = \frac{(s^2 + 1) - 2}{(s^2 + 1)^2} = \frac{1}{s^2 + 1} - 2\frac{1}{(s^2 + 1)^2}$$

so that

$$\mathcal{L}^{-1}\left\{\frac{1}{(s^2 + 1)^2}\right\} = \frac{1}{2}\sin t - \frac{1}{2}t \cos t.$$

It follows that the solution of the given initial value problem is

$$y(t) = \mathcal{L}^{-1}\left\{\frac{s - 1}{s^2 + 1}\right\} + \mathcal{L}^{-1}\left\{\frac{s + 1}{(s^2 + 1)^2}\right\} = \left(1 - \frac{1}{2}t\right)\cos t - \frac{1}{2}(1 - t)\sin t.$$

The same result can also be obtained by using a computer algebra system to invert the Laplace transform (22).

▶ **Laplace Transforms of Systems of Differential Equations** The Laplace transform method easily extends to systems of equations. Consider the initial value problem

$$y_1' = a_{11}y_1 + a_{12}y_2 + f_1(t), \quad y_1(0) = y_{10},$$

$$y_2' = a_{21}y_1 + a_{22}y_2 + f_2(t), \quad y_2(0) = y_{20}. \tag{23}$$

Taking the Laplace transform of each equation in the system (23) gives

$$sY_1 - y_1(0) = a_{11}Y_1 + a_{12}Y_2 + F_1(s),$$

$$sY_2 - y_2(0) = a_{21}Y_1 + a_{22}Y_2 + F_2(s). \tag{24}$$

The pair of equations (24) may be rewritten in the form

$$(s - a_{11})Y_1 - a_{12}Y_2 = y_{10} + F_1(s),$$

$$-a_{21}Y_1 + (s - a_{22})Y_2 = y_{20} + F_2(s), \tag{25}$$

where we have also substituted for the initial conditions specified in Eqs. (23). Using matrix notation, the algebraic system (25) is conveniently expressed by

$$(s\mathbf{I} - \mathbf{A})\mathbf{Y} = \mathbf{y}_0 + \mathbf{F}(s) \tag{26}$$

where

$$\mathbf{Y} = \begin{pmatrix} Y_1 \\ Y_2 \end{pmatrix}, \quad \mathbf{A} = \begin{pmatrix} a_{11} & a_{12} \\ a_{21} & a_{22} \end{pmatrix}, \quad \mathbf{y}_0 = \begin{pmatrix} y_{10} \\ y_{20} \end{pmatrix}, \quad \mathbf{F}(s) = \begin{pmatrix} F_1(s) \\ F_2(s) \end{pmatrix}$$

and $\mathbf{I}$ is the 2×2 identity matrix. It follows that

$$\mathbf{Y} = (s\mathbf{I} - \mathbf{A})^{-1}\mathbf{y}_0 + (s\mathbf{I} - \mathbf{A})^{-1}\mathbf{F}(s) \tag{27}$$

where, using the formula for the inverse of a 2×2 matrix given by Eq. (14) in Section 3.1,

$$(s\mathbf{I} - \mathbf{A})^{-1} = \frac{1}{|s\mathbf{I} - \mathbf{A}|} \begin{pmatrix} s - a_{22} & a_{12} \\ a_{21} & s - a_{11} \end{pmatrix}.$$

We note that $Z(s) = |s\mathbf{I} - \mathbf{A}| = s^2 - (a_{11} + a_{22})s + a_{11}a_{22} - a_{12}a_{21}$ is the characteristic polynomial of $\mathbf{A}$. Inverting each component of $\mathbf{Y}$ in Eq. (27) then yields the solution of Eqs. (23).

EXAMPLE 4

Use the Laplace transform to solve the system

$$y_1' = -3y_1 + 4y_2 + \sin t,$$
$$y_2' = -2y_1 + 3y_2 + t \tag{28}$$

subject to the initial conditions

$$y_1(0) = 0, \quad y_2(0) = 1. \tag{29}$$

Taking the Laplace transform of each equation in the system (28) and substituting the initial conditions (29) yields the pair of algebraic equations

$$sY_1 - 0 = -3Y_1 + 4Y_2 + 1/(s^2 + 1),$$
$$sY_2 - 1 = -2Y_1 + 3Y_2 + 1/s^2.$$

Rewriting this system in the matrix form (26) gives

$$\begin{pmatrix} s+3 & -4 \\ 2 & s-3 \end{pmatrix} \begin{pmatrix} Y_1 \\ Y_2 \end{pmatrix} = \begin{pmatrix} 0 \\ 1 \end{pmatrix} + \begin{pmatrix} 1/(s^2+1) \\ 1/s^2 \end{pmatrix} = \begin{pmatrix} 1/(s^2+1) \\ (s^2+1)/s^2 \end{pmatrix}. \tag{30}$$

The solution of Eq. (30) is

$$Y(s) = \frac{1}{s^2(s^2-1)(s^2+1)} \left(\frac{4s^4 + s^3 + 5s^2 + 4}{s^5 + 3s^4 + 2s^3 + 4s^2 + s + 3} \right). \tag{31}$$

Inverting the partial fraction decomposition of each component in the vector (31) then yields

$$y_1(t) = \frac{7}{2} e^t - 3 e^{-t} + \frac{3}{2} \sin t - \frac{1}{2} \cos t - 4t,$$

and

$$y_2(t) = \frac{7}{2} e^t - \frac{3}{2} e^{-t} + \sin t - 3t - 1.$$

A computer algebra system can be used to facilitate the calculations.

The method of Laplace transforms can also be used to solve coupled systems of mixed order equations without necessarily converting them to a system of first order equations. The primary requirement is that the equations have constant coefficients.

EXAMPLE 5

Use the Laplace transform to solve the system

$$x'' + y' + 2x = 0,$$
$$2x' - y' = \cos t, \tag{32}$$

subject to the initial conditions

$$x(0) = 0, \qquad x'(0) = 0, \qquad y(0) = 0. \tag{33}$$

Letting $X(s) = \mathcal{L}\{x(t)\}$ and $Y(s) = \mathcal{L}\{y(t)\}$ and taking the Laplace transform of each equation in the system (32) gives

$$s^2 X(s) - sx(0) - x'(0) + sY(s) - y(0) + 2X(s) = 0$$
$$2s X(s) - 2x(0) - s Y(s) + y(0) = \frac{s}{s^2+1}. \tag{34}$$

Substituting the prescribed initial values for $x(0)$, $x'(0)$, and $y(0)$, we write the resultant system in the form

$$(s^2 + 2)X + sY = 0$$
$$2s X - s Y = \frac{s}{s^2+1}. \tag{35}$$

Solving for X we find

$$X(s) = \frac{s}{(s^2+1)[(s+1)^2+1]}$$
$$= \frac{1}{5} \frac{s}{s^2+1} + \frac{2}{5} \frac{1}{s^2+1} - \frac{1}{5} \frac{s+1}{(s+1)^2+1} - \frac{3}{5} \frac{1}{(s+1)^2+1}$$

where the last expression is the partial fraction expansion for X. Thus

$$x(t) = \frac{1}{5} \cos t + \frac{2}{5} \sin t - \frac{1}{5} e^{-t} \cos t - \frac{3}{5} e^{-t} \sin t.$$

Solving the second equation in system (35) for Y gives $Y(s) = 2X(s) - 1/(s^2 + 1)$ so $y(t) = \mathcal{L}^{-1}\{Y(s)\} = 2\mathcal{L}^{-1}\{X(s)\} - \sin t = 2x(t) - \sin t$, that is,

$$y(t) = \frac{2}{5}\cos t - \frac{1}{5}\sin t - \frac{2}{5}e^{-t}\cos t - \frac{6}{5}e^{-t}\sin t.$$

PROBLEMS

In each of Problems 1 through 13, use the Laplace transform to solve the given initial value problem.

1. $y'' - y' - 6y = 0$; $y(0) = 1$, $y'(0) = -1$

2. $y'' + 3y' + 2y = t$; $y(0) = 1$, $y'(0) = 0$

3. $y'' - 2y' + 2y = 0$; $y(0) = 0$, $y'(0) = 1$

4. $y'' - 4y' + 4y = 0$; $y(0) = 1$, $y'(0) = 1$

5. $y'' - 2y' + 4y = 0$; $y(0) = 2$, $y'(0) = 0$

6. $y'' + 2y' + 5y = e^{-t}\sin 2t$;
   $y(0) = 2$, $y'(0) = -1$

7. $y'' + \omega^2 y = \cos 2t$, $\omega^2 \ne 4$;
   $y(0) = 1$, $y'(0) = 0$

8. $y'' - 2y' + 2y = \cos t$;
   $y(0) = 1$, $y'(0) = 0$

9. $y'' - 2y' + 2y = e^{-t}$;
   $y(0) = 0$, $y'(0) = 1$

10. $y'' + 2y' + y = 4e^{-t}$;
    $y(0) = 2$, $y'(0) = -1$

11. $y^{(4)} - 4y''' + 6y'' - 4y' + y = 0$; $y(0) = 0$,
    $y'(0) = 1$, $y''(0) = 0$, $y'''(0) = 1$

12. $y^{(4)} - y = 0$; $y(0) = 1$, $y'(0) = 0$,
    $y''(0) = 1$, $y'''(0) = 0$

13. $y^{(4)} - 4y = 0$; $y(0) = 1$, $y'(0) = 0$,
    $y''(0) = -2$, $y'''(0) = 0$

In each of Problems 14 through 19, use the Laplace transform to solve the given initial value problem.

14. $\mathbf{y}' = \begin{pmatrix} -3 & 4 \\ -2 & 3 \end{pmatrix}\mathbf{y}$, $\mathbf{y}(0) = \begin{pmatrix} 1 \\ 0 \end{pmatrix}$

15. $\mathbf{y}' = \begin{pmatrix} 5 & -2 \\ 6 & -2 \end{pmatrix}\mathbf{y}$, $\mathbf{y}(0) = \begin{pmatrix} 1 \\ 0 \end{pmatrix}$

16. $\mathbf{y}' = \begin{pmatrix} 4 & -4 \\ 5 & -4 \end{pmatrix}\mathbf{y}$, $\mathbf{y}(0) = \begin{pmatrix} 1 \\ 0 \end{pmatrix}$

17. $\mathbf{y}' = \begin{pmatrix} 0 & 1 \\ -1 & 0 \end{pmatrix}\mathbf{y}$, $\mathbf{y}(0) = \begin{pmatrix} 1 \\ 1 \end{pmatrix}$

18. $\mathbf{y}' = \begin{pmatrix} -4 & -1 \\ 1 & -2 \end{pmatrix}\mathbf{y}$, $\mathbf{y}(0) = \begin{pmatrix} 1 \\ 0 \end{pmatrix}$

19. $\mathbf{y}' = \begin{pmatrix} 5 & -7 \\ 7 & -9 \end{pmatrix}\mathbf{y}$, $\mathbf{y}(0) = \begin{pmatrix} 0 \\ 1 \end{pmatrix}$

In each of Problems 20 through 24, use a computer algebra system to assist in solving the given initial value problem by the method of Laplace transforms. CAS

20. $\mathbf{y}' = \begin{pmatrix} -4 & -1 \\ 1 & -2 \end{pmatrix}\mathbf{y} + \begin{pmatrix} e^t \\ \sin 2t \end{pmatrix}$,
    $\mathbf{y}(0) = \begin{pmatrix} 1 \\ 2 \end{pmatrix}$

21. $\mathbf{y}' = \begin{pmatrix} 5 & -1 \\ 1 & 3 \end{pmatrix}\mathbf{y} + \begin{pmatrix} 2e^{-t} \\ e^t \end{pmatrix}$, $\mathbf{y}(0) = \begin{pmatrix} -3 \\ 2 \end{pmatrix}$

22. $\mathbf{y}' = \begin{pmatrix} -1 & -5 \\ 1 & 3 \end{pmatrix}\mathbf{y} + \begin{pmatrix} 3 \\ 5\cos t \end{pmatrix}$,
    $\mathbf{y}(0) = \begin{pmatrix} 1 \\ -1 \end{pmatrix}$

23. $\mathbf{y}' = \begin{pmatrix} -2 & 1 \\ 1 & -2 \end{pmatrix}\mathbf{y} + \begin{pmatrix} 0 \\ \sin t \end{pmatrix}$, $\mathbf{y}(0) = \begin{pmatrix} 0 \\ 0 \end{pmatrix}$

24. $\mathbf{y}' = \begin{pmatrix} 0 & 1 & -1 \\ 1 & 0 & 1 \\ 1 & 1 & 0 \end{pmatrix}\mathbf{y} + \begin{pmatrix} 0 \\ -e^{-t} \\ e^t \end{pmatrix}$,

    $\mathbf{y}(0) = \begin{pmatrix} 1 \\ 2 \\ 3 \end{pmatrix}$

25 Use the Laplace transform to solve the system CAS
$$x'' - y'' + x - 4y = 0,$$
$$x' + y' = \cos t,$$
$$x(0) = 0 \quad x'(0) = 1, \quad y(0) = 0, \quad y'(0) = 2.$$

CAS 26 A radioactive substance R_1 having decay rate k_1 disintegrates into a second radioactive substance R_2 having decay rate $k_2 \neq k_1$. Substance R_2 disintegrates into R_3, which is stable. If $m_i(t)$ represents the mass of substance R_i at time t, $i = 1, 2, 3$, the applicable equations are

$$m_1' = -k_1 m_1$$
$$m_2' = k_1 m_1 - k_2 m_2$$
$$m_3' = k_2 m_2.$$

Use the Laplace transform to solve this system under the conditions

$$m_1(0) = m_0 \quad m_2(0) = 0, \quad m_3(0) = 0.$$

5.5 Discontinuous Functions and Periodic Functions

In Section 5.4, we presented the general procedure used to solve initial value problems by means of the Laplace transform. Some of the most interesting elementary applications of the transform method occur in the solution of linear differential equations with discontinuous, periodic, or impulsive forcing functions. Equations of this type frequently arise in the analysis of the flow of current in electric circuits or the vibrations of mechanical systems. In this section and the following ones, we develop some additional properties of the Laplace transform that are useful in the solution of such problems. Unless a specific statement is made to the contrary, all functions appearing below are assumed to be piecewise continuous and of exponential order, so that their Laplace transforms exist, at least for s sufficiently large.

► **The Unit Step Function**   To deal effectively with functions having jump discontinuities, it is very helpful to introduce a function known as the **unit step function** or **Heaviside function**. This function is defined by

$$u(t) = \begin{cases} 0, & t < 0, \\ 1, & t \geq 0. \end{cases} \tag{1}$$

In applications, the Heaviside function often represents a force, voltage, current, or signal that is turned on at time $t = 0$, and left on thereafter. Translations of the Heaviside function are used to turn such functions on at times other than 0. For a real number c, we define

$$u_c(t) = \begin{cases} 0, & t < c, \\ 1, & t \geq c. \end{cases} \tag{2}$$

The graph of $y = u_c(t)$ is shown at the top of Figure 5.5.1. It is often necessary to turn a signal or function on at time $t = c$ and then turn it off at time $t = d > c$. This can be accomplished by using an **indicator function**, $u_{cd}(t)$, for the interval $[c, d)$ defined by

$$u_{cd}(t) = u_c(t) - u_d(t) = \begin{cases} 0, & t < c \text{ or } t \geq d, \\ 1, & c \leq t < d. \end{cases} \tag{3}$$

The graph of $y = u_{cd}(t)$ is shown at the bottom of Figure 5.5.1. Note that $u_{0d}(t) = u_0(t) - u_d(t) = 1 - u_d(t)$ for $t \geq 0$ has a negative step at $t = d$ and is used to turn off, at time $t = d$, a function that is initially turned on at $t = 0$.

Remark. Note that the definition of $u_c(t)$ at the point where the jump discontinuity occurs is immaterial. In fact, $u_c(t)$ need not be defined at all at $t = c$. This is consistent with comments regarding piecewise continuous functions that appear in the remark immediately following Theorem 5.3.1. For example, we will use the model $\hat{f}(t) = 3u_{24}(t) = 3[u_2(t) - u_4(t)]$ to describe

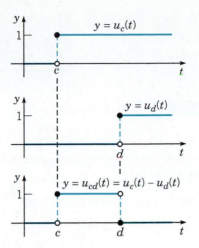

FIGURE 5.5.1 The function $u_{cd}(t) = u_c(t) - u_d(t)$ used to model the indicator function for the intervals $[c, d)$, $(c, d]$, (c, d), and $[c, d]$.

the function

$$f(t) = \begin{cases} 0, & t < 2, \\ 3, & 2 < t < 4, \\ 0, & 4 < t, \end{cases}$$

even though f is not defined at $t = 2$ and $t = 4$ while $\hat{f}(2) = 3$ and $\hat{f}(4) = 0$.

EXAMPLE 1

Use the unit step function to give a representation of the piecewise continuous function

$$f(t) = \begin{cases} t, & 0 < t < 2, \\ 1, & 2 \le t < 3, \\ e^{-2t}, & 3 \le t. \end{cases}$$

This can be accomplished by:

(i) turning on the function t at $t = 0$ and turning it off at $t = 2$,

(ii) turning on the constant function 1 at $t = 2$ and then turning it off at $t = 3$, and

(iii) turning on e^{-2t} at $t = 3$ and leaving it on.

Thus

$$f(t) = t u_{02}(t) + 1 u_{23}(t) + e^{-2t} u_3(t)$$

$$= t[1 - u_2(t)] + 1[u_2(t) - u_3(t)] + e^{-2t} u_3(t)$$

$$= t - (t - 1) u_2(t) + (e^{-2t} - 1) u_3(t), \ t > 0.$$

EXAMPLE 2

Use unit step functions to represent the function described by the graph in Figure 5.5.2.

The increasing portion of the triangular pulse is a straight line segment with slope 1 that passes through the point $(1, 0)$ in the ty-plane. It is therefore described by $t - 1$ for $1 \le t \le 2$.

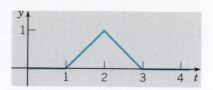

FIGURE 5.5.2 A triangular pulse.

The decreasing portion of the pulse is a straight line segment with slope -1 that passes through the point $(3, 0)$ in the ty-plane. It is described by $3 - t$ for $2 \le t \le 3$. Using indicator functions, we have

$$f(t) = (t - 1)u_{12}(t) + (3 - t)u_{23}(t)$$

$$= (t - 1)[u_1(t) - u_2(t)] + (3 - t)[u_2(t) - u_3(t)]$$

$$= (t - 1)u_1(t) - 2(t - 2)u_2(t) + (t - 3)u_3(t).$$

Using the definition of the unit step function and noting where the step function changes value, the last line can also be written in the form

$$f(t) = \begin{cases} 0, & t < 1, \\ t - 1, & 1 \le t < 2, \\ 3 - t, & 2 \le t < 3, \\ 0, & 3 \le t, \end{cases}$$

thus confirming the representation in terms of unit step functions.

▶ **The Laplace Transform of the Unit Step Function** The Laplace transform of u_c with $c \ge 0$ is

$$\mathcal{L}\{u_c(t)\} = \frac{e^{-cs}}{s}, \qquad s > 0, \tag{4}$$

since, for $s > 0$

$$\mathcal{L}\{u_c(t)\} = \int_0^\infty e^{-st} u_c(t)\, dt = \int_c^\infty e^{-st}\, dt$$

$$= \frac{e^{-cs}}{s}.$$

In particular, note that $\mathcal{L}\{u(t)\} = \mathcal{L}\{u_0(t)\} = 1/s$ which is identical to the Laplace transform of the function $f(t) = 1$. This is because $f(t) = 1$ and $u(t)$ are identical functions on $[0, \infty)$. For the indicator function $u_{cd}(t) = u_c(t) - u_d(t)$, the linearity of $\mathcal{L}$ gives

$$\mathcal{L}\{u_{cd}(t)\} = \mathcal{L}\{u_c(t)\} - \mathcal{L}\{u_d(t)\} = \frac{e^{-cs} - e^{-ds}}{s}, \qquad s > 0. \tag{5}$$

▶ **Laplace Transforms of Time-Shifted Functions** For a given function f defined for $t \ge 0$, we will often want to consider the related function g defined by

$$y = g(t) = \begin{cases} 0, & t < c, \\ f(t - c), & t \ge c, \end{cases}$$

which represents a translation of f a distance c in the positive t direction (see Figure 5.5.3). In terms of the unit step function we can write $g(t)$ in the convenient form

$$g(t) = u_c(t)f(t - c).$$

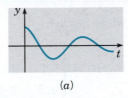
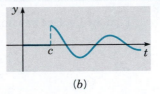

(a)   (b)

FIGURE 5.5.3   A translation of the given function. (a) $y = f(t)$; (b) $y = u_c(t)f(t - c)$.

The unit step function is particularly important in transform use because of the following relation between the transform of $f(t)$ and that of its translation $u_c(t)f(t - c)$.

THEOREM 5.5.1

If $F(s) = \mathcal{L}\{f(t)\}$ exists for $s > a \geq 0$, and if c is a positive constant, then

$$\mathcal{L}\{u_c(t)f(t - c)\} = e^{-cs}\mathcal{L}\{f(t)\} = e^{-cs}F(s), \qquad s > a. \qquad (6)$$

Conversely, if $f(t) = \mathcal{L}^{-1}\{F(s)\}$, then

$$u_c(t)f(t - c) = \mathcal{L}^{-1}\{e^{-cs}F(s)\}. \qquad (7)$$

Theorem 5.5.1 simply states that the translation of $f(t)$ a distance c in the positive t direction corresponds to the multiplication of $F(s)$ by e^{-cs}.

Proof

To prove Theorem 5.5.1, it is sufficient to compute the transform of $u_c(t)f(t - c)$:

$$\mathcal{L}\{u_c(t)f(t - c)\} = \int_0^\infty e^{-st}u_c(t)f(t - c)\, dt$$

$$= \int_c^\infty e^{-st}f(t - c)\, dt.$$

Introducing a new integration variable $\xi = t - c$, we have

$$\mathcal{L}\{u_c(t)f(t - c)\} = \int_0^\infty e^{-(\xi + c)s}f(\xi)\, d\xi = e^{-cs}\int_0^\infty e^{-s\xi}f(\xi)\, d\xi$$

$$= e^{-cs}F(s).$$

Thus Eq. (6) is established; Eq. (7) follows by taking the inverse transform of both sides of Eq. (6). ∎

A simple example of this theorem occurs if we take $f(t) = 1$. Recalling that $\mathcal{L}\{1\} = 1/s$, we immediately have $\mathcal{L}\{u_c(t)\} = e^{-cs}/s$ from Eq. (6). This result agrees with that of Eq. (4).

Remark. In order to use Theorem 5.5.1, the translations of u and f have to be identical, that is, their product must be of the form $f(t - c)u_c(t) = f(t - c)u(t - c)$. If, as is usually the case, a term involving a step function is of the form $g(t)u_c(t)$, a systematic way to deduce f is simply to set $f(t - c) = g(t)$ and then substitute $t + c$ for t in the resulting equation to obtain $f(t + c - c) = g(t + c)$ or $f(t) = g(t + c)$.

EXAMPLE 3

Find the Laplace transform of the function

$$f(t) = \begin{cases} t, & 0 < t < 2, \\ 1, & 2 \le t < 3, \\ e^{-2t}, & 3 \le t. \end{cases}$$

In Example 1, we computed the following representation of f in terms of step functions,

$$f(t) = t - (t - 1)u_2(t) + (e^{-2t} - 1)u_3(t), \ t > 0.$$

If we set $f_1(t) = t$, then $\mathcal{L}\{f_1(t)\} = 1/s^2$. Following the above remark, we set $f_2(t - 2) = t - 1$ and substitute $t + 2$ for t so that $f_2(t) = t + 1$, and $\mathcal{L}\{f_2(t)\} = 1/s^2 + 1/s = (1 + s)/s^2$. Similarly, set $f_3(t - 3) = e^{-2t} - 1$ and substitute $t + 3$ for t to obtain $f_3(t) = e^{-2(t+3)} - 1 = e^{-6}e^{-2t} - 1$. Thus $\mathcal{L}\{f_3(t)\} = e^{-6}/(s + 2) - 1/s$. Using the linearity of $\mathcal{L}$ and Theorem 5.5.1 gives

$$\mathcal{L}\{f(t)\} = \mathcal{L}\{f_1(t)\} - \mathcal{L}\{f_2(t - 2)u_2(t)\} + \mathcal{L}\{f_3(t - 3)u_2(t)\}$$

$$= \mathcal{L}\{f_1(t)\} - e^{-2s}\mathcal{L}\{f_2(t)\} + e^{-3s}\mathcal{L}\{f_3(t)\}$$

$$= \frac{1}{s^2} - e^{-2s}\frac{1 + s}{s^2} + e^{-3s}\left[\frac{e^{-6}}{s + 2} - \frac{1}{s}\right].$$

EXAMPLE 4

Find the Laplace transform of

$$f(t) = \begin{cases} 0, & t < 1, \\ t - 1, & 1 \le t < 2, \\ 3 - t, & 2 \le t < 3, \\ 0, & 3 \le t. \end{cases}$$

This is the triangular pulse of Example 2 which can be represented by

$$f(t) = (t - 1)u_1(t) - 2(t - 2)u_2(t) + (t - 3)u_3(t)$$

and is already nearly in the form to which Theorem 5.5.1 is directly applicable. If we set $f_1(t - 1) = t - 1$, $f_2(t - 2) = 2(t - 2)$, and $f_3 = t - 3$ we see that $f_1(t) = t$, $f_2(t) = 2t$, and $f_3(t) = t$. It follows that

$$\mathcal{L}\{f(t)\} = \frac{e^{-s}}{s^2} - \frac{2e^{-2s}}{s^2} + \frac{e^{-3s}}{s^2}.$$

EXAMPLE 5

Find the inverse transform of

$$F(s) = \frac{1 - e^{-2s}}{s^2}.$$

From the linearity of the inverse transform we have

$$f(t) = \mathcal{L}^{-1}\{F(s)\} = \mathcal{L}^{-1}\left\{\frac{1}{s^2}\right\} - \mathcal{L}^{-1}\left\{\frac{e^{-2s}}{s^2}\right\}$$

$$= t - u_2(t)(t-2).$$

The function f may also be written as

$$f(t) = \begin{cases} t, & 0 \le t < 2, \\ 2, & t \ge 2. \end{cases}$$

▶ **Periodic Functions**

DEFINITION 5.5.2	A function $f(t)$ is said to be **periodic with period** $T > 0$ if $$f(t+T) = f(t)$$ for all t in the domain of f.

A periodic function can be defined by indicating the length of its period and specifying its values over a single period. For example, the periodic function shown in Figure 5.5.4 can be expressed as

$$f(t) = \begin{cases} 1 - t, & 0 \le t < 1, \\ 0, & 1 \le t < 2, \end{cases} \qquad \text{and } f(t) \text{ has period 2.} \tag{8}$$

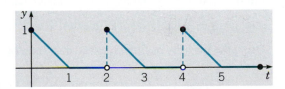

FIGURE 5.5.4 The periodic function described in Eq. (8).

More generally, in discussing a periodic function $f(t)$, it is convenient to introduce a **window function** $f_T(t)$ defined by

$$f_T(t) = f(t)[1 - u_T(t)] = \begin{cases} f(t), & 0 \le t \le T, \\ 0, & \text{otherwise,} \end{cases} \tag{9}$$

and its Laplace transform $F_T(s)$ given by

$$F_T(s) = \int_0^\infty e^{-st} f_T(t)\, dt = \int_0^T e^{-st} f(t)\, dt. \tag{10}$$

The window function specifies values of $f(t)$ over a single period. A replication of $f_T(t)$ shifted k periods to the right, that is, a distance kT, can be represented by

$$f_T(t - kT)u_{kT}(t) = \begin{cases} f(t - kT), & kT \le t < (k+1)T, \\ 0, & \text{otherwise.} \end{cases} \tag{11}$$

Summing the time shifted replications $f_T(t - kT)u_{kT}(t)$, $k = 0, \cdots, n - 1$, gives $f_{nT}(t)$, the periodic extension of $f_T(t)$ to the interval $[0, nT]$,

$$f_{nT}(t) = \sum_{k=0}^{n-1} f_T(t - kT)u_{kT}(t). \tag{12}$$

Thus $f_{nT}(t)$ consists of exactly n periods of the window function $f_T(t)$ as shown in Figure 5.5.5.

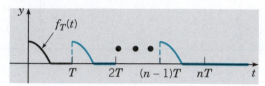

FIGURE 5.5.5   The periodic extension $f_{nT}(t)$ of the window function $f_T(t)$ to the interval $[0, nT]$, $f_{nT}(t) = \sum_{k=0}^{n-1} f_T(t - kT)u_{kT}(t)$.

The entire periodic function is then represented by

$$f(t) = \sum_{n=0}^{\infty} f_T(t - nT)u_{nT}(t). \tag{13}$$

THEOREM 5.5.3

If f is periodic with period T and is piecewise continuous on $[0, T]$, then

$$\mathcal{L}\{f(t)\} = \frac{F_T(s)}{1 - e^{-sT}} = \frac{\int_0^T e^{-st} f(t)\, dt}{1 - e^{-sT}}. \tag{14}$$

Proof

From Theorem 5.5.1 and Eq. (10), we know that for each $k \geq 0$,

$$\mathcal{L}\{f_T(t - kT)u_{kT}(t)\} = e^{-kTs}\mathcal{L}\{f_T(t)\} = e^{-kTs}F_T(s).$$

Using the linearity of $\mathcal{L}$ the Laplace transform of $f_{nT}(t)$ in (12) is therefore

$$F_{nT}(s) = \int_0^{nT} e^{-st} f(t)\, dt = \sum_{k=0}^{n-1} \mathcal{L}\{f_T(t - kT)u_{kT}(t)\}$$

$$= \sum_{k=0}^{n-1} e^{-kTs} F_T(s) = F_T(s) \sum_{k=0}^{n-1} \left(e^{-sT}\right)^k = F_T(s)\frac{1 - (e^{-sT})^n}{1 - e^{-sT}}, \tag{15}$$

where the last equality arises from the formula for the sum of the first n terms of a geometric series. Since $e^{-sT} < 1$ for $sT > 0$, it follows from (15) that

$$F(s) = \lim_{n \to \infty} \int_0^{nT} e^{-sT} f(t)\, dt = \lim_{n \to \infty} F_T(s)\frac{1 - (e^{-sT})^n}{1 - e^{-sT}} = \frac{F_T(s)}{1 - e^{-sT}}. \quad \blacksquare$$

EXAMPLE 6

Find the Laplace transform of the periodic function described by the graph in Figure 5.5.6. The sawtooth waveform can be expressed as

$$f(t) = \begin{cases} t, & 0 < t < 1, \\ 0, & 1 < t < 2, \end{cases} \qquad \text{and } f(t) \text{ has period 2.}$$

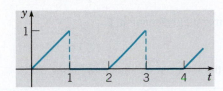

FIGURE 5.5.6   A periodic sawtooth signal.

Using $T = 2$, $F_T(s) = \displaystyle\int_0^2 e^{-st} f(t)\,dt = \int_0^1 e^{-st} t\,dt = \dfrac{1-e^{-s}}{s^2} - \dfrac{e^{-s}}{s}$, and Theorem 5.5.3 gives

$$F(s) = \frac{1-e^{-s}}{s^2(1-e^{-2s})} - \frac{e^{-s}}{s(1-e^{-2s})}.$$

EXAMPLE 7

Find the inverse Laplace transform of

$$F(s) = \frac{1-e^{-s}}{s(1-e^{-2s})}. \tag{16}$$

According to Eq. (14), the presence of $1 - e^{-2s}$ in the denominator suggests that the inverse transform is a periodic function of period 2 and that $F_2(s) = (1 - e^{-s})/s$. Since $\mathcal{L}^{-1}\{1/s\} = 1$ and $\mathcal{L}^{-1}\{e^{-s}/s\} = u_1(t)$ it follows that

$$f_2(t) = 1 - u_1(t) = \begin{cases} 1, & 0 \le t < 1, \\ 0, & 1 \le t < 2. \end{cases}$$

Hence,

$$f(t) = \begin{cases} 1, & 0 \le t < 1, \\ 0, & 1 \le t < 2, \end{cases} \qquad \text{and } f(t) \text{ has period 2.}$$

The graph of $f(t)$, a periodic square wave, is shown in Figure 5.5.7.

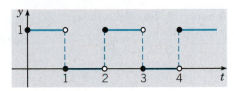

FIGURE 5.5.7   A periodic square wave.

Alternatively, note that $F(s)$ in Eq. (16) can be expressed as

$$F(s) = \frac{1}{s(1+e^{-s})} = \frac{1}{s}\left(1 - e^{-s} + e^{-2s} - e^{-3s} + \cdots\right),$$

where the last equality arises from using the geometric series representation of $1/(1 + e^{-s})$. A term-by-term inversion of this infinite series using the linearity of $\mathcal{L}^{-1}$ yields the time domain representation

$$f(t) = 1 - u_1(t) + u_2(t) - u_3(t) + \cdots$$

of the square wave shown in Figure 5.5.7.

PROBLEMS

In each of Problems 1 through 6, sketch the graph of the given function on the interval $t \geq 0$.

1. $u_1(t) + 2u_3(t) - 6u_4(t)$

2. $(t - 3)u_2(t) - (t - 2)u_3(t)$

3. $f(t - \pi)u_\pi(t)$,   where $f(t) = t^2$

4. $f(t - 3)u_3(t)$,   where $f(t) = \sin t$

5. $f(t - 1)u_2(t)$,   where $f(t) = 2t$

6. $(t - 1)u_1(t) - 2(t - 2)u_2(t) + (t - 3)u_3(t)$

In each of Problems 7 through 12, find the Laplace transform of the given function.

7. $f(t) = \begin{cases} 0, & t < 2 \\ (t - 2)^2, & t \geq 2 \end{cases}$

8. $f(t) = \begin{cases} 0, & t < 1 \\ t^2 - 2t + 2, & t \geq 1 \end{cases}$

9. $f(t) = \begin{cases} 0, & t < \pi \\ t - \pi, & \pi \leq t < 2\pi \\ 0, & t \geq 2\pi \end{cases}$

10. $f(t) = u_1(t) + 2u_3(t) - 6u_4(t)$

11. $f(t) = (t - 3)u_2(t) - (t - 2)u_3(t)$

12. $f(t) = t - u_1(t)(t - 1)$, $t \geq 0$

In each of Problems 13 through 18, find the inverse Laplace transform of the given function.

13. $F(s) = \dfrac{3!e^{-s}}{(s - 2)^4}$

14. $F(s) = \dfrac{e^{-2s}}{s^2 + s - 2}$

15. $F(s) = \dfrac{2(s - 1)e^{-2s}}{s^2 - 2s + 2}$

16. $F(s) = \dfrac{2e^{-2s}}{s^2 - 4}$

17. $F(s) = \dfrac{(s - 2)e^{-s}}{s^2 - 4s + 3}$

18. $F(s) = \dfrac{e^{-s} + e^{-2s} - e^{-3s} - e^{-4s}}{s}$

In each of Problems 19 through 21, find the Laplace transform of the given function.

19. $f(t) = \begin{cases} 1, & 0 \leq t < 1 \\ 0, & t \geq 1 \end{cases}$

20. $f(t) = \begin{cases} 1, & 0 \leq t < 1 \\ 0, & 1 \leq t < 2 \\ 1, & 2 \leq t < 3 \\ 0, & t \geq 3 \end{cases}$

21. $f(t) = 1 - u_1(t) + \cdots + u_{2n}(t) - u_{2n+1}(t) = 1 + \displaystyle\sum_{k=1}^{2n+1} (-1)^k u_k(t)$

In each of Problems 22 through 24, find the Laplace transform of the periodic function.

22. $f(t) = \begin{cases} 1, & 0 \leq t < 1, \\ -1, & 1 \leq t < 2; \end{cases}$ and $f(t)$ has period 2. See Figure 5.5.8.

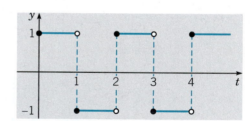

FIGURE 5.5.8   A periodic square wave.

23. $f(t) = t$, $0 \leq t < 1$; and $f(t)$ has period 1. See Figure 5.5.9.

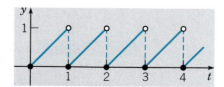

FIGURE 5.5.9   A sawtooth wave.

24. $f(t) = \sin t$, $0 \leq t < \pi$; and $f(t)$ has period π. See Figure 5.5.10.

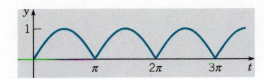

FIGURE 5.5.10 A rectified sine wave.

25. (a) If $f(t) = 1 - u_1(t)$, find $\mathcal{L}\{f(t)\}$; compare with Problem 19. Sketch the graph of $y = f(t)$.

(b) Let $g(t) = \displaystyle\int_0^t f(\xi)\, d\xi$, where the function f is defined in part **(a)**. Sketch the graph of $y = g(t)$ and find $\mathcal{L}\{g(t)\}$.

(c) Let $h(t) = g(t) - u_1(t)g(t-1)$, where g is defined in part **(b)**. Sketch the graph of $y = h(t)$ and find $\mathcal{L}\{h(t)\}$.

26. Consider the function p defined by

$$p(t) = \begin{cases} t, & 0 \le t < 1, \\ 2 - t, & 1 \le t < 2; \end{cases} \quad \text{and } p(t) \text{ has period 2.}$$

(a) Sketch the graph of $y = p(t)$.

(b) Find $\mathcal{L}\{p(t)\}$ by noting that p is the periodic extension of the function h in Problem 25(c) and then using the result of Theorem 5.5.3.

(c) Find $\mathcal{L}\{p(t)\}$ by noting that

$$p(t) = \int_0^t f(\xi)\, d\xi,$$

where f is the function in Problem 22, and then using Theorem 5.2.2.

5.6 Differential Equations with Discontinuous Forcing Functions

In this section, we turn our attention to some examples in which the nonhomogeneous term, or forcing function, of a differential equation is modeled by a discontinuous function. This is often done even if in the actual physical system the forcing function is continuous, but changes rapidly over a very short time interval. In many applications, systems are often tested by subjecting them to discontinuous forcing functions. For example, engineers frequently wish to know how a system responds to a step input, a common situation where the input to a system suddenly changes from one constant level to another.

While not immediately obvious, solutions of the constant coefficient equation

$$ay'' + by' + cy = g(t),$$

are in fact continuous whenever the input g is piecewise continuous. To understand how discontinuous inputs can yield continuous outputs, we consider the simple initial value problem

$$y''(t) = u_c(t), \qquad y(0) = 0, \quad y'(0) = 0 \tag{1}$$

where $c > 0$. Integrating both sides of Eq. (1) from 0 to t and using the initial condition $y'(0) = 0$ yields

$$y'(t) - y'(0) = \int_0^t u_c(\tau)\, d\tau = \begin{cases} 0, & 0 \le t < c, \\ t - c, & t \ge c, \end{cases} \tag{2}$$

or

$$y'(t) = (t - c)u_c(t). \tag{3}$$

Integrating both sides of Eq. (3) from 0 to t then yields

$$y(t) - y(0) = \int_0^t (\tau - c)u_c(\tau)\, d\tau = \begin{cases} 0, & 0 \le t < c, \\ \dfrac{(t - c)^2}{2}, & t \ge c, \end{cases} \tag{4}$$

or, since $y(0) = 0$,

$$y(t) = \frac{(t - c)^2}{2}\, u_c(t). \tag{5}$$

The graphs of $y''(t)$, $y'(t)$, and $y(t)$ are shown in Figure 5.6.1.

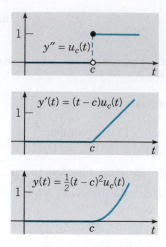

FIGURE 5.6.1   The smoothing effect of integration.

While $y''(t)$ has a jump discontinuity of magnitude 1 at $t = c$, $y'(t)$ is continuous at $t = c$ as can be seen from Eq. (3) where $\lim_{t \to c} y'(t) = 0$. From Eq. (5) we see that $y(t)$ is not only continuous for all values of $t > 0$ but its derivative $y'(t)$ exists and is continuous for all values of $t > 0$. Each integration yields a function that is smoother than the function being integrated. The smoothing effect of integration generalizes to variable coefficient equations and higher order equations. Consider

$$y'' + p(t)y' + q(t)y = g(t), \tag{6}$$

where p and q are continuous on some interval $\alpha < t < \beta$, but g is only piecewise continuous there. If $y = \psi(t)$ is a solution of Eq. (6), then ψ and ψ' are continuous on $\alpha < t < \beta$, but ψ'' has jump discontinuities at the same points as g. Similar remarks apply to higher order equations; the highest derivative of the solution appearing in the differential equation has jump discontinuities at the same points as the forcing function, but the solution itself and its lower order derivatives are continuous even at those points.

EXAMPLE
1

Describe the qualitative nature of the solution of the initial value problem

$$y'' + 4y = g(t), \tag{7}$$

$$y(0) = 0, \qquad y'(0) = 0, \tag{8}$$

where

$$g(t) = \begin{cases} 0, & 0 \le t < 5, \\ (t - 5)/5, & 5 \le t < 10, \\ 1, & t \ge 10, \end{cases} \tag{9}$$

and then find the solution.

In this example, the forcing function has the graph shown in Figure 5.6.2 and is known as ramp loading.

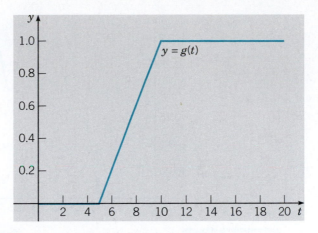

FIGURE 5.6.2   Ramp loading: $y = g(t)$ from Eq. (9).

It is relatively easy to identify the general form of the solution. For $t < 5$ the solution is simply $y = 0$. On the other hand, for $t > 10$ the solution has the form

$$y = c_1 \cos 2t + c_2 \sin 2t + 1/4. \tag{10}$$

The constant $1/4$ is a particular solution of the nonhomogeneous equation, while the other two terms are the general solution of the corresponding homogeneous equation. Thus the solution (10) is a simple harmonic oscillation about $y = 1/4$. Similarly, in the intermediate range $5 < t < 10$, the solution is an oscillation about a certain linear function. In an engineering context, for example, we might be interested in knowing the amplitude of the eventual steady oscillation.

To solve the problem it is convenient to write

$$g(t) = \left[u_5(t)(t-5) - u_{10}(t)(t-10) \right] /5, \tag{11}$$

as you may verify. Then we take the Laplace transform of the differential equation and use the initial conditions, thereby obtaining

$$(s^2 + 4)Y(s) = (e^{-5s} - e^{-10s})/5s^2,$$

or

$$Y(s) = (e^{-5s} - e^{-10s})H(s)/5, \tag{12}$$

where

$$H(s) = \frac{1}{s^2(s^2 + 4)}. \tag{13}$$

Thus the solution of the initial value problem (7), (8), (9) is

$$y = \phi(t) = \left[u_5(t)h(t-5) - u_{10}(t)h(t-10) \right] /5, \tag{14}$$

where $h(t)$ is the inverse transform of $H(s)$. The partial fraction expansion of $H(s)$ is

$$H(s) = \frac{1/4}{s^2} - \frac{1/4}{s^2 + 4}, \tag{15}$$

and it then follows from lines 3 and 5 of Table 5.3.1 that

$$h(t) = \frac{1}{4}t - \frac{1}{8}\sin 2t. \tag{16}$$

The graph of $y = \phi(t)$ is shown in Figure 5.6.3.

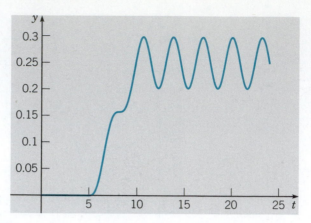

FIGURE 5.6.3   Solution of the initial value problem (7), (8), (9).

Observe that it has the qualitative form that we indicated earlier. To find the amplitude of the eventual steady oscillation, it is sufficient to locate one of the maximum or minimum points for $t > 10$. Setting the derivative of the solution (14) equal to zero, we find that the first maximum is located approximately at (10.642, 0.2979), so the amplitude of the oscillation is approximately 0.0479. Note that in this example the forcing function g is continuous but g' is discontinuous at $t = 5$ and $t = 10$. It follows that the solution ϕ and its first two derivatives are continuous everywhere, but ϕ''' has discontinuities at $t = 5$ and at $t = 10$ that match the discontinuities in g' at those points.

**EXAMPLE
2**

Resonance. Solve the initial value problem

$$y'' + \pi^2 y = f(t), \qquad y(0) = 0, \quad y'(0) = 0 \tag{17}$$

where $f(t)$ is the square wave in Example 7, Section 5.5.

Taking the Laplace transform of the differential equation in the initial value problem (17) and using the initial conditions gives

$$(s^2 + \pi^2)Y = \frac{1}{s\,(1 + e^{-s})}, \tag{18}$$

where we have used the representation $1/s\,(1 + e^{-s})$ for the Laplace transform of $f(t)$. Thus

$$Y(s) = \frac{1}{s\,(s^2 + \pi^2)}\,\frac{1}{1 + e^{-s}}.$$

Using the partial fraction expansion

$$\frac{1}{s\,(s^2 + \pi^2)} = \frac{1}{\pi^2}\left[\frac{1}{s} - \frac{s}{s^2 + \pi^2}\right]$$

gives

$$\mathcal{L}^{-1}\left\{\frac{1}{s\,(s^2 + \pi^2)}\right\} = \frac{1}{\pi^2}\left[1 - \cos \pi t\right].$$

Using the geometric series representation $1/(1 + e^{-s}) = \sum_{k=0}^{\infty}(-1)^k e^{-ks}$ we find that

$$y(t) = \frac{1}{\pi^2} \sum_{k=0}^{\infty} (-1)^k \left\{ 1 - \cos\left[\pi(t-k)\right] \right\} u_k(t). \tag{19}$$

Graphs of y and the square wave input are shown in Figure 5.6.4. The graph of y suggests that the system is exhibiting resonance, a fact that is not immediately apparent from the representation (19). However, the trigonometric identity $\cos\pi(t-k) = \cos\pi t \cos\pi k = (-1)^k \cos\pi t$ allows us to express the sum (19) as

$$y(t) = \frac{1}{\pi^2} \sum_{k=0}^{\infty} \left[(-1)^k - \cos\pi t\right] u_k(t). \tag{20}$$

If we evaluate Eq. (20) at the integers $t = n$, then using the facts that $u_k(n) = 0$ if $k > n$ and $\cos\pi n = (-1)^n$ we find that

$$y(n) = -\frac{n}{\pi^2} \quad \text{if } n \text{ is an even integer,}$$

and

$$y(n) = \frac{n+1}{\pi^2} \quad \text{if } n \text{ is an odd integer.}$$

The graphs of $y(n)$ versus n for $n = 1, 3, 5, \ldots$ and $y(n)$ versus n for $n = 0, 2, 4, \ldots$, form upper and lower envelopes for the graph of $y(t)$, $t \geq 0$ respectively and are shown as dashed lines in Figure 5.6.4. From the graphs of y and the square wave input f, we see that $f(t) = 1$, a force in the positive y direction, during time intervals when $y(t)$ is increasing and $f(t) = 0$ during time intervals when $y(t)$ is decreasing. Thus phase synchronization between input $f(t)$ and response $y(t)$ gives rise to the phenomenon of resonance.

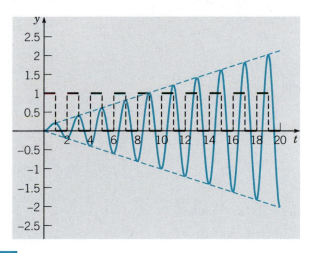

FIGURE 5.6.4   Resonance due to a square wave input.

PROBLEMS

ODEA In each of Problems 1 through 13, find the solution of the given initial value problem. Draw the graphs of the solution and of the forcing function; explain how they are related.

1. $y'' + y = f(t); \qquad y(0) = 0, \quad y'(0) = 1;$

$$f(t) = \begin{cases} 1, & 0 \leq t < \pi/2 \\ 0, & \pi/2 \leq t < \infty \end{cases}$$

2. $y'' + 2y' + 2y = h(t)$; $y(0) = 0$, $y'(0) = 1$;

$$h(t) = \begin{cases} 1, & \pi \le t < 2\pi \\ 0, & 0 \le t < \pi \quad \text{and} \quad t \ge 2\pi \end{cases}$$

3. $y'' + 4y = \sin t - u_{2\pi}(t) \sin(t - 2\pi)$;

   $y(0) = 0$, $y'(0) = 0$

4. $y'' + 4y = \sin t + u_\pi(t) \sin(t - \pi)$;

   $y(0) = 0$, $y'(0) = 0$

5. $y'' + 3y' + 2y = f(t)$; $y(0) = 0$, $y'(0) = 0$;

$$f(t) = \begin{cases} 1, & 0 \le t < 10 \\ 0, & t \ge 10 \end{cases}$$

6. $y'' + 3y' + 2y = u_2(t)$;

   $y(0) = 0$, $y'(0) = 1$

7. $y'' + y = u_{3\pi}(t)$;

   $y(0) = 1$, $y'(0) = 0$

8. $y'' + y' + 1.25y = t - u_{\pi/2}(t)(t - \pi/2)$;

   $y(0) = 0$, $y'(0) = 0$

9. $y'' + y = g(t)$; $y(0) = 0$, $y'(0) = 1$;

$$g(t) = \begin{cases} t/2, & 0 \le t < 6 \\ 3, & t \ge 6 \end{cases}$$

10. $y'' + y' + 1.25y = g(t)$;

    $y(0) = 0$, $y'(0) = 0$;

$$g(t) = \begin{cases} \sin t, & 0 \le t < \pi \\ 0, & t \ge \pi \end{cases}$$

11. $y'' + 4y = u_\pi(t) - u_{3\pi}(t)$;

    $y(0) = 0$, $y'(0) = 0$

12. $y^{(4)} - y = u_1(t) - u_2(t)$; $y(0) = 0$,

    $y'(0) = 0$, $y''(0) = 0$, $y'''(0) = 0$

13. $y^{(4)} + 5y'' + 4y = 1 - u_\pi(t)$; $y(0) = 0$,

    $y'(0) = 0$, $y''(0) = 0$, $y'''(0) = 0$

14. Find an expression involving $u_c(t)$ for a function f that ramps up from zero at $t = t_0$ to the value h at $t = t_0 + k$.

15. Find an expression involving $u_c(t)$ for a function g that ramps up from zero at $t = t_0$ to the value h at $t = t_0 + k$ and then ramps back down to zero at $t = t_0 + 2k$.

16. A certain spring–mass system satisfies the initial value problem

$$u'' + 0.25u' + u = kg(t), \quad u(0) = 0, \quad u'(0) = 0,$$

where $g(t)$ is of unit magnitude over the time interval $3/2 < t < 5/2$ and $k > 0$ is a parameter.

(a) Sketch the graph of $g(t)$.

(b) Solve the initial value problem.

(c) Plot the solution for $k = 1/2$, $k = 1$, and $k = 2$. ODEA
Describe the principal features of the solution and how they depend on k.

(d) Find, to two decimal places, the smallest value ODEA of k for which the solution $u(t)$ reaches the value 2.

(e) Suppose $k = 2$. Find the time τ after which ODEA $|u(t)| < 0.1$ for all $t > \tau$.

17. Modify the problem in Example 1 of this section by replacing the given forcing function $g(t)$ by

$$f(t) = \left[u_5(t)(t - 5) - u_{5+k}(t)(t - 5 - k) \right] / k.$$

(a) Sketch the graph of $f(t)$ and describe how it depends on k. For what value of k is $f(t)$ identical to $g(t)$ in the example?

(b) Solve the initial value problem

$$y'' + 4y = f(t), \qquad y(0) = 0, \quad y'(0) = 0.$$

(c) The solution in part (b) depends on k, but for ODEA sufficiently large t the solution is always a simple harmonic oscillation about $y = 1/4$. Try to decide how the amplitude of this eventual oscillation depends on k. Then confirm your conclusion by plotting the solution for a few different values of k.

18. Consider the initial value problem

$$y'' + \frac{1}{3}y' + 4y = f_k(t), \quad y(0) = 0, \quad y'(0) = 0,$$

where

$$f_k(t) = \begin{cases} 1/2k, & 4 - k \le t < 4 + k \\ 0, & 0 \le t < 4 - k \quad \text{and} \quad t \ge 4 + k \end{cases}$$

and $0 < k < 4$.

(a) Sketch the graph of $f_k(t)$. Observe that the area under the graph is independent of k. If $f_k(t)$ represents a force, this means that the product of the magnitude of the force and the time interval during which it acts does not depend on k.

(b) Write $f_k(t)$ in terms of the unit step function and then solve the given initial value problem.

(c) Plot the solution for $k = 2$, $k = 1$, and $k = 1/2$. Describe how the solution depends on k.

Resonance and Beats. In Section 4.7, we observed that an undamped harmonic oscillator (such as a spring–mass system) with a sinusoidal forcing term experiences resonance if the frequency of the forcing term is the same as the natural frequency. If the forcing frequency is slightly different from the natural frequency, then the system exhibits a beat. In Problems 19 through 23, we explore the effect of some nonsinusoidal periodic forcing functions.

19. Consider the initial value problem

$$y'' + y = f(t), \qquad y(0) = 0, \quad y'(0) = 0,$$

where

$$f(t) = 1 + 2\sum_{k=1}^{n}(-1)^k u_{k\pi}(t).$$

(a) Find the solution of the initial value problem.

(b) Let $n = 15$ and plot the graph of $f(t)$ and $y(t)$ for $0 \le t \le 60$ on the same set of coordinate axes. Describe the solution and explain why it behaves as it does.

(c) Investigate how the solution changes as n increases. What happens as $n \to \infty$?

20. Consider the initial value problem

$$y'' + 0.1y' + y = f(t), \quad y(0) = 0, \quad y'(0) = 0,$$

where $f(t)$ is the same as in Problem 19.

(a) Find the solution of the initial value problem.

(b) Plot the graph of $f(t)$ and $y(t)$ on the same set of coordinate axes. Use a large enough value of n and a long enough t-interval so that the transient part of the solution has become negligible and the steady state is clearly shown.

(c) Estimate the amplitude and frequency of the steady-state part of the solution.

(d) Compare the results of part (b) with those from Section 4.7 for a sinusoidally forced oscillator.

21. Consider the initial value problem

$$y'' + y = g(t), \qquad y(0) = 0, \quad y'(0) = 0,$$

where

$$g(t) = 1 + \sum_{k=1}^{n}(-1)^k u_{k\pi}(t).$$

(a) Draw the graph of $g(t)$ on an interval such as $0 \le t \le 6\pi$. Compare the graph with that of $f(t)$ in Problem 19(a).

(b) Find the solution of the initial value problem.

(c) Let $n = 15$ and plot the graph of the solution for $0 \le t \le 60$. Describe the solution and explain why it behaves as it does. Compare it with the solution of Problem 19.

(d) Investigate how the solution changes as n increases. What happens as $n \to \infty$?

22. Consider the initial value problem

$$y'' + 0.1y' + y = g(t), \quad y(0) = 0, \quad y'(0) = 0,$$

where $g(t)$ is the same as in Problem 21.

(a) Plot the graph of the solution. Use a large enough value of n and a long enough t-interval so that the transient part of the solution has become negligible and the steady state is clearly shown.

(b) Estimate the amplitude and frequency of the steady-state part of the solution.

(c) Compare the results of part (b) with those from Problem 20 and from Section 4.7 for a sinusoidally forced oscillator.

23. Consider the initial value problem

$$y'' + y = f(t), \qquad y(0) = 0, \quad y'(0) = 0,$$

where

$$f(t) = 1 + 2\sum_{k=1}^{n}(-1)^k u_{11k/4}(t).$$

Observe that this problem is identical to Problem 19 except that the frequency of the forcing term has been increased somewhat.

(a) Find the solution of this initial value problem.

(b) Let $n \ge 33$ and plot $y(t)$ and $f(t)$ for $0 \le t \le 90$ or longer on the same set of coordinate axes. Your plot should show a clearly recognizable beat.

(c) From the graph in part (b) estimate the "slow period" and the "fast period" for this oscillator.

(d) For a sinusoidally forced oscillator, it was shown in Section 4.7 that the "slow frequency" is given by $|\omega - \omega_0|/2$, where ω_0 is the natural frequency of the system and ω is the forcing frequency. Similarly, the "fast frequency" is $(\omega + \omega_0)/2$. Use these expressions to calculate the "fast period" and the "slow period" for the oscillator in this problem. How well do the results compare with your estimates from part (c)?

5.7 Impulse Functions

In some applications, it is necessary to deal with phenomena of an impulsive nature—for example, voltages or forces of large magnitude that act over very short time intervals. Such problems often lead to differential equations of the form

$$ay'' + by' + cy = g(t), \tag{1}$$

where $g(t)$ is large during a short time interval $t_0 \leq t < t_0 + \epsilon$ and is otherwise zero.

The integral $I(\epsilon)$, defined by

$$I(\epsilon) = \int_{t_0}^{t_0+\epsilon} g(t)\, dt, \tag{2}$$

or, since $g(t) = 0$ outside of the interval $[t_0, t_0 + \epsilon)$,

$$I(\epsilon) = \int_{-\infty}^{\infty} g(t)\, dt, \tag{3}$$

is a measure of the strength of the input function. For example, the initial value problem for a mass-spring system at equilibrium during $0 \leq t < t_0$, suddenly set into motion by striking the mass with a hammer at time t_0, is

$$my'' + \gamma y' + ky = g(t), \qquad y(0) = 0, \quad y'(0) = 0, \tag{4}$$

where the nonhomogeneous term $g(t)$ represents a force of large magnitude and short duration as shown in Figure 5.7.1. However, its detailed behavior may be difficult or impossible to ascertain.

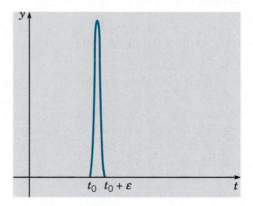

FIGURE 5.7.1   A force of large magnitude active during the short time interval $[t_0, t_0 + \epsilon)$.

In applications where $g(t)$ represents a force, $I(\epsilon)$ is referred to as the total **impulse** of the force over the time interval $[t_0, t_0 + \epsilon)$ and has the physical units of momentum (force $\times$ time). Prior to time t_0 the mass has zero momentum, but during the time interval $[t_0, t_0 + \epsilon)$ an amount of momentum equal to $I(\epsilon)$ is transferred to the mass. The idea that we wish to quantify here is that the dominant contribution to the system response $y(t)$ for times $t \geq t_0 + \epsilon$ is primarily determined by the magnitude of the impulse $I(\epsilon)$ rather than the detailed behavior of the forcing function $g(t)$. This is illustrated by the following example.

EXAMPLE 1

Let I_0 be a real number and let ε be a small positive constant. Suppose that t_0 is zero and that $g(t)$ is given by $g(t) = I_0 \delta_\epsilon(t)$ where

$$\delta_\epsilon(t) = \frac{1 - u_\epsilon(t)}{\epsilon} = \begin{cases} \dfrac{1}{\epsilon}, & 0 \le t < \epsilon, \\[2mm] 0, & t < 0 \quad \text{or} \quad t \ge \epsilon. \end{cases} \tag{5}$$

Graphs of δ_ϵ for various small values of ε are shown in Figure 5.7.2.

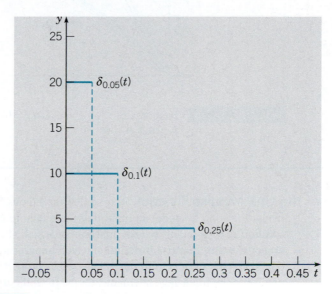

FIGURE 5.7.2 Graphs of $\delta_\epsilon(t)$ for $\epsilon = .25, .1,$ and $.05$.

Since the area beneath the graph of $\delta_\epsilon(t)$ is equal to 1 for every $\epsilon \ne 0$, Eq. (3) implies that $I(\epsilon) = I_0$ for any $\epsilon \ne 0$. Now consider the system described by the initial value problem

$$y'' + y = I_0 \delta_\epsilon(t), \qquad y(0) = 0, \quad y'(0) = 0. \tag{6}$$

The response, easily obtained by using Laplace transforms, is

$$y_\epsilon(t) = \frac{I_0}{\epsilon} \{1 - \cos(t) - u_\epsilon(t)[1 - \cos(t - \epsilon)]\}$$

$$= \begin{cases} \dfrac{I_0}{\epsilon}[1 - \cos(t)], & 0 \le t < \epsilon, \\[3mm] \dfrac{I_0}{\epsilon}[\cos(t - \epsilon) - \cos(t)], & t \ge \epsilon. \end{cases} \tag{7}$$

For any $t > 0$, $y_0(t) = \lim_{\epsilon \to 0} y_\epsilon(t)$ can be determined by using L'Hospital's rule,

$$y_0(t) = \lim_{\epsilon \to 0} I_0 \frac{\cos(t - \epsilon) - \cos(t)}{\epsilon} = I_0 \lim_{\epsilon \to 0} \sin(t - \epsilon) = I_0 \sin t. \tag{8}$$

The graphs of y_ϵ for $I_0 = 1, 2$ and $\epsilon = 0.2, 0.1, 0$ in Figure 5.7.3 show that the response is relatively insensitive to variation in small values of ε. Figure 5.7.3 also illustrates that the amplitude of the response is determined by I_0, a fact readily apparent in the form of expressions (7) and (8).

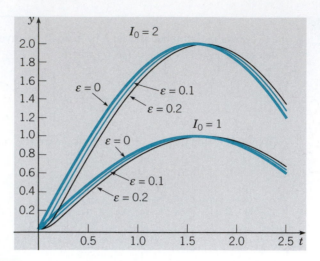

FIGURE 5.7.3   Response of $y'' + y = I_0\delta_\epsilon(t)$, $y(0) = 0$, $y'(0) = 0$ for $I_0 = 1, 2$, and $\epsilon = 0.2, 0.1, 0$.

▶ **Definition of the Unit Impulse Function**   It is somewhat tedious first to solve an initial value problem using $\delta_\epsilon(t - t_0)$ to model an input pulse and then to compute the limiting behavior of the solution as $\epsilon \to 0$. This type of analysis can usually be avoided by introducing an idealized object, the **unit impulse function** δ, which imparts an impulse of magnitude one at $t = t_0$, but is zero for all values of t other than t_0. The properties that define the "function" δ, made precise in the following statements, derive from the limiting behavior of δ_ϵ as $\epsilon \to 0$:

1.

$$\delta(t - t_0) = \lim_{\epsilon \to 0} \delta_\epsilon(t - t_0) = 0, \qquad t \neq t_0, \tag{9}$$

2. for any function continuous on an interval $a \le t_0 < b$ containing t_0,

$$\int_a^b f(t)\delta(t - t_0)dt = \lim_{\epsilon \to 0} \int_a^b f(t)\delta_\epsilon(t - t_0)\,dt = f(t_0). \tag{10}$$

The first property is due to the fact that the interval on which $\delta_\epsilon(t - t_0)$ is nonzero is $[t_0, t_0 + \epsilon)$ so any $t \neq t_0$ must lie outside the interval for sufficiently small values of ϵ. To understand the second property, note that if $\epsilon < b - t_0$, $[t_0, t_0 + \epsilon) \subset [a, b)$ and therefore

$$\int_a^b f(t)\delta_\epsilon(t - t_0)\,dt = \frac{1}{\epsilon} \int_{t_0}^{t_0+\epsilon} f(t)\,dt = \frac{1}{\epsilon} \cdot \epsilon \cdot f(t^*) = f(t^*)$$

where the second equality follows from the mean value theorem for integrals. Since $t_0 \le t^* \le t_0 + \epsilon$, $t^* \to t_0$ as $\epsilon \to 0$. Consequently $f(t^*) \to f(t_0)$ as $\epsilon \to 0$ since f is continuous at t_0. Note that if $f(t) = 1$, Eqs. (9) and (10) imply that

$$\int_a^b \delta\,(t - t_0)\,dt = \begin{cases} 1, & \text{if } a \le t_0 < b, \\ 0, & \text{if } t_0 \notin [a, b). \end{cases}$$

There is no ordinary function of the kind studied in elementary calculus that satisfies both of Eqs. (9) and (10). The "function" δ, defined by those equations, is an example of what are known as generalized functions; it is usually called the Dirac **delta function**.

▶ The Laplace Transform of $\delta(t - t_0)$ Since e^{-st} is a continuous function of t for all t and s, property (10) gives

$$\mathcal{L}\{\delta(t - t_0)\} = \int_0^\infty e^{-st}\delta(t - t_0)\,dt = e^{-st_0} \tag{11}$$

for all $t_0 \geq 0$. In the important case when $t_0 = 0$,

$$\mathcal{L}\{\delta(t)\} = \int_0^\infty e^{-st}\delta(t)\,dt = 1. \tag{12}$$

It is often convenient to introduce the delta function when working with impulse problems and to operate formally on it as though it were a function of the ordinary kind. This is illustrated in the examples below. It is important to realize, however, that the ultimate justification of such procedures must rest on a careful analysis of the limiting operations involved. Such a rigorous mathematical theory has been developed, but we do not discuss it here.

EXAMPLE 2

In the initial value problem (6) of Example 1, we replace the model of the input pulse δ_ϵ by δ,
$$y'' + y = I_0\delta(t), \qquad y(0) = 0, \quad y'(0) = 0.$$

Since $\mathcal{L}\{\delta(t)\} = 1$, Laplace transforming the initial value problem gives $(s^2 + 1)Y = I_0$ and therefore $Y(s) = I_0/(s^2 + 1)$. Thus, $y(t) = \mathcal{L}^{-1}\{Y(s\} = I_0 \sin t$ in agreement with $y_0(t) = \lim\limits_{\epsilon \to 0} y_\epsilon(t) = I_0 \sin t$ of Example 1.

EXAMPLE 3

Find the solution of the initial value problem

$$2y'' + y' + 2y = \delta(t - 5), \tag{13}$$

$$y(0) = 0, \qquad y'(0) = 0. \tag{14}$$

To solve the given problem, we take the Laplace transform of the differential equation and use the initial conditions, obtaining

$$(2s^2 + s + 2)Y(s) = e^{-5s}.$$

Thus

$$Y(s) = \frac{e^{-5s}}{2s^2 + s + 2} = \frac{e^{-5s}}{2}\,\frac{1}{\left(s + \dfrac{1}{4}\right)^2 + \dfrac{15}{16}}. \tag{15}$$

By Theorem 5.2.1 or from line 9 of Table 5.3.1,

$$\mathcal{L}^{-1}\left\{\frac{1}{\left(s+\dfrac{1}{4}\right)^2+\dfrac{15}{16}}\right\} = \frac{4}{\sqrt{15}}\,e^{-t/4}\sin\frac{\sqrt{15}}{4}\,t. \tag{16}$$

Hence, by Theorem 5.5.1, we have

$$y = \mathcal{L}^{-1}\{Y(s)\} = \frac{2}{\sqrt{15}}\,u_5(t)e^{-(t-5)/4}\sin\frac{\sqrt{15}}{4}\,(t-5), \tag{17}$$

which is the formal solution of the given problem. It is also possible to write y in the form

$$y = \begin{cases} 0, & t < 5, \\[2mm] \dfrac{2}{\sqrt{15}}\,e^{-(t-5)/4}\sin\dfrac{\sqrt{15}}{4}\,(t-5), & t \ge 5. \end{cases} \tag{18}$$

The graph of the solution (17), or (18), is shown in Figure 5.7.4.

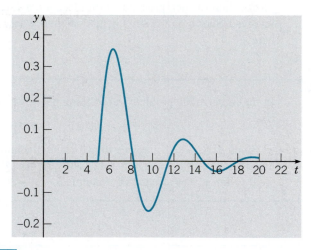

FIGURE 5.7.4   Solution of the initial value problem (13), (14).

Since the initial conditions at $t=0$ are homogeneous and there is no external excitation until $t=5$, there is no response in the interval $0 < t < 5$. The impulse at $t=5$ produces a decaying oscillation that persists indefinitely. The response is continuous at $t=5$ despite the singularity in the forcing function at that point. However, the first derivative of the solution has a jump discontinuity at $t=5$, and the second derivative has an infinite discontinuity there. This is required by the differential equation (13), since a singularity on one side of the equation must be balanced by a corresponding singularity on the other side.

▶ **$\delta(t-t_0)$ as the Derivative of $u(t-t_0)$**   In view of the definition of $\delta(t-t_0)$ given in Eqs. (9) and (10)

$$\int_{-\infty}^{t}\delta(\tau-t_0)\,d\tau = \begin{cases} 0, & t < t_0, \\ 1, & t > t_0, \end{cases} \tag{19}$$

or

$$\int_{-\infty}^{t} \delta(\tau - t_0)\,d\tau = u(t - t_0) \tag{20}$$

for all $t \neq t_0$. Formally differentiating both sides of Eq. (20) with respect to t and treating $\delta(t - t_0)$ as an ordinary function such that the indefinite integral on the left-hand side of Eq. (20) actually satisfies the fundamental theorem of calculus, we find that

$$\delta(t - t_0) = u'(t - t_0), \tag{21}$$

that is, the delta function is the derivative of the unit step function. In the context of the theory of **generalized functions**, or **distributions**, this is indeed true in a rigorous sense.

We note that the statement in Eq. (21) is consistent with an equivalent statement in the s-domain. If we formally apply the result $\mathcal{L}\{f'(t)\} = sF(s) - f(0)$ to $u'(t - t_0)$ for $t_0 > 0$, we find that $\mathcal{L}\{u'(t - t_0)\} = se^{-st_0}/s - u(-t_0) = e^{-st_0} = \mathcal{L}\{\delta(t - t_0)\}$ in agreement with Eq. (11). Letting $t_0 \to 0$ extends the result to $\mathcal{L}\{u'(t)\} = 1 = \mathcal{L}\{\delta(t)\}$.

PROBLEMS

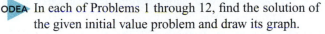

ODEA In each of Problems 1 through 12, find the solution of the given initial value problem and draw its graph.

1. $y'' + 2y' + 2y = \delta(t - \pi)$;
   $y(0) = 1, \quad y'(0) = 0$

2. $y'' + 4y = \delta(t - \pi) - \delta(t - 2\pi)$;
   $y(0) = 0, \quad y'(0) = 0$

3. $y'' + 3y' + 2y = \delta(t - 5) + u_{10}(t)$;
   $y(0) = 0, \quad y'(0) = 1/2$

4. $y'' - y = -20\delta(t - 3)$;
   $y(0) = 1, \quad y'(0) = 0$

5. $y'' + 2y' + 3y = \sin t + \delta(t - 3\pi)$;
   $y(0) = 0, \quad y'(0) = 0$

6. $y'' + 4y = \delta(t - 4\pi)$;
   $y(0) = 1/2, \quad y'(0) = 0$

7. $y'' + y = \delta(t - 2\pi)\cos t$;
   $y(0) = 0, \quad y'(0) = 1$

8. $y'' + 4y = 2\delta(t - \pi/4)$;
   $y(0) = 0, \quad y'(0) = 0$

9. $y'' + y = u_{\pi/2}(t) + 3\delta(t - 3\pi/2) - u_{2\pi}(t)$;
   $y(0) = 0, \quad y'(0) = 0$

10. $2y'' + y' + 4y = \delta(t - \pi/6)\sin t$;
    $y(0) = 0, \quad y'(0) = 0$

11. $y'' + 2y' + 2y = \cos t + \delta(t - \pi/2)$;
    $y(0) = 0, \quad y'(0) = 0$

12. $y^{(4)} - y = \delta(t - 1)$; $\quad y(0) = 0, \quad y'(0) = 0,$
    $y''(0) = 0, \quad y'''(0) = 0$

13. Consider again the system in Example 3 of this section, in which an oscillation is excited by a unit impulse at $t = 5$. Suppose that it is desired to bring the system to rest again after exactly one cycle—that is, when the response first returns to equilibrium moving in the positive direction.

    (a) Determine the impulse $k\delta(t - t_0)$ that should be applied to the system in order to accomplish this objective. Note that k is the magnitude of the impulse and t_0 is the time of its application.

    (b) Solve the resulting initial value problem and ODEA plot its solution to confirm that it behaves in the specified manner.

14. Consider the initial value problem ODEA

$$y'' + \gamma y' + y = \delta(t - 1), \quad y(0) = 0, \quad y'(0) = 0,$$

where γ is the damping coefficient (or resistance).

(a) Let $\gamma = 1/2$. Find the solution of the initial value problem and plot its graph.

(b) Find the time t_1 at which the solution attains its maximum value. Also find the maximum value y_1 of the solution.

(c) Let $\gamma = 1/4$ and repeat parts (a) and (b).

(d) Determine how t_1 and y_1 vary as γ decreases. What are the values of t_1 and y_1 when $\gamma = 0$?

15. Consider the initial value problem

$$y'' + \gamma y' + y = k\delta(t - 1), \quad y(0) = 0, \quad y'(0) = 0,$$

where k is the magnitude of an impulse at $t = 1$ and γ is the damping coefficient (or resistance).

(a) Let $\gamma = 1/2$. Find the value of k for which the response has a peak value of 2; call this value k_1.

(b) Repeat part (a) for $\gamma = 1/4$.

(c) Determine how k_1 varies as γ decreases. What is the value of k_1 when $\gamma = 0$?

16. Consider the initial value problem

$$y'' + y = f_k(t), \quad y(0) = 0, \quad y'(0) = 0,$$

where $\quad f_k(t) = [u_{4-k}(t) - u_{4+k}(t)]/2k \quad$ with $0 < k \le 1$.

(a) Find the solution $y = \phi(t, k)$ of the initial value problem.

(b) Calculate $\lim_{k \to 0} \phi(t, k)$ from the solution found in part (a).

(c) Observe that $\lim_{k \to 0} f_k(t) = \delta(t - 4)$. Find the solution $\phi_0(t)$ of the given initial value problem with $f_k(t)$ replaced by $\delta(t - 4)$. Is it true that $\phi_0(t) = \lim_{k \to 0} \phi(t, k)$?

ODEA (d) Plot $\phi(t,1/2)$, $\phi(t,1/4)$, and $\phi_0(t)$ on the same axes. Describe the relation between $\phi(t,k)$ and $\phi_0(t)$.

ODEA Problems 17 through 22 deal with the effect of a sequence of impulses on an undamped oscillator. Suppose that

$$y'' + y = f(t), \quad y(0) = 0, \quad y'(0) = 0.$$

For each of the following choices for $f(t)$:

(a) Try to predict the nature of the solution without solving the problem.

(b) Test your prediction by finding the solution and drawing its graph.

(c) Determine what happens after the sequence of impulses ends.

17. $f(t) = \sum_{k=1}^{20} \delta(t - k\pi)$

18. $f(t) = \sum_{k=1}^{20} (-1)^{k+1}\delta(t - k\pi)$

19. $f(t) = \sum_{k=1}^{20} \delta(t - k\pi/2)$

20. $f(t) = \sum_{k=1}^{20} (-1)^{k+1}\delta(t - k\pi/2)$

21. $f(t) = \sum_{k=1}^{15} \delta[t - (2k - 1)\pi]$

22. $f(t) = \sum_{k=1}^{40} (-1)^{k+1}\delta(t - 11k/4)$

23. The position of a certain lightly damped oscillator satisfies the initial value problem

$$y'' + 0.1y' + y = \sum_{k=1}^{20} (-1)^{k+1}\delta(t - k\pi),$$

$$y(0) = 0, \quad y'(0) = 0.$$

Observe that, except for the damping term, this problem is the same as Problem 18.

(a) Try to predict the nature of the solution without solving the problem.

(b) Test your prediction by finding the solution ODEA and drawing its graph.

(c) Determine what happens after the sequence of ODEA impulses ends.

24. Proceed as in Problem 23 for the oscillator ODEA satisfying

$$y'' + 0.1y' + y = \sum_{k=1}^{15} \delta[t - (2k - 1)\pi],$$

$$y(0) = 0, \quad y'(0) = 0.$$

Observe that, except for the damping term, this problem is the same as Problem 21.

25. (a) By the method of variation of parameters, show that the solution of the initial value problem

$$y'' + 2y' + 2y = f(t); \quad y(0) = 0,$$
$$y'(0) = 0$$

is

$$y = \int_0^t e^{-(t-\tau)} f(\tau) \sin(t - \tau) \, d\tau.$$

(b) Show that if $f(t) = \delta(t - \pi)$, then the solution of part (a) reduces to

$$y = u_\pi(t)e^{-(t-\pi)} \sin(t - \pi).$$

(c) Use a Laplace transform to solve the given initial value problem with $f(t) = \delta(t - \pi)$ and confirm that the solution agrees with the result of part (b).

5.8 Convolution Integrals and Their Applications

Consider the initial value problem

$$y'' + y = g(t), \qquad y(0) = 0, \quad y'(0) = 0, \tag{1}$$

where the input $g(t)$ is assumed to be piecewise continuous and of exponential order for $t \geq 0$. Using the method of variation of parameters (see Section 4.8), the solution of (1) can be represented by the integral

$$y(t) = \int_0^t \sin(t - \tau)g(\tau)\,d\tau. \tag{2}$$

The integral operation involving $\sin t$ and $g(t)$ that appears on the right-hand side of Eq. (2), termed a convolution integral, is an example of the type of operation that arises naturally in representing the response or output $y(t)$ of a linear constant coefficient equation to an input function $g(t)$ in the t-domain. Other places where convolution integrals arise are in various applications in which the behavior of the system at time t depends not only on its state at time t but also on its past history. Systems of this kind are sometimes called hereditary systems and occur in such diverse fields as neutron transport, viscoelasticity, and population dynamics. A formal definition of the operation exhibited in Eq. (2) follows.

▶ Definition and Properties of Convolution

DEFINITION 5.8.1	Let $f(t)$ and $g(t)$ be piecewise continuous functions on $[0, \infty)$. The **convolution of f and g** is defined by

$$h(t) = \int_0^t f(t - \tau)g(\tau)\,d\tau. \tag{3}$$

The integral in Eq. (3) is known as a **convolution integral**. It is conventional to emphasize that the convolution integral can be thought of as a "generalized product" by writing

$$h(t) = (f * g)(t). \tag{4}$$

The convolution $f * g$ has several of the properties of ordinary multiplication. These properties are summarized in the following theorem.

THEOREM 5.8.2

$$
\begin{aligned}
f * g &= g * f && \text{(commutative law)} & (5)\\
f * (g_1 + g_2) &= f * g_1 + f * g_2 && \text{(distributive law)} & (6)\\
(f * g) * h &= f * (g * h) && \text{(associative law)} & (7)\\
f * 0 &= 0 * f = 0. && & (8)
\end{aligned}
$$

Proof

The proof of Eq. (5) begins with the definition of $f * g$,

$$(f * g)(t) = \int_0^t f(t - \tau)g(\tau)\,d\tau.$$

Changing the variable of integration from τ to $u = t - \tau$ we get

$$(f * g)(t) = -\int_t^0 f(u)g(t-u)\,du = \int_0^t g(t-u)f(u)\,du = (g * f)(t). \qquad \blacksquare$$

The proofs of properties (6) and (7) are left as exercises. Equation (8) follows from the fact that $f(t - \tau) \cdot 0 = 0$.

There are other properties of ordinary multiplication that the convolution integral does not have. For example, it is not true in general that $f * 1$ is equal to f. To see this, note that

$$(f * 1)(t) = \int_0^t f(t-\tau) \cdot 1\,d\tau = \int_0^t f(u)\,du,$$

where $u = t - \tau$. Thus, if $(f * 1)(t) = f(t)$, it must be the case that $f'(t) = f(t)$ and therefore $f(t) = ce^t$. Since $f(0) = (f * 1)(0) = 0$, it follows that $c = 0$. Therefore the only f satisfying $(f * 1)(t) = f(t)$ is the zero function. Similarly, $f * f$ is not necessarily nonnegative. See Problem 2 for an example.

▶ **The Convolution Theorem.** Let us again consider the initial value problem (1). By Theorem 4.2.1, Eq. (2) is the unique solution of the initial value problem (1) for $t \geq 0$. On the other hand, Laplace transforming the initial value problem (1) and solving for $Y(s) = \mathcal{L}\{y(t)\}$ shows that

$$Y(s) = \frac{1}{1+s^2}\,G(s)$$

where $G = \mathcal{L}\{g\}$. By Theorem 5.3.1, there is one and only one continuous version of the inverse Laplace transform of $Y(s)$. We must therefore conclude that

$$\int_0^t \sin(t-\tau)g(\tau)\,d\tau = \mathcal{L}^{-1}\left\{\frac{1}{1+s^2}\,G(s)\right\}, \qquad (9)$$

or equivalently,

$$\mathcal{L}\left\{\int_0^t \sin(t-\tau)g(\tau)\,d\tau\right\} = \frac{1}{1+s^2}\,G(s) = \mathcal{L}\{\sin t\}\mathcal{L}\{g(t)\}. \qquad (10)$$

Equality between the Laplace transform of the convolution of two functions and the product of the Laplace transforms of the two functions involved in the convolution operation, as expressed in Eq. (10), is an instance of the following general result.

THEOREM 5.8.3

Convolution Theorem. If $F(s) = \mathcal{L}\{f(t)\}$ and $G(s) = \mathcal{L}\{g(t)\}$ both exist for $s > a \geq 0$, then

$$H(s) = F(s)G(s) = \mathcal{L}\{h(t)\}, \qquad s > a, \qquad (11)$$

where

$$h(t) = \int_0^t f(t-\tau)g(\tau)\,d\tau = \int_0^t f(\tau)g(t-\tau)\,d\tau. \qquad (12)$$

Proof

We note first that if

$$F(s) = \int_0^\infty e^{-s\xi} f(\xi)\,d\xi$$

and

$$G(s) = \int_0^\infty e^{-s\tau} g(\tau)\,d\tau,$$

then

$$F(s)G(s) = \int_0^\infty e^{-s\xi} f(\xi)\,d\xi \int_0^\infty e^{-s\tau} g(\tau)\,d\tau. \tag{13}$$

Since the integrand of the first integral does not depend on the integration variable of the second, we can write $F(s)G(s)$ as an iterated integral,

$$F(s)G(s) = \int_0^\infty e^{-s\tau} g(\tau)\left[\int_0^\infty e^{-s\xi} f(\xi)\,d\xi\right] d\tau$$

$$= \int_0^\infty g(\tau)\left[\int_0^\infty e^{-s(\xi+\tau)} f(\xi)\,d\xi\right] d\tau. \tag{14}$$

The latter integral can be put into a more convenient form by introducing a change of variable. Let $\xi = t - \tau$, for fixed τ, so that $d\xi = dt$. Further, $\xi = 0$ corresponds to $t = \tau$ and $\xi = \infty$ corresponds to $t = \infty$. Then, the integral with respect to ξ in Eq. (14) is transformed into one with respect to t:

$$F(s)G(s) = \int_0^\infty g(\tau)\left[\int_\tau^\infty e^{-st} f(t-\tau)\,dt\right] d\tau. \tag{15}$$

The iterated integral on the right side of Eq. (15) is carried out over the shaded wedge-shaped region extending to infinity in the $t\tau$-plane shown in Figure 5.8.1.

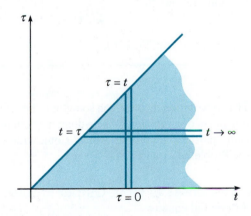

FIGURE 5.8.1

Assuming that the order of integration can be reversed, we finally obtain

$$F(s)G(s) = \int_0^\infty e^{-st}\left[\int_0^t f(t-\tau)g(\tau)\,d\tau\right] dt, \tag{16}$$

or

$$F(s)G(s) = \int_0^\infty e^{-st} h(t)\,dt$$

$$= \mathcal{L}\{h(t)\}, \tag{17}$$

where $h(t)$ is defined by Eq. (3). ∎

Test Theorem 5.8.3 by finding the convolution of $f(t) = t$ and $g(t) = e^{-2t}$ in two different ways:

(i) by direct evaluation of the convolution integral (3) in Definition 5.8.1, and

(ii) by directly computing the inverse Laplace transform of $F(s)G(s)$, the product of the Laplace transforms of $f(t)$ and $g(t)$.

By using integration by parts, the convolution of $f(t) = t$ and $g(t) = e^{-2t}$ is found to be

$$(f * g)(t) = \int_0^t (t - \tau)e^{-2\tau}\, d\tau$$

$$= -\frac{1}{2}(t - \tau)e^{-2\tau}\Big|_{\tau=0}^{\tau=t} - \frac{1}{2}\int_0^t e^{-2\tau}\, d\tau$$

$$= \frac{t}{2} - \frac{1}{4} + \frac{1}{4}e^{-2t}. \tag{18}$$

On the other hand, since $F(s) = 1/s^2$ and $G(s) = 1/(s + 2)$,

$$F(s)G(s) = \frac{1}{s^2(s + 2)}.$$

Using partial fractions, we can express $F(s)G(s)$ in the form

$$F(s)G(s) = \frac{1}{2}\frac{1}{s^2} - \frac{1}{4}\frac{1}{s} + \frac{1}{4}\frac{1}{s + 2}.$$

It readily follows that

$$\mathcal{L}^{-1}\{F(s)G(s)\} = \frac{t}{2} - \frac{1}{4} + \frac{1}{4}e^{-2t},$$

in agreement with Eq. (18).

The convolution theorem can sometimes be effectively used to compute the inverse Laplace transform of a function in the s-domain when it may not be readily apparent how to proceed using partial fraction expansions along with a table of Laplace transforms. For example, see Problems 11 and 12.

▶ **Free and Forced Responses of Input-Output Problems.** Consider the problem consisting of the differential equation

$$ay'' + by' + cy = g(t), \tag{19}$$

where a, b, and c are real constants and g is a given function, together with the initial conditions

$$y(0) = y_0, \qquad y'(0) = y_1. \tag{20}$$

The initial value problem (19), (20) is often referred to as an **input-output problem**. The coefficients a, b, and c describe the properties of some physical system, $g(t)$ is the input to the system, and the values y_0 and y_1 describe the initial state of the system. The solution of Eqs. (19) and (20), the total response, can be separated into two parts: the forced response and the free response. We now show how representations of these responses are found in both the s-domain and the t-domain.

By taking the Laplace transform of Eq. (19) and using the initial conditions (20), we obtain

$$(as^2 + bs + c)Y(s) - (as + b)y_0 - ay_1 = G(s).$$

If we set

$$H(s) = \frac{1}{as^2 + bs + c}, \tag{21}$$

then we can express $Y(s)$ as

$$Y(s) = H(s)[(as + b)y_0 + ay_1] + H(s)G(s). \tag{22}$$

Consequently,

$$y(t) = \mathcal{L}^{-1}\{H(s)[(as + b)y_0 + ay_1]\} + \int_0^t h(t - \tau)g(\tau)\,d\tau \tag{23}$$

where we have used the convolution theorem to represent $\mathcal{L}^{-1}\{H(s)G(s)\}$ as a convolution integral. Observe that the first term on the right-hand side of Eq. (23) is the solution of the initial value problem

$$ay'' + by' + cy = 0, \qquad y(0) = y_0, \quad y'(0) = y_1, \tag{24}$$

obtained from Eqs. (19) and (20) by setting $g(t)$ equal to zero. The solution of the initial value problem (24) is called the **free response** of the system in the t-domain. Its representation in the s-domain is the first term on the right-hand side of Eq. (22). The second term on the right-hand side of Eq. (23) is the solution of

$$ay'' + by' + cy = g(t), \qquad y(0) = 0, \quad y'(0) = 0, \tag{25}$$

in which the initial values y_0 and y_1 are each replaced by zero. The solution of the initial value problem (25) is called the **forced response** of the system. Its counterpart in the s-domain is given by the second term on the right-hand side of Eq. (22), the two representations being related by Theorem 5.8.3.

Using the time-domain methods of Chapter 4, an alternative representation of the free response is

$$y(t) = \alpha_1 y_1(t) + \alpha_2 y_2(t)$$

where $y_1(t)$ and $y_2(t)$ comprise a fundamental set of solutions of $ay'' + by' + cy = 0$, and α_1 and α_2 are determined by solving the linear system

$$\alpha_1 y_1(0) + \alpha_2 y_2(0) = y_0,$$
$$\alpha_1 y_1'(0) + \alpha_2 y_2'(0) = y_1.$$

Thus

$$\alpha_1 = \frac{y_0 y_2'(0) - y_1 y_2(0)}{y_1(0)y_2'(0) - y_1'(0)y_2(0)} \quad \text{and} \quad \alpha_2 = \frac{y_1 y_1(0) - y_0 y_1'(0)}{y_1(0)y_2'(0) - y_1'(0)y_2(0)}. \tag{26}$$

By the uniqueness of solutions of the initial value problem (Theorem 4.2.1) and the one to one correspondence between piecewise continuous functions of exponential order and their Laplace transforms (Theorem 5.3.1), we must conclude that, with the values of α_1 and α_2 specified in Eqs. (26),

$$\mathcal{L}\{\alpha_1 y_1(t) + \alpha_2 y_2(t)\} = H(s)[(as + b)y_0 + ay_1].$$

Results of the above discussion are summarized in Table 5.8.1.

TABLE 5.8.1

The total response of $ay'' + by' + cy = g(t)$, $y(0) = y_0$, $y'(0) = y_1$ as a sum of the free response and the forced response in both the s-domain and the t-domain.

	Total Response		Free Response		Forced Response
s-domain:	$Y(s)$	$=$	$H(s)[(as + b)y_0 + ay_1]$	$+$	$H(s)G(s)$
t-domain:	$y(t)$	$=$	$\alpha_1 y_1(t) + \alpha_2 y_2(t)$	$+$	$\int_0^t h(t - \tau)g(\tau)\,d\tau$

▶ Transfer Functions and Impulse Responses. There are many applications where the dominant component of the total response is the forced response and the free response is of little importance. For example, consider an automobile suspension system consisting of a quarter of the automobile's mass, a wheel, a spring, and a shock absorber. If the input to the system arises from the vertical motion of the wheel rolling over a rough surface, it is intuitively clear that the initial vertical displacement and velocity of the automobile's mass are of little importance and it is the forced response that best represents the smoothness of the ride. Mathematically speaking, if the real parts of the roots of the characteristic equation, $a\lambda^2 + b\lambda + c = 0$, for $ay'' + by' + cy = 0$ are negative, due to damping for example, then the free response of the system is transient and often undetectable after a short period of time; it is the forced response that is of primary interest. Unless otherwise stated, when we refer to the output of an input-output system in either the t-domain or the s-domain, we mean the forced response.

DEFINITION 5.8.4	The **transfer function** of the input-output problem (19), (20) is the ratio of the forced response to the input in the s-domain. Equivalently, the transfer function is the factor in the equation for $Y(s)$ multiplying the Laplace transform of the input, $G(s)$.

Obviously, the transfer function for the input-output problem (19), (20) is $H(s)$ defined in Eq. (21),

$$H(s) = \frac{1}{as^2 + bs + c}.$$

Note that the transfer function $H(s)$ completely characterizes the system (19) since it contains all of the information concerning the coefficients of the system (19). On the other hand, $G(s)$ depends only on the external excitation $g(t)$ that is applied to the system.

The s-domain forced response $H(s)G(s)$ reduces to the transfer function $H(s)$ when the input $G(s) = 1$. This means that the corresponding input in the time domain is $g(t) = \delta(t)$ since $\mathcal{L}\{\delta(t)\} = 1$. Consequently, the inverse Laplace transform of the transfer function, $h(t) = \mathcal{L}^{-1}\{H(s)\}$, is the solution of the initial value problem

$$ay'' + by' + cy = \delta(t), \qquad y(0) = 0, \quad y'(0) = 0, \tag{27}$$

obtained from the initial value problem (25) by replacing $g(t)$ by $\delta(t)$. Thus $h(t)$ is the response of the system to a unit impulse applied at $t = 0$ under the condition that all initial conditions are zero. It is natural to call $h(t)$ the **impulse response** of the system. The forced response in the t-domain is the convolution of the impulse response and the forcing function. The block diagrams in Figure 5.8.2 represent the mapping of inputs to outputs in both the t-domain and s-domain.

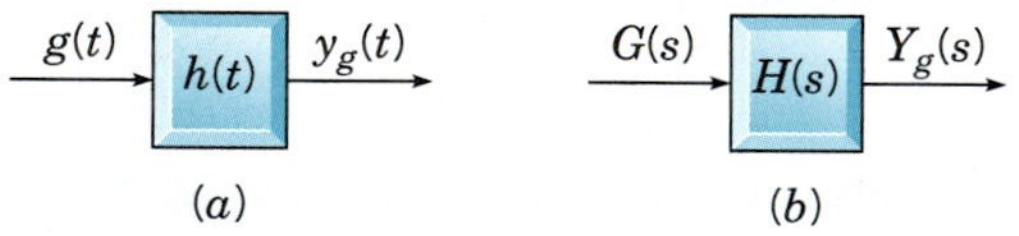

FIGURE 5.8.2 Block diagram of forced response of $ay'' + by' + cy = g(t)$ in (a) the t-domain, and (b) the s-domain.

In the t-domain, the system is characterized by the impulse response $h(t)$ as shown in Figure 5.8.2(a). The response or output $y_g(t)$ due to the input $g(t)$ is represented by the convolution integral

$$y_g(t) = \int_0^t h(t - \tau)g(\tau)\,d\tau. \tag{28}$$

In the Laplace or s-domain, the system is characterized by the transfer function $H(s) = \mathcal{L}\{h(t)\}$. The output $Y_g(s)$ due to the input $G(s)$ is represented by the product of $H(s)$ and $G(s)$,

$$Y_g(s) = H(s)G(s). \qquad (29)$$

The representation (29) results from taking the Laplace transform of Eq. (28) and using the convolution theorem. In summary, to obtain the system output in the t-domain:

1. Find the transfer function $H(s)$.
2. Find the Laplace transform of the input, $G(s)$.
3. Construct the output in the s-domain, a simple algebraic operation $Y_g(s) = H(s)G(s)$.
4. Compute the output in the t-domain, $y_g(t) = \mathcal{L}^{-1}\{Y_g(s)\}$.

EXAMPLE 2

Consider the input-output system

$$y'' + 2y' + 5y = g(t) \qquad (30)$$

where $g(t)$ is any piecewise continuous function of exponential order.

1. Find the transfer function and the impulse response.
2. Using a convolution integral to represent the forced response, find the general solution of Eq. (30).
3. Find the total response if the initial state of the system is prescribed by $y(0) = 1$, $y'(0) = -3$.
4. Compute the forced response when $g(t) = t$.

   (i) The transfer function $H(s)$ is the Laplace transform of the solution of

$$y'' + 2y' + 5y = \delta(t), \qquad y(0) = 0, \quad y'(0) = 0.$$

Taking the Laplace transform of the differential equation and using the zero-valued initial conditions gives $(s^2 + 2s + 5)Y(s) = 1$. Thus, the transfer function is

$$H(s) = Y(s) = \frac{1}{s^2 + 2s + 5}.$$

To find the impulse response, $h(t)$, we compute the inverse transform of $H(s)$. Completing the square in the denominator of $H(s)$ yields

$$H(s) = \frac{1}{(s+1)^2 + 4} = \frac{1}{2} \frac{2}{(s+1)^2 + 4}.$$

Thus

$$h(t) = \mathcal{L}^{-1}\{H(s)\} = \frac{1}{2}e^{-t}\sin 2t.$$

   (ii) A particular solution of Eq. (30) is provided by the forced response

$$y_g(t) = \int_0^t h(t-\tau)g(\tau)\,d\tau = \frac{1}{2}\int_0^t e^{-(t-\tau)}\sin 2(t-\tau)g(\tau)\,d\tau.$$

The characteristic equation of $y'' + 2y' + 5y = 0$ is $\lambda^2 + 2\lambda + 5 = 0$; it has the complex solutions $\lambda = -1 \pm 2i$. Therefore a fundamental set of real-valued solutions of $y'' + 2y' + 5y = 0$ is $\{e^{-t}\cos 2t, e^{-t}\sin 2t\}$. It follows that the general solution of Eq. (30) is

$$y(t) = c_1 e^{-t}\cos 2t + c_2 e^{-t}\sin 2t + \frac{1}{2}\int_0^t e^{-(t-\tau)}\sin 2(t-\tau)g(\tau)\,d\tau.$$

   (iii) The free response is obtained by requiring the complementary solution $y_c(t) = c_1 e^{-t}\cos 2t + c_2 e^{-t}\sin 2t$ to satisfy the initial conditions $y(0) = 1$ and

$y'(0) = -3$. This leads to the system $c_1 = 1$ and $-c_1 + 2c_2 = -3$, so $c_2 = -1$. Consequently, the total response is represented by

$$y(t) = e^{-t}\cos 2t - e^{-t}\sin 2t + \frac{1}{2}\int_0^t e^{-(t-\tau)}\sin 2(t-\tau)g(\tau)\,d\tau.$$

Remark. It is worth noting that the particular solution represented by the forced response does not affect the values of the constants c_1 and c_2 since $y_g(0) = 0$ and $y_g'(0) = 0$.

(iv) If $g(t) = t$, the forced response may be calculated either by evaluating the convolution integral

$$y_g(t) = \frac{1}{2}\int_0^t \tau e^{-(t-\tau)}\sin 2(t-\tau)\,d\tau$$

or by computing the inverse Laplace transform of the product of $H(s)$ and the Laplace transform of t, $1/s^2$. We choose the latter method. Using partial fractions we find that

$$Y_g(s) = \frac{1}{s^2[(s+1)^2+4]} = \frac{1}{5s^2} - \frac{2}{25s} + \frac{2}{25}\frac{s+1}{(s+1)^2+4} - \frac{3}{50}\frac{2}{(s+1)^2+4},$$

and therefore

$$y_g(t) = \frac{1}{5}t - \frac{2}{25} + \frac{2}{25}e^{-t}\cos 2t - \frac{3}{50}e^{-t}\sin 2t.$$

PROBLEMS

1. Establish the distributive and associative properties of the convolution integral.
   (a) $f * (g_1 + g_2) = f * g_1 + f * g_2$
   (b) $f * (g * h) = (f * g) * h$

2. Show, by means of the example $f(t) = \sin t$, that $f * f$ is not necessarily nonnegative.

In each of Problems 3 through 6, find the Laplace transform of the given function.

3. $f(t) = \displaystyle\int_0^t (t-\tau)^2 \cos 2\tau\, d\tau$

4. $f(t) = \displaystyle\int_0^t e^{-(t-\tau)}\sin \tau\, d\tau$

5. $f(t) = \displaystyle\int_0^t (t-\tau)e^\tau\, d\tau$

6. $f(t) = \displaystyle\int_0^t \sin(t-\tau)\cos \tau\, d\tau$

In each of Problems 7 through 12, find the inverse Laplace transform of the given function by using the convolution theorem.

7. $F(s) = \dfrac{1}{s^4(s^2+1)}$

8. $F(s) = \dfrac{s}{(s+1)(s^2+4)}$

9. $F(s) = \dfrac{1}{(s+1)^2(s^2+4)}$

10. $F(s) = \dfrac{G(s)}{s^2+1}$

11. $F(s) = \dfrac{1}{(s^2+1)^2}$

12. $F(s) = \dfrac{s}{(s^2+1)^2}$

13. **(a)** If $f(t) = t^m$ and $g(t) = t^n$, where m and n are positive integers, show that

$$(f * g)(t) = t^{m+n+1}\int_0^1 u^m(1-u)^n\, du.$$

   (b) Use the convolution theorem to show that

$$\int_0^1 u^m(1-u)^n\, du = \frac{m!\,n!}{(m+n+1)!}.$$

   (c) Extend the result of part **(b)** to the case where m and n are positive numbers but not necessarily integers.

In each of Problems 14 through 21, express the total response of the given initial value problem using a convolution integral to represent the forced response.

14 $y'' + \omega^2 y = g(t);$ $\quad y(0) = 0,$ $\quad y'(0) = 1$

15. $y'' + 2y' + 2y = \sin \alpha t$;
$y(0) = 0, \quad y'(0) = 0$

16. $4y'' + 4y' + 17y = g(t)$;
$y(0) = 0, \quad y'(0) = 0$

17. $y'' + y' + 1.25y = 1 - u_\pi(t)$;
$y(0) = 1, \quad y'(0) = -1$

18. $y'' + 4y' + 4y = g(t)$;
$y(0) = 2, \quad y'(0) = -3$

19. $y'' + 3y' + 2y = \cos \alpha t$;
$y(0) = 1, \quad y'(0) = 0$

20. $y^{(4)} - y = g(t); \quad y(0) = 0,$
$y'(0) = 0, \quad y''(0) = 0, \quad y'''(0) = 0$

21. $y^{(4)} + 5y'' + 4y = g(t); \quad y(0) = 1,$
$y'(0) = 0, \quad y''(0) = 0, \quad y'''(0) = 0$

22. **Unit Step Responses**. The unit step response of a system is the output $y(t)$ when the input is the unit step function $u(t)$ and all initial conditions are zero.

(a) Show that the derivative of the unit step response is the impulse response.

ODEA

(b) Plot and graph on the same set of coordinates the unit step response of

$$y'' + 2\delta y' + y = u(t)$$

for $\delta = .2, .3, .4, .5, .6, .8,$ and 1.

23. Consider the equation

$$\phi(t) + \int_0^t k(t - \xi)\phi(\xi)\,d\xi = f(t),$$

in which f and k are known functions, and ϕ is to be determined. Since the unknown function ϕ appears under an integral sign, the given equation is called an **integral equation**; in particular, it belongs to a class of integral equations known as Volterra integral equations. Take the Laplace transform of the given integral equation and obtain an expression for $\mathcal{L}\{\phi(t)\}$ in terms of the transforms $\mathcal{L}\{f(t)\}$ and $\mathcal{L}\{k(t)\}$ of the given functions f and k. The inverse transform of $\mathcal{L}\{\phi(t)\}$ is the solution of the original integral equation.

24. Consider the Volterra integral equation (see Problem 23)

$$\phi(t) + \int_0^t (t - \xi)\phi(\xi)\,d\xi = \sin 2t. \qquad (i)$$

(a) Solve the integral equation (i) by using the Laplace transform.

(b) By differentiating Eq. (i) twice, show that $\phi(t)$ satisfies the differential equation

$$\phi''(t) + \phi(t) = -4 \sin 2t.$$

Show also that the initial conditions are

$$\phi(0) = 0, \qquad \phi'(0) = 2.$$

(c) Solve the initial value problem in part (b) and verify that the solution is the same as the one in part (a).

In each of Problems 25 through 27:

(a) Solve the given Volterra integral equation by using the Laplace transform.

(b) Convert the integral equation into an initial value problem, as in Problem 24(b).

(c) Solve the initial value problem in part (b) and verify that the solution is the same as the one in part (a).

25. $\phi(t) + \displaystyle\int_0^t (t - \xi)\phi(\xi)\,d\xi = 1$

26. $\phi(t) - \displaystyle\int_0^t (t - \xi)\phi(\xi)\,d\xi = 1$

27. $\phi(t) + 2\displaystyle\int_0^t \cos(t - \xi)\phi(\xi)\,d\xi = e^{-t}$

There are also equations, known as **integro-differential equations**, in which both derivatives and integrals of the unknown function appear. In each of Problems 28 through 30:

(a) Solve the given integro-differential equation by using the Laplace transform.

(b) By differentiating the integro-differential equation a sufficient number of times, convert it into an initial value problem.

(c) Solve the initial value problem in part (b) and verify that the solution is the same as the one in part (a).

28. $\phi'(t) + \displaystyle\int_0^t (t - \xi)\phi(\xi)\,d\xi = t,$
$\phi(0) = 0$

29. $\phi'(t) - \dfrac{1}{2}\displaystyle\int_0^t (t - \xi)^2\phi(\xi)\,d\xi = -t,$
$\phi(0) = 1$

30. $\phi'(t) + \phi(t) - \displaystyle\int_0^t \sin(t - \xi)\phi(\xi)\,d\xi,$
$\phi(0) = 1$

31. **The Tautochrone**. A problem of interest in the history of mathematics is that of finding the

tautochrone[2]—the curve down which a particle will slide freely under gravity alone, reaching the bottom in the same time regardless of its starting point on the curve. This problem arose in the construction of a clock pendulum whose period is independent of the amplitude of its motion. The tautochrone was found by Christian Huygens in 1673 by geometrical methods, and later by Leibniz and Jakob Bernoulli using analytical arguments. Bernoulli's solution (in 1690) was one of the first occasions on which a differential equation was explicitly solved.

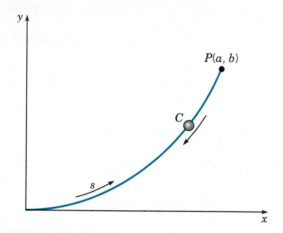

FIGURE 5.8.3 The tautochrone.

The geometric configuration is shown in Figure 5.8.3. The starting point $P(a,b)$ is joined to the terminal point $(0, 0)$ by the arc C. Arc length s is measured from the origin, and $f(y)$ denotes the

rate of change of s with respect to y:

$$f(y) = \frac{ds}{dy} = \left[1 + \left(\frac{dx}{dy}\right)^2\right]^{1/2}. \quad (i)$$

Then it follows from the principle of conservation of energy that the time $T(b)$ required for a particle to slide from P to the origin is

$$T(b) = \frac{1}{\sqrt{2g}} \int_0^b \frac{f(y)}{\sqrt{b - y}}\, dy. \quad (ii)$$

(a) Assume that $T(b) = T_0$, a constant, for each b. By taking the Laplace transform of Eq. (ii) in this case, and using the convolution theorem, show that

$$F(s) = \sqrt{\frac{2g}{\pi}} \frac{T_0}{\sqrt{s}}; \quad (iii)$$

then show that

$$f(y) = \frac{\sqrt{2g}}{\pi} \frac{T_0}{\sqrt{y}}. \quad (iv)$$

Hint: See Problem 37 of Section 5.1.

(b) Combining Eqs. (i) and (iv), show that

$$\frac{dx}{dy} = \sqrt{\frac{2\alpha - y}{y}}, \quad (v)$$

where $\alpha = gT_0^2/\pi^2$.

(c) Use the substitution $y = 2\alpha \sin^2(\theta/2)$ to solve Eq. (v), and show that

$$x = \alpha(\theta + \sin\theta), \quad y = \alpha(1 - \cos\theta). \quad (vi)$$

Equations (vi) can be identified as parametric equations of a cycloid. Thus the tautochrone is an arc of a cycloid.

5.9 Linear Systems and Feedback Control

The development of methods and theory of feedback control is one of the most important achievements of modern engineering. Instances of feedback control systems are common: thermostatic control of heating and cooling systems, automobile cruise control, airplane autopilots, industrial robots, and so on. In this section, we discuss the application of the Laplace transform to input-output problems and feedback control of systems that can be modeled by linear differential equations with constant coefficients. The subject area is extensive. Thus we limit our presentation to the introduction of only a few important methods and concepts.

[2]The word "tautochrone" comes from the Greek words *tauto*, meaning same, and *chronos*, meaning time.

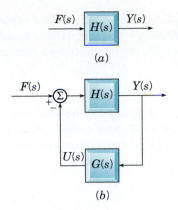

FIGURE 5.9.1 (a) Open-loop system. (b) Closed-loop system.

▶ Feedback Control. Control systems are classified into two general categories: *open-loop* and *closed-loop* systems. An **open-loop** control system is one in which the control action is independent of the output. An example of an open-loop system is a toaster that is controlled by a timer. Once a slice of bread is inserted and the timer is set (the input), the toasting process continues until the timer expires. The input is not affected by the degree to which the slice of bread is toasted (the output). A **closed-loop** control system is one in which the control action depends on the output in some manner, for example, a home heating and cooling system controlled by a thermostat. A block diagram of a simple open-loop control system with transfer function $H(s)$ is shown in Figure 5.9.1(a). The output is specified by $Y(s) = H(s)F(s)$ where $F(s)$ represents the control input. In Figure 5.9.1(b), a feedback control loop has been added to the open-loop system. The output is used as an input to an element, represented by a transfer function $G(s)$, called a **controller**.[3] The output of the controller $U(s) = G(s)Y(s)$ is then added to (positive feedback) or subtracted from (negative feedback) the external input $F(s)$ at the junction point represented by the circle at the left side of Figure 5.9.1(b).[4] In the case of negative feedback, as indicated by the minus sign at the junction point in Figure 5.9.1(b), this yields an algebraic equation

$$Y(s) = H(s)[F(s) - U(s)] = H(s)[F(s) - G(s)Y(s)] \tag{1}$$

in which $Y(s)$ appears on both sides of the equal sign. Solving Eq. (1) for $Y(s)$ yields

$$Y(s) = \frac{H(s)F(s)}{1 + G(s)H(s)}. \tag{2}$$

Thus the transfer function for the closed-loop system is

$$H_G(s) = \frac{Y(s)}{F(s)} = \frac{H(s)}{1 + G(s)H(s)}. \tag{3}$$

---

[3] The physical hardware associated with the quantity being controlled is often called the **plant**. The plant is usually considered to be fixed and unalterable. The elements added to implement control are referred to as the **controller**. The combined entity of the plant and the controller is called the **closed-loop system**, or simply the **system**.

[4] Note that there is some imprecision in this convention since if the transfer function of the controller were represented by $-G(s)$ instead of $G(s)$, the effect of adding and subtracting the output of the controller would be reversed. In the end, what is really important is achieving a desired performance in the closed-loop system.

It is often desirable to introduce explicitly a parameter associated with the **gain** (output amplification or attenuation) of a transfer function. For example, if the open loop transfer function $H(s)$ is multiplied by a gain parameter K, then Eq. (3) becomes

$$H_G(s) = \frac{KH(s)}{1 + KG(s)H(s)}. \tag{4}$$

Block diagrams can be used to express relationships between inputs and outputs of components of more complex systems from which the overall transfer function may be deduced by algebraic operations. For example, from the block diagram in Figure 5.9.2 we have

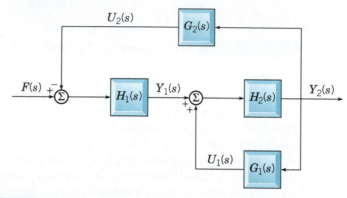

FIGURE 5.9.2   Block diagram of a feedback control system with both positive and negative feedback loops.

$$Y_1(s) = H_1(s)[F(s) - U_2(s)]$$
$$Y_2(s) = H_2(s)[Y_1(s) + U_1(s)]$$
$$U_1(s) = G_1(s)Y_2(s)$$
$$U_2(s) = G_2(s)Y_2(s).$$

Eliminating $U_1(s)$, $U_2(s)$, and $Y_1(s)$ and solving for $Y_2(s)$ yields

$$Y_2(s) = \frac{H_2(s)H_1(s)F(s)}{1 - H_2(s)G_1(s) + H_2(s)H_1(s)G_2(s)}.$$

Thus the overall transfer function of the system described by Figure 5.9.2 is

$$H(s) = \frac{Y_2(s)}{F(s)} = \frac{H_2(s)H_1(s)}{1 - H_2(s)G_1(s) + H_2(s)H_1(s)G_2(s)}. \tag{5}$$

If the transfer functions for each of the components are rational functions, then the closed-loop transfer function will also be a rational function. For example, if $G_1(s) = 1/(s + 1)$, $G_2(s) = 1/(s^2 + 2s + 2)$, $H_1(s) = 1/(s + 2)$, and $H_2(s) = 1/(s + 3)$, then Eq. (5) can be written as

$$H(s) = \frac{\dfrac{1}{(s + 3)(s + 2)}}{1 - \dfrac{1}{(s + 1)(s + 3)} + \dfrac{1}{(s^2 + 2s + 2)(s + 3)(s + 2)}}$$

$$= \frac{s^3 + 3s^2 + 4s + 2}{s^5 + 8s^4 + 24s^3 + 36s^2 + 29s + 9}. \tag{6}$$

▶ **Poles, Zeros, and Stability.** There are typically several application dependent performance characteristics that a closed-loop system must satisfy. Achieving these characteristics is part of the design problem of control theory. However, independent of these characteristics, all feedback control systems must satisfy the conditions that we now define for stability.

DEFINITION 5.9.1	A function $f(t)$ defined on $0 \leq t < \infty$ is said to be **bounded** if there is a number M such that $$|f(t)| \leq M \qquad \text{for all } t \geq 0.$$

DEFINITION 5.9.2	A system is said to be **bounded-input bounded-output** (BIBO) **stable** if every bounded input results in a bounded output.

We wish to characterize BIBO stable systems that can be represented by rational transfer functions. We consider transfer functions having the general form

$$H(s) = \frac{P(s)}{Q(s)} = \frac{b_m s^m + b_{m-1} s^{m-1} + \cdots + b_0}{s^n + a_{n-1} s^{n-1} + \cdots + a_0} \tag{7}$$

where the polynomial in the numerator is $P(s) = b_m s^m + b_{m-1} s^{m-1} + \cdots + b_0$ and the polynomial in the denominator is $Q(s) = s^n + a_{n-1} s^{n-1} + \cdots + a_0$. The transfer function (7) is said to have m zeros and n poles. The reason for this is clear if we represent $H(s)$ as

$$H(s) = b_m \frac{(s - \zeta_1)(s - \zeta_2) \cdots (s - \zeta_m)}{(s - \lambda_1)(s - \lambda_2) \cdots (s - \lambda_n)}$$

where we have expressed the numerator polynomial $P(s)$ and denominator polynomial $Q(s)$ in their factored forms

$$P(s) = b_m(s - \zeta_1)(s - \zeta_2) \cdots (s - \zeta_m)$$

and

$$Q(s) = (s - \lambda_1)(s - \lambda_2) \cdots (s - \lambda_n).$$

The m values of s, namely $\zeta_1, \zeta_2, \ldots, \zeta_m$, that make $P(s)$ zero are called the **zeros** of $H(s)$. The n values of s, that is $\lambda_1, \lambda_2, \ldots, \lambda_n$, that make $Q(s)$ zero are called the **poles** of $H(s)$. The zeros and poles may be real or complex numbers. Since the coefficients of the polynomials are assumed to be real numbers, complex poles or zeros must occur in complex conjugate pairs. We assume that all transfer functions have more poles than zeros, that is, $n > m$. Such transfer functions are called **strictly proper**. If $H(s)$ is strictly proper, then $\lim_{s \to \infty} H(s) = 0$. Recall from Corollary 5.1.7 that this is necessary in order for $H(s)$ to be the Laplace transform of a piecewise continuous function of exponential order.

THEOREM 5.9.3	An input-output system with strictly proper transfer function $$H(s) = \frac{P(s)}{Q(s)} = \frac{b_m s^m + b_{m-1} s^{m-1} + \cdots + b_0}{s^n + a_{n-1} s^{n-1} + \cdots + a_0}$$ is BIBO stable if and only if all of the poles have negative real parts.

Proof

The output $Y(s)$ is related to input $F(s)$ in the s-domain by

$$Y(s) = H(s)F(s).$$

The convolution theorem gives the corresponding relation in the time domain,

$$y(t) = \int_0^t h(t - \tau)f(\tau)\,d\tau.$$

If all of the poles of $H(s)$ have negative real parts, the partial fraction expansion of $H(s)$ is a linear combination of terms of the form

$$\frac{1}{(s - \alpha)^m}, \quad \frac{(s - \alpha)}{\left[(s - \alpha)^2 + \beta^2\right]^m}, \quad \frac{\beta}{\left[(s - \alpha)^2 + \beta^2\right]^m}.$$

where $m \geq 1$ and $\alpha < 0$. Consequently, $h(t)$ will consist only of a linear combination of terms of the form

$$t^k e^{\alpha t}, \quad t^k e^{\alpha t} \cos \beta t, \quad t^k e^{\alpha t} \sin \beta t, \tag{8}$$

where $k \leq m - 1$. Assuming that there exists a constant M_1 such that $|f(t)| \leq M_1$ for all $t \geq 0$, it follows that

$$|y(t)| = \left| \int_0^t h(t - \tau)f(\tau)\,d\tau \right| \leq \int_0^t |h(t - \tau)|\,|f(\tau)|\,d\tau$$

$$\leq M_1 \int_0^t |h(t - \tau)|\,d\tau = M_1 \int_0^t |h(\tau)|\,d\tau \leq M_1 \int_0^\infty |h(\tau)|\,d\tau.$$

Integrating by parts shows that integrals of any terms of the form exhibited in the list (8) are bounded by $k!/|\alpha|^{k+1}$. Consequently, the output $y(t)$ is bounded, which proves the first part of the theorem.

To prove the second part, we now suppose that every bounded input results in a bounded response. We first show that this implies that $\int_0^\infty |h(\tau)|\,d\tau$ is bounded. Consider the bounded input defined by

$$f(t - \tau) = \begin{cases} 1, & \text{if } h(\tau) \geq 0 \text{ and } \tau < t \\ -1, & \text{if } h(\tau) < 0 \text{ and } \tau < t \\ 0, & \text{if } t < \tau. \end{cases}$$

Then

$$y(t) = \int_0^t h(\tau)f(t - \tau)\,d\tau = \int_0^t |h(\tau)|\,d\tau.$$

Since $y(t)$ is bounded, there exists a constant M_2 such that $|y(t)| \leq M_2$ for all $t \geq 0$. Then from the preceding equation we have

$$|y(t)| = \left| \int_0^t |h(\tau)|\,d\tau \right| = \int_0^t |h(\tau)|\,d\tau \leq M_2$$

for all $t \geq 0$. Thus $\int_0^t |h(\tau)|\,d\tau$ is bounded for all $t \geq 0$ and consequently, $\int_0^\infty |h(\tau)|\,d\tau$ is also bounded, as was to be shown. Since $H(s)$ is the Laplace transform of $h(t)$,

$$H(s) = \int_0^\infty e^{-st} h(t)\,dt.$$

If $\text{Re}(s) \geq 0$, then $|e^{-st}| = e^{-\text{Re}(s)t} \leq 1$ for all $t \geq 0$ so

$$|H(s)| = \left| \int_0^\infty e^{-st} h(t)\,dt \right| \leq \int_0^\infty |e^{-st}|\,|h(t)|\,dt \leq \int_0^\infty |h(t)|\,dt \leq M_2.$$

Since the rational function $|H(s)|$ is bounded on the set $\{s : \text{Re}(s) \geq 0\}$, all poles of $H(s)$ must lie in the left half of the complex s-plane. ∎

▶ **Root-Locus Analysis.** Theorem 5.9.3 points out the importance of pole locations with regard to the linear system stability problem. For rational transfer functions, the mathematical problem is to determine whether all of the roots of the polynomial equation

$$s^n + a_{n-1}s^{n-1} + \cdots + a_0 = 0 \tag{9}$$

have negative real parts.[5] Given numerical values of the polynomial coefficients, powerful computer programs can then be used to find the roots. It is often the case that one or more of the coefficients of the polynomial depend on a parameter. Then it is of major interest to understand how the locations of the roots in the complex s-plane change as the parameter is varied. The graph of all possible roots of Eq. (9) relative to some particular parameter is called a **root locus**. The design technique based on this graph is called the **root-locus method** of analysis.

EXAMPLE 1

Suppose the open-loop transfer function for a plant is given by

$$H(s) = \frac{K}{s^2 + 2s + 2}$$

and that it is desired to synthesize a closed loop system with negative feedback using the controller

$$G(s) = \frac{1}{s + 0.1}.$$

Since the poles of $H(s)$ are $s = -1 \pm i$ while $G(s)$ has a single pole at $s = -0.1$, Theorem 5.9.3 implies that open-loop input-output systems described by $H(s)$ and $G(s)$ are BIBO stable. The transfer function for the closed loop system corresponding to Figure 5.9.1(b) is

$$H_G(s) = \frac{H}{1 + GH} = \frac{K(s + 0.1)}{s^3 + 2.1s^2 + 2.2s + K + 0.2}.$$

The root-locus diagram for $H_G(s)$ as K varies from 1 to 10 is shown in Figure 5.9.3.

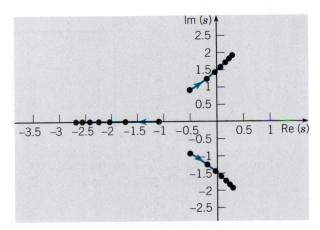

FIGURE 5.9.3 The root locus of the polynomial equation $s^3 + 2.1s^2 + 2.2s + K + 0.2 = 0$ as K varies from 1 to 10.

---

[5] A polynomial with real coefficients and roots which are either negative or pairwise conjugate with negative real parts is called a **Hurwitz polynomial**.

The value of K for which the complex roots cross the imaginary axis is found by assuming for that value of K the roots of the polynomial, α and $\pm i\beta$, must satisfy the equation

$$(s - \alpha)(s^2 + \beta^2) = s^3 - \alpha s^2 + \beta^2 s - \alpha\beta^2 = s^3 + 2.1s^2 + 2.2s + K + 0.2.$$

Matching the coefficients of the last two polynomials on the right,

$$\alpha = -2.1, \quad \beta^2 = 2.2, \quad -\alpha\beta^2 = K + 0.2,$$

yields $K = 4.42$. Thus, for $K < 4.42$, the real parts of all poles of $H_G(s)$ are negative and the closed-loop system is BIBO stable. For $K = 4.42$ two poles are pure imaginary and for $K > 4.42$ the real parts of two of the poles are positive. Thus the closed-loop system is unstable, that is, not BIBO stable, for gains $K \geq 4.42$.

▶ **The Routh Stability Criterion.** In view of the importance of knowing polynomial root locations for the purpose of establishing BIBO stability, a valuable tool is a criterion for establishing whether the roots of a polynomial have negative real parts in terms of the coefficients of the polynomial.[6] For example, the real parts of the roots of

$$Q(s) = s^2 + a_1 s + a_0$$

are negative if and only if $a_1 > 0$ and $a_0 > 0$, a result easily obtained by comparing with a quadratic function of s with roots $\alpha \pm i\beta$,

$$(s - \alpha + i\beta)(s - \alpha - i\beta) = s^2 - 2\alpha s + \alpha^2 + \beta^2.$$

Thus $a_1 = -2\alpha$ and $a_0 = \alpha^2 + \beta^2$. Clearly, the real parts of the roots are negative, that is, $\alpha < 0$, if and only if $a_1 > 0$ and $a_0 > 0$.

We now present, without proof, a classical result for nth degree polynomials that provides a test for determining whether all of the roots of the polynomial (9) have negative real parts in terms of the coefficients of the polynomial. It is useful if the degree n of Eq. (9) is not too large and the coefficients depend on one or more parameters. The criterion is applied using a **Routh table** defined by

s^n	1	a_{n-2}	a_{n-4}	$\cdots$
s^{n-1}	a_{n-1}	a_{n-3}	a_{n-5}	$\cdots$
$\cdot$	b_1	b_2	b_3	$\cdots$
$\cdot$	c_1	c_2	c_3	$\cdots$
$\cdot$	$\cdot$	$\cdot$	$\cdot$	

where $1, a_{n-1}, a_{n-2}, \ldots, a_0$ are the coefficients of the polynomial (9) and $b_1, b_2, \ldots, c_1, c_2, \ldots$ are defined by the quotients

$$b_1 = -\frac{\begin{vmatrix} 1 & a_{n-2} \\ a_{n-1} & a_{n-3} \end{vmatrix}}{a_{n-1}}, \quad b_2 = -\frac{\begin{vmatrix} 1 & a_{n-4} \\ a_{n-1} & a_{n-5} \end{vmatrix}}{a_{n-1}}, \quad \ldots$$

---

[6]More generally, such a criterion is valuable for answering stability questions about linear constant coefficient systems, $\mathbf{x}' = \mathbf{A}\mathbf{x}$, discussed in Chapter 6. The original formulation of polynomial root location problems dates at least back to Cauchy in 1831.

$$c_1 = -\frac{\begin{vmatrix} a_{n-1} & a_{n-3} \\ b_1 & b_2 \end{vmatrix}}{b_1}, \quad c_2 = -\frac{\begin{vmatrix} a_{n-1} & a_{n-5} \\ b_1 & b_3 \end{vmatrix}}{b_1}, \quad \ldots$$

The table is continued horizontally and vertically until only zeros are obtained. Any row of the table may be multiplied by a positive constant before the next row is computed without altering the properties of the table.

THEOREM 5.9.4

The Routh Criterion. All of the roots of the polynomial equation

$$s^n + a_{n-1}s^{n-1} + \cdots + a_0 = 0$$

have negative real parts if and only if the elements of the first column of the Routh table have the same sign. Otherwise, the number of roots with positive real parts is equal to the number of changes of sign.

EXAMPLE 2

The Routh table for the denominator polynomial in the transfer function

$$H_G(s) = \frac{K(s + 0.1)}{s^3 + 2.1s^2 + 2.2s + K + 0.2}$$

of Example 1 is

s^3	1	2.2	0
s^2	2.1	$K + 0.2$	0
s	$\dfrac{4.42 - K}{2.1}$	0	
s^0	$K + 0.2$		

No sign changes occur in the first column if $K < 4.42$ and $K > -0.2$. Thus we conclude that the closed-loop system is BIBO stable if $-0.2 < K < 4.42$. If $K > 4.42$, there are two roots with positive real parts. If $K < -0.2$, there is one root with positive real part.

PROBLEMS

1. Find the transfer function of the system shown in Figure 5.9.4.

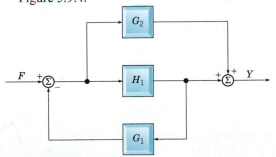

FIGURE 5.9.4 Block diagram for Problem 1.

2. Find the transfer function of the system shown in Figure 5.9.5.

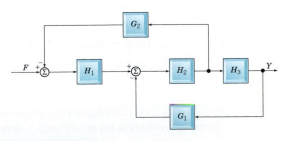

FIGURE 5.9.5 Block diagram for Problem 2.

3. If $h(t)$ is any one of the functions, $t^k e^{\alpha t}$, $t^k e^{\alpha t} \cos \beta t$, or $t^k e^{\alpha t} \sin \beta t$ and $\alpha < 0$, show that

$$\int_0^\infty |h(t)|\, dt \le \frac{k!}{|\alpha|^{k+1}}.$$

4. Show that, if all of the real roots of a polynomial are negative and the real parts of all of the complex roots are negative, then the coefficients of the polynomial are necessarily positive. (Thus if any coefficients are negative or zero, the polynomial must have roots with nonnegative real parts. This provides a useful initial check on whether there may be roots in the right half plane.)

5. Use Routh's criterion to find necessary and sufficient conditions on a_2, a_1, and a_0 that guarantee all roots of $s^3 + a_2 s^2 + a_1 s + a_0$ lie in the left half plane.

CAS For each of the characteristic functions in Problems 6 through 11, use Routh's stability criterion to determine the number of roots with positive real parts. Confirm your answers by using a computer to find the roots.

6. $s^3 + 2s^2 + 3s + 7$

7. $s^3 + 5s^2 + 9s + 5$

8. $s^4 + 9s^3 + 23s^2 + s - 34$

9. $s^4 + 5s^3 - 5s^2 - 35s + 34$

10. $s^4 + 14s^3 + 78s^2 + 206s + 221$

11. $s^4 + 8s^3 + 24s^2 + 32s + 100$

In each of Problems 12 through 15, use the Routh Criterion to find the range of K for which all the poles of the transfer functions are in the left half of the complex plane. Then use a computer program to verify your answers by plotting the poles in the s-plane for various values of K. CAS

12. $\dfrac{K(s+1)}{s^3 + 5s^2 + (K-6)s + K}$

13. $\dfrac{K(s+1)(s+2)}{s^3 + 4s^2 + 6s + 4 + K}$

14. $\dfrac{K(s+1)}{s^3 + 3s^2 + 10s + 8 + K}$

15. $\dfrac{(s^2 + 2s + 2)K}{s^4 + 4s^3 + 14s^2 + 20s + 16 + K}$

CHAPTER SUMMARY

For constant coefficient linear differential equations, the Laplace transform converts initial value problems in the t-domain to algebraic problems in the s-domain. The Laplace transform is used to study input-output behavior of linear systems, feedback control systems, and electric circuits.

Section 5.1 The Laplace transform is an **integral transform**.

▶ If f is **piecewise continuous** and of **exponential order**, the Laplace transform is defined by

$$\mathcal{L}\{f\}(s) = F(s) = \int_0^\infty e^{-st} f(t)\, dt.$$

The **domain** of F is the set of all s for which the **improper integral** converges.

▶ $\mathcal{L}$ is a **linear operator**, $\mathcal{L}\{c_1 f_1 + c_2 f_2\} = c_1 \mathcal{L}\{f_1\} + c_2 \mathcal{L}\{f_2\}$.

Section 5.2 Some properties of $\mathcal{L}$: if $\mathcal{L}\{f(t)\} = F(s)$, then

▶ $\mathcal{L}\{e^{at} f(t)\}(s) = F(s-a)$,

▶ $\mathcal{L}\{f^{(n)}(t)\}(s) = s^n F(s) - s^{n-1} f(0) - \cdots - f^{(n-1)}(0)$,

▶ $\mathcal{L}\{t^n f(t)\}(s) = (-1)^n F^{(n)}(s)$.

▶ If f is piecewise continuous and of exponential order, $\lim_{s\to\infty} F(s) = 0$.

Section 5.3 If f is piecewise continuous and of exponential order and $\mathcal{L}\{f\} = F$, then f is the **inverse transform** of F and is denoted by $f = \mathcal{L}^{-1}\{F\}$.

▶ $\mathcal{L}^{-1}$ is a **linear operator**, $\mathcal{L}^{-1}\{c_1 F_1 + c_2 F_2\} = c_1 \mathcal{L}^{-1}\{F_1\} + c_2 \mathcal{L}^{-1}\{F_2\}$.

▶ Inverse transforms of many functions can be found using (i) the linearity of $\mathcal{L}^{-1}$, (ii) **partial fraction decompositions**, and (iii) a table of Laplace transforms. Labor can be greatly reduced, or eliminated, by using a computer algebra system.

Section 5.4 The Laplace transform is used to solve initial value problems for linear constant coefficient differential equations and systems of linear constant coefficient differential equations.

Section 5.5 Discontinuous functions are modeled using the **unit step function**

$$u_c(t) = \begin{cases} 0, & t < c, \\ 1, & t \geq c. \end{cases}$$

▶ $\mathcal{L}\{u_c(t)f(t-c)\} = e^{-cs}\mathcal{L}\{f(t)\} = e^{-cs}F(s).$

▶ If f is **periodic with period** T and is piecewise continuous on $[0, T]$, then

$$\mathcal{L}\{f(t)\} = \frac{\int_0^T e^{-st}f(t)\,dt}{1 - e^{-sT}}.$$

Section 5.6 The Laplace transform is convenient for solving constant coefficient differential equations and systems with discontinuous forcing functions.

Section 5.7 Large magnitude inputs of short duration are modeled by the **unit impulse**, or **Dirac delta function** defined by

$$\delta(t - t_0) = 0 \text{ if } t \neq t_0, \text{ and}$$

$$\int_a^b f(t)\delta(t - t_0)\,dt = f(t_0) \text{ if } a \leq t_0 < b \text{ and } f \text{ is continuous.}$$

▶ $\mathcal{L}\{\delta(t - t_0)\} = e^{-st_0}$

▶ $u'(t - t_0) = \delta(t - t_0)$ in a generalized sense.

Section 5.8 The **convolution** of f and g is defined by $(f * g)(t) = \int_0^t f(t - \tau)g(\tau)\,d\tau.$

▶ **The Convolution Theorem.** $\mathcal{L}\{(f * g)(t)\} = F(s)G(s)$

▶ The **transfer function** for $ay'' + by' + cy = g(t)$ is $H(s) = 1/(as^2 + bs + c)$ and the corresponding **impulse response** is $h(t) = \mathcal{L}^{-1}\{H(s)\}.$

▶ The **forced response** for $ay'' + by' + cy = g(t)$ is $H(s)G(s)$ in the s-domain and $\int_0^t h(t - \tau)g(\tau)\,d\tau$ in the t-domain.

Section 5.9 Transfer functions for many feedback control systems modeled by linear constant coefficient differential equations are of the form

$$H(s) = \frac{P(s)}{Q(s)} = \frac{b_m s^m + b_{m-1}s^{m-1} + \cdots + b_0}{s^n + a_{n-1}s^{n-1} + \cdots + a_0}$$

where $m < n$ (**strictly rational transfer functions**). The roots of $Q(s)$ are the **poles** of the transfer function.

▶ An input-output system with a strictly rational transfer function is **bounded input-bounded output stable** if and only if all of the poles have negative real parts.

▶ The **Routh criterion** can be used to determine whether the real parts of the roots of a polynomial are negative in terms of the coefficients of the polynomial.

PROJECTS

Project 1 An Electric Circuit Problem

Assume the following values for the elements in the electric circuit shown in Figure 5.P.1:

$$R = 0.01 \text{ ohm}$$
$$L_1 = 1 \text{ henry}$$
$$L_2 = 1 \text{ henry}$$
$$C = 1 \text{ farad}$$

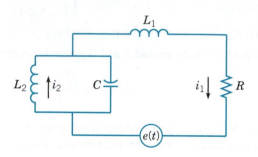

FIGURE 5.P.1

Project 1 PROBLEMS

1. Find $G(s)$ such that the Laplace transform of the charge on the capacitor, $Q(s) = \mathcal{L}\{q(t)\}$, can be expressed as $Q(s) = G(s)E(s)$ where $E(s) = \mathcal{L}\{e(t)\}$ is the Laplace transform of the impressed voltage. Assume that at time $t = 0$ the charge on the capacitor is zero and the currents i_1 and i_2 are zero.

 2. Suppose that the impressed voltage is prescribed by the square wave

$$e(t) = \begin{cases} 1, & 0 \le t < 0.6\pi, \\ -1, & 0.6\pi \le t < 1.2\pi; \end{cases}$$

and $e(t)$ has period 1.2π.

Assuming zero initial conditions as stated in Problem 1, find an expression for $q(t)$ and plot the graphs of $e(t)$ and $q(t)$ on the same set of coordinate axes. Explain the behavior of $q(t)$.

3. Suppose that the impressed voltage is prescribed CAS by the square wave

$$e(t) = \begin{cases} 1, & 0 \le t < \pi/\sqrt{2}, \\ -1, & \pi/\sqrt{2} \le t < \sqrt{2}\pi; \end{cases}$$

and $e(t)$ has period $\sqrt{2}\pi$.

Assuming zero initial conditions as stated in Problem 1, find an expression for $q(t)$ and plot the graphs of $e(t)$ and $q(t)$ on the same set of coordinate axes. Explain the behavior of $q(t)$.

Project 2 Effects of Pole Locations on Step Responses of Second Order Systems

A technique commonly used in engineering system design is to analyze the unit step response, that is, the response of the system to a unit step input function. This provides information about transient behavior as the system changes from one operating point to another. The transient behavior is usually described by a number of performance characteristics associated with the unit step response. The purpose of this project is to present some of these performance characteristics and draw relationships between them and pole locations of the transfer function in the context of a second order system expressed in the following form convenient for analysis,

$$y'' + 2\zeta\omega_0 y' + \omega_0^2 y = \omega_0^2 u(t), \qquad y(0) = 0, \quad y'(0) = 0. \tag{1}$$

The parameter ω_0 is the **undamped natural frequency** of the system and ζ is the **damping ratio**. We assume that $\omega_0 > 0$ and $\zeta \geq 0$. The equation in the initial value problem (1) could be a model for a damped mass–spring system, an RLC circuit, or a closed loop system where the parameters ζ and ω_0 are related to gain constants that appear in the feed forward and feedback loops. In the latter case, the parameters are normally chosen to achieve some set of performance specifications. Not only is the second order model relatively easy to analyze, performance specifications for the second order case serve as a point of reference for higher order systems (see Problem 9). The transfer function of the differential equation in the initial value problem (1) is

$$H(s) = \frac{\omega_0^2}{s^2 + 2\zeta\omega_0 s + \omega_0^2}, \tag{2}$$

and the corresponding unit step response is

$$Y(s) = \frac{1}{s}H(s) = \frac{\omega_0^2}{s(s^2 + 2\zeta\omega_0 s + \omega_0^2)}. \tag{3}$$

However, transient performance characteristics are more naturally described in the time domain. Two commonly used performance characteristics (see Figure 5.P.2) associated with the unit step response are:

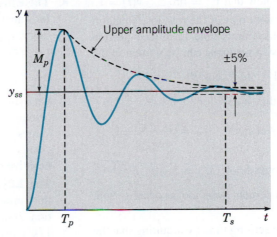

FIGURE 5.P.2 The unit step response for an underdamped $(0 < \zeta < 1)$ second order system.

1. **Overshoot** M_p. The maximum difference between the transient and steady state solutions for a unit step input. It is often represented as a percentage of the final steady-state value, y_{ss}, of the output,

$$\text{percent overshoot} = \frac{M_p - y_{ss}}{y_{ss}} \times 100.$$

2. **Settling Time** T_s. The time required for the unit step response to reach and remain within a specified percentage of its final value. A typical nominal value is 5%.

We restrict attention to the underdamped case, $0 < \zeta < 1$. Then the transfer function $H(s)$ has a pair of conjugate complex poles and no zeros in the finite s-plane. In the following problems, we study the relationship between pole locations of $H(s)$ and transient behavior characterized by M_p and T_s.

Project 2 PROBLEMS

1. If the dimensionless time $\tau = \omega_0 t$ is introduced in Eq. (1), show that the only parameter left in the problem is ζ. Thus, changing the value of ω_0 in Eq. (1) is tantamount to changing the time scale of the response.

CAS 2. Assume that the poles of the transfer function (2) are given by $s = -\alpha \pm i\beta$. Show that $\alpha = \zeta \omega_0$ and $\beta = \omega_0 \sqrt{1 - \zeta^2}$. Then find the polar coordinate representation of the poles as a function of ζ. For a fixed value of ω_0, describe the locus plot of the poles as ζ varies from 0 to 1. For a fixed value of ζ, describe the locus plot of the poles for ω_0 increasing from 0.

3. For $0 < \zeta < 1$ find the unit step response in the time domain, $y(t)$, by calculating the inverse transform of $Y(s)$ in Eq. (3). Show that $y_{ss} = \lim_{t \to \infty} y(t) = 1$.

CAS 4. Express the transient part of the response found in Exercise 3 in phase amplitude form (see Section 4.5) so that the step response can be written in the form

$$y(t) = 1 - Ae^{-\alpha t}\cos(\beta t - \phi).$$

Then show that the

time to maximum overshoot, $\quad T_p = \dfrac{2\pi}{\omega_0 \sqrt{1 - \zeta^2}}$,

and the maximum overshoot, $M_p = 1 + e^{-\zeta \pi / \sqrt{1 - \zeta^2}}$.

Plot the graph of percent overshoot as ζ varies from 0 to 1.

5. Approximate the settling time by requiring that the amplitude envelope $Ae^{-\alpha t}$ of the transient part of the solution at time T_s is less than .05 times its original value. Thus show that

$$T_s = \frac{3}{\zeta \omega_0}.$$

CAS 6. Holding the value of ω_0 equal to 1, plot the poles of $H(s)$ in the complex s-plane for each of the following values of ζ: 0.1, 0.3, 0.5, $1/\sqrt{2}$, and 0.9. Then for each value of ζ, plot the corresponding unit step response $y(t)$ on a single set of coordinate

axes. From the graphs explain in terms of maximum overshoot and settling time the changes observed in the step response as the corresponding pole locations change. Explain why $\zeta = 1/\sqrt{2}$ may be a good practical choice for the damping ratio if it is desired to minimize overshoot and settling time, simultaneously.

7. Holding the value of ζ fixed at 0.5, plot the poles CAS of $H(s)$ in the complex s-plane for each of the following values of ω_0: 0.5, 1.0, 1.5, and 2.0. Then for each value of ω_0, plot the corresponding unit step response $y(t)$ on a single set of coordinate axes. From the graphs, explain in terms of maximum overshoot and settling time the changes observed in the step responses as the corresponding pole locations change.

8. **The Effect of an Added Zero**. Consider the trans- CAS fer function $H_1(s)$ obtained from $H(s)$ by setting $\zeta = 1/\sqrt{2}$, $\omega_0 = \sqrt{2}$ and including a zero at $s = -z_0$,

$$H_1(s) = \frac{2s/z_0 + 2}{s^2 + 2s + 2}.$$

On a single set of coordinate axes, plot the unit step response, $y_1(t)$ of $H_1(s)$ for each of the following values of z_0: 0.5, 1.0, 3.0, and 5.0. Describe the change in the response as the zero moves along the negative real axis toward the origin.

9. **The Effect of an Additional Pole**. Consider the CAS third order system represented by the transfer function $H_2(s)$ obtained from $H(s)$ by setting $\zeta = 1/\sqrt{2}$, $\omega_0 = \sqrt{2}$ and including an additional pole at $s = -\lambda_0$,

$$H_2(s) = \frac{2\lambda_0}{s(s + \lambda_0)(s^2 + 2s + 2)}.$$

On a single set of coordinate axes, plot the unit step response $y_2(t)$ of $H_2(s)$ for each of the following values of λ_0: 0.5, 1.0, 3.0, and 5.0. Verify that the step response of $H_2(s)$ is very well approximated by the step response of the second order system $H(s)$ provided that the pole located at $-\lambda_0$ lies sufficiently far to the left of the real parts of the poles of $H(s)$.

Project 3 The Watt Governor, Feedback Control, and Stability

In the latter part of the 18th century, James Watt designed and built a steam engine with a rotary output motion (see Figure 5.P.3).

It was highly desirable to maintain a uniform rotational speed for powering various types of machinery, but fluctuations in steam pressure and work load on the engine caused the rotational speed to vary. At first, the speed was controlled manually by using a throttle valve to vary the flow of steam to the engine inlet. Then, using principles observed in a device for controlling the speed of the grinding stone in a wind-driven flour mill, Watt designed a **flyball** or **centrifugal governor**, based on the motion of a pair of centrifugal pendulums, to regulate the angular velocity of the steam engine's flywheel. A sketch of the essential components of the governor and its mechanical linkage to the throttle valve is shown in Figure 5.P.4.

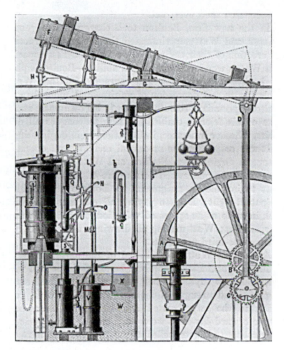

FIGURE 5.P.3 The Watt steam engine, 1781.

To understand how the mechanism automatically controls the angular velocity of the steam engine's flywheel and drive shaft assembly, assume that the engine is operating under a constant load at a desired equilibrium speed or operating point. The engine, via a belt and pulley assembly, also rotates a straight vertical shaft to which a pair of flyballs are connected. If the load on the engine is decreased or the steam pressure driving the engine increases, the engine speed increases. A corresponding increase in the rotational speed of the flyball shaft simultaneously causes the flyballs to swing outward due to an increase in centrifugal force. This motion, in turn, causes a sliding collar on the vertical shaft to move downward. A lever arm connected to the sliding collar on one end and a throttle control rod on the other end then partially closes the throttle valve, the steam flow to the engine is reduced, and the engine returns to its operating point. Adjustable elements in the linkage allow a desirable engine speed to be set during the startup phase of operation.

A Problem with Instability. While some governors on some steam engines were successful in maintaining constant rotational speed, others would exhibit an anomalous rhythmic

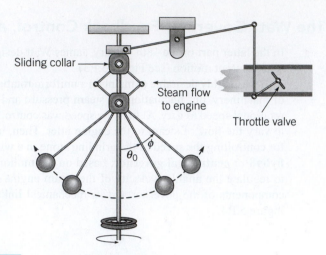

FIGURE 5.P.4   The centrifugal governor.

oscillation in which the system appeared to "hunt" unsuccessfully about its equilibrium value for a constant speed. The amplitude of oscillations would increase until limited by mechanical constraints on the motion of the flyballs or the throttle valve.

The purpose of this project is to develop a relatively simple mathematical model of the feedback control system and to gain an understanding of the underlying cause of the unstable behavior. We will need a differential equation for the angular velocity of the steam engine flywheel and drive shaft and a differential equation for the angle of deflection between the flyball connecting arms and the vertical shaft about which the flyballs revolve. Furthermore, the equations need to be coupled to account for the mechanical linkage between the governor and the throttle valve.

The Flyball Motion. The model for the flyball governor follows from taking into account all of the forces acting on the flyball and applying Newton's law, $ma = F$. The angle between the flyball connecting arm and the vertical shaft about which the flyballs revolve will be denoted by θ. Assuming that the angular velocity of the vertical shaft and the rotational speed of the engine have the same value, Ω, there is a centrifugal acceleration acting on the flyballs in the outward direction of magnitude $\Omega^2 L \sin\theta$ (see Figure 5.P.5).

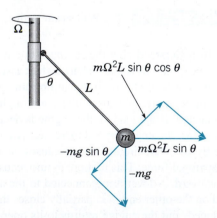

FIGURE 5.P.5   The angle of deflection of the centrifugal pendulum is determined by the opposing components of gravitational force and centrifugal force.

Recall from calculus that the magnitude of this acceleration is the curvature of the motion, $1/L \sin\theta$, times the square of the tangential velocity, $\Omega^2 L^2 \sin^2\theta$. Taking into account the force due to gravitational acceleration and assuming a damping force of magnitude $\gamma\theta'$, we obtain the following equation for θ,

$$mL\theta'' = -\gamma\theta' - mg\sin\theta + m\Omega^2 L \sin\theta\cos\theta. \tag{1}$$

Eq. (1) results from equating components of inertial forces to impressed forces parallel to the line tangent to the circular arc along which the flyball moves in the vertical plane determined by the pendulum arm and the vertical shaft (Problem 1).

Angular Velocity of the Steam Engine Flywheel and Drive Shaft. The equation for the rotational speed of the flywheel and drive shaft assembly of the steam engine is assumed to be

$$J\frac{d\Omega}{dt} = -\beta\Omega + \tau \tag{2}$$

where J is the moment of inertia of the flywheel, the first term on the right is the torque due to the load, and the second term on the right is the steam generated torque referred to the drive shaft.

Linearization About the Operating Point. In order to use the feedback control concepts of Section 5.9, it is necessary to determine the linear equations that are good approximations to Eqs. (1) and (2) when θ and Ω are near their equilibrium operating points, a mathematical technique known as **linearization**. The equilibrium operating point of the steam engine will be denoted by Ω_0, the equilibrium angle that the flyball connecting arm makes with the vertical will be denoted by θ_0, and the equilibrium torque delivered to the engine drive shaft will be denoted by τ_0. Note that in the equilibrium state, corresponding to $\theta'' = 0$, $\theta' = 0$, and $\Omega' = 0$, Eqs. (1) and (2) imply that

$$g = L\Omega_0^2 \cos\theta_0. \tag{3}$$

and

$$\tau_0 = \beta\Omega_0. \tag{4}$$

To linearize Eqs. (1) and (2) about θ_0, Ω_0, and τ_0, we assume that $\theta = \theta_0 + \phi$, $\Omega = \Omega_0 + y$, and $\tau = \tau_0 + u$ where ϕ, y, and u are perturbations that are small relative to θ_0, Ω_0, and τ_0, respectively. Note that ϕ represents the **error** in deflection of the flyball connecting arm from its desired value θ_0, in effect, measuring or sensing the error in the rotational speed of the engine. If $\phi > 0$, the engine is rotating too rapidly and must be slowed down; if $\phi < 0$, the engine is rotating too slowly and must be sped up. Substituting the expressions for θ, Ω, and τ into Eqs. (1) and (2), retaining only linear terms in ϕ, y, and u, and making use of Eqs. (3) and (4) yields

$$\phi'' + 2\delta\phi' + \omega_0^2\phi = \alpha_0 y \tag{5}$$

and

$$Jy' = -\beta y + u \tag{6}$$

where $\delta = \gamma/2mL$, $\omega_0^2 = \Omega_0^2 \sin^2\theta_0$, and $\alpha_0 = \Omega_0 \sin 2\theta_0$ (Problem 3).

The Closed Loop System. Regarding the error in rotational speed, y, as the input, the transfer function associated with Eq. (5) is easily seen to be

$$G(s) = \frac{\alpha_0}{s^2 + 2\delta s + \omega_0^2}.$$

Thus $\Phi(s) = G(s)Y(s)$ where $\Phi(s) = \mathcal{L}\{\phi(t)\}$ and $Y(s) = \mathcal{L}\{y(t)\}$. Regarding u as the input, the transfer function associated with Eq. (6) is

$$H(s) = \frac{1}{Js + \beta}$$

so that $Y(s) = H(s)U(s)$ where $U(s) = \mathcal{L}\{u(t)\}$. The closed loop system is synthesized by subtracting $\Phi(s)$ at the summation point as shown in Figure 5.P.6.

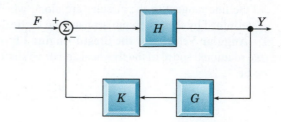

FIGURE 5.P.6   Block diagram of the feedback control system corresponding to the Watt governor linearized about the operating point.

Thus a positive error in deflection of the flyball connecting arm causes a decrease in steam generated torque and vice-versa. A proportional gain constant K has been inserted into the feedback loop to model the sensitivity of the change in steam generated torque to the error $\Phi(s)$ in the deflection of the flyball connecting arm. Physically, K can be altered by changing the location of the pivot point of the lever arm that connects the sliding collar on the governor to the vertical rod attached to the throttle valve. Note that any external input affecting engine speed is represented by $F(s)$.

Project 3 PROBLEMS

In the following problems, we ask the reader to supply some of the details left out of the above discussion, to analyze the closed loop system for stability properties, and to conduct a numerical simulation of the nonlinear system.

1. Work out the details leading to Eq. (1).
2. Give physical explanations for the meaning of Eqs. (3) and (4).
3. Derive the linearized system (5) and (6).
4. Show that the transfer function of the closed loop system linearized about the operating point is

$$H_K(s) = \frac{s^2 + 2\delta s + \omega_0^2}{(Js + \beta)(s^2 + 2\delta s + \omega_0^2) + K}.$$

5. Use the Routh Criterion to show that if the gain factor K is sufficiently large, $H_K(s)$ will have two poles with positive real parts and the corresponding closed loop system is therefore unstable. Derive an expression for K_c, that value of K at which the real parts of a pair of conjugate complex poles of $H_K(s)$ are equal to 0.

6. **The Nonlinear Feedback Control System.** Using the relations (3) and (4), show that Eqs. (1) and (2) can be expressed in terms of ϕ and y as

$$mL\phi'' = -\gamma\phi - m\Omega_0^2 L \cos(\theta_0)\sin(\theta_0 + \phi)$$
$$+ m(\Omega_0 + y)^2 L \sin(\theta_0 + \phi)\cos(\theta_0 + \phi)$$

$$\text{(i)}$$

and

$$Jy' = -\beta y - K\phi \tag{ii}$$

where the negative feedback loop has been incorporated into the nonlinear system.

7. **Simulations.** Consider the following parameter values expressed in SI units

$$m = 12, \quad L = 1/2, \quad g = 9.8, \quad \gamma = .01,$$
$$J = 400, \quad \beta = 20, \quad \theta_0 = \pi/6,$$

with Ω_0 determined by the relation (3).

(a) Using the above parameter values, construct CAS a root-locus plot of the poles of $H_K(s)$ as K varies over an interval containing K_c. Verify that a pair of poles cross from the left half plane to the right half plane as K increases through K_c.

(b) Conduct computer simulations of the system (i), (ii) in Problem 6 using the above parameter values. Do this for various values of K less than K_c and greater than K_c while experimenting with different values of $y(0) \neq 0$ to represent departures from the equilibrium operating point. For simplicity, you may always assume that $\phi(0) = \phi'(0) = 0$. Plot graphs of the functions $\phi(t)$ and $y(t)$ generated from the simulations. Note that if $\phi(t)$ wanders outside the interval $[-\theta_0, \pi - \theta_0]$, the results are nonphysical (Why?). Are the results of your simulations consistent with your theoretical predictions? Discuss the results of your computer experiments addressing such issues as whether the feedback control strategy actually works, good choices for K, stability and instability, and the "hunting" phenomenon discussed above.

Systems of First Order Linear Equations

I n this chapter, we build on elementary theory and solution techniques for first order linear systems introduced in a two-dimensional setting in Chapters 3 and 4. Science and engineering applications that possess even a modest degree of complexity often lead to systems of differential equations of dimension $n > 2$. The language, concepts, and tools of linear algebra, combined with certain elements of calculus, are essential for the study of these systems. The required mathematical background from matrices and linear algebra is presented in Appendix A. A reader with little or no previous exposure to this material will find it necessary to study some or all of Appendix A before proceeding on to Chapter 6. Alternatively, the reader may take up the study of Chapter 6 straightaway, drawing on necessary results from Appendix A as needed.

A large portion of Chapter 6 generalizes the eigenvalue method presented in Chapter 3 to constant coefficient linear systems of dimension $n > 2$. Since carrying out eigenvalue calculations by hand for matrices of dimension greater than 2 is time consuming, tedious, and susceptible to error, we strongly recommend using a computer or a calculator to perform the required calculations, once an understanding of the general theory and methodology has been acquired.

6.1 Definitions and Examples

First order linear systems of dimension 2,

$$x' = p_{11}(t)x + p_{12}(t)y + g_1(t),$$
$$y' = p_{21}(t)x + p_{22}(t)y + g_2(t),$$

(1)

were first introduced in Section 3.2. We now discuss the general mathematical framework for first order linear systems of dimension n, followed by additional examples from science and engineering. The matrix algebra required to understand the general framework is presented in Appendix A.1.

▶ **Matrix-Valued Functions.** The principal mathematical objects involved in the study of linear systems of differential equations are **matrix-valued functions**, or simply **matrix functions**. These objects are vectors or matrices whose elements are functions of t. We write

$$\mathbf{x}(t) = \begin{pmatrix} x_1(t) \\ \vdots \\ x_n(t) \end{pmatrix}, \qquad \mathbf{P}(t) = \begin{pmatrix} p_{11}(t) & \cdots & p_{1n}(t) \\ \vdots & & \vdots \\ p_{n1}(t) & \cdots & p_{nn}(t) \end{pmatrix},$$

respectively.

The matrix $\mathbf{P} = \mathbf{P}(t)$ is said to be continuous at $t = t_0$ or on an interval $I = (\alpha, \beta)$ if each element of $\mathbf{P}$ is a continuous function at the given point or on the given interval. Similarly, $\mathbf{P}(t)$ is said to be differentiable if each of its elements is differentiable, and its derivative $d\mathbf{P}/dt$ is defined by

$$\frac{d\mathbf{P}}{dt} = \left(\frac{dp_{ij}}{dt} \right).$$

(2)

In other words, each element of $d\mathbf{P}/dt$ is the derivative of the corresponding element of $\mathbf{P}$. Similarly, the integral of a matrix function is defined as

$$\int_a^b \mathbf{P}(t)\, dt = \left(\int_a^b p_{ij}(t)\, dt \right).$$

(3)

Many of the results of elementary calculus extend easily to matrix functions, in particular,

$$\frac{d}{dt}(\mathbf{CP}) = \mathbf{C}\frac{d\mathbf{P}}{dt}, \quad \text{where } \mathbf{C} \text{ is a constant matrix,}$$

(4)

$$\frac{d}{dt}(\mathbf{P} + \mathbf{Q}) = \frac{d\mathbf{P}}{dt} + \frac{d\mathbf{Q}}{dt},$$

(5)

$$\frac{d}{dt}(\mathbf{PQ}) = \mathbf{P}\frac{d\mathbf{Q}}{dt} + \frac{d\mathbf{P}}{dt}\mathbf{Q}.$$

(6)

In Eqs. (4) and (6), care must be taken in each term to avoid interchanging the order of multiplication. The definitions expressed by Eqs. (2) and (3) also apply as special cases to vectors.

First Order Linear Systems: General Framework

The general form of a first order linear system of dimension n is

$$
\begin{aligned}
x_1' &= p_{11}(t)x_1 + p_{12}(t)x_2 + \cdots + p_{1n}(t)x_n + g_1(t), \\
x_2' &= p_{21}(t)x_1 + p_{22}(t)x_2 + \cdots + p_{2n}(t)x_n + g_2(t), \\
&\;\;\vdots \\
x_n' &= p_{n1}(t)x_1 + p_{n2}(t)x_2 + \cdots + p_{nn}(t)x_n + g_n(t),
\end{aligned}
\tag{7}
$$

or, using matrix notation,

$$
\mathbf{x}' = \mathbf{P}(t)\mathbf{x} + \mathbf{g}(t)
\tag{8}
$$

where

$$
\mathbf{P}(t) =
\begin{pmatrix}
p_{11}(t) & \cdots & p_{1n}(t) \\
\vdots & & \vdots \\
p_{n1}(t) & \cdots & p_{nn}(t)
\end{pmatrix}
$$

is referred to as the **matrix of coefficients** of the system (7) and

$$
\mathbf{g}(t) =
\begin{pmatrix}
g_1(t) \\
\vdots \\
g_n(t)
\end{pmatrix}
$$

is referred to as the **nonhomogeneous** term of the system. We will assume that $\mathbf{P}(t)$ and $\mathbf{g}(t)$ are continuous on an interval $I = (\alpha, \beta)$. If $\mathbf{g}(t) = \mathbf{0}$ for all $t \in I$, then the system (7) or (8) is said to be **homogeneous**, otherwise the system is said to be **nonhomogeneous**. The function $\mathbf{g}(t)$ is often referred to as the **input**, or **forcing function**, to the system (8) and provides a means for modeling interaction between the physical system represented by $\mathbf{x}' = \mathbf{P}(t)\mathbf{x}$ and the world external to the system.

The system (8) is said to have a solution on the interval I if there exists a vector

$$
\mathbf{x} = \boldsymbol{\phi}(t)
\tag{9}
$$

with n components that is differentiable at all points in the interval I and satisfies Eq. (8) at all points in this interval. In addition to the system of differential equations, there may also be given an initial condition of the form

$$
\mathbf{x}(t_0) = \mathbf{x}_0
\tag{10}
$$

where t_0 is a specified value of t in I and $\mathbf{x}_0$ is a given constant vector with n components. The system (8) and the initial condition (10) together form an **initial value problem**.

The initial value problem (8), (10) generalizes the framework for 2-dimensional systems presented in Chapter 3 to systems of dimension n. A solution (9) can be viewed as a set of parametric equations in an n-dimensional space. For a given value of t, Eq. (9) gives values of the coordinates $x_1, x_2, \ldots, x_n$ of a point $\mathbf{x}$ in n-dimensional space, $\mathbf{R}^n$. As t changes, the coordinates in general also change. The collection of points corresponding to $\alpha < t < \beta$ form a curve in the space. As with 2-dimensional systems, it is often helpful to think of the curve as the trajectory, or path, of a particle moving in accordance with the system (8). The initial condition (10) determines the starting point of the moving particle. The components of $\mathbf{x}$ are again referred to as **state variables** and the vector $\mathbf{x} = \boldsymbol{\phi}(t)$ is referred to as the **state of the system** at time t. The initial condition (10) prescribes the state of the system at time t_0. Given the initial condition (10), the differential equation (8) is a rule for advancing the state of the system through time.

▶ **Linear *n*th Order Equations.** Single equations of higher order can always be transformed into systems of first order equations. This is usually required if a numerical approach is planned, because almost all codes for generating numerical approximations to solutions of differential equations are written for systems of first order equations. An *n*th order linear differential equation in standard form is given by

$$\frac{d^n y}{dt^n} + p_1(t)\frac{d^{n-1}y}{dt^{n-1}} + \cdots + p_{n-1}(t)\frac{dy}{dt} + p_n(t)y = g(t). \tag{11}$$

We will assume that the functions $p_1, \ldots, p_n$ and g are continuous real-valued functions on some interval $I = (\alpha, \beta)$. Since Eq. (11) involves the *n*th derivative of y with respect to t, it will, so to speak, require n integrations to solve Eq. (11). Each of these integrations introduces an arbitrary constant. Hence, we can expect that, to obtain a unique solution, it is necessary to specify n initial conditions,

$$y(t_0) = y_0, \quad y'(t_0) = y_1, \quad \ldots, \quad y^{(n-1)}(t_0) = y_{n-1}. \tag{12}$$

To transform Eq. (11) into a system of n first order equations, we introduce the variables $x_1, x_2, \ldots, x_n$ defined by

$$x_1 = y, \quad x_2 = y', \quad x_3 = y'', \quad \ldots, \quad x_n = y^{(n-1)}. \tag{13}$$

It then follows immediately that

$$\begin{aligned} x_1' &= x_2, \\ x_2' &= x_3, \\ &\vdots \\ x_{n-1}' &= x_n, \end{aligned} \tag{14}$$

and, from Eq. (11),

$$x_n' = -p_n(t)x_1 - p_{n-1}(t)x_2 - \cdots - p_1(t)x_n + g(t). \tag{15}$$

Using matrix notation, the system (14), (15) can be written as

$$\mathbf{x}' = \begin{pmatrix} 0 & 1 & 0 & 0 & \cdots & 0 \\ 0 & 0 & 1 & 0 & \cdots & 0 \\ \vdots & \vdots & \vdots & \vdots & & \vdots \\ 0 & 0 & 0 & 0 & \cdots & 1 \\ -p_n(t) & -p_{n-1}(t) & -p_{n-2}(t) & -p_{n-3}(t) & \cdots & -p_1(t) \end{pmatrix} \mathbf{x} + \begin{pmatrix} 0 \\ 0 \\ \vdots \\ 0 \\ g(t) \end{pmatrix} \tag{16}$$

where $\mathbf{x} = (x_1, \ldots, x_n)^T$. Using the definitions for the state variables in the list (13), the initial condition for Eq. (16) is expressed by

$$\mathbf{x}(t_0) = \begin{pmatrix} y_0 \\ y_1 \\ \vdots \\ y_{n-1} \end{pmatrix}. \tag{17}$$

Applications Modeled by First Order Linear Systems

In previous chapters, we have primarily encountered first order systems of dimension 2 although systems of higher dimension appear in some of the projects. We now present several additional examples of applications that illustrate how higher dimensional linear systems can arise.

▶ Coupled Mass-Spring Systems. Interconnected systems of masses and springs are often used as a starting point in modeling flexible mechanical structures or other systems idealized as an assortment of elastically coupled bodies (for an example, see Project 2 at the end of this chapter). As a consequence of Newton's second law of motion, the mathematical description of the dynamics results in a coupled system of second order equations. Using the same technique that was demonstrated above to convert a single higher order equation into a system of first order equations, we show how a system of second order equations is easily converted to a first order system. Consider two masses, m_1 and m_2, connected to three springs in the arrangement shown in Figure 6.1.1(a).

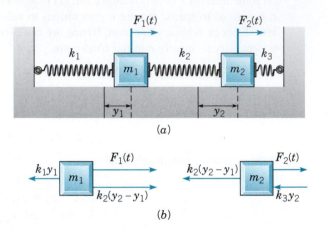

(a)

(b)

FIGURE 6.1.1 (a) A two-mass, three-spring system. (b) Free body diagrams of the forces acting on each of the masses.

The masses are constrained to move only in the horizontal direction on a frictionless surface under the influence of external forces $F_1(t)$ and $F_2(t)$. The springs, assumed to obey Hooke's Law, have spring constants k_1, k_2, and k_3 and, when the system is at equilibrium, we assume that the springs are at their rest lengths. Displacements of m_1 and m_2 from their equilibrium positions are denoted by y_1 and y_2, respectively. For simplicity, we neglect the effects of friction. Aided by the free body diagrams in Figure 6.1.1(b), the following differential equations result from equating the inertial force of each mass (mass times acceleration) to the totality of external forces acting on the mass,

$$
\begin{aligned}
m_1 \frac{d^2 y_1}{dt^2} &= k_2(y_2 - y_1) - k_1 y_1 + F_1(t) \\
&= -(k_1 + k_2)y_1 + k_2 y_2 + F_1(t), \\
m_2 \frac{d^2 y_2}{dt^2} &= -k_3 y_2 - k_2(y_2 - y_1) + F_2(t) \\
&= k_2 y_1 - (k_2 + k_3)y_2 + F_2(t).
\end{aligned}
\tag{18}
$$

For example, if $y_2 > y_1 > 0$ as shown in Figure 6.1.1(a), then the force ky_1 exerted on m_1 by the left-most spring points in the negative direction since that spring is in an elongated state while the force $k_2(y_2 - y_1)$ exerted on m_1 by the middle spring points in the positive direction since that spring is also in an elongated state. The second equation follows by applying analogous reasoning to m_2. Considering instantaneous configurations other than $y_2 > y_1 > 0$ yields the same set of equations. Specifying the displacement and velocity of each mass at time $t = 0$ provides initial conditions for the system (18),

$$
y_1(0) = y_{10}, \quad y_2(0) = y_{20}, \quad y_1'(0) = v_{10}, \quad y_2'(0) = v_{20}.
\tag{19}
$$

The second order system (18) is subsumed within the framework of first order systems by introducing the state variables

$$x_1 = y_1, \quad x_2 = y_2, \quad x_3 = y_1', \quad x_4 = y_2'.$$

Then

$$
\begin{aligned}
x_1' &= x_3, \\
x_2' &= x_4, \\
x_3' &= -\frac{k_1 + k_2}{m_1}x_1 + \frac{k_2}{m_1}x_2 + \frac{1}{m_1}F_1(t), \\
x_4' &= \frac{k_2}{m_2}x_1 - \frac{k_2 + k_3}{m_2}x_2 + \frac{1}{m_2}F_2(t),
\end{aligned}
\tag{20}
$$

where we have used Eqs. (18) to obtain the last two equations. Using matrix notation the system (20) is expressed as

$$
\mathbf{x}' =
\begin{pmatrix}
0 & 0 & 1 & 0 \\
0 & 0 & 0 & 1 \\
-(k_1 + k_2)/m_1 & k_2/m_1 & 0 & 0 \\
k_2/m_2 & -(k_2 + k_3)/m_2 & 0 & 0
\end{pmatrix}
\mathbf{x} +
\begin{pmatrix}
0 \\
0 \\
F_1(t)/m_1 \\
F_2(t)/m_2
\end{pmatrix}
\tag{21}
$$

where $\mathbf{x} = (x_1, x_2, x_3, x_4)^T$. The initial condition for the first order system (21) is then $\mathbf{x}(0) = (y_{10}, y_{20}, v_{10}, v_{20})^T$. Using the parameter values $m_1 = m_2 = k_1 = k_2 = k_3 = 1$ and under the condition of zero inputs, Figure 6.1.2 shows component plots of the solution of Eq. (21) subject to the initial condition $\mathbf{x}(0) = (-3, 2, 0, 0)^T$, that is, the left mass is pulled 3 units to the left and released with zero initial velocity while the right mass is simultaneously pulled 2 units to the right and released with zero initial velocity.

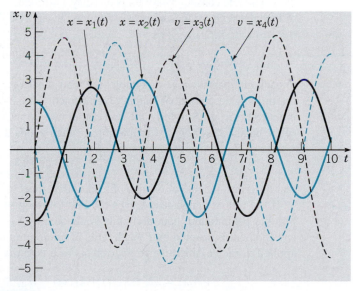

FIGURE 6.1.2   The response (displacements and velocities of the masses) of the system (21) subject to the initial conditions $x_1(0) = -3$, $x_2(0) = 2$, $x_3(0) = x_4(0) = 0$; $m_1 = m_2 = k_1 = k_2 = k_3 = 1$ and $F_1(t) = F_2(t) = 0$.

Solutions of Eq. (21) are analyzed in Section 6.4 using the eigenvalue method.

▶ Compartment Models. A frequently used modeling paradigm idealizes the physical, biological, or economic system as a collection of subunits, or **compartments**, which exchange contents (matter, energy, capital, and so forth) with one another. The amount of material in each compartment is represented by a component of a state vector and differential equations are used to describe transport of material between compartments. For example, a model for lead in the human body views the body as composed of blood (Compartment 1), soft tissue (Compartment 2), and bones (Compartment 3) as shown in Figure 6.1.3.

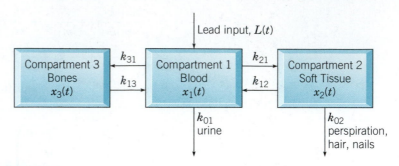

FIGURE 6.1.3 A three-compartment model of lead in the human body.

Lead enters the blood from the environment through the gastrointestinal tract and lungs and is taken up rapidly by the liver and kidneys. It is absorbed less rapidly by other soft tissues, and very slowly by the bones. Lead is then eliminated from the blood via urine, feces, skin, hair, and nails.

Denote the amount of lead in Compartment i by $x_i(t)$ and the rate at which lead moves from Compartment i to Compartment j by the rate constant k_{ji}. In this example, the amount of lead is measured in micrograms (1 microgram $= 10^{-6}$ gram) and time is measured in days. The rate constants have units of 1/day. Differential equations describing the exchange of lead between compartments are obtained by applying the mass balance law,

$$\frac{dx}{dt} = \text{input rate} - \text{output rate}, \tag{22}$$

to each compartment. For example, the lead input rate to Compartment 3 is obtained by multiplying the amount of lead in Compartment 1, x_1, by the rate constant k_{31}. The resulting input rate is $k_{31}x_1$. Similarly, the lead output rate from Compartment 3 is $k_{13}x_3$. Application of the balance law (22) to Compartment 3 then yields the differential equation

$$x_3' = k_{31}x_1 - k_{13}x_3$$

that describes the rate of change of lead with respect to time in Compartment 3.

If the rate at which lead leaves the body through urine is denoted by k_{01}, the rate of loss via perspiration, hair, and nails is denoted by k_{02}, and exposure level to lead in the environment is denoted by L (with units of microgram/day), the principle of mass balance applied to each of the three compartments results in the following system of equations,

$$
\begin{aligned}
x_1' &= (L + k_{12}x_2 + k_{13}x_3) - (k_{21} + k_{31} + k_{01})x_1, \\
x_2' &= k_{21}x_1 - (k_{02} + k_{12})x_2, \\
x_3' &= k_{31}x_1 - k_{13}x_3.
\end{aligned}
\tag{23}
$$

Note that these equations assume that there is no transfer of lead between soft tissue and bones.

We consider a case in which the amount of lead in each compartment is initially zero but the lead-free body is then subjected to a constant level of exposure over a 365 day time period. This is followed by complete removal of the source to see how quickly the amount of lead in each compartment decreases. In Problem 12, Section 6.6, you are asked to solve the initial value problem for the system (23) that yields the component plots shown in Figure 6.1.4.

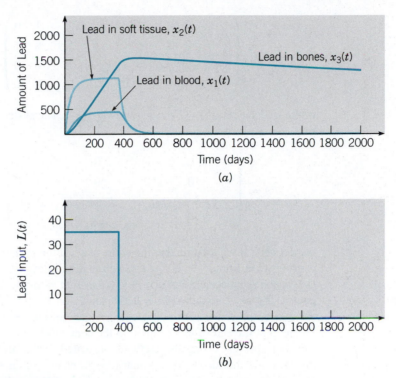

FIGURE 6.1.4 (a) Amounts of lead in body compartments during and after an exposure period of one year. (b) Lead input to Compartment 1 (blood).

Note that once the source is removed, lead is eliminated from the blood and soft tissues fairly rapidly, but persists in the bones for a much longer time. Mathematically, the slow decay in the amount of lead in the bones is due to the fact that the rate constant k_{13} is very small relative to the other rate constants.

▶ **Linear Control Systems.** There are many physical, biological, and economic systems in which it is desirable to **control**, or **steer**, the state of the system to some desired value or along some desired path in the state space. A standard mathematical model for linear control systems consists of the pair of equations

$$\mathbf{x}' = \mathbf{A}\mathbf{x} + \mathbf{B}\mathbf{u}(t) \tag{24}$$

$$\mathbf{y} = \mathbf{C}\mathbf{x} \tag{25}$$

where **A** is an $n \times n$ **system matrix**, **B** is an $n \times m$ **input matrix**, and **C** is an $r \times n$ **output matrix**. Eq. (24) is referred to as the **plant equation** while the linear algebraic equation (25) is referred to as the **output equation**. The output **y**, a vector function with r components, allows for the possibility that some components of **x** may not be directly observable, or for the possibility that only certain linear combinations of the state variables are observed or measured. The $m \times 1$ vector function $\mathbf{u}(t)$ is the **plant input**. Note that $\mathbf{B}\mathbf{u}(t)$, an $n \times 1$ vector, is simply a nonhomogeneous term in Eq. (24).

A common type of control problem is to choose or design the input $\mathbf{u}(t)$ in order to achieve some desired objective. As an example, consider again the vibration problem consisting of two masses and three springs shown in Figure 6.1.1. Given an initial state (initial position and velocity of each mass), suppose the objective is to bring the entire system to equilibrium during a specified time interval $[0, T]$ by applying a suitable forcing function only to the mass on the left. We may write the system of equations (21) in the form

$$\mathbf{x}' = \begin{pmatrix} 0 & 0 & 1 & 0 \\ 0 & 0 & 0 & 1 \\ -(k_1 + k_2)/m_1 & k_2/m_1 & 0 & 0 \\ k_2/m_2 & -(k_2 + k_3)/m_2 & 0 & 0 \end{pmatrix} \mathbf{x} + \begin{pmatrix} 0 \\ 0 \\ 1 \\ 0 \end{pmatrix} u(t) \tag{26}$$

where we have set $u(t) = F_1(t)/m_1$ and $F_2(t) = 0$. The system and input matrices in Eq. (26) are

$$\mathbf{A} = \begin{pmatrix} 0 & 0 & 1 & 0 \\ 0 & 0 & 0 & 1 \\ -(k_1 + k_2)/m_1 & k_2/m_1 & 0 & 0 \\ k_2/m_2 & -(k_2 + k_3)/m_2 & 0 & 0 \end{pmatrix}, \quad \text{and} \quad \mathbf{B} = \begin{pmatrix} 0 \\ 0 \\ 1 \\ 0 \end{pmatrix},$$

respectively. If we assume that the entire state of the system is observable, then the output matrix is $\mathbf{C} = \mathbf{I}_4$ and $\mathbf{y} = \mathbf{I}_4\mathbf{x} = \mathbf{x}$. Given an initial state $\mathbf{x}(0) = \mathbf{x}_0$, the control problem then is to specify an acceleration $u(t) = F_1(t)/m_1$ that is to be applied to mass m_1 over the time interval $0 \le t \le T$ so that $\mathbf{x}(T) = \mathbf{0}$. Obviously, an essential first step is to ascertain whether the desired objective can be achieved. A general answer to this question is provided, in Project 3 at the end of this chapter, in the context of an analogous system of three masses and four springs. Using the methods presented in the aforementioned project, an input function $u(t)$ that drives the system (26) from the initial state $\mathbf{x}(0) = (-3, 2, 0, 0)^T$ to equilibrium over the time interval $0 \le t \le 10$ is shown in Figure 6.1.5.

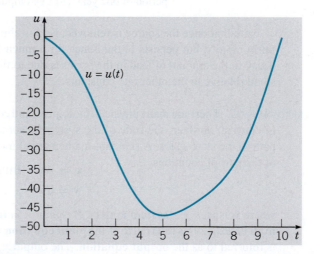

FIGURE 6.1.5   The acceleration $u(t)$, $0 \le t \le T$ applied to the left mass in Figure 6.1.1 drives the system from its initial state $\mathbf{x}(0) = (-3, 2, 0, 0)$ to equilibrium at time T (see Figure 6.1.6); $m_1 = m_2 = k_1 = k_2 = k_3 = 1$, $u(t) = F_1(t)/m_1$, $F_2(t) = 0$, and $T = 10$.

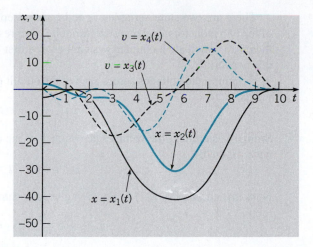

FIGURE 6.1.6 The response (displacements and velocities of the masses) of the system (26) subject to the initial conditions $x_1(0) = -3$, $x_2(0) = 2$, $x_3(0) = x_4(0) = 0$, $m_1 = m_2 = k_1 = k_2 = k_3 = 1$ and the control input $u(t)$ shown in Figure 6.1.5.

Graphs of the resulting mass displacements and velocities are shown in Figure 6.1.6. While the initial motion is similar to that of the unforced response shown in Figure 6.1.2, subsequent motion is greatly modified as the input acceleration forces the system to a state of rest.

▶ **The State Variable Approach to Circuit Analysis.** Energy storage elements in an electrical network are the capacitors and the inductors. The energy in a charged capacitor is due to the separation of charge on its plates. The energy in an inductor is stored in its magnetic field. The state variables of the circuit are the set of variables associated with the energy of the energy storage elements, namely, the voltage of each capacitor and the current in each inductor. Application of Kirchhoff's current law and Kirchhoff's voltage law yields a first order system of differential equations for these state variables. Then, given the initial conditions of these variables, the complete response of the circuit to a forcing function (such as an impressed voltage source) is completely determined. Obviously, electrical networks consisting of hundreds or thousands of storage elements lead to first order systems of correspondingly large dimension.

To illustrate the state variable approach, we consider the electric circuit shown in Figure 6.1.7.

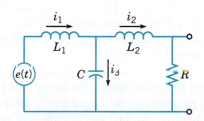

FIGURE 6.1.7 An electrical network composed of two loops.

There are two state variables, i_1 and i_2, representing the currents passing through each of the two inductors, and one state variable, v, representing the voltage on the capacitor. We therefore anticipate a coupled system of three first order differential equations describing this network.

Summing the voltages across each of the circuit elements in the left loop of the network yields

$$L_1 i_1' + v = e(t). \tag{27}$$

Similarly, summing the voltages across each of the circuit elements in the right loop of the network yields

$$L_2 i_2' + R i_2 - v = 0. \tag{28}$$

Note that adding Eq. (27) to Eq. (28) gives the sum of the voltage drops around the outer loop,

$$L_1 i_1' + L_2 i_2' + R i_2 = e(t),$$

a redundant equation since it can be obtained from Eqs. (27) and (28). Next, Kirchhoff's current law applied to the node connecting L_1, L_2, and C yields

$$i_1 = i_2 + i_3$$

or $i_3 = i_1 - i_2$. Since the voltage on the capacitor is given by $v = q/C$ where q is the charge on the capacitor, it follows that

$$Cv' = q' = i_3 = i_1 - i_2. \tag{29}$$

Equations (27) through (29) are efficiently represented in matrix notation by

$$\begin{pmatrix} i_1' \\ i_2' \\ v' \end{pmatrix} = \begin{pmatrix} 0 & 0 & -1/L_1 \\ 0 & -R/L_2 & 1/L_2 \\ 1/C & -1/C & 0 \end{pmatrix} \begin{pmatrix} i_1 \\ i_2 \\ v \end{pmatrix} + \begin{pmatrix} 1/L_1 \\ 0 \\ 0 \end{pmatrix} e(t), \tag{30}$$

where we have expressed the system in a form analogous to Eq. (24). Given initial conditions for each of the state variables, the complete response of the network shown in Figure 6.1.7 to the input $e(t)$ can be found by solving the system (30) subject to the prescribed initial conditions. The output voltage across the resistor, $v_R = R i_2$, can be conveniently expressed in the form of the output equation (25),

$$v_R = \begin{pmatrix} 0 & R & 0 \end{pmatrix} \begin{pmatrix} i_1 \\ i_2 \\ v \end{pmatrix}.$$

If we assume zero initial conditions for the state variables and circuit parameter values $L_1 = 3/2$, $L_2 = 1/2$, $C = 4/3$, and $R = 1$, Figure 6.1.8 shows the output voltage $v_R(t) = R i_R(t)$ across the resistor due to a harmonic input $e(t) = \sin \omega t$ with $\omega = 1/2$ (Figure 6.1.8(a)) and $\omega = 2$ (Figure 6.1.8(b)).

At the lower frequency the amplitude of the steady state response is approximately equal to 1 while at the higher frequency the steady state response is greatly attenuated. With the given parameter values, the circuit acts as a **low-pass filter**. Low frequency input signals $e(t)$ pass through the circuit easily while high frequency signals are greatly attenuated. In Section 6.6, we find the amplitude of the steady-state output $v_R(t)$ of the circuit for all frequencies.

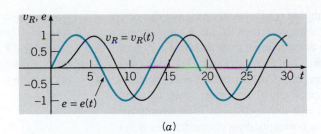

(a)

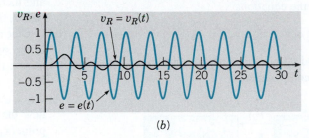

(b)

FIGURE 6.1.8 Output voltage $v_R(t) = Ri_R(t)$ across the resistor of the circuit in Figure 6.1.7 due to the input $e(t) = \sin \omega t$: (a) $\omega = 1/2$, (b) $\omega = 2$.

PROBLEMS

1. If $\mathbf{A}(t) = \begin{pmatrix} e^t & 2e^{-t} & e^{2t} \\ 2e^t & e^{-t} & -e^{2t} \\ -e^t & 3e^{-t} & 2e^{2t} \end{pmatrix}$

and

$\mathbf{B}(t) = \begin{pmatrix} 2e^t & e^{-t} & 3e^{2t} \\ -e^t & 2e^{-t} & e^{2t} \\ 3e^t & -e^{-t} & -e^{2t} \end{pmatrix}$ find

(a) $\mathbf{A} + 3\mathbf{B}$ (b) $\mathbf{AB}$
(c) $d\mathbf{A}/dt$ (d) $\int_0^1 \mathbf{A}(t)\,dt$

2. Verify that $\mathbf{x} = e^{-t} \begin{pmatrix} 6 \\ -8 \\ -4 \end{pmatrix} + 2e^{2t} \begin{pmatrix} 0 \\ 1 \\ -1 \end{pmatrix}$ satisfies

$\mathbf{x}' = \begin{pmatrix} 1 & 1 & 1 \\ 2 & 1 & -1 \\ 0 & -1 & 1 \end{pmatrix} \mathbf{x}$.

3. Verify that $\boldsymbol{\Psi} = \begin{pmatrix} e^t & e^{-2t} & e^{3t} \\ -4e^t & -e^{-2t} & 2e^{3t} \\ -e^t & -e^{-2t} & e^{3t} \end{pmatrix}$

satisfies the matrix differential equation

$\boldsymbol{\Psi}' = \begin{pmatrix} 1 & -1 & 4 \\ 3 & 2 & -1 \\ 2 & 1 & -1 \end{pmatrix} \boldsymbol{\Psi}$.

In each of Problems 4 through 9, transform the equation into an equivalent first order system.

4. $y^{(4)} + 4y''' + 3y = t$
5. $ty''' + (\sin t)y'' + 3y = \cos t$
6. $t(t-1)y^{(4)} + e^t y'' + 4t^2 y = 0$
7. $y''' + ty'' + t^2 y' + t^2 y = \ln t$
8. $(x-1)y^{(4)} + (x+1)y'' + (\tan x)y = 0$
9. $(x^2 - 4)y^{(6)} + x^2 y'' + 9y = 0$

10. Derive the differential equations for $x_1(t)$ and $x_2(t)$ in the system (23) by applying the balance law (22) to the compartment model illustrated in Figure 6.1.3.

11. Determine the matrix $\mathbf{K}$ and input $\mathbf{g}(t)$ if the system (23) is to be expressed using matrix notation, $\mathbf{x}' = \mathbf{Kx} + \mathbf{g}(t)$.

12. Find a system of first order linear differential equations for the four state variables of the circuit shown in Figure 6.1.9.

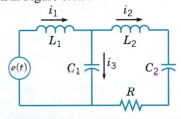

FIGURE 6.1.9

13. An initial amount α of a tracer (such as a dye or a radioactive isotope) is injected into Compartment 1 of the two compartment system shown in Figure 6.1.10. At time $t > 0$ let $x_1(t)$ and $x_2(t)$ denote the amount of tracer in Compartment 1 and Compartment 2, respectively. Thus under the conditions stated, $x_1(0) = \alpha$ and $x_2(0) = 0$. The amounts are related to the corresponding concentrations $\rho_1(t)$ and $\rho_2(t)$ by the equations

$$x_1 = \rho_1 V_1 \quad \text{and} \quad x_2 = \rho_2 V_2 \qquad (i)$$

where V_1 and V_2 are the constant respective volumes of the compartments.

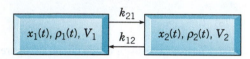

FIGURE 6.1.10 A closed two-compartment system.

The differential equations that describe the exchange of tracer between the compartments are

$$\frac{dx_1}{dt} = -k_{21}\rho_1 + k_{12}\rho_2$$
$$\frac{dx_2}{dt} = k_{21}\rho_1 - k_{12}\rho_2, \qquad (ii)$$

or, using the relations in (i),

$$\frac{dx_1}{dt} = -L_{21}x_1 + L_{12}x_2$$
$$\frac{dx_2}{dt} = L_{21}x_1 - L_{12}x_2, \qquad (iii)$$

where $L_{21} = k_{21}/V_1$ is the **fractional turnover rate** of Compartment 1 with respect to 2 and $L_{12} = k_{12}/V_2$ is the **fractional turnover rate** of Compartment 2 with respect to 1.

(a) Use Eqs. (iii) to show that

$$\frac{d}{dt}[x_1(t) + x_2(t)] = 0$$

and therefore $x_1(t) + x_2(t) = \alpha$ for all $t \geq 0$, that is, the tracer is conserved.

(b) Use the eigenvalue method to find the solution of the system (iii) subject to the initial conditions $x_1(0) = \alpha$ and $x_2(0) = 0$.

(c) What are the limiting values $\bar{x}_1 = \lim_{t\to\infty} x_1(t)$ and $\bar{x}_2 = \lim_{t\to\infty} x_2(t)$? Explain how the rate of approach to the equilibrium point $(\bar{x}_1, \bar{x}_2)$ depends on L_{12} and L_{21}.

(d) Give a qualitative sketch of the phase portrait for the system (iii).

(e) Plot the graphs of $\bar{x}_1/\alpha$ and $\bar{x}_2/\alpha$ as a function of $L_{21}/L_{12} \geq 0$ on the same set of coordinates and explain the meaning of the graphs.

14. (a) Using matrix notation, show that the system of second order equations (18) for the displacements $\mathbf{y} = (y_1, y_2)^T$ of the masses can be written in the form

$$\mathbf{y}'' = \mathbf{K}\mathbf{y} + \mathbf{f}(t).$$

(b) Under what conditions on k_1, k_2, k_3, m_1, and m_2 is $\mathbf{K}$ a symmetric matrix?

15. Consider the plant equation (26) for the control system consisting of two masses and three springs. Determine a suitable output matrix $\mathbf{C}$ for an output equation (25) if only the displacement of m_2 is observable.

6.2 Basic Theory of First Order Linear Systems

The theory presented in this section generalizes to higher dimensions the theory for two dimensional linear systems introduced in Sections 3.2 and 4.2. The extension of this theory requires definitions for linearly dependent and linearly independent sets of vectors (see Appendix A.3) as well as the use of the determinant (i) to determine whether the column vectors of an $n \times n$ matrix are linearly dependent or linearly independent, and (ii) to test whether a system of n linear algebraic equations in n unknowns has a unique solution (see Theorems A.3.6 and A.3.7 in Appendix A.3).

Using the matrix notation of Section 6.1, the first order linear system of dimension n is

$$\mathbf{x}' = \mathbf{P}(t)\mathbf{x} + \mathbf{g}(t). \tag{1}$$

Conditions under which a unique solution to the initial value problem associated with Eq. (1) exists are provided by the following generalization of Theorem 3.2.1.

THEOREM 6.2.1

(**Existence and Uniqueness for First Order Linear Systems**). If $\mathbf{P}(t)$ and $\mathbf{g}(t)$ are continuous on an open interval $I = (\alpha, \beta)$, then there exists a unique solution $\mathbf{x} = \boldsymbol{\phi}(t)$ to the initial value problem

$$\mathbf{x}' = \mathbf{P}(t)\mathbf{x} + \mathbf{g}(t), \qquad \mathbf{x}(t_0) = \mathbf{x}_0, \tag{2}$$

where t_0 is any point in I, and $\mathbf{x}_0$ is any constant vector with n components. Moreover, the solution exists throughout the interval I.

The proof of this theorem, too difficult to include here, can be found in more advanced books on differential equations. However, just as in the two dimensional case, the theorem is easy to apply. If the functions $p_{11}, p_{12}, \ldots, p_{nn}, g_1, \ldots, g_n$ are continuous on the interval I containing the point t_0, then we are assured that one and only one solution exists on I, irrespective of the vector $\mathbf{x}_0$ of prescribed initial values. A commonly occurring special case of the initial value problem (2) is

$$\mathbf{x}' = \mathbf{A}\mathbf{x}, \qquad \mathbf{x}(0) = \mathbf{x}_0 \tag{3}$$

where $\mathbf{A}$ is a constant $n \times n$ matrix. Since the coefficients of $\mathbf{A}$ are continuous for all values of t, Theorem 6.2.1 guarantees that a solution exists and is unique on the entire t-axis. The study of solutions of the constant coefficient initial value problem (3), based on eigenvalue methods, is taken up in the next section.

In this section, we discuss properties and structure of solutions of the homogeneous equation

$$\mathbf{x}' = \mathbf{P}(t)\mathbf{x} \tag{4}$$

obtained from Eq. (1) by setting $\mathbf{g}(t) = \mathbf{0}$. The solution of the nonhomogeneous equation, $\mathbf{x}' = \mathbf{P}(t)\mathbf{x} + \mathbf{g}(t)$, is dealt with in Section 6.6.

We use the notation

$$\mathbf{x}_1(t) = \begin{pmatrix} x_{11}(t) \\ x_{21}(t) \\ \vdots \\ x_{n1}(t) \end{pmatrix}, \quad \ldots, \quad \mathbf{x}_k(t) = \begin{pmatrix} x_{1k}(t) \\ x_{2k}(t) \\ \vdots \\ x_{nk}(t) \end{pmatrix}, \quad \ldots$$

to designate specific solutions of the system (4). Note that $x_{ij}(t)$ refers to the ith component of the jth solution $\mathbf{x}_j(t)$. An expression of the form

$$c_1\mathbf{x}_1 + \cdots + c_k\mathbf{x}_k$$

where $c_1, \ldots, c_k$ are arbitrary constants is called a **linear combination** of solutions.

The following result, a generalization of Theorem 3.3.1, is the principal distinguishing property of linear systems. This property is not shared by nonlinear systems.

THEOREM 6.2.2

(Principle of Superposition). If $x_1, x_2, \ldots, x_k$ are solutions of the homogeneous linear system

$$x' = P(t)x \tag{5}$$

on the interval $I = (\alpha, \beta)$, then the linear combination

$$c_1 x_1 + c_2 x_2 + \cdots + c_k x_k$$

is also a solution of Eq. (5) on I.

Proof

Let $x = c_1 x_1 + c_2 x_2 + \cdots + c_k x_k$. The result follows from the linear operations of matrix multiplication and differentiation,

$$\begin{aligned}
P(t)x &= P(t)\left[c_1 x_1 + \cdots + c_k x_k\right] \\
&= c_1 P(t)x_1 + \cdots + c_k P(t)x_k \\
&= c_1 x_1' + \cdots + c_k x_k' = x'.
\end{aligned}$$

∎

We use Theorem 6.2.2 to enlarge a finite set of solutions $\{x_1, x_2, \ldots, x_k\}$ to a k-fold infinite family of solutions $c_1 x_1 + c_2 x_2 + \cdots + c_k x_k$ parameterized by $c_1, \ldots, c_k$. In the discussion that follows, we show that all solutions of Eq. (5) are contained in an n-parameter family $c_1 x_1 + c_2 x_2 + \cdots + c_n x_n$ provided that the n solutions $x_1, \ldots, x_n$ are distinct in a sense made precise by the following definition.

DEFINITION 6.2.3

The n vector functions $x_1, \ldots, x_n$ are said to be **linearly independent on an interval I** if the only constants $c_1, c_2, \ldots, c_n$ such that

$$c_1 x_1(t) + \cdots + c_n x_n(t) = 0 \tag{6}$$

for all $t \in I$ are $c_1 = c_2 = \cdots = c_n = 0$. If there exist constants $c_1, c_2, \ldots, c_n$, *not all zero*, such that Eq. (6) is true for all $t \in I$, the vector functions are said to be **linearly dependent** on I.

EXAMPLE 1

Show that the vector functions

$$x_1(t) = \begin{pmatrix} e^{-2t} \\ 0 \\ -e^{-2t} \end{pmatrix} \quad \text{and} \quad x_2(t) = \begin{pmatrix} e^t \\ e^t \\ e^t \end{pmatrix}$$

are linearly independent on $I = (-\infty, \infty)$.

To prove independence we assume that

$$c_1 x_1(t) + c_2 x_2(t) = 0 \tag{7}$$

is true for all $t \in I$. Setting $t = 0$ in Eq. (7) leads to the requirement

$$c_1 \begin{pmatrix} 1 \\ 0 \\ -1 \end{pmatrix} + c_2 \begin{pmatrix} 1 \\ 1 \\ 1 \end{pmatrix} = \begin{pmatrix} 0 \\ 0 \\ 0 \end{pmatrix},$$

which is equivalent to the system of linear equations

$$c_1 + c_2 = 0,$$
$$c_2 = 0, \tag{8}$$
$$-c_1 + c_2 = 0.$$

Since the only solution of the system (8) is $c_1 = c_2 = 0$, we conclude that the vector functions $\mathbf{x}_1$ and $\mathbf{x}_2$ are linearly independent on any interval containing $t = 0$ and in particular, are linearly independent on $I = (-\infty, \infty)$.

EXAMPLE 2

Show that the vector functions

$$\mathbf{x}_1(t) = \begin{pmatrix} 1+t \\ t \\ 1-t \end{pmatrix}, \quad \mathbf{x}_2(t) = \begin{pmatrix} 3 \\ t+2 \\ t \end{pmatrix}, \quad \text{and} \quad \mathbf{x}_3(t) = \begin{pmatrix} 1-2t \\ 2-t \\ 3t-2 \end{pmatrix}$$

are linearly dependent on $I = (-\infty, \infty)$.

We begin by assuming that

$$c_1\mathbf{x}_1(t) + c_2\mathbf{x}_2(t) + c_3\mathbf{x}_3(t) = \mathbf{0} \quad \text{for all } t \in I, \tag{9}$$

and attempt to determine constants c_1, c_2, and c_3, not all of which are zero, such that condition (9) is a true statement. Candidate values for c_1, c_2, and c_3 are found by evaluating the equation in condition (9) at a particular value of $t \in I$. For example, choosing $t = 0$ yields the system of equations

$$c_1\begin{pmatrix} 1 \\ 0 \\ 1 \end{pmatrix} + c_2\begin{pmatrix} 3 \\ 2 \\ 0 \end{pmatrix} + c_3\begin{pmatrix} 1 \\ 2 \\ -2 \end{pmatrix} = \begin{pmatrix} 0 \\ 0 \\ 0 \end{pmatrix}, \tag{10}$$

which is equivalent to the system of linear equations

$$c_1 + 3c_2 + c_3 = 0,$$
$$2c_2 + 2c_3 = 0, \tag{11}$$
$$c_1 - 2c_3 = 0.$$

Using Gaussian elimination, we find the general solution of the system (11) to be

$$\mathbf{c} = \alpha \begin{pmatrix} 2 \\ -1 \\ 1 \end{pmatrix}$$

where α is arbitrary. Setting $\alpha = 1$ gives us $c_1 = 2$, $c_2 = -1$, and $c_3 = 1$. Thus Eq. (10) is true for these values of c_1, c_2, and c_3, that is, the three vectors $(1, 0, 1)^T$, $(3, 2, 0)^T$, and $(1, 2, -2)^T$, are linearly dependent. We still need to verify that the statement (9) is true using the values $c_1 = 2$, $c_2 = -1$, and $c_3 = 1$. Since

$$2\begin{pmatrix} 1+t \\ t \\ 1-t \end{pmatrix} - 1\begin{pmatrix} 3 \\ t+2 \\ t \end{pmatrix} + 1\begin{pmatrix} 1-2t \\ 2-t \\ 3t-2 \end{pmatrix} = \begin{pmatrix} 2+2t-3+1-2t \\ 2t-t-2+2-t \\ 2-2t-t+3t-2 \end{pmatrix} = \begin{pmatrix} 0 \\ 0 \\ 0 \end{pmatrix}$$

for all $t \in I$, we conclude that $\mathbf{x}_1$, $\mathbf{x}_2$, and $\mathbf{x}_3$ are linearly dependent vector functions on I.

In Section 3.3, the Wronskian of two vector functions with two components was defined. We now extend the definition to sets of n vector functions of length n.

DEFINITION 6.2.4

Let $\mathbf{x}_1, \ldots, \mathbf{x}_n$ be n solutions of Eq. (4) and let $\mathbf{X}(t)$ be the $n \times n$ matrix whose jth column is $\mathbf{x}_j(t)$, $j = 1, \ldots, n$,

$$\mathbf{X}(t) = \begin{pmatrix} x_{11}(t) & \cdots & x_{1n}(t) \\ \vdots & & \vdots \\ x_{n1}(t) & \cdots & x_{nn}(t) \end{pmatrix}. \tag{12}$$

The **Wronskian** $W = W[\mathbf{x}_1, \ldots, \mathbf{x}_n]$ of the n solutions $\mathbf{x}_1, \ldots, \mathbf{x}_n$ is defined by

$$W[\mathbf{x}_1, \ldots, \mathbf{x}_n](t) = \det \mathbf{X}(t). \tag{13}$$

The next theorem shows how W is used to test whether a set of n solutions of $\mathbf{x}' = \mathbf{P}(t)\mathbf{x}$ is linearly independent or linearly dependent on an interval I.

THEOREM 6.2.5

Let $\mathbf{x}_1, \ldots, \mathbf{x}_n$ be solutions of $\mathbf{x}' = \mathbf{P}(t)\mathbf{x}$ on an interval $I = (\alpha, \beta)$ in which $\mathbf{P}(t)$ is continuous.

(i) If $\mathbf{x}_1, \ldots, \mathbf{x}_n$ are linearly independent on I, then $W[\mathbf{x}_1, \ldots, \mathbf{x}_n](t) \neq 0$ at every point in I,

(ii) If $\mathbf{x}_1, \ldots, \mathbf{x}_n$ are linearly dependent on I, then $W[\mathbf{x}_1, \ldots, \mathbf{x}_n](t) = 0$ at every point in I.

Proof

Assume first that $\mathbf{x}_1, \ldots, \mathbf{x}_n$ are linearly independent on I. We then want to show that $W[\mathbf{x}_1, \ldots, \mathbf{x}_n](t) \neq 0$ throughout I. To do this, we assume the contrary, that is, there is a point $t_0 \in I$ such that $W[\mathbf{x}_1, \ldots, \mathbf{x}_n](t_0) = 0$. This means that the column vectors $\{\mathbf{x}_1(t_0), \ldots, \mathbf{x}_n(t_0)\}$ are linearly dependent (Theorem A.3.6) so that there exist constants $\hat{c}_1, \ldots, \hat{c}_n$, not all zero, such that $\hat{c}_1 \mathbf{x}_1(t_0) + \cdots + \hat{c}_n \mathbf{x}_n(t_0) = \mathbf{0}$. Then Theorem 6.2.2 implies that $\boldsymbol{\phi}(t) = \hat{c}_1 \mathbf{x}_1(t) + \cdots + \hat{c}_n \mathbf{x}_n(t)$ is a solution of $\mathbf{x}' = \mathbf{P}(t)\mathbf{x}$ that satisfies the initial condition $\mathbf{x}(t_0) = \mathbf{0}$. The zero solution also satisfies the same initial value problem. The uniqueness part of Theorem 6.2.1 therefore implies that $\boldsymbol{\phi}$ is the zero solution, that is, $\boldsymbol{\phi}(t) = \hat{c}_1 \mathbf{x}_1(t) + \cdots + \hat{c}_n \mathbf{x}_n(t) = \mathbf{0}$ for every $t \in (\alpha, \beta)$, contradicting our original assumption that $\mathbf{x}_1, \ldots, \mathbf{x}_n$ are linearly independent on I. This proves (i).

To prove (ii), assume that $\mathbf{x}_1, \ldots, \mathbf{x}_n$ are linearly dependent on I. Then there exists $\alpha_1, \ldots, \alpha_n$, not all zero, such that $\alpha_1 \mathbf{x}_1(t) + \cdots + \alpha_n \mathbf{x}_n(t) = \mathbf{0}$ for every $t \in I$. Consequently, for each $t \in I$, the vectors $\mathbf{x}_1(t), \ldots, \mathbf{x}_n(t)$ are linearly dependent. Thus $W[\mathbf{x}_1, \ldots, \mathbf{x}_n](t) = 0$ at every point in I (Theorem A.3.6). ∎

Thus $\mathbf{x}_1, \ldots, \mathbf{x}_n$ are linearly independent solutions of $\mathbf{x}' = \mathbf{P}(t)\mathbf{x}$ on I if and only if $W[\mathbf{x}_1, \ldots, \mathbf{x}_n](t) \neq 0$ for every $t \in I$. The next theorem shows that all solutions of $\mathbf{x}' = \mathbf{P}(t)\mathbf{x}$ are contained in the n-fold infinite family $c_1 \mathbf{x}_1 + \cdots + c_n \mathbf{x}_n$ provided that $\mathbf{x}_1, \ldots, \mathbf{x}_n$ are linearly independent on I.

THEOREM 6.2.6

Let $\mathbf{x}_1, \ldots, \mathbf{x}_n$ be solutions of

$$\mathbf{x}' = \mathbf{P}(t)\mathbf{x} \tag{14}$$

on the interval $\alpha < t < \beta$ such that, for some point $t_0 \in (\alpha, \beta)$, the Wronskian is nonzero, $W[\mathbf{x}_1, \ldots, \mathbf{x}_n](t_0) \neq 0$. Then each solution $\mathbf{x} = \boldsymbol{\phi}(t)$ of Eq. (14) can be expressed as a linear combination of $\mathbf{x}_1, \ldots, \mathbf{x}_n$,

$$\boldsymbol{\phi}(t) = \hat{c}_1 \mathbf{x}_1(t) + \cdots + \hat{c}_n \mathbf{x}_n(t), \tag{15}$$

where the constants $\hat{c}_1, \ldots, \hat{c}_n$ are uniquely determined.

Proof

Let $\boldsymbol{\phi}(t)$ be a given solution of Eq. (14). If we set $\mathbf{x}_0 = \boldsymbol{\phi}(t_0)$, then the vector function $\boldsymbol{\phi}$ is a solution of the initial value problem

$$\mathbf{x}' = \mathbf{P}(t)\mathbf{x}, \qquad \mathbf{x}(t_0) = \mathbf{x}_0. \tag{16}$$

By the Principle of Superposition, the linear combination $\boldsymbol{\psi}(t) = c_1 \mathbf{x}_1(t) + \cdots + c_n \mathbf{x}_n(t)$ is also a solution of (14) for any choice of constants $c_1, \ldots, c_n$. The requirement $\boldsymbol{\psi}(t_0) = \mathbf{x}_0$ leads to the linear algebraic system

$$\mathbf{X}(t_0)\mathbf{c} = \mathbf{x}_0 \tag{17}$$

where $\mathbf{X}(t)$ is defined by Eq. (12). Since $W[\mathbf{x}_1, \ldots, \mathbf{x}_n](t_0) \neq 0$, the linear algebraic system (17) has a unique solution (see Theorem A.3.7) which we denote by $\hat{c}_1, \ldots, \hat{c}_n$. Thus the particular member $\hat{\boldsymbol{\psi}}(t) = \hat{c}_1 \mathbf{x}_1(t) + \cdots + \hat{c}_n \mathbf{x}_n(t)$ of the n-parameter family represented by $\boldsymbol{\psi}(t)$ also satisfies the initial value problem (16). By the uniqueness part of Theorem 6.2.1, it follows that $\boldsymbol{\phi} = \hat{\boldsymbol{\psi}} = \hat{c}_1 \mathbf{x}_1 + \cdots + \hat{c}_n \mathbf{x}_n$. Since $\boldsymbol{\phi}$ is arbitrary, the result holds (with different constants of course) for every solution of Eq. (14). ∎

Remark. It is customary to call the n-parameter family

$$c_1 \mathbf{x}_1(t) + \cdots + c_n \mathbf{x}_n(t)$$

the **general solution** of Eq. (14) if $W[\mathbf{x}_1, \ldots, \mathbf{x}_n](t_0)$ is nonzero for some $t_0 \in (\alpha, \beta)$. Theorem 6.2.6 guarantees that the general solution includes all possible solutions of Eq. (14). Any set of solutions $\mathbf{x}_1, \ldots, \mathbf{x}_n$ of Eq. (14), which are linearly independent on an interval $\alpha < t < \beta$, is said to be a **fundamental set of solutions** for that interval.

EXAMPLE 3

If

$$\mathbf{A} = \begin{pmatrix} 0 & -1 & 2 \\ 2 & -3 & 2 \\ 3 & -3 & 1 \end{pmatrix},$$

show that

$$\mathbf{x}_1(t) = e^{-2t} \begin{pmatrix} 1 \\ 0 \\ -1 \end{pmatrix}, \quad \mathbf{x}_2(t) = e^{-t} \begin{pmatrix} 1 \\ 1 \\ 0 \end{pmatrix}, \quad \text{and} \quad \mathbf{x}_3(t) = e^{t} \begin{pmatrix} 1 \\ 1 \\ 1 \end{pmatrix}$$

is a fundamental set for

$$\mathbf{x}' = \mathbf{A}\mathbf{x}. \tag{18}$$

Then solve the initial value problem consisting of Eq. (18) subject to the initial condition $\mathbf{x}(0) = (1, -2, 1)^T$.

Substituting each of x_1, x_2, and x_3 into Eq. (18) and verifying that the equation reduces to an identity shows that x_1, x_2, and x_3 are solutions of Eq. (18). In the next section we will show you how to find x_1, x_2, and x_3 yourself. Since

$$W[x_1, x_2, x_3](0) = \begin{vmatrix} 1 & 1 & 1 \\ 0 & 1 & 1 \\ -1 & 0 & 1 \end{vmatrix} = 1 \neq 0,$$

it follows from Theorem 6.2.6 that x_1, x_2, and x_3 are a fundamental set for Eq. (18) and that the general solution is $x = c_1 x_1(t) + c_2 x_2(t) + c_3 x_3(t)$. Substituting the general solution into the initial condition $x(0) = (1, -2, 1)^T$ yields the system

$$\begin{pmatrix} 1 & 1 & 1 \\ 0 & 1 & 1 \\ -1 & 0 & 1 \end{pmatrix} \begin{pmatrix} c_1 \\ c_2 \\ c_3 \end{pmatrix} = \begin{pmatrix} 1 \\ -2 \\ 1 \end{pmatrix}. \tag{19}$$

Solving Eq. (19) by Gaussian elimination gives $c_1 = 3$, $c_2 = -6$, and $c_3 = 4$. Thus the solution of Eq. (18) satisfying $x(0) = (1, -2, 1)^T$ is

$$x = 3e^{-2t} \begin{pmatrix} 1 \\ 0 \\ -1 \end{pmatrix} - 6e^{-t} \begin{pmatrix} 1 \\ 1 \\ 0 \end{pmatrix} + 4e^{t} \begin{pmatrix} 1 \\ 1 \\ 1 \end{pmatrix}.$$

The next theorem states that the system (4) always has at least one fundamental set of solutions.

THEOREM 6.2.7

Let

$$e_1 = \begin{pmatrix} 1 \\ 0 \\ 0 \\ \vdots \\ 0 \end{pmatrix}, \quad e_2 = \begin{pmatrix} 0 \\ 1 \\ 0 \\ \vdots \\ 0 \end{pmatrix}, \quad \dots, \quad e_n = \begin{pmatrix} 0 \\ 0 \\ 0 \\ \vdots \\ 1 \end{pmatrix};$$

further, let $x_1, \dots, x_n$ be solutions of Eq. (4) that satisfy the initial conditions

$$x_1(t_0) = e_1, \quad \dots, \quad x_n(t_0) = e_n,$$

respectively, where t_0 is any point in $\alpha < t < \beta$. Then $x_1, \dots, x_n$ form a fundamental set of solutions of the system (4).

Proof

Existence and uniqueness of the solutions $x_1, \dots, x_n$ are guaranteed by Theorem 6.2.1 and it is easy to see that $W[x_1, \dots, x_n](t_0) = \det I_n = 1$. It then follows from Theorem 6.2.6 that $x_1, \dots, x_n$ are a fundamental set of solutions for the system (4). ∎

Once one fundamental set of solutions has been found, other sets can be generated by forming (independent) linear combinations of the first set. For theoretical purposes, the set given by Theorem 6.2.7 is usually the simplest.

To summarize, any set of n linearly independent solutions of the system (4) constitutes a fundamental set of solutions. Under the conditions given in this section, such fundamental sets always exist, and every solution of the system (4) can be represented as a linear combination of any fundamental set of solutions.

▶ **Linear nth Order Equations.** Recall from Section 6.1 that by introducing the variables

$$x_1 = y, \quad x_2 = y', \quad x_3 = y'', \quad \dots, \quad x_n = y^{(n-1)}, \tag{20}$$

the initial value problem for the linear nth order equation,

$$\frac{d^n y}{dt^n} + p_1(t)\frac{d^{n-1}y}{dt^{n-1}} + \cdots + p_{n-1}(t)\frac{dy}{dt} + p_n(t)y = g(t), \tag{21}$$

$$y(t_0) = y_0, \quad y'(t_0) = y_1, \quad \dots, \quad y^{(n-1)}(t_0) = y_{n-1} \tag{22}$$

can be expressed as an initial value problem for a first order system,

$$\mathbf{x}' = \begin{pmatrix} 0 & 1 & 0 & 0 & \cdots & 0 \\ 0 & 0 & 1 & 0 & \cdots & 0 \\ \vdots & \vdots & \vdots & \vdots & & \vdots \\ 0 & 0 & 0 & 0 & \cdots & 1 \\ -p_n(t) & -p_{n-1}(t) & -p_{n-2}(t) & -p_{n-3}(t) & \cdots & -p_1(t) \end{pmatrix} \mathbf{x} + \begin{pmatrix} 0 \\ 0 \\ \vdots \\ 0 \\ g(t) \end{pmatrix}, \tag{23}$$

$$\mathbf{x}(t_0) = \begin{pmatrix} y_0 \\ y_1 \\ \vdots \\ y_{n-1} \end{pmatrix}, \tag{24}$$

where $\mathbf{x} = (x_1, \dots, x_n)^T$. Theorem 6.2.1 then provides sufficient conditions for existence and uniqueness of a solution to the initial value problem (21), (22). These conditions are stated in the following corollary to Theorem 6.2.1.

COROLLARY 6.2.8	If the functions $p_1(t), p_2(t), \dots, p_n(t)$, and $g(t)$ are continuous on the open interval $I = (\alpha, \beta)$, then there exists exactly one solution $y = \phi(t)$ of the differential equation (21) that also satisfies the initial conditions (22). This solution exists throughout the interval I.

Proof

Under the stated conditions, the matrix of coefficients and nonhomogeneous term in Eq. (23) are continuous on I. By Theorem 6.2.1, a unique solution $\mathbf{x} = \boldsymbol{\phi}(t) = (\phi_1(t), \dots, \phi_n(t))^T$ to the initial value problem (23), (24) exists throughout I. The definitions for the state variables in the list (20) then show that $y = \phi(t) = \phi_1(t)$ is the unique solution of the initial value problem (21), (22). ■

We now restrict our attention to the homogeneous equation associated with Eq. (21),

$$\frac{d^n y}{dt^n} + p_1(t)\frac{d^{n-1}y}{dt^{n-1}} + \cdots + p_{n-1}(t)\frac{dy}{dt} + p_n(t)y = 0, \tag{25}$$

and the corresponding homogeneous system associated with Eq. (23)

$$\mathbf{x}' = \begin{pmatrix} 0 & 1 & 0 & 0 & \cdots & 0 \\ 0 & 0 & 1 & 0 & \cdots & 0 \\ \vdots & \vdots & \vdots & \vdots & & \vdots \\ 0 & 0 & 0 & 0 & \cdots & 1 \\ -p_n(t) & -p_{n-1}(t) & -p_{n-2}(t) & -p_{n-3}(t) & \cdots & -p_1(t) \end{pmatrix} \mathbf{x}. \tag{26}$$

Using the relations (20), the scalar functions $y_1, \ldots, y_n$ are solutions of Eq. (25) if and only if the vectors $\mathbf{x}_1 = (y_1, y_1', \ldots, y_1^{(n-1)})^T, \ldots, \mathbf{x}_n = (y_n, y_n', \ldots, y_n^{(n-1)})^T$ are solutions of Eq. (26). In accordance with Eq. (13), we will define the Wronskian of the scalar functions $y_1, \ldots, y_n$ by

$$W[y_1, \ldots, y_n](t) = \begin{vmatrix} y_1(t) & y_2(t) & \cdots & y_n(t) \\ y_1'(t) & y_2'(t) & \cdots & y_n'(t) \\ \vdots & \vdots & & \vdots \\ y_1^{(n-1)}(t) & y_2^{(n-1)}(t) & & y_n^{(n-1)}(t) \end{vmatrix}. \tag{27}$$

Theorem 6.2.6 then allows us to conclude that if $y_1, \ldots, y_n$ are solutions of Eq. (25) on an interval $I = (\alpha, \beta)$ and

$$W[y_1, \ldots, y_n](t_0) \neq 0$$

for some $t_0 \in (\alpha, \beta)$, then each solution $y = \phi(t)$ of Eq. (25) can be written as a linear combination of $y_1, \ldots, y_n$,

$$\phi(t) = \hat{c}_1 y_1(t) + \cdots + \hat{c}_n y_n(t),$$

where the constants $\hat{c}_1, \ldots, \hat{c}_n$ are uniquely determined. We state this result in the following corollary to Theorem 6.2.6.

COROLLARY 6.2.9

Let $y_1, \ldots, y_n$ be solutions of

$$\frac{d^n y}{dt^n} + p_1(t)\frac{d^{n-1} y}{dt^{n-1}} + \cdots + p_{n-1}(t)\frac{dy}{dt} + p_n(t)y = 0 \tag{28}$$

on an interval $I = (\alpha, \beta)$ in which $p_1, \ldots, p_n$ are continuous. If for some point $t_0 \in I$ these solutions satisfy

$$W[y_1, \ldots, y_n](t_0) \neq 0,$$

then each solution $y = \phi(t)$ of Eq. (28) can be expressed as a linear combination of $y_1, \ldots, y_n$,

$$\phi(t) = \hat{c}_1 y_1(t) + \cdots + \hat{c}_n y_n(t),$$

where the constants $\hat{c}_1, \ldots, \hat{c}_n$ are uniquely determined.

The terminology for solutions of the nth order scalar equation is identical to that used for solutions of $\mathbf{x}' = \mathbf{P}(t)\mathbf{x}$. A set of solutions $y_1, \ldots, y_n$ such that $W[y_1, \ldots, y_n](t_0) \neq 0$ for some $t_0 \in (\alpha, \beta)$ is called a **fundamental set of solutions** for Eq. (28) and the n-parameter family represented by the linear combination

$$y = c_1 y_1(t) + \cdots + c_n y_n(t),$$

where $c_1, \ldots, c_n$ are arbitrary constants, is called the **general solution** of Eq. (28). Corollary 6.2.9 guarantees that each solution of Eq. (28) corresponds to some member of this n-parameter family of solutions.

EXAMPLE 4

Show that $y_1(x) = x$, $y_2(x) = x^{-1}$, and $y_3(x) = x^2$ constitute a fundamental set of solutions for

$$x^3 y''' + x^2 y'' - 2xy' + 2y = 0 \tag{29}$$

on $I = (0, \infty)$.

Substituting each of y_1, y_2, and y_3 into Eq. (29) shows that they are solutions on the given interval. To verify that the three functions are linearly independent on I, we compute the Wronskian

$$W[y_1, y_2, y_3](x) = \begin{vmatrix} x & x^{-1} & x^2 \\ 1 & -x^{-2} & 2x \\ 0 & 2x^{-3} & 2 \end{vmatrix} = -6x^{-1}.$$

Since $W[y_1, y_2, y_3](x) < 0$ on I, Corollary 6.2.9 implies that x, x^{-1}, and x^2 are a fundamental set for Eq. (29). Note that it is only necessary to confirm that $W[y_1, y_2, y_3]$ is nonzero at one point in I; for example, showing that $W[y_1, y_2, y_3](1) = -6$ would have sufficed.

PROBLEMS

In each of Problems 1 through 6, determine intervals in which solutions are sure to exist.

1. $y^{(4)} + 4y''' + 3y = t$
2. $ty''' + (\sin t)y'' + 3y = \cos t$
3. $t(t-1)y^{(4)} + e^t y'' + 4t^2 y = 0$
4. $y''' + ty'' + t^2 y' + t^3 y = \ln t$
5. $(x-1)y^{(4)} + (x+1)y'' + (\tan x)y = 0$
6. $(x^2 - 4)y^{(6)} + x^2 y'' + 9y = 0$
7. Consider the vectors

$$\mathbf{x}_1(t) = \begin{pmatrix} e^t \\ 2e^t \\ -e^t \end{pmatrix}, \quad \mathbf{x}_2(t) = \begin{pmatrix} e^{-t} \\ -2e^{-t} \\ e^{-t} \end{pmatrix},$$

$$\mathbf{x}_3(t) = \begin{pmatrix} 2e^{4t} \\ 2e^{4t} \\ -8e^{4t} \end{pmatrix}$$

and let $\mathbf{X}(t)$ be the matrix whose columns are the vectors $\mathbf{x}_1(t)$, $\mathbf{x}_2(t)$, $\mathbf{x}_3(t)$. Compare the amounts of work between the following two methods for obtaining $W[\mathbf{x}_1, \mathbf{x}_2, \mathbf{x}_3](0)$:
Method 1: First find $|\mathbf{X}(t)|$ and then set $t = 0$.
Method 2: Evaluate $\mathbf{X}(t)$ at $t = 0$ and then find $|\mathbf{X}(0)|$.

8. Determine whether

$$\mathbf{x}_1(t) = e^{-t} \begin{pmatrix} 1 \\ 0 \\ -1 \end{pmatrix}, \quad \mathbf{x}_2(t) = e^{-t} \begin{pmatrix} 1 \\ -2 \\ 1 \end{pmatrix},$$

$$\mathbf{x}_3(t) = e^{-t} \begin{pmatrix} 0 \\ 2 \\ -2 \end{pmatrix}$$

is a fundamental set of solutions for

$$\mathbf{x}' = \begin{pmatrix} 0 & 1 & 1 \\ 1 & 0 & 1 \\ 1 & 1 & 0 \end{pmatrix} \mathbf{x}.$$

9. Determine whether

$$\mathbf{x}_1(t) = e^{-t} \begin{pmatrix} 1 \\ 0 \\ -1 \end{pmatrix}, \quad \mathbf{x}_2(t) = e^{-t} \begin{pmatrix} 1 \\ -4 \\ 1 \end{pmatrix},$$

$$\mathbf{x}_3(t) = e^{8t} \begin{pmatrix} 2 \\ 1 \\ 2 \end{pmatrix}$$

is a fundamental set of solutions for

$$\mathbf{x}' = \begin{pmatrix} 3 & 2 & 4 \\ 2 & 0 & 2 \\ 4 & 2 & 3 \end{pmatrix} \mathbf{x}.$$

10. In Section 4.2, it was shown that if $\mathbf{x}_1$ and $\mathbf{x}_2$ are solutions of

$$\mathbf{x}' = \begin{pmatrix} p_{11}(t) & p_{12}(t) \\ p_{21}(t) & p_{22}(t) \end{pmatrix} \mathbf{x},$$

on an interval I, then the Wronskian W of $\mathbf{x}_1$ and $\mathbf{x}_2$ satisfies the differential equation $W' = (p_{11} + p_{22})W$. A generalization of that proof shows that if $\mathbf{x}_1, \ldots, \mathbf{x}_n$ are solutions of Eq. (1) on I, then the Wronskian of $\mathbf{x}_1, \ldots, \mathbf{x}_n$, denoted by W, satisfies the differential equation

$$W' = (p_{11} + p_{22} + \cdots + p_{nn})W = \text{tr}(\mathbf{P}(t))W. \tag{i}$$

(a) Explain why Eq. (i) also implies that W is either identically zero or else never vanishes on I in accordance with Theorem 6.2.5.

(b) If $y_1, \ldots, y_n$ are solutions of Eq. (28), find a counterpart to (i) satisfied by $W = W[y_1, \ldots, y_n](t)$.

In each of Problems 11 through 16, verify that the given functions are solutions of the differential equations, and determine their Wronskian.

11. $y''' + y' = 0$,   1, $\cos t$, $\sin t$

12. $y^{(4)} + y'' = 0$,   1, t, $\cos t$, $\sin t$

13. $y''' + 2y'' - y' - 2y = 0$, e^t, e^{-t}, e^{-2t}

14. $y^{(4)} + 2y''' + y'' = 0$,   1, t, e^{-t}, te^{-t}

15. $xy''' - y'' = 0$,   1, x, x^3

16. $x^3 y''' + x^2 y'' - 2xy' + 2y = 0$, x, x^2, $1/x$

17. Verify that the differential operator defined by

$$L[y] = y^{(n)} + p_1(t)y^{(n-1)} + \cdots + p_n(t)y$$

is a linear operator. That is, show that

$$L[c_1 y_1 + c_2 y_2] = c_1 L[y_1] + c_2 L[y_2],$$

where y_1 and y_2 are n times differentiable functions and c_1 and c_2 are arbitrary constants. Hence, show that if $y_1, y_2, \ldots, y_n$ are solutions of $L[y] = 0$, then the linear combination $c_1 y_1 + \cdots + c_n y_n$ is also a solution of $L[y] = 0$.

6.3 Homogeneous Linear Systems with Constant Coefficients

In Chapter 3, the eigenvalue method was used to find fundamental solution sets for linear constant coefficient systems of dimension 2,

$$\mathbf{x}' = \begin{pmatrix} a_{11} & a_{12} \\ a_{21} & a_{22} \end{pmatrix} \mathbf{x}. \tag{1}$$

In this section, we extend the eigenvalue method to the system

$$\mathbf{x}' = \mathbf{A}\mathbf{x} \tag{2}$$

where $\mathbf{A}$ is a real constant $n \times n$ matrix. As in Chapter 3, we assume solutions of the form

$$\mathbf{x} = e^{\lambda t}\mathbf{v} \tag{3}$$

where the scalar λ and the constant $n \times 1$ vector $\mathbf{v}$ are to be determined. The steps leading to the eigenvalue problem are identical to the 2-dimensional case. Substituting from Eq. (3) into Eq. (2) we find that

$$\lambda e^{\lambda t}\mathbf{v} = e^{\lambda t}\mathbf{A}\mathbf{v}, \tag{4}$$

where we have used the fact that $\mathbf{x}' = \lambda e^{\lambda t}\mathbf{v}$. Since $e^{\lambda t}$ is nonzero, Eq. (4) reduces to

$$\lambda \mathbf{v} = \mathbf{A}\mathbf{v},$$

or

$$(\mathbf{A} - \lambda \mathbf{I}_n)\mathbf{v} = \mathbf{0} \tag{5}$$

where $\mathbf{I}_n$ is the $n \times n$ identity matrix. Given a square matrix $\mathbf{A}$, recall that the problem of

(i) finding values of λ for which Eq. (5) has nontrivial solution vectors $\mathbf{v}$, and

(ii) finding the corresponding nontrivial solutions,

is known as the **eigenvalue problem** for $\mathbf{A}$ (see Appendix A.4). We distinguish the following three cases:

1. $\mathbf{A}$ has a complete set of n linearly independent eigenvectors and all of the eigenvalues of $\mathbf{A}$ are real,

2. $\mathbf{A}$ has a complete set of n linearly independent eigenvectors and one or more complex conjugate pairs of eigenvalues,

3. A is defective, that is, there are one or more eigenvalues of **A** for which the geometric multiplicity is less than the algebraic multiplicity (see Appendix A.4).

In the rest of this section, we analyze the first case while the second and third cases are dealt with in Sections 6.4 and 6.7, respectively.

The Matrix A Is Nondefective With Real Eigenvalues

THEOREM 6.3.1

Let $(\lambda_1, \mathbf{v}_1), \ldots, (\lambda_n, \mathbf{v}_n)$ be eigenpairs for the real, $n \times n$ constant matrix **A**. Assume that the eigenvalues $\lambda_1, \ldots, \lambda_n$ are real and that the corresponding eigenvectors $\mathbf{v}_1, \ldots, \mathbf{v}_n$ are linearly independent. Then

$$\left\{ e^{\lambda_1 t} \mathbf{v}_1, \ldots, e^{\lambda_n t} \mathbf{v}_n \right\} \tag{6}$$

is a fundamental set of solutions to $\mathbf{x}' = \mathbf{A}\mathbf{x}$ on the interval $(-\infty, \infty)$. The general solution of $\mathbf{x}' = \mathbf{A}\mathbf{x}$ is therefore given by

$$\mathbf{x}(t) = c_1 e^{\lambda_1 t} \mathbf{v}_1 + \cdots + c_n e^{\lambda_n t} \mathbf{v}_n \tag{7}$$

where $c_1, \ldots, c_n$ are arbitrary constants.

Remark. The eigenvalues need not be distinct. All that is required is that for each eigenvalue λ_j, $g_j = m_j$, where g_j and m_j are the geometric and algebraic multiplicities, respectively, of λ_j.

Proof

Let

$$\mathbf{x}_1(t) = e^{\lambda_1 t} \mathbf{v}_1, \ldots, \mathbf{x}_n(t) = e^{\lambda_n t} \mathbf{v}_n.$$

We have $\mathbf{x}'_j = e^{\lambda_j t} \lambda_j \mathbf{v}_j = e^{\lambda_j t} \mathbf{A} \mathbf{v}_j = \mathbf{A} \mathbf{x}_j$, so $\mathbf{x}_j$ is a solution of $\mathbf{x}' = \mathbf{A}\mathbf{x}$ for each $j = 1, \ldots, n$. To show that $\mathbf{x}_1, \ldots, \mathbf{x}_n$ comprise a fundamental set of solutions, we evaluate the Wronskian,

$$W[\mathbf{x}_1, \ldots, \mathbf{x}_n](t) = \det[e^{\lambda_1 t} \mathbf{v}_1, \cdots, e^{\lambda_n t} \mathbf{v}_n] = e^{(\lambda_1 + \cdots + \lambda_n)t} \det[\mathbf{v}_1, \cdots, \mathbf{v}_n]. \tag{8}$$

The exponential function is never zero and since the eigenvectors $\mathbf{v}_1, \ldots, \mathbf{v}_n$ are linearly independent, the determinant in the last term is nonzero. Therefore the Wronskian is nonzero and the result follows. ∎

EXAMPLE 1

Find the general solution of

$$\mathbf{x}' = \begin{pmatrix} -4/5 & -1/5 & 4/5 \\ -1/5 & -4/5 & -4/5 \\ 2/5 & -2/5 & 3/5 \end{pmatrix} \mathbf{x}. \tag{9}$$

The characteristic polynomial of the matrix **A** of coefficients is

$$|\mathbf{A} - \lambda \mathbf{I}| = \begin{vmatrix} -4/5 - \lambda & -1/5 & 4/5 \\ -1/5 & -4/5 - \lambda & -4/5 \\ 2/5 & -2/5 & 3/5 - \lambda \end{vmatrix} = -\lambda^3 - \lambda^2 + \lambda + 1 = -(\lambda + 1)^2 (\lambda - 1)$$

so the eigenvalues of $\mathbf{A}$ are $\lambda_1 = -1$ and $\lambda_2 = 1$ with algebraic multiplicities 2 and 1, respectively. To find the eigenvector(s) belonging to λ_1, we set $\lambda = -1$ in $(\mathbf{A} - \lambda\mathbf{I})\mathbf{v} = \mathbf{0}$. This gives the linear algebraic system

$$\begin{pmatrix} 1/5 & -1/5 & 4/5 \\ -1/5 & 1/5 & -4/5 \\ 2/5 & -2/5 & 8/5 \end{pmatrix} \begin{pmatrix} v_1 \\ v_2 \\ v_3 \end{pmatrix} = \begin{pmatrix} 0 \\ 0 \\ 0 \end{pmatrix}. \tag{10}$$

By using elementary row operations we reduce the system (10) to

$$\begin{pmatrix} 1 & -1 & 4 \\ 0 & 0 & 0 \\ 0 & 0 & 0 \end{pmatrix} \begin{pmatrix} v_1 \\ v_2 \\ v_3 \end{pmatrix} = \begin{pmatrix} 0 \\ 0 \\ 0 \end{pmatrix}$$

so the only constraint on the components of $\mathbf{v}$ is $v_1 - v_2 + 4v_3 = 0$. Setting $v_2 = a_1$ and $v_3 = a_2$, where a_1 and a_2 are arbitrary constants, and then solving for v_1 gives $v_1 = a_1 - 4a_2$. Consequently the general solution of (10) can be represented by

$$\mathbf{v} = \begin{pmatrix} a_1 - 4a_2 \\ a_1 \\ a_2 \end{pmatrix} = a_1 \begin{pmatrix} 1 \\ 1 \\ 0 \end{pmatrix} + a_2 \begin{pmatrix} -4 \\ 0 \\ 1 \end{pmatrix}. \tag{11}$$

First, setting $a_1 = 2$ and $a_2 = 0$ and then setting $a_1 = 2$ and $a_2 = 1$ yields a pair of linearly independent eigenvectors associated with $\lambda_1 = -1$,

$$\mathbf{v}_1 = \begin{pmatrix} 2 \\ 2 \\ 0 \end{pmatrix} \quad \text{and} \quad \mathbf{v}_2 = \begin{pmatrix} -2 \\ 2 \\ 1 \end{pmatrix}.$$

Remark. Any choices for a_1 and a_2 that give us a pair of linearly independent eigenvectors for λ_1 would suffice. For example, we could have first set $a_1 = 1$ and $a_2 = 0$ and then set $a_1 = 0$ and $a_2 = 1$.

To find the eigenvector associated with $\lambda_2 = 1$ we set $\lambda = 1$ in $(\mathbf{A} - \lambda\mathbf{I})\mathbf{v} = \mathbf{0}$ to obtain the system

$$\begin{pmatrix} -9/5 & -1/5 & 4/5 \\ -1/5 & -9/5 & -4/5 \\ 2/5 & -2/5 & -2/5 \end{pmatrix} \begin{pmatrix} v_1 \\ v_2 \\ v_3 \end{pmatrix} = \begin{pmatrix} 0 \\ 0 \\ 0 \end{pmatrix}. \tag{12}$$

Using elementary row operations reduces the system (12) to

$$\begin{pmatrix} 1 & 0 & -1/2 \\ 0 & 1 & 1/2 \\ 0 & 0 & 0 \end{pmatrix} \begin{pmatrix} v_1 \\ v_2 \\ v_3 \end{pmatrix} = \begin{pmatrix} 0 \\ 0 \\ 0 \end{pmatrix}. \tag{13}$$

Equation (12) is therefore equivalent to the two equations $v_1 - v_3/2 = 0$ and $v_2 + v_3/2 = 0$. The general solution of this pair of equations can be expressed as

$$\mathbf{v} = b_1 \begin{pmatrix} 1 \\ -1 \\ 2 \end{pmatrix} \tag{14}$$

where b_1 is arbitrary. Choosing $b_1 = 1$ yields the eigenvector

$$\mathbf{v}_3 = \begin{pmatrix} 1 \\ -1 \\ 2 \end{pmatrix}$$

belonging to the eigenvalue $\lambda_2 = 1$. The general solution of Eq. (9) is therefore given by

$$\mathbf{x}(t) = c_1 e^{-t} \begin{pmatrix} 2 \\ 2 \\ 0 \end{pmatrix} + c_2 e^{-t} \begin{pmatrix} -2 \\ 2 \\ 1 \end{pmatrix} + c_3 e^{t} \begin{pmatrix} 1 \\ -1 \\ 2 \end{pmatrix}. \tag{15}$$

To help understand the qualitative behavior of all solutions of Eq. (9), we introduce the subset S of $\mathbf{R}^3$ spanned by the eigenvectors $\mathbf{v}_1$ and $\mathbf{v}_2$,

$$S = \{ \mathbf{v} : \mathbf{v} = a_1 \mathbf{v}_1 + a_2 \mathbf{v}_2, \ -\infty < a_1, a_2 < \infty \}. \tag{16}$$

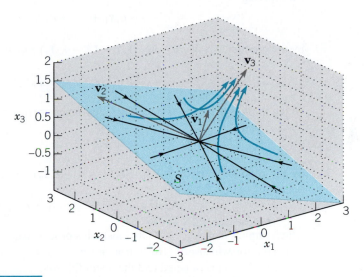

FIGURE 6.3.1 Solution trajectories for the system (9).

Geometrically, S corresponds to a plane passing through the origin as shown in Figure 6.3.1. From Eq. (15) we see that if $\mathbf{x}(0) \in S$, then $c_3 = 0$ and consequently $\mathbf{x}(t) \in S$ for all $t \geq 0$. Furthermore $\mathbf{x}(t) \to (0, 0, 0)^T$ as $t \to \infty$ due to the fact that the eigenvalue $\lambda_1 = -1 < 0$. The straight line trajectories lying in S in Figure 6.3.1 correspond to solutions $\mathbf{x}(t)$ of Eq. (9) where the initial conditions are such that $\mathbf{x}(0) = \mathbf{x}_0 \in S$. Also shown in Figure 6.3.1 are the graphs of four trajectories (heavy curves) represented by Eq. (15) in which c_3 is chosen to be slightly positive. For each of these trajectories, the initial condition $\mathbf{x}(0) \notin S$, and it is clear from Eq. (15) and Figure 6.3.1 that each trajectory in this case must asymptotically approach the line passing through the origin with direction vector $\mathbf{v}_3$, parametrically expressed by $\mathbf{x} = \tau \mathbf{v}_3$, $-\infty < \tau < \infty$, as $t \to \infty$.

▶ **Real and Distinct Eigenvalues.** A frequently occurring special case for which the general solution of $\mathbf{x}' = \mathbf{A}\mathbf{x}$ is always of the form (7) is given by the following corollary to Theorem 6.3.1.

COROLLARY 6.3.2

Suppose that the matrix $\mathbf{A}$ has n eigenpairs $(\lambda_1, \mathbf{v}_1), \ldots, (\lambda_n, \mathbf{v}_n)$ with the property that the eigenvalues $\lambda_1, \ldots, \lambda_n$ are real and distinct. Then

$$\left\{ e^{\lambda_1 t} \mathbf{v}_1, \ldots, e^{\lambda_n t} \mathbf{v}_n \right\}$$

is a fundamental solution set for the homogeneous system $\mathbf{x}' = \mathbf{A}\mathbf{x}$.

Proof

Since the eigenvalues are distinct, the eigenvectors $\mathbf{v}_1, \ldots, \mathbf{v}_n$ are linearly independent. To see why this is so, note that

$$(\mathbf{A} - \lambda_i \mathbf{I}_n)\mathbf{v}_j = \mathbf{A}\mathbf{v}_j - \lambda_i \mathbf{I}_n \mathbf{v}_j = \lambda_j \mathbf{v}_j - \lambda_i \mathbf{v}_j = \begin{cases} \mathbf{0}, & i = j \\ (\lambda_j - \lambda_i)\mathbf{v}_j, & i \neq j. \end{cases} \tag{17}$$

Multiplying

$$c_1 \mathbf{v}_1 + \cdots + c_n \mathbf{v}_n = \mathbf{0} \tag{18}$$

by the product matrix $(\mathbf{A} - \lambda_{n-1}\mathbf{I}_n) \cdots (\mathbf{A} - \lambda_1 \mathbf{I}_n)$ and using Eqs. (17) yields the equation

$$c_n (\lambda_n - \lambda_1) \cdots (\lambda_n - \lambda_{n-1}) \mathbf{v}_n = \mathbf{0}$$

which implies that $c_n = 0$ since the eigenvalues are distinct. Next, multiplying

$$c_1 \mathbf{v}_1 + \cdots + c_{n-1} \mathbf{v}_{n-1} = \mathbf{0}$$

by the product matrix $(\mathbf{A} - \lambda_{n-2}\mathbf{I}_n) \cdots (\mathbf{A} - \lambda_1 \mathbf{I}_n)$ and again using Eqs. (17) yields the equation

$$c_{n-1}(\lambda_{n-1} - \lambda_1) \cdots (\lambda_{n-1} - \lambda_{n-2}) \mathbf{v}_{n-1} = \mathbf{0}$$

which implies that $c_{n-1} = 0$. Obviously this process can be continued to show that the only constants for which Eq. (18) is true are $c_1 = c_2 = \cdots = c_n = 0$. Thus the hypothesis of Theorem (6.3.1) is satisfied and the result follows. ∎

EXAMPLE 2

Find the general solution of

$$\mathbf{x}' = \begin{pmatrix} -1 & -1 & 1 & 1 \\ -3 & -4 & -3 & 6 \\ 0 & -3 & -2 & 3 \\ -3 & -5 & -3 & 7 \end{pmatrix} \mathbf{x}. \tag{19}$$

The characteristic polynomial of the matrix of coefficients is

$$\begin{vmatrix} -1-\lambda & -1 & 1 & 1 \\ -3 & -4-\lambda & -3 & 6 \\ 0 & -3 & -2-\lambda & 3 \\ -3 & -5 & -3 & 7-\lambda \end{vmatrix} = \lambda^4 - 5\lambda^2 + 4 = (\lambda+2)(\lambda+1)(\lambda-1)(\lambda-2).$$

Thus the eigenvalues $\lambda_1 = -2$, $\lambda_2 = -1$, $\lambda_3 = 1$, $\lambda_4 = 2$ are distinct and Corollary 6.3.2 is applicable. The respective eigenvectors are found to be

$$\mathbf{v}_1 = \begin{pmatrix} 1 \\ 0 \\ -1 \\ 0 \end{pmatrix}, \quad \mathbf{v}_2 = \begin{pmatrix} 1 \\ 1 \\ 0 \\ 1 \end{pmatrix}, \quad \mathbf{v}_3 = \begin{pmatrix} 1 \\ 0 \\ 1 \\ 1 \end{pmatrix}, \quad \mathbf{v}_4 = \begin{pmatrix} 0 \\ 1 \\ 0 \\ 1 \end{pmatrix}.$$

Of course, the eigenvalues and eigenvectors can be found easily by using a computer or calculator. A fundamental set of solutions of Eq. (19) is therefore

$$\left\{ e^{-2t} \begin{pmatrix} 1 \\ 0 \\ -1 \\ 0 \end{pmatrix}, \quad e^{-t} \begin{pmatrix} 1 \\ 1 \\ 0 \\ 1 \end{pmatrix}, \quad e^{t} \begin{pmatrix} 1 \\ 0 \\ 1 \\ 1 \end{pmatrix}, \quad e^{2t} \begin{pmatrix} 0 \\ 1 \\ 0 \\ 1 \end{pmatrix} \right\}$$

and the general solution is

$$\mathbf{x} = c_1 e^{-2t} \begin{pmatrix} 1 \\ 0 \\ -1 \\ 0 \end{pmatrix} + c_2 e^{-t} \begin{pmatrix} 1 \\ 1 \\ 0 \\ 1 \end{pmatrix} + c_3 e^{t} \begin{pmatrix} 1 \\ 0 \\ 1 \\ 1 \end{pmatrix} + c_4 e^{2t} \begin{pmatrix} 0 \\ 1 \\ 0 \\ 1 \end{pmatrix}.$$

From the general solution we can deduce the behavior of solutions for large t: (i) if $c_3 = c_4 = 0$, $\lim_{t \to \infty} \mathbf{x}(t) = (0, 0, 0, 0)^T$, (ii) if $c_3 \neq 0$ and $c_4 = 0$, $\mathbf{x}(t)$ asymptotically approaches the line passing through the origin in $\mathbf{R}^4$ with direction vector $\mathbf{v}_3$, (iii) if $c_4 \neq 0$, $\mathbf{x}(t)$ asymptotically approaches the line passing through the origin in $\mathbf{R}^4$ with direction vector $\mathbf{v}_4$.

▶ **Symmetric Matrices.** Even though the matrix in Example 1 has an eigenvalue ($\lambda = -1$) with algebraic multiplicity 2, we were able to find two linearly independent eigenvectors $\mathbf{v}_1$ and $\mathbf{v}_2$, and, as a consequence, were able to construct the general solution. In general, given a matrix $\mathbf{A}$, the geometric multiplicity of each eigenvalue is less than or equal to its algebraic multiplicity. If for one or more eigenvalues of $\mathbf{A}$ the geometric multiplicity is less than the algebraic multiplicity, then Eq. (2) will not have a fundamental set of solutions of the form (6). However, if $\mathbf{A}$ belongs to the class of real and symmetric matrices, then Eq. (2) will always have a fundamental set of solutions of the form (6) with the general solution given by Eq. (7). This is a consequence of the following properties of the eigenvalues and eigenvectors of a real symmetric matrix $\mathbf{A}$ (see Appendix A.4):

1. All the eigenvalues $\lambda_1, \ldots, \lambda_n$ of $\mathbf{A}$ are real.
2. $\mathbf{A}$ has a complete set of n real and linearly independent eigenvectors $\mathbf{v}_1, \ldots, \mathbf{v}_n$. Furthermore, eigenvectors corresponding to different eigenvalues are orthogonal to one another and all eigenvectors belonging to the same eigenvalue can be chosen to be orthogonal to one another.

The following continuous time, discrete space model of particle diffusion in one dimension leads to a first order system of linear differential equations $\mathbf{x}' = \mathbf{A}\mathbf{x}$ where $\mathbf{A}$ is a symmetric matrix.

EXAMPLE

3

Diffusion on a One-Dimensional Lattice with Reflecting Boundaries. Consider particles that can occupy any of n equally spaced points lying along the real line (a one-dimensional lattice) as shown in Figure 6.3.2.

FIGURE 6.3.2 Diffusion on a one-dimensional lattice.

If $x_j(t)$ is the number of particles residing at the jth lattice point at time t, we make the following assumptions governing the movement of particles from site to site:

(i) particle transitions to site j are permitted only from nearest neighbor sites, and

(ii) particles move from more populated sites to less populated sites with the rate of transition proportional to the difference between the numbers of particles at adjacent sites.

Then the differential equation describing the rate of change in the number of particles at the jth interior point of the lattice is

$$\frac{dx_j}{dt} = k(x_{j-1} - x_j) + k(x_{j+1} - x_j) = k(x_{j-1} - 2x_j + x_{j+1}), \quad j = 2, \ldots, n-1 \tag{20}$$

where k is a rate constant. The rate equations for the numbers of particles at the left and right endpoints are

$$\frac{dx_1}{dt} = k(x_2 - x_1) \tag{21}$$

and

$$\frac{dx_n}{dt} = k(x_{n-1} - x_n), \tag{22}$$

respectively. The left and right endpoints are referred to as reflecting boundaries since Eqs. (21) and (22) do not permit particles to escape from the set of lattice points. This will be made evident in the discussion that follows.

Using Eqs. (20) through (22), the system describing diffusion on a lattice consisting of $n = 3$ points is expressed in matrix notation as

$$\mathbf{x}' = k\mathbf{A}\mathbf{x} \tag{23}$$

where $\mathbf{x} = (x_1, x_2, x_3)^T$ and

$$\mathbf{A} = \begin{pmatrix} -1 & 1 & 0 \\ 1 & -2 & 1 \\ 0 & 1 & -1 \end{pmatrix}. \tag{24}$$

The eigenvalues of the symmetric matrix $\mathbf{A}$ are $\lambda_1 = 0$, $\lambda_2 = -1$, and $\lambda_3 = -3$, and the corresponding eigenvectors are

$$\mathbf{v}_1 = \begin{pmatrix} 1 \\ 1 \\ 1 \end{pmatrix}, \quad \mathbf{v}_2 = \begin{pmatrix} 1 \\ 0 \\ -1 \end{pmatrix}, \quad \mathbf{v}_3 = \begin{pmatrix} 1 \\ -2 \\ 1 \end{pmatrix}. \tag{25}$$

The eigenvectors are mutually orthogonal and are frequently referred to as **normal modes**. Plots of the components of the eigenvectors are shown in Figure 6.3.3.

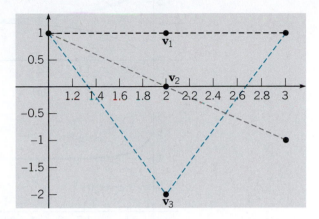

FIGURE 6.3.3 The components of the eigenvectors, or normal modes, of the symmetric matrix A.

Thus the general solution of Eq. (23) is

$$\mathbf{x}(t) = c_1 \begin{pmatrix} 1 \\ 1 \\ 1 \end{pmatrix} + c_2 e^{-kt} \begin{pmatrix} 1 \\ 0 \\ -1 \end{pmatrix} + c_3 e^{-3kt} \begin{pmatrix} 1 \\ -2 \\ 1 \end{pmatrix}. \tag{26}$$

From Eq. (26) it is clear that

$$\lim_{t \to \infty} \mathbf{x}(t) = c_1 \mathbf{v}_1 = \begin{pmatrix} c_1 \\ c_1 \\ c_1 \end{pmatrix}, \tag{27}$$

that is, all solutions approach an equilibrium state in which the numbers of particles at each of the lattice sites are identical, that is, a uniform distribution. The components on both sides of Eq. (23) may be summed by multiplying both sides of Eq. (23) by the row vector $(1, 1, 1)$. This yields the equation $(d/dt)(x_1 + x_2 + x_3) = 0$ and therefore $x_1(t) + x_2(t) + x_3(t)$ is constant for $t > 0$, a statement that the total number of particles in the system is conserved. Since $\lim_{t \to 0}[x_1(t) + x_2(t) + x_3(t)] = x_{10} + x_{20} + x_{30}$, the initial total number of particles, and Eq. (27) implies that $\lim_{t \to \infty}[x_1(t) + x_2(t) + x_3(t)] = 3c_1$, it follows that $3c_1 = x_{10} + x_{20} + x_{30}$. Consequently $c_1 = (x_{10} + x_{20} + x_{30})/3$, the average value of the initial total number of particles.

By choosing $\mathbf{c} = (1, 1, 0)^T$ in the general solution (26), corresponding to the initial condition $\mathbf{x}(0) = (2, 1, 0)^T$, decay towards equilibrium that involves only the eigenvectors $\mathbf{v}_1$ and $\mathbf{v}_2$ is illustrated in Figure 6.3.4(a) where we have also set the rate constant $k = 1$. In this case, the decay rate is controlled by the eigenvalue $\lambda_2 = -1$. Furthermore, the components of $\mathbf{x}(t)$ are antisymmetric about the equilibrium solution due to the fact that $\mathbf{v}_2$ is antisymmetric about its middle component (see Figure 6.3.3).

Similarly, by choosing $\mathbf{c} = (1, 0, 1/2)^T$ in (26), corresponding to the initial condition $\mathbf{x}(0) = (3/2, 0, 3/2)^T$, decay towards equilibrium that involves only the eigenvectors $\mathbf{v}_1$ and $\mathbf{v}_3$ is shown in Figure 6.3.4(b) where again $k = 1$. In this case, the rate of decay toward equilibrium is faster since it is controlled by the eigenvalue $\lambda_3 = -3$. Note also that the

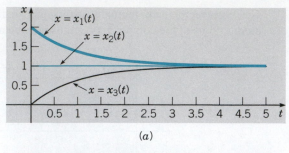

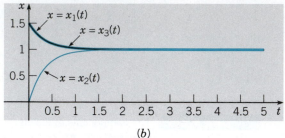

FIGURE 6.3.4 (a) Component plots of Eq. (26) with $\mathbf{c} = (1, 1, 0)^T$ corresponding to the initial condition $\mathbf{x}(0) = (2, 1, 0)^T$. (b) Component plots of Eq. (26) with $\mathbf{c} = (1, 0, 1/2)^T$ corresponding to the initial condition $\mathbf{x}(0) = (3/2, 0, 3/2)$. In both cases $k = 1$.

components of $\mathbf{x}(t)$ are symmetric about the equilibrium solution due to the fact that $\mathbf{v}_3$ is symmetric about its middle component (see Figure 6.3.3).

In Figure 6.3.5, we show component plots of (26) with $\mathbf{c} = (1, 3/2, 1/2)^T$, corresponding to the initial condition $\mathbf{x}(0) = (3, 0, 0)^T$ in which all of the particles are initially located at the first lattice point, again with $k = 1$. In this case, all three eigenvectors are required to represent the solution of the initial value problem.

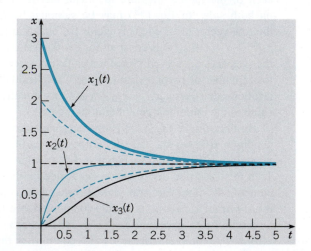

FIGURE 6.3.5 Component plots of Eq. (26) with $\mathbf{c} = (1, 3/2, 1/2)^T$ corresponding to the initial condition $\mathbf{x}(0) = (3, 0, 0)^T$. The dashed curves are the solutions shown in Figure 6.3.4(a); $k = 1$.

The contribution to the solution from $e^{-3t}\mathbf{v}_3$ decays rapidly while the long term decay rate towards equilibrium is controlled by the eigenvalue $\lambda_2 = -1$. Since the component $x_2(t) = 1 - e^{-3t}$ contains no e^{-t} term, it converges more quickly than $x_1(t)$ and $x_2(t)$. In the long run, the component plots closely match those shown in Figure 6.3.4(a) since the total numbers of initial particles used in each of the two initial value problems are the same.

PROBLEMS

In each of Problems 1 through 8, find the general solution of the given system of equations.

1. $\begin{aligned} x_1' &= -4x_1 + x_2 \\ x_2' &= x_1 - 5x_2 + x_3 \\ x_3' &= x_2 - 4x_3 \end{aligned}$

2. $\begin{aligned} x_1' &= x_1 + 4x_2 + 4x_3 \\ x_2' &= 3x_2 + 2x_3 \\ x_3' &= 2x_2 + 3x_3 \end{aligned}$

3. $\begin{aligned} x_1' &= 2x_1 - 4x_2 + 2x_3 \\ x_2' &= -4x_1 + 2x_2 - 2x_3 \\ x_3' &= 2x_1 - 2x_2 - x_3 \end{aligned}$

4. $\begin{aligned} x_1' &= -2x_1 + 2x_2 - x_3 \\ x_2' &= -2x_1 + 3x_2 - 2x_3 \\ x_3' &= -2x_1 + 4x_2 - 3x_3 \end{aligned}$

5. $\mathbf{x}' = \begin{pmatrix} 1 & 1 & 2 \\ 1 & 2 & 1 \\ 2 & 1 & 1 \end{pmatrix} \mathbf{x}$ 6. $\mathbf{x}' = \begin{pmatrix} 3 & 2 & 4 \\ 2 & 0 & 2 \\ 4 & 2 & 3 \end{pmatrix} \mathbf{x}$

7. $\mathbf{x}' = \begin{pmatrix} 1 & 1 & 1 \\ 2 & 1 & -1 \\ -8 & -5 & -3 \end{pmatrix} \mathbf{x}$

8. $\mathbf{x}' = \begin{pmatrix} 1 & -1 & 4 \\ 3 & 2 & -1 \\ 2 & 1 & -1 \end{pmatrix} \mathbf{x}$

ODEA In each of Problems 9 through 12, solve the given initial value problem and plot the graph of the solution in $\mathbf{R}^3$. Describe the behavior of the solution as $t \to \infty$.

9. $\mathbf{x}' = \begin{pmatrix} 1 & 1 & 2 \\ 0 & 2 & 2 \\ -1 & 1 & 3 \end{pmatrix} \mathbf{x}, \qquad \mathbf{x}(0) = \begin{pmatrix} 2 \\ 0 \\ 1 \end{pmatrix}$

10. $\mathbf{x}' = \begin{pmatrix} 0 & 0 & -1 \\ 2 & 0 & 0 \\ -1 & 2 & 4 \end{pmatrix} \mathbf{x}, \qquad \mathbf{x}(0) = \begin{pmatrix} 7 \\ 5 \\ 5 \end{pmatrix}$

11. $\mathbf{x}' = \begin{pmatrix} -1 & 0 & 3 \\ 0 & -2 & 0 \\ 3 & 0 & -1 \end{pmatrix} \mathbf{x}, \qquad \mathbf{x}(0) = \begin{pmatrix} 2 \\ -1 \\ -2 \end{pmatrix}$

12. $\mathbf{x}' = \begin{pmatrix} 1/2 & -1 & -3/2 \\ 3/2 & -2 & -3/2 \\ -2 & 2 & 1 \end{pmatrix} \mathbf{x}, \qquad \mathbf{x}(0) = \begin{pmatrix} 2 \\ 1 \\ 1 \end{pmatrix}$

13. Using the rate equations (20) through (22), ex- CAS press, in matrix notation, the system of differential equations that describe diffusion on a one-dimensional lattice with reflecting boundaries consisting of $n = 4$ lattice sites. Find the general solution of the resulting system and plot the components of each of the eigenvectors of the matrix $\mathbf{A}$ on the same set of coordinate axes. Which eigenvalue controls the long term rate of decay towards equilibrium?

14. **Diffusion on a one-dimensional lattice with an absorbing boundary.** Consider a one-dimensional lattice consisting of $n = 4$ lattice points as shown in Figure 6.3.6.

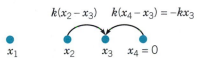

$k(x_2 - x_3)$ $k(x_4 - x_3) = -kx_3$

x_1 x_2 x_3 $x_4 = 0$

FIGURE 6.3.6 An absorbing boundary at the right endpoint.

Assume that the left endpoint is a reflecting boundary so that the rate equation for the number of particles $x_1(t)$ occupying that site is given by Eq. (21). Further, assume that the sites 2 and 3 are interior points with rate equations given by Eq. (20) but with the condition that $x_4(t) = 0$ for all $t > 0$.

(a) Find the system of differential equations that describe the rate equations for $\mathbf{x} = (x_1, x_2, x_3)^T$ and express the equations in matrix notation, $\mathbf{x}' = \mathbf{Ax}$.

(b) Site 4 is referred to as an absorbing boundary because particles that land on that site from site 3 are removed from the diffusion process. Show that

$$\frac{d}{dt}(x_1 + x_2 + x_3) = -kx_3$$

and explain the meaning of this equation.

CAS (c) Find the eigenvalues and eigenvectors for the matrix $\mathbf{A}$ in part (a) and plot the components of the eigenvectors. Then compare the eigenvalues and eigenvectors with those found in Example 3 of this section.

(d) Find the general solution of the system of equations found in part (a). Explain the asymptotic behavior of $\mathbf{x}(t)$ as $t \to \infty$.

CAS 15. Find constant 3×1 vectors $\mathbf{u}_1$ and $\mathbf{u}_2$ such that the solution of the initial value problem

$$\mathbf{x}' = \begin{pmatrix} 2 & -4 & -3 \\ 3 & -5 & -3 \\ -2 & 2 & 1 \end{pmatrix} \mathbf{x}, \qquad \mathbf{x}(0) = \mathbf{x}_0$$

tends to $(0, 0, 0)^T$ as $t \to \infty$ for any $\mathbf{x}_0 \in S$ where

$$S = \{\mathbf{u} : \mathbf{u} = a_1\mathbf{u}_1 + a_2\mathbf{u}_2, \ -\infty < a_1, a_2 < \infty\}.$$

In $\mathbf{R}^3$, plot solutions to the initial value problem for several different choices of $\mathbf{x}_0 \in S$ overlaying the trajectories on a graph of the plane determined by $\mathbf{u}_1$ and $\mathbf{u}_2$. Describe the behavior of solutions as $t \to \infty$ if $\mathbf{x}_0 \notin S$.

16. Find constant 4×1 vectors $\mathbf{u}_1$, $\mathbf{u}_2$, and $\mathbf{u}_3$ such that the solution of the initial value problem

$$\mathbf{x}' = \begin{pmatrix} 1 & 5 & 3 & -5 \\ 2 & 3 & 2 & -4 \\ 0 & -1 & -2 & 1 \\ 2 & 4 & 2 & -5 \end{pmatrix} \mathbf{x}, \qquad \mathbf{x}(0) = \mathbf{x}_0$$

tends to $(0, 0, 0, 0)^T$ as $t \to \infty$ for any $\mathbf{x}_0 \in S$ where

$$S = \{\mathbf{u} : \mathbf{u} = a_1\mathbf{u}_1 + a_2\mathbf{u}_2 + a_3\mathbf{u}_3,$$

$$-\infty < a_1, a_2, a_3 < \infty\}.$$

17. A radioactive substance R_1 having decay rate k_1 disintegrates into a second radioactive substance R_2 having decay rate k_2. Substance R_2 disintegrates into R_3, which is stable. If $m_i(t)$ represents the mass of substance R_i at time t, $i = 1, 2, 3$, the

applicable equations are

$$m_1' = -k_1 m_1$$
$$m_2' = k_1 m_1 - k_2 m_2$$
$$m_3' = k_2 m_2.$$

Draw a block diagram of a compartment model of the overall reaction. Label the directed arrows that represent the mass flows between compartments with the appropriate radioactive decay rate constants. Use the eigenvalue method to solve the above system under the conditions

$$m_1(0) = m_0 \quad m_2(0) = 0, \quad m_3(0) = 0.$$

For each of the matrices in Problems 18 through 23, CAS use a computer to assist in finding a fundamental set of solutions to the system $\mathbf{x}' = \mathbf{A}\mathbf{x}$.

18. $\mathbf{A} = \begin{pmatrix} -5 & 1 & -4 & -1 \\ 0 & -3 & 0 & 0 \\ 1 & -1 & 0 & 1 \\ 2 & -1 & 2 & -2 \end{pmatrix}$

19. $\mathbf{A} = \begin{pmatrix} 2 & 2 & 0 & -1 \\ 2 & -1 & 0 & 2 \\ 0 & 0 & 3 & 0 \\ -1 & 2 & 0 & 2 \end{pmatrix}$

20. $\mathbf{A} = \begin{pmatrix} 1 & 8 & 5 & 3 \\ 2 & 16 & 10 & 6 \\ 5 & -14 & -11 & -3 \\ -1 & -8 & -5 & -3 \end{pmatrix}$

21. $\mathbf{A} = \begin{pmatrix} -2 & 2 & 0 & -2 \\ -1 & 3 & -1 & 1 \\ -2 & -2 & -4 & 2 \\ -7 & 1 & -7 & 3 \end{pmatrix}$

22. $\mathbf{A} = \begin{pmatrix} -5 & -2 & -1 & 2 & 3 \\ 0 & -3 & 0 & 0 & 0 \\ 1 & 0 & -1 & 0 & -1 \\ 2 & 1 & 0 & -4 & -2 \\ -3 & -2 & -1 & 2 & 1 \end{pmatrix}$

23. $\mathbf{A} = \begin{pmatrix} 0 & -3 & -2 & 3 & 2 \\ 8 & 6 & 4 & -8 & -16 \\ -8 & -8 & -6 & 8 & 16 \\ 8 & 7 & 4 & -9 & -16 \\ -3 & -5 & -3 & 5 & 7 \end{pmatrix}$

6.4 Nondefective Matrices with Complex Eigenvalues

In this section, we again consider a system of n linear homogeneous equations with constant coefficients,

$$\mathbf{x}' = \mathbf{A}\mathbf{x}, \tag{1}$$

where the coefficient matrix $\mathbf{A}$ is real-valued, nondefective, and has one or more complex eigenvalues. Since $\mathbf{A}$ is real, the coefficients of the characteristic equation

$$p(\lambda) = |\mathbf{A} - \lambda\mathbf{I}_n| = 0 \tag{2}$$

are real. Consequently, complex eigenvalues must occur in conjugate pairs. If $\lambda = \mu + i\nu$ and $(\lambda, \mathbf{v})$ is an eigenpair of $\mathbf{A}$, then so is $(\bar{\lambda}, \bar{\mathbf{v}})$ where $\bar{\lambda}$ and $\bar{\mathbf{v}}$ are the complex conjugates of λ and $\mathbf{v}$, respectively. It follows that the corresponding complex conjugates

$$\mathbf{u}(t) = e^{\lambda t}\mathbf{v} \qquad \bar{\mathbf{u}}(t) = e^{\bar{\lambda}t}\bar{\mathbf{v}} \tag{3}$$

are solutions of Eq. (1). By an argument identical to that in Section 3.4, we obtain real-valued solutions of Eq. (1) by taking the real and imaginary parts of $\mathbf{u}(\mathbf{t})$ or $\bar{\mathbf{u}}(t)$. If we let $\mathbf{v} = \mathbf{a} + i\mathbf{b}$ where $\mathbf{a}$ and $\mathbf{b}$ are real constant $n \times 1$ vectors, then the vectors

$$\mathbf{x}_1(t) = \text{Re }\mathbf{u}(t) = e^{\mu t}(\mathbf{a}\cos\nu t - \mathbf{b}\sin\nu t)$$
$$\mathbf{x}_2(t) = \text{Im }\mathbf{u}(t) = e^{\mu t}(\mathbf{a}\sin\nu t + \mathbf{b}\cos\nu t) \tag{4}$$

are real-valued solutions of Eq. (1). It is possible to show that $\mathbf{x}_1$ and $\mathbf{x}_2$ are linearly independent solutions (see Problem 11).

If all of the eigenvectors of $\mathbf{A}$, real and complex, are linearly independent, then a fundamental set of real solutions of Eq. (1) consists of solutions of the form (4) associated with complex eigenvalues and solutions of the form $e^{\lambda_j t}\mathbf{v}_j$ associated with real eigenvalues. For example, suppose that $\lambda_1 = \mu + i\nu$, $\lambda_2 = \mu - i\nu$, and that $\lambda_3, \ldots, \lambda_n$ are all real and distinct. Let the corresponding eigenvectors be $\mathbf{v}_1 = \mathbf{a} + i\mathbf{b}$, $\mathbf{v}_2 = \mathbf{a} - i\mathbf{b}$, $\mathbf{v}_3, \ldots, \mathbf{v}_n$. Then the general solution of Eq. (1) is

$$\mathbf{x} = c_1\mathbf{x}_1(t) + c_2\mathbf{x}_2(t) + c_3 e^{\lambda_3 t}\mathbf{v}_3 + \cdots + c_n e^{\lambda_n t}\mathbf{v}_n,$$

where $\mathbf{x}_1(t)$ and $\mathbf{x}_2(t)$ are given by Eqs. (4). We emphasize that this analysis applies only if the coefficient matrix $\mathbf{A}$ in Eq. (1) is real, for it is only then that complex eigenvalues and eigenvectors occur in conjugate pairs.

EXAMPLE 1

Find a fundamental set of real-valued solutions of the system

$$\mathbf{x}' = \begin{pmatrix} -4 & 5 & -3 \\ -17/3 & 4/3 & 7/3 \\ 23/3 & -25/3 & -4/3 \end{pmatrix} \mathbf{x} \tag{5}$$

and describe the solution trajectories.

To find a fundamental set of solutions, we assume that $\mathbf{x} = e^{\lambda t}\mathbf{v}$ and obtain the set of linear algebraic equations

$$\begin{pmatrix} -4 - \lambda & 5 & -3 \\ -17/3 & 4/3 - \lambda & 7/3 \\ 23/3 & -25/3 & -4/3 - \lambda \end{pmatrix} \mathbf{v} = \mathbf{0}. \tag{6}$$

The characteristic equation of the matrix of coefficients in Eq. (5) is

$$
\begin{vmatrix}
-4 - \lambda & 5 & -3 \\
-17/3 & 4/3 - \lambda & 7/3 \\
23/3 & -25/3 & -4/3 - \lambda
\end{vmatrix}
= -(\lambda^3 + 4\lambda^2 + 69\lambda + 130)
$$

$$
= -(\lambda + 2)(\lambda^2 + 2\lambda + 65)
$$

$$
= -(\lambda + 2)(\lambda + 1 - 8i)(\lambda + 1 + 8i)
$$

and therefore the eigenvalues are $\lambda_1 = -2$, $\lambda_2 = -1 + 8i$, and $\lambda_3 = -1 - 8i$. Substituting $\lambda = -2$ into Eq. (6) and using elementary row operations yields the reduced system

$$
\begin{pmatrix}
1 & 0 & -1 \\
0 & 1 & -1 \\
0 & 0 & 0
\end{pmatrix} \mathbf{v} = \mathbf{0}.
$$

The general solution of this system is $\mathbf{v} = a_1(1, 1, 1)^T$ and taking $a_1 = 1$ gives the eigenvector associated with $\lambda_1 = -2$,

$$
\mathbf{v}_1 = \begin{pmatrix} 1 \\ 1 \\ 1 \end{pmatrix}.
$$

Substituting $\lambda = -1 + 8i$ into Eq. (6) gives the system

$$
\begin{pmatrix}
-3 - 8i & 5 & -3 \\
-17/3 & 7/3 - 8i & 7/3 \\
23/3 & -25/3 & 1/3 - 8i
\end{pmatrix}
\begin{pmatrix} v_1 \\ v_2 \\ v_3 \end{pmatrix}
= \begin{pmatrix} 0 \\ 0 \\ 0 \end{pmatrix}.
$$

Using complex arithmetic and elementary row operations, we reduce this system to

$$
\begin{pmatrix}
1 & 0 & (1 - i)/2 \\
0 & 1 & (1 + i)/2 \\
0 & 0 & 0
\end{pmatrix}
\begin{pmatrix} v_1 \\ v_2 \\ v_3 \end{pmatrix}
= \begin{pmatrix} 0 \\ 0 \\ 0 \end{pmatrix}.
$$

The general solution is represented by

$$
\mathbf{v} = a_1 \begin{pmatrix} 1/2 \\ i/2 \\ -(1 + i)/2 \end{pmatrix}
$$

where a_1 is an arbitrary constant. Taking $a_1 = 2$ gives the eigenvector belonging to λ_2,

$$
\mathbf{v}_2 = \begin{pmatrix} 1 \\ i \\ -1 - i \end{pmatrix} = \mathbf{a} + i\mathbf{b}
$$

where

$$
\mathbf{a} = \begin{pmatrix} 1 \\ 0 \\ -1 \end{pmatrix} \quad \text{and} \quad \mathbf{b} = \begin{pmatrix} 0 \\ 1 \\ -1 \end{pmatrix}. \tag{7}
$$

The eigenvector belonging to $\lambda_3 = -1 - 8i = \bar{\lambda}_2$ is the complex conjugate of $\mathbf{v}_2$,

$$
\mathbf{v}_3 = \mathbf{a} - \mathbf{b}i = \begin{pmatrix} 1 \\ -i \\ -1 + i \end{pmatrix}.
$$

Hence a fundamental set of solutions of the system (5) is

$$\mathbf{x}_1(t) = e^{-2t} \begin{pmatrix} 1 \\ 1 \\ 1 \end{pmatrix}, \quad \mathbf{u}(t) = e^{(-1+8i)t} \begin{pmatrix} 1 \\ i \\ -1-i \end{pmatrix},$$

and

$$\bar{\mathbf{u}}(t) = e^{(-1-8i)t} \begin{pmatrix} 1 \\ -i \\ -1+i \end{pmatrix}.$$

To obtain a fundamental set of real-valued solutions, we must find the real and imaginary parts of either $\mathbf{u}$ or $\bar{\mathbf{u}}$. A direct calculation using complex arithmetic gives

$$\mathbf{u}(t) = e^{-t}(\cos 8t + i \sin 8t) \begin{pmatrix} 1 \\ i \\ -1-i \end{pmatrix}$$

$$= e^{-t} \begin{pmatrix} \cos 8t \\ -\sin 8t \\ -\cos 8t + \sin 8t \end{pmatrix} + i e^{-t} \begin{pmatrix} \sin 8t \\ \cos 8t \\ -\sin 8t - \cos 8t \end{pmatrix}.$$

Hence

$$\mathbf{x}_2(t) = e^{-t} \begin{pmatrix} \cos 8t \\ -\sin 8t \\ -\cos 8t + \sin 8t \end{pmatrix} \quad \text{and} \quad \mathbf{x}_3(t) = e^{-t} \begin{pmatrix} \sin 8t \\ \cos 8t \\ -\sin 8t - \cos 8t \end{pmatrix}$$

are real-valued solutions of Eq. (5). To verify that $\mathbf{x}_1(t)$, $\mathbf{x}_2(t)$, and $\mathbf{x}_3(t)$ are linearly independent, we compute their Wronskian at $t = 0$,

$$W[\mathbf{x}_1, \mathbf{x}_2, \mathbf{x}_3](0) = \begin{vmatrix} 1 & 1 & 0 \\ 1 & 0 & 1 \\ 1 & -1 & -1 \end{vmatrix} = 3.$$

Since $W[\mathbf{x}_1, \mathbf{x}_2, \mathbf{x}_3](0) \neq 0$, it follows that $\mathbf{x}_1(t)$, $\mathbf{x}_2(t)$, and $\mathbf{x}_3(t)$ constitute a fundamental set of (real-valued) solutions of the system (5). Therefore the general solution of Eq. (5) is

$$\mathbf{x}(t) = c_1 e^{-2t} \begin{pmatrix} 1 \\ 1 \\ 1 \end{pmatrix} + c_2 e^{-t} \begin{pmatrix} \cos 8t \\ -\sin 8t \\ -\cos 8t + \sin 8t \end{pmatrix}$$

$$+ c_3 e^{-t} \begin{pmatrix} \sin 8t \\ \cos 8t \\ -\sin 8t - \cos 8t \end{pmatrix} \tag{8}$$

or, using $\mathbf{a}$ and $\mathbf{b}$ defined in Eqs. (7),

$$\mathbf{x}(t) = c_1 e^{-2t} \mathbf{v}_1 + c_2 e^{-t}(\mathbf{a} \cos 8t - \mathbf{b} \sin 8t) + c_3 e^{-t}(\mathbf{a} \sin 8t + \mathbf{b} \cos 8t). \tag{9}$$

Plots of the components of $\mathbf{x}$ corresponding to $c_1 = 10$, $c_2 = 30$, and $c_3 = 20$ are shown in Figure 6.4.1.

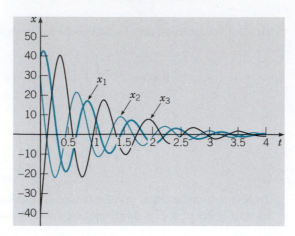

FIGURE 6.4.1 Plots of the components of the solution vector (8) using $c_1 = 10$, $c_2 = 30$, and $c_3 = 20$.

Geometric understanding of the solution (8) or (9) is facilitated by rewriting Eq. (9) in the form

$$\mathbf{x}(t) = c_1 e^{-2t}\mathbf{v}_1 + e^{-t}(c_2 \cos 8t + c_3 \sin 8t)\mathbf{a} + e^{-t}(c_3 \cos 8t - c_2 \sin 8t)\mathbf{b}. \tag{10}$$

Since the real parts of all eigenvalues are negative and appear in the exponential (decay) factors multiplying each term in Eqs. (8) through (10), all solutions approach $\mathbf{0}$ as $t \to \infty$. If $c_1 \neq 0$, since $\lambda_1 = -2$ is less than $\text{Re }\lambda_2 = \text{Re }\lambda_3 = -1$, solutions decay toward $\mathbf{0}$ in the direction parallel to $\mathbf{v}_1$ at a faster rate than their simultaneous spiral toward $\mathbf{0}$ in the plane S determined by $\mathbf{a}$ and $\mathbf{b}$,

$$S = \{\mathbf{v}: \mathbf{v} = d_1\mathbf{a} + d_2\mathbf{b}, \ -\infty < d_1, d_2 < \infty\}.$$

Solution trajectories decay toward the plane S parallel to the direction $\mathbf{v}_1$ while simultaneously spiraling toward $\mathbf{0}$ in the plane S. A typical solution trajectory for this case is shown in Figure 6.4.2. If $c_1 = 0$, solutions start in S and spiral toward $\mathbf{0}$ as $t \to \infty$, remaining in S all the while.

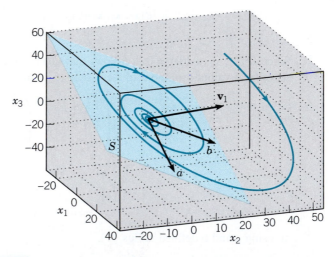

FIGURE 6.4.2 A typical solution trajectory for Eq. (5) that is not in S at time $t = 0$ spirals toward S as $t \to \infty$. The lengths of $\mathbf{v}_1$, $\mathbf{a}$, and $\mathbf{b}$ have been scaled to enhance visualization.

▶ **Natural Frequencies and Principal Modes of Vibration.** Consider again the system of two masses and three springs shown in Figure 6.1.1. If we assume that there are no external forces, then $F_1(t) = 0$, $F_2(t) = 0$ and Eq. (21) in Section 6.1 reduces to the homogeneous system

$$\mathbf{x}' = \begin{pmatrix} 0 & 0 & 1 & 0 \\ 0 & 0 & 0 & 1 \\ -(k_1 + k_2)/m_1 & k_2/m_1 & 0 & 0 \\ k_2/m_2 & -(k_2 + k_3)/m_2 & 0 & 0 \end{pmatrix} \mathbf{x}. \tag{11}$$

This system is said to have **two degrees of freedom** since there are two independent coordinates, x_1 and x_2, necessary to describe the motion of the system. Under certain conditions, both masses will undergo harmonic motion at the same frequency. In such a case, both masses will attain their maximum displacements at the same times even if they do not both move in the same direction. When this kind of motion occurs, the frequency is called a **natural frequency** of the system, and the motion is called a **principal mode of vibration**. In general, the number of natural frequencies and principal modes possessed by a vibrating system of masses is equal to the number of degrees of freedom of the system. Thus, for the two degrees of freedom system under discussion, there will be two natural frequencies and two principal modes of vibration. The principal mode of vibration corresponding to the lowest natural frequency is referred to as the **first mode**. The principal mode of vibration corresponding to the next higher frequency is called the **second mode** and so on if there are more than two degrees of freedom. The following example illustrates the important relationship between eigenvalues and eigenvectors and natural frequencies and principal modes of vibration.

EXAMPLE 2

Suppose that $m_1 = 2$, $m_2 = 9/4$, $k_1 = 1$, $k_2 = 3$, $k_3 = 15/4$ in Eq. (11),

$$\mathbf{x}' = \begin{pmatrix} 0 & 0 & 1 & 0 \\ 0 & 0 & 0 & 1 \\ -2 & 3/2 & 0 & 0 \\ 4/3 & -3 & 0 & 0 \end{pmatrix} \mathbf{x} = \mathbf{Ax}. \tag{12}$$

Find and describe the principal modes of vibration and the associated natural frequencies for the spring-mass system described by Eq. (12).

Keep in mind that x_1 and x_2 are the positions of m_1 and m_2, relative to their equilibrium positions, and that x_3 and x_4 are their corresponding velocities. Employing the eigenvalue method of solution, we assume, as usual, that $\mathbf{x} = e^{\lambda t}\mathbf{v}$, where $(\lambda, \mathbf{v})$ must be an eigenpair of the matrix $\mathbf{A}$. It is possible, though a bit tedious, to find the eigenvalues and eigenvectors of $\mathbf{A}$ by hand, but it is easy with appropriate computer software. The characteristic polynomial of $\mathbf{A}$ is

$$\lambda^4 + 5\lambda^2 + 4 = (\lambda^2 + 1)(\lambda^2 + 4)$$

so the eigenvalues are $\lambda_1 = i$, $\lambda_2 = -i$, $\lambda_3 = 2i$, and $\lambda_4 = -2i$. The corresponding eigenvectors are

$$\mathbf{v}_1 = \begin{pmatrix} 3 \\ 2 \\ 3i \\ 2i \end{pmatrix}, \quad \mathbf{v}_2 = \begin{pmatrix} 3 \\ 2 \\ -3i \\ -2i \end{pmatrix}, \quad \mathbf{v}_3 = \begin{pmatrix} 3 \\ -4 \\ 6i \\ -8i \end{pmatrix}, \quad \mathbf{v}_4 = \begin{pmatrix} 3 \\ -4 \\ -6i \\ 8i \end{pmatrix}.$$

The complex-valued solutions $e^{it}\mathbf{v}_1$ and $e^{-it}\mathbf{v}_2$ are complex conjugates, so two real-valued solutions can be found by finding the real and imaginary parts of either of them. For instance, we have

$$e^{it}\mathbf{v}_1 = (\cos t + i\sin t)\begin{pmatrix} 3 \\ 2 \\ 3i \\ 2i \end{pmatrix}$$

$$= \begin{pmatrix} 3\cos t \\ 2\cos t \\ -3\sin t \\ -2\sin t \end{pmatrix} + i\begin{pmatrix} 3\sin t \\ 2\sin t \\ 3\cos t \\ 2\cos t \end{pmatrix} = \mathbf{x}_1(t) + i\mathbf{x}_2(t).$$

In a similar way, we obtain

$$e^{2it}\mathbf{v}_3 = (\cos 2t + i\sin 2t)\begin{pmatrix} 3 \\ -4 \\ 6i \\ -8i \end{pmatrix}$$

$$= \begin{pmatrix} 3\cos 2t \\ -4\cos 2t \\ -6\sin 2t \\ 8\sin 2t \end{pmatrix} + i\begin{pmatrix} 3\sin 2t \\ -4\sin 2t \\ 6\cos 2t \\ -8\cos 2t \end{pmatrix} = \mathbf{x}_3(t) + i\mathbf{x}_4(t).$$

We leave it to you to verify that $\mathbf{x}_1$, $\mathbf{x}_2$, $\mathbf{x}_3$, and $\mathbf{x}_4$ are linearly independent and therefore form a fundamental set of solutions. Thus the general solution of Eq. (12) is

$$\mathbf{x} = c_1\begin{pmatrix} 3\cos t \\ 2\cos t \\ -3\sin t \\ -2\sin t \end{pmatrix} + c_2\begin{pmatrix} 3\sin t \\ 2\sin t \\ 3\cos t \\ 2\cos t \end{pmatrix} + c_3\begin{pmatrix} 3\cos 2t \\ -4\cos 2t \\ -6\sin 2t \\ 8\sin 2t \end{pmatrix} + c_4\begin{pmatrix} 3\sin 2t \\ -4\sin 2t \\ 6\cos 2t \\ -8\cos 2t \end{pmatrix} \tag{13}$$

where c_1, c_2, c_3, and c_4 are arbitrary constants.

The state space for this system is four-dimensional, and each solution, obtained by a particular set of values for $c_1, \ldots, c_4$ in Eq. (13), corresponds to a trajectory in this space. Since each solution, given by Eq. (13), is periodic with period 2π, each trajectory is a closed curve. No matter where the trajectory starts in $\mathbf{R}^4$ at $t = 0$, it returns to that point at $t = 2\pi$, $t = 4\pi$, and so forth, repeatedly traversing the same curve in each time interval of length 2π. Thus, even though we cannot graph solutions in $\mathbf{R}^4$, we infer from Eq. (13) that the trajectories are closed curves in $\mathbf{R}^4$.

Note that the only two vibration frequencies present in Eq. (13) are $\omega_1 = \text{Im } \lambda_1 = 1$ and $\omega_2 = \text{Im } \lambda_3 = 2$ radians per second. The corresponding periods are 2π and π. In accordance with the defining property of the first mode of vibration given above, both masses will vibrate at the lowest frequency, $\omega_1 = 1$, if we choose $c_3 = c_4 = 0$ in Eq. (13). The first pure mode of vibration occurs for any c_1 and c_2, not both zero, whenever $x_1(0) = 3c_1$, $x_2(0) = 2c_1$, $x_3(0) = 3c_2$, and $x_4(0) = 2c_2$. The displacements of m_1 and m_2 are then given by

$$x_1(t) = 3(c_1 \cos t + c_2 \sin t) = 3A_1 \cos(t - \phi_1)$$

and

$$x_2(t) = 2(c_1 \cos t + c_2 \sin t) = 2A_1 \cos(t - \phi_1)$$

where A_1 and ϕ_1 are determined by the relationships $A_1 = \sqrt{c_1^2 + c_2^2}$, $A_1 \cos \phi_1 = c_1$, and $A_1 \sin \phi_1 = c_2$. Thus, $x_2 = (2/3)x_1$, that is, the displacement of the right hand mass is always in the same direction as the displacement of the left hand mass, but only $2/3$ as much. In particular, the amplitude of the motion of m_2 is $2/3$ that of the amplitude of motion of m_1. Thus the motions of both masses are exactly in phase with one another, albeit with different amplitudes. Component plots of the displacements of m_1 and m_2 using $c_1 = 1$, $c_2 = 0$, $c_3 = 0$, and $c_4 = 0$ are shown in Figure 6.4.3. Time histories of the positions of the masses are depicted by phantom images.

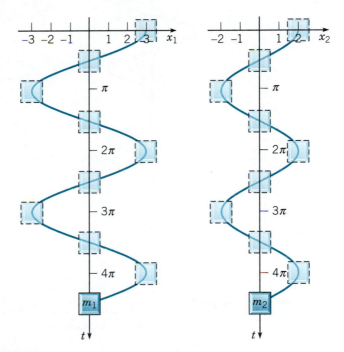

FIGURE 6.4.3 The first principal mode of vibration; the motions of m_1 and m_2 are in phase but have different amplitudes.

The second pure mode of vibration, corresponding to the frequency $\omega_2 = 2$, arises by choosing $c_1 = c_2 = 0$ in Eq. (13). This mode can be realized by choosing initial conditions $x_1(0) = 3c_3$, $x_2(0) = -4c_3$, $x_3(0) = 6c_4$, $x_4(0) = -8c_4$ for any c_3 and c_4, not both zero. The displacements of m_1 and m_2 in this case are given by

$$x_1(t) = 3(c_3 \cos 2t + c_4 \sin 2t) = 3A_2 \cos(2t - \phi_2)$$

and

$$x_2(t) = -4(c_3 \cos 2t + c_4 \sin 2t) = -4A_2 \cos(2t - \phi_2),$$

where $A_2 = \sqrt{c_3^2 + c_4^2}$, $A_2 \cos\phi_2 = c_3$, and $A_2 \sin\phi_2 = c_4$. Consequently, $x_2(t) = -(4/3)x_1$, that is, the displacement of m_2 is always in a direction opposite to the direction of displacement of m_1 and greater by a factor of 4/3. The motions of the masses are 180 degrees out of phase with each other and have different amplitudes. Component plots of the displacements of m_1 and m_2 using $c_1 = 0$, $c_2 = 0$, $c_3 = 1$, and $c_4 = 0$ are shown in Figure 6.4.4.

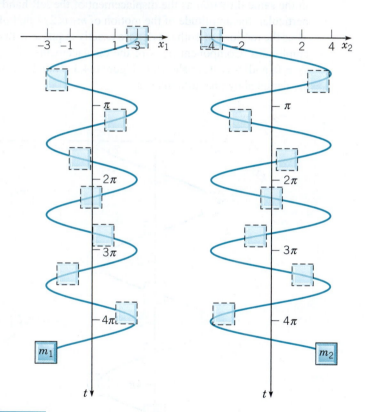

FIGURE 6.4.4 The second principal mode of vibration; the motions of m_1 and m_2 are 180 degrees out of phase and have different amplitudes.

Each of the pure modes of vibration is realized by one or the other special sets of initial conditions discussed above. Equation (13) shows that the motion of m_1 and m_2 for any other choice of initial conditions will be a superposition of both principal modes of vibration. For example, if initially m_1 is at equilibrium and m_2 is released with zero velocity from a position 2 units to the right of its equilibrium position so that the initial condition is given by $\mathbf{x}(0) = (0, 2, 0, 0)^T$, then $c_1 = 1/3$, $c_2 = 0$, $c_3 = -1/3$ and $c_4 = 0$. Thus both modes of vibration are required to describe the motion. Component plots of the displacements of m_1 and m_2 for this case are shown in Figure 6.4.5. Note that the period of the vibrations is 2π.

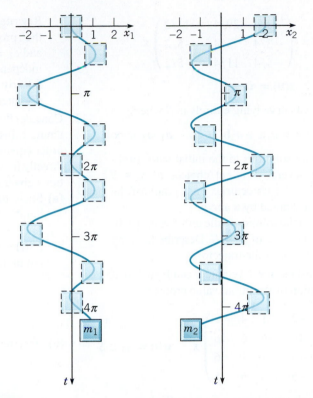

FIGURE 6.4.5 Component plots of the solution of Eq. (12) corresponding to the initial condition $\mathbf{x}(0) = (0, 2, 0, 0)^T$.

PROBLEMS

In each of Problems 1 through 8, express the general solution of the given system of equations in terms of real-valued functions.

1. $x_1' = -2x_1 + 2x_2 + x_3$
   $x_2' = -2x_1 + 2x_2 + 2x_3$
   $x_3' = 2x_1 - 3x_2 - 3x_3$

2. $x_1' = 2x_1 - 4x_2 - x_3$
   $x_2' = x_1 + x_2 + 3x_3$
   $x_3' = 3x_1 - 4x_2 - 2x_3$

3. $x_1' = -2x_2 - x_3$
   $x_2' = x_1 - x_2 + x_3$
   $x_3' = x_1 - 2x_2 - 2x_3$

4. $x_1' = -4x_1 + 2x_2 - x_3$
   $x_2' = -6x_1 - 3x_3$
   $x_3' = (8/3)x_2 - 2x_3$

5. $\mathbf{x}' = \begin{pmatrix} -7 & 6 & -6 \\ -9 & 5 & -9 \\ 0 & -1 & -1 \end{pmatrix} \mathbf{x}$

6. $\mathbf{x}' = \begin{pmatrix} 1/3 & 1/3 & -5/3 \\ -7/3 & -1/3 & 5/3 \\ 1 & -1 & -1 \end{pmatrix} \mathbf{x}$

7. $\mathbf{x}' = \begin{pmatrix} 1 & 1 & 1 \\ 2 & 1 & -1 \\ -8 & -5 & -3 \end{pmatrix} \mathbf{x}$

8. $\mathbf{x}' = \begin{pmatrix} 1 & -1 & 4 \\ 3 & 2 & -1 \\ 2 & 1 & -1 \end{pmatrix} \mathbf{x}$

9. **(a)** Find constant 3×1 vectors $\mathbf{a}$ and $\mathbf{b}$ such that the solution of the initial value problem

$$x' = \begin{pmatrix} 3/4 & 29/4 & -11/2 \\ -3/4 & 3/4 & -5/2 \\ 5/4 & 11/4 & -5/2 \end{pmatrix} x,$$

$$x(0) = x_0 \in S$$

is a closed curve lying entirely in S where

$$S = \{u : u = a_1 a + a_2 b, \ -\infty < a_1, a_2 < \infty\}.$$

CAS (b) In $\mathbf{R}^3$, plot solutions to the initial value problem for several different choices of $x_0 \in S$ overlaying the trajectories on a graph of the plane determined by a and b.

CAS (c) In $\mathbf{R}^3$, plot the solution to the initial value problem for a choice of $x_0 \notin S$. Describe the long time behavior of the trajectory.

10. (a) Find constant 4×1 vectors a and b such that the solution of the initial value problem

$$x' = \begin{pmatrix} -2 & -1 & 4 & 2 \\ -19 & -6 & 6 & 16 \\ -9 & -1 & 1 & 6 \\ -5 & -3 & 6 & 5 \end{pmatrix} x, \quad x(0) = x_0 \in S$$

is a closed curve lying entirely in S where

$$S = \{u : u = a_1 a + a_2 b, \ -\infty < a_1, a_2 < \infty\}.$$

(b) Describe the long time behavior of any solutions for which $x_0 \notin S$.

11. In this problem, we indicate how to show that $x_1(t)$ and $x_2(t)$, as given by Eqs. (4), are linearly independent. Let $\lambda_1 = \mu + i\nu$ and $\bar{\lambda}_1 = \mu - i\nu$ be a pair of conjugate eigenvalues of the coefficient matrix A of Eq. (1); let $v_1 = a + ib$ and $\bar{v}_1 = a - ib$ be the corresponding eigenvectors. Recall from Corollary 6.3.2 that if the eigenvalues of a real matrix are real and distinct, then the eigenvectors are linearly independent. The proof of Corollary 6.3.2 holds equally well if some of the eigenvalues are complex. Thus if $\nu \neq 0$ so that $\lambda_1 \neq \bar{\lambda}_1$, then v_1 and $\bar{v}_1$ are linearly independent.

(a) First we show that a and b are linearly independent. Consider the equation $c_1 a + c_2 b = 0$. Express a and b in terms of v_1 and $\bar{v}_1$, and then show that $(c_1 - ic_2)v_1 + (c_1 + ic_2)\bar{v}_1 = 0$.

(b) Show that $c_1 - ic_2 = 0$ and $c_1 + ic_2 = 0$ and then that $c_1 = 0$ and $c_2 = 0$. Consequently, a and b are linearly independent.

(c) To show that $x_1(t)$ and $x_2(t)$ are linearly independent, consider the equation $c_1 x_1(t_0) + c_2 x_2(t_0) = 0$, where t_0 is an arbitrary point.

Rewrite this equation in terms of a and b, and then proceed, as in part (b), to show that $c_1 = 0$ and $c_2 = 0$. Hence $x_1(t)$ and $x_2(t)$ are linearly independent at the point t_0. Therefore they are linearly independent at every point and on every interval.

12. Consider the two-mass, three-spring system of Example 2. Instead of solving the system of four first order equations, we indicate here how to proceed directly from the system of two second order equations given in Eq. (18) of Section 6.1.

(a) Show that using the parameter values of Example 2, $m_1 = 2$, $m_2 = 9/4$, $k_1 = 1$, $k_2 = 3$, $k_3 = 15/4$ and assuming that $F_1(t) = 0$ and $F_2(t) = 0$, Eqs. (18) in Section 6.1 can be written in the form

$$y'' = \begin{pmatrix} -2 & 3/2 \\ 4/3 & -3 \end{pmatrix} y = By.$$

(b) Assume that $y = e^{\omega t} u$ and show that

$$(B - \omega^2 I) u = 0.$$

Note that ω^2 is an eigenvalue of B corresponding to the eigenvector u.

(c) Find the eigenvalues and eigenvectors of B.

(d) Write down the expressions for y_1 and y_2. There should be four arbitrary constants in these expressions.

(e) By differentiating the results from part (d), write down expressions for y_1' and y_2'. Your results from parts (d) and (e) should agree with Eq. (13) in the text.

13. Consider the two-mass, three-spring system whose equations of motion are Eqs. (11) in the text. Let $m_1 = m_2 = 1$ and $k_1 = k_2 = k_3 = 1$.

(a) Using the values of the parameters given in the previous statement, find the eigenvalues and eigenvectors of the matrix A.

(b) Write down the general solution of the system.

(c) For each fundamental mode draw graphs of x_1 versus t and x_2 versus t and describe the fundamental modes of vibration.

(d) Consider the initial conditions $x(0) = (-1, 3, 0, 0)^T$. Evaluate the arbitrary constants in the general solution in part (b) and plot graphs of x_1 versus t and x_2 versus t.

(e) Consider other initial conditions of your own choice, and plot graphs of x_1 versus t and x_2 versus t in each case.

14. Consider the two-mass, three-spring system whose equations of motion are Eqs. (11) in the text. Let $m_1 = m_2 = m$ and $k_1 = k_2 = k_3 = k$. Find the general solution in terms of the parameter $\omega = \sqrt{k/m}$. In terms of ω, what are the principal modes of vibration and the corresponding natural frequencies?

For each of the matrices in Problems 15 through 18, use a computer to assist in finding a fundamental set of real solutions to the system $\mathbf{x}' = \mathbf{A}\mathbf{x}$.

 15. $\mathbf{A} = \begin{pmatrix} -3 & 6 & 2 & -2 \\ 2 & -3 & -6 & 2 \\ -4 & 8 & 3 & -4 \\ 2 & -2 & -6 & 1 \end{pmatrix}$

16. $\mathbf{A} = \begin{pmatrix} -3 & -4 & 5 & 9 \\ -2 & -5 & 4 & 12 \\ -2 & 0 & -1 & 2 \\ 0 & -2 & 2 & 3 \end{pmatrix}$ CAS

17. $\mathbf{A} = \begin{pmatrix} -3 & -5 & 8 & 14 \\ -6 & -8 & 11 & 27 \\ -6 & -4 & 7 & 17 \\ 0 & -2 & 2 & 4 \end{pmatrix}$ CAS

18. $\mathbf{A} = \begin{pmatrix} 0 & 3 & 0 & -2 \\ -1/2 & 1 & -3 & -5/2 \\ 0 & 3 & -5 & -3 \\ 1 & 3 & 0 & -3 \end{pmatrix}$ CAS

6.5 Fundamental Matrices and the Exponential of a Matrix

Fundamental Matrices

The structure of the solutions of linear differential equations can be further illuminated by introducing the idea of a fundamental matrix. Suppose that $\mathbf{x}_1(t), \ldots, \mathbf{x}_n(t)$ form a fundamental set of solutions for the equation

$$\mathbf{x}' = \mathbf{P}(t)\mathbf{x} \qquad (1)$$

on some interval $\alpha < t < \beta$. Then the matrix

$$\mathbf{X}(t) = [\mathbf{x}_1(t), \mathbf{x}_2(t), \ldots, \mathbf{x}_n(t)] = \begin{pmatrix} x_{11}(t) & \cdots & x_{1n}(t) \\ \vdots & & \vdots \\ x_{n1}(t) & \cdots & x_{nn}(t) \end{pmatrix}, \qquad (2)$$

whose columns are the vectors $\mathbf{x}_1(t), \ldots, \mathbf{x}_n(t)$, is said to be a **fundamental matrix** for the system (1). Note that a fundamental matrix is nonsingular since its columns are linearly independent vectors (see Theorems 6.2.5, 6.2.6 and the remark following Theorem 6.2.6).

EXAMPLE 1

Find a fundamental matrix for the system

$$\mathbf{x}' = \begin{pmatrix} 1 & 1 \\ 4 & 1 \end{pmatrix}\mathbf{x}. \qquad (3)$$

Using the eigenvalue method, we find the pair of linearly independent solutions

$$\mathbf{x}_1(t) = \begin{pmatrix} e^{3t} \\ 2e^{3t} \end{pmatrix}, \qquad \mathbf{x}_2(t) = \begin{pmatrix} e^{-t} \\ -2e^{-t} \end{pmatrix},$$

of Eq. (3). Thus a fundamental matrix for the system (3) is

$$\mathbf{X}(t) = \begin{pmatrix} e^{3t} & e^{-t} \\ 2e^{3t} & -2e^{-t} \end{pmatrix}. \tag{4}$$

The solution of an initial value problem can be written very compactly in terms of a fundamental matrix. The general solution of Eq. (1) is

$$\mathbf{x} = c_1\mathbf{x}_1(t) + \cdots + c_n\mathbf{x}_n(t) \tag{5}$$

or, in terms of $\mathbf{X}(t)$,

$$\mathbf{x} = \mathbf{X}(t)\mathbf{c}, \tag{6}$$

where $\mathbf{c}$ is a constant vector with arbitrary components $c_1, \ldots, c_n$. For an initial value problem consisting of the differential equation (1) and the initial condition

$$\mathbf{x}(t_0) = \mathbf{x}_0 \tag{7}$$

where t_0 is a given point in $\alpha < t < \beta$ and $\mathbf{x}_0$ is a given initial vector, it is only necessary to choose the vector $\mathbf{c}$ in Eq. (6) so as to satisfy the initial condition (7). Hence $\mathbf{c}$ must satisfy

$$\mathbf{X}(t_0)\mathbf{c} = \mathbf{x}_0. \tag{8}$$

Therefore, since $\mathbf{X}(t_0)$ is nonsingular,

$$\mathbf{c} = \mathbf{X}^{-1}(t_0)\mathbf{x}_0$$

and

$$\mathbf{x} = \mathbf{X}(t)\mathbf{X}^{-1}(t_0)\mathbf{x}_0 \tag{9}$$

is the solution of the initial value problem (1), (7). We emphasize, however, that to solve a given initial value problem, one would ordinarily solve Eq. (8) by row reduction and then substitute for $\mathbf{c}$ in Eq. (6), rather than compute $\mathbf{X}^{-1}(t_0)$ and use Eq. (9).

Recall that each column of the fundamental matrix $\mathbf{X}$ is a solution of Eq. (1). It follows that $\mathbf{X}$ satisfies the matrix differential equation

$$\mathbf{X}' = \mathbf{P}(t)\mathbf{X}. \tag{10}$$

This relation is readily confirmed by comparing the two sides of Eq. (10) column by column,

$$\underbrace{\left[\mathbf{x}_1', \cdots, \mathbf{x}_n'\right]}_{\mathbf{X}'} = \underbrace{\left[\mathbf{P}(t)\mathbf{x}_1, \cdots, \mathbf{P}(t)\mathbf{x}_n\right]}_{\mathbf{P}(t)\mathbf{X}}.$$

Sometimes it is convenient to make use of the special fundamental matrix, denoted by $\mathbf{\Phi}(t)$, whose columns are the vectors $\mathbf{x}_1(t), \ldots, \mathbf{x}_n(t)$ designated in Theorem 6.2.7. Besides the differential equation (1), for each $j = 1, \ldots, n$ these vectors satisfy the initial condition

$$\mathbf{x}_j(t_0) = \mathbf{e}_j = \begin{pmatrix} 0 \\ \vdots \\ 0 \\ 1 \\ 0 \\ \vdots \\ 0 \end{pmatrix} \leftarrow j\text{th position.}$$

Thus $\Phi(t)$ has the property that

$$\Phi(t_0) = \begin{pmatrix} 1 & 0 & \cdots & 0 \\ 0 & 1 & \cdots & 0 \\ \vdots & \vdots & \ddots & \vdots \\ 0 & 0 & \cdots & 1 \end{pmatrix} = \mathbf{I}_n. \tag{11}$$

We will always reserve the symbol Φ to denote the fundamental matrix satisfying the initial condition (11) and use $\mathbf{X}$ when an arbitrary fundamental matrix is intended. In terms of $\Phi(t)$, the solution of the initial value problem (1), (7) is even simpler in appearance; since $\Phi^{-1}(t_0) = \mathbf{I}_n$, it follows from Eq. (9) that

$$\mathbf{x} = \Phi(t)\mathbf{x}_0. \tag{12}$$

Remark. Although the fundamental matrix $\Phi(t)$ is often more complicated than $\mathbf{X}(t)$, it is especially helpful if the same system of differential equations is to be solved repeatedly subject to many different initial conditions. This corresponds to a given physical system that can be started from many different initial states. If the fundamental matrix $\Phi(t)$ has been determined, then the solution for each set of initial conditions can be found simply by matrix multiplication, as indicated by Eq. (12). The matrix $\Phi(t)$ thus represents a transformation of the initial conditions $\mathbf{x}_0$ into the solution $\mathbf{x}(t)$ at an arbitrary time t. Comparing Eqs. (9) and (12) makes it clear that $\Phi(t) = \mathbf{X}(t)\mathbf{X}^{-1}(t_0)$.

EXAMPLE 2

For the system (3) of Example 1, find the fundamental matrix Φ such that $\Phi(0) = \mathbf{I}_2$.

The columns of Φ are solutions of Eq. (3) that satisfy the initial conditions

$$\mathbf{x}_1(0) = \begin{pmatrix} 1 \\ 0 \end{pmatrix}, \qquad \mathbf{x}_2(0) = \begin{pmatrix} 0 \\ 1 \end{pmatrix}.$$

Since the general solution of Eq. (3) is

$$\mathbf{x} = c_1 e^{3t} \begin{pmatrix} 1 \\ 2 \end{pmatrix} + c_2 e^{-t} \begin{pmatrix} 1 \\ -2 \end{pmatrix},$$

we can find the solution satisfying the first set of these initial conditions by choosing $c_1 = c_2 = 1/2$; similarly, we obtain the solution satisfying the second set of initial conditions by choosing $c_1 = 1/4$ and $c_2 = -1/4$. Hence

$$\Phi(t) = \begin{pmatrix} \frac{1}{2}e^{3t} + \frac{1}{2}e^{-t} & \frac{1}{4}e^{3t} - \frac{1}{4}e^{-t} \\ e^{3t} - e^{-t} & \frac{1}{2}e^{3t} + \frac{1}{2}e^{-t} \end{pmatrix}.$$

Equivalently, $\Phi(t)$ can be found by computing the matrix product

$$\Phi(t) = \mathbf{X}(t)\mathbf{X}^{-1}(0) = \begin{pmatrix} e^{3t} & e^{-t} \\ 2e^{3t} & -2e^{-t} \end{pmatrix} \begin{pmatrix} \frac{1}{2} & \frac{1}{4} \\ \frac{1}{2} & -\frac{1}{4} \end{pmatrix}$$

$$= \begin{pmatrix} \frac{1}{2}e^{3t} + \frac{1}{2}e^{-t} & \frac{1}{4}e^{3t} - \frac{1}{4}e^{-t} \\ e^{3t} - e^{-t} & \frac{1}{2}e^{3t} + \frac{1}{2}e^{-t} \end{pmatrix}.$$

Note that the elements of $\Phi(t)$ are more complicated than those of the fundamental matrix $\mathbf{X}(t)$ given by Eq. (4); however, it is now easy to determine the solution corresponding to any set of initial conditions.

The Matrix Exponential Function e^{At}

Recall that the solution of the scalar initial value problem

$$x' = ax, \qquad x(0) = x_0, \tag{13}$$

where a is a constant, is

$$x = x_0 e^{at}.$$

Now consider the corresponding initial value problem for an $n \times n$ system, namely

$$\mathbf{x}' = \mathbf{A}\mathbf{x}, \qquad \mathbf{x}(0) = \mathbf{x}_0, \tag{14}$$

where $\mathbf{A}$ is a constant matrix. Applying the results of this section to the problem (14), we can write its solution as

$$\mathbf{x} = \mathbf{\Phi}(t)\mathbf{x}_0, \tag{15}$$

where $\mathbf{\Phi}(0) = \mathbf{I}_n$. Comparing the problems (13) and (14), and their solutions, suggests that the matrix $\mathbf{\Phi}(t)$ might have an exponential character. We now explore this possibility.

The scalar exponential function e^{at} can be represented by the power series

$$e^{at} = 1 + at + \frac{1}{2!}a^2 t^2 + \frac{1}{3!}a^3 t^3 + \cdots = \sum_{k=0}^{\infty} \frac{a^k t^k}{k!}, \tag{16}$$

which converges for all t. By analogy with the power series (16), we now define the symbolic expression $e^{\mathbf{A}t}$.

DEFINITION 6.5.1

Let $\mathbf{A}$ be an $n \times n$ constant matrix. The **matrix exponential function**, denoted by $e^{\mathbf{A}t}$, is defined to be

$$e^{\mathbf{A}t} = \mathbf{I}_n + \mathbf{A}t + \frac{1}{2!}\mathbf{A}^2 t^2 + \frac{1}{3!}\mathbf{A}^3 t^3 + \cdots = \sum_{k=0}^{\infty} \mathbf{A}^k \frac{t^k}{k!}. \tag{17}$$

Remark. The powers of $\mathbf{A}$ that appear in Eq. (17) symbolically represent matrix products, with $\mathbf{A}^0 = \mathbf{I}_n$ by convention, $\mathbf{A}^2 = \mathbf{A}\mathbf{A}$, $\mathbf{A}^3 = \mathbf{A}\mathbf{A}\mathbf{A}$, and so forth. Since $\mathbf{A}$ is an $n \times n$ matrix, each term in the series is an $n \times n$ matrix as is any partial sum $\mathbf{S}_N(t) = \sum_{k=0}^{N} \mathbf{A}^k t^k / k!$. Given any $n \times n$ matrix $\mathbf{A}$, it can be shown that each elemental sequence (that is, the sequence associated with each entry) in the sequence of $n \times n$ matrices $\mathbf{S}_N(t)$ converges for all t as $N \to \infty$. Thus $e^{\mathbf{A}t}$ is an $n \times n$ matrix equal to the convergent sequence of partial sums, $\lim_{N \to \infty} \mathbf{S}_N(t)$. Furthermore, the convergence is sufficiently rapid that all operations performed on the series (17), such as term-by-term differentiation, are justified.

EXAMPLE 3

Use Definition 6.5.1 to find $e^{\mathbf{A}t}$ if

$$\mathbf{A} = \begin{pmatrix} 1 & 1 \\ 0 & 1 \end{pmatrix}.$$

Noting that

$$\mathbf{A}^2 = \begin{pmatrix} 1 & 2 \\ 0 & 1 \end{pmatrix}, \qquad \mathbf{A}^3 = \begin{pmatrix} 1 & 3 \\ 0 & 1 \end{pmatrix}, \qquad \cdots, \qquad \mathbf{A}^k = \begin{pmatrix} 1 & k \\ 0 & 1 \end{pmatrix}, \cdots$$

and using Eq. (17), we find that

$$
\begin{aligned}
e^{\mathbf{A}t} &= \begin{pmatrix} 1 & 0 \\ 0 & 1 \end{pmatrix} + \frac{t}{1!}\begin{pmatrix} 1 & 1 \\ 0 & 1 \end{pmatrix} + \frac{t^2}{2!}\begin{pmatrix} 1 & 2 \\ 0 & 1 \end{pmatrix} + \cdots + \frac{t^k}{k!}\begin{pmatrix} 1 & k \\ 0 & 1 \end{pmatrix} + \cdots \\
&= \begin{pmatrix} \sum_{k=0}^{\infty} t^k/k! & \sum_{k=1}^{\infty} kt^k/k! \\ 0 & \sum_{k=0}^{\infty} t^k/k! \end{pmatrix} = \begin{pmatrix} \sum_{k=0}^{\infty} t^k/k! & t\sum_{k=0}^{\infty} t^k/k! \\ 0 & \sum_{k=0}^{\infty} t^k/k! \end{pmatrix} \\
&= \begin{pmatrix} e^t & te^t \\ 0 & e^t \end{pmatrix}
\end{aligned}
$$

where we have used the series (16) with $a = 1$.

In general, it is not possible to identify the entries of $e^{\mathbf{A}t}$ in terms of elementary functions from their infinite series representations, even if the infinite series can be found. However, one family of matrices for which the entries on the right-hand side of Eq. (17) are identifiable from their infinite series representation is the class of diagonal matrices, say

$$
\mathbf{D} = \begin{pmatrix} \lambda_1 & 0 & \cdots & 0 \\ 0 & \lambda_2 & \cdots & 0 \\ \vdots & \vdots & \ddots & \vdots \\ 0 & 0 & \cdots & \lambda_n \end{pmatrix}. \tag{18}
$$

In this case, the powers of $\mathbf{D}$ are easy to calculate,

$$
\mathbf{D}^2 = \begin{pmatrix} \lambda_1^2 & 0 & \cdots & 0 \\ 0 & \lambda_2^2 & \cdots & 0 \\ \vdots & \vdots & \ddots & \vdots \\ 0 & 0 & \cdots & \lambda_n^2 \end{pmatrix}, \ldots, \mathbf{D}^k = \begin{pmatrix} \lambda_1^k & 0 & \cdots & 0 \\ 0 & \lambda_2^k & \cdots & 0 \\ \vdots & \vdots & \ddots & \vdots \\ 0 & 0 & \cdots & \lambda_n^k \end{pmatrix}, \ldots.
$$

It follows that

$$
\begin{aligned}
e^{\mathbf{D}t} = \sum_{k=0}^{\infty} \mathbf{D}^k \frac{t^k}{k!} &= \begin{pmatrix} \sum_{k=0}^{\infty} \lambda_1^k t^k/k! & 0 & \cdots & 0 \\ 0 & \sum_{k=0}^{\infty} \lambda_2^k t^k/k! & \cdots & 0 \\ \vdots & \vdots & & \vdots \\ 0 & 0 & \cdots & \sum_{k=0}^{\infty} \lambda_n^k t^k/k! \end{pmatrix} \\
&= \begin{pmatrix} e^{\lambda_1 t} & 0 & \cdots & 0 \\ 0 & e^{\lambda_2 t} & \cdots & 0 \\ \vdots & \vdots & & \vdots \\ 0 & 0 & : & e^{\lambda_n t} \end{pmatrix}. \tag{19}
\end{aligned}
$$

The next theorem shows the equivalency of $e^{\mathbf{A}t}$ and the fundamental matrix $\mathbf{\Phi}(t)$ for the system $\mathbf{x}' = \mathbf{A}\mathbf{x}$.

THEOREM 6.5.2

If $\mathbf{A}$ is an $n \times n$ constant matrix, then

$$\boxed{e^{\mathbf{A}t} = \mathbf{\Phi}(t).} \qquad (20)$$

Consequently, the solution to the initial value problem $\mathbf{x}' = \mathbf{A}\mathbf{x}$, $\mathbf{x}(0) = \mathbf{x}_0$, is $\mathbf{x} = e^{\mathbf{A}t}\mathbf{x}_0$.

Proof

Differentiating the series (17) term by term we obtain

$$\frac{d}{dt}e^{\mathbf{A}t} = \frac{d}{dt}\left(\mathbf{I}_n + \mathbf{A}t + \frac{1}{2!}\mathbf{A}^2t^2 + \frac{1}{3!}\mathbf{A}^3t^3 + \cdots\right)$$

$$= \left(\mathbf{A} + \frac{1}{1!}\mathbf{A}^2t + \frac{1}{2!}\mathbf{A}^3t^2 + \cdots\right)$$

$$= \mathbf{A}\left(\mathbf{I}_n + \mathbf{A}t + \frac{1}{2!}\mathbf{A}^2t^2 + \cdots\right)$$

$$= \mathbf{A}e^{\mathbf{A}t}.$$

Thus $e^{\mathbf{A}t}$ satisfies the differential equation

$$\frac{d}{dt}e^{\mathbf{A}t} = \mathbf{A}e^{\mathbf{A}t}. \qquad (21)$$

Further, when $t = 0$, $e^{\mathbf{A}t}$ satisfies the initial condition

$$e^{\mathbf{A}t}\big|_{t=0} = \mathbf{I}_n. \qquad (22)$$

The fundamental matrix $\mathbf{\Phi}$ satisfies the same initial value problem as $e^{\mathbf{A}t}$, namely,

$$\mathbf{\Phi}' = \mathbf{A}\mathbf{\Phi}, \quad \mathbf{\Phi}(0) = \mathbf{I}_n.$$

By the uniqueness part of Theorem 6.2.1 (extended to matrix differential equations), we conclude that $e^{\mathbf{A}t}$ and the fundamental matrix $\mathbf{\Phi}(t)$ are identical. Thus we can write the solution of the initial value problem (14) in the form

$$\mathbf{x} = e^{\mathbf{A}t}\mathbf{x}_0. \qquad \blacksquare$$

Equations (21) and (22) show that two properties of e^{at}, namely $d(e^{at})/dt = ae^{at}$ and $e^{a\cdot 0} = 1$, generalize to $e^{\mathbf{A}t}$. The following theorem summarizes some additional properties of e^{at} that are also shared by $e^{\mathbf{A}t}$.

THEOREM 6.5.3

Let $\mathbf{A}$ and $\mathbf{B}$ be $n \times n$ constant matrices and t, τ be real or complex numbers. Then,

(a) $e^{\mathbf{A}(t+\tau)} = e^{\mathbf{A}t}e^{\mathbf{A}\tau}$

(b) $\mathbf{A}$ commutes with $e^{\mathbf{A}t}$, that is, $\mathbf{A}e^{\mathbf{A}t} = e^{\mathbf{A}t}\mathbf{A}$.

(c) $\left(e^{\mathbf{A}t}\right)^{-1} = e^{-\mathbf{A}t}$.

(d) $e^{(\mathbf{A}+\mathbf{B})t} = e^{\mathbf{A}t}e^{\mathbf{B}t}$ if $\mathbf{AB} = \mathbf{BA}$, that is, if $\mathbf{A}$ and $\mathbf{B}$ commute.

Proof

(a) Multiplying the series for $e^{\mathbf{A}t}$ and $e^{\mathbf{A}\tau}$ in a manner analogous to multiplying polynomial expressions and regrouping terms appropriately yields

$$e^{\mathbf{A}t}e^{\mathbf{A}\tau} =$$

$$= \left(\mathbf{I}_n + \mathbf{A}t + \frac{1}{2!}\mathbf{A}^2 t^2 + \frac{1}{3!}\mathbf{A}^3 t^3 + \cdots\right)\left(\mathbf{I}_n + \mathbf{A}\tau + \frac{1}{2!}\mathbf{A}^2 \tau^2 + \frac{1}{3!}\mathbf{A}^3 \tau^3 + \cdots\right)$$

$$= \mathbf{I}_n + \mathbf{A}(t + \tau) + \frac{1}{2!}\mathbf{A}^2(t + \tau)^2 + \frac{1}{3!}\mathbf{A}^3(t + \tau)^3 + \cdots$$

$$= e^{\mathbf{A}(t+\tau)}.$$

(b) From the series (17)

$$\mathbf{A}e^{\mathbf{A}t} = \mathbf{A}\left(\mathbf{I}_n + \mathbf{A}t + \frac{1}{2!}\mathbf{A}^2 t^2 + \frac{1}{3!}\mathbf{A}^3 t^3 + \cdots\right)$$

$$= \left(\mathbf{A} + \mathbf{A}^2 t + \frac{1}{2!}\mathbf{A}^3 t^2 + \frac{1}{3!}\mathbf{A}^4 t^3 + \cdots\right)$$

$$= \left(\mathbf{I}_n + \mathbf{A}t + \frac{1}{2!}\mathbf{A}^2 t^2 + \frac{1}{3!}\mathbf{A}^3 t^3 + \cdots\right)\mathbf{A} = e^{\mathbf{A}t}\mathbf{A}.$$

(c) Setting $\tau = -t$ in result (a) gives $e^{\mathbf{A}(t+(-t))} = e^{\mathbf{A}t}e^{\mathbf{A}(-t)}$, or $\mathbf{I}_n = e^{\mathbf{A}t}e^{-\mathbf{A}t}$. Thus $e^{-\mathbf{A}t} = \left(e^{\mathbf{A}t}\right)^{-1}$

(d) Let $\mathbf{\Phi}(t) = e^{(\mathbf{A}+\mathbf{B})t}$ and $\hat{\mathbf{\Phi}}(t) = e^{\mathbf{A}t}e^{\mathbf{B}t}$. Then $\mathbf{\Phi}$ is the unique solution to

$$\mathbf{\Phi}' = (\mathbf{A} + \mathbf{B})\mathbf{\Phi}, \qquad \mathbf{\Phi}(0) = \mathbf{I}_n. \tag{23}$$

Since we are assuming that $\mathbf{A}$ and $\mathbf{B}$ commute, it follows that $e^{\mathbf{A}t}\mathbf{B} = \mathbf{B}e^{\mathbf{A}t}$. The argument makes use of the series representation (17) for $e^{\mathbf{A}t}$ and is identical to that used in the proof of part (b). Then

$$\hat{\mathbf{\Phi}}' = \mathbf{A}e^{\mathbf{A}t}e^{\mathbf{B}t} + e^{\mathbf{A}t}\mathbf{B}e^{\mathbf{B}t} = \mathbf{A}e^{\mathbf{A}t}e^{\mathbf{B}t} + \mathbf{B}e^{\mathbf{A}t}e^{\mathbf{B}t} = (\mathbf{A}+\mathbf{B})e^{\mathbf{A}t}e^{\mathbf{B}t} = (\mathbf{A}+\mathbf{B})\hat{\mathbf{\Phi}}.$$

Since $\hat{\mathbf{\Phi}}(0) = \mathbf{I}_n$, $\hat{\mathbf{\Phi}}$ is also a solution of the initial value problem (23). The result follows from the uniqueness part of Theorem 6.2.1. ∎

Methods for Constructing $e^{\mathbf{A}t}$

Given an $n \times n$ constant matrix $\mathbf{A}$, we now discuss relationships between different representations of the fundamental matrix $e^{\mathbf{A}t}$ for the system

$$\mathbf{x}' = \mathbf{A}\mathbf{x} \tag{24}$$

that provide practical methods for its construction.

If a fundamental set of solutions,

$$\mathbf{x}_1(t) = \begin{pmatrix} x_{11}(t) \\ x_{21}(t) \\ \vdots \\ x_{n1}(t) \end{pmatrix}, \quad \mathbf{x}_2(t) = \begin{pmatrix} x_{12}(t) \\ x_{22}(t) \\ \vdots \\ x_{n2}(t) \end{pmatrix}, \dots, \mathbf{x}_n(t) = \begin{pmatrix} x_{1n}(t) \\ x_{2n}(t) \\ \vdots \\ x_{nn}(t) \end{pmatrix},$$

of Eq. (24) is available, then by Theorem 6.5.2

$$e^{\mathbf{A}t} = \mathbf{X}(t)\mathbf{X}^{-1}(0) \tag{25}$$

where

$$\mathbf{X}(t) = [\mathbf{x}_1(t), \mathbf{x}_2(t), \dots, \mathbf{x}_n(t)] = \begin{pmatrix} x_{11}(t) & \cdots & x_{1n}(t) \\ \vdots & & \vdots \\ x_{n1}(t) & \cdots & x_{nn}(t) \end{pmatrix}.$$

EXAMPLE 4

Find $e^{\mathbf{A}t}$ for the system

$$\mathbf{x}' = \begin{pmatrix} 1 & -1 \\ 1 & 3 \end{pmatrix} = \mathbf{A}\mathbf{x}. \tag{26}$$

Then find the inverse $\left(e^{\mathbf{A}t}\right)^{-1}$ of $e^{\mathbf{A}t}$.

A fundamental set of solutions to Eq. (26) was found in Example 3 of Section 3.5. Since the characteristic equation of $\mathbf{A}$ is $(\lambda - 2)^2$, the only eigenvalue of $\mathbf{A}$, $\lambda_1 = 2$, has algebraic multiplicity 2. However, the geometric multiplicity of λ_1 is 1 since the only eigenvector that can be found is $\mathbf{v}_1 = (1, -1)^T$. Thus one solution of Eq. (26) is $\mathbf{x}_1(t) = e^{2t}(1, -1)^T$. A second linearly independent solution,

$$\mathbf{x}_2(t) = te^{2t}\begin{pmatrix} 1 \\ -1 \end{pmatrix} + e^{2t}\begin{pmatrix} 0 \\ -1 \end{pmatrix} = \begin{pmatrix} te^{2t} \\ -(1+t)e^{2t} \end{pmatrix},$$

is found using the method of generalized eigenvectors for systems of dimension 2 discussed in Section 3.5. Thus a fundamental matrix for Eq. (26) is

$$\mathbf{X}(t) = [\mathbf{x}_1(t), \mathbf{x}_2(t)] = \begin{pmatrix} e^{2t} & te^{2t} \\ -e^{2t} & -(1+t)e^{2t} \end{pmatrix}.$$

Using (25) gives

$$e^{\mathbf{A}t} = \mathbf{X}(t)\mathbf{X}^{-1}(0)$$

$$= \begin{pmatrix} e^{2t} & te^{2t} \\ -e^{2t} & -(1+t)e^{2t} \end{pmatrix}\begin{pmatrix} 1 & 0 \\ -1 & -1 \end{pmatrix}$$

$$= \begin{pmatrix} (1-t)e^{2t} & -te^{2t} \\ te^{2t} & (1+t)e^{2t} \end{pmatrix}. \tag{27}$$

By part (c) of Theorem 6.5.3, the inverse of $e^{\mathbf{A}t}$ is obtained by replacing t by $-t$ in the matrix (27),

$$\left(e^{\mathbf{A}t}\right)^{-1} = e^{-\mathbf{A}t} = \begin{pmatrix} (1+t)e^{-2t} & te^{-2t} \\ -te^{-2t} & (1-t)e^{-2t} \end{pmatrix}.$$

▶ **$e^{\mathbf{A}t}$ When A Is Nondefective.** In the case that $\mathbf{A}$ has n linearly independent eigenvectors $\{\mathbf{v}_1, \dots, \mathbf{v}_n\}$, then, by Theorem 6.3.1, a fundamental set for Eq. (24) is $\{e^{\lambda_1 t}\mathbf{v}_1, e^{\lambda_2 t}\mathbf{v}_2, \dots, e^{\lambda_n t}\mathbf{v}_n\}$. In this case,

$$\mathbf{X}(t) = \left[e^{\lambda_1 t} \mathbf{v}_1, e^{\lambda_2 t} \mathbf{v}_2, \ldots, e^{\lambda_n t} \mathbf{v}_n \right]$$

$$= \begin{pmatrix} e^{\lambda_1 t} v_{11} & e^{\lambda_2 t} v_{12} & \cdots & e^{\lambda_n t} v_{1n} \\ e^{\lambda_1 t} v_{21} & e^{\lambda_2 t} v_{22} & \cdots & e^{\lambda_n t} v_{2n} \\ \vdots & \vdots & & \vdots \\ e^{\lambda_1 t} v_{n1} & e^{\lambda_2 t} v_{n2} & \cdots & e^{\lambda_n t} v_{nn} \end{pmatrix} = \mathbf{T} e^{\mathbf{D} t} \tag{28}$$

where

$$\mathbf{T} = [\mathbf{v}_1, \mathbf{v}_2, \ldots, \mathbf{v}_n] = \begin{pmatrix} v_{11} & v_{12} & \cdots & v_{1n} \\ v_{21} & v_{22} & \cdots & v_{2n} \\ \vdots & \vdots & & \vdots \\ v_{n1} & v_{n2} & \cdots & v_{nn} \end{pmatrix}$$

and $e^{\mathbf{D}t}$ is defined in Eq. (19). The matrix $\mathbf{T}$ is nonsingular since its columns are the linearly independent eigenvectors of $\mathbf{A}$. Substituting $\mathbf{T}e^{\mathbf{D}t}$ for $\mathbf{X}(t)$ in Eq. (25) and noting that $\mathbf{X}(0) = \mathbf{T}$ and $\mathbf{X}^{-1}(0) = \mathbf{T}^{-1}$ gives us

$$e^{\mathbf{A}t} = \underbrace{\mathbf{T} e^{\mathbf{D}t}}_{\mathbf{X}(t)} \underbrace{\mathbf{T}^{-1}}_{\mathbf{X}^{-1}(0)}. \tag{29}$$

Thus, in the case that $\mathbf{A}$ has n linearly independent eigenvectors, Eq. (29) expresses $e^{\mathbf{A}t}$ directly in terms of the eigenpairs of $\mathbf{A}$. Note that if $\mathbf{A}$ is real, $e^{\mathbf{A}t}$ must be real and therefore the right-hand sides of Eqs. (25) and (29) must produce real matrix functions even if some of the eigenvalues and eigenvectors are complex (see Problem 23 for an alternative derivation of Eq. (29)).

EXAMPLE 5

Consider again the system of differential equations

$$\mathbf{x}' = \mathbf{A}\mathbf{x}$$

where $\mathbf{A}$ is given by Eq. (3). Use Eq. (29) to obtain $e^{\mathbf{A}t}$.

Since the eigenpairs of $\mathbf{A}$ are $(\lambda_1, \mathbf{v}_1)$ and $(\lambda_2, \mathbf{v}_2)$ with $\lambda_1 = 3$, $\lambda_2 = -1$, $\mathbf{v}_1 = (1, 2)^T$, and $\mathbf{v}_2 = (1, -2)^T$,

$$\mathbf{T} = \begin{pmatrix} 1 & 1 \\ 2 & -2 \end{pmatrix}, \quad \mathbf{D} = \begin{pmatrix} 3 & 0 \\ 0 & -1 \end{pmatrix}, \quad \text{and} \quad e^{\mathbf{D}t} = \begin{pmatrix} e^{3t} & 0 \\ 0 & e^{-t} \end{pmatrix}.$$

Therefore

$$e^{\mathbf{A}t} = \underbrace{\begin{pmatrix} 1 & 1 \\ 2 & -2 \end{pmatrix}}_{\mathbf{T}} \underbrace{\begin{pmatrix} e^{3t} & 0 \\ 0 & e^{-t} \end{pmatrix}}_{e^{\mathbf{D}t}} \underbrace{\begin{pmatrix} \frac{1}{2} & \frac{1}{4} \\ \frac{1}{2} & -\frac{1}{4} \end{pmatrix}}_{\mathbf{T}^{-1}} = \begin{pmatrix} \frac{1}{2}e^{3t} + \frac{1}{2}e^{-t} & \frac{1}{4}e^{3t} - \frac{1}{4}e^{-t} \\ e^{3t} - e^{-t} & \frac{1}{2}e^{3t} + \frac{1}{2}e^{-t} \end{pmatrix}$$

in agreement with the results of Example 2. This is expected since in Example 2, $\mathbf{X}(t) = \mathbf{T}e^{\mathbf{D}t}$ and $\mathbf{X}^{-1}(0) = \mathbf{T}^{-1}$.

▶ **Using the Laplace Transform to Find $e^{\mathbf{A}t}$.** Recall that the method of Laplace transforms was applied to constant coefficient, first order linear systems of dimension 2 in Section 5.4. The method generalizes in a natural way to first order linear systems of dimension n. Here we apply the method to the matrix initial value problem

$$\mathbf{\Phi}' = \mathbf{A}\mathbf{\Phi}, \qquad \mathbf{\Phi}(0) = \mathbf{I}_n, \tag{30}$$

which is equivalent to n initial value problems for the individual columns of $\mathbf{\Phi}$, each a first order system with coefficient matrix $\mathbf{A}$ and initial condition prescribed by the corresponding column of $\mathbf{I}_n$. We denote the Laplace transform of $\mathbf{\Phi}(t)$ by

$$\hat{\mathbf{\Phi}}(s) = \mathcal{L}\left\{\mathbf{\Phi}(t)\right\}(s) = \int_0^\infty e^{-st}\mathbf{\Phi}(t)\,dt.$$

Taking the Laplace transform of the differential equation in the initial value problem (30) yields

$$s\hat{\mathbf{\Phi}} - \mathbf{\Phi}(0) = \mathbf{A}\hat{\mathbf{\Phi}},$$

or

$$(s\mathbf{I}_n - \mathbf{A})\hat{\mathbf{\Phi}}(s) = \mathbf{I}_n,$$

where we have used the initial condition in (30) and rearranged terms. It follows that

$$\hat{\mathbf{\Phi}}(s) = (s\mathbf{I}_n - \mathbf{A})^{-1}. \tag{31}$$

We can then recover $\mathbf{\Phi}(t) = e^{\mathbf{A}t}$ by taking the inverse Laplace transform of the expression on the right-hand side of Eq. (31),

$$e^{\mathbf{A}t} = \mathcal{L}^{-1}\left\{(s\mathbf{I}_n - \mathbf{A})^{-1}\right\}(t). \tag{32}$$

Provided that we can find the inverse Laplace transform of each entry of the matrix $(s\mathbf{I}_n - \mathbf{A})^{-1}$, the inversion formula (32) will yield $e^{\mathbf{A}t}$ whether $\mathbf{A}$ is defective or nondefective.

EXAMPLE 6

Find $e^{\mathbf{A}t}$ for the system

$$\mathbf{x}' = \begin{pmatrix} 4 & 6 & 6 \\ 1 & 3 & 2 \\ -1 & -5 & -2 \end{pmatrix} \mathbf{x} = \mathbf{A}\mathbf{x}.$$

Then find the inverse $(e^{\mathbf{A}t})^{-1} = e^{-\mathbf{A}t}$.

The eigenvalues of $\mathbf{A}$ are $\lambda_1 = 1$ with algebraic multiplicity 1 and $\lambda_2 = 2$ with algebraic multiplicity 2 and geometric multiplicity 1 so $\mathbf{A}$ is defective. Thus we attempt to find $e^{\mathbf{A}t}$ using the Laplace transform. The matrix $s\mathbf{I}_3 - \mathbf{A}$ is given by

$$s\mathbf{I}_3 - \mathbf{A} = \begin{pmatrix} s-4 & -6 & -6 \\ -1 & s-3 & -2 \\ 1 & 5 & s+2 \end{pmatrix}.$$

Using a computer algebra system, we find that

$$(s\mathbf{I}_3 - \mathbf{A})^{-1} = \frac{1}{(s-1)(s-2)^2} \begin{pmatrix} s^2 - s + 4 & 6(s-3) & 6(s-1) \\ s & s^2 - 2s - 2 & 2(s-1) \\ -(s+2) & -(5s-14) & (s-6)(s-1) \end{pmatrix},$$

and then taking the inverse Laplace transform gives,

$$
e^{At} = \begin{pmatrix} (6t-3)e^{2t}+4e^t & -12e^t+6(2-t)e^{2t} & 6te^{2t} \\ (2t-1)e^{2t}+e^t & (4-2t)e^{2t}-3e^t & 2te^{2t} \\ (3-4t)e^{2t}-3e^t & (-9+4t)e^{2t}+9e^t & (1-4t)e^{2t} \end{pmatrix}.
$$

Replacing t by $-t$ in the last result easily provides us with the inverse of e^{At},

$$
e^{-At} = \begin{pmatrix} -(6t+3)e^{-2t}+4e^{-t} & -12e^{-t}+6(2+t)e^{-2t} & -6te^{-2t} \\ -(2t+1)e^{-2t}+e^{-t} & (4+2t)e^{-2t}-3e^{-t} & -2te^{-2t} \\ (3+4t)e^{-2t}-3e^{-t} & -(9+4t)e^{-2t}+9e^{-t} & (1+4t)e^{-2t} \end{pmatrix}.
$$

PROBLEMS

In each of Problems 1 through 14, find a fundamental matrix for the given system of equations. In each case also find the fundamental matrix e^{At}.

1. $\mathbf{x}' = \begin{pmatrix} 3 & -2 \\ 2 & -2 \end{pmatrix} \mathbf{x}$ 2. $\mathbf{x}' = \begin{pmatrix} -\frac{3}{4} & \frac{1}{2} \\ \frac{1}{8} & -\frac{3}{4} \end{pmatrix} \mathbf{x}$

3. $\mathbf{x}' = \begin{pmatrix} 3 & -4 \\ 1 & -1 \end{pmatrix} \mathbf{x}$ 4. $\mathbf{x}' = \begin{pmatrix} 4 & -2 \\ 8 & -4 \end{pmatrix} \mathbf{x}$

5. $\mathbf{x}' = \begin{pmatrix} 2 & -5 \\ 1 & -2 \end{pmatrix} \mathbf{x}$ 6. $\mathbf{x}' = \begin{pmatrix} -1 & -4 \\ 1 & -1 \end{pmatrix} \mathbf{x}$

7. $\mathbf{x}' = \begin{pmatrix} 5 & -1 \\ 3 & 1 \end{pmatrix} \mathbf{x}$ 8. $\mathbf{x}' = \begin{pmatrix} 1 & -1 \\ 5 & -3 \end{pmatrix} \mathbf{x}$

9. $\mathbf{x}' = \begin{pmatrix} 2 & -1 \\ 3 & -2 \end{pmatrix} \mathbf{x}$ 10. $\mathbf{x}' = \begin{pmatrix} 1 & 1 \\ 4 & -2 \end{pmatrix} \mathbf{x}$

11. $\mathbf{x}' = \begin{pmatrix} -\frac{3}{2} & 1 \\ -\frac{1}{4} & -\frac{1}{2} \end{pmatrix} \mathbf{x}$ 12. $\mathbf{x}' = \begin{pmatrix} -3 & \frac{5}{2} \\ -\frac{5}{2} & 2 \end{pmatrix} \mathbf{x}$

13. $\mathbf{x}' = \begin{pmatrix} 1 & 1 & 1 \\ 2 & 1 & -1 \\ -8 & -5 & -3 \end{pmatrix} \mathbf{x}$

14. $\mathbf{x}' = \begin{pmatrix} 1 & -1 & 4 \\ 3 & 2 & -1 \\ 2 & 1 & -1 \end{pmatrix} \mathbf{x}$

15. Solve the initial value problem

$$
\mathbf{x}' = \begin{pmatrix} -1 & -4 \\ 1 & -1 \end{pmatrix} \mathbf{x}, \quad \mathbf{x}(0) = \begin{pmatrix} 3 \\ 1 \end{pmatrix}
$$

by using the fundamental matrix e^{At} found in Problem 6.

16. Solve the initial value problem

$$
\mathbf{x}' = \begin{pmatrix} 2 & -1 \\ 3 & -2 \end{pmatrix} \mathbf{x}, \quad \mathbf{x}(0) = \begin{pmatrix} 2 \\ -1 \end{pmatrix}
$$

by using the fundamental matrix e^{At} found in Problem 9.

In each of Problems 17 through 20, use the method of Laplace transforms to find the fundamental matrix e^{At}.

17. $\mathbf{x}' = \begin{pmatrix} -4 & -1 \\ 1 & -2 \end{pmatrix} \mathbf{x}$

18. $\mathbf{x}' = \begin{pmatrix} 5 & -1 \\ 1 & 3 \end{pmatrix} \mathbf{x}$

19. $\mathbf{x}' = \begin{pmatrix} -1 & -5 \\ 1 & 3 \end{pmatrix} \mathbf{x}$

20. $\mathbf{x}' = \begin{pmatrix} 0 & 1 & -1 \\ 1 & 0 & 1 \\ 1 & 1 & 0 \end{pmatrix} \mathbf{x}$

21. Consider an oscillator satisfying the initial value problem

$$
u'' + \omega^2 u = 0, \qquad u(0) = u_0, \quad u'(0) = v_0. \quad (i)
$$

(a) Let $x_1 = u$, $x_2 = u'$, and transform Eqs. (i) into the form

$$
\mathbf{x}' = \mathbf{Ax}, \qquad \mathbf{x}(0) = \mathbf{x}_0. \quad (ii)
$$

(b) By using the series (17), show that

$$
e^{At} = \mathbf{I}_2 \cos \omega t + \mathbf{A} \frac{\sin \omega t}{\omega}. \quad (iii)
$$

(c) Find the solution of the initial value problem (ii).

CAS **22.** The matrix of coefficients for the system of differential equations describing the radioactive decay process in Problem 17, Section 6.3 is

$$\mathbf{A} = \begin{pmatrix} -k_1 & 0 & 0 \\ k_1 & -k_2 & 0 \\ 0 & k_2 & 0 \end{pmatrix}.$$

Use a computer algebra system to find the fundamental matrix $e^{\mathbf{A}t}$ and use the result to solve the initial value problem $\mathbf{m}' = \mathbf{A}\mathbf{m}$, $\mathbf{m}(0) = (m_0, 0, 0)^T$ where the components of $\mathbf{m} = (m_1, m_2, m_3)^T$ are the amounts of each of the three substances R_1, R_2, and R_3, the first two of which are radioactive while the third is stable.

23. Assume that the real $n \times n$ matrix $\mathbf{A}$ has n linearly independent eigenvectors $\mathbf{v}_1, \ldots, \mathbf{v}_n$ corresponding to the (possibly repeated and possibly complex) eigenvalues $\lambda_1, \ldots, \lambda_n$. If $\mathbf{T}$ is the matrix whose columns are the eigenvectors of $\mathbf{A}$,

$$\mathbf{T} = [\mathbf{v}_1, \ldots, \mathbf{v}_n],$$

it is shown in Appendix A.4 that

$$\mathbf{T}^{-1}\mathbf{A}\mathbf{T} = \mathbf{D} = \begin{pmatrix} \lambda_1 & 0 & \cdots & 0 \\ 0 & \lambda_2 & \cdots & 0 \\ \vdots & & & \vdots \\ 0 & 0 & \cdots & \lambda_n \end{pmatrix}, \tag{i}$$

that is, $\mathbf{A}$ is similar to the diagonal matrix $\mathbf{D} = \text{diag}(\lambda_1, \ldots, \lambda_n)$.

(a) Use the relation in Eq. (i) to show that

$$\mathbf{A}^n = \mathbf{T}\mathbf{D}^n\mathbf{T}^{-1}$$

for each $n = 0, 1, 2, \ldots$.

(b) Use the results in part **(a)**, Eq. (17) in Definition 6.5.1, and Eq. (19) to show that

$$e^{\mathbf{A}t} = \mathbf{T}e^{\mathbf{D}t}\mathbf{T}^{-1}.$$

24. The Method of Successive Approximations. Consider the initial value problem

$$\mathbf{x}' = \mathbf{A}\mathbf{x}, \qquad \mathbf{x}(0) = \mathbf{x}_0, \tag{i}$$

where $\mathbf{A}$ is a constant matrix and $\mathbf{x}_0$ is a prescribed vector.

(a) Assuming that a solution $\mathbf{x} = \boldsymbol{\phi}(t)$ exists, show that it must satisfy the integral equation

$$\boldsymbol{\phi}(t) = \mathbf{x}_0 + \int_0^t \mathbf{A}\boldsymbol{\phi}(s)\,ds. \tag{ii}$$

(b) Start with the initial approximation $\boldsymbol{\phi}^{(0)} = \mathbf{x}_0$. Substitute this expression for $\boldsymbol{\phi}(t)$ in the right-hand side of Eq. (ii) and obtain a new approximation $\boldsymbol{\phi}^{(1)}(t)$. Show that

$$\boldsymbol{\phi}^{(1)}(t) = (\mathbf{I}_n + \mathbf{A}t)\mathbf{x}_0. \tag{iii}$$

(c) Repeat the process and thereby obtain a sequence of approximations $\boldsymbol{\phi}^{(0)}, \boldsymbol{\phi}^{(1)}, \boldsymbol{\phi}^{(2)}, \ldots, \boldsymbol{\phi}^{(n)}, \ldots$. Use an inductive argument to show that

$$\boldsymbol{\phi}^{(n)}(t) = \left(\mathbf{I}_n + \mathbf{A}t + \mathbf{A}^2\frac{t^2}{2!} + \cdots + \mathbf{A}^n\frac{t^n}{n!}\right)\mathbf{x}_0. \tag{iv}$$

(d) Let $n \to \infty$ and show that the solution of the initial value problem (i) is

$$\boldsymbol{\phi}(t) = e^{\mathbf{A}t}\mathbf{x}_0. \tag{v}$$

6.6 Nonhomogeneous Linear Systems

Variation of Parameters

In this section, we turn to the nonhomogeneous system

$$\mathbf{x}' = \mathbf{P}(t)\mathbf{x} + \mathbf{g}(t), \tag{1}$$

where the $n \times n$ matrix $\mathbf{P}(t)$ and the $n \times 1$ vector $\mathbf{g}(t)$ are continuous for $\alpha < t < \beta$. Assume that a fundamental matrix $\mathbf{X}(t)$ for the corresponding homogeneous system

$$\mathbf{x}' = \mathbf{P}(t)\mathbf{x} \tag{2}$$

has been found. We use the method of variation of parameters to construct a particular solution, and hence the general solution, of the nonhomogeneous system (1).

Since the general solution of the homogeneous system (2) is $\mathbf{X}(t)\mathbf{c}$, it is natural to proceed as in Section 4.8 and to seek a solution of the nonhomogeneous system (1) by replacing the constant vector $\mathbf{c}$ by a vector function $\mathbf{u}(t)$. Thus we assume that

$$\mathbf{x} = \mathbf{X}(t)\mathbf{u}(t), \tag{3}$$

where $\mathbf{u}(t)$ is a vector to be determined. Upon differentiating $\mathbf{x}$ as given by Eq. (3) and requiring that Eq. (1) be satisfied, we obtain

$$\mathbf{X}'(t)\mathbf{u}(t) + \mathbf{X}(t)\mathbf{u}'(t) = \mathbf{P}(t)\mathbf{X}(t)\mathbf{u}(t) + \mathbf{g}(t). \tag{4}$$

Since $\mathbf{X}(t)$ is a fundamental matrix, $\mathbf{X}'(t) = \mathbf{P}(t)\mathbf{X}(t)$ so that the terms involving $\mathbf{u}(t)$ drop out; hence, Eq. (4) reduces to

$$\mathbf{X}(t)\mathbf{u}'(t) = \mathbf{g}(t). \tag{5}$$

Recall that $\mathbf{X}(t)$ is nonsingular on any interval where $\mathbf{P}$ is continuous. Hence $\mathbf{X}^{-1}(t)$ exists, and therefore

$$\mathbf{u}'(t) = \mathbf{X}^{-1}(t)\mathbf{g}(t). \tag{6}$$

Thus for $\mathbf{u}(t)$ we can select any vector from the class of vectors that satisfy Eq. (6); these vectors are determined only up to an arbitrary additive constant vector. Therefore we denote $\mathbf{u}(t)$ by

$$\mathbf{u}(t) = \int \mathbf{X}^{-1}(t)\mathbf{g}(t)dt + \mathbf{c}, \tag{7}$$

where the constant vector $\mathbf{c}$ is arbitrary. If the integral in Eq. (7) can be evaluated, then the general solution of the system (1) is found by substituting for $\mathbf{u}(t)$ from Eq. (7) in Eq. (3). However, even if the integral cannot be evaluated, we can still write the general solution of Eq. (1) in the form

$$\mathbf{x} = \mathbf{X}(t)\mathbf{c} + \mathbf{X}(t)\int_{t_1}^{t} \mathbf{X}^{-1}(s)\mathbf{g}(s)ds, \tag{8}$$

where t_1 is any point in the interval (α, β). Observe that the first term on the right side of Eq. (8) is the general solution of the corresponding homogeneous system (2), and the second term is a particular solution of Eq. (1) (see Problem 1).

Now let us consider the initial value problem consisting of the differential equation (1) and the initial condition

$$\mathbf{x}(t_0) = \mathbf{x}_0. \tag{9}$$

We can find the solution of this problem most conveniently if we choose the lower limit of integration in Eq. (8) to be the initial point t_0. Then the general solution of the differential equation is

$$\mathbf{x} = \mathbf{X}(t)\mathbf{c} + \mathbf{X}(t)\int_{t_0}^{t} \mathbf{X}^{-1}(s)\mathbf{g}(s)ds. \tag{10}$$

For $t = t_0$, the integral in Eq. (10) is zero, so the initial condition (9) is also satisfied if we choose

$$\mathbf{c} = \mathbf{X}^{-1}(t_0)\mathbf{x}_0. \tag{11}$$

Therefore

$$\mathbf{x} = \mathbf{X}(t)\mathbf{X}^{-1}(t_0)\mathbf{x}_0 + \mathbf{X}(t)\int_{t_0}^{t} \mathbf{X}^{-1}(s)\mathbf{g}(s)ds. \tag{12}$$

is the solution of the given initial value problem. Again, although it is helpful to use $\mathbf{X}^{-1}$ to write the solutions (8) and (12), it is usually better in particular cases to solve the necessary equations by row reduction than to calculate $\mathbf{X}^{-1}(t)$ and substitute into Eqs. (8) and (12).

The solution (12) takes a slightly simpler form if we use the fundamental matrix $\Phi(t)$ satisfying $\Phi(t_0) = \mathbf{I}_n$. In this case, we have

$$\mathbf{x} = \Phi(t)\mathbf{x}_0 + \Phi(t) \int_{t_0}^{t} \Phi^{-1}(s)\mathbf{g}(s)\,ds. \tag{13}$$

▶ **The Case of Constant P.** If the coefficient matrix $\mathbf{P}(t)$ in Eq. (1) is a constant matrix, $\mathbf{P}(t) = \mathbf{A}$, it is natural and convenient to use the fundamental matrix $\Phi(t) = e^{\mathbf{A}t}$ to represent solutions to

$$\mathbf{x}' = \mathbf{A}\mathbf{x} + \mathbf{g}(t). \tag{14}$$

Since $(e^{\mathbf{A}t})^{-1} = e^{-\mathbf{A}t}$, the general solution (10) takes the form

$$\mathbf{x} = e^{\mathbf{A}t}\mathbf{c} + e^{\mathbf{A}t} \int_{t_0}^{t} e^{-\mathbf{A}s}\mathbf{g}(s)\,ds. \tag{15}$$

If an initial condition is prescribed at $t = t_0$ as in Eq. (9), then $\mathbf{c} = e^{-\mathbf{A}t_0}\mathbf{x}_0$ and from Eq. (15) we get

$$\mathbf{x} = e^{\mathbf{A}(t-t_0)}\mathbf{x}_0 + e^{\mathbf{A}t} \int_{t_0}^{t} e^{-\mathbf{A}s}\mathbf{g}(s)\,ds. \tag{16}$$

where we have used the property $e^{\mathbf{A}t}e^{-\mathbf{A}t_0} = e^{\mathbf{A}(t-t_0)}$. If the initial condition is prescribed at $t = t_0 = 0$, Eq. (16) reduces to

$$\mathbf{x} = e^{\mathbf{A}t}\mathbf{x}_0 + e^{\mathbf{A}t} \int_{0}^{t} e^{-\mathbf{A}s}\mathbf{g}(s)\,ds. \tag{17}$$

EXAMPLE 1

Use the method of variation of parameters to find the solution of the initial value problem

$$\mathbf{x}' = \begin{pmatrix} -3 & 4 \\ -2 & 3 \end{pmatrix}\mathbf{x} + \begin{pmatrix} \sin t \\ t \end{pmatrix} = \mathbf{A}\mathbf{x} + \mathbf{g}(t), \quad \mathbf{x}(0) = \begin{pmatrix} 0 \\ 1 \end{pmatrix}. \tag{18}$$

The calculations may be performed by hand. However, it is highly recommended to use a computer algebra system to facilitate the operations. The eigenvalues of $\mathbf{A}$ are $\lambda_1 = 1$ and $\lambda_2 = -1$ with eigenvectors $\mathbf{v}_1 = (1, 1)^T$ and $\mathbf{v}_2 = (2, 1)^T$, respectively. A fundamental matrix for the homogeneous equation $\mathbf{x}' = \mathbf{A}\mathbf{x}$ is therefore

$$\mathbf{X}(t) = \begin{pmatrix} e^t & 2e^{-t} \\ e^t & e^{-t} \end{pmatrix}$$

and it follows that

$$e^{\mathbf{A}t} = \mathbf{X}(t)\mathbf{X}^{-1}(0) = \begin{pmatrix} 2e^{-t} - e^t & 2e^t - 2e^{-t} \\ e^{-t} - e^t & 2e^t - e^{-t} \end{pmatrix}.$$

A particular solution of the differential equation (18) is

$$\mathbf{x}_p(t) = e^{\mathbf{A}t} \int_{0}^{t} e^{-\mathbf{A}s}\mathbf{g}(s)\,ds = \begin{pmatrix} -4t + \frac{3}{2}e^t - e^{-t} - \frac{1}{2}\cos t + \frac{3}{2}\sin t \\ -1 - 3t + \frac{3}{2}e^t - \frac{1}{2}e^{-t} + \sin t \end{pmatrix}.$$

Thus the solution of the initial value problem is

$$\mathbf{x}(t) = e^{\mathbf{A}t}\mathbf{x}_0 + \mathbf{x}_p(t)$$

$$= \begin{pmatrix} 2e^t - 2e^{-t} \\ 2e^t - e^{-t} \end{pmatrix} + \begin{pmatrix} -4t + \frac{3}{2}e^t - e^{-t} - \frac{1}{2}\cos t + \frac{3}{2}\sin t \\ -1 - 3t + \frac{3}{2}e^t - \frac{1}{2}e^{-t} + \sin t \end{pmatrix}$$

$$= \begin{pmatrix} \frac{7}{2}e^t - 3e^{-t} + \frac{3}{2}\sin t - \frac{1}{2}\cos t - 4t \\ \frac{7}{2}e^t - \frac{3}{2}e^{-t} + \sin t - 3t - 1 \end{pmatrix}$$

in agreement with the solution found using the method of Laplace transforms in Example 4 of Section 5.4.

Undetermined Coefficients and Frequency Response

The method of undetermined coefficients, discussed in Section 4.6, can be used to find a particular solution of

$$\mathbf{x}' = \mathbf{A}\mathbf{x} + \mathbf{g}(t)$$

if $\mathbf{A}$ is an $n \times n$ constant matrix and the entries of $\mathbf{g}(t)$ consist of polynomials, exponential functions, sines and cosines, or finite sums and products of these functions. The methodology described in Section 4.6 extends in a natural way to these types of problems and is discussed in the exercises (see Problems 14–16). In the next example, we illustrate the method of undetermined coefficients in the special but important case of determining the frequency response and gain function for a first order, constant coefficient linear system when there is a single, or one-dimensional, input and the real parts of the eigenvalues of the system matrix are all negative. In Project 2 at the end of this chapter, an analogous problem arises in the analysis of the response of tall buildings to earthquake induced seismic vibrations.

EXAMPLE 2

Consider the circuit shown in Figure 6.6.1 that was discussed in Section 6.1. Using the circuit parameter values $L_1 = 3/2$, $L_2 = 1/2$, $C = 4/3$, and $R = 1$, find the frequency response and plot a graph of the gain function for the output voltage $v_R = Ri_2(t)$ across the resistor in the circuit.

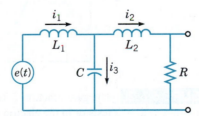

FIGURE 6.6.1 An electric circuit with input $e = e(t)$ and output $v_R = Ri_2(t)$.

By setting $\mathbf{x} = \begin{pmatrix} i_1 & i_2 & v \end{pmatrix}^T$, we may express Eq. (30) in Section 6.1 as

$$\mathbf{x}' = \mathbf{A}\mathbf{x} + \mathbf{B}e(t) \tag{19}$$

where

$$\mathbf{A} = \begin{pmatrix} 0 & 0 & -1/L_1 \\ 0 & -R/L_2 & 1/L_2 \\ 1/C & -1/C & 0 \end{pmatrix} \quad \text{and} \quad \mathbf{B} = \begin{pmatrix} 1/L_1 \\ 0 \\ 0 \end{pmatrix}.$$

The frequency response is found in a manner analogous to that for finding the frequency response of a damped spring-mass system in Section 4.7. For the given values of the parameters, the eigenvalues of the system matrix $\mathbf{A}$ are $\lambda_1 = -1$, $\lambda_2 = -1/2 + i\sqrt{3}/2$, and $\lambda_3 = \bar{\lambda}_2 = -1/2 - i\sqrt{3}/2$.[1] Since the real parts of the eigenvalues are all negative, the transient part of any solution will die out leaving, in the long term, only the steady state response of the system to the harmonic input $e(t) = e^{i\omega t}$. The frequency response for the state variables is then found by assuming a steady state solution of the form

$$\mathbf{x} = \mathbf{G}(i\omega)e^{i\omega t} \tag{20}$$

where $\mathbf{G}(i\omega)$ is a 3×1 frequency response vector. Substituting the right-hand side of Eq. (20) for $\mathbf{x}$ in Eq. (19) and canceling the $e^{i\omega t}$, which appears in every term, yields the equation

$$i\omega\mathbf{G} = \mathbf{A}\mathbf{G} + \mathbf{B}. \tag{21}$$

Solving Eq. (21) for $\mathbf{G}$ we get

$$\mathbf{G}(i\omega) = -(\mathbf{A} - i\omega\mathbf{I}_3)^{-1}\mathbf{B}. \tag{22}$$

It follows from Eq. (22) that the frequency response of the output voltage $v_R = Ri_2(t)$ is given by

$$RG_2(i\omega) = \begin{pmatrix} 0 & R & 0 \end{pmatrix} \mathbf{G}(i\omega).$$

Using a computer, we can solve Eq. (21) for each ω on a sufficiently fine grid. The gain function $R|G_2(i\omega)|$ evaluated on such a grid is shown in Figure 6.6.2.

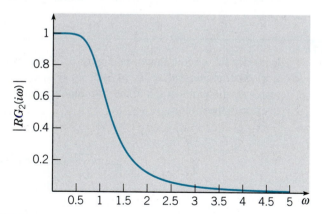

FIGURE 6.6.2 The gain function $R|G_2(i\omega)|$ of the output voltage $v_R = Rv(t)$ across the resistor in the electric circuit shown in Figure 6.6.1.

---

[1]The characteristic polynomial of $\mathbf{A}$ can be shown to be $p(\lambda) = -[\lambda^3 + (R/L_2)\lambda^2 + (1/CL_1 + 1/CL_2)\lambda + R/(CL_1L_2)]$. The Routh stability criterion, discussed in Section 5.9, can then be used to show that the real parts of all the roots of p are negative whenever L_1, L_2, R, and C are positive.

With the specified values of the circuit parameters, the circuit in Figure 6.6.1 is a **low-pass filter**, that is, a circuit offering easy passage to low-frequency signals and difficult passage to high-frequency signals.

PROBLEMS

1. Assuming that $\mathbf{X}(t)$ is a fundamental matrix for $\mathbf{x}' = \mathbf{P}(t)\mathbf{x}$, show directly that

$$\mathbf{x}_p = \mathbf{X}(t) \int_{t_1}^{t} \mathbf{X}^{-1}(s)\mathbf{g}(s)\,ds,$$

is a particular solution of Eq. (1), $\mathbf{x}' = \mathbf{P}(t)\mathbf{x} + \mathbf{g}(t)$.

CAS In each of Problems 2 through 9, find the general solution of the given system. Calculations are greatly facilitated by using a computer algebra system.

2. $\mathbf{x}' = \begin{pmatrix} 2 & -1 \\ 3 & -2 \end{pmatrix} \mathbf{x} + \begin{pmatrix} e^t \\ t \end{pmatrix}$

3. $\mathbf{x}' = \begin{pmatrix} 1 & \sqrt{3} \\ \sqrt{3} & -1 \end{pmatrix} \mathbf{x} + \begin{pmatrix} e^t \\ \sqrt{3}e^{-t} \end{pmatrix}$

4. $\mathbf{x}' = \begin{pmatrix} 2 & -5 \\ 1 & -2 \end{pmatrix} \mathbf{x} + \begin{pmatrix} -\cos t \\ \sin t \end{pmatrix}$

5. $\mathbf{x}' = \begin{pmatrix} 1 & 1 \\ 4 & -2 \end{pmatrix} \mathbf{x} + \begin{pmatrix} e^{-2t} \\ -2e^t \end{pmatrix}$

6. $\mathbf{x}' = \begin{pmatrix} 0 & -1 & 1 \\ 0 & 2 & 0 \\ -2 & -1 & 3 \end{pmatrix} \mathbf{x} + \begin{pmatrix} 1 \\ t \\ e^{-t} \end{pmatrix}$

7. $\mathbf{x}' = \begin{pmatrix} -\frac{1}{2} & \frac{1}{2} & -\frac{1}{2} \\ -1 & -2 & 1 \\ \frac{1}{2} & \frac{1}{2} & -\frac{3}{2} \end{pmatrix} \mathbf{x} + \begin{pmatrix} 1 \\ t \\ e^{-3t} \end{pmatrix}$

8. $\mathbf{x}' = \begin{pmatrix} -4 & 1 & 3 \\ 0 & -2 & 0 \\ -2 & 1 & 1 \end{pmatrix} \mathbf{x} + \begin{pmatrix} t \\ 0 \\ \cos t \end{pmatrix}$

9. $\mathbf{x}' = \begin{pmatrix} -\frac{1}{2} & 1 & \frac{1}{2} \\ 1 & -1 & 1 \\ \frac{1}{2} & 1 & -\frac{1}{2} \end{pmatrix} \mathbf{x} + \begin{pmatrix} 0 \\ \sin t \\ 0 \end{pmatrix}$

ODEA 10. Diffusion of particles on a lattice with reflecting boundaries was described in Example 3, Section 6.3. In this problem, we introduce a known particle source $x_0(t) = f(t)$ at the left end of the lattice as shown in Figure 6.6.3. Treating locations 1 and 2 as interior points and location 3 as a reflecting

boundary, the system of differential equations for $\mathbf{x} = (x_1, x_2, x_3)$, the number of particles at lattice points 1, 2, and 3, respectively, is

$$\mathbf{x}' = k \begin{pmatrix} -2 & 1 & 0 \\ 1 & -2 & 1 \\ 0 & 1 & -1 \end{pmatrix} \mathbf{x} + k \begin{pmatrix} f(t) \\ 0 \\ 0 \end{pmatrix}. \qquad (i)$$

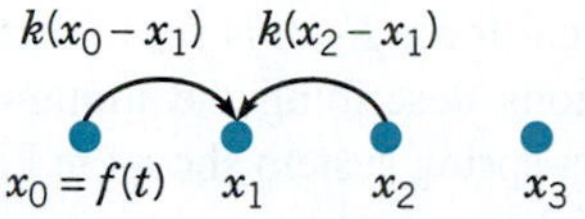

$$k(x_0 - x_1) \quad k(x_2 - x_1)$$

$$x_0 = f(t) \quad x_1 \qquad x_2 \qquad x_3$$

FIGURE 6.6.3 Diffusion on a lattice with a particle source at the left endpoint.

(a) Find a numerical approximation to the solution of Eq. (i) subject to the initial condition $\mathbf{x}(0) = (0, 0, 0)^T$ if $f(t) = 1$ and $k = 1$. Draw component plots of the solution and determine $\lim_{t \to \infty} \mathbf{x}(t)$.

(b) Find a numerical approximation to the solution of Eq. (i) subject to the initial condition $\mathbf{x}(0) = (0, 0, 0)^T$ if $k = 1$ and $f(t) = 1 - \cos \omega t$ for each of the cases $\omega = 1$ and $\omega = 4$. Draw component plots of the solutions in each of the two cases.

11. Find numerical approximations to the initial value ODEA problems posed in Problem 10 under the assumption that the right boundary is an absorbing boundary (see Problem 14, Section 6.3).

12. The equations presented in Section 6.1 for mod- ODEA eling lead uptake, subsequent exchange between tissue compartments, and removal from the human body are

$$x_1' = (L + k_{12}x_2 + k_{13}x_3) - (k_{21} + k_{31} + k_{01})x_1$$
$$x_2' = k_{21}x_1 - (k_{02} + k_{12})x_2$$
$$x_3' = k_{31}x_1 - k_{13}x_3.$$

Employing the methods of Section 6.6, find the solution of this system subject to the initial conditions $x_1(0) = 0$, $x_2(0) = 0$, and $x_3(0) = 0$ using the parameter values $k_{21} = .011$, $k_{12} = .012$,

$k_{31} = .0039$, $k_{13} = .000035$, $k_{01} = .021$, $k_{02} = .016$, and input function

$$L(t) = \begin{cases} 35, & 0 \le t \le 365, \\ 0, & t > 365. \end{cases}$$

Solve the problem in two stages, the first over the time interval $0 \le t \le 365$ and the second over the time interval $t \ge 365$. Evaluate the solution for the first stage at $t = 365$ to provide initial conditions for the second stage of the problem. You may wish to use a computer to assist your calculations. Check your results by comparing graphs of the solutions with the graphs in Figure 6.1.4 or by comparing with an approximation obtained directly by using an initial value problem solver.

CAS **13.** When viscous damping forces are included and the external force $F_2(t)$ is set to zero, the system of equations describing the motion of the coupled mass-spring system shown in Figure 6.1.1 of Section 6.1 is

$$\mathbf{x'} = \begin{pmatrix} 0 & 0 & 1 & 0 \\ 0 & 0 & 0 & 1 \\ -(k_1+k_2)/m_1 & k_2/m_1 & -\gamma/m_1 & 0 \\ k_2/m_2 & -(k_2+k_3)/m_2 & 0 & -\gamma/m_2 \end{pmatrix} \mathbf{x}$$

$$+ \begin{pmatrix} 0 \\ 0 \\ F_1(t)/m_1 \\ 0 \end{pmatrix},$$

where γ is the damping coefficient for both m_1 and m_2. Assume a harmonic input $F_1(t)/m_1 = e^{i\omega t}$ and use the parameter values $m_1 = 2$, $m_2 = 9/4$, $k_1 = 1$, $k_2 = 3$, $k_3 = 15/4$ of Example 2, Section 6.4, to compute and plot the graphs of the gain functions $|G_1(i\omega)|$ and $|G_2(i\omega)|$ for the displacements x_1 and x_2, respectively. Compare the graphs of the gain functions with the natural frequencies of the undamped system for each of the damping coefficients $\gamma = 1, 0.5$, and 0.1.

Undetermined Coefficients. For each of the nonhomogeneous terms specified in Problems 14 through 16, use the method of undetermined coefficients to find a particular solution of

$$\mathbf{x'} = \begin{pmatrix} -2 & 1 \\ 1 & -2 \end{pmatrix} \mathbf{x} + \mathbf{g}(t) = \mathbf{A}\mathbf{x} + \mathbf{g}(t) \qquad \text{(ii)}$$

given that the general solution of the corresponding homogeneous system $\mathbf{x'} = \mathbf{A}\mathbf{x}$ is

$$\mathbf{x} = c_1 e^{-3t} \begin{pmatrix} 1 \\ -1 \end{pmatrix} + c_2 e^{-t} \begin{pmatrix} 1 \\ 1 \end{pmatrix}.$$

14. $\mathbf{g}(t) = \begin{pmatrix} 0 \\ 3t \end{pmatrix} = \begin{pmatrix} 0 \\ 0 \end{pmatrix} + t \begin{pmatrix} 0 \\ 3 \end{pmatrix}$.

Since the entries of $\mathbf{g}(t)$ are linear functions of t, substitute an expression of the form

$$\mathbf{x}_p(t) = t\mathbf{a} + \mathbf{b} = t \begin{pmatrix} a_1 \\ a_2 \end{pmatrix} + \begin{pmatrix} b_1 \\ b_2 \end{pmatrix}$$

into Eq. (ii) and match the coefficients of the vector polynomial on both sides of the resulting equation to obtain the two systems

$$\mathbf{Aa} = -\begin{pmatrix} 0 \\ 3 \end{pmatrix}, \qquad \mathbf{Ab} = \mathbf{a}.$$

Solve the first equation for $\mathbf{a}$, substitute the result into the second equation and solve for $\mathbf{b}$.

15. $\mathbf{g}(t) = \begin{pmatrix} \sin t \\ 0 \end{pmatrix} = \sin t \begin{pmatrix} 1 \\ 0 \end{pmatrix}$.

Since an entry of $\mathbf{g}(t)$ contains a sine function, substitute an expression of the form

$$\mathbf{x}_p(t) = (\cos t)\mathbf{a} + (\sin t)\mathbf{b} =$$

$$\cos t \begin{pmatrix} a_1 \\ a_2 \end{pmatrix} + \sin t \begin{pmatrix} b_1 \\ b_2 \end{pmatrix}$$

into Eq. (ii) and match the coefficients of the sine function and the cosine function on both sides of the resulting equation to obtain the coupled systems

$$\mathbf{Aa} = \mathbf{b}, \qquad \mathbf{Ab} = -\mathbf{a} - \begin{pmatrix} 1 \\ 0 \end{pmatrix}.$$

Show that $(\mathbf{A}^2 + \mathbf{I}_2)\mathbf{a} = -\begin{pmatrix} 1 \\ 0 \end{pmatrix}$ and solve for $\mathbf{a}$. Then substitute this result into the second equation above and solve for $\mathbf{b}$.

16. $\mathbf{g}(t) = \begin{pmatrix} 2e^{-t} \\ 0 \end{pmatrix} = e^{-t} \begin{pmatrix} 2 \\ 0 \end{pmatrix}$.

Since the homogeneous equation has a solution of the form $\mathbf{x}_p(t) = e^{-t}\mathbf{a}$, it is necessary to include both $te^{-t}\mathbf{a}$ and $e^{-t}\mathbf{b}$ in the assumed form for the particular solution,

$$\mathbf{x}_p(t) = e^{-t}\mathbf{b} + te^{-t}\mathbf{a}.$$

Substitute this expression into Eq. (ii) using the given $\mathbf{g}$ and show that $\mathbf{a}$ and $\mathbf{b}$ must be selected to satisfy

$$(\mathbf{A} + \mathbf{I}_2)\mathbf{a} = \mathbf{0}, \qquad (\mathbf{A} + \mathbf{I}_2)\mathbf{b} = \mathbf{a} - \begin{pmatrix} 2 \\ 0 \end{pmatrix}.$$

Then find $\mathbf{a}$ and $\mathbf{b}$ that satisfy both of these equations.

6.7 Defective Matrices

We conclude the chapter on systems of first order linear equations with a discussion of the case in which the real, constant matrix **A** is defective, that is, **A** has one or more defective eigenvalues. Recall that an eigenvalue is defective if its geometric multiplicity g is less than its algebraic multiplicity m. In other words, there are fewer than m linearly independent eigenvectors associated with this eigenvalue. We include this section for completeness but it may be regarded as optional for the following reasons:

1. In the majority of applications, **A** has a complete set of linearly independent eigenvectors. Thus the theory and methods presented in the preceding sections are adequate to handle these problems.

2. The repeated eigenvalue problem for linear constant coefficient systems of dimension 2 was treated in Section 3.5.

3. For linear systems of dimension larger than, say 4, the payoff utility of finding analytic representations of fundamental sets of solutions is questionable since the complexity of the representations often requires that graphs of the components be constructed in order to help understand their behavior. Thus most scientists and engineers resort directly to numerical approximations in order to obtain the same information.

4. If it is necessary to find analytical representations of fundamental sets of solutions for systems of dimension greater than 2 in the case that **A** is defective, then a computer algebra system or the method of Laplace transforms discussed in Section 6.5 together with a computer algebra system may be used to find an analytic representation of $e^{\mathbf{A}t}$ in terms of elementary functions.

▶ **Fundamental Sets for Defective Matrices.** Let λ be an eigenvalue of **A** with algebraic multiplicity m. The following theorem shows how to find m linearly independent solutions of $\mathbf{x}' = \mathbf{A}\mathbf{x}$ associated with λ.

THEOREM 6.7.1

Suppose **A** is a real $n \times n$ matrix and λ is an eigenvalue of **A** with algebraic multiplicity m. Let $\mathbf{v}_1, \ldots, \mathbf{v}_m$ be linearly independent solutions of $(\mathbf{A} - \lambda\mathbf{I}_n)^m\mathbf{v} = \mathbf{0}$. Then

$$\mathbf{x}_k = e^{\lambda t}\left[\mathbf{v}_k + \frac{t}{1!}(\mathbf{A} - \lambda\mathbf{I}_n)\mathbf{v}_k + \cdots + \frac{t^{m-1}}{(m-1)!}(\mathbf{A} - \lambda\mathbf{I}_n)^{m-1}\mathbf{v}_k\right], \qquad (1)$$

$k = 1, \ldots m$, are linearly independent solutions of $\mathbf{x}' = \mathbf{A}\mathbf{x}$.

Proof

We rewrite $\mathbf{x}' = \mathbf{A}\mathbf{x}$ in the form

$$\mathbf{x}' = (\mathbf{A} - \lambda\mathbf{I}_n)\mathbf{x} + \lambda\mathbf{x} \qquad (2)$$

and show that $\mathbf{x}_k$ defined by Eq. (1) satisfies Eq. (2). Differentiating the expression (1) yields

$$\frac{d\mathbf{x}_k}{dt} = e^{\lambda t}\left[(\mathbf{A} - \lambda\mathbf{I}_n)\mathbf{v}_k + \cdots + \frac{t^{m-2}}{(m-2)!}(\mathbf{A} - \lambda\mathbf{I}_n)^{m-1}\mathbf{v}_k\right] + \lambda\mathbf{x}_k. \qquad (3)$$

On the other hand, multiplying $\mathbf{x}_k$ by $(\mathbf{A} - \lambda\mathbf{I}_n)$ gives

$$(\mathbf{A} - \lambda\mathbf{I}_n)\mathbf{x}_k = e^{\lambda t}\left[(\mathbf{A} - \lambda\mathbf{I}_n)\mathbf{v}_k + \cdots + \frac{t^{m-1}}{(m-1)!}(\mathbf{A} - \lambda\mathbf{I}_n)^m\mathbf{v}_k\right]. \tag{4}$$

Since $(\mathbf{A} - \lambda\mathbf{I}_n)^m\mathbf{v}_k = \mathbf{0}$, the last term within the brackets on the right-hand side of Eq. (4) equals $\mathbf{0}$. Therefore Eq. (4) reduces to

$$(\mathbf{A} - \lambda\mathbf{I}_n)\mathbf{x}_k = e^{\lambda t}\left[(\mathbf{A} - \lambda\mathbf{I}_n)\mathbf{v}_k + \cdots + \frac{t^{m-2}}{(m-2)!}(\mathbf{A} - \lambda\mathbf{I}_n)^{m-1}\mathbf{v}_k\right]. \tag{5}$$

From Eqs. (3) and (5) we see that $\dfrac{d\mathbf{x}_k}{dt} = (\mathbf{A} - \lambda\mathbf{I}_n)\mathbf{x}_k + \lambda\mathbf{x}_k$.

The set of solutions $\{\mathbf{x}_1, \ldots, \mathbf{x}_m\}$ is linearly independent on $I = (-\infty, \infty)$ since setting $t = 0$ in the statement

$$c_1\mathbf{x}_1(t) + \cdots + c_m\mathbf{x}_m(t) = \mathbf{0} \quad \text{for every } t \in I$$

yields the equation

$$c_1\mathbf{v}_1 + \cdots + c_m\mathbf{v}_m = \mathbf{0}.$$

Since $\mathbf{v}_1, \ldots, \mathbf{v}_m$ are linearly independent, the constants $c_1, \ldots, c_m$ are necessarily zero. ∎

Remark

1. Theorem 6.7.1 applies to both defective and nondefective eigenvalues.
2. For $\mathbf{A}$, as stated in the theorem, the existence of m linearly independent solutions to $(\mathbf{A} - \lambda\mathbf{I}_n)^m\mathbf{v} = \mathbf{0}$ is guaranteed by Theorem A.4.2 in Appendix A.4.
3. Assuming that $\mathbf{A}$ has r eigenvalues $\lambda_1, \ldots, \lambda_r$ with corresponding algebraic multiplicities $m_1, \ldots, m_r$ such that $m_1 + \cdots + m_r = n$, it can be shown that the union of the solution sets of $(\mathbf{A} - \lambda_j\mathbf{I}_n)^{m_j}\mathbf{v} = \mathbf{0}$ for $j = 1, \ldots r$, say $\left\{\mathbf{v}_1^{(1)}, \ldots, \mathbf{v}_{m_1}^{(1)}, \ldots, \mathbf{v}_1^{(r)}, \ldots, \mathbf{v}_{m_r}^{(r)}\right\}$, is linearly independent. Thus, the union of the sets of solutions obtained by applying Theorem 6.7.1 to each eigenvalue, $\left\{\mathbf{x}_1^{(1)}(t), \ldots, \mathbf{x}_{m_1}^{(1)}(t), \ldots, \mathbf{x}_1^{(r)}(t), \ldots, \mathbf{x}_{m_r}^{(r)}(t)\right\}$, is a fundamental set of solutions of $\mathbf{x}' = \mathbf{A}\mathbf{x}$.
4. That there exist solutions to $\mathbf{x}' = \mathbf{A}\mathbf{x}$ of the form (1) is made plausible by noting that $e^{\mathbf{A}t}\mathbf{v}$ is a solution of $\mathbf{x}' = \mathbf{A}\mathbf{x}$ for any vector $\mathbf{v}$. Therefore $e^{\mathbf{A}t}\mathbf{v}_k$ for $\mathbf{v}_k$ as in Theorem 6.7.1 is a solution of $\mathbf{x}' = \mathbf{A}\mathbf{x}$ and can be expressed in the form $e^{\lambda t}e^{(\mathbf{A}-\lambda\mathbf{I}_n)t}\mathbf{v}_k$. Theorem 6.7.1 shows that a series expansion of $e^{(\mathbf{A}-\lambda\mathbf{I}_n)t}\mathbf{v}_k$ based on Eq. (17) in Section 6.5 (with $\mathbf{A}$ replaced by $\mathbf{A} - \lambda\mathbf{I}_n$) is guaranteed to consist of no more than m nonzero terms.
5. If the λ in Theorem 6.7.1 is complex, $\lambda = \mu + i\nu$, $\nu \neq 0$, then $2m$ linearly independent, real-valued solutions of $\mathbf{x}' = \mathbf{A}\mathbf{x}$ associated with λ are $\{\text{Re } \mathbf{x}_1, \ldots, \text{Re } \mathbf{x}_m\}$ and $\{\text{Im } \mathbf{x}_1, \ldots, \text{Im } \mathbf{x}_m\}$.

EXAMPLE 1

Find a fundamental set of solutions for

$$\mathbf{x}' = \begin{pmatrix} 1 & -1 \\ 1 & 3 \end{pmatrix} = \mathbf{A}\mathbf{x}. \tag{6}$$

The characteristic equation of $\mathbf{A}$ is $p(\lambda) = \lambda^2 - 4\lambda + 4 = (\lambda - 2)^2$ so the only eigenvalue of $\mathbf{A}$ is $\lambda_1 = 2$ with algebraic multiplicity 2. Thus,

$$(\mathbf{A} - \lambda_1 \mathbf{I}_2) = \begin{pmatrix} -1 & -1 \\ 1 & 1 \end{pmatrix}, \quad (\mathbf{A} - \lambda_1 \mathbf{I}_2)^2 = \begin{pmatrix} 0 & 0 \\ 0 & 0 \end{pmatrix}.$$

Two linearly independent solutions of $(\mathbf{A} - \lambda_1 \mathbf{I}_2)^2 \mathbf{v} = \mathbf{0}$ are $\mathbf{v}_1 = (1, 0)^T$ and $\mathbf{v}_2 = (0, 1)^T$. Following Theorem 6.7.1 we find the fundamental set

$$\hat{\mathbf{x}}_1(t) = e^{\lambda_1 t} [\mathbf{v}_1 + t(\mathbf{A} - \lambda_1 \mathbf{I}_2)\mathbf{v}_1]$$

$$= e^{2t} \left[\begin{pmatrix} 1 \\ 0 \end{pmatrix} + t \begin{pmatrix} -1 \\ 1 \end{pmatrix} \right] = e^{2t} \begin{pmatrix} 1 - t \\ t \end{pmatrix}$$

and

$$\hat{\mathbf{x}}_2(t) = e^{\lambda_1 t} [\mathbf{v}_2 + t(\mathbf{A} - \lambda_1 \mathbf{I}_2)\mathbf{v}_2]$$

$$= e^{2t} \left[\begin{pmatrix} 0 \\ 1 \end{pmatrix} + t \begin{pmatrix} -1 \\ 1 \end{pmatrix} \right] = e^{2t} \begin{pmatrix} -t \\ 1 + t \end{pmatrix}.$$

Comparing with the fundamental set for Eq. (6) found in Example 4, Section 6.5,

$$\mathbf{x}_1(t) = e^{2t} \begin{pmatrix} 1 \\ -1 \end{pmatrix}, \quad \mathbf{x}_2(t) = e^{2t} \begin{pmatrix} t \\ -1 - t \end{pmatrix}$$

we see that $\hat{\mathbf{x}}_1(t) = \mathbf{x}_1(t) - \mathbf{x}_2(t)$ and $\hat{\mathbf{x}}_2(t) = -\mathbf{x}_2(t)$. Since the fundamental matrix $\hat{\mathbf{X}}(t) = [\hat{\mathbf{x}}_1(t), \hat{\mathbf{x}}_2(t)]$ satisfies $\hat{\mathbf{X}}(0) = \mathbf{I}_2$, it follows that $\hat{\mathbf{X}}(t) = e^{\mathbf{A}t}$.

EXAMPLE 2

Find a fundamental set of solutions for

$$\mathbf{x}' = \begin{pmatrix} 4 & 6 & 6 \\ 1 & 3 & 2 \\ -1 & -5 & -2 \end{pmatrix} \mathbf{x} = \mathbf{A}\mathbf{x}. \tag{7}$$

The characteristic polynomial of $\mathbf{A}$ is $p(\lambda) = -(\lambda^3 - 5\lambda^2 + 8\lambda - 4) = -(\lambda - 1)(\lambda - 2)^2$ so the eigenvalues of $\mathbf{A}$ are the simple eigenvalue $\lambda_1 = 1$ and $\lambda_2 = 2$ with algebraic multiplicity 2. Since the eigenvector associated with λ_1 is $\mathbf{v}_1 = (4, 1, -3)^T$, one solution of Eq. (7) is

$$\mathbf{x}_1(t) = e^t \begin{pmatrix} 4 \\ 1 \\ -3 \end{pmatrix}. \tag{8}$$

Next we compute

$$\mathbf{A} - \lambda_2 \mathbf{I}_3 = \begin{pmatrix} 2 & 6 & 6 \\ 1 & 1 & 2 \\ -1 & -5 & -4 \end{pmatrix}$$

and

$$(\mathbf{A} - \lambda_2 \mathbf{I}_3)^2 = \begin{pmatrix} 4 & -12 & 0 \\ 1 & -3 & 0 \\ -3 & 9 & 0 \end{pmatrix}.$$

Reducing $(\mathbf{A} - \lambda_2\mathbf{I}_3)^2$ to echelon form

$$\begin{pmatrix} 1 & -3 & 0 \\ 0 & 0 & 0 \\ 0 & 0 & 0 \end{pmatrix}$$

using elementary row operations confirms that rank$[(\mathbf{A} - \lambda_2\mathbf{I}_3)^2] = 1$. Linearly independent solutions of $(\mathbf{A} - \lambda_2\mathbf{I}_3)^2\mathbf{v} = \mathbf{0}$ are $\mathbf{v}_1^{(2)} = (3, 1, 0)^T$ and $\mathbf{v}_2^{(2)} = (0, 0, 1)^T$. Therefore two linearly independent solutions of Eq. (7) associated with λ_2 are

$$\mathbf{x}_2(t) = e^{\lambda_2 t}\left[\mathbf{v}_1^{(2)} + t(\mathbf{A} - \lambda_2\mathbf{I}_3)\mathbf{v}_1^{(2)}\right]$$

$$= e^{2t}\left[\begin{pmatrix} 3 \\ 1 \\ 0 \end{pmatrix} + t\begin{pmatrix} 12 \\ 4 \\ -8 \end{pmatrix}\right] = e^{2t}\begin{pmatrix} 3 + 12t \\ 1 + 4t \\ -8t \end{pmatrix}, \tag{9}$$

and

$$\mathbf{x}_3(t) = e^{\lambda_2 t}\left[\mathbf{v}_2^{(2)} + t(\mathbf{A} - \lambda_2\mathbf{I}_3)\mathbf{v}_2^{(2)}\right]$$

$$= e^{2t}\left[\begin{pmatrix} 0 \\ 0 \\ 1 \end{pmatrix} + t\begin{pmatrix} 6 \\ 2 \\ -4 \end{pmatrix}\right] = e^{2t}\begin{pmatrix} 6t \\ 2t \\ 1 - 4t \end{pmatrix}. \tag{10}$$

A fundamental set for Eq. (7) consists of $\mathbf{x}_1(t)$, $\mathbf{x}_2(t)$, and $\mathbf{x}_3(t)$ given by Eqs. (8), (9), and (10), respectively.

EXAMPLE 3

Find the general solution of

$$\mathbf{x}' = \begin{pmatrix} -3 & -1 & -6 \\ -2 & -1 & -4 \\ 1 & 0 & 1 \end{pmatrix}\mathbf{x} = \mathbf{A}\mathbf{x}. \tag{11}$$

The characteristic polynomial of $\mathbf{A}$ is $p(\lambda) = -(\lambda^3 + 3\lambda^2 + 3\lambda + 1) = -(\lambda + 1)^3$ so $\lambda_1 = -1$ is the only eigenvalue. The algebraic multiplicity of this eigenvalue is $m_1 = 3$. Since

$$(\mathbf{A} - \lambda_1\mathbf{I}_3)^3 = \begin{pmatrix} 0 & 0 & 0 \\ 0 & 0 & 0 \\ 0 & 0 & 0 \end{pmatrix},$$

three linearly independent solutions of $(\mathbf{A} - \lambda_1\mathbf{I}_3)^3 = 0$ are

$$\mathbf{v}_1 = \begin{pmatrix} 1 \\ 0 \\ 0 \end{pmatrix}, \quad \mathbf{v}_2 = \begin{pmatrix} 0 \\ 1 \\ 0 \end{pmatrix}, \quad \mathbf{v}_3 = \begin{pmatrix} 0 \\ 0 \\ 1 \end{pmatrix}.$$

Using

$$\mathbf{A} - \lambda_1\mathbf{I}_3 = \begin{pmatrix} -2 & -1 & -6 \\ -2 & 0 & -4 \\ 1 & 0 & 2 \end{pmatrix}, \quad (\mathbf{A} - \lambda_1\mathbf{I}_3)^2 = \begin{pmatrix} 0 & 2 & 4 \\ 0 & 2 & 4 \\ 0 & -1 & -2 \end{pmatrix}$$

we find that a fundamental set of solutions of Eq. (11) is comprised of

$$\mathbf{x}_1(t) = e^{-t}\left[\mathbf{v}_1 + t(\mathbf{A} - \lambda_1\mathbf{I}_3)\mathbf{v}_1 + \frac{t^2}{2!}(\mathbf{A} - \lambda_1\mathbf{I}_3)^2\mathbf{v}_1\right]$$

$$= e^{-t}\left[\begin{pmatrix} 1 \\ 0 \\ 0 \end{pmatrix} + t\begin{pmatrix} -2 \\ -2 \\ 1 \end{pmatrix}\right] = e^{-t}\begin{pmatrix} 1 - 2t \\ -2t \\ t \end{pmatrix},$$

$$\mathbf{x}_2(t) = e^{-t}\left[\mathbf{v}_2 + t(\mathbf{A} - \lambda_1\mathbf{I}_3)\mathbf{v}_2 + \frac{t^2}{2!}(\mathbf{A} - \lambda_1\mathbf{I}_3)^2\mathbf{v}_2\right]$$

$$= e^{-t}\left[\begin{pmatrix} 0 \\ 1 \\ 0 \end{pmatrix} + t\begin{pmatrix} -1 \\ 0 \\ 0 \end{pmatrix} + \frac{t^2}{2!}\begin{pmatrix} 2 \\ 2 \\ -1 \end{pmatrix}\right] = e^{-t}\begin{pmatrix} t^2 - t \\ 1 + t^2 \\ -t^2/2 \end{pmatrix},$$

and

$$\mathbf{x}_3(t) = e^{-t}\left[\mathbf{v}_3 + t(\mathbf{A} - \lambda_1\mathbf{I}_3)\mathbf{v}_3 + \frac{t^2}{2!}(\mathbf{A} - \lambda_1\mathbf{I}_3)^2\mathbf{v}_3\right]$$

$$= e^{-t}\left[\begin{pmatrix} 0 \\ 0 \\ 1 \end{pmatrix} + t\begin{pmatrix} -6 \\ -4 \\ 2 \end{pmatrix} + \frac{t^2}{2!}\begin{pmatrix} 4 \\ 4 \\ -2 \end{pmatrix}\right] = e^{-t}\begin{pmatrix} 2t^2 - 6t \\ 2t^2 - 4t \\ 1 + 2t - t^2 \end{pmatrix}.$$

Note that the fundamental matrix $\mathbf{X}(t) = [\mathbf{x}_1(t), \mathbf{x}_2(t), \mathbf{x}_3(t)]$ is in fact equal to $e^{\mathbf{A}t}$ since $\mathbf{X}(0) = \mathbf{I}_3$. Thus the general solution of Eq. (11) is given by

$$\mathbf{x}(t) = \mathbf{X}(t)\mathbf{c} = e^{\mathbf{A}t}\mathbf{c}.$$

EXAMPLE 4

Use the eigenvalue method to find a fundamental set of solutions of the 4th order equation

$$\frac{d^4y}{dt^4} + 2\frac{d^2y}{dt^2} + y = 0. \tag{12}$$

If we define $\mathbf{x} = (y, y', y'', y''')^T$, the dynamical system equivalent to Eq. (12) is

$$\mathbf{x}' = \begin{pmatrix} 0 & 1 & 0 & 0 \\ 0 & 0 & 1 & 0 \\ 0 & 0 & 0 & 1 \\ -1 & 0 & -2 & 0 \end{pmatrix} \mathbf{x} = \mathbf{A}\mathbf{x}. \tag{13}$$

Since the characteristic polynomial of $\mathbf{A}$ is $p(\lambda) = \lambda^4 + 2\lambda^2 + 1 = (\lambda - i)^2(\lambda + i)^2$, the only eigenvalues of $\mathbf{A}$ are $\lambda_1 = i$ and $\lambda_2 = -i$ with corresponding algebraic multiplicities $m_1 = 2$ and $m_2 = 2$. Using Gaussian elimination, we find two linearly independent solutions of

$$(\mathbf{A} - \lambda_1\mathbf{I}_4)^2\mathbf{v} = \begin{pmatrix} -1 & -2i & 1 & 0 \\ 0 & -1 & -2i & 1 \\ -1 & 0 & -3 & -2i \\ 2i & -1 & 4i & -3 \end{pmatrix} \mathbf{v} = \mathbf{0},$$

$$\mathbf{v}_1 = \begin{pmatrix} -3 \\ -2i \\ 1 \\ 0 \end{pmatrix} \quad \text{and} \quad \mathbf{v}_2 = \begin{pmatrix} -2i \\ 1 \\ 0 \\ 1 \end{pmatrix}.$$

Two complex-valued, linearly independent solutions of Eq. (13) are therefore

$$\hat{\mathbf{x}}_1(t) = e^{\lambda_1 t}\,[\mathbf{v}_1 + t(\mathbf{A} - \lambda_1 \mathbf{I}_4)\mathbf{v}_1]$$

$$= e^{it}\left[\begin{pmatrix} -3 \\ -2i \\ 1 \\ 0 \end{pmatrix} + t\begin{pmatrix} i \\ -1 \\ -i \\ 1 \end{pmatrix}\right]$$

and

$$\hat{\mathbf{x}}_2(t) = e^{\lambda_1 t}\,[\mathbf{v}_2 + t(\mathbf{A} - \lambda_1 \mathbf{I}_4)\mathbf{v}_2]$$

$$= e^{it}\left[\begin{pmatrix} -2i \\ 1 \\ 0 \\ 1 \end{pmatrix} + t\begin{pmatrix} -1 \\ -i \\ 1 \\ i \end{pmatrix}\right].$$

Using complex arithmetic and Euler's formula, we find the following fundamental set of four real-valued solutions of Eq. (13),

$$\mathbf{x}_1(t) = \operatorname{Re}\hat{\mathbf{x}}_1(t) = \begin{pmatrix} -3\cos t - t\sin t \\ -t\cos t + 2\sin t \\ \cos t + t\sin t \\ t\cos t \end{pmatrix},$$

$$\mathbf{x}_2(t) = \operatorname{Im}\hat{\mathbf{x}}_1(t) = \begin{pmatrix} t\cos t - 3\sin t \\ -2\cos t - t\sin t \\ -t\cos t + \sin t \\ t\sin t \end{pmatrix},$$

$$\mathbf{x}_3(t) = \operatorname{Re}\hat{\mathbf{x}}_2(t) = \begin{pmatrix} -t\cos t + 2\sin t \\ \cos t + t\sin t \\ t\cos t \\ \cos t - t\sin t \end{pmatrix},$$

and

$$\mathbf{x}_4(t) = \operatorname{Im}\hat{\mathbf{x}}_2(t) = \begin{pmatrix} -t\sin t - 2\cos t \\ \sin t - t\cos t \\ t\sin t \\ \sin t + t\cos t \end{pmatrix}.$$

The first components of $\mathbf{x}_1$, $\mathbf{x}_2$, $\mathbf{x}_3$, and $\mathbf{x}_4$ provide us with a fundamental set for Eq. (12), $y_1(t) = -3\cos t - t\sin t$, $y_2(t) = t\cos t - 3\sin t$, $y_3(t) = -t\cos t + 2\sin t$, and $y_4(t) = -t\sin t - 2\cos t$. We note that each of these solutions is a linear combination of a simpler fundamental set for Eq. (12) consisting of $\cos t$, $\sin t$, $t\cos t$, and $t\sin t$.

PROBLEMS

In each of Problems 1 through 8, find a fundamental matrix for the given system.

1. $\mathbf{x}' = \begin{pmatrix} 3 & -4 \\ 1 & -1 \end{pmatrix} \mathbf{x}$

2. $\mathbf{x}' = \begin{pmatrix} 4 & -2 \\ 8 & -4 \end{pmatrix} \mathbf{x}$

3. $\mathbf{x}' = \begin{pmatrix} 1 & 1 & 1 \\ 2 & 1 & -1 \\ -3 & 2 & 4 \end{pmatrix} \mathbf{x}$

4. $\mathbf{x}' = \begin{pmatrix} 5 & -3 & -2 \\ 8 & -5 & -4 \\ -4 & 3 & 3 \end{pmatrix} \mathbf{x}$

5. $\mathbf{x}' = \begin{pmatrix} -7 & 9 & -6 \\ -8 & 11 & -7 \\ -2 & 3 & -1 \end{pmatrix} \mathbf{x}$

6. $\mathbf{x}' = \begin{pmatrix} 5 & 6 & 2 \\ -2 & -2 & -1 \\ -2 & -3 & 0 \end{pmatrix} \mathbf{x}$

7. $\mathbf{x}' = \begin{pmatrix} -2 & -7 & -7 & -5 \\ -3 & -8 & -7 & -7 \\ 1 & 1 & 0 & 1 \\ 2 & 8 & 8 & 6 \end{pmatrix} \mathbf{x}$

8. $\mathbf{x}' = \begin{pmatrix} 1 & -1 & -2 & 3 \\ 2 & -\frac{3}{2} & -1 & \frac{7}{2} \\ -1 & \frac{1}{2} & 0 & -\frac{3}{2} \\ -2 & \frac{3}{2} & 3 & -\frac{7}{2} \end{pmatrix} \mathbf{x}$

In each of Problems 9 and 10, find the solution of the ODEA given initial value problem. Draw the corresponding trajectory in x_1x_2-space and also draw the graph of x_1 versus t.

9. $\mathbf{x}' = \begin{pmatrix} 1 & -4 \\ 4 & -7 \end{pmatrix} \mathbf{x}, \quad \mathbf{x}(0) = \begin{pmatrix} 3 \\ 2 \end{pmatrix}$

10. $\mathbf{x}' = \begin{pmatrix} 3 & -4 \\ 1 & -1 \end{pmatrix} \mathbf{x}, \quad \mathbf{x}(0) = \begin{pmatrix} -2 \\ 4 \end{pmatrix}$

Ineach of Problems 11 and 12, find the solution of the ODEA given initial value problem. Draw the corresponding trajectory in $x_1x_2x_3$-space and also draw the graph of x_1 versus t.

11. $\mathbf{x}' = \begin{pmatrix} 4 & 1 & 3 \\ 6 & 4 & 6 \\ -5 & -2 & -4 \end{pmatrix} \mathbf{x}, \quad \mathbf{x}(0) = \begin{pmatrix} 1 \\ -1 \\ 1 \end{pmatrix}$

12. $\mathbf{x}' = \begin{pmatrix} 1 & 1 & 0 \\ -14 & -5 & 1 \\ 15 & 5 & -2 \end{pmatrix} \mathbf{x}, \quad \mathbf{x}(0) = \begin{pmatrix} 5 \\ 7 \\ -4 \end{pmatrix}$

CHAPTER SUMMARY

Section 6.1 Many science and engineering problems are modeled by systems of differential equations of dimension $n > 2$: vibrating systems with two or more degrees of freedom; compartment models arising in biology, ecology, pharmacokinetics, transport theory, and chemical reactor systems; linear control systems; and electrical networks.

Section 6.2

▶ If $\mathbf{P}(t)$ and $\mathbf{g}(t)$ are continuous on I, a unique solution to the initial value problem

$$\mathbf{x}' = \mathbf{P}(t)\mathbf{x} + \mathbf{g}(t), \quad \mathbf{x}(t_0) = \mathbf{x}_0, \quad t_0 \in I$$

exists throughout I.

▶ A set of n solutions $\mathbf{x}_1, \ldots, \mathbf{x}_n$ to the homogeneous equation

$$\mathbf{x}' = \mathbf{P}(t)\mathbf{x}, \quad \mathbf{P} \text{ continuous on } I$$

is a **fundamental set** on I if their **Wronskian** $W[\mathbf{x}_1, \ldots, \mathbf{x}_n](t)$ is nonzero for some (and hence all) $t \in I$. If $\mathbf{x}_1, \ldots, \mathbf{x}_n$ is a fundamental set of solutions to the homogeneous equation, then a **general solution** is

$$\mathbf{x} = c_1 \mathbf{x}_1(t) + \cdots + c_n \mathbf{x}_n(t)$$

where $c_1, \ldots, c_n$ are arbitrary constants. The n solutions are **linearly independent** on I if and only if their Wronskian is nonzero on I.

▶ The theory of nth order linear equations

$$\frac{d^n y}{dt^n} + p_1(t)\frac{d^{n-1} y}{dt^{n-1}} + \cdots + p_{n-1}(t)\frac{dy}{dt} + p_n(t)y = g(t),$$

follows from the theory of first order linear systems of dimension n.

Section 6.3

▶ If $\mathbf{v}_1, \ldots, \mathbf{v}_n$ are linearly independent eigenvectors of the real, constant $n \times n$ matrix $\mathbf{A}$ and the corresponding eigenvalues $\lambda_1, \ldots, \lambda_n$ are real, then a real general solution of $\mathbf{x}' = \mathbf{A}\mathbf{x}$ is

$$\mathbf{x}(t) = c_1 e^{\lambda_1 t}\mathbf{v}_1 + \cdots + c_n e^{\lambda_n t}\mathbf{v}_n$$

where $c_1, \ldots, c_n$ are arbitrary real constants.

▶ Two classes of $n \times n$ matrices which have n linearly independent eigenvectors are (i) matrices with n distinct eigenvalues and (ii) **symmetric matrices**, that is, matrices satisfying $\mathbf{A}^T = \mathbf{A}$.

Section 6.4
If $\mathbf{A}$ is real, constant, and nondefective, each pair of complex conjugate eigenvalues $\mu \pm i\nu$ with corresponding eigenvectors $\mathbf{v} = \mathbf{a} \pm i\mathbf{b}$ yields two linearly independent, real vector solutions Re $\{\exp[(\mu + i\nu)t][\mathbf{a} + i\mathbf{b}]\} = \exp(\mu t)(\cos \nu t \, \mathbf{a} - \sin \nu t \, \mathbf{b})$ and Im $\{\exp[(\mu + i\nu)t][\mathbf{a} + i\mathbf{b}]\} = \exp(\mu t)(\sin \nu t \, \mathbf{a} + \cos \nu t \, \mathbf{b})$.

Section 6.5

▶ If $\mathbf{x}_1, \ldots, \mathbf{x}_n$ is a fundamental set for $\mathbf{x}' = \mathbf{P}(t)\mathbf{x}$, the $n \times n$ matrix $\mathbf{X}(t) = [\mathbf{x}_1, \ldots, \mathbf{x}_n]$ is called a **fundamental matrix** and satisfies $\mathbf{X}' = \mathbf{P}(t)\mathbf{X}$. In addition, the fundamental matrix $\boldsymbol{\Phi}(t) = \mathbf{X}(t)\mathbf{X}^{-1}(t_0)$ satisfies $\boldsymbol{\Phi}(t_0) = \mathbf{I}_n$.

▶ If $\mathbf{A}$ is a real, constant $n \times n$ matrix, the matrix exponential function

$$e^{\mathbf{A}t} = \mathbf{I}_n + \mathbf{A}t + \frac{1}{2!}\mathbf{A}^2 t^2 + \cdots + \frac{1}{n!}\mathbf{A}^n t^n + \cdots$$

is a fundamental matrix satisfying $e^{\mathbf{A}0} = \mathbf{I}_n$ and $(e^{\mathbf{A}t})^{-1} = e^{-\mathbf{A}t}$. Methods for computing $e^{\mathbf{A}t}$:

▶ If $\mathbf{X}(t)$ is any fundamental matrix for $\mathbf{x}' = \mathbf{A}\mathbf{x}$, $e^{\mathbf{A}t} = \mathbf{X}(t)\mathbf{X}^{-1}(0)$.

▶ If $\mathbf{A}$ is **diagonalizable**, that is $\mathbf{T}^{-1}\mathbf{A}\mathbf{T} = \mathbf{D}$, then $e^{\mathbf{A}t} = \mathbf{T}e^{\mathbf{D}t}\mathbf{T}^{-1}$.

▶ Use Laplace transforms to find the solution $\boldsymbol{\phi}(t)$ to $\mathbf{X}' = \mathbf{A}\mathbf{X}$, $\mathbf{X}(0) = \mathbf{I}_n$. Then $e^{\mathbf{A}t} = \boldsymbol{\phi}(t)$.

Section 6.6 If $\mathbf{X}(t)$ is a fundamental matrix for the homogeneous system, the general solution of $\mathbf{x}' = \mathbf{P}(t)\mathbf{x} + \mathbf{g}(t)$, $\mathbf{x}(t_0) = \mathbf{x}_0$ is $\mathbf{x} = \mathbf{X}(t)\mathbf{X}^{-1}(t_0)\mathbf{c} + \mathbf{x}_p(t)$ where a particular solution $\mathbf{x}_p(t)$ is given by the **variation of parameters formula**

$$\mathbf{x}_p(t) = \mathbf{X}(t)\int \mathbf{X}^{-1}(t)\mathbf{g}(t)\,dt.$$

Section 6.7

▶ If $\mathbf{A}$ is a real, constant $n \times n$ matrix and λ is a real eigenvalue of $\mathbf{A}$ with algebraic multiplicity m, then m linearly independent solutions of $\mathbf{x}' = \mathbf{A}\mathbf{x}$ associated with λ are

$$\mathbf{x}_k = e^{\lambda t}\left[\mathbf{v}_k + \frac{t}{1!}(\mathbf{A} - \lambda\mathbf{I}_n)\mathbf{v}_k + \cdots + \frac{t^{m-1}}{(m-1)!}(\mathbf{A} - \lambda\mathbf{I}_n)^{m-1}\mathbf{v}_k\right],$$

$k = 1, \ldots, m$ where $\mathbf{v}_1, \ldots, \mathbf{v}_m$ are linearly independent solutions of $(\mathbf{A} - \lambda\mathbf{I}_n)^m\mathbf{v} = \mathbf{0}$. If λ is complex, $\lambda = \mu + i\nu$, $\nu \neq 0$, then $2m$ linearly independent, real-valued solutions of $\mathbf{x}' = \mathbf{A}\mathbf{x}$ associated with λ are $\{\text{Re }\mathbf{x}_1, \ldots, \text{Re }\mathbf{x}_m\}$ and $\{\text{Im }\mathbf{x}_1, \ldots, \text{Im }\mathbf{x}_m\}$.

▶ The union of the sets of solutions generated by this method comprise a fundamental set for $\mathbf{x}' = \mathbf{A}\mathbf{x}$.

PROJECTS

Project 1 A Compartment Model of Heat Flow in a Rod

This project shows how, in principle, to estimate the thermal diffusivity of a homogeneous material from experimental data. Consider a metal rod of length L and cross-sectional area A which is insulated around its lateral boundary and its right end as shown in Figure 6.P.1.

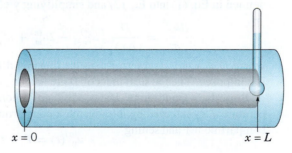

$x = 0$ $x = L$

FIGURE 6.P.1 A metal rod with its entire surface insulated except for the left end.

A thermometer is also placed in thermal contact with the right end of the rod. The rod is initially brought to a uniform temperature of zero degrees by placing the left end of the rod in thermal contact with a heat reservoir held at constant temperature zero and waiting a sufficiently long period of time. The left end is then quickly placed into contact with a heat reservoir which is maintained at a constant temperature $T > 0$. The temperature readings of the thermometer are then recorded at several different points in time. The arrival of the wave of heat at the right end of the rod is therefore observed and it is intuitively obvious that information is being gathered on the rate at which heat energy is being transported through the material of which the rod is composed. We will assume that the thermometer has negligible effect on the rate at which the rod heats.

▶ **A Compartment Model for Heat Transport.** We imagine the rod as consisting of n subsections or compartments, each of width $\Delta x = L/n$ where Δx is small enough so that the temperature at time t in the mth compartment can be represented by a single number, $u_m^{(n)}(t)$, for each $m = 1, \ldots, n$.

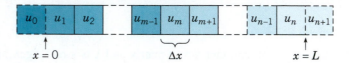

$x = 0 \qquad \Delta x \qquad x = L$

FIGURE 6.P.2   A compartment model for heat transport in a finite rod.

The rate of heat flow from compartment $m - 1$ to compartment m is given by a discrete version of the Newton-Fourier law of cooling

$$J_m^{(n)} = -\kappa \frac{u_m^{(n)} - u_{m-1}^{(n)}}{\Delta x}, \tag{1}$$

where κ is the **thermal conductivity** of the material of which the rod is composed. Thus if $u_{m-1}^{(n)}(t) > u_m^{(n)}(t)$, $J_m^{(n)}$ is positive and heat flows from compartment $m - 1$ to compartment m.

Conservation of energy requires that the time rate of change of heat energy in compartment m equals the rate at which heat enters (or exits) across the interfaces between each of the adjacent compartments. Applying the law to each interior compartment m results in the system of equations

$$\frac{d}{dt}\left(\rho c A \Delta x u_m^{(n)}\right) = A\left(J_m^{(n)} - J_{m+1}^{(n)}\right), \quad m = 2, \ldots, n - 1, \tag{2}$$

where ρ and c are, respectively, the mass density and specific heat of the rod. Note that $-J_{m+1}^{(n)} = \kappa\left(u_{m+1}^{(n)} - u_m^{(n)}\right)/\Delta x$ is positive if $u_{m+1}^{(n)}(t) > u_m^{(n)}(t)$. Substituting $J_m^{(n)}$ and $J_{m+1}^{(n)}$ defined in Eq. (1) into Eq. (2) and simplifying yields the system

$$\frac{du_m^{(n)}}{dt} = \frac{D}{(\Delta x)^2}\left(u_{m-1}^{(n)} - 2u_m^{(n)} + u_{m+1}^{(n)}\right), \quad m = 2, \ldots n - 1. \tag{3}$$

where $D = \kappa/(\rho c)$ is referred to as the **thermal diffusivity** of the material of which the rod is composed. The physical units of D are length^2/time.

Boundary Conditions. The condition of holding the left end of the rod at a fixed temperature T is achieved by introducing a compartment labeled 0 immediately to the left of compartment 1 and setting

$$u_0^{(n)}(t) = T, \quad t > 0. \tag{4}$$

Setting $m = 1$ in Eqs. (2) and using the boundary condition (4) gives a nonhomogeneous equation for compartment 1,

$$\frac{du_1^{(n)}}{dt} = \frac{D}{(\Delta x)^2}\left(-2u_1^{(n)} + u_2^{(n)}\right) + \frac{D}{(\Delta x)^2}T. \tag{5}$$

The insulated condition at the right end of the rod is achieved by introducing a virtual compartment, labeled by $n + 1$, satisfying

$$u_{n+1}^{(n)}(t) = u_n^{(n)}(t), \quad t > 0. \tag{6}$$

From Eq. (1), we see that $J_{n+1}^{(n)} = 0$, that is, the flow of heat across the interface at the right end of the rod is zero, and is consistent with our notion of an insulated boundary. Setting $m = n$ in Eq. (3) and using the boundary condition (6) gives the equation for the temperature of the nth compartment,

$$\frac{du_n^{(n)}}{dt} = \frac{D}{(\Delta x)^2} \left(u_{n-1}^{(n)} - u_n^{(n)} \right). \tag{7}$$

Project 1 PROBLEMS

1. Show that in terms of the vectors

$$\mathbf{u}^{(n)} = \begin{pmatrix} u_1^{(n)} \\ u_2^{(n)} \\ \vdots \\ u_n^{(n)} \end{pmatrix}, \quad \mathbf{g}^{(n)} = \frac{n^2}{L^2} \begin{pmatrix} T \\ 0 \\ \vdots \\ 0 \end{pmatrix}$$

and the matrix

$$\mathbf{A}^{(n)} = \frac{n^2}{L^2} \begin{pmatrix} -2 & 1 & 0 & 0 & \cdots & 0 & 0 \\ 1 & -2 & 1 & 0 & \cdots & 0 & 0 \\ 0 & 1 & -2 & 1 & \vdots & 0 & 0 \\ 0 & 0 & \ddots & \ddots & \vdots & & \vdots \\ \vdots & \vdots & & & & & \\ 0 & 0 & & \cdots & & -2 & 1 \\ 0 & 0 & & \cdots & & 1 & -1 \end{pmatrix},$$

the system of equations (3), (5), and (7) can be written in the form

$$\frac{d\mathbf{u}^{(n)}}{dt} = D\mathbf{A}^{(n)}\mathbf{u}^{(n)} + D\mathbf{g}^{(n)} \tag{i}$$

where the superscript n indexes the number of compartments into which the rod has been partitioned.

2. **Conservation of Energy.** Show that the boundary condition on the right end of the rod is correct in the sense that energy is conserved, that is, the time rate of change in heat energy in the rod satisfies the equation

$$\frac{d}{dt} \sum_{m=1}^{n} \rho c A \Delta x u_m^{(n)} = \kappa A \frac{T - u_1^{(n)}}{\Delta x}. \tag{i}$$

Give an explanation of the physical significance of Eq. (i).

3. The following pairs of time and temperature observations were recorded at the right end of a rod of length $L = 10$ cm subsequent to the instant that the left end was placed in thermal contact with a heating element kept at a constant temperature $T = 100$ degrees.

TABLE 6.P.1 Temperatures recorded at the right end of the rod in Figure 6.P.1

Time	Temperature	Time	Temperature
10	3.2	60	63.57
20	16.65	70	70.33
30	32.03	80	76.03
40	44.81	90	80.53
50	55.15	100	84.05

From the family of model equations in Eq. (i), parameterized by D, we wish to choose a value of D that minimizes the function

$$V(D) = \sum_{k=1}^{10} \left[u_n^{(n)}(t_k; D) - \theta_k \right]^2 \tag{i}$$

where (t_k, θ_k), $k = 1, \ldots, 10$ denote the time and temperature measurements in Table 6.P.1 and $u_n^{(n)}(t; D)$ is the nth component of the solution $\mathbf{u}^{(n)}(t) = \mathbf{u}^{(n)}(t; D)$ of the initial value problem

$$\frac{d\mathbf{u}^{(n)}}{dt} = D\mathbf{A}^{(n)}\mathbf{u}^{(n)} + D\mathbf{g}^{(n)}, \quad \mathbf{u}^{(n)}(0) = \mathbf{0}. \tag{ii}$$

Estimate a value of D that minimizes $V(D)$ by finding numerical approximations to the initial value problem (ii) for a number of different values of D, say $D_1, \ldots, D_m$, and plotting the graph of $V(D_j)$ versus D_j for $j = 1, \ldots, m$.

Test the sensitivity of your estimate to a few different, but sensible, choices for the number n of compartments in your model.

CAS 4. Denote the estimate of D found in the preceding problem by $\hat{D}$. Test the accuracy of $\hat{D}$ by solving the initial value problem (ii) in Problem 3 numerically using $D = \hat{D}$ and then plotting $u_n^{(n)}(t; \hat{D})$

versus t and the data in Table 6.P.1 on the same set of coordinate axes. Examine the sensitivity of the plots to the choice of n.

5. Plot compartment temperature profiles, $u_j^{(n)}(t; \hat{D})$ CAS versus j, $j = 1, \ldots, n$, for several different values of t. Examine the sensitivity of the profiles to the choice of n.

Project 2 Earthquakes and Tall Buildings

Simplistic differential equation models are often used to introduce concepts and principles which are important for understanding the dynamic behavior of complex physical systems. In this project, we employ such a model to study the response of a tall building due to horizontal seismic motion at the foundation generated by an earthquake.

Figure 6.P.3 is an illustration of a building idealized as a collection of n floors, each of mass m, connected together by vertical walls. If we neglect gravitation and restrict motion to the horizontal direction, the displacements of the floors, relative to a fixed frame of reference, are denoted by $x_1, x_2, \ldots, x_n$. At equilibrium, all of the displacements and velocities are zero and the floors are in perfect vertical alignment.

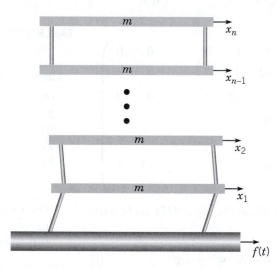

FIGURE 6.P.3 A building consisting of floors of mass m connected by stiff but flexible vertical walls.

When adjacent floors are not in alignment, we assume that the walls exert a flexural restoring force proportional to the difference in displacements between the floors with proportionality constant k. Thus the equation of motion for the j th floor is

$$mx_j'' = -k(x_j - x_{j-1}) - k(x_j - x_{j+1}), \qquad j = 2, \ldots, n-1, \tag{1}$$

while the equations for the first floor and nth, or top, floor are

$$mx_1'' = -k(x_1 - f(t)) - k(x_1 - x_2), \tag{2}$$

and

$$mx_n'' = -k(x_n - x_{n-1}), \tag{3}$$

respectively. The horizontal motion of the foundation generated by the earthquake is described by the input function $f(t)$.

Project 2 PROBLEMS

1. **The Undamped Building**.
   (a) Show that Eqs. (1) through (3) can be expressed in matrix notation as

   $$\mathbf{x}'' + \omega_0^2 \mathbf{K}\mathbf{x} = \omega_0^2 f(t)\mathbf{z} \qquad (i)$$

   where

   $$\omega_0^2 = k/m, \qquad \mathbf{x} = (x_1, x_2, \ldots, x_n)^T,$$

   $$\mathbf{z} = (1, 0, \ldots, 0)^T,$$

   and

   $$\mathbf{K} = \begin{pmatrix} 2 & -1 & 0 & 0 & \cdots & 0 \\ -1 & 2 & -1 & 0 & \cdots & 0 \\ 0 & -1 & 2 & -1 & \cdots & 0 \\ \vdots & & \ddots & \ddots & \ddots & \vdots \\ 0 & \cdots & & -1 & 2 & -1 \\ 0 & \cdots & & 0 & -1 & 1 \end{pmatrix}. \qquad (ii)$$

   (b) A real $n \times n$ matrix $\mathbf{A}$ is said to be **positive definite** if $\mathbf{x}^T \mathbf{A}\mathbf{x} > 0$ for every real n-vector $\mathbf{x} \neq \mathbf{0}$. Show that $\mathbf{K}$ in (ii) satisfies

   $$\mathbf{x}^T \mathbf{K}\mathbf{x} = \sum_{j=1}^{n-1} \left[x_j^2 + (x_j - x_{j+1})^2 \right],$$

   and is therefore positive definite.

   (c) Eigenvalues and eigenvectors of real symmetric matrices are real (see Appendix A.4). Show that if $\mathbf{K}$ is positive definite and λ and $\mathbf{u}$ are an eigenvalue-eigenvector pair for $\mathbf{K}$, then

   $$\lambda = \frac{\mathbf{u}^T \mathbf{K}\mathbf{u}}{\mathbf{u}^T \mathbf{u}} > 0.$$

   Thus all eigenvalues of $\mathbf{K}$ in (ii) are real and positive.

   (d) For the cases $n = 5, 10,$ and 20, demonstrate numerically that the eigenvalues of $\mathbf{K}$, $\lambda_j = \omega_j^2$, $j = 1, \ldots n$ can be ordered as follows, CAS

   $$0 < \omega_1^2 < \omega_2^2 < \cdots < \omega_n^2.$$

   (e) Since K is real and symmetric, it possesses a set of n orthogonal eigenvectors, $\{\mathbf{u}_1, \mathbf{u}_2, \ldots, \mathbf{u}_n\}$, that is, $\mathbf{u}_i^T \mathbf{u}_j = 0$ if $i \neq j$ (see Appendix A.4). These eigenvectors can be used to construct a **normal mode representation**,

   $$\mathbf{x} = a_1(t)\mathbf{u}_1 + \cdots + a_n(t)\mathbf{u}_n, \qquad (iii)$$

   of the solution of

   $$\mathbf{x}'' + \omega_0^2 \mathbf{K}\mathbf{x} = \omega_0^2 f(t)\mathbf{z},$$
   $$\mathbf{x}(0) = \mathbf{x}_0, \qquad (iv)$$
   $$\mathbf{x}'(0) = \mathbf{v}_0.$$

   Substitute the representation (iii) into the differential equation and initial conditions in Eqs. (iv) and use the fact that $\mathbf{K}\mathbf{u}_j = \omega_j^2 \mathbf{u}_j$, $j = 1, \ldots, n$ and the orthogonality of $\mathbf{u}_1, \mathbf{u}_2, \ldots, \mathbf{u}_n$ to show that for each $i = 1, \ldots n$, the **mode amplitude** $a_i(t)$ satisfies the initial value problem

   $$a_i'' + \omega_i^2 \omega_0^2 a_i = \omega_0^2 f(t)z_i, \qquad a_i(0) = \alpha_i,$$

   $$a_i'(0) = \beta_i, \qquad (v)$$

   where

   $$z_i = \frac{\mathbf{u}_i^T \mathbf{z}}{\mathbf{u}_i^T \mathbf{u}_i}, \qquad \alpha_i = \frac{\mathbf{u}_i^T \mathbf{x}_0}{\mathbf{u}_i^T \mathbf{u}_i}, \qquad \beta_i = \frac{\mathbf{u}_i^T \mathbf{v}_0}{\mathbf{u}_i^T \mathbf{u}_i}.$$

   (f) An unforced pure mode of vibration, say the jth mode, can be realized by solving the initial value problem (iv) subject to the initial conditions $\mathbf{x}(0) = A_j \mathbf{u}_j$, where A_j is a mode amplitude factor, and $\mathbf{x}'(0) = \mathbf{0}$ with zero input, $f(t) = 0$. Show that, in this case, the normal mode solution of the initial value problem (iv) consists of a single term, CAS

   $$\mathbf{x}^{(j)}(t) = A_j \cos(\omega_0 \omega_j t) \mathbf{u}_j.$$

   Thus the natural frequency of the jth mode of vibration is $\omega_0 \omega_j$ and the corresponding period is $2\pi/(\omega_0 \omega_j)$. Assuming that $A_1 = 1$, $\omega_0 = 41$, and $n = 20$ plot a graph of the components (floor displacement versus floor number) of the first mode $\mathbf{x}^{(1)}(t)$ for several values of t over an entire cycle. Then generate analogous graphs for the second and third pure modes of vibration. Describe and compare the modes of vibration and their relative frequencies.

2. **The Building with Damping Devices**. In addition to flexural restoring forces, all buildings possess intrinsic internal friction, due to structural and nonstructural elements, that causes the amplitude of vibrations to decay as these elements absorb vibrational energy. One of several techniques employed in modern earthquake resistant design uses added damping devices such as shock absorbers between adjacent floors to artificially increase the internal

damping of a building and improve its resistance to earthquake induced motion (see Figure 6.P.4).

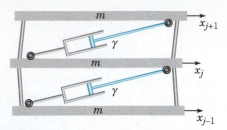

FIGURE 6.P.4 Shock absorbers are used to dampen the amplitude of relative motion between adjacent floors.

The effects of intrinsic and artificial damping are accounted for by including damping forces of the form

$$F_d^{(j)} = -\gamma(x_j' - x_{j-1}') - \gamma(x_j' - x_{j+1}'),$$
$$j = 2, \ldots, n-1,$$

$$F_d^{(1)} = -\gamma(x_1' - f') - \gamma(x_1' - x_2'),$$

and

$$F_d^{(n)} = -\gamma(x_n' - x_{n-1}')$$

on the right-hand sides of Eqs. (1), (2), and (3) respectively.

(a) Show that including damping devices between each pair of adjacent floors changes the system (i) in Problem 1

$$\mathbf{x}'' + 2\delta\mathbf{K}\mathbf{x}' + \omega_0^2\mathbf{K}\mathbf{x} = \left[\omega_0^2 f(t) + 2\delta f'(t)\right]\mathbf{z} \tag{i}$$

where $\delta = \dfrac{\gamma}{2m}$.

(b) Assume an input to Eq. (i) in part (a) of the form $f(t) = e^{i\omega t}$ and show that the n-floor frequency response vector $\mathbf{G}(i\omega)$ (see Sections 4.7 and 6.6) satisfies the equation

$$\left[(\omega_0^2 + i2\delta\omega)\mathbf{K} - \omega^2\mathbf{I}_n\right]\mathbf{G}(i\omega) = (\omega_0^2 + i2\delta\omega)\mathbf{z}. \tag{ii}$$

Thus the gain function for the frequency response of the jth floor is the absolute value of the jth component of $\mathbf{G}(i\omega)$, $|G_j(i\omega)|$. On the same set of coordinate axes, plot the graphs of $|G_j(i\omega)|$ versus $\omega \in [0, 6\pi]$, $j = 5, 10, 20$ using the parameter values $n = 20$, $\omega_0 = 41$, and $\delta = 10$. Repeat using $\delta = 50$ and interpret the results of your graphs.

(c) Numerically approximate the solution of Eq. (i) in part (a) subject to the initial conditions $\mathbf{x}(0) = \mathbf{0}$, $\mathbf{x}'(0) = \mathbf{0}$ using the parameter values $n = 20$, $\omega_0 = 41$, and $\delta = 10$. For the input function, use $f(t) = A \sin \omega_r t$ where $A = 2$ inches and ω_r is the lowest resonant frequency of the system, that is, the value of ω at which $|G_{20}(i\omega)|$ in part (b) attains its maximum value. Plot the graphs of $x_j(t)$ versus t, $j = 5, 10, 20$ on the same set of coordinate axes. What is the amplitude of oscillatory displacement of each of these floors once the steady state response is attained? Repeat the simulation using $\delta = 50$ and compare your results with the case $\delta = 10$.

3. A majority of the buildings which collapsed during the Mexico City earthquake of September 19, 1985 were around 20 stories tall. The predominant period of the earthquake ground motion, around 2.0 seconds, was close to the natural period of vibration of these buildings. Other buildings, of different heights and with different vibrational characteristics, were often found undamaged even though they were located right next to the damaged 20 story buildings. Explain whether the model described in this project provides a possible explanation for these observations. Your argument should be supported by computer simulations and graphs.

Project 3 Controlling a Spring-Mass System to Equilibrium

Consider a system of three masses coupled together by springs or elastic bands as depicted in Figure 6.P.5. We will assume that the magnitudes of the three masses are all equal to m and the spring constants are all equal to k. If $x_j(t)$, $j = 1, 2, 3$, represent the displacements of the masses from their equilibrium positions at time t, then the differential equations describing the motion of the masses are

$$m\frac{d^2x_1}{dt^2} = -2kx_1 + kx_2 + \hat{u}_1(t),$$

$$m\frac{d^2x_2}{dt^2} = kx_1 - 2kx_2 + kx_3 + \hat{u}_2(t), \tag{1}$$

$$m\frac{d^2x_3}{dt^2} = kx_2 - 2kx_3 + \hat{u}_3(t).$$

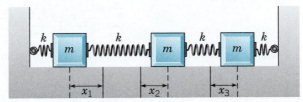

FIGURE 6.P.5   A three mass, four spring system.

The functions $\hat{u}_j(t)$ $j = 1, 2, 3$ represent externally applied forces acting on each of the masses. Dividing both sides of the set of equations (1) by m and using matrix notation, the system can be expressed as

$$\frac{d^2\mathbf{x}}{dt^2} = \mathbf{G}\mathbf{x} + \mathbf{u}(t) \tag{2}$$

where

$$\mathbf{G} = \omega_0^2 \begin{pmatrix} -2 & 1 & 0 \\ 1 & -2 & 1 \\ 0 & 1 & -2 \end{pmatrix}, \qquad \mathbf{u}(t) = \begin{pmatrix} u_1(t) \\ u_2(t) \\ u_3(t) \end{pmatrix} = \frac{1}{m}\begin{pmatrix} \hat{u}_1(t) \\ \hat{u}_2(t) \\ \hat{u}_3(t) \end{pmatrix}, \tag{3}$$

and $\omega_0^2 = k/m$. In addition to Eq. (2), we need to specify initial positions for the masses,

$$\mathbf{x}(0) = \mathbf{x}_0, \tag{4}$$

and initial velocities,

$$\frac{d\mathbf{x}}{dt}(0) = \mathbf{v}_0. \tag{5}$$

Given the system (2) and initial conditions (4) and (5), the control problem of interest is to choose the external forcing function $\mathbf{u}(t)$ in such a way as to steer or drive the system to the equilibrium state at some prescribed time $T > 0$. We are particularly interested in whether this can be done by applying external forces to only one of the masses. For example, we could require $u_2(t) \equiv 0$ and $u_3(t) \equiv 0$ and try to control the system to equilibrium by selecting an appropriate control function $u_1(t)$, $0 \le t \le T$.

▶ **Normal Mode Solutions.** Prior to addressing the control problem, we will discuss normal mode solutions of the initial value problem (2), (4), and (5). Normal mode analysis is a method for solving equations where an orthogonal basis is generated by a matrix or operator that appears in the equation. Let us consider Eqs. (2), (4), and (5) in the n-dimensional case, that is, $\mathbf{x}(t)$, $\mathbf{u}(t)$, $\mathbf{x}_0$, and $\mathbf{v}_0$ are n-vectors while $\mathbf{G}$ is an $n \times n$ symmetric matrix. Since $\mathbf{G}$ is symmetric, it has n real eigenvalues, $\lambda_1, \ldots, \lambda_n$, with a corresponding set of orthogonal eigenvectors $S = \{\mathbf{v}_1, \ldots, \mathbf{v}_n\}$. Thus S provides us with an orthogonal basis for $\mathbf{R}^n$. Since $\mathbf{x}(t) \in \mathbf{R}^n$, it is natural to assume the following normal mode representation for $\mathbf{x}(t)$,

$$\mathbf{x}(t) = a_1(t)\mathbf{v}_1 + \cdots + a_n(t)\mathbf{v}_n. \tag{6}$$

Each vector $\mathbf{v}_j$ in the sum (6) is called a normal mode while the coefficient $a_j(t)$ is the corresponding mode amplitude function. Equation (6) expresses the solution of the initial value problem (2), (4), and (5) at any time t as a superposition of the normal modes. In order to determine how the amplitude functions vary with time, we substitute the representation (6) into each of the equations (2), (4), and (5), and then use the orthogonality property of the normal modes to deduce differential equations and initial conditions for the amplitude functions. For example, substituting the sum (6) into Eq. (2) yields the equation

$$\sum_{j=1}^{n} \frac{d^2 a_j}{dt^2} \mathbf{v}_j = \sum_{j=1}^{n} a_j \mathbf{G} \mathbf{v}_j + \mathbf{u}(t). \tag{7}$$

Using the fact that $\mathbf{G}\mathbf{v}_j = \lambda_j \mathbf{v}_j$, $j = 1, \ldots, n$ permits Eq. (7) to be written as

$$\sum_{j=1}^{n} \left[\frac{d^2 a_j}{dt^2} - \lambda_j a_j \right] \mathbf{v}_j = \mathbf{u}(t). \tag{8}$$

Equation (8) may be viewed as an expansion of $\mathbf{u}(t)$ in terms of the orthogonal system S. Thus the coefficients $\dfrac{d^2 a_j}{dt^2} - \lambda_j a_j$ in the sum on the left hand side of Eq. (8) must satisfy

$$\frac{d^2 a_i}{dt^2} - \lambda_i a_i = \frac{\langle \mathbf{v}_i, \mathbf{u}(t) \rangle}{||\mathbf{v}_i||^2}, \qquad i = 1, \ldots, n \tag{9}$$

where the inner product $\langle , \rangle$ and norm $|| \cdot ||$ in Eqs. (9) refer to the Euclidean inner product and norm on $\mathbf{R}^n$, respectively. Equations (9) are easily derived by taking the inner product of both sides of Eq. (8) with $\mathbf{v}_i$ and using the fact that

$$\langle \mathbf{v}_i, \mathbf{v}_j \rangle = \begin{cases} ||\mathbf{v}_i||^2, & j = i \\ 0, & j \neq i. \end{cases} \tag{10}$$

In a similar manner, substituting the sum (6) into Eqs. (4) and (5) yields initial conditions for each of the differential equations (9),

$$a_i(0) = \frac{\langle \mathbf{v}_i, \mathbf{x}_0 \rangle}{||\mathbf{v}_i||^2}, \qquad i = 1, \ldots, n, \tag{11}$$

and

$$\frac{da_i}{dt}(0) = \frac{\langle \mathbf{v}_i, \mathbf{v}_0 \rangle}{||\mathbf{v}_i||^2}, \qquad i = 1, \ldots, n. \tag{12}$$

Thus the mode amplitudes can be obtained by solving the initial value problems defined by Eqs. (9), (11), and (12). For our particular problem of interest, it can be shown that the eigenvalues of $\mathbf{G}$ are strictly negative (Problem 8), a fact which can be made explicit by setting $\lambda_j = -\omega_j^2$, $j = 1, \ldots, n$. Thus Eqs. (9) may be written as

$$\frac{d^2 a_i}{dt^2} + \omega_i^2 a_i = \frac{\langle \mathbf{v}_i, \mathbf{u}(t) \rangle}{||\mathbf{v}_i||^2}, \qquad i = 1, \ldots, n. \tag{13}$$

It is of particular interest to observe solutions of the initial value problems (13), (11), and (12) for the special case where the initial position is taken to be the jth eigenvector, $\mathbf{x}_0 = \mathbf{v}_j$, the initial velocity is taken to be zero, $\mathbf{v}_0 = \mathbf{0}$, and the forcing function is set to zero, $\mathbf{u}(t) \equiv \mathbf{0}$. Then the solutions of the initial value problems (13), (11), and (12) are given by

$$a_i(t) = \begin{cases} 0, & i \neq j \\ \cos \omega_j t, & i = j \end{cases}, \tag{14}$$

and therefore the motion consists of a single pure mode with time varying amplitude of angular frequency ω_j,

$$\mathbf{x}(t) = \cos\omega_j t \ \ \mathbf{v}_j. \tag{15}$$

This observation has the following important implication for the control problem. If the jth mode has a zero component (which we may call a nodal point), for example, the kth component is zero, then a single control function applied to the kth mass will not be able to force vibrations of the jth mode to the equilibrium state. This can be shown mathematically by setting $\mathbf{u}(t) = u_k(t)\mathbf{e}_k$ and noting that $\langle\mathbf{v}_j, \mathbf{u}(t)\rangle = u_k(t)\langle\mathbf{v}_j, \mathbf{e}_k\rangle = 0$. Here $\mathbf{e}_k$ is the n-vector with 1 in the kth entry and 0's elsewhere. Consequently, $\mathbf{u}(t) = u_k(t)\mathbf{e}_k$ cannot influence the amplitude of the jth mode.

▶ **Linear Constant Coefficient Control Problems.** We wish to recast the second order system (2) as a system of first order differential equations in order to take advantage of existing classical theory of linear control. This can be accomplished in many ways. We choose to define

$$\mathbf{y} = \begin{pmatrix} \mathbf{x} \\ \mathbf{x}' \end{pmatrix}.$$

Thus the first three components of $\mathbf{y}$ refer to the positions of the three masses while the last three components refer to the velocities of the masses. Then the system (2) can be written as

$$\frac{d\mathbf{y}}{dt} = \mathbf{Ay} + \mathbf{Bu}(t) \tag{16}$$

where

$$\mathbf{A} = \begin{pmatrix} \mathbf{0} & \mathbf{I}_3 \\ \mathbf{G} & \mathbf{0} \end{pmatrix}.$$

The input matrix $\mathbf{B}$ depends on the number of controls and to what components of the system they will be applied. In the case of a scalar control $u(t)$ applied to the first mass, system (16) will be

$$\frac{d\mathbf{y}}{dt} = \mathbf{Ay} + \begin{pmatrix} 0 \\ 0 \\ 0 \\ 1 \\ 0 \\ 0 \end{pmatrix} u(t), \tag{17}$$

that is,

$$\mathbf{B} = \begin{pmatrix} 0 \\ 0 \\ 0 \\ 1 \\ 0 \\ 0 \end{pmatrix}.$$

In the case of a scalar control applied to the second mass,

$$\mathbf{B} = \begin{pmatrix} 0 \\ 0 \\ 0 \\ 0 \\ 1 \\ 0 \end{pmatrix}.$$

The initial condition for Eq. (16) obtained from Eqs. (4) and (5) is

$$\mathbf{y}(0) = \mathbf{y}_0 = \begin{pmatrix} \mathbf{x}_0 \\ \mathbf{v}_0 \end{pmatrix}. \tag{18}$$

More generally, a linear constant coefficient control system is expressed as

$$\frac{d\mathbf{y}}{dt} = \mathbf{A}\mathbf{y} + \mathbf{B}\mathbf{u}(t) \tag{19}$$

where $\mathbf{y}$ is $n \times 1$, $\mathbf{A}$ is $n \times n$, $\mathbf{B}$ is $n \times m$, and the control vector $\mathbf{u}(t)$ is $m \times 1$.

DEFINITION 6.P.1

The system (19) is said to be **completely controllable** at time $T > 0$ if there exists a control $\mathbf{u}(t)$ such that any arbitrary state $\mathbf{y}(0) = \mathbf{y}_0$ can be driven to the zero state $\mathbf{y}(T) = 0$.

In order to arrive at a test for controllability, we consider the following representation for the solution of the initial value problem (16) and (18),

$$\mathbf{y}(t) = e^{\mathbf{A}t}\mathbf{y}_0 + \int_0^t e^{\mathbf{A}(t-s)}\mathbf{B}\mathbf{u}(s)ds. \tag{20}$$

Setting $\mathbf{y}(T) = \mathbf{0}$ yields an integral equation for $\mathbf{u}(t)$,

$$\int_0^T e^{\mathbf{A}(T-s)}\mathbf{B}\mathbf{u}(s)ds = -e^{\mathbf{A}T}\mathbf{y}_0. \tag{21}$$

The Cayley-Hamilton theorem from linear algebra states that if

$$p(\lambda) = \det(\lambda\mathbf{I}_n - \mathbf{A}) = \lambda^n + p_{n-1}\lambda^{n-1} + \cdots + p_1\lambda + p_0 \tag{22}$$

is the characteristic polynomial of $\mathbf{A}$, then

$$p(\mathbf{A}) = \mathbf{A}^n + p_{n-1}\mathbf{A}^{n-1} + \cdots + p_1\mathbf{A} + p_0\mathbf{I}_n = \mathbf{0}. \tag{23}$$

Remark. The characteristic polynomial of $\mathbf{A}$ defined in Appendix A.4, $\det(\mathbf{A} - \lambda\mathbf{I}_n)$, differs from the definition given in Eq. (22) by the largely irrelevant factor $(-1)^n$ since $\det(\mathbf{A} - \lambda\mathbf{I}_n) = (-1)^n \det(\lambda\mathbf{I}_n - \mathbf{A})$.

If the Cayley-Hamilton theorem is applied to the matrix exponential function $\exp[\mathbf{A}(T - s)]$, then the left-hand side of Eq. (21) may be expressed in the form

$$\mathbf{B}\int_0^T q_0(T-s)\mathbf{u}(s)ds + \mathbf{A}\mathbf{B}\int_0^T q_1(T-s)\mathbf{u}(s)ds + \cdots +$$
$$\mathbf{A}^{n-1}\mathbf{B}\int_0^T q_{n-1}(T-s)\mathbf{u}(s)ds = -e^{\mathbf{A}T}\mathbf{y}_0. \tag{24}$$

If the system (19) is to be controllable at time T, according to Eq. (24) we must be able to express any vector in $\mathbf{R}^n$ (the right-hand side of Eq. (24)) as a linear combination of the columns of the matrix $[\mathbf{B}, \mathbf{AB}, \ldots, \mathbf{A}^{n-1}\mathbf{B}]$. This leads to the *rank test for controllability*,

---

THEOREM 6.P.2

The system (19) is *completely controllable at time $T > 0$ if and only if the matrix*

$$\Psi = [\mathbf{B}, \mathbf{AB}, \ldots, \mathbf{A}^{n-1}\mathbf{B}] \qquad (25)$$

has rank n.

The matrix Ψ is called the *controllability matrix* for the system (19).

▶ **Construction of the Control Function.** Given that the system (19) is completely controllable, we now address the problem of determining a control that will do the job. We will restrict attention to the case where $\mathbf{B}$ is $n \times 1$ and $\mathbf{u}(t)$ is a scalar control but the method easily extends to the case where $\mathbf{B}$ is $n \times m$ and $\mathbf{u}$ is $m \times 1$. We choose a set of functions $E = \{\phi_1(t), \phi_2(t), \ldots\}$ which are linearly independent over the time interval $[0, T]$. For example, we might choose $\phi_j(t) = \sin j\pi t/T$, $j = 1, 2, 3, \ldots$. We will represent the scalar control function $u(t) = u_N(t)$ as a superposition of a finite number of functions from E,

$$u_N(t) = \sum_{j=1}^{N} c_j \phi_j(t) \qquad (26)$$

where the coefficients c_j are to be determined. The general idea is to permit construction of a sufficiently diverse class of inputs so at least one will do the job of controlling the system to equilibrium. In order to generate the vector on the right-hand side of Eq. (21), we numerically solve the initial value problem

$$\frac{d\mathbf{y}}{dt} = \mathbf{Ay}, \quad \mathbf{y}(0) = \mathbf{y}_0 \qquad (27)$$

over the time interval $[0, T]$. The right-hand side of Eq. (21) is the solution of the initial value problem (27) evaluated at time T and multiplied by -1,

$$-\exp[\mathbf{A}T]\mathbf{y}_0 = -\mathbf{y}(T).$$

Next, solve each of the N initial value problems

$$\frac{d\mathbf{y}}{dt} = \mathbf{Ay} + \mathbf{B}\phi_j(t), \quad \mathbf{y}(0) = \mathbf{0}, \quad j = 1, \ldots, N \qquad (28)$$

over the time interval $[0, T]$. The solutions of the initial value problems (28) will be denoted by $\mathbf{z}_j(t)$. By the *principle of superposition* of solutions for linear nonhomogeneous differential equations, the solution $\mathbf{y}_N(t)$ of

$$\frac{d\mathbf{y}}{dt} = \mathbf{Ay} + \mathbf{B}u_N(t), \quad \mathbf{y}(0) = \mathbf{0} \qquad (29)$$

is simply

$$\mathbf{y}_N(t) = \int_0^t e^{\mathbf{A}(t-s)}\mathbf{B}u_N(s)\,ds = \sum_{j=1}^{N} c_j \mathbf{z}_j(t). \qquad (30)$$

Setting $\mathbf{y}_N(T) = -\mathbf{y}(T)$ yields a linear algebraic system to be solved for $\mathbf{c} = (c_1, \ldots, c_N)^T$. If $N = 6$, the system to be solved is

$$\overbrace{[\mathbf{z}_1(T), \ldots, \mathbf{z}_6(T)]}^{\mathbf{Z}(T)} \overbrace{\begin{bmatrix} c_1 \\ c_2 \\ c_3 \\ c_4 \\ c_5 \\ c_6 \end{bmatrix}}^{\mathbf{c}} = -\mathbf{y}(T),$$

or, using matrix notation,

$$\mathbf{Z}(T)\mathbf{c} = -\mathbf{y}(T). \tag{31}$$

How large should N be? Since the right-hand side of Eq. (21) may be any vector in $\mathbf{R}^n$, the matrix $\mathbf{Z} = [\mathbf{z}_1(T), \mathbf{z}_2(T), \cdots, \mathbf{z}_N(T)]$ must be of full rank, that is, rank $(\mathbf{Z}) = n$. Therefore, we conclude that the number of inputs N must be greater than or equal to n. It is conceivable that even though the inputs $\phi_j(t)$ are linearly independent functions, the outputs $\mathbf{z}_j(T)$, $j = 1, \ldots, N = n$, due to errors in numerical approximation or due to the effects of the system on the inputs, may not be sufficiently diverse so that rank $(\mathbf{Z}) < n$. Thus we take a common sense approach. Construct $\mathbf{Z}$ with $N = n$ columns and then check to see if rank $(\mathbf{Z}) = n$. If so, then solve Eq. (31). If not, include more functions in the sum (26) (that is, increase the value of N) until rank $(\mathbf{Z}) = n$. If $N > n$, then Eq. (31) may have many solutions, any one of which may be used to realize $u_N(t)$ in Eq. (26). The apparent freedom in the choice of the control function is exploited in the field of optimal control theory.

Project 3 PROBLEMS

1. Derive the system of equations (1) by applying Newton's second law, $ma = F$, to each of the masses. Assume that the springs follow Hooke's Law: *the force exerted by a spring on a mass is proportional to the length of its departure from its equilibrium length.*

CAS 2. Find the eigenvalues and eigenvectors of the matrix $\mathbf{G}$ in (3). Assume that $\omega_0^2 = 1$. Plot the components of the eigenvectors. Verbally, and with the aid of sketches, describe the normal modes of vibration in terms of the motion of the masses relative to one another.

3. From the normal mode representation of the solution of the initial value problem (2), (4), and (5), explain why the system cannot be completely controllable by applying a control function only to the center mass.

4. Repeat Problem 2 for a system of 4 masses connected by springs. Give a physical interpretation of the normal modes of vibration. Does the normal mode representation rule out complete controllability for the case of a scalar control applied to any single mass? Explain.

5. Find the rank of the controllability matrix for the CAS three mass system (i) in the case that a single control is applied to the first mass and (ii) in the case that a single control is applied to the second mass.

6. Find the rank of the controllability matrix for the CAS four mass system in all of the cases that a single control is applied to each of the four masses and compare your results with the conclusions obtained in Problem 4. (Note that by symmetry considerations, it is only necessary to consider one of the masses on the end and one of the interior masses.)

7. Prove the Cayley-Hamilton Theorem for the special case that $\mathbf{A}$ is diagonalizable.

8. A symmetric matrix G is said to be *negative definite* if $\mathbf{x}^T G \mathbf{x} < 0$ for every $\mathbf{x} \neq 0$. Prove that a

symmetric matrix $\mathbf{G}$ is negative definite if and only if its eigenvalues are all negative. Use this result to show that $\mathbf{G}$ in (3) has negative eigenvalues.

CAS **9.** For the three mass system, find a scalar control $u(t)$ which, when applied to the first mass, will drive the system (16) from the initial state $\mathbf{y}(0) = [-1, 1, 2, 0, 0, 0]^T$ to equilibrium at time $T = 10$. Plot the graphs of $u(t)$ and $y_j(t)$, $j = 1, \ldots, 6$ to confirm that the system is actually driven to equilibrium.

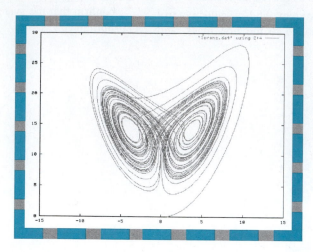

CHAPTER SEVEN

Nonlinear Differential Equations and Stability

I n this chapter, we take up the investigation of systems of nonlinear equations. Such systems can be solved by analytical methods only in rare instances. Numerical approximation methods provide one means of dealing with nonlinear systems. Another approach, presented in this chapter, is geometrical in character and leads to a qualitative understanding of the behavior of solutions rather than to detailed quantitative information. A combination of methods is often needed to achieve optimal results.

7.1 Autonomous Systems and Stability

We first introduced two-dimensional systems of the form

$$\frac{dx}{dt} = F(x, y), \quad \frac{dy}{dt} = G(x, y) \tag{1}$$

in Section 3.6. Recall that the system (1) is called **autonomous** because the functions F and G do not depend on the independent variable t. In Chapter 3, we were mainly concerned with showing how to find the solutions of homogeneous linear systems, and we presented only a few examples of nonlinear systems. Now we want to focus on the analysis of two-dimensional nonlinear systems of the form (1). Unfortunately, it is only in exceptional cases that solutions can be found by analytical methods. One alternative is to use numerical methods to approximate solutions. Software packages often include one or more algorithms, such as the Runge-Kutta method discussed in Section 3.7, for this purpose. Detailed quantitative information about solutions usually requires the use of such methods and we employed them in producing many of the figures in this chapter. Another alternative is to consider what qualitative information can be obtained about solutions without actually solving the equations. The main purpose of this chapter is to show that a good deal of information about the qualitative behavior of solutions can often be obtained with relatively little effort. The questions that we consider in this chapter are associated with the idea of stability of a solution, and the methods that we employ are basically geometrical. The numerical and geometrical approaches complement each other rather well: the numerical methods provide detailed information about a single solution,[1] while the geometrical methods yield qualitative information about all solutions simultaneously.

Both the concept of stability and the use of geometrical methods were introduced in Chapter 1 and have been used repeatedly in later chapters as well: in Section 2.5 for first order autonomous equations

$$dy/dt = f(y), \tag{2}$$

and in Chapters 3 and 6 for linear systems with constant coefficients. In this chapter, we refine the ideas and extend the discussion to nonlinear autonomous systems of equations.

▶ **Autonomous Systems.** We are concerned with systems of two simultaneous differential equations of the form (1). We assume that the functions F and G are continuous and have continuous partial derivatives in some domain D of the xy-plane. If (x_0, y_0) is a point in this domain, then by Theorem 3.6.1 there exists a unique solution $x = \phi(t)$, $y = \psi(t)$ of the system (1) satisfying the initial conditions

$$x(t_0) = x_0, \qquad y(t_0) = y_0. \tag{3}$$

The solution is defined in some time interval I that contains the point t_0.

Frequently, we will write the initial value problem (1), (3) in the vector form

$$d\mathbf{x}/dt = \mathbf{f}(\mathbf{x}), \qquad \mathbf{x}(t_0) = \mathbf{x}_0, \tag{4}$$

where $\mathbf{x} = x\mathbf{i} + y\mathbf{j}$, $\mathbf{f}(\mathbf{x}) = F(x, y)\mathbf{i} + G(x, y)\mathbf{j}$, and $\mathbf{x}_0 = x_0\mathbf{i} + y_0\mathbf{j}$. In this case, the solution is expressed as $\mathbf{x} = \boldsymbol{\phi}(t)$, where $\boldsymbol{\phi}(t) = \phi(t)\mathbf{i} + \psi(t)\mathbf{j}$. As usual, we interpret a solution $\mathbf{x} = \boldsymbol{\phi}(t)$ as a curve traced by a moving point in the xy-plane, the phase plane.

The autonomous system (1) has an associated direction field that is independent of time. Consequently, there is only one trajectory passing through each point (x_0, y_0) in the phase plane. In other words, all solutions that satisfy an initial condition of the form (3) lie on the same trajectory, regardless of the time t_0 at which they pass through (x_0, y_0). Thus, just as for the constant coefficient linear system

$$d\mathbf{x}/dt = \mathbf{A}\mathbf{x}, \tag{5}$$

a single phase portrait simultaneously displays important qualitative information about all solutions of the system (1). We will see this fact confirmed repeatedly in this chapter.

Autonomous systems occur frequently in applications. Physically, an autonomous system is one whose configuration, including physical parameters and external forces or effects, is independent of time. The response of the system to given initial conditions is then independent of the time at which the conditions are imposed.

▶ **Stability and Instability.** The concepts of stability, asymptotic stability, and instability have already been mentioned several times in this book. It is now time to give a precise mathematical definition

---

[1] Of course, they can be used repeatedly with different initial conditions to approximate more than one solution.

of these concepts, at least for autonomous systems of the form

$$\mathbf{x}' = \mathbf{f}(\mathbf{x}). \tag{6}$$

In the following definitions, and elsewhere, we use the notation $||\mathbf{x}||$ to designate the length, or magnitude, of the vector $\mathbf{x}$. If $\mathbf{x} = x\mathbf{i} + y\mathbf{j}$, then $||\mathbf{x}|| = \sqrt{x^2 + y^2}$.

The points, if any, where $\mathbf{f}(\mathbf{x}) = \mathbf{0}$ are called **critical points** of the autonomous system (6). At such points $\mathbf{x}' = \mathbf{0}$ also, so critical points correspond to constant, or equilibrium, solutions of the system of differential equations.

Roughly speaking, we have seen that a critical point is stable if all trajectories that start close to the critical point remain close to it for all future time. A critical point is asymptotically stable if all nearby trajectories not only remain nearby but actually approach the critical point as $t \to \infty$. A critical point is unstable if at least some nearby trajectories do not remain close to the critical point as t increases.

More precisely, a critical point $\mathbf{x}_0$ of the system (6) is said to be **stable** if, given any $\epsilon > 0$, there is a $\delta > 0$ such that every solution $\mathbf{x} = \boldsymbol{\phi}(t)$ of the system (1), which at $t = 0$ satisfies

$$||\boldsymbol{\phi}(0) - \mathbf{x}_0|| < \delta, \tag{7}$$

exists for all positive t and satisfies

$$||\boldsymbol{\phi}(t) - \mathbf{x}_0|| < \epsilon \tag{8}$$

for all $t \geq 0$. A critical point that is not stable is said to be **unstable**.

Stability of a critical point is illustrated geometrically in Figures 7.1.1(a) and 7.1.1(b). The mathematical statements (7) and (8) say that all solutions that start "sufficiently close" (that is, within the distance δ) to $\mathbf{x}_0$ stay "close" (within the distance ϵ) to $\mathbf{x}_0$. Note that in Figure 7.1.1(a) the trajectory is within the circle $||\mathbf{x} - \mathbf{x}_0|| = \delta$ at $t = 0$ and, although it soon passes outside of this circle, it remains within the circle $||\mathbf{x} - \mathbf{x}_0|| = \epsilon$ for all $t \geq 0$. In fact, it eventually approaches $\mathbf{x}_0$. However, this limiting behavior is not necessary for stability, as illustrated in Figure 7.1.1(b).

A critical point $\mathbf{x}_0$ is said to be **asymptotically stable** if it is stable and if there exists a $\delta_0 > 0$ such that, if a solution $\mathbf{x} = \boldsymbol{\phi}(t)$ satisfies

$$||\boldsymbol{\phi}(0) - \mathbf{x}_0|| < \delta_0, \tag{9}$$

then

$$\lim_{t \to \infty} \boldsymbol{\phi}(t) = \mathbf{x}_0. \tag{10}$$

Thus trajectories that start "sufficiently close" to $\mathbf{x}_0$ not only must stay "close" but must eventually approach $\mathbf{x}_0$ as $t \to \infty$. This is the case for the trajectory in Figure 7.1.1(a) but not for the one in Figure 7.1.1(b). Note that asymptotic stability is a stronger property

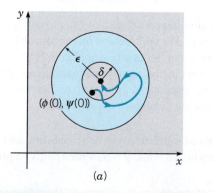
(a)

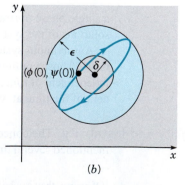
(b)

FIGURE 7.1.1 (a) Asymptotic stability. (b) Stability.

than stability, since a critical point must be stable before we can even talk about whether it might be asymptotically stable. On the other hand, the limit condition (10), which is an essential feature of asymptotic stability, does not by itself imply even ordinary stability. Indeed, examples can be constructed in which all the trajectories approach x_0 as $t \to \infty$, but for which x_0 is not a stable critical point. Geometrically, all that is needed is a family of trajectories having members that start arbitrarily close to x_0, and then depart an arbitrarily large distance before eventually approaching x_0 as $t \to \infty$.

In this chapter, we are concentrating on two-dimensional systems, but the definitions just given are independent of the dimension of the system. If you interpret the vectors in Eqs. (6) through (10) as n-dimensional, then the definitions of stability, asymptotic stability, and instability apply also to n-dimensional systems. These definitions can be made more concrete by interpreting them in terms of a specific physical problem.

▶ **The Oscillating Pendulum.** The concepts of asymptotic stability, stability, and instability can be easily visualized in terms of an oscillating pendulum. Consider the configuration shown in Figure 7.1.2, in which a mass m is attached to one end of a rigid, but weightless, rod of length L. The other end of the rod is supported at the origin O, and the rod is free to rotate in the plane of the paper. The position of the pendulum is described by the angle θ between the rod and the downward vertical direction, with the counterclockwise direction taken as positive. The gravitational force mg acts downward, while the damping force $c|d\theta/dt|$, where c is positive, is always opposite to the direction of motion. We assume that θ and $d\theta/dt$ are both positive. The equation of motion can be quickly derived from the principle of angular momentum,[2] which states that the time rate of change of angular momentum about any point is equal to the moment of the resultant force about that point. The angular momentum about the origin, $mL^2(d\theta/dt)$, is the product of the mass m, the moment arm L, and the velocity $L d\theta/dt$. Thus the equation of motion is

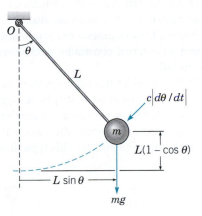

FIGURE 7.1.2   An oscillating pendulum.

$$mL^2\frac{d^2\theta}{dt^2} = -cL\frac{d\theta}{dt} - mgL\sin\theta. \tag{11}$$

The factors L and $L\sin\theta$ on the right side of Eq. (11) are the moment arms of the resistive force and of the gravitational force, respectively; the minus signs are due to the fact that the two forces tend to make the pendulum rotate in the clockwise (negative) direction. You

[2]Recall that in Section 4.1 we derived the pendulum equation by using Newton's second law of motion.

should verify, as an exercise, that the same equation is obtained for the other three possible sign combinations of θ and $d\theta/dt$.

By straightforward algebraic operations, we can write Eq. (11) in the standard form

$$\frac{d^2\theta}{dt^2} + \frac{c}{mL}\frac{d\theta}{dt} + \frac{g}{L}\sin\theta = 0, \tag{12}$$

or

$$\frac{d^2\theta}{dt^2} + \gamma\frac{d\theta}{dt} + \omega^2\sin\theta = 0, \tag{13}$$

where $\gamma = c/mL$ and $\omega^2 = g/L$. To convert Eq. (13) to a system of two first order equations, we let $x = \theta$ and $y = d\theta/dt$; then

$$\frac{dx}{dt} = y, \qquad \frac{dy}{dt} = -\omega^2\sin x - \gamma y. \tag{14}$$

Since γ and ω^2 are constants, the system (14) is an autonomous system of the form (1).

The critical points of Eqs. (14) are found by solving the equations

$$y = 0, \qquad -\omega^2\sin x - \gamma y = 0. \tag{15}$$

We obtain $y = 0$ and $x = \pm n\pi$, where n is an integer. These points correspond to two physical equilibrium positions, one with the mass directly below the point of support ($\theta = 0$) and the other with the mass directly above the point of support ($\theta = \pi$). Our intuition suggests that the first is stable and the second is unstable.

More precisely, if the mass is slightly displaced from the lower equilibrium position, it will oscillate back and forth with gradually decreasing amplitude, eventually approaching the equilibrium position as the initial potential energy is dissipated by the damping force. This type of motion illustrates *asymptotic stability*.

On the other hand, if the mass is slightly displaced from the upper equilibrium position, it will rapidly fall, under the influence of gravity, and will ultimately approach the lower equilibrium position in this case also. This type of motion illustrates *instability*. In practice, it is impossible to maintain the pendulum in its upward equilibrium position for very long without an external constraint mechanism, since the slightest perturbation will cause the mass to fall.

Finally, consider the ideal situation in which the damping coefficient c (or γ) is zero. In this case, if the mass is displaced slightly from its lower equilibrium position, it will oscillate indefinitely with constant amplitude about the equilibrium position. Since there is no dissipation in the system, the mass will remain near the equilibrium position but will not approach it asymptotically. This type of motion is *stable* but not asymptotically stable. In

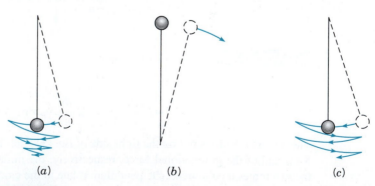

(a)   (b)   (c)

FIGURE 7.1.3 Qualitative motion of a pendulum. (*a*) With air resistance. (*b*) With or without air resistance. (*c*) Without air resistance.

general, this motion is impossible to achieve experimentally, because the slightest degree of air resistance or friction at the point of support will eventually cause the pendulum to approach its rest position.

These three types of motion are illustrated schematically in Figure 7.1.3. Solutions of the pendulum equations are discussed in more detail in the next section.

▶ **The Importance of Critical Points.** Critical points correspond to equilibrium solutions, that is, solutions in which $x(t)$ and $y(t)$ are constant. For such a solution, the system described by x and y is not changing; it remains in its initial state forever. It might seem reasonable to conclude that such points are not very interesting. However, recall that in Section 2.5 and later in Chapter 3, we found that the behavior of solutions in the *neighborhood* of a critical point has important implications for the behavior of solutions farther away. We will find that the same is true for nonlinear systems of the form (1). Consider the following examples.

EXAMPLE 1

Undamped Pendulum

An oscillating pendulum without damping is described by the equations

$$\frac{dx}{dt} = y, \qquad \frac{dy}{dt} = -\omega^2 \sin x, \tag{16}$$

obtained by setting γ equal to zero in Eq. (14). The critical points for the system (16) are $(\pm n\pi, 0)$; even values of n, including zero, correspond to the downward equilbrium position and odd values of n to the upward equilibrium position. Let $\omega = 2$ and draw a phase portrait for this system. From the phase portrait, describe the behavior of solutions near each critical point and relate this behavior to the overall motion of the pendulum.

A direction field and phase portrait for the system (16) with $\omega = 2$ are shown in Figure 7.1.4. Looking first at a fairly small rectangle centered at the origin, we see that the trajectories there are closed curves resembling ellipses. These correspond to periodic solutions, or to a periodic oscillation of the pendulum about its stable downward equilibrium position. The same behavior occurs near other critical points corresponding to even values of n.

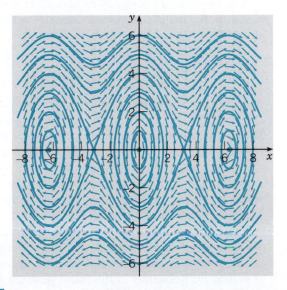

FIGURE 7.1.4 Direction field and phase portrait for the system (16) with $\omega = 2$.

The situation is different near the critical points $(\pm\pi, 0)$. In a small rectangle centered on each of these points, the trajectories display the pattern associated with saddle points. Observe that one trajectory appears to connect the points $(\pi, 0)$ and $(-\pi, 0)$, passing from the former to the latter in the lower half plane $(y < 0)$. More precisely, a solution on this trajectory approaches $(\pi, 0)$ as $t \to -\infty$ and approaches $(-\pi, 0)$ as $t \to \infty$. Similarly, there is a trajectory on which $y > 0$ that leaves $(-\pi, 0)$ as $t \to -\infty$ and approaches $(\pi, 0)$ as $t \to \infty$. These two trajectories enclose a region about the origin within which periodic motions occur. As indicated in Figure 7.1.4, a similar situation exists between each successive pair of saddle points.

Outside of these regions where periodic motions take place, the pendulum experiences a different kind of motion. If the angular velocity y is sufficiently large in magnitude, then the pendulum exhibits a whirling motion in which the angular position x steadily increases or decreases, depending on whether y is positive or negative. In the phase portrait, these are shown by the wavy curves in the upper and lower parts of the figure. The curves that "connect" pairs of saddle points are called **separatrices** because they separate the regions of periodic motions from the regions of whirling motions in the phase plane.

Finally, note that if you choose any initial point close to the origin, or close to one of the other stable equilibria, then a small oscillation will ensue. However, if you choose an initial point near one of the saddle points, there are two possibilities. If you are within the separatrices associated with that point, the motion will be a large oscillation; the pendulum will swing from near its vertical position almost all the way around, and then back and forth. However, if you start at a point just outside the separatrices, then a whirling motion will result.

EXAMPLE 2

Consider the system

$$\frac{dx}{dt} = -(2 + y)(x + y), \qquad \frac{dy}{dt} = -y(1 - x). \tag{17}$$

Find all of the critical points for this system. Then draw a direction field and phase portrait on a rectangle large enough to include all of the critical points. From your plot, classify each critical point as to type and determine whether it is asymptotically stable, stable, or unstable.

The critical points are found by solving the equations

$$-(2 + y)(x + y) = 0, \quad -y(1 - x) = 0. \tag{18}$$

One way to satisfy the first equation is to choose $y = -2$. Then the second equation becomes $2(1 - x) = 0$, so $x = 1$. The first of Eqs. (18) can also be satisfied by choosing $x = -y$. Then the second equation becomes $y(1 + y) = 0$, so either $y = 0$ or else $y = -1$. Since $x = -y$, this leads to the two additional critical points $(0, 0)$ and $(1, -1)$. Thus the system (17) has three critical points: $(1, -2)$, $(0, 0)$, and $(1, -1)$. Alternatively, you can start with the second of Eqs. (18), which can be satisfied by choosing $y = 0$ or $x = 1$. By substituting each of these values in the first of Eqs. (18), you obtain the same three critical points.

Figure 7.1.5 shows a direction field and phase portrait for the system (17). If we look at the immediate neighborhood of each of the critical points, it should be clear that $(1, -2)$ is a spiral point, $(0, 0)$ is a node, and $(1, -1)$ is a saddle point. The direction of motion on the trajectories can be inferred from the underlying direction field. Thus the direction of motion on the spirals in the fourth quadrant is clockwise. The trajectories approach $(1, -2)$, albeit quite slowly, so this point is asymptotically stable. Similarly, the direction

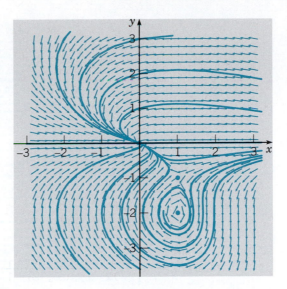

FIGURE 7.1.5   Direction field and phase portrait for the system (17).

field shows that trajectories near the node (0, 0) are approaching this point; thus it is also asymptotically stable. Finally, the saddle point at $(1, -1)$ is unstable, since this is true of all saddle points. Two trajectories enter the saddle point with slopes of approximately 0.6. All other trajectories eventually leave the neighborhood of this point.

Observe that the line $y = 0$ is a trajectory; on this line $y' = 0$ and x' is positive or negative depending on whether x is less than or greater than 0. Thus no other trajectories can cross this line.

Suppose that an autonomous system has at least one asymptotically stable critical point, as in Example 2. Then it is often of interest to determine where in the phase plane the trajectories lie that ultimately approach this critical point. If a point P in the xy-plane has the property that a trajectory that starts at P approaches the critical point as $t \to \infty$, then this trajectory is said to be attracted by the critical point. The set of all such points P is called the **basin of attraction** or the **region of asymptotic stability** of the critical point. A trajectory that bounds a basin of attraction is called a **separatrix** because it separates trajectories that approach a particular critical point from other trajectories that do not do so. Determination of basins of attraction is important in understanding the large-scale behavior of the solutions of a given autonomous system.

EXAMPLE 3

For the system (17) in Example 2, describe the basins of attraction for the node and spiral point.

In Figure 7.1.6, we look more closely at the region of the phase plane containing the saddle and spiral points. The two pairs of nearby trajectories bracket the two trajectories that enter the saddle point as $t \to \infty$. In the region between these two trajectories, every trajectory approaches the spiral point. Thus this is the basin of attraction for the spiral point. It consists of an oval region surrounding the point $(1, -2)$ together with a thin tail extending to the right just below the x-axis. The two bounding trajectories, or separatrices

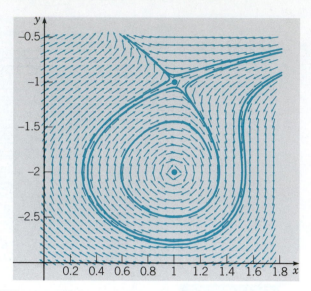

FIGURE 7.1.6 A direction field and phase portrait for the system (17) near the saddle and spiral points.

separate the trajectories that approach the spiral point from those that approach the node. It is often the case that separatrices are trajectories that enter a saddle point; consequently these trajectories may be of greater than average significance. From Figures 7.1.5 and 7.1.6, it appears that all trajectories that do not approach the spiral point or the saddle point ultimately approach the asymptotically stable node at the origin.

In the two preceding examples, the pattern of the trajectories near each critical point is essentially the same as one of the patterns that occur for homogeneous linear systems with constant coefficients. This situation is typical of a very large class of two-dimensional nonlinear autonomous systems. Further, the behavior of solutions of an autonomous system near its critical points is one of the major factors influencing the overall behavior of the solutions throughout the xy-plane. In the next section, we take up the question of finding a linear system that approximates a nonlinear system near a critical point.

PROBLEMS

ODEA For each of the systems in Problems 1 through 18:

(a) Find all the critical points (equilibrium solutions).

(b) Use a computer to draw a direction field and phase portrait for the system.

(c) From the plot(s) in part (b), determine whether each critical point is asymptotically stable, stable, or unstable, and classify it as to type.

(d) Describe the basin of attraction for each asymptotically stable critical point.

1. $dx/dt = -y + xy$, $dy/dt = x + 2xy$
2. $dx/dt = 1 + 2y$, $dy/dt = 1 - 3x^2$

3. $dx/dt = 2x - x^2 - xy$,
   $dy/dt = 3y - 2y^2 - 3xy$

4. $dx/dt = -(x - y)(1 - x - y)$, $dy/dt = -x(2+y)$

5. $dx/dt = x(2 - x - y)$, $dy/dt = -x + 3y - 2xy$

6. $dx/dt = (2 - x)(y - x)$, $dy/dt = y(2 - x - x^2)$

7. $dx/dt = (2 - y)(x - y)$, $dy/dt = (1 + x)(x + y)$

8. $dx/dt = x(2 - x - y)$, $dy/dt = (1 - y)(2 + x)$

9. $dx/dt = (2 + x)(3y - x)$,
   $dy/dt = (3 - x)(y - x)$

10. $dx/dt = x - 2y - x^2$, $\qquad dy/dt = (1 + y)(1 - x)$

11. $dx/dt = -y$, $\qquad dy/dt = -3y - x(x - 1)(x - 2)$

12. $dx/dt = (2 + x)(1 - x + y)$,
    $dy/dt = (y - 1)(1 + x + 2y)$

13. $dx/dt = x(2 - x - 3y)$,
    $dy/dt = y(3 - x)(2 + x)$

14. $dx/dt = (x - 2y + 1)(y + 1)$,
    $dy/dt = -y(3 - 2x - y)$

15. $dx/dt = (y - 1)(x - 1)(x + 2)$,
    $dy/dt = x(2 - y)$

16. $dx/dt = 9 - y^2$, $\qquad dy/dt = (1 - x)(y + x)$

17. $dx/dt = y$, $\qquad dy/dt = x - \frac{1}{6}x^3$

18. $dx/dt = -x + y + x^2$, $\qquad dy/dt = y - 2xy$

19. (a) Consider the equations of motion of an un-
    damped pendulum,

$$x' = y, \qquad y' = -\omega^2 \sin x; \qquad (i)$$

see Example 1. Convert this system into a single equation for dy/dx, and then show that the
trajectories of the undamped pendulum satisfy
the equation

$$\frac{1}{2}y^2 + \omega^2(1 - \cos x) = c, \qquad (ii)$$

where c is a constant of integration.

(b) Multiply Eq. (ii) by mL^2 and recall that
$\omega^2 = g/L$. Then, by expressing Eq. (ii) in terms
of θ, obtain

$$\frac{1}{2}mL^2 \left(\frac{d\theta}{dt}\right)^2 + mgL(1 - \cos \theta) = E,$$
$$(iii)$$

where $E = mL^2 c$.

(c) Show that the first term in Eq. (iii) is the kinetic energy of the pendulum and that the second term is the potential energy due to gravity. Thus the total energy E of the pendulum is
constant along any trajectory; in other words,
the undamped pendulum satisfies the principle of conservation of energy. The value of E
is determined by the initial conditions.

ODEA 20. The motion of a certain undamped pendulum is
described by the equations

$$dx/dt = y, \qquad dy/dt = -4 \sin x.$$

If the pendulum is set in motion with an angular
displacement A and no initial velocity, then the
initial conditions are $x(0) = A$, $y(0) = 0$.

(a) Let $A = 0.25$ and plot x versus t. From the
graph, estimate the amplitude R and period T
of the resulting motion of the pendulum.

(b) Repeat part (a) for $A = 0.5, 1.0, 1.5$, and 2.0.

(c) How do the amplitude and period of the pendulum's motion depend on the initial position A?
Draw a graph to show each of these relationships. Can you say anything about the limiting
value of the period as $A \to 0$?

(d) Let $A = 4$ and plot x versus t. Explain why this
graph differs from those in parts (a) and (b).
For what value of A does the transition take
place?

21. Consider the pendulum equations   ODEA

$$dx/dt = y, \qquad dy/dt = -6 \sin x.$$

If the pendulum is set in motion from its downward equilibrium position with angular velocity v, then the initial conditions are $x(0) = 0$,
$y(0) = v$.

(a) Plot x versus t for $v = 3$ and also for $v = 6$.
Explain the differing motions of the pendulum
that these two graphs represent.

(b) There is a critical value of v, which we denote
by v_c, such that one type of motion occurs for
$v < v_c$ and a qualitatively different type of motion occurs for $v > v_c$. Determine the value of
v_c to two decimal places.

22. In this problem, we derive a formula for the natural
period of an undamped nonlinear pendulum. The
equation of motion is

$$mL^2 \frac{d^2\theta}{dt^2} + mgL \sin \theta = 0$$

obtained by setting $c = 0$ in Eq. (12). Suppose that
the bob is pulled through a positive angle α and
then released with zero velocity.

(a) We usually think of θ and $d\theta/dt$ as functions
of t. However, if we reverse the roles of t and
θ, we can regard t as a function of θ and, consequently, we can also think of $d\theta/dt$ as a function of θ. Then derive the following sequence
of equations:

$$\frac{1}{2}mL^2 \frac{d}{d\theta}\left[\left(\frac{d\theta}{dt}\right)^2\right] = -mgL \sin \theta,$$

$$\frac{1}{2}m\left(L\frac{d\theta}{dt}\right)^2 = mgL(\cos \theta - \cos \alpha),$$

$$dt = -\sqrt{\frac{L}{2g}} \frac{d\theta}{\sqrt{\cos\theta - \cos\alpha}}.$$

Why was the negative square root chosen in the last equation?

(b) If T is the natural period of oscillation, derive the formula

$$\frac{T}{4} = -\sqrt{\frac{L}{2g}} \int_\alpha^0 \frac{d\theta}{\sqrt{\cos\theta - \cos\alpha}}.$$

(c) By using the identities $\cos\theta = 1 - 2\sin^2(\theta/2)$ and $\cos\alpha = 1 - 2\sin^2(\alpha/2)$, followed by the change of variable $\sin(\theta/2) = k\sin\phi$ with $k = \sin(\alpha/2)$, show that

$$T = 4\sqrt{\frac{L}{g}} \int_0^{\pi/2} \frac{d\phi}{\sqrt{1 - k^2\sin^2\phi}}.$$

The integral is called the **elliptic integral** of the first kind. Note that the period depends on the ratio L/g and also on the initial displacement α through $k = \sin(\alpha/2)$.

(d) By evaluating the integral in the expression for T, obtain values for T that you can compare with the graphical estimates you obtained in Problem 20.

23. Given that $x = \phi(t)$, $y = \psi(t)$ is a solution of the autonomous system

$$dx/dt = F(x, y), \qquad dy/dt = G(x, y)$$

for $\alpha < t < \beta$, show that $x = \Phi(t) = \phi(t - s)$, $y = \Psi(t) = \psi(t - s)$ is a solution for $\alpha + s < t < \beta + s$ for any real number s.

24. Prove that, for the system

$$dx/dt = F(x, y), \qquad dy/dt = G(x, y),$$

there is at most one trajectory passing through a given point (x_0, y_0).
Hint: Let C_0 be the trajectory generated by the solution $x = \phi_0(t)$, $y = \psi_0(t)$, with $\phi_0(t_0) = x_0$, $\psi_0(t_0) = y_0$, and let C_1 be the trajectory generated by the solution $x = \phi_1(t)$, $y = \psi_1(t)$, with $\phi_1(t_1) = x_0$, $\psi_1(t_1) = y_0$. Use the fact that the system is autonomous, and also the existence and uniqueness theorem, to show that C_0 and C_1 are the same.

25. Prove that if a trajectory starts at a noncritical point of the system

$$dx/dt = F(x, y), \qquad dy/dt = G(x, y),$$

then it cannot reach a critical point (x_0, y_0) in a finite length of time.
Hint: Assume the contrary, that is, assume that the solution $x = \phi(t)$, $y = \psi(t)$ satisfies $\phi(a) = x_0$, $\psi(a) = y_0$. Then use the fact that $x = x_0$, $y = y_0$ is a solution of the given system satisfying the initial condition $x = x_0$, $y = y_0$ at $t = a$.

26. Assuming that the trajectory corresponding to a solution $x = \phi(t)$, $y = \psi(t)$, $-\infty < t < \infty$, of an autonomous system is closed, show that the solution is periodic.
Hint: Since the trajectory is closed, there exists at least one point (x_0, y_0) such that $\phi(t_0) = x_0$, $\psi(t_0) = y_0$ and a number $T > 0$ such that $\phi(t_0 + T) = x_0$, $\psi(t_0 + T) = y_0$. Show that $x = \Phi(t) = \phi(t + T)$ and $y = \Psi(t) = \psi(t + T)$ is a solution, and then use the existence and uniqueness theorem to show that $\Phi(t) = \phi(t)$ and $\Psi(t) = \psi(t)$ for all t.

7.2 Almost Linear Systems

In Chapter 3 we investigated solutions of the two dimensional linear homogeneous system with constant coefficients,

$$\mathbf{x}' = \mathbf{A}\mathbf{x}. \tag{1}$$

If we assume that $\det(\mathbf{A}) \neq 0$, which is equivalent to assuming that zero is not an eigenvalue of $\mathbf{A}$, then $\mathbf{x} = \mathbf{0}$ is the only critical point (equilibrium solution) of the system (1). The stability properties of this critical point depend on the eigenvalues of $\mathbf{A}$. The results were summarized in Table 3.5.1, which is repeated here as Table 7.2.1 for convenience. We can also restate these results in the following theorem.

	Stability Properties of Linear Systems $\mathbf{x}' = \mathbf{A}\mathbf{x}$ with $\det(\mathbf{A}-\lambda\mathbf{I}) = 0$ and $\det \mathbf{A} \neq 0$.

TABLE 7.2.1

Eigenvalues	Type of Critical Point	Stability
$\lambda_1 > \lambda_2 > 0$	Node	Unstable
$\lambda_1 < \lambda_2 < 0$	Node	Asymptotically stable
$\lambda_2 < 0 < \lambda_1$	Saddle point	Unstable
$\lambda_1 = \lambda_2 > 0$	Proper or improper node	Unstable
$\lambda_1 = \lambda_2 < 0$	Proper or improper node	Asymptotically stable
$\lambda_1, \lambda_2 = \mu \pm i\nu$	Spiral point	
$\quad \mu > 0$		Unstable
$\quad \mu < 0$		Asymptotically stable
$\lambda_1 = i\nu \,, \lambda_2 = -i\nu$	Center	Stable

THEOREM 7.2.1

The critical point $\mathbf{x} = \mathbf{0}$ of the linear system (1) is asymptotically stable if the eigenvalues λ_1, λ_2 are real and negative or are complex with negative real part; stable, but not asymptotically stable, if λ_1 and λ_2 are pure imaginary; unstable if λ_1 and λ_2 are real and either is positive, or if they are complex with positive real part.

▶ **Effect of Small Perturbations.** It is apparent from this theorem or from Table 7.2.1 that the eigenvalues λ_1, λ_2 of the coefficient matrix $\mathbf{A}$ determine the type of critical point at $\mathbf{x} = \mathbf{0}$ and its stability characteristics. In turn, the values of λ_1 and λ_2 depend on the coefficients in the system (1). When such a system arises in some applied field, the values of the coefficients usually result from measurements of certain physical quantities. Such measurements are often subject to small random errors, so it is of interest to investigate whether small changes (perturbations) in the coefficients can affect the stability or instability of a critical point and/or significantly alter the pattern of trajectories.

The eigenvalues λ_1, λ_2 are the roots of the polynomial equation

$$\det(\mathbf{A} - \lambda\mathbf{I}) = 0. \tag{2}$$

It is possible to show that *small* perturbations in some or all of the coefficients are reflected in *small* perturbations in the eigenvalues. The most sensitive situation occurs when $\lambda_1 = i\nu$ and $\lambda_2 = -i\nu$, that is, when the critical point is a center and the trajectories are closed curves surrounding it. If a slight change is made in the coefficients, then the eigenvalues λ_1 and λ_2 will take on new values $\lambda_1' = \mu' + i\nu'$ and $\lambda_2' = \mu' - i\nu'$, where μ' is small in magnitude and $\nu' \cong \nu$ (see Figure 7.2.1). It is possible that $\mu' = 0$, in which case the critical point remains a center. However, in most cases $\mu' \neq 0$, and then the trajectories of the perturbed system are spirals rather than closed curves. The system is asymptotically stable if $\mu' < 0$ but unstable if $\mu' > 0$; see Problem 25.

Another slightly less sensitive case occurs if the eigenvalues λ_1 and λ_2 are equal; in this case, the critical point is a node. Small perturbations in the coefficients will normally cause the two equal roots to separate (bifurcate). If the separated roots are real, then the critical point of the perturbed system remains a node, but if the separated roots are complex conjugates, then the critical point becomes a spiral point. These two possibilities are shown schematically in Figure 7.2.2. In this case, the stability or instability of the system is not affected by small perturbations in the coefficients, but the trajectories may be altered considerably (see Problem 26).

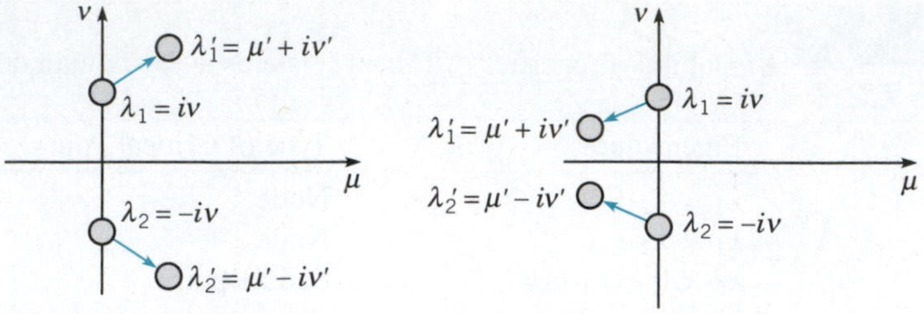

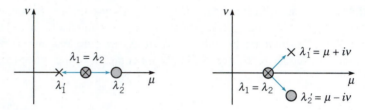

(Figure at top of page)

FIGURE 7.2.1 Schematic perturbation of $\lambda_1 = iv$, $\lambda_2 = -iv$.

In all other cases, the stability or instability of the system is not changed, nor is the type of critical point altered, by sufficiently small perturbations in the coefficients of the system. For example, if λ_1 and λ_2 are real, negative, and unequal, then a *small* change in the coefficients will neither change the sign of λ_1 and λ_2 nor allow them to coalesce. Thus the critical point remains an asymptotically stable node.

FIGURE 7.2.2 Schematic perturbation of $\lambda_1 = \lambda_2$.

▶ **Linear Approximations to Nonlinear Systems.** Now let us consider a nonlinear two-dimensional autonomous system

$$\mathbf{x}' = \mathbf{f}(\mathbf{x}). \tag{3}$$

Our main object is to investigate the behavior of trajectories of the system (3) near a critical point $\mathbf{x}_0$. We will seek to do this by approximating the nonlinear system (3) by an appropriate linear system, whose trajectories are easy to describe. The crucial question is whether the trajectories of the linear system are good approximations to those of the nonlinear system. Of course, we also need to know how to find the approximating linear system.

It is convenient to choose the critical point to be the origin. This involves no loss of generality, since if $\mathbf{x}_0 \neq \mathbf{0}$, it is always possible to make the substitution $\mathbf{u} = \mathbf{x} - \mathbf{x}_0$ in Eq. (3). Then $\mathbf{u}$ will satisfy an autonomous system with a critical point at the origin.

First, let us consider what it means for a nonlinear system (3) to be "close" to a linear system (1). Accordingly, suppose that

$$\mathbf{x}' = \mathbf{A}\mathbf{x} + \mathbf{g}(\mathbf{x}) \tag{4}$$

and that $\mathbf{x} = \mathbf{0}$ is an **isolated** critical point of the system (4). This means that there is some circle about the origin within which there are no other critical points. In addition, we assume that $\det \mathbf{A} \neq 0$, so that $\mathbf{x} = \mathbf{0}$ is also an isolated critical point of the linear system $\mathbf{x}' = \mathbf{A}\mathbf{x}$. For the nonlinear system (4) to be close to the linear system $\mathbf{x}' = \mathbf{A}\mathbf{x}$, we must assume that $\mathbf{g}(\mathbf{x})$ is small. More precisely, we assume that the components of $\mathbf{g}$ have continuous first partial derivatives and satisfy the limit condition

$$\|\mathbf{g}(\mathbf{x})\|/\|\mathbf{x}\| \to 0 \qquad \text{as} \quad \mathbf{x} \to \mathbf{0}, \tag{5}$$

that is, $\|\mathbf{g}\|$ is small in comparison to $\|\mathbf{x}\|$ itself near the origin. Such a system is called an **almost linear system** in the neighborhood of the critical point $\mathbf{x} = \mathbf{0}$.

It may be helpful to express the condition (5) in scalar form. If we let $\mathbf{x}^T = (x,y)$, then $||\mathbf{x}|| = (x^2 + y^2)^{1/2} = r$. Similarly, if $\mathbf{g}^T(\mathbf{x}) = (g_1(x,y), g_2(x,y))$, then $||\mathbf{g}(\mathbf{x})|| = [g_1^2(x, y) + g_2^2(x, y)]^{1/2}$. Then it follows that condition (5) is satisfied if and only if

$$g_1(x, y)/r \to 0, \qquad g_2(x, y)/r \to 0 \qquad \text{as} \quad r \to 0. \tag{6}$$

EXAMPLE 1

Determine whether the system

$$\begin{pmatrix} x \\ y \end{pmatrix}' = \begin{pmatrix} 1 & 0 \\ 0 & 0.5 \end{pmatrix} \begin{pmatrix} x \\ y \end{pmatrix} + \begin{pmatrix} -x^2 - xy \\ -0.75xy - 0.25y^2 \end{pmatrix} \tag{7}$$

is almost linear in the neighborhood of the origin.

Observe that the system (7) is of the form (4), that $(0, 0)$ is a critical point, and that $\det \mathbf{A} \neq 0$. It is not hard to show that the other critical points of Eqs. (7) are $(0, 2)$, $(1, 0)$, and $(0.5, 0.5)$; consequently, the origin is an isolated critical point. In checking the condition (6), it is convenient to introduce polar coordinates by letting $x = r \cos \theta$, $y = r \sin \theta$. Then

$$\frac{g_1(x, y)}{r} = \frac{-x^2 - xy}{r} = \frac{-r^2 \cos^2 \theta - r^2 \sin \theta \cos \theta}{r}$$

$$= -r(\cos^2 \theta + \sin \theta \cos \theta) \to 0$$

as $r \to 0$. In a similar way you can show that $g_2(x, y)/r \to 0$ as $r \to 0$. Hence the system (7) is almost linear near the origin.

EXAMPLE 2

The motion of a pendulum is described by the system (see Eq. (14) of Section 7.1)

$$\frac{dx}{dt} = y, \qquad \frac{dy}{dt} = -\omega^2 \sin x - \gamma y. \tag{8}$$

The critical points are $(0, 0)$, $(\pm \pi, 0)$, $(\pm 2\pi, 0)$, ..., so the origin is an isolated critical point of this system. Show that the system is almost linear near the origin.

To compare Eqs. (8) with Eq. (4), we must rewrite the former so that the linear and nonlinear terms are clearly identified. If we write $\sin x = x + (\sin x - x)$ and substitute this expression in the second of Eqs. (8), we obtain the equivalent system

$$\begin{pmatrix} x \\ y \end{pmatrix}' = \begin{pmatrix} 0 & 1 \\ -\omega^2 & -\gamma \end{pmatrix} \begin{pmatrix} x \\ y \end{pmatrix} - \omega^2 \begin{pmatrix} 0 \\ \sin x - x \end{pmatrix}. \tag{9}$$

Comparing Eqs. (9) and (4), we see that $g_1(x, y) = 0$ and $g_2(x, y) = -\omega^2 (\sin x - x)$. The Taylor series for $\sin x$ implies that $\sin x - x$ behaves like $-x^3/3! = -(r^3 \cos^3 \theta)/3!$ when x is small. Consequently, $(\sin x - x)/r \to 0$ as $r \to 0$. Thus the conditions (6) are satisfied and the system (9) is almost linear near the origin.

Let us now return to the general nonlinear system (3), which we write in the scalar form

$$x' = F(x, y), \qquad y' = G(x, y). \tag{10}$$

We assume that (x_0, y_0) is an isolated critical point of this system. The system (10) is almost linear in the neighborhood of (x_0, y_0) whenever the functions F and G have continuous partial derivatives up to order 2. To show this, we use Taylor expansions about the point (x_0, y_0) to write $F(x, y)$ and $G(x, y)$ in the form

$$F(x, y) = F(x_0, y_0) + F_x(x_0, y_0)(x - x_0) + F_y(x_0, y_0)(y - y_0) + \eta_1(x, y),$$

$$G(x, y) = G(x_0, y_0) + G_x(x_0, y_0)(x - x_0) + G_y(x_0, y_0)(y - y_0) + \eta_2(x, y),$$

where $\eta_1(x, y)/[(x - x_0)^2 + (y - y_0)^2]^{1/2} \to 0$ as $(x, y) \to (x_0, y_0)$, and similarly for η_2. Note that $F(x_0, y_0) = G(x_0, y_0) = 0$; also $dx/dt = d(x - x_0)/dt$ and $dy/dt = d(y - y_0)/dt$. Then the system (10) reduces to

$$\frac{d}{dt} \begin{pmatrix} x - x_0 \\ y - y_0 \end{pmatrix} = \begin{pmatrix} F_x(x_0, y_0) & F_y(x_0, y_0) \\ G_x(x_0, y_0) & G_y(x_0, y_0) \end{pmatrix} \begin{pmatrix} x - x_0 \\ y - y_0 \end{pmatrix} + \begin{pmatrix} \eta_1(x, y) \\ \eta_2(x, y) \end{pmatrix}, \tag{11}$$

or, in vector notation,

$$\frac{d\mathbf{u}}{dt} = \frac{d\mathbf{f}}{d\mathbf{x}}(\mathbf{x}_0)\mathbf{u} + \boldsymbol{\eta}(\mathbf{x}), \tag{12}$$

where $\mathbf{u} = (x - x_0, y - y_0)^T$ and $\boldsymbol{\eta} = (\eta_1, \eta_2)^T$.

The significance of this result is twofold. First, if the functions F and G are twice differentiable, then the system (10) is almost linear, and it is unnecessary to resort to the limiting process used in Examples 1 and 2. Second, the linear system that corresponds to the nonlinear system (10) near (x_0, y_0) is given by the linear part of Eqs. (11) or (12):

$$\frac{d}{dt} \begin{pmatrix} u_1 \\ u_2 \end{pmatrix} = \begin{pmatrix} F_x(x_0, y_0) & F_y(x_0, y_0) \\ G_x(x_0, y_0) & G_y(x_0, y_0) \end{pmatrix} \begin{pmatrix} u_1 \\ u_2 \end{pmatrix}, \tag{13}$$

where $u_1 = x - x_0$ and $u_2 = y - y_0$. Equation (13) provides a simple and general method for finding the linear system corresponding to an almost linear system near a given critical point.

The matrix
$$\mathbf{J} = \begin{pmatrix} F_x & F_y \\ G_x & G_y \end{pmatrix}, \tag{14}$$

which appears as the coefficient matrix in Eq. (13), is called the **Jacobian matrix** of the functions F and G with respect to the variables x and y. We need to assume that $\det(\mathbf{J})$ is not zero at (x_0, y_0) so that this point is also an isolated critical point of the linear system (13).

EXAMPLE 3

Use Eq. (13) to find the linear system corresponding to the pendulum equations (8) near the origin; near the critical point $(\pi, 0)$.

In this case, we have, from Eq. (8),

$$F(x, y) = y, \qquad G(x, y) = -\omega^2 \sin x - \gamma y. \tag{15}$$

Since these functions are differentiable as many times as necessary, the system (8) is almost linear near each critical point. The Jacobian matrix for the system (8) is

$$\begin{pmatrix} F_x & F_y \\ G_x & G_y \end{pmatrix} = \begin{pmatrix} 0 & 1 \\ -\omega^2 \cos x & -\gamma \end{pmatrix}. \tag{16}$$

Thus, at the origin, the corresponding linear system is

$$\frac{d}{dt}\begin{pmatrix} x \\ y \end{pmatrix} = \begin{pmatrix} 0 & 1 \\ -\omega^2 & -\gamma \end{pmatrix}\begin{pmatrix} x \\ y \end{pmatrix}, \tag{17}$$

which agrees with Eq. (9).

Similarly, evaluating the Jacobian matrix at $(\pi, 0)$, we obtain

$$\frac{d}{dt}\begin{pmatrix} u \\ w \end{pmatrix} = \begin{pmatrix} 0 & 1 \\ \omega^2 & -\gamma \end{pmatrix}\begin{pmatrix} u \\ w \end{pmatrix}, \tag{18}$$

where $u = x - \pi$, $w = y$. This is the linear system corresponding to Eqs. (8) near the point $(\pi, 0)$.

We now return to the almost linear system (4). Since the nonlinear term $\mathbf{g}(\mathbf{x})$ is small compared to the linear term $\mathbf{Ax}$ when $\mathbf{x}$ is small, it is reasonable to hope that the trajectories of the linear system (1) are good approximations to those of the nonlinear system (4), at least near the origin. This turns out to be true in many (but not all) cases, as the following theorem states.

THEOREM 7.2.2

Let λ_1 and λ_2 be the eigenvalues of the linear system (1),

$$\mathbf{x}' = \mathbf{Ax},$$

corresponding to the almost linear system (4),

$$\mathbf{x}' = \mathbf{Ax} + \mathbf{g}(\mathbf{x}).$$

Assume that $\mathbf{x} = \mathbf{0}$ is an isolated critical point of both of these systems. Then the type and stability of $\mathbf{x} = \mathbf{0}$ for the linear system (1) and for the almost linear system (4) are as shown in Table 7.2.2.

The proof of Theorem 7.2.2 is beyond the scope of this book, so we will use the results without proof. Essentially, Theorem 7.2.2 says that, for small $\mathbf{x}$ (or $\mathbf{x} - \mathbf{x}_0$), the nonlinear terms are also small and do not affect the stability and type of critical point as determined by the linear terms except in two sensitive cases: λ_1 and λ_2 pure imaginary, and λ_1 and λ_2 real and equal. Recall that, earlier in this section, we stated that small perturbations in the coefficients of the linear system (1), and hence in the eigenvalues λ_1 and λ_2, can alter the type and stability of the critical point only in these two sensitive cases. It is reasonable to expect that the small nonlinear term in Eq. (4) might have a similar effect, at least in these two sensitive cases. This is so, but the main significance of Theorem 7.2.2 is that, in *all other cases*, the small nonlinear term does not alter the type or stability of the critical point. Thus, except in the two sensitive cases, the type and stability of the critical point of the nonlinear system (4) can be determined from a study of the much simpler linear system (1).

In a small neighborhood of the critical point, the trajectories of the nonlinear system are also similar to those of the approximating linear system. In particular, the slopes at which trajectories "enter" or "leave" the critical point are given correctly by the linear equations. Farther away, the nonlinear terms become dominant and the trajectories of the nonlinear system usually bear no resemblance to those of the linear system.

TABLE 7.2.2   Stability and Instability Properties of Linear and Almost Linear Systems.

	Linear System		Almost Linear System	
λ_1, λ_2	Type	Stability	Type	Stability
$\lambda_1 > \lambda_2 > 0$	N	Unstable	N	Unstable
$\lambda_1 < \lambda_2 < 0$	N	Asymptotically stable	N	Asymptotically stable
$\lambda_2 < 0 < \lambda_1$	SP	Unstable	SP	Unstable
$\lambda_1 = \lambda_2 > 0$	PN or IN	Unstable	N or SpP	Unstable
$\lambda_1 = \lambda_2 < 0$	PN or IN	Asymptotically stable	N or SpP	Asymptotically stable
$\lambda_1, \lambda_2 = \mu \pm i\nu$				
$\mu > 0$	SpP	Unstable	SpP	Unstable
$\mu < 0$	SpP	Asymptotically stable	SpP	Asymptotically stable
$\lambda_1 = i\nu, \lambda_2 = -i\nu$	C	Stable	C or SpP	Indeterminate

Note: N, node; IN, improper node; PN, proper node; SP, saddle point; SpP, spiral point; C, center.

▶ **Damped Pendulum.** We continue the discussion of the damped pendulum begun in Examples 2 and 3. Near the origin the nonlinear equations (8) are approximated by the linear system (17). The characteristic equation for this system is

$$\lambda^2 + \gamma\lambda + \omega^2 = 0, \tag{19}$$

so the eigenvalues are

$$\lambda_1, \lambda_2 = \frac{-\gamma \pm \sqrt{\gamma^2 - 4\omega^2}}{2}. \tag{20}$$

The nature of the solutions of Eqs. (8) and (17) depends on the sign of $\gamma^2 - 4\omega^2$ as follows:

1. If $\gamma^2 - 4\omega^2 > 0$, then the eigenvalues are real, unequal, and negative. The critical point $(0, 0)$ is an asymptotically stable node of the linear system (17) and of the almost linear system (8).

2. If $\gamma^2 - 4\omega^2 = 0$, then the eigenvalues are real, equal, and negative. The critical point $(0, 0)$ is an asymptotically stable (proper or improper) node of the linear system (17). It may be either an asymptotically stable node or spiral point of the almost linear system (8).

3. If $\gamma^2 - 4\omega^2 < 0$, then the eigenvalues are complex with negative real part. The critical point $(0, 0)$ is an asymptotically stable spiral point of the linear system (17) and of the almost linear system (8).

Thus the critical point $(0, 0)$ is a spiral point of the system (8) if the damping is small and a node if the damping is large enough. In either case, the origin is asymptotically stable.

Let us now consider the case $\gamma^2 - 4\omega^2 < 0$, corresponding to small damping, in more detail. The direction of motion on the spirals near $(0, 0)$ can be obtained directly from Eqs. (8). Consider the point at which a spiral intersects the positive y-axis ($x = 0$ and $y > 0$). At such a point it follows from Eqs. (8) that $dx/dt > 0$. Thus the point (x, y) on the trajectory is moving to the right, so the direction of motion on the spirals is clockwise.

The behavior of the pendulum near the critical points $(\pm\, n\pi, 0)$, with n even, is the same as its behavior near the origin. We expect this on physical grounds since all these critical points correspond to the downward equilibrium position of the pendulum. The conclusion can be confirmed by repeating the analysis carried out above for the origin. Figure 7.2.3 shows the clockwise spirals at a few of these critical points.

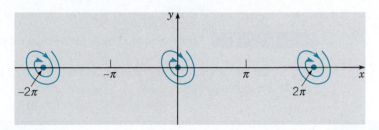

FIGURE 7.2.3 Asymptotically stable spiral points for the damped pendulum.

Now let us consider the critical point $(\pi, 0)$. Here the nonlinear equations (8) are approximated by the linear system (18), whose eigenvalues and eigenvectors are determined from

$$\begin{pmatrix} -\lambda & 1 \\ \omega^2 & -\gamma - \lambda \end{pmatrix} \begin{pmatrix} v_1 \\ v_2 \end{pmatrix} = \begin{pmatrix} 0 \\ 0 \end{pmatrix}. \tag{21}$$

The characteristic equation of this system is

$$\lambda^2 + \gamma\lambda - \omega^2 = 0, \tag{22}$$

so the eigenvalues are

$$\lambda_1, \lambda_2 = \frac{-\gamma \pm \sqrt{\gamma^2 + 4\omega^2}}{2}. \tag{23}$$

One eigenvalue (λ_1) is positive and the other (λ_2) is negative. Therefore, regardless of the amount of damping, the critical point $x = \pi, y = 0$ is an (unstable) saddle point both of the linear system (18) and of the almost linear system (8).

To examine the behavior of trajectories near the saddle point $(\pi, 0)$ in more detail, we also need the eigenvectors. From the first row of Eq. (21) we have

$$-\lambda v_1 + v_2 = 0.$$

Therefore the eigenvectors are $(1, \lambda_1)$ and $(1, \lambda_2)$ corresponding to the eigenvalues λ_1 and λ_2 respectively. Consequently, the general solution of Eqs. (18) is

$$\begin{pmatrix} u \\ w \end{pmatrix} = C_1 e^{\lambda_1 t} \begin{pmatrix} 1 \\ \lambda_1 \end{pmatrix} + C_2 e^{\lambda_2 t} \begin{pmatrix} 1 \\ \lambda_2 \end{pmatrix}, \tag{24}$$

where C_1 and C_2 are arbitrary constants. Since $\lambda_1 > 0$ and $\lambda_2 < 0$, the first solution becomes unbounded and the second tends to zero as $t \to \infty$. Hence the trajectories that "enter" the saddle point are obtained by setting $C_1 = 0$. As they approach the saddle point, the entering trajectories are tangent to the line having slope $\lambda_2 < 0$. Thus one lies in the second quadrant ($C_2 < 0$), and the other lies in the fourth quadrant ($C_2 > 0$). For $C_2 = 0$, we obtain the pair of trajectories "exiting" from the saddle point. These trajectories have slope $\lambda_1 > 0$; one lies in the first quadrant ($C_1 > 0$), and the other lies in the third quadrant ($C_1 < 0$).

The situation is the same at other critical points $(n\pi, 0)$ with n odd. These all correspond to the upward equilibrium position of the pendulum, so we expect them to be unstable. The analysis at $(\pi, 0)$ can be repeated to show that they are saddle points oriented in the same

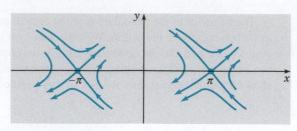

FIGURE 7.2.4 Unstable saddle points for the damped pendulum.

way as the one at $(\pi, 0)$. Diagrams of the trajectories in the neighborhood of two saddle points are shown in Figure 7.2.4.

EXAMPLE 4

The equations of motion of a certain pendulum are

$$dx/dt = y, \qquad dy/dt = -9\sin x - 0.2y, \qquad (25)$$

where $x = \theta$ and $y = d\theta/dt$. Draw a phase portrait for this system and explain how it shows the possible motions of the pendulum.

By plotting the trajectories starting at various initial points in the phase plane, we obtain the phase portrait shown in Figure 7.2.5. As we have seen, the critical points (equilibrium solutions) are the points $(n\pi, 0)$, where $n = 0, \pm 1, \pm 2, \dots$. Even values of n, including zero, correspond to the downward position of the pendulum, while odd values of n correspond to the upward position. Near each of the asymptotically stable critical points, the trajectories are clockwise spirals that represent a decaying oscillation about the downward equilibrium position. The wavy horizontal portions of the trajectories that occur for larger values of $|y|$ represent whirling motions of the pendulum. Note that a whirling motion cannot continue indefinitely, no matter how large $|y|$ is. Eventually the angular velocity is so much reduced by the damping term that the pendulum can no longer go over the top, and instead begins to oscillate about its downward position.

Each of the asymptotically stable critical points has its own basin of attraction, that is, those points from which trajectories ultimately approach the given critical point. The basin

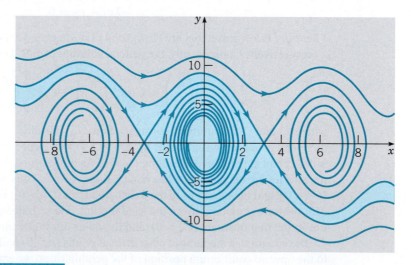

FIGURE 7.2.5 Phase portrait for the damped pendulum of Example 4.

of attraction for the origin is shown in color in Figure 7.2.5 and consists of a roughly oval region surrounding the origin together with two thin tails that extend infinitely far to either side. It is bounded by the trajectories (separatrices) that enter the saddle points at $(-\pi, 0)$ and $(\pi, 0)$. The basins of attraction for the other asymptotically stable critical points have the same shape as for the origin; they are simply shifted right or left by multiples of 2π. Note that it is mathematically possible (but physically unrealizable) to choose initial conditions on a separatrix so that the resulting motion leads to a balanced pendulum in a vertically upward position of unstable equilibrium.

An important difference between linear and nonlinear autonomous systems is illustrated by the pendulum equations. Recall that the linear system (1) has only the single critical point $\mathbf{x} = \mathbf{0}$ if $\det \mathbf{A} \neq 0$. Thus, if the origin is asymptotically stable, then not only do trajectories that start close to the origin approach it, but, in fact, every trajectory approaches the origin. In this case, the critical point $\mathbf{x} = \mathbf{0}$ is said to be **globally asymptotically stable**. This property of linear systems is not, in general, true for nonlinear systems. For nonlinear systems, it is important to determine (or to estimate) the basin of attraction for each asymptotically stable critical point.

PROBLEMS

ODEA In each of Problems 1 through 20:

(a) Determine all critical points of the given system of equations.

(b) Find the corresponding linear system near each critical point.

(c) Find the eigenvalues of each linear system. What conclusions can you then draw about the nonlinear system?

(d) Draw a phase portrait of the nonlinear system to confirm your conclusions, or to extend them in those cases where the linear system does not provide definite information about the nonlinear system.

(e) Draw a sketch of, or describe in words, the basin of attraction of each asymptotically stable critical point.

1. $dx/dt = -2x + y, \qquad dy/dt = x^2 - y$

2. $dx/dt = x - y, \qquad dy/dt = x - 2y + xy - 2$

3. $dx/dt = x + y^2, \qquad dy/dt = x + 2y$

4. $dx/dt = x - y^2, \qquad dy/dt = x - 2y + x^2$

5. $dx/dt = (2 + x)(y - x), \qquad dy/dt = (4 - x)(y + x)$

6. $dx/dt = x - x^2 - xy, \qquad dy/dt = 3y - xy - 2y^2$

7. $dx/dt = 1 - y, \qquad dy/dt = x^2 - y^2$

8. $dx/dt = x - x^2 - 2xy, \qquad dy/dt = -y(x + 1)$

9. $dx/dt = -(x - y)(1 - x - y),$
$dy/dt = x(2 + y)$

10. $dx/dt = x + x^2 + y^2, \qquad dy/dt = y - xy$

11. $dx/dt = 2x + y + xy^3, \qquad dy/dt = x - 2y - xy$

12. $dx/dt = (1 + x)\sin y, \qquad dy/dt = 1 - x - \cos y$

13. $dx/dt = x - y^2, \qquad dy/dt = y - x^2$

14. $dx/dt = 1 - xy, \qquad dy/dt = x - y^3$

15. $dx/dt = -2x - y - x(x^2 + y^2),$
$dy/dt = x - y + y(x^2 + y^2)$

16. $dx/dt = y + x(1 - x^2 - y^2),$
$dy/dt = -x + y(1 - x^2 - y^2)$

17. $dx/dt = (2 + y)(y - 0.5x),$
$dy/dt = (2 - x)(y + 0.5x)$

18. $dx/dt = 4 - y^2, \qquad dy/dt = (1.5 + x)(y - x)$

19. $dx/dt = (1 - y)(2x - y),$
$dy/dt = (2 + x)(x - 2y)$

20. $dx/dt = 2x^2y - 3x^2 - 4y,$
$dy/dt = -2xy^2 + 6xy$

21. Consider the autonomous system

$$dx/dt = y, \qquad dy/dt = x + 2x^3.$$

(a) Show that the origin is the only critical point.

(b) Find the approximating linear system near the origin. Write down the corresponding equation for dy/dx and solve it. Sketch the trajectories of the linear system and observe that the origin is a saddle point. From the parametric form of the solution show that the trajectories that enter and leave the origin lie on the lines $y = -x$ and $y = x$, respectively.

(c) Now consider the original nonlinear equation. Write down the corresponding equation for dy/dx and solve it. Sketch the trajectories for the nonlinear system that correspond to $y = -x$ and $y = x$ for the linear system.

22. Consider the autonomous system

$$dx/dt = x, \qquad dy/dt = -2y + x^3.$$

(a) Show that the origin is the only critical point.

(b) Find the approximating linear system near the origin. Write down the corresponding equation for dy/dx and solve it. Sketch the trajectories of the linear system and observe that the origin is a saddle point. From the parametric form of the solution, show that the trajectories that enter and leave the origin lie on the y-axis and the x-axis, respectively.

(c) Now consider the original nonlinear equation. Write down the corresponding equation for dy/dx and solve it. Sketch the trajectories for the nonlinear system that correspond to $x = 0$ and $y = 0$ for the linear system.

ODEA 23. The equations of motion of a certain nonlinear damped pendulum are

$$dx/dt = y, \qquad dy/dt = -6\sin x - \gamma y,$$

where γ is the damping coefficient. Suppose that the initial conditions are $x(0) = 0$, $y(0) = v$.

(a) For $\gamma = 1/4$, plot x versus t for $v = 3$ and for $v = 6$. Explain these plots in terms of the motions of the pendulum that they represent. Also explain how they relate to the corresponding graphs in Problem 21 of Section 7.1.

(b) Let v_c be the critical value of the initial velocity where the transition from one type of motion to the other occurs. Determine v_c to two decimal places.

(c) Repeat part (b) for other values of γ and determine how v_c depends on γ.

24. Theorem 7.2.2 provides no information about the stability of a critical point of an almost linear system if that point is a center of the corresponding linear system. The systems

$$dx/dt = y + x(x^2 + y^2),$$
$$dy/dt = -x + y(x^2 + y^2) \tag{i}$$

and

$$dx/dt = y - x(x^2 + y^2),$$
$$dy/dt = -x - y(x^2 + y^2) \tag{ii}$$

show that this must be so.

(a) Show that $(0, 0)$ is a critical point of each system and, furthermore, is a center of the corresponding linear system.

(b) Show that each system is almost linear.

(c) Let $r^2 = x^2 + y^2$, and note that $x\,dx/dt + y\,dy/dt = r\,dr/dt$. For system (ii), show that $dr/dt < 0$ and that $r \to 0$ as $t \to \infty$; hence the critical point is asymptotically stable. For system (i), show that the solution of the initial value problem for r with $r = r_0$ at $t = 0$ becomes unbounded as $t \to 1/2r_0^2$; hence the critical point is unstable.

25. In this problem, we show how small changes in the coefficients of a system of linear equations can affect a critical point that is a center. Consider the system

$$\mathbf{x}' = \begin{pmatrix} 0 & 1 \\ -1 & 0 \end{pmatrix} \mathbf{x}.$$

Show that the eigenvalues are $\pm i$ so that $(0, 0)$ is a center. Now consider the system

$$\mathbf{x}' = \begin{pmatrix} \epsilon & 1 \\ -1 & \epsilon \end{pmatrix} \mathbf{x},$$

where $|\epsilon|$ is arbitrarily small. Show that the eigenvalues are $\epsilon \pm i$. Thus no matter how small $|\epsilon| \neq 0$ is, the center becomes a spiral point. If $\epsilon < 0$, the spiral point is asymptotically stable; if $\epsilon > 0$, the spiral point is unstable.

26. In this problem, we show how small changes in the coefficients of a system of linear equations can affect the nature of a critical point when the eigenvalues are equal. Consider the system

$$\mathbf{x}' = \begin{pmatrix} -1 & 1 \\ 0 & -1 \end{pmatrix} \mathbf{x}.$$

Show that the eigenvalues are $\lambda_1 = -1$, $\lambda_2 = -1$ so that the critical point $(0, 0)$ is an asymptotically stable node. Now consider the system

$$x' = \begin{pmatrix} -1 & 1 \\ -\epsilon & -1 \end{pmatrix} x,$$

where $|\epsilon|$ is arbitrarily small. Show that if $\epsilon > 0$, then the eigenvalues are $-1 \pm i\sqrt{\epsilon}$, so that the asymptotically stable node becomes an asymptotically stable spiral point. If $\epsilon < 0$, then the eigenvalues are $-1 \pm \sqrt{|\epsilon|}$, and the critical point remains an asymptotically stable node.

27. A generalization of the damped pendulum equation discussed in the text, or a damped spring–mass system, is the Liénard equation

$$\frac{d^2x}{dt^2} + c(x)\frac{dx}{dt} + g(x) = 0.$$

If $c(x)$ is a constant and $g(x) = kx$, then this equation has the form of the linear pendulum equation (replace $\sin\theta$ with θ in Eq. (13) of Section 7.1); otherwise, the damping force $c(x)\,dx/dt$ and the restoring force $g(x)$ are nonlinear. Assume that c is continuously differentiable, g is twice continuously differentiable, and $g(0) = 0$.

(a) Write the Liénard equation as a system of two first order equations by introducing the variable $y = dx/dt$.

(b) Show that $(0, 0)$ is a critical point and that the system is almost linear in the neighborhood of $(0, 0)$.

(c) Show that if $c(0) > 0$ and $g'(0) > 0$, then the critical point is asymptotically stable, and that if $c(0) < 0$ or $g'(0) < 0$, then the critical point is unstable.

Hint: Use Taylor series to approximate c and g in the neighborhood of $x = 0$.

7.3 Competing Species

In this section, we use phase plane methods to investigate some problems involving competition for scarce resources. We will express the equations in terms of two species competing for the same food supply. However, the same or similar models have also been used to study other competitive situations, for example, businesses competing in the same economic markets.

The problems that we consider here involve two interacting populations and are extensions of those discussed in Section 2.5, which dealt with a single population. Although the models described here are extremely simple compared to the very complex relationships that exist in nature, it is still possible to acquire some insight into ecological principles from a study of these idealized problems.

Suppose that in some closed environment there are two similar species competing for a limited food supply—for example, two species of fish in a pond that do not prey on each other but do compete for the available food. Let x and y be the populations of the two species at time t. As discussed in Section 2.5, we assume that the population of each of the species, in the absence of the other, is governed by a logistic equation. Thus

$$dx/dt = x(\epsilon_1 - \sigma_1 x), \tag{1a}$$

$$dy/dt = y(\epsilon_2 - \sigma_2 y), \tag{1b}$$

respectively, where ϵ_1 and ϵ_2 are the growth rates of the two populations, and ϵ_1/σ_1 and ϵ_2/σ_2 are their saturation levels. However, when both species are present, each will impinge on the available food supply for the other. In effect, they reduce each other's growth rates and saturation populations. The simplest way to reduce the growth rate of species x due to the presence of species y is to replace the growth rate factor $\epsilon_1 - \sigma_1 x$ in Eq. (1a) by $\epsilon_1 - \sigma_1 x - \alpha_1 y$, where α_1 is a measure of the degree to which species y interferes with species x. Similarly, in Eq. (1b) we replace $\epsilon_2 - \sigma_2 y$ by $\epsilon_2 - \sigma_2 y - \alpha_2 x$. Thus we have the system of equations

$$dx/dt = x(\epsilon_1 - \sigma_1 x - \alpha_1 y),$$
$$dy/dt = y(\epsilon_2 - \sigma_2 y - \alpha_2 x). \tag{2}$$

The values of the positive constants ϵ_1, σ_1, α_1, ϵ_2, σ_2, and α_2 depend on the particular species under consideration and, in general, must be determined from observations. We are interested in solutions of Eqs. (2) for which x and y are nonnegative. In the following two examples, we discuss two typical problems in some detail. At the end of the section, we return to the general equations (2).

EXAMPLE 1

Discuss the qualitative behavior of solutions of the system

$$dx/dt = x(1 - x - y),$$
$$dy/dt = y(0.75 - y - 0.5x). \tag{3}$$

We find the critical points by solving the system of algebraic equations

$$x(1 - x - y) = 0, \qquad y(0.75 - y - 0.5x) = 0. \tag{4}$$

The first equation can be satisfied by choosing $x = 0$; then the second equation requires that $y = 0$ or $y = 0.75$. Similarly, the second equation can be satisfied by choosing $y = 0$, and then the first equation requires that $x = 0$ or $x = 1$. Thus we have found three critical points, namely $(0, 0)$, $(0, 0.75)$, and $(1, 0)$. Equations (4) are also satisfied by solutions of the system

$$1 - x - y = 0, \qquad 0.75 - y - 0.5x = 0,$$

which leads to a fourth critical point $(0.5, 0.5)$. The four critical points correspond to equilibrium solutions of the system (3). The first three of these points involve the extinction of one or both species; only the last corresponds to the long-term survival of both species. Other solutions are represented as curves or trajectories in the xy-plane that describe the evolution of the populations in time. To begin to discover their qualitative behavior we can proceed in the following way.

First, observe that the coordinate axes are themselves trajectories. This follows directly from Eqs. (3) since $dx/dt = 0$ on the y-axis (where $x = 0$) and, similarly, $dy/dt = 0$ on the x-axis. Thus no other trajectories can cross the coordinate axes. For a population problem, only nonnegative values of x and y are significant, and we conclude that any trajectory that starts in the first quadrant must remain there for all time.

A direction field for the system (3) in the positive quadrant is shown in Figure 7.3.1. The heavy dots in this figure are the critical points or equilibrium solutions. Based on the direction field, it appears that the point $(0.5, 0.5)$ attracts other solutions and is therefore asymptotically stable, while the other three critical points are unstable. To confirm these conclusions, we can look at the linear approximations near each critical point.

The system (3) is almost linear in the neighborhood of each critical point. There are two ways to obtain the linear system near a critical point (X, Y). First, we can use the substitution $x = X + u$, $y = Y + w$ in Eqs. (3), retaining only the terms that are linear in u and w. Alternatively, we can evaluate the Jacobian matrix

$$\mathbf{J} = \begin{pmatrix} F_x(X, Y) & F_y(X, Y) \\ G_x(X, Y) & G_y(X, Y) \end{pmatrix} \tag{5}$$

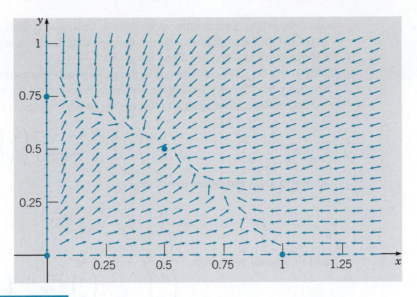

FIGURE 7.3.1   Critical points and direction field for the system (3).

to find the coefficient matrix of the approximating linear system. When several critical points are to be investigated, it is usually more efficient to use the Jacobian matrix. For the system (3),

$$F(x, y) = x(1 - x - y), \qquad G(x, y) = y(0.75 - y - 0.5x), \tag{6}$$

so by evaluating **J** we obtain the linear system

$$\frac{d}{dt}\begin{pmatrix} u \\ w \end{pmatrix} = \begin{pmatrix} 1 - 2X - Y & -X \\ -0.5Y & 0.75 - 2Y - 0.5X \end{pmatrix}\begin{pmatrix} u \\ w \end{pmatrix}, \tag{7}$$

where $u = x - X$ and $w = y - Y$. The system (7) is the approximate linear system near the critical point (X, Y). We will now examine each of the four critical points in turn.

x = 0, y = 0. This critical point corresponds to a state in which both species die as a result of their competition. By rewriting the system (3) in the form

$$\frac{d}{dt}\begin{pmatrix} x \\ y \end{pmatrix} = \begin{pmatrix} 1 & 0 \\ 0 & 0.75 \end{pmatrix}\begin{pmatrix} x \\ y \end{pmatrix} - \begin{pmatrix} x^2 + xy \\ 0.5xy + y^2 \end{pmatrix}, \tag{8}$$

or by setting $X = Y = 0$ in Eq. (7), we see that near the origin the corresponding linear system is

$$\frac{d}{dt}\begin{pmatrix} x \\ y \end{pmatrix} = \begin{pmatrix} 1 & 0 \\ 0 & 0.75 \end{pmatrix}\begin{pmatrix} x \\ y \end{pmatrix}. \tag{9}$$

The eigenvalues and eigenvectors of the system (9) are

$$\lambda_1 = 1, \qquad \mathbf{v}_1 = \begin{pmatrix} 1 \\ 0 \end{pmatrix}; \qquad \lambda_2 = 0.75, \qquad \mathbf{v}_2 = \begin{pmatrix} 0 \\ 1 \end{pmatrix}, \tag{10}$$

so the general solution of the system is

$$\begin{pmatrix} x \\ y \end{pmatrix} = c_1 e^t \begin{pmatrix} 1 \\ 0 \end{pmatrix} + c_2 e^{0.75t} \begin{pmatrix} 0 \\ 1 \end{pmatrix}. \tag{11}$$

Thus the origin is an unstable node of both the linear system (9) and the nonlinear system (8) or (3). In the neighborhood of the origin, all trajectories are tangent to the y-axis except for one trajectory that lies along the x-axis.

$x = 1, y = 0$. This corresponds to a state in which species x survives the competition but species y does not. By evaluating the Jacobian matrix in the system (7) for $X = 1, Y = 0$, we find that the linear system coresponding to the system (3) near the critical point $(1, 0)$ is

$$\frac{d}{dt}\begin{pmatrix} u \\ w \end{pmatrix} = \begin{pmatrix} -1 & -1 \\ 0 & 0.25 \end{pmatrix} \begin{pmatrix} u \\ w \end{pmatrix}. \tag{12}$$

Its eigenvalues and eigenvectors are

$$\lambda_1 = -1, \quad \mathbf{v}_1 = \begin{pmatrix} 1 \\ 0 \end{pmatrix}; \quad \lambda_2 = 0.25, \quad \mathbf{v}_2 = \begin{pmatrix} 4 \\ -5 \end{pmatrix}, \tag{13}$$

and its general solution is

$$\begin{pmatrix} u \\ w \end{pmatrix} = c_1 e^{-t}\begin{pmatrix} 1 \\ 0 \end{pmatrix} + c_2 e^{0.25t}\begin{pmatrix} 4 \\ -5 \end{pmatrix}. \tag{14}$$

Since the eigenvalues have opposite signs, the point $(1, 0)$ is a saddle point, and hence is an unstable equilibrium point of the linear system (12) and of the nonlinear system (3). The behavior of the trajectories near $(1, 0)$ can be seen from Eq. (14). If $c_2 = 0$, then there is one pair of trajectories that approaches the critical point along the x-axis. If $c_1 = 0$, then another pair of trajectories leaves the critical point (more precisely, approaches it as $t \to -\infty$) tangent to the line of slope $-5/4$. All other trajectories also depart from the neighborhood of $(1, 0)$.

$x = 0, y = 0.75$. In this case, species y survives but x does not. The analysis is similar to that for the point $(1, 0)$. The corresponding linear system is

$$\frac{d}{dt}\begin{pmatrix} u \\ w \end{pmatrix} = \begin{pmatrix} 0.25 & 0 \\ -0.375 & -0.75 \end{pmatrix} \begin{pmatrix} u \\ w \end{pmatrix}. \tag{15}$$

The eigenvalues and eigenvectors are

$$\lambda_1 = 0.25, \quad \mathbf{v}_1 = \begin{pmatrix} 8 \\ -3 \end{pmatrix}; \quad \lambda_2 = -0.75, \quad \mathbf{v}_2 = \begin{pmatrix} 0 \\ 1 \end{pmatrix}, \tag{16}$$

so the general solution of Eq. (15) is

$$\begin{pmatrix} u \\ w \end{pmatrix} = c_1 e^{0.25t}\begin{pmatrix} 8 \\ -3 \end{pmatrix} + c_2 e^{-0.75t}\begin{pmatrix} 0 \\ 1 \end{pmatrix}. \tag{17}$$

Thus the point $(0, 0.75)$ is also a saddle point. One pair of trajectories approaches the critical point along the y-axis and another departs tangent to the line with slope $-3/8$. All other trajectories also leave the neighborhood of the critical point.

$x = 0.5, y = 0.5$. This critical point corresponds to a mixed equilibrium state, or coexistence, in the competition between the two species. The eigenvalues and eigenvectors of the corresponding linear system

$$\frac{d}{dt}\begin{pmatrix} u \\ w \end{pmatrix} = \begin{pmatrix} -0.5 & -0.5 \\ -0.25 & -0.5 \end{pmatrix} \begin{pmatrix} u \\ w \end{pmatrix} \tag{18}$$

are

$$\lambda_1 = (-2 + \sqrt{2})/4 \cong -0.146, \qquad \mathbf{v}_1 = \begin{pmatrix} \sqrt{2} \\ -1 \end{pmatrix};$$

$$\lambda_2 = (-2 - \sqrt{2})/4 \cong -0.854, \qquad \mathbf{v}_2 = \begin{pmatrix} \sqrt{2} \\ 1 \end{pmatrix}.$$ (19)

Therefore the general solution of Eq. (18) is

$$\begin{pmatrix} u \\ w \end{pmatrix} = c_1 e^{-0.146t} \begin{pmatrix} \sqrt{2} \\ -1 \end{pmatrix} + c_2 e^{-0.854t} \begin{pmatrix} \sqrt{2} \\ 1 \end{pmatrix}.$$ (20)

Since both eigenvalues are negative, the critical point (0.5, 0.5) is an asymptotically stable node of the linear system (18) and of the nonlinear system (3). All trajectories approach the critical point as $t \to \infty$. One pair of trajectories approaches the critical point along the line with slope $\sqrt{2}/2$ determined from the eigenvector $\mathbf{v}_2$. All other trajectories approach the critical point tangent to the line with slope $-\sqrt{2}/2$ determined from the eigenvector $\mathbf{v}_1$.

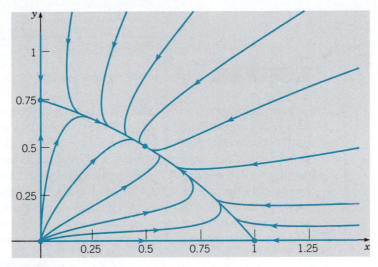

FIGURE 7.3.2   A phase portrait of the system (3).

A phase portrait for the system (3) is shown in Figure 7.3.2. By looking closely at the trajectories near each critical point, you can see that they behave in the manner predicted by the linear system near that point. In addition, note that the quadratic terms on the right side of Eqs. (3) are all negative. Since for x and y large and positive these terms are the dominant ones, it follows that far from the origin in the first quadrant both x' and y' are negative, that is, the trajectories are directed inward. Thus all trajectories that start at a point (x_0, y_0) with $x_0 > 0$ and $y_0 > 0$ eventually approach the point (0.5, 0.5).

EXAMPLE 2

Discuss the qualitative behavior of the solutions of the system

$$dx/dt = x(1 - x - y),$$
$$dy/dt = y(0.8 - 0.6y - x),$$ (21)

when x and y are nonnegative. Observe that this system is also a special case of the system (2) for two competing species.

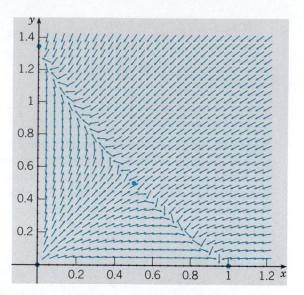

FIGURE 7.3.3 Critical points and direction field for the system (21).

Once again, there are four critical points, namely $(0, 0)$, $(1, 0)$, $(0, 4/3)$, and $(0.5, 0.5)$, corresponding to equilibrium solutions of the system (21). Figure 7.3.3 shows a direction field for the system (21), together with the four critical points. From the direction field, it appears that the mixed equilibrium solution $(0.5, 0.5)$ is a saddle point, and therefore unstable, while the points $(1, 0)$ and $(0, 4/3)$ are asymptotically stable. Thus, for the competition described by Eqs. (21), one species will eventually overwhelm the other and drive it to extinction. The surviving species is determined by the initial state of the system. To confirm these conclusions, we can look at the linear approximations near each critical point. For later reference, we record the Jacobian matrix for the system (21) evaluated at a critical point (X, Y):

$$\begin{pmatrix} F_x(X, Y) & F_y(X, Y) \\ G_x(X, Y) & G_y(X, Y) \end{pmatrix} = \begin{pmatrix} 1 - 2X - Y & -X \\ -Y & 0.8 - 1.2Y - X \end{pmatrix}. \tag{22}$$

x = 0, y = 0. Neglecting the nonlinear terms in Eqs. (21), or setting $X = 0$ and $Y = 0$ in the Jacobian matrix, we obtain the linear system

$$\frac{d}{dt} \begin{pmatrix} x \\ y \end{pmatrix} = \begin{pmatrix} 1 & 0 \\ 0 & 0.8 \end{pmatrix} \begin{pmatrix} x \\ y \end{pmatrix}, \tag{23}$$

which is valid near the origin. The eigenvalues and eigenvectors of the system (23) are

$$\lambda_1 = 1, \quad \mathbf{v}_1 = \begin{pmatrix} 1 \\ 0 \end{pmatrix}; \quad \lambda_2 = 0.8, \quad \mathbf{v}_2 = \begin{pmatrix} 0 \\ 1 \end{pmatrix}, \tag{24}$$

so the general solution is

$$\begin{pmatrix} x \\ y \end{pmatrix} = c_1 e^t \begin{pmatrix} 1 \\ 0 \end{pmatrix} + c_2 e^{0.8t} \begin{pmatrix} 0 \\ 1 \end{pmatrix}. \tag{25}$$

Therefore the origin is an unstable node of the linear system (23) and also of the nonlinear system (21). All trajectories leave the origin tangent to the y-axis except for one trajectory that lies along the x-axis.

x = 1, y = 0. The corresponding linear system is

$$\frac{d}{dt}\begin{pmatrix} u \\ w \end{pmatrix} = \begin{pmatrix} -1 & -1 \\ 0 & -0.2 \end{pmatrix}\begin{pmatrix} u \\ w \end{pmatrix}. \tag{26}$$

Its eigenvalues and eigenvectors are

$$\lambda_1 = -1, \quad \mathbf{v}_1 = \begin{pmatrix} 1 \\ 0 \end{pmatrix}; \quad \lambda_2 = -0.2, \quad \mathbf{v}_2 = \begin{pmatrix} 5 \\ -4 \end{pmatrix}, \tag{27}$$

and its general solution is

$$\begin{pmatrix} u \\ w \end{pmatrix} = c_1 e^{-t}\begin{pmatrix} 1 \\ 0 \end{pmatrix} + c_2 e^{-0.2t}\begin{pmatrix} 5 \\ -4 \end{pmatrix}. \tag{28}$$

The point $(1, 0)$ is an asymptotically stable node of the linear system (26) and of the nonlinear system (21). If the initial values of x and y are sufficiently close to $(1, 0)$, then the interaction process will lead ultimately to that state, that is, to the survival of species x and the extinction of species y. There is one pair of trajectories that approaches the critical point along the x-axis. All other trajectories approach $(1, 0)$ tangent to the line with slope $-4/5$ that is determined by the eigenvector $\mathbf{v}_2$.

x = 0, y = 4/3. The analysis in this case is similar to that for the point $(1, 0)$. The appropriate linear system is

$$\frac{d}{dt}\begin{pmatrix} u \\ w \end{pmatrix} = \begin{pmatrix} -1/3 & 0 \\ -4/3 & -4/5 \end{pmatrix}\begin{pmatrix} u \\ w \end{pmatrix}. \tag{29}$$

The eigenvalues and eigenvectors of this system are

$$\lambda_1 = -1/3, \quad \mathbf{v}_1 = \begin{pmatrix} 1 \\ -20/7 \end{pmatrix}; \quad \lambda_2 = -4/5, \quad \mathbf{v}_2 = \begin{pmatrix} 0 \\ 1 \end{pmatrix}, \tag{30}$$

and its general solution is

$$\begin{pmatrix} u \\ w \end{pmatrix} = c_1 e^{-t/3}\begin{pmatrix} 1 \\ -20/7 \end{pmatrix} + c_2 e^{-4t/5}\begin{pmatrix} 0 \\ 1 \end{pmatrix}. \tag{31}$$

Thus the critical point $(0, 4/3)$ is an asymptotically stable node of both the linear system (29) and the nonlinear system (21). All nearby trajectories approach the critical point tangent to the line with slope $-20/7$ except for one trajectory that lies along the y-axis.

x = 0.5, y = 0.5. The corresponding linear system is

$$\frac{d}{dt}\begin{pmatrix} u \\ w \end{pmatrix} = \begin{pmatrix} -0.5 & -0.5 \\ -0.5 & -0.3 \end{pmatrix}\begin{pmatrix} u \\ w \end{pmatrix}. \tag{32}$$

The eigenvalues and eigenvectors are

$$\lambda_1 = \frac{-4 + \sqrt{26}}{10} \cong 0.1099, \quad \mathbf{v}_1 = \begin{pmatrix} 5 \\ -1 - \sqrt{26} \end{pmatrix} \cong \begin{pmatrix} 5 \\ -6.0990 \end{pmatrix},$$

$$\lambda_2 = \frac{-4 - \sqrt{26}}{10} \cong -0.9099, \quad \mathbf{v}_2 = \begin{pmatrix} 5 \\ -1 + \sqrt{26} \end{pmatrix} \cong \begin{pmatrix} 5 \\ 4.0990 \end{pmatrix}, \tag{33}$$

so the general solution is

$$\begin{pmatrix} u \\ w \end{pmatrix} = c_1 e^{0.1099t} \begin{pmatrix} 5 \\ -6.0990 \end{pmatrix} + c_2 e^{-0.9099t} \begin{pmatrix} 5 \\ 4.0990 \end{pmatrix}. \tag{34}$$

Since the eigenvalues are of opposite sign, the critical point $(0.5, 0.5)$ is a saddle point and therefore is unstable, as we had surmised earlier. One pair of trajectories approaches the critical point as $t \to \infty$; the others depart from it. As they approach the critical point, the entering trajectories are tangent to the line with slope $(\sqrt{26} - 1)/5 \cong 0.8198$ determined from the eigenvector $\mathbf{v}_2$.

A phase portrait for the system (21) is shown in Figure 7.3.4. Near each of the critical points the trajectories of the nonlinear system behave as predicted by the corresponding linear approximation. Of particular interest is the pair of trajectories that enter the saddle point. These trajectories form a separatrix that divides the first quadrant into two basins of attraction. Trajectories starting above the separatrix ultimately approach the node at $(0, 4/3)$, while trajectories starting below the separatrix approach the node at $(1, 0)$. If the initial state lies precisely on the separatrix, then the solution (x, y) will approach the saddle point as $t \to \infty$. However, the slightest perturbation as one follows this trajectory will dislodge the point (x, y) from the separatrix and cause it to approach one of the nodes instead. Thus, in practice, one species will survive the competition and the other will not.

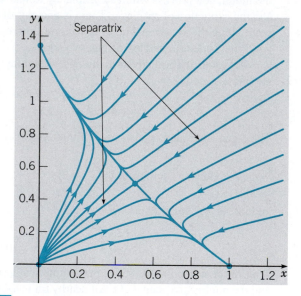

FIGURE 7.3.4 A phase portrait of the system (21).

Examples 1 and 2 show that in some cases the competition between two species leads to an equilibrium state of coexistence, while in other cases the competition results in the eventual extinction of one of the species. To understand more clearly how and why this happens, and to learn how to predict which situation will occur, it is useful to look again at the general system (2). There are four cases to be considered, depending on the relative orientation of the lines

$$\epsilon_1 - \sigma_1 x - \alpha_1 y = 0 \quad \text{and} \quad \epsilon_2 - \sigma_2 y - \alpha_2 x = 0, \tag{35}$$

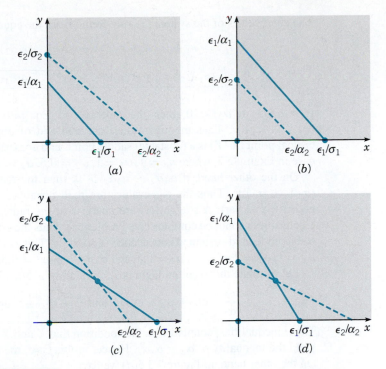

FIGURE 7.3.5   The various cases for the competing species system (2).

as shown in Figure 7.3.5. These lines are called the x and y **nullclines**, respectively, because x' is zero on the first and y' is zero on the second. In each part of Figure 7.3.5, the, x-nullcline is the solid line and the y-nullcline is the dashed line.

Let (X, Y) denote any critical point in any one of the four cases. As in Examples 1 and 2, the system (2) is almost linear in the neighborhood of this point because the right side of each differential equation is a quadratic polynomial. To study the system (2) in the neighborhood of this critical point, we can look at the corresponding linear system

$$\frac{d}{dt}\begin{pmatrix} u \\ w \end{pmatrix} = \begin{pmatrix} \epsilon_1 - 2\sigma_1 X - \alpha_1 Y & -\alpha_1 X \\ -\alpha_2 Y & \epsilon_2 - 2\sigma_2 Y - \alpha_2 X \end{pmatrix}\begin{pmatrix} u \\ w \end{pmatrix}. \tag{36}$$

We now use Eq. (36) to determine the conditions under which the model described by Eqs. (2) permits the coexistence of the two species x and y. Of the four possible cases shown in Figure 7.3.5, coexistence is possible only in cases (c) and (d). In these cases, the nonzero values of X and Y are readily obtained by solving the algebraic equations (35); the result is

$$X = \frac{\epsilon_1 \sigma_2 - \epsilon_2 \alpha_1}{\sigma_1 \sigma_2 - \alpha_1 \alpha_2}, \qquad Y = \frac{\epsilon_2 \sigma_1 - \epsilon_1 \alpha_2}{\sigma_1 \sigma_2 - \alpha_1 \alpha_2}. \tag{37}$$

Further, since $\epsilon_1 - \sigma_1 X - \alpha_1 Y = 0$ and $\epsilon_2 - \sigma_2 Y - \alpha_2 X = 0$, Eq. (36) immediately reduces to

$$\frac{d}{dt}\begin{pmatrix} u \\ w \end{pmatrix} = \begin{pmatrix} -\sigma_1 X & -\alpha_1 X \\ -\alpha_2 Y & -\sigma_2 Y \end{pmatrix}\begin{pmatrix} u \\ w \end{pmatrix}. \tag{38}$$

The eigenvalues of the system (38) are found from the equation

$$\lambda^2 + (\sigma_1 X + \sigma_2 Y)\lambda + (\sigma_1\sigma_2 - \alpha_1\alpha_2)XY = 0. \tag{39}$$

Thus

$$\lambda_{1,2} = \frac{-(\sigma_1 X + \sigma_2 Y) \pm \sqrt{(\sigma_1 X + \sigma_2 Y)^2 - 4(\sigma_1\sigma_2 - \alpha_1\alpha_2)XY}}{2}. \tag{40}$$

If $\sigma_1\sigma_2 - \alpha_1\alpha_2 < 0$, then the radicand of Eq. (40) is positive and greater than $(\sigma_1 X + \sigma_2 Y)^2$. Thus the eigenvalues are real and of opposite sign. Consequently, the critical point (X, Y) is a (unstable) saddle point, and coexistence is not possible. This is the case in Example 2, where $\sigma_1 = 1$, $\alpha_1 = 1$, $\sigma_2 = 0.6$, $\alpha_2 = 1$, and $\sigma_1\sigma_2 - \alpha_1\alpha_2 = -0.4$.

On the other hand, if $\sigma_1\sigma_2 - \alpha_1\alpha_2 > 0$, then the radicand of Eq. (40) is less than $(\sigma_1 X + \sigma_2 Y)^2$. Thus the eigenvalues are real, negative, and unequal, or complex with negative real part. A straightforward analysis of the radicand of Eq. (40) shows that the eigenvalues cannot be complex (see Problem 7). Thus the critical point is an asymptotically stable node, and sustained coexistence is possible. This is illustrated by Example 1, where $\sigma_1 = 1$, $\alpha_1 = 1$, $\sigma_2 = 1$, $\alpha_2 = 0.5$, and $\sigma_1\sigma_2 - \alpha_1\alpha_2 = 0.5$.

Let us relate this result to Figures 7.3.5(c) and 7.3.5(d). In Figure 7.3.5(c) we have

$$\frac{\epsilon_1}{\sigma_1} > \frac{\epsilon_2}{\alpha_2} \quad \text{or} \quad \epsilon_1\alpha_2 > \epsilon_2\sigma_1 \quad \text{and} \quad \frac{\epsilon_2}{\sigma_2} > \frac{\epsilon_1}{\alpha_1} \quad \text{or} \quad \epsilon_2\alpha_1 > \epsilon_1\sigma_2. \tag{41}$$

These inequalities, coupled with the condition that X and Y given by Eqs. (37) be positive, yield the inequality $\sigma_1\sigma_2 < \alpha_1\alpha_2$. Hence, in this case, the critical point is a saddle point. On the other hand, in Figure 7.3.5(d) we have

$$\frac{\epsilon_1}{\sigma_1} < \frac{\epsilon_2}{\alpha_2} \quad \text{or} \quad \epsilon_1\alpha_2 < \epsilon_2\sigma_1 \quad \text{and} \quad \frac{\epsilon_2}{\sigma_2} < \frac{\epsilon_1}{\alpha_1} \quad \text{or} \quad \epsilon_2\alpha_1 < \epsilon_1\sigma_2. \tag{42}$$

Now the condition that X and Y be positive yields $\sigma_1\sigma_2 > \alpha_1\alpha_2$. Hence the critical point is asymptotically stable. For this case, we can also show that the other critical points $(0, 0)$, $(\epsilon_1/\sigma_1, 0)$, and $(0, \epsilon_2/\sigma_2)$ are unstable. Thus for any positive initial values of x and y, the two populations approach the equilibrium state of coexistence given by Eqs. (37).

Equations (2) provide the biological interpretation of the result that coexistence occurs or not depending on whether $\sigma_1\sigma_2 - \alpha_1\alpha_2$ is positive or negative. The σ's are a measure of the inhibitory effect that the growth of each population has on itself, while the α's are a measure of the inhibiting effect that the growth of each population has on the other species. Thus, when $\sigma_1\sigma_2 > \alpha_1\alpha_2$, interaction (competition) is "weak" and the species can coexist; when $\sigma_1\sigma_2 < \alpha_1\alpha_2$, interaction (competition) is "strong" and the species cannot coexist—one must die out.

PROBLEMS

ODEA Each of Problems 1 through 6 can be interpreted as describing the interaction of two species with populations x and y. In each of these problems, carry out the following steps.

(a) Draw a direction field and describe how solutions seem to behave.

(b) Find the critical points.

(c) For each critical point, find the corresponding linear system. Find the eigenvalues and eigenvectors of the linear system, classify each critical point as to type, and determine whether it is asymptotically stable, stable, or unstable.

(d) Sketch the trajectories in the neighborhood of each critical point.

(e) Compute and plot enough trajectories of the given system to show clearly the behavior of the solutions.

(f) Determine the limiting behavior of x and y as $t \to \infty$, and interpret the results in terms of the populations of the two species.

1. $dx/dt = x(1.5 - x - 0.5y)$
   $dy/dt = y(2 - y - 0.75x)$

2. $dx/dt = x(1.5 - x - 0.5y)$
   $dy/dt = y(2 - 0.5y - 1.5x)$

3. $dx/dt = x(1.5 - 0.5x - y)$
   $dy/dt = y(2 - y - 1.125x)$

4. $dx/dt = x(1.5 - 0.5x - y)$
   $dy/dt = y(0.75 - y - 0.125x)$

5. $dx/dt = x(1 - x - y)$
   $dy/dt = y(1.5 - y - x)$

6. $dx/dt = x(1 - x + 0.5y)$
   $dy/dt = y(2.5 - 1.5y + 0.25x)$

7. Show that

$$(\sigma_1 X + \sigma_2 Y)^2 - 4(\sigma_1\sigma_2 - \alpha_1\alpha_2)XY$$
$$= (\sigma_1 X - \sigma_2 Y)^2 + 4\alpha_1\alpha_2 XY.$$

Hence conclude that the eigenvalues given by Eq. (40) can never be complex.

8. Consider the system (2) in the text, and assume that $\sigma_1\sigma_2 - \alpha_1\alpha_2 = 0$.
   (a) Find all the critical points of the system. Observe that the result depends on whether $\sigma_1\epsilon_2 - \alpha_2\epsilon_1$ is zero.
   (b) If $\sigma_1\epsilon_2 - \alpha_2\epsilon_1 > 0$, classify each critical point and determine whether it is asymptotically stable, stable, or unstable. Note that Problem 5 is of this type. Then do the same if $\sigma_1\epsilon_2 - \alpha_2\epsilon_1 < 0$.
   (c) Analyze the nature of the trajectories when $\sigma_1\epsilon_2 - \alpha_2\epsilon_1 = 0$.

9. Consider the system (3) in Example 1 of the text. Recall that this system has an asymptotically stable critical point at $(0.5, 0.5)$, corresponding to the stable coexistence of the two population species. Now suppose that immigration or emigration occurs at the constant rates of δa and δb for the species x and y, respectively. In this case, Eqs. (3) are replaced by

$$dx/dt = x(1 - x - y) + \delta a,$$
$$dy/dt = y(0.75 - y - 0.5x) + \delta b. \tag{i}$$

The question is what effect this has on the location of the stable equilibrium point.
   (a) To find the new critical point we must solve the equations

$$x(1 - x - y) + \delta a = 0,$$
$$y(0.75 - y - 0.5x) + \delta b = 0. \tag{ii}$$

One way to proceed is to assume that x and y are given by power series in the parameter δ; thus

$$x = x_0 + x_1\delta + \cdots, \quad y = y_0 + y_1\delta + \cdots. \tag{iii}$$

Substitute Eqs. (iii) into Eqs. (ii) and collect terms according to powers of δ.
   (b) From the constant terms (the terms not involving δ), show that $x_0 = 0.5$ and $y_0 = 0.5$, thus confirming that, in the absence of immigration or emigration, the critical point is $(0.5, 0.5)$.
   (c) From the terms that are linear in δ, show that

$$x_1 = 4a - 4b, \quad y_1 = -2a + 4b. \tag{iv}$$

   (d) Suppose that $a > 0$ and $b > 0$ so that immigration occurs for both species. Show that the resulting equilibrium solution may represent an increase in both populations, or an increase in one but a decrease in the other. Explain intuitively why this is a reasonable result.

10. The system

$$x' = -y, \quad y' = -\gamma y - x(x - 0.15)(x - 2)$$

results from an approximation to the Hodgkin–Huxley equations, which model the transmission of neural impulses along an axon.
   (a) Find the critical points and classify them by investigating the approximate linear system near each one.
   (b) Draw phase portraits for $\gamma = 0.8$ and for $\gamma = 1.5$.
   (c) Consider the trajectory that leaves the critical point $(2, 0)$. Find the value of γ for which this trajectory ultimately approaches the origin as $t \to \infty$. Draw a phase portrait for this value of γ.

Bifurcation Points. Consider the system

$$x' = F(x, y, \alpha), \quad y' = G(x, y, \alpha), \tag{i}$$

where α is a parameter. The equations

$$F(x, y, \alpha) = 0, \quad G(x, y, \alpha) = 0 \tag{ii}$$

determine the x and y nullclines, respectively. Any point where an x nullcline and a y nullcline intersect is a critical point. As α varies and the configuration of the nullclines changes, it may well happen that, at a certain value of α, two critical points coalesce into one. For further variations in α, the critical point may disappear altogether, or two critical points may reappear, often with stability characteristics different than before they coalesced. Of course, the process may occur in the reverse order. For a certain value of α, two formerly nonintersecting nullclines may come together, creating a critical point, which, for further changes in α, may split into two. A value of α at which critical points coalesce, and possibly are lost or gained, is a bifurcation point. Since a phase portrait of a system is very dependent on the location and nature of the critical points, an understanding of bifurcations is essential to an understanding of the global behavior of the system's solutions. Problems 11 through 17 illustrate some possibilities that involve bifurcations.

ODEA In each of Problems 11 through 14:

(a) Sketch the nullclines and describe how the critical points move as α increases.

(b) Find the critical points.

(c) Let $\alpha = 2$. Classify each critical point by investigating the corresponding approximate linear system. Draw a phase portrait in a rectangle containing the critical points.

(d) Find the bifurcation point α_0 at which the critical points coincide. Locate this critical point and find the eigenvalues of the approximate linear system. Draw a phase portrait.

(e) For $\alpha > \alpha_0$, there are no critical points. Choose such a value of α and draw a phase portrait.

11. $x' = -4x + y + x^2, \qquad y' = \frac{3}{2}\alpha - y$

12. $x' = \frac{3}{2}\alpha - y, \qquad y' = -4x + y + x^2$

13. $x' = -4x + y + x^2, \qquad y' = -\alpha - x + y$

14. $x' = -\alpha - x + y, y' = -4x + y + x^2$

In each of Problems 15 and 16:

(a) Find the critical points.

(b) Determine the value of α, denoted by α_0, where two critical points coincide.

(c) By finding the approximating linear systems and their eigenvalues, determine how the stability properties of these two critical points change as α passes through the bifurcation point α_0.

(d) Draw phase portraits for values of α near α_0 to ODEA show how the transition through the bifurcation point occurs.

15. $x' = (2 + x)(1 - x + y),$
$y' = (y - 1)(1 + x + \alpha y); \alpha > 0$

16. $x' = y(\alpha - 2x + 3y), \qquad y' = (2 - x)(3 + y)$

17. Suppose that a certain pair of competing species are described by the system

$$dx/dt = x(4 - x - y),$$
$$dy/dt = y(2 + 2\alpha - y - \alpha x),$$

where $\alpha > 0$ is a parameter.

(a) Find the critical points. Note that $(2, 2)$ is a critical point for all values of α.

(b) Determine the nature of the critical point $(2, 2)$ for $\alpha = 0.75$ and for $\alpha = 1.25$. There is a value of α between 0.75 and 1.25 where the nature of the critical point changes abruptly. Denote this value by α_0; it is also called a bifurcation point.

(c) Find the approximate linear system near the point $(2, 2)$ in terms of α.

(d) Find the eigenvalues of the linear system in part (c) as functions of α. Then determine α_0.

(e) Draw phase portraits near $(2, 2)$ for $\alpha = \alpha_0$ and ODEA for values of α slightly less than, and slightly greater than, α_0. Explain how the transition in the phase portrait takes place as α passes through α_0.

7.4 Predator–Prey Equations

In the preceding section, we discussed a model of two species that interact by competing for a common food supply or other natural resource. In this section, we investigate the situation in which one species (the predator) preys on the other species (the prey), while the prey lives on a different source of food. For example, consider foxes and rabbits in a closed forest. The foxes prey on the rabbits, and the rabbits live on the vegetation in the forest. Other examples are bass in a lake as predators and redear as prey, or ladybugs as predators and aphids as prey. We emphasize again that a model involving only two species cannot fully describe the complex relationships among species that actually occur in nature. Nevertheless, the study of simple models is the first step toward an understanding of more complicated phenomena.

We will denote by x and y the populations of the prey and predator, respectively, at time t. In constructing a model of the interaction of the two species, we make the following assumptions:

1. In the absence of the predator, the prey grows at a rate proportional to the current population; thus $dx/dt = ax$, $a > 0$, when $y = 0$.

2. In the absence of the prey, the predator dies out; thus $dy/dt = -cy$, $c > 0$, when $x = 0$.

3. The number of encounters between predator and prey is proportional to the product of their populations. Each such encounter tends to promote the growth of the predator and to inhibit the growth of the prey. Thus the growth rate of the predator includes a term of the form γxy, while the growth rate of the prey includes a term of the form $-\alpha xy$, where γ and α are positive constants.

As a consequence of these assumptions, we are led to the equations

$$dx/dt = ax - \alpha xy = x(a - \alpha y),$$
$$dy/dt = -cy + \gamma xy = y(-c + \gamma x).$$
$$(1)$$

The constants a, c, α, and γ are all positive; a and c are the growth rate of the prey and the death rate of the predator, respectively, and α and γ are measures of the effect of the interaction between the two species. In specific situations values for a, α, c, and γ must be determined by observation. Equations (1) are known as the Lotka–Volterra equations. Although these are rather simple equations, they have been used as a preliminary model for many predator-prey relationships. Ways of making them more realistic are discussed at the end of this section and in the problems. Our goal here is to determine the qualitative behavior of the solutions (trajectories) of the system (1) for arbitrary positive initial values of x and y. We do this first for a specific example and then return to the general equations (1) at the end of the section.

**EXAMPLE
1**

Discuss the solutions of the system

$$dx/dt = x(1 - 0.5y) = x - 0.5xy = F(x, y),$$
$$dy/dt = y(-0.75 + 0.25x) = -0.75y + 0.25xy = G(x, y)$$
$$(2)$$

for x and y positive.

The critical points of this system are the solutions of the algebraic equations

$$x(1 - 0.5y) = 0, \qquad y(-0.75 + 0.25x) = 0. \qquad (3)$$

Thus the critical points are $(0, 0)$ and $(3, 2)$. Figure 7.4.1 shows the critical points and a direction field for the system (2). From this figure, it appears that trajectories in the first quadrant encircle the critical point $(3, 2)$. Whether the trajectories are actually closed curves, or whether they spiral slowly in or out cannot be definitely determined from the direction field. The origin appears to be a saddle point. Just as for the competition equations in Section 7.3, the coordinate axes are trajectories of Eqs. (1) or (2). Consequently, no other trajectory can cross a coordinate axis, which means that every solution starting in the first quadrant remains there for all time.

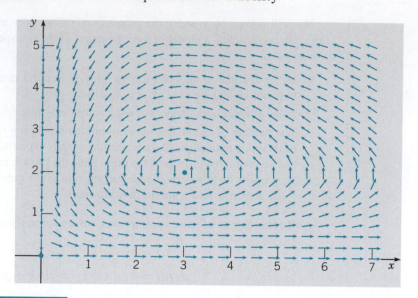

FIGURE 7.4.1   Critical points and direction field for the predator–prey system (2).

Next we examine the local behavior of solutions near each critical point. Near the origin we can neglect the nonlinear terms in Eqs. (2) to obtain the corresponding linear system

$$\frac{d}{dt}\begin{pmatrix} x \\ y \end{pmatrix} = \begin{pmatrix} 1 & 0 \\ 0 & -0.75 \end{pmatrix}\begin{pmatrix} x \\ y \end{pmatrix}. \tag{4}$$

The eigenvalues and eigenvectors of Eq. (4) are

$$\lambda_1 = 1, \quad \mathbf{v}_1 = \begin{pmatrix} 1 \\ 0 \end{pmatrix}; \qquad \lambda_2 = -0.75, \quad \mathbf{v}_2 = \begin{pmatrix} 0 \\ 1 \end{pmatrix}, \tag{5}$$

so its general solution is

$$\begin{pmatrix} x \\ y \end{pmatrix} = c_1 e^t \begin{pmatrix} 1 \\ 0 \end{pmatrix} + c_2 e^{-0.75t} \begin{pmatrix} 0 \\ 1 \end{pmatrix}. \tag{6}$$

Thus the origin is a saddle point both of the linear system (4) and of the nonlinear system (2), and therefore is unstable. One pair of trajectories enters the origin along the y-axis and another leaves the origin along the x-axis. All other trajectories also depart from the neighborhood of the origin.

To examine the critical point (3, 2) let us first find the Jacobian matrix

$$\begin{pmatrix} F_x & F_y \\ G_x & G_y \end{pmatrix} = \begin{pmatrix} 1 - 0.5y & -0.5x \\ 0.25y & -0.75 + 0.25x \end{pmatrix}. \tag{7}$$

Evaluating this matrix at the point (3, 2), we obtain the linear system

$$\frac{d}{dt}\begin{pmatrix} u \\ w \end{pmatrix} = \begin{pmatrix} 0 & -1.5 \\ 0.5 & 0 \end{pmatrix}\begin{pmatrix} u \\ w \end{pmatrix}, \tag{8}$$

where $u = x - 3$ and $w = y - 2$. The eigenvalues and eigenvectors of this system are

$$\lambda_1 = \frac{\sqrt{3}\,i}{2}, \quad \mathbf{v}_1 = \begin{pmatrix} 1 \\ -i/\sqrt{3} \end{pmatrix}; \qquad \lambda_2 = -\frac{\sqrt{3}\,i}{2}, \quad \mathbf{v}_2 = \begin{pmatrix} 1 \\ i/\sqrt{3} \end{pmatrix}. \tag{9}$$

Since the eigenvalues are imaginary, the critical point $(3, 2)$ is a center of the linear system (8) and is therefore a stable critical point for that system. Recall from Section 7.2 that this is one of the cases in which the behavior of the linear system may or may not carry over to the nonlinear system, so the nature of the point $(3, 2)$ for the nonlinear system (2) cannot be determined from this information. The simplest way to find the trajectories of the linear system (8) is to divide the second of Eqs. (8) by the first so as to obtain the differential equation

$$\frac{dw}{du} = \frac{dw/dt}{du/dt} = \frac{0.5u}{-1.5w} = -\frac{u}{3w},$$

or

$$u\, du + 3w\, dw = 0. \tag{10}$$

Consequently,

$$u^2 + 3w^2 = k, \tag{11}$$

where k is an arbitrary nonnegative constant of integration. Thus the trajectories of the linear system (8) lie on ellipses centered at the critical point and elongated somewhat in the horizontal direction.

Now let us return to the nonlinear system (2). Dividing the second of Eqs. (2) by the first, we obtain

$$\frac{dy}{dx} = \frac{y(-0.75 + 0.25x)}{x(1 - 0.5y)}. \tag{12}$$

Equation (12) is a separable equation and can be put in the form

$$\frac{1 - 0.5y}{y}\, dy = \frac{-0.75 + 0.25x}{x}\, dx,$$

from which it follows that

$$0.75 \ln x + \ln y - 0.5y - 0.25x = c, \tag{13}$$

where c is a constant of integration. Although by using only elementary functions we cannot solve Eq. (13) explicitly for either variable in terms of the other, it is possible to show that the graph of the equation for a fixed value of c is a closed curve surrounding the critical point $(3, 2)$. Thus the critical point is also a center of the nonlinear system (2), and the predator and prey populations exhibit a cyclic variation.

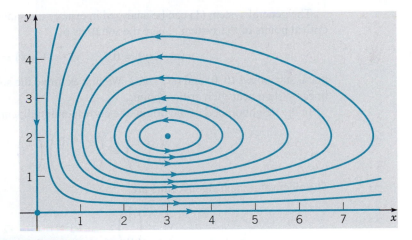

FIGURE 7.4.2 A phase portrait of the system (2).

Figure 7.4.2 shows a phase portrait of the system (2). For some initial conditions, the trajectory represents small variations in x and y about the critical point, and is almost elliptical in shape, as the linear analysis suggests. For other initial conditions, the oscillations in x and y are more pronounced, and the shape of the trajectory is significantly different from an ellipse. Observe that the trajectories are traversed in the counterclockwise direction. The dependence of x and y on t for a typical set of initial conditions is shown in Figure 7.4.3. Note that x and y are periodic functions of t, as they must be since the trajectories are closed curves. Further, the oscillation of the predator population lags behind that of the prey. Starting from a state in which both predator and prey populations are relatively small, the prey first increase because there is little predation. Then the predators, with abundant food, increase in population also. This causes heavier predation, and the prey tend to decrease. Finally, with a diminished food supply, the predator population also decreases, and the system returns to the original state.

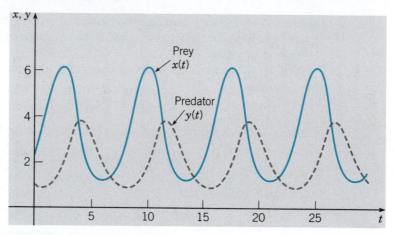

FIGURE 7.4.3   Variations of the prey and predator populations with time for the system (2).

The general system (1) can be analyzed in exactly the same way as in the example. The critical points of the system (1) are the solutions of

$$x(a - \alpha y) = 0, \qquad y(-c + \gamma x) = 0,$$

that is, the points $(0, 0)$ and $(c/\gamma, a/\alpha)$. We first examine the solutions of the corresponding linear system near each critical point.

In the neighborhood of the origin, the corresponding linear system is

$$\frac{d}{dt} \begin{pmatrix} x \\ y \end{pmatrix} = \begin{pmatrix} a & 0 \\ 0 & -c \end{pmatrix} \begin{pmatrix} x \\ y \end{pmatrix}. \tag{14}$$

The eigenvalues and eigenvectors are

$$\lambda_1 = a, \quad \mathbf{v}_1 = \begin{pmatrix} 1 \\ 0 \end{pmatrix}; \qquad \lambda_2 = -c, \quad \mathbf{v}_2 = \begin{pmatrix} 0 \\ 1 \end{pmatrix}, \tag{15}$$

so the general solution is

$$\begin{pmatrix} x \\ y \end{pmatrix} = c_1 e^{at} \begin{pmatrix} 1 \\ 0 \end{pmatrix} + c_2 e^{-ct} \begin{pmatrix} 0 \\ 1 \end{pmatrix}. \tag{16}$$

Thus the origin is a saddle point and hence unstable. Entrance to the saddle point is along the y-axis; all other trajectories depart from the neighborhood of the critical point.

Next consider the critical point $(c/\gamma, a/\alpha)$. The Jacobian matix is

$$\begin{pmatrix} F_x & F_y \\ G_x & G_y \end{pmatrix} = \begin{pmatrix} a - \alpha y & -\alpha x \\ \gamma y & -c + \gamma x \end{pmatrix},$$

Evaluating this matrix at $(c/\gamma, a/\alpha)$, we obtain the approximate linear system

$$\frac{d}{dt}\begin{pmatrix} u \\ w \end{pmatrix} = \begin{pmatrix} 0 & -\alpha c/\gamma \\ \gamma a/\alpha & 0 \end{pmatrix}\begin{pmatrix} u \\ w \end{pmatrix}, \tag{17}$$

where $u = x - c/\gamma$ and $w = y - a/\alpha$. The eigenvalues of the system (17) are $\lambda = \pm i\sqrt{ac}$, so the critical point is a (stable) center of the linear system. To find the trajectories of the system (17), we can divide the second equation by the first to obtain

$$\frac{dw}{du} = \frac{dw/dt}{du/dt} = -\frac{(\gamma a/\alpha)u}{(\alpha c/\gamma)w}, \tag{18}$$

or

$$\gamma^2 au\, du + \alpha^2 cw\, dw = 0. \tag{19}$$

Consequently,

$$\gamma^2 au^2 + \alpha^2 cw^2 = k, \tag{20}$$

where k is a nonnegative constant of integration. Thus the trajectories of the linear system (17) are ellipses, just as in the example.

Returning briefly to the nonlinear system (1), observe that it can be reduced to the single equation

$$\frac{dy}{dx} = \frac{dy/dt}{dx/dt} = \frac{y(-c + \gamma x)}{x(a - \alpha y)}. \tag{21}$$

Equation (21) is separable and has the solution

$$a \ln y - \alpha y + c \ln x - \gamma x = C, \tag{22}$$

where C is a constant of integration. Again it is possible to show that, for fixed C, the graph of Eq. (22) is a closed curve surrounding the critical point $(c/\gamma, a/\alpha)$. Thus this critical point is also a center for the general nonlinear system (1).

The cyclic variation of the predator and prey populations can be analyzed in more detail when the deviations from the point $(c/\gamma, a/\alpha)$ are small and the linear system (17) can be used. The solution of the system (17) can be written in the form

$$u = \frac{c}{\gamma}K \cos(\sqrt{ac}\, t + \phi), \qquad w = \frac{a}{\alpha}\sqrt{\frac{c}{a}}K \sin(\sqrt{ac}\, t + \phi), \tag{23}$$

where the constants K and ϕ are determined by the initial conditions. Thus

$$x = \frac{c}{\gamma} + \frac{c}{\gamma}K \cos(\sqrt{ac}\, t + \phi),$$

$$y = \frac{a}{\alpha} + \frac{a}{\alpha}\sqrt{\frac{c}{a}}K \sin(\sqrt{ac}\, t + \phi). \tag{24}$$

These equations are good approximations for the nearly elliptical trajectories close to the critical point $(c/\gamma, a/\alpha)$. We can use them to draw several conclusions about the cyclic variation of the predator and prey on such trajectories.

1. The sizes of the predator and prey populations vary sinusoidally with period $2\pi/\sqrt{ac}$. This period of oscillation is independent of the initial conditions.

2. The predator and prey populations are out of phase by one-quarter of a cycle. The prey leads and the predator lags, as explained in the example.

3. The amplitudes of the oscillations are Kc/γ for the prey and $a\sqrt{c}K/\alpha\sqrt{a}$ for the predator and hence depend on the initial conditions as well as on the parameters of the problem.

4. The average populations of prey and predator over one complete cycle are c/γ and a/α, respectively. These are the same as the equilibrium populations (see Problem 10).

Cyclic variations of predator and prey as predicted by Eqs. (1) have been observed in nature. One striking example is described by Odum (pp. 191–192). Based on the records of the Hudson Bay Company of Canada, the abundance of lynx and snowshoe hare, as indicated by the number of pelts turned in over the period 1845–1935, shows a distinct periodic variation with period of 9 to 10 years. The peaks of abundance are followed by very rapid declines, and the peaks of abundance of the lynx and hare are out of phase, with that of the hare preceding that of the lynx by a year or more.

Since the critical point $(c/\gamma, a/\alpha)$ is a center, we expect that small perturbations of the Lotka–Volterra equations may well lead to significant changes in the solutions. Put another way, unless the Lotka–Volterra equations exactly describe a given predator–prey relationship, the actual fluctuations of the populations may differ substantially from those predicted by the Lotka–Volterra equations, due to small inaccuracies in the model equations. This situation has led to many attempts[3] to replace the Lotka–Volterra equations by other systems that are less susceptible to the effects of small perturbations. Problem 13 introduces one such alternative model.

Another criticism of the Lotka–Volterra equations is that in the absence of the predator, the prey will grow without bound. This can be corrected by allowing for the natural inhibiting effect that an increasing population has on the growth rate of the population. For example, the first of Eqs. (1) can be modified so that when $y = 0$, it reduces to a logistic equation for x. The effects of this modification are explored in Problems 11 and 12. Problems 14 through 16 deal with harvesting in a predator–prey relationship. The results may seem rather counterintuitive.

Finally, we repeat a warning stated earlier: relationships among species in the natural world are often complex and subtle. You should not expect too much of a simple system of two differential equations in describing such relationships. Even if you are convinced that the general form of the equations is sound, the determination of numerical values for the coefficients may present serious difficulties.

PROBLEMS

ODEA Each of Problems 1 through 5 can be interpreted as describing the interaction of two species with population densities x and y. In each of these problems, carry out the following steps.

(a) Draw a direction field and describe how solutions seem to behave.

(b) Find the critical points.

(c) For each critical point, find the corresponding linear system. Find the eigenvalues and eigenvectors of the linear system. Classify each critical point as to type, and determine whether it is asymptotically stable, stable, or unstable.

(d) Sketch the trajectories in the neighborhood of each critical point.

---

[3] See the book by Brauer and Castillo-Chávez, listed in the references, for an extensive discussion of alternative models for predator–prey relationships.

(e) Draw a phase portrait for the system.

(f) Determine the limiting behavior of x and y as $t \to \infty$ and interpret the results in terms of the populations of the two species.

1. $dx/dt = x\,(1.5 - 0.5y)$
   $dy/dt = y\,(-0.5 + x)$

2. $dx/dt = x\,(1 - 0.5y)$
   $dy/dt = y\,(-0.25 + 0.5x)$

3. $dx/dt = x\,(1 - 0.5x - 0.5y)$
   $dy/dt = y\,(-0.25 + 0.5x)$

4. $dx/dt = x\,(1.125 - x - 0.5y)$
   $dy/dt = y\,(-1 + x)$

5. $dx/dt = x\,(-1 + 2.5x - 0.3y - x^2)$
   $dy/dt = y\,(-1.5 + x)$

6. In this problem, we examine the phase difference between the cyclic variations of the predator and prey populations as given by Eqs. (24) of this section. Suppose we assume that $K > 0$ and that t is measured from the time that the prey population (x) is a maximum; then $\phi = 0$. Show that the predator population (y) is a maximum at $t = \pi/(2\sqrt{ac}) = T/4$, where T is the period of the oscillation. When is the prey population increasing most rapidly? decreasing most rapidly? a minimum? Answer the same questions for the predator population. Draw a typical elliptic trajectory enclosing the point $(c/\gamma, a/\alpha)$, and mark these points on it.

ODEA 7. (a) Find the ratio of the amplitudes of the oscillations of the prey and predator populations about the critical point $(c/\gamma, a/\alpha)$, using the approximation (24), which is valid for small oscillations. Observe that the ratio is independent of the initial conditions.

(b) Evaluate the ratio found in part (a) for the system (2).

(c) Estimate the amplitude ratio for the solution of the nonlinear system (2) shown in Figure 7.4.3. Does the result agree with that obtained from the linear approximation?

(d) Determine the prey–predator amplitude ratio for other solutions of the system (2), that is, for solutions satisfying other initial conditions. Is the ratio independent of the initial conditions?

ODEA 8. (a) Find the period of the oscillations of the prey and predator populations, using the approximation (24), which is valid for small oscillations. Note that the period is independent of the amplitude of the oscillations.

(b) For the solution of the nonlinear system (2) shown in Figure 7.4.3, estimate the period as well as possible. Is the result the same as for the linear approximation?

(c) Calculate other solutions of the system (2), that is, solutions satisfying other initial conditions, and determine their periods. Is the period the same for all initial conditions?

9. Consider the system ODEA

$$dx/dt = ax[1 - (y/2)], \quad dy/dt = by[-1 + (x/3)],$$

where a and b are positive constants. Observe that this system is the same as in the example in the text if $a = 1$ and $b = 0.75$. Suppose the initial conditions are $x(0) = 5$ and $y(0) = 2$.

(a) Let $a = 1$ and $b = 1$. Plot the trajectory in the phase plane and determine (or estimate) the period of the oscillation.

(b) Repeat part (a) for $a = 3$ and $a = 1/3$, with $b = 1$.

(c) Repeat part (a) for $b = 3$ and $b = 1/3$, with $a = 1$.

(d) Describe how the period and the shape of the trajectory depend on a and b.

10. The average sizes of the prey and predator populations are defined as

$$\bar{x} = \frac{1}{T} \int_A^{A+T} x(t)\,dt, \qquad \bar{y} = \frac{1}{T} \int_A^{A+T} y(t)\,dt,$$

respectively, where T is the period of a full cycle, and A is any nonnegative constant.

(a) Using the approximation (24), which is valid near the critical point, show that $\bar{x} = c/\gamma$ and $\bar{y} = a/\alpha$.

(b) For the solution of the nonlinear system (2) shown in Figure 7.4.3, estimate $\bar{x}$ and $\bar{y}$ as well as you can. Try to determine whether $\bar{x}$ and $\bar{y}$ are given by c/γ and a/α, respectively, in this case.

Hint: Consider how you might estimate the value of an integral even though you do not have a formula for the integrand.

(c) Calculate other solutions of the system (2), that CAS is, solutions satisfying other initial conditions, and determine $\bar{x}$ and $\bar{y}$ for these solutions. Are the values of $\bar{x}$ and $\bar{y}$ the same for all solutions?

In Problems 11 and 12, we consider the effect of modifying the equation for the prey x by including a term $-\sigma x^2$ so that this equation reduces to a logistic equation in the absence of the predator y. Problem 11 deals with a specific system of this kind and Problem 12 takes up this modification to the general Lotka–Volterra system. The systems in Problems 3 and 4 are other examples of this type.

11. Consider the system

$$x' = x(1 - \sigma x - 0.5y), \qquad y' = y(-0.75 + 0.25x),$$

where $\sigma > 0$. Observe that this system is a modification of the system (2) in Example 1.
(a) Find all of the critical points. How does their location change as σ increases from zero? Observe that there is a critical point in the interior of the first quadrant only if $\sigma < 1/3$.
(b) Determine the type and stability property of each critical point. Find the value $\sigma_1 < 1/3$ where the nature of the critical point in the interior of the first quadrant changes. Describe the change that takes place in this critical point as σ passes through σ_1.
(c) Draw a direction field and phase portrait for a value of σ between zero and σ_1; for a value of σ between σ_1 and $1/3$.
(d) Describe the effect on the two populations as σ increases from zero to $1/3$.

12. Consider the system

$$dx/dt = x(a - \sigma x - \alpha y), \quad dy/dt = y(-c + \gamma x),$$

where a, σ, α, c, and γ are positive constants.
(a) Find all critical points of the given system. How does their location change as σ increases from zero? Assume that $a/\sigma > c/\gamma$, or that $\sigma < a\gamma/c$. Why is this assumption necessary?
(b) Determine the nature and stability characteristics of each critical point.
(c) Show that there is a value of σ between zero and $a\gamma/c$ where the critical point in the interior of the first quadrant changes from a spiral point to a node.
(d) Describe the effect on the two populations as σ increases from zero to $a\gamma/c$.

13. In the Lotka–Volterra equations, the interaction between the two species is modeled by terms proportional to the product xy of the respective populations. If the prey population is much larger than the predator population, this may overstate the interaction; for example, a predator may hunt only when it is hungry, and ignore the prey at other times. In this problem, we consider an alterna-

tive model of a type proposed by Rosenzweig and MacArthur.[4]
(a) Consider the system

$$x' = x\left(1 - 0.2x - \frac{2y}{x+6}\right),$$

$$y' = y\left(-0.25 + \frac{x}{x+6}\right).$$

Find all of the critical points of this system.
(b) Determine the type and stability characteristics of each critical point.
(c) Draw a direction field and phase portrait for this system.

Harvesting in a Predator–Prey Relationship. In a predator–prey situation it may happen that one or perhaps both species are valuable sources of food (for example). Or, the prey may be regarded as a pest, leading to efforts to reduce their number. In a constant-effort model of harvesting, we introduce a term $-E_1 x$ in the prey equation and a term $-E_2 y$ in the predator equation. A constant-yield model of harvesting is obtained by including the term $-H_1$ in the prey equation and the term $-H_2$ in the predator equation. The constants E_1, E_2, H_1, and H_2 are always nonnegative. Problems 14 and 15 deal with constant-effort harvesting, and Problem 16 with constant-yield harvesting.

14. Applying a constant-effort model of harvesting to the Lotka–Volterra equations (1), we obtain the system

$$x' = x(a - \alpha y - E_1), \quad y' = y(-c + \gamma x - E_2).$$

When there is no harvesting, the equilibrium solution is $(c/\gamma, a/\alpha)$.
(a) Before doing any mathematical analysis, think about the situation intuitively. How do you think the populations will change if the prey alone is harvested? if the predator alone is harvested? if both are harvested?
(b) How does the equilibrium solution change if the prey is harvested, but not the predator ($E_1 > 0$, $E_2 = 0$)?
(c) How does the equilibrium solution change if the predator is harvested, but not the prey ($E_1 = 0$, $E_2 > 0$)?
(d) How does the equilibrium solution change if both are harvested ($E_1 > 0$, $E_2 > 0$)?

---

[4]See the book by Brauer and Castillo-Chávez for further details.

15. If we modify the Lotka–Volterra equations by including a self-limiting term $-\sigma x^2$ in the prey equation, and then assume constant-effort harvesting, we obtain the equations

$$x' = x(a - \sigma x - \alpha y - E_1),$$
$$y' = y(-c + \gamma x - E_2).$$

In the absence of harvesting, an equilibrium solution is $x = c/\gamma$, $y = (a/\alpha) - (\sigma c)/(\alpha \gamma)$.

(a) How does this equilibrium solution change if the prey is harvested, but not the predator $(E_1 > 0, E_2 = 0)$?

(b) How does this equilibrium solution change if the predator is harvested, but not the prey $(E_1 = 0, E_2 > 0)$?

(c) How does this equilibrium solution change if both are harvested $(E_1 > 0, E_2 > 0)$?

16. In this problem, we apply a constant-yield model of harvesting to the situation in Example 1. Consider the system

$$x' = x(1 - 0.5y) - H_1,$$
$$y' = y(-0.75 + 0.25x) - H_2,$$

where H_1 and H_2 are nonnegative constants. Recall that if $H_1 = H_2 = 0$, then $(3, 2)$ is an equilibrium solution for this system.

(a) Before doing any mathematical analysis, think about the situation intuitively. How do you think the populations will change if the prey alone is harvested? if the predator alone is harvested? if both are harvested?

(b) How does the equilibrium solution change if the prey is harvested, but not the predator $(H_1 > 0, H_2 = 0)$?

(c) How does the equilibrium solution change if the predator is harvested, but not the prey $(H_1 = 0, H_2 > 0)$?

(d) How does the equilibrium solution change if both are harvested $(H_1 > 0, H_2 > 0)$?

7.5 Periodic Solutions and Limit Cycles

In this section, we discuss further the possible existence of periodic solutions of two-dimensional autonomous systems

$$x' = f(x). \tag{1}$$

Such solutions satisfy the relation

$$x(t + T) = x(t) \tag{2}$$

for all t and for some nonnegative constant T called the period. The corresponding trajectories are *closed curves* in the phase plane. Periodic solutions often play an important role in physical problems because they represent phenomena that occur repeatedly. In many situations, a periodic solution represents a "final state" that is approached by all "neighboring" solutions as the transients due to the initial conditions die out.

A special case of a periodic solution is a constant solution $x = x_0$, which corresponds to a critical point of the autonomous system. Such a solution is clearly periodic with any period. In this section, when we speak of a periodic solution, we mean a nonconstant periodic solution. In this case, the period T is positive and is usually chosen as the smallest positive number for which Eq. (2) is valid.

Recall that the solutions of the linear autonomous system

$$x' = Ax \tag{3}$$

are periodic if and only if the eigenvalues of A are pure imaginary. In this case the critical point at the origin is a center. We emphasize that if the eigenvalues of A are pure imaginary, then every solution of the linear system (3) is periodic, while if the eigenvalues are not pure imaginary, then there are no (nonconstant) periodic solutions. The predator–prey equations discussed in Section 7.4, although nonlinear, behave similarly; all solutions in the first quadrant are periodic. The following example illustrates a different way in which periodic solutions of nonlinear autonomous systems can occur.

EXAMPLE 1

Discuss the solutions of the system

$$\begin{pmatrix} x \\ y \end{pmatrix}' = \begin{pmatrix} x + y - x(x^2 + y^2) \\ -x + y - y(x^2 + y^2) \end{pmatrix}. \tag{4}$$

It is not difficult to show that $(0, 0)$ is the only critical point of the system (4), and the system is almost linear in the neighborhood of the origin. The corresponding linear system

$$\begin{pmatrix} x \\ y \end{pmatrix}' = \begin{pmatrix} 1 & 1 \\ -1 & 1 \end{pmatrix} \begin{pmatrix} x \\ y \end{pmatrix} \tag{5}$$

has eigenvalues $1 \pm i$. Therefore the origin is an unstable spiral point for both the linear system (5) and the nonlinear system (4). Thus any solution that starts near the origin in the phase plane will spiral away from the origin. Since there are no other critical points, we might think that all solutions of Eqs. (4) correspond to trajectories that spiral out to infinity. However, we now show that this is incorrect, because far away from the origin the trajectories are directed inward.

It is convenient to introduce polar coordinates r and θ, where

$$x = r \cos \theta, \qquad y = r \sin \theta, \tag{6}$$

and $r \geq 0$. If we multiply the first of Eqs. (4) by x, the second by y, and add, we then obtain

$$x \frac{dx}{dt} + y \frac{dy}{dt} = (x^2 + y^2) - (x^2 + y^2)^2. \tag{7}$$

Since $r^2 = x^2 + y^2$ and $r(dr/dt) = x(dx/dt) + y(dy/dt)$, it follows from Eq. (7) that

$$r \frac{dr}{dt} = r^2(1 - r^2). \tag{8}$$

This is similar to the equations discussed in Section 2.5. The critical points (for $r \geq 0$) are the origin and the point $r = 1$, which corresponds to the unit circle in the phase plane. From Eq. (8), it follows that $dr/dt > 0$ if $r < 1$ and $dr/dt < 0$ if $r > 1$. Thus inside the unit circle the trajectories are directed outward, while outside the unit circle they are directed inward. Apparently, the circle $r = 1$ is a limiting trajectory for this system.

To determine an equation for θ, we multiply the first of Eqs. (4) by y, the second by x, and subtract, obtaining

$$y \frac{dx}{dt} - x \frac{dy}{dt} = x^2 + y^2. \tag{9}$$

Upon calculating dx/dt and dy/dt from Eqs. (6), we find that the left side of Eq. (9) is $-r^2 (d\theta/dt)$, so Eq. (9) reduces to

$$\frac{d\theta}{dt} = -1. \tag{10}$$

The system of equations (8), (10) for r and θ is equivalent to the original system (4). One solution of the system (8), (10) is

$$r = 1, \qquad \theta = -t + t_0, \tag{11}$$

where t_0 is an arbitrary constant. As t increases, a point satisfying Eqs. (11) moves clockwise around the unit circle. Thus the autonomous system (4) has a periodic solution. Other solutions can be obtained by solving Eq. (8) by separation of variables; if $r \neq 0$ and $r \neq 1$, then

$$\frac{dr}{r(1 - r^2)} = dt. \tag{12}$$

Equation (12) can be solved by using partial fractions to rewrite the left side and then integrating. By performing these calculations, we find that the solution of Eqs. (10) and (12) is

$$r = \frac{1}{\sqrt{1 + c_0 e^{-2t}}}, \qquad \theta = -t + t_0, \tag{13}$$

where c_0 and t_0 are arbitrary constants. The solution (13) also contains the solution (11), which is obtained by setting $c_0 = 0$ in the first of Eqs. (13).

The solution satisfying the initial conditions $r = \rho$, $\theta = \alpha$ at $t = 0$ is given by

$$r = \frac{1}{\sqrt{1 + [(1/\rho^2) - 1]e^{-2t}}}, \qquad \theta = -(t - \alpha). \tag{14}$$

If $\rho < 1$, then $r \to 1$ from the inside as $t \to \infty$; if $\rho > 1$, then $r \to 1$ from the outside as $t \to \infty$. Thus, in all cases, the trajectories spiral toward the circle $r = 1$ as $t \to \infty$. Several trajectories are shown in Figure 7.5.1.

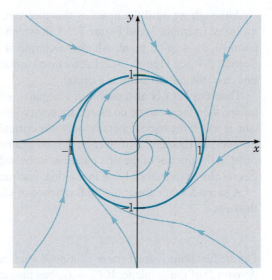

FIGURE 7.5.1 Trajectories of the system (4); a limit cycle.

In this example, the circle $r = 1$ not only corresponds to periodic solutions of the system (4), but it also attracts other nonclosed trajectories that spiral toward it as $t \to \infty$. In general, a closed trajectory in the phase plane such that other nonclosed trajectories spiral toward it, either from the inside or the outside, as $t \to \infty$, is called a **limit cycle**. Thus the circle $r = 1$ is a limit cycle for the system (4). If all trajectories that start near a closed trajectory (both inside and outside) spiral toward the closed trajectory as $t \to \infty$, then the limit cycle is **asymptotically stable**. Since the limiting trajectory is itself a periodic orbit rather than an equilibrium point, this type of stability is often called **orbital stability**. If the trajectories on one side spiral toward the closed trajectory, while those on the other side spiral away as $t \to \infty$, then the limit cycle is said to be **semistable**. If the trajectories on both sides of the closed trajectory spiral away as $t \to \infty$, then the closed trajectory is **unstable**. It is also possible to have closed trajectories that other trajectories neither approach nor depart from—for example, the periodic solutions of the predator–prey equations in Section 7.4. In this case, the closed trajectory is **stable**.

In Example 1, the existence of an asymptotically stable limit cycle was established by solving the equations explicitly. Unfortunately, this is usually not possible, so it is worthwhile to know general theorems concerning the existence or nonexistence of limit cycles of nonlinear autonomous systems. In discussing these theorems, it is convenient to rewrite the system (1) in the scalar form

$$dx/dt = F(x, y), \qquad dy/dt = G(x, y). \tag{15}$$

THEOREM 7.5.1

Let the functions F and G have continuous first partial derivatives in a domain D of the xy-plane. A closed trajectory of the system (15) must necessarily enclose at least one critical (equilibrium) point. If it encloses only one critical point, the critical point cannot be a saddle point.

Although we omit the proof of this theorem, it is easy to show examples of it. One is given by Example 1 and Figure 7.5.1 in which the closed trajectory encloses the critical point $(0, 0)$, a spiral point. Another example is the system of predator–prey equations in Section 7.4; see Figure 7.4.2. Each closed trajectory surrounds the critical point $(3, 2)$; in this case, the critical point is a center.

Theorem 7.5.1 is also useful in a negative sense. If a given region contains no critical points, then there can be no closed trajectory lying entirely in the region. The same conclusion is true if the region contains only one critical point, and this point is a saddle point. For instance, in Example 2 of Section 7.3, an example of competing species, the only critical point in the interior of the first quadrant is the saddle point $(0.5, 0.5)$. Therefore this system has no closed trajectory lying in the first quadrant.

A second result about the nonexistence of closed trajectories is given in the following theorem.

THEOREM 7.5.2

Let the functions F and G have continuous first partial derivatives in a simply connected domain D of the xy-plane. If $F_x + G_y$ has the same sign throughout D, then there is no closed trajectory of the system (15) lying entirely in D.

A simply connected two-dimensional domain is one with no holes. Theorem 7.5.2 is a straightforward consequence of Green's theorem in the plane; see Problem 13. Note that if $F_x + G_y$ changes sign in the domain, then no conclusion can be drawn; there may or may not be closed trajectories in D.

To illustrate Theorem 7.5.2, consider the system (4). A routine calculation shows that

$$F_x(x, y) + G_y(x, y) = 2 - 4(x^2 + y^2) = 2(1 - 2r^2), \tag{16}$$

where, as usual, $r^2 = x^2 + y^2$. Hence $F_x + G_y$ is positive for $0 \le r < 1/\sqrt{2}$, so there is no closed trajectory in this circular disk. Of course, we showed in Example 1 that there is no closed trajectory in the larger region $r < 1$. This illustrates that the information given by Theorem 7.5.2 may not be the best possible result. Again referring to Eq. (16), note that $F_x + G_y < 0$ for $r > 1\sqrt{2}$. However, the theorem is not applicable in this case because this annular region is not simply connected. Indeed, as shown in Example 1, it does contain a limit cycle.

The following theorem gives conditions that guarantee the existence of a closed trajectory.

THEOREM 7.5.3	**(Poincaré–Bendixson Theorem)** Let the functions F and G have continuous first partial derivatives in a domain D of the xy-plane. Let D_1 be a bounded subdomain in D, and let R be the region that consists of D_1 plus its boundary (all points of R are in D). Suppose that R contains no critical point of the system (15). If there exists a constant t_0 such that $x = \phi(t), y = \psi(t)$ is a solution of the system (15) that exists and stays in R for all $t \geq t_0$, then either $x = \phi(t), y = \psi(t)$ is a periodic solution (closed trajectory), or $x = \phi(t), y = \psi(t)$ spirals toward a closed trajectory as $t \to \infty$. In either case, the system (15) has a periodic solution in R.

Note that if R does contain a closed trajectory, then necessarily, by Theorem 7.5.1, this trajectory must enclose a critical point. However, this critical point cannot be in R. Thus R cannot be simply connected; it must have a hole.

As an application of the Poincaré–Bendixson theorem, consider again the system (4). Since the origin is a critical point, it must be excluded. For instance, we can consider the region R defined by $0.5 \leq r \leq 2$. Next, we must show that there is a solution whose trajectory stays in R for all t greater than or equal to some t_0. This follows immediately from Eq. (8). For $r = 0.5$, $dr/dt > 0$, so r increases, while for $r = 2$, $dr/dt < 0$, so r decreases. Thus any trajectory that crosses the boundary of R is entering R. Consequently, any solution of Eqs. (4) that starts in R at $t = t_0$ cannot leave but must stay in R for $t > t_0$. Of course, other numbers could be used instead of 0.5 and 2; all that is important is that $r = 1$ is included.

One should not infer from this discussion of the preceding theorems that it is easy to determine whether a given nonlinear autonomous system has periodic solutions; often it is not a simple matter at all. Theorems 7.5.1 and 7.5.2 are frequently inconclusive, and for Theorem 7.5.3 it is often difficult to determine a region R and a solution that always remains within it.

We close this section with another example of a nonlinear system that has a limit cycle.

EXAMPLE 2	The van der Pol equation

$$u'' - \mu(1 - u^2)u' + u = 0, \qquad (17)$$

where μ is a nonnegative constant, describes the current u in a triode oscillator. Discuss the solutions of this equation.

If $\mu = 0$, Eq. (17) reduces to $u'' + u = 0$, whose solutions are sine or cosine waves of period 2π. For $\mu > 0$, the second term on the left side of Eq. (17) must also be considered. This is the resistance term, proportional to u', with a coefficient $-\mu(1-u^2)$ that depends on u. For large u, this term is positive and acts as usual to reduce the amplitude of the response. However, for small u, the resistance term is negative and so causes the response to grow. This suggests that perhaps there is a solution of intermediate size that other solutions approach as t increases.

To analyze Eq. (17) more carefully, we write it as a two-dimensional system by introducing the variables $x = u, y = u'$. Then it follows that

$$x' = y, \qquad y' = -x + \mu(1 - x^2)y. \qquad (18)$$

The only critical point of the system (18) is the origin. Near the origin the corresponding linear system is

$$\begin{pmatrix} x \\ y \end{pmatrix}' = \begin{pmatrix} 0 & 1 \\ -1 & \mu \end{pmatrix} \begin{pmatrix} x \\ y \end{pmatrix}, \tag{19}$$

whose eigenvalues are $(\mu \pm \sqrt{\mu^2 - 4})/2$. Thus the origin is an unstable spiral point for $0 < \mu < 2$ and an unstable node for $\mu \geq 2$. In all cases, a solution that starts near the origin grows as t increases.

With regard to periodic solutions, Theorems 7.5.1 and 7.5.2 provide only partial information. From Theorem 7.5.1 we conclude that if there are closed trajectories, they must enclose the origin. Next we calculate $F_x(x, y) + G_y(x, y)$, with the result that

$$F_x(x, y) + G_y(x, y) = \mu(1 - x^2). \tag{20}$$

Then it follows from Theorem 7.5.2 that closed trajectories, if there are any, are not contained in the strip $|x| < 1$ where $F_x + G_y > 0$.

The application of the Poincaré–Bendixson theorem to this problem is not nearly as simple as for Example 1. If we introduce polar coordinates, we find that the equation for the radial variable r is

$$r' = \mu(1 - r^2 \cos^2 \theta) r \sin^2 \theta. \tag{21}$$

Again, consider an annular region R given by $r_1 \leq r \leq r_2$, where r_1 is small and r_2 is large. When $r = r_1$, the linear term on the right side of Eq. (21) dominates, and $r' > 0$ except on the x-axis, where $\sin \theta = 0$ and consequently $r' = 0$ also. Thus trajectories are entering R at every point on the circle $r = r_1$, except possibly for those on the x-axis, where the trajectories are tangent to the circle. When $r = r_2$, the cubic term on the right side of Eq. (21) is the dominant one. Thus $r' < 0$, except for points on the x-axis where $r' = 0$ and for points near the y-axis where $r^2 \cos^2 \theta < 1$ and the linear term makes $r' > 0$. Thus, no matter how large a circle is chosen, there will be points on it (namely, the points on or near the y-axis) where trajectories are leaving R. Therefore, the Poincaré–Bendixson theorem is not applicable unless we consider more complicated regions.

It is possible to show, by a more intricate analysis, that the van der Pol equation does have a unique limit cycle. However, we will not follow this line of argument further. We turn instead to a different approach in which we plot numerically computed solutions. Experimental

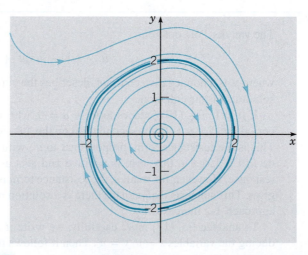

FIGURE 7.5.2 Trajectories of the van der Pol equation (17) for $\mu = 0.2$.

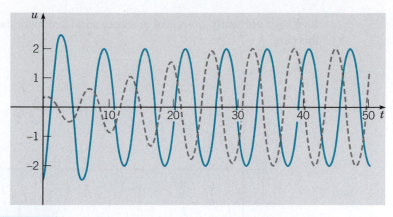

Plots of u versus t for the trajectories in Figure 7.5.2.

observations indicate that the van der Pol equation has an asymptotically stable periodic solution whose period and amplitude depend on the parameter μ. By looking at graphs of trajectories in the phase plane and of u versus t, we can gain some understanding of this periodic behavior.

Figure 7.5.2 shows two trajectories of the van der Pol equation in the phase plane for $\mu = 0.2$. The trajectory starting near the origin spirals outward in the clockwise direction. This is consistent with the behavior of the linear approximation near the origin. The other trajectory passes through $(-3, 2)$ and spirals inward, again in the clockwise direction. Both trajectories approach a closed curve that corresponds to an asymptotically stable periodic solution. In Figure 7.5.3, we show the plots of u versus t for the solutions corresponding to the trajectories in Figure 7.5.2. The solution that is initially smaller gradually increases in amplitude, while the larger solution gradually decays. Both solutions approach a stable periodic motion that corresponds to the limit cycle. Figure 7.5.3 also shows that there is a phase difference between the two solutions as they approach the limit cycle. The plots of

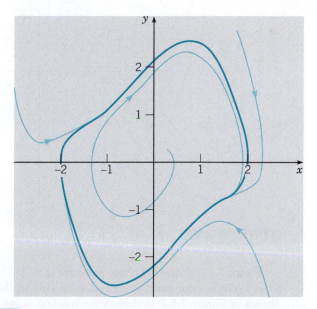

Trajectories of the van der Pol equation (17) for $\mu = 1$.

u versus *t* are nearly sinusoidal in shape, consistent with the nearly circular limit cycle in this case.

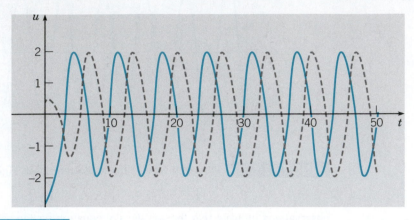

FIGURE 7.5.5 Plots of *u* versus *t* for the trajectories in Figure 7.5.4.

Figures 7.5.4 and 7.5.5 show similar plots for the case $\mu = 1$. Trajectories again move clockwise in the phase plane, but the limit cycle is considerably different from a circle. The plots of *u* versus *t* tend more rapidly to the limiting oscillation, and again show a phase difference. The oscillations are somewhat less symmetric in this case, rising somewhat more steeply than they fall.

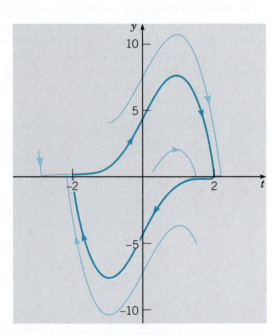

FIGURE 7.5.6 Trajectories of the van der Pol equation (17) for $\mu = 5$.

Figure 7.5.6 shows the phase plane for $\mu = 5$. The motion remains clockwise, and the limit cycle is even more elongated, especially in the *y* direction. A plot of *u* versus *t* is shown in Figure 7.5.7. Although the solution starts far from the limit cycle, the limiting oscillation is virtually reached in a fraction of a period. Starting from one of its extreme

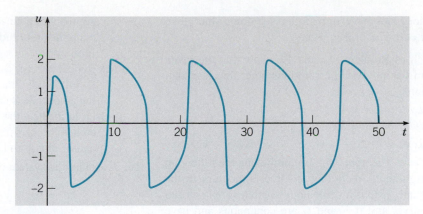

FIGURE 7.5.7   Plot of u versus t for the outward spiralling trajectory in Figure 7.5.6.

values on the x-axis in the phase plane, the solution moves toward the other extreme position slowly at first, but once a certain point on the trajectory is reached, the remainder of the transition is completed very swiftly. The process is then repeated in the opposite direction. The waveform of the limit cycle, as shown in Figure 7.5.7, is quite different from a sine wave.

These graphs clearly show that, in the absence of external excitation, the van der Pol oscillator has a certain characteristic mode of vibration for each value of μ. The graphs of u versus t show that the amplitude of this oscillation changes very little with μ, but the period increases as μ increases. At the same time, the waveform changes from one that is very nearly sinusoidal to one that is much less smooth.

The presence of a single periodic motion that attracts all (nearby) solutions, that is, an asymptotically stable limit cycle, is one of the characteristic phenomena associated with nonlinear differential equations.

PROBLEMS

In each of Problems 1 through 6, an autonomous system is expressed in polar coordinates. Determine all periodic solutions, all limit cycles, and determine their stability characteristics.

1. $dr/dt = r^2 (1 - r^2), \qquad d\theta/dt = 1$

2. $dr/dt = r (1 - r)^2, \qquad d\theta/dt = -1$

3. $dr/dt = r (r - 1)(r - 3), \qquad d\theta/dt = 1$

4. $dr/dt = r (1 - r)(r - 2), \qquad d\theta/dt = -1$

5. $dr/dt = \sin \pi r, \qquad d\theta/dt = 1$

6. $dr/dt = r \, |r - 2| \, (r - 3), \qquad d\theta/dt = -1$

7. If $x = r \cos\theta$, $y = r \sin\theta$, show that
   $$y (dx/dt) - x (dy/dt) = -r^2 (d\theta/dt).$$

8. **(a)** Show that the system
$$dx/dt = -y + xf(r)/r, \qquad dy/dt = x + yf(r)/r$$

has periodic solutions corresponding to the zeros of $f(r)$. What is the direction of motion on the closed trajectories in the phase plane?

   (b) Let $f(r) = r (r - 2)^2 (r^2 - 4r + 3)$. Determine all periodic solutions and determine their stability characteristics.

9. Determine the periodic solutions, if any, of the system
$$\frac{dx}{dt} = y + \frac{x}{\sqrt{x^2 + y^2}}(x^2 + y^2 - 2),$$
$$\frac{dy}{dt} = -x + \frac{y}{\sqrt{x^2 + y^2}}(x^2 + y^2 - 2).$$

10. Using Theorem 7.5.2, show that the linear autonomous system
$$dx/dt = a_{11}x + a_{12}y, \qquad dy/dt = a_{21}x + a_{22}y$$

does not have a periodic solution (other than $x = 0, y = 0$) if $a_{11} + a_{22} \neq 0$.

In each of Problems 11 and 12, show that the given system has no periodic solutions other than constant solutions.

11. $dx/dt = x + y + x^3 - y^2$,
    $dy/dt = -x + 2y + x^2 y + y^3/3$

12. $dx/dt = -2x - 3y - xy^2$, $\quad dy/dt = y + x^3 - x^2 y$

13. Prove Theorem 7.5.2 by completing the following argument. According to Green's theorem in the plane, if C is a sufficiently smooth simple closed curve, and if F and G are continuous and have continuous first partial derivatives, then

$$\int_C [F(x, y)\, dy - G(x, y)\, dx]$$

$$= \iint_R \left[F_x(x, y) + G_y(x, y) \right] dA,$$

where C is traversed counterclockwise and R is the region enclosed by C. Assume that $x = \phi(t)$, $y = \psi(t)$ is a solution of the system (15) that is periodic with period T. Let C be the closed curve given by $x = \phi(t), y = \psi(t)$ for $0 \leq t \leq T$. Show that the line integral is zero for this curve. Then show that the conclusion of Theorem 7.5.2 must follow.

ODEA 14. (a) By examining the graphs of u versus t in Figures 7.5.3, 7.5.5, and 7.5.7, estimate the period T of the van der Pol oscillator in these cases.

(b) Calculate and plot the graphs of solutions of the van der Pol equation for other values of the parameter μ. Estimate the period T in these cases also.

(c) Plot the estimated values of T versus μ. Describe how T depends on μ.

ODEA 15. The equation

$$u'' - \mu(1 - \tfrac{1}{3}u'^2)\, u' + u = 0$$

is often called the Rayleigh equation.

(a) Write the Rayleigh equation as a system of two first order equations.

(b) Show that the origin is the only critical point of this system. Determine its type and whether it is asymptotically stable, stable, or unstable.

(c) Let $\mu = 1$. Choose initial conditions and compute the corresponding solution of the system on an interval such as $0 \leq t \leq 20$ or longer. Plot u versus t and also plot the trajectory in the phase plane. Observe that the trajectory

approaches a closed curve (limit cycle). Estimate the amplitude A and the period T of the limit cycle.

(d) Repeat part (c) for other values of μ, such as $\mu = 0.2, 0.5, 2$, and 5. In each case, estimate the amplitude A and the period T.

(e) Describe how the limit cycle changes as μ increases. For example, make a table of values and/or plot A and T as functions of μ.

16. Consider the system of equations

$$x' = \mu x + y - x(x^2 + y^2),$$

$$y' = -x + \mu y - y(x^2 + y^2), \tag{i}$$

where μ is a parameter of unspecified sign. Observe that this system is the same as the one in Example 1, except for the introduction of μ.

(a) Show that the origin is the only critical point.

(b) Find the linear system that approximates Eqs. (i) near the origin and find its eigenvalues. Determine the type and stability of the critical point at the origin. How does this classification depend on μ?

(c) Referring to Example 1 if necessary, rewrite Eqs. (i) in polar coordinates.

(d) Show that when $\mu > 0$ there is a periodic solution $r = \sqrt{\mu}$. By solving the system found in part (c), or by plotting numerically computed solutions, conclude that this periodic solution attracts all other nonzero solutions.

Note: As the parameter μ increases through the value zero, the previously asymptotically stable critical point at the origin loses its stability, and simultaneously a new asymptotically stable solution (the limit cycle) emerges. Thus the point $\mu = 0$ is a bifurcation point; this type of bifurcation is called a **Hopf bifurcation**.

17. Consider the van der Pol system   ODEA

$$x' = y, \quad y' = -x + \mu(1 - x^2)y,$$

where we now allow the parameter μ to be any real number.

(a) Show that the origin is the only critical point. Determine its type and stability property, and how these depend on μ.

(b) Let $\mu = -1$; draw a phase portrait and conclude that there is a periodic solution that surrounds the origin. Observe that this periodic solution is unstable. Compare your plot with Figure 7.5.4.

(c) Draw a phase portrait for a few other negative values of μ. Describe how the shape of the periodic solution changes with μ.

(d) Consider small positive or negative values of μ. By drawing phase portraits, determine how the periodic solution changes as $\mu \to 0$. Compare the behavior of the van der Pol system as μ increases through zero with the behavior of the system in Problem 16.

Problems 18 and 19 extend the consideration of the Rosenzweig–MacArthur predator-prey model introduced in Problem 13 of Section 7.4.

ODEA 18. Consider the system

$$x' = x\left(2.4 - 0.2x - \frac{2y}{x+6}\right),$$

$$y' = y\left(-0.25 + \frac{x}{x+6}\right).$$

Observe that this system differs from that in Problem 13 of Section 7.4 only in the growth rate for the prey.

(a) Find all of the critical points.

(b) Determine the type and stability of each critical point.

(c) Draw a phase portrait in the first quadrant and conclude that there is an asymptotically stable limit cycle. Thus this model predicts a stable long-term oscillation of the prey and predator populations.

ODEA 19. Consider the system

$$x' = x\left(a - 0.2x - \frac{2y}{x+6}\right),$$

$$y' = y\left(-0.25 + \frac{x}{x+6}\right),$$

where a is a positive parameter. Observe that this system includes the one in Problem 18 above and also the one in Problem 13 in Section 7.4.

(a) Find all of the critical points.

(b) Consider the critical point in the interior of the first quadrant. Find the eigenvalues of the approximate linear system. Determine the value a_0 where this critical point changes from asymptotically stable to unstable.

(c) Draw a phase portrait for a value of a slightly greater than a_0. Observe that a limit cycle has appeared. How does the limit cycle change as a increases further?

20. There are certain chemical reactions in which the ODEA constituent concentrations oscillate periodically over time. The system

$$x' = 1 - (b+1)x + x^2 y/4, \qquad y' = bx - x^2 y/4$$

is a special case of a model, known as the Brusselator, of this kind of reaction. Assume that b is a positive parameter, and consider solutions in the first quadrant of the xy-plane.

(a) Show that the only critical point is $(1, 4b)$.

(b) Find the eigenvalues of the approximate linear system at the critical point.

(c) Classify the critical point as to type and stability. How does the classification depend on b?

(d) As b increases through a certain value b_0, the critical point changes from asymptotically stable to unstable. What is that value b_0?

(e) Plot trajectories in the phase plane for values of b slightly less than and slightly greater than b_0. Observe the limit cycle when $b > b_0$; the Brusselator has a Hopf bifurcation point at b_0.

(f) Plot trajectories for several values of $b > b_0$ and observe how the limit cycle deforms as b increases.

21. The system ODEA

$$x' = 3(x + y - \tfrac{1}{3}x^3 - k),$$

$$y' = -\tfrac{1}{3}(x + 0.8y - 0.7)$$

is a special case of the Fitzhugh–Nagumo equations, which model the transmission of neural impulses along an axon. The parameter k is the external stimulus.

(a) Show that the system has one critical point regardless of the value of k.

(b) Find the critical point for $k = 0$ and show that it is an asymptotically stable spiral point. Repeat the analysis for $k = 0.5$ and show that the critical point is now an unstable spiral point. Draw a phase portrait for the system in each case.

(c) Find the value k_0 where the critical point changes from asymptotically stable to unstable. Find the critical point and draw a phase portrait for the system for $k = k_0$.

(d) For $k > k_0$ the system exhibits an asymptotically stable limit cycle; the system has a Hopf bifurcation point at k_0. Draw a phase portrait for $k = 0.4, 0.5$, and 0.6. Observe that the limit cycle is not small when k is near k_0. Also plot x versus t and estimate the period T in each case.

(e) As k increases further, there is a value k_1 at which the critical point again becomes asymptotically stable and the limit cycle vanishes. Find k_1.

7.6 Chaos and Strange Attractors: The Lorenz Equations

In principle, the methods described in this chapter for two-dimensional autonomous systems can be applied to higher dimensional systems as well. In practice, several difficulties arise when we try to do this. One problem is that there is simply a greater number of possible cases that can occur, and the number increases with the dimension of the system and its phase space. Another problem is the difficulty of graphing trajectories accurately in a phase space of more than two dimensions. Even in three dimensions, it may not be easy to construct a clear and understandable plot of the trajectories, and it becomes more difficult as the number of variables increases. Finally, and this has been clearly realized only in the last thirty years or so, there are different and very complex phenomena that can occur, and frequently do occur, in systems of three or more dimensions that are not present in two-dimensional systems. Our goal in this section is to provide a brief introduction to some of these phenomena by discussing one particular three-dimensional autonomous system that has been intensively studied.

An important problem in meteorology, and in other applications of fluid dynamics, concerns the motion of a layer of fluid, such as the earth's atmosphere, that is warmer at the bottom than at the top; see Figure 7.6.1. If the vertical temperature difference ΔT is small, then there is a linear variation of temperature with altitude, but no significant motion of the fluid layer. However, if ΔT is large enough, then the warmer air rises, displacing the cooler air above it, and a steady convective motion results. If the temperature difference increases further, then eventually the steady convective flow breaks up and a more complex and turbulent motion ensues.

FIGURE 7.6.1 A layer of fluid heated from below.

While investigating this phenomenon, Edward N. Lorenz was led (by a process too involved to describe here) to the nonlinear autonomous three-dimensional system

$$
\begin{aligned}
dx/dt &= \sigma(-x + y), \\
dy/dt &= rx - y - xz, \\
dz/dt &= -bz + xy.
\end{aligned}
\tag{1}
$$

Equations (1) are now commonly referred to as the Lorenz equations.[5] Observe that the second and third equations involve quadratic nonlinearities. However, except for being a

---

[5] A very thorough treatment of the Lorenz equations appears in the book by Sparrow listed in the references.

three-dimensional system, superficially the Lorenz equations appear no more complicated than the competing species or predator–prey equations discussed in Sections 7.3 and 7.4. The variable x in Eqs. (1) is related to the intensity of the fluid motion, while the variables y and z are related to the temperature variations in the horizontal and vertical directions. The Lorenz equations also involve three parameters σ, r, and b, all of which are real and positive. The parameters σ and b depend on the material and geometrical properties of the fluid layer. For the earth's atmosphere, reasonable values of these parameters are $\sigma = 10$ and $b = 8/3$; they will be assigned these values in much of what follows in this section. The parameter r, on the other hand, is proportional to the temperature difference ΔT, and our purpose is to investigate how the nature of the solutions of Eqs. (1) changes with r.

The first step in analyzing the Lorenz equations is to locate the critical points by solving the algebraic system

$$\sigma x - \sigma y = 0,$$
$$rx - y - xz = 0, \tag{2}$$
$$-bz + xy = 0.$$

From the first equation, we have $y = x$. Then, eliminating y from the second and third equations, we obtain

$$x(r - 1 - z) = 0, \tag{3}$$
$$-bz + x^2 = 0. \tag{4}$$

One way to satisfy Eq. (3) is to choose $x = 0$. Then it follows that $y = 0$ and, from Eq. (4), $z = 0$. Alternatively, we can satisfy Eq. (3) by choosing $z = r - 1$. Then Eq. (4) requires that $x = \pm\sqrt{b(r-1)}$ and then $y = \pm\sqrt{b(r-1)}$ also. Observe that these expressions for x and y are real only when $r \geq 1$. Thus $(0, 0, 0)$, which we will denote by P_1, is a critical point for all values of r, and it is the only critical point for $r < 1$. However, when $r > 1$, there are also two other critical points, namely $(\sqrt{b(r-1)}, \sqrt{b(r-1)}, r-1)$, and $(-\sqrt{b(r-1)}, -\sqrt{b(r-1)}, r-1)$. We will denote the latter two points by P_2 and P_3, respectively. Note that all three critical points coincide when $r = 1$. As r increases through the value 1, the critical point P_1 at the origin *bifurcates*, and the critical points P_2 and P_3 come into existence.

Next we will determine the local behavior of solutions in the neighborhood of each critical point. Although much of the following analysis can be carried out for arbitrary values of σ and b, we will simplify our work by using the values $\sigma = 10$ and $b = 8/3$. Near the origin (the critical point P_1) the approximating linear system is

$$\begin{pmatrix} x \\ y \\ z \end{pmatrix}' = \begin{pmatrix} -10 & 10 & 0 \\ r & -1 & 0 \\ 0 & 0 & -8/3 \end{pmatrix} \begin{pmatrix} x \\ y \\ z \end{pmatrix}. \tag{5}$$

The eigenvalues are determined from the equation

$$\begin{vmatrix} -10 - \lambda & 10 & 0 \\ r & -1 - \lambda & 0 \\ 0 & 0 & -8/3 - \lambda \end{vmatrix} = -(8/3 + \lambda)[\lambda^2 + 11\lambda - 10(r - 1)] = 0. \tag{6}$$

Therefore

$$\lambda_1 = -\frac{8}{3}, \qquad \lambda_2 = \frac{-11 - \sqrt{81 + 40r}}{2}, \qquad \lambda_3 = \frac{-11 + \sqrt{81 + 40r}}{2}. \tag{7}$$

Note that all three eigenvalues are negative for $r < 1$. For example, when $r = 1/2$, the eigenvalues are $\lambda_1 = -8/3$, $\lambda_2 = -10.52494$, $\lambda_3 = -0.47506$. Hence the origin is asymptotically stable for this range of r, both for the linear approximation (5) and for the original system (1). However, λ_3 changes sign when $r = 1$ and is positive for $r > 1$. The value $r = 1$ corresponds to the initiation of convective flow in the physical problem described earlier. The origin is unstable for $r > 1$. All solutions starting near the origin tend to grow, except for those lying precisely in the plane determined by the eigenvectors associated with λ_1 and λ_2 (or, for the nonlinear system (1), in a certain surface tangent to this plane at the origin).

Next, let us consider the neighborhood of the critical point $P_2(\sqrt{8(r-1)/3}, \sqrt{8(r-1)/3}, r-1)$ for $r > 1$. If u, v, and w are the perturbations from the critical point in the x, y, and z directions, respectively, then the approximating linear system is

$$\begin{pmatrix} u \\ v \\ w \end{pmatrix}' = \begin{pmatrix} -10 & 10 & 0 \\ 1 & -1 & -\sqrt{8(r-1)/3} \\ \sqrt{8(r-1)/3} & \sqrt{8(r-1)/3} & -8/3 \end{pmatrix} \begin{pmatrix} u \\ v \\ w \end{pmatrix}. \tag{8}$$

The eigenvalues of the coefficient matrix of Eq. (8) are determined from the equation

$$3\lambda^3 + 41\lambda^2 + 8(r+10)\lambda + 160(r-1) = 0, \tag{9}$$

which is obtained by straightforward algebraic steps that are omitted here. The solutions of Eq. (9) depend on r in the following way:

For $1 < r < r_1 \cong 1.3456$, there are three negative real eigenvalues.

For $r_1 < r < r_2 \cong 24.737$, there is one negative real eigenvalue and two complex eigenvalues with negative real part.

For $r_2 < r$, there is one negative real eigenvalue and two complex eigenvalues with positive real part.

The same results are obtained for the critical point P_3. Thus there are several different situations.

For $0 < r < 1$, the only critical point is P_1 and it is asymptotically stable. All solutions approach this point (the origin) as $t \to \infty$.

For $1 < r < r_1$, the critical points P_2 and P_3 are asymptotically stable and P_1 is unstable. All nearby solutions approach one or the other of the points P_2 and P_3 exponentially.

For $r_1 < r < r_2$, the critical points P_2 and P_3 are asymptotically stable and P_1 is unstable. All nearby solutions approach one or the other of the points P_2 and P_3; most of them spiral inward to the critical point.

For $r_2 < r$, all three critical points are unstable. Most solutions near P_2 or P_3 spiral away from the critical point.

However, this is by no means the end of the story. Let us consider solutions for r somewhat greater than r_2. In this case, P_1 has one positive eigenvalue and each of P_2 and P_3 has a pair of complex eigenvalues with positive real part. A trajectory can approach any one of the critical points only on certain highly restricted paths. The slightest deviation from these paths causes the trajectory to depart from the critical point. Since none of the critical points is asymptotically stable, one might expect that most trajectories would approach infinity for large t. However, it can be shown that all solutions remain bounded as $t \to \infty$; see Problem 4. In fact, it can be shown that all solutions ultimately approach a certain limiting set of points that has zero volume. Indeed, this is true, not only for $r > r_2$, but for all positive values of r.

A plot of computed values of x versus t for a typical solution with $r > r_2$ is shown in Figure 7.6.2. Note that the solution oscillates back and forth between positive and negative values in a rather erratic manner. Indeed, the graph of x versus t resembles a random vibration, although the Lorenz equations are entirely deterministic and the solution is completely determined by the initial conditions. Nevertheless, the solution also exhibits a certain *regularity* in that the frequency and amplitude of the oscillations are essentially constant in time.

The solutions of the Lorenz equations are also extremely sensitive to perturbations in the initial conditions. Figure 7.6.3 shows the graphs of computed values of x versus t for the two solutions whose initial points are (5, 5, 5) and (5.01, 5, 5). The dashed graph is the same as the one in Figure 7.6.2, while the solid graph starts at a nearby point. The two solutions remain close until t is near 10, after which they are quite different and, indeed, seem to have no relation to each other. It was this property that particularly attracted the attention of Lorenz in his original study of these equations, and caused him to conclude that detailed long-range weather predictions are probably not possible.

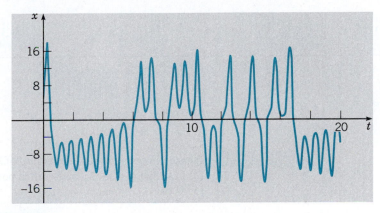

FIGURE 7.6.2 A plot of x versus t for the Lorenz equations (1) with $r = 28$; the initial point is (5, 5, 5).

The attracting set in this case, although of zero volume, has a rather complicated structure and is called a **strange attractor**. The term **chaotic** has come into general use to describe solutions such as those shown in Figures 7.6.2 and 7.6.3.

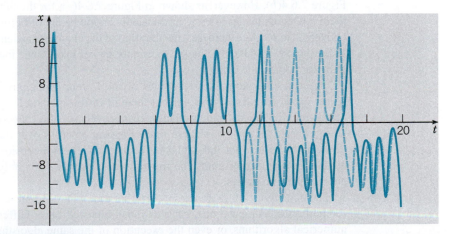

FIGURE 7.6.3 Plots of x versus t for two initially nearby solutions of Lorenz equations with $r = 28$; the initial point is (5, 5, 5) for the dashed curve and is (5.01, 5, 5) for the solid curve.

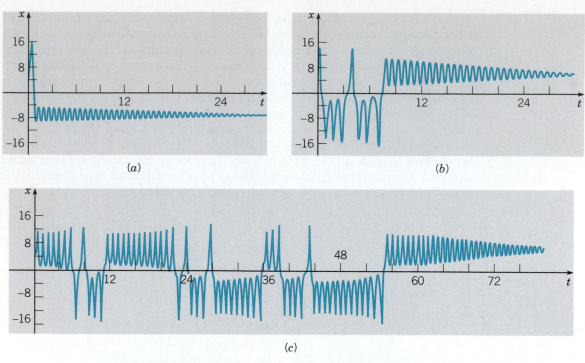

FIGURE 7.6.4 Plots of x versus t for three solutions of Lorenz equations with $r = 21$. (a) Initial point is (3, 8, 0). (b) Initial point is (5, 5, 5). (c) Initial point is (5, 5, 10).

To determine how and when the strange attractor is created, it is illuminating to investigate solutions for smaller values of r. For $r = 21$, solutions starting at three different initial points are shown in Figure 7.6.4. For the initial point (3, 8, 0), the solution begins to converge to the point P_3 almost at once; see Figure 7.6.4(a). For the second initial point (5, 5, 5), there is a fairly short interval of transient behavior, after which the solution converges to P_2; see Figure 7.6.4(b). However, as shown in Figure 7.6.4(c), for the third initial point (5, 5, 10), there is a much longer interval of transient chaotic behavior before the solution eventually converges to P_2. As r increases, the duration of the chaotic transient behavior also increases. When $r = r_3 \cong 24.06$, the chaotic transients appear to continue indefinitely, and the strange attractor comes into being.

One can also show the trajectories of the Lorenz equations in the three-dimensional phase space, or at least projections of them in various planes. Figures 7.6.5 and 7.6.6 show projections in the xy- and xz-planes, respectively, of the trajectory starting at (5, 5, 5). Observe that the graphs in these figures appear to cross over themselves repeatedly, but this cannot be true for the actual trajectories in three-dimensional space because of the general uniqueness theorem. The apparent crossings are due wholly to the two-dimensional character of the figures.

The sensitivity of solutions to perturbations of the initial data also has implications for numerical computations, such as those reported here. Different step sizes, different numerical algorithms, or even the execution of the same algorithm on different machines will introduce small differences in the computed solution, which eventually lead to large deviations. For example, the exact sequence of positive and negative loops in the calculated solution depends strongly on the precise numerical algorithm and its implementation, as

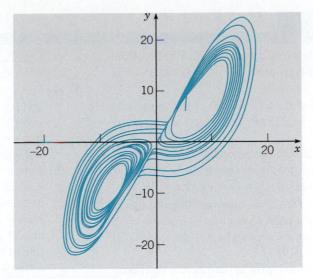

FIGURE 7.6.5   Projections of a trajectory of the Lorenz equations (with $r = 28$) in the xy-plane.

well as on the initial conditions. However, the general appearance of the solution and the structure of the attracting set are independent of all these factors.

Solutions of the Lorenz equations for other parameter ranges exhibit other interesting types of behavior. For example, for certain values of r greater than r_2, intermittent bursts of chaotic behavior separate long intervals of apparently steady periodic oscillation. For other ranges of r, solutions show a period-doubling property. Some of these features are taken up in the problems.

Since about 1975, the Lorenz equations and other higher dimensional autonomous systems have been studied intensively, and this is one of the most active areas of current mathematical research. Chaotic behavior of solutions appears to be much more common than was suspected at first, and many questions remain unanswered. Some of these are mathematical in nature, while others relate to the physical applications or interpretations of solutions.

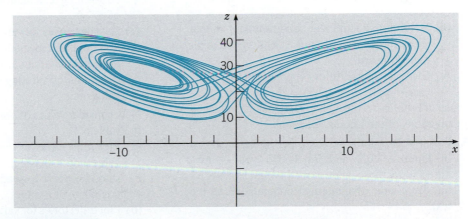

FIGURE 7.6.6   Projections of a trajectory of the Lorenz equations (with $r = 28$) in the xz-plane.

PROBLEMS

Problems 1 through 3 ask you to fill in some of the details of the analysis of the Lorenz equations in this section.

1. **(a)** Show that the eigenvalues of the linear system (5), valid near the origin, are given by Eq. (7).
   (b) Determine the corresponding eigenvectors.
   (c) Determine the eigenvalues and eigenvectors of the system (5) in the case where $r = 28$.

2. **(a)** Show that the linear approximation valid near the critical point P_2 is given by Eq. (8).
   (b) Show that the eigenvalues of the system (8) satisfy Eq. (9).
   (c) For $r = 28$, solve Eq. (9) and thereby determine the eigenvalues of the system (8).

CAS 3. **(a)** By solving Eq. (9) numerically, show that the real part of the complex roots changes sign when $r \cong 24.737$.
   (b) Show that a cubic polynomial $x^3 + Ax^2 + Bx + C$ has one real zero and two pure imaginary zeros only if $AB = C$.
   (c) By applying the result of part **(b)** to Eq. (9), show that the real part of the complex roots changes sign when $r = 470/19$.

4. Consider the ellipsoid

$$V(x, y, z) = rx^2 + \sigma y^2 + \sigma(z - 2r)^2 = c > 0.$$

   (a) Calculate

$$\frac{dV}{dt} = \frac{\partial V}{\partial x}\frac{dx}{dt} + \frac{\partial V}{\partial y}\frac{dy}{dt} + \frac{\partial V}{\partial z}\frac{dz}{dt}$$

   along trajectories of the Lorenz equations (1).
   (b) Determine a sufficient condition on c so that every trajectory crossing $V(x, y, z) = c$ is directed inward.
   (c) Evaluate the condition found in part **(b)** for the case $\sigma = 10$, $b = 8/3$, $r = 28$.

In each of Problems 5 through 7, carry out the indicated investigations of the Lorenz equations.

CAS 5. For $r = 28$, plot x versus t for the cases shown in Figures 7.6.2 and 7.6.3. Do your graphs agree with those shown in the figures? Recall the discussion of numerical computation in the text.

CAS 6. For $r = 28$, plot the projections in the xy- and xz-planes, respectively, of the trajectory starting at the point (5, 5, 5). Do the graphs agree with those in Figures 7.6.5 and 7.6.6?

CAS 7. **(a)** For $r = 21$, plot x versus t for the solutions starting at the initial points (3, 8, 0), (5, 5, 5), and (5, 5, 10). Use a t interval of at least $0 \le t \le 30$. Compare your graphs with those in Figure 7.6.4.
   (b) Repeat the calculation in part **(a)** for $r = 22$, $r = 23$, and $r = 24$. Increase the t interval as necessary so that you can determine when each solution begins to converge to one of the critical points. Record the approximate duration of the chaotic transient in each case. Describe how this quantity depends on the value of r.
   (c) Repeat the calculations in parts **(a)** and **(b)** for values of r slightly greater than 24. Try to estimate the value of r for which the duration of the chaotic transient approaches infinity.

8. For certain r intervals, or windows, the Lorenz CAS equations exhibit a period-doubling property. Careful calculations may reveal this phenomenon.
   (a) One period-doubling window contains the value $r = 100$. Let $r = 100$ and plot the trajectory starting at (5, 5, 5) or some other initial point of your choice. Does the solution appear to be periodic? What is the period?
   (b) Repeat the calculation in part **(a)** for slightly smaller values of r. When $r \cong 99.98$, you may be able to observe that the period of the solution doubles. Try to observe this result by performing calculations with nearby values of r.
   (c) As r decreases further, the period of the solution doubles repeatedly. The next period doubling occurs at about $r = 99.629$. Try to observe this by plotting trajectories for nearby values of r.

9. Now consider values of r slightly larger than those CAS in Problem 8.
   (a) Plot trajectories of the Lorenz equations for values of r between 100 and 100.78. You should observe a steady periodic solution for this range of r values.
   (b) Plot trajectories for values of r between 100.78 and 100.8. Determine, as best you can, how and when the periodic trajectory breaks up.

CHAPTER SUMMARY

Nonlinear two-dimensional autonomous systems have the form

$$dx/dt = F(x, y), \qquad dy/dt = G(x, y),$$

or, in vector notation,

$$dx/dt = f(x).$$

The first four sections in this chapter deal mainly with approximating a nonlinear system by a linear one. The last two sections introduce phenomena that occur only in nonlinear systems.

Section 7.1

▶ **Critical points** of the system $x' = f(x)$ satisfy $f(x) = 0$.

▶ Formal definitions of **stability, asymptotic stability**, and **instability** of critical points are given. Stability and asymptotic stability are illustrated by an undamped and a damped simple pendulum, respectively, about its downward equilibrium position. Instability is illustrated by a pendulum, damped or undamped, about its upward equilibrium position.

▶ Examples illustrate **basins of attraction** and their boundaries, called **separatrices.**

Section 7.2

▶ If F and G are twice differentiable, then the nonlinear autonomous system $x' = f(x)$ can be approximated near a critical point x_0 by a linear system $u' = Au$, where $u = x - x_0$. The coefficient matrix A is the Jacobian matrix J evaluated at x_0. Thus

$$A = J(x_0) = \begin{pmatrix} F_x(x_0, y_0) & F_y(x_0, y_0) \\ G_x(x_0, y_0) & G_y(x_0, y_0) \end{pmatrix}.$$

▶ Theorem 7.2.2 states that the trajectories of the nonlinear system locally resemble those of the linear approximation, except possibly in the cases where the eigenvalues of the linear system are either pure imaginary or real and equal. Thus, in most cases, the linear system is a good local approximation to the nonlinear system.

Section 7.3   Application: Competing Species

▶ The equations

$$dx/dt = x(\epsilon_1 - \sigma_1 x - \alpha_1 y), \qquad dy/dt = y(\epsilon_2 - \sigma_2 y - \alpha_2 x)$$

are often used as a model of competition, such as between two species in nature or perhaps between two businesses.

▶ Examples show that sometimes the two competitors can coexist in a stable manner, but sometimes one will overwhelm the other and drive it to extinction. The analysis in this section explains why this happens and enables you to predict which outcome will occur for a given system.

Section 7.4   Application: Predator–Prey

▶ The predator-prey, or **Lotka–Volterra**, equations

$$dx/dt = x(a - \alpha y), \qquad dy/dt = y(-c + \gamma x)$$

are a starting point for the study of the relation between a prey x and its predator y.

▶ The solutions of this system exhibit a cyclic variation about a critical point (a center) in the first quadrant. This type of behavior has sometimes been observed in nature.

Section 7.5 Nonlinear systems, unlike linear systems, sometimes have periodic solutions, or **limit cycles**, that attract other nearby solutions.

▶ Several theorems specify conditions under which limit cycles do, or do not, exist.

▶ The **van der Pol equation** (written in system form)

$$x' = y, \qquad y' = -x + \mu \left(1 - x^2\right) y$$

is an important equation that illustrates the occurrence of a limit cycle.

Section 7.6 In three or more dimensions there is the possibility that solutions may be **chaotic**. In addition to critical points and limit cycles, solutions may converge to sets of points known as **strange attractors**.

▶ The **Lorenz equations**, arising in a study of the atmosphere,

$$dx/dt = \sigma \left(-x + y\right), \qquad dy/dt = rx - y - xz, \qquad dz/dt = -bz + xy$$

provide an example of the occurence of chaos in a relatively simple three-dimensional nonlinear system.

PROJECTS

Project 1 Modeling of Epidemics

Infectious disease is disease caused by a biological agent (virus, bacterium, or parasite) that can be spread directly or indirectly from one organism to another. A sudden outbreak of infectious disease which spreads rapidly and affects a large number of people, animals, or plants in a particular area for a limited period of time is referred to as an *epidemic*. Mathematical models are used to help understand the dynamics of an epidemic, to design treatment and control strategies (such as a vaccination program or quarantine policy), and to help forecast whether an epidemic will occur. In this project, we consider two simple models which highlight some important principles of epidemics.

The SIR Model. Most mathematical models of disease assume that the population is subdivided into a set of distinct compartments, or classes. The class in which an individual resides at time t depends on that individual's experience with respect to the disease. The simplest of these models classifies individuals as either susceptible, infectious, or removed from the population following the infectious period (see Figure 7.P.1).

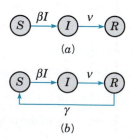

FIGURE 7.P.1 (*a*) The SIR epidemic model, and (*b*) the SIRS epidemic model.

Accordingly, we define the state variables

$S(t)$ = number of susceptible individuals at time t,
$I(t)$ = number of infected individuals at time t,
$R(t)$ = number of post-infective individuals removed from the
population at time t (due to immunity, quarantine, or death).

Susceptible individuals are able to catch the disease, after which they move into the infectious class. Infectious individuals spread the disease to susceptibles, and remain in the infectious class for a period of time (the infectious period) before moving into the removed class. Individuals in the removed class consist of those who can no longer acquire or spread the disease. The mathematical model (referred to as the SIR model) describing the temporal evolution of the sizes of the classes is based on the following assumptions:

1. The rate at which susceptibles become infected is proportional to the number of encounters between susceptible and infected individuals, which in turn is proportional to the product of the two populations, βSI. Larger values of β correspond to higher contact rates between infecteds and susceptibles.

2. The rate of transition from class I to class R is proportional to I, that is, νI. The biological meaning of ν is that $1/\nu$ is the average length of the infectious period.

3. During the time period over which the disease evolves there is no immigration, emigration, births, or deaths except possibly from the disease.

With these assumptions, the differential equations that describe the number of individuals in the three classes are

$$\begin{aligned} S' &= -\beta I S, \\ I' &= \beta I S - \nu I, \\ R' &= \nu I. \end{aligned} \tag{1}$$

It is convenient to restrict analysis to the first two equations in Eq. (1) since they are independent of R,

$$\begin{aligned} S' &= -\beta I S, \\ I' &= \beta I S - \gamma I. \end{aligned} \tag{2}$$

The SIRS Model. A slight variation in the SIR model results by assuming that individuals in the R class are temporarily immune, say for an average length of time $1/\gamma$, after which they rejoin the class of susceptibles. The governing equations in this scenario, referred to as the SIRS model, are

$$\begin{aligned} S' &= -\beta I S + \gamma R, \\ I' &= \beta I S - \nu I, \\ R' &= \nu I - \gamma R. \end{aligned} \tag{3}$$

Project 1 PROBLEMS

1. Assume that $S(0) + I(0) + R(0) = N$, that is, the total size of the population at time $t = 0$ is N. Show that $S(t) + I(t) + R(t) = N$ for all $t > 0$ for both the SIR and SIRS models.

2. The triangular region $\Gamma = \{(S, I): 0 \leq S + I \leq N\}$ in the SI-plane is depicted in Figure 7.P.2. Use an analysis based strictly on direction fields to show that no solution of the system (2) can leave the set Γ. More precisely, show that each point on the boundary of Γ is either a critical point of the system (2), or else the direction field vectors point toward

the interior of Γ or are parallel to the boundary of Γ.

3. If epidemics are identified with solution trajectories in which the number of infected individuals initially increases, reaches a maximum, and then decreases, use a nullcline analysis to show that an epidemic occurs if and only if $S(0) > \rho = \nu/\beta$. Assume that $\nu/\beta < 1$. Thus, $\rho = \nu/\beta$ is, in effect, a threshold value of susceptibles separating Γ into an epidemic region and a nonepidemic region. Explain how the size of the nonepidemic

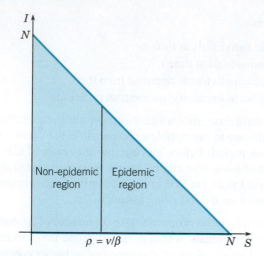

FIGURE 7.P.2 The state variables S and I for the SIR and SIRS models must lie in the region $\Gamma = \{(S, I) : 0 \le S + I \le N\}$.

region depends on contact rate and length of infection period.

ODEA **4.** Find an equation of the form $H(S, I) = c$ satisfied by the solutions of Eq. (2). Then construct a phase portrait within Γ for the system (2) consisting of

trajectories emanating from points along the upper boundary $S + I = N$ of Γ corresponding to initial states in which $R(0) = 0$.

5. In the SIR system (1), describe qualitatively the asymptotic behavior of S, I, and R as $t \to \infty$. In particular, answer the question, "Does everyone get infected?" Then explain the statement, "The epidemic does not die out due to the lack of susceptibles, but rather due to a lack of infectives."

6. Vaccinated individuals are protected from acquiring the disease and are, in effect, removed from participating in the transmission of the disease. Explain how an epidemic can be avoided by vaccinating a sufficiently large fraction p of the population, but it is not necessary to vaccinate the entire population.

7. Use the equation $S + I + R = N$ to reduce the SIRS ODEA model (3) to a system of dimension 2. Then use the qualitative methods of Chapter 7 and numerical simulations to discover as much as you can about the properties of solutions of the system (3). Compare and contrast your findings with the properties of solutions of the SIR model.

Project 2 Harvesting in a Competitive Environment

Consider again the system (Eq. (2) of Section 7.3)

$$dx/dt = x(\epsilon_1 - \sigma_1 x - \alpha_1 y),$$
$$dy/dt = y(\epsilon_2 - \sigma_2 y - \alpha_2 x),$$
(1)

which models competition between two species. To be specific, suppose that x and y are the populations of two species of fish in a pond, lake, or ocean. Suppose further that species x is a good source of nourishment, so that it is desirable to harvest members of x for food. Intuitively, it may seem reasonable to believe that if x is harvested too aggressively, then its numbers may be reduced to the point where it is no longer able to survive the competition with y and will decline to possible extinction. So the policy issue is how to determine a harvest rate that will provide useful food without threatening the long-term survival of the species. There are two simple models that have been used to investigate harvesting in a competitive situation, a constant-effort model and a constant-yield model. The first of these is described in Problems 1 through 3, and the second in Problem 4.

Project 2 PROBLEMS

1. Consider again the system

$$dx/dt = x(1 - x - y),$$
$$dy/dt = y(0.75 - y - 0.5x),$$
(i)

which appeared in Example 1 of Section 7.3. A constant-effort model, applied to the species x alone, assumes that the rate of growth of x is altered by including the term $-Ex$, where E is a positive constant measuring the effort invested in harvesting members of species x. This assumption means that, for a given effort E, the rate of catch is proportional to the population x, and that for a given population x the rate of catch is proportional to the effort E. Based on this assumption, Eqs. (i) are replaced by

$$dx/dt = x(1 - x - y) - Ex = x(1 - E - x - y),$$
$$dy/dt = y(0.75 - y - 0.5x). \qquad \text{(ii)}$$

(a) For $E = 0$, the critical points of Eqs. (ii) are as in Example 1 of Section 7.3. As E increases, some critical points move while others remain fixed. Which ones move and how?

(b) For a certain value of E, denoted by E_0, the asymptotically stable node, originally at the point (0.5, 0.5), coincides with the saddle point (0, 0.75). Find the value of E_0.

ODEA (c) Draw a direction field and/or a phase portrait for $E = E_0$ and for values of E slightly less than and slightly greater than E_0.

(d) How does the nature of the critical point (0, 0.75) change as E passes through E_0?

(e) What happens to the species x for $E > E_0$?

2. Consider the system

$$dx/dt = x(1 - x - y),$$
$$dy/dt = y(0.8 - 0.6y - x), \qquad \text{(iii)}$$

which appeared in Example 2 of Section 7.3. If constant-effort harvesting is applied to species x, then the modified equations are

$$dx/dt = x(1 - x - y) - Ex = x(1 - E - x - y),$$
$$dy/dt = y(0.8 - 0.6y - x). \qquad \text{(iv)}$$

(a) For $E = 0$, the critical points of Eqs. (iv) are as in Example 2 of Section 7.3. As E increases, some critical points move while others remain fixed. Which ones move and how?

(b) For a certain value of E, denoted by E_0, the saddle point originally at (0.5, 0.5), coincides with the asymptotically stable node originally at (1, 0). Find the value of E_0.

ODEA (c) Draw a direction field and/or a phase portrait for $E = E_0$ and for values of E slightly less than, and slightly greater than, E_0. Estimate the basin of attraction for each asymptotically stable critical point.

(d) Consider the asymptotically stable node originally at (1, 0). How does the nature of this critical point change as E passes through E_0?

(e) What happens to the species x for $E > E_0$?

3. Consider the system (i) in Problem 1, and assume now that both x and y are harvested, with efforts E_1 and E_2, respectively. Then the modified equations are

$$dx/dt = x(1 - E_1 - x - y),$$
$$dy/dt = y(0.75 - E_2 - y - 0.5x). \qquad \text{(v)}$$

(a) When $E_1 = E_2 = 0$ there is an asymptotically stable node at (0.5, 0.5). Find conditions on E_1 and E_2 that permit the continued long-term survival of both species.

(b) Use the conditions found in part (a) to sketch the region in the E_1E_2-plane that corresponds to the long-term survival of both species. Also identify regions where one species survives but not the other, and a region where both decline to extinction.

4. A constant-yield model, applied to species x, assumes that dx/dt is reduced by a positive constant H, the yield rate. For the situation described by Eqs. (i), the modified equations are

$$dx/dt = x(1 - x - y) - H,$$
$$dy/dt = y(0.75 - y - 0.5x). \qquad \text{(vi)}$$

(a) For $H = 0$, the x-nullclines are the lines $x = 0$ and $x + y = 1$. For $H > 0$ show that the x-nullcline is a hyperbola whose asymptotes are $x = 0$ and $x + y = 1$.

(b) How do the critical points move as H increases from zero?

(c) For a certain value of H, denoted by H_c, the asymptotically stable node originally at (0.5, 0.5) coincides with the saddle point originally at (0, 0.75). Determine the value of H_c. Also determine the values of x and y where the two critical points coincide.

(d) Where are the critical points for $H > H_c$? Classify them as to type.

(e) What happens to species x for $H > H_c$? What happens to species y?

(f) Draw a direction field and/or phase portrait for ODEA $H = H_c$ and for values of H slightly less than, and slightly greater than, H_c.

Project 3 The Rössler System

The system

$$x' = -y - z, \quad y' = x + ay, \quad z' = b + z(x - c), \tag{1}$$

where a, b, and c are positive parameters, is known as the Rössler[6] system. It is a relatively simple system, consisting of two linear equations and a third equation with a single quadratic nonlinearity. In the following problems, we ask you to carry out some numerical investigations of this system, with the goal of exploring its period-doubling property. To simplify matters set $a = 0.25$, $b = 0.5$, and let $c > 0$ remain arbitrary.

Project 3 PROBLEMS

ODEA **1. (a)** Show that there are no critical points when $c < \sqrt{0.5}$, one critical point for $c = \sqrt{0.5}$, and two critical points when $c > \sqrt{0.5}$.

(b) Find the critical point(s) and determine the eigenvalues of the associated Jacobian matrix when $c = \sqrt{0.5}$ and when $c = 1$.

(c) How do you think trajectories of the system will behave for $c = 1$? Plot the trajectory starting at the origin. Does it behave the way that you expected?

(d) Choose one or two other initial points and plot the corresponding trajectories. Do these plots agree with your expectations?

ODEA **2. (a)** Let $c = 1.3$. Find the critical points and the corresponding eigenvalues. What conclusions, if any, can you draw from this information?

(b) Plot the trajectory starting at the origin. What is the limiting behavior of this trajectory? To see the limiting behavior clearly, you may wish to choose a t-interval for your plot so that the initial transients are eliminated.

(c) Choose one or two other initial points and plot the corresponding trajectories. Are the limiting behavior(s) the same as in part **(b)**?

(d) Observe that there is a limit cycle whose basin of attraction is fairly large (although not all of xyz-space). Draw a plot of x, y, or z versus t and estimate the period of motion around the limit cycle.

3. The limit cycle found in Problem 2 comes into existence as a result of a Hopf bifurcation at a value c_1 of c between 1 and 1.3. Determine, or at least estimate more precisely, the value of c_1.

There are several ways in which you might do this.

(a) Draw plots of trajectories for different values of c. ODEA

(b) Calculate eigenvalues at critical points for different values of c.

(c) Use the result of Problem 3**(b)** in Section 7.6.

4. (a) Let $c = 3$. Find the critical points and the corresponding eigenvalues. ODEA

(b) Plot the trajectory starting at the point $(1, 0, -2)$. Observe that the limit cycle now consists of two loops before it closes; it is often called a 2-cycle.

(c) Plot x, y, or z versus t and show that the period of motion on the 2-cycle is very nearly double the period of the simple limit cycle in Problem 2. There has been a period-doubling bifurcation of cycles for a certain value of c between 1.3 and 3.

5. (a) Let $c = 3.8$. Find the critical points and the corresponding eigenvalues. ODEA

(b) Plot the trajectory starting at the point $(1, 0, -2)$. Observe that the limit cycle is now a 4-cycle. Find the period of motion. Another period-doubling bifurcation has occurred for c between 3 and 3.8.

(c) For $c = 3.85$ show that the limit cycle is an 8-cycle. Verify that its period is very close to eight times the period of the simple limit cycle in Problem 2.

Note: As c increases further, there is an accelerating cascade of period-doubling bifurcations. The bifurcation values of c converge to a limit, which marks the onset of chaos.

---

[6]See the book by Strogatz for a more extensive discussion and further references.

Series Solutions of Second Order Linear Equations

T he general solution of a linear second order equation

$$P(x)y'' + Q(x)y' + R(x)y = 0$$

is

$$y = c_1 y_1(x) + c_2 y_2(x),$$

where y_1 and y_2 are a fundamental set of solutions of the differential equation. So far, we have given a systematic procedure for constructing fundamental solutions only if the equation has constant coefficients:

$$ay'' + by' + cy = 0.$$

Furthermore, y_1 and y_2 can be expressed in closed form in terms of elementary functions. In Section 4.4 we were able to extend this technique to finding the general solution of the Cauchy–Euler equation:

$$ax^2 y'' + bxy' + cy = 0.$$

The special form of the variable coefficients in this equation allows us to transform it into a constant coefficient equation by using the substitution $z = \ln x$. We will briefly revisit the Cauchy–Euler equation in Section 8.4.

   To deal with equations that have general nonconstant coefficients, it is necessary to consider alternative solution techniques. For some applications we may find that approximations using an initial value problem solver are satisfactory for our needs. However, there are some variable coefficient equations that frequently recur in applications, and it is either convenient or necessary to represent their

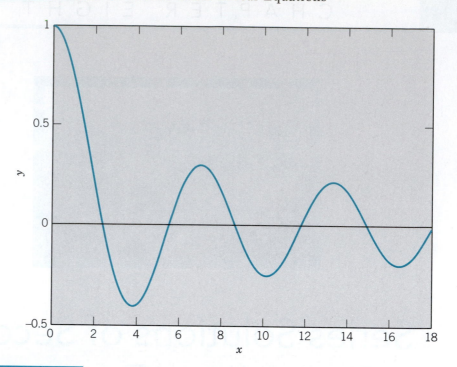

FIGURE 8.0.1 The graph of $J_0(x)$ represented by the power series (1).

solutions by infinite series. For instance, a solution of

$$x^2 y'' + xy' + x^2 y = 0, \qquad x > 0,$$

is given by the power series

$$y(x) = \sum_{n=0}^{\infty} (-1)^n \frac{x^{2n}}{2^{2n}(n!)^2} = 1 - \frac{x^2}{2^2} + \frac{x^4}{2^4(2!)^2} - \frac{x^6}{2^6(3!)^2} + \cdots, \tag{1}$$

as we shall see later in this chapter. The function given by the power series, commonly denoted by $J_0(x)$, is an oscillatory function with an infinite number of zeros on the positive axis. Not only are there tables listing its values and the location of many of its zeros, it can be evaluated by an appropriate function call in a computer algebra system in the same way that elementary functions such as e^x or $\sin x$ are evaluated. The graph of $J_0(x)$ is shown in Figure 8.0.1. The function $J_0(x)$, called the **Bessel function of the first kind of order zero**, is used to represent solutions to problems in heat transport, mechanical vibrations, acoustics, and electrodynamics. For example, the shape of one of the radially symmetric modes of vibration of a circular elastic membrane is a surface of revolution generated by revolving $J_0(x)$ about the z-axis as shown in Figure 8.0.2.

If this mode is excited, its amplitude oscillates at a pure frequency determined by the tension, radius, and planar mass density of the membrane.

The function $J_0(x)$ is an example of a **special function**. Special functions refer to mathematical functions that have acquired conventional names and notations due to their widespread use in applied mathematics, physics, and engineering. There are several different approaches to defining and analyzing special functions. For many, a natural and direct approach is to define them to be solutions of certain variable coefficient differential equations, the focus of this chapter. Often, many properties of a special function, in addition to

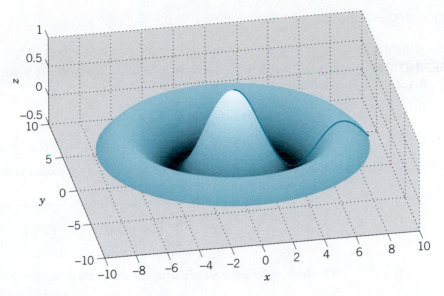

FIGURE 8.0.2 A radially symmetric vibrational mode of a circular elastic membrane obtained by revolving $J_0(x)$ around the z-axis.

its representation in terms of an infinite series, can then be deduced from the differential equation itself.

8.1 Review of Power Series

In this chapter we discuss the use of power series to construct fundamental sets of solutions of second order linear differential equations whose coefficients are nonconstant functions of the independent variable. We begin by briefly summarizing the pertinent results about power series that we need. Readers who are familiar with power series may go on to Section 8.2. Those who need more details than are presented here should consult a book on calculus.

DEFINITION 8.1.1

A **power series** is an infinite series of the form

$$\sum_{n=0}^{\infty} a_n(x - x_0)^n = a_0 + a_1(x - x_0) + a_2(x - x_0)^2 + \cdots. \tag{1}$$

The constants $a_0, a_1, a_2 \ldots$ are called the **coefficients** of the series, the constant x_0 is called the **center** of the series, and x is a variable. Setting $x_0 = 0$ in Eq. (1) gives us a **power series centered at $x_0 = 0$**:

$$\sum_{n=0}^{\infty} a_n x^n = a_0 + a_1 x + a_2 x^2 + \cdots.$$

Convergence Concepts

DEFINITION 8.1.2	A power series $\sum_{n=0}^{\infty} a_n(x - x_0)^n$ is said to **converge** at a point x if the sequence of **partial sums**

$$S_m(x) = \sum_{n=0}^{m} a_n(x - x_0)^n$$

$$= a_0 + a_1(x - x_0) + a_2(x - x_0)^2 + \cdots + a_m(x - x_0)^m$$

converges as $m \to \infty$. The sum of the series at the point x is defined to be the limit of the sequence of partial sums, and we write

$$\sum_{n=0}^{\infty} a_n(x - x_0)^n = \lim_{m \to \infty} S_m(x).$$

If the limit of the sequence of partial sums does not exist, then the series is said to **diverge** at x.

Remark. The series (1) certainly converges for $x = x_0$; it may converge for all x, or it may converge for some values of x and not for others.

EXAMPLE 1

Use Definition 8.1.2 to show that the infinite **geometric series**

$$a + ax + ax^2 + \cdots = \sum_{n=0}^{\infty} ax^n$$

converges to $a/(1 - x)$ if $|x| < 1$ and diverges if $|x| \geq 1$ provided that $a \neq 0$.

Let $S_m(x) = a + ax + ax^2 + \cdots + ax^m$. Subtracting $x S_m(x) = ax + ax^2 + \cdots + ax^m + ax^{m+1}$ from $S_m(x)$ gives $S_m(x) - x S_m(x) = (1 - x)S_m(x) = a - ax^{m+1}$, or

$$S_m(x) = \frac{a(1 - x^{m+1})}{1 - x}.$$

If $|x| < 1$, then $\lim_{m \to \infty} S_m(x) = \lim_{m \to \infty} \frac{a(1 - x^{m+1})}{1 - x} = \frac{a}{1 - x}$. On the other hand, if $|x| \geq 1$ and $a \neq 0$, the nth term of the series, ax^n, does not approach zero as $n \to \infty$, a necessary condition for convergence of an infinite series. Thus the series diverges if $|x| \geq 1$ and $a \neq 0$.

DEFINITION 8.1.3	The series $\sum_{n=0}^{\infty} a_n(x - x_0)^n$ is said to **converge absolutely** at a point x if the series

$$\sum_{n=0}^{\infty} |a_n(x - x_0)^n| = \sum_{n=0}^{\infty} |a_n||x - x_0|^n \qquad (2)$$

converges.

There are several tests from calculus that can be used to determine whether an infinite series of *positive* terms converges. Combining such a test with the following theorem, which states that **absolute convergence implies convergence**, gives us a valuable tool for determining values of x for which a power series converges.

THEOREM 8.1.4

If the power series $\sum_{n=0}^{\infty} |a_n(x - x_0)^n|$ converges at x, then so does the series $\sum_{n=0}^{\infty} a_n(x - x_0)^n$.

Remark. The converse of Theorem 8.1.4 is false.

▶ **The Ratio Test.** One of the most useful tests for the absolute convergence of a power series is the ratio test.

THEOREM 8.1.5

The Ratio Test. If $a_n \neq 0$, and if, for a fixed value of x,

$$\lim_{n \to \infty} \left| \frac{a_{n+1}(x - x_0)^{n+1}}{a_n(x - x_0)^n} \right| = |x - x_0| \lim_{n \to \infty} \left| \frac{a_{n+1}}{a_n} \right| = |x - x_0|L,$$

then the power series converges absolutely at that value of x if $|x - x_0|L < 1$ and diverges if $|x - x_0|L > 1$. If $|x - x_0|L = 1$, the test is inconclusive.

EXAMPLE 2

For which values of x does the power series

$$\sum_{n=1}^{\infty} (-1)^{n+1} n (x - 2)^n$$

converge?

To test for convergence, we use the ratio test. We have

$$\lim_{n \to \infty} \left| \frac{(-1)^{n+2}(n + 1)(x - 2)^{n+1}}{(-1)^{n+1} n (x - 2)^n} \right| = |x - 2| \lim_{n \to \infty} \frac{n + 1}{n} = |x - 2|.$$

According to Theorem 8.1.5, the series converges absolutely for $|x - 2| < 1$, or $1 < x < 3$, and diverges for $|x - 2| > 1$. The values of x corresponding to $|x - 2| = 1$ are $x = 1$ and $x = 3$. The series diverges for each of these values of x since the nth term of the series does not approach zero as $n \to \infty$.

▶ **Radius of Convergence.** For each power series in x there is a nonnegative number ρ such that $\sum_{n=0}^{\infty} a_n(x - x_0)^n$ converges absolutely for $|x - x_0| < \rho$ and diverges for $|x - x_0| > \rho$. The number ρ is called the **radius of convergence** and the interval $|x - x_0| < \rho$ is called the **interval of convergence**; it is indicated by the shaded region in Figure 8.1.1. If the series

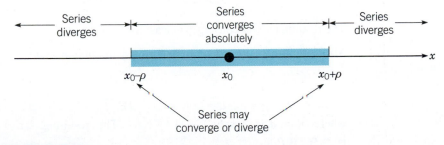

FIGURE 8.1.1   The interval of convergence of a power series.

converges only at $x = x_0$, we set $\rho = 0$. If the series converges for all x, we set $\rho = \infty$. Using these conventions, we can state that *each power series has a radius of convergence ρ, where $0 \le \rho \le \infty$. If $0 < \rho < \infty$, the series may either converge or diverge when $|x - x_0| = \rho$.*

EXAMPLE 3

Determine the radius of convergence of the power series

$$\sum_{n=1}^{\infty} \frac{(x+1)^n}{n2^n}.$$

We apply the ratio test:

$$\lim_{n\to\infty} \left| \frac{(x+1)^{n+1}}{(n+1)2^{n+1}} \frac{n2^n}{(x+1)^n} \right| = \frac{|x+1|}{2} \lim_{n\to\infty} \frac{n}{n+1} = \frac{|x+1|}{2}.$$

Thus the series converges absolutely for $|x+1| < 2$, or $-3 < x < 1$, and diverges for $|x+1| > 2$. The radius of convergence of the power series is $\rho = 2$. Finally, we check the endpoints of the interval of convergence. At $x = 1$ the series becomes the harmonic series

$$\sum_{n=1}^{\infty} \frac{1}{n},$$

which diverges. At $x = -3$ we have

$$\sum_{n=1}^{\infty} \frac{(-3+1)^n}{n2^n} = \sum_{n=1}^{\infty} \frac{(-1)^n}{n},$$

which converges but does not converge absolutely. The series is said to **converge conditionally** at $x = -3$. To summarize, the given power series converges for $-3 \le x < 1$ and diverges otherwise. It converges absolutely for $-3 < x < 1$ and has a radius of convergence 2.

Algebraic Operations on Power Series

A power series defines a function at those points x where the series converges. Suppose that the series $f(x) = \sum_{n=0}^{\infty} a_n(x - x_0)^n$ and $g(x) = \sum_{n=0}^{\infty} b_n(x - x_0)^n$ converge for $|x - x_0| < \rho$ where $\rho > 0$. The algebraic operations of addition, subtraction, and multiplication of power series are similar to the corresponding operations for polynomials, but with the additional requirement that convergence issues must be addressed.

Addition and subtraction. The series can be added or subtracted termwise, and

$$f(x) \pm g(x) = \sum_{n=0}^{\infty} (a_n \pm b_n)(x - x_0)^n = a_0 \pm b_0 + (a_1 \pm b_1)(x - x_0) + \cdots.$$

The resulting series converges at least for $|x - x_0| < \rho$.

Multiplication. The series can be formally multiplied, and

$$f(x)g(x) = \left[\sum_{n=0}^{\infty} a_n(x - x_0)^n \right]\left[\sum_{n=0}^{\infty} b_n(x - x_0)^n \right] = \sum_{n=0}^{\infty} c_n(x - x_0)^n,$$

where $c_n = a_0 b_n + a_1 b_{n-1} + \cdots + a_n b_0$. The resulting series converges at least for $|x - x_0| < \rho$.

The next theorem is a tool that is often used to help find the coefficients of a power series.

THEOREM 8.1.6	**The Identity Principle.** If $\sum_{n=0}^{\infty} a_n(x - x_0)^n = \sum_{n=0}^{\infty} b_n(x - x_0)^n$ for each x in some open interval with center x_0, then $a_n = b_n$ for $n = 0, 1, 2, 3, \ldots$. In particular, if $\sum_{n=0}^{\infty} a_n(x - x_0)^n = 0$ for each such x, then $a_0 = a_1 = \cdots = a_n = \cdots = 0$.

As a simple application of Theorem 8.1.6, we use it to help compute the quotient of two power series.

Division. If $g(x_0) \neq 0$, the series can be formally divided, and

$$\frac{f(x)}{g(x)} = \sum_{n=0}^{\infty} d_n(x - x_0)^n.$$

In most cases the coefficients d_n can be most easily obtained by equating coefficients in the equivalent relation

$$\sum_{n=0}^{\infty} a_n(x - x_0)^n = \left[\sum_{n=0}^{\infty} d_n(x - x_0)^n\right]\left[\sum_{n=0}^{\infty} b_n(x - x_0)^n\right]$$

$$= \sum_{n=0}^{\infty} \left(\sum_{k=0}^{n} d_k b_{n-k}\right)(x - x_0)^n.$$

Using Theorem 8.1.6, we then set

$$a_n = \sum_{k=0}^{n} d_k b_{n-k} \text{ for each } n = 0, 1, 2, \ldots.$$

This gives

$$a_0 = d_0 b_0, \qquad a_1 = d_0 b_1 + d_1 b_0, \qquad a_2 = d_0 b_2 + d_1 b_1 + d_2 b_0, \ldots,$$

which can be solved sequentially for $d_0, d_1, d_2, \ldots$. In the case of division the radius of convergence of the resulting power series may be less than ρ. For example, $\rho = \infty$ for $f(x) = 1$ and $g(x) = 1 - x$, but $\rho = 1$ for $f(x)/g(x) = 1/(1 - x) = \sum_{n=0}^{\infty} x^n$.

Taylor Series and Analytic Functions

DEFINITION 8.1.7	Let f be a function with derivatives of all orders throughout the interval $|x - x_0| < \rho$, where $\rho > 0$. Then the **Taylor series of f at x_0** is the power series

$$\sum_{n=0}^{\infty} \frac{f^{(n)}(x_0)}{n!}(x - x_0)^n = f(x_0) + \frac{f'(x_0)}{1!}(x - x_0) + \frac{f''(x_0)}{2!}(x - x_0) + \cdots.$$

If $x \neq x_0$, the Taylor series of f may not converge at x, or, if it does converge, it may not converge to the function value $f(x)$. For example, the Taylor series of $f(x) = \exp(-1/x^2)$ at $x_0 = 0$ is the zero function since $f^{(n)}(0) = 0$ for each $n = 0, 1, \ldots$. But obviously, $\exp(-1/x^2) \neq 0$. The functions that are useful to us are those that can be represented by a power series with a nonzero radius of convergence. Such functions are said to be **analytic**. With respect to the algebraic operations of addition, subtraction, and multiplication, analytic functions generalize the notion of polynomials.

DEFINITION 8.1.8

A function f that has a power series expansion of the form

$$f(x) = \sum_{n=0}^{\infty} a_n(x - x_0)^n$$

with a radius of convergence $\rho > 0$ is said to be **analytic** at x_0.

Suppose that f is analytic at x_0. Then f is continuous and has derivatives of all orders for $|x - x_0| < \rho$. Further, f', f'', ... can be computed by differentiating the series termwise; that is,

$$f'(x) = a_1 + 2a_2(x - x_0) + \cdots + na_n(x - x_0)^{n-1} + \cdots$$

$$= \sum_{n=1}^{\infty} na_n(x - x_0)^{n-1},$$

$$f''(x) = 2a_2 + 6a_3(x - x_0) + \cdots + n(n - 1)a_n(x - x_0)^{n-2} + \cdots$$

$$= \sum_{n=2}^{\infty} n(n - 1)a_n(x - x_0)^{n-2},$$

and so forth, and each of the series converges absolutely for $|x - x_0| < \rho$. Setting $x = x_0$ in each of the successive differentiated series yields the following formula relating the a_n's to f and its derivatives:

$$a_n = \frac{f^{(n)}(x_0)}{n!}.$$

Consequently, *if f is analytic at x_0, then its power series must coincide with its Taylor series.* Thus

$$f(x) = \sum_{n=0}^{\infty} a_n(x - x_0)^n = \sum_{n=0}^{\infty} \frac{f^{(n)}(x_0)}{n!}(x - x_0)^n$$

for $|x - x_0| < \rho$. All of the familiar functions of calculus are analytic except perhaps at certain easily recognized points. For example, $\sin x$ and e^x are analytic everywhere, $1/x$ is analytic except at $x = 0$, and $\tan x$ is analytic except at odd multiples of $\pi/2$. If f and g are analytic at x_0, then the algebraic properties discussed earlier imply that $f \pm g$, $f \cdot g$, and f/g [provided that $g(x_0) \neq 0$] are also analytic at $x = x_0$.

Remark. In many respects the natural context for the use of power series is the complex plane. The methods and results of this chapter nearly always can be directly extended to differential equations in which the independent and dependent variables are complex-valued.

Shift of Index of Summation

Shifting the index of summation in a power series is analogous to changing the variable of integration in an integral. This operation, in conjunction with Theorem 8.1.6, is a useful tool for computing power series solutions of differential equations.

The index of summation in an infinite series is a dummy parameter just as the integration variable in a definite integral is a dummy variable. Thus it is immaterial which letter is used for the index of summation. For example,

$$\sum_{n=0}^{\infty} \frac{2^n x^n}{n!} = \sum_{j=0}^{\infty} \frac{2^j x^j}{j!}.$$

Just as we make changes of the variable of integration in a definite integral, we find it convenient to make changes of summation indices in calculating series solutions of differential equations. We illustrate by several examples how to shift the summation index.

EXAMPLE 4

Write $\sum_{n=2}^{\infty} a_n x^n$ as a series whose first term corresponds to $n = 0$ rather than $n = 2$.

Let $m = n - 2$; then $n = m + 2$, and $n = 2$ corresponds to $m = 0$. Hence

$$\sum_{n=2}^{\infty} a_n x^n = \sum_{m=0}^{\infty} a_{m+2} x^{m+2}. \tag{3}$$

By writing out the first few terms of each of these series, you can verify that they contain precisely the same terms. Finally, in the series on the right side of Eq. (3), we can replace the dummy index m by n, obtaining

$$\sum_{n=2}^{\infty} a_n x^n = \sum_{n=0}^{\infty} a_{n+2} x^{n+2}.$$

In effect, we have shifted the index upward by 2 and have compensated by starting to count at a level 2 lower than originally.

EXAMPLE 5

Write the series

$$\sum_{n=2}^{\infty} (n + 2)(n + 1) a_n (x - x_0)^{n-2} \tag{4}$$

as a series whose generic term involves $(x - x_0)^n$ rather than $(x - x_0)^{n-2}$.

Again, we shift the index by 2 so that n is replaced by $n + 2$ and start counting 2 lower. We obtain

$$\sum_{n=0}^{\infty} (n + 4)(n + 3) a_{n+2} (x - x_0)^n. \tag{5}$$

You can readily verify that the terms in the series (4) and (5) are exactly the same.

EXAMPLE 6

Write the expression

$$x^2 \sum_{n=0}^{\infty} (r + n) a_n x^{r+n-1} \tag{6}$$

as a series whose generic term involves x^{r+n}.

First, take the x^2 inside the summation, obtaining

$$\sum_{n=0}^{\infty} (r + n) a_n x^{r+n+1}.$$

Next, shift the index down by 1 and start counting 1 higher. Thus

$$\sum_{n=0}^{\infty} (r+n)a_n x^{r+n+1} = \sum_{n=1}^{\infty} (r+n-1)a_{n-1} x^{r+n}. \tag{7}$$

Again, you can easily verify that the two series in Eq. (7) are identical and that both are exactly the same as the expression (6).

EXAMPLE 7

Assume that

$$\sum_{n=1}^{\infty} na_n x^{n-1} = \sum_{n=0}^{\infty} a_n x^n \tag{8}$$

for all x, and determine what this assumption implies about the coefficients a_n.

We want to use Theorem 8.1.6 to equate corresponding coefficients in the two series. In order to do this, we must first rewrite Eq. (8) so that the series display the same power of x in their generic terms. For instance, in the series on the left side of Eq. (8), we can replace n by $n + 1$ and start counting 1 lower. Thus Eq. (8) becomes

$$\sum_{n=0}^{\infty} (n+1)a_{n+1} x^n = \sum_{n=0}^{\infty} a_n x^n. \tag{9}$$

According to Theorem 8.1.6, we conclude that

$$(n+1)a_{n+1} = a_n, \qquad n = 0, 1, 2, 3, \ldots$$

or

$$a_{n+1} = \frac{a_n}{n+1}, \qquad n = 0, 1, 2, 3, \ldots. \tag{10}$$

Hence, choosing successive values of n in Eq. (10), we have

$$a_1 = a_0, \qquad a_2 = \frac{a_1}{2} = \frac{a_0}{2!}, \qquad a_3 = \frac{a_2}{3} = \frac{a_0}{3!},$$

and so forth. In general,

$$a_n = \frac{a_0}{n!}, \qquad n = 1, 2, 3, \ldots. \tag{11}$$

Thus the relation (10) determines all the following coefficients in terms of a_0. Finally, using the coefficients given by Eq. (11), we obtain

$$\sum_{n=0}^{\infty} a_n x^n = a_0 \sum_{n=0}^{\infty} \frac{x^n}{n!} = a_0 e^x,$$

where we have followed the usual convention that $0! = 1$.

PROBLEMS

In each of Problems 1 through 8 determine the radius of convergence of the given power series.

1. $\displaystyle\sum_{n=0}^{\infty} (x-3)^n$

2. $\displaystyle\sum_{n=0}^{\infty} \frac{n}{2^n} x^n$

3. $\displaystyle\sum_{n=0}^{\infty} \frac{x^{2n}}{n!}$

4. $\displaystyle\sum_{n=0}^{\infty} 2^n x^n$

5. $\displaystyle\sum_{n=1}^{\infty} \frac{(2x+1)^n}{n^2}$

6. $\displaystyle\sum_{n=1}^{\infty} \frac{(x-x_0)^n}{n}$

7. $\displaystyle\sum_{n=1}^{\infty} \frac{(-1)^n n^2 (x+2)^n}{3^n}$

8. $\displaystyle\sum_{n=1}^{\infty} \frac{n! x^n}{n^n}$

In each of Problems 9 through 16 determine the Taylor series about the point x_0 for the given function. Also determine the radius of convergence of the series.

9. $\sin x$, $x_0 = 0$

10. e^x, $x_0 = 0$

11. x, $x_0 = 1$

12. x^2, $x_0 = -1$

13. $\ln x$, $x_0 = 1$

14. $\dfrac{1}{1+x}$, $x_0 = 0$

15. $\dfrac{x}{x^4+9} = \dfrac{x}{9} \cdot \dfrac{1}{1+(x^4/9)}$, $x_0 = 0$

16. $\dfrac{1}{1-x} = -\dfrac{1}{1+(x-2)}$, $x_0 = 2$

17. Given that $y = \sum_{n=0}^{\infty} nx^n$, compute y' and y'' and write out the first four terms of each series as well as the coefficient of x^n in the general term.

18. Given that $y = \sum_{n=0}^{\infty} a_n x^n$, compute y' and y'' and write out the first four terms of each series as well as the coefficient of x^n in the general term. Show that if $y'' = y$, then the coefficients a_0 and a_1 are arbitrary, and determine a_2 and a_3 in terms of a_0 and a_1. Show that $a_{n+2} = a_n/(n+2)(n+1)$, $n = 0, 1, 2, 3, \ldots$.

In each of Problems 19 and 20 verify the given equation.

19. $\sum_{n=0}^{\infty} a_n(x-1)^{n+1} = \sum_{n=1}^{\infty} a_{n-1}(x-1)^n$

20. $\sum k = 0^{\infty} a_{k+1} x^k + \sum_{k=0}^{\infty} a_k x^{k+1}$
$= a_1 + \sum_{k=1}^{\infty} (a_{k+1} + a_{k-1}) x^k$

In each of Problems 21 through 27 rewrite the given expression as a sum whose generic term involves x^n.

21. $\sum_{n=2}^{\infty} n(n-1)a_n x^{n-2}$ 22. $\sum_{n=0}^{\infty} a_n x^{n+2}$

23. $x \sum_{n=1}^{\infty} na_n x^{n-1} + \sum_{k=0}^{\infty} a_k x^k$

24. $(1-x^2)\sum_{n=2}^{\infty} n(n-1)a_n x^{n-2}$

25. $\sum_{m=2}^{\infty} m(m-1)a_m x^{m-2} + x\sum_{k=1}^{\infty} ka_k x^{k-1}$

26. $\sum_{n=1}^{\infty} na_n x^{n-1} + x\sum_{n=0}^{\infty} a_n x^n$

27. $x\sum_{n=2}^{\infty} n(n-1)a_n x^{n-2} + \sum_{n=0}^{\infty} a_n x^n$

28. Determine the a_n so that the equation

$$\sum_{n=1}^{\infty} na_n x^{n-1} + 2\sum_{n=0}^{\infty} a_n x^n = 0$$

is satisfied. Try to identify the function represented by the series $\sum_{n=0}^{\infty} a_n x^n$.

8.2 Series Solutions Near an Ordinary Point, Part I

We now consider methods of solving second order linear equations when the coefficients are nonconstant functions of the independent variable. It is sufficient to consider the homogeneous equation

$$P(x)\frac{d^2y}{dx^2} + Q(x)\frac{dy}{dx} + R(x)y = 0, \tag{1}$$

since the procedure for the corresponding nonhomogeneous equation is similar. The following table lists several variable coefficient equations that frequently occur in applied mathematics, engineering, and physics:

Airy equation	$y'' - xy = 0$	acoustics, fiber optics
Bessel equation	$x^2 y'' + xy' + (x^2 - \nu^2)y = 0$	acoustics, electrodynamics
Chebyshev equation	$(1 - x^2)y'' - xy' + \alpha^2 y = 0$	approximation theory
Hermite equation	$y'' - 2xy' + \lambda y = 0$	quantum mechanics
Laguerre equation	$xy'' + (1 - x)y' + \lambda y = 0$	approximation theory
Legendre equation	$(1 - x^2)y'' - 2xy' + \alpha(\alpha + 1)y = 0$	heat flow, electrodynamics

The solutions of these equations are well-known special functions of mathematical physics. In this chapter we discuss several of the properties of these equations and their solutions, and in the projects at the end of this chapter we study applications that require certain solutions of the Bessel equation and the Hermite equation. In most applications the independent variable refers to a spatial coordinate, so we denote the independent variable by x. Since the coefficients that appear in each of the above equations are polynomials, we emphasize the case in which the functions P, Q, and R in Eq. (1) are polynomials and have no common factors. However, the series method of solution is also applicable when the coefficients are general analytic functions.

Ordinary and Singular Points

In this section we consider solutions of Eq. (1) in intervals around a point x_0 where P, Q, and R are analytic and $P(x_0) \neq 0$.

DEFINITION
8.2.1

A point x_0 is said to be an **ordinary point** of Eq. (1) if the coefficients P, Q, and R are analytic at x_0, and $P(x_0) \neq 0$. If x_0 is not an ordinary point, it is called a **singular point** of the equation.

If x_0 is an ordinary point, then we can divide Eq. (1) by $P(x)$ to obtain

$$y'' + p(x)y' + q(x)y = 0, \qquad (2)$$

where $p(x) = Q(x)/P(x)$ and $q(x) = R(x)/P(x)$ are analytic at x_0, and therefore continuous in an interval around x_0. Hence, according to the existence and uniqueness Theorem 4.2.1, there exists in that interval a unique solution of Eq. (2) or (1) that also satisfies the initial conditions $y(x_0) = y_0$, $y'(x_0) = y_1$ for arbitrary values of y_0 and y_1. On the other hand, if x_0 is a singular point of Eq. (1), then at least one of $Q(x_0)$ and $R(x_0)$ is not zero. Consequently, at least one of the coefficients p and q in Eq. (2) becomes unbounded as $x \rightarrow x_0$, and therefore Theorem 4.2.1 does not apply in this case.

We note that $x = 0$ is a regular point for the Airy, Chebyshev, Hermite, and Legendre equations, but that it is a singular point for the Bessel and Laguerre equations. Sections 8.4 through 8.7 deal with finding solutions of Eq. (1) in the neighborhood of a singular point.

We now take up the problem of solving Eq. (1) in the neighborhood of an ordinary point x_0. We look for solutions of the form

$$y = a_0 + a_1(x - x_0) + \cdots + a_n(x - x_0)^n + \cdots = \sum_{n=0}^{\infty} a_n(x - x_0)^n, \qquad (3)$$

and assume that the series converges in the interval $|x - x_0| < \rho$ for some $\rho > 0$. Although at first sight it may appear unattractive to seek a solution in the form of a power series, this is actually a convenient and useful form for a solution. Within their intervals of convergence, power series behave very much like polynomials and are easy to manipulate both analytically and numerically. Indeed, even if we can obtain a solution in terms of elementary functions, such as exponential or trigonometric functions, we are likely to need a power series or some equivalent expression if we want to evaluate them numerically or to plot their graphs.

The most practical way to determine the coefficients a_n is to substitute the series (3) and its derivatives for y, y', and y'' in Eq. (1). The following example illustrates this process. The operations, such as differentiation, that are involved in the procedure are justified so long as we stay within the interval of convergence.

EXAMPLE 1

Find a series solution of the equation

$$y'' + y = 0, \qquad -\infty < x < \infty. \tag{4}$$

As we know, two linearly independent solutions of this equation are $\sin x$ and $\cos x$, so series methods are not needed to solve this equation. However, this example illustrates the use of power series in a relatively simple case. For Eq. (4), $P(x) = 1$, $Q(x) = 0$, and $R(x) = 1$; hence every point is an ordinary point.

We look for a solution in the form of a power series about $x_0 = 0$:

$$y = a_0 + a_1 x + a_2 x^2 + \cdots + a_n x^n + \cdots = \sum_{n=0}^{\infty} a_n x^n \tag{5}$$

and assume that the series converges in some interval $|x| < \rho$.

Differentiating Eq. (5) term by term yields

$$y' = a_1 + 2a_2 x + \cdots + n a_n x^{n-1} + \cdots = \sum_{n=1}^{\infty} n a_n x^{n-1}, \tag{6}$$

$$y'' = 2a_2 + \cdots + n(n-1)a_n x^{n-2} + \cdots = \sum_{n=2}^{\infty} n(n-1) a_n x^{n-2}. \tag{7}$$

Substituting the series (5) and (7) for y and y'' in Eq. (4) gives

$$\sum_{n=2}^{\infty} n(n-1) a_n x^{n-2} + \sum_{n=0}^{\infty} a_n x^n = 0.$$

To combine the two series, we need to rewrite at least one of them so that both series display the same generic term. Thus, in the first sum, we shift the index of summation by replacing n by $n + 2$ and starting the sum at 0 rather than 2. We obtain

$$\sum_{n=0}^{\infty} (n+2)(n+1) a_{n+2} x^n + \sum_{n=0}^{\infty} a_n x^n = 0$$

or

$$\sum_{n=0}^{\infty} [(n+2)(n+1) a_{n+2} + a_n] x^n = 0.$$

If we require that this equation be satisfied for all x in a neighborhood of $x_0 = 0$, then Theorem 8.1.16, the Identity Principle, implies that the coefficient of each power of x must be zero. Hence we conclude that

$$(n+2)(n+1) a_{n+2} + a_n = 0, \qquad n = 0, 1, 2, 3, \ldots. \tag{8}$$

Equation (8) is referred to as a **recurrence relation**. The successive coefficients can be evaluated one by one by writing the recurrence relation first for $n = 0$, then for $n = 1$, and so forth. In this example Eq. (8) relates each coefficient to the second one before it. Thus the even-numbered coefficients ($a_0, a_2, a_4, \dots$) and the odd-numbered ones ($a_1, a_3, a_5, \dots$) are determined separately. For the even-numbered coefficients we have

$$a_2 = -\frac{a_0}{2 \cdot 1} = -\frac{a_0}{2!}, \qquad a_4 = -\frac{a_2}{4 \cdot 3} = +\frac{a_0}{4!}, \qquad a_6 = -\frac{a_4}{6 \cdot 5} = -\frac{a_0}{6!}, \dots$$

These results suggest that in general, if $n = 2k$, then

$$a_n = a_{2k} = \frac{(-1)^k}{(2k)!} a_0, \qquad k = 1, 2, 3, \dots . \tag{9}$$

We can prove Eq. (9) by mathematical induction. First, observe that it is true for $k = 1$. Next, assume that it is true for an arbitrary value of k and consider the case $k + 1$. We have

$$a_{2k+2} = -\frac{a_{2k}}{(2k+2)(2k+1)} = -\frac{(-1)^k}{(2k+2)(2k+1)(2k)!} a_0 = \frac{(-1)^{k+1}}{(2k+2)!} a_0.$$

Hence Eq. (9) is also true for $k + 1$, and consequently, it is true for all positive integers k.

   Similarly, for the odd-numbered coefficients

$$a_3 = -\frac{a_1}{2 \cdot 3} = -\frac{a_1}{3!}, \qquad a_5 = -\frac{a_3}{5 \cdot 4} = +\frac{a_1}{5!}, \qquad a_7 = -\frac{a_5}{7 \cdot 6} = -\frac{a_1}{7!}, \dots,$$

and in general, if $n = 2k + 1$, then[1]

$$a_n = a_{2k+1} = \frac{(-1)^k}{(2k+1)!} a_1, \qquad k = 1, 2, 3, \dots . \tag{10}$$

Substituting these coefficients into Eq. (5), we have

$$y = a_0 + a_1 x - \frac{a_0}{2!} x^2 - \frac{a_1}{3!} x^3 + \frac{a_0}{4!} x^4 + \frac{a_1}{5!} x^5$$

$$+ \cdots + \frac{(-1)^n a_0}{(2n)!} x^{2n} + \frac{(-1)^n a_1}{(2n+1)!} x^{2n+1} + \cdots$$

$$= a_0 \left[1 - \frac{x^2}{2!} + \frac{x^4}{4!} + \cdots + \frac{(-1)^n}{(2n)!} x^{2n} + \cdots \right]$$

$$+ a_1 \left[x - \frac{x^3}{3!} + \frac{x^5}{5!} + \cdots + \frac{(-1)^n}{(2n+1)!} x^{2n+1} + \cdots \right]$$

$$= a_0 \sum_{n=0}^{\infty} \frac{(-1)^n}{(2n)!} x^{2n} + a_1 \sum_{n=0}^{\infty} \frac{(-1)^n}{(2n+1)!} x^{2n+1}. \tag{11}$$

   Now that we have formally obtained two series solutions of Eq. (4), we can test them for convergence. Using the ratio test, it is easy to show that each of the series in Eq. (11) converges for all x, and this justifies retroactively all the steps used in obtaining the solutions. Indeed, we recognize that the first series in Eq. (11) is exactly the Taylor series for $\cos x$ about $x = 0$ and that the second is the Taylor series for $\sin x$ about $x = 0$. Thus, as expected, we obtain the solution $y = a_0 \cos x + a_1 \sin x$.

---

[1]The result given in Eq. (10) and other similar formulas in this chapter can be proved by an induction argument resembling the one just given for Eq. (9). We assume that the results are plausible and omit the inductive argument hereafter.

Notice that no conditions are imposed on a_0 and a_1; hence they are arbitrary. From Eqs. (5) and (6) we see that y and y' evaluated at $x = 0$ are a_0 and a_1, respectively. Since the initial conditions $y(0)$ and $y'(0)$ can be chosen arbitrarily, it follows that a_0 and a_1 should be arbitrary until specific initial conditions are stated.

The Truncated Power Series for Sin(x)

In Example 1 we knew from the start that $\sin x$ and $\cos x$ form a fundamental set of solutions of Eq. (4). However, if we had not known this and had simply solved Eq. (4) using series methods, we would still have obtained the solution (11). In recognition of the fact that the differential equation (4) often occurs in applications, we might decide to give the two solutions of Eq. (11) special names, perhaps

$$C(x) = \sum_{n=0}^{\infty} \frac{(-1)^n}{(2n)!} x^{2n}, \quad S(x) = \sum_{n=0}^{\infty} \frac{(-1)^n}{(2n+1)!} x^{2n+1}.$$

Then we might ask what properties these functions have. For instance, it follows at once from the series expansions that $C(0) = 1$, $S(0) = 0$, $C(-x) = C(x)$, and $S(-x) = -S(x)$. It is also easy to show that

$$S'(x) = C(x), \qquad C'(x) = -S(x).$$

Moreover, by calculating with the infinite series,[2] we can show that the functions $C(x)$ and $S(x)$ have all the usual analytical and algebraic properties of the cosine and sine functions, respectively. If we use these properties, it follows that knowing values of $S(x)$ in the interval $0 \le x \le \pi/2$ suffices to determine the value for any other elementary trigonometric function at any x in its domain; see Problem 29(a). Consequently, if we wish to compute a numerical approximation of a trigonometric function at any point in its domain, we could use a truncated power series to approximate $S(x)$:

$$S_n(x) = \sum_{k=0}^{n} \frac{(-1)^k}{(2k+1)!} x^{2k+1} \tag{12}$$

at one or two appropriate values of $x \in [0, \pi/2]$. In general, the number of terms required to attain a prescribed level of accuracy in a truncated power series increases as the distance between x and the center of the series increases. The graph in Figure 8.2.1 shows that by using $n = 7$ in Eq. (12), a polynomial of degree 15, the absolute error $|S(x) - S_7(x)|$ is less than 10^{-11} uniformly for $0 \le x \le \pi/2$; see Problem 29(b).

Thus the properties of, and relationships between, the elementary trigonometric functions make the truncated power series for $S(x)$ an accurate and efficient numerical method for approximating their values at all points in their domains. By contrast, to approximate $S(100)$ by directly substituting $x = 100$ into Eq. (12) requires that $n = 145$, a polynomial of degree 291, to attain the same level of accuracy; see Problem 29(c). Furthermore, to avoid underflows and overflows, we would have to use sensible numerical procedures for evaluating and summing the terms $x^{2k+1}/(2k+1)!$ for large values of x and k.

The foregoing discussion, with regard to the elementary trigonometric functions, is indicative of a general theme involving the use of series methods for approximating values of special functions that arise as solutions of linear differential equations. The series

---

[2] Such an analysis is given in Section 24 of K. Knopp, *Theory and Applications of Infinite Series* (New York: Hafner, 1951).

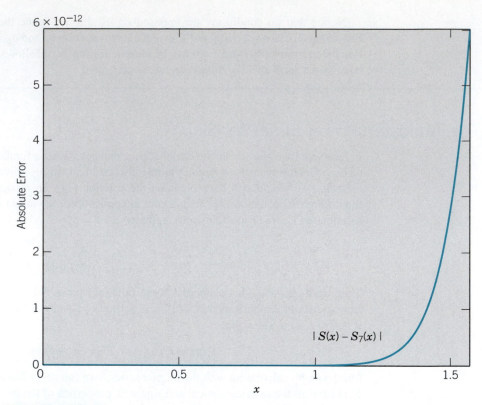

FIGURE 8.2.1 The absolute error $|S(x) - S_7(x)|$ between the sine function $S(x) = \sin x$ and its truncated Taylor series, $S_7(x) = \sum_{k=0}^{7}(-1)^k x^{2k+1}/(2k+1)!$, over the interval $0 \leq x \leq \pi/2$.

representation plays an important role, but it is often the case that a truncated series does not provide a viable approximation for desired values of the special function. In such cases, properties of the special function, derived either from a series representation, directly from the differential equation, or from an alternative representation such as an integral, are used to obtain accurate approximations in an efficient manner.

EXAMPLE 2

Airy Equation. Find a series solution in powers of x of Airy's equation

$$y'' - xy = 0, \qquad -\infty < x < \infty. \tag{13}$$

For this equation $P(x) = 1$, $Q(x) = 0$, and $R(x) = -x$; hence every point is an ordinary point. We assume that

$$y = \sum_{n=0}^{\infty} a_n x^n \tag{14}$$

and that the series converges in some interval $|x| < \rho$. The series for y'' is given by Eq. (7); as explained in the preceding example, we can rewrite it as

$$y'' = \sum_{n=0}^{\infty}(n+2)(n+1)a_{n+2}x^n. \tag{15}$$

Substituting the series (14) and (15) for y and y'' in Eq. (13), we obtain

$$\sum_{n=0}^{\infty}(n+2)(n+1)a_{n+2}x^n = x\sum_{n=0}^{\infty}a_n x^n = \sum_{n=0}^{\infty}a_n x^{n+1}. \tag{16}$$

Next, we shift the index of summation in the series on the right side of this equation by replacing n by $n-1$ and starting the summation at 1 rather than zero. Thus we have

$$2\cdot 1a_2 + \sum_{n=1}^{\infty}(n+2)(n+1)a_{n+2}x^n = \sum_{n=1}^{\infty}a_{n-1}x^n.$$

Again, for this equation to be satisfied for all x, it is necessary that the coefficients of like powers of x be equal; hence $a_2 = 0$, and we obtain the recurrence relation

$$(n+2)(n+1)a_{n+2} = a_{n-1} \qquad \text{for} \quad n = 1, 2, 3, \ldots. \tag{17}$$

Since a_{n+2} is given in terms of a_{n-1}, the a's are determined in steps of three. Thus a_0 determines a_3, which in turn determines $a_6, \ldots$; a_1 determines a_4, which in turn determines $a_7, \ldots$; and a_2 determines a_5, which in turn determines $a_8, \ldots$. Since $a_2 = 0$, we immediately conclude that $a_5 = a_8 = a_{11} = \cdots = 0$.

For the sequence $a_0, a_3, a_6, a_9, \ldots$ we set $n = 1, 4, 7, 10, \ldots$ in the recurrence relation:

$$a_3 = \frac{a_0}{2\cdot 3}, \qquad a_6 = \frac{a_3}{5\cdot 6} = \frac{a_0}{2\cdot 3\cdot 5\cdot 6}, \qquad a_9 = \frac{a_6}{8\cdot 9} = \frac{a_0}{2\cdot 3\cdot 5\cdot 6\cdot 8\cdot 9}, \ldots.$$

These results suggest the general formula

$$a_{3n} = \frac{a_0}{2\cdot 3\cdot 5\cdot 6\cdots(3n-1)(3n)}, \qquad n \geq 4.$$

For the sequence $a_1, a_4, a_7, a_{10}, \ldots$, we set $n = 2, 5, 8, 11, \ldots$ in the recurrence relation:

$$a_4 = \frac{a_1}{3\cdot 4}, \qquad a_7 = \frac{a_4}{6\cdot 7} = \frac{a_1}{3\cdot 4\cdot 6\cdot 7}, \qquad a_{10} = \frac{a_7}{9\cdot 10} = \frac{a_1}{3\cdot 4\cdot 6\cdot 7\cdot 9\cdot 10}, \ldots.$$

In general, we have

$$a_{3n+1} = \frac{a_1}{3\cdot 4\cdot 6\cdot 7\cdots(3n)(3n+1)}, \qquad n \geq 4.$$

Thus the general solution of Airy's equation is

$$y = a_0\underbrace{\left[1 + \frac{x^3}{2\cdot 3} + \frac{x^6}{2\cdot 3\cdot 5\cdot 6} + \cdots + \frac{x^{3n}}{2\cdot 3\cdots(3n-1)(3n)} + \cdots\right]}_{y_1(x)}$$

$$+ a_1\underbrace{\left[x + \frac{x^4}{3\cdot 4} + \frac{x^7}{3\cdot 4\cdot 6\cdot 7} + \cdots + \frac{x^{3n+1}}{3\cdot 4\cdots(3n)(3n+1)} + \cdots\right]}_{y_2(x)}. \tag{18}$$

Having obtained these two series solutions, we can now investigate their convergence. Because of the rapid growth of the denominators of the terms in the series (18), we might expect these series to have a large radius of convergence. Indeed, it is easy to use the ratio test to show that both these series converge for all x; see Problem 20.

Assuming for the moment that the series do converge for all x, let y_1 and y_2 denote the functions defined by the expressions in the first and second sets of brackets, respectively, in Eq. (18). Then, by choosing first $a_0 = 1, a_1 = 0$ and then $a_0 = 0, a_1 = 1$, it follows that y_1 and y_2 are individually solutions of Eq. (13). Notice that y_1 satisfies the initial conditions $y_1(0) = 1, y_1'(0) = 0$ and that y_2 satisfies the initial conditions $y_2(0) = 0, y_2'(0) = 1$. Thus

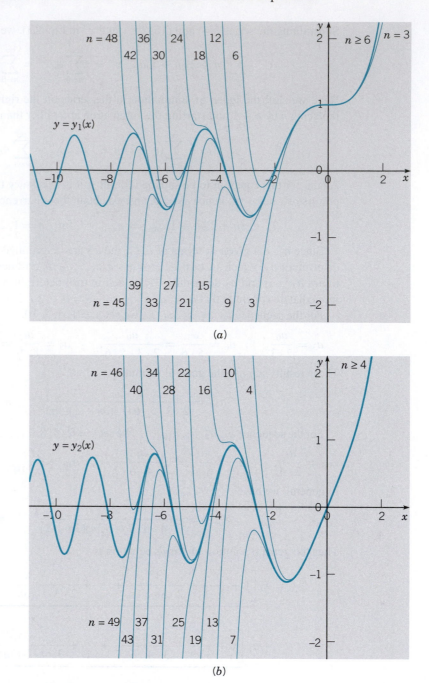

FIGURE 8.2.2 (a) Polynomial approximations to the solution $y_1(x)$ of Airy's equation. (b) Polynomial approximations to the solution $y_2(x)$ of Airy's equation. In each case the value of n is the degree of the approximating polynomial.

$W[y_1, y_2](0) = 1 \neq 0$, and consequently, y_1 and y_2 are linearly independent. Hence the general solution of Airy's equation is

$$y = a_0 y_1(x) + a_1 y_2(x), \qquad -\infty < x < \infty.$$

In Figure 8.2.2 we show the graphs of the solutions y_1 and y_2 of Airy's equation, as well as graphs of several partial sums of the two series in Eq. (18).

The partial sums provide local approximations to the solutions in a neighborhood of the origin. Although the quality of the approximation improves as the number of terms increases, no polynomial can adequately represent y_1 and y_2 for large $|x|$. A practical way to estimate the interval in which a given partial sum is reasonably accurate is to compare the graphs of that partial sum and the next one, obtained by including one more term. As soon as the graphs begin to separate noticeably, we can be confident that the original partial sum is no longer accurate. For example, in Figure 8.2.2(a) the graphs for $n = 24$ and $n = 27$ begin to separate at about $x = -9/2$. Thus, beyond this point, the partial sum of degree 24 is worthless as an approximation to the solution.

The Airy Functions Ai(x) and Bi(x)

Since Airy's equation is linear, we can form other fundamental sets of solutions by taking pairs of linear combinations of y_1 and y_2 in Eq. (18), provided that the resultant pairs of functions are linearly independent. Using the gamma function, denoted by $\Gamma(p)$ (see Problem 36, Section 5.1) and defined by the integral

$$\Gamma(p + 1) = \int_0^\infty e^{-x} x^p \, dx, \qquad p > 1,$$

we obtain the following conventional definitions for a fundamental set of solutions to Airy's equation:

$$\text{Ai}(x) = a_1 y_1(x) - a_2 y_2(x) \tag{19}$$

$$\text{Bi}(x) = \sqrt{3}[a_1 y_1(x) + a_2 y_2(x)] \tag{20}$$

where $a_1 = 3^{-2/3}/\Gamma(2/3) \cong 0.355028053887817$ and $a_2 = 3^{-1/3}/\Gamma(1/3) \cong 0.258819403792807$. It can be verified that the Wronskian $W[\text{Ai}, \text{Bi}](x) = \pi^{-1}$ and, therefore, Ai and Bi are indeed a fundamental set of solutions for Airy's equation.

Because of their importance in optics, acoustics, and electrodynamics, the Airy functions have been extensively studied, and their properties are well known to applied mathematicians and scientists. The definitions of Ai(x) and Bi(x) are primarily dictated by their behavior for large positive values of x. A mathematical technique known as **asymptotic analysis**[3] yields the following approximations to Ai(x) and Bi(x) for large values of $|x|$:

$$\text{Ai}(x) \cong \begin{cases} \pi^{-1/2}(-x)^{-1/4} \sin\left[\frac{2}{3}(-x)^{3/2} + \frac{1}{4}\pi\right], & x \to -\infty \\ \frac{1}{2}\pi^{-1/2}x^{-1/4}e^{-2x^{3/2}/3}, & x \to \infty, \end{cases} \tag{21}$$

and

$$\text{Bi}(x) \cong \begin{cases} \pi^{-1/2}(-x)^{-1/4} \cos\left[\frac{2}{3}(-x)^{3/2} + \frac{1}{4}\pi\right], & x \to -\infty \\ \pi^{-1/2}x^{-1/4}e^{2x^{3/2}/3}, & x \to \infty, \end{cases} \tag{22}$$

so Ai(x) decays exponentially while Bi(x) grows exponentially as $x \to \infty$. Graphs of the Airy functions and the asymptotic approximations (21) and (22) are shown in Figure 8.2.3.

The point $x = 0$ is referred to as a **turning point** because the qualitative character of the solutions changes from oscillatory for $x < 0$ to exponential for $x > 0$. It is evident

---

[3] Asymptotic analysis includes a variety of mathematical methods used to find and describe the limiting behavior of functions. Asymptotic approximations for Ai(x) and Bi(x) and many other special functions appear in the handbook by Abramowitz and Stegun listed in the References.

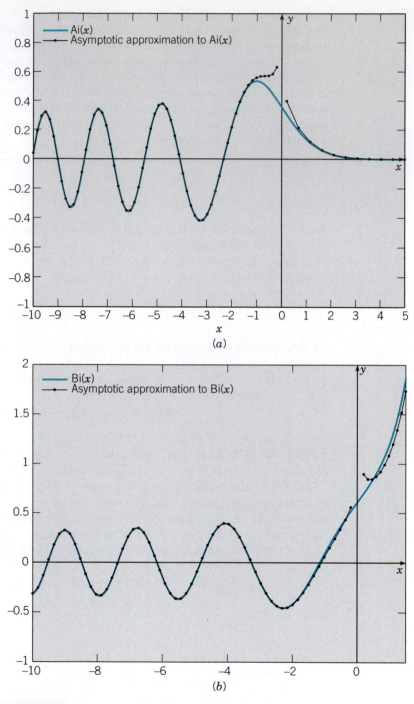

FIGURE 8.2.3 (a) The graphs of Ai(x) and the asymptotic approximations in Eq. (21) for values of $|x| \geq 0.2$. (b) The graphs of Bi(x) and the asymptotic approximations in Eq. (22) for values of $|x| \geq 0.2$.

from the graphs in Figure 8.2.3 and from the asymptotic approximations (21) and (22) that the oscillations for $x < 0$ are not uniform but, rather, decay in amplitude and increase in frequency as the distance from the origin increases.

In contrast to Example 1, the solutions $Ai(x)$ and $Bi(x)$ of Airy's equation are not elementary functions that you have already encountered in calculus. We also note that truncated power series of Ai and Bi can be obtained by substituting truncated power series of y_1 and y_2 from Eq. (18) into Ai and Bi defined in Eqs. (19) and (20). The truncated power series approximations, which are most practical and accurate for small values of $|x|$, are complementary to the asymptotic approximations (21) and (22), which are most practical and accurate for large values of $|x|$.

Numerical Evaluation of Special Functions

In this chapter we use infinite series as a starting point for representing solutions of variable coefficient equations. Particularly important are equations, such as those listed at the beginning of this section, that give rise to special functions which frequently occur in applications. Although series methods occupy an important place in the study of special functions, we do not want the reader to infer that in order to find an accurate approximation of a special function at a given value of x, all that needs to be done is to sum a sufficiently large number of terms in an infinite series. Even if this is a viable option, the operation can be problematic unless proper numerical procedures are employed to speed convergence, and to avoid excessive round-off error, overflows, and underflows.

In general, the problem of accurately computing special functions is a nontrivial task; much effort has been devoted to the development of reliable software packages for this purpose. In order to cope with the problem of evaluating special functions for different ranges of the independent variable and different ranges of parameters that appear in the special functions, the numerical algorithms employed in these packages rely on a variety of mathematical approximation methods, including (truncated) infinite series, other types of polynomial approximations, rational function approximations, recurrence formulas, and asymptotic approximations. As a consequence, in most cases the numerical value of a special function generated by an appropriate function call in a CAS can be considered as highly accurate and reliable.

PROBLEMS

In each of Problems 1 through 14 solve the given differential equation by means of a power series about the given point x_0. Find the recurrence relation; also find the first four terms in each of two linearly independent solutions (unless the series terminates sooner). If possible, find the general term in each solution.

1. $y'' - y = 0,$ $\quad x_0 = 0$
2. $y'' - xy' - y = 0,$ $\quad x_0 = 0$
3. $y'' - xy' - y = 0,$ $\quad x_0 = 1$
4. $y'' + k^2 x^2 y = 0,$ $\quad x_0 = 0,$ $\quad k$ a constant
5. $(1 - x)y'' + y = 0,$ $\quad x_0 = 0$
6. $(2 + x^2)y'' - xy' + 4y = 0,$ $\quad x_0 = 0$
7. $y'' + xy' + 2y = 0,$ $\quad x_0 = 0$
8. $xy'' + y' + xy = 0,$ $\quad x_0 = 1$

9. $(1 + x^2)y'' - 4xy' + 6y = 0,$ $\quad x_0 = 0$
10. $(4 - x^2)y'' + 2y = 0,$ $\quad x_0 = 0$
11. $(3 - x^2)y'' - 3xy' - y = 0,$ $\quad x_0 = 0$
12. $(1 - x)y'' + xy' - y = 0,$ $\quad x_0 = 0$
13. $2y'' + xy' + 3y = 0,$ $\quad x_0 = 0$
14. $2y'' + (x + 1)y' + 3y = 0,$ $\quad x_0 = 2$

In each of Problems 15 through 18:
(a) Find the first five nonzero terms in the solution of the given initial value problem.
(b) Plot the four-term and the five-term approximations to the solution on the same axes.
(c) From the plot in part (b), estimate the interval in which the four-term approximation is reasonably accurate.

CAS **15.** $y'' - xy' - y = 0$, $\quad y(0) = 2$, $\quad y'(0) = 1$
(see Problem 2)

CAS **16.** $(2 + x^2)y'' - xy' + 4y = 0$, $\quad y(0) = -1$,
$y'(0) = 3$ (see Problem 6)

CAS **17.** $y'' + xy' + 2y = 0$, $\quad y(0) = 4$, $\quad y'(0) = -1$
(see Problem 7)

CAS **18.** $(1 - x)y'' + xy' - y = 0$, $\quad y(0) = -3$,
$y'(0) = 2$ (see Problem 12)

19. By making the change of variable $x - 1 = t$ and assuming that y has a Taylor series in powers of t, find two linearly independent series solutions of

$$y'' + (x - 1)^2 y' + (x^2 - 1)y = 0$$

in powers of $x - 1$. Show that you obtain the same result directly by assuming that y has a Taylor series in powers of $x - 1$ and also expressing the coefficient $x^2 - 1$ in powers of $x - 1$.

20. Show directly, using the ratio test, that the two series solutions of Airy's equation about $x = 0$ converge for all x; see Eq. (18) of the text.

21. The Hermite Equation. The equation

$$y'' - 2xy' + \lambda y = 0, \qquad -\infty < x < \infty,$$

where λ is a constant, is known as the Hermite equation. It is an important equation in mathematical physics (see Project 2 at the end of this chapter).

(a) Find the first four terms in each of two linearly independent solutions about $x = 0$.

(b) Observe that if λ is a nonnegative even integer, then one or the other of the series solutions terminates and becomes a polynomial. Find the polynomial solutions for $\lambda = 0, 2, 4, 6, 8$, and 10. Note that each polynomial is determined only up to a multiplicative constant.

(c) The Hermite polynomial $H_n(x)$ is defined as the polynomial solution of the Hermite equation with $\lambda = 2n$ for which the coefficient of x^n is 2^n. Find $H_0(x), \ldots, H_5(x)$.

22. Consider the initial value problem $y' = \sqrt{1 - y^2}$, $y(0) = 0$.

(a) Show that $y = \sin x$ is the solution of this initial value problem.

(b) Look for a solution of the initial value problem in the form of a power series about $x = 0$. Find the coefficients up to the term in x^3 in this series.

In each of Problems 23 through 28 plot several partial sums in a series solution of the given initial value problem about $x = 0$, thereby obtaining graphs analogous to those in Figure 8.2.2.

23. $y'' - xy' - y = 0$, $\quad y(0) = 1$, $\quad y'(0) = 0$ CAS
(see Problem 2)

24. $(2 + x^2)y'' - xy' + 4y = 0$, $\quad y(0) = 1$, CAS
$y'(0) = 0$ (see Problem 6)

25. $y'' + xy' + 2y = 0$, $\quad y(0) = 0$, $\quad y'(0) = 1$ CAS
(see Problem 7)

26. $(4 - x^2)y'' + 2y = 0$, $\quad y(0) = 0$, $\quad y'(0) = 1$ CAS
(see Problem 10)

27. $y'' + x^2 y = 0$, $\quad y(0) = 1$, $\quad y'(0) = 0$ CAS
(see Problem 4)

28. $(1 - x)y'' + xy' - 2y = 0$, $\quad y(0) = 0$, CAS
$y'(0) = 1$

29. (a) Use the algebraic and analytic properties of the sine and cosine functions to show that any elementary trigonometric function can be evaluated at any point in its domain by evaluating $y = \sin x$ at one or two appropriate values of $x \in [0, \pi/2]$.

(b) Use the facts that the terms in the series (12) are alternating in sign and decreasing in absolute value for $0 \le x \le \pi/2$ to show that $|S_7(x) - S(x)| < 10^{-11}$ for all $x \in [0, \pi/2]$.

(c) Show that in order to achieve $|S(100) - S_n(100)| < 10^{-11}$ using Eq. (12), we must use $n = 145$, a polynomial of degree 291.

8.3 Series Solutions Near an Ordinary Point, Part II

In the preceding section we considered the problem of finding solutions of

$$P(x)y'' + Q(x)y' + R(x)y = 0, \tag{1}$$

where P, Q, and R are polynomials, in the neighborhood of an ordinary point x_0. Assuming that Eq. (1) does have a solution $y = \phi(x)$ and that ϕ has a Taylor series

$$y = \phi(x) = \sum_{n=0}^{\infty} a_n(x - x_0)^n, \tag{2}$$

which converges for $|x - x_0| < \rho$, where $\rho > 0$, we found that the a_n can be determined by directly substituting the series (2) for y in Eq. (1).

Let us now consider how we might justify the statement that if x_0 is an ordinary point of Eq. (1), then there exist solutions of the form (2). We also consider the question of the radius of convergence of such a series.

General Solutions in Neighborhoods of Ordinary Points

Suppose, then, that there is a solution of Eq. (1) of the form (2). By differentiating Eq. (2) m times and setting x equal to x_0, we obtain

$$m!a_m = \phi^{(m)}(x_0).$$

Hence, to compute a_n in the series (2), we must show that we can determine $\phi^{(n)}(x_0)$ for $n = 0, 1, 2, \ldots$ from the differential equation (1).

Suppose that $y = \phi(x)$ is a solution of Eq. (1) satisfying the initial conditions $y(x_0) = y_0$, $y'(x_0) = y_0'$. Then $a_0 = y_0$ and $a_1 = y_0'$. If we are solely interested in finding a solution of Eq. (1) without specifying any initial conditions, then a_0 and a_1 remain arbitrary. To determine $\phi^{(n)}(x_0)$ and the corresponding a_n for $n = 2, 3, \ldots$, we turn to Eq. (1). Since ϕ is a solution of Eq. (1), we have

$$P(x)\phi''(x) + Q(x)\phi'(x) + R(x)\phi(x) = 0.$$

For the interval about x_0 for which P is nonvanishing, we can write this equation in the form

$$\phi''(x) = -p(x)\phi'(x) - q(x)\phi(x), \tag{3}$$

where $p(x) = Q(x)/P(x)$ and $q(x) = R(x)/P(x)$. Setting x equal to x_0 in Eq. (3) gives

$$\phi''(x_0) = -p(x_0)\phi'(x_0) - q(x_0)\phi(x_0).$$

Hence a_2 is given by

$$2!a_2 = \phi''(x_0) = -p(x_0)a_1 - q(x_0)a_0. \tag{4}$$

To determine a_3, we differentiate Eq. (3) and then set x equal to x_0, obtaining

$$3!a_3 = \phi'''(x_0) = -[p\phi'' + (p' + q)\phi' + q'\phi]\big|_{x=x_0}$$
$$= -2!p(x_0)a_2 - [p'(x_0) + q(x_0)]a_1 - q'(x_0)a_0. \tag{5}$$

Substituting for a_2 from Eq. (4) gives a_3 in terms of a_1 and a_0. Since P, Q, and R are polynomials and $P(x_0) \neq 0$, all the derivatives of p and q exist at x_0. Hence we can continue to differentiate Eq. (3) indefinitely, determining after each differentiation the successive coefficients a_4, a_5, $\ldots$ by setting x equal to x_0.

Notice that the important property that we use in determining the a_n is that we can compute infinitely many derivatives of the functions p and q. Of course, this is possible if the functions p and q are *analytic* at x_0; that is, they have Taylor series expansions that converge in some interval about the point x_0:

$$p(x) = p_0 + p_1(x - x_0) + \cdots + p_n(x - x_0)^n + \cdots = \sum_{n=0}^{\infty} p_n(x - x_0)^n, \qquad (6)$$

$$q(x) = q_0 + q_1(x - x_0) + \cdots + q_n(x - x_0)^n + \cdots = \sum_{n=0}^{\infty} q_n(x - x_0)^n. \qquad (7)$$

Now let us turn to the question of the interval of convergence of the series solution. One possibility is to compute the series solution for each problem and then to apply one of the tests for convergence of an infinite series to determine its radius of convergence. However, the question can be answered at once for a wide class of problems by the following theorem.

THEOREM 8.3.1

If x_0 is an ordinary point of the differential equation (1),

$$P(x)y'' + Q(x)y' + R(x)y = 0,$$

that is, if $p = Q/P$ and $q = R/P$ are analytic at x_0, then the general solution of Eq. (1) is

$$y = \sum_{n=0}^{\infty} a_n(x - x_0)^n = a_0 y_1(x) + a_1 y_2(x), \qquad (8)$$

where a_0 and a_1 are arbitrary, and y_1 and y_2 are two power series solutions that are analytic at x_0. The solutions y_1 and y_2 form a fundamental set of solutions. Further, the radius of convergence for each of the series solutions y_1 and y_2 is at least as large as the minimum of the radii of convergence of the series for p and q.

To see that y_1 and y_2 are a fundamental set of solutions, note that they have the form $y_1(x) = 1 + b_2(x - x_0)^2 + \cdots$ and $y_2(x) = (x - x_0) + c_2(x - x_0)^2 + \cdots$, where $b_2 + c_2 = a_2$. Hence y_1 satisfies the initial conditions $y_1(x_0) = 1$, $y_1'(x_0) = 0$, and y_2 satisfies the initial conditions $y_2(x_0) = 0$, $y_2'(x_0) = 1$. Thus $W[y_1, y_2](x_0) = 1$.

Also note that although calculating the coefficients by successively differentiating the differential equation is excellent in theory, it is usually not a practical computational procedure. Rather, one should substitute the series (2) for y in the differential equation (1) and determine the coefficients so that the differential equation is satisfied, as in the examples in the preceding section.

We will not prove Theorem 8.3.1. What is important for our purposes is that there is a series solution of the form (2) and that the radius of convergence of the series solution cannot be less than the smaller of the radii of convergence of the series for p and q; hence we need only determine these.

This can be done in either of two ways. Again, one possibility is simply to compute the power series for p and q and then to determine the radii of convergence by using one of the convergence tests for infinite series. However, there is an easier way when P, Q, and R are polynomials. It is shown in the theory of functions of a complex variable that the ratio of two polynomials, say, Q/P, has a convergent power series expansion about a point $x = x_0$ if $P(x_0) \neq 0$. Furthermore, if we assume that any factors common to Q and P have been canceled, then the radius of convergence of the power series for Q/P about the point x_0 is precisely the distance from x_0 to the nearest zero of P. In determining this distance, we must remember that $P(x) = 0$ may have complex roots, and these must also be considered.

EXAMPLE 1

What is the radius of convergence of the Taylor series for $(1 + x^2)^{-1}$ about $x = 0$?

One way to proceed is to find the Taylor series in question, namely

$$\frac{1}{1 + x^2} = 1 - x^2 + x^4 - x^6 + \cdots + (-1)^n x^{2n} + \cdots.$$

Then it can be verified by the ratio test that $\rho = 1$. Another approach is to note that the zeros of $1 + x^2$ are $x = \pm i$. Since the distance in the complex plane from 0 to i or to $-i$ is 1, the radius of convergence of the power series about $x = 0$ is 1.

EXAMPLE 2

What is the radius of convergence of the Taylor series for $(x^2 - 2x + 2)^{-1}$ about $x = 0$? About $x = 1$?

First notice that

$$x^2 - 2x + 2 = 0$$

has solutions $x = 1 \pm i$. The distance in the complex plane from $x = 0$ to either $x = 1 + i$ or $x = 1 - i$ is $\sqrt{2}$; hence the radius of convergence of the Taylor series expansion $\sum_{n=0}^{\infty} a_n x^n$ about $x = 0$ is $\sqrt{2}$.

The distance in the complex plane from $x = 1$ to either $x = 1 + i$ or $x = 1 - i$ is 1; hence the radius of convergence of the Taylor series expansion $\sum_{n=0}^{\infty} b_n (x - 1)^n$ about $x = 1$ is 1.

In Example 2 of the previous section we found series solutions of the Airy equation, $y'' - xy = 0$. Since $P(x) = 1$ while $Q(x) = 0$ and $R(x) = -x$ have infinite radii of convergence, if follows from Theorem 8.3.1 that the series converge for all values of x.

A series solution may converge for a wider range of x than indicated by Theorem 8.3.1, so the theorem actually gives only a lower bound on the radius of convergence of the series solution. This is illustrated by the Legendre polynomial solution of the Legendre equation given in the next example.

EXAMPLE 3

Determine a lower bound for the radius of convergence of series solutions about $x = 0$ for the Legendre equation

$$(1 - x^2)y'' - 2xy' + \alpha(\alpha + 1)y = 0,$$

where α is a constant.

Note that $P(x) = 1 - x^2$, $Q(x) = -2x$, and $R(x) = \alpha(\alpha + 1)$ are polynomials, and that the zeros of P, namely $x = \pm 1$, are a distance 1 from $x = 0$. Hence a series solution of the form $\sum_{n=0}^{\infty} a_n x^n$ converges for $|x| < 1$ at least, and possibly for larger values of x. Indeed, it can be shown that if α is a positive integer, one of the series solutions terminates after a finite number of terms and hence converges not just for $|x| < 1$ but for all x. For example, if $\alpha = 1$, the polynomial solution is $y = x$. See Problems 22 through 29 at the end of this section for a more comprehensive discussion of the Legendre equation.

EXAMPLE 4

Determine a lower bound for the radius of convergence of series solutions of the differential equation

$$(1 + x^2)y'' + 2xy' + 4x^2y = 0 \tag{9}$$

about the point $x = 0$ and about the point $x = -\frac{1}{2}$.

Again P, Q, and R are polynomials, and P has zeros at $x = \pm i$. The distance in the complex plane from 0 to $\pm i$ is 1, and from $-\frac{1}{2}$ to $\pm i$ is $\sqrt{1 + \frac{1}{4}} = \sqrt{5}/2$. Hence in the first case the series $\sum_{n=0}^{\infty} a_n x^n$ converges at least for $|x| < 1$, and in the second case the series $\sum_{n=0}^{\infty} b_n (x + \frac{1}{2})^n$ converges at least for $|x + \frac{1}{2}| < \sqrt{5}/2$.

An interesting observation that we can make about Eq. (9) follows from Theorems 4.2.1 and 8.3.1. Suppose that initial conditions $y(0) = y_0$ and $y'(0) = y_1$ are given. Since $1 + x^2 \neq 0$ for all x, we know from Theorem 4.2.1 that there exists a unique solution of the initial value problem on $-\infty < x < \infty$. On the other hand, Theorem 8.3.1 only guarantees a series solution of the form $\sum_{n=0}^{\infty} a_n x^n$ (with $a_0 = y_0$, $a_1 = y_1$) for $-1 < x < 1$. The unique solution on the interval $-\infty < x < \infty$ may not have a power series about $x = 0$ that converges for all x.

EXAMPLE 5

Can we determine a series solution about $x = 0$ for the differential equation

$$y'' + (\sin x)y' + (1 + x^2)y = 0,$$

and if so, what is the radius of convergence?

For this differential equation, $p(x) = \sin x$ and $q(x) = 1 + x^2$. Recall from calculus that $\sin x$ has a Taylor series expansion about $x = 0$ that converges for all x. Furthermore, q also has a Taylor series expansion about $x = 0$, namely $q(x) = 1 + x^2$, that converges for all x. Thus there is a series solution of the form $y = \sum_{n=0}^{\infty} a_n x^n$ with a_0 and a_1 arbitrary, and the series converges for all x.

PROBLEMS

In each of Problems 1 through 4 determine $\phi''(x_0)$, $\phi'''(x_0)$, and $\phi^{(4)}(x_0)$ for the given point x_0 if $y = \phi(x)$ is a solution of the given initial value problem.

1. $y'' + xy' + y = 0$; $y(0) = 1$, $y'(0) = 0$

2. $y'' + (\sin x)y' + (\cos x)y = 0$; $y(0) = 0$, $y'(0) = 1$

3. $x^2y'' + (1 + x)y' + 3(\ln x)y = 0$; $y(1) = 2$, $y'(1) = 0$

4. $y'' + x^2y' + (\sin x)y = 0$; $y(0) = a_0$, $y'(0) = a_1$

In each of Problems 5 through 8 determine a lower bound for the radius of convergence of series solutions about each given point x_0 for the given differential equation.

5. $y'' + 4y' + 6xy = 0$; $x_0 = 0$, $x_0 = 4$

6. $(x^2 - 2x - 3)y'' + xy' + 4y = 0$; $x_0 = 4$, $x_0 = -4$, $x_0 = 0$

7. $(1 + x^3)y'' + 4xy' + y = 0$; $x_0 = 0$, $x_0 = 2$

8. $xy'' + y = 0$; $x_0 = 1$

9. Determine a lower bound for the radius of convergence of series solutions about the given x_0 for each of the differential equations in Problems 1 through 14 of Section 8.2.

10. **The Chebyshev Equation.** The Chebyshev differential equation is

$$(1 - x^2)y'' - xy' + \alpha^2 y = 0,$$

where α is a constant.

(a) Determine two linearly independent solutions in powers of x for $|x| < 1$.

(b) Show that if α is a nonnegative integer n, then there is a polynomial solution of degree n. These polynomials, when properly normalized, are called the Chebyshev polynomials. They are very useful in problems that require a polynomial approximation to a function defined on $-1 \le x \le 1$.

(c) Find a polynomial solution for each of the cases $\alpha = n = 0, 1, 2, 3$.

For each of the differential equations in Problems 11 through 14 find the first four nonzero terms in each of two linearly independent power series solutions about the origin. What do you expect the radius of convergence to be for each solution?

11. $y'' + (\sin x)y = 0$

12. $e^x y'' + xy = 0$

13. $(\cos x)y'' + xy' - 2y = 0$

14. $e^{-x}y'' + \ln(1 + x)y' - xy = 0$

15. Let x and x^2 be solutions of a differential equation $P(x)y'' + Q(x)y' + R(x)y = 0$. Can you say whether the point $x = 0$ is an ordinary point or a singular point? Prove your answer.

First Order Equations. The series methods discussed in this section are directly applicable to the first order linear differential equation $P(x)y' + Q(x)y = 0$ at a point x_0, if the function $p = Q/P$ has a Taylor series expansion about that point. Such a point is called an ordinary point, and further, the radius of convergence of the series $y = \sum_{n=0}^{\infty} a_n(x - x_0)^n$ is at least as large as the radius of convergence of the series for Q/P. In each of Problems 16 through 21 solve the given differential equation by a series in powers of x and verify that a_0 is arbitrary in each case. Problems 20 and 21 involve nonhomogeneous differential equations to which series methods can be easily extended. Where possible, compare the series solution with the solution obtained by using the methods described in Chapter 2.

16. $y' - y = 0$

17. $y' - xy = 0$

18. $y' = e^{x^2}y$,   three terms only

19. $(1 - x)y' = y$

20. $y' - y = x^2$

21. $y' + xy = 1 + x$

The Legendre Equation. Problems 22 through 29 deal with the Legendre equation

$$(1 - x^2)y'' - 2xy' + \alpha(\alpha + 1)y = 0.$$

As indicated in Example 3, the point $x = 0$ is an ordinary point of this equation, and the distance from the origin to the nearest zero of $P(x) = 1 - x^2$ is 1. Hence the radius of convergence of series solutions about $x = 0$ is at least 1. Also notice that we need to consider only $\alpha \ge -1/2$ since $\beta = \alpha(\alpha + 1) = (\alpha + 1/2)^2 - 1/4$ maps $\alpha \in [-1/2, \infty)$ onto the interval $[-1/4, \infty)$, the set of all possible values that β can assume.

22. Show that two solutions of the Legendre equation for $|x| < 1$ are

$y_1(x) =$

$$1 - \frac{\alpha(\alpha + 1)}{2!}x^2 + \frac{\alpha(\alpha - 2)(\alpha + 1)(\alpha + 3)}{4!}x^4$$
$$+ \sum_{m=3}^{\infty}(-1)^m \frac{\alpha \cdots (\alpha - 2m + 2)(\alpha + 1) \cdots (\alpha + 2m - 1)}{(2m)!}x^{2m},$$

$y_2(x) =$

$$x - \frac{(\alpha - 1)(\alpha + 2)}{3!}x^3 + \frac{(\alpha - 1)(\alpha - 3)(\alpha + 2)(\alpha + 4)}{5!}x^5$$
$$+ \sum_{m=3}^{\infty}(-1)^m \frac{(\alpha - 1) \cdots (\alpha - 2m + 1)(\alpha + 2) \cdots (\alpha + 2m)}{(2m + 1)!}x^{2m+1}.$$

23. Show that if α is zero or a positive even integer $2n$, the series solution y_1 reduces to a polynomial of degree $2n$ containing only even powers of x. Find the polynomials corresponding to $\alpha = 0, 2,$ and 4. Show that if α is a positive odd integer $2n + 1$, the series solution y_2 reduces to a polynomial of degree $2n + 1$ containing only odd powers of x. Find the polynomials corresponding to $\alpha = 1, 3,$ and 5.

24. The Legendre polynomial $P_n(x)$ is defined as the CAS polynomial solution of the Legendre equation with $\alpha = n$ that also satisfies the condition $P_n(1) = 1$.

(a) Using the results of Problem 23, find the Legendre polynomials $P_0(x), \ldots, P_5(x)$.

(b) Plot the graphs of $P_0(x), \ldots, P_5(x)$ for $-1 \le x \le 1$.

(c) Find the zeros of $P_0(x), \ldots, P_5(x)$.

25. It can be shown that the general formula for $P_n(x)$ is

$$P_n(x) = \frac{1}{2^n}\sum_{k=0}^{[n/2]} \frac{(-1)^k(2n - 2k)!}{k!(n - k)!(n - 2k)!}x^{n-2k},$$

where $[n/2]$ denotes the greatest integer less than or equal to $n/2$. By observing the form of $P_n(x)$ for n even and n odd, show that $P_n(-1) = (-1)^n$.

26. The Legendre polynomials play an important role in mathematical physics. For example, in solving Laplace's equation (the potential equation) in

spherical coordinates, we encounter the equation

$$\frac{d^2 F(\varphi)}{d\varphi^2} + \cot\varphi \frac{dF(\varphi)}{d\varphi} + n(n+1)F(\varphi) = 0,$$

$$0 < \varphi < \pi,$$

where n is a positive integer. Show that the change of variable $x = \cos\varphi$ leads to the Legendre equation with $\alpha = n$ for $y = f(x) = F(\arccos x)$.

27. Show that for $n = 0, 1, 2, 3$, the corresponding Legendre polynomial is given by

$$P_n(x) = \frac{1}{2^n n!} \frac{d^n}{dx^n}(x^2 - 1)^n.$$

This formula, known as Rodrigues's formula, is true for all positive integers n.

28. Show that the Legendre equation can also be written as

$$[(1 - x^2)y']' = -\alpha(\alpha + 1)y.$$

Then it follows that

$$[(1 - x^2)P_n'(x)]' = -n(n+1)P_n(x) \quad \text{and}$$
$$[(1 - x^2)P_m'(x)]' = -m(m+1)P_m(x).$$

By multiplying the first equation by $P_m(x)$ and the second equation by $P_n(x)$, integrating by parts, and then subtracting one equation from the other, show that

$$\int_{-1}^{1} P_n(x)P_m(x)\,dx = 0 \qquad \text{if} \quad n \neq m.$$

This property of the Legendre polynomials is known as the orthogonality property. If $m = n$, it can be shown that the value of the preceding integral is $2/(2n + 1)$.

29. Given a polynomial f of degree n, it is possible to express f as a linear combination of $P_0, P_1, P_2, \ldots, P_n$:

$$f(x) = \sum_{k=0}^{n} a_k P_k(x).$$

Using the result of Problem 28, show that

$$a_k = \frac{2k + 1}{2} \int_{-1}^{1} f(x)P_k(x)\,dx.$$

8.4 Regular Singular Points

In Sections 8.4 through 8.7 we consider the nonconstant coefficient equation

$$P(x)y'' + Q(x)y' + R(x)y = 0, \tag{1}$$

where x_0 is a singular point. This means that $P(x_0) = 0$ and at least one of Q and R is not zero at x_0.

If we attempt to use the power series method of the preceding section to solve Eq. (1) in the neighborhood of a singular point x_0, we find that the method fails. This is because the solution of Eq. (1) is often not analytic at x_0 and consequently cannot be represented by a Taylor series in powers of $x - x_0$. Therefore, to have any chance of solving Eq. (1) in the neighborhood of a singular point, we must use a more general type of series expansion.

Since the singular points of a differential equation are usually few in number, we might ask whether we can simply ignore them, especially since we already know how to construct solutions about ordinary points. However, this is not feasible because the singular points determine the principal features of the solution to a much larger extent than you might at first suspect. In the neighborhood of a singular point the solution often becomes large, or experiences rapid changes in magnitude.

Cauchy–Euler Equations Revisited

To illustrate the preceding statement, we consider a special case of Eq. (1), the Cauchy–Euler equation $ax^2 y'' + bxy' + cy = 0$, where $a \neq 0$, b, and c are real constants. Dividing through by a, we can write the Cauchy–Euler equation in the form

$$L[y] = x^2 y'' + \alpha x y' + \beta y = 0, \tag{2}$$

where $\alpha = b/a$ and $\beta = c/a$ are real constants. In this case $P(x) = x^2$, so $x = 0$ is the only singular point for Eq. (2); all other points are ordinary points. In Section 4.4 we were able to find fundamental sets of solutions for Eq. (2) on the interval $(0, \infty)$ by introducing a new independent variable $z = \ln x$, thereby transforming Eq. (2) into the constant coefficient equation

$$y'' + (\alpha - 1)y' + \beta y = 0 \tag{3}$$

for $y = y(z)$. Equation (3) always has at least one solution of the form $y = e^{rz}$, where r is one of the roots of the characteristic equation for Eq. (3):

$$F(r) = r(r - 1) + \alpha r + \beta = 0. \tag{4}$$

If we substitute $\ln x$ for z in $y = e^{rz}$, it follows that Eq. (2) always has at least one solution of the form $y = e^{r \ln x} = x^r$ if $x > 0$. This suggests that we can skip the intermediate step of changing the independent variable from x to $z = \ln x$, and directly seek solutions of Eq. (2) in the form $y = x^r$. Observing that $(x^r)' = rx^{r-1}$ and $(x^r)'' = r(r - 1)x^{r-2}$, we obtain

$$L[x^r] = x^2(x^r)'' + \alpha x(x^r)' + \beta x^r$$
$$= x^r[r(r - 1) + \alpha r + \beta] = x^r F(r), \tag{5}$$

where $F(r)$ is the characteristic polynomial defined in Eq. (4). This shows that if r is a root of Eq. (4), then $L[x^r]$ is zero, and $y = x^r$ is a solution of Eq. (2).

Let us denote the roots of Eq. (4) by

$$r_1, r_2 = \frac{-(\alpha - 1) \pm \sqrt{(\alpha - 1)^2 - 4\beta}}{2}. \tag{6}$$

Thus $F(r) = (r - r_1)(r - r_2)$. General solutions of Eq. (2) on both $x < 0$ and $x > 0$ consisting of real-valued functions are, of course, identical to those given in Section 4.4 for each of the cases in which the roots of Eq. (4) are real and different, real but equal, and complex conjugates. For convenience, we list these solutions here in Table 8.4.1. Below however, we recount some of the ideas used to solve Eq. (2) in the context of seeking solutions directly in the form x^r.

In the case that the roots r_1 and r_2 are equal, we know from Section 4.4 that $y_1(x) = x^{r_1}$ and $y_2(x) = x^{r_1} \ln x$ are a fundamental set of solutions to Eq. (2) in the interval $x > 0$. Although y_2 can be obtained by the method of reduction of order (see Problem 40), we exhibit here an alternative method that will be used to simplify a calculation which arises in Section 8.6. Since $r_1 = r_2$, it follows that $F(r) = (r - r_1)^2$. In this case not only does

TABLE 8.4.1

General solutions of the Cauchy–Euler equation $x^2 y'' + \alpha x y' + \beta y = 0$ for $x \neq 0$. The roots r_1 and r_2 are solutions of the characteristic equation $r(r - 1) + \alpha r + \beta = 0$.

Roots	General Solution
r_1 and r_2 real, $r_1 \neq r_2$	$y = c_1 |x|^{r_1} + c_2 |x|^{r_2}, \ x \neq 0$
r_1 and r_2 real, $r_1 = r_2$	$y = c_1 |x|^{r_1} + c_2 |x|^{r_1} \ln|x|, \ x \neq 0$
$r_1 = \mu + i\nu, r_2 = \mu - i\nu$	$y = |x|^{\mu}[c_1 \cos(\nu \ln|x|) + c_2 \sin(\nu \ln|x|)], \ x \neq 0$

$F(r_1) = 0$ but also $F'(r_1) = 0$. This suggests differentiating Eq. (5) with respect to r and then setting r equal to r_1. Differentiating Eq. (5) with respect to r gives

$$\frac{\partial}{\partial r} L[x^r] = \frac{\partial}{\partial r}[x^r F(r)].$$

Substituting $(r - r_1)^2$ for $F(r)$, interchanging differentiation with respect to x and with respect to r, and noting that $\partial(x^r)/\partial r = x^r \ln x$, we obtain

$$L[x^r \ln x] = (r - r_1)^2 x^r \ln x + 2(r - r_1)x^r. \tag{7}$$

The right side of Eq. (7) is zero for $r = r_1$; consequently,

$$y_2(x) = x^{r_1}\ln x, \qquad x > 0 \tag{8}$$

is a second solution of Eq. (2).

To obtain real-valued solutions of Eq. (2) in the interval $x < 0$, we make the change of variable $\hat{x} = -x$, where $\hat{x} > 0$, and let $y = y(\hat{x})$. The transformed equation, in terms of the independent variable $\hat{x} \in (0, \infty)$, is (see Problem 46, Section 4.4)

$$\hat{x}^2 \frac{d^2 y}{d\hat{x}^2} + \alpha \hat{x}\frac{dy}{d\hat{x}} + \beta y = 0, \qquad \hat{x} > 0.$$

Since this equation is exactly the same as Eq. (2), its general solutions are identical to those of Eq. (2) in $x > 0$, but expressed in terms of the independent variable $\hat{x}$. To get back to the general solution of Eq. (2) in $I = (-\infty, 0)$, we then substitute $-x$ for $\hat{x}$. We can then combine the results for $x > 0$ and $x < 0$ by recalling that $|x| = x$ when $x > 0$ and that $|x| = -x$ when $x < 0$.

Solutions of Cauchy–Euler equations serve to illustrate the potentially interesting behavior of a physical system modeled by a differential equation with a singular point, in the neighborhood of that singular point. Often geometric singularities in a physical problem, such as corners or sharp edges, lead to singular points in the corresponding differential equation. Thus, although at first sight we might want to avoid the few points where a differential equation is singular, it is necessary to study the solution most carefully in neighborhoods of these points. To illustrate, we briefly refer to the examples of Cauchy–Euler equations presented in Section 4.4.

▶ In Example 4 of Section 4.4, we found that the general solution of the Cauchy–Euler equation $2x^2 y'' + 3xy' - 7 = 0$ on $x > 0$ is $y = c_1 x^{1/2} + c_2 x^{-1}$. Thus all nontrivial solutions are nonanalytic at $x = 0$, and any solution for which $c_2 \neq 0$ becomes unbounded as $x \to \infty$ (see Figure 4.4.4).

▶ In Example 5 of Section 4.4, we found that the general solution of $y'' + 5y'/x + 4y/x^2 = 0$ on the interval $x < 0$ is $y = c_1 x^{-2} + c_2 x^{-2}\ln(-x)$. All solutions, other than the trivial solution $y = 0$, are unbounded as $x \to 0$ (see Figure 4.4.5).

▶ And in Example 6 of Section 4.4, we found that the general solution of $4x^2 y'' + 8xy' + 65y = 0$ on the interval $x > 0$ is $y = x^{-1/2}[c_1 \cos(4\ln x) + c_2 \sin(4\ln x)]$. Thus nontrivial solutions rapidly oscillate with ever-increasing amplitude and frequency as $x \to 0$ (see Figure 4.4.6).

The solutions of a Cauchy–Euler equation of the form

$$(x - x_0)^2 y'' + \alpha(x - x_0)y' + \beta y = 0 \tag{9}$$

are similar. If one looks for solutions of the form $y = (x - x_0)^r$, then the general solution is given in Table 8.4.1 with x replaced by $x - x_0$. Alternatively, we can reduce Eq. (9) to the form of Eq. (2) by making the change of independent variable $t = x - x_0$.

Regular and Irregular Singular Points

Without any additional information about the behavior of Q/P and R/P in the neighborhood of the singular point, it is impossible to describe the behavior of the solutions of Eq. (1) near $x = x_0$. It may be that there are two distinct solutions of Eq. (1) that remain bounded as $x \rightarrow x_0$; or there may be only one, with the other becoming unbounded as $x \rightarrow x_0$; or they may both become unbounded as $x \rightarrow x_0$. If Eq. (1) has solutions that become unbounded as $x \rightarrow x_0$, it is often important to determine the precise functional nature of these solutions as $x \rightarrow x_0$. For example, does $y \rightarrow \infty$ in the same way as $(x - x_0)^{-1}$ or $|x - x_0|^{-1/2}$, or in some other manner?

Our goal is to extend the method already developed for solving Eq. (1) near an ordinary point so that it also applies to the neighborhood of a singular point x_0. To do this in a reasonably simple manner, it is necessary to restrict ourselves to cases in which the singularities in the functions Q/P and R/P at $x = x_0$ are not too severe—that is, to what we might call "weak singularities." At this stage it is not clear exactly what is an acceptable singularity. However, in the case that P, Q, and R are polynomials, as we develop the method of solution, you will see that the appropriate conditions (see also Problem 21, Section 8.6) to distinguish "weak singularities" are

$$\lim_{x \to x_0} (x - x_0) \frac{Q(x)}{P(x)} \quad \text{is finite} \tag{10}$$

and

$$\lim_{x \to x_0} (x - x_0)^2 \frac{R(x)}{P(x)} \quad \text{is finite.} \tag{11}$$

This means that the singularity in Q/P can be no worse than $(x - x_0)^{-1}$ and the singularity in R/P can be no worse than $(x - x_0)^{-2}$. Such a point is called a regular singular point of Eq. (1). The definition of a regular singular point for equations with more general coefficients than polynomials follows.

DEFINITION 8.4.1

The point x_0 is a **regular singular point** of Eq. (1):

$$P(x)y'' + Q(x)y' + R(x)y = 0,$$

if it is a singular point and if both[4]

$$(x - x_0) \frac{Q(x)}{P(x)} \quad \text{and} \quad (x - x_0)^2 \frac{R(x)}{P(x)} \tag{12}$$

have convergent Taylor series about x_0—that is, if the functions in Eq. (12) are analytic at $x = x_0$. Any singular point of Eq. (1) that is not a regular singular point is called an **irregular singular point** of Eq. (1).

Observe that Eqs. (10) and (11) imply that the functions in Eq. (12) are analytic if P, Q, and R are polynomials. In particular, note that the conditions in Eqs. (10) and (11) are

---

[4]The functions given in Eq. (12) may not be defined at x_0, in which case their values at x_0 are to be assigned as their limits as $x \rightarrow x_0$.

satisfied by the Cauchy–Euler equation (2). Thus the singularity in a Cauchy–Euler equation is a regular singular point. Indeed, we will see that all equations of the form (1) behave very much like Cauchy–Euler equations near a regular singular point. That is, solutions near a regular singular point may include powers of x with negative or nonintegral exponents, logarithms, or sines or cosines of logarithmic arguments.

In the following sections we discuss how to solve Eq. (1) in the neighborhood of a regular singular point. A discussion of the solutions of differential equations in the neighborhood of irregular singular points is more complicated and may be found in more advanced books.

As an alternative to analytical methods, one can consider the use of a numerical method, such as the Runge–Kutta method discussed in Section 3.7. However, numerical methods are ill suited to the study of solutions near a singular point. Thus, even if one adopts a numerical approach, it is advantageous to combine it with the analytical methods of this chapter in order to examine the behavior of solutions near singular points.

EXAMPLE 1

Determine the singular points of the Legendre equation

$$(1 - x^2)y'' - 2xy' + \alpha(\alpha + 1)y = 0 \tag{13}$$

and determine whether they are regular or irregular.

In this case $P(x) = 1 - x^2$, so the singular points are $x = 1$ and $x = -1$. Observe that when we divide Eq. (13) by $1 - x^2$, the coefficients of y' and y are $-2x/(1 - x^2)$ and $\alpha(\alpha + 1)/(1 - x^2)$, respectively. We consider the point $x = 1$ first. Thus, from Eqs. (10) and (11), we calculate

$$\lim_{x \to 1}(x - 1)\frac{-2x}{1 - x^2} = \lim_{x \to 1}\frac{(x - 1)(-2x)}{(1 - x)(1 + x)} = \lim_{x \to 1}\frac{2x}{1 + x} = 1$$

and

$$\lim_{x \to 1}(x - 1)^2\frac{\alpha(\alpha + 1)}{1 - x^2} = \lim_{x \to 1}\frac{(x - 1)^2\alpha(\alpha + 1)}{(1 - x)(1 + x)}$$

$$= \lim_{x \to 1}\frac{(x - 1)(-\alpha)(\alpha + 1)}{1 + x} = 0.$$

Since these limits are finite, the point $x = 1$ is a regular singular point. It can be shown in a similar manner that $x = -1$ is also a regular singular point.

EXAMPLE 2

Determine the singular points of the differential equation

$$2x(x - 2)^2 y'' + 3xy' + (x - 2)y = 0$$

and classify them as regular or irregular.

Dividing the differential equation by $2x(x - 2)^2$, we have

$$y'' + \frac{3}{2(x - 2)^2}y' + \frac{1}{2x(x - 2)}y = 0,$$

so $p(x) = \dfrac{Q(x)}{P(x)} = \dfrac{3}{2(x-2)^2}$ and $q(x) = \dfrac{R(x)}{P(x)} = \dfrac{1}{2x(x-2)}$. The singular points are $x = 0$ and $x = 2$. Consider $x = 0$. We have

$$\lim_{x \to 0} xp(x) = \lim_{x \to 0} x \dfrac{3}{2(x-2)^2} = 0,$$

$$\lim_{x \to 0} x^2 q(x) = \lim_{x \to 0} x^2 \dfrac{1}{2x(x-2)} = 0.$$

Since these limits are finite, $x = 0$ is a regular singular point. For $x = 2$ we have

$$\lim_{x \to 2}(x-2)p(x) = \lim_{x \to 2}(x-2)\dfrac{3}{2(x-2)^2} = \lim_{x \to 2}\dfrac{3}{2(x-2)},$$

so the limit does not exist; hence $x = 2$ is an irregular singular point.

EXAMPLE 3

Determine the singular points of

$$\left(x - \frac{\pi}{2}\right)^2 y'' + (\cos x)y' + (\sin x)y = 0$$

and classify them as regular or irregular.

The only singular point is $x = \pi/2$. To study it, we consider the functions

$$\left(x - \frac{\pi}{2}\right)p(x) = \left(x - \frac{\pi}{2}\right)\frac{Q(x)}{P(x)} = \frac{\cos x}{x - \pi/2}$$

and

$$\left(x - \frac{\pi}{2}\right)^2 q(x) = \left(x - \frac{\pi}{2}\right)^2 \frac{R(x)}{P(x)} = \sin x.$$

Since

$$\cos x = \cos(x - \pi/2 + \pi/2) = -\sin(x - \pi/2)$$

$$= -\frac{x - \pi/2}{1!} + \frac{(x - \pi/2)^3}{3!} - \frac{(x - \pi/2)^5}{5!} + \cdots,$$

the Taylor series for $\cos x$ about $x = \pi/2$ is

$$\frac{\cos x}{x - \pi/2} = -1 + \frac{(x - \pi/2)^2}{3!} - \frac{(x - \pi/2)^4}{5!} + \cdots,$$

which converges for all x. Similarly, $\sin x$ is analytic at $x = \pi/2$. Therefore, we conclude that $\pi/2$ is a regular singular point for this equation.

PROBLEMS

In each of Problems 1 through 12 determine the general solution of the given differential equation that is valid in any interval not including the singular point.

1. $x^2 y'' + 4xy' + 2y = 0$
2. $(x + 1)^2 y'' + 3(x + 1)y' + 0.75y = 0$
3. $x^2 y'' - 3xy' + 4y = 0$
4. $x^2 y'' + 3xy' + 5y = 0$
5. $x^2 y'' - xy' + y = 0$
6. $(x - 1)^2 y'' + 8(x - 1)y' + 12y = 0$
7. $x^2 y'' + 6xy' - y = 0$

8. $2x^2y'' - 4xy' + 6y = 0$

9. $x^2y'' - 5xy' + 9y = 0$

10. $(x-2)^2y'' + 5(x-2)y' + 8y = 0$

11. $x^2y'' + 2xy' + 4y = 0$

12. $x^2y'' - 4xy' + 4y = 0$

In each of Problems 13 through 16 find the solution of the given initial value problem. Plot the graph of the solution and describe how the solution behaves as $x \to 0$.

CAS 13. $2x^2y'' + xy' - 3y = 0, \quad y(1) = 1, \quad y'(1) = 4$

CAS 14. $4x^2y'' + 8xy' + 17y = 0, \quad y(1) = 2, \quad y'(1) = -3$

CAS 15. $x^2y'' - 3xy' + 4y = 0, \quad y(-1) = 2, \quad y'(-1) = 3$

CAS 16. $x^2y'' + 3xy' + 5y = 0, \quad y(1) = 1, \quad y'(1) = -1$

In each of Problems 17 through 34 find all singular points of the given equation and determine whether each one is regular or irregular.

17. $xy'' + (1-x)y' + xy = 0$

18. $x^2(1-x)^2y'' + 2xy' + 4y = 0$

19. $x^2(1-x)y'' + (x-2)y' - 3xy = 0$

20. $x^2(1-x^2)y'' + (2/x)y' + 4y = 0$

21. $(1-x^2)^2y'' + x(1-x)y' + (1+x)y = 0$

22. $x^2y'' + xy' + (x^2 - v^2)y = 0, \quad$ Bessel equation

23. $(x+3)y'' - 2xy' + (1-x^2)y = 0$

24. $x(1-x^2)^3y'' + (1-x^2)^2y' + 2(1+x)y = 0$

25. $(x+2)^2(x-1)y'' + 3(x-1)y' - 2(x+2)y = 0$

26. $x(3-x)y'' + (x+1)y' - 2y = 0$

27. $(x^2 + x - 2)y'' + (x+1)y' + 2y = 0$

28. $xy'' + e^xy' + (3\cos x)y = 0$

29. $y'' + (\ln|x|)y' + 3xy = 0$

30. $x^2y'' + 2(e^x - 1)y' + (e^{-x}\cos x)y = 0$

31. $x^2y'' - 3(\sin x)y' + (1+x^2)y = 0$

32. $xy'' + y' + (\cot x)y = 0$

33. $(\sin x)y'' + xy' + 4y = 0$

34. $(x\sin x)y'' + 3y' + xy = 0$

35. Find all values of α for which all solutions of $x^2y'' + \alpha xy' + (5/2)y = 0$ approach zero as $x \to 0$.

36. Find all values of β for which all solutions of $x^2y'' + \beta y = 0$ approach zero as $x \to 0$.

37. Find γ so that the solution of the initial value problem $x^2y'' - 2y = 0, \ y(1) = 1, \ y'(1) = \gamma$ is bounded as $x \to 0$.

38. Find all values of α for which all solutions of $x^2y'' + \alpha xy' + (5/2)y = 0$ approach zero as $x \to \infty$.

39. Consider the Cauchy–Euler equation $x^2y'' + \alpha xy' + \beta y = 0$. Find conditions on α and β so that:
   (a) All solutions approach zero as $x \to 0$.
   (b) All solutions are bounded as $x \to 0$.
   (c) All solutions approach zero as $x \to \infty$.
   (d) All solutions are bounded as $x \to \infty$.
   (e) All solutions are bounded both as $x \to 0$ and as $x \to \infty$.

40. Using the method of reduction of order, show that if r_1 is a repeated root of

$$r(r-1) + \alpha r + \beta = 0,$$

then x^{r_1} and $x^{r_1}\ln x$ are solutions of $x^2y'' + \alpha xy' + \beta y = 0$ for $x > 0$.

In each of Problems 41 and 42 show that the point $x = 0$ is a regular singular point. In each problem try to find solutions of the form $\sum_{n=0}^{\infty} a_n x^n$. Show that there is only one nonzero solution of this form in Problem 41 and that there are no nonzero solutions of this form in Problem 42. Thus in neither case can the general solution be found in this manner. This is typical of equations with singular points.

41. $2xy'' + 3y' + xy = 0$

42. $2x^2y'' + 3xy' - (1+x)y = 0$

43. **Singularities at Infinity.** The definitions of an ordinary point and a regular singular point given in the preceding sections apply only if the point x_0 is finite. In more advanced work in differential equations it is often necessary to consider the point at infinity. This is done by making the change of variable $\xi = 1/x$ and studying the resulting equation at $\xi = 0$. Show that, for the differential equation

$$P(x)y'' + Q(x)y' + R(x)y = 0,$$

the point at infinity is an ordinary point if

$$\frac{1}{P(1/\xi)}\left[\frac{2P(1/\xi)}{\xi} - \frac{Q(1/\xi)}{\xi^2}\right] \quad \text{and} \quad \frac{R(1/\xi)}{\xi^4 P(1/\xi)}$$

have Taylor series expansions about $\xi = 0$. Show also that the point at infinity is a regular singular point if at least one of the above functions does not have a Taylor series expansion, but both

$$\frac{\xi}{P(1/\xi)}\left[\frac{2P(1/\xi)}{\xi} - \frac{Q(1/\xi)}{\xi^2}\right] \quad \text{and} \quad \frac{R(1/\xi)}{\xi^2 P(1/\xi)}$$

do have such expansions.

In each of Problems 44 through 49 use the results of Problem 43 to determine whether the point at infinity is an ordinary point, a regular singular point, or an irregular singular point of the given differential equation.

44. $y'' + y = 0$

45. $x^2y'' + xy' - 4y = 0$

46. $(1 - x^2)y'' - 2xy' + \alpha(\alpha + 1)y = 0$, Legendre equation

47. $x^2y'' + xy' + (x^2 - \nu^2)y = 0$, Bessel equation

48. $y'' - 2xy' + \lambda y = 0$, Hermite equation

49. $y'' - xy = 0$, Airy equation

8.5 Series Solutions Near a Regular Singular Point, Part I

We now consider the question of solving the general second order linear equation

$$P(x)y'' + Q(x)y' + R(x)y = 0 \tag{1}$$

in the neighborhood of a regular singular point $x = x_0$. For convenience we assume that $x_0 = 0$. If $x_0 \neq 0$, we can transform the equation into one for which the regular singular point is at the origin by letting $x - x_0$ equal t.

The fact that $x = 0$ is a regular singular point of Eq. (1) means that $xQ(x)/P(x) = xp(x)$ and $x^2R(x)/P(x) = x^2q(x)$ have finite limits as $x \to 0$ and are analytic at $x = 0$. Thus they have convergent power series expansions of the form

$$xp(x) = \sum_{n=0}^{\infty} p_n x^n, \qquad x^2q(x) = \sum_{n=0}^{\infty} q_n x^n, \tag{2}$$

on some interval $|x| < \rho$ about the origin, where $\rho > 0$. To make the quantities $xp(x)$ and $x^2q(x)$ appear in Eq. (1), it is convenient to divide Eq. (1) by $P(x)$ and then to multiply by x^2, obtaining

$$x^2y'' + x[xp(x)]y' + [x^2q(x)]y = 0 \tag{3}$$

or

$$x^2y'' + x(p_0 + p_1x + \cdots + p_nx^n + \cdots)y' + (q_0 + q_1x + \cdots + q_nx^n + \cdots)y = 0. \tag{4}$$

If all of the coefficients p_n and q_n are zero, except possibly

$$p_0 = \lim_{x \to 0} \frac{xQ(x)}{P(x)} \quad \text{and} \quad q_0 = \lim_{x \to 0} \frac{x^2R(x)}{P(x)}, \tag{5}$$

then Eq. (4) reduces to the Cauchy–Euler equation

$$x^2y'' + p_0xy' + q_0y = 0, \tag{6}$$

which was discussed in the preceding section and in Section 4.4. In general, of course, some of the p_n and q_n, $n \geq 1$, are not zero. However, the essential character of solutions of Eq. (4) is identical to that of solutions of the Cauchy–Euler equation (6). The presence of the terms $p_1 x + \cdots + p_n x^n + \cdots$ and $q_1 x + \cdots + q_n x^n + \cdots$ merely complicates the calculations.

We restrict our discussion primarily to the interval $x > 0$. The interval $x < 0$ can be treated, just as for the Cauchy–Euler equation, by making the change of variable $x = -\xi$ and then solving the resulting equation for $\xi > 0$.

Since the coefficients in Eq. (4) are "Cauchy–Euler coefficients" times power series, it is natural to seek solutions in the form of "Cauchy–Euler solutions" times power series. Thus we assume that

$$y = x^r (a_0 + a_1 x + \cdots + a_n x^n + \cdots) = x^r \sum_{n=0}^{\infty} a_n x^n = \sum_{n=0}^{\infty} a_n x^{r+n}, \tag{7}$$

where $a_0 \neq 0$. In other words, r is the exponent of the first term in the series, and a_0 is its coefficient. As part of the solution, we have to determine:

1. The values of r for which Eq. (1) has a solution of the form (7).
2. The recurrence relation for the coefficients a_n.
3. The radius of convergence of the series $\sum_{n=0}^{\infty} a_n x^n$.

The general theory was constructed by Frobenius and is fairly complicated. Rather than trying to present this theory, we simply assume in this and the next two sections that there does exist a solution of the stated form. In particular, we assume that any power series in an expression for a solution has a nonzero radius of convergence and concentrate on showing how to determine the coefficients in such a series. To illustrate the method of Frobenius, we first consider an example.

**EXAMPLE
1**

Solve the differential equation

$$2x^2 y'' - xy' + (1+x)y = 0. \tag{8}$$

It is easy to show that $x = 0$ is a regular singular point of Eq. (8). Furthermore, $xp(x) = -1/2$ and $x^2 q(x) = (1+x)/2$. Thus $p_0 = -1/2$, $q_0 = 1/2$, $q_1 = 1/2$, and all other p's and q's are zero. Noting that Eq. (6), depends only on p_0 and q_0, we find the Cauchy–Euler equation corresponding to Eq. (8):

$$2x^2 y'' - xy' + y = 0. \tag{9}$$

To solve Eq. (8), we assume that there is a solution of the form (7). Then y' and y'' are given by

$$y' = \sum_{n=0}^{\infty} a_n (r+n) x^{r+n-1} \tag{10}$$

and

$$y'' = \sum_{n=0}^{\infty} a_n (r+n)(r+n-1) x^{r+n-2}. \tag{11}$$

By substituting the expressions for y, y', and y'' in Eq. (8), we obtain

$$2x^2 y'' - xy' + (1+x)y = \sum_{n=0}^{\infty} 2a_n(r+n)(r+n-1)x^{r+n}$$

$$- \sum_{n=0}^{\infty} a_n(r+n)x^{r+n} + \sum_{n=0}^{\infty} a_n x^{r+n} + \sum_{n=0}^{\infty} a_n x^{r+n+1}. \quad (12)$$

The last term in Eq. (12) can be written as $\sum_{n=1}^{\infty} a_{n-1} x^{r+n}$, so by combining the terms in Eq. (12), we obtain

$$2x^2 y'' - xy' + (1+x)y = a_0[2r(r-1) - r + 1]x^r$$

$$+ \sum_{n=1}^{\infty} \{[2(r+n)(r+n-1) - (r+n) + 1]a_n + a_{n-1}\}x^{r+n} = 0. \quad (13)$$

If Eq. (13) is to be satisfied for all x, the coefficient of each power of x in Eq. (13) must be zero. From the coefficient of x^r we obtain, since $a_0 \neq 0$,

$$2r(r-1) - r + 1 = 2r^2 - 3r + 1 = (r-1)(2r-1) = 0. \quad (14)$$

Equation (14) is called the **indicial equation** for Eq. (8). Note that it is exactly the polynomial equation we would obtain for the Cauchy–Euler equation (9) associated with Eq. (8). The roots of the indicial equation are

$$r_1 = 1, \qquad r_2 = 1/2. \quad (15)$$

These values of r are called the **exponents at the singularity** for the regular singular point $x = 0$. They determine the qualitative behavior of the solution (7) in the neighborhood of the singular point.

Now we return to Eq. (13) and set the coefficient of x^{r+n} equal to zero. This gives the relation

$$[2(r+n)(r+n-1) - (r+n) + 1]a_n + a_{n-1} = 0, \qquad n \geq 1, \quad (16)$$

or

$$a_n = -\frac{a_{n-1}}{2(r+n)^2 - 3(r+n) + 1}$$

$$= -\frac{a_{n-1}}{[(r+n)-1][2(r+n)-1]}, \qquad n \geq 1. \quad (17)$$

For each root r_1 and r_2 of the indicial equation, we use the recurrence relation (17) to determine a set of coefficients $a_1, a_2, \ldots$. For $r = r_1 = 1$, Eq. (17) becomes

$$a_n = -\frac{a_{n-1}}{(2n+1)n}, \qquad n \geq 1.$$

Thus

$$a_1 = -\frac{a_0}{3 \cdot 1},$$

$$a_2 = -\frac{a_1}{5 \cdot 2} = \frac{a_0}{(3 \cdot 5)(1 \cdot 2)},$$

and

$$a_3 = -\frac{a_2}{7 \cdot 3} = -\frac{a_0}{(3 \cdot 5 \cdot 7)(1 \cdot 2 \cdot 3)}.$$

In general, we have

$$a_n = \frac{(-1)^n}{[3 \cdot 5 \cdot 7 \cdots (2n+1)]n!} a_0, \qquad n \geq 4. \quad (18)$$

Multiplying the numerator and denominator of the right side of Eq. (18) by $2 \cdot 4 \cdot 6 \cdots 2n = 2^n n!$, we can rewrite a_n as

$$a_n = \frac{(-1)^n 2^n}{(2n+1)!} a_0, \qquad n \geq 1.$$

Hence, if we omit the constant multiplier a_0, one solution of Eq. (8) is

$$y_1(x) = x \left[1 + \sum_{n=1}^{\infty} \frac{(-1)^n 2^n}{(2n+1)!} x^n \right], \qquad x > 0. \tag{19}$$

To determine the radius of convergence of the series in Eq. (19), we use the ratio test:

$$\lim_{n \to \infty} \left| \frac{a_{n+1} x^{n+1}}{a_n x^n} \right| = \lim_{n \to \infty} \frac{2|x|}{(2n+2)(2n+3)} = 0$$

for all x. Thus the series converges for all x.

Corresponding to the second root $r = r_2 = \frac{1}{2}$, we proceed similarly. From Eq. (17) we have

$$a_n = -\frac{a_{n-1}}{2n(n - \frac{1}{2})} = -\frac{a_{n-1}}{n(2n-1)}, \qquad n \geq 1.$$

Hence

$$a_1 = -\frac{a_0}{1 \cdot 1},$$

$$a_2 = -\frac{a_1}{2 \cdot 3} = \frac{a_0}{(1 \cdot 2)(1 \cdot 3)},$$

$$a_3 = -\frac{a_2}{3 \cdot 5} = -\frac{a_0}{(1 \cdot 2 \cdot 3)(1 \cdot 3 \cdot 5)},$$

and, in general,

$$a_n = \frac{(-1)^n}{n![1 \cdot 3 \cdot 5 \cdots (2n-1)]} a_0, \qquad n \geq 4. \tag{20}$$

Just as in the case of the first root r_1, we multiply the numerator and denominator by $2 \cdot 4 \cdot 6 \cdots 2n = 2^n n!$. Then we have

$$a_n = \frac{(-1)^n 2^n}{(2n)!} a_0, \qquad n \geq 1.$$

Again omitting the constant multiplier a_0, we obtain the second solution

$$y_2(x) = x^{1/2} \left[1 + \sum_{n=1}^{\infty} \frac{(-1)^n 2^n}{(2n)!} x^n \right], \qquad x > 0. \tag{21}$$

As before, we can show that the series in Eq. (21) converges for all x. Since the series solutions y_1 and y_2 are not a constant multiple of each other, they form a fundamental set of solutions. Hence the general solution of Eq. (8) is

$$y = c_1 y_1(x) + c_2 y_2(x), \qquad x > 0.$$

The preceding example illustrates that if $x = 0$ is a regular singular point, then sometimes there are two solutions of the form (7) in the neighborhood of this point. Similarly, if there is a regular singular point at $x = x_0$, then there may be two solutions of the form

$$y = (x - x_0)^r \sum_{n=0}^{\infty} a_n(x - x_0)^n \qquad (22)$$

that are valid near $x = x_0$. However, just as a Cauchy–Euler equation may not have two solutions of the form $y = x^r$, so a more general equation with a regular singular point may not have two solutions of the form (7) or (22). In particular, we show in the next section that if the roots r_1 and r_2 of the indicial equation are equal, or differ by an integer, then the second solution normally has a more complicated structure. In all cases, though, it is possible to find at least one solution of the form (7) or (22); if r_1 and r_2 differ by an integer, this solution corresponds to the larger value of r. If there is only one such solution, then the second solution involves a logarithmic term, just as for the Cauchy–Euler equation when the roots of the characteristic equation are equal. The method of reduction of order or some other procedure can be invoked to determine the second solution in such cases. This is discussed in Sections 8.6 and 8.7.

If the roots of the indicial equation are complex, then they cannot be equal or differ by an integer, so there are always two solutions of the form (7) or (22). Of course, these solutions are complex-valued functions of x. However, as for the Cauchy–Euler equation, it is possible to obtain real-valued solutions by taking the real and imaginary parts of the complex solutions.

Finally, we mention a practical point. If P, Q, and R are polynomials, it is often much better to work directly with Eq. (1) than with Eq. (3). This avoids the necessity of expressing $xQ(x)/P(x)$ and $x^2 R(x)/P(x)$ as power series. For example, it is more convenient to consider the equation

$$x(1 + x)y'' + 2y' + xy = 0$$

than to write it in the form

$$x^2 y'' + \frac{2x}{1 + x} y' + \frac{x^2}{1 + x} y = 0,$$

which would entail expanding $2x/(1 + x)$ and $x^2/(1 + x)$ in power series.

PROBLEMS

In each of Problems 1 through 10:

(a) Show that the given differential equation has a regular singular point at $x = 0$.

(b) Determine the indicial equation, the recurrence relation, and the roots of the indicial equation.

(c) Find the series solution ($x > 0$) corresponding to the larger root.

(d) If the roots are unequal and do not differ by an integer, find the series solution corresponding to the smaller root also.

1. $2xy'' + y' + xy = 0$

2. $x^2 y'' + xy' + (x^2 - \frac{1}{9})y = 0$

3. $xy'' + y = 0$

4. $xy'' + y' - y = 0$

5. $3x^2 y'' + 2xy' + x^2 y = 0$

6. $x^2 y'' + xy' + (x - 2)y = 0$

7. $xy'' + (1 - x)y' - y = 0$

8. $2x^2 y'' + 3xy' + (2x^2 - 1)y = 0$

9. $x^2 y'' - x(x + 3)y' + (x + 3)y = 0$

10. $x^2 y'' + (x^2 + \frac{1}{4})y = 0$

11. The Legendre equation of order α is

$$(1 - x^2)y'' - 2xy' + \alpha(\alpha + 1)y = 0.$$

The solution of this equation near the ordinary point $x = 0$ was discussed in Problems 22 and 23 of Section 8.3. In Example 1 of Section 8.4 it was shown that $x = \pm 1$ are regular singular points.

(a) Determine the indicial equation and its roots for the point $x = 1$.

(b) Find a series solution in powers of $x - 1$ for $x - 1 > 0$.

Hint: Write $1 + x = 2 + (x - 1)$ and $x = 1 + (x - 1)$. Alternatively, make the change of variable $x - 1 = t$ and determine a series solution in powers of t.

12. The Chebyshev equation is

$$(1 - x^2)y'' - xy' + \alpha^2 y = 0,$$

where α is a constant; see Problem 10 in Section 8.3.

(a) Show that $x = 1$ and $x = -1$ are regular singular points, and find the exponents at each of these singularities.

(b) Find two solutions about $x = 1$.

13. The Laguerre differential equation is

$$xy'' + (1 - x)y' + \lambda y = 0.$$

(a) Show that $x = 0$ is a regular singular point.

(b) Determine the indicial equation, its roots, and the recurrence relation.

(c) Find one solution $(x > 0)$. Show that if $\lambda = m$, a positive integer, this solution reduces to a polynomial. When properly normalized, this polynomial is known as the Laguerre polynomial, $L_m(x)$.

14. The Bessel equation of order zero is

$$x^2 y'' + xy' + x^2 y = 0.$$

(a) Show that $x = 0$ is a regular singular point.

(b) Show that the roots of the indicial equation are $r_1 = r_2 = 0$.

(c) Show that one solution for $x > 0$ is

$$J_0(x) = 1 + \sum_{n=1}^{\infty} \frac{(-1)^n x^{2n}}{2^{2n}(n!)^2}.$$

(d) Show that the series for $J_0(x)$ converges for all x. The function J_0 is known as the Bessel function of the first kind of order zero.

15. Referring to Problem 14, use the method of reduction of order to show that the second solution of the Bessel equation of order zero contains a logarithmic term.

Hint: If $y_2(x) = J_0(x)v(x)$, then

$$y_2(x) = J_0(x) \int \frac{dx}{x[J_0(x)]^2}.$$

Find the first term in the series expansion of $1/x[J_0(x)]^2$.

16. The Bessel equation of order one is

$$x^2 y'' + xy' + (x^2 - 1)y = 0.$$

(a) Show that $x = 0$ is a regular singular point.

(b) Show that the roots of the indicial equation are $r_1 = 1$ and $r_2 = -1$.

(c) Show that one solution for $x > 0$ is

$$J_1(x) = \frac{x}{2} \sum_{n=0}^{\infty} \frac{(-1)^n x^{2n}}{(n+1)! \, n! \, 2^{2n}}.$$

(d) Show that the series for $J_1(x)$ converges for all x. The function J_1 is known as the Bessel function of the first kind of order one.

(e) Show that it is impossible to determine a second solution of the form

$$x^{-1} \sum_{n=0}^{\infty} b_n x^n, \qquad x > 0.$$

8.6 Series Solutions Near a Regular Singular Point, Part II

Now let us consider the general problem of determining a solution of the equation

$$L[y] = x^2 y'' + x[xp(x)]y' + [x^2 q(x)]y = 0, \tag{1}$$

where

$$xp(x) = \sum_{n=0}^{\infty} p_n x^n, \qquad x^2 q(x) = \sum_{n=0}^{\infty} q_n x^n, \tag{2}$$

and both series converge in an interval $|x| < \rho$ for some $\rho > 0$. The point $x = 0$ is a regular singular point, and the corresponding Cauchy–Euler equation is

$$x^2 y'' + p_0 x y' + q_0 y = 0. \tag{3}$$

We seek a solution of Eq. (1) for $x > 0$ and assume that it has the form

$$y = \phi(r, x) = x^r \sum_{n=0}^{\infty} a_n x^n = \sum_{n=0}^{\infty} a_n x^{r+n}, \tag{4}$$

where $a_0 \neq 0$, and we have written $y = \phi(r, x)$ to emphasize that ϕ depends on r as well as x. It follows that

$$y' = \sum_{n=0}^{\infty} (r + n) a_n x^{r+n-1}, \qquad y'' = \sum_{n=0}^{\infty} (r + n)(r + n - 1) a_n x^{r+n-2}. \tag{5}$$

Then, substituting from Eqs. (2), (4), and (5) in Eq. (1) gives

$$a_0 r(r-1) x^r + a_1(r+1) r x^{r+1} + \cdots + a_n(r+n)(r+n-1) x^{r+n} + \cdots$$
$$+ (p_0 + p_1 x + \cdots + p_n x^n + \cdots)$$
$$\times [a_0 r x^r + a_1(r+1) x^{r+1} + \cdots + a_n(r+n) x^{r+n} + \cdots]$$
$$+ (q_0 + q_1 x + \cdots + q_n x^n + \cdots)$$
$$\times (a_0 x^r + a_1 x^{r+1} + \cdots + a_n x^{r+n} + \cdots) = 0.$$

Multiplying the infinite series together and then collecting terms, we obtain

$$a_0 F(r) x^r + [a_1 F(r+1) + a_0(p_1 r + q_1)] x^{r+1}$$
$$+ \{a_2 F(r+2) + a_0(p_2 r + q_2) + a_1[p_1(r+1) + q_1]\} x^{r+2}$$
$$+ \cdots + \{a_n F(r+n) + a_0(p_n r + q_n) + a_1[p_{n-1}(r+1) + q_{n-1}]$$
$$+ \cdots + a_{n-1}[p_1(r+n-1) + q_1]\} x^{r+n} + \cdots = 0,$$

or, in a more compact form,

$$L[\phi](r, x) = a_0 F(r) x^r$$
$$+ \sum_{n=1}^{\infty} \left\{ F(r+n) a_n + \sum_{k=0}^{n-1} a_k[(r+k) p_{n-k} + q_{n-k}] \right\} x^{r+n} = 0, \tag{6}$$

where

$$F(r) = r(r-1) + p_0 r + q_0. \tag{7}$$

For Eq. (6) to be satisfied for all $x > 0$, the coefficient of each power of x must be zero.

Since $a_0 \neq 0$, the term involving x^r yields the equation $F(r) = 0$. This equation is called the **indicial equation**; note that it is exactly the equation we would obtain in looking for solutions $y = x^r$ of the Cauchy–Euler equation (3). Let us denote the roots of the indicial equation by r_1 and r_2 with $r_1 \geq r_2$ if the roots are real. If the roots are complex, the designation of the roots is immaterial. Only for these values of r can we expect to find solutions of Eq. (1) of the form (4). The roots r_1 and r_2 are called the **exponents at the singularity**; they determine the qualitative nature of the solution in the neighborhood of the singular point.

Setting the coefficient of x^{r+n} in Eq. (6) equal to zero gives the **recurrence relation**

$$F(r+n)a_n + \sum_{k=0}^{n-1} a_k[(r+k)p_{n-k} + q_{n-k}] = 0, \qquad n \geq 1. \tag{8}$$

Unequal Roots

Equation (8) shows that, in general, a_n depends on the value of r and all the preceding coefficients $a_0, a_1, \ldots, a_{n-1}$. It also shows that we can successively compute $a_1, a_2, \ldots, a_n, \ldots$ in terms of a_0 and the coefficients in the series for $xp(x)$ and $x^2q(x)$, provided that $F(r+1), F(r+2), \ldots, F(r+n), \ldots$ are not zero. The only values of r for which $F(r) = 0$ are $r = r_1$ and $r = r_2$; since $r_1 \geq r_2$, it follows that $r_1 + n$ is not equal to r_1 or r_2 for $n \geq 1$. Consequently, $F(r_1 + n) \neq 0$ for $n \geq 1$. Hence we can always determine one solution of Eq. (1) in the form (4), namely

$$y_1(x) = x^{r_1}\left[1 + \sum_{n=1}^{\infty} a_n(r_1)x^n\right], \qquad x > 0. \tag{9}$$

Here we have introduced the notation $a_n(r_1)$ to indicate that a_n has been determined from Eq. (8) with $r = r_1$. To specify the arbitrary constant in the solution, we have taken a_0 to be 1.

If r_2 is not equal to r_1, and $r_1 - r_2$ is not a positive integer, then $r_2 + n$ is not equal to r_1 for any value of $n \geq 1$; hence $F(r_2 + n) \neq 0$, and we can also obtain a second solution

$$y_2(x) = x^{r_2}\left[1 + \sum_{n=1}^{\infty} a_n(r_2)x^n\right], \qquad x > 0. \tag{10}$$

Just as for the series solutions about ordinary points discussed in Section 8.3, the series in Eqs. (9) and (10) converge at least in the interval $|x| < \rho$ where the series for both $xp(x)$ and $x^2q(x)$ converge. Within their radii of convergence, the power series $1 + \sum_{n=1}^{\infty} a_n(r_1)x^n$ and $1 + \sum_{n=1}^{\infty} a_n(r_2)x^n$ define functions that are analytic at $x = 0$. Thus the singular behavior, if there is any, of the solutions y_1 and y_2 is due to the factors x^{r_1} and x^{r_2} that multiply these two analytic functions. Next, to obtain real-valued solutions for $x < 0$, we can make the substitution $x = -\xi$ with $\xi > 0$. As we might expect from our discussion of the Cauchy–Euler equation, it turns out that we need only replace x^{r_1} in Eq. (9) and x^{r_2} in Eq. (10) by $|x|^{r_1}$ and $|x|^{r_2}$, respectively. Finally, note that if r_1 and r_2 are complex numbers, then they are necessarily complex conjugates and $r_2 \neq r_1 + N$. Thus, in this case we can always find two series solutions of the form (4); however, they are complex-valued functions of x. Real-valued solutions can be obtained by taking the real and imaginary parts of the complex-valued solutions. The exceptional cases in which $r_1 = r_2$ or $r_1 - r_2 = N$, where N is a positive integer, require more discussion and will be considered later in this section.

It is important to realize that r_1 and r_2, the exponents at the singular point, are easy to find and that they determine the qualitative behavior of the solutions. To calculate r_1 and r_2, it is only necessary to solve the quadratic indicial equation

$$r(r-1) + p_0 r + q_0 = 0, \tag{11}$$

whose coefficients are given by

$$p_0 = \lim_{x \to 0} xp(x), \qquad q_0 = \lim_{x \to 0} x^2 q(x). \tag{12}$$

Note that these are exactly the limits that must be evaluated in order to classify the singularity as a regular singular point; thus they have usually been determined at an earlier stage of the investigation.

Further, if $x = 0$ is a regular singular point of the equation

$$P(x)y'' + Q(x)y' + R(x)y = 0, \tag{13}$$

where the functions P, Q, and R are polynomials, then $xp(x) = xQ(x)/P(x)$ and $x^2 q(x) = x^2 R(x)/P(x)$. Thus

$$p_0 = \lim_{x \to 0} x\frac{Q(x)}{P(x)}, \qquad q_0 = \lim_{x \to 0} x^2 \frac{R(x)}{P(x)}. \tag{14}$$

Finally, the radii of convergence for the series in Eqs. (9) and (10) are at least equal to the distance from the origin to the nearest zero of P other than $x = 0$ itself.

EXAMPLE 1

Discuss the nature of the solutions of the equation

$$2x(1 + x)y'' + (3 + x)y' - xy = 0$$

near the singular points.

This equation is of the form (13) with $P(x) = 2x(1 + x)$, $Q(x) = 3 + x$, and $R(x) = -x$. The points $x = 0$ and $x = -1$ are the only singular points. The point $x = 0$ is a regular singular point, since

$$\lim_{x \to 0} x\frac{Q(x)}{P(x)} = \lim_{x \to 0} x\frac{3 + x}{2x(1 + x)} = \frac{3}{2},$$

$$\lim_{x \to 0} x^2 \frac{R(x)}{P(x)} = \lim_{x \to 0} x^2 \frac{-x}{2x(1 + x)} = 0.$$

Further, from Eq. (14), $p_0 = \frac{3}{2}$ and $q_0 = 0$. Thus the indicial equation is $r(r - 1) + \frac{3}{2}r = 0$, and the roots are $r_1 = 0$, $r_2 = -\frac{1}{2}$. Since these roots are not equal and do not differ by an integer, there are two solutions of the form

$$y_1(x) = 1 + \sum_{n=1}^{\infty} a_n(0)x^n \quad \text{and} \quad y_2(x) = |x|^{-1/2}\left[1 + \sum_{n=1}^{\infty} a_n\left(-\frac{1}{2}\right)x^n\right]$$

for $0 < |x| < \rho$. A lower bound for the radius of convergence of each series is 1, the distance from $x = 0$ to $x = -1$, the other zero of $P(x)$. Note that the solution y_1 is bounded as $x \to 0$, indeed is analytic there, and that the second solution y_2 is unbounded as $x \to 0$.

The point $x = -1$ is also a regular singular point, since

$$\lim_{x \to -1} (x + 1)\frac{Q(x)}{P(x)} = \lim_{x \to -1} \frac{(x + 1)(3 + x)}{2x(1 + x)} = -1,$$

$$\lim_{x \to -1} (x + 1)^2 \frac{R(x)}{P(x)} = \lim_{x \to -1} \frac{(x + 1)^2(-x)}{2x(1 + x)} = 0.$$

In this case $p_0 = -1$, $q_0 = 0$, so the indicial equation is $r(r - 1) - r = 0$. The roots of the indicial equation are $r_1 = 2$ and $r_2 = 0$. Corresponding to the larger root there is a solution of the form

$$y_1(x) = (x + 1)^2 \left[1 + \sum_{n=1}^{\infty} a_n(2)(x + 1)^n \right].$$

The series converges at least for $|x + 1| < 1$, and y_1 is an analytic function there. Since the two roots differ by a positive integer, there may or may not be a second solution of the form

$$y_2(x) = 1 + \sum_{n=1}^{\infty} a_n(0)(x + 1)^n.$$

We cannot say more without further analysis.

Observe that no complicated calculations were required to discover the information about the solutions presented in this example. All that was needed was to evaluate a few limits and solve two quadratic equations.

We now consider the cases in which the roots of the indicial equation are equal, or differ by a positive integer, $r_1 - r_2 = N$. As we have shown earlier, there is always one solution of the form (9) corresponding to the larger root r_1 of the indicial equation. By analogy with the Cauchy–Euler equation, we might expect that if $r_1 = r_2$, then the second solution contains a logarithmic term. This may also be true if the roots differ by an integer.

Equal Roots

The method of finding the second solution is essentially the same as the one we used to find the second solution of the Cauchy–Euler equation (see Section 8.4) when the roots of the indicial equation were equal. We consider r to be a continuous variable and determine a_n as a function of r by solving the recurrence relation (8). For this choice of $a_n(r)$ for $n \geq 1$, Eq. (6) reduces to

$$L[\phi](r, x) = a_0 F(r) x^r = a_0 (r - r_1)^2 x^r, \tag{15}$$

since r_1 is a repeated root of $F(r)$. Setting $r = r_1$ in Eq. (15), we find that $L[\phi](r_1, x) = 0$; hence, as we already know, $y_1(x)$ given by Eq. (9) is one solution of Eq. (1). But more important, it also follows from Eq. (15), just as for the Cauchy–Euler equation, that

$$L\left[\frac{\partial \phi}{\partial r} \right](r_1, x) = a_0 \frac{\partial}{\partial r} [x^r (r - r_1)^2] \Big|_{r=r_1}$$

$$= a_0 [(r - r_1)^2 x^r \ln x + 2(r - r_1)x^r] \Big|_{r=r_1} = 0. \tag{16}$$

Hence a second solution of Eq. (1) is

$$y_2(x) = \frac{\partial \phi(r, x)}{\partial r} \Big|_{r=r_1} = \frac{\partial}{\partial r} \left\{ x^r \left[a_0 + \sum_{n=1}^{\infty} a_n(r)x^n \right] \right\} \Big|_{r=r_1}$$

$$= (x^{r_1} \ln x) \left[a_0 + \sum_{n=1}^{\infty} a_n(r_1)x^n \right] + x^{r_1} \sum_{n=1}^{\infty} a_n'(r_1)x^n$$

$$= y_1(x) \ln x + x^{r_1} \sum_{n=1}^{\infty} a_n'(r_1)x^n, \qquad x > 0, \tag{17}$$

where $a_n'(r_1)$ denotes da_n/dr evaluated at $r = r_1$.

It may turn out that it is difficult to determine $a_n(r)$ as a function of r from the recurrence relation (8) and then to differentiate the resulting expression with respect to r. An alternative is simply to assume that y has the *form* of Eq. (17). That is, assume that

$$y = y_1(x) \ln x + x^{r_1} \sum_{n=1}^{\infty} b_n x^n, \qquad x > 0, \tag{18}$$

where $y_1(x)$ has already been found. The coefficients b_n are calculated, as usual, by substituting into the differential equation, collecting terms, and setting the coefficient of each power of x equal to zero. A third possibility is to use the method of reduction of order to find $y_2(x)$ once $y_1(x)$ is known.

Roots Differing by an Integer

For this case the derivation of the second solution is considerably more complicated and will not be given here. The form of this solution is stated in Eq. (24) in the following theorem. The coefficients $c_n(r_2)$ in Eq. (24) are given by

$$c_n(r_2) = \frac{d}{dr}[(r - r_2)a_n(r)]\Big|_{r=r_2}, \qquad n = 1, 2, \ldots, \tag{19}$$

where $a_n(r)$ is determined from the recurrence relation (8) with $a_0 = 1$. Further, the coefficient a in Eq. (24) is

$$a = \lim_{r \to r_2} (r - r_2)a_N(r). \tag{20}$$

If $a_N(r_2)$ is finite, then $a = 0$ and there is no logarithmic term in y_2. A full derivation of formulas (19) and (20) may be found in Coddington (Chapter 4).

In practice, the best way to determine whether a is zero in the second solution is simply to try to compute the a_n corresponding to the root r_2 and to see whether it is possible to determine $a_N(r_2)$. If so, there is no further problem. If not, we must use the form (24) with $a \neq 0$.

When $r_1 - r_2 = N$, there are again three ways to find a second solution. First, we can calculate a and $c_n(r_2)$ directly by substituting the expression (24) for y in Eq. (1). Second, we can calculate $c_n(r_2)$ and a of Eq. (24) using the formulas (19) and (20). If this is the planned procedure, then in calculating the solution corresponding to $r = r_1$, be sure to obtain the general formula for $a_n(r)$ rather than just $a_n(r_1)$. The third alternative is to use the method of reduction of order.

THEOREM 8.6.1	Consider the differential equation (1)

$$x^2 y'' + x[xp(x)]y' + [x^2 q(x)]y = 0,$$

where $x = 0$ is a regular singular point. Then $xp(x)$ and $x^2 q(x)$ are analytic at $x = 0$ with convergent power series expansions

$$xp(x) = \sum_{n=0}^{\infty} p_n x^n, \qquad x^2 q(x) = \sum_{n=0}^{\infty} q_n x^n$$

for $|x| < \rho$, where $\rho > 0$ is the minimum of the radii of convergence of the power series for $xp(x)$ and $x^2 q(x)$. Let r_1 and r_2 be the roots of the indicial equation

$$F(r) = r(r - 1) + p_0 r + q_0 = 0,$$

with $r_1 \geq r_2$ if r_1 and r_2 are real. Then in either the interval $-\rho < x < 0$ or the interval $0 < x < \rho$, there exists a solution of the form

$$y_1(x) = |x|^{r_1} \left[1 + \sum_{n=1}^{\infty} a_n(r_1) x^n \right], \tag{21}$$

where the $a_n(r_1)$ are given by the recurrence relation (8) with $a_0 = 1$ and $r = r_1$.

If $r_1 - r_2$ is not zero or a positive integer, then in either the interval $-\rho < x < 0$ or the interval $0 < x < \rho$, there exists a second solution of the form

$$y_2(x) = |x|^{r_2} \left[1 + \sum_{n=1}^{\infty} a_n(r_2) x^n \right]. \tag{22}$$

The $a_n(r_2)$ are also determined by the recurrence relation (8) with $a_0 = 1$ and $r = r_2$. The power series in Eqs. (21) and (22) converge at least for $|x| < \rho$.

If $r_1 = r_2$, then the second solution is

$$y_2(x) = y_1(x) \ln |x| + |x|^{r_1} \sum_{n=1}^{\infty} b_n(r_1) x^n. \tag{23}$$

If $r_1 - r_2 = N$, a positive integer, then

$$y_2(x) = a y_1(x) \ln |x| + |x|^{r_2} \left[1 + \sum_{n=1}^{\infty} c_n(r_2) x^n \right]. \tag{24}$$

The coefficients $a_n(r_1), b_n(r_1), c_n(r_2)$, and the constant a can be determined by substituting the form of the series solutions for y in Eq. (1). The constant a may turn out to be zero, in which case there is no logarithmic term in the solution (24). Each of the series in Eqs. (23) and (24) converges at least for $|x| < \rho$ and defines a function that is analytic in some neighborhood of $x = 0$.

In all three cases the two solutions $y_1(x)$ and $y_2(x)$ form a fundamental set of solutions of the given differential equation.

PROBLEMS

In each of Problems 1 through 12:
(a) Find all the regular singular points of the given differential equation.
(b) Determine the indicial equation and the exponents at the singularity for each regular singular point.

1. $xy'' + 2xy' + 6e^x y = 0$
2. $x^2 y'' - x(2 + x)y' + (2 + x^2)y = 0$
3. $x(x - 1)y'' + 6x^2 y' + 3y = 0$
4. $y'' + 4xy' + 6y = 0$
5. $x^2 y'' + 3(\sin x)y' - 2y = 0$
6. $2x(x + 2)y'' + y' - xy = 0$
7. $x^2 y'' + \frac{1}{2}(x + \sin x)y' + y = 0$
8. $(x + 1)^2 y'' + 3(x^2 - 1)y' + 3y = 0$
9. $x^2(1 - x)y'' - (1 + x)y' + 2xy = 0$
10. $(x - 2)^2(x + 2)y'' + 2xy' + 3(x - 2)y = 0$
11. $(4 - x^2)y'' + 2xy' + 3y = 0$
12. $x(x + 3)^2 y'' - 2(x + 3)y' - xy = 0$

In each of Problems 13 through 17:
(a) Show that $x = 0$ is a regular singular point of the given differential equation.
(b) Find the exponents at the singular point $x = 0$.
(c) Find the first three nonzero terms in each of two solutions about $x = 0$.

13. $xy'' + y' - y = 0$
14. $xy'' + 2xy' + 6e^x y = 0$ (see Problem 1)
15. $x(x - 1)y'' + 6x^2 y' + 3y = 0$ (see Problem 3)
16. $xy'' + y = 0$
17. $x^2 y'' + (\sin x)y' - (\cos x)y = 0$

18. (a) Show that
$$(\ln x)y'' + \frac{1}{2}y' + y = 0$$
has a regular singular point at $x = 1$.
(b) Determine the roots of the indicial equation at $x = 1$.
(c) Determine the first three nonzero terms in the series $\sum_{n=0}^{\infty} a_n(x - 1)^{r+n}$ corresponding to the larger root. Take $x - 1 > 0$.
(d) What would you expect the radius of convergence of the series to be?

19. In several problems in mathematical physics, it is necessary to study the differential equation
$$x(1 - x)y'' + [\gamma - (1 + \alpha + \beta)x]y' - \alpha\beta y = 0, \quad (i)$$
where α, β, and γ are constants. This equation is known as the **hypergeometric equation**.

(a) Show that $x = 0$ is a regular singular point and that the roots of the indicial equation are 0 and $1 - \gamma$.
(b) Show that $x = 1$ is a regular singular point and that the roots of the indicial equation are 0 and $\gamma - \alpha - \beta$.
(c) Assuming that $1 - \gamma$ is not a positive integer, show that, in the neighborhood of $x = 0$, one solution of Eq. (i) is
$$y_1(x) = 1 + \frac{\alpha\beta}{\gamma \cdot 1!}x + \frac{\alpha(\alpha + 1)\beta(\beta + 1)}{\gamma(\gamma + 1)2!}x^2 + \cdots.$$
What would you expect the radius of convergence of this series to be?
(d) Assuming that $1 - \gamma$ is not an integer or zero, show that a second solution for $0 < x < 1$ is
$$y_2(x) = x^{1-\gamma}\left[1 + \frac{(\alpha - \gamma + 1)(\beta - \gamma + 1)}{(2 - \gamma)1!}x\right.$$
$$\left. + \frac{(\alpha - \gamma + 1)(\alpha - \gamma + 2)(\beta - \gamma + 1)(\beta - \gamma + 2)}{(2 - \gamma)(3 - \gamma)2!}x^2 + \cdots\right].$$
(e) Show that the point at infinity is a regular singular point and that the roots of the indicial equation are α and β. See Problem 43 in Section 8.4.

20. Consider the differential equation
$$x^3 y'' + \alpha xy' + \beta y = 0,$$
where α and β are real constants and $\alpha \neq 0$.
(a) Show that $x = 0$ is an irregular singular point.
(b) By attempting to determine a solution of the form $\sum_{n=0}^{\infty} a_n x^{r+n}$, show that the indicial equation for r is linear and that, consequently, there is only one formal solution of the assumed form.
(c) Show that if $\beta/\alpha = -1, 0, 1, 2, \ldots$, then the formal series solution terminates and therefore is an actual solution. For other values of β/α, show that the formal series solution has a zero radius of convergence and so does not represent an actual solution in any interval.

21. Consider the differential equation
$$y'' + \frac{\alpha}{x^s}y' + \frac{\beta}{x^t}y = 0 \quad (i)$$
where $\alpha \neq 0$ and $\beta \neq 0$ are real numbers, and s and t are positive integers that for the moment are arbitrary.

(a) Show that if $s > 1$ or $t > 2$, then the point $x = 0$ is an irregular singular point.

(b) Try to find a solution of Eq. (i) of the form

$$y = \sum_{n=0}^{\infty} a_n x^{r+n}, \qquad x > 0. \qquad \text{(ii)}$$

Show that if $s = 2$ and $t = 2$, then there is only one possible value of r for which a formal solution of Eq. (i) of the form (ii) exists.

(c) Show that if $s = 1$ and $t = 3$, then there are no solutions of Eq. (i) of the form (ii).

(d) Show that the maximum values of s and t for which the indicial equation is quadratic in r [and hence we can hope to find two solutions of the form (ii)] are $s = 1$ and $t = 2$. These are precisely the conditions that distinguish a "weak singularity," or a regular singular point, from an irregular singular point, as we defined in Section 8.4.

As a note of caution, we point out that although it is sometimes possible to obtain a formal series solution of the form (ii) at an irregular singular point, the series may not have a positive radius of convergence. See Problem 20 for an example.

8.7 Bessel's Equation

In this section we illustrate the discussion in Section 8.6 by considering three special cases of **Bessel's equation**:

$$x^2 y'' + xy' + (x^2 - v^2)y = 0, \qquad (1)$$

where v is a constant. It is easy to show that $x = 0$ is a regular singular point of Eq. (1). We have

$$p_0 = \lim_{x \to 0} x \frac{Q(x)}{P(x)} = \lim_{x \to 0} x \frac{1}{x} = 1,$$

$$q_0 = \lim_{x \to 0} x^2 \frac{R(x)}{P(x)} = \lim_{x \to 0} x^2 \frac{x^2 - v^2}{x^2} = -v^2.$$

Thus the indicial equation is

$$F(r) = r(r-1) + p_0 r + q_0 = r(r-1) + r - v^2 = r^2 - v^2 = 0,$$

with the roots $r = \pm v$. We will consider the three cases $v = 0$, $v = \frac{1}{2}$, and $v = 1$ for the interval $x > 0$.

Bessel Equation of Order Zero

In this case $v = 0$, so Eq. (1) reduces to

$$L[y] = x^2 y'' + xy' + x^2 y = 0, \qquad (2)$$

and the roots of the indicial equation are equal: $r_1 = r_2 = 0$. Substituting

$$y = \phi(r, x) = a_0 x^r + \sum_{n=1}^{\infty} a_n x^{r+n}, \qquad (3)$$

in Eq. (2), we obtain

$$L[\phi](r, x) = \sum_{n=0}^{\infty} a_n[(r + n)(r + n - 1) + (r + n)]x^{r+n} + \sum_{n=0}^{\infty} a_n x^{r+n+2}$$

$$= a_0[r(r - 1) + r]x^r + a_1[(r + 1)r + (r + 1)]x^{r+1}$$

$$+ \sum_{n=2}^{\infty} \{a_n[(r + n)(r + n - 1) + (r + n)] + a_{n-2}\}x^{r+n} = 0. \qquad (4)$$

As we have already noted, the roots of the indicial equation $F(r) = r(r - 1) + r = 0$ are $r_1 = 0$ and $r_2 = 0$. The recurrence relation is

$$a_n(r) = -\frac{a_{n-2}(r)}{(r + n)(r + n - 1) + (r + n)} = -\frac{a_{n-2}(r)}{(r + n)^2}, \qquad n \geq 2. \qquad (5)$$

To determine $y_1(x)$, we set r equal to 0. Then, from Eq. (4), it follows that for the coefficient of x^{r+1} to be zero, we must choose $a_1 = 0$. Hence, from Eq. (5), $a_3 = a_5 = a_7 = \cdots = 0$. Furthermore,

$$a_n(0) = -a_{n-2}(0)/n^2, \qquad n = 2, 4, 6, 8, \ldots,$$

or, letting $n = 2m$, we obtain

$$a_{2m}(0) = -a_{2m-2}(0)/(2m)^2, \qquad m = 1, 2, 3, \ldots.$$

Thus

$$a_2(0) = -\frac{a_0}{2^2}, \qquad a_4(0) = \frac{a_0}{2^4 2^2}, \qquad a_6(0) = -\frac{a_0}{2^6(3 \cdot 2)^2},$$

and, in general,

$$a_{2m}(0) = \frac{(-1)^m a_0}{2^{2m}(m!)^2}, \qquad m = 1, 2, 3, \ldots. \qquad (6)$$

Hence

$$y_1(x) = a_0 \left[1 + \sum_{m=1}^{\infty} \frac{(-1)^m x^{2m}}{2^{2m}(m!)^2} \right], \qquad x > 0. \qquad (7)$$

The function in brackets is known as the **Bessel function of the first kind of order zero** and is denoted by $J_0(x)$. It follows from Theorem 8.6.1 that the series converges for all x and that J_0 is analytic at $x = 0$. Some of the important properties of J_0 are discussed in the problems.

To determine $y_2(x)$, we will calculate $a_n'(0)$.[5] First we note from the coefficient of x^{r+1} in Eq. (4) that $(r + 1)^2 a_1(r) = 0$. Thus $a_1(r) = 0$ for all r near zero, so not only does $a_1(0) = 0$ but also $a_1'(0) = 0$. From the recurrence relation (5) it follows that $a_3'(0) = a_5'(0) = \cdots = a_{2n+1}'(0) = \cdots = 0$; hence we need only compute $a_{2m}'(0)$, $m = 1, 2, 3,$. From Eq. (5) we have

$$a_{2m}(r) = -a_{2m-2}(r)/(r + 2m)^2, \qquad m = 1, 2, 3, \ldots.$$

[5]Problem 10 outlines an alternative procedure, in which we simply substitute the form (23) of Section 8.6 in Eq. (2) and then determine the b_n.

By solving this recurrence relation, we obtain

$$a_2(r) = -\frac{a_0}{(r+2)^2}, \qquad a_4(r) = \frac{a_0}{(r+2)^2(r+4)^2},$$

and, in general,

$$a_{2m}(r) = \frac{(-1)^m a_0}{(r+2)^2 \cdots (r+2m)^2}, \qquad m \geq 3. \tag{8}$$

The computation of $a'_{2m}(r)$ can be carried out most conveniently by noting that if

$$f(x) = (x - \alpha_1)^{\beta_1}(x - \alpha_2)^{\beta_2}(x - \alpha_3)^{\beta_3} \cdots (x - \alpha_n)^{\beta_n},$$

and if x is not equal to $\alpha_1, \alpha_2, \ldots, \alpha_n$, then

$$\frac{f'(x)}{f(x)} = \frac{\beta_1}{x - \alpha_1} + \frac{\beta_2}{x - \alpha_2} + \cdots + \frac{\beta_n}{x - \alpha_n}.$$

Applying this result to $a_{2m}(r)$ from Eq. (8), we find that

$$\frac{a'_{2m}(r)}{a_{2m}(r)} = -2\left(\frac{1}{r+2} + \frac{1}{r+4} + \cdots + \frac{1}{r+2m}\right),$$

and setting r equal to 0, we obtain

$$a'_{2m}(0) = -2\left[\frac{1}{2} + \frac{1}{4} + \cdots + \frac{1}{2m}\right]a_{2m}(0).$$

Substituting for $a_{2m}(0)$ from Eq. (6) and letting

$$H_m = 1 + \frac{1}{2} + \frac{1}{3} + \cdots + \frac{1}{m}, \tag{9}$$

we obtain, finally,

$$a'_{2m}(0) = -H_m \frac{(-1)^m a_0}{2^{2m}(m!)^2}, \qquad m = 1, 2, 3, \ldots.$$

The second solution of the Bessel equation of order zero is found by setting $a_0 = 1$ and substituting for $y_1(x)$ and $a'_{2m}(0) = b_{2m}(0)$ in Eq. (23) of Section 8.6. We obtain

$$y_2(x) = J_0(x) \ln x + \sum_{m=1}^{\infty} \frac{(-1)^{m+1} H_m}{2^{2m}(m!)^2} x^{2m}, \qquad x > 0. \tag{10}$$

Instead of y_2, the second solution is usually taken to be a certain linear combination of J_0 and y_2. It is known as the **Bessel function of the second kind of order zero** and is denoted by Y_0. Following Copson (Chapter 12), we define[6]

$$Y_0(x) = \frac{2}{\pi}[y_2(x) + (\gamma - \ln 2)J_0(x)]. \tag{11}$$

Here γ is a constant known as the Euler–Máscheroni constant; it is defined by the equation

$$\gamma = \lim_{n \to \infty}(H_n - \ln n) \cong 0.5772. \tag{12}$$

[6]Other authors use other definitions for Y_0. The present choice for Y_0 is also known as the Weber function.

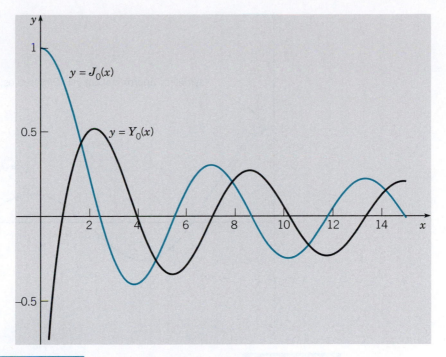

FIGURE 8.7.1 The Bessel functions J_0 and Y_0.

Substituting for $y_2(x)$ in Eq. (11), we obtain

$$Y_0(x) = \frac{2}{\pi}\left[\left(\gamma + \ln\frac{x}{2}\right)J_0(x) + \sum_{m=1}^{\infty}\frac{(-1)^{m+1}H_m}{2^{2m}(m!)^2}x^{2m}\right], \qquad x > 0. \tag{13}$$

The general solution of the Bessel equation of order zero for $x > 0$ is

$$y = c_1 J_0(x) + c_2 Y_0(x).$$

Note that $J_0(x) \to 1$ as $x \to 0$ and that $Y_0(x)$ has a logarithmic singularity at $x = 0$; that is, $Y_0(x)$ behaves as $(2/\pi)\ln x$ when $x \to 0$ through positive values. Thus, if we are interested in solutions of Bessel's equation of order zero that are finite at the origin, which is often the case, we must discard Y_0. The graphs of the functions J_0 and Y_0 are shown in Figure 8.7.1.

It is interesting to note from Figure 8.7.1 that for x large, both $J_0(x)$ and $Y_0(x)$ are oscillatory. Such a behavior might be anticipated from the original equation; indeed it is true for the solutions of the Bessel equation of order ν. If we divide Eq. (1) by x^2, we obtain

$$y'' + \frac{1}{x}y' + \left(1 - \frac{\nu^2}{x^2}\right)y = 0.$$

For x very large it is reasonable to conjecture that the terms $(1/x)y'$ and $(\nu^2/x^2)y$ are small and hence can be neglected. If this is true, then the Bessel equation of order ν can be approximated by

$$y'' + y = 0.$$

The solutions of this equation are $\sin x$ and $\cos x$; thus we might anticipate that the solutions of Bessel's equation for large x are similar to linear combinations of $\sin x$ and $\cos x$. This

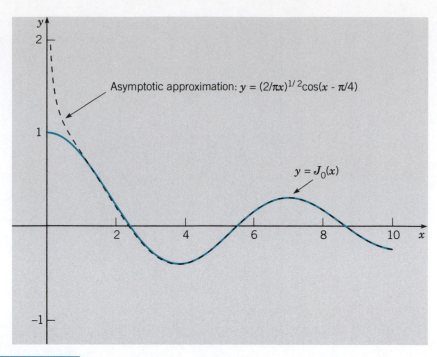

FIGURE 8.7.2   Asymptotic approximation to $J_0(x)$.

is correct insofar as the Bessel functions are oscillatory; however, it is only partly correct. For x large the functions J_0 and Y_0 also decay as x increases; thus the equation $y'' + y = 0$ does not provide an adequate approximation to the Bessel equation for large x, and a more delicate analysis is required. In fact, it is possible to show that

$$J_0(x) \cong \left(\frac{2}{\pi x}\right)^{1/2} \cos\left(x - \frac{\pi}{4}\right) \qquad \text{as} \quad x \to \infty, \tag{14}$$

and that

$$Y_0(x) \cong \left(\frac{2}{\pi x}\right)^{1/2} \sin\left(x - \frac{\pi}{4}\right) \qquad \text{as} \quad x \to \infty. \tag{15}$$

These **asymptotic approximations**, as $x \to \infty$, are actually very good. For example, Figure 8.7.2 shows that the asymptotic approximation (14) to $J_0(x)$ is reasonably accurate for all $x \geq 1$. Thus to approximate $J_0(x)$ over the entire range from zero to infinity, one can use two or three terms of the series (7) for $x \leq 1$ and the asymptotic approximation (14) for $x \geq 1$.

Bessel Equation of Order One-Half

This case illustrates the situation in which the roots of the indicial equation differ by a positive integer but there is no logarithmic term in the second solution. Setting $\nu = \frac{1}{2}$ in Eq. (1) gives

$$L[y] = x^2 y'' + xy' + \left(x^2 - \frac{1}{4}\right) y = 0. \tag{16}$$

When we substitute the series (3) for $y = \phi(r, x)$, we obtain

$$L[\phi](r, x) = \sum_{n=0}^{\infty} \left[(r + n)(r + n - 1) + (r + n) - \frac{1}{4} \right] a_n x^{r+n} + \sum_{n=0}^{\infty} a_n x^{r+n+2}$$

$$= \left(r^2 - \frac{1}{4} \right) a_0 x^r + \left[(r + 1)^2 - \frac{1}{4} \right] a_1 x^{r+1}$$

$$+ \sum_{n=2}^{\infty} \left\{ \left[(r + n)^2 - \frac{1}{4} \right] a_n + a_{n-2} \right\} x^{r+n} = 0. \tag{17}$$

The roots of the indicial equation are $r_1 = \frac{1}{2}, r_2 = -\frac{1}{2}$; hence the roots differ by an integer. The recurrence relation is

$$\left[(r + n)^2 - \frac{1}{4} \right] a_n = -a_{n-2}, \qquad n \geq 2. \tag{18}$$

Corresponding to the larger root $r_1 = \frac{1}{2}$, we find from the coefficient of x^{r+1} in Eq. (17) that $a_1 = 0$. Hence, from Eq. (18), $a_3 = a_5 = \cdots = a_{2n+1} = \cdots = 0$. Further, for $r = \frac{1}{2}$,

$$a_n = -\frac{a_{n-2}}{n(n + 1)}, \qquad n = 2, 4, 6 \ldots,$$

or, letting $n = 2m$, we obtain

$$a_{2m} = -\frac{a_{2m-2}}{2m(2m + 1)}, \qquad m = 1, 2, 3, \ldots.$$

By solving this recurrence relation, we find that

$$a_2 = -\frac{a_0}{3!}, \qquad a_4 = \frac{a_0}{5!}, \cdots$$

and, in general,

$$a_{2m} = \frac{(-1)^m a_0}{(2m + 1)!}, \qquad m = 1, 2, 3, \ldots.$$

Hence, taking $a_0 = 1$, we obtain

$$y_1(x) = x^{1/2} \left[1 + \sum_{m=1}^{\infty} \frac{(-1)^m x^{2m}}{(2m + 1)!} \right] = x^{-1/2} \sum_{m=0}^{\infty} \frac{(-1)^m x^{2m+1}}{(2m + 1)!}, \qquad x > 0. \tag{19}$$

The second power series in Eq. (19) is precisely the Taylor series for $\sin x$; hence one solution of the Bessel equation of order one-half is $x^{-1/2} \sin x$. The **Bessel function of the first kind of order one-half**, $J_{1/2}$, is defined as $(2/\pi)^{1/2} y_1$. Thus

$$J_{1/2}(x) = \left(\frac{2}{\pi x} \right)^{1/2} \sin x, \qquad x > 0. \tag{20}$$

Corresponding to the root $r_2 = -\frac{1}{2}$, it is possible that we may have difficulty in computing a_1 since $N = r_1 - r_2 = 1$. However, from Eq. (17) for $r = -\frac{1}{2}$, the coefficients of x^r and x^{r+1} are both zero regardless of the choice of a_0 and a_1. Hence a_0 and a_1 can be chosen arbitrarily. From the recurrence relation (18), we obtain a set of even-numbered coefficients corresponding to a_0 and a set of odd-numbered coefficients corresponding to a_1. Thus no logarithmic term is needed to obtain a second solution in this case. It is left as an exercise to show that, for $r = -\frac{1}{2}$,

$$a_{2n} = \frac{(-1)^n a_0}{(2n)!}, \qquad a_{2n+1} = \frac{(-1)^n a_1}{(2n+1)!}, \qquad n = 1, 2, \ldots.$$

Hence

$$y_2(x) = x^{-1/2}\left[a_0 \sum_{n=0}^{\infty} \frac{(-1)^n x^{2n}}{(2n)!} + a_1 \sum_{n=0}^{\infty} \frac{(-1)^n x^{2n+1}}{(2n+1)!} \right]$$

$$= a_0 \frac{\cos x}{x^{1/2}} + a_1 \frac{\sin x}{x^{1/2}}, \qquad x > 0. \tag{21}$$

The constant a_1 simply introduces a multiple of $y_1(x)$. The second solution of the Bessel equation of order one-half is usually taken to be the solution for which $a_0 = (2/\pi)^{1/2}$ and $a_1 = 0$. It is denoted by $J_{-1/2}$. Then

$$J_{-1/2}(x) = \left(\frac{2}{\pi x} \right)^{1/2} \cos x, \qquad x > 0. \tag{22}$$

The general solution of Eq. (16) is $y = c_1 J_{1/2}(x) + c_2 J_{-1/2}(x)$.

By comparing Eqs. (20) and (22) with Eqs. (14) and (15), we see that, except for a phase shift of $\pi/4$, the functions $J_{-1/2}$ and $J_{1/2}$ resemble J_0 and Y_0, respectively, for large x. The graphs of $J_{1/2}$ and $J_{-1/2}$ are shown in Figure 8.7.3.

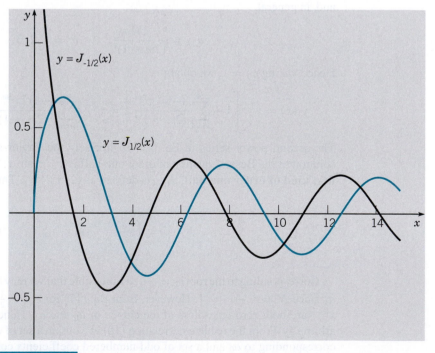

FIGURE 8.7.3 The Bessel functions $J_{1/2}$ and $J_{-1/2}$.

Bessel Equation of Order One

This case illustrates the situation in which the roots of the indicial equation differ by a positive integer and the second solution involves a logarithmic term. Setting $\nu = 1$ in Eq. (1) gives

$$L[y] = x^2 y'' + xy' + (x^2 - 1)y = 0. \tag{23}$$

If we substitute the series (3) for $y = \phi(r, x)$ and collect terms as in the preceding cases, we obtain

$$L[\phi](r, x) = a_0(r^2 - 1)x^r + a_1[(r + 1)^2 - 1]x^{r+1}$$

$$+ \sum_{n=2}^{\infty} \{[(r + n)^2 - 1]a_n + a_{n-2}\}x^{r+n} = 0. \tag{24}$$

The roots of the indicial equation are $r_1 = 1$ and $r_2 = -1$. The recurrence relation is

$$[(r + n)^2 - 1]a_n(r) = -a_{n-2}(r), \qquad n \geq 2. \tag{25}$$

Corresponding to the larger root $r = 1$, the recurrence relation becomes

$$a_n = -\frac{a_{n-2}}{(n + 2)n}, \qquad n = 2, 3, 4, \ldots.$$

We also find from the coefficient of x^{r+1} in Eq. (24) that $a_1 = 0$; hence from the recurrence relation, $a_3 = a_5 = \cdots = 0$. For even values of n, let $n = 2m$; then

$$a_{2m} = -\frac{a_{2m-2}}{(2m + 2)(2m)} = -\frac{a_{2m-2}}{2^2(m + 1)m}, \qquad m = 1, 2, 3, \ldots.$$

By solving this recurrence relation, we obtain

$$a_{2m} = \frac{(-1)^m a_0}{2^{2m}(m + 1)!m!}, \qquad m = 1, 2, 3, \ldots. \tag{26}$$

The Bessel function of the first kind of order one, denoted by J_1, is obtained by choosing $a_0 = 1/2$. Hence

$$J_1(x) = \frac{x}{2} \sum_{m=0}^{\infty} \frac{(-1)^m x^{2m}}{2^{2m}(m + 1)!m!}. \tag{27}$$

The series converges absolutely for all x, so the function J_1 is analytic everywhere.

In determining a second solution of Bessel's equation of order one, we illustrate the method of direct substitution. The calculation of the general term in Eq. (28) below is rather complicated, but the first few coefficients can be found fairly easily. According to Theorem 8.6.1, we assume that

$$y_2(x) = aJ_1(x) \ln x + x^{-1}\left[1 + \sum_{n=1}^{\infty} c_n x^n\right], \qquad x > 0. \tag{28}$$

Computing $y_2'(x)$, $y_2''(x)$, substituting in Eq. (23), and making use of the fact that J_1 is a solution of Eq. (23), we obtain

$$2axJ_1'(x) + \sum_{n=0}^{\infty} [(n - 1)(n - 2)c_n + (n - 1)c_n - c_n]x^{n-1} + \sum_{n=0}^{\infty} c_n x^{n+1} = 0, \tag{29}$$

where $c_0 = 1$. Substituting for $J_1(x)$ from Eq. (27), shifting the indices of summation in the two series, and carrying out several steps of algebra, we arrive at

$$-c_1 + [0 \cdot c_2 + c_0]x + \sum_{n=2}^{\infty} [(n^2 - 1)c_{n+1} + c_{n-1}]x^n$$

$$= -a\left[x + \sum_{m=1}^{\infty} \frac{(-1)^m (2m+1)x^{2m+1}}{2^{2m}(m+1)! \, m!}\right]. \tag{30}$$

From Eq. (30) we observe first that $c_1 = 0$, and $a = -c_0 = -1$. Furthermore, since there are only odd powers of x on the right, the coefficient of each even power of x on the left must be zero. Thus, since $c_1 = 0$, we have $c_3 = c_5 = \cdots = 0$. Corresponding to the odd powers of x, we obtain the recurrence relation [let $n = 2m + 1$ in the series on the left side of Eq. (30)]

$$[(2m+1)^2 - 1]c_{2m+2} + c_{2m} = \frac{(-1)^m (2m+1)}{2^{2m}(m+1)! \, m!}, \qquad m = 1, 2, 3, \dots . \tag{31}$$

When we set $m = 1$ in Eq. (31), we obtain

$$(3^2 - 1)c_4 + c_2 = (-1)3/(2^2 \cdot 2!).$$

Notice that c_2 can be selected *arbitrarily*, and then this equation determines c_4. Also notice that in the equation for the coefficient of x, c_2 appeared to be multiplied by 0, and that equation was used to determine a. That c_2 is arbitrary is not surprising, since c_2 is the coefficient of x in the expression $x^{-1}\left[1 + \sum_{n=1}^{\infty} c_n x^n\right]$. Consequently, c_2 simply generates a multiple of J_1, and y_2 is determined only up to an additive multiple of J_1. In accordance with the usual practice, we choose $c_2 = 1/2^2$. Then we obtain

$$c_4 = \frac{-1}{2^4 \cdot 2}\left[\frac{3}{2} + 1\right] = \frac{-1}{2^4 2!}\left[\left(1 + \frac{1}{2}\right) + 1\right]$$

$$= \frac{(-1)}{2^4 \cdot 2!}(H_2 + H_1).$$

It is possible to show that the solution of the recurrence relation (31) is

$$c_{2m} = \frac{(-1)^{m+1}(H_m + H_{m-1})}{2^{2m}m!(m-1)!}, \qquad m = 1, 2, \dots$$

with the understanding that $H_0 = 0$. Thus

$$y_2(x) = -J_1(x)\ln x + \frac{1}{x}\left[1 - \sum_{m=1}^{\infty} \frac{(-1)^m (H_m + H_{m-1})}{2^{2m}m!(m-1)!}x^{2m}\right], \qquad x > 0. \tag{32}$$

The calculation of $y_2(x)$ using the alternative procedure [see Eqs. (19) and (20) of Section 8.6] in which we determine the $c_n(r_2)$ is slightly easier. In particular, the latter procedure yields the general formula for c_{2m} without the necessity of solving a recurrence relation of the form (31) (see Problem 11). In this regard, you may also wish to compare the calculations of the second solution of Bessel's equation of order zero in the text and in Problem 10.

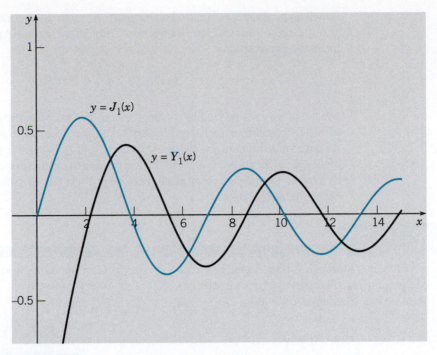

FIGURE 8.7.4 The Bessel functions J_1 and Y_1.

The second solution of Eq. (23), the Bessel function of the second kind of order one, Y_1, is usually taken to be a certain linear combination of J_1 and y_2. Following Copson (Chapter 12), Y_1 is defined as

$$Y_1(x) = \frac{2}{\pi}[-y_2(x) + (\gamma - \ln 2)J_1(x)], \qquad (33)$$

where γ is defined in Eq. (12). The general solution of Eq. (23) for $x > 0$ is

$$y = c_1 J_1(x) + c_2 Y_1(x).$$

Notice that although J_1 is analytic at $x = 0$, the second solution Y_1 becomes unbounded in the same manner as $1/x$ as $x \to 0$. The graphs of J_1 and Y_1 are shown in Figure 8.7.4.

Numerical Evaluation of Bessel Functions

In this section we have shown how to obtain infinite series solutions of Bessel's equation of orders zero, one-half, and one. In applications it is not unusual to require Bessel functions of other orders. For example, the problem of the vibrating circular elastic membrane requires Bessel functions of the first kind of order n, $J_n(x)$, for $n = 0, 1, 2, \ldots$. In most instances, accurate numerical approximations can be obtained by appropriate function calls in a computer algebra system in the same way that approximations of elementary functions such as $\sin x$ and e^x are found. However, this does not obviate the need for understanding some of the properties of these functions, such as their behavior near the singular point, which dictates that we use only the $J_n(x)$, but not the Bessel functions of the second kind, $Y_n(x)$, which have a logarithmic singularity, and therefore are unbounded as $x \to 0$, the center of the membrane. It is also the case that in the course of solving a problem, Bessel functions may appear disguised as integrals, or some other form. For example, in Project 1 at the end of this chapter, Diffraction Through a Circular Aperture, we encounter the integral representation

$$J_0(x) = \frac{1}{2\pi} \int_0^{2\pi} e^{-ix \cos \theta}\, d\theta,$$

a fact that may not be readily apparent to you. Not only can the infinite series for $J_0(x)$ be extracted from this integral (see Problem 4 in Project 1), but also asymptotic approximations for large values of x such as the one shown in Eq. (14). In applications that require Bessel, or other special functions, it is sometimes necessary to examine one or more of the references devoted to the subject, such as the books by Hochstadt or Abramowitz and Stegun, that appear in the References.

PROBLEMS

In each of Problems 1 through 4 show that the given differential equation has a regular singular point at $x = 0$, and determine two solutions for $x > 0$.

1. $x^2 y'' + 2xy' + xy = 0$
2. $x^2 y'' + 3xy' + (1 + x)y = 0$
3. $x^2 y'' + xy' + 2xy = 0$
4. $x^2 y'' + 4xy' + (2 + x)y = 0$
5. Find two solutions of the Bessel equation of order $\frac{3}{2}$:

$$x^2 y'' + xy' + \left(x^2 - \tfrac{9}{4}\right)y = 0, \qquad x > 0.$$

6. Show that the Bessel equation of order one-half:

$$x^2 y'' + xy' + \left(x^2 - \tfrac{1}{4}\right)y = 0, \qquad x > 0,$$

can be reduced to the equation

$$v'' + v = 0$$

by the change of dependent variable $y = x^{-1/2} v(x)$. From this conclude that $y_1(x) = x^{-1/2} \cos x$ and $y_2(x) = x^{-1/2} \sin x$ are solutions of the Bessel equation of order one-half.

7. Show directly that the series for $J_0(x)$, Eq. (7), converges absolutely for all x.
8. Show directly that the series for $J_1(x)$, Eq. (27), converges absolutely for all x and that $J_0'(x) = -J_1(x)$.
9. Consider the Bessel equation of order ν

$$x^2 y'' + xy' + (x^2 - \nu^2)y = 0, \qquad x > 0,$$

where ν is real and positive.

(a) Show that $x = 0$ is a regular singular point and that the roots of the indicial equation are ν and $-\nu$.

(b) Corresponding to the larger root ν, show that one solution is

$$y_1(x) = x^\nu \left[1 - \frac{1}{1!(1+\nu)}\left(\frac{x}{2}\right)^2 + \frac{1}{2!(1+\nu)(2+\nu)}\left(\frac{x}{2}\right)^4 \right.$$
$$\left. + \sum_{m=3}^{\infty} \frac{(-1)^m}{m!(1+\nu)\cdots(m+\nu)}\left(\frac{x}{2}\right)^{2m} \right].$$

(c) If 2ν is not an integer, show that a second solution is

$$y_2(x) = x^{-\nu} \left[1 - \frac{1}{1!(1-\nu)}\left(\frac{x}{2}\right)^2 + \frac{1}{2!(1-\nu)(2-\nu)}\left(\frac{x}{2}\right)^4 \right.$$
$$\left. + \sum_{m=3}^{\infty} \frac{(-1)^m}{m!(1-\nu)\cdots(m-\nu)}\left(\frac{x}{2}\right)^{2m} \right].$$

Note that $y_1(x) \to 0$ as $x \to 0$, and that $y_2(x)$ is unbounded as $x \to 0$.

(d) Verify by direct methods that the power series in the expressions for $y_1(x)$ and $y_2(x)$ converge absolutely for all x. Also verify that y_2 is a solution, provided only that ν is not an integer.

10. In this section we showed that one solution of Bessel's equation of order zero

$$L[y] = x^2 y'' + xy' + x^2 y = 0,$$

is J_0, where $J_0(x)$ is given by Eq. (7) with $a_0 = 1$. According to Theorem 8.6.1, a second solution has the form $(x > 0)$

$$y_2(x) = J_0(x)\ln x + \sum_{n=1}^{\infty} b_n x^n.$$

(a) Show that

$$L[y_2](x) = \sum_{n=2}^{\infty} n(n-1)b_n x^n + \sum_{n=1}^{\infty} nb_n x^n$$

$$+ \sum_{n=1}^{\infty} b_n x^{n+2} + 2xJ_0'(x). \qquad \text{(i)}$$

(b) Substituting the series representation for $J_0(x)$ in Eq. (i), show that

$$b_1 x + 2^2 b_2 x^2 + \sum_{n=3}^{\infty} (n^2 b_n + b_{n-2})x^n$$

$$= -2\sum_{n=1}^{\infty} \frac{(-1)^n 2nx^{2n}}{2^{2n}(n!)^2}. \qquad \text{(ii)}$$

(c) Note that only even powers of x appear on the right side of Eq. (ii). Show that $b_1 = b_3 = b_5 = \cdots = 0$, $b_2 = 1/2^2(1!)^2$, and that

$$(2n)^2 b_{2n} + b_{2n-2} = -2(-1)^n (2n)/2^{2n}(n!)^2,$$

$$n = 2, 3, 4, \ldots.$$

Deduce that

$$b_4 = -\frac{1}{2^2 \, 4^2}\left(1 + \frac{1}{2}\right) \qquad \text{and}$$

$$b_6 = \frac{1}{2^2 \, 4^2 \, 6^2}\left(1 + \frac{1}{2} + \frac{1}{3}\right).$$

The general solution of the recurrence relation is $b_{2n} = (-1)^{n+1} H_n/2^{2n}(n!)^2$. Substituting for b_n in the expression for $y_2(x)$, we obtain the solution given in Eq. (10).

11. Find a second solution of Bessel's equation of order one by computing the $c_n(r_2)$ and a of Eq. (24) of Section 8.5 according to the formulas (19) and (20) given in that section. Some guidelines along the way of this calculation are the following. First, use Eq. (24) of this section to show that $a_1(-1)$ and $a_1'(-1)$ are 0. Then show that $c_1(-1) = 0$ and, from the recurrence relation, that $c_n(-1) = 0$ for $n = 3, 5, \ldots$. Finally, use Eq. (25) to show that

$$a_2(r) = -\frac{a_0}{(r+1)(r+3)},$$

$$a_4(r) = \frac{a_0}{(r+1)(r+3)(r+3)(r+5)},$$

and that

$$a_{2m}(r)$$

$$= \frac{(-1)^m a_0}{(r+1)\cdots(r+2m-1)(r+3)\cdots(r+2m+1)},$$

$$m \geq 3.$$

Then show that

$$c_{2m}(-1) = (-1)^{m+1}(H_m + H_{m-1})/2^{2m}m!(m-1)!,$$

$$m \geq 1.$$

12. By a suitable change of variables it is sometimes possible to transform another differential equation into a Bessel equation. For example, show that a solution of

$$x^2 y'' + \left(\alpha^2 \beta^2 x^{2\beta} + \tfrac{1}{4} - \nu^2 \beta^2\right)y = 0, \qquad x > 0$$

is given by $y = x^{1/2} f(\alpha x^\beta)$, where $f(\xi)$ is a solution of the Bessel equation of order ν.

13. Using the result of Problem 12, show that the general solution of the Airy equation

$$y'' - xy = 0, \qquad x > 0$$

is $y = x^{1/2}[c_1 f_1(\tfrac{2}{3}ix^{3/2}) + c_2 f_2(\tfrac{2}{3}ix^{3/2})]$, where $f_1(\xi)$ and $f_2(\xi)$ are a fundamental set of solutions of the Bessel equation of order one-third.

14. It can be shown that J_0 has infinitely many zeros for $x > 0$. In particular, the first three zeros are approximately 2.405, 5.520, and 8.653 (see Figure 5.7.1). Let λ_j, $j = 1, 2, 3, \ldots$, denote the zeros of J_0; it follows that

$$J_0(\lambda_j x) = \begin{cases} 1, & x = 0, \\ 0, & x = 1. \end{cases}$$

Verify that $y = J_0(\lambda_j x)$ satisfies the differential equation

$$y'' + \frac{1}{x}y' + \lambda_j^2 y = 0, \qquad x > 0.$$

Hence show that

$$\int_0^1 xJ_0(\lambda_i x)J_0(\lambda_j x)\, dx = 0 \qquad \text{if} \quad \lambda_i \neq \lambda_j.$$

This important property of $J_0(\lambda_i x)$, known as the orthogonality property, is useful in solving boundary value problems.

Hint: Write the differential equation for $J_0(\lambda_i x)$. Multiply it by $xJ_0(\lambda_j x)$ and subtract it from $xJ_0(\lambda_i x)$ times the differential equation for $J_0(\lambda_j x)$. Then integrate from 0 to 1.

CHAPTER SUMMARY

Section 8.1 Review of Power Series

▶ Every **power series** $\sum_{n=0}^{\infty} a_n(x - x_0)^n$ has a **radius of convergence** ρ, $0 \leq \rho \leq \infty$, such that the series converges absolutely for $|x - x_0| < \rho$ and diverges for $|x - x_0| > \rho$.

▶ **The Ratio Test.** If $\lim_{n \to \infty} \dfrac{|a_{n+1}|}{|a_n|} = L$, then the radius of convergence is given by $\rho = 1/L$.

▶ A function f that has a power series expansion at x_0 with a radius of convergence $\rho > 0$ is said to be **analytic** at x_0.

▶ If f is analytic at x_0, then its power series expansion at x_0 is the **Taylor series**

$$f(x) = \sum_{n=0}^{\infty} \frac{f^{(n)}(x_0)}{n!}(x - x_0)^n.$$

▶ If f and g are analytic at x_0, then $f \pm g$, fg, and f/g [provided $g(x_0) \neq 0$] are analytic at x_0.

Sections 8.2 and 8.3 Series Solutions Near an Ordinary Point, Parts I and II

▶ The point x_0 is an **ordinary point** of the linear equation

$$y'' + p(x)y' + q(x)y = 0$$

if p and q are analytic at x_0.

▶ If x_0 is an ordinary point of

$$y'' + p(x)y' + q(x)y = 0,$$

then the general solution has a representation of the form

$$y = \sum_{n=0}^{\infty} a_n(x - x_0)^n = a_0 y_1(x) + a_1 y_2(x),$$

where a_0 and a_1 are arbitrary, and y_1 and y_2 are linearly independent series solutions that are analytic at x_0.

▶ The series solutions y_1 and y_2 can be found by formally substituting $y = \sum_{n=0}^{\infty} a_n(x - x_0)^n$ into $y'' + p(x)y' + q(x)y = 0$, grouping coefficients of like powers of $x - x_0$, and setting each coefficient in the resulting series equal to zero. This yields a **recurrence formula** that, in general, gives a_{n+2} in terms of $a_0, a_1, \ldots, a_{n+1}$ for each $n \geq 0$. Starting with arbitrary a_0 and a_1, we can compute, in order, a_2, a_3, a_4, and so on. Choosing $a_0 = 1$ and $a_1 = 0$ then gives y_1, while choosing $a_0 = 0$ and $a_1 = 1$ gives y_2. In some cases, a general formula can be found for the a_n.

▶ The radius of convergence for each of the series solutions y_1 and y_2 is greater than, or equal to, the minimum of the radii of convergence of the series for p and q.

Section 8.4 Regular Singular Points

▶ The point x_0 is a **singular point** of

$$y'' + p(x)y' + q(x)y = 0$$

if either p or q fails to be analytic at x_0.

▶ If x_0 is a singular point of $y'' + p(x)y' + q(x)y = 0$ for which $(x - x_0)p(x)$ and $(x - x_0)^2 q(x)$ are both analytic at x_0, then x_0 is a **regular singular point**; otherwise, x_0 is an **irregular singular point**.

▶ The **Cauchy–Euler equation**

$$x^2 y'' + \alpha x y' + \beta y = 0$$

has a regular singular point at $x = 0$. If the roots of the **indicial equation** $r(r-1) + \alpha r + \beta = 0$ are r_1 and r_2, then general solutions of the Cauchy–Euler equation on either the interval $x < 0$, or the interval $x > 0$, are

$$y = c_1 |x|^{r_1} + c_2 |x|^{r_2}, \qquad \text{if } r_1 \text{ and } r_2 \text{ are real and not equal,}$$
$$y = (c_1 + c_2 \ln |x|)|x|^{r_1}, \qquad \text{if } r_1 \text{ and } r_2 \text{ are real and equal,}$$
$$y = |x|^\mu [c_1 \cos(\nu \ln |x|) + c_2 \sin(\nu \ln |x|)], \qquad \text{if } r_1, r_2 = \mu \pm i\nu \text{ and } \nu \neq 0.$$

Sections 8.5 and 8.6   Series Solutions Near a Regular Singular Point, Parts I and II

▶ If each of the power series $xp(x) = \sum_{n=0}^\infty p_n x^n$ and $x^2 q(x) = \sum_{n=0}^\infty q_n x^n$ converge for $|x| < \rho$, where $\rho > 0$, then $x = 0$ is a regular singular point of

$$x^2 y'' + x[xp(x)]y' + [x^2 q(x)]y = 0.$$

▶ The **Method of Frobenius.** Substituting a series of the form

$$y = \phi(r, x) = x^r \sum_{n=0}^\infty a_n x^n = \sum_{n=0}^\infty a_n x^{r+n}$$

into the equation yields the **indicial equation**

$$F(r) = r(r-1) + p_0 r + q_0 = 0,$$

and a **recurrence relation**

$$F(r+n)a_n + \sum_{k=0}^{n-1} a_k[(r+k)p_{n-k} + q_{n-k}] = 0, \qquad n \geq 1.$$

If the roots r_1 and r_2 of $F(r) = 0$ are real, with $r_1 \geq r_2$, then the general solution on either $-\rho < x < 0$ or $0 < x < \rho$ is

$$y = a_1 y_1(x) + a_2 y_2(x)$$

where

$$y_1(x) = |x|^{r_1}\left[1 + \sum_{n=1}^\infty a_n(r_1)x^n\right]$$

and

$$y_2(x) = \begin{cases} |x|^{r_2}[1 + \sum_{n=1}^\infty a_n(r_2)x^n], & \text{if } r_1 - r_2 \text{ is not zero or a positive integer,} \\ y_1(x)\ln|x| + |x|^{r_1}\sum_{n=1}^\infty b_n(r_1)x^n, & \text{if } r_1 = r_2, \\ ay_1(x)\ln|x| + |x|^{r_2}\sum_{n=1}^\infty c_n(r_2)x^n, & \text{if } r_1 - r_2 = N, \text{ a positive integer.} \end{cases}$$

All of the coefficients can be determined by substituting the form of the series solution into $x^2 y'' + x[xp(x)]y' + [x^2 q(x)]y = 0$; the a_n will necessarily satisfy the recurrence relation $F(r+n)a_n + \sum_{k=0}^{n-1} a_k[(r+k)p_{n-k} + q_{n-k}] = 0$. All of the power series appearing in the solutions converge at least for $|x| < \rho$.

Section 8.7 Bessel's Equation

▶ **Bessel's equation of order** ν

$$x^2 y'' + xy' + (x^2 - \nu^2)y = 0,$$

has a regular singular point at $x = 0$.

▶ The general solution of **Bessel's equation of order zero** is $y = c_1 J_0(x) + c_2 Y_0(x)$, $x > 0$, where

$$J_0(x) = 1 + \sum_{m=1}^{\infty} \frac{(-1)^m x^{2m}}{2^{2m}(m!)^2}$$

and

$$Y_0(x) = \frac{2}{\pi}\left[\left(\gamma + \ln\frac{x}{2}\right)J_0(x) + \sum_{m=1}^{\infty} \frac{(-1)^{m+1} H_m}{2^{2m}(m!)^2}x^{2m}\right].$$

▶ The general solution of **Bessel's equation of order one-half** is $y = c_1 J_{1/2}(x) + c_2 J_{-1/2}(x)$, $x > 0$, where

$$J_{1/2}(x) = \left(\frac{2}{\pi x}\right)^{1/2} \cos x \qquad \text{and} \qquad J_{-1/2}(x) = \left(\frac{2}{\pi x}\right)^{1/2} \sin x.$$

▶ The general solution of **Bessel's equation of order one** is $y = c_1 J_1(x) + c_2 Y_1(x)$, $x > 0$, where

$$J_1(x) = \frac{x}{2}\sum_{m=0}^{\infty} \frac{(-1)^m x^{2m}}{2^{2m}(m+1)!m!} \qquad \text{and} \qquad Y_1(x) = \frac{2}{\pi}[-y_2(x) + (\gamma - \ln 2)J_1(x)].$$

PROJECTS

Project 1 Diffraction Through a Circular Aperture

In this project Bessel functions are used to describe the wave field that results from monochromatic light waves[7] passing through a small circular aperture (see Figure 8.P.1). The tendency of light to bend around objects, or when passing through slits and apertures, is called **diffraction**. The theory that light travels in straight lines (**geometrical optics**) fails to describe in an adequate manner certain effects that are observed when the physical dimensions of the objects, slits, or apertures are comparable with the wavelength of the light. In particular, if the wavelength of the light is denoted by λ, and the radius of the aperture is denoted by a, diffraction effects are important when $a \cong \lambda$, and become more pronounced as $a/\lambda \to 0$.

Since divergent waves travel different distances, some move out of phase and begin to interfere with one another. The intensity of the light increases in regions where the amplitudes of the waves add, and decreases in regions where the amplitudes cancel. The resulting interference pattern of bright and dark regions can be seen on an observation screen that is located at an appropriate distance from the aperture screen.

In addition to light, all other types of waves, such as sound, water, and matter waves, also exhibit diffraction. We restrict our attention to the simplest case of monochromatic waves

---

[7]Electromagnetic radiation consisting of a single wavelength.

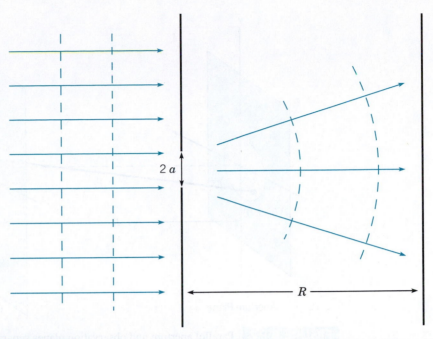

FIGURE 8.P.1 Plane waves tend to diverge and interfere with one another after passing through a small circular aperture.

that can be represented by a real, scalar valued function $u(x, y, z)$ satisfying the **Helmholtz equation**[8]

$$\frac{\partial^2 u}{\partial x^2} + \frac{\partial^2 u}{\partial y^2} + \frac{\partial^2 u}{\partial z^2} + k^2 u = 0, \tag{1}$$

where $k = 2\pi/\lambda$ is called the **wave number** of the wave and, as mentioned above, λ is the wavelength of the wave. In the case of light waves, u could be one of the Cartesian components of the electric field vector. For acoustic waves, u is the amplitude of a pressure disturbance corresponding to a sound wave.

To solve the problem of diffraction of plane waves passing through a circular aperture, we require only very simple solutions of Eq. (1). For example, plane waves of amplitude 1 propagating in the positive z direction, and incident from the left on the aperture plane in Figure 8.P.2, are represented by

$$u_I = e^{ikz}. \tag{2}$$

It is customary, and convenient, to use complex notation to represent wave functions. If we wish to know the physical meaning of a result, we simply take the real part.[9]

---

[8] The Helmholtz, or steady-state, wave equation arises from assuming time harmonic solutions $\psi = ue^{-i\omega t}$ to the time-dependent wave equation

$$\frac{\partial^2 \psi}{\partial t^2} = c^2 \left(\frac{\partial^2 \psi}{\partial x^2} + \frac{\partial^2 \psi}{\partial y^2} + \frac{\partial^2 \psi}{\partial z^2} \right),$$

where c is the wave's phase velocity. Substituting $ue^{-i\omega t}$ for ψ in this equation yields the Helmholtz equation (1) and the relation $k = \omega/c$, or $c/\lambda = \omega/(2\pi)$.

[9] Note that including the time-dependent factor $e^{-i\omega t}$ in the incident wave yields $e^{-i\omega t} u_I = e^{ik(z-ct)} = \cos k(z - ct) + i \sin k(z - ct)$, the real part of which is a cosine wave moving in the positive z-direction with velocity c.

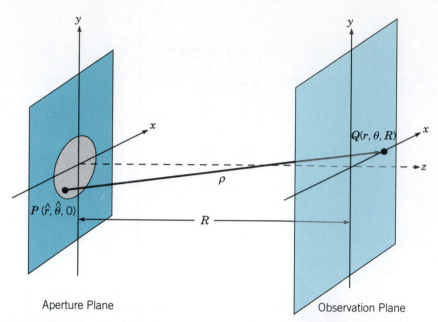

FIGURE 8.P.2 Parallel aperture and observation planes separated by a distance R. The cylindrical coordinates of generic source and field points in the planes are denoted by $P(\hat{r}, \hat{\theta}, 0)$ and $Q(r, \theta, R)$, respectively.

Another solution of Eq. (1), representing spherically symmetric waves propagating outward from a source point (ξ, η, ζ), is

$$H^1(\rho) = e^{ik\rho}/\rho, \tag{3}$$

where $\rho = \sqrt{(x - \xi)^2 + (y - \eta)^2 + (z - \zeta)^2}$ is the distance between the source point and the point (x, y, z) in the wave field. This is easily verified (see Problem 1) if we express the Helmholtz equation in spherical coordinates and assume spherical symmetry:

$$\frac{\partial^2 u}{\partial \rho^2} + \frac{2}{\rho}\frac{\partial u}{\partial \rho} + k^2 u = 0, \qquad \rho > 0. \tag{4}$$

▶ **Simplifying Approximations.** Assume that the aperture plane coincides with the xy-coordinate plane and that the observation plane is located at a distance $z = R$ to the right of the aperture plane. If we use cylindrical coordinates, points in the aperture are represented by $(\hat{r}, \hat{\theta}, 0)$, $\hat{r} \le a$, while points lying outside the aperture, field points, are represented by (r, θ, z). In particular, points in the observation plane have coordinates (r, θ, R).

▶ **Huygen's Principle.** In order to obtain an approximate mathematical description of the diffraction pattern on the observation screen, we invoke the following form of **Huygen's Principle:** *All points on a wave front in the incident wave field can be considered as a point source for the production of secondary, spherical, outward-propagating wavelets.* Assuming that the magnitude of the effect of the incident wave $u_I = e^{ikz}$ in the aperture is a constant α across the aperture, we write the amplitude of the wave field in the region $z > 0$ contributed by the elemental plane wavelet at the point $(\hat{r}, \hat{\theta}, 0)$ in the aperture by

$$dA = \alpha\frac{e^{ik\rho}}{\rho}\hat{r}\,d\hat{\theta}\,d\hat{r}, \tag{5}$$

where

$$\rho = \rho(r, \theta, z; \hat{r}, \hat{\theta}, 0) = \sqrt{(r \cos \theta - \hat{r} \cos \hat{\theta})^2 + (r \sin \theta - \hat{r} \sin \hat{\theta})^2 + z^2}$$

$$= \sqrt{r^2 + \hat{r}^2 - 2r\hat{r} \cos(\hat{\theta} - \theta) + z^2} \tag{6}$$

is the distance between $(\hat{r}, \hat{\theta}, 0)$ and (r, θ, z). "Summing" the wavelets in Eq. (5) across the aperture yields the following integral representation for the wave field in the region $z > 0$:

$$u(r, \theta, z) = \alpha \int_0^a \int_0^{2\pi} \frac{e^{ik\rho}}{\rho} \hat{r} d\hat{\theta} d\hat{r}. \tag{7}$$

Remark. It can be shown that the exact formula for the wave field in the region $z \geq 0$ satisfying the boundary condition $\partial u / \partial z = 0$ in the region $r > a$ of the aperture plane is

$$u(r, \theta, z) = -\frac{1}{2\pi} \int_0^a \int_0^{2\pi} \frac{e^{ik\rho}}{\rho} \frac{\partial u}{\partial z}(\hat{r}, \hat{\theta}, 0) \hat{r} d\hat{\theta} d\hat{r}. \tag{8}$$

Since the values of $\partial u / \partial z$ are not known in the aperture, Eq. (8) represents an equation that must be solved for $u(r, \theta, z)$ on $z \geq 0$. A simplifying approximation made by Kirchhoff is to replace the unknown quantity $\partial u / \partial z$ in the aperture by that of the incident wave, $\partial u_I / \partial z|_{z=0} = ike^{ikz}|_{z=0} = ik$. Substituting this result into the right-hand side of Eq. (8) yields the approximation

$$u(r, \theta, z) \cong \frac{k}{2\pi i} \int_0^a \int_0^{2\pi} \frac{e^{ik\rho}}{\rho} \hat{r} d\hat{\theta} d\hat{r}, \tag{9}$$

which agrees with Eq. (7) if we set $\alpha = \frac{k}{2\pi i}$. In Problem 2 we ask the reader to show that Eq. (9) [and hence Eq. (7)] satisfies the Helmholtz equation (1) in the region $z > 0$ and the boundary condition, $\dfrac{\partial u}{\partial z} = 0$, at $z = 0$ in the region $r > 0$ outside the aperture.

▶ **The Far Field Approximation.** We now assume that $kR >> 1$, read as "kR is large compared to unity." Since $kR = 2\pi R / \lambda$, this means that the distance R between the aperture and observation planes is large relative to the wavelength λ, so that R consists of a large number of wave periods. Since $\lambda \cong a$, $\hat{r}/R << 1$ for all $\hat{r} \leq a$. Thus we can approximate the distance between source points in the aperture and field points at $z = R$ by

$$\rho = \rho(r, \theta, R; \hat{r}, \hat{\theta}, 0)$$

$$= R\sqrt{1 + \frac{r^2 + \hat{r}^2}{R^2} - \frac{2r\hat{r}}{R^2} \cos(\hat{\theta} - \theta)} \cong R + \frac{r^2}{2R} - \frac{r}{R}\hat{r} \cos(\hat{\theta} - \theta). \tag{10}$$

This approximation permits us to approximate the outward-propagating spherical wave by

$$\frac{e^{ik\rho}}{\rho} \cong \frac{1}{R} e^{ik(R + r^2/2R)} e^{-(ikr\hat{r}/R)\cos(\hat{\theta} - \theta)}, \tag{11}$$

where we have retained terms in the complex exponent that account for the variation in phase of waves due to variation in the location of the source points $(\hat{r}, \hat{\theta}, 0)$, but have simply approximated the amplitude factor $1/\rho$ by $1/R$. Note that the term $\exp[ik(R + r^2/2R)]/R$ represents an amplitude factor and phase shift that is uniform over the aperture. Substituting the right side of Eq. (11) for $e^{ik\rho}/\rho$ in Eq. (7) yields the approximation

$$u(r, R) \cong \alpha \frac{e^{ik(R+r^2/2R)}}{R} \int_0^a \int_0^{2\pi} e^{-(ikr\hat{r}/R)\cos\hat{\theta}} \hat{r}\, d\hat{\theta}\, d\hat{r}. \quad [10]$$

(12)

Since the integral of $\exp[-(ikr\hat{r}/R)\cos(\hat{\theta} - \theta)]$ over any interval of length 2π is independent of θ, $\cos(\hat{\theta} - \theta)$ has been replaced by $\cos\hat{\theta}$ and we have dropped the dependence of u on θ. As expected, the wave field is symmetric about the z-axis.

When we use the integral representation $J_0(x) = \frac{1}{2\pi} \int_0^{2\pi} e^{-ix\cos\phi}\, dx$ (Problem 4), it follows that

$$u(r, R) \cong 2\pi\alpha \frac{e^{ik(R+r^2/2R)}}{R} \int_0^a \hat{r} J_0(kr\hat{r}/R)\, d\hat{r}.$$

(13)

Finally, using the identity $\int_0^a x J_0(x)\, dx = aJ_1(a)$ (Problem 5), we obtain

$$u(r, R) \cong 2\pi a^2\alpha \frac{e^{ik(R+r^2/2R)}}{R} \frac{J_1(kra/R)}{(kra/R)}.$$

(14)

What is observed in the observation plane is not the amplitude of the wave but rather its intensity

$$I(r, R) = |u(r, R)|^2 = I_0 \left[\frac{2J_1(kar/R)}{kar/R} \right]^2 = I_0 \left[\frac{2J_1(ka\sin\phi)}{ka\sin\phi} \right]^2,$$

(15)

FIGURE 8.P.3 A computer image of the Airy pattern generated from the irradiance function in Eq. (15). The color scale intensities have been adjusted to enhance the brightness of the outer rings of the Airy pattern.

---

[10]The far field approximation in Eq. (12) in which spherical waves are approximated by plane waves is known as **Fraunhoffer diffraction**.

where I_0 is the intensity at the center of the diffraction pattern, and ϕ is the angle, in spherical coordinates, of the position vector of any point on a circle of radius r centered at the z-axis in the observation screen. Note that since $J_1(x)/x \to 1/2$ as $x \to 0$, $2J_1(ka\sin\phi)/(ka\sin\phi) \to 1$ as $\phi \to 0$. The amplitude of the function in Eq. (15) in the observation plane is shown in Figure 8.P.3.

The resulting diffraction pattern has a bright region in the center, known as the **Airy disk**. The pattern of concentric rings of alternating high and low intensities is called the **Airy pattern**. The diameter of the Airy disk is a function of the wavelength of the illuminating light and the size of the circular aperture. Since the zeros of $J_1(x)$ are at $0, 3.8317, 7.0156, \dots$, it follows that the distance from the center of the Airy disk to the intensity minimum at the midpoint of the first dark ring on the screen is given by (see Problem 6)

$$r_1 = R\sin\phi_1 = 1.22\frac{R\lambda}{2a}. \tag{16}$$

Project 1 PROBLEMS

1. Show that $H^1(\rho) = e^{ik\rho}/\rho$ and $H^2(\rho) = e^{-ik\rho}/\rho$ (spherically symmetric waves propagating in from infinity toward the source point) are a fundamental set of solutions for Eq. (4) in $\rho > 0$.

2. Assuming that derivatives with respect to x, y, and z can be taken inside the double integral, show that Eq. (9) satisfies the Helmholtz equation (1) in the region $z > 0$ and the boundary condition, $\dfrac{\partial u}{\partial z} = 0$, at $z = 0$ in the region $r > 0$ outside the aperture. [*Hint:* Use Cartesian coordinates $\rho = \sqrt{(x-\hat{x})^2 + (y-\hat{y})^2 + z^2}$ and $\rho_x = (x - \hat{x})\rho^{-1}$, etc. to simplify calculations.]

3. Verify the approximation (10).

4. Verify the integral representations

$$J_0(x) = \frac{1}{2\pi}\int_0^{2\pi} e^{-ix\cos\theta}\,d\theta = \frac{1}{2\pi}\int_0^{2\pi}\cos(x\cos\theta)\,d\theta$$

by carrying out the following steps.

(a) Use integration by parts to derive the identity $n\int_0^{2\pi}\cos^n\theta\,d\theta = (n-1)\int_0^{2\pi}\cos^{(n-2)}\theta\,d\theta$ for all $n \geq 2$.

(b) Integrate term-by-term in each of the series

$$\cos(x\cos\theta) = \sum_{n=0}^{\infty}\frac{(-1)^n}{(2n)!}x^{2n}\cos^{2n}\theta \quad \text{and}$$

$$\sin(x\cos\theta) = \sum_{n=0}^{\infty}\frac{(-1)^{(n+1)}}{(2n+1)!}x^{2n+1}\cos^{2n+1}\theta$$

using the identity from Problem 4(a), in addition to the facts that $\int_0^{2\pi} 1\,d\theta = 2\pi$, $\int_0^{2\pi}\cos\theta\,d\theta = 0$, and $\int_0^{2\pi}\cos^2\theta\,d\theta = \pi$.

5. Use series representations to show that $[xJ_1(x)]' = xJ_0(x)$ and, therefore,

$$aJ_1(a) = \int_0^a xJ_0(x)\,dx.$$

Use this result to obtain Eq. (14) from Eq. (13).

6. Verify Eq. (16).

Project 2 Hermite Polynomials and the Quantum Mechanical Harmonic Oscillator

Recall from Chapter 4 that a particle of mass m constrained to move along the x-axis, and bound to the equilibrium position $x = 0$ by a restoring force $-kx$, satisfies, in the absence of damping forces, the equation of motion

$$mx'' = -kx. \tag{1}$$

If we multiply Eq. (1) by x', the resulting equation can be written as

$$\frac{d}{dt}\left[\frac{1}{2}m(x')^2 + \frac{1}{2}kx^2\right] = 0. \tag{2}$$

It follows from Eq. (2) that

$$\frac{1}{2}m(x')^2 + \frac{1}{2}kx^2 = E, \text{ a constant.} \tag{3}$$

In Eq. (3) the terms $\frac{1}{2}m(x')^2$, $\frac{1}{2}kx^2$, and E are, respectively, the kinetic, potential, and total energies of the system. If we denote the natural frequency of the system (1) by $\omega = \sqrt{k/m}$, Eq. (3) can be expressed in the form

$$\frac{1}{2}m(x')^2 + \frac{1}{2}m\omega^2 x^2 = E, \tag{4}$$

where the potential energy is now represented by the term $\frac{1}{2}m\omega^2 x^2$. If initial conditions $x(0) = x_0$ and $x'(0) = v_0$ are prescribed for the system (1), then by evaluating Eq. (4) at $t = 0$, it follows that the total energy of the system is $E = \frac{1}{2}mv_0^2 + \frac{1}{2}m\omega^2 x_0^2$, and, depending on the initial conditions, can take on any nonnegative value.

In quantum mechanics, the (steady-state) **Schrödinger wave equation** corresponding to a one-dimensional problem is the ordinary differential equation

$$-\frac{\hbar^2}{2\mu}\frac{d^2\psi}{dx^2} + V(x)\psi = E\psi, \tag{5}$$

where $\hbar$ is Planck's constant[11] divided by 2π, E is the total energy of the quantum mechanical system, and $V(x)$ is the potential function for the system. For example, the potential energy function for the distance between atoms in a diatomic molecule, oscillating in the neighborhood of a stable equilibrium position, may be approximated by

$$V(x) = \frac{1}{2}\mu\omega^2 x^2, \tag{6}$$

where ω is loosely called the classical frequency of the harmonic oscillator, and μ is the reduced mass[12] of the system. Substituting the right-hand side of Eq. (6) for $V(x)$ in Eq. (5) gives the Schrodinger equation for the linear harmonic oscillator:

$$-\frac{\hbar^2}{2\mu}\frac{d^2\psi}{dx^2} + \frac{1}{2}\mu\omega^2 x^2\psi = E\psi. \tag{7}$$

The function $\rho(x) = |\psi(x)|^2$ is interpreted as a probability density function for the position of a particle in the system. Thus $\rho(x)dx = |\psi(x)|^2 dx$ is the probability, that upon a measurement of its position, the particle will be found in an interval of width dx about the point x. It follows that physically admissible solutions $\psi(x)$, known as **Schrödinger wave functions**, are required to satisfy

$$\psi \to 0 \quad \text{as} \quad |x| \to \infty \quad \text{and} \quad \int_{-\infty}^{\infty} |\psi(x)|^2 \, dx = 1. \tag{8}$$

In this project, we ask the reader to show that solutions of Eq. (7) satisfying the conditions (8) occur only for certain discrete values of E, called **eigenvalues**. The corresponding solutions of Eq. (7)—the Schrödinger wave functions—are called **eigenfunctions** of the problem.

---

[11]Planck's constant is $h = 6.6255 \times 10^{-27}$ erg s, whereas $\hbar = h/(2\pi) = 1.0545 \times 10^{-27}$ erg s.

[12]The reduced mass μ of a system of two bodies is defined by $\mu = (1/m_1 + 1/m_2)^{-1} = m_1 m_2/(m_1 + m_2)$, where m_1 and m_2 are the masses of the bodies. When considering the vibration of a diatomic molecule, using the reduced mass (i) assures us that we are viewing the motion from a framework (the center of mass of the system) that is truly stationary, and (ii) allows the two-body problem to be solved as if it were a one-body problem.

Project 2 PROBLEMS

1. Show that changing the independent variable to $\xi = \sqrt{\mu\omega/\hbar}\, x$ transforms Eq. (7) into

$$\frac{d^2\psi}{d\xi^2} + (\lambda + 1 - \xi^2)\psi = 0, \qquad (i)$$

where $\lambda + 1 = 2E/(\hbar\omega)$.

2. Show that if we substitute $\psi = e^{-\xi^2/2} y(\xi)$ into Eq. (i), then $y(\xi)$ must satisfy the Hermite equation

$$y'' - 2\xi y' + \lambda y = 0. \qquad (ii)$$

3. Find power series solutions $y_1(\xi; \lambda)$ and $y_2(\xi; \lambda)$ for Eq. (ii) such that $y_1(0; \lambda) = 1$ and $y_1'(0; \lambda) = 0$, while $y_2(0; \lambda) = 0$ and $y_2'(0; \lambda) = 1$. What is the radius of convergence for each of these two series? (Also see Problem 21, Section 8.2.)

4. Carry out the following steps to show that solutions of Eq. (7) satisfying the conditions (8) exist only for a discrete set of values of E.

   (a) Show that $y_1(\xi; 2n)$ is a polynomial of degree n for each $n = 0, 2, 4, \ldots$, and that $y_2(\xi; 2n)$ is a polynomial of degree n for each $n = 1, 3, 5 \ldots$.

   (b) The Hermite polynomial $H_n(\xi)$ is defined as the degree n polynomial solution of the Hermite equation for which the coefficient of ξ^n is 2^n:

$$H_n(\xi) = \begin{cases} (-1)^{n/2} \dfrac{n!}{(n/2)!}\, y_1(\xi; 2n), & n = 0, 2, 4 \ldots, \\[2mm] (-1)^{(n-1)/2} \dfrac{2(n!)}{[(n-1)/2]!}\, y_2(\xi; 2n), & n = 1, 3, 5, \ldots. \end{cases}$$

Find $H_0(\xi), \ldots, H_5(\xi)$.

   (c) Explain why $\psi_n(\xi) = c_n e^{-\xi^2/2} H_n(\xi)$, $n = 0, 1, 2, \ldots$, with appropriate normalization constants c_n, all satisfy the conditions (8).

   (d) Plot the graphs of the (unnormalized) eigenfunctions $e^{-\xi^2/2} H_0(\xi), \ldots, e^{-\xi^2/2} H_5(\xi)$.

   (e) From the recursion formula for the coefficients in the series for $y_1(\xi; \lambda)$, show that for values of $\lambda \neq 2n$, $n = 0, 2, 4, \ldots$, all coefficients starting with a certain one have the same sign, and that the ratio of the coefficient of ξ^n to the coefficient of ξ^{n-2} is

$$\frac{2n - 2 - \lambda}{n(n-1)} \sim \frac{2}{n} \quad \text{as} \quad n \to \infty.$$

Compare this result with the ratio of successive coefficients in the power series for e^{ξ^2} to conclude that $e^{-\xi^2/2} y_1(\xi; \lambda)$ satisfies the conditions (8) only for $\lambda = 2n$, $n = 0, 2, 4, \ldots$. Similarly, show that the tail of the series for $e^{-\xi^2/2} y_2(\xi; \lambda)$ for values of $\lambda \neq 2n$, $n = 1, 3, 5, \ldots$ is asymptotically similar to ξe^{ξ^2}. Conclude that $\psi_n(\xi) = c_n e^{-\xi^2/2} H_n(\xi)$, $n = 0, 1, 2, \ldots$ (with $\xi = \sqrt{\mu\omega/\hbar}\, x$) are the only solutions of Eq. (7) that satisfy the conditions (8).

   (f) Conclude that E can only assume the discrete values $E_n = \hbar\omega(n + \tfrac{1}{2})$ or $E = \tfrac{1}{2}\hbar\omega$, $E = \tfrac{3}{2}\hbar\omega$, $E = \tfrac{5}{2}\hbar\omega, \ldots$. The lowest energy level is called the **ground state** of the system. All discrete higher energy levels are classified as **excited states**.

Project 3 Perturbation Methods

Perturbation theory consists of a collection of methods used to find an approximate solution of a differential equation that is "close" to a nearby problem which is exactly solvable. The distance between the two problems is usually quantified by a small parameter that appears in the differential equation. The systematic calculations used in the most basic perturbation method are identical to those used to construct power series solutions of differential equations with one important difference—the series expansions are carried out, not with respect to the independent variable, but with respect to the small parameter that appears in the differential equation. We present the method in the context of motion of an object launched vertically upward from the surface of the earth.

▶ **A Projectile Motion Problem.** Assume that the earth is a sphere of radius R with total mass M, and that the mass density depends only on the radial distance from the earth's center. According to Newton's theory of universal gravitation, the magnitude of the earth's acceleration field at a height r above the earth's surface is

$$\frac{GM}{(R+r)^2}, \tag{1}$$

where the constant of universal gravitation $G \cong 6.6732 \times 10^{-11}$ N-m^2/kg^2. By using Newton's second law of motion, $ma = F$, and neglecting air resistance, it follows that the differential equation for motion of an object of mass m, projected radially outward from the earth's surface, is

$$mr'' = -\frac{mMG}{(R+r)^2}, \tag{2}$$

where r is the distance between the projectile's center of mass and the surface of the earth (see Figure 8.P.4).

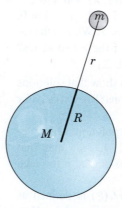

FIGURE 8.P.4   Radial distance r from the earth's surface of an object of mass m.

If the projectile is launched from the surface of the earth with velocity v_0, then initial conditions for Eq. (2) are

$$r(0) = 0, \qquad r'(0) = v_0. \tag{3}$$

Using g to denote the value of gravitational acceleration at the surface of the earth, we set $r = 0$ in Eq. (1) to get $g = MG/R^2$. Then the differential equation in the initial value problem (2) may be written (see also Example 4 in Section 2.3) as

$$r'' = -\frac{gR^2}{(R+r)^2} = -\frac{g}{(1+r/R)^2}. \tag{4}$$

If $r/R < 1$, then Eq. (4) may be expressed as

$$r'' = -g[1 - 2(r/R) + 3(r/R)^2 - \cdots], \tag{5}$$

where we have used a geometric power series to represent the right-hand side of Eq. (4). If the initial velocity v_0 is relatively small, then we expect that r/R, the ratio of the object's height to the radius of the earth, will remain small relative to 1 during the entire time period of the projectiles flight. This would certainly be a correct assumption if we had an estimate of the maximum attained altitude, say, r_m, such that $r_m/R << 1$ (read as "r_m/R is small compared to unity"). In this case we might neglect all terms involving r/R to reduce Eq. (5) to the approximate equation

$$r'' = -g, \tag{6}$$

which describes the motion of the object acted on by constant acceleration g. Solving Eq. (6) subject to the initial conditions (3) then gives

$$r = v_0 t - gt^2/2. \tag{7}$$

But suppose that we wished to study the effect, on the motion of the projectile, of the neglected terms involving r/R in Eq. (5). Suppose that r/R is small relative to 1, but not necessarily negligible for the purpose of accurately predicting the maximum attained altitude and flight time of the projectile for the problem at hand. Let us rewrite Eq. (5) in the form

$$r'' = -g + f(r), \tag{8}$$

where $f(r) = g[2(r/R) - 3(r/R)^2 + \cdots]$. If r/R is small relative to 1, so that $f(r)$ is small relative to g, then the term $f(r)$ is referred to as a **perturbation** to the equation $r'' = -g$. In this project, and in the problems, we introduce a **perturbation method** for systematically obtaining approximate solutions to initial value problems that contain perturbations in the equations or the initial conditions. The calculations used to obtain these approximations are formally identical to the calculations required to obtain series solutions of linear variable coefficient equations, except that the series are developed in terms of a small parameter that appears in the problem instead of the independent variable.

▶ **Setting Up the Problem: Scaling.** In order to apply perturbation methods in a sensible and systematic way, we first introduce a technique commonly used by applied mathematicians to set up a differential equation so that the magnitude of a perturbation, relative to other terms in the equation, is represented by a small, dimensionless, multiplicative parameter. Note that the time to reach the approximate maximum height, as computed from Eq. (7), is v_0/g. The corresponding height is $v_0^2/(2g)$. These quantities, although approximations obtained from consideration of the unperturbed problem, provide us with a natural, or characteristic, unit of time T, and a natural, or characteristic, unit of length L, defined by

$$T = \frac{v_0}{g} \quad \text{and} \quad L = \frac{v_0^2}{g}, \tag{9}$$

where we have neglected the irrelevant factor of $1/2$ in the definition of L. The units T and L are referred to as **intrinsic time** and **length scales**, respectively. We now use these scales to define the dimensionless independent and dependent variables

$$\tau = \frac{t}{T} \quad \text{and} \quad y = \frac{r}{L}.$$

Thus $y = 1$ corresponds to (twice) the approximate maximum attained height, whereas $\tau = 1$ corresponds to the unit of time required to reach maximum height, no matter what fundamental physical units of time and length are used to specify g and v_0. We now want to express Eq. (4) and the initial conditions (3) in terms of the dimensionless variables y and τ. Since $r(t) = Ly(\tau(t))$, we use the chain rule from calculus to get

$$\frac{dr}{dt} = L \frac{dy}{d\tau} \frac{d\tau}{dt} = \frac{L}{T} \frac{dy}{d\tau} = v_0 \frac{dy}{d\tau} \tag{10}$$

and

$$\frac{d^2r}{dt^2} = \frac{L}{T} \frac{d^2y}{d\tau^2} \frac{d\tau}{dt} = \frac{L}{T^2} \frac{d^2y}{d\tau^2} = g \frac{d^2y}{d\tau^2}. \tag{11}$$

Substituting Ly for r and the right-hand side of Eq. (11) for d^2r/dt^2, in Eq. (4), we obtain

$$\frac{d^2y}{d\tau^2} = -\frac{1}{(1+\varepsilon y)^2}, \tag{12}$$

where $\varepsilon = v_0^2/(gR) = (v_0^2/g)/R$ is a dimensionless parameter equal to the ratio of (twice) the approximate maximum attained height to the radius of the earth. The initial condition $r(0) = Ly(0) = 0$ implies that $y(0) = 0$. Using Eq. (10) in the initial condition $r'(0) = v_0$ yields $y'(0) = 1$. Thus the initial conditions for Eq. (12) are

$$y(0) = 0, \qquad \frac{dy}{d\tau}(0) = 1. \tag{13}$$

There are two advantages to the dimensionless formulation consisting of Eqs. (12) and (13) over Eqs. (4) and (3): (i) the number of parameters that appear in the problem has been reduced from 3 to 1, and (ii) under the assumed experimental conditions, the nominal maximum magnitude of the perturbation εy is, in fact, represented by the dimensionless parameter ε, which we assume to be small relative to 1.

Remark. In this problem, the parameters R and g are physical constants, so it would appear that the reduction in the number of parameters from 3 to 1 is of little consequence, since we are only interested in varying the initial velocity v_0. However, if we wish to consider projectiles launched from the surface of large heavenly bodies with different values of R and g, then the gain in reducing the number of parameters from 3 to 1 is significant if the purpose is to conduct a parametric study of the effect of different values of v_0, R, and g on the projectile's motion. In this case, for each fixed value of ε, the projectile motion is identical for all points on the surface $v_0 = \sqrt{\varepsilon g R}$ in the positive octant of three-dimensional gRv_0-space. By varying ε, this one-parameter family of surfaces will contain all parameter values of interest in the octant. In differential equation models, the process of **nondimensionalization**, that is, putting the problem into dimensionless form, normally leads to dimensionless combinations of parameters, called **dimensionless groups**, that are fewer in number than the dimensional parameters that appear in the original formulation of the problem. **Scaling** is nondimensionalizing in such a way that a dimensionless multiplicative factor indicates the relative magnitude of each term in the equation.

▶ **A Basic Perturbation Method.** We now illustrate the most basic perturbation method on the initial value problem

$$y'' = -\frac{1}{(1+\varepsilon y)^2}, \qquad y(0) = 0, \quad y'(0) = 1, \tag{14}$$

where ε is assumed to be a small positive number, relative to 1. The solution of this initial value problem depends on both the independent variable τ and the parameter ε, so we denote it by $y(\tau, \varepsilon)$. We begin by assuming a perturbation expansion for $y(\tau, \varepsilon)$ in the form of a power series in ε:

$$y(\tau, \varepsilon) = y_0(\tau) + y_1(\tau)\varepsilon + y_2(\tau)\varepsilon^2 + \cdots. \tag{15}$$

Note that the coefficients in the expansion are functions of the independent variable τ. In practical applications, it is normally the case that only a few terms in the expansion are actually computed. The **zeroth order** perturbation approximation to the initial value problem is $y_0(\tau)$. This corresponds to solving the **unperturbed problem**, obtained by setting $\varepsilon = 0$ in Eq. (14). The **first order** approximation to the initial value problem is $y_0(\tau) + y_1(\tau)\varepsilon$; the **second order** approximation is $y_0(\tau) + y_1(\tau)\varepsilon + y_2(\tau)\varepsilon^2$, and so on.

The general idea of this perturbation method is to substitute the expansion (15) into the differential equation, insuring that both the left- and right-hand sides of the equation are expressed as a power series in ε. By equating coefficients of like powers of ε from both sides of the equation (see Theorem 8.1.6, the Identity Principle), we obtain differential equations for each of the coefficient functions $y_0, y_1, y_2, \ldots$. Initial conditions for each of the equations are obtained by carrying out analogous calculations for each of the prescribed initial conditions.

Thus, we begin by expanding the right-hand side of the differential equation in terms of a power series in ε:

$$y'' = -\frac{1}{(1 + \varepsilon y)^2} = -1 + 2y\varepsilon - 3y^2\varepsilon^2 + \cdots. \tag{16}$$

Substituting the expansion (15) into Eq. (16) gives

$$\begin{aligned}
y_0'' + y_1''\varepsilon + y_2''\varepsilon^2 + \cdots &= -1 + 2(y_0 + y_1\varepsilon + y_2\varepsilon^2 + \cdots)\varepsilon \\
&\quad - 3(y_0 + y_1\varepsilon + y_2\varepsilon^2 + \cdots)^2\varepsilon^2 + \cdots \\
&= -1 + 2y_0\varepsilon + (2y_1 - 3y_0^2)\varepsilon^2 + \cdots.
\end{aligned} \tag{17}$$

Similarly, we substitute the expansion (15) into each of the initial conditions in the initial value problem (14) to get

$$y_0(0) + y_1(0)\varepsilon + y_2(0)\varepsilon^2 + \cdots = 0 = 0 \cdot \varepsilon^0 + 0 \cdot \varepsilon^1 + 0 \cdot \varepsilon^2 + \cdots \tag{18}$$

and

$$y_0'(0) + y_1'(0)\varepsilon + y_2'(0)\varepsilon^2 + \cdots = 1 = 1 \cdot \varepsilon^0 + 0 \cdot \varepsilon^1 + 0 \cdot \varepsilon^2 + \cdots. \tag{19}$$

Equating coefficients of like powers of ε in each of Eqs. (17–19) gives us the following sequence of initial value problems:

Coefficients of ε^0: $\quad y_0'' = -1, \qquad y_0(0) = 0, \qquad y_0'(0) = 1,$ (20)

Coefficients of ε^1: $\quad y_1'' = 2y_0, \qquad y_1(0) = 0, \qquad y_1'(0) = 0,$ (21)

Coefficients of ε^2: $\quad y_2'' = 2y_1 - 3y_0^2, \qquad y_2(0) = 0, \qquad y_2'(0) = 0.$ (22)

The solutions of the initial value problems (20–22) are

$$y_0 = \tau - \tau^2/2, \qquad y_1 = \tau^3/3 - \tau^4/12, \qquad y_2 = -\tau^4/4 + 11\tau^5/60 - 11\tau^6/360.$$

Substituting these expressions into the expansion (15) gives the second-order perturbation approximation to the initial value problem:

$$y(\tau, \varepsilon) \cong \tau - \tau^2/2 + (\tau^3/3 - \tau^4/12)\varepsilon + (-\tau^4/4 + 11\tau^5/60 - 11\tau^6/360)\varepsilon^2. \tag{23}$$

Equation (23) can now be used to find a perturbation approximation for the time $\tau_m = \tau_m(\varepsilon)$ at which the projectile attains maximum height. Since the velocity of the projectile must equal zero at this time, τ_m must satisfy the equation $y'(\tau_m, \varepsilon) = 0$. Setting the derivative of the right-hand side of Eq. (23) equal to zero gives the algebraic equation

$$1 - \tau_m + \left(\tau_m^2 - \tau_m^3/3\right)\varepsilon + \left(-\tau_m^3 + 11\tau_m^4/12 - 11\tau_m^5/60\right)\varepsilon^2 = 0. \tag{24}$$

Since we seek the root of Eq. (24) near 1 (why?), we assume an expansion for τ_m of the form

$$\tau_m = 1 + a_1\varepsilon + a_2\varepsilon^2 + \cdots. \tag{25}$$

If we substitute the expansion (25) into Eq. (24) and retain all terms up to second order in ε, we obtain a pair of algebraic equations from which the coefficients a_1 and a_2 can be computed. The second order expansion in Eq. (25) can then be substituted back into Eq. (23) to get the maximum height out to second order in ε (see Problem 1).

▶ **Regular and Singular Perturbation Problems.** In this project we have introduced the most basic perturbation method. Perturbation theory consists of a collection of methods used to find approximate solutions to problems that are "close" to exactly solvable (unperturbed) problems, with the distance between the perturbed and unperturbed problems often quantified by a small parameter. We were able to find the perturbation expansion for $y(\tau, \varepsilon)$ in Eq. (15) because we were able to find a closed form solution of the unperturbed problem.

It is often the case that an expansion such as

$$y_0(t) + y_1(t)\varepsilon + y_2(t)\varepsilon^2 + \cdots$$

is not a good approximation to the solution over the entire domain of the problem, or that for one or more values of $n \geq 1$, $y_n(t)\varepsilon^n$ is not small compared to the preceding term $y_{n-1}(t)\varepsilon^{n-1}$ for all values of t in the domain of the problem. For example, the two-term expansion

$$1 - t\varepsilon$$

is not a good approximation of $e^{-\varepsilon t}$ for $t \in (-\infty, \infty)$ since εt is small relative to 1 only if t is small relative to $1/\varepsilon$. On the other hand, any finite number of terms retained in the expansion

$$\sin(t + \varepsilon) = \sin t \cos \varepsilon + \cos t \sin \varepsilon$$

$$= \sin t\left(1 - \frac{\varepsilon^2}{2!} + \frac{\varepsilon^4}{4!} + \cdots\right) + \cos t\left(\varepsilon - \frac{\varepsilon^3}{3!} + \frac{\varepsilon^5}{5!} - \cdots\right)$$

$$= \sin t + \varepsilon \cos t - \frac{\sin t}{2!}\varepsilon^2 - \frac{\cos t}{3!}\varepsilon^3 + \frac{\sin t}{4!}\varepsilon^4 + \cdots$$

are uniformly valid on $(-\infty, \infty)$ for small values of ε since the coefficients of all powers of ε are bounded for all values of t. The expansion (23) is uniformly valid for τ in the range $0 < \tau \ll 1/\varepsilon$, which includes the range of physical interest since the projectile hits the ground well before $1/\varepsilon$ if $\varepsilon \ll 1$.

It is the rule rather than the exception that perturbation approximations break down in some subsets of the domain called *regions of nonuniformity*. Infinite or semi-infinite domains are but one source leading to such regions. Nonuniformity may also arise if there is a small parameter multiplying the highest derivative in the differential equation or if the perturbation expansion has singularities in the region of interest. When the straightforward perturbation solution is uniformly valid over the entire domain of the problem, we refer to it as a **regular perturbation problem**. If the straightforward perturbation approximation is not uniformly valid throughout the problem domain, we refer to it as a **singular perturbation problem**. Numerous special techniques have been devised to obtain perturbation approximations that are uniformly valid over entire problem domains. For an introduction to dimensional analysis, scaling, regular, and singular perturbation methods, see the book by Lin and Segel in the References. A more advanced treatment of perturbation methods can be found in the book by Holmes.

Project 3 PROBLEMS

1. Calculate the coefficients a_1 and a_2 in Eq. (25). Then find the second order perturbation approximation of $y_m = y(\tau_m, \varepsilon)$. Explain why the approximations for τ_m and y_m make sense in terms of the physics.

2. For each $\varepsilon = 0.1$, 0.25, and 0.5, examine the ac- CAS curacy of the perturbation expansion (23) by plotting, on the same set of coordinate axes, the zeroth, first, and second order perturbation approximation

of $y(\tau, \varepsilon)$, along with the numerical approximation of the initial value problem (14) obtained by using an initial value problem solver.

3. (a) Calculate the first order perturbation approximation to the initial value problem

$$y'' + (1 + \varepsilon \cos t)y = 0, \qquad y(0) = 1, \quad y'(0) = 0.$$

CAS (b) For each $\varepsilon = 0.1, 0.25, 0.5,$ and 1.0, compare the graphs of the first order approximation with the numerical approximation obtained by using an initial value problem solver.

4. Consider the initial value problem for the displacement $\tilde{y}(\tilde{t})$ from equilibrium of a mass m in a damped harmonic oscillator

$$m\tilde{y}'' + 2\gamma \tilde{y}' + k\tilde{y} = 0, \qquad \tilde{y}(0) = y_0, \quad \tilde{y}'(0) = 0.$$

(a) Using the time scale $T = \sqrt{m/k}$ and the length scale $L = y_0$, show that the initial value problem for $y(t) = \tilde{y}(\tilde{t}(t))/L$, where $\tilde{t}(t) = tT$ is

$$y'' + 2\varepsilon y' + y = 0, \qquad y(0) = 1, \ y'(0) = 0,$$

where $\varepsilon = \gamma/\sqrt{mk}$. Why are these appropriate time and length scales if the damping force is assumed to be small relative to inertial and spring forces?

(b) Assume that $0 < \varepsilon << 1$ and calculate first and second order perturbation approximations to the initial value problem in (a). Is the expansion uniform for $t > 0$?

CAS (c) For each $\varepsilon = 0.1, 0.25,$ and 0.5, compare the graphs of the first and second order approximations with the graph of the exact solution.

5. Consider the initial value problem

$$y' + y = \varepsilon y^2, \qquad y(0) = 1.$$

(a) Determine a second order perturbation approximation for small ε.

(b) Show that the exact solution is

$$y = \frac{e^{-t}}{1 + \varepsilon(e^{-t} - 1)}.$$

(c) Expand this exact solution for small ε and compare the result with (a). Is this expansion valid for all t?

6. Consider the oscillations of a mass connected to a nonlinear spring (see Duffing's equation, Problem 17, Section 4.2) described by

$$y'' + y + \varepsilon y^3 = 0, \qquad y(0) = 1, \quad y'(0) = 0,$$

where ε is a small positive number.

(a) Multiply the equation by y' and integrate to get

$$\frac{1}{2}(y')^2 + \frac{1}{2}y^2 + \frac{1}{4}\varepsilon y^4 = E$$

and then determine E from the initial conditions. Explain why this equation shows that y is bounded for all times when ε is positive.

(b) Find a first order perturbation approximation to the initial value problem for small ε. Explain why the approximation is valid only for $\varepsilon t << 1$.

(c) For each $\varepsilon = 0.1, 0.25,$ and 0.5, compare the CAS graphs of the first order approximation with the numerical approximation obtained by using an initial value problem solver.

7. Consider the initial value problem describing the motion of a small mass to which is delivered a large impulse at time $t = 0$:

$$\varepsilon y'' + y' + y = 0, \qquad y(0) = 0, \qquad \varepsilon y'(0) = 1. \quad \text{(i)}$$

(a) Find the exact solution $y(t, \varepsilon)$ to the initial CAS value problem (i). Then, for each $\varepsilon = 0.01, 0.1,$ and 0.2, plot the graphs of $y(t, \varepsilon)$ on the same set of coordinate axes. The small interval near $x = 0$ where y changes very rapidly is called a **boundary layer**.

(b) Show that to lowest order in powers of ε, the CAS roots of the characteristic equation $\varepsilon \lambda^2 + \lambda + 1 = 0$ for Eq. (i) are $\lambda_1 \cong -1$ and $\lambda_2 \cong -1/\varepsilon$, suggesting that $y_U(t) = e^{-t} - e^{-t/\varepsilon}$ may be a good approximation to $y(t, \varepsilon)$. Use a computer to demonstrate that $|y(t, \varepsilon) - y_U(t)| < K\varepsilon$ for small values of ε, where K is a constant; therefore, y_U appears to be an approximation that is uniformly valid in $[0, \infty)$.

(c) Using the stretching transformation $\tau = t/\varepsilon$ to blow up the boundary layer, show that $Y(\tau, \varepsilon) = y(\varepsilon\tau, \varepsilon)$ satisfies the initial value problem

$$Y'' + Y' + \varepsilon Y = 0, \qquad Y(0) = 0, \quad Y'(0) = 1.$$

Then show that the zeroth order approximation to this initial value problem is $Y_I(\tau) = 1 - e^{-\tau}$ (the inner solution) Explain why $Y_I(t/\varepsilon)$ cannot be a uniformly valid approximation to $y(t, \varepsilon)$ in $[0, \infty)$.

(d) Show that $y_O(t) = Ce^{-t}$ (the outer solution) is the general solution of the unperturbed equation in the initial value problem (i). Use the "matching condition"

$$\lim_{t \to 0} y_O(t) = \lim_{\tau \to \infty} Y_I(\tau)$$

(the inner limit of the outer solution equals the outer limit of the inner solution) to show that

$C = 1$. Explain why y_O cannot be a uniformly valid approximation to $y(t, \varepsilon)$ on $[0, \infty)$.

(e) Show that the composite approximation

$$y_U(t) = y_O(t) + Y_I(t/\varepsilon) - \lim_{t \to 0} y_O(t)$$

agrees with (b).

CHAPTER NINE

Partial Differential Equations and Fourier Series

n many important physical problems there are two or more independent variables, so the corresponding mathematical models involve partial, rather than ordinary, differential equations. This chapter treats one important method for solving partial differential equations, a method known as separation of variables. Its essential feature is the replacement of the partial differential equation by a set of ordinary differential equations, which must be solved subject to given initial or boundary conditions. The first section of this chapter deals with some basic properties of boundary value problems for ordinary differential equations. The desired solution of the partial differential equation is then expressed as a sum, usually an infinite series, formed from solutions of the ordinary differential equations. In many cases we ultimately need to deal with a series of sines and/or cosines, so part of the chapter is devoted to a discussion of such series, which are known as Fourier series. With the necessary mathematical background in place, we then illustrate the use of separation of variables in a variety of problems arising from heat conduction, wave propagation, and potential theory.

9.1 Two-Point Boundary Value Problems

Up to this point in the book we have dealt with initial value problems, consisting of a differential equation together with suitable initial conditions at a given point. A typical example, which was discussed at length in Chapter 4, is the differential equation

$$y'' + p(t)y' + q(t)y = g(t), \tag{1}$$

with the initial conditions

$$y(t_0) = y_0, \qquad y'(t_0) = y_1. \tag{2}$$

Physical applications often lead to another type of problem, one in which the value of the dependent variable y or its derivative is specified at two *different* points. Such conditions are called **boundary conditions** to distinguish them from initial conditions that specify the value of y and y' at the *same* point. A differential equation and suitable boundary conditions form a **two-point boundary value problem**. A typical example is the differential equation

$$y'' + p(x)y' + q(x)y = g(x) \tag{3}$$

with the boundary conditions

$$y(\alpha) = y_0, \qquad y(\beta) = y_1. \tag{4}$$

The natural occurrence of boundary value problems usually involves a space coordinate as the independent variable, so we have used x rather than t in Eqs. (3) and (4). To solve the boundary value problem (3), (4), we need to find a function $y = \phi(x)$ that satisfies the differential equation (3) in the interval $\alpha < x < \beta$ and that takes on the specified values y_0 and y_1 at the endpoints of the interval. Usually, we first seek the general solution of the differential equation and then use the boundary conditions to determine the values of the arbitrary constants.

Boundary value problems can also be posed for nonlinear differential equations, but we will restrict ourselves to a consideration of linear equations only. An important classification of linear boundary value problems is whether they are homogeneous or nonhomogeneous. If the function g has the value zero for each x, and if the boundary values y_0 and y_1 are also zero, then the problem (3), (4) is called **homogeneous**. Otherwise, the problem is **nonhomogeneous**.

Although the initial value problem (1), (2) and the boundary value problem (3), (4) may superficially appear to be quite similar, their solutions differ in some very important ways. Under mild conditions on the coefficients initial value problems are certain to have a unique solution. On the other hand, boundary value problems under similar conditions may have a unique solution, but they may also have no solution or, in some cases, infinitely many solutions. In this respect, linear boundary value problems resemble systems of linear algebraic equations.

Let us recall some facts (see Appendices A.2 and A.3) about the system

$$\mathbf{Ax} = \mathbf{b}, \tag{5}$$

where $\mathbf{A}$ is a given $n \times n$ matrix, $\mathbf{b}$ is a given $n \times 1$ vector, and $\mathbf{x}$ is an $n \times 1$ vector to be determined. If $\mathbf{A}$ is nonsingular, then the system (5) has a unique solution for any $\mathbf{b}$. However, if $\mathbf{A}$ is singular, then the system (5) has no solution unless $\mathbf{b}$ satisfies a certain additional condition, in which case the system has infinitely many solutions. Now consider the corresponding homogeneous system

$$\mathbf{Ax} = \mathbf{0}, \tag{6}$$

obtained from the system (5) when $\mathbf{b} = \mathbf{0}$. The homogeneous system (6) always has the solution $\mathbf{x} = \mathbf{0}$, which is often referred to as the trivial solution. If $\mathbf{A}$ is nonsingular, then this is the only solution, but if $\mathbf{A}$ is singular, then there are infinitely many nonzero, or nontrivial, solutions. Note that it is impossible for the homogeneous system to have no solution. These results can also be stated in the following way: the nonhomogeneous system (5) has a unique solution if and only if the homogeneous system (6) has only the solution $\mathbf{x} = \mathbf{0}$, and the nonhomogeneous system (5) has either no solution or infinitely many solutions if and only if the homogeneous system (6) has nonzero solutions.

We now turn to some examples of linear boundary value problems that illustrate very similar behavior. A more general discussion of linear boundary value problems appears in Chapter 10.

EXAMPLE 1

Solve the boundary value problem

$$y'' + 2y = 0, \qquad y(0) = 1, \quad y(\pi) = 0. \tag{7}$$

The general solution of the differential equation (7) is

$$y = c_1 \cos \sqrt{2}\, x + c_2 \sin \sqrt{2}\, x. \tag{8}$$

The first boundary condition requires that $c_1 = 1$. The second boundary condition implies that $c_1 \cos \sqrt{2}\, \pi + c_2 \sin \sqrt{2}\, \pi = 0$, so $c_2 = -\cot \sqrt{2}\, \pi \cong -0.2762$. Thus the solution of the boundary value problem (7) is

$$y = \cos \sqrt{2}\, x - \cot \sqrt{2}\, \pi \sin \sqrt{2}\, x. \tag{9}$$

This example illustrates the case of a nonhomogeneous boundary value problem with a unique solution.

EXAMPLE 2

Solve the boundary value problem

$$y'' + y = 0, \qquad y(0) = 1, \quad y(\pi) = a, \tag{10}$$

where a is a given number.

The general solution of this differential equation is

$$y = c_1 \cos x + c_2 \sin x, \tag{11}$$

and from the first boundary condition we find that $c_1 = 1$. The second boundary condition now requires that $-c_1 = a$. These two conditions on c_1 are incompatible if $a \neq -1$, so the problem has no solution in that case. However, if $a = -1$, then both boundary conditions are satisfied provided that $c_1 = 1$, regardless of the value of c_2. In this case there are infinitely many solutions of the form

$$y = \cos x + c_2 \sin x, \tag{12}$$

where c_2 remains arbitrary. This example illustrates that a nonhomogeneous boundary value problem may have no solution, and also that under special circumstances it may have infinitely many solutions.

Corresponding to the nonhomogeneous boundary value problem (3), (4) is the homogeneous problem consisting of the differential equation

$$y'' + p(x)y' + q(x)y = 0 \tag{13}$$

and the boundary conditions

$$y(\alpha) = 0, \qquad y(\beta) = 0. \tag{14}$$

Observe that this problem has the solution $y = 0$ for all x, regardless of the coefficients $p(x)$ and $q(x)$. This (trivial) solution is rarely of interest. What we usually want to know is whether the problem has other, nonzero, solutions. Consider the following two examples.

EXAMPLE 3

Solve the boundary value problem

$$y'' + 2y = 0, \qquad y(0) = 0, \quad y(\pi) = 0. \tag{15}$$

The general solution of the differential equation is again given by Eq. (8):

$$y = c_1 \cos \sqrt{2}\, x + c_2 \sin \sqrt{2}\, x.$$

The first boundary condition requires that $c_1 = 0$, and the second boundary condition leads to $c_2 \sin \sqrt{2}\, \pi = 0$. Since $\sin \sqrt{2}\, \pi \neq 0$, it follows that $c_2 = 0$ also. Consequently, $y = 0$ for all x is the only solution of the problem (15). This example illustrates that a homogeneous boundary value problem may have only the trivial solution $y = 0$.

EXAMPLE 4

Solve the boundary value problem

$$y'' + y = 0, \qquad y(0) = 0, \quad y(\pi) = 0. \tag{16}$$

The general solution is given by Eq. (11):

$$y = c_1 \cos x + c_2 \sin x,$$

and the first boundary condition requires that $c_1 = 0$. Since $\sin \pi = 0$, the second boundary condition is also satisfied when $c_1 = 0$, regardless of the value of c_2. Thus the solution of the problem (16) is $y = c_2 \sin x$, where c_2 remains arbitrary. This example illustrates that a homogeneous boundary value problem may have infinitely many solutions.

Examples 1 through 4 illustrate (but of course do not prove) that there is the same relationship between homogeneous and nonhomogeneous linear boundary value problems as there is between homogeneous and nonhomogeneous linear algebraic systems. A nonhomogeneous boundary value problem (Example 1) has a unique solution, and the corresponding homogeneous problem (Example 3) has only the trivial solution. Further, a nonhomogeneous problem (Example 2) has either no solution or infinitely many, and the corresponding homogeneous problem (Example 4) has nontrivial solutions.

Eigenvalue Problems

Recall the matrix equation

$$\mathbf{Ax} = \lambda \mathbf{x}, \tag{17}$$

which is discussed in Appendix A.4. Equation (17) has the solution $\mathbf{x} = \mathbf{0}$ for every value of λ, but for certain values of λ, called eigenvalues, there are also nonzero solutions, called eigenvectors. The situation is similar for boundary value problems.

Consider the problem consisting of the differential equation

$$y'' + \lambda y = 0, \tag{18}$$

together with the boundary conditions

$$y(0) = 0, \qquad y(\pi) = 0. \tag{19}$$

Observe that the problem (18), (19) is the same as the problems in Examples 3 and 4 if $\lambda = 2$ and $\lambda = 1$, respectively. Recalling the results of these examples, we note that for $\lambda = 2$, Eqs. (18), (19) have only the trivial solution $y = 0$, while for $\lambda = 1$, the problem (18), (19) has other, nontrivial, solutions. By extension of the terminology associated with Eq. (17), the values of λ for which nontrivial solutions of (18), (19) occur are called **eigenvalues**, and the nontrivial solutions themselves are called **eigenfunctions**. Restating the results of Examples 3 and 4, we have found that $\lambda = 1$ is an eigenvalue of the problem (18), (19) and that $\lambda = 2$ is not. Further, any nonzero multiple of $\sin x$ is an eigenfunction corresponding to the eigenvalue $\lambda = 1$.

Let us now turn to the problem of finding other eigenvalues and eigenfunctions of the problem (18), (19). We need to consider separately the cases $\lambda > 0, \lambda = 0$, and $\lambda < 0$, since the form of the solution of Eq. (18) is different in each of these cases. Suppose first that $\lambda > 0$. To avoid the frequent appearance of radical signs, it is convenient to let $\lambda = \mu^2$ and to rewrite Eq. (18) as

$$y'' + \mu^2 y = 0. \tag{20}$$

The characteristic polynomial equation for Eq. (20) is $r^2 + \mu^2 = 0$ with roots $r = \pm i\mu$, so the general solution is

$$y = c_1 \cos \mu x + c_2 \sin \mu x. \tag{21}$$

Note that μ is nonzero (since $\lambda > 0$) and there is no loss of generality if we also assume that μ is positive. The first boundary condition requires that $c_1 = 0$, and then the second boundary condition reduces to

$$c_2 \sin \mu\pi = 0. \tag{22}$$

We are seeking nontrivial solutions so we must require that $c_2 \neq 0$. Consequently, $\sin \mu\pi$ must be zero, and our task is to choose μ so that this will occur. We know that the sine function has the value zero at every integer multiple of π, so we can choose μ to be any (positive) integer. The corresponding values of λ are the squares of the positive integers, so we have determined that

$$\lambda_1 = 1, \quad \lambda_2 = 4, \quad \lambda_3 = 9, \quad \ldots, \quad \lambda_n = n^2, \quad \ldots \tag{23}$$

are eigenvalues of the problem (18), (19). The eigenfunctions are given by Eq. (21) with $c_1 = 0$, so they are just multiples of the functions $\sin nx$ for $n = 1, 2, 3, \ldots$. Observe that the constant c_2 in Eq. (21) is never determined, so eigenfunctions are determined only up to an arbitrary multiplicative constant [just as the eigenvectors of the matrix problem (17) are]. We will usually choose the multiplicative constant to be 1 and write the eigenfunctions as

$$y_1(x) = \sin x, \quad y_2(x) = \sin 2x, \quad \ldots, \quad y_n(x) = \sin nx, \quad \ldots, \tag{24}$$

remembering that multiples of these functions are also eigenfunctions.

Now let us suppose that $\lambda < 0$. If we let $\lambda = -\mu^2$, then Eq. (18) becomes

$$y'' - \mu^2 y = 0. \tag{25}$$

The characteristic equation for Eq. (25) is $r^2 - \mu^2 = 0$ with roots $r = \pm\mu$, so its general solution can be written as

$$y = c_1 \cosh \mu x + c_2 \sinh \mu x. \tag{26}$$

We have chosen the hyperbolic functions $\cosh \mu x$ and $\sinh \mu x$, rather than the exponential functions $\exp(\mu x)$ and $\exp(-\mu x)$, as a fundamental set of solutions for convenience in applying the boundary conditions. The first boundary condition requires that $c_1 = 0$, and then the second boundary condition gives $c_2 \sinh \mu\pi = 0$. Since $\mu \neq 0$, it follows that $\sinh \mu\pi \neq 0$, and therefore we must have $c_2 = 0$. Consequently, $y = 0$ and there are no nontrivial solutions for $\lambda < 0$. In other words, the problem (18), (19) has no negative eigenvalues.

Finally, consider the possibility that $\lambda = 0$. Then Eq. (18) becomes

$$y'' = 0, \tag{27}$$

and its general solution is

$$y = c_1 x + c_2. \tag{28}$$

The boundary conditions (19) can be satisfied only by choosing $c_1 = 0$ and $c_2 = 0$, so there is only the trivial solution $y = 0$ in this case as well. That is, $\lambda = 0$ is not an eigenvalue.

To summarize our results: we have shown that the problem (18), (19) has an infinite sequence of positive eigenvalues $\lambda_n = n^2$ for $n = 1, 2, 3, \ldots$ and that the corresponding eigenfunctions are proportional to $\sin nx$. Further, there are no other real eigenvalues. There remains the possibility that there might be some complex eigenvalues; recall that a matrix with real elements may very well have complex eigenvalues. In Problem 23 we outline an argument showing that the particular problem (18), (19) cannot have complex eigenvalues. Later, in Section 10.2, we discuss an important class of boundary value problems that includes (18), (19). One of the useful properties of this class of problems is that all their eigenvalues are real.

In later sections of this chapter, we will often encounter the problem

$$y'' + \lambda y = 0, \qquad y(0) = 0, \quad y(L) = 0, \tag{29}$$

which differs from the problem (18), (19) only in that the second boundary condition is imposed at an arbitrary point $x = L$ rather than at $x = \pi$. The solution process for $\lambda > 0$ is exactly the same as before up to the step where the second boundary condition is applied. For the problem (29) this condition requires that

$$c_2 \sin \mu L = 0 \tag{30}$$

rather than Eq. (22), as in the former case. Consequently, μL must be an integer multiple of π, so $\mu = n\pi/L$, where n is a positive integer. Hence the eigenvalues and eigenfunctions of the problem (29) are given by

$$\lambda_n = n^2 \pi^2 / L^2, \qquad y_n(x) = \sin(n\pi x/L), \qquad n = 1, 2, 3, \ldots. \tag{31}$$

As usual, the eigenfunctions $y_n(x)$ are determined only up to an arbitrary multiplicative constant. In the same way as for the problem (18), (19), you can show that the problem (29) has no eigenvalues or eigenfunctions other than those in Eq. (31).

The problems following this section explore to some extent the effect of different boundary conditions on the eigenvalues and eigenfunctions. A more systematic discussion of two-point boundary and eigenvalue problems appears in Chapter 10.

PROBLEMS

In each of Problems 1 through 13 either solve the given boundary value problem or else show that it has no solution.

1. $y'' + y = 0$, $y(0) = 0$, $y'(\pi) = 1$
2. $y'' + 2y = 0$, $y'(0) = 1$, $y'(\pi) = 0$
3. $y'' + y = 0$, $y(0) = 0$, $y(L) = 0$
4. $y'' + y = 0$, $y'(0) = 1$, $y(L) = 0$
5. $y'' + y = x$, $y(0) = 0$, $y(\pi) = 0$

6. $y'' + 2y = x$, $y(0) = 0$, $y(\pi) = 0$
7. $y'' + 4y = \cos x$, $y(0) = 0$, $y(\pi) = 0$
8. $y'' + 4y = \sin x$, $y(0) = 0$, $y(\pi) = 0$
9. $y'' + 4y = \cos x$, $y'(0) = 0$, $y'(\pi) = 0$
10. $y'' + 3y = \cos x$, $y'(0) = 0$, $y'(\pi) = 0$
11. $x^2 y'' - 2xy' + 2y = 0$, $y(1) = -1$, $y(2) = 1$
12. $x^2 y'' + 3xy' + y = x^2$, $y(1) = 0$, $y(e) = 0$

13. $x^2 y'' + 5xy' + (4 + \pi^2)y = \ln x, \quad y(1) = 0,$
$y(e) = 0$

In each of Problems 14 through 20 find the eigenvalues and eigenfunctions of the given boundary value problem. Assume that all eigenvalues are real.

14. $y'' + \lambda y = 0, \qquad y(0) = 0, \quad y'(\pi) = 0$

15. $y'' + \lambda y = 0, \qquad y'(0) = 0, \quad y(\pi) = 0$

16. $y'' + \lambda y = 0, \qquad y'(0) = 0, \quad y'(\pi) = 0$

17. $y'' + \lambda y = 0, \qquad y'(0) = 0, \quad y(L) = 0$

18. $y'' + \lambda y = 0, \qquad y'(0) = 0, \quad y'(L) = 0$

19. $y'' - \lambda y = 0, \qquad y(0) = 0, \quad y'(L) = 0$

20. $x^2 y'' - xy' + \lambda y = 0, \qquad y(1) = 0,$
$y(L) = 0, \quad L > 1$

21. The axially symmetric laminar flow of a viscous incompressible fluid through a long straight tube of circular cross section under a constant axial pressure gradient is known as Poiseuille flow. The axial velocity w is a function of the radial variable r only and satisfies the boundary value problem

$$w'' + \frac{1}{r}w' = -\frac{G}{\mu}, \qquad w(R) = 0,$$

$$w(r) \text{ bounded for } 0 < r < R,$$

where R is the radius of the tube, G is the pressure gradient, and μ is the coefficient of viscosity of the fluid.

(a) Find the velocity profile $w(r)$.

(b) By integrating $w(r)$ over a cross section, show that the total flow rate Q is given by

$$Q = \pi R^4 G / 8\mu.$$

Since Q, R, and G can be measured, this result provides a practical way to determine the viscosity μ.

(c) Suppose that R is reduced to $3/4$ of its original value. What is the corresponding reduction in

Q? This result has implications for blood flow through arteries constricted by plaque.

22. Consider a horizontal metal beam of length L subject to a vertical load $f(x)$ per unit length. The resulting vertical displacement in the beam $y(x)$ satisfies the differential equation

$$EI \frac{d^4 y}{dx^4} = f(x),$$

where E is Young's modulus and I is the moment of inertia of the cross section about an axis through the centroid perpendicular to the xy-plane. Suppose that $f(x)/EI$ is a constant k. For each of the boundary conditions given below solve for the displacement $y(x)$, and plot y versus x in the case that $L = 1$ and $k = -1$.

(a) Simply supported at both ends: $y(0) = y''(0) = y(L) = y''(L) = 0$

(b) Clamped at both ends: $y(0) = y'(0) = y(L) = y'(L) = 0$

(c) Clamped at $x = 0$, free at $x = L$: $y(0) = y'(0) = y''(L) = y'''(L) = 0$

23. In this problem we outline a proof that the eigenvalues of the boundary value problem (18), (19) are real.

(a) Write the solution of Eq. (18) as $y = k_1 \exp(i\mu x) + k_2 \exp(-i\mu x)$, where $\lambda = \mu^2$, and impose the boundary conditions (19). Show that nontrivial solutions exist if and only if

$$\exp(i\mu\pi) - \exp(-i\mu\pi) = 0. \tag{i}$$

(b) Let $\mu = \nu + i\sigma$ and use Euler's relation $\exp(i\nu\pi) = \cos(\nu\pi) + i\sin(\nu\pi)$ to determine the real and imaginary parts of Eq. (i).

(c) By considering the equations found in part (b), show that ν is an integer and that $\sigma = 0$. Consequently, μ is real and so is λ.

9.2 Fourier Series

Later in this chapter you will find that you can solve many important problems involving partial differential equations, provided that you can express a given function as an infinite sum of sines and/or cosines. In this and the following two sections we explain in detail how this can be done. These trigonometric series are called **Fourier series**; they are somewhat analogous to Taylor series in that both types of series provide a means of expressing quite complicated functions in terms of certain familiar elementary functions.

We begin with a series of the form

$$\frac{a_0}{2} + \sum_{m=1}^{\infty}\left(a_m \cos\frac{m\pi x}{L} + b_m \sin\frac{m\pi x}{L}\right).$$ (1)

On the set of points where the series (1) converges, it defines a function f, whose value at each point is the sum of the series for that value of x. In this case the series (1) is said to be the Fourier series for f. Our immediate goals are to determine what functions can be represented as a sum of a Fourier series and to find some means of computing the coefficients in the series corresponding to a given function. The first term in the series (1) is written as $a_0/2$ rather than simply as a_0 to simplify a formula for the coefficients that we derive below. Besides their association with the method of separation of variables and partial differential equations, Fourier series are also useful in various other ways, such as in the analysis of mechanical or electrical systems acted on by periodic external forces.

Periodicity of the Sine and Cosine Functions

To discuss Fourier series, it is necessary to develop certain properties of the trigonometric functions $\sin(m\pi x/L)$ and $\cos(m\pi x/L)$, where m is a positive integer. The first property is their periodic character. A function f is said to be **periodic** with period $T > 0$ if the domain of f contains $x + T$ whenever it contains x, and if

$$f(x + T) = f(x)$$ (2)

for every value of x. An example of a periodic function is shown in Figure 9.2.1. It follows immediately from the definition that if T is a period of f, then $2T$ is also a period, and so indeed is any integral multiple of T. The smallest value of T for which Eq. (2) holds is called the **fundamental period** of f. A constant function is a periodic function with an arbitrary period but no fundamental period.

If f and g are any two periodic functions with common period T, then their product fg and any linear combination $c_1 f + c_2 g$ are also periodic with period T. To prove the latter statement, let $F(x) = c_1 f(x) + c_2 g(x)$; then for any x

$$F(x + T) = c_1 f(x + T) + c_2 g(x + T) = c_1 f(x) + c_2 g(x) = F(x).$$ (3)

Moreover, it can be shown that the sum of any finite number, or even the sum of a convergent infinite series, of functions of period T is also periodic with period T.

In particular, the functions $\sin(m\pi x/L)$ and $\cos(m\pi x/L)$, $m = 1, 2, 3, \ldots$, are periodic with fundamental period $T = 2L/m$. To see this, recall that $\sin x$ and $\cos x$ have

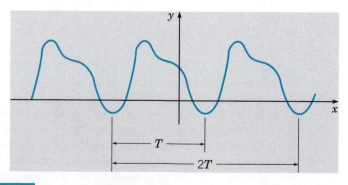

FIGURE 9.2.1 A periodic function.

fundamental period 2π and that $\sin \alpha x$ and $\cos \alpha x$ have fundamental period $2\pi/\alpha$. If we choose $\alpha = m\pi/L$, then the period T of $\sin(m\pi x/L)$ and $\cos(m\pi x/L)$ is given by $T = 2\pi L/m\pi = 2L/m$.

Note also that, since every positive integral multiple of a period is also a period, each of the functions $\sin(m\pi x/L)$ and $\cos(m\pi x/L)$ has the common period $2L$.

Orthogonality of the Sine and Cosine Functions

To describe a second essential property of the functions $\sin(m\pi x/L)$ and $\cos(m\pi x/L)$, we generalize the concept of orthogonality of vectors (see Appendix A.4). The standard **inner product** (u, v) of two real-valued functions u and v on the interval $\alpha \le x \le \beta$ is defined by

$$(u, v) = \int_{\alpha}^{\beta} u(x)v(x)\, dx. \tag{4}$$

The functions u and v are said to be **orthogonal** on $\alpha \le x \le \beta$ if their inner product is zero—that is, if

$$\int_{\alpha}^{\beta} u(x)v(x)\, dx = 0. \tag{5}$$

A set of functions is said to be **mutually orthogonal** if each distinct pair of functions in the set is orthogonal.

The functions $\sin(m\pi x/L)$ and $\cos(m\pi x/L)$, $m = 1, 2, \ldots$ form a mutually orthogonal set of functions on the interval $-L \le x \le L$. In fact, they satisfy the following orthogonality relations:

$$\int_{-L}^{L} \cos\frac{m\pi x}{L} \cos\frac{n\pi x}{L}\, dx = \begin{cases} 0, & m \ne n, \\ L, & m = n; \end{cases} \tag{6}$$

$$\int_{-L}^{L} \cos\frac{m\pi x}{L} \sin\frac{n\pi x}{L}\, dx = 0, \qquad \text{all } m, n; \tag{7}$$

$$\int_{-L}^{L} \sin\frac{m\pi x}{L} \sin\frac{n\pi x}{L}\, dx = \begin{cases} 0, & m \ne n, \\ L, & m = n. \end{cases} \tag{8}$$

These results can be obtained by direct integration. For example, to derive Eq. (8), note that

$$\int_{-L}^{L} \sin\frac{m\pi x}{L} \sin\frac{n\pi x}{L}\, dx = \frac{1}{2}\int_{-L}^{L}\left[\cos\frac{(m-n)\pi x}{L} - \cos\frac{(m+n)\pi x}{L}\right] dx$$

$$= \frac{1}{2}\frac{L}{\pi}\left\{\frac{\sin[(m-n)\pi x/L]}{m-n} - \frac{\sin[(m+n)\pi x/L]}{m+n}\right\}\Bigg|_{-L}^{L}$$

$$= 0$$

as long as $m + n$ and $m - n$ are not zero. Since m and n are positive, $m + n \ne 0$. On the other hand, if $m - n = 0$, then $m = n$, and the integral must be evaluated in a different way.

In this case

$$\int_{-L}^{L} \sin \frac{m\pi x}{L} \sin \frac{n\pi x}{L} \, dx = \int_{-L}^{L} \left(\sin \frac{m\pi x}{L} \right)^2 dx$$

$$= \frac{1}{2} \int_{-L}^{L} \left[1 - \cos \frac{2m\pi x}{L} \right] dx$$

$$= \frac{1}{2} \left\{ x - \frac{\sin (2m\pi x/L)}{2m\pi/L} \right\} \Big|_{-L}^{L}$$

$$= L.$$

This establishes Eq. (8); Eqs. (6) and (7) can be verified by similar computations.

The Euler–Fourier Formulas

Now let us suppose that a series of the form (1) converges, and let us call its sum $f(x)$:

$$f(x) = \frac{a_0}{2} + \sum_{m=1}^{\infty} \left(a_m \cos \frac{m\pi x}{L} + b_m \sin \frac{m\pi x}{L} \right). \tag{9}$$

The coefficients a_m and b_m can be related to $f(x)$ as a consequence of the orthogonality conditions (6), (7), and (8). First multiply Eq. (9) by $\cos (n\pi x/L)$, where n is a *fixed* positive integer $(n > 0)$, and integrate with respect to x from $-L$ to L. Assuming that the integration can be legitimately carried out term by term,[1] we obtain

$$\int_{-L}^{L} f(x) \cos \frac{n\pi x}{L} \, dx = \frac{a_0}{2} \int_{-L}^{L} \cos \frac{n\pi x}{L} \, dx + \sum_{m=1}^{\infty} a_m \int_{-L}^{L} \cos \frac{m\pi x}{L} \cos \frac{n\pi x}{L} \, dx$$

$$+ \sum_{m=1}^{\infty} b_m \int_{-L}^{L} \sin \frac{m\pi x}{L} \cos \frac{n\pi x}{L} \, dx. \tag{10}$$

Keeping in mind that n is fixed whereas m ranges over the positive integers, it follows from the orthogonality relations (6) and (7) that the only nonzero term on the right side of Eq. (10) is the one for which $m = n$ in the first summation. Hence

$$\int_{-L}^{L} f(x) \cos \frac{n\pi x}{L} \, dx = La_n, \qquad n = 1, 2, \dots. \tag{11}$$

To determine a_0, we can integrate Eq. (9) from $-L$ to L, obtaining

$$\int_{-L}^{L} f(x) \, dx = \frac{a_0}{2} \int_{-L}^{L} dx + \sum_{m=1}^{\infty} a_m \int_{-L}^{L} \cos \frac{m\pi x}{L} \, dx + \sum_{m=1}^{\infty} b_m \int_{-L}^{L} \sin \frac{m\pi x}{L} \, dx$$

$$= La_0, \tag{12}$$

since each integral involving a trigonometric function is zero. Thus

$$a_n = \frac{1}{L} \int_{-L}^{L} f(x) \cos \frac{n\pi x}{L} \, dx, \qquad n = 0, 1, 2, \dots. \tag{13}$$

[1] This is a nontrivial assumption, since not all convergent series with variable terms can be so integrated. For the special case of Fourier series, however, term-by-term integration can always be justified.

By writing the constant term in Eq. (9) as $a_0/2$, it is possible to compute all the a_n from Eq. (13). Otherwise, a separate formula would have to be used for a_0.

A similar expression for b_n may be obtained by multiplying Eq. (9) by $\sin(n\pi x/L)$, integrating termwise from $-L$ to L, and using the orthogonality relations (7) and (8); thus

$$b_n = \frac{1}{L}\int_{-L}^{L} f(x)\sin\frac{n\pi x}{L}\,dx, \qquad n = 1, 2, 3, \ldots. \tag{14}$$

Equations (13) and (14) are known as the Euler–Fourier formulas for the coefficients in a Fourier series. Hence, if the series (9) converges to $f(x)$, and if the series can be integrated term by term, then the coefficients *must be given* by Eqs. (13) and (14).

Note that Eqs. (13) and (14) are explicit formulas for a_n and b_n in terms of f, and that the determination of any particular coefficient is independent of all the other coefficients. Of course, the difficulty in evaluating the integrals in Eqs. (13) and (14) depends very much on the particular function f involved.

Note also that the formulas (13) and (14) depend only on the values of $f(x)$ in the interval $-L \le x \le L$. Since each of the terms in the Fourier series (9) is periodic with period $2L$, the series converges for all x whenever it converges in $-L \le x \le L$, and its sum is also a periodic function with period $2L$. Hence $f(x)$ is determined for all x by its values in the interval $-L \le x \le L$.

It is possible to show (see Problem 27) that if g is periodic with period T, then every integral of g over an interval of length T has the same value. If we apply this result to the Euler–Fourier formulas (13) and (14), it follows that the interval of integration, $-L \le x \le L$, can be replaced, if it is more convenient to do so, by any other interval of length $2L$.

EXAMPLE 1

Assume that there is a Fourier series converging to the function f defined by

$$f(x) = \begin{cases} -x, & -2 \le x < 0, \\ x, & 0 \le x < 2; \end{cases} \tag{15}$$

$$f(x+4) = f(x).$$

Determine the coefficients in this Fourier series.

This function represents a triangular wave (see Figure 9.2.2) and is periodic with period 4. Thus in this case $L = 2$, and the Fourier series has the form

$$f(x) = \frac{a_0}{2} + \sum_{m=1}^{\infty}\left(a_m \cos\frac{m\pi x}{2} + b_m \sin\frac{m\pi x}{2}\right), \tag{16}$$

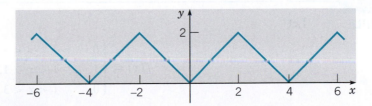

FIGURE 9.2.2 Triangular wave.

where the coefficients are computed from Eqs. (13) and (14) with $L = 2$. Substituting for $f(x)$ in Eq. (13) with $m = 0$, we have

$$a_0 = \frac{1}{2} \int_{-2}^{0} (-x)\, dx + \frac{1}{2} \int_{0}^{2} x\, dx = 1 + 1 = 2. \tag{17}$$

For $m > 0$, Eq. (13) yields

$$a_m = \frac{1}{2} \int_{-2}^{0} (-x) \cos \frac{m\pi x}{2}\, dx + \frac{1}{2} \int_{0}^{2} x \cos \frac{m\pi x}{2}\, dx.$$

These integrals can be evaluated through integration by parts, with the result that

$$
\begin{aligned}
a_m &= \frac{1}{2} \left[-\frac{2}{m\pi} x \sin \frac{m\pi x}{2} - \left(\frac{2}{m\pi} \right)^2 \cos \frac{m\pi x}{2} \right]\Bigg|_{-2}^{0} \\
&\quad + \frac{1}{2} \left[\frac{2}{m\pi} x \sin \frac{m\pi x}{2} + \left(\frac{2}{m\pi} \right)^2 \cos \frac{m\pi x}{2} \right]\Bigg|_{0}^{2} \\
&= \frac{1}{2} \left[-\left(\frac{2}{m\pi} \right)^2 + \left(\frac{2}{m\pi} \right)^2 \cos m\pi + \left(\frac{2}{m\pi} \right)^2 \cos m\pi - \left(\frac{2}{m\pi} \right)^2 \right] \\
&= \frac{4}{(m\pi)^2} (\cos m\pi - 1), \qquad m = 1, 2, \ldots \\
&= \begin{cases} -8/(m\pi)^2, & m \text{ odd}, \\ 0, & m \text{ even}. \end{cases}
\end{aligned}
\tag{18}
$$

Finally, from Eq. (14) it follows in a similar way that

$$b_m = 0, \qquad m = 1, 2, \ldots. \tag{19}$$

By substituting the coefficients from Eqs. (17), (18), and (19) in the series (16), we obtain the Fourier series for f:

$$
\begin{aligned}
f(x) &= 1 - \frac{8}{\pi^2} \left(\cos \frac{\pi x}{2} + \frac{1}{3^2} \cos \frac{3\pi x}{2} + \frac{1}{5^2} \cos \frac{5\pi x}{2} + \cdots \right) \\
&= 1 - \frac{8}{\pi^2} \sum_{m=1,3,5,\ldots}^{\infty} \frac{\cos (m\pi x / 2)}{m^2} \\
&= 1 - \frac{8}{\pi^2} \sum_{n=1}^{\infty} \frac{\cos (2n-1)\pi x / 2}{(2n-1)^2}.
\end{aligned}
\tag{20}
$$

EXAMPLE 2

Let

$$
f(x) = \begin{cases} 0, & -3 < x < -1, \\ 1, & -1 < x < 1, \\ 0, & 1 < x < 3 \end{cases}
\tag{21}
$$

and suppose that $f(x + 6) = f(x)$; see Figure 9.2.3. Find the coefficients in the Fourier series for f.

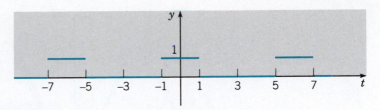

FIGURE 9.2.3   Graph of $f(x)$ in Example 2.

Note that $f(x)$ is not assigned a value at the points of discontinuity, such as $x = -1$ and $x = 1$. This has no effect on the values of the Fourier coefficients, because they result from the evaluation of integrals, and the value of an integral is not affected by the value of the integrand at a single point, or at a finite number of points. Thus the coefficients are the same regardless of what value, if any, $f(x)$ is assigned at a point of discontinuity.

Since f has period 6, it follows that $L = 3$ in this problem. Consequently, the Fourier series for f has the form

$$f(x) = \frac{a_0}{2} + \sum_{n=1}^{\infty}\left(a_n \cos\frac{n\pi x}{3} + b_n \sin\frac{n\pi x}{3}\right), \tag{22}$$

where the coefficients a_n and b_n are given by Eqs. (13) and (14) with $L = 3$. We have

$$a_0 = \frac{1}{3}\int_{-3}^{3} f(x)\,dx = \frac{1}{3}\int_{-1}^{1} dx = \frac{2}{3}. \tag{23}$$

Similarly,

$$a_n = \frac{1}{3}\int_{-1}^{1}\cos\frac{n\pi x}{3}\,dx = \frac{1}{n\pi}\sin\frac{n\pi x}{3}\Big|_{-1}^{1} = \frac{2}{n\pi}\sin\frac{n\pi}{3}, \quad n = 1, 2, \ldots, \tag{24}$$

and

$$b_n = \frac{1}{3}\int_{-1}^{1}\sin\frac{n\pi x}{3}\,dx = -\frac{1}{n\pi}\cos\frac{n\pi x}{3}\Big|_{-1}^{1} = 0, \quad n = 1, 2, \ldots. \tag{25}$$

Thus the Fourier series for f is

$$f(x) = \frac{1}{3} + \sum_{n=1}^{\infty}\frac{2}{n\pi}\sin\frac{n\pi}{3}\cos\frac{n\pi x}{3}$$

$$= \frac{1}{3} + \frac{\sqrt{3}}{\pi}\left[\cos(\pi x/3) + \frac{\cos(2\pi x/3)}{2} - \frac{\cos(4\pi x/3)}{4} - \frac{\cos(5\pi x/3)}{5} + \cdots\right]. \tag{26}$$

EXAMPLE 3

Consider again the function in Example 1 and its Fourier series (20). Investigate the speed with which the series converges. In particular, determine how many terms are needed so that the error is no greater than 0.01 for all x.

The mth partial sum in this series,

$$s_m(x) = 1 - \frac{8}{\pi^2}\sum_{n=1}^{m}\frac{\cos(2n-1)\pi x/2}{(2n-1)^2}, \tag{27}$$

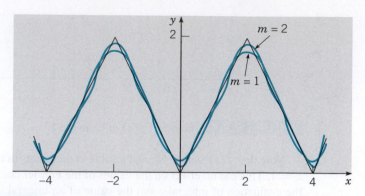

FIGURE 9.2.4   Partial sums in the Fourier series, Eq. (20), for the triangular wave.

can be used to approximate the function f. The coefficients diminish as $(2n - 1)^{-2}$, so the series converges fairly rapidly. This is borne out by Figure 9.2.4, where the partial sums for $m = 1$ and $m = 2$ are plotted. To investigate the convergence in more detail, we can consider the error $e_m(x) = f(x) - s_m(x)$. Figure 9.2.5 shows a plot of $|e_6(x)|$ versus x for $0 \leq x \leq 2$. Observe that $|e_6(x)|$ is greatest at the points $x = 0$ and $x = 2$ where the graph of $f(x)$ has corners. It is more difficult for the series to approximate the function near these points, resulting in a larger error there for a given m. Similar graphs are obtained for other values of m.

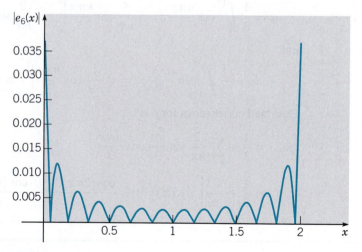

FIGURE 9.2.5   Plot of $|e_6(x)|$ versus x for the triangular wave.

Once you realize that the maximum error always occurs at $x = 0$ or $x = 2$, you can obtain a *uniform* error bound for each m simply by evaluating $|e_m(x)|$ at one of these points. For example, for $m = 6$ we have $e_6(2) = 0.03370$, so $|e_6(x)| < 0.034$ for $0 \leq x \leq 2$ and consequently for all x. Table 9.2.1 shows corresponding data for other values of m; these data are plotted in Figure 9.2.6. From this information you can begin to estimate the number of terms that are needed in the series in order to achieve a given level of accuracy in the approximation. For example, to guarantee that $|e_m(x)| \leq 0.01$, we need to choose $m = 21$.

TABLE 9.2.1	Values of the error $e_m(2)$ for the triangular wave.	

m	$e_m(2)$
2	0.09937
4	0.05040
6	0.03370
10	0.02025
15	0.01350
20	0.01013
25	0.00810

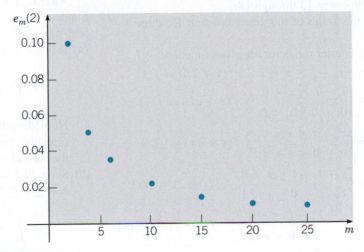

FIGURE 9.2.6 Plot of $e_m(2)$ versus m for the triangular wave.

In this book Fourier series appear mainly as a means of solving certain problems in partial differential equations. However, such series have much wider application in science and engineering and, in general, are valuable tools in the investigation of periodic phenomena. A basic problem is resolving an incoming signal into its harmonic components, which amounts to constructing its Fourier series representation. In some frequency ranges the separate terms correspond to different colors or to different audible tones. The magnitude of the coefficient determines the amplitude of each component. This process is referred to as spectral analysis.

PROBLEMS

In each of Problems 1 through 8 determine whether the given function is periodic. If so, find its fundamental period.

1. $\sin 5x$

2. $\cos 2\pi x$

3. $\sinh 2x$

4. $\sin \pi x / L$

5. $\tan \pi x$

6. x^2

7. $f(x) = \begin{cases} 0, & 2n-1 \le x < 2n, \\ 1, & 2n \le x < 2n+1; \end{cases}$ $n = 0, \pm 1, \pm 2, \ldots$

8. $f(x) = \begin{cases} (-1)^n, & 2n-1 \le x < 2n, \\ 1, & 2n \le x < 2n+1; \end{cases}$ $n = 0, \pm 1, \pm 2, \ldots$

9. If $f(x) = -x$ for $-L < x < L$, and if $f(x+2L) = f(x)$, find a formula for $f(x)$

in the interval $L < x < 2L$; in the interval $-3L < x < -2L$.

10. If $f(x) = \begin{cases} x+1, & -1 < x < 0, \\ x, & 0 < x < 1, \end{cases}$ and if $f(x+2) = f(x)$, find a formula for $f(x)$ in the interval $1 < x < 2$; in the interval $8 < x < 9$.

11. If $f(x) = L - x$ for $0 < x < 2L$, and if $f(x+2L) = f(x)$, find a formula for $f(x)$ in the interval $-L < x < 0$.

12. Verify Eqs. (6) and (7) in this section by direct integration.

In each of Problems 13 through 18:
(a) Sketch the graph of the given function for three periods.
(b) Find the Fourier series for the given function.

13. $f(x) = -x$, $-L \le x < L$; $f(x+2L) = f(x)$

14. $f(x) = \begin{cases} 1, & -L \le x < 0, \\ 0, & 0 \le x < L; \end{cases}$ $f(x+2L) = f(x)$

15. $f(x) = \begin{cases} x, & -\pi \le x < 0, \\ 0, & 0 \le x < \pi; \end{cases}$ $f(x+2\pi) = f(x)$

16. $f(x) = \begin{cases} x+1, & -1 \le x < 0, \\ 1-x, & 0 \le x < 1; \end{cases}$ $f(x+2) = f(x)$

17. $f(x) = \begin{cases} x+L, & -L \le x \le 0, \\ L, & 0 < x < L; \end{cases}$ $f(x+2L) = f(x)$

18. $f(x) = \begin{cases} 0, & -2 \le x \le -1, \\ x, & -1 < x < 1, \\ 0, & 1 \le x < 2; \end{cases}$ $f(x+4) = f(x)$

In each of Problems 19 through 24:
(a) Sketch the graph of the given function for three periods.
(b) Find the Fourier series for the given function.
(c) Plot $s_m(x)$ versus x for $m = 5$, 10, and 20.
(d) Describe how the Fourier series seems to be converging.

CAS 19. $f(x) = \begin{cases} -1, & -2 \le x < 0, \\ 1, & 0 \le x < 2; \end{cases}$ $f(x+4) = f(x)$

CAS 20. $f(x) = x$, $-1 \le x < 1$; $f(x+2) = f(x)$

CAS 21. $f(x) = x^2/2$, $-2 \le x \le 2$; $f(x+4) = f(x)$

CAS 22. $f(x) = \begin{cases} x+2, & -2 \le x < 0, \\ 2-2x, & 0 \le x < 2; \end{cases}$ $f(x+4) = f(x)$

CAS 23. $f(x) = \begin{cases} -\dfrac{1}{2}x, & -2 \le x < 0, \\ 2x - \dfrac{1}{2}x^2, & 0 \le x < 2; \end{cases}$ $f(x+4) = f(x)$

CAS 24. $f(x) = \begin{cases} 0, & -3 \le x \le 0, \\ x^2(3-x), & 0 < x < 3; \end{cases}$ $f(x+6) = f(x)$

25. Consider the function f defined in Problem 21 and let $e_m(x) = f(x) - s_m(x)$.
(a) Plot $|e_m(x)|$ versus x for $0 \le x \le 2$ for several values of m. CAS
(b) Find the smallest value of m for which $|e_m(x)| \le 0.01$ for all x.

26. Consider the function f defined in Problem 24 and let $e_m(x) = f(x) - s_m(x)$.
(a) Plot $|e_m(x)|$ versus x for $0 \le x \le 3$ for several values of m. CAS
(b) Find the smallest value of m for which $|e_m(x)| \le 0.1$ for all x.

27. Suppose that g is an integrable periodic function with period T.
(a) If $0 \le a \le T$, show that

$$\int_0^T g(x)\,dx = \int_a^{a+T} g(x)\,dx.$$

Hint: Show first that $\int_0^a g(x)\,dx = \int_T^{a+T} g(x)\,dx$. Consider the change of variable $s = x - T$ in the second integral.

(b) Show that for any value of a, not necessarily in $0 \le a \le T$,

$$\int_0^T g(x)\,dx = \int_a^{a+T} g(x)\,dx.$$

(c) Show that for any values of a and b,

$$\int_a^{a+T} g(x)\,dx = \int_b^{b+T} g(x)\,dx.$$

28. If f is differentiable and is periodic with period T, show that f' is also periodic with period T. Determine whether

$$F(x) = \int_0^x f(t)\,dt$$

is always periodic.

29. In this problem we indicate certain similarities between three-dimensional geometric vectors and Fourier series.
(a) Let v_1, v_2, and v_3 be a set of mutually orthogonal vectors in three dimensions, and let u be any three-dimensional vector. Show that

$$u = a_1 v_1 + a_2 v_2 + a_3 v_3, \qquad \text{(i)}$$

where

$$a_i = \frac{u \cdot v_i}{v_i \cdot v_i}, \qquad i = 1, 2, 3. \qquad \text{(ii)}$$

Show that a_i can be interpreted as the projection of $\mathbf{u}$ in the direction of $\mathbf{v}_i$ divided by the length of $\mathbf{v}_i$.

(b) Define the inner product (u, v) by

$$(u, v) = \int_{-L}^{L} u(x)v(x)\, dx. \qquad \text{(iii)}$$

Also let

$$\phi_n(x) = \cos(n\pi x/L), \quad n = 0, 1, 2, \ldots;$$

$$\psi_n(x) = \sin(n\pi x/L), \quad n = 1, 2, \ldots. \qquad \text{(iv)}$$

Show that Eq. (10) can be written in the form

$$(f, \phi_n) = \frac{a_0}{2}(\phi_0, \phi_n) + \sum_{m=1}^{\infty} a_m(\phi_m, \phi_n)$$

$$+ \sum_{m=1}^{\infty} b_m(\psi_m, \phi_n). \qquad \text{(v)}$$

(c) Use Eq. (v) and the corresponding equation for (f, ψ_n), together with the orthogonality

relations, to show that

$$a_n = \frac{(f, \phi_n)}{(\phi_n, \phi_n)}, \quad n = 0, 1, 2, \ldots;$$

$$b_n = \frac{(f, \psi_n)}{(\psi_n, \psi_n)}, \quad n = 1, 2, \ldots. \qquad \text{(vi)}$$

Note the resemblance between Eqs. (vi) and Eq. (ii). The functions ϕ_n and ψ_n play a role for functions similar to that of the orthogonal vectors $\mathbf{v}_1$, $\mathbf{v}_2$, and $\mathbf{v}_3$ in three-dimensional space. The coefficients a_n and b_n can be interpreted as projections of the function f onto the base functions ϕ_n and ψ_n.

Observe also that any vector in three dimensions can be expressed as a linear combination of three mutually orthogonal vectors. In a somewhat similar way, any sufficiently smooth function defined on $-L \leq x \leq L$ can be expressed as a linear combination of the mutually orthogonal functions $\cos(n\pi x/L)$ and $\sin(n\pi x/L)$, that is, as a Fourier series.

9.3 The Fourier Convergence Theorem

In the preceding section we showed that if the Fourier series

$$\frac{a_0}{2} + \sum_{m=1}^{\infty} \left(a_m \cos \frac{m\pi x}{L} + b_m \sin \frac{m\pi x}{L} \right) \qquad (1)$$

converges and thereby defines a function f, then f is periodic with period $2L$, and the coefficients a_m and b_m are related to $f(x)$ by the Euler–Fourier formulas:

$$a_m = \frac{1}{L} \int_{-L}^{L} f(x) \cos \frac{m\pi x}{L}\, dx, \quad m = 0, 1, 2, \ldots; \qquad (2)$$

$$b_m = \frac{1}{L} \int_{-L}^{L} f(x) \sin \frac{m\pi x}{L}\, dx, \quad m = 1, 2, \ldots. \qquad (3)$$

In this section we suppose that a function f is given. If this function is periodic with period $2L$ and integrable on the interval $[-L, L]$, then a set of coefficients a_m and b_m can be computed from Eqs. (2) and (3), and a series of the form (1) can be formally constructed. The question is whether this series converges for each value of x and, if so, whether its sum is $f(x)$. Examples have been discovered showing that the Fourier series corresponding to a function f may not converge to $f(x)$ or may even diverge. Functions whose Fourier series do not converge to the value of the function at isolated points are easily constructed, and

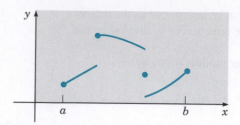

FIGURE 9.3.1 A piecewise continuous function.

examples will be presented later in this section. Functions whose Fourier series diverge at one or more points are more pathological, and we will not consider them in this book.

To guarantee convergence of a Fourier series to the function from which its coefficients were computed, it is essential to place additional conditions on the function. From a practical point of view, such conditions should be broad enough to cover all situations of interest, yet simple enough to be easily checked for particular functions. Through the years several sets of conditions have been devised to serve this purpose.

Before stating a convergence theorem for Fourier series, we define a term that appears in the theorem. A function f is said to be **piecewise continuous** on an interval $a \le x \le b$ if the interval can be partitioned by a finite number of points $a = x_0 < x_1 < \cdots < x_n = b$ so that

1. f is continuous on each open subinterval $x_{i-1} < x < x_i$.

2. f approaches a finite limit as the endpoints of each subinterval are approached from within the subinterval.

The graph of a piecewise continuous function is shown in Figure 9.3.1.

The notation $f(c+)$ is used to denote the limit of $f(x)$ as $x \to c$ from the right; similarly, $f(c-)$ denotes the limit of $f(x)$ as x approaches c from the left.

Note that it is not essential that the function even be defined at the partition points x_i. For example, in the following theorem we assume that f' is piecewise continuous; but certainly f' does not exist at those points where f itself is discontinuous. It is also not essential that the interval be closed; it may also be open, or open at one end and closed at the other.

THEOREM 9.3.1

Suppose that f and f' are piecewise continuous on the interval $-L \le x < L$. Further, suppose that f is defined outside the interval $-L \le x < L$ so that it is periodic with period $2L$. Then f has a Fourier series

$$f(x) = \frac{a_0}{2} + \sum_{m=1}^{\infty} \left(a_m \cos \frac{m\pi x}{L} + b_m \sin \frac{m\pi x}{L} \right), \tag{2}$$

whose coefficients are given by Eqs. (2) and (3). The Fourier series converges to $f(x)$ at all points where f is continuous, and to $[f(x+) + f(x-)]/2$ at all points where f is discontinuous.

Note that $[f(x+) + f(x-)]/2$ is the mean value of the right- and left-hand limits at the point x. At any point where f is continuous, $f(x+) = f(x-) = f(x)$. Thus it is correct to say that the Fourier series converges to $[f(x+) + f(x-)]/2$ at all points. Whenever we say that a Fourier series converges to a function f, we always mean that it converges in this sense.

It should be emphasized that the conditions given in this theorem are only sufficient for the convergence of a Fourier series; they are by no means necessary. Nor are they the most general sufficient conditions that have been discovered. In spite of this, the proof of the theorem is fairly difficult and we do not discuss it here.[2] Under more restrictive conditions a much simpler convergence proof is possible; see Problem 18.

To obtain a better understanding of the content of the theorem, it is helpful to consider some classes of functions that fail to satisfy the assumed conditions. Functions that are not included in the theorem are primarily those with infinite discontinuities in the interval $[-L, L]$, such as $1/x^2$ as $x \to 0$, or $\ln|x - L|$ as $x \to L$. Functions having an infinite number of jump discontinuities in this interval are also excluded; however, such functions are rarely encountered.

It is noteworthy that a Fourier series may converge to a sum that is not differentiable, or even continuous, in spite of the fact that each term in the series (4) is continuous, and even differentiable infinitely many times. The example below is an illustration of this, as is Example 2 in Section 9.2.

EXAMPLE 1

Let

$$f(x) = \begin{cases} 0, & -L < x < 0, \\ L, & 0 < x < L, \end{cases} \tag{3}$$

and let f be defined outside this interval so that $f(x + 2L) = f(x)$ for all x. We will temporarily leave open the definition of f at the points $x = 0, \pm L$. Find the Fourier series for this function and determine where it converges.

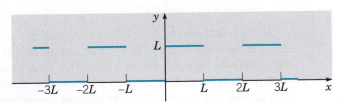

FIGURE 9.3.2   Square wave.

The equation $y = f(x)$ has the graph shown in Figure 9.3.2, extended to infinity in both directions. It can be thought of as representing a square wave. The interval $[-L, L]$ can be partitioned to give the two open subintervals $(-L, 0)$ and $(0, L)$. In $(0, L)$, $f(x) = L$ and $f'(x) = 0$. Clearly, both f and f' are continuous and furthermore have limits as $x \to 0$ from the right and as $x \to L$ from the left. The situation in $(-L, 0)$ is similar. Consequently, both f and f' are piecewise continuous on $[-L, L)$, so f satisfies the conditions of Theorem 9.3.1. If the coefficients a_m and b_m are computed from Eqs. (2) and (3), the convergence of the resulting Fourier series to $f(x)$ is ensured at all points where f is continuous. Note that the values of a_m and b_m are the same regardless of the definition of f at its points of discontinuity. This is true because the value of an integral is unaffected by changing the

---

[2]Proofs of the convergence of a Fourier series can be found in most books on advanced calculus. See, for example, Kaplan or Buck in the References.

value of the integrand at a finite number of points. From Eq. (2),

$$a_0 = \frac{1}{L} \int_{-L}^{L} f(x)\, dx = \int_0^L dx = L;$$

$$a_m = \frac{1}{L} \int_{-L}^{L} f(x) \cos \frac{m\pi x}{L}\, dx = \int_0^L \cos \frac{m\pi x}{L}\, dx$$

$$= 0, \quad m \neq 0.$$

Similarly, from Eq. (3),

$$b_m = \frac{1}{L} \int_{-L}^{L} f(x) \sin \frac{m\pi x}{L}\, dx = \int_0^L \sin \frac{m\pi x}{L}\, dx$$

$$= \frac{L}{m\pi}(1 - \cos m\pi)$$

$$= \begin{cases} 0, & m \text{ even}; \\ 2L/m\pi, & m \text{ odd}. \end{cases}$$

Hence

$$f(x) = \frac{L}{2} + \frac{2L}{\pi}\left(\sin \frac{\pi x}{L} + \frac{1}{3} \sin \frac{3\pi x}{L} + \frac{1}{5} \sin \frac{5\pi x}{L} + \cdots \right)$$

$$= \frac{L}{2} + \frac{2L}{\pi} \sum_{m=1,3,5,\ldots}^{\infty} \frac{\sin(m\pi x/L)}{m}$$

$$= \frac{L}{2} + \frac{2L}{\pi} \sum_{n=1}^{\infty} \frac{\sin(2n-1)\pi x/L}{2n-1}. \tag{4}$$

At the points $x = 0, \pm nL$, where the function f in the example is not continuous, all terms in the series after the first vanish and the sum is $L/2$. This is the mean value of the limits from the right and left, as it should be. Thus we might as well define f at these points to have the value $L/2$. If we choose to define it otherwise, the series still gives the value $L/2$ at these points, since all of the preceding calculations remain valid. The series simply does not converge to the function at those points unless f is defined to have the value $L/2$. This illustrates the possibility that the Fourier series corresponding to a function may not converge to it at points of discontinuity unless the function is suitably defined at such points.

The manner in which the partial sums

$$s_n(x) = \frac{L}{2} + \frac{2L}{\pi}\left(\sin \frac{\pi x}{L} + \cdots + \frac{1}{2n-1} \sin \frac{(2n-1)\pi x}{L} \right), \qquad n = 1, 2, \ldots$$

of the Fourier series (6) converge to $f(x)$ is indicated in Figure 9.3.3, where L has been chosen to be 1 and the graph of $s_8(x)$ is plotted. The figure suggests that at points where f is continuous the partial sums do approach $f(x)$ as n increases. However, in the neighborhood of points of discontinuity, such as $x = 0$ and $x = L$, the partial sums do not converge smoothly to the mean value. Instead they tend to overshoot the mark at each end of the jump, as though they cannot quite accommodate themselves to the sharp turn required at this point. This behavior is typical of Fourier series at points of discontinuity and is known as the **Gibbs phenomenon**.

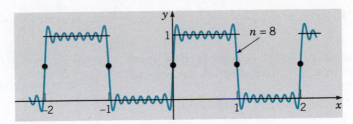

FIGURE 9.3.3 The partial sum $s_8(x)$ in the Fourier series, Eq. (6), for the square wave.

Additional insight is attained by considering the error $e_n(x) = f(x) - s_n(x)$. Figure 9.3.4 shows a plot of $|e_n(x)|$ versus x for $n = 8$ and for $L = 1$. The least upper bound of $|e_8(x)|$ is 0.5 and is approached as $x \to 0$ and as $x \to 1$. As n increases, the error decreases in the interior of the interval [where $f(x)$ is continuous], but the least upper bound does not diminish with increasing n. Thus we cannot uniformly reduce the error throughout the interval by increasing the number of terms.

Figures 9.3.3 and 9.3.4 also show that the series in this example converges more slowly than the one in Example 1 in Section 9.2. This is due to the fact that the coefficients in the series (6) are proportional only to $1/(2n - 1)$.

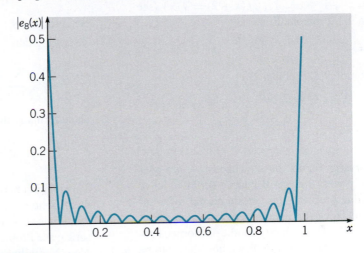

FIGURE 9.3.4 A plot of the error $|e_8(x)|$ versus x for the square wave.

PROBLEMS

In each of Problems 1 through 6 assume that the given function is periodically extended outside the original interval.

(a) Find the Fourier series for the extended function.
(b) Sketch the graph of the function to which the series converges for three periods.

1. $f(x) = \begin{cases} -1, & -1 \leq x < 0, \\ 1, & 0 \leq x < 1 \end{cases}$

2. $f(x) = \begin{cases} 0, & -\pi \leq x < 0, \\ x, & 0 \leq x < \pi \end{cases}$

3. $f(x) = \begin{cases} L + x, & -L \leq x < 0, \\ L - x, & 0 \leq x < L \end{cases}$

4. $f(x) = 1 - x^2, \quad -1 \leq x < 1$

5. $f(x) = \begin{cases} 0, & -\pi \leq x < -\pi/2, \\ 1, & -\pi/2 \leq x < \pi/2, \\ 0, & \pi/2 \leq x < \pi \end{cases}$

6. $f(x) = \begin{cases} 0, & -1 \leq x < 0, \\ x^2, & 0 \leq x < 1 \end{cases}$

In each of Problems 7 through 12 assume that the given function is periodically extended outside the original interval.

(a) Find the Fourier series for the given function.

(b) Let $e_n(x) = f(x) - s_n(x)$. Find the least upper bound or the maximum value (if it exists) of $|e_n(x)|$ for $n = 10, 20$, and 40.

(c) If possible, find the smallest n for which $|e_n(x)| \leq 0.01$ for all x.

CAS 7. $f(x) = \begin{cases} x, & -\pi \leq x < 0, \\ 0, & 0 \leq x < \pi; \end{cases}$ $\quad f(x + 2\pi) = f(x)$

(see Section 9.2, Problem 15)

CAS 8. $f(x) = \begin{cases} x + 1, & -1 \leq x < 0, \\ 1 - x, & 0 \leq x < 1; \end{cases}$ $\quad f(x + 2) = f(x)$

(see Section 9.2, Problem 16)

CAS 9. $f(x) = x, \ -1 \leq x < 1; \quad f(x + 2) = f(x)$ (see Section 9.2, Problem 20)

CAS 10. $f(x) = \begin{cases} x + 2, & -2 \leq x < 0, \\ 2 - 2x, & 0 \leq x < 2; \end{cases}$ $\quad f(x + 4) = f(x)$

(see Section 9.2, Problem 22)

CAS 11. $f(x) = \begin{cases} 0, & -1 \leq x < 0, \\ x^2, & 0 \leq x < 1; \end{cases}$ $\quad f(x + 2) = f(x)$

(see Problem 6)

CAS 12. $f(x) = x - x^3, \ -1 \leq x < 1; \quad f(x + 2) = f(x)$

Periodic Forcing Terms. In this chapter we are concerned mainly with the use of Fourier series to solve boundary value problems for certain partial differential equations. However, Fourier series are also useful in many other situations where periodic phenomena occur. Problems 13 through 16 indicate how they can be employed to solve initial value problems with periodic forcing terms.

13. Find the solution of the initial value problem

$$y'' + \omega^2 y = \sin nt, \qquad y(0) = 0, \quad y'(0) = 0,$$

where n is a positive integer and $\omega^2 \neq n^2$. What happens if $\omega^2 = n^2$?

14. Find the formal solution of the initial value problem

$$y'' + \omega^2 y = \sum_{n=1}^{\infty} b_n \sin nt, \qquad y(0) = 0, \quad y'(0) = 0,$$

where $\omega > 0$ is not equal to a positive integer. How is the solution altered if $\omega = m$, where m is a positive integer?

15. Find the formal solution of the initial value problem

$$y'' + \omega^2 y = f(t), \qquad y(0) = 0, \quad y'(0) = 0,$$

where f is periodic with period 2π and

$$f(t) = \begin{cases} 1, & 0 < t < \pi; \\ 0, & t = 0, \pi, 2\pi; \\ -1, & \pi < t < 2\pi. \end{cases}$$

See Problem 1.

16. Find the formal solution of the initial value problem

$$y'' + \omega^2 y = f(t), \qquad y(0) = 1, \quad y'(0) = 0,$$

where f is periodic with period 2 and

$$f(t) = \begin{cases} 1 - t, & 0 \leq t < 1; \\ -1 + t, & 1 \leq t < 2. \end{cases}$$

See Problem 8.

17. Assuming that

$$f(x) = \frac{a_0}{2} + \sum_{n=1}^{\infty} \left(a_n \cos \frac{n\pi x}{L} + b_n \sin \frac{n\pi x}{L} \right), \tag{i}$$

show formally that

$$\frac{1}{L} \int_{-L}^{L} [f(x)]^2 \, dx = \frac{a_0^2}{2} + \sum_{n=1}^{\infty} (a_n^2 + b_n^2).$$

This relation between a function f and its Fourier coefficients is known as **Parseval's equation**. This relation is very important in the theory of Fourier series; see Problem 9 in Section 10.6.

Hint: Multiply Eq. (i) by $f(x)$, integrate from $-L$ to L, and use the Euler–Fourier formulas.

18. This problem indicates a proof of convergence of a Fourier series under conditions more restrictive than those in Theorem 9.3.1.

(a) If f and f' are piecewise continuous on $-L \leq x < L$, and if f is periodic with period $2L$, show that na_n and nb_n are bounded as $n \to \infty$.

Hint: Use integration by parts.

(b) If f is continuous on $-L \leq x \leq L$ and periodic with period $2L$, and if f' and f'' are piecewise continuous on $-L \leq x < L$, show that $n^2 a_n$ and $n^2 b_n$ are bounded as $n \to \infty$. If f is continuous on the *closed* interval, then it is continuous for all x. Why is this important?

Hint: Again, use integration by parts.

(c) Using the result of part (b), show that $\sum_{n=1}^{\infty}|a_n|$

and $\sum_{n=1}^{\infty}|b_n|$ converge.

(d) From the result in part (c), show that the Fourier series (4) converges absolutely[3] for all x.

Acceleration of Convergence. In the next problem we show how it is sometimes possible to improve the speed of convergence of a Fourier series.

19. Suppose that we wish to calculate values of the function g, where

$$g(x) = \sum_{n=1}^{\infty} \frac{(2n-1)}{1+(2n-1)^2} \sin(2n-1)\pi x. \qquad \text{(i)}$$

It is possible to show that this series converges, albeit rather slowly. However, observe that for large n the terms in the series (i) are approximately

equal to $[\sin(2n-1)\pi x]/(2n-1)$ and that the latter terms are similar to those in the example in the text, Eq. (6).

(a) Show that

$$\sum_{n=1}^{\infty} [\sin(2n-1)\pi x]/(2n-1) = (\pi/2)\left[f(x) - \frac{1}{2}\right], \qquad \text{(ii)}$$

where f is the square wave in the example with $L=1$.

(b) Subtract Eq. (ii) from Eq. (i) and show that

$$g(x) = \frac{\pi}{2}\left[f(x) - \frac{1}{2}\right] - \sum_{n=1}^{\infty} \frac{\sin(2n-1)\pi x}{(2n-1)[1+(2n-1)^2]}. \qquad \text{(iii)}$$

The series (iii) converges much faster than the series (i) and thus provides a better way to calculate values of $g(x)$.

9.4 Even and Odd Functions

Before looking at further examples of Fourier series, it is useful to distinguish two classes of functions for which the Euler–Fourier formulas can be simplified. These are even and odd functions, which are characterized geometrically by the property of symmetry with respect to the y-axis and the origin, respectively (see Figure 9.4.1).

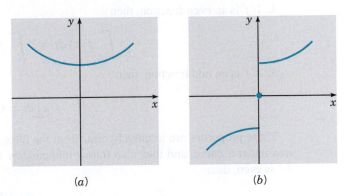

(a)   (b)

FIGURE 9.4.1   (a) An even function. (b) An odd function.

Analytically, f is an **even function** if its domain contains the point $-x$ whenever it contains the point x, and if

$$f(-x) = f(x) \qquad (1)$$

---

[3] It also converges uniformly; for an explanation of what this means, see a book on advanced calculus or analysis.

for each x in the domain of f. Similarly, f is an **odd function** if its domain contains $-x$ whenever it contains x, and if

$$f(-x) = -f(x) \tag{2}$$

for each x in the domain of f. Examples of even functions are 1, x^2, $\cos nx$, $|x|$, and x^{2n}. The functions x, x^3, $\sin nx$, and x^{2n+1} are examples of odd functions. Note that according to Eq. (2), $f(0)$ must be zero if f is an odd function whose domain contains the origin. Most functions are neither even nor odd, for instance, e^x. Only one function, f identically zero, is both even and odd.

Elementary properties of even and odd functions include the following:

1. The sum (difference) and product (quotient) of two even functions are even.
2. The sum (difference) of two odd functions is odd; the product (quotient) of two odd functions is even.
3. The sum (difference) of an odd function and an even function is neither even nor odd; the product (quotient) of two such functions is odd.[4]

The proofs of all these assertions are simple and follow directly from the definitions. For example, if both f_1 and f_2 are odd, and if $g(x) = f_1(x) + f_2(x)$, then

$$g(-x) = f_1(-x) + f_2(-x) = -f_1(x) - f_2(x)$$
$$= -[f_1(x) + f_2(x)] = -g(x), \tag{3}$$

so $f_1 + f_2$ is an odd function also. Similarly, if $h(x) = f_1(x)f_2(x)$, then

$$h(-x) = f_1(-x)f_2(-x) = [-f_1(x)][-f_2(x)] = f_1(x)f_2(x) = h(x), \tag{4}$$

so that $f_1 f_2$ is even.

Also of importance are the following two integral properties of even and odd functions:

4. If f is an even function, then

$$\int_{-L}^{L} f(x)\, dx = 2 \int_{0}^{L} f(x)\, dx. \tag{5}$$

5. If f is an odd function, then

$$\int_{-L}^{L} f(x)\, dx = 0. \tag{6}$$

These properties are intuitively clear from the interpretation of an integral in terms of area under a curve, and they also follow immediately from the definitions. For example, if f is even, then

$$\int_{-L}^{L} f(x)\, dx = \int_{-L}^{0} f(x)\, dx + \int_{0}^{L} f(x)\, dx.$$

Letting $x = -s$ in the first term on the right side and using Eq. (1), we obtain

$$\int_{-L}^{L} f(x)\, dx = -\int_{L}^{0} f(s)\, ds + \int_{0}^{L} f(x)\, dx = 2 \int_{0}^{L} f(x)\, dx.$$

The proof of the corresponding property for odd functions is similar.

Even and odd functions are particularly important in applications of Fourier series since their Fourier series have special forms, which occur frequently in physical problems.

---

[4]These statements may need to be modified if either function vanishes identically.

Cosine Series

Suppose that f and f' are piecewise continuous on $-L \le x < L$ and that f is an even periodic function with period $2L$. Then it follows from properties 1 and 3 that $f(x) \cos(n\pi x/L)$ is even and $f(x) \sin(n\pi x/L)$ is odd. As a consequence of Eqs. (5) and (6), the Fourier coefficients of f are then given by

$$
\begin{aligned}
a_n &= \frac{2}{L} \int_0^L f(x) \cos \frac{n\pi x}{L}\, dx, \qquad n = 0, 1, 2, \ldots; \\
b_n &= 0, \qquad n = 1, 2, \ldots.
\end{aligned}
\tag{7}
$$

Thus f has the Fourier series

$$
f(x) = \frac{a_0}{2} + \sum_{n=1}^{\infty} a_n \cos \frac{n\pi x}{L}.
$$

In other words, the Fourier series of any even function consists only of the even trigonometric functions $\cos(n\pi x/L)$ and the constant term; it is natural to call such a series a **Fourier cosine series**. From a computational point of view, observe that only the coefficients a_n, for $n = 0, 1, 2, \ldots$, need to be calculated from the integral formula (7). Each of the b_n, for $n = 1, 2, \ldots$, is automatically zero for any even function and so does not need to be calculated by integration.

Sine Series

Suppose that f and f' are piecewise continuous on $-L \le x < L$ and that f is an odd periodic function of period $2L$. Then it follows from properties 2 and 3 that $f(x) \cos(n\pi x/L)$ is odd and $f(x) \sin(n\pi x/L)$ is even. In this case the Fourier coefficients of f are

$$
\begin{aligned}
a_n &= 0, \qquad n = 0, 1, 2, \ldots, \\
b_n &= \frac{2}{L} \int_0^L f(x) \sin \frac{n\pi x}{L}\, dx, \qquad n = 1, 2, \ldots,
\end{aligned}
\tag{8}
$$

and the Fourier series for f is of the form

$$
f(x) = \sum_{n=1}^{\infty} b_n \sin \frac{n\pi x}{L}.
$$

Thus the Fourier series for any odd function consists only of the odd trigonometric functions $\sin(n\pi x/L)$; such a series is called a **Fourier sine series**. Again observe that only half of the coefficients need to be calculated by integration, since each a_n, for $n = 0, 1, 2, \ldots$, is zero for any odd function.

EXAMPLE 1

Let $f(x) = x$, $-L < x < L$, and let $f(-L) = f(L) = 0$. Let f be defined elsewhere so that it is periodic of period $2L$ (see Figure 9.4.2). The function defined in this manner is known as a sawtooth wave. Find the Fourier series for this function.

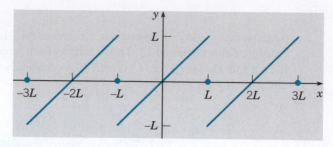

FIGURE 9.4.2 Sawtooth wave.

Since f is an odd function, its Fourier coefficients are, according to Eq. (8),

$$a_n = 0, \qquad n = 0, 1, 2, \ldots;$$

$$b_n = \frac{2}{L} \int_0^L x \sin \frac{n\pi x}{L} \, dx$$

$$= \frac{2}{L} \left(\frac{L}{n\pi} \right)^2 \left\{ \sin \frac{n\pi x}{L} - \frac{n\pi x}{L} \cos \frac{n\pi x}{L} \right\} \Bigg|_0^L$$

$$= \frac{2L}{n\pi} (-1)^{n+1}, \qquad n = 1, 2, \ldots.$$

Hence the Fourier series for f, the sawtooth wave, is

$$f(x) = \frac{2L}{\pi} \sum_{n=1}^{\infty} \frac{(-1)^{n+1}}{n} \sin \frac{n\pi x}{L}. \tag{9}$$

Observe that the periodic function f is discontinuous at the points $\pm L, \pm 3L, \ldots$, as shown in Figure 9.4.2. At these points the series (9) converges to the mean value of the left and right limits, namely, zero. The partial sum of the series (9) for $n = 9$ is shown in Figure 9.4.3. The Gibbs phenomenon (mentioned in Section 9.3) again occurs near the points of discontinuity.

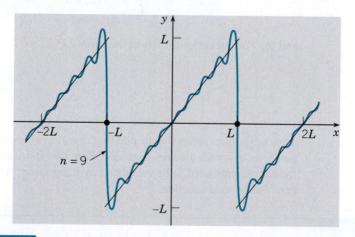

FIGURE 9.4.3 A partial sum in the Fourier series, Eq. (9), for the sawtooth wave.

Note that in this example $f(-L) = f(L) = 0$, as well as $f(0) = 0$. This is required if the function f is to be both odd and periodic with period $2L$. When we speak of constructing a sine series for a function defined on $0 \leq x \leq L$, it is understood that, if necessary, we must first redefine the function to be zero at $x = 0$ and $x = L$.

It is worthwhile to observe that the triangular wave function (Example 1 of Section 9.2) and the sawtooth wave function just considered are identical on the interval $0 \leq x < L$. Therefore, their Fourier series converge to the same function, $f(x) = x$, on this interval. Thus, if it is required to represent the function $f(x) = x$ on $0 \leq x < L$ by a Fourier series, it is possible to do this by *either a cosine series or a sine series*. In the former case f is extended as an *even* function into the interval $-L < x < 0$ and elsewhere periodically (the triangular wave). In the latter case f is extended into $-L < x < 0$ as an *odd* function and elsewhere periodically (the sawtooth wave). If f is extended in any other way, the resulting Fourier series will still converge to x in $0 \leq x < L$ but will involve both sine and cosine terms.

In solving problems in differential equations, it is often useful to expand in a Fourier series of period $2L$ a function f originally defined only on the interval $[0, L]$. As indicated previously for the function $f(x) = x$, several alternatives are available. Explicitly, we can

1. Define a function g of period $2L$ so that

$$g(x) = \begin{cases} f(x), & 0 \leq x \leq L, \\ f(-x), & -L < x < 0. \end{cases} \tag{10}$$

The function g is thus the **even periodic extension** of f. Its Fourier series, which is a cosine series, represents f on $[0, L]$.

2. Define a function h of period $2L$ so that

$$h(x) = \begin{cases} f(x), & 0 < x < L, \\ 0, & x = 0, L, \\ -f(-x), & -L < x < 0. \end{cases} \tag{11}$$

The function h is thus the **odd periodic extension** of f. Its Fourier series, which is a sine series, also represents f on $(0, L)$.

3. Define a function k of period $2L$ so that

$$k(x) = f(x), \qquad 0 \leq x \leq L, \tag{12}$$

and let $k(x)$ be defined for $(-L, 0)$ in any way consistent with the conditions of Theorem 9.3.1. Sometimes it is convenient to define $k(x)$ to be zero for $-L < x < 0$. The Fourier series for k, which involves both sine and cosine terms, also represents f on $[0, L]$, regardless of the manner in which $k(x)$ is defined in $(-L, 0)$. Thus there are infinitely many such series, all of which converge to $f(x)$ in the original interval.

Usually, the form of the expansion to be used will be dictated (or at least suggested) by the purpose for which it is needed. However, if there is a choice as to the kind of Fourier series to be used, the selection can sometimes be based on the rapidity of convergence. For example, the cosine series for the triangular wave [Eq. (20) of Section 9.2] converges more rapidly than the sine series for the sawtooth wave [Eq. (9) in this section], although both converge to the same function for $0 \leq x < L$. This is because the triangular wave is a smoother function than the sawtooth wave and is therefore easier to approximate.

In general, the more continuous derivatives possessed by a function over the entire interval $-\infty < x < \infty$, the faster its Fourier series will converge. See Problem 18 of Section 9.3.

EXAMPLE 2

Suppose that

$$f(x) = \begin{cases} 1 - x, & 0 < x \le 1, \\ 0, & 1 < x \le 2. \end{cases} \qquad (13)$$

As indicated previously, we can represent f either by a cosine series or by a sine series. Sketch the graph of the sum of each of these series for $-6 \le x \le 6$.

In this example $L = 2$, so the cosine series for f converges to the even periodic extension of f of period 4, whose graph is sketched in Figure 9.4.4.

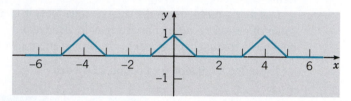

FIGURE 9.4.4   Even periodic extension of $f(x)$ given by Eq. (13).

Similarly, the sine series for f converges to the odd periodic extension of f of period 4. The graph of this function is shown in Figure 9.4.5.

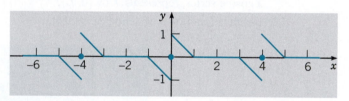

FIGURE 9.4.5   Odd periodic extension of $f(x)$ given by Eq. (13).

PROBLEMS

In each of Problems 1 through 6 determine whether the given function is even, odd, or neither.

1. $x^3 - 2x$

2. $x^3 - 2x + 1$

3. $\tan 2x$

4. $\sec x$

5. $|x|^3$

6. e^{-x}

In each of Problems 7 through 12 a function f is given on an interval of length L. In each case sketch the graphs of the even and odd extensions of f of period $2L$.

7. $f(x) = \begin{cases} x, & 0 \le x < 2, \\ 1, & 2 \le x < 3 \end{cases}$

8. $f(x) = \begin{cases} 0, & 0 \le x < 1, \\ x - 1, & 1 \le x < 2 \end{cases}$

9. $f(x) = 2 - x, \quad 0 < x < 2$

10. $f(x) = x - 3, \quad 0 < x < 4$

11. $f(x) = \begin{cases} 0, & 0 \le x < 1, \\ 1, & 1 \le x < 2 \end{cases}$

12. $f(x) = 4 - x^2, \quad 0 < x < 1$

13. Prove that any function can be expressed as the sum of two other functions, one of which is even and the other odd. That is, for any function f,

whose domain contains $-x$ whenever it contains x, show that there are an even function g and an odd function h such that $f(x) = g(x) + h(x)$. *Hint:* What can you say about $f(x) + f(-x)$?

14. Find the coefficients in the cosine and sine series described in Example 2.

In each of Problems 15 through 22:
(a) Find the required Fourier series for the given function.
(b) Sketch the graph of the function to which the series converges over three periods.

15. $f(x) = \begin{cases} 1, & 0 < x < 1, \\ 0, & 1 < x < 2; \end{cases}$ cosine series, period 4
Compare with Example 1 and Problem 5 of Section 9.3.

16. $f(x) = \begin{cases} x, & 0 \le x < 1, \\ 1, & 1 \le x < 2; \end{cases}$ sine series, period 4

17. $f(x) = 1$, $0 \le x \le \pi$; cosine series, period 2π

18. $f(x) = 1$, $0 < x < \pi$; sine series, period 2π

19. $f(x) = \begin{cases} 0, & 0 < x < \pi, \\ 1, & \pi < x < 2\pi, \\ 2, & 2\pi < x < 3\pi; \end{cases}$ sine series, period 6π

20. $f(x) = x$, $0 \le x < 1$; series of period 1

21. $f(x) = L - x$, $0 \le x \le L$; cosine series, period $2L$
Compare with Example 1 of Section 9.2.

22. $f(x) = L - x$, $0 < x < L$; sine series, period $2L$

In each of Problems 23 through 26:
(a) Find the required Fourier series for the given function.
(b) Sketch the graph of the function to which the series converges for three periods.
(c) Plot one or more partial sums of the series.

CAS 23. $f(x) = \begin{cases} x, & 0 < x < \pi, \\ 0, & \pi < x < 2\pi; \end{cases}$ cosine series, period 4π

CAS 24. $f(x) = -x$, $-\pi < x < 0$; sine series, period 2π

CAS 25. $f(x) = 2 - x^2$, $0 < x < 2$; sine series, period 4

CAS 26. $f(x) = x^2 - 2x$, $0 < x < 4$; cosine series, period 8

In each of Problems 27 through 30 a function is given on an interval $0 < x < L$.
(a) Sketch the graphs of the even extension $g(x)$ and the odd extension $h(x)$ of the given function of period $2L$ over three periods.

(b) Find the Fourier cosine and sine series for the given function.
(c) Plot a few partial sums of each series.
(d) For each series investigate the dependence on n of the maximum error on $[0, L]$.

27. $f(x) = 3 - x$, $0 < x < 3$ CAS

28. $f(x) = \begin{cases} x, & 0 < x < 1, \\ 0, & 1 < x < 2 \end{cases}$ CAS

29. $f(x) = (4x^2 - 4x - 3)/4$, $0 < x < 2$ CAS

30. $f(x) = x^3 - 5x^2 + 5x + 1$, $0 < x < 3$ CAS

31. Prove that if f is an odd function, then

$$\int_{-L}^{L} f(x)\, dx = 0.$$

32. Prove properties 2 and 3 of even and odd functions, as stated in the text.

33. Prove that the derivative of an even function is odd and that the derivative of an odd function is even.

34. Let $F(x) = \int_0^x f(t)\, dt$. Show that if f is even, then F is odd, and that if f is odd, then F is even.

35. From the Fourier series for the square wave in Example 1 of Section 9.3, show that

$$\frac{\pi}{4} = 1 - \frac{1}{3} + \frac{1}{5} - \frac{1}{7} + \cdots = \sum_{n=0}^{\infty} \frac{(-1)^n}{2n + 1}.$$

This relation between π and the odd positive integers was discovered by Leibniz in 1674.

36. From the Fourier series for the triangular wave (Example 1 of Section 9.2), show that

$$\frac{\pi^2}{8} = 1 + \frac{1}{3^2} + \frac{1}{5^2} + \cdots = \sum_{n=0}^{\infty} \frac{1}{(2n + 1)^2}.$$

37. Assume that f has a Fourier sine series

$$f(x) = \sum_{n=1}^{\infty} b_n \sin(n\pi x/L), \qquad 0 \le x \le L.$$

(a) Show formally that

$$\frac{2}{L} \int_0^L [f(x)]^2\, dx = \sum_{n=1}^{\infty} b_n^2.$$

Compare this result (Parseval's equation) with that of Problem 17 in Section 9.3. What is the corresponding result if f has a cosine series?

(b) Apply the result of part (a) to the series for the sawtooth wave given in Eq. (9), and thereby show that

$$\frac{\pi^2}{6} = 1 + \frac{1}{2^2} + \frac{1}{3^2} + \cdots = \sum_{n=1}^{\infty} \frac{1}{n^2}.$$

This relation was discovered by Euler about 1735.
More Specialized Fourier Series. Let f be a function originally defined on $0 \le x \le L$ and satisfying there the continuity conditions of Theorem 9.3.1. In this section we have shown that it is possible to represent f by either a sine series or a cosine series by constructing odd or even periodic extensions of f, respectively. Problems 38 through 40 concern some other, more specialized Fourier series that converge to the given function f on $(0, L)$.

38. Let f be extended into $(L, 2L]$ in an arbitrary manner. Then extend the resulting function into $(-2L, 0)$ as an odd function and elsewhere as a periodic function of period $4L$ (see Figure 9.4.6). Show that this function has a Fourier sine series in terms of the functions $\sin(n\pi x/2L)$, $n = 1, 2, 3, \ldots$; that is,

$$f(x) = \sum_{n=1}^{\infty} b_n \sin(n\pi x/2L),$$

where

$$b_n = \frac{1}{L} \int_0^{2L} f(x) \sin(n\pi x/2L)\, dx.$$

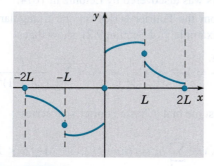

FIGURE 9.4.6 Graph of the function in Problem 38.

This series converges to the original function on $(0, L)$.

39. Let f first be extended into $(L, 2L)$ so that it is symmetric about $x = L$; that is, so as to satisfy $f(2L - x) = f(x)$ for $0 \le x < L$. Let the resulting function be extended into $(-2L, 0)$ as an odd function and elsewhere as a periodic function of period $4L$ (see Figure 9.4.7). Show that this function has a Fourier series in terms of the functions $\sin(\pi x/2L)$, $\sin(3\pi x/2L)$, $\sin(5\pi x/2L), \ldots$; that is,

$$f(x) = \sum_{n=1}^{\infty} b_n \sin\frac{(2n-1)\pi x}{2L},$$

where

$$b_n = \frac{2}{L} \int_0^L f(x) \sin\frac{(2n-1)\pi x}{2L}\, dx.$$

This series converges to the original function on $(0, L]$.

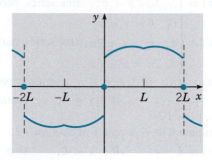

FIGURE 9.4.7 Graph of the function in Problem 39.

40. (a) How should f, originally defined on $[0, L]$, be extended so as to obtain a Fourier series involving only the functions $\cos(\pi x/2L)$, $\cos(3\pi x/2L)$, $\cos(5\pi x/2L), \ldots$? Refer to Problems 38 and 39.

(b) If $f(x) = x$ for $0 \le x \le L$, sketch the function to which the Fourier series converges for $-4L \le x \le 4L$.

9.5 Separation of Variables; Heat Conduction in a Rod

The basic partial differential equations of heat conduction, wave propagation, and potential theory that we discuss in this chapter are associated with three distinct types of physical phenomena: diffusive processes, oscillatory processes, and time-independent or steady processes. Consequently, they are of fundamental importance in many branches of physics. They are also of considerable significance from a mathematical point of view. The partial differential equations whose theory is best developed and whose applications are most significant and varied are the linear equations of second order. All such equations can be classified into one of three categories: the heat conduction equation, the wave equation, and the potential equation, respectively, are prototypes of each of these categories. Thus a study of these three equations yields much information about more general second order linear partial differential equations.

During the last two centuries several methods have been developed for solving partial differential equations. The method of separation of variables is the oldest systematic method, having been used by D'Alembert, Daniel Bernoulli, and Euler about 1750 in their investigations of waves and vibrations. It has been considerably refined and generalized in the meantime, and it remains a method of great importance and frequent use today. To show how the method of separation of variables works, we consider first a basic problem of heat conduction in a solid body. The mathematical study of heat conduction originated about 1800, and it continues to command the attention of modern scientists. For example, analysis of the dissipation and transfer of heat away from its sources in high-speed machinery is frequently an important technological problem.

Let us now consider a heat conduction problem for a straight bar of uniform cross section and homogeneous material. Let the x-axis be chosen to lie along the axis of the bar, and let $x = 0$ and $x = L$ denote the ends of the bar (see Figure 9.5.1). Suppose further that the sides of the bar are perfectly insulated so that no heat passes through them. We also assume that the cross-sectional dimensions are so small that the temperature u can be considered constant on any given cross section. Then u is a function only of the axial coordinate x and the time t.

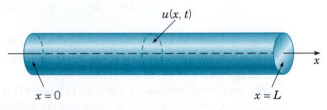

FIGURE 9.5.1   A heat-conducting solid bar.

▶ **The Heat Equation.**   The variation of temperature in the bar is governed by a partial differential equation whose derivation appears in Appendix 9.A at the end of this chapter. The equation is called the **heat conduction equation** and has the form

$$\alpha^2 u_{xx} = u_t, \qquad 0 < x < L, \quad t > 0, \tag{1}$$

where α^2 is a constant known as the **thermal diffusivity**. The parameter α^2 depends only on the material from which the bar is made and is defined by

$$\alpha^2 = \kappa/\rho s, \tag{2}$$

where κ is the thermal conductivity, ρ is the density, and s is the specific heat of the material in the bar. The units of α^2 are (length)^2/time. Typical values of α^2 are given in Table 9.5.1.

TABLE 9.5.1	Values of the thermal diffusivity for some common materials.

Material	$\alpha^2(\text{cm}^2/\text{s})$
Silver	1.71
Copper	1.14
Aluminum	0.86
Cast iron	0.12
Granite	0.011
Brick	0.0038
Water	0.00144

▶ **Initial Condition.** In addition, we assume that the initial temperature distribution in the bar is given; thus

$$u(x, 0) = f(x), \qquad 0 \le x \le L, \tag{3}$$

where f is a given function.

▶ **Boundary Conditions.** Finally, we assume that the ends of the bar are held at fixed temperatures: the temperature T_1 at $x = 0$ and the temperature T_2 at $x = L$. However, it turns out that we need only consider the case where $T_1 = T_2 = 0$. We show in Section 9.6 how to reduce the more general problem to this special case. Thus in this section we will assume that u is always zero when $x = 0$ or $x = L$:

$$u(0, t) = 0, \qquad u(L, t) = 0, \quad t > 0. \tag{4}$$

The fundamental problem of heat conduction is to find $u(x, t)$ that satisfies the differential equation (1) for $0 < x < L$ and for $t > 0$, the initial condition (3) when $t = 0$, and the boundary conditions (4) at $x = 0$ and $x = L$.

The problem described by Eqs. (1), (3), and (4) is an initial value problem in the time variable t; an initial condition is given and the differential equation governs what happens later. However, with respect to the space variable x, the problem is a boundary value problem; boundary conditions are imposed at each end of the bar and the differential equation describes the evolution of the temperature in the interval between them. Alternatively, we can consider the problem as a boundary value problem in the xt-plane (see Figure 9.5.2). The solution $u(x, t)$ of Eq. (1) is sought in the semi-infinite strip $0 < x < L, t > 0$, subject to the requirement that $u(x, t)$ must assume a prescribed value at each point on the boundary of this strip.

The Method of Separation of Variables

The heat conduction problem (1), (3), (4) is *linear* since u appears only to the first power throughout. The differential equation and boundary conditions are also *homogeneous*. This suggests that we might approach the problem by seeking solutions of the differential equation and boundary conditions, and then superposing them to satisfy the initial condition. The remainder of this section describes how this plan can be implemented.

One solution of the differential equation (1) that satisfies the boundary conditions (4) is the function $u(x, t) = 0$, but this solution does not satisfy the initial condition (3) except in

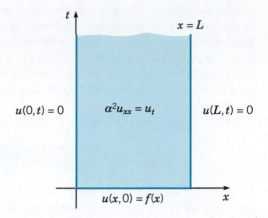

Boundary value problem for the heat conduction equation.

the trivial case in which $f(x)$ is also zero. Thus our goal is to find other, nonzero solutions of the differential equation and boundary conditions. To find the needed solutions, we start by making a basic assumption about the form of the solutions that has far-reaching, and perhaps unforeseen, consequences. The assumption is that $u(x, t)$ is a product of two functions, one depending only on x and the other depending only on t; thus

$$u(x, t) = X(x)T(t). \tag{5}$$

Substituting from Eq. (5) for u in the differential equation (1) yields

$$\alpha^2 X''T = XT', \tag{6}$$

where primes refer to ordinary differentiation with respect to the independent variable, whether x or t. Equation (6) is equivalent to

$$\frac{X''}{X} = \frac{1}{\alpha^2}\frac{T'}{T}, \tag{7}$$

in which the variables are separated; that is, the left side depends only on x and the right side only on t.

It is now crucial to realize that for Eq. (7) to be valid for $0 < x < L, t > 0$, it is necessary that both sides of Eq. (7) must be equal to the same constant. Otherwise, if one independent variable (say, x) were kept fixed and the other were allowed to vary, one side (the left in this case) of Eq. (7) would remain unchanged while the other varied, thus violating the equality. If we call this separation constant $-\lambda$, then Eq. (7) becomes

$$\frac{X''}{X} = \frac{1}{\alpha^2}\frac{T'}{T} = -\lambda. \tag{8}$$

Hence we obtain the following two ordinary differential equations for $X(x)$ and $T(t)$:

$$X'' + \lambda X = 0, \tag{9}$$
$$T' + \alpha^2 \lambda T = 0. \tag{10}$$

We denote the separation constant by $-\lambda$ (rather than λ) because it turns out that it must be negative, and it is convenient to exhibit the minus sign explicitly.

The assumption (5) has led to the replacement of the partial differential equation (1) by the two ordinary differential equations (9) and (10). Each of these equations is linear and homogeneous, with constant coefficients, and so can be readily solved for *any* value of λ. The product of two solutions of Eq. (9) and (10), respectively, provides a solution of the partial differential equation (1). However, we are interested only in those solutions of Eq. (1) that also satisfy the boundary conditions (4). As we now show, this severely restricts the possible values of λ.

Substituting for $u(x, t)$ from Eq. (5) in the boundary condition at $x = 0$, we obtain

$$u(0, t) = X(0)T(t) = 0. \tag{11}$$

If Eq. (11) is satisfied by choosing $T(t)$ to be zero for all t, then $u(x, t)$ is zero for all x and t, and we have already rejected this possibility. Therefore, Eq. (11) must be satisfied by requiring that

$$X(0) = 0. \tag{12}$$

Similarly, the boundary condition at $x = L$ requires that

$$X(L) = 0. \tag{13}$$

We now want to consider Eq. (9) subject to the boundary conditions (12) and (13). This is an eigenvalue problem and, in fact, is the same problem that we discussed in detail at the end of Section 9.1; see especially the paragraph following Eq. (29) in that section. The only difference is that the dependent variable there was called y rather than X. If we refer to the results obtained earlier [Eq. (31) of Section 9.1], the only nontrivial solutions of Eqs. (9), (12), and (13) are the eigenfunctions

$$X_n(x) = \sin(n\pi x/L), \qquad n = 1, 2, 3, \ldots \tag{14}$$

associated with the eigenvalues

$$\lambda_n = n^2\pi^2/L^2, \qquad n = 1, 2, 3, \ldots. \tag{15}$$

Turning now to Eq. (10) for $T(t)$ and substituting $n^2\pi^2/L^2$ for λ, we have

$$T' + (n^2\pi^2\alpha^2/L^2)T = 0. \tag{16}$$

Thus $T(t)$ is proportional to $\exp(-n^2\pi^2\alpha^2 t/L^2)$. Hence multiplying solutions of Eqs. (9) and (10) together, and neglecting arbitrary constants of proportionality, we conclude that the functions

$$u_n(x, t) = e^{-n^2\pi^2\alpha^2 t/L^2} \sin(n\pi x/L), \qquad n = 1, 2, 3, \ldots \tag{17}$$

satisfy the partial differential equation (1) and the boundary conditions (4) for each positive integer value of n. The functions u_n are sometimes called fundamental solutions of the heat conduction problem (1), (3), and (4).

It remains only to satisfy the initial condition (3)

$$u(x, 0) = f(x), \qquad 0 \le x \le L. \tag{18}$$

Recall that we have often solved initial value problems by forming linear combinations of a set of fundamental solutions and then choosing the coefficients to satisfy the initial conditions. The analogous step in the present problem is to form a linear combination of the functions (17) and then to choose the coefficients to satisfy Eq. (18). The main difference from earlier problems is that there are infinitely many functions (17), so a general linear combination of them is an infinite series. Thus we assume that

$$u(x, t) = \sum_{n=1}^{\infty} c_n u_n(x, t) = \sum_{n=1}^{\infty} c_n e^{-n^2 \pi^2 \alpha^2 t / L^2} \sin \frac{n\pi x}{L}, \tag{19}$$

where the coefficients c_n are as yet undetermined. The individual terms in the series (19) satisfy the differential equation (1) and boundary conditions (4). We will assume that the infinite series of Eq. (19) converges and also satisfies Eqs. (1) and (4). To satisfy the initial condition (3), we must have

$$u(x, 0) = \sum_{n=1}^{\infty} c_n \sin \frac{n\pi x}{L} = f(x). \tag{20}$$

In other words, we need to choose the coefficients c_n so that the series of sine functions in Eq. (20) converges to the initial temperature distribution $f(x)$ for $0 \le x \le L$. The series in Eq. (20) is just the Fourier sine series for f; according to Eq. (8) of Section 9.4, its coefficients are given by

$$c_n = \frac{2}{L} \int_0^L f(x) \sin \frac{n\pi x}{L} dx. \tag{21}$$

Hence the solution of the heat conduction problem of Eqs. (1), (3), and (4) is given by the series in Eq. (19) with the coefficients computed from Eq. (21).

EXAMPLE 1

Find the temperature $u(x, t)$ at any time in a metal rod 50 cm long, insulated on the sides, which initially has a uniform temperature of $20°C$ throughout and whose ends are maintained at $0°C$ for all $t > 0$.

The temperature in the rod satisfies the heat conduction problem (1), (3), (4) with $L = 50$ and $f(x) = 20$ for $0 < x < 50$. Thus, from Eq. (19), the solution is

$$u(x, t) = \sum_{n=1}^{\infty} c_n e^{-n^2 \pi^2 \alpha^2 t / 2500} \sin \frac{n\pi x}{50}, \tag{22}$$

where, from Eq. (21),

$$c_n = \frac{4}{5} \int_0^{50} \sin \frac{n\pi x}{50} dx$$

$$= \frac{40}{n\pi}(1 - \cos n\pi) = \begin{cases} 80/n\pi, & n \text{ odd}; \\ 0, & n \text{ even}. \end{cases} \tag{23}$$

Finally, by substituting for c_n in Eq. (22), we obtain

$$u(x, t) = \frac{80}{\pi} \sum_{n=1,3,5,\ldots}^{\infty} \frac{1}{n} e^{-n^2 \pi^2 \alpha^2 t / 2500} \sin \frac{n\pi x}{50}. \tag{24}$$

The expression (24) for the temperature is moderately complicated, but the negative exponential factor in each term of the series causes the series to converge quite rapidly, except for small values of t or α^2. Therefore accurate results can usually be obtained by using only a few terms of the series.

In order to display quantitative results, let us measure t in seconds; then α^2 has the units of cm²/s. If we choose $\alpha^2 = 1$ for convenience, this corresponds to a rod of a material whose thermal properties are somewhere between copper and aluminum. The behavior of the solution can be seen from the graphs in Figures 9.5.3 through 9.5.5. In Figure 9.5.3 we show the temperature distribution in the bar at several different times. Observe that the temperature diminishes steadily as heat in the bar is lost through the end points. The way in which the temperature decays at a given point in the bar is indicated in Figure 9.5.4, where temperature is plotted against time for a few selected points in the bar. Finally, Figure 9.5.5 is a three-dimensional plot of u versus both x and t. Observe that we obtain the graphs in Figures 9.5.3 and 9.5.4 by intersecting the surface in Figure 9.5.5 by planes on which either t or x is constant. The slight waviness in Figure 9.5.5 at $t = 0$ results from using only a finite number of terms in the series for $u(x, t)$ and from the slow convergence of the series for $t = 0$.

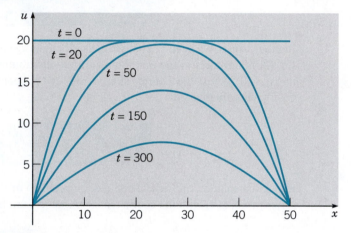

FIGURE 9.5.3 Temperature distributions at several times for the heat conduction problem of Example 1.

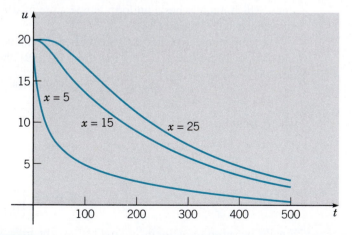

FIGURE 9.5.4 Dependence of temperature on time at several locations for the heat conduction problem of Example 1.

A problem with possible practical implications is to determine the time τ at which the entire bar has cooled to a specified temperature. For example, when is the temperature in the entire bar no greater than $1°C$? Because of the symmetry of the initial temperature distribution and the boundary conditions, the warmest point in the bar is always the center. Thus τ is found by solving $u(25, t) = 1$ for t. Using one term in the series expansion (24), we obtain

$$\tau = \frac{2500}{\pi^2} \ln(80/\pi) \cong 820 \text{ s}.$$

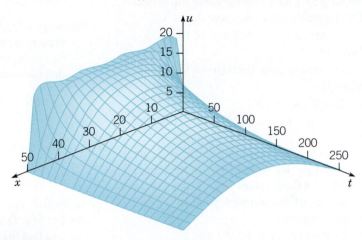

FIGURE 9.5.5   Plot of temperature u versus x and t for the heat conduction problem of Example 1.

PROBLEMS

In each of Problems 1 through 6 determine whether the method of separation of variables can be used to replace the given partial differential equation by a pair of ordinary differential equations. If so, find the equations.

1. $xu_{xx} + u_t = 0$
2. $tu_{xx} + xu_t = 0$
3. $u_{xx} + u_{xt} + u_t = 0$
4. $[p(x)u_x]_x - r(x)u_{tt} = 0$
5. $u_{xx} + (x + y)u_{yy} = 0$
6. $u_{xx} + u_{yy} + xu = 0$

7. Find the solution of the heat conduction problem

$$100u_{xx} = u_t, \qquad 0 < x < 1, \quad t > 0;$$
$$u(0, t) = 0, \qquad u(1, t) = 0, \quad t > 0;$$
$$u(x, 0) = \sin 2\pi x - \sin 5\pi x, \qquad 0 \le x \le 1.$$

8. Find the solution of the heat conduction problem

$$u_{xx} = 4u_t, \qquad 0 < x < 2, \quad t > 0;$$
$$u(0, t) = 0, \qquad u(2, t) = 0, \quad t > 0;$$
$$u(x, 0) = 2\sin(\pi x/2) - \sin \pi x + 4 \sin 2\pi x,$$
$$0 \le x \le 2.$$

Consider the conduction of heat in a rod 40 cm in length whose ends are maintained at $0°C$ for all $t > 0$. In each of Problems 9 through 12 find an expression for the tem-

perature $u(x, t)$ if the initial temperature distribution in the rod is the given function. Suppose that $\alpha^2 = 1$.

9. $u(x, 0) = 50, \qquad 0 < x < 40$

10. $u(x, 0) = \begin{cases} x, & 0 \le x < 20, \\ 40 - x, & 20 \le x \le 40 \end{cases}$

11. $u(x, 0) = \begin{cases} 0, & 0 \le x < 10, \\ 50, & 10 \le x \le 30, \\ 0, & 30 < x \le 40 \end{cases}$

12. $u(x, 0) = x, \qquad 0 < x < 40$

13. Consider again the rod in Problem 9. For $t = 5$ and $x = 20$ determine how many terms are needed to find the solution correct to three decimal places. A reasonable way to do this is to find n so that including one more term does not change the first three decimal places of $u(20, 5)$. Repeat for $t = 20$ and $t = 80$. Form a conclusion about the speed of convergence of the series for $u(x, t)$. **CAS**

14. For the rod in Problem 9:
    (a) Plot u versus x for $t = 5, 10, 20, 40, 100,$ and 200. Put all of the graphs on the same set of axes and thereby obtain a picture of the way **CAS**

in which the temperature distribution changes with time.

(b) Plot u versus t for $x = 5, 10, 15$, and 20.

(c) Draw a three-dimensional plot of u versus x and t.

(d) How long does it take for the entire rod to cool off to a temperature of no more than $1°C$?

CAS **15.** Follow the instructions in Problem 14 for the rod in Problem 10.

CAS **16.** Follow the instructions in Problem 14 for the rod in Problem 11.

CAS **17.** For the rod in Problem 12:

(a) Plot u versus x for $t = 5, 10, 20, 40, 100$, and 200.

(b) For each value of t used in part **(a)** estimate the value of x for which the temperature is greatest. Plot these values versus t to see how the location of the warmest point in the rod changes with time.

(c) Plot u versus t for $x = 10, 20$, and 30.

(d) Draw a three-dimensional plot of u versus x and t.

(e) How long does it take for the entire rod to cool off to a temperature of no more than $1°C$?

CAS **18.** Let a metallic rod 20 cm long be heated to a uniform temperature of $100°C$. Suppose that at $t = 0$ the ends of the bar are plunged into an ice bath at $0°C$, and thereafter maintained at this temperature, but that no heat is allowed to escape through the lateral surface. Find an expression for the temperature at any point in the bar at any later time. Determine the temperature at the center of the bar at time $t = 30$ s if the bar is made of (a) silver, (b) aluminum, or (c) cast iron.

CAS **19.** For the rod of Problem 18 find the time that will elapse before the center of the bar cools to a temperature of $5°C$ if the bar is made of (a) silver, (b) aluminum, or (c) cast iron.

20. In solving differential equations, the computations can almost always be simplified by the use of **dimensionless variables**.

(a) Show that if the dimensionless variable $\xi = x/L$ is introduced, the heat conduction equation becomes

$$\frac{\partial^2 u}{\partial \xi^2} = \frac{L^2}{\alpha^2} \frac{\partial u}{\partial t}, \qquad 0 < \xi < 1, \quad t > 0.$$

(b) Since L^2/α^2 has the units of time, it is convenient to use this quantity to define a dimensionless time variable $\tau = (\alpha^2/L^2)t$. Then show that the heat conduction equation reduces to

$$\frac{\partial^2 u}{\partial \xi^2} = \frac{\partial u}{\partial \tau}, \qquad 0 < \xi < 1, \quad \tau > 0.$$

21. Consider the equation

$$au_{xx} - bu_t + cu = 0, \qquad \text{(i)}$$

where a, b, and c are constants.

(a) Let $u(x, t) = e^{\delta t} w(x, t)$, where δ is constant, and find the corresponding partial differential equation for w.

(b) If $b \neq 0$, show that δ can be chosen so that the partial differential equation found in part **(a)** has no term in w. Thus, by a change of dependent variable, it is possible to reduce Eq. (i) to the heat conduction equation.

22. The heat conduction equation in two space dimensions is

$$\alpha^2(u_{xx} + u_{yy}) = u_t.$$

Assuming that $u(x, y, t) = X(x)Y(y)T(t)$, find ordinary differential equations that are satisfied by $X(x)$, $Y(y)$, and $T(t)$.

23. The heat conduction equation in two space dimensions may be expressed in terms of polar coordinates as

$$\alpha^2[u_{rr} + (1/r)u_r + (1/r^2)u_{\theta\theta}] = u_t.$$

Assuming that $u(r, \theta, t) = R(r)\Theta(\theta)T(t)$, find ordinary differential equations that are satisfied by $R(r)$, $\Theta(\theta)$, and $T(t)$.

9.6 Other Heat Conduction Problems

In Section 9.5 we considered the problem consisting of the heat conduction equation

$$\alpha^2 u_{xx} = u_t, \qquad 0 < x < L, \quad t > 0, \qquad \text{(1)}$$

the boundary conditions

$$u(0, t) = 0, \qquad u(L, t) = 0, \qquad t > 0, \tag{2}$$

and the initial condition

$$u(x, 0) = f(x), \qquad 0 \le x \le L. \tag{3}$$

We found the solution to be

$$u(x, t) = \sum_{n=1}^{\infty} c_n e^{-n^2 \pi^2 \alpha^2 t / L^2} \sin \frac{n \pi x}{L}, \tag{4}$$

where the coefficients c_n are the same as in the series

$$f(x) = \sum_{n=1}^{\infty} c_n \sin \frac{n \pi x}{L}. \tag{5}$$

The series in Eq. (5) is just the Fourier sine series for f; according to Section 9.4, its coefficients are given by

$$c_n = \frac{2}{L} \int_0^L f(x) \sin \frac{n \pi x}{L} \, dx. \tag{6}$$

Hence the solution of the heat conduction problem, Eqs. (1) to (3), is given by the series in Eq. (4) with the coefficients computed from Eq. (6).

We emphasize that at this stage the solution (4) must be regarded as a *formal* solution; that is, we obtained it without rigorous justification of the limiting processes involved. Such a justification is beyond the scope of this book. However, once the series (4) has been obtained, it is possible to show that in $0 < x < L, t > 0$ it converges to a continuous function, that the derivatives u_{xx} and u_t can be computed by differentiating the series (4) term by term, and that the heat conduction equation (1) is indeed satisfied. The argument relies heavily on the fact that each term of the series (4) contains a negative exponential factor, and this results in relatively rapid convergence of the series. A further argument establishes that the function u given by Eq. (4) also satisfies the boundary and initial conditions; this completes the justification of the formal solution.

It is interesting to note that although f satisfies the conditions of the Fourier convergence theorem (Theorem 9.3.1), it may have points of discontinuity. In this case the initial temperature distribution $u(x, 0) = f(x)$ is discontinuous at one or more points. Nevertheless, the solution $u(x, t)$ is continuous for arbitrarily small values of $t > 0$. This illustrates the fact that heat conduction is a diffusive process that instantly smooths out any discontinuities that may be present in the initial temperature distribution. Finally, since f is bounded, it follows from Eq. (6) that the coefficients c_n are also bounded. Consequently, the presence of the negative exponential factor in each term of the series (4) guarantees that

$$\lim_{t \to \infty} u(x, t) = 0 \tag{7}$$

for all x regardless of the initial condition. This is in accord with the result expected from physical intuition.

We now consider two other problems of one-dimensional heat conduction that can be handled by the method developed in Section 9.5.

Nonhomogeneous Boundary Conditions

Suppose now that one end of the bar is held at a constant temperature T_1 and the other is maintained at a constant temperature T_2. Then the boundary conditions are

$$u(0, t) = T_1, \qquad u(L, t) = T_2, \quad t > 0. \tag{8}$$

The differential equation (1) and the initial condition (3) remain unchanged.

This problem is only slightly more difficult, because of the nonhomogeneous boundary conditions, than the one in Section 9.5. We can solve it by reducing it to a problem having homogeneous boundary conditions, which can then be solved as in Section 9.5. The technique for doing this is suggested by the following physical argument.

▶ **The Steady State Solution.** After a long time—that is, as $t \to \infty$—we anticipate that a steady temperature distribution $v(x)$ will be reached, which is independent of the time t and the initial conditions. Since $v(x)$ must satisfy the equation of heat conduction (1), we have

$$v''(x) = 0, \qquad 0 < x < L. \tag{9}$$

Hence the steady state temperature distribution is a linear function of x. Further, $v(x)$ must satisfy the boundary conditions

$$v(0) = T_1, \qquad v(L) = T_2, \tag{10}$$

which are valid even as $t \to \infty$. The solution of Eq. (9) satisfying Eqs. (10) is

$$v(x) = (T_2 - T_1)\frac{x}{L} + T_1. \tag{11}$$

▶ **The Transient Solution.** Returning to the original problem, Eqs. (1), (3), and (8), we will try to express $u(x, t)$ as the sum of the steady state temperature distribution $v(x)$ and another (transient) temperature distribution $w(x, t)$; thus we write

$$u(x, t) = v(x) + w(x, t). \tag{12}$$

Since $v(x)$ is given by Eq. (11), the problem will be solved, provided that we can determine $w(x, t)$. The boundary value problem for $w(x, t)$ is found by substituting the expression in Eq. (12) for $u(x, t)$ in Eqs. (1), (3), and (8).

From Eq. (1) we have

$$\alpha^2(v + w)_{xx} = (v + w)_t;$$

it follows that

$$\alpha^2 w_{xx} = w_t, \tag{13}$$

since $v_{xx} = 0$ and $v_t = 0$. Similarly, from Eqs. (12), (8), and (10),

$$
\begin{aligned}
w(0, t) &= u(0, t) - v(0) = T_1 - T_1 = 0, \\
w(L, t) &= u(L, t) - v(L) = T_2 - T_2 = 0.
\end{aligned}
\tag{14}
$$

Finally, from Eqs. (12) and (3),

$$
w(x, 0) = u(x, 0) - v(x) = f(x) - v(x),
\tag{15}
$$

where $v(x)$ is given by Eq. (11). Thus the transient part of the solution to the original problem is found by solving the problem consisting of Eqs. (13), (14), and (15). This latter problem is precisely the one solved in Section 9.5, provided that $f(x) - v(x)$ is now regarded as the initial temperature distribution. Hence

$$
u(x, t) = (T_2 - T_1)\frac{x}{L} + T_1 + \sum_{n=1}^{\infty} c_n e^{-n^2\pi^2\alpha^2 t/L^2} \sin \frac{n\pi x}{L},
\tag{16}
$$

where

$$
c_n = \frac{2}{L} \int_0^L \left[f(x) - (T_2 - T_1)\frac{x}{L} - T_1 \right] \sin \frac{n\pi x}{L}\, dx.
\tag{17}
$$

This is another case in which a more difficult problem is solved by reducing it to a simpler problem that has already been solved. The technique of reducing a problem with nonhomogeneous boundary conditions to one with homogeneous boundary conditions by subtracting the steady state solution has wide application.

**EXAMPLE
1**

Consider the heat conduction problem

$$
u_{xx} = u_t, \qquad 0 < x < 30, \quad t > 0,
\tag{18}
$$

$$
u(0, t) = 20, \qquad u(30, t) = 50, \quad t > 0,
\tag{19}
$$

$$
u(x, 0) = 60 - 2x, \qquad 0 < x < 30.
\tag{20}
$$

Find the steady state temperature distribution and the boundary value problem that determines the transient distribution.

The steady state temperature satisfies $v''(x) = 0$ and the boundary conditions $v(0) = 20$ and $v(30) = 50$. Thus $v(x) = 20 + x$. The transient distribution $w(x, t)$ satisfies the heat conduction equation

$$
w_{xx} = w_t,
\tag{21}
$$

the homogeneous boundary conditions

$$
w(0, t) = 0, \qquad w(30, t) = 0,
\tag{22}
$$

and the modified initial condition

$$w(x, 0) = 60 - 2x - (20 + x) = 40 - 3x. \tag{23}$$

Note that this problem is of the form (1), (2), (3) with $f(x) = 40 - 3x$, $\alpha^2 = 1$, and $L = 30$. Thus the solution is given by Eqs. (4) and (6).

Figure 9.6.1 shows a plot of the initial temperature distribution $60 - 2x$, the final temperature distribution $20 + x$, and the temperature at three intermediate times found by solving Eqs. (21) through (23). Note that the intermediate temperature satisfies the boundary conditions (19) for any $t > 0$. As t increases, the effect of the boundary conditions gradually moves from the ends of the bar toward its center.

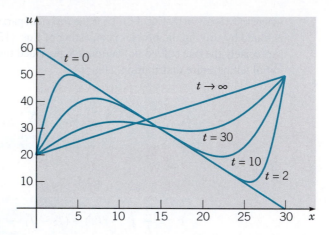

FIGURE 9.6.1   Temperature distributions at several times for the heat conduction problem of Example 1.

Bar with Insulated Ends

A slightly different problem occurs if the ends of the bar are insulated so that there is no passage of heat through them. According to Eq. (2) in Appendix 9.A, the rate of flow of heat across a cross section is proportional to the rate of change of temperature in the x direction. Thus, in the case of no heat flow, the boundary conditions are

$$u_x(0, t) = 0, \qquad u_x(L, t) = 0, \quad t > 0. \tag{24}$$

The problem posed by Eqs. (1), (3), and (24) can also be solved by the method of separation of variables. If we let

$$u(x, t) = X(x)T(t), \tag{25}$$

and substitute for u in Eq. (1), then it follows, as in Section 9.5, that

$$\frac{X''}{X} = \frac{1}{\alpha^2} \frac{T'}{T} = -\lambda, \tag{26}$$

where λ is a constant. Thus we obtain again the two ordinary differential equations

$$X'' + \lambda X = 0, \tag{27}$$

$$T' + \alpha^2 \lambda T = 0. \tag{28}$$

For any value of λ a product of solutions of Eqs. (27) and (28) is a solution of the partial differential equation (1). However, we are interested only in those solutions that also satisfy the boundary conditions (24).

If we substitute for $u(x, t)$ from Eq. (25) in the boundary condition at $x = 0$, we obtain $X'(0)T(t) = 0$. We cannot permit $T(t)$ to be zero for all t, since then $u(x, t)$ would also be zero for all t. Hence we must have

$$X'(0) = 0. \tag{29}$$

Proceeding in the same way with the boundary condition at $x = L$, we find that

$$X'(L) = 0. \tag{30}$$

Thus we wish to solve Eq. (27) subject to the boundary conditions (29) and (30). It is possible to show that nontrivial solutions of this problem can exist only if λ is real. One way to show this is indicated in Problem 18; alternatively, we can appeal to a more general theory to be discussed in Section 10.2. We will assume that λ is real and will consider, in turn, the three cases $\lambda < 0$, $\lambda = 0$, and $\lambda > 0$.

If $\lambda < 0$, it is convenient to let $\lambda = -\mu^2$, where μ is real and positive. Then Eq. (27) becomes $X'' - \mu^2 X = 0$, and its general solution is

$$X(x) = k_1 \sinh \mu x + k_2 \cosh \mu x. \tag{31}$$

In this case the boundary conditions can be satisfied only by choosing $k_1 = k_2 = 0$. Since this is unacceptable, it follows that λ cannot be negative; in other words, the problem (27), (29), (30) has no negative eigenvalues.

If $\lambda = 0$, then Eq. (27) is $X'' = 0$, and therefore

$$X(x) = k_1 x + k_2. \tag{32}$$

The boundary conditions (29) and (30) require that $k_1 = 0$ but do not determine k_2. Thus $\lambda = 0$ is an eigenvalue, corresponding to the eigenfunction $X(x) = 1$. For $\lambda = 0$ it follows from Eq. (28) that $T(t)$ is also a constant, which can be combined with k_2. Hence, for $\lambda = 0$, we obtain the constant solution $u(x, t) = k_2$.

Finally, if $\lambda > 0$, let $\lambda = \mu^2$, where μ is real and positive. Then Eq. (27) becomes $X'' + \mu^2 X = 0$, and consequently,

$$X(x) = k_1 \sin \mu x + k_2 \cos \mu x. \tag{33}$$

The boundary condition (29) requires that $k_1 = 0$, and the boundary condition (30) requires that $\mu = n\pi/L$ for $n = 1, 2, 3, \ldots$ but leaves k_2 arbitrary. Thus the problem (27), (29), (30) has an infinite sequence of positive eigenvalues $\lambda = n^2\pi^2/L^2$ with the corresponding eigenfunctions $X(x) = \cos(n\pi x/L)$. For these values of λ the solutions $T(t)$ of Eq. (28) are proportional to $\exp(-n^2\pi^2\alpha^2 t/L^2)$.

Combining all these results, we have the following fundamental solutions for the problem (1), (3), and (24):

$$
\begin{aligned}
u_0(x, t) &= 1, \\
u_n(x, t) &= e^{-n^2\pi^2\alpha^2 t/L^2} \cos \frac{n\pi x}{L}, \qquad n = 1, 2, \ldots,
\end{aligned}
\tag{34}
$$

where arbitrary constants of proportionality have been dropped. Each of these functions satisfies the differential equation (1) and the boundary conditions (24). Because both the differential equation and the boundary conditions are linear and homogeneous, any finite linear combination of the fundamental solutions satisfies them. We will assume that this is true for convergent infinite linear combinations of fundamental solutions as well. Thus, to satisfy the initial condition (3), we assume that $u(x, t)$ has the form

$$
\begin{aligned}
u(x, t) &= \frac{c_0}{2} u_0(x, t) + \sum_{n=1}^{\infty} c_n u_n(x, t) \\
&= \frac{c_0}{2} + \sum_{n=1}^{\infty} c_n e^{-n^2\pi^2\alpha^2 t/L^2} \cos \frac{n\pi x}{L}.
\end{aligned}
\tag{35}
$$

The coefficients c_n are determined by the requirement that

$$
u(x, 0) = \frac{c_0}{2} + \sum_{n=1}^{\infty} c_n \cos \frac{n\pi x}{L} = f(x).
\tag{36}
$$

Thus the unknown coefficients in Eq. (35) must be the coefficients in the Fourier cosine series of period $2L$ for f. Hence

$$
c_n = \frac{2}{L} \int_0^L f(x) \cos \frac{n\pi x}{L}\, dx, \qquad n = 0, 1, 2, \ldots.
\tag{37}
$$

With this choice of the coefficients $c_0, c_1, c_2, \ldots$, the series (35) provides the solution to the heat conduction problem for a rod with insulated ends, Eqs. (1), (3), and (24).

It is worth observing that the solution (35) can also be thought of as the sum of a steady state temperature distribution (given by the constant $c_0/2$), which is independent of time t, and a transient distribution (given by the rest of the infinite series) that vanishes in the limit as t approaches infinity. That the steady state is a constant is consistent with the expectation that the process of heat conduction will gradually smooth out the initial temperature distribution in the bar as long as no heat is allowed to enter from, or to escape to, the outside. The physical interpretation of the term

$$
\frac{c_0}{2} = \frac{1}{L} \int_0^L f(x)\, dx
\tag{38}
$$

is that it is the mean value of the original temperature distribution.

EXAMPLE 2

Find the temperature $u(x, t)$ in a metal rod of length 25 cm that is insulated on the ends as well as on the sides and whose initial temperature distribution is $u(x, 0) = x$ for $0 < x < 25$.

The temperature in the rod satisfies the heat conduction problem (1), (3), (24) with $L = 25$. Thus, from Eq. (35), the solution is

$$u(x, t) = \frac{c_0}{2} + \sum_{n=1}^{\infty} c_n e^{-n^2 \pi^2 \alpha^2 t / 625} \cos \frac{n \pi x}{25}, \tag{39}$$

where the coefficients are determined from Eq. (37). We have

$$c_0 = \frac{2}{25} \int_0^{25} x \, dx = 25 \tag{40}$$

and, for $n \geq 1$,

$$
\begin{aligned}
c_n &= \frac{2}{25} \int_0^{25} x \cos \frac{n \pi x}{25} \, dx \\
&= 50(\cos n\pi - 1)/(n\pi)^2 = \begin{cases} -100/(n\pi)^2, & n \text{ odd;} \\ 0, & n \text{ even.} \end{cases}
\end{aligned}
\tag{41}
$$

Thus

$$u(x, t) = \frac{25}{2} - \frac{100}{\pi^2} \sum_{n=1,3,5,\ldots}^{\infty} \frac{1}{n^2} e^{-n^2 \pi^2 \alpha^2 t / 625} \cos (n\pi x / 25) \tag{42}$$

is the solution of the given problem.

For $\alpha^2 = 1$, Figure 9.6.2 shows plots of the temperature distribution in the bar at several times. Again the convergence of the series is rapid so that only a relatively few terms are needed to generate the graphs.

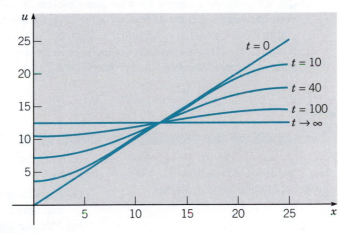

FIGURE 9.6.2 Temperature distributions at several times for the heat conduction problem of Example 2.

More General Problems

The method of separation of variables can also be used to solve heat conduction problems with other boundary conditions than those given by Eqs. (8) and Eqs. (24). For example, the left end of the bar might be held at a fixed temperature T while the other end is insulated. In this case the boundary conditions are

$$u(0, t) = T, \qquad u_x(L, t) = 0, \quad t > 0. \tag{43}$$

The first step in solving this problem is to reduce the given boundary conditions to homogeneous ones by subtracting the steady state solution. The resulting problem is solved by essentially the same procedure as in the problems previously considered. However, the extension of the initial function f outside of the interval $[0, L]$ is somewhat different from that in any case considered so far (see Problem 15).

A more general type of boundary condition occurs when the rate of heat flow through the end of the bar is proportional to the temperature. It is shown in Appendix 9.A that the boundary conditions in this case are of the form

$$u_x(0, t) - h_1 u(0, t) = 0, \qquad u_x(L, t) + h_2 u(L, t) = 0, \quad t > 0, \tag{44}$$

where h_1 and h_2 are nonnegative constants. If we apply the method of separation of variables to the problem consisting of Eqs. (1), (3), and (44), we find that $X(x)$ must be a solution of

$$X'' + \lambda X = 0, \qquad X'(0) - h_1 X(0) = 0, \qquad X'(L) + h_2 X(L) = 0, \tag{45}$$

where λ is the separation constant. Once again it is possible to show that nontrivial solutions can exist only for certain nonnegative real values of λ, the eigenvalues, but these values are not given by a simple formula (see Problem 20). It is also possible to show that the corresponding solutions of Eqs. (45), the eigenfunctions, satisfy an orthogonality relation and that we can satisfy the initial condition (3) by superposing solutions of Eqs. (45). However, the resulting series is not included in the discussion of this chapter. There is more general theory that covers such problems, and it is outlined in Chapter 10.

PROBLEMS

In each of Problems 1 through 8 find the steady-state solution of the heat conduction equation $\alpha^2 u_{xx} = u_t$ that satisfies the given set of boundary conditions.

1. $u(0, t) = 10, \quad u(50, t) = 40$
2. $u(0, t) = 30, \quad u(40, t) = -20$
3. $u_x(0, t) = 0, \quad u(L, t) = 0$
4. $u_x(0, t) = 0, \quad u(L, t) = T$
5. $u(0, t) = 0, \quad u_x(L, t) = 0$
6. $u(0, t) = T, \quad u_x(L, t) = 0$
7. $u_x(0, t) - u(0, t) = 0, \quad u(L, t) = T$
8. $u(0, t) = T, \quad u_x(L, t) + u(L, t) = 0$

CAS 9. Let an aluminum rod of length 20 cm be initially at the uniform temperature of $25°$C. Suppose that at time $t = 0$ the end $x = 0$ is cooled to $0°$C while the end $x = 20$ is heated to $60°$C, and

both are thereafter maintained at those temperatures.

(a) Find the temperature distribution in the rod at any time t.

(b) Plot the initial temperature distribution, the final (steady-state) temperature distribution, and the temperature distributions at two representative intermediate times on the same set of axes.

(c) Plot u versus t for $x = 5, 10$, and 15.

(d) Determine how much time must elapse before the temperature at $x = 5$ cm comes (and remains) within 1% of its steady-state value.

10. (a) Let the ends of a copper rod 100 cm long be CAS maintained at $0°$C. Suppose that the center of the bar is heated to $100°$C by an external heat

source and that this situation is maintained until a steady state results. Find this steady-state temperature distribution.

(b) At a time $t = 0$ [after the steady state of part (a) has been reached], let the heat source be removed. At the same instant let the end $x = 0$ be placed in thermal contact with a reservoir at $20°C$, while the other end remains at $0°C$. Find the temperature as a function of position and time.

(c) Plot u versus x for several values of t. Also plot u versus t for several values of x.

(d) What limiting value does the temperature at the center of the rod approach after a long time? How much time must elapse before the center of the rod cools to within $1°$ of its limiting value?

CAS 11. Consider a rod of length 30 for which $\alpha^2 = 1$. Suppose the initial temperature distribution is given by $u(x, 0) = x(60 - x)/30$ and that the boundary conditions are $u(0, t) = 30$ and $u(30, t) = 0$.

(a) Find the temperature in the rod as a function of position and time.

(b) Plot u versus x for several values of t. Also plot u versus t for several values of x.

(c) Plot u versus t for $x = 12$. Observe that u initially decreases, then increases for a while, and finally decreases to approach its steady-state value. Explain physically why this behavior occurs at this point.

CAS 12. Consider a uniform rod of length L with an initial temperature given by $u(x, 0) = \sin(\pi x/L)$, $0 \le x \le L$. Assume that both ends of the bar are insulated.

(a) Find the temperature $u(x, t)$.

(b) What is the steady-state temperature as $t \to \infty$?

(c) Let $\alpha^2 = 1$ and $L = 40$. Plot u versus x for several values of t. Also plot u versus t for several values of x.

(d) Describe briefly how the temperature in the rod changes as time progresses.

CAS 13. Consider a bar of length 40 cm whose initial temperature is given by $u(x, 0) = x(60 - x)/30$. Suppose that $\alpha^2 = 1/4$ cm^2/s and that both ends of the bar are insulated.

(a) Find the temperature $u(x, t)$.

(b) Plot u versus x for several values of t. Also plot u versus t for several values of x.

(c) Determine the steady-state temperature in the bar.

(d) Determine how much time must elapse before the temperature at $x = 40$ comes within $1°$ of its steady-state value.

14. Consider a bar 30 cm long that is made of a material for which $\alpha^2 = 1$ and whose ends are insulated. Suppose that the initial temperature is zero except for the interval $5 < x < 10$, where the initial temperature is $25°C$.

(a) Find the temperature $u(x, t)$.

(b) Plot u versus x for several values of t. Also plot u versus t for several values of x.

(c) Plot $u(4, t)$ and $u(11, t)$ versus t. Observe that the points $x = 4$ and $x = 11$ are symmetrically located with respect to the initial temperature pulse, yet their temperature plots are significantly different. Explain physically why this is so.

15. Consider a uniform bar of length L having an initial temperature distribution given by $f(x)$, $0 \le x \le L$. Assume that the temperature at the end $x = 0$ is held at $0°C$, while the end $x = L$ is insulated so that no heat passes through it.

(a) Show that the fundamental solutions of the partial differential equation and boundary conditions are

$$u_n(x, t) = e^{-(2n-1)^2\pi^2\alpha^2 t/4L^2} \sin[(2n-1)\pi x/2L],$$
$$n = 1, 2, 3, \ldots.$$

(b) Find a formal series expansion for the temperature $u(x, t)$

$$u(x, t) = \sum_{n=1}^{\infty} c_n u_n(x, t)$$

that also satisfies the initial condition $u(x, 0) = f(x)$.

Hint: Even though the fundamental solutions involve only the odd sines, it is still possible to represent f by a Fourier series involving only these functions. See Problem 39 of Section 9.4.

16. In the bar of Problem 15 suppose that $L = 30$, $\alpha^2 = 1$, and the initial temperature distribution is $f(x) = 30 - x$ for $0 < x < 30$.

(a) Find the temperature $u(x, t)$.

(b) Plot u versus x for several values of t. Also plot u versus t for several values of x.

(c) How does the location x_m of the warmest point in the bar change as t increases? Draw a graph of x_m versus t.

(d) Plot the maximum temperature in the bar versus t.

CAS **17.** Suppose that the conditions are as in Problems 15 and 16 except that the boundary condition at $x = 0$ is $u(0, t) = 40$.

(a) Find the temperature $u(x, t)$.

(b) Plot u versus x for several values of t. Also plot u versus t for several values of x.

(c) Compare the plots you obtained in this problem with those from Problem 16. Explain how the change in the boundary condition at $x = 0$ causes the observed differences in the behavior of the temperature in the bar.

18. Consider the problem

$$X'' + \lambda X = 0, \quad X'(0) = 0, \quad X'(L) = 0. \quad \text{(i)}$$

Let $\lambda = \mu^2$, where $\mu = \nu + i\sigma$ with ν and σ real. Show that if $\sigma \neq 0$, then the only solution to Eqs. (i) is the trivial solution $X(x) = 0$.

Hint: Use an argument similar to that in Problem 23 of Section 9.1.

19. The right end of a bar of length a with thermal conductivity κ_1 and cross-sectional area A_1 is joined to the left end of a bar of thermal conductivity κ_2 and cross-sectional area A_2. The composite bar has a total length L. Suppose that the end $x = 0$ is held at temperature zero, while the end $x = L$ is held at temperature T. Find the steady-state temperature in the composite bar, assuming that the temperature and rate of heat flow are continuous at $x = a$.

Hint: See Eq. (2) in Appendix 9.A.

20. Consider the problem

$$\alpha^2 u_{xx} = u_t \quad 0 < x < L, \quad t > 0;$$
$$u(0, t) = 0, \quad u_x(L, t) + \gamma u(L, t) = 0, \quad t > 0; \quad \text{(i)}$$
$$u(x, 0) = f(x), \quad 0 \leq x \leq L.$$

(a) Let $u(x, t) = X(x)T(t)$ and show that

$$X'' + \lambda X = 0, \quad X(0) = 0, \quad X'(L) + \gamma X(L) = 0, \quad \text{(ii)}$$

and

$$T' + \lambda \alpha^2 T = 0,$$

where λ is the separation constant.

(b) Assume that λ is real, and show that problem (ii) has no nontrivial solutions if $\lambda \leq 0$.

(c) If $\lambda > 0$, let $\lambda = \mu^2$ with $\mu > 0$. Show that problem (ii) has nontrivial solutions only if μ

is a solution of the equation

$$\mu \cos \mu L + \gamma \sin \mu L = 0. \quad \text{(iii)}$$

(d) Rewrite Eq. (iii) as $\tan \mu L = -\mu/\gamma$. Then, by drawing the graphs of $y = \tan \mu L$ and $y = -\mu L/\gamma L$ for $\mu > 0$ on the same set of axes, show that Eq. (iii) is satisfied by infinitely many positive values of μ; denote these by $\mu_1, \mu_2, \ldots, \mu_n, \ldots$, ordered in increasing size.

(e) Determine the set of fundamental solutions $u_n(x, t)$ corresponding to the values μ_n found in part **(d)**.

An External Heat Source. Consider the heat conduction problem in a bar that is in thermal contact with an external heat source or sink. Then the modified heat conduction equation is

$$u_t = \alpha^2 u_{xx} + s(x), \quad \text{(i)}$$

where the term $s(x)$ describes the effect of the external agency; $s(x)$ is positive for a source and negative for a sink. Suppose that the boundary conditions are

$$u(0, t) = T_1, \quad u(L, t) = T_2 \quad \text{(ii)}$$

and the initial condition is

$$u(x, 0) = f(x). \quad \text{(iii)}$$

Problems 21 through 23 deal with this kind of problem.

21. Write $u(x, t) = v(x) + w(x, t)$, where v and w are the steady state and transient parts of the solution, respectively. State the boundary value problems that $v(x)$ and $w(x, t)$, respectively, satisfy. Observe that the problem for w is the fundamental heat conduction problem discussed in Section 9.5, with a modified initial temperature distribution.

22. **(a)** Suppose that $\alpha^2 = 1$ and $s(x) = k$, a constant, CAS in Eq. (i). Find $v(x)$.

(b) Assume that $T_1 = 0$, $T_2 = 0$, $L = 20$, $k = 1/5$, and that $f(x) = 0$ for $0 < x < L$. Determine $w(x, t)$. Then plot $u(x, t)$ versus x for several values of t; on the same axes also plot the steady-state part of the solution $v(x)$.

23. **(a)** Let $\alpha^2 = 1$ and $s(x) = kx/L$, where k is a con- CAS stant, in Eq. (i). Find $v(x)$.

(b) Assume that $T_1 = 10$, $T_2 = 30$, $L = 20$, $k = 1/2$, and that $f(x) = 0$ for $0 < x < L$. Determine $w(x, t)$. Then plot $u(x, t)$ versus x for several values of t; on the same axes also plot the steady-state part of the solution $v(x)$.

9.7 The Wave Equation: Vibrations of an Elastic String

A second partial differential equation that occurs frequently in applied mathematics is the wave equation. Some form of this equation, or a generalization of it, almost inevitably arises in any mathematical analysis of phenomena involving the propagation of waves in a continuous medium. For example, the studies of acoustic waves, water waves, electromagnetic waves, and seismic waves are all based on this equation.

Perhaps the easiest situation to visualize occurs in the investigation of mechanical vibrations. Suppose that an elastic string of length L is tightly stretched between two supports at the same horizontal level, so that the x-axis lies along the string (see Figure 9.7.1).

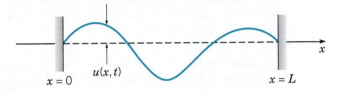

$x = 0$ $u(x, t)$ $x = L$

FIGURE 9.7.1   A vibrating string.

The elastic string may be thought of as a violin string, a guy wire, or possibly an electric power line. Suppose that the string is set in motion (by plucking, for example) so that it vibrates in a vertical plane, and let $u(x, t)$ denote the vertical displacement experienced by the string at the point x at time t. If damping effects, such as air resistance, are neglected, and if the amplitude of the motion is not too large, then $u(x, t)$ satisfies the partial differential equation

$$a^2 u_{xx} = u_{tt} \tag{1}$$

in the domain $0 < x < L$, $t > 0$. Equation (1) is known as the one-dimensional **wave equation** and is derived in Appendix 9.B at the end of the chapter. The constant coefficient a^2 appearing in Eq. (1) is given by

$$a^2 = T/\rho, \tag{2}$$

where T is the tension (force) in the string, and ρ is the mass per unit length of the string material. It follows that a has the units of length/time—that is, of velocity. In Problem 14 it is shown that a is the velocity of propagation of waves along the string.

▶ **Boundary and Initial Conditions.** To describe the motion of the string completely, it is necessary also to specify suitable initial and boundary conditions for the displacement $u(x, t)$. The ends are assumed to remain fixed, and therefore the boundary conditions are

$$u(0, t) = 0, \qquad u(L, t) = 0, \qquad t \geq 0. \tag{3}$$

Since the differential equation (1) is of second order with respect to t, it is plausible to prescribe two initial conditions. These are the initial position of the string

$$u(x, 0) = f(x), \qquad 0 \leq x \leq L \tag{4}$$

and its initial velocity

$$u_t(x, 0) = g(x), \qquad 0 \leq x \leq L, \tag{5}$$

where f and g are given functions. In order for Eqs. (3), (4), and (5) to be consistent, it is also necessary to require that

$$f(0) = f(L) = 0, \qquad g(0) = g(L) = 0. \tag{6}$$

The mathematical problem then is to determine the solution of the wave equation (1) that also satisfies the boundary conditions (3) and the initial conditions (4) and (5). Like the heat conduction problem of Sections 9.5 and 9.6, this problem is an initial value problem in the time variable t and a boundary value problem in the space variable x. Alternatively, it can be considered as a boundary value problem in the semi-infinite strip $0 < x < L, t > 0$ of the xt-plane (see Figure 9.7.2). One condition is imposed at each point on the semi-infinite sides, and two are imposed at each point on the finite base.

It is important to realize that Eq. (1) governs a large number of other wave problems besides the transverse vibrations of an elastic string. For example, it is only necessary to interpret the function u and the constant a appropriately to have problems dealing with water waves in an ocean, acoustic or electromagnetic waves in the atmosphere, or elastic waves in a solid body. If more than one space dimension is significant, then Eq. (1) must be slightly generalized. The two-dimensional wave equation is

$$a^2(u_{xx} + u_{yy}) = u_{tt}. \tag{7}$$

This equation would arise, for example, if we considered the motion of a thin elastic sheet, such as a drumhead. Similarly, in three dimensions the wave equation is

$$a^2(u_{xx} + u_{yy} + u_{zz}) = u_{tt}. \tag{8}$$

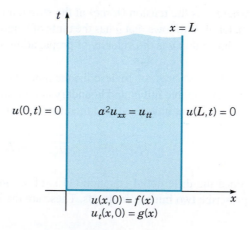

$$u(0, t) = 0 \qquad a^2 u_{xx} = u_{tt} \qquad u(L, t) = 0$$

$$u(x, 0) = f(x)$$
$$u_t(x, 0) = g(x)$$

FIGURE 9.7.2   Boundary value problem for the wave equation.

In connection with the latter two equations, the boundary and initial conditions must also be suitably generalized.

We now solve some typical boundary value problems involving the one-dimensional wave equation.

▶ **Elastic String with Nonzero Initial Displacement.** First suppose that the string is disturbed from its equilibrium position and then released at time $t = 0$ with zero velocity to vibrate freely. Then the vertical displacement $u(x, t)$ must satisfy the wave equation (1)

$$a^2 u_{xx} = u_{tt}, \qquad 0 < x < L, \quad t > 0;$$

the boundary conditions (3)

$$u(0, t) = 0, \qquad u(L, t) = 0, \quad t \geq 0;$$

and the initial conditions

$$u(x, 0) = f(x), \qquad u_t(x, 0) = 0, \quad 0 \leq x \leq L, \tag{9}$$

where f is a given function describing the configuration of the string at $t = 0$.

The method of separation of variables can be used to obtain the solution of Eqs. (1), (3), and (9). Assuming that

$$u(x, t) = X(x)T(t) \tag{10}$$

and substituting for u in Eq. (1), we obtain

$$\frac{X''}{X} = \frac{1}{a^2} \frac{T''}{T} = -\lambda, \tag{11}$$

where λ is a separation constant. Thus we find that $X(x)$ and $T(t)$ satisfy the ordinary differential equations

$$X'' + \lambda X = 0, \tag{12}$$

$$T'' + a^2 \lambda T = 0. \tag{13}$$

Further, by substituting from Eq. (10) for $u(x, t)$ in the boundary conditions (3), we find that $X(x)$ must satisfy the boundary conditions

$$X(0) = 0, \qquad X(L) = 0. \tag{14}$$

Finally, by substituting from Eq. (10) into the second of the initial conditions (9), we also find that $T(t)$ must satisfy the initial condition

$$T'(0) = 0. \tag{15}$$

Our next task is to determine $X(x)$, $T(t)$, and λ by solving Eq. (12) subject to the boundary conditions (14) and Eq. (13) subject to the initial condition (15).

The problem of solving the differential equation (12) subject to the boundary conditions (14) is *precisely the same problem* that arose in Section 9.5 in connection with the heat conduction equation. Thus we can use the results obtained there and at the end of Section 9.1: the problem (12), (14) has nontrivial solutions if and only if λ is an eigenvalue

$$\lambda = n^2\pi^2/L^2, \qquad n = 1, 2, \ldots, \tag{16}$$

and $X(x)$ is proportional to the corresponding eigenfunction $\sin(n\pi x/L)$.

Using the values of λ given by Eq. (16) in Eq. (13), we obtain

$$T'' + \frac{n^2\pi^2 a^2}{L^2}T = 0. \tag{17}$$

Therefore

$$T(t) = k_1 \cos\frac{n\pi at}{L} + k_2 \sin\frac{n\pi at}{L}, \tag{18}$$

where k_1 and k_2 are arbitrary constants. The initial condition (15) requires that $k_2 = 0$, so $T(t)$ must be proportional to $\cos(n\pi at/L)$.

Thus the functions

$$u_n(x, t) = \sin\frac{n\pi x}{L} \cos\frac{n\pi at}{L}, \qquad n = 1, 2, \ldots \tag{19}$$

satisfy the partial differential equation (1), the boundary conditions (3), and the second initial condition (9). These functions are the fundamental solutions for the given problem.

To satisfy the remaining (nonhomogeneous) initial condition (9), we will consider a superposition of the fundamental solutions (19) with properly chosen coefficients. Thus we assume that $u(x, t)$ has the form

$$u(x, t) = \sum_{n=1}^{\infty} c_n u_n(x, t) = \sum_{n=1}^{\infty} c_n \sin\frac{n\pi x}{L} \cos\frac{n\pi at}{L}, \tag{20}$$

where the constants c_n remain to be chosen. The initial condition $u(x, 0) = f(x)$ requires that

$$u(x, 0) = \sum_{n=1}^{\infty} c_n \sin\frac{n\pi x}{L} = f(x). \tag{21}$$

Consequently, the coefficients c_n must be the coefficients in the Fourier sine series of period $2L$ for f; hence

$$c_n = \frac{2}{L}\int_0^L f(x) \sin\frac{n\pi x}{L}\, dx, \qquad n = 1, 2, \ldots. \tag{22}$$

Thus the formal solution of the problem of Eqs. (1), (3), and (9) is given by Eq. (20) with the coefficients calculated from Eq. (22).

For a fixed value of n the expression $\sin(n\pi x/L)\cos(n\pi at/L)$ in Eq. (19) is periodic in time t with the period $2L/na$; it therefore represents a vibratory motion of the string having this period, or having the frequency $n\pi a/L$. The quantities $n\pi a/L$ for $n = 1, 2, \ldots$ are the **natural frequencies** of the string—that is, the frequencies at which the string will freely vibrate. The factor $\sin(n\pi x/L)$ represents the displacement pattern occurring in the string when it is executing vibrations of the given frequency. Each displacement pattern is called a **natural mode** of vibration and is periodic in the space variable x; the spatial period $2L/n$ is called the **wavelength** of the mode of frequency $n\pi a/L$. Thus the eigenvalues $n^2\pi^2/L^2$ of the problem (12), (14) are proportional to the squares of the natural frequencies, and the eigenfunctions $\sin(n\pi x/L)$ give the natural modes. The first three natural modes are sketched in Figure 9.7.3. The total motion of the string, given by the function $u(x, t)$ of Eq. (20), is thus a combination of the natural modes of vibration and is also a periodic function of time with period $2L/a$.

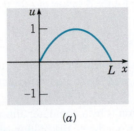
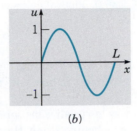
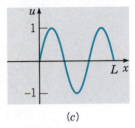

(a)   (b)   (c)

FIGURE 9.7.3   First three fundamental modes of vibration of an elastic string. (a) Frequency $= \pi a/L$, wavelength $= 2L$; (b) frequency $= 2\pi a/L$, wavelength $= L$; (c) frequency $= 3\pi a/L$, wavelength $= 2L/3$.

EXAMPLE 1

Consider a vibrating string of length $L = 30$ that satisfies the wave equation

$$4u_{xx} = u_{tt}, \qquad 0 < x < 30, \quad t > 0. \tag{23}$$

Assume that the ends of the string are fixed and that the string is set in motion with no initial velocity from the initial position

$$u(x, 0) = f(x) = \begin{cases} x/10, & 0 \le x \le 10, \\ (30 - x)/20, & 10 < x \le 30. \end{cases} \tag{24}$$

Find the displacement $u(x, t)$ of the string and describe its motion through one period.

The solution is given by Eq. (20) with $a = 2$ and $L = 30$; that is,

$$u(x, t) = \sum_{n=1}^{\infty} c_n \sin \frac{n\pi x}{30} \cos \frac{2n\pi t}{30}, \tag{25}$$

where c_n is calculated from Eq. (22). Substituting from Eq. (24) into Eq. (22), we obtain

$$c_n = \frac{2}{30} \int_0^{10} \frac{x}{10} \sin \frac{n\pi x}{30}\, dx + \frac{2}{30} \int_{10}^{30} \frac{30 - x}{20} \sin \frac{n\pi x}{30}\, dx. \tag{26}$$

By evaluating the integrals in Eq. (26), we find that

$$c_n = \frac{9}{n^2\pi^2} \sin \frac{n\pi}{3}, \qquad n = 1, 2, \ldots. \tag{27}$$

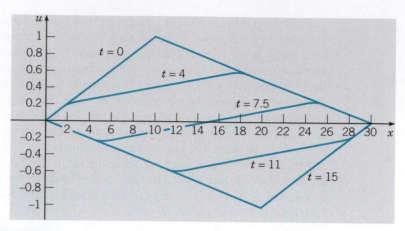

FIGURE 9.7.4   Plots of u versus x for fixed values of t for the string in Example 1.

The solution (25), (27) gives the displacement of the string at any point x at any time t. The motion is periodic in time with period 30, so it is sufficient to analyze the solution for $0 \le t \le 30$.

The best way to visualize the solution is by a computer animation showing the dynamic behavior of the vibrating string. Here we indicate the motion of the string in Figures 9.7.4, 9.7.5, and 9.7.6. Plots of u versus x for $t = 0, 4, 7.5, 11$, and 15 are shown in Figure 9.7.4. Observe that the maximum initial displacement is positive and occurs at $x = 10$, while at $t = 15$, a half-period later, the maximum displacement is negative and occurs at $x = 20$. The string then retraces its motion and returns to its original configuration at $t = 30$. Figure 9.7.5 shows the behavior of the points $x = 10, 15$, and 20 by plots of u versus t for these fixed values of x. The plots confirm that the motion is indeed periodic with period 30. Observe also that each interior point on the string is motionless for one-third of each period. Figure 9.7.6 shows a three-dimensional plot of u versus both x and t, from which the overall nature of the solution is apparent. Of course, the curves in Figures 9.7.4 and 9.7.5 lie on the surface shown in Figure 9.7.6.

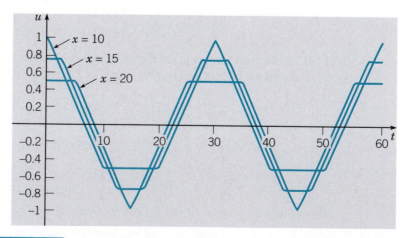

FIGURE 9.7.5   Plots of u versus t for fixed values of x for the string in Example 1.

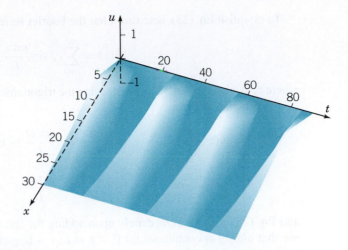

FIGURE 9.7.6 Plot of u versus x and t for the string in Example 1.

▶ **Justification of the Solution.** As in the heat conduction problem considered earlier, Eq. (20) with the coefficients c_n given by Eq. (22) is only a *formal* solution of Eqs. (1), (3), and (9). To ascertain whether Eq. (20) *actually* represents the solution of the given problem requires some further investigation. As in the heat conduction problem, it is tempting to try to show this directly by substituting Eq. (20) for $u(x, t)$ in Eqs. (1), (3), and (9). However, upon formally computing u_{xx}, for example, we obtain

$$u_{xx}(x, t) = -\sum_{n=1}^{\infty} c_n \left(\frac{n\pi}{L}\right)^2 \sin\frac{n\pi x}{L} \cos\frac{n\pi at}{L};$$

due to the presence of the n^2 factor in the numerator, this series may not converge. This would not necessarily mean that the series (20) for $u(x, t)$ is incorrect, but only that the series (20) cannot be used to calculate u_{xx} and u_{tt}. A basic difference between solutions of the wave equation and those of the heat conduction equation is that the latter contain negative exponential terms that approach zero very rapidly with increasing n, which ensures the convergence of the series solution and its derivatives. In contrast, series solutions of the wave equation contain only oscillatory terms that do not decay with increasing n.

However, there is an alternative way to establish the validity of Eq. (20) indirectly. At the same time, we will gain additional information about the structure of the solution. First we will show that Eq. (20) is equivalent to

$$u(x, t) = \frac{1}{2}[h(x - at) + h(x + at)], \tag{28}$$

where h is the function obtained by extending the initial data f into $(-L, 0)$ as an odd function, and to other values of x as a periodic function of period $2L$. That is,

$$h(x) = \begin{cases} f(x), & 0 \leq x \leq L, \\ -f(-x), & -L < x < 0; \end{cases} \tag{29}$$

$$h(x + 2L) = h(x).$$

To establish Eq. (28), note that h has the Fourier series

$$h(x) = \sum_{n=1}^{\infty} c_n \sin \frac{n\pi x}{L}, \tag{30}$$

where c_n is given by Eq. (22). Then, using the trigonometric identities for the sine of a sum or difference, we obtain

$$h(x - at) = \sum_{n=1}^{\infty} c_n \left(\sin \frac{n\pi x}{L} \cos \frac{n\pi at}{L} - \cos \frac{n\pi x}{L} \sin \frac{n\pi at}{L} \right),$$

$$h(x + at) = \sum_{n=1}^{\infty} c_n \left(\sin \frac{n\pi x}{L} \cos \frac{n\pi at}{L} + \cos \frac{n\pi x}{L} \sin \frac{n\pi at}{L} \right),$$

and Eq. (28) follows immediately upon adding the last two equations. From Eq. (28) we see that $u(x, t)$ is continuous for $0 < x < L$, $t > 0$, provided that h is continuous on the interval $(-\infty, \infty)$. This requires f to be continuous on the original interval $[0, L]$. Similarly, u is twice continuously differentiable with respect to either variable in $0 < x < L$, $t > 0$, provided that h is twice continuously differentiable on $(-\infty, \infty)$. This requires f' and f'' to be continuous on $[0, L]$. Furthermore, since h'' is the odd extension of f'', we must also have $f''(0) = f''(L) = 0$. However, since h' is the even extension of f', no further conditions are required on f'. Provided that these conditions are met, u_{xx} and u_{tt} can be computed from Eq. (28), and it is an elementary exercise to show that these derivatives satisfy the wave equation. Some of the details of the argument just indicated are given in Problems 19 and 20.

If some of the continuity requirements stated in the last paragraph are not met, then u is not differentiable at some points in the semi-infinite strip $0 < x < L$, $t > 0$, and thus is a solution of the wave equation only in a somewhat restricted sense. An important physical consequence of this observation is that if there are any discontinuities present in the initial data f, then they will be preserved in the solution $u(x, t)$ for all time. In contrast, in heat conduction problems, initial discontinuities are instantly smoothed out (Section 9.6). Suppose that the initial displacement f has a jump discontinuity at $x = x_0$, $0 \le x_0 \le L$. Since h is a periodic extension of f, the same discontinuity is present in $h(\xi)$ at $\xi = x_0 + 2nL$ and at $\xi = -x_0 + 2nL$, where n is any integer. Thus $h(x - at)$ is discontinuous when $x - at = x_0 + 2nL$, or when $x - at = -x_0 + 2nL$. For a fixed x in $[0, L]$ the discontinuity that was originally at x_0 will reappear in $h(x - at)$ at the times $t = (x \pm x_0 - 2nL)/a$. Similarly, $h(x + at)$ is discontinuous at the point x at the times $t = (-x \pm x_0 + 2mL)/a$, where m is any integer. If we refer to Eq. (28), it then follows that the solution $u(x, t)$ is also discontinuous at the given point x at these times. Since the physical problem is posed for $t > 0$, only those values of m and n that yield positive values of t are of interest.

▶ **General Problem for the Elastic String.** Let us modify the problem just considered by supposing that the string is set in motion from its equilibrium position with a given velocity. Then the vertical displacement $u(x, t)$ must satisfy the wave equation (1)

$$a^2 u_{xx} = u_{tt}, \qquad 0 < x < L, \quad t > 0;$$

the boundary conditions (3)

$$u(0, t) = 0, \qquad u(L, t) = 0, \qquad t \ge 0;$$

and the initial conditions

$$u(x, 0) = 0, \qquad u_t(x, 0) = g(x), \qquad 0 \le x \le L, \tag{31}$$

where $g(x)$ is the initial velocity at the point x of the string.

The solution of this new problem can be obtained by following the procedure described above for the problem (1), (3), and (9). Upon separating variables, we find that the problem for $X(x)$ is exactly the same as before. Thus, once again, $\lambda = n^2 \pi^2 / L^2$ and $X(x)$ is proportional to $\sin(n\pi x/L)$. The differential equation for $T(t)$ is again Eq. (17), but the associated initial condition is now

$$T(0) = 0, \tag{32}$$

corresponding to the first of the initial conditions (31). The general solution of Eq. (17) is given by Eq. (18), but now the initial condition (32) requires that $k_1 = 0$. Therefore, $T(t)$ is now proportional to $\sin(n\pi a t/L)$ and the fundamental solutions for the problem (1), (3), and (31) are

$$u_n(x, t) = \sin \frac{n\pi x}{L} \sin \frac{n\pi a t}{L}, \qquad n = 1, 2, 3, \ldots. \tag{33}$$

Each of the functions $u_n(x, t)$ satisfies the wave equation (1), the boundary conditions (3), and the first of the initial conditions (31). The main consequence of using the initial conditions (31) rather than (9) is that the time-dependent factor in $u_n(x, t)$ involves a sine rather than a cosine.

To satisfy the remaining (nonhomogeneous) initial condition, we assume that $u(x, t)$ can be expressed as a linear combination of the fundamental solutions (33); that is,

$$u(x, t) = \sum_{n=1}^{\infty} k_n u_n(x, t) = \sum_{n=1}^{\infty} k_n \sin \frac{n\pi x}{L} \sin \frac{n\pi a t}{L}. \tag{34}$$

To determine the values of the coefficients k_n, we differentiate Eq. (34) with respect to t, set $t = 0$, and use the second initial condition (31); this gives the equation

$$u_t(x, 0) = \sum_{n=1}^{\infty} \frac{n\pi a}{L} k_n \sin \frac{n\pi x}{L} = g(x). \tag{35}$$

Hence the quantities $(n\pi a/L)k_n$ are the coefficients in the Fourier sine series of period $2L$ for g. Therefore,

$$\frac{n\pi a}{L} k_n = \frac{2}{L} \int_0^L g(x) \sin \frac{n\pi x}{L} \, dx, \qquad n = 1, 2, \ldots. \tag{36}$$

Thus Eq. (34), with the coefficients given by Eq. (36), constitutes a formal solution to the problem of Eqs. (1), (3), and (31). The validity of this formal solution can be established by arguments similar to those previously outlined for the solution of Eqs. (1), (3), and (9).

Finally, we turn to the problem consisting of the wave equation (1), the boundary conditions (3), and the general initial conditions (4), (5):

$$u(x, 0) = f(x), \qquad u_t(x, 0) = g(x), \quad 0 < x < L, \tag{37}$$

where $f(x)$ and $g(x)$ are the given initial position and velocity, respectively, of the string. Although this problem can be solved by separating variables, as in the cases discussed previously, it is important to note that it can also be solved simply by adding together the two solutions that we obtained above. To show that this is true, let $v(x, t)$ be the solution of the problem (1), (3), and (9), and let $w(x, t)$ be the solution of the problem (1), (3), and (31). Thus $v(x, t)$ is given by Eqs. (20) and (22), and $w(x, t)$ is given by Eqs. (34) and (36). Now let $u(x, t) = v(x, t) + w(x, t)$; what problem does $u(x, t)$ satisfy? First, observe that

$$a^2 u_{xx} - u_{tt} = (a^2 v_{xx} - v_{tt}) + (a^2 w_{xx} - w_{tt}) = 0 + 0 = 0, \tag{38}$$

so $u(x, t)$ satisfies the wave equation (1). Next, we have

$$u(0, t) = v(0, t) + w(0, t) = 0 + 0 = 0, \qquad u(L, t) = v(L, t) + w(L, t) = 0 + 0 = 0, \tag{39}$$

so $u(x, t)$ also satisfies the boundary conditions (3). Finally, we have

$$u(x, 0) = v(x, 0) + w(x, 0) = f(x) + 0 = f(x) \tag{40}$$

and

$$u_t(x, 0) = v_t(x, 0) + w_t(x, 0) = 0 + g(x) = g(x). \tag{41}$$

Thus $u(x, t)$ satisfies the general initial conditions (37).

We can restate the result we have just obtained in the following way. To solve the wave equation with the general initial conditions (37), you can solve instead the somewhat simpler problems with the initial conditions (9) and (31), respectively, and then add together the two solutions. This is another use of the principle of superposition.

PROBLEMS

Consider an elastic string of length L whose ends are held fixed. The string is set in motion with no initial velocity from an initial position $u(x, 0) = f(x)$. In each of Problems 1 through 4 carry out the following steps. Let $L = 10$ and $a = 1$ in parts (b) through (d).
(a) Find the displacement $u(x, t)$ for the given initial position $f(x)$.
(b) Plot $u(x, t)$ versus x for $0 \le x \le 10$ and for several values of t between $t = 0$ and $t = 20$.
(c) Plot $u(x, t)$ versus t for $0 \le t \le 20$ and for several values of x.
(d) Construct an animation of the solution in time for at least one period.
(e) Describe the motion of the string in a few sentences.

CAS **1.** $f(x) = \begin{cases} 2x/L, & 0 \le x \le L/2, \\ 2(L-x)/L, & L/2 < x \le L \end{cases}$

2. $f(x) = \begin{cases} 4x/L, & 0 \le x \le L/4, \\ 1, & L/4 < x < 3L/4, \\ 4(L-x)/L, & 3L/4 \le x \le L \end{cases}$ CAS

3. $f(x) = 8x(L-x)^2/L^3$ CAS

4. $f(x) = \begin{cases} 1, & L/2 - 1 < x < L/2 + 1 \ (L > 2), \\ 0, & \text{otherwise} \end{cases}$ CAS

Consider an elastic string of length L whose ends are held fixed. The string is set in motion from its equilibrium position with an initial velocity $u_t(x, 0) = g(x)$. In each of Problems 5 through 8 carry out the following steps. Let $L = 10$ and $a = 1$ in parts (b) through (d).
(a) Find the displacement $u(x, t)$ for the given $g(x)$.
(b) Plot $u(x, t)$ versus x for $0 \le x \le 10$ and for several CAS values of t between $t = 0$ and $t = 20$.

(c) Plot $u(x, t)$ versus t for $0 \le t \le 20$ and for several values of x.

(d) Construct an animation of the solution in time for at least one period.

(e) Describe the motion of the string in a few sentences.

CAS **5.** $g(x) = \begin{cases} 2x/L, & 0 \le x \le L/2, \\ 2(L-x)/L, & L/2 < x \le L \end{cases}$

CAS **6.** $g(x) = \begin{cases} 4x/L, & 0 \le x \le L/4, \\ 1, & L/4 < x < 3L/4, \\ 4(L-x)/L, & 3L/4 \le x \le L \end{cases}$

CAS **7.** $g(x) = 8x(L-x)^2/L^3$

CAS **8.** $g(x) = \begin{cases} 1, & L/2 - 1 < x < L/2 + 1 \quad (L > 2), \\ 0, & \text{otherwise} \end{cases}$

9. If an elastic string is free at one end, the boundary condition to be satisfied there is that $u_x = 0$. Find the displacement $u(x, t)$ in an elastic string of length L, fixed at $x = 0$ and free at $x = L$, set in motion with no initial velocity from the initial position $u(x, 0) = f(x)$, where f is a given function.

Hint: Show that the fundamental solutions for this problem, satisfying all conditions except the nonhomogeneous initial condition, are

$$u_n(x, t) = \sin \lambda_n x \cos \lambda_n a t,$$

where $\lambda_n = (2n - 1)\pi/2L, n = 1, 2, \ldots$. Compare this problem with Problem 15 of Section 9.6; pay particular attention to the extension of the initial data out of the original interval $[0, L]$.

CAS **10.** Consider an elastic string of length L. The end $x = 0$ is held fixed, while the end $x = L$ is free; thus the boundary conditions are $u(0, t) = 0$ and $u_x(L, t) = 0$. The string is set in motion with no initial velocity from the initial position $u(x, 0) = f(x)$, where

$$f(x) = \begin{cases} 1, & L/2 - 1 < x < L/2 + 1 \quad (L > 2), \\ 0, & \text{otherwise.} \end{cases}$$

(a) Find the displacement $u(x, t)$.

(b) With $L = 10$ and $a = 1$, plot u versus x for $0 \le x \le 10$ and for several values of t. Pay particular attention to values of t between 3 and 7. Observe how the initial disturbance is reflected at each end of the string.

(c) With $L = 10$ and $a = 1$, plot u versus t for several values of x.

(d) Construct an animation of the solution in time for at least one period.

(e) Describe the motion of the string in a few sentences.

11. Suppose that the string in Problem 10 CAS is started instead from the initial position $f(x) = 8x(L-x)^2/L^3$. Follow the instructions in Problem 10 for this new problem.

12. Dimensionless variables can be introduced into the wave equation $a^2 u_{xx} = u_{tt}$ in the following manner:

(a) Let $s = x/L$ and show that the wave equation becomes

$$a^2 u_{ss} = L^2 u_{tt}.$$

(b) Show that L/a has the dimensions of time and thus can be used as the unit on the time scale. Thus, let $\tau = at/L$ and show that the wave equation then reduces to

$$u_{ss} = u_{\tau\tau}.$$

Problems 13 and 14 indicate the form of the general solution of the wave equation and the physical significance of the constant a.

13. (a) Show that the wave equation

$$a^2 u_{xx} = u_{tt}$$

can be reduced to the form $u_{\xi\eta} = 0$ by the change of variables $\xi = x - at, \eta = x + at$.

(b) Show that $u(x, t)$ can be written as

$$u(x, t) = \phi(x - at) + \psi(x + at),$$

where ϕ and ψ are arbitrary functions.

14. (a) Plot the value of $\phi(x - at)$ for $t = 0, 1/a, 2/a$, CAS and t_0/a if $\phi(s) = \sin s$. Note that for any $t \ne 0$ the graph of $y = \phi(x - at)$ is the same as that of $y = \phi(x)$ when $t = 0$, but displaced a distance at in the positive x direction. Thus a represents the velocity at which a disturbance moves along the string.

(b) What is the interpretation of $\phi(x + at)$?

15. A steel wire 5 ft in length is stretched by a tensile force of 50 lb. The wire has a weight per unit length of 0.026 lb/ft.

(a) Find the velocity of propagation of transverse waves in the wire.

(b) Find the natural frequencies of vibration.

(c) If the tension in the wire is increased, how are the natural frequencies changed? Are the natural modes also changed?

16. Consider the wave equation

$$a^2 u_{xx} = u_{tt}$$

in an infinite one-dimensional medium subject to the initial conditions

$$u(x, 0) = f(x), \quad u_t(x, 0) = 0, \quad -\infty < x < \infty.$$

(a) Using the form of the solution obtained in Problem 13, show that ϕ and ψ must satisfy

$$\phi(x) + \psi(x) = f(x),$$
$$-\phi'(x) + \psi'(x) = 0.$$

(b) Solve the equations of part (a) for ϕ and ψ, and thereby show that

$$u(x, t) = \frac{1}{2}[f(x - at) + f(x + at)].$$

This form of the solution was obtained by D'Alembert in 1746.

Hint: Note that the equation $\psi'(x) = \phi'(x)$ is solved by choosing $\psi(x) = \phi(x) + c$.

(c) Let

$$f(x) = \begin{cases} 2, & -1 < x < 1, \\ 0, & \text{otherwise.} \end{cases}$$

Show that

$$f(x - at) = \begin{cases} 2, & -1 + at < x < 1 + at, \\ 0, & \text{otherwise.} \end{cases}$$

Also determine $f(x + at)$.

(d) Sketch the solution found in part (b) at $t = 0$, $t = 1/2a$, $t = 1/a$, and $t = 2/a$, obtaining the results shown in Figure 9.7.7. Observe that an initial displacement produces two waves moving in opposite directions away from the original location; each wave consists of one-half of the initial displacement.

17. Consider the wave equation

$$a^2 u_{xx} = u_{tt}$$

in an infinite one-dimensional medium subject to the initial conditions

$$u(x, 0) = 0, \quad u_t(x, 0) = g(x), \quad -\infty < x < \infty.$$

(a) Using the form of the solution obtained in Problem 13, show that

$$\phi(x) + \psi(x) = 0,$$
$$-a\phi'(x) + a\psi'(x) = g(x).$$

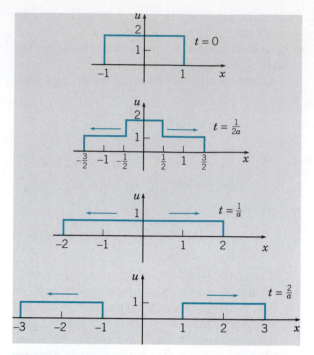

FIGURE 9.7.7 Propagation of initial disturbance in an infinite one-dimensional medium.

(b) Use the first equation of part (a) to show that $\psi'(x) = -\phi'(x)$. Then use the second equation to show that $-2a\phi'(x) = g(x)$ and therefore that

$$\phi(x) = -\frac{1}{2a} \int_{x_0}^x g(\xi)\, d\xi + \phi(x_0),$$

where x_0 is arbitrary. Finally, determine $\psi(x)$.

(c) Show that

$$u(x, t) = \frac{1}{2a} \int_{x-at}^{x+at} g(\xi)\, d\xi.$$

18. By combining the results of Problems 16 and 17, show that the solution of the problem

$$a^2 u_{xx} = u_{tt},$$
$$u(x, 0) = f(x), \quad u_t(x, 0) = g(x), \quad -\infty < x < \infty$$

is given by

$$u(x, t) = \frac{1}{2}[f(x - at) + f(x + at)]$$

$$+ \frac{1}{2a} \int_{x-at}^{x+at} g(\xi)\, d\xi.$$

Problems 19 and 20 indicate how the formal solution (20), (22) of Eqs. (1), (3), and (9) can be shown to constitute the actual solution of that problem.

19. By using the trigonometric identity $\sin A \cos B = \frac{1}{2}[\sin(A + B) + \sin(A - B)]$, show that the solution (20) of the problem of Eqs. (1), (3), and (9) can be written in the form (28).

20. Let $h(\xi)$ represent the initial displacement in $[0, L]$, extended into $(-L, 0)$ as an odd function and extended elsewhere as a periodic function of period $2L$. Assuming that h, h', and h'' are continuous, show by direct differentiation that $u(x, t)$ as given in Eq. (28) satisfies the wave equation (1) and also the initial conditions (9). Note also that since Eq. (20) clearly satisfies the boundary conditions (3), the same is true of Eq. (28). Comparing Eq. (28) with the solution of the corresponding problem for the infinite string (Problem 16), we see that they have the same form, provided that the initial data for the finite string, defined originally only on the interval $0 \leq x \leq L$, are extended in the given manner over the entire x-axis. If this is done, the solution for the infinite string is also applicable to the finite one.

21. The motion of a circular elastic membrane, such as a drumhead, is governed by the two-dimensional wave equation in polar coordinates

$$u_{rr} + (1/r)u_r + (1/r^2)u_{\theta\theta} = a^{-2}u_{tt}.$$

Assuming that $u(r, \theta, t) = R(r)\Theta(\theta)T(t)$, find ordinary differential equations satisfied by $R(r)$, $\Theta(\theta)$, and $T(t)$.

22. The total energy $E(t)$ of the vibrating string is given as a function of time by

$$E(t) = \int_0^L \left[\frac{1}{2}\rho u_t^2(x, t) + \frac{1}{2}T u_x^2(x, t)\right] dx; \quad \text{(i)}$$

the first term is the kinetic energy due to the motion of the string, and the second term is the potential energy created by the displacement of the string away from its equilibrium position.

For the displacement $u(x, t)$ given by Eq. (20)—that is, for the solution of the string problem with zero initial velocity—show that

$$E(t) = \frac{\pi^2 T}{4L}\sum_{n=1}^{\infty} n^2 c_n^2. \quad \text{(ii)}$$

Note that the right side of Eq. (ii) does not depend on t. Thus the total energy E is a constant

and therefore is *conserved* during the motion of the string.

Hint: Use Parseval's equation (Problem 37 of Section 9.4 and Problem 17 of Section 9.3), and recall that $a^2 = T/\rho$.

23. Dispersive Waves. Consider the modified wave equation

$$a^{-2}u_{tt} + \gamma^2 u = u_{xx}, \quad 0 < x < L, \quad t > 0 \quad \text{(i)}$$

with the boundary conditions

$$u(0, t) = 0, \quad u(L, t) = 0, \quad t > 0 \quad \text{(ii)}$$

and the initial conditions

$$u(x, 0) = f(x), \quad u_t(x, 0) = 0, \quad 0 < x < L. \quad \text{(iii)}$$

(a) Show that the solution can be written as

$$u(x, t) = \sum_{n=1}^{\infty} c_n \cos\left(\sqrt{\frac{n^2\pi^2}{L^2} + \gamma^2}\, at\right)\sin\frac{n\pi x}{L},$$

where

$$c_n = \frac{2}{L}\int_0^L f(x)\sin\frac{n\pi x}{L}\, dx.$$

(b) By using trigonometric identities, rewrite the solution as

$$u(x, t) = \frac{1}{2}\sum_{n=1}^{\infty} c_n\left[\sin\frac{n\pi}{L}(x + a_n t) + \sin\frac{n\pi}{L}(x - a_n t)\right].$$

Determine a_n, the speed of wave propagation.

(c) Observe that a_n, found in part **(b)**, depends on n. This means that components of different wave lengths (or frequencies) are propagated at different speeds, resulting in a distortion of the original wave form over time. This phenomenon is called **dispersion**. Find the condition under which a_n is independent of n—that is, there is no dispersion.

24. Consider the situation in Problem 23 with $a^2 = 1$, CAS $L = 10$, and

$$f(x) = \begin{cases} x - 4, & 4 \leq x \leq 5, \\ 6 - x, & 5 \leq x \leq 6, \\ 0, & \text{otherwise.} \end{cases}$$

(a) Determine the coefficients c_n in the solution of Problem 23(a).

(b) Plot

$$\sum_{n=1}^{N} c_n \sin \frac{n\pi x}{10} \quad \text{for} \quad 0 \le x \le 10,$$

choosing N large enough so that the plot accurately displays the graph of $f(x)$. Use this value of N for the remaining plots called for in this problem.

(c) Let $\gamma = 0$. Plot $u(x, t)$ versus x for $t = 60$.

(d) Let $\gamma = 1/8$. Plot $u(x, t)$ versus x for $t = 20$, 40, and 60.

(e) Let $\gamma = 1/4$. Plot $u(x, t)$ versus x for $t = 20$, 40, and 60.

9.8 Laplace's Equation

Laplace's equation in two dimensions is

$$u_{xx} + u_{yy} = 0, \tag{1}$$

and in three dimensions

$$u_{xx} + u_{yy} + u_{zz} = 0. \tag{2}$$

Laplace's equation describes many types of physical systems that are in an equilibrium state. For example, in a two-dimensional heat conduction problem, the temperature $u(x, y, t)$ must satisfy the differential equation

$$\alpha^2 (u_{xx} + u_{yy}) = u_t, \tag{3}$$

where α^2 is the thermal diffusivity. If a steady state exists, u is a function of x and y only, and the time derivative vanishes; in this case Eq. (3) reduces to Eq. (1). Similarly, for the steady-state heat conduction problem in three dimensions, the temperature must satisfy the three-dimensional form of Laplace's equation. Equations (1) and (2) also occur in other branches of mathematical physics. In the consideration of electrostatic fields, the electric potential function in a dielectric medium containing no electric charges must satisfy either Eq. (1) or Eq. (2), depending on the number of space dimensions involved. Similarly, the potential function of a particle in free space acted on only by gravitational forces satisfies the same equations. Consequently, Laplace's equation is often referred to as the **potential equation**. Another example arises in the study of the steady (time-independent), two-dimensional, inviscid, irrotational motion of an incompressible fluid. This study centers on two functions, known as the velocity potential function and the stream function, both of which satisfy Eq. (1). In elasticity, the displacements that occur when a perfectly elastic bar is twisted are described in terms of the so-called warping function, which also satisfies Eq. (1).

Since there is no time dependence in any of the problems mentioned previously, there are no initial conditions to be satisfied by the solutions of Eq. (1) or (2). They must, however, satisfy certain boundary conditions on the bounding curve or surface of the region in which the differential equation is to be solved. Since Laplace's equation is of second order, it might be plausible to expect that two boundary conditions would be required to determine the solution completely. This, however, is not the case. Recall that in the heat conduction problem for the finite bar (Sections 9.5 and 9.6) it was necessary to prescribe one condition at each end of the bar—that is, *one condition at each point of the boundary*. If we generalize this observation to multidimensional problems, it is then natural to prescribe one condition on the function u at each point on the boundary of the region in which a solution of

Eq. (1) or (2) is sought. The most common boundary condition occurs when the value of u is specified at each boundary point; in terms of the heat conduction problem, this corresponds to prescribing the temperature on the boundary. In some problems the value of the derivative, or rate of change, of u in the direction normal to the boundary is specified instead; the condition on the boundary of a thermally insulated body, for example, is of this type. It is entirely possible for more complicated boundary conditions to occur; for example, u might be prescribed on part of the boundary and its normal derivative specified on the remainder. The problem of finding a solution of Laplace's equation that takes on given boundary values is known as a **Dirichlet problem**. In contrast, if the values of the normal derivative are prescribed on the boundary, the problem is said to be a **Neumann problem**. The Dirichlet and Neumann problems are also known as the first and second boundary value problems of potential theory, respectively.

Physically, it is plausible to expect that the types of boundary conditions just mentioned will be sufficient to determine the solution completely. Indeed, it is possible to establish the existence and uniqueness of the solution of Laplace's equation under the boundary conditions mentioned, provided that the shape of the boundary and the functions appearing in the boundary conditions satisfy certain very mild requirements. However, the proofs of these theorems, and even their accurate statement, are beyond the scope of the present book. Our only concern will be solving some typical problems by means of separation of variables and Fourier series.

While the problems chosen as examples have interesting physical interpretations (in terms of electrostatic potentials or steady state temperature distributions, for instance), our purpose here is primarily to point out some of the features that may occur during their mathematical solution. It is also worth noting again that more complicated problems can sometimes be solved by expressing the solution as the sum of solutions of several simpler problems (see Problems 3 and 4).

Dirichlet Problem for a Rectangle.

Consider the mathematical problem of finding the function u satisfying Laplace's equation (1)

$$u_{xx} + u_{yy} = 0,$$

in the rectangle $0 < x < a, 0 < y < b$, and also satisfying the boundary conditions

$$\begin{aligned} u(x,0) &= 0, & u(x,b) = 0, & \quad 0 < x < a, \\ u(0,y) &= 0, & u(a,y) = f(y), & \quad 0 \le y \le b, \end{aligned} \tag{4}$$

where f is a given function on $0 \le y \le b$ (see Figure 9.8.1).

To solve this problem, we wish to construct a fundamental set of solutions satisfying the partial differential equation and the homogeneous boundary conditions; then we will superpose these solutions so as to satisfy the remaining boundary condition. Let us assume that

$$u(x,y) = X(x)Y(y) \tag{5}$$

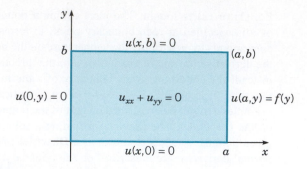

FIGURE 9.8.1 Dirichlet problem for a rectangle.

and substitute for u in Eq. (1). This yields

$$\frac{X''}{X} = -\frac{Y''}{Y} = \lambda,$$

where λ is the separation constant. Thus we obtain the two ordinary differential equations

$$X'' \quad - \quad \lambda X = 0, \tag{6}$$

$$Y'' \quad + \quad \lambda Y = 0. \tag{7}$$

If we now substitute for u from Eq. (5) in each of the homogeneous boundary conditions, we find that

$$X(0) = 0 \tag{8}$$

and

$$Y(0) = 0, \qquad Y(b) = 0. \tag{9}$$

We will first determine the solution of the differential equation (7) subject to the boundary conditions (9). However, this problem is essentially identical to one encountered previously in Sections 9.1, 9.5, and 9.7. We conclude that there are nontrivial solutions if and only if λ is an eigenvalue, namely,

$$\lambda = (n\pi/b)^2, \qquad n = 1, 2, \ldots, \tag{10}$$

and $Y(y)$ is proportional to the corresponding eigenfunction $\sin(n\pi y/b)$. Next, we substitute from Eq. (10) for λ in Eq. (6) and solve this equation subject to the boundary condition (8). It is convenient to write the general solution of Eq. (6) as

$$X(x) = k_1 \cosh(n\pi x/b) + k_2 \sinh(n\pi x/b), \tag{11}$$

and the boundary condition (8) then requires that $k_1 = 0$. Therefore $X(x)$ must be proportional to $\sinh(n\pi x/b)$. Thus we obtain the fundamental solutions

$$u_n(x, y) = \sinh\frac{n\pi x}{b} \sin\frac{n\pi y}{b}, \qquad n = 1, 2, \ldots. \tag{12}$$

These functions satisfy the differential equation (1) and all the homogeneous boundary conditions for each value of n.

To satisfy the remaining nonhomogeneous boundary condition at $x = a$, we assume, as usual, that we can represent the solution $u(x, y)$ in the form

$$u(x, y) = \sum_{n=1}^{\infty} c_n u_n(x, y) = \sum_{n=1}^{\infty} c_n \sinh \frac{n\pi x}{b} \sin \frac{n\pi y}{b}. \tag{13}$$

The coefficients c_n are determined by the boundary condition

$$u(a, y) = \sum_{n=1}^{\infty} c_n \sinh \frac{n\pi a}{b} \sin \frac{n\pi y}{b} = f(y). \tag{14}$$

Therefore, the quantities $c_n \sinh (n\pi a/b)$ must be the coefficients in the Fourier sine series of period $2b$ for f and are given by

$$c_n \sinh \frac{n\pi a}{b} = \frac{2}{b} \int_0^b f(y) \sin \frac{n\pi y}{b}\, dy. \tag{15}$$

Thus the solution of the partial differential equation (1) satisfying the boundary conditions (4) is given by Eq. (13) with the coefficients c_n computed from Eq. (15).

From Eqs. (13) and (15) we see that the solution contains the factor $\sinh (n\pi x/b)/ \sinh (n\pi a/b)$. To estimate this quantity for large n, we can use the approximation $\sinh \xi \cong e^{\xi}/2$, and thereby obtain

$$\frac{\sinh (n\pi x/b)}{\sinh (n\pi a/b)} \cong \frac{\frac{1}{2}\exp (n\pi x/b)}{\frac{1}{2}\exp (n\pi a/b)} = \exp\left[- n\pi(a - x)/b\right].$$

Thus this factor has the character of a negative exponential; consequently, the series (13) converges quite rapidly unless $a - x$ is very small.

EXAMPLE 1

To illustrate these results, let $a = 3$, $b = 2$, and

$$f(y) = \begin{cases} y, & 0 \le y \le 1, \\ 2 - y, & 1 \le y \le 2. \end{cases} \tag{16}$$

By evaluating c_n from Eq. (15), we find that

$$c_n = \frac{8 \sin (n\pi/2)}{n^2\pi^2 \sinh (3n\pi/2)}. \tag{17}$$

Then $u(x, y)$ is given by Eq. (13). Keeping 20 terms in the series, we can plot u versus x and y, as shown in Figure 9.8.2. Alternatively, we can construct a contour plot showing level curves of $u(x, y)$; Figure 9.8.3 is such a plot, with an increment of 0.1 between adjacent curves.

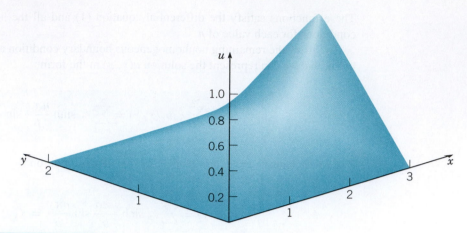

FIGURE 9.8.2 Plot of u versus x and y for Example 1.

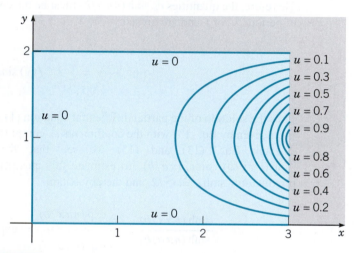

FIGURE 9.8.3 Level curves of $u(x, y)$ for Example 1.

Dirichlet Problem for a Circle.

Consider the problem of solving Laplace's equation in a circular region $r < a$ subject to the boundary condition

$$u(a, \theta) = f(\theta), \tag{18}$$

where f is a given function on $0 \le \theta < 2\pi$ (see Figure 9.8.4). In polar coordinates Laplace's equation has the form

$$u_{rr} + \frac{1}{r} u_r + \frac{1}{r^2} u_{\theta\theta} = 0. \tag{19}$$

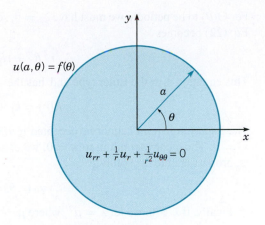

FIGURE 9.8.4 Dirichlet problem for a circle.

To complete the statement of the problem, we note that for $u(r, \theta)$ to be single-valued, it is necessary that u be periodic in θ with period 2π. Moreover, we state explicitly that $u(r, \theta)$ must be bounded for $r \leq a$, since this will become important later.

To apply the method of separation of variables to this problem, we assume that

$$u(r, \theta) = R(r)\Theta(\theta), \tag{20}$$

and substitute for u in the differential equation (19). This yields

$$R''\Theta + \frac{1}{r}R'\Theta + \frac{1}{r^2}R\Theta'' = 0,$$

or

$$r^2 \frac{R''}{R} + r\frac{R'}{R} = -\frac{\Theta''}{\Theta} = \lambda, \tag{21}$$

where λ is the separation constant. Thus we obtain the two ordinary differential equations

$$r^2 R'' + rR' - \lambda R = 0, \tag{22}$$

$$\Theta'' + \lambda \Theta = 0. \tag{23}$$

In this problem there are no homogeneous boundary conditions; recall, however, that solutions must be bounded and also periodic in θ with period 2π. It is possible to show (Problem 9) that the periodicity condition requires λ to be real. We will consider in turn the cases in which λ is negative, zero, and positive.

If $\lambda < 0$, let $\lambda = -\mu^2$, where $\mu > 0$. Then Eq. (23) becomes $\Theta'' - \mu^2\Theta = 0$, and consequently,

$$\Theta(\theta) = c_1 e^{\mu\theta} + c_2 e^{-\mu\theta}. \tag{24}$$

Thus $\Theta(\theta)$ can be periodic only if $c_1 = c_2 = 0$, and we conclude that λ cannot be negative. If $\lambda = 0$, then Eq. (23) becomes $\Theta'' = 0$, and thus

$$\Theta(\theta) = c_1 + c_2\theta. \tag{25}$$

For $\Theta(\theta)$ to be periodic we must have $c_2 = 0$, so that $\Theta(\theta)$ is a constant. Further, for $\lambda = 0$, Eq. (22) becomes

$$r^2 R'' + r R' = 0. \tag{26}$$

This equation is of the Euler type and has the solution

$$R(r) = k_1 + k_2 \ln r. \tag{27}$$

The logarithmic term cannot be accepted if $u(r, \theta)$ is to remain bounded as $r \to 0$; hence $k_2 = 0$. Thus, corresponding to $\lambda = 0$, we conclude that $u(r, \theta)$ must be a constant—that is, proportional to the solution

$$u_0(r, \theta) = 1. \tag{28}$$

Finally, if $\lambda > 0$, we let $\lambda = \mu^2$, where $\mu > 0$. Then Eqs. (22) and (23) become

$$r^2 R'' + r R' - \mu^2 R = 0 \tag{29}$$

and

$$\Theta'' + \mu^2 \Theta = 0, \tag{30}$$

respectively. Equation (29) is an Euler equation and has the solution

$$R(r) = k_1 r^\mu + k_2 r^{-\mu}, \tag{31}$$

while Eq. (30) has the solution

$$\Theta(\theta) = c_1 \sin \mu\theta + c_2 \cos \mu\theta. \tag{32}$$

In order for Θ to be periodic with period 2π, it is necessary for μ to be a positive integer n. With $\mu = n$ it follows that the solution $r^{-\mu}$ in Eq. (31) must be discarded since it becomes unbounded as $r \to 0$. Consequently, $k_2 = 0$ and the appropriate solutions of Eq. (19) are

$$u_n(r, \theta) = r^n \cos n\theta, \qquad v_n(r, \theta) = r^n \sin n\theta, \qquad n = 1, 2, \ldots . \tag{33}$$

These functions, together with $u_0(r, \theta) = 1$, form a set of fundamental solutions for the present problem.

In the usual way we now assume that u can be expressed as a linear combination of the fundamental solutions; that is,

$$u(r, \theta) = \frac{c_0}{2} + \sum_{n=1}^{\infty} r^n (c_n \cos n\theta + k_n \sin n\theta). \tag{34}$$

The boundary condition (18) then requires that

$$u(a, \theta) = \frac{c_0}{2} + \sum_{n=1}^{\infty} a^n (c_n \cos n\theta + k_n \sin n\theta) = f(\theta) \tag{35}$$

for $0 \le \theta < 2\pi$. The function f may be extended outside this interval so that it is periodic with period 2π and therefore has a Fourier series of the form (35). Since the extended function has period 2π, we may compute its Fourier coefficients by integrating over any period of the function. In particular, it is convenient to use the original interval $(0, 2\pi)$; then

$$a^n c_n = \frac{1}{\pi} \int_0^{2\pi} f(\theta) \cos n\theta \, d\theta, \qquad n = 0, 1, 2, \ldots; \tag{36}$$

$$a^n k_n = \frac{1}{\pi} \int_0^{2\pi} f(\theta) \sin n\theta \, d\theta, \qquad n = 1, 2, \ldots. \tag{37}$$

With this choice of the coefficients, Eq. (34) represents the solution of the boundary value problem of Eqs. (18) and (19). Note that in this problem we needed both sine and cosine terms in the solution. This is because the boundary data were given on $0 \le \theta < 2\pi$ and have period 2π. As a consequence, the full Fourier series is required, rather than sine or cosine terms alone.

PROBLEMS

CAS 1. (a) Find the solution $u(x, y)$ of Laplace's equation in the rectangle $0 < x < a$, $0 < y < b$, that satisfies the boundary conditions

$$u(0, y) = 0, \quad u(a, y) = 0, \qquad 0 < y < b,$$
$$u(x, 0) = 0, \quad u(x, b) = g(x), \qquad 0 \le x \le a.$$

(b) Find the solution if

$$g(x) = \begin{cases} x, & 0 \le x \le a/2, \\ a - x, & a/2 \le x \le a. \end{cases}$$

(c) For $a = 3$ and $b = 1$ plot u versus x for several values of y and also plot u versus y for several values of x.

(d) Plot u versus both x and y in three dimensions. Also draw a contour plot showing several level curves of $u(x, y)$ in the xy-plane.

2. Find the solution $u(x, y)$ of Laplace's equation in the rectangle $0 < x < a$, $0 < y < b$, that satisfies the boundary conditions

$$u(0, y) = 0, \qquad u(a, y) = 0, \qquad 0 < y < b,$$
$$u(x, 0) = h(x), \qquad u(x, b) = 0, \qquad 0 \le x \le a.$$

CAS 3. (a) Find the solution $u(x, y)$ of Laplace's equation in the rectangle $0 < x < a$, $0 < y < b$, that satisfies the boundary conditions

$$u(0, y) = 0, \qquad u(a, y) = f(y), \qquad 0 < y < b,$$
$$u(x, 0) = h(x), \quad u(x, b) = 0, \qquad 0 \le x \le a.$$

Hint: Consider the possibility of adding the solutions of two problems, one with

homogeneous boundary conditions except for $u(a, y) = f(y)$, and the other with homogeneous boundary conditions except for $u(x, 0) = h(x)$.

(b) Find the solution if $h(x) = (x/a)^2$ and $f(y) = 1 - (y/b)$.

(c) Let $a = 2$ and $b = 2$. Plot the solution in several ways: u versus x, u versus y, u versus both x and y, and a contour plot.

4. Show how to find the solution $u(x, y)$ of Laplace's equation in the rectangle $0 < x < a$, $0 < y < b$, that satisfies the boundary conditions

$$u(0, y) = k(y), \quad u(a, y) = f(y), \qquad 0 < y < b,$$
$$u(x, 0) = h(x), \quad u(x, b) = g(x), \qquad 0 \le x \le a.$$

Hint: See Problem 3.

5. Find the solution $u(r, \theta)$ of Laplace's equation

$$u_{rr} + (1/r)u_r + (1/r^2)u_{\theta\theta} = 0$$

outside the circle $r = a$, that satisfies the boundary condition

$$u(a, \theta) = f(\theta), \qquad 0 \le \theta < 2\pi,$$

on the circle. Assume that $u(r, \theta)$ is single-valued and bounded for $r > a$.

6. (a) Find the solution $u(r, \theta)$ of Laplace's equation CAS in the semicircular region $r < a$, $0 < \theta < \pi$, that satisfies the boundary conditions

$$u(r, 0) = 0, \qquad u(r, \pi) = 0, \quad 0 \le r < a,$$
$$u(a, \theta) = f(\theta), \qquad 0 \le \theta \le \pi.$$

Assume that u is single-valued and bounded in the given region.

(b) Find the solution if $f(\theta) = \theta(\pi - \theta)$.

(c) Let $a = 2$ and plot the solution in several ways: u versus r, u versus θ, u versus both r and θ, and a contour plot.

7. Find the solution $u(r, \theta)$ of Laplace's equation in the circular sector $0 < r < a, 0 < \theta < \alpha$, that satisfies the boundary conditions

$$u(r, 0) = 0, \qquad u(r, \alpha) = 0, \qquad 0 \le r < a,$$
$$u(a, \theta) = f(\theta), \qquad 0 \le \theta \le \alpha.$$

Assume that u is single-valued and bounded in the sector.

CAS 8. (a) Find the solution $u(x, y)$ of Laplace's equation in the semi-infinite strip $0 < x < a, y > 0$, that satisfies the boundary conditions

$$u(0, y) = 0, \qquad u(a, y) = 0, \qquad y > 0,$$
$$u(x, 0) = f(x), \qquad 0 \le x \le a$$

and the additional condition that $u(x, y) \to 0$ as $y \to \infty$.

(b) Find the solution if $f(x) = x(a - x)$.

(c) Let $a = 5$. Find the smallest value of y_0 for which $u(x, y) \le 0.1$ for all $y \ge y_0$.

9. Show that Eq. (23) has periodic solutions only if λ is real.
Hint: Let $\lambda = -\mu^2$, where $\mu = \nu + i\sigma$ with ν and σ real.

10. Consider the problem of finding a solution $u(x, y)$ of Laplace's equation in the rectangle $0 < x < a$, $0 < y < b$, that satisfies the boundary conditions

$$u_x(0, y) = 0, \qquad u_x(a, y) = f(y), \qquad 0 < y < b,$$
$$u_y(x, 0) = 0, \qquad u_y(x, b) = 0, \qquad 0 \le x \le a.$$

This is an example of a Neumann problem.

(a) Show that Laplace's equation and the homogeneous boundary conditions determine the fundamental set of solutions

$$u_0(x, y) = c_0,$$
$$u_n(x, y) = c_n \cosh(n\pi x/b) \cos(n\pi y/b),$$
$$n = 1, 2, 3, \ldots.$$

(b) By superposing the fundamental solutions of part (a), formally determine a function u satisfying the nonhomogeneous boundary condition $u_x(a, y) = f(y)$. Note that when $u_x(a, y)$ is calculated, the constant term in $u(x, y)$ is eliminated, and there is no condition from which to determine c_0. Furthermore, it must

be possible to express f by means of a Fourier cosine series of period $2b$, which does not have a constant term. This means that

$$\int_0^b f(y)\, dy = 0$$

is a necessary condition for the given problem to be solvable. Finally, note that c_0 remains arbitrary, and hence the solution is determined only up to this additive constant. This is a property of all Neumann problems.

11. Find a solution $u(r, \theta)$ of Laplace's equation inside the circle $r = a$ that satisfies the boundary condition on the circle

$$u_r(a, \theta) = g(\theta), \qquad 0 \le \theta < 2\pi.$$

Note that this is a Neumann problem and that its solution is determined only up to an arbitrary additive constant. State a necessary condition on $g(\theta)$ for this problem to be solvable by the method of separation of variables (see Problem 10).

12. (a) Find the solution $u(x, y)$ of Laplace's equa- CAS tion in the rectangle $0 < x < a, 0 < y < b$, that satisfies the boundary conditions

$$u(0, y) = 0, \quad u(a, y) = 0, \qquad 0 < y < b,$$
$$u_y(x, 0) = 0, \quad u(x, b) = g(x), \qquad 0 \le x \le a.$$

Note that this is neither a Dirichlet nor a Neumann problem, but a mixed problem in which u is prescribed on part of the boundary and its normal derivative on the rest.

(b) Find the solution if

$$g(x) = \begin{cases} x, & 0 \le x \le a/2, \\ a - x, & a/2 \le x \le a. \end{cases}$$

(c) Let $a = 3$ and $b = 1$. By drawing suitable plots, compare this solution with the solution of Problem 1.

13. (a) Find the solution $u(x, y)$ of Laplace's equa- CAS tion in the rectangle $0 < x < a, 0 < y < b$, that satisfies the boundary conditions

$$u(0, y) = 0, \quad u(a, y) = f(y), \qquad 0 < y < b,$$
$$u(x, 0) = 0, \quad u_y(x, b) = 0, \qquad 0 \le x \le a.$$

Hint: Eventually, it will be necessary to expand $f(y)$ in a series that makes use of the functions $\sin(\pi y/2b), \sin(3\pi y/2b), \sin(5\pi y/2b), \ldots$ (see Problem 39 of Section 9.4).

(b) Find the solution if $f(y) = y(2b - y)$.

(c) Let $a = 3$ and $b = 2$; plot the solution in several ways.

CAS **14. (a)** Find the solution $u(x, y)$ of Laplace's equation in the rectangle $0 < x < a$, $0 < y < b$, that satisfies the boundary conditions

$$u_x(0, y) = 0, \quad u_x(a, y) = 0, \qquad 0 < y < b,$$
$$u(x, 0) = 0, \quad u(x, b) = g(x), \quad 0 \leq x \leq a.$$

(b) Find the solution if $g(x) = 1 + x^2(x - a)^2$.

(c) Let $a = 3$ and $b = 2$; plot the solution in several ways.

15. By writing Laplace's equation in cylindrical coordinates r, θ, and z and then assuming that the solution is axially symmetric (no dependence on θ), we obtain the equation

$$u_{rr} + (1/r)u_r + u_{zz} = 0.$$

Assuming that $u(r, z) = R(r)Z(z)$, show that R and Z satisfy the equations

$$rR'' + R' + \lambda^2 rR = 0, \qquad Z'' - \lambda^2 Z = 0.$$

The equation for R is Bessel's equation of order zero with independent variable λr.

CAS **16. Flow in an Aquifer.** Consider the flow of water in a porous medium, such as sand, in an aquifer. The flow is driven by the hydraulic head, a measure of the potential energy of the water above the aquifer. Let $R : 0 < x < a, 0 < z < b$ be a vertical section of an aquifer. In a uniform, homogeneous medium the hydraulic head $u(x, z)$ satisfies Laplace's equation

$$u_{xx} + u_{zz} = 0 \text{ in } R. \qquad \text{(i)}$$

If water cannot flow through the sides and bottom of R, then the boundary conditions there are

$$u_x(0, z) = 0, \quad u_x(a, z) = 0, \quad 0 \leq z \leq b \qquad \text{(ii)}$$
$$u_z(x, 0) = 0, \qquad 0 \leq x \leq a. \qquad \text{(iii)}$$

Finally, suppose that the boundary condition at $z = b$ is

$$u(x, b) = b + \alpha x, \qquad 0 \leq x \leq a, \qquad \text{(iv)}$$

where α is the slope of the water table.

(a) Solve the given boundary value problem for $u(x, z)$.

(b) Let $a = 1000$, $b = 500$, and $\alpha = 0.1$. Draw a contour plot of the solution in R; that is, plot some level curves of $u(x, z)$.

(c) Water flows along paths in R that are orthogonal to the level curves of $u(x, z)$ in the direction of decreasing u. Plot some of the flow paths.

CHAPTER SUMMARY

Section 9.1   Two-Point Boundary Value Problems

▶ **The two-point boundary value problem**

$$y'' + p(x)y' + q(x)y = g(x), \quad \alpha < x < \beta, \qquad y(\alpha) = y_1, \quad y(\beta) = y_2$$

is **nonhomogeneous** if at least one of the **boundary values** y_1 and y_2 is nonzero, or if the function g is not everywhere zero. If both of the boundary values y_1 and y_2 are zero, and g is the zero function, then the resulting two-point boundary value problem

$$y'' + p(x)y' + q(x)y = 0, \quad \alpha < x < \beta, \qquad y(\alpha) = 0, \quad y(\beta) = 0$$

is **homogeneous**.

  ▶ If $y = 0$ is the unique solution of the homogeneous two-point boundary value problem, then the nonhomogeneous two-point boundary value problem has a unique solution.

  ▶ If the solution $y = 0$ of the homogeneous two-point boundary value problem is not unique, then the nonhomogeneous two-point boundary value problem may have no solution or infinitely many solutions.

▶ The problem

$$y'' + \lambda y = 0, \quad \alpha < x < \beta, \qquad y(\alpha) = 0, \quad y(\beta) = 0$$

containing the parameter λ is an example of a special type of two-point boundary value problem, called an **eigenvalue problem**. Values of λ for which nonzero solutions exist are called **eigenvalues**, and the corresponding nonzero solutions are called **eigenfunctions**.

Section 9.2 Fourier Series

Section 9.3 The Fourier Convergence Theorem

▶ If f and f' are piecewise continuous on the interval $[-L, L]$, then the **Fourier series** of f is the **trigonometric series**

$$\frac{a_0}{2} + \sum_{n=1}^{\infty}\left(a_n \cos \frac{n\pi x}{L} + b_n \sin \frac{n\pi x}{L}\right),$$

where the **Fourier coefficients** a_n and b_n are given by the **Euler–Fourier formulas**

$$a_n = \frac{1}{L}\int_{-L}^{L} f(x)\cos \frac{n\pi x}{L}\, dx, \qquad n = 0, 1, 2, \ldots,$$

$$b_n = \frac{1}{L}\int_{-L}^{L} f(x)\sin \frac{n\pi x}{L}\, dx, \qquad n = 1, 2, 3, \ldots.$$

▶ If F denotes the periodic extension of f to the entire real line, then the Fourier series of f converges to $F(x)$ when x is a point of continuity of F, and to $[F(x+) + F(x-)]/2$ when x is a point of discontinuity of F.

▶ If we let $\phi_n(x) = \cos(n\pi x/L)$, $n = 0, 1, 2, \ldots$ and $\psi_n(x) = \sin(n\pi x/L)$, $n = 1, 2, \ldots$, then in terms of the inner product (u, v) defined by $(u, v) = \int_{-L}^{L} u(x)v(x)\, dx$, the Fourier coefficients are given by

$$a_n = \frac{(f, \phi_n)}{(\phi_n, \phi_n)}, \qquad n = 0, 1, 2, \ldots; \qquad b_n = \frac{(f, \psi_n)}{(\psi_n, \psi_n)}, \qquad n = 1, 2, \ldots.$$

Section 9.4 Even and Odd Functions

▶ If f and f' are piecewise continuous on $[0, L]$, then the **Fourier cosine series** of f is the Fourier series of the **even extension** of f to the interval $[-L, L]$,

$$\frac{a_0}{2} + \sum_{n=1}^{\infty} a_n \cos \frac{n\pi x}{L},$$

where

$$a_n = \frac{2}{L}\int_{0}^{L} f(x)\cos \frac{n\pi x}{L}\, dx, \qquad n = 0, 1, 2, \ldots.$$

▶ If f and f' are piecewise continuous on $[0, L]$, then the **Fourier sine series** of f is the Fourier series of the **odd extension** of f to the interval $[-L, L]$:

$$\sum_{n=1}^{\infty} b_n \sin \frac{n\pi x}{L},$$

where

$$b_n = \frac{2}{L}\int_{0}^{L} f(x)\sin \frac{n\pi x}{L}\, dx, \qquad n = 1, 2, \ldots.$$

Section 9.5   Separation of Variables; Heat Conduction in a Rod

Section 9.6   Other Heat Conduction Problems

▶ The classical equation for unidirectional heat conduction in a rod of finite length L is

$$\alpha^2 u_{xx} = u_t, \qquad 0 < x < L, \quad t > 0,$$

where $u(x, t)$ represents the temperature at time t and position x along the rod.

▶ The initial temperature profile in the rod is specified by an **initial condition** $u(x, 0) = f(x), 0 < x < L$.

▶ One of the following homogeneous **boundary conditions** is commonly imposed at $x = 0$:

$$u(0, t) = 0, \quad t > 0 \qquad \text{Left end held at temperature zero}$$
$$u_x(0, t) = 0, \quad t > 0 \qquad \text{Left end insulated}$$
$$u_x(0, t) - h_1 u(0, t) = 0, \quad t > 0 \qquad \text{Heat flow at left end proportional to}$$
$$\text{temperature}$$

A boundary condition analogous to one of the above is also imposed at $x = L$.

▶ Nonhomogeneous boundary conditions may be accommodated by finding a function $v(x)$ (often a steady-state solution of the heat equation) that satisfies the boundary conditions, assuming that $u(x, t) = v(x) + w(x, t)$, and finding the initial boundary value problem with homogeneous boundary conditions satisfied by $w(x, t)$.

▶ The classical technique known as the **method of separation of variables** used to solve the homogeneous heat equation with homogeneous boundary conditions consists of the following steps:

1. Assume there exists a solution of the heat equation of the form $u(x, t) = X(x)T(t)$.

2. Find an eigenvalue problem satisfied by X by substituting $u = XT$ into the heat equation and imposing the boundary conditions specified for the heat equation.

3. Find the eigenvalues and eigenfunctions to obtain fundamental solutions of the heat equation in the form $u_n(x, t) = X_n(x)T_n(t)$ that satisfy the heat equation and the boundary condtions.

4. Assume a solution of the heat conduction problem of the form

$$u(x, t) = \sum_n c_n u_n(x, t) = \sum_n c_n X_n(x) T_n(t).$$

5. Determine the coefficients c_n by requiring that $u(x, 0) = \sum_n c_n X_n(x) T_n(0) = f(x)$ and using the orthogonality properties of the eigenfunctions X_n.

▶ The method of separation of variables can be extended to solve heat conduction problems in which the partial differential equation contains a nonhomogeneous term (heat source or sink) or in which the boundary conditions are nonhomogeneous.

Section 9.7   The Wave Equation: Vibrations of an Elastic String

▶ The classical wave equation that describes the vibrations of an elastic string of length L is

$$a^2 u_{xx} = u_{tt}, \qquad 0 < x < L, \quad t > 0,$$

where $u(x, t)$ represents the string's displacement from equilibrium at time t and position x along the string.

▶ The initial displacement and initial velocity at each point along the string are specified by the **initial conditions** $u(x, 0) = f(x)$, $0 < x < L$, and $u_t(x, 0) = g(x)$, $0 < x < L$, respectively.

▶ **Boundary conditions** for the case in which the left and right ends are fixed are

$$u(0, t) = 0, \qquad u(L, t) = 0, \quad t > 0.$$

▶ The method of separation of variables yields a formal solution of the form

$$u(x, t) = \sum_{n=1}^{\infty} \left[c_n \cos \frac{n\pi a}{L} t + k_n \sin \frac{n\pi a}{L} t \right] \sin \frac{n\pi x}{L},$$

in which the coefficients c_n and k_n are determined by substituting the solution into the initial conditions:

$$f(x) = \sum_{n=1}^{\infty} c_n \sin \frac{n\pi x}{L},$$

$$g(x) = \sum_{n=1}^{\infty} k_n \left(\frac{n\pi a}{L} \right) \sin \frac{n\pi x}{L}.$$

Thus, the c_n and $k_n(n\pi a/L)$ are the coefficients in the Fourier sine series of f and g, respectively, on the interval $[0, L]$.

▶ The nth term in the series solution

$$\left[c_n \cos \frac{n\pi a}{L} t + k_n \sin \frac{n\pi a}{L} t \right] \sin \frac{n\pi x}{L} = A_n \cos \left(\frac{n\pi a}{L} t - \delta_n \right) \sin \frac{n\pi x}{L}$$

is a standing wave with **modal shape** $\sin n\pi x/L$ that oscillates at a **natural frequency** $n\pi a/L$.

▶ The two- and three-dimensional wave equations in Cartesian coordinates are

$$a^2(u_{xx} + u_{yy}) = u_{tt} \qquad \text{and} \qquad a^2(u_{xx} + u_{yy} + u_{zz}) = u_{tt},$$

respectively.

Section 9.8 Laplace's Equation
Laplace's equation arises in the study of steady-state heat conduction problems and in electric and gravitational field theory.

▶ In Cartesian coordinates, Laplace's equation in $n = 1, 2$, and 3 space dimensions, respectively, is

$$\begin{aligned} u_{xx}(x) &= 0, & \text{for } n = 1, \\ u_{xx}(x, y) + u_{yy}(x, y) &= 0, & \text{for } n = 2, \\ u_{xx}(x, y, z) + u_{yy}(x, y, z) + u_{zz}(x, y, z) &= 0, & \text{for } n = 3. \end{aligned}$$

▶ A problem involving Laplace's equation in a domain where the value of the dependent variable is specified around the entire boundary of the domain is called a **Dirichlet problem**, and the boundary condition is called a **Dirichlet condition**.

▶ **The Dirichlet problem in a rectangle** $D = \{(x, y) \mid 0 < x < a \text{ and } 0 < y < b\}$ is

$$u_{xx} + u_{yy} = 0, \qquad (x, y) \in D,$$

subject to the boundary conditions

$$\begin{aligned} u(x, 0) &= f_1(x), & u(x, b) &= f_2(x), & 0 &\leq x \leq a, \\ u(0, y) &= g_1(y), & u(a, y) &= g_2(y), & 0 &< y < b. \end{aligned}$$

▶ The **Dirichlet problem in a circular region** $D = \{(r, \theta) \mid 0 \le r < a \text{ and } 0 \le \theta < 2\pi\}$ is

$$u_{rr} + \frac{1}{r}u_r + \frac{1}{r^2}u_{\theta\theta} = 0, \qquad (r, \theta) \in D,$$

Laplace's equation in polar coordinates, subject to the boundary condition

$$u(a, \theta) = f(\theta), \qquad 0 \le \theta < 2\pi,$$

where f is 2π-periodic, and the requirement that $|u(r, \theta)|$ be bounded on D.

▶ If **n** denotes the vector of unit length normal to the boundary of a domain, a problem involving Laplace's equation in which the directional derivative $\partial u/\partial \mathbf{n}$ is specified around the entire boundary of the domain, except at points where **n** is not defined, is called a **Neumann problem**, and the boundary condition is referred to as a **Neumann condition**.

Appendix 9.A Derivation of the Heat Equation

In this appendix we derive the differential equation that, to a first approximation at least, governs the conduction of heat in solids. It is important to understand that the mathematical analysis of a physical situation or process such as this ultimately rests on a foundation of empirical knowledge of the phenomenon involved. The mathematician must have a place to start, so to speak, and this place is furnished by experience. Consider a uniform rod insulated on the lateral surfaces so that heat can flow only in the axial direction. It has been demonstrated many times that if two parallel cross sections of the same area A and different temperatures T_1 and T_2, respectively, are separated by a small distance d, an amount of heat per unit time will pass from the warmer section to the cooler one. Moreover, this amount of heat is proportional to the area A and to the temperature difference $|T_2 - T_1|$, and is inversely proportional to the separation distance d. Thus

$$\text{Amount of heat per unit time} = \kappa A|T_2 - T_1|/d, \tag{5}$$

where the positive proportionality factor κ is called the thermal conductivity and depends primarily on the material[5] of the rod. The relation (1) is often called Fourier's law of heat conduction. We repeat that Eq. (1) is an empirical, not a theoretical, result and that it can be, and has often been, verified by careful experiment. It is the basis of the mathematical theory of heat conduction.

Now consider a straight rod of uniform cross section and homogeneous material, oriented so that the x-axis lies along the axis of the rod (see Figure 9.A.1). Let $x = 0$ and $x = L$ designate the ends of the bar.

We will assume that the sides of the bar are perfectly insulated so that there is no passage of heat through them. We will also assume that the temperature u depends only on the axial position x and the time t, and not on the lateral coordinates y and z. In other words, we assume that the temperature remains constant on any cross section of the bar. This assumption is usually satisfactory when the lateral dimensions of the rod are small compared to its length.

---

[5] Actually, κ also depends on the temperature, but if the temperature range is not too great, it is satisfactory to assume that κ is independent of temperature.

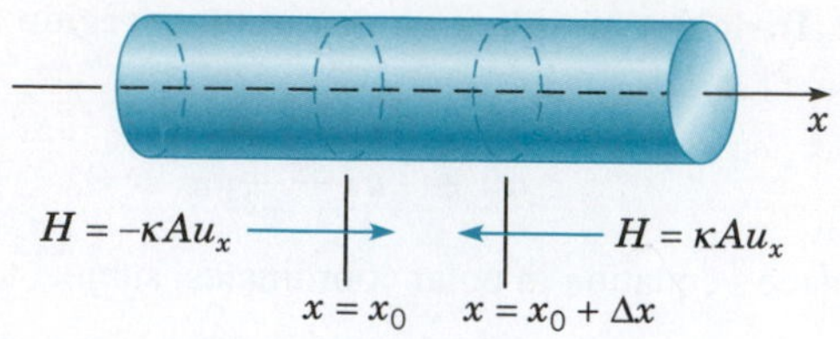

FIGURE 9.A.1 Conduction of heat in an element of a rod.

The differential equation governing the temperature in the bar is an expression of a fundamental physical balance; the rate at which heat flows into any portion of the bar is equal to the rate at which heat is absorbed in that portion of the bar. The terms in the equation are called the flux (flow) term and the absorption term, respectively.

We will first calculate the flux term. Consider an element of the bar lying between the cross sections $x = x_0$ and $x = x_0 + \Delta x$, where x_0 is arbitrary and Δx is small. The instantaneous rate of heat transfer $H(x_0, t)$ from left to right across the cross section $x = x_0$ is given by

$$H(x_0, t) = -\lim_{d \to 0} \kappa A \frac{u(x_0 + d/2, t) - u(x_0 - d/2, t)}{d}$$
$$= -\kappa A u_x(x_0, t). \tag{6}$$

The minus sign appears in this equation because there will be a positive flow of heat from left to right only if the temperature is greater to the left of $x = x_0$ than to the right; in this case $u_x(x_0, t)$ is negative. In a similar manner, the rate at which heat passes from left to right through the cross section $x = x_0 + \Delta x$ is given by

$$H(x_0 + \Delta x, t) = -\kappa A u_x(x_0 + \Delta x, t). \tag{7}$$

The net rate at which heat flows into the segment of the bar between $x = x_0$ and $x = x_0 + \Delta x$ is thus given by

$$Q = H(x_0, t) - H(x_0 + \Delta x, t) = \kappa A[u_x(x_0 + \Delta x, t) - u_x(x_0, t)], \tag{8}$$

and the amount of heat entering this bar element in time Δt is

$$Q \Delta t = \kappa A[u_x(x_0 + \Delta x, t) - u_x(x_0, t)] \Delta t. \tag{9}$$

Let us now calculate the absorption term. The average change in temperature Δu, in the time interval Δt, is proportional to the amount of heat $Q \Delta t$ introduced and inversely proportional to the mass Δm of the element. Thus

$$\Delta u = \frac{1}{s} \frac{Q \Delta t}{\Delta m} = \frac{Q \Delta t}{s \rho A \Delta x}, \tag{10}$$

where the constant of proportionality s is known as the specific heat of the material of the bar, and ρ is its density.[6] The average temperature change Δu in the bar element under consideration is the actual temperature change at some intermediate point $x = x_0 + \theta \Delta x$, where $0 < \theta < 1$. Thus Eq. (6) can be written as

$$u(x_0 + \theta \Delta x, t + \Delta t) - u(x_0 + \theta \Delta x, t) = \frac{Q \Delta t}{s \rho A \Delta x}, \tag{11}$$

---

[6]The dependence of the density and specific heat on temperature is relatively small and will be neglected. Thus both ρ and s will be considered as constants.

or as

$$Q\Delta t = [u(x_0 + \theta\Delta x, t + \Delta t) - u(x_0 + \theta\Delta x, t)]s\rho A\Delta x. \tag{12}$$

To balance the flux and absorption terms, we equate the two expressions for $Q\Delta t$:

$$\kappa A[u_x(x_0 + \Delta x, t) - u_x(x_0, t)]\Delta t$$
$$= s\rho A[u(x_0 + \theta\Delta x, t + \Delta t) - u(x_0 + \theta\Delta x, t)]\Delta x. \tag{13}$$

Upon dividing Eq. (9) by $\Delta x\,\Delta t$ and then letting $\Delta x \to 0$ and $\Delta t \to 0$, we obtain the *heat conduction* or *diffusion* equation

$$\alpha^2 u_{xx} = u_t. \tag{14}$$

The quantity α^2 defined by

$$\alpha^2 = \kappa/\rho s \tag{15}$$

is called the *thermal diffusivity* and is a parameter that depends only on the material of the bar. The units of α^2 are (length)^2/time. Typical values of α^2 are given in Table 9.5.1.

Several relatively simple conditions may be imposed at the ends of the bar. For example, the temperature at an end may be maintained at some constant value T. This might be accomplished by placing the end of the bar in thermal contact with some reservoir of sufficient size so that any heat that flows between the bar and the reservoir does not appreciably alter the temperature of the reservoir. At an end where this is done the boundary condition is

$$u = T. \tag{16}$$

Another simple boundary condition occurs if the end is insulated so that no heat passes through it. Recalling the expression (2) for the amount of heat crossing any cross section of the bar, we conclude that the condition of insulation is that this quantity vanish. Thus

$$u_x = 0 \tag{17}$$

is the boundary condition at an insulated end.

A more general type of boundary condition occurs if the rate of flow of heat through an end of the bar is proportional to the temperature there. Let us consider the end $x = 0$, where the rate of flow of heat from left to right is given by $-\kappa A u_x(0, t)$; see Eq. (2). Hence the rate of heat flow out of the bar (from right to left) at $x = 0$ is $\kappa A u_x(0, t)$. If this quantity is proportional to the temperature $u(0, t)$, then we obtain the boundary condition

$$u_x(0, t) - h_1 u(0, t) = 0, \qquad t > 0, \tag{18}$$

where h_1 is a nonnegative constant of proportionality. Note that $h_1 = 0$ corresponds to an insulated end and that $h_1 \to \infty$ corresponds to an end held at zero temperature.

If heat flow is taking place at the right end of the bar ($x = L$), then in a similar way we obtain the boundary condition

$$u_x(L, t) + h_2 u(L, t) = 0, \qquad t > 0, \tag{19}$$

where again h_2 is a nonnegative constant of proportionality.

Finally, to determine completely the flow of heat in the bar, it is necessary to state the temperature distribution at one fixed instant, usually taken as the initial time $t = 0$. This initial condition is of the form

$$u(x, 0) = f(x), \qquad 0 \le x \le L. \tag{20}$$

The problem then is to determine the solution of the differential equation (10) subject to one of the boundary conditions (12) to (15) at each end and to the initial condition (16) at $t = 0$.

Several generalizations of the heat equation (10) also occur in practice. First, the bar material may be nonuniform and the cross section may not be constant along the length of the bar. In this case, the parameters κ, ρ, s, and A may depend on the axial variable x. Going back to Eq. (2), we see that the rate of heat transfer from left to right across the cross section at $x = x_0$ is now given by

$$H(x_0, t) = -\kappa(x_0)A(x_0)u_x(x_0, t) \tag{21}$$

with a similar expression for $H(x_0 + \Delta x, t)$. If we introduce these quantities into Eq. (4) and eventually into Eq. (9), and proceed as before, we obtain the partial differential equation

$$(\kappa A u_x)_x = s\rho A u_t. \tag{22}$$

We will usually write Eq. (18) in the form

$$r(x)u_t = [p(x)u_x]_x, \tag{23}$$

where $p(x) = \kappa(x)A(x)$ and $r(x) = s(x)\rho(x)A(x)$. Note that both of these quantities are intrinsically positive.

A second generalization occurs if there are other ways in which heat enters or leaves the bar. Suppose that there is a *source* that adds heat to the bar at a rate $G(x, t, u)$ per unit time per unit length, where $G(x, t, u) > 0$. In this case we must add the term $G(x, t, u) \Delta x \Delta t$ to the left side of Eq. (9), and this leads to the differential equation

$$r(x)u_t = [p(x)u_x]_x + G(x, t, u). \tag{24}$$

If $G(x, t, u) < 0$, then we speak of a *sink* that removes heat from the bar at the rate $G(x, t, u)$ per unit time per unit length. To make the problem tractable, we must restrict the form of the function G. In particular, we assume that G is linear in u and that the coefficient of u does not depend on t. Thus we write

$$G(x, t, u) = F(x, t) - q(x)u. \tag{25}$$

The minus sign in Eq. (21) has been introduced so that certain equations that appear later will have their customary forms. Substituting from Eq. (21) into Eq. (20), we obtain

$$r(x)u_t = [p(x)u_x]_x - q(x)u + F(x, t). \tag{26}$$

This equation is sometimes called the generalized heat conduction equation. Boundary value problems for Eq. (22) will be discussed to some extent in Chapter 10.

Finally, if instead of a one-dimensional bar, we consider a body with more than one significant space dimension, then the temperature is a function of two or three space coordinates rather than of x alone. Considerations similar to those leading to Eq. (10) can be employed to derive the heat conduction equation in two dimensions

$$\alpha^2(u_{xx} + u_{yy}) = u_t, \tag{27}$$

or in three dimensions

$$\alpha^2(u_{xx} + u_{yy} + u_{zz}) = u_t. \tag{28}$$

The boundary conditions corresponding to Eqs. (12) and (13) for multidimensional problems correspond to a prescribed temperature distribution on the boundary, or to an insulated boundary. Similarly, the initial temperature distribution will in general be a function of x and y for Eq. (23) and a function of x, y, and z for Eq. (24).

Appendix 9.B Derivation of the Wave Equation

In this appendix we derive the wave equation in one space dimension as it applies to the transverse vibrations of an elastic string, or cable; the elastic string may be thought of as a violin string, a guy wire, or possibly an electric power line. The same equation, however, with the variables properly interpreted, occurs in many other wave problems having only one significant space variable.

Consider a perfectly flexible elastic string stretched tightly between supports fixed at the same horizontal level (see Figure 9.B.1a). Let the x-axis lie along the string with the endpoints located at $x = 0$ and $x = L$. If the string is set in motion at some initial time $t = 0$ (by plucking, for example) and is thereafter left undisturbed, it will vibrate freely in a vertical plane, provided that damping effects, such as air resistance, are neglected. To determine the differential equation governing this motion, we will consider the forces acting on a small element of the string of length Δx lying between the points x and $x + \Delta x$ (see Figure 9.B.1b). We assume that the motion of the string is small and that, as a consequence, each point on the string moves solely in a vertical line. We denote by $u(x, t)$ the vertical displacement of the point x at the time t. Let the tension in the string, which always acts in the tangential direction, be denoted by $T(x, t)$, and let ρ denote the mass per unit length of the string.

Newton's law, as it applies to the element Δx of the string, states that the net external force, due to the tension at the ends of the element, must be equal to the product of the mass of the element and the acceleration of its mass center. Since there is no horizontal acceleration, the horizontal components must satisfy

$$T(x + \Delta x, t)\cos(\theta + \Delta\theta) - T(x, t)\cos\theta = 0. \tag{29}$$

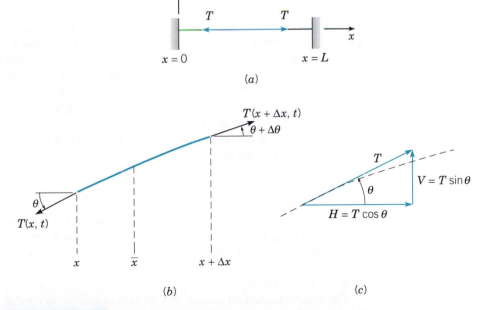

(a)

(b)   (c)

FIGURE 9.B.1   (a) An elastic string under tension. (b) An element of the displaced string. (c) Resolution of the tension T into components.

If we denote the horizontal component of the tension (see Figure 9.B.1c) by H, then Eq. (1) states that H is independent of x.

On the other hand, the vertical components satisfy

$$T(x + \Delta x, t) \sin(\theta + \Delta\theta) - T(x, t) \sin\theta = \rho \, \Delta x \, u_{tt}(x, t), \tag{30}$$

where $\bar{x}$ is the coordinate of the center of mass of the element of the string under consideration. Clearly, $\bar{x}$ lies in the interval $x < \bar{x} < x + \Delta x$. The weight of the string, which acts vertically downward, is assumed to be negligible and has been neglected in Eq. (2).

If the vertical component of T is denoted by V, then Eq. (2) can be written as

$$\frac{V(x + \Delta x, t) - V(x, t)}{\Delta x} = \rho u_{tt}(\bar{x}, t).$$

Passing to the limit as $\Delta x \to 0$ gives

$$V_x(x, t) = \rho u_{tt}(x, t). \tag{31}$$

To express Eq. (3) entirely in terms of u, we note that

$$V(x, t) = H(t) \tan\theta = H(t) u_x(x, t).$$

Hence Eq. (3) becomes

$$(Hu_x)_x = \rho u_{tt},$$

or, since H is independent of x,

$$Hu_{xx} = \rho u_{tt}. \tag{32}$$

For small motions of the string it is permissible to replace $H = T \cos\theta$ by T. Then Eq. (4) takes its customary form

$$a^2 u_{xx} = u_{tt}, \tag{33}$$

where

$$a^2 = T/\rho. \tag{34}$$

We will assume further that a^2 is a constant, although this is not required in our derivation, even for small motions. Equation (5) is called the wave equation for one space dimension. Since T has the dimension of force, and ρ that of mass/length, it follows that the constant a has the dimension of velocity. It is possible to identify a as the velocity with which a small disturbance (wave) moves along the string. According to Eq. (6), the wave velocity a varies directly with the tension in the string, but inversely with the density of the string material. These facts are in agreement with experience.

As in the case of the heat conduction equation, there are various generalizations of the wave equation (5). One important equation is known as the **telegraph equation** and has the form

$$u_{tt} + cu_t + ku = a^2 u_{xx} + F(x, t), \tag{35}$$

where c and k are nonnegative constants. The terms cu_t, ku, and $F(x, t)$ arise from a viscous damping force, an elastic restoring force, and an external force, respectively. Note the similarity of Eq. (7), except for the term $a^2 u_{xx}$, with the equation for the spring–mass system derived in Section 4.1; the additional term $a^2 u_{xx}$ arises from a consideration of internal elastic forces.

The telegraph equation also governs the flow of voltage, or current, in a transmission line (hence its name); in this case the coefficients are related to electrical parameters in the line.

For a vibrating system with more than one significant space coordinate, it may be necessary to consider the wave equation in two dimensions

$$a^2(u_{xx} + u_{yy}) = u_{tt},\tag{36}$$

or in three dimensions

$$a^2(u_{xx} + u_{yy} + u_{zz}) = u_{tt}.\tag{37}$$

PROJECTS

Project 1 Estimating the Diffusion Coefficient in the Heat Equation[7]

A mathematical model usually contains one or more parameters that need to be determined before the model can be used in a specific design, prediction or control problem. In this project we show how, in principle, to estimate the thermal diffusivity of a homogeneous material from experimental data. Consider a metal rod of uniform cross section and length L that is insulated around its lateral boundary, and at its right end, as shown in Figure 9.P.1.

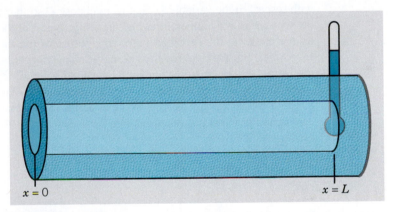

$x = 0$ ‎ ‎ ‎ $x = L$

FIGURE 9.P.1 A metal rod with its entire surface insulated except for the left end.

A thermometer is also placed in thermal contact with the right end of the rod. The rod is initially brought to a uniform temperature of $0°C$ by placing its left end in thermal contact with a heat reservoir held at a constant temperature of $0°C$ and waiting a sufficiently long period of time. The left end is then quickly placed into contact with a heat reservoir that is maintained at a constant temperature $T > 0$. Temperature readings of the thermometer are subsequently recorded at several different points in time. The arrival of the wave of heat at the right end of the rod is therefore partially observed, and it is intuitively obvious that information is being gathered about the rate at which heat energy is being transported through the material of which the rod is composed. We will assume that the thermometer has negligible effect on the rate at which the rod heats. If we denote the temperature in the rod at position x and time t by $u(x, t)$, then the initial boundary value problem for u,

[7]This problem is also the subject of Project 1, Chapter 6, where a compartment model is used to describe the transport of heat.

corresponding to this experiment, is

$$u_t = Du_{xx}, \qquad 0 < x < L, \quad t > 0, \tag{1}$$

$$u(0, t) = T, \quad u_x(L, t) = 0, \qquad t > 0, \tag{2}$$

$$u(x, 0) = 0, \qquad 0 \le x \le L. \tag{3}$$

We assume that we can easily measure the length L of the rod, and that we can precisely control the temperature T at the left end of the rod, so that the only parameter that needs to be determined is the thermal diffusivity, D. Since the solution of Eqs. (1–3) depends on D as well as x and t, we explicitly indicate this dependence by using the notation $u(x, t, D)$. In a **parameter estimation** problem such as this, it is useful to think of the model equations (1–3) as a family of models parameterized by D. The object then is to select a particular member of this family that best agrees with the observational data.

Suppose that the temperature measurements made at the right end of the rod at times $\{t_1, t_2, \ldots, t_m\}$ are denoted by $\{\theta_1, \theta_2, \ldots \theta_m\}$. The **output least squares** method for estimating the thermal diffusivity D is to minimize the sum of squares

$$V(D) = \sum_{j=1}^{m} [u(L, t_j, D) - \theta_j]^2 \tag{4}$$

with respect to the variable D, that is, to choose D so that the sum of squared differences between measured temperatures and the temperatures at the right end of the rod predicted by the model at the specified times (the output of the model) are as small as possible. In general, Eq. (4) requires that we have a way to accurately approximate $u(x, t, D)$ at $x = L$, as well as a systematic method for finding the value of D that minimizes $V(D)$. As shown below, these problems are slightly simplified by taking advantage of the fact that the high-frequency modes in the Fourier series representation of the solution of the heat conduction problem decay at much faster rates than the amplitudes of the low-frequency modes.

Project 1 PROBLEMS

1. Find the steady-state solution $v(x)$ of the problem defined by Eqs. (1) and (2) by setting $u_t = 0$ in Eq. (1), and then substituting the general solution of the resulting equation into the boundary conditions to determine the constants of integration.

2. If the solution of the problem (1–3) is expressed as $u(x, t, D) = v(x) + w(x, t, D)$, find the complete problem (the partial differential equation, boundary conditions, and initial condition) that the transient part of the solution, $w(x, t, D)$, must satisfy.

3. Solve the initial boundary value problem, found in Problem 2, for $w(x, t, D)$. Conclude that the solution of the problem (1–3) is

$u(x, t, D)$

$$= T - \frac{4T}{\pi} \sum_{n=1}^{\infty} \frac{1}{2n-1} \exp\left[-\left(\frac{(2n-1)\pi}{2L}\right)^2 Dt\right] \sin\left(\frac{(2n-1)\pi x}{2L}\right).$$

(i)

4. Denote by

$u_N(x, t, D)$

$$= T - \frac{4T}{\pi} \sum_{n=1}^{N} \frac{1}{2n-1} \exp\left[-\left(\frac{(2n-1)\pi}{2L}\right)^2 Dt\right] \sin\left(\frac{(2n-1)\pi x}{2L}\right)$$

the approximation of $u(t, x, D)$ obtained by truncating the series in Eq. (i), Problem 3. Use the *alternating series estimation theorem* from calculus to show that

$$|u(L, t, D) - u_N(L, t, D)| \le \frac{4T}{\pi(2N+1)} \exp\left[-\left(\frac{(2N+1)\pi}{2L}\right)^2 Dt\right].$$

Suppose we believe that D is much larger than 0.1 cm^2/s but much smaller than 100 cm^2/s. Assuming that $L = 10$ cm and $T = 100°$C, and that a nominal value for D is 1 cm^2/s, determine how large t must be such that $|u(L, t) - u_1(L, t)| \le 0.001$ degrees.

CAS 5. Table 9.P.1 lists pairs of time and (slightly noisy) temperature observations recorded at the right end of a rod of length $L = 10$ cm subsequent to the instant that the left end was placed in thermal contact with a heating element kept at a constant temperature $T = 100°C$. Assuming that $D \cong 1$ cm²/s, use the result from Problem 4 to justify minimizing

$$V(D) \cong \sum_j [u_1(L, t_j, D) - \theta_j]^2 \qquad \text{(i)}$$

using only values of t_j such that $|u(L, t_j, D) - u_1(L, t_j, D)| \leq 0.001$ degrees. Under this restriction,

estimate a value of D that minimizes $V(D)$ in Eq. (i) by plotting the graph of $V(D_k)$ versus a suitable set of values D_k, $k = 1, \ldots, L$.

6. Denote by $\hat{D}$ the estimate of D found in Problem 5. CAS Test the accuracy of $\hat{D}$ by plotting $u_N(L, t; \hat{D})$ versus t and the data in Table 9.P.1 on the same set of coordinate axes. Examine the sensitivity of the plots to the choice of N.

7. Plot temperature profiles, $u_N(x, t; \hat{D})$ versus x for CAS several different values of t. Examine the sensitivity of the profiles to the choice of N.

TABLE 9.P.1	Temperatures recorded at the right end of the rod in Figure 9.P.1.		
Time	Temperature (°C)	Time	Temperature (°C)
10	3.2	60	63.57
20	16.65	70	70.33
30	32.03	80	76.03
40	44.81	90	80.53
50	55.15	100	84.05

Project 2 The Transmission Line Problem

Electric cables, or transmission lines, are used to transmit either power or information. An electric power cable transmits electrical energy from the point where the energy is generated to the point where it is used. An electric communication cable transmits information, by electrical means, from one point to another. Transmission lines are constructed from conducting materials and insulating materials taking into account design factors such as voltage and current loads, amount of information to be transmitted, frequencies, and environmental conditions. Common examples of transmission lines are parallel wire lines and coaxial cables.

▶ The Transmission Line and Telegraph Equations. We conceptualize the uniform transmission line as a distributed circuit represented by a cascade, or ladder, of interconnected identical cells of very short length Δx, as shown in Figure 9.P.2.

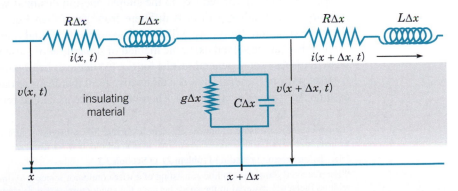

FIGURE 9.P.2   The currents and voltages in a segment of a transmission line.

Each cell consists of a pair of conductors separated by an insulating, or dielectric, material. Across each cell, the conductors have a series resistance and inductance denoted by $R\Delta x$ and $L\Delta x$, respectively, while the insulating material has a parallel shunt capacitance and shunt conductance[8] denoted by $C\Delta x$ and $g\Delta x$, respectively.

We will denote the voltage between the two conductors at position x (measured from the beginning of the line) and time t by $v(x,t)$ and the current in the "hot" wire by $i(x,t)$. Kirchoff's voltage and current laws[9] will then tell us what happens to v and i between two adjacent points, say, x and $x+\Delta x$, along the transmission line. The change in voltage due to the resistance and inductance across the element of length Δx gives us the equation

$$-[v(x+\Delta x, t) - v(x,t)] = [Ri(x,t) + L\frac{\partial i}{\partial t}(x,t)]\Delta x. \tag{1}$$

Since the shunt current across the dielectric is $gv(x+\Delta x, t)\Delta x + C\frac{\partial v}{\partial t}(x+\Delta x, t)\Delta x$, the requirement that the sum of the currents at the node located at $x+\Delta x$ in the hot wire be equal to zero gives us the equation

$$-[i(x+\Delta x, t) - i(x,t)] = [gv(x+\Delta x, t) + C\frac{\partial v}{\partial t}(x+\Delta x, t)]\Delta x. \tag{2}$$

Dividing each of Eqs. (1) and (2) by Δx and letting $\Delta x \to 0$ yield the system of first order partial differential equations

$$-\frac{\partial v}{\partial x} = Ri + L\frac{\partial i}{\partial t}, \tag{3}$$

$$-\frac{\partial i}{\partial x} = gv + C\frac{\partial v}{\partial t}, \tag{4}$$

for the two unknowns v and i. These equations are known as the **transmission line equations**. If Eq. (4) is differentiated with respect to t and Eq. (3) is differentiated with respect to x, one can then use algebraic elimination to obtain the **telegraph equation** (see Problem 1)

$$\sigma^{-2}v_{tt} + 2\gamma v_t + \omega^2 v = v_{xx}, \tag{5}$$

where

$$\sigma = \frac{1}{\sqrt{LC}}, \qquad \gamma = (RC + gL)/2, \qquad \omega = \sqrt{Rg}. \tag{6}$$

Once the voltage v is known, the current i can be found by solving Eq. (3).

▶ **Dissipation, Dispersion, and Diffusion.** If resistance R and conductance g are set to zero, then $\gamma = \omega = 0$ and Eq. (5) reduces to the simple one-dimensional wave equation with phase velocity[10] σ. If either γ or ω is different from zero, then Eq. (5) is a generalization of the wave equation. Below we will formulate a specific initial boundary value problem for Eq. (5) that can be solved using the method of separation of variables. Before doing so, we wish to discuss some general results for Eq. (5) that can be obtained with very little effort.

Dissipation. The term $2\gamma v_t$ is a dissipative term that leads to damping. To see this, let us assume that $v(x,t) = z(t)y(x,t)$. Then first derivatives can be removed from Eq. (5) if we

[8]Conductance, a measure of how easily electric current flows through a substance, is the reciprocal of resistance.

[9]See the discussion preceding Problem 21 in Section 3.2.

[10]For a coaxial transmission line consisting of a wire conductor inside a cylindrical outer conductor, and with no dielectric material in the space between the conductors, it can be shown that the phase velocity $\sigma = c$, the speed of light.

choose z so that $\sigma^{-2}z' + \gamma z = 0$, that is, $z = e^{-\gamma\sigma^2 t}$ (see Problem 2). It follows that $y(x, t)$ must then satisfy

$$\sigma^{-2}y_{tt} + ky = y_{xx}, \tag{7}$$

where

$$k = \omega^2 - \gamma^2\sigma^2. \tag{8}$$

Hence, the solution of the telegraph equation (5) can be expressed in the form

$$v(x, t) = e^{-\gamma\sigma^2 t}y(x, t), \tag{9}$$

where y satisfies Eq. (7).

Dispersion. We will now show that the term ky in Eq. (7) leads to the phenomenon of dispersion.[11] Note that if $k = 0$, then Eq. (7) is the simple wave equation with a general solution of the form $y(x, t) = \phi(x - \sigma t) + \psi(x + \sigma t)$ (see Problem 13, Section 9.7). From Eq. (9) it follows that the general solution to Eq. (5), in the case that $k = 0$, is

$$v(x, t) = e^{-\gamma\sigma^2 t}[\phi(x - \sigma t) + \psi(x + \sigma t)]. \tag{10}$$

Equation (10) shows that, in the absence of the term ky, the waves retain their shape except for an amplitude attenuation factor that depends on time. Such waves are called **relatively undistorted**.

However, if $k \neq 0$ and we look for progressive wave solutions to Eq. (7) of the form $y(x, t) = \phi(x - \sigma t)$, we find on substitution that $\phi'' + k\phi = \phi''$, or $k\phi = 0$. Therefore, if $k \neq 0$, there are no nontrivial solutions of the form $\phi(x - \sigma t)$. The same is true for solutions of the form $\psi(x + \sigma t)$. If instead we seek solutions to Eq. (7) of the form $\phi(x - \alpha t)$, where $\alpha \neq \sigma$, then we find that

$$\phi'' + \frac{k\sigma^2}{\sigma^2 - \alpha^2}\phi = 0. \tag{11}$$

Equation (11) has nontrivial solutions consisting of sines and cosines if $\alpha < \sigma$ and real exponential solutions if $\alpha > \sigma$. This equation implies that not all propagating waves travel with the same velocity. If, for example, $\phi(x) = \sin \lambda x$, then the dependence of the phase velocity α on spatial frequency λ is given by (see Problem 3)

$$\alpha = \sigma\sqrt{1 - k/\lambda^2}, \tag{12}$$

which shows that not all waves propagate with the same velocity. This property is referred to as **dispersion**. Due to the presence of the term ku, Eq. (7) is a special case of a **dispersive wave equation**. Below, we will examine dispersion in the context of a specific initial boundary value problem for Eq. (5).

Diffusion. Finally, we note that there is a diffusion equation embedded in Eq. (5). If inductance and conductance are ignored ($L = \omega = 0$), then Eq. (5) reduces to the simple one dimensional diffusion equation with diffusion coefficient $1/RC$. For frequencies on the order of 100 Hz and realistic values of R and C, the diffusion coefficient is very large and the diffusion speed is about $c/6$.

▶ **The Cable Charging Problem.** We now consider a cable of length l that is initially in an uncharged state described by the initial conditions

$$v(x, 0) = 0, \qquad v_t(x, 0) = 0, \qquad 0 < x < l. \tag{13}$$

[11]The solution of an initial-boundary value problem for Eq. (7) is discussed in Problem 23, Section 9.7

If at time $t = 0$ a battery with constant voltage V is connected to the left end of the cable, then boundary conditions for Eq. (5) are given by

$$v(0, t) = V, \qquad v(l, t) = 0, \quad t > 0. \tag{14}$$

The *cable charging problem* consists of Eq. (5), the initial conditions (13), and the boundary conditions (14). Due to the presence of damping in the cable, we anticipate that as $t \to \infty$, a steady-state voltage distribution $\hat{v}(x)$ will be reached [see Problem 4(a)]. If the solution of (5), (13), and (14) is expressed as the difference between $\hat{v}(x)$ and a transient solution $u(x, t)$,

$$v(x, t) = \hat{v}(x) - u(x, t), \tag{15}$$

it follows that $u(x, t)$ satisfies the initial boundary value problem [see Problem 4(b)]

$$\sigma^{-2} u_{tt} + 2\gamma u_t + \omega^2 u = u_{xx}, \qquad 0 < x < l, \quad t > 0, \tag{16}$$

$$u(0, t) = 0, \qquad u(l, t) = 0, \quad t > 0 \tag{17}$$

and

$$u(x, 0) = \hat{v}(x), \qquad u_t(x, 0) = 0, \quad 0 < x < l. \tag{18}$$

Using the method of separation of variables, we find that the solution of Eqs. (16–18) is (see Problem 5)

$$
\begin{aligned}
u(x, t) = \, & e^{-\gamma \sigma^2 t} \sum_{n=1}^{N} \left[A_n \cosh \sigma \sqrt{(\sigma \gamma)^2 - \lambda_n^2 - \omega^2}\, t \right. \\
& \left. + B_n \sinh \sigma \sqrt{(\sigma \gamma)^2 - \lambda_n^2 - \omega^2}\, t \right] \sin \lambda_n x \\
& + e^{-\gamma \sigma^2 t} \sum_{n=N+1}^{\infty} \left[A_n \cos \sigma \sqrt{\lambda_n^2 + \omega^2 - (\sigma \gamma)^2}\, t \right. \\
& \left. + B_n \sin \sigma \sqrt{\lambda_n^2 + \omega^2 - (\sigma \gamma)^2}\, t \right] \sin \lambda_n x,
\end{aligned}
\tag{19}
$$

where

$$0 \leq N < \frac{l}{\pi} \sqrt{(\sigma \gamma)^2 - \omega^2} < N + 1,$$

$$\lambda_n = \frac{n\pi}{l}, \qquad 1 \leq n \leq \infty,$$

$$A_n = \frac{2}{l} \int_0^l \hat{v}(x) \sin \lambda_n x, \qquad 1 \leq n \leq \infty, \tag{20}$$

and

$$
B_n = \begin{cases}
\dfrac{\gamma \sigma^2 A_n}{\sqrt{(\sigma \gamma)^2 - \lambda_n^2 - \omega^2}}, & n \leq N, \\[4mm]
\dfrac{\gamma \sigma^2 A_n}{\sqrt{\lambda_n^2 + \omega^2 - (\sigma \gamma)^2}}, & n > N.
\end{cases}
\tag{21}
$$

Note that we are assuming that $l\sqrt{(\sigma\gamma)^2 - \omega^2}/\pi$ is not equal to an integer. If $N \geq 1$, then $u(x, t)$ is a superposition of N modes with amplitudes that decay exponentially in time plus an infinite number of damped standing waves.

Distortionless Transmission. We wish to continue the above discussion of dispersion in the context of the initial value problem (16–18) using initial conditions different from Eq. (18) that allow us to easily see the effects of dispersion on a particular waveform (see Problem 9). Toward this end, it is useful to exhibit $u(x, t)$ as a superposition of propagating modes. Trigonometric identities and the relations $\sin ix = i \sinh x$ and $\cos ix = \cosh x$ can be used (see Problem 6) to write $u(x, t)$ as

$$
u(x, t) = \frac{1}{2}e^{-\gamma\sigma^2 t} \sum_{n=1}^{N} \{A_n[\sin\lambda_n(x + i\sigma_n t) + \sin\lambda_n(x - i\sigma_n t)]
$$

$$
+ i B_n[\cos\lambda_n(x + i\sigma_n t) - \cos\lambda_n(x - i\sigma_n t)]\}
$$

$$
+ \frac{1}{2}e^{-\gamma\sigma^2 t} \sum_{n=N+1}^{\infty} \{A_n[\sin\lambda_n(x + \sigma_n t) + \sin\lambda_n(x - \sigma_n t)]
$$

$$
- B_n[\cos\lambda_n(x + \sigma_n t) - \cos\lambda_n(x - \sigma_n t)]\}, \tag{22}
$$

where

$$
\sigma_n = \begin{cases} \sigma\sqrt{(\sigma\gamma/\lambda_n)^2 - (\omega/\lambda_n)^2 - 1}, & 1 \leq n \leq N, \\ \sigma\sqrt{1 + (\omega/\lambda_n)^2 - (\sigma\gamma/\lambda_n)^2}, & n \geq N + 1. \end{cases} \tag{23}
$$

Equation (22) represents $u(x, t)$ as a superposition of modes traveling in the positive and negative directions along the x-axis with complex phase velocities $i\sigma_n$ (*diffusing modes*) for $1 \leq n \leq N$ and real phase velocities σ_n for $n \geq N + 1$ (*propagating modes*). Observe that the real phase velocities σ_n decrease with λ_n if $\omega > \sigma\gamma$ and increase with λ_n if $\omega < \sigma\gamma$. As mentioned above, the dependence of phase velocity on spatial frequency or wavelength is called **dispersion** or **phase distortion**. The dependence of attenuation on spatial frequency due to the complex phase velocities in the first N modes is called **amplitude distortion**. Signal transmission is called **undistorted transmission** if the shape of the signal at the beginning of the line is the same as that at the end. The condition for distortionless transmission (see Problem 7) is

$$
\frac{L}{R} = \frac{C}{g}. \tag{24}
$$

Note that Eq. (24) is true if and only if $k = 0$ in Eq. (8). In this case $N = 0$ so there is no amplitude distortion; all the modes propagate. The relation (24) is a transmission line design criterion. The ratio L/R is usually smaller than C/g unless special measures are taken to achieve approximate equality between the two terms. Decreasing resistance R by using wires of large diameter is expensive. Increasing the conductance g of the insulation also increases the attenuation of the line. The most effective and economical method of obtaining approximate equality in Eq. (24) is to explicitly increase the inductance by connecting induction coils in the line at fixed intervals or by using a cable, the conductors of which are bound by a thin tape made of a material with a high magnetic permeability.

Project 2 PROBLEMS

1. Derive the telegraph equation (5) from the transmission line equations (3) and (4).

2. Show that if we look for solutions of Eq. (5) in the form $v(x, t) = z(t)y(x, t)$, then first derivative terms can be removed if we choose z to satisfy $\sigma^{-2}z' + \gamma z = 0$, and consequently, $y(x, t)$ must satisfy Eq. (7).

3. Show that if $\phi(x) = \sin \lambda x$ is a solution of Eq. (11), then the dependence of phase velocity on spatial frequency λ is given by Eq. (12).

4. (a) Find the steady-state solution $\hat{v}(x)$ of Eq. (5) subject to the boundary conditions in Eq. (14) by setting $v_t = v_{tt} = 0$ in Eq. (5), and then substituting the general solution of the resulting equation into the boundary conditions to determine the constants of integration. Use Eq. (3) to then find the steady-state current in the cable.

   (b) Show that the transient term $u(x, t)$ in Eq. (15) must satisfy the initial boundary value problem defined by Eqs. (16–18).

5. Use the method of separation of variables to solve the inital boundary value problem defined by Eqs. (16–18).

6. Use the relations $\sin ia = i \sinh a$ and $\cos ia = \cosh a$ and the trigonometric identities

$$\sin(a \pm b) = \sin a \cos b \pm \cos a \sin b \quad \text{and}$$
$$\cos(ab) = \cos a \cos b \pm \sin a \sin b$$

to derive the propagating mode representation (22) from Eq. (19).

7. Derive the condition (24) for distortionless transmission and show that in this case, $N = 0$, so that all modes propagate.

8. In order to examine the effects of dispersion,[12] we CAS consider the problem defined by Eqs. (16–17) on a transmission line of length $l = 5$, using for initial conditions the square wave pulse

$$u(x, 0) = f(x) = \begin{cases} 1, & 2 < x < 3 \\ 0, & \text{otherwise}, \end{cases}$$

and $u_t(x, 0) = 0$, $0 \le x \le 5$,

and the following sets of unrealistic, but computationally convenient, parameter values:
(i) $\sigma = 1/2$, $\gamma = 1/2$, $\omega = 1/4$; (ii) $\sigma = 1/2$, $\gamma = 1/2$, $\omega = 1/2$; (iii) $\sigma = 1/2$, $\gamma = 1/2$, $\omega = 1$. Using the coefficients

$$A_n = \frac{2}{5} \int_0^5 f(x) \sin \lambda_n x \quad \text{and}$$

$$B_n = \gamma \sigma^2 A_n / \sqrt{\lambda_n^2 + \omega^2 - (\sigma \gamma)^2},$$

plot and compare the graphs of

$$e^{\gamma \sigma^2 t} u(x, t) = \sum_{n=1}^{1000} \left[A_n \cos \sigma \sqrt{\lambda_n^2 + \omega^2 - (\sigma \gamma)^2} t \right.$$
$$\left. + B_n \sin \sigma \sqrt{\lambda_n^2 + \omega^2 - (\sigma \gamma)^2} t \right] \sin \lambda_n x$$

versus x, $0 \le x \le 5$, for each of the cases (i–iii) on a single set of coordinate axes at the time $t = 20$. Repeat the experiment for the time $t = 40$.

Project 3 Solving Poisson's Equation by Finite Differences

Laplace's equation (see Section 9.8), or the **potential equation**, is a partial differential equation used to model physical phenomena such as the steady-state temperature distribution in an isotropic and homogeneous material or the electric potential field generated by a distribution of electric charge. It also arises in gravitational potential field theory and in fluid mechanics. In two-dimensional rectangular coordinates, Laplace's equation is

$$\nabla^2 u = \frac{\partial^2 u}{\partial x^2} + \frac{\partial^2 u}{\partial y^2} = 0. \tag{1}$$

---

[12]Using Fourier transforms, we can perform the numerical simulation in this problem on a transmission line of infinite length. We have chosen instants of time where both halves of the initial wave pulse have coalesced at the center of the finite interval following an even number of reflections from the endpoints. The shape of the pulse at these times is identical to that of the same initial pulse after it has propagated along an infinite transmission line for equivalent periods of time.

The operator

$$\nabla^2 = \frac{\partial^2}{\partial x^2} + \frac{\partial^2}{\partial y^2}$$

is called the **Laplacian** operator or just the Laplacian. It is often the case that Laplace's equation has a nonhomogeneous term, say, $f(x, y)$, representing a distributed source[13] ($f > 0$) or sink ($f < 0$):

$$\nabla^2 u = \frac{\partial^2 u}{\partial x^2} + \frac{\partial^2 u}{\partial y^2} = -f(x, y). \tag{2}$$

Equation (2), Laplace's equation with a nonhomogeneous term, is referred to as **Poisson's equation**. A well-posed problem involving Laplace's or Poisson's equation requires boundary conditions. If the value of the dependent variable u is specified on the boundary, then the boundary conditions are called **Dirichlet conditions** and the resulting problem is called a **Dirichlet problem**. For example, the Dirichlet problem for Poisson's equation in a rectangular region (see Figure 9.P.3)

$$R = \{(x, y): 0 < x < a, 0 < y < b\} \tag{3}$$

is prescribed by

$$\frac{\partial^2 u}{\partial x^2} + \frac{\partial^2 u}{\partial y^2} = -f(x, y), \qquad (x, y) \in R, \tag{4}$$

$$u(x, 0) = p(x), \qquad u(x, b) = q(x), \quad 0 < x < a, \tag{5}$$

$$u(0, y) = r(y), \qquad u(a, y) = s(y), \quad 0 < y < b. \tag{6}$$

Equations (5) specify the values of u along the bottom and top edges of the rectangle, while Eqs. (6) specify the values of u along the left and right edges of the rectangle.

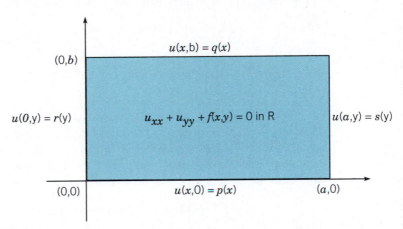

FIGURE 9.P.3   The boundary value problem described by Eqs. (4–6).

A numerical method for approximating the solution of the Dirichlet problem (4–6) is the finite difference method. The finite difference method is based on approximating derivatives

---

[13] To see this, for example, in the case of heat transport in a flat plate, the time-dependent heat equation with a distributed heat source in the interior of the plate is $u_t = u_{xx} + u_{yy} + f(x, y)$, where $f > 0$. Thus, the steady-state equation with a heat source in the interior is $u_{xx} + u_{yy} = -f(x, y)$.

by difference quotients. For example, a centered difference formula for the second derivative of a function $F(x)$ at the point x_0 is given by

$$\frac{F(x_0 - \Delta x) - 2F(x_0) + F(x_0 + \Delta x)}{(\Delta x)^2} \cong \frac{\partial^2 F}{\partial x^2}(x_0), \tag{7}$$

where Δx is typically a small increment (see Problem 1). We partition the region R into rectangular elements with mesh point coordinates defined by

$$x_m = m \Delta x, \qquad m = 0, 1, 2, \ldots, M + 1, \tag{8}$$

and

$$y_n = n \Delta y, \qquad n = 0, 1, 2, \ldots, N + 1, \tag{9}$$

where

$$\Delta x = \frac{a}{M + 1} \qquad \text{and} \qquad \Delta y = \frac{b}{N + 1}.$$

It is convenient to introduce the following notation to represent values of $u, f, p, q, r,$ and s at the grid points:

$$u_{mn} = u(x_m, y_n), \quad f_{mn} = f(x_m, y_n), \qquad 0 \le m \le M + 1, \quad 0 \le n \le N + 1, \tag{10}$$

$$p_m = p(x_m), \quad q_m = q(x_m), \qquad 0 \le m \le M + 1, \tag{11}$$

$$r_n = r(y_n), \quad s_n = s(y_n), \qquad 0 \le n \le N + 1. \tag{12}$$

Applying finite difference approximations to the partial derivatives appearing in Eq. (4) at the grid point (x_m, y_n) yields, using the notation in Eq. (10),

$$\frac{u_{m-1,n} - 2u_{mn} + u_{m+1,n}}{(\Delta x)^2} + \frac{u_{m,n-1} - 2u_{mn} + u_{m,n+1}}{(\Delta y)^2} = -f_{mn}. \tag{13}$$

If we choose $\Delta x = \Delta y = h$, then Eq. (13) can be written as

$$u_{m-1,n} - 4u_{mn} + u_{m+1,n} + u_{m,n-1} + u_{m,n+1} = -h^2 f_{mn}. \tag{14}$$

Observe that if the point (x_m, y_n) at which the discretization is centered is an interior point adjacent to the boundary, then one or two of the terms on the left-hand side of Eq. (14) are known from the boundary conditions. For example, the finite difference approximation applied at the point (x_1, y_1) may be written as

$$-4u_{11} + u_{21} + u_{12} = -h^2 f(x_1, y_1) - p_1 - r_1, \tag{15}$$

since $u_{01} = r_1$ and $u_{10} = p_1$. Only the MN values of u_{mn}, $m = 1, \ldots, M$, $n = 1, \ldots, N$, at the interior grid points are regarded as unknowns since the values of grid points on the boundary of R are prescribed by the boundary conditions (5–6).

We associate with Eq. (14) the computational molecule shown in Figure 9.P.4, where the weights 1, 1, 1, 1 are assigned to the four grid points (x_{m-1}, y_n), (x_{m+1}, y_n), (x_m, y_{n-1}), and (x_m, y_{n+1}) adjacent to (x_m, y_n), which has a weight of -4. This computational molecule makes it easy to write down directly, using matrices, the system of equations satisfied by the unknowns. For example, using the computational molecule in Figure 9.P.4 on the 3×2

grid of interior points shown in Figure 9.P.5, we can write down the following system of six equations for the unknown values u_{mn}, $m = 1, 2, 3$, $n = 1, 2$ at the interior nodes:

$$
\begin{bmatrix}
-4 & 1 & 0 & 1 & 0 & 0 \\
1 & -4 & 1 & 0 & 1 & 0 \\
0 & 1 & -4 & 0 & 0 & 1 \\
1 & 0 & 0 & -4 & 1 & 0 \\
0 & 1 & 0 & 1 & -4 & 1 \\
0 & 0 & 1 & 0 & -1 & 4
\end{bmatrix}
\begin{bmatrix}
u_{11} \\ u_{21} \\ u_{31} \\ u_{12} \\ u_{22} \\ u_{32}
\end{bmatrix}
= -h^2
\begin{bmatrix}
f_{11} \\ f_{21} \\ f_{31} \\ f_{12} \\ f_{22} \\ f_{32}
\end{bmatrix}
-
\begin{bmatrix}
p_1 \\ p_2 \\ p_3 \\ 0 \\ 0 \\ 0
\end{bmatrix}
-
\begin{bmatrix}
0 \\ 0 \\ 0 \\ q_1 \\ q_2 \\ q_3
\end{bmatrix}
-
\begin{bmatrix}
r_1 \\ 0 \\ 0 \\ r_2 \\ 0 \\ 0
\end{bmatrix}
-
\begin{bmatrix}
0 \\ 0 \\ s_1 \\ 0 \\ 0 \\ s_2
\end{bmatrix},
$$
(16)

or, multiplying both sides of Eq. (16) by -1,

$$
\begin{bmatrix}
4 & -1 & 0 & -1 & 0 & 0 \\
-1 & 4 & -1 & 0 & -1 & 0 \\
0 & -1 & 4 & 0 & 0 & -1 \\
-1 & 0 & 0 & 4 & -1 & 0 \\
0 & -1 & 0 & -1 & 4 & -1 \\
0 & 0 & -1 & 0 & -1 & 4
\end{bmatrix}
\begin{bmatrix}
u_{11} \\ u_{21} \\ u_{31} \\ u_{12} \\ u_{22} \\ u_{32}
\end{bmatrix}
= h^2
\begin{bmatrix}
f_{11} \\ f_{21} \\ f_{31} \\ f_{12} \\ f_{22} \\ f_{32}
\end{bmatrix}
+
\begin{bmatrix}
p_1 \\ p_2 \\ p_3 \\ 0 \\ 0 \\ 0
\end{bmatrix}
+
\begin{bmatrix}
0 \\ 0 \\ 0 \\ q_1 \\ q_2 \\ q_3
\end{bmatrix}
+
\begin{bmatrix}
r_1 \\ 0 \\ 0 \\ r_2 \\ 0 \\ 0
\end{bmatrix}
+
\begin{bmatrix}
0 \\ 0 \\ s_1 \\ 0 \\ 0 \\ s_2
\end{bmatrix}.
$$
(17)

Using obvious matrix notation, we can write Eq. (17) as

$$\mathbf{Au} = h^2\mathbf{f} + \mathbf{p} + \mathbf{q} + \mathbf{r} + \mathbf{s}. \tag{18}$$

We infer from Eq. (17) the pattern of entries that allow us to easily define, using a computer algebra system, the required matrix entries for a more densely defined grid:

▶ The $MN \times MN$ matrix $\mathbf{A}$ is **symmetric** and **banded** with 4's along the main diagonal, -1's along the first subdiagonal above and below the main diagonal except for every Mth entry which is zero, and -1's along the Mth subdiagonal above and below the main diagonal. All other entries of $\mathbf{A}$ are zero.

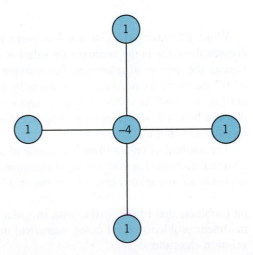

FIGURE 9.P.4   Computational molecule for two-dimensional Laplace operator.

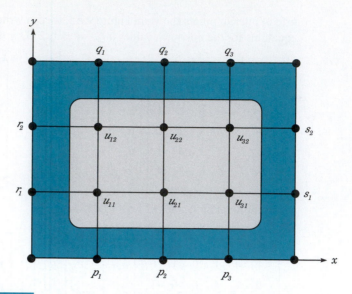

FIGURE 9.P.5 A 5 × 4 grid with 3 × 2 interior points.

▶ Beginning with the first entry, each set of M consecutive entries of the vector $\mathbf{f}$ corresponds to values of $f(x, y)$ at interior nodal points along the rows of the computational grid.

▶ The first M entries of the $\mathbf{p}$ vector correspond to $p(x)$ evaluated at the interior bottom row of grid points, with the remaining entries equal to 0.

▶ The last M entries of the $\mathbf{q}$ vector correspond to $q(x)$ evaluated at the interior top row of grid points, with the remaining entries equal to 0.

▶ Beginning with the first entry, each Mth consecutive entry of the $\mathbf{r}$ vector is equal to $r(y)$ evaluated at a grid point along the left boundary of R.

▶ Beginning with the Mth entry, each Mth consecutive entry of the $\mathbf{s}$ vector is equal to $s(y)$ evaluated at a grid point along the right boundary of R.

While decreasing the distance h between nodes in the finite difference algorithm decreases the error in the approximate solution of the boundary value problem (4–6), it increases the number of unknowns. For example, decreasing internodal distances by a factor of 1/2 increases the number of unknowns by 4. However, understanding the pattern of the entries in $\mathbf{A}$ and the vectors $\mathbf{f}$, $\mathbf{p}$, $\mathbf{q}$, $\mathbf{r}$, and $\mathbf{s}$ allows us to construct a system of equations that can be easily solved using a computer algebra system, provided that the dimension of the system is not too large.

The method of finite differences is one of several different types of algorithms used to compute numerical approximations to solutions of partial differential equations. Numerical approximation methods enjoy three major advantages over the method of separation of variables: (1) they generalize to equations with variable coefficients, (2) they can be used on problems that have domains with irregular geometries, and (3) they can be applied to nonlinear problems. In all cases, numerical linear algebra plays an important role in the solution algorithms.

Project 3 PROBLEMS

1. If a function $F(x)$ has n continuous derivatives, the Taylor series expansion of $F(x_0 + \Delta x)$ with remainder about the point x_0 is

$$F(x_0 + \Delta x) = F(x_0) + \frac{1}{1!}F^{(1)}(x_0)\Delta x + \frac{1}{2!}F^{(2)}(x_0)(\Delta x)^2$$
$$+ \cdots + \frac{1}{(n-1)!}F^{(n-1)}(x_0)(\Delta x)^{n-1}$$
$$+ \frac{1}{n!}F^{(n)}(x_1)(\Delta x)^n,$$

(i)

where $|x_1 - x_0| < |\Delta x|$. Use Eq. (i) to show that if $F(x)$ has four continuous derivatives, then the local error between the second order centered finite difference approximation at x_0 and $\frac{\partial^2 F}{\partial x^2}(x_0)$ satisfies

$$\left| \frac{F(x_0 - \Delta x) - 2F(x_0) + F(x_0 + \Delta x)}{(\Delta x)^2} - \frac{\partial^2 F}{\partial x^2}(x_0) \right| \leq C(\Delta x)^2$$

where C is a constant. If the increment size Δx is reduced by a factor of $1/2$, by what factor is the local error reduced?

CAS 2. Consider the following Dirichlet problem:

$$\frac{\partial^2 u}{\partial x^2} + \frac{\partial^2 u}{\partial y^2} = -6 - xy, \quad 0 < x < 2, \quad 0 < y < 1,$$

(i)

$$u(x, 0) = x^2 - 2x, \quad u(x, 1) = x(-13/6),$$
$$0 < x < 2,$$

(ii)

$$u(0, y) = 4y(1 - y), \quad u(2, y) = 4y(1 - y - y^2/12),$$
$$0 < y < 1.$$

(iii)

(a) In the case that the increment size $h = 1/3$, sketch the computational grid and write out by

hand the matrices $\mathbf{A}, \mathbf{f}, \mathbf{p}, \mathbf{q}, \mathbf{r}$ and $\mathbf{s}$ analogous to those appearing in Eq. (18).

(b) Use a computer algebra system to solve the resulting system of equations obtained in part (a).

(c) Verify that $u(x, y) = x^2 - 2x - 4y^2 + 4y - xy^3/6$ satisfies Eqs. (i–iii).

(d) At each of the computational grid points, compute the difference between the approximation obtained in part (b) and the exact solution.

3. Using a computer algebra system, construct the matrix $\mathbf{A}$ and the vectors $\mathbf{f}, \mathbf{p}, \mathbf{q}, \mathbf{r},$ and $\mathbf{s}$ and then solve Eq. (18) for each of the following increment sizes: (i) $h = 0.2$, (ii) $h = 0.1$, and (iii) $h = 0.05$. In each case, compute the maximum error between the numerical approximation and the exact solution at each of the grid points: **CAS**

$$\max_{1 \leq m \leq M, \ 1 \leq n \leq N} |u_{mn} - u(x_m, y_n)|$$

Does it appear that $\max_{1 \leq m \leq M, \ 1 \leq n \leq N} |u_{mn} - u(x_m, y_n)| \leq \text{constant} \cdot h$ as $h \to 0$?

4. How many unknowns are there in the resulting linear system and how many entries does the full matrix $\mathbf{A}$ (counting zero and nonzero entries) have in the case $h = 0.025$? How many nonzero entries does $\mathbf{A}$ have in this case? What percentage of the entries of $\mathbf{A}$ are nonzero in this case?

5. Draw a surface plot of the numerical approximation to the solution of Eqs. (21–23) for the case $h = 0.04$. **CAS**

CHAPTER TEN

Boundary Value Problems and Sturm–Liouville Theory

As a result of separating variables in a partial differential equation in Chapter 9, we repeatedly encountered the differential equation

$$X'' + \lambda X = 0, \qquad 0 < x < L,$$

with the boundary conditions

$$X(0) = 0, \qquad X(L) = 0.$$

This boundary value problem is the prototype of a large class of problems that are important in applied mathematics. These problems are known as Sturm–Liouville boundary value problems. In this chapter we discuss the major properties of Sturm–Liouville problems and their solutions; in the process we are able to generalize somewhat the method of separation of variables for partial differential equations.

10.1 The Occurrence of Two-Point Boundary Value Problems

In Chapter 9 we described the method of separation of variables as a means of solving certain problems involving partial differential equations. The heat conduction problem consisting of the partial differential equation

$$\alpha^2 u_{xx} = u_t, \qquad 0 < x < L, \quad t > 0, \tag{1}$$

subject to the boundary conditions

$$u(0, t) = 0, \qquad u(L, t) = 0, \qquad t > 0, \tag{2}$$

and the initial condition

$$u(x, 0) = f(x), \qquad 0 \le x \le L, \tag{3}$$

is typical of the problems considered there. A crucial part of the process of solving such problems is to find the eigenvalues and eigenfunctions of the differential equation

$$X'' + \lambda X = 0, \qquad 0 < x < L, \tag{4}$$

with the boundary conditions

$$X(0) = 0, \qquad X(L) = 0, \tag{5}$$

or perhaps

$$X'(0) = 0, \qquad X'(L) = 0. \tag{6}$$

The sine or cosine functions that result from solving Eq. (4) subject to the boundary conditions (5) or (6) are used to expand the initial temperature distribution $f(x)$ in a Fourier series.

In this chapter we extend and generalize the results of Chapter 9. Our main goal is to show how the method of separation of variables can be used to solve problems somewhat more general than that of Eqs. (1), (2), and (3). We are interested in three types of generalizations.

First, we wish to consider more general partial differential equations—for example, the equation

$$r(x)u_t = [p(x)u_x]_x - q(x)u + F(x, t). \tag{7}$$

This equation can arise, as indicated in Appendix 9.A of Chapter 9, in the study of heat conduction in a bar of variable material properties in the presence of heat sources. If p and r are constants, and if the source terms qu and F are zero, then Eq. (7) reduces to Eq. (1). The partial differential equation (7) also occurs in the investigation of other phenomena of a diffusive character.

A second generalization is to allow more general boundary conditions. In particular, we wish to consider boundary conditions of the form

$$u_x(0, t) - h_1 u(0, t) = 0, \qquad u_x(L, t) + h_2 u(L, t) = 0. \tag{8}$$

Such conditions occur when the rate of heat flow through an end of the bar is proportional to the temperature there. Usually, h_1 and h_2 are nonnegative constants, but in some cases they may be negative or depend on t. The boundary conditions (2) are obtained in the limit as $h_1 \to \infty$ and $h_2 \to \infty$. The other important limiting case, $h_1 = h_2 = 0$, gives the boundary conditions for insulated ends.

The final generalization that we discuss in this chapter concerns the geometry of the region in which the problem is posed. The results of Chapter 9 are adequate only for a rather restricted class of problems, chiefly those in which the region of interest is rectangular or, in a few cases, circular. Later in this chapter we consider certain problems posed in a few other geometrical regions.

Let us consider the equation

$$r(x)u_t = [p(x)u_x]_x - q(x)u \tag{9}$$

obtained by setting the term $F(x, t)$ in Eq. (7) equal to zero. To separate the variables, we assume that

$$u(x, t) = X(x)T(t), \tag{10}$$

and substitute for u in Eq. (9). We obtain

$$r(x)XT' = [p(x)X']'T - q(x)XT \tag{11}$$

or, upon dividing by $r(x)XT$,

$$\frac{T'}{T} = \frac{[p(x)X']'}{r(x)X} - \frac{q(x)}{r(x)} = -\lambda. \tag{12}$$

We have denoted the separation constant by $-\lambda$ in anticipation of the fact that usually it will turn out to be real and negative. From Eq. (12) we obtain the following two ordinary differential equations for X and T:

$$
\begin{aligned}
[p(x)X']' - q(x)X + \lambda r(x)X &= 0, & (13) \\
T' + \lambda T &= 0. & (14)
\end{aligned}
$$

If we substitute from Eq. (10) for u in Eqs. (8) and assume that h_1 and h_2 are constants, then we obtain the boundary conditions

$$X'(0) - h_1 X(0) = 0, \qquad X'(L) + h_2 X(L) = 0. \tag{15}$$

To proceed further, we need to solve Eq. (13) subject to the boundary conditions (15). Although this is a more general linear homogeneous two-point boundary value problem than the problem consisting of the differential equation (4) and the boundary conditions (5) or (6), the solutions behave in very much the same way. For every value of λ, the problem (13), (15) has the trivial solution $X(x) = 0$. For certain values of λ, called **eigenvalues**, there are also other, nontrivial, solutions called **eigenfunctions**. These eigenfunctions form the basis for series solutions of a variety of problems in partial differential equations, such as the generalized heat conduction equation (9) subject to the boundary conditions (8) and the initial condition (3).

In this chapter we discuss some of the properties of solutions of two-point boundary value problems for second order linear equations. Sometimes we consider the general linear homogeneous equation

$$P(x)y'' + Q(x)y' + R(x)y = 0, \tag{16}$$

investigated in Chapter 4. However, for most purposes it is better to discuss equations in which the first and second derivative terms are related as in Eq. (13). It is always possible to transform the general equation (16) so that the derivative terms appear as in Eq. (13) (see Problem 11).

Boundary value problems with higher order differential equations can also occur; in them the number of boundary conditions must equal the order of the differential equation. As a rule, the order of the differential equation is even, and half the boundary conditions are given at each end of the interval. It is also possible for a single boundary condition to involve values of the solution and/or its derivatives at both boundary points; for example,

$$y(0) - y(L) = 0. \tag{17}$$

The following example involves one boundary condition of the form (15) and is therefore more complicated than the problems in Section 9.1.

EXAMPLE 1

Find the eigenvalues and the corresponding eigenfunctions of the boundary value problem

$$y'' + \lambda y = 0, \tag{18}$$

$$y(0) = 0, \qquad y'(1) + y(1) = 0. \tag{19}$$

One place where this problem occurs is in the heat conduction problem in a bar of unit length. The boundary condition at $x = 0$ corresponds to a zero temperature there. The boundary condition at $x = 1$ corresponds to a rate of heat flow that is proportional to the temperature there, and units are chosen so that the constant of proportionality is 1 (see Appendix 9.A of Chapter 9).

The solution of the differential equation may have one of several forms, depending on λ, so it is necessary to consider several cases. First, if $\lambda = 0$, the general solution of the differential equation is

$$y = c_1 x + c_2. \tag{20}$$

The two boundary conditions require that

$$c_2 = 0, \qquad 2c_1 + c_2 = 0, \tag{21}$$

respectively. The only solution of Eqs. (21) is $c_1 = c_2 = 0$, so the boundary value problem has no nontrivial solution in this case. Hence $\lambda = 0$ is not an eigenvalue.

If $\lambda > 0$, then the general solution of the differential equation (18) is

$$y = c_1 \sin\sqrt{\lambda}\, x + c_2 \cos\sqrt{\lambda}\, x, \tag{22}$$

where $\sqrt{\lambda} > 0$. The boundary condition at $x = 0$ requires that $c_2 = 0$; from the boundary condition at $x = 1$ we then obtain the equation

$$c_1(\sin\sqrt{\lambda} + \sqrt{\lambda}\cos\sqrt{\lambda}) = 0.$$

For a nontrivial solution y we must have $c_1 \neq 0$, and thus λ must satisfy

$$\sin\sqrt{\lambda} + \sqrt{\lambda}\cos\sqrt{\lambda} = 0. \tag{23}$$

Note that if λ is such that $\cos\sqrt{\lambda} = 0$, then $\sin\sqrt{\lambda} \neq 0$, and Eq. (23) is not satisfied. Hence we may assume that $\cos\sqrt{\lambda} \neq 0$; dividing Eq. (23) by $\cos\sqrt{\lambda}$, we obtain

$$\sqrt{\lambda} = -\tan\sqrt{\lambda}. \tag{24}$$

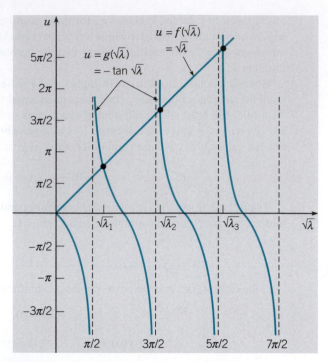

FIGURE 10.1.1 Graphical solution of $\sqrt{\lambda} = -\tan\sqrt{\lambda}$.

The solutions of Eq. (24) can be determined numerically. They can also be found approximately by sketching the graphs of $f(\sqrt{\lambda}) = \sqrt{\lambda}$ and $g(\sqrt{\lambda}) = -\tan\sqrt{\lambda}$ for $\sqrt{\lambda} > 0$ on the same set of axes, and identifying the points of intersection of the two curves (see Figure 10.1.1). The point $\sqrt{\lambda} = 0$ is specifically excluded from this argument because the solution (22) is valid only for $\sqrt{\lambda} \neq 0$. Despite the fact that the curves intersect there, $\lambda = 0$ is not an eigenvalue, as we have already shown. The first three positive solutions of Eq. (24) are $\sqrt{\lambda_1} \cong 2.029$, $\sqrt{\lambda_2} \cong 4.913$, and $\sqrt{\lambda_3} \cong 7.979$. As can be seen from Figure 10.1.1, the other roots are given with reasonable accuracy by $\sqrt{\lambda_n} \cong (2n-1)\pi/2$ for $n = 4, 5, \ldots$, the precision of this estimate improving as n increases. Hence the eigenvalues are

$$\lambda_1 \cong 4.116, \qquad \lambda_2 \cong 24.14,$$
$$\lambda_3 \cong 63.66, \qquad \lambda_n \cong (2n-1)^2\pi^2/4 \qquad \text{for } n = 4, 5, \ldots. \tag{25}$$

Finally, since $c_2 = 0$, the eigenfunction corresponding to the eigenvalue λ_n is

$$\phi_n(x, \lambda_n) = k_n \sin\sqrt{\lambda_n}\,x; \qquad n = 1, 2, \ldots, \tag{26}$$

where the constant k_n remains arbitrary.

Next consider $\lambda < 0$. In this case it is convenient to let $\lambda = -\mu$ so that $\mu > 0$. Then Eq. (14) becomes

$$y'' - \mu y = 0, \tag{27}$$

and its general solution is

$$y = c_1 \sinh\sqrt{\mu}\,x + c_2 \cosh\sqrt{\mu}\,x, \tag{28}$$

where $\sqrt{\mu} > 0$. Proceeding as in the previous case, we find that μ must satisfy the equation

$$\sqrt{\mu} = -\tanh \sqrt{\mu}. \tag{29}$$

From Figure 10.1.2 it is clear that the graphs of $f(\sqrt{\mu}) = \sqrt{\mu}$ and $g(\sqrt{\mu}) = -\tanh \sqrt{\mu}$ intersect only at the origin. Hence there are no positive values of $\sqrt{\mu}$ that satisfy Eq. (29), and hence the boundary value problem (18), (19) has no negative eigenvalues.

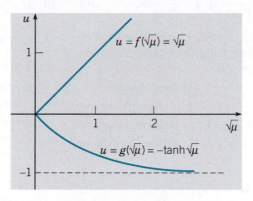

FIGURE 10.1.2 Graphical solution of $\sqrt{\mu} = -\tanh \sqrt{\mu}$.

Finally, it is necessary to consider the possibility that λ may be complex. It is possible to show by direct calculation that the problem (18), (19) has no complex eigenvalues. However, in Section 10.2 we consider in more detail a large class of problems that includes this example. One of the things we show there is that every problem in this class has only real eigenvalues. Therefore we omit the discussion of the nonexistence of complex eigenvalues here. Thus we conclude that all the eigenvalues and eigenfunctions of the problem (18), (19) are given by Eqs. (25) and (26).

PROBLEMS

In each of Problems 1 through 6 state whether the given boundary value problem is homogeneous or nonhomogeneous.

1. $y'' + 4y = 0$, $\quad y(-1) = 0$, $\quad y(1) = 0$
2. $[(1 + x^2)y']' + 4y = 0$, $\quad y(0) = 0$, $\quad y(1) = 1$
3. $y'' + 4y = \sin x$, $\quad y(0) = 0$, $\quad y(1) = 0$
4. $-y'' + x^2 y = \lambda y$, $\quad y'(0) - y(0) = 0$, $y'(1) + y(1) = 0$
5. $-[(1 + x^2)y']' = \lambda y + 1$, $\quad y(-1) = 0$, $y(1) = 0$
6. $-y'' = \lambda(1 + x^2)y$, $\quad y(0) = 0$, $y'(1) + 3y(1) = 0$

In each of Problems 7 through 10 assume that all eigenvalues are real.

(a) Determine the form of the eigenfunctions and the determinantal equation satisfied by the nonzero eigenvalues.

(b) Determine whether $\lambda = 0$ is an eigenvalue.

(c) Find approximate values for λ_1 and λ_2, the nonzero eigenvalues of smallest absolute value.

(d) Estimate λ_n for large values of n.

7. $y'' + \lambda y = 0$, $\quad y(0) = 0$, $\quad y(\pi) + y'(\pi) = 0$
8. $y'' + \lambda y = 0$, $\quad y'(0) = 0$, $\quad y(1) + y'(1) = 0$
9. $y'' + \lambda y = 0$, $\quad y(0) - y'(0) = 0$, $y(1) + y'(1) = 0$
10. $y'' - \lambda y = 0$, $\quad y(0) + y'(0) = 0$, $\quad y(1) = 0$
11. Consider the general linear homogeneous second order equation

$$P(x)y'' + Q(x)y' + R(x)y = 0. \tag{i}$$

We seek an integrating factor $\mu(x)$ such that, upon multiplying Eq. (i) by $\mu(x)$, we can write the resulting equation in the form

$$[\mu(x)P(x)y']' + \mu(x)R(x)y = 0. \tag{ii}$$

(a) By equating coefficients of y' in Eqs. (i) and (ii), show that μ must be a solution of

$$P\mu' = (Q - P')\mu. \qquad \text{(iii)}$$

(b) Solve Eq. (iii) and thereby show that

$$\mu(x) = \frac{1}{P(x)} \exp \int_{x_0}^x \frac{Q(s)}{P(s)}\, ds. \qquad \text{(iv)}$$

In each of Problems 12 through 15 use the method of Problem 11 to transform the given equation into the form $[p(x)y']' + q(x)y = 0$.

12. $y'' - 2xy' + \lambda y = 0$,   Hermite equation
13. $x^2 y'' + xy' + (x^2 - v^2)y = 0$,   Bessel equation
14. $xy'' + (1 - x)y' + \lambda y = 0$,   Laguerre equation
15. $(1 - x^2)y'' - xy' + \alpha^2 y = 0$,   Chebyshev equation
16. The equation

$$u_{tt} + cu_t + ku = a^2 u_{xx} + F(x, t), \qquad \text{(i)}$$

where $a^2 > 0$, $c \geq 0$, and $k \geq 0$ are constants, is known as the **telegraph equation**. It arises in the study of an elastic string under tension (see Appendix 9.B of Chapter 9). Equation (i) also occurs in other applications. Assuming that $F(x, t) = 0$, let $u(x, t) = X(x)T(t)$, separate the variables in Eq. (i), and derive ordinary differential equations for X and T.

17. Consider the boundary value problem

$$y'' - 2y' + (1 + \lambda)y = 0, \quad y(0) = 0, \ y(1) = 0.$$

(a) Introduce a new dependent variable u by the relation $y = s(x)u$. Determine $s(x)$ so that the differential equation for u has no u' term.
(b) Solve the boundary value problem for u and thereby determine the eigenvalues and eigenfunctions of the original problem. Assume that all eigenvalues are real.
(c) Also solve the given problem directly (without introducing u).

18. Consider the boundary value problem

$$y'' + 4y' + (4 + 9\lambda)y = 0, \qquad y(0) = 0, \quad y'(L) = 0.$$

(a) Determine, at least approximately, the real eigenvalues and the corresponding eigenfunctions by proceeding as in Problem 17(**a, b**).
(b) Also solve the given problem directly (without introducing a new variable).
Hint: In part (**a**) be sure to pay attention to the boundary conditions as well as the differential equation.

The differential equations in Problems 19 and 20 differ from those in previous problems in that the parameter λ multiplies the y' term as well as the y term. In each of these problems determine the real eigenvalues and the corresponding eigenfunctions.

19. $y'' + y' + \lambda(y' + y) = 0$, $y'(0) = 0$, $y(1) = 0$
20. $x^2 y'' - \lambda(xy' - y) = 0$, $y(1) = 0$,
     $y(2) - y'(2) = 0$

21. Consider the problem

$$y'' + \lambda y = 0, \qquad 2y(0) + y'(0) = 0, \quad y(1) = 0.$$

(a) Find the determinantal equation satisfied by the positive eigenvalues.
(b) Show that there is an infinite sequence of such eigenvalues.
(c) Find λ_1 and λ_2. Then show that $\lambda_n \cong [(2n + 1)\pi/2]^2$ for large n.
(d) Find the determinantal equation satisfied by the negative eigenvalues.
(e) Show that there is exactly one negative eigenvalue and find its value.

22. Consider the problem

$$y'' + \lambda y = 0, \qquad \alpha y(0) + y'(0) = 0, \quad y(1) = 0,$$

where α is a given constant.
(a) Show that for all values of α there is an infinite sequence of positive eigenvalues.
(b) If $\alpha < 1$, show that all (real) eigenvalues are positive. Show that the smallest eigenvalue approaches zero as α approaches 1 from below.
(c) Show that $\lambda = 0$ is an eigenvalue only if $\alpha = 1$.
(d) If $\alpha > 1$, show that there is exactly one negative eigenvalue and that this eigenvalue decreases as α increases.

23. Consider the problem

$$y'' + \lambda y = 0, \qquad y(0) = 0, \quad y'(L) = 0.$$

Show that if ϕ_m and ϕ_n are eigenfunctions corresponding to the eigenvalues λ_m and λ_n, respectively, with $\lambda_m \neq \lambda_n$, then

$$\int_0^L \phi_m(x)\phi_n(x)\, dx = 0.$$

Hint: Note that

$$\phi_m'' + \lambda_m \phi_m = 0, \qquad \phi_n'' + \lambda_n \phi_n = 0.$$

Multiply the first of these equations by ϕ_n, the second by ϕ_m, and integrate from 0 to L, using integration by parts. Finally, subtract one equation from the other.

24. In this problem we consider a higher order eigenvalue problem. The analysis of transverse vibrations of a uniform elastic bar is based on the differential equation

$$y^{(4)} - \lambda y = 0,$$

where y is the transverse displacement and $\lambda = m\omega^2/EI$; m is the mass per unit length of the rod, E is Young's modulus, I is the moment of inertia of the cross section about an axis through the centroid perpendicular to the plane of vibration, and ω is the frequency of vibration. Thus for a bar whose material and geometrical properties are given, the eigenvalues determine the natural frequencies of vibration. Boundary conditions at each end are usually one of the following types:

$$
\begin{aligned}
y = y' &= 0, & \text{clamped end,}\\
y = y'' &= 0, & \text{simply supported or hinged end,}\\
y'' = y''' &= 0, & \text{free end.}
\end{aligned}
$$

For each of the following three cases find the form of the eigenfunctions and the equation satisfied by the eigenvalues of this fourth order boundary value problem. Determine λ_1 and λ_2, the two eigenvalues of smallest magnitude. Assume that the eigenvalues are real and positive.

(a) $y(0) = y''(0) = 0$, $y(L) = y''(L) = 0$
(b) $y(0) = y''(0) = 0$, $y(L) = y'(L) = 0$
(c) $y(0) = y'(0) = 0$, $y''(L) = y'''(L) = 0$
(cantilevered bar)

25. This problem illustrates that the eigenvalue parameter sometimes appears in the boundary conditions as well as in the differential equation. Consider the longitudinal vibrations of a uniform straight elastic bar of length L. It can be shown that the axial displacement $u(x, t)$ satisfies the partial differential equation

$$(E/\rho)u_{xx} = u_{tt}; \qquad 0 < x < L, \quad t > 0, \qquad \text{(i)}$$

where E is Young's modulus and ρ is the mass per unit volume. If the end $x = 0$ is fixed, then the boundary condition there is

$$u(0, t) = 0, \qquad t > 0. \qquad \text{(ii)}$$

Suppose that the end $x = L$ is rigidly attached to a mass m but is otherwise unrestrained. We can obtain the boundary condition here by writing Newton's law for the mass. From the theory of elasticity it can be shown that the force exerted by the bar on the mass is given by $-EAu_x(L, t)$. Hence the boundary condition is

$$EAu_x(L, t) + mu_{tt}(L, t) = 0, \qquad t > 0. \qquad \text{(iii)}$$

(a) Assume that $u(x, t) = X(x)T(t)$, and show that $X(x)$ and $T(t)$ satisfy the differential equations

$$
\begin{aligned}
X'' + \lambda X &= 0, & \text{(iv)}\\
T'' + \lambda(E/\rho)T &= 0. & \text{(v)}
\end{aligned}
$$

(b) Show that the boundary conditions are

$$X(0) = 0, \qquad X'(L) - \gamma\lambda LX(L) = 0, \qquad \text{(vi)}$$

where $\gamma = m/\rho AL$ is a dimensionless parameter that gives the ratio of the end mass to the mass of the bar.
Hint: Use the differential equation for $T(t)$ in simplifying the boundary condition at $x = L$.

(c) Determine the form of the eigenfunctions and the equation satisfied by the real eigenvalues of Eqs. (iv) and (vi).

(d) Find the first two eigenvalues λ_1 and λ_2 if $\gamma = 0.5$.

10.2 Sturm–Liouville Boundary Value Problems

We now consider a class of two-point boundary value problems of the type obtained in Section 10.1 by separating the variables in a heat conduction problem for a bar of variable material properties, and with a source term proportional to the temperature. The class consists of a differential equation of the form

$$[p(x)y']' - q(x)y + \lambda r(x)y = 0 \qquad \text{(1)}$$

on the interval $0 < x < 1$, together with the boundary conditions

$$\alpha_1 y(0) + \alpha_2 y'(0) = 0, \qquad \beta_1 y(1) + \beta_2 y'(1) = 0 \tag{2}$$

at the endpoints. This class of problems also occurs in many other applications. It is often convenient to introduce the linear homogeneous differential operator L defined by

$$L[y] = -[p(x)y']' + q(x)y. \tag{3}$$

Then the differential equation (1) can be written as

$$L[y] = \lambda r(x)y. \tag{4}$$

We assume that the functions p, p', q, and r are continuous on the interval $0 \le x \le 1$ and, further, that $p(x) > 0$ and $r(x) > 0$ at all points in $0 \le x \le 1$. These assumptions are necessary to render the theory as simple as possible while retaining considerable generality. It turns out that these conditions are satisfied in many significant problems in mathematical physics. For example, the equation $y'' + \lambda y = 0$, which arose repeatedly in the preceding chapter, is of the form (1) with $p(x) = 1$, $q(x) = 0$, and $r(x) = 1$. The boundary conditions (2) are said to be **separated**; that is, each involves only one of the boundary points. These are the most general separated boundary conditions that are possible for a second order differential equation.

Lagrange's Identity

Before proceeding to establish some of the properties of the Sturm–Liouville problem (1), (2), it is necessary to derive an identity, known as **Lagrange's identity**, which is basic to the study of linear boundary value problems. Let u and v be functions having continuous second derivatives on the interval $0 \le x \le 1$. Then[1]

$$\int_0^1 L[u]v \, dx = \int_0^1 [-(pu')'v + quv] \, dx.$$

Integrating the first term on the right side twice by parts, we obtain

$$\int_0^1 L[u]v \, dx = -p(x)u'(x)v(x)\Big|_0^1 + p(x)u(x)v'(x)\Big|_0^1 + \int_0^1 [-u(pv')' + uqv] \, dx$$

$$= -p(x)[u'(x)v(x) - u(x)v'(x)]\Big|_0^1 + \int_0^1 uL[v] \, dx.$$

Hence, upon transposing the integral on the right side, we have

$$\int_0^1 \{L[u]v - uL[v]\} \, dx = -p(x)[u'(x)v(x) - u(x)v'(x)]\Big|_0^1, \tag{5}$$

which is Lagrange's identity.

---

[1] For brevity we sometimes use the notation $\int_0^1 f \, dx$ rather than $\int_0^1 f(x) \, dx$ in this chapter.

Now let us suppose that the functions u and v in Eq. (5) also satisfy the boundary conditions (2). Then, if we assume that $\alpha_2 \neq 0$ and $\beta_2 \neq 0$, the right side of Eq. (5) becomes

$$-p(x)\left[u'(x)v(x) - u(x)v'(x)\right]\Big|_0^1$$

$$= -p(1)[u'(1)v(1) - u(1)v'(1)] + p(0)[u'(0)v(0) - u(0)v'(0)]$$

$$= -p(1)\left[-\frac{\beta_1}{\beta_2}u(1)v(1) + \frac{\beta_1}{\beta_2}u(1)v(1)\right] + p(0)\left[-\frac{\alpha_1}{\alpha_2}u(0)v(0) + \frac{\alpha_1}{\alpha_2}u(0)v(0)\right]$$

$$= 0.$$

The same result holds if either α_2 or β_2 is zero; the proof in this case is even simpler and is left for you. Thus, if the differential operator L is defined by Eq. (3), and if the functions u and v satisfy the boundary conditions (2), then Lagrange's identity reduces to

$$\int_0^1 \{L[u]v - uL[v]\}\, dx = 0. \tag{6}$$

Let us now write Eq. (6) in a slightly different way. In Eq. (4) of Section 9.2 we introduced the inner product (u, v) of two real-valued functions u and v on a given interval; using the interval $0 \leq x \leq 1$, we have

$$(u, v) = \int_0^1 u(x)v(x)\, dx. \tag{7}$$

In this notation Eq. (6) becomes

$$(L[u], v) - (u, L[v]) = 0. \tag{8}$$

In proving Theorem 10.2.1 below, it is necessary to deal with complex-valued functions. The proof requires an inner product that generalizes Eq. (7) to handle such functions. If u and v are complex-valued functions on $0 \leq x \leq 1$, we define the complex inner product of u and v by

$$(u, v) = \int_0^1 u(x)\bar{v}(x)\, dx, \tag{9}$$

where $\bar{v}$ is the complex conjugate of v. Clearly, Eq. (9) coincides with Eq. (7) if $u(x)$ and $v(x)$ are real. It is important to know that Eq. (8) remains valid under the stated conditions if u and v are complex-valued functions and if the inner product (9) is used. To see this, one can start with the quantity $\int_0^1 L[u]\bar{v}\, dx$ and retrace the steps leading to Eq. (6), making use of the fact that $p(x)$, $q(x)$, α_1, α_2, β_1, and β_2 are all real quantities (see Problem 22).

We now consider some of the implications of Eq. (8) for the Sturm–Liouville boundary value problem (1), (2). We assume without proof[2] that this problem actually has eigenvalues and eigenfunctions. In Theorems 10.2.1 to 10.2.4 below, we state several of their important, but relatively elementary, properties. Each of these properties is illustrated by the basic

[2]The proof may be found, for example, in the references by Sagan (Chapter 5) or Birkhoff and Rota (Chapter 10).

Sturm–Liouville problem

$$y'' + \lambda y = 0, \qquad y(0) = 0, \quad y(1) = 0, \tag{10}$$

whose eigenvalues are $\lambda_n = n^2 \pi^2$, with the corresponding eigenfunctions $\phi_n(x) = \sin n\pi x$.

Properties of Eigenvalues and Eigenvectors of Sturm-Liouville Problems

THEOREM 10.2.1	All the eigenvalues of the Sturm–Liouville problem (1), (2) are real.

To prove this theorem, let us suppose that λ is a (possibly complex) eigenvalue of the problem (1), (2) and that ϕ is a corresponding eigenfunction, also possibly complex-valued. Let us write $\lambda = \mu + i\nu$ and $\phi(x) = U(x) + iV(x)$, where μ, ν, $U(x)$, and $V(x)$ are real. Then, if we let $u = \phi$ and also $v = \phi$ in Eq. (8), we have

$$(L[\phi], \phi) = (\phi, L[\phi]). \tag{11}$$

However, we know that $L[\phi] = \lambda r \phi$, so Eq. (11) becomes

$$(\lambda r\phi, \phi) = (\phi, \lambda r\phi). \tag{12}$$

Writing out Eq. (12) in full, using the definition (9) of the inner product, we obtain

$$\int_0^1 \lambda r(x)\phi(x)\overline{\phi}(x)\, dx = \int_0^1 \phi(x)\overline{\lambda}\overline{r}(x)\overline{\phi}(x)\, dx. \tag{13}$$

Since $r(x)$ is real, Eq. (13) reduces to

$$(\lambda - \overline{\lambda}) \int_0^1 r(x)\phi(x)\overline{\phi}(x)\, dx = 0,$$

or

$$(\lambda - \overline{\lambda}) \int_0^1 r(x)[U^2(x) + V^2(x)]\, dx = 0. \tag{14}$$

The integrand in Eq. (14) is nonnegative and not identically zero. Since the integrand is also continuous, it follows that the integral is positive. Therefore, the factor $\lambda - \overline{\lambda} = 2i\nu$ must be zero. Hence $\nu = 0$ and λ is real, so the theorem is proved.

An important consequence of Theorem 10.2.1 is that in finding eigenvalues and eigenfunctions of a Sturm–Liouville boundary value problem, we need look only for real eigenvalues. Recall that this is what we did in Chapter 9. It is also possible to show that the eigenfunctions of the boundary value problem (1), (2) are real. A proof is sketched in Problem 23.

THEOREM 10.2.2

If ϕ_1 and ϕ_2 are two eigenfunctions of the Sturm–Liouville problem (1), (2) corresponding to eigenvalues λ_1 and λ_2, respectively, and if $\lambda_1 \neq \lambda_2$, then

$$\int_0^1 r(x)\phi_1(x)\phi_2(x)\,dx = 0. \tag{15}$$

This theorem expresses the property of **orthogonality** of the eigenfunctions with respect to the weight function r. To prove the theorem, we note that ϕ_1 and ϕ_2 satisfy the differential equations

$$L[\phi_1] = \lambda_1 r \phi_1 \tag{16}$$

and

$$L[\phi_2] = \lambda_2 r \phi_2, \tag{17}$$

respectively. If we let $u = \phi_1, v = \phi_2$, and substitute for $L[u]$ and $L[v]$ in Eq. (8), we obtain

$$(\lambda_1 r \phi_1, \phi_2) - (\phi_1, \lambda_2 r \phi_2) = 0,$$

or, using Eq. (9),

$$\lambda_1 \int_0^1 r(x)\phi_1(x)\overline{\phi}_2(x)\,dx - \overline{\lambda}_2 \int_0^1 \phi_1(x)\overline{r}(x)\overline{\phi}_2(x)\,dx = 0.$$

Because $\lambda_2, r(x)$, and $\phi_2(x)$ are real, this equation becomes

$$(\lambda_1 - \lambda_2)\int_0^1 r(x)\phi_1(x)\phi_2(x)\,dx = 0. \tag{18}$$

Since by hypothesis $\lambda_1 \neq \lambda_2$, it follows that ϕ_1 and ϕ_2 must satisfy Eq. (15), and the theorem is proved.

THEOREM 10.2.3

The eigenvalues of the Sturm–Liouville problem (1), (2) are all simple; that is, to each eigenvalue there corresponds only one linearly independent eigenfunction. Further, the eigenvalues form an infinite sequence and can be ordered according to increasing magnitude so that

$$\lambda_1 < \lambda_2 < \lambda_3 < \cdots < \lambda_n < \cdots.$$

Moreover,

$$\lambda_n \to \infty \text{ as } n \to \infty.$$

The proof of this theorem is somewhat more advanced than those of the two previous theorems and will be omitted. However, a proof that the eigenvalues are simple is outlined in Problem 20.

Again we note that all the properties stated in Theorems 10.2.1 to 10.2.3 are exemplified by the eigenvalues $\lambda_n = n^2\pi^2$ and eigenfunctions $\phi_n(x) = \sin n\pi x$ of the example problem

(10). Clearly, the eigenvalues are real. The eigenfunctions satisfy the orthogonality relation

$$\int_0^1 \phi_m(x)\phi_n(x)\,dx = \int_0^1 \sin m\pi x \sin n\pi x\,dx = 0, \qquad m \neq n, \tag{19}$$

which was established in Section 10.2 by direct integration. Further, the eigenvalues can be ordered so that $\lambda_1 < \lambda_2 < \cdots$, and $\lambda_n \to \infty$ as $n \to \infty$. Finally, to each eigenvalue there corresponds a single linearly independent eigenfunction.

We will now assume that the eigenvalues of the Sturm–Liouville problem (1), (2) are ordered as indicated in Theorem 10.2.3. Associated with the eigenvalue λ_n is a corresponding eigenfunction ϕ_n, determined up to a multiplicative constant. It is often convenient to choose the arbitrary constant multiplying each eigenfunction so as to satisfy the condition

$$\int_0^1 r(x)\phi_n^2(x)\,dx = 1, \qquad n = 1, 2, \ldots. \tag{20}$$

Equation (20) is called a normalization condition, and eigenfunctions satisfying this condition are said to be **normalized**. Indeed, in this case, the eigenfunctions are said to form an **orthonormal set** (with respect to the weight function r) since they already satisfy the orthogonality relation (15). It is sometimes useful to combine Eqs. (15) and (20) into a single equation. To this end we introduce the symbol δ_{mn}, known as the **Kronecker delta** and defined by

$$\delta_{mn} = \begin{cases} 0, & \text{if } m \neq n, \\ 1, & \text{if } m = n. \end{cases} \tag{21}$$

Making use of the Kronecker delta, we can write Eqs. (15) and (20) as

$$\int_0^1 r(x)\phi_m(x)\phi_n(x)\,dx = \delta_{mn}. \tag{22}$$

EXAMPLE 1

Determine the normalized eigenfunctions of the problem (10):

$$y'' + \lambda y = 0, \qquad y(0) = 0, \quad y(1) = 0.$$

The eigenvalues of this problem are $\lambda_1 = \pi^2$, $\lambda_2 = 4\pi^2$, ..., $\lambda_n = n^2\pi^2$, ..., and the corresponding eigenfunctions are $k_1 \sin \pi x$, $k_2 \sin 2\pi x$, ..., $k_n \sin n\pi x$, ..., respectively. In this case the weight function is $r(x) = 1$. To satisfy Eq. (20), we must choose k_n so that

$$\int_0^1 (k_n \sin n\pi x)^2\,dx = 1 \tag{23}$$

for each value of n. Since

$$k_n^2 \int_0^1 \sin^2 n\pi x\,dx = k_n^2 \int_0^1 \left(\frac{1}{2} - \frac{1}{2}\cos 2n\pi x \right) dx = \frac{1}{2}k_n^2,$$

Eq. (23) is satisfied if k_n is chosen to be $\sqrt{2}$ for each value of n. Hence the normalized eigenfunctions of the given boundary value problem are

$$\phi_n(x) = \sqrt{2} \sin n\pi x, \qquad n = 1, 2, 3, \ldots. \tag{24}$$

EXAMPLE 2

Determine the normalized eigenfunctions of the problem

$$y'' + \lambda y = 0, \qquad y(0) = 0, \quad y'(1) + y(1) = 0. \tag{25}$$

In Example 1 of Section 10.1 we found that the eigenvalues λ_n satisfy the equation

$$\sin \sqrt{\lambda_n} + \sqrt{\lambda_n} \cos \sqrt{\lambda_n} = 0 \tag{26}$$

and that the corresponding eigenfunctions are

$$\phi_n(x) = k_n \sin \sqrt{\lambda_n}\, x, \tag{27}$$

where k_n is arbitrary. We can determine k_n from the normalization condition (20). Since $r(x) = 1$ in this problem, we have

$$
\begin{aligned}
\int_0^1 \phi_n^2(x)\, dx &= k_n^2 \int_0^1 \sin^2 \sqrt{\lambda_n}\, x\, dx \\
&= k_n^2 \int_0^1 \left(\frac{1}{2} - \frac{1}{2} \cos 2\sqrt{\lambda_n}\, x \right) dx = k_n^2 \left(\frac{x}{2} - \frac{\sin 2\sqrt{\lambda_n}\, x}{4\sqrt{\lambda_n}} \right) \Bigg|_0^1 \\
&= k_n^2 \frac{2\sqrt{\lambda_n} - \sin 2\sqrt{\lambda_n}}{4\sqrt{\lambda_n}} = k_n^2 \frac{\sqrt{\lambda_n} - \sin \sqrt{\lambda_n} \cos \sqrt{\lambda_n}}{2\sqrt{\lambda_n}} \\
&= k_n^2 \frac{1 + \cos^2 \sqrt{\lambda_n}}{2},
\end{aligned}
$$

where in the last step we have used Eq. (26). Hence, to normalize the eigenfunctions ϕ_n, we must choose

$$k_n = \left(\frac{2}{1 + \cos^2 \sqrt{\lambda_n}} \right)^{1/2}. \tag{28}$$

The normalized eigenfunctions of the given problem are

$$\phi_n(x) = \frac{\sqrt{2} \sin \sqrt{\lambda_n}\, x}{(1 + \cos^2 \sqrt{\lambda_n})^{1/2}}; \qquad n = 1, 2, \ldots. \tag{29}$$

Generalized Fourier Series

We now turn to the question of expressing a given function f as a series of eigenfunctions of the Sturm–Liouville problem (1), (2). We have already seen examples of such expansions in Sections 10.2 to 10.4. For example, it was shown there that if f is continuous and has a piecewise continuous derivative on $0 \le x \le 1$, and satisfies the boundary conditions $f(0) = f(1) = 0$, then f can be expanded in a Fourier sine series of the form

$$f(x) = \sum_{n=1}^{\infty} b_n \sin n\pi x. \tag{30}$$

The functions $\sin n\pi x, n = 1, 2, \ldots$, are precisely the eigenfunctions of the boundary value problem (10). The coefficients b_n are given by

$$b_n = 2 \int_0^1 f(x) \sin n\pi x \, dx \tag{31}$$

and the series (30) converges for each x in $0 \leq x \leq 1$. In a similar way f can be expanded in a Fourier cosine series using the eigenfunctions $\cos n\pi x, n = 0, 1, 2, \ldots$, of the boundary value problem $y'' + \lambda y = 0$, $y'(0) = 0$, $y'(1) = 0$.

Now suppose that a given function f, satisfying suitable conditions, can be expanded in an infinite series of eigenfunctions of the more general Sturm–Liouville problem (1), (2). If this can be done, then we have

$$f(x) = \sum_{n=1}^{\infty} c_n \phi_n(x), \tag{32}$$

where the functions $\phi_n(x)$ satisfy Eqs. (1), (2) and also the orthogonality condition (22). To compute the coefficients in the series (32), we multiply Eq. (32) by $r(x)\phi_m(x)$, where m is a fixed positive integer, and integrate from $x = 0$ to $x = 1$. Assuming that the series can be integrated term by term, we obtain

$$\int_0^1 r(x)f(x)\phi_m(x) \, dx = \sum_{n=1}^{\infty} c_n \int_0^1 r(x)\phi_m(x)\phi_n(x) \, dx = \sum_{n=1}^{\infty} c_n \delta_{mn}. \tag{33}$$

Hence, using the definition of δ_{mn}, we have

$$c_m = \int_0^1 r(x)f(x)\phi_m(x) \, dx = (f, r\phi_m), \qquad m = 1, 2, \ldots. \tag{34}$$

The coefficients in the series (32) have thus been formally determined. Equation (34) has the same structure as the Euler–Fourier formulas for the coefficients in a Fourier series, and the eigenfunction series (32) also has convergence properties similar to those of Fourier series. The following theorem is analogous to Theorem 9.3.1.

**THEOREM
10.2.4**

Let $\phi_1, \phi_2, \ldots, \phi_n, \ldots$ be the normalized eigenfunctions of the Sturm–Liouville problem (1), (2):

$$[p(x)y']' - q(x)y + \lambda r(x)y = 0,$$

$$\alpha_1 y(0) + \alpha_2 y'(0) = 0, \qquad \beta_1 y(1) + \beta_2 y'(1) = 0.$$

Let f and f' be piecewise continuous on $0 \leq x \leq 1$. Then the series (32) whose coefficients c_m are given by Eq. (34) converges to $[f(x+) + f(x-)]/2$ at each point in the open interval $0 < x < 1$.

If f satisfies further conditions, then a stronger conclusion can be established. Suppose that, in addition to the hypotheses of Theorem 10.2.4, the function f is continuous on $0 \leq x \leq 1$. If $\alpha_2 = 0$ in the first of Eqs. (2) [so that $\phi_n(0) = 0$], then assume that $f(0) = 0$.

Similarly, if $\beta_2 = 0$ in the second of Eqs. (2), assume that $f(1) = 0$. Otherwise no boundary conditions need be prescribed for f. Then the series (32) converges to $f(x)$ at each point in the closed interval $0 \leq x \leq 1$.

EXAMPLE 3

Expand the function

$$f(x) = x, \qquad 0 \leq x \leq 1 \tag{35}$$

in terms of the normalized eigenfunctions $\phi_n(x)$ of the problem (25).

In Example 2 we found the normalized eigenfunctions to be

$$\phi_n(x) = k_n \sin \sqrt{\lambda_n}\, x, \tag{36}$$

where k_n is given by Eq. (28) and λ_n satisfies Eq. (26). To find the expansion for f in terms of the eigenfunctions ϕ_n, we write

$$f(x) = \sum_{n=1}^{\infty} c_n \phi_n(x), \tag{37}$$

where the coefficients are given by Eq. (34). Thus

$$c_n = \int_0^1 f(x)\phi_n(x)\, dx = k_n \int_0^1 x \sin \sqrt{\lambda_n}\, x \, dx.$$

Integrating by parts, we obtain

$$c_n = k_n \left(\frac{\sin \sqrt{\lambda_n}}{\lambda_n} - \frac{\cos \sqrt{\lambda_n}}{\sqrt{\lambda_n}} \right) = k_n \frac{2 \sin \sqrt{\lambda_n}}{\lambda_n},$$

where we have used Eq. (26) in the last step. Upon substituting for k_n from Eq. (28), we obtain

$$c_n = \frac{2\sqrt{2} \sin \sqrt{\lambda_n}}{\lambda_n (1 + \cos^2 \sqrt{\lambda_n})^{1/2}}. \tag{38}$$

Thus

$$f(x) = 4 \sum_{n=1}^{\infty} \frac{\sin \sqrt{\lambda_n} \sin \sqrt{\lambda_n}\, x}{\lambda_n (1 + \cos^2 \sqrt{\lambda_n})}. \tag{39}$$

Observe that although the right side of Eq. (39) is a series of sines, it is not included in the discussion of Fourier sine series in Section 9.4.

Self-Adjoint Problems

Sturm–Liouville boundary value problems are of great importance in their own right, but they can also be viewed as belonging to a much more extensive class of problems that have many of the same properties. For example, there are many similarities between Sturm–Liouville problems and the algebraic system

$$\mathbf{A}\mathbf{x} = \lambda \mathbf{x}, \tag{40}$$

where the $n \times n$ matrix $\mathbf{A}$ is symmetric. Comparing the results mentioned in Appendix A.4 with those of this section, we note that in both cases the eigenvalues are real and the eigenfunctions or eigenvectors form an orthogonal set. Further, the eigenfunctions or eigenvectors can be used as the basis for expressing an essentially arbitrary function or vector, respectively, as a sum. The most important difference is that a matrix has only a finite number of eigenvalues and eigenvectors, while a Sturm–Liouville problem has infinitely many. It is interesting and of fundamental importance in mathematics that these seemingly different problems—the matrix problem (40) and the Sturm–Liouville problem (1), (2)—which arise in different ways, are actually parts of a single underlying theory. This theory is usually referred to as linear operator theory and is part of the subject of functional analysis.

We now point out some ways in which Sturm–Liouville problems can be generalized, while still preserving the main results of Theorems 10.2.1 to 10.2.4—the existence of a sequence of real eigenvalues tending to infinity, the orthogonality of the eigenfunctions, and the possibility of expressing an arbitrary function as a series of eigenfunctions. These generalizations depend on the continued validity of the crucial relation (8).

Let us consider the boundary value problem consisting of the differential equation

$$L[y] = \lambda r(x)y, \qquad 0 < x < 1, \tag{41}$$

where

$$L[y] = P_n(x)\frac{d^n y}{dx^n} + \cdots + P_1(x)\frac{dy}{dx} + P_0(x)y, \tag{42}$$

and n linear homogeneous boundary conditions at the endpoints. If Eq. (8) is valid for every pair of sufficiently differentiable functions that satisfy the boundary conditions, then the given problem is said to be **self-adjoint**. It is important to observe that Eq. (8) involves restrictions on both the differential equation and the boundary conditions. The differential operator L must be such that the same operator appears in both terms of Eq. (8). This requires L to be of even order. Further, a second order operator must have the form (3), a fourth order operator must have the form

$$L[y] = [p(x)y'']'' - [q(x)y']' + s(x)y, \tag{43}$$

and higher order operators must have an analogous structure. In addition, the boundary conditions must be such as to eliminate the boundary terms that arise during the integration by parts used in deriving Eq. (8). For example, in a second order problem this is true for the separated boundary conditions (2) and also in certain other cases, one of which is given in Example 4 below.

Let us suppose that we have a self-adjoint boundary value problem for Eq. (41), where $L[y]$ is given now by Eq. (43). We assume that p, q, r, and s are continuous on $0 \le x \le 1$ and that the derivatives of p and q indicated in Eq. (43) are also continuous. If in addition $p(x) > 0$ and $r(x) > 0$ for $0 \le x \le 1$, then there is an infinite sequence of real eigenvalues tending to $+\infty$, the eigenfunctions are orthogonal with respect to the weight function r, and an arbitrary function can be expressed as a series of eigenfunctions. However, the eigenvalues may not be simple in these more general problems.

We turn now to the relation between Sturm–Liouville problems and Fourier series. We have noted previously that Fourier sine and cosine series can be obtained by using the eigenfunctions of certain Sturm–Liouville problems involving the differential equation $y'' + \lambda y = 0$. This raises the question of whether we can obtain a full Fourier series, including both sine and cosine terms, by choosing a suitable set of boundary conditions. The answer is provided by the following example, which also serves to illustrate the occurrence of nonseparated boundary conditions.

EXAMPLE 4

Find the eigenvalues and eigenfunctions of the boundary value problem

$$y'' + \lambda y = 0, \tag{44}$$

$$y(-L) - y(L) = 0, \qquad y'(-L) - y'(L) = 0. \tag{45}$$

This is not a Sturm–Liouville problem because the boundary conditions are not separated. The boundary conditions (45) are called **periodic boundary conditions** since they require that y and y' assume the same values at $x = L$ as at $x = -L$. Nevertheless, it is straight-forward to show that the problem (44), (45) is self-adjoint. A simple calculation establishes that $\lambda_0 = 0$ is an eigenvalue and that the corresponding eigenfunction is $\phi_0(x) = 1$. Further, there are additional eigenvalues $\lambda_1 = (\pi/L)^2$, $\lambda_2 = (2\pi/L)^2$, ..., $\lambda_n = (n\pi/L)^2$, ... To each of these nonzero eigenvalues there correspond *two* linearly independent eigenfunc-tions; for example, corresponding to λ_n are the two eigenfunctions $\phi_n(x) = \cos(n\pi x/L)$ and $\psi_n(x) = \sin(n\pi x/L)$. This illustrates that the eigenvalues may not be simple when the boundary conditions are not separated. Further, if we seek to expand a given function f of period $2L$ in a series of eigenfunctions of the problem (44), (45), we obtain the series

$$f(x) = \frac{a_0}{2} + \sum_{n=1}^{\infty} \left(a_n \cos \frac{n\pi x}{L} + b_n \sin \frac{n\pi x}{L} \right),$$

which is just the Fourier series for f.

We will not give further consideration to problems that have nonseparated boundary conditions, nor will we deal with problems of higher than second order, except in a few problems. There is, however, one other kind of generalization that we do wish to discuss. That is the case in which the coefficients p, q, and r in Eq. (1) do not quite satisfy the rather strict continuity and positivity requirements laid down at the beginning of this sec-tion. Such problems are called singular Sturm–Liouville problems and are the subject of Section 10.4.

PROBLEMS

In each of Problems 1 through 5 determine the normal-ized eigenfunctions of the given problem.

1. $y'' + \lambda y = 0$, $\qquad y(0) = 0$, $\quad y'(1) = 0$
2. $y'' + \lambda y = 0$, $\qquad y'(0) = 0$, $\quad y(1) = 0$
3. $y'' + \lambda y = 0$, $\qquad y'(0) = 0$, $\quad y'(1) = 0$
4. $y'' + \lambda y = 0$, $\qquad y'(0) = 0$, $\quad y'(1) + y(1) = 0$
   (see Section 10.1, Problem 8.)
5. $y'' - 2y' + (1 + \lambda)y = 0$, $\qquad y(0) = 0$,
   $y(1) = 0$ (see Section 10.1, Problem 17.)

In each of Problems 6 through 9 find the coefficients in the eigenfunction expansion $\sum_{n=1}^{\infty} a_n \phi_n(x)$ of the given function, using the normalized eigenfunctions of Prob-lem 1.

6. $f(x) = 1$, $\qquad 0 \le x \le 1$
7. $f(x) = x$, $\qquad 0 \le x \le 1$

8. $f(x) = \begin{cases} 1, & 0 \le x < \frac{1}{2} \\ 0, & \frac{1}{2} \le x \le 1 \end{cases}$

9. $f(x) = \begin{cases} 2x, & 0 \le x < \frac{1}{2} \\ 1, & \frac{1}{2} \le x \le 1 \end{cases}$

In each of Problems 10 through 13 find the coefficients in the eigenfunction expansion $\sum_{n=1}^{\infty} a_n \phi_n(x)$ of the given function, using the normalized eigenfunctions of Problem 4.

10. $f(x) = 1$, $\qquad 0 \le x \le 1$
11. $f(x) = x$, $\qquad 0 \le x \le 1$
12. $f(x) = 1 - x$, $\qquad 0 \le x \le 1$

13. $f(x) = \begin{cases} 1, & 0 \le x < \frac{1}{2} \\ 0, & \frac{1}{2} \le x \le 1 \end{cases}$

In each of Problems 14 through 18 determine whether the given boundary value problem is self-adjoint.

14. $y'' + y' + 2y = 0$, $y(0) = 0$, $y(1) = 0$

15. $(1 + x^2)y'' + 2xy' + y = 0$, $y'(0) = 0$,
$y(1) + 2y'(1) = 0$

16. $y'' + y = \lambda y$, $y(0) - y'(1) = 0$,
$y'(0) - y(1) = 0$

17. $(1 + x^2)y'' + 2xy' + y = \lambda(1 + x^2)y$,
$y(0) - y'(1) = 0$, $y'(0) + 2y(1) = 0$

18. $y'' + \lambda y = 0$, $y(0) = 0$, $y(\pi) + y'(\pi) = 0$

19. Show that if the functions u and v satisfy Eqs. (2), and either $\alpha_2 = 0$ or $\beta_2 = 0$, or both, then

$$p(x)\left[u'(x)v(x) - u(x)v'(x)\right]\Big|_0^1 = 0.$$

20. In this problem we outline a proof of the first part of Theorem 10.2.3: that the eigenvalues of the Sturm–Liouville problem (1), (2) are simple. The proof is by contradiction.

  (a) Suppose that a given eigenvalue λ is not simple. Then there exist two corresponding eigenfunctions ϕ_1 and ϕ_2 that are linearly independent, that is, not multiples of each other.

  (b) Compute the Wronskian $W[\phi_1, \phi_2](x)$ and use the boundary conditions (2) to show that $W[\phi_1, \phi_2](0) = 0$.

  (c) Use the fact that the vectors $\mathbf{x}_1 = (\phi_1, \phi_1')^T$ and $\mathbf{x}_2 = (\phi_2, \phi_2')^T$ are linearly independent solutions of

$$\mathbf{x}' = \begin{pmatrix} 0 & 1 \\ (q - \lambda r)/p & -p'/p \end{pmatrix}\mathbf{x}$$

and Theorem 6.2.5 to reach a contradiction. This establishes that the eigenvalues must be simple, as asserted in Theorem 10.2.3.

21. Consider the Sturm–Liouville problem

$$-[p(x)y']' + q(x)y = \lambda r(x)y,$$
$$\alpha_1 y(0) + \alpha_2 y'(0) = 0, \quad \beta_1 y(1) + \beta_2 y'(1) = 0,$$

where p, q, and r satisfy the conditions stated in the text.

  (a) Show that if λ is an eigenvalue and ϕ a corresponding eigenfunction, then

$$\lambda \int_0^1 r\phi^2\, dx = \int_0^1 (p\phi'^2 + q\phi^2)\, dx$$
$$+ \frac{\beta_1}{\beta_2}p(1)\phi^2(1) - \frac{\alpha_1}{\alpha_2}p(0)\phi^2(0),$$

provided that $\alpha_2 \neq 0$ and $\beta_2 \neq 0$. How must this result be modified if $\alpha_2 = 0$ or $\beta_2 = 0$?

  (b) Show that if $q(x) \geq 0$ and if β_1/β_2 and $-\alpha_1/\alpha_2$ are nonnegative, then the eigenvalue λ is non-negative.

  (c) Under the conditions of part (b) show that the eigenvalue λ is strictly positive unless $\alpha_1 = \beta_1 = 0$ and $q(x) = 0$ for each x in $0 \leq x \leq 1$.

22. Derive Eq. (8) using the inner product (9) and assuming that u and v are complex-valued functions. *Hint:* Consider the quantity $\int_0^1 L[u]\bar{v}\, dx$, split u and v into real and imaginary parts, and proceed as in the text.

23. In this problem we outline a proof that the eigenfunctions of the Sturm–Liouville problem (1), (2) are real.

  (a) Let λ be an eigenvalue and ϕ a corresponding eigenfunction. Let $\phi(x) = U(x) + iV(x)$, and show that U and V are also eigenfunctions corresponding to λ.

  (b) Using Theorem 10.2.3, or the result of Problem 20, show that U and V are linearly dependent.

  (c) Show that ϕ must be real, apart from an arbitrary multiplicative constant that may be complex.

24. Consider the problem

$$x^2 y'' = \lambda(xy' - y), \quad y(1) = 0, \quad y(2) = 0.$$

Note that λ appears as a coefficient of y' as well as of y itself. It is possible to extend the definition of self-adjointness to this type of problem and to show that this particular problem is not self-adjoint. Show that the problem has eigenvalues but that none of them is real. This illustrates that in general nonself-adjoint problems may have eigenvalues that are not real.

Buckling of an Elastic Column. An investigation of the buckling of a uniform elastic column of length L by an axial load P (Figure 10.2.1a) leads to the differential equation

$$y^{(4)} + \lambda y'' = 0, \quad 0 < x < L. \quad\quad \text{(i)}$$

The parameter λ is equal to P/EI, where E is Young's modulus and I is the moment of inertia of the cross section about an axis through the centroid perpendicular to the xy-plane. The boundary conditions at $x = 0$ and $x = L$ depend on how the ends of the column are supported. Typical boundary conditions are

$$y = y' = 0, \qquad \text{clamped end};$$
$$y = y'' = 0, \qquad \text{simply supported (hinged) end}.$$

The bar shown in Figure 10.2.1a is simply supported at $x = 0$ and clamped at $x = L$. It is desired to determine the eigenvalues and eigenfunctions of Eq. (i) subject to suitable boundary conditions. In particular, the smallest eigenvalue λ_1 gives the load at which the column buckles, or can assume a curved equilibrium position, as shown in Figure 10.2.1b. The corresponding eigenfunction describes the configuration of the buckled column. Note that the differential equation (i) does not fall within the theory discussed in this section. It is possible to show, however, that in each of the cases given here all the eigenvalues are real and positive. Problems 25 and 26 deal with column buckling problems.

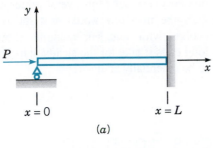

(a)

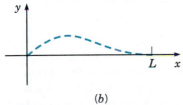

(b)

FIGURE 10.2.1   (a) A column under compression. (b) Shape of the buckled column.

25. For each of the following boundary conditions find the smallest eigenvalue (the buckling load) of $y^{(4)} + \lambda y'' = 0$, and also find the corresponding eigenfunction (the shape of the buckled column).
    (a) $y(0) = y''(0) = 0, \qquad y(L) = y''(L) = 0$
    (b) $y(0) = y''(0) = 0, \qquad y(L) = y'(L) = 0$
    (c) $y(0) = y'(0) = 0, \qquad y(L) = y'(L) = 0$

26. In some buckling problems the eigenvalue parameter appears in the boundary conditions as well as in the differential equation. One such case occurs when one end of the column is clamped and the other end is free. In this case the differential equation $y^{(4)} + \lambda y'' = 0$ must be solved subject to the boundary conditions

$$y(0) = 0, \qquad y'(0) = 0,$$
$$y''(L) = 0, \qquad y'''(L) + \lambda y'(L) = 0.$$

Find the smallest eigenvalue and the corresponding eigenfunction.

27. Solutes in an aquifer are transported by two sep- CAS arate mechanisms. The process by which a solute is transported by the bulk motion of the flowing groundwater is called **advection**. In addition, the solute is spread by small-scale fluctuations in the groundwater velocity along the tortuous flow paths within individual pores, a process called **mechanical dispersion**. The one-dimensional form of the advection-dispersion equation for a nonreactive dissolved solute in a saturated, homogeneous, isotropic porous medium under steady, uniform flow is

$$c_t + v c_x = D c_{xx}, \qquad 0 < x < L, \quad t > 0, \quad \text{(i)}$$

where $c(x, t)$ is the concentration of the solute, v is the average linear groundwater velocity, D is the coefficient of hydrodynamic dispersion, and L is the length of the aquifer. Suppose that the boundary conditions are

$$c(0, t) = 0, \qquad c_x(L, t) = 0, \qquad t > 0 \quad \text{(ii)}$$

and that the initial condition is

$$c(x, 0) = f(x), \qquad 0 < x < L, \quad \text{(iii)}$$

where $f(x)$ is the given initial concentration of the solute.

(a) Assume that $c(x, t) = X(x)T(t)$, use the method of separation of variables, and find the equations satisfied by $X(x)$ and $T(t)$, respectively. Show that the problem for $X(x)$ can be written in the Sturm-Liouville form

$$[p(x)X']' + \lambda r(x)X = 0, \quad 0 < x < L, \quad \text{(iv)}$$
$$X(0) = 0, \qquad X'(L) = 0, \quad \text{(v)}$$

where $p(x) = r(x) = \exp(-vx/D)$. Hence the eigenvalues are real and the eigenfunctions are orthogonal with respect to the weight function $r(x)$.

(b) Let $\mu^2 = \lambda - (v^2/4D^2)$. Show that the eigenfunctions are

$$X_n(x) = e^{vx/2D} \sin \mu_n x, \quad \text{(vi)}$$

where μ_n satisfies the equation

$$\tan \mu L = -2D\mu/v. \quad \text{(vii)}$$

(c) Show graphically that Eq. (vii) has an infinite sequence of positive roots and that $\mu_n \cong (2n-1)\pi/2L$ for large n.

(d) Show that

$$\int_0^L r(x)X_n^2(x)\,dx = \frac{L}{2} + \frac{v}{4D\mu_n^2}\sin^2\mu_n L.$$

(e) Find a formal solution of the problem (i), (ii), (iii) in terms of a series of the eigenfunctions $X_n(x)$.

(f) Let $v = 1$, $D = 0.5$, $L = 10$, and $f(x) = \delta(x-3)$, where δ is the Dirac delta[3] function. Using the solution found in part (e), plot $c(x,t)$ versus x for several values of t, such as $t = 0.5, 1, 3, 6$, and 10. Also plot $c(x,t)$ versus t for several values of x. Note that the number of terms that are needed to obtain an accurate plot depends strongly on the values of t and x.

(g) Describe in a few words how the solution evolves as time advances.

CAS 28. A nonreactive tracer at concentration c_0 is continuously introduced into a steady flow at the upstream end of a column of length L packed with a homogeneous granular medium. If we assume that the tracer concentration in the column is initially zero, the boundary value problem that models this process is

$$c_t + vc_x = Dc_{xx}, \qquad 0 < x < L, \quad t > 0,$$
$$c(0,t) = c_0, \qquad c_x(L,t) = 0, \qquad t > 0,$$
$$c(x,0) = 0, \qquad 0 < x < L,$$

where $c(x,t)$, v, and D are as in Problem 27.

(a) Assume that $c(x,t) = c_0 + u(x,t)$ and find the boundary value problem satisfied by $u(x,t)$.

(b) Proceeding as in Problem 27, find $u(x,t)$ in terms of an eigenfunction expansion.

(c) Let $v = 1$, $D = 0.5$, $c_0 = 1$, and $L = 10$. Plot $c(x,t)$ versus x for several values of t, and also plot $c(x,t)$ versus t for several values of x.

(d) Describe in a few words how the solution evolves with time. For example, about how long does it take for the steady-state solution to be essentially attained?

10.3 Nonhomogeneous Boundary Value Problems

In this section we discuss how to solve nonhomogeneous boundary value problems for both ordinary and partial differential equations. Most of our attention is directed toward problems in which the differential equation alone is nonhomogeneous, while the boundary conditions are homogeneous. We assume that the solution can be expanded in a series of eigenfunctions of a related homogeneous problem, and then we determine the coefficients in this series so that the nonhomogeneous problem is satisfied. We first describe this method as it applies to boundary value problems for second order linear ordinary differential equations. Later we illustrate its use for partial differential equations by solving a heat conduction problem in a bar with variable material properties and in the presence of source terms.

Nonhomogeneous Sturm–Liouville Problems

Consider the boundary value problem consisting of the nonhomogeneous differential equation

$$L[y] = -[p(x)y']' + q(x)y = \mu r(x)y + f(x), \tag{1}$$

[3]See Section 5.7, especially Eq. (10) of that section.

where μ is a given constant and f is a given function on $0 \le x \le 1$, and the boundary conditions are

$$\alpha_1 y(0) + \alpha_2 y'(0) = 0, \qquad \beta_1 y(1) + \beta_2 y'(1) = 0. \tag{2}$$

As in Section 10.2, we assume that p, p', q, and r are continuous on $0 \le x \le 1$ and that $p(x) > 0$ and $r(x) > 0$ there. We will solve the problem (1), (2) by making use of the eigenfunctions of the corresponding homogeneous problem consisting of the differential equation

$$L[y] = \lambda r(x) y \tag{3}$$

and the boundary conditions (2). Let $\lambda_1 < \lambda_2 < \cdots < \lambda_n < \cdots$ be the eigenvalues of this problem, and let $\phi_1, \phi_2, \ldots, \phi_n, \ldots$ be the corresponding normalized eigenfunctions.

We now assume that the solution $y = \phi(x)$ of the nonhomogeneous problem (1), (2) can be expressed as a series of the form

$$\phi(x) = \sum_{n=1}^{\infty} b_n \phi_n(x). \tag{4}$$

From Eq. (34) of Section 10.2 we know that

$$b_n = \int_0^1 r(x) \phi(x) \phi_n(x)\, dx, \qquad n = 1, 2, \ldots. \tag{5}$$

However, since we do not know $\phi(x)$, we cannot use Eq. (5) to calculate b_n. Instead, we will try to determine b_n so that the problem (1), (2) is satisfied and then use Eq. (4) to find $\phi(x)$. Note first that ϕ as given by Eq. (4) always satisfies the boundary conditions (2) since each ϕ_n does.

Now consider the differential equation that ϕ must satisfy. This is just Eq. (1) with y replaced by ϕ:

$$L[\phi](x) = \mu r(x) \phi(x) + f(x). \tag{6}$$

We substitute the series (4) into the differential equation (6) and attempt to determine b_n so that the differential equation is satisfied. The term on the left side of Eq. (6) becomes

$$L[\phi](x) = L\left[\sum_{n=1}^{\infty} b_n \phi_n\right](x) = \sum_{n=1}^{\infty} b_n L[\phi_n](x)$$

$$= \sum_{n=1}^{\infty} b_n \lambda_n r(x) \phi_n(x), \tag{7}$$

where we have assumed that we can interchange the operations of summation and differentiation.

Note that the function r appears in Eq. (7) and also in the term $\mu r(x)\phi(x)$ in Eq. (6). This suggests that we rewrite the nonhomogeneous term in Eq. (6) as $r(x)[f(x)/r(x)]$ so that $r(x)$ also appears as a multiplier in this term. If the function f/r satisfies the conditions of

Theorem 10.2.4, then

$$\frac{f(x)}{r(x)} = \sum_{n=1}^{\infty} c_n \phi_n(x), \tag{8}$$

where, using Eq. (5) with ϕ replaced by f/r,

$$c_n = \int_0^1 r(x) \frac{f(x)}{r(x)} \phi_n(x)\, dx = \int_0^1 f(x) \phi_n(x)\, dx, \qquad n = 1, 2, \ldots. \tag{9}$$

Upon substituting for $\phi(x)$, $L[\phi](x)$, and $f(x)$ in Eq. (6) from Eqs. (4), (7), and (8), respectively, we find that

$$\sum_{n=1}^{\infty} b_n \lambda_n r(x) \phi_n(x) = \mu r(x) \sum_{n=1}^{\infty} b_n \phi_n(x) + r(x) \sum_{n=1}^{\infty} c_n \phi_n(x).$$

After collecting terms and canceling the common nonzero factor $r(x)$, we have

$$\sum_{n=1}^{\infty} [(\lambda_n - \mu)b_n - c_n] \phi_n(x) = 0. \tag{10}$$

If Eq. (10) is to hold for each x in the interval $0 \le x \le 1$, then the coefficient of $\phi_n(x)$ must be zero for each n; see Problem 14 for a proof of this fact. Hence

$$(\lambda_n - \mu)b_n - c_n = 0, \qquad n = 1, 2, \ldots. \tag{11}$$

We must now distinguish two main cases, one of which also has two subcases.

First suppose that $\mu \ne \lambda_n$ for $n = 1, 2, 3, \ldots$; that is, μ is not equal to any eigenvalue of the corresponding homogeneous problem. Then

$$b_n = \frac{c_n}{\lambda_n - \mu}, \qquad n = 1, 2, 3, \ldots, \tag{12}$$

and

$$y = \phi(x) = \sum_{n=1}^{\infty} \frac{c_n}{\lambda_n - \mu} \phi_n(x). \tag{13}$$

Equation (13), with c_n given by Eq. (9), is a formal solution of the nonhomogeneous boundary value problem (1), (2). Our argument does not prove that the series (13) converges. However, any solution of the boundary value problem (1), (2) clearly satisfies the conditions of Theorem 10.2.4; indeed, it satisfies the more stringent conditions given in the paragraph following that theorem. Thus it is reasonable to expect that the series (13) does converge at each point, and this fact can be established, provided, for example, that f is continuous.

Now suppose that μ is equal to one of the eigenvalues of the corresponding homogeneous problem, say, $\mu = \lambda_m$; then the situation is quite different. In this event, for $n = m$, Eq. (11) has the form $0 \cdot b_m - c_m = 0$. Again we must consider two cases.

If $\mu = \lambda_m$ and $c_m \ne 0$, then there is no value of b_m that satisfies Eq. (11), and therefore the nonhomogeneous problem (1), (2) has no solution.

If $\mu = \lambda_m$ and $c_m = 0$, then Eq. (11) is satisfied regardless of the value of b_m; in other words, b_m remains arbitrary. In this case the boundary value problem (1), (2) does have a solution, but it is not unique, since it contains an arbitrary multiple of the eigenfunction ϕ_m.

Since c_m is given by Eq. (9), the condition $c_m = 0$ means that

$$\int_0^1 f(x)\phi_m(x)\,dx = 0. \tag{14}$$

Thus, if $\mu = \lambda_m$, the nonhomogeneous boundary value problem (1), (2) can be solved only if f is orthogonal to the eigenfunction corresponding to the eigenvalue λ_m.

The results we have formally obtained are summarized in the following theorem.

THEOREM 10.3.1

The nonhomogeneous boundary value problem (1), (2) has a unique solution for each continuous f whenever μ is different from all the eigenvalues of the corresponding homogeneous problem; the solution is given by Eq. (13), and the series converges for each x in $0 \le x \le 1$. If μ is equal to an eigenvalue λ_m of the corresponding homogeneous problem, then the nonhomogeneous boundary value problem has no solution unless f is orthogonal to ϕ_m, that is, unless the condition (14) holds. In that case, the solution is not unique and contains an arbitrary multiple of $\phi_m(x)$.

The main part of Theorem 10.3.1 is sometimes stated in the following way:

THEOREM 10.3.2

For a given value of μ, either the nonhomogeneous problem (1), (2) has a unique solution for each continuous f (if μ is not equal to any eigenvalue λ_m of the corresponding homogeneous problem), or else the homogeneous problem (3), (2) has a nontrivial solution (the eigenfunction corresponding to λ_m).

This latter form of the theorem is known as the Fredholm alternative theorem. This is one of the basic theorems of mathematical analysis and occurs in many different contexts.

EXAMPLE 1

Solve the boundary value problem

$$y'' + 2y = -x, \tag{15}$$

$$y(0) = 0, \qquad y(1) + y'(1) = 0. \tag{16}$$

This particular problem can be solved directly in an elementary way and has the solution

$$y = \frac{\sin\sqrt{2}\,x}{\sin\sqrt{2} + \sqrt{2}\cos\sqrt{2}} - \frac{x}{2}. \tag{17}$$

The method of solution described below illustrates the use of eigenfunction expansions, a method that can be employed in many problems not accessible by elementary procedures.

We begin by rewriting Eq. (15) as

$$-y'' = 2y + x \tag{18}$$

so that it will have the same form as Eq. (1). We seek the solution of the given problem as a series of normalized eigenfunctions ϕ_n of the corresponding homogeneous problem

$$y'' + \lambda y = 0, \qquad y(0) = 0, \quad y(1) + y'(1) = 0. \tag{19}$$

These eigenfunctions were found in Example 2 of Section 10.2 and are

$$\phi_n(x) = k_n \sin \sqrt{\lambda_n}\, x, \tag{20}$$

where

$$k_n = \left(\frac{2}{1 + \cos^2 \sqrt{\lambda_n}} \right)^{1/2} \tag{21}$$

and λ_n satisfies

$$\sin \sqrt{\lambda_n} + \sqrt{\lambda_n} \cos \sqrt{\lambda_n} = 0. \tag{22}$$

Recall that in Example 1 of Section 10.1 we found that

$$\lambda_1 \cong 4.116, \qquad \lambda_2 \cong 24.14,$$

$$\lambda_3 \cong 63.66, \qquad \lambda_n \cong (2n-1)^2 \pi^2/4 \qquad \text{for } n = 4, 5, \ldots.$$

We assume that y is given by Eq. (4)

$$y = \sum_{n=1}^{\infty} b_n \phi_n(x),$$

and it follows that the coefficients b_n are found from Eq. (12)

$$b_n = \frac{c_n}{\lambda_n - 2},$$

where the c_n are the expansion coefficients of the nonhomogeneous term $f(x) = x$ in Eq. (18) in terms of the eigenfunctions ϕ_n. These coefficients were found in Example 3 of Section 10.2 and are

$$c_n = \frac{2\sqrt{2} \sin \sqrt{\lambda_n}}{\lambda_n (1 + \cos^2 \sqrt{\lambda_n})^{1/2}}. \tag{23}$$

Putting everything together, we finally obtain the solution

$$y = 4 \sum_{n=1}^{\infty} \frac{\sin \sqrt{\lambda_n}}{\lambda_n (\lambda_n - 2)(1 + \cos^2 \sqrt{\lambda_n})} \sin \sqrt{\lambda_n}\, x. \tag{24}$$

Although Eqs. (17) and (24) are quite different in appearance, they are actually two different expressions for the same function. This follows from the uniqueness part of Theorem 10.3.1 or 10.3.2 since $\lambda = 2$ is not an eigenvalue of the homogeneous problem (19). Alternatively, you can show the equivalence of Eqs. (17) and (24) by expanding the right side of Eq. (17) in terms of the eigenfunctions $\phi_n(x)$. For this problem it is fairly obvious that Eq. (17) is a more convenient expression for the solution than Eq. (24). However, we emphasize again that in other problems we may not be able to obtain the solution except by series (or numerical) methods.

Nonhomogeneous Heat Conduction Problems

To show how eigenfunction expansions can be used to solve nonhomogeneous problems for partial differential equations, let us consider the generalized heat conduction equation

$$r(x)u_t = [p(x)u_x]_x - q(x)u + F(x, t) \tag{25}$$

with the boundary conditions

$$u_x(0, t) - h_1 u(0, t) = 0, \qquad u_x(1, t) + h_2 u(1, t) = 0 \tag{26}$$

and the initial condition

$$u(x, 0) = f(x). \tag{27}$$

This problem was previously discussed in Appendix 9.A of Chapter 9 and in Section 10.1. In the latter section we let $u(x, t) = X(x)T(t)$ in the homogeneous equation obtained by setting $F(x, t) = 0$, and showed that $X(x)$ must be a solution of the boundary value problem

$$-[p(x)X']' + q(x)X = \lambda r(x)X, \tag{28}$$

$$X'(0) - h_1 X(0) = 0, \qquad X'(1) + h_2 X(1) = 0. \tag{29}$$

If we assume that p, q, and r satisfy the proper continuity requirements and that $p(x)$ and $r(x)$ are always positive, the problem (28), (29) is a Sturm–Liouville problem as discussed in Section 10.2. Thus we obtain a sequence of eigenvalues $\lambda_1 < \lambda_2 < \cdots < \lambda_n < \cdots$ and corresponding normalized eigenfunctions $\phi_1(x), \phi_2(x), \ldots, \phi_n(x), \ldots$.

We will solve the given nonhomogeneous boundary value problem (25) to (27) by assuming that $u(x, t)$ can be expressed as a series of eigenfunctions

$$u(x, t) = \sum_{n=1}^{\infty} b_n(t)\phi_n(x), \tag{30}$$

and then showing how to determine the coefficients $b_n(t)$. The procedure is basically the same as that used in the problem (1), (2) considered earlier, although it is more complicated in certain respects. For instance, the coefficients b_n must now depend on t, because otherwise u would be a function of x only. Note that the boundary conditions (26) are automatically satisfied by an expression of the form (30) because each $\phi_n(x)$ satisfies the boundary conditions (29).

Next we substitute from Eq. (30) for u in Eq. (25). From the first two terms on the right side of Eq. (25) we formally obtain

$$[p(x)u_x]_x - q(x)u = \frac{\partial}{\partial x}\left[p(x)\sum_{n=1}^{\infty} b_n(t)\phi_n'(x)\right] - q(x)\sum_{n=1}^{\infty} b_n(t)\phi_n(x)$$

$$= \sum_{n=1}^{\infty} b_n(t)\{[p(x)\phi_n'(x)]' - q(x)\phi_n(x)\}. \tag{31}$$

Since $[p(x)\phi_n'(x)]' - q(x)\phi_n(x) = -\lambda_n r(x)\phi_n(x)$, we obtain finally

$$[p(x)u_x]_x - q(x)u = -r(x)\sum_{n=1}^{\infty} b_n(t)\lambda_n \phi_n(x). \tag{32}$$

Now consider the term on the left side of Eq. (25). We have

$$
\begin{aligned}
r(x)u_t &= r(x)\frac{\partial}{\partial t}\sum_{n=1}^{\infty} b_n(t)\phi_n(x) \\[2mm]
&= r(x)\sum_{n=1}^{\infty} b_n'(t)\phi_n(x).
\end{aligned}
\tag{33}
$$

We must also express the nonhomogeneous term in Eq. (25) as a series of eigenfunctions. Once again, it is convenient to look at the ratio $F(x, t)/r(x)$ and to write

$$\frac{F(x, t)}{r(x)} = \sum_{n=1}^{\infty} \gamma_n(t)\phi_n(x), \tag{34}$$

where the coefficients are given by

$$\gamma_n(t) = \int_0^1 r(x)\frac{F(x, t)}{r(x)}\phi_n(x)\,dx = \int_0^1 F(x, t)\phi_n(x)\,dx, \qquad n = 1, 2, \dots. \tag{35}$$

Since $F(x, t)$ is given, we can consider the functions $\gamma_n(t)$ to be known.

Gathering all these results together, we substitute from Eqs. (32), (33), and (34) in Eq. (25), and find that

$$r(x)\sum_{n=1}^{\infty} b_n'(t)\phi_n(x) = -r(x)\sum_{n=1}^{\infty} b_n(t)\lambda_n \phi_n(x) + r(x)\sum_{n=1}^{\infty} \gamma_n(t)\phi_n(x). \tag{36}$$

To simplify Eq. (36), we cancel the common nonzero factor $r(x)$ from all terms and write everything in one summation:

$$\sum_{n=1}^{\infty}[b_n'(t) + \lambda_n b_n(t) - \gamma_n(t)]\phi_n(x) = 0. \tag{37}$$

Once again, if Eq. (37) is to hold for all x in $0 < x < 1$, it is necessary for the quantity in square brackets to be zero for each n (again see Problem 14). Hence $b_n(t)$ is a solution of the first order linear ordinary differential equation

$$b_n'(t) + \lambda_n b_n(t) = \gamma_n(t), \qquad n = 1, 2, \dots, \tag{38}$$

where $\gamma_n(t)$ is given by Eq. (35). To determine $b_n(t)$ completely, we must have an initial condition

$$b_n(0) = B_n, \qquad n = 1, 2, \dots \tag{39}$$

for Eq. (38). This we obtain from the initial condition (27). Setting $t = 0$ in Eq. (30) and using Eq. (27), we have

$$u(x, 0) = \sum_{n=1}^{\infty} b_n(0)\phi_n(x) = \sum_{n=1}^{\infty} B_n \phi_n(x) = f(x). \tag{40}$$

Thus the initial values B_n are the coefficients in the eigenfunction expansion for $f(x)$. Therefore

$$B_n = \int_0^1 r(x)f(x)\phi_n(x)\,dx, \qquad n = 1, 2, \ldots. \tag{41}$$

Note that everything on the right side of Eq. (41) is known, so we can consider B_n as known.

The initial value problem (38), (39) is solved by the methods of Section 2.1. The integrating factor is $\mu(t) = \exp(\lambda_n t)$, and it follows that

$$b_n(t) = B_n e^{-\lambda_n t} + \int_0^t e^{-\lambda_n(t-s)}\gamma_n(s)\,ds, \qquad n = 1, 2, \ldots. \tag{42}$$

The details of this calculation are left to you. Note that the first term on the right side of Eq. (42) depends on the function f through the coefficients B_n, while the second depends on the nonhomogeneous term F through the coefficients $\gamma_n(s)$.

Thus an explicit solution of the boundary value problem (25) to (27) is given by Eq. (30)

$$u(x, t) = \sum_{n=1}^{\infty} b_n(t)\phi_n(x),$$

where the coefficients $b_n(t)$ are determined from Eq. (42). The quantities B_n and $\gamma_n(s)$ in Eq. (42) are found in turn from Eqs. (41) and (35), respectively.

Summarizing, to use this method to solve a boundary value problem such as that given by Eqs. (25) to (27), we must:

1. Find the eigenvalues λ_n and the normalized eigenfunctions ϕ_n of the homogeneous problem (28), (29).
2. Calculate the coefficients B_n and $\gamma_n(t)$ from Eqs. (41) and (35), respectively.
3. Evaluate the integral in Eq. (42) to determine $b_n(t)$.
4. Sum the infinite series (30).

Since any or all of these steps may be difficult, the entire process can be quite formidable. One redeeming feature is that often the series (30) converges rapidly, in which case only a very few terms may be needed to obtain an adequate approximation to the solution.

EXAMPLE 2

Find the solution of the heat conduction problem

$$u_t = u_{xx} + xe^{-t}, \tag{43}$$

$$u(0, t) = 0, \qquad u_x(1, t) + u(1, t) = 0, \tag{44}$$

$$u(x, 0) = 0. \tag{45}$$

Again we use the normalized eigenfunctions ϕ_n of the problem (19) and assume that u is given by Eq. (30)

$$u(x, t) = \sum_{n=1}^{\infty} b_n(t)\phi_n(x).$$

The coefficients b_n are determined from the differential equation

$$b_n' + \lambda_n b_n = \gamma_n(t), \tag{46}$$

where λ_n is the nth eigenvalue of problem (19) and $\gamma_n(t)$ is the nth expansion coefficient of the nonhomogeneous term xe^{-t} in terms of the eigenfunctions ϕ_n. Thus we have

$$
\begin{aligned}
\gamma_n(t) &= \int_0^1 xe^{-t}\phi_n(x)\,dx = e^{-t}\int_0^1 x\phi_n(x)\,dx \\
&= c_n e^{-t}, \tag{47}
\end{aligned}
$$

where $c_n = \int_0^1 x\phi_n(x)\,dx$ is given by Eq. (23). The initial condition for Eq. (46) is

$$b_n(0) = 0 \tag{48}$$

since the initial temperature distribution (45) is zero everywhere. The solution of the initial value problem (46), (48) is

$$
\begin{aligned}
b_n(t) &= e^{-\lambda_n t}\int_0^t e^{\lambda_n s} c_n e^{-s}\,ds = c_n e^{-\lambda_n t}\frac{e^{(\lambda_n - 1)t} - 1}{\lambda_n - 1} \\
&= \frac{c_n}{\lambda_n - 1}(e^{-t} - e^{-\lambda_n t}). \tag{49}
\end{aligned}
$$

Thus the solution of the heat conduction problem (43) to (45) is given by

$$u(x, t) = 4\sum_{n=1}^\infty \frac{(\sin\sqrt{\lambda_n})(e^{-t} - e^{-\lambda_n t})\sin\sqrt{\lambda_n}\,x}{\lambda_n(\lambda_n - 1)(1 + \cos^2\sqrt{\lambda_n})}. \tag{50}$$

The solution given by Eq. (50) is exact but complicated. To judge whether a satisfactory approximation to the solution can be obtained by using only a few terms in this series, we must estimate its speed of convergence. First we split the right side of Eq. (50) into two parts:

$$u(x, t) = 4e^{-t}\sum_{n=1}^\infty \frac{\sin\sqrt{\lambda_n}\,\sin\sqrt{\lambda_n}\,x}{\lambda_n(\lambda_n - 1)(1 + \cos^2\sqrt{\lambda_n})} - 4\sum_{n=1}^\infty \frac{e^{-\lambda_n t}\sin\sqrt{\lambda_n}\,\sin\sqrt{\lambda_n}\,x}{\lambda_n(\lambda_n - 1)(1 + \cos^2\sqrt{\lambda_n})}. \tag{51}$$

Recall from Example 1 in Section 10.1 that the eigenvalues λ_n are very nearly proportional to n^2. In the first series on the right side of Eq. (51) the trigonometric factors are all bounded as $n \to \infty$; therefore this series converges similarly to the series $\sum_{n=1}^\infty \lambda_n^{-2}$ or $\sum_{n=1}^\infty n^{-4}$. Hence at most two or three terms are required for us to obtain an excellent approximation to this part of the solution. The second series contains the additional factor $e^{-\lambda_n t}$, so its convergence is even more rapid for $t > 0$; all terms after the first are almost surely negligible.

Further Discussion

Eigenfunction expansions can be used to solve a much greater variety of problems than the preceding discussion and examples may suggest. For example, time-independent non-homogeneus boundary conditions can be handled much as in Section 9.6. To reduce the problem to one with homogeneous boundary conditions, subtract from u a function v that is chosen to satisfy the given boundary conditions. Then the difference $w = u - v$ satisfies a problem with homogeneous boundary conditions, but with a modified forcing term and initial condition. This problem can be solved by the procedure described in this section.

One potential difficulty in using eigenfunction expansions is that the normalized eigenfunctions of the corresponding homogeneous problem must be found. For a differential equation with variable coefficients this may be difficult, if not impossible. In such a case it is sometimes possible to use other functions, such as eigenfunctions of a simpler problem, that satisfy the same boundary conditions. For instance, if the boundary conditions are

$$u(0, t) = 0, \qquad u(1, t) = 0, \tag{52}$$

then it may be convenient to replace the functions $\phi_n(x)$ in Eq. (30) by $\sin n\pi x$. These functions at least satisfy the correct boundary conditions, although in general they are not solutions of the corresponding homogeneous differential equation. Next we expand the nonhomogeneous term $F(x, t)$ in a series of the form (34), again with $\phi_n(x)$ replaced by $\sin n\pi x$, and then substitute for both u and F in Eq. (25). Upon collecting the coefficients of $\sin n\pi x$ for each n, we have an infinite set of linear first order differential equations from which to determine $b_1(t), b_2(t), \ldots$. The essential difference between this case and the one considered earlier is that now the equations for the functions $b_n(t)$ are *coupled*. Thus they cannot be solved one by one, as before, but must be dealt with simultaneously. In practice, the infinite system is replaced by an approximating finite system, from which approximations to a finite number of coefficients are calculated.

Boundary value problems for equations of higher than second order can also often be solved by eigenfunction expansions. In some cases the procedure parallels almost exactly that for second order problems. However, a variety of complications can also arise.

Finally, we emphasize that the discussion in this section has been purely formal. Separate and sometimes elaborate arguments must be used to establish convergence of eigenfunction expansions or to justify some of the steps used, such as term-by-term differentiation of eigenfunction series.

There are also other, altogether different, methods for solving nonhomogeneous boundary value problems. One of these leads to a solution expressed as a definite integral rather than as an infinite series. This approach involves certain functions known as Green's functions and, for ordinary differential equations, is the subject of Problems 28 through 36.

PROBLEMS

In each of Problems 1 through 5 solve the given problem by means of an eigenfunction expansion.

1. $y'' + 2y = -x, \qquad y(0) = 0, \quad y(1) = 0$

2. $y'' + 2y = -x, \qquad y(0) = 0, \quad y'(1) = 0$
   (see Section 10.2, Problem 7)

3. $y'' + 2y = -x, \qquad y'(0) = 0, \quad y'(1) = 0$
   (see Section 10.2, Problem 3)

4. $y'' + 2y = -x, \quad y'(0) = 0, \quad y'(1) + y(1) = 0$
   (see Section 10.2, Problem 11)

5. $y'' + 2y = -1 + |1 - 2x|, \qquad y(0) = 0,$
   $y(1) = 0$

In each of Problems 6 through 9 determine a formal eigenfunction series expansion for the solution of the given problem. Assume that f satisfies the conditions of Theorem 10.3.1. State the values of μ for which the solution exists.

6. $y'' + \mu y = -f(x), \qquad y(0) = 0, \quad y'(1) = 0$

7. $y'' + \mu y = -f(x), \qquad y'(0) = 0, \quad y(1) = 0$

8. $y'' + \mu y = -f(x), \qquad y'(0) = 0, \quad y'(1) = 0$

9. $y'' + \mu y = -f(x), \qquad y'(0) = 0,$
   $y'(1) + y(1) = 0$

In each of Problems 10 through 13 determine whether there is any value of the constant a for which the problem has a solution. Find the solution for each such value.

10. $y'' + \pi^2 y = a + x, \qquad y(0) = 0, \quad y(1) = 0$

11. $y'' + 4\pi^2 y = a + x, \qquad y(0) = 0, \quad y(1) = 0$

12. $y'' + \pi^2 y = a, \qquad y'(0) = 0, \quad y'(1) = 0$

13. $y'' + \pi^2 y = a - \cos \pi x, \qquad y(0) = 0,$
    $y(1) = 0$

14. Let $\phi_1, \ldots, \phi_n, \ldots$ be the normalized eigenfunctions of the differential equation (3) subject to the boundary conditions (2). If $\sum_{n=1}^{\infty} c_n \phi_n(x)$

converges to $f(x)$, where $f(x) = 0$ for each x in $0 \le x \le 1$, show that $c_n = 0$ for each n.

Hint: Multiply by $r(x)\phi_m(x)$, integrate, and use the orthogonality property of the eigenfunctions.

15. Let L be a second order linear differential operator. Show that the solution $y = \phi(x)$ of the problem

$$L[y] = f(x),$$

$$\alpha_1 y(0) + \alpha_2 y'(0) = a, \quad \beta_1 y(1) + \beta_2 y'(1) = b$$

can be written as $y = u + v$, where $u = \phi_1(x)$ and $v = \phi_2(x)$ are solutions of the problems

$$L[u] = 0,$$

$$\alpha_1 u(0) + \alpha_2 u'(0) = a, \quad \beta_1 u(1) + \beta_2 u'(1) = b$$

and

$$L[v] = f(x),$$

$$\alpha_1 v(0) + \alpha_2 v'(0) = 0, \quad \beta_1 v(1) + \beta_2 v'(1) = 0,$$

respectively.

16. Show that the problem

$$y'' + \pi^2 y = \pi^2 x, \quad y(0) = 1, \quad y(1) = 0$$

has the solution

$$y = c_1 \sin \pi x + \cos \pi x + x.$$

Also show that this solution cannot be obtained by splitting the problem as suggested in Problem 15, since neither of the two subsidiary problems can be solved in this case.

17. Consider the problem

$$y'' + p(x)y' + q(x)y = 0, \quad y(0) = a, \quad y(1) = b.$$

Let $y = u + v$, where v is any twice differentiable function satisfying the boundary conditions (but not necessarily the differential equation). Show that u is a solution of the problem

$$u'' + p(x)u' + q(x)u = g(x), \quad u(0) = 0,$$
$$u(1) = 0,$$

where $g(x) = -[v'' + p(x)v' + q(x)v]$ and is known once v is chosen. Thus nonhomogeneities can be transferred from the boundary conditions to the differential equation. Find a function v for this problem.

18. Using the method of Problem 17, transform the problem

$$y'' + 2y = 2 - 4x, \quad y(0) = 1,$$
$$y(1) + y'(1) = -2$$

into a new problem in which the boundary conditions are homogeneous. Solve the latter problem by reference to Example 1 of the text.

In each of Problems 19 through 22 use eigenfunction expansions to find the solution of the given boundary value problem.

19. $u_t = u_{xx} - x, \quad u(0, t) = 0, \quad u_x(1, t) = 0,$
$u(x, 0) = \sin(\pi x/2);$ (see Problem 2)

20. $u_t = u_{xx} + e^{-t}, \quad u_x(0, t) = 0,$
$u_x(1, t) + u(1, t) = 0, \quad u(x, 0) = 1 - x;$ (see Section 10.2, Problems 10 and 12)

21. $u_t = u_{xx} + 1 - |1 - 2x|, \quad u(0, t) = 0,$
$u(1, t) = 0, \quad u(x, 0) = 0;$ (see Problem 5)

22. $u_t = u_{xx} + e^{-t}(1 - x), \quad u(0, t) = 0,$
$u_x(1, t) = 0, \quad u(x, 0) = 0;$ (see Section 10.2, Problems 6 and 7)

23. Consider the boundary value problem

$$r(x)u_t = [p(x)u_x]_x - q(x)u + F(x),$$

$$u(0, t) = T_1, \quad u(1, t) = T_2, \quad u(x, 0) = f(x).$$

(a) Let $v(x)$ be a solution of the problem

$$[p(x)v']' - q(x)v = -F(x), \quad v(0) = T_1,$$
$$v(1) = T_2.$$

If $w(x, t) = u(x, t) - v(x)$, find the boundary value problem satisfied by w. Note that this problem can be solved by the method of this section.

(b) Generalize the procedure of part (a) to the case where u satisfies the boundary conditions

$$u_x(0, t) - h_1 u(0, t) = T_1,$$
$$u_x(1, t) + h_2 u(1, t) = T_2.$$

In each of Problems 24 and 25 use the method indicated in Problem 23 to solve the given boundary value problem.

24. $u_t = u_{xx} - 2,$
$u(0, t) = 1, \quad u(1, t) = 0,$
$u(x, 0) = x^2 - 2x + 2$

25. $u_t = u_{xx} - \pi^2 \cos \pi x,$
$u_x(0, t) = 0, \quad u(1, t) = 1,$
$u(x, 0) = \cos(3\pi x/2) - \cos \pi x$

26. The method of eigenfunction expansions is often useful for nonhomogeneous problems related to the wave equation or its generalizations. Consider the problem

$$r(x)u_{tt} = [p(x)u_x]_x - q(x)u + F(x, t), \quad \text{(i)}$$

$$u_x(0, t) - h_1 u(0, t) = 0,$$
$$u_x(1, t) + h_2 u(1, t) = 0, \qquad \text{(ii)}$$

$$u(x, 0) = f(x), \qquad u_t(x, 0) = g(x). \qquad \text{(iii)}$$

This problem can arise in connection with generalizations of the telegraph equation (Problem 16 in Section 10.1) or the longitudinal vibrations of an elastic bar (Problem 25 in Section 10.1).

(a) Let $u(x, t) = X(x)T(t)$ in the homogeneous equation corresponding to Eq. (i), and show that $X(x)$ satisfies Eqs. (28) and (29) of the text. Let λ_n and $\phi_n(x)$ denote the eigenvalues and normalized eigenfunctions of this problem.

(b) Assume that $u(x, t) = \sum_{n=1}^{\infty} b_n(t)\phi_n(x)$, and show that $b_n(t)$ must satisfy the initial value problem

$$b_n''(t) + \lambda_n b_n(t) = \gamma_n(t), \qquad b_n(0) = \alpha_n,$$
$$b_n'(0) = \beta_n,$$

where α_n, β_n, and $\gamma_n(t)$ are the expansion coefficients for $f(x)$, $g(x)$, and $F(x, t)/r(x)$ in terms of the eigenfunctions $\phi_1(x), \ldots,$ $\phi_n(x), \ldots$.

27. In this problem we explore a little further the analogy between Sturm–Liouville boundary value problems and symmetric matrices. Let $\mathbf{A}$ be an $n \times n$ symmetric matrix with eigenvalues $\lambda_1, \ldots, \lambda_n$ and corresponding orthogonal eigenvectors $\boldsymbol{\xi}^{(1)}, \ldots, \boldsymbol{\xi}^{(n)}$.

Consider the nonhomogeneous system of equations

$$\mathbf{Ax} - \mu\mathbf{x} = \mathbf{b}, \qquad \text{(i)}$$

where μ is a given real number and $\mathbf{b}$ is a given vector. We will point out a way of solving Eq. (i) that is analogous to the method presented in the text for solving Eqs. (1) and (2).

(a) Show that $\mathbf{b} = \sum_{i=1}^{n} b_i \boldsymbol{\xi}^{(i)}$, where $b_i = (\mathbf{b}, \boldsymbol{\xi}^{(i)})$.

(b) Assume that $\mathbf{x} = \sum_{i=1}^{n} a_i \boldsymbol{\xi}^{(i)}$ and show that for Eq. (i) to be satisfied, it is necessary that $a_i = b_i/(\lambda_i - \mu)$. Thus

$$\mathbf{x} = \sum_{i=1}^{n} \frac{(\mathbf{b}, \boldsymbol{\xi}^{(i)})}{\lambda_i - \mu} \boldsymbol{\xi}^{(i)}, \qquad \text{(ii)}$$

provided that μ is not one of the eigenvalues of $\mathbf{A}$, $\mu \neq \lambda_i$ for $i = 1, \ldots, n$. Compare this result with Eq. (13).

Green's Functions. Consider the nonhomogeneous system of algebraic equations

$$\mathbf{Ax} - \mu\mathbf{x} = \mathbf{b}, \qquad \text{(i)}$$

where $\mathbf{A}$ is an $n \times n$ symmetric matrix, μ is a given real number, and $\mathbf{b}$ is a given vector. Instead of using an eigenvector expansion as in Problem 27, we can solve Eq. (i) by computing the inverse matrix $(\mathbf{A} - \mu\mathbf{I})^{-1}$, which exists if μ is not an eigenvalue of $\mathbf{A}$. Then

$$\mathbf{x} = (\mathbf{A} - \mu\mathbf{I})^{-1}\mathbf{b}. \qquad \text{(ii)}$$

Problems 28 through 36 indicate a way of solving nonhomogeneous boundary value problems that is analogous to using the inverse matrix for a system of linear algebraic equations. The Green's function plays a part similar to the inverse of the matrix of coefficients. This method leads to solutions expressed as definite integrals rather than as infinite series. Except in Problem 35, we will assume that $\mu = 0$ for simplicity.

28. (a) Show by the method of variation of parameters that the general solution of the differential equation

$$-y'' = f(x)$$

can be written in the form

$$y = \phi(x) = c_1 + c_2 x - \int_0^x (x - s)f(s)\, ds,$$

where c_1 and c_2 are arbitrary constants.

(b) Let $y = \phi(x)$ also be required to satisfy the boundary conditions $y(0) = 0$, $y(1) = 0$. Show that in this case

$$c_1 = 0, \qquad c_2 = \int_0^1 (1 - s)f(s)\, ds.$$

(c) Show that, under the conditions of parts (a) and (b), $\phi(x)$ can be written in the form

$$\phi(x) = \int_0^x s(1 - x)f(s)\, ds + \int_x^1 x(1 - s)f(s)\, ds.$$

(d) Defining

$$G(x, s) = \begin{cases} s(1 - x), & 0 \leq s \leq x, \\ x(1 - s), & x \leq s \leq 1, \end{cases}$$

show that the solution can be written as

$$\phi(x) = \int_0^1 G(x, s)f(s)\, ds.$$

The function $G(x, s)$ appearing under the integral sign is a Green's function. The usefulness of a Green's function solution rests on the fact that the Green's function is independent of the nonhomogeneous term in the differential equation. Thus, once the Green's function is determined, the solution of the boundary value problem for any nonhomogeneous term $f(x)$ is obtained by a single integration. Note further that no determination of arbitrary constants is required, since $\phi(x)$ as given by the Green's function integral formula automatically satisfies the boundary conditions.

29. By a procedure similar to that in Problem 28 show that the solution of the boundary value problem

$$-(y'' + y) = f(x), \qquad y(0) = 0, \quad y(1) = 0$$

is

$$y = \phi(x) = \int_0^1 G(x, s) f(s)\, ds,$$

where

$$G(x, s) = \begin{cases} \dfrac{\sin s \sin (1 - x)}{\sin 1}, & 0 \le s \le x, \\[2ex] \dfrac{\sin x \sin (1 - s)}{\sin 1}, & x \le s \le 1. \end{cases}$$

30. It is possible to show that the Sturm–Liouville problem

$$L[y] = -[p(x)y']' + q(x)y = f(x), \qquad (i)$$

$$\alpha_1 y(0) + \alpha_2 y'(0) = 0, \qquad \beta_1 y(1) + \beta_2 y'(1) = 0 \tag{ii}$$

has a Green's function solution

$$y = \phi(x) = \int_0^1 G(x, s) f(s)\, ds, \tag{iii}$$

provided that $\lambda = 0$ is not an eigenvalue of $L[y] = \lambda y$ subject to the boundary conditions (ii). Further, $G(x, s)$ is given by

$$G(x, s) = \begin{cases} -y_1(s)y_2(x)/p(x)W[y_1, y_2](x), & 0 \le s \le x, \\ -y_1(x)y_2(s)/p(x)W[y_1, y_2](x), & x \le s \le 1, \end{cases} \tag{iv}$$

where y_1 is a solution of $L[y] = 0$ satisfying the boundary condition at $x = 0$, y_2 is a solution of $L[y] = 0$ satisfying the boundary condition at $x = 1$, and $W[y_1, y_2]$ is the Wronskian of y_1 and y_2.

(a) Verify that the Green's function obtained in Problem 28 is given by formula (iv).

(b) Verify that the Green's function obtained in Problem 29 is given by formula (iv).

(c) Show that $p(x)W[y_1, y_2](x)$ is a constant by showing that its derivative is zero.

(d) Using Eq. (iv) and the result of part (c), show that $G(x, s) = G(s, x)$.

(e) Verify that $y = \phi(x)$ from Eq. (iii) with $G(x, s)$ given by Eq. (iv) satisfies the differential equation (i) and the boundary conditions (ii).

In each of Problems 31 through 34 solve the given boundary value problem by determining the appropriate Green's function and expressing the solution as a definite integral. Use Eqs. (i) to (iv) of Problem 30.

31. $-y'' = f(x), \qquad y'(0) = 0, \quad y(1) = 0$
32. $-y'' = f(x), \qquad y(0) = 0, \quad y(1) + y'(1) = 0$
33. $-(y'' + y) = f(x), \qquad y'(0) = 0, \quad y(1) = 0$
34. $-y'' = f(x), \qquad y(0) = 0, \quad y'(1) = 0$

35. Consider the boundary value problem

$$L[y] = -[p(x)y']' + q(x)y = \mu r(x)y + f(x), \tag{i}$$

$$\alpha_1 y(0) + \alpha_2 y'(0) = 0, \qquad \beta_1 y(1) + \beta_2 y'(1) = 0. \tag{ii}$$

According to the text, the solution $y = \phi(x)$ is given by Eq. (13), where c_n is defined by Eq. (9), provided that μ is not an eigenvalue of the corresponding homogeneous problem. In this case it can also be shown that the solution is given by a Green's function integral of the form

$$y = \phi(x) = \int_0^1 G(x, s, \mu) f(s)\, ds. \tag{iii}$$

Note that in this problem the Green's function also depends on the parameter μ.

(a) Show that if these two expressions for $\phi(x)$ are to be equivalent, then

$$G(x, s, \mu) = \sum_{i=1}^{\infty} \frac{\phi_i(x)\phi_i(s)}{\lambda_i - \mu}, \tag{iv}$$

where λ_i and ϕ_i are the eigenvalues and eigenfunctions, respectively, of Eqs. (3), (2) of the text. Again we see from Eq. (iv) that μ cannot be equal to any eigenvalue λ_i.

(b) Derive Eq. (iv) directly by assuming that $G(x, s, \mu)$ has the eigenfunction expansion

$$G(x, s, \mu) = \sum_{i=1}^{\infty} a_i(x, \mu)\phi_i(s). \tag{v}$$

Determine $a_i(x, \mu)$ by multiplying Eq. (v) by $r(s)\phi_j(s)$ and integrating with respect to s from $s = 0$ to $s = 1$.

Hint: Show first that λ_i and ϕ_i satisfy the equation

$$\phi_i(x) = (\lambda_i - \mu) \int_0^1 G(x, s, \mu)r(s)\phi_i(s)\, ds.$$

(vi)

36. Consider the boundary value problem

$$-d^2y/ds^2 = \delta(s - x), \qquad y(0) = 0, \quad y(1) = 0,$$

where s is the independent variable, $s = x$ is a definite point in the interval $0 < s < 1$, and δ is the Dirac delta function (see Section 5.7). Show that the solution of this problem is the Green's function $G(x, s)$ obtained in Problem 28.

In solving the given problem, note that $\delta(s - x) = 0$ in the intervals $0 \leq s < x$ and $x < s \leq 1$. Note further that $-dy/ds$ experiences a jump of magnitude 1 as s passes through the value x.

This problem illustrates a general property, namely, that the Green's function $G(x, s)$ can be identified as the response at the point s to a unit impulse at the point x. A more general nonhomogeneous term f on $0 \leq x \leq 1$ can be regarded as a continuous distribution of impulses with magnitude $f(x)$ at the point x. The solution of a nonhomogeneous boundary value problem in terms of a Green's function integral can then be interpreted as the result of superposing the responses to the set of impulses represented by the nonhomogeneous term $f(x)$.

10.4 Singular Sturm–Liouville Problems

In the preceding sections of this chapter we considered Sturm–Liouville boundary value problems: the differential equation

$$L[y] = -[p(x)y']' + q(x)y = \lambda r(x)y, \qquad 0 < x < 1,$$

(1)

together with boundary conditions of the form

$$\alpha_1 y(0) + \alpha_2 y'(0) = 0,$$

(2)

$$\beta_1 y(1) + \beta_2 y'(1) = 0.$$

(3)

Until now, we have always assumed that the problem is **regular**; that is, p is differentiable, q and r are continuous, and $p(x) > 0$ and $r(x) > 0$ at all points in the *closed* interval. However, there are also equations of physical interest in which some of these conditions are not satisfied.

For example, suppose that we wish to study Bessel's equation of order ν on the interval $0 < x < 1$. This equation is sometimes written in the form[4]

$$-(xy')' + \frac{\nu^2}{x}y = \lambda xy$$

(4)

so that $p(x) = x$, $q(x) = \nu^2/x$, and $r(x) = x$. Thus $p(0) = 0$, $r(0) = 0$, and $q(x)$ is unbounded and hence discontinuous as $x \to 0$. However, the conditions imposed on regular Sturm–Liouville problems are met elsewhere in the interval.

Similarly, for Legendre's equation we have

$$-[(1 - x^2)y']' = \lambda y, \qquad -1 < x < 1,$$

(5)

---

[4] The substitution $t = \sqrt{\lambda}\, x$ reduces Eq. (4) to the standard form $t^2 y'' + t y' + (t^2 - \nu^2)y = 0$.

where $\lambda = \alpha(\alpha + 1)$, $p(x) = 1 - x^2$, $q(x) = 0$, and $r(x) = 1$. Here the required conditions on p, q, and r are satisfied in the interval $0 \leq x \leq 1$ except at $x = 1$, where p is zero.

We use the term **singular Sturm–Liouville problem** to refer to a certain class of boundary value problems for the differential equation (1) in which the functions p, q, and r satisfy the conditions stated earlier on the open interval $0 < x < 1$, but at least one of these functions fails to satisfy them at one or both of the boundary points. We also prescribe suitable separated boundary conditions of a kind to be described in more detail later in this section. Singular problems also occur if the interval is unbounded, for example, $0 \leq x < \infty$. We do not consider this latter kind of singular problem in this book.

As an example of a singular problem on a finite interval, consider the equation

$$xy'' + y' + \lambda xy = 0, \tag{6}$$

or

$$-(xy')' = \lambda xy, \tag{7}$$

on the interval $0 < x < 1$, and suppose that $\lambda > 0$. This equation arises in the study of free vibrations of a circular elastic membrane and is discussed further in Section 10.5. If we introduce the new independent variable t defined by $t = \sqrt{\lambda}\, x$, then

$$\frac{dy}{dx} = \sqrt{\lambda}\,\frac{dy}{dt}, \qquad \frac{d^2y}{dx^2} = \lambda\,\frac{d^2y}{dt^2}.$$

Hence Eq. (6) becomes

$$\frac{t}{\sqrt{\lambda}}\,\lambda\,\frac{d^2y}{dt^2} + \sqrt{\lambda}\,\frac{dy}{dt} + \lambda\frac{t}{\sqrt{\lambda}}y = 0,$$

or, if we cancel the common factor $\sqrt{\lambda}$ in each term,

$$t\frac{d^2y}{dt^2} + \frac{dy}{dt} + ty = 0. \tag{8}$$

Equation (8) is Bessel's equation of order zero (see Section 8.7). The general solution of Eq. (8) for $t > 0$ is

$$y = c_1 J_0(t) + c_2 Y_0(t);$$

hence the general solution of Eq. (7) for $x > 0$ is

$$y = c_1 J_0(\sqrt{\lambda}\,x) + c_2 Y_0(\sqrt{\lambda}\,x), \tag{9}$$

where J_0 and Y_0 denote the Bessel functions of the first and second kinds of order zero. From Eqs. (7) and (13) of Section 8.7 we have

$$J_0(\sqrt{\lambda}\,x) = 1 + \sum_{m=1}^{\infty} \frac{(-1)^m \lambda^m x^{2m}}{2^{2m}(m!)^2}, \qquad x > 0, \tag{10}$$

$$Y_0(\sqrt{\lambda}\,x) = \frac{2}{\pi}\left[\left(\gamma + \ln\frac{\sqrt{\lambda}\,x}{2}\right) J_0(\sqrt{\lambda}\,x) + \sum_{m=1}^{\infty} \frac{(-1)^{m+1} H_m \lambda^m x^{2m}}{2^{2m}(m!)^2}\right], \qquad x > 0, \tag{11}$$

where $H_m = 1 + (1/2) + \cdots + (1/m)$ and $\gamma = \lim_{m \to \infty}(H_m - \ln m)$. The graphs of $y = J_0(x)$ and $y = Y_0(x)$ are given in Figure 8.7.1.

Suppose that we seek a solution of Eq. (7) that also satisfies the boundary conditions

$$y(0) \;=\; 0, \tag{12}$$

$$y(1) \;=\; 0, \tag{13}$$

which are typical of those we have met in other problems in this chapter. Since $J_0(0) = 1$ and $Y_0(x) \to -\infty$ as $x \to 0$, the condition $y(0) = 0$ can be satisfied only by choosing $c_1 = c_2 = 0$ in Eq. (9). Thus the boundary value problem (7), (12), (13) has only the trivial solution.

One interpretation of this result is that the boundary condition (12) at $x = 0$ is too restrictive for the differential equation (7). This illustrates the general situation, namely, that at a singular boundary point it is necessary to consider a modified type of boundary condition. In the present problem, suppose that we require only that the solution (9) and its derivative remain bounded. In other words, we take as the boundary condition at $x = 0$ the requirement

$$y, \ y' \text{ bounded as } x \to 0. \tag{14}$$

This condition can be satisfied by choosing $c_2 = 0$ in Eq. (9), so as to eliminate the unbounded solution Y_0. The second boundary condition, $y(1) = 0$, then yields

$$J_0(\sqrt{\lambda}) = 0. \tag{15}$$

It is possible to show[5] that Eq. (15) has an infinite set of discrete positive roots, which yield the eigenvalues $0 < \lambda_1 < \lambda_2 < \cdots < \lambda_n < \cdots$ of the given problem. The corresponding eigenfunctions are

$$\phi_n(x) = J_0(\sqrt{\lambda_n}\, x), \tag{16}$$

determined only up to a multiplicative constant. The boundary value problem (7), (13), and (14) is an example of a singular Sturm–Liouville problem. This example illustrates that if the boundary conditions are relaxed in an appropriate way, then a singular Sturm–Liouville problem may have an infinite sequence of eigenvalues and eigenfunctions, just as a regular Sturm–Liouville problem does.

Because of their importance in applications, it is worthwhile to investigate singular boundary value problems a little further. There are two main questions that are of concern:

1. Precisely what type of boundary conditions can be allowed in a singular Sturm–Liouville problem?

2. To what extent do the eigenvalues and eigenfunctions of a singular problem share the properties of eigenvalues and eigenfunctions of regular Sturm–Liouville problems? In particular, are the eigenvalues real, are the eigenfunctions orthogonal, and can a given function be expanded as a series of eigenfunctions?

Both these questions can be answered by a study of the identity

$$\int_0^1 \{L[u]v - uL[v]\}\, dx = 0, \tag{17}$$

which played an essential part in the development of the theory of regular Sturm–Liouville problems. We therefore investigate the conditions under which this relation holds for singular problems, where the integral in Eq. (17) may now have to be examined as an improper integral. To be definite, we consider the differential equation (1) and assume that $x = 0$ is a singular boundary point but that $x = 1$ is not. The boundary condition (3) is imposed at the

---

[5] The function J_0 is well tabulated; the roots of Eq. (15) can be found in various tables, for example, those in Jahnke and Emde or Abramowitz and Stegun. You can also use a computer algebra system to compute them quickly. The first three roots of Eq. (15) are $\sqrt{\lambda} = 2.405, 5.520$, and 8.654, respectively, to four significant figures; $\sqrt{\lambda_n} \cong (n - 1/4)\pi$ for large n.

nonsingular boundary point $x = 1$, but we leave unspecified, for the moment, the boundary condition at $x = 0$. Indeed, our principal objective is to determine what kinds of boundary conditions are allowable at a singular boundary point if Eq. (17) is to hold.

Since the boundary value problem under investigation is singular at $x = 0$, we choose $\epsilon > 0$ and consider the integral $\int_\epsilon^1 L[u]v \, dx$, instead of $\int_0^1 L[u]v \, dx$, as in Section 10.2. Afterwards we let ϵ approach zero. Assuming that u and v have at least two continuous derivatives on $\epsilon \leq x \leq 1$, and integrating twice by parts, we find that

$$\int_\epsilon^1 \{L[u]v - uL[v]\} \, dx = -p(x)\left[u'(x)v(x) - u(x)v'(x)\right]\Big|_\epsilon^1. \tag{18}$$

The boundary term at $x = 1$ is again eliminated if both u and v satisfy the boundary condition (3), and thus

$$\int_\epsilon^1 \{L[u]v - uL[v]\} \, dx = p(\epsilon)[u'(\epsilon)v(\epsilon) - u(\epsilon)v'(\epsilon)]. \tag{19}$$

Taking the limit as $\epsilon \to 0$ yields

$$\int_0^1 \{L[u]v - uL[v]\} \, dx = \lim_{\epsilon \to 0} p(\epsilon)[u'(\epsilon)v(\epsilon) - u(\epsilon)v'(\epsilon)]. \tag{20}$$

Hence Eq. (17) holds if and only if, in addition to the assumptions stated previously,

$$\lim_{\epsilon \to 0} p(\epsilon)[u'(\epsilon)v(\epsilon) - u(\epsilon)v'(\epsilon)] = 0 \tag{21}$$

for every pair of functions u and v in the class under consideration. Equation (21) is therefore the criterion that determines what boundary conditions are allowable at $x = 0$ if that point is a singular boundary point. A similar condition applies at $x = 1$ if that boundary point is singular, namely,

$$\lim_{\epsilon \to 0} p(1 - \epsilon)[u'(1 - \epsilon)v(1 - \epsilon) - u(1 - \epsilon)v'(1 - \epsilon)] = 0. \tag{22}$$

In summary, as in Section 10.2, a singular boundary value problem for Eq. (1) is said to be **self-adjoint** if Eq. (17) is valid, possibly as an improper integral, for each pair of functions u and v with the following properties: they are twice continuously differentiable on the open interval $0 < x < 1$, they satisfy a boundary condition of the form (2) at each regular boundary point, and they satisfy a boundary condition sufficient to ensure Eq. (21) if $x = 0$ is a singular boundary point, or Eq. (22) if $x = 1$ is a singular boundary point. If at least one boundary point is singular, then the differential equation (1), together with two boundary conditions of the type just described, are said to form a **singular Sturm–Liouville problem**.

For example, for Eq. (7) we have $p(x) = x$. If both u and v satisfy the boundary condition (14) at $x = 0$, it is clear that Eq. (21) will hold. Hence the singular boundary value problem, consisting of the differential equation (7), the boundary condition (14) at $x = 0$, and any boundary condition of the form (3) at $x = 1$, is self-adjoint.

The most striking difference between regular and singular Sturm–Liouville problems is that in a singular problem the eigenvalues may not be discrete. That is, the problem may have nontrivial solutions for every value of λ, or for every value of λ in some interval. In such a case the problem is said to have a **continuous spectrum**. It may happen that a singular problem has a mixture of discrete eigenvalues and also a continuous spectrum. Finally, it is possible that only a discrete set of eigenvalues exists, just as in the regular case discussed in Section 10.2. For example, this is true of the problem consisting of Eqs. (7), (13), and (14). In general, it may be difficult to determine which case actually occurs in a given problem.

A systematic discussion of singular Sturm–Liouville problems is quite sophisticated[6] indeed, requiring a substantial extension of the methods presented in this book. We restrict ourselves to some examples related to physical applications; in each of these examples it is known that there is an infinite set of discrete eigenvalues.

If a singular Sturm–Liouville problem does have only a discrete set of eigenvalues and eigenfunctions, then Eq. (17) can be used, just as in Section 10.2, to prove that the eigenvalues of such a problem are real and that the eigenfunctions are orthogonal with respect to the weight function r. The expansion of a given function in terms of a series of eigenfunctions then follows as in Section 10.2.

Such expansions are useful, as in the regular case, for solving nonhomogeneous boundary value problems. The procedure is very similar to that described in Section 10.3. Some examples for ordinary differential equations are indicated in Problems 1 to 4, and some problems for partial differential equations appear in Section 10.5.

For instance, the eigenfunctions $\phi_n(x) = J_0(\sqrt{\lambda_n}\, x)$ of the singular Sturm–Liouville problem

$$-(xy')' = \lambda xy, \qquad 0 < x < 1,$$

$$y, \ y' \text{ bounded as } x \to 0, \qquad y(1) = 0$$

satisfy the orthogonality relation

$$\int_0^1 x\phi_m(x)\phi_n(x)\,dx = 0, \qquad m \neq n \tag{23}$$

with respect to the weight function $r(x) = x$. Then, if f is a given function, we assume that

$$f(x) = \sum_{n=1}^{\infty} c_n J_0(\sqrt{\lambda_n}\, x). \tag{24}$$

Multiplying Eq. (24) by $xJ_0(\sqrt{\lambda_m}\, x)$ and integrating term by term from $x = 0$ to $x = 1$ yield

$$\int_0^1 x f(x) J_0(\sqrt{\lambda_m}\, x)\,dx = \sum_{n=1}^{\infty} c_n \int_0^1 x J_0(\sqrt{\lambda_m}\, x) J_0(\sqrt{\lambda_n}\, x)\,dx. \tag{25}$$

Because of the orthogonality condition (23), the right side of Eq. (25) collapses to a single term; hence

$$c_m = \frac{\displaystyle\int_0^1 x f(x) J_0(\sqrt{\lambda_m}\, x)\,dx}{\displaystyle\int_0^1 x J_0^2(\sqrt{\lambda_m}\, x)\,dx}, \tag{26}$$

which determines the coefficients in the series (24).

The convergence of the series (24) is established by an extension of Theorem 10.2.4 to cover this case. This theorem can also be shown to hold for other sets of Bessel functions, which are solutions of appropriate boundary value problems, for Legendre polynomials,

---

[6]See, for example, Chapter 5 of the book by Yosida listed in the References at the end of this book.

and for solutions of a number of other singular Sturm–Liouville problems of considerable interest.

It must be emphasized that the singular problems mentioned here are not necessarily typical. In general, singular boundary value problems are characterized by continuous spectra, rather than by discrete sets of eigenvalues. The corresponding sets of eigenfunctions are therefore not denumerable, and series expansions of the type described in Theorem 10.2.4 do not exist. They are replaced by appropriate integral representations.

PROBLEMS

1. Find a formal solution of the nonhomogeneous boundary value problem

$$-(xy')' = \mu xy + f(x),$$

$$y, \ y' \text{ bounded as } x \to 0, \qquad y(1) = 0,$$

where f is a given continuous function on $0 \le x \le 1$, and μ is not an eigenvalue of the corresponding homogeneous problem.

Hint: Use a series expansion similar to those in Section 10.3.

2. Consider the boundary value problem

$$-(xy')' = \lambda xy,$$

$$y, \ y' \text{ bounded as } x \to 0, \qquad y'(1) = 0.$$

(a) Show that $\lambda_0 = 0$ is an eigenvalue of this problem corresponding to the eigenfunction $\phi_0(x) = 1$. If $\lambda > 0$, show formally that the eigenfunctions are given by $\phi_n(x) = J_0(\sqrt{\lambda_n}\, x)$, where $\sqrt{\lambda_n}$ is the nth positive root (in increasing order) of the equation $J_0'(\sqrt{\lambda}) = 0$. It is possible to show that there is an infinite sequence of such roots.

(b) Show that if $m, \ n = 0, 1, 2, \ldots$, then

$$\int_0^1 x\phi_m(x)\phi_n(x)\, dx = 0, \qquad m \ne n.$$

(c) Find a formal solution to the nonhomogeneous problem

$$-(xy')' = \mu xy + f(x),$$

$$y, \ y' \text{ bounded as } x \to 0, \qquad y'(1) = 0,$$

where f is a given continuous function on $0 \le x \le 1$, and μ is not an eigenvalue of the corresponding homogeneous problem.

3. Consider the problem

$$-(xy')' + (k^2/x)y = \lambda xy,$$

$$y, \ y' \text{ bounded as } x \to 0, \qquad y(1) = 0,$$

where k is a positive integer.

(a) Using the substitution $t = \sqrt{\lambda}\, x$, show that the given differential equation reduces to Bessel's equation of order k (see Problem 9 of Section 8.7). One solution is $J_k(t)$; a second linearly independent solution, denoted by $Y_k(t)$, is unbounded as $t \to 0$.

(b) Show formally that the eigenvalues $\lambda_1, \lambda_2, \ldots$ of the given problem are the squares of the positive zeros of $J_k(\sqrt{\lambda})$ and that the corresponding eigenfunctions are $\phi_n(x) = J_k(\sqrt{\lambda_n}\, x)$. It is possible to show that there is an infinite sequence of such zeros.

(c) Show that the eigenfunctions $\phi_n(x)$ satisfy the orthogonality relation

$$\int_0^1 x\phi_m(x)\phi_n(x)\, dx = 0, \qquad m \ne n.$$

(d) Determine the coefficients in the formal series expansion

$$f(x) = \sum_{n=1}^{\infty} a_n \phi_n(x).$$

(e) Find a formal solution of the nonhomogeneous problem

$$-(xy')' + (k^2/x)y = \mu xy + f(x),$$

$$y, \ y' \text{ bounded as } x \to 0, \qquad y(1) = 0,$$

where f is a given continuous function on $0 \le x \le 1$, and μ is not an eigenvalue of the corresponding homogeneous problem.

4. Consider Legendre's equation (see Problems 22 through 24 in Section 8.3)

$$-[(1 - x^2)y']' = \lambda y$$

subject to the boundary conditions

$$y(0) = 0, \qquad y, \ y' \text{ bounded as } x \to 1.$$

The eigenfunctions of this problem are the odd Legendre polynomials $\phi_1(x) = P_1(x) = x$, $\phi_2(x) = P_3(x) = (5x^3 - 3x)/2, \ldots, \phi_n(x) = P_{2n-1}(x), \ldots$

corresponding to the eigenvalues $\lambda_1 = 2$, $\lambda_2 = 4 \cdot 3, \ldots, \lambda_n = 2n(2n - 1), \ldots$.

(a) Show that

$$\int_0^1 \phi_m(x)\phi_n(x)\,dx = 0, \qquad m \neq n.$$

(b) Find a formal solution of the nonhomogeneous problem

$$-[(1 - x^2)y']' = \mu y + f(x),$$
$$y(0) = 0, \qquad y, \ y' \text{ bounded as } x \to 1,$$

where f is a given continuous function on $0 \leq x \leq 1$, and μ is not an eigenvalue of the corresponding homogeneous problem.

5. The equation

$$(1 - x^2)y'' - xy' + \lambda y = 0 \qquad \text{(i)}$$

is Chebyshev's equation; see Problem 10 in Section 8.3.

(a) Show that Eq. (i) can be written in the form

$$-[(1 - x^2)^{1/2}y']' = \lambda(1 - x^2)^{-1/2}y,$$
$$-1 < x < 1. \qquad \text{(ii)}$$

(b) Consider the boundary conditions

$$y, \ y' \text{ bounded as } x \to -1,$$

$$y, \ y' \text{ bounded as } x \to 1. \qquad \text{(iii)}$$

Show that the boundary value problem (ii), (iii) is self-adjoint.

(c) It can be shown that the boundary value problem (ii), (iii) has the eigenvalues $\lambda_0 = 0$, $\lambda_1 = 1$, $\lambda_2 = 4, \ldots, \lambda_n = n^2, \ldots$. The corresponding eigenfunctions are the Chebyshev polynomials $T_n(x)$: $T_0(x) = 1$, $T_1(x) = x$, $T_2(x) = 1 - 2x^2, \ldots$. Show that

$$\int_{-1}^1 \frac{T_m(x)T_n(x)}{(1 - x^2)^{1/2}}\,dx = 0, \qquad m \neq n. \qquad \text{(iv)}$$

Note that this is a convergent improper integral.

10.5 Further Remarks on the Method of Separation of Variables: A Bessel Series Expansion

In this chapter we are interested in extending the method of separation of variables developed in Chapter 9 to a larger class of problems—to problems involving more general differential equations, more general boundary conditions, or different geometrical regions. We indicated in Section 10.3 how to deal with a class of more general differential equations or boundary conditions. Here we concentrate on problems posed in various geometrical regions, with emphasis on those leading to singular Sturm–Liouville problems when the variables are separated.

Because of its relative simplicity, as well as the considerable physical significance of many problems to which it is applicable, the method of separation of variables merits its important place in the theory and application of partial differential equations. However, this method does have certain limitations that should not be forgotten. In the first place, the problem must be linear so that the principle of superposition can be invoked to construct additional solutions by forming linear combinations of the fundamental solutions of an appropriate homogeneous problem.

As a practical matter, we must also be able to solve the ordinary differential equations, obtained after separating the variables, in a reasonably convenient manner. In some problems to which the method of separation of variables can be applied in principle, it is of very limited practical value due to a lack of information about the solutions of the ordinary differential equations that appear.

Furthermore, the geometry of the region involved in the problem is subject to rather severe restrictions. On the one hand, a coordinate system must be employed in which the variables can be separated, and the partial differential equation replaced by a set of ordinary differential equations. For Laplace's equation there are about a dozen such coordinate systems; only rectangular, circular cylindrical, and spherical coordinates are likely to be familiar to most readers of this book. On the other hand, the boundary of the region of interest must consist of coordinate curves or surfaces—that is, curves or surfaces on which one variable remains constant. Thus, at an elementary level, one is limited to regions bounded by straight lines or circular arcs in two dimensions, or by planes, circular cylinders, circular cones, or spheres in three dimensions.

In three-dimensional problems the separation of variables in Laplace's operator $u_{xx} + u_{yy} + u_{zz}$ leads to the equation $X'' + \lambda X = 0$ in rectangular coordinates, to Bessel's equation in cylindrical coordinates, and to Legendre's equation in spherical coordinates. It is this fact that is largely responsible for the intensive study that has been made of these equations and the functions defined by them. It is also noteworthy that two of the three most important situations lead to singular, rather than regular, Sturm–Liouville problems. Thus singular problems are by no means exceptional and may be of even greater interest than regular ones. The remainder of this section is devoted to an example involving an expansion of a given function as a series of Bessel functions.

The Vibrations of a Circular Elastic Membrane

In Section 9.7 [Eq. (7)] we noted that the transverse vibrations of a thin elastic membrane are governed by the two-dimensional wave equation

$$a^2(u_{xx} + u_{yy}) = u_{tt}. \tag{1}$$

To study the motion of a circular membrane, it is convenient to write Eq. (1) in polar coordinates:

$$a^2\left(u_{rr} + \frac{1}{r}u_r + \frac{1}{r^2}u_{\theta\theta}\right) = u_{tt}. \tag{2}$$

We will assume that the membrane has unit radius, that it is fixed securely around its circumference, and that initially it occupies a displaced position independent of the angular variable θ, from which it is released at time $t = 0$. Because of the circular symmetry of the initial and boundary conditions, it is natural to assume also that u is independent of θ; that is, u is a function of r and t only. In this event the differential equation (2) becomes

$$a^2\left(u_{rr} + \frac{1}{r}u_r\right) = u_{tt}, \qquad 0 < r < 1, \quad t > 0. \tag{3}$$

The boundary condition at $r = 1$ is

$$u(1, t) = 0, \qquad t \geq 0, \tag{4}$$

and the initial conditions are

$$u(r, 0) = f(r), \qquad 0 \le r \le 1, \tag{5}$$

$$u_t(r, 0) = 0, \qquad 0 \le r \le 1, \tag{6}$$

where $f(r)$ describes the initial configuration of the membrane. For consistency we also require that $f(1) = 0$. Finally, we state explicitly the requirement that $u(r, t)$ be bounded for $0 \le r \le 1$.

Assuming that $u(r, t) = R(r)T(t)$, and substituting for $u(r, t)$ in Eq. (3), we obtain

$$\frac{R'' + (1/r)R'}{R} = \frac{1}{a^2}\frac{T''}{T} = -\lambda^2. \tag{7}$$

We have anticipated that the separation constant must be negative by writing it as $-\lambda^2$ with $\lambda > 0$.[7] Then Eq. (7) yields the following two ordinary differential equations:

$$r^2 R'' + rR' + \lambda^2 r^2 R = 0, \tag{8}$$

$$T'' + \lambda^2 a^2 T = 0. \tag{9}$$

Thus, from Eq. (9),

$$T(t) = k_1 \sin \lambda a t + k_2 \cos \lambda a t. \tag{10}$$

Introducing the new independent variable $\xi = \lambda r$ into Eq. (8), we obtain

$$\xi^2 \frac{d^2 R}{d\xi^2} + \xi \frac{dR}{d\xi} + \xi^2 R = 0, \tag{11}$$

which is Bessel's equation of order zero. Thus

$$R = c_1 J_0(\xi) + c_2 Y_0(\xi), \tag{12}$$

where J_0 and Y_0 are Bessel functions of the first and second kinds, respectively, of order zero (see Section 10.4). In terms of r we have

$$R = c_1 J_0(\lambda r) + c_2 Y_0(\lambda r). \tag{13}$$

The boundedness condition on $u(r, t)$ requires that R remain bounded as $r \to 0$. Since $Y_0(\lambda r) \to -\infty$ as $r \to 0$, we must choose $c_2 = 0$. The boundary condition (4) then requires that

$$J_0(\lambda) = 0. \tag{14}$$

Consequently, the allowable values of the separation constant are obtained from the roots of the transcendental equation (14). Recall from Section 10.4 that $J_0(\lambda)$ has an infinite set of discrete positive zeros, which we denote by $\lambda_1, \lambda_2, \lambda_3, \ldots, \lambda_n, \ldots$, ordered in increasing magnitude. Further, the functions $J_0(\lambda_n r)$ are the eigenfunctions of a singular Sturm–Liouville problem and can be used as the basis of a series expansion for the given function

---

[7]By denoting the separation constant by $-\lambda^2$, rather than simply by $-\lambda$, we avoid the appearance of numerous radical signs in the following discussion.

f. The fundamental solutions of this problem, satisfying the partial differential equation (3), the boundary condition (4), and boundedness condition, are

$$u_n(r, t) = J_0(\lambda_n r) \sin \lambda_n at, \qquad n = 1, 2, \ldots, \tag{15}$$

$$v_n(r, t) = J_0(\lambda_n r) \cos \lambda_n at, \qquad n = 1, 2, \ldots. \tag{16}$$

Next we assume that $u(r, t)$ can be expressed as an infinite linear combination of the fundamental solutions (15), (16):

$$\begin{aligned} u(r, t) &= \sum_{n=1}^{\infty} [k_n u_n(r, t) + c_n v_n(r, t)] \\ &= \sum_{n=1}^{\infty} [k_n J_0(\lambda_n r) \sin \lambda_n at + c_n J_0(\lambda_n r) \cos \lambda_n at]. \end{aligned} \tag{17}$$

The initial conditions require that

$$u(r, 0) = \sum_{n=1}^{\infty} c_n J_0(\lambda_n r) = f(r) \tag{18}$$

and

$$u_t(r, 0) = \sum_{n=1}^{\infty} \lambda_n a k_n J_0(\lambda_n r) = 0. \tag{19}$$

From Eq. (26) of Section 10.4 we obtain

$$k_n = 0, \qquad c_n = \frac{\displaystyle\int_0^1 r f(r) J_0(\lambda_n r) \, dr}{\displaystyle\int_0^1 r [J_0(\lambda_n r)]^2 \, dr}; \qquad n = 1, 2, \ldots. \tag{20}$$

Thus the solution of the partial differential equation (3) satisfying the boundary condition (4) and the initial conditions (5) and (6) is given by

$$u(r, t) = \sum_{n=1}^{\infty} c_n J_0(\lambda_n r) \cos \lambda_n at \tag{21}$$

with the coefficients c_n defined by Eq. (20).

PROBLEMS

1. Consider Laplace's equation $u_{xx} + u_{yy} = 0$ in the parallelogram whose vertices are $(0, 0)$, $(2, 0)$, $(3, 2)$, and $(1, 2)$. Suppose that on the side $y = 2$ the boundary condition is $u(x, 2) = f(x)$ for $1 \leq x \leq 3$, and that on the other three sides $u = 0$ (see Figure 10.5.1).

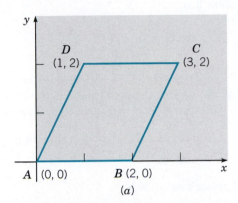

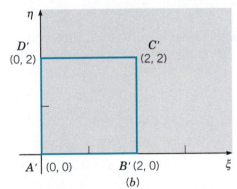

FIGURE 10.5.1 The region in Problem 1.

(a) Show that there are no nontrivial solutions of the partial differential equation of the form $u(x, y) = X(x)Y(y)$ that also satisfy the homogeneous boundary conditions.

(b) Let $\xi = x - \frac{1}{2}y$, $\eta = y$. Show that the given parallelogram in the xy-plane transforms into the square $0 \leq \xi \leq 2$, $0 \leq \eta \leq 2$ in the $\xi\eta$-plane. Show that the differential equation transforms into

$$\frac{5}{4}u_{\xi\xi} - u_{\xi\eta} + u_{\eta\eta} = 0.$$

How are the boundary conditions transformed?

(c) Show that in the $\xi\eta$-plane the differential equation possesses no solution of the form

$$u(\xi, \eta) = U(\xi)V(\eta).$$

Thus in the xy-plane the shape of the boundary precludes a solution by the method of the separation of variables, while in the $\xi\eta$-plane the region is acceptable but the variables in the differential equation can no longer be separated.

2. Find the displacement $u(r, t)$ in a vibrating circular elastic membrane of radius 1 that satisfies the boundary condition

$$u(1, t) = 0, \qquad t \geq 0,$$

and the initial conditions

$$u(r, 0) = 0, \qquad u_t(r, 0) = g(r), \qquad 0 \leq r \leq 1,$$

where $g(1) = 0$.
Hint: The differential equation to be satisfied is Eq. (3) of this section.

3. Find the displacement $u(r, t)$ in a vibrating circular elastic membrane of radius 1 that satisfies the boundary condition

$$u(1, t) = 0, \qquad t \geq 0,$$

and the initial conditions

$$u(r, 0) = f(r), \qquad u_t(r, 0) = g(r), \qquad 0 \leq r \leq 1,$$

where $f(1) = g(1) = 0$.

4. The wave equation in polar coordinates is

$$u_{rr} + (1/r)u_r + (1/r^2)u_{\theta\theta} = a^{-2}u_{tt}.$$

Show that if $u(r, \theta, t) = R(r)\Theta(\theta)T(t)$, then R, Θ, and T satisfy the ordinary differential equations

$$
\begin{aligned}
r^2 R'' + rR' + (\lambda^2 r^2 - n^2)R &= 0, \\
\Theta'' + n^2\Theta &= 0, \\
T'' + \lambda^2 a^2 T &= 0.
\end{aligned}
$$

5. In the circular cylindrical coordinates r, θ, z defined by

$$x = r\cos\theta, \qquad y = r\sin\theta, \qquad z = z,$$

Laplace's equation is

$$u_{rr} + (1/r)u_r + (1/r^2)u_{\theta\theta} + u_{zz} = 0.$$

(a) Show that if $u(r, \theta, z) = R(r)\Theta(\theta)Z(z)$, then R, Θ, and Z satisfy the ordinary differential equations

$$r^2 R'' + rR' + (\lambda^2 r^2 - n^2)R = 0,$$
$$\Theta'' + n^2\Theta = 0,$$
$$Z'' - \lambda^2 Z = 0.$$

(b) Show that if $u(r, \theta, z)$ is independent of θ, then the first equation in part **(a)** becomes

$$r^2 R'' + rR' + \lambda^2 r^2 R = 0,$$

the second is omitted altogether, and the third is unchanged.

6. Find the steady-state temperature in a semi-infinite rod $0 < z < \infty$, $0 \leq r < 1$, if the temperature is independent of θ and approaches zero as $z \to \infty$. Assume that the temperature $u(r, z)$ satisfies the boundary conditions

$$u(1, z) = 0, \qquad z > 0,$$
$$u(r, 0) = f(r), \qquad 0 \leq r \leq 1.$$

Hint: Refer to Problem 5.

7. The equation

$$v_{xx} + v_{yy} + k^2 v = 0$$

is a generalization of Laplace's equation and is sometimes called the Helmholtz equation.

(a) In polar coordinates the Helmholtz equation is

$$v_{rr} + (1/r)v_r + (1/r^2)v_{\theta\theta} + k^2 v = 0.$$

If $v(r, \theta) = R(r)\Theta(\theta)$, show that R and Θ satisfy the ordinary differential equations

$$r^2 R'' + rR' + (k^2 r^2 - \lambda^2)R = 0,$$
$$\Theta'' + \lambda^2\Theta = 0.$$

(b) Consider the Helmholtz equation in the disk $r < c$. Find the solution that remains bounded at all points in the disk, that is periodic in θ with period 2π, and that satisfies the boundary condition $v(c, \theta) = f(\theta)$, where f is a given function on $0 \leq \theta < 2\pi$.
Hint: The equation for R is a Bessel equation. See Problem 3 in Section 10.4.

8. Consider the flow of heat in an infinitely long cylinder of radius 1: $0 \leq r < 1$, $0 \leq \theta < 2\pi$, $-\infty < z < \infty$. Let the surface of the cylinder be held at temperature zero, and let the initial temperature distribution be a function of the radial variable r only. Then the temperature u is a function

of r and t only, and satisfies the heat conduction equation

$$\alpha^2[u_{rr} + (1/r)u_r] = u_t, \qquad 0 < r < 1, \quad t > 0,$$

and the following initial and boundary conditions:

$$u(r, 0) = f(r), \qquad 0 \leq r \leq 1,$$
$$u(1, t) = 0, \qquad t > 0.$$

Show that

$$u(r, t) = \sum_{n=1}^{\infty} c_n J_0(\lambda_n r)e^{-\alpha^2\lambda_n^2 t},$$

where $J_0(\lambda_n) = 0$. Find a formula for c_n.

9. In the spherical coordinates ρ, θ, ϕ ($\rho > 0$, $0 \leq \theta < 2\pi$, $0 \leq \phi \leq \pi$) defined by the equations

$$x = \rho \cos\theta \sin\phi, \qquad y = \rho \sin\theta \sin\phi,$$
$$z = \rho \cos\phi,$$

Laplace's equation is

$$\rho^2 u_{\rho\rho} + 2\rho u_\rho + (\csc^2\phi)u_{\theta\theta} + u_{\phi\phi} + (\cot\phi)u_\phi = 0.$$

(a) Show that if $u(\rho, \theta, \phi) = P(\rho)\Theta(\theta)\Phi(\phi)$, then P, Θ, and Φ satisfy ordinary differential equations of the form

$$\rho^2 P'' + 2\rho P' - \mu^2 P = 0,$$
$$\Theta'' + \lambda^2\Theta = 0,$$
$$(\sin^2\phi)\Phi'' + (\sin\phi\cos\phi)\Phi'$$
$$+ (\mu^2\sin^2\phi - \lambda^2)\Phi = 0.$$

The first of these equations is of the Euler type, while the third is related to Legendre's equation.

(b) Show that if $u(\rho, \theta, \phi)$ is independent of θ, then the first equation in part **(a)** is unchanged, the second is omitted, and the third becomes

$$(\sin^2\phi)\Phi'' + (\sin\phi\cos\phi)\Phi' + (\mu^2\sin^2\phi)\Phi = 0.$$

(c) Show that if a new independent variable is defined by $s = \cos\phi$, then the equation for Φ in part **(b)** becomes

$$(1 - s^2)\frac{d^2\Phi}{ds^2} - 2s\frac{d\Phi}{ds} + \mu^2\Phi = 0,$$

$$-1 \leq s \leq 1.$$

Note that this is Legendre's equation.

10. Find the steady-state temperature $u(\rho, \phi)$ in a sphere of unit radius if the temperature is independent of θ and satisfies the boundary condition

$$u(1, \phi) = f(\phi), \qquad 0 \leq \phi \leq \pi.$$

Hint: Refer to Problem 9 and to Problems 22 through 29 in Section 8.3. Use the fact that the only solutions of Legendre's equation that are finite at both ± 1 are the Legendre polynomials.

10.6 Series of Orthogonal Functions: Mean Convergence

In Section 10.2 we stated that, under certain restrictions, a given function f can be expanded in a series of eigenfunctions of a Sturm–Liouville boundary value problem, the series converging to $[f(x+) + f(x-)]/2$ at each point in the open interval. Under somewhat more restrictive conditions the series converges to $f(x)$ at each point in the closed interval. This type of convergence is referred to as pointwise convergence. In this section we describe a different kind of convergence that is especially useful for series of orthogonal functions, such as eigenfunctions.

Suppose that we are given the set of functions $\phi_1, \phi_2, \ldots, \phi_n$, that are continuous on the interval $0 \leq x \leq 1$ and satisfy the orthonormality condition

$$\int_0^1 r(x)\phi_i(x)\phi_j(x)\,dx = \begin{cases} 0, & i \neq j, \\ 1, & i = j, \end{cases} \tag{1}$$

where r is a nonnegative weight function. Suppose also that we wish to approximate a given function f, defined on $0 \leq x \leq 1$, by a linear combination of $\phi_1, \ldots, \phi_n$. That is, if

$$S_n(x) = \sum_{i=1}^n a_i \phi_i(x), \tag{2}$$

we wish to choose the coefficients $a_1, \ldots, a_n$ so that the function S_n will best approximate f on $0 \leq x \leq 1$. The first problem that we must face in doing this is to state precisely what we mean by "best approximate f on $0 \leq x \leq 1$." There are several reasonable meanings that can be attached to this phrase.

1. We can choose n points $x_1, \ldots, x_n$ in the interval $0 \leq x \leq 1$ and require that $S_n(x)$ have the same value as $f(x)$ at each of these points. The coefficients $a_1, \ldots, a_n$ are found by solving the set of linear algebraic equations

$$\sum_{i=1}^n a_i \phi_i(x_j) = f(x_j), \qquad j = 1, \ldots, n. \tag{3}$$

This procedure is known as the **method of collocation**. It has the advantage that it is very easy to write down Eqs. (3); one needs only to evaluate the functions involved at the points $x_1, \ldots, x_n$. If these points are well chosen, and if n is fairly large, then presumably $S_n(x)$ will not only be equal to $f(x)$ at the chosen points but will be reasonably close to it at other points as well. However, collocation has several deficiencies. One is that if one more base function ϕ_{n+1} is added, then one more point x_{n+1} is required, and *all* the coefficients must be recomputed. Thus it is inconvenient to improve the accuracy of a collocation approximation by including additional terms. Further, the coefficients a_i depend on the location of the points $x_1, \ldots, x_n$, and it is not obvious how best to select these points.

2. Alternatively, we can consider the difference $|f(x) - S_n(x)|$ and try to make it as small as possible. The trouble here is that $|f(x) - S_n(x)|$ is a function of x as well as of the coefficients $a_1, \ldots, a_n$, and it is not clear how to calculate a_i. The choice of a_i that makes $|f(x) - S_n(x)|$ small at one point may make it large at another. One way to proceed is to consider instead the least upper bound[8] of $|f(x) - S_n(x)|$ for x in $0 \le x \le 1$, and then to choose $a_1, \ldots, a_n$ so as to make this quantity as small as possible. That is, if

$$E_n(a_1, \ldots, a_n) = \operatorname*{lub}_{0 \le x \le 1} |f(x) - S_n(x)|, \tag{4}$$

then choose $a_1, \ldots, a_n$ so as to minimize E_n. This approach is intuitively appealing and is often used in theoretical calculations. In practice, however, it is usually very hard, if not impossible, to write down an explicit formula for $E_n(a_1, \ldots, a_n)$. Further, this procedure also shares one of the disadvantages of collocation: upon adding an additional term to $S_n(x)$, one must recompute all the preceding coefficients. Thus it is not often useful in practical problems.

3. Another way to proceed is to consider

$$I_n(a_1, \ldots, a_n) = \int_0^1 r(x) |f(x) - S_n(x)| \, dx. \tag{5}$$

If $r(x) = 1$, then I_n is the area between the graphs of $y = f(x)$ and $y = S_n(x)$ (see Figure 10.6.1). We can then determine the coefficients a_i so as to minimize I_n. To avoid the complications resulting from calculations with absolute values, it is more convenient to consider instead

$$R_n(a_1, \ldots, a_n) = \int_0^1 r(x) [f(x) - S_n(x)]^2 \, dx \tag{6}$$

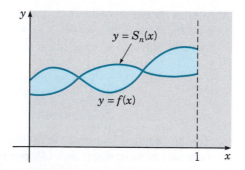

FIGURE 10.6.1   Approximation of $f(x)$ by $S_n(x)$.

---

[8]The least upper bound (lub) is an upper bound that is smaller than any other upper bound. The lub of a bounded function always exists and is equal to the function's maximum if it has one.

as our measure of the quality of approximation of the linear combination $S_n(x)$ to $f(x)$. Although R_n is clearly similar in some ways to I_n, it lacks the simple geometric interpretation of the latter. Nevertheless, it is much easier mathematically to deal with R_n than with I_n. The quantity R_n is called the **mean square error** of the approximation S_n to f. If $a_1, \ldots, a_n$ are chosen so as to minimize R_n, then S_n is said to approximate f in the mean square sense.

To choose $a_1, \ldots, a_n$ so as to minimize R_n, we must satisfy the necessary conditions

$$\partial R_n/\partial a_i = 0, \qquad i = 1, \ldots, n. \tag{7}$$

Writing out Eq. (7), and noting that $\partial S_n(x; a_1, \ldots, a_n)/\partial a_i$ is equal to $\phi_i(x)$, we obtain

$$-\frac{\partial R_n}{\partial a_i} = 2 \int_0^1 r(x)[f(x) - S_n(x)]\phi_i(x)\, dx = 0. \tag{8}$$

Substituting for $S_n(x)$ from Eq. (2) and making use of the orthogonality relation (1), we find that

$$a_i = \int_0^1 r(x)f(x)\phi_i(x)\, dx, \qquad i = 1, \ldots, n. \tag{9}$$

The coefficients defined by Eq. (9) are called the Fourier coefficients of f with respect to the orthonormal set $\phi_1, \phi_2, \ldots, \phi_n$ and the weight function r. Since the conditions (7) are only necessary and not sufficient for R_n to be a minimum, a separate argument is required to show that R_n is actually minimized if the a_i are chosen by Eq. (9). This argument is outlined in Problem 5.

Note that the coefficients (9) are the same as those in the eigenfunction series whose convergence, under certain conditions, was stated in Theorem 10.2.4. Thus $S_n(x)$ is the nth partial sum in this series and constitutes the best mean square approximation to $f(x)$ that is possible with the functions $\phi_1, \ldots, \phi_n$. We will assume hereafter that the coefficients a_i in $S_n(x)$ are given by Eq. (9).

Equation (9) is noteworthy in two other important respects. In the first place, it gives a formula for each a_i *separately*, rather than a set of linear algebraic equations for $a_1, \ldots, a_n$ as in the method of collocation, for example. This is due to the orthogonality of the base functions $\phi_1, \ldots, \phi_n$. Further, the formula for a_i is *independent* of n, the number of terms in $S_n(x)$. The practical significance of this is as follows. Suppose that, to obtain a better approximation to f, we desire to use an approximation with more terms—say, k terms, where $k > n$. It is then unnecessary to recompute the first n coefficients in $S_k(x)$. All that is required is to compute, from Eq. (9), the coefficients $a_{n+1}, \ldots, a_k$ arising from the additional base functions $\phi_{n+1}, \ldots, \phi_k$. Of course, if f, r, and the ϕ_n are complicated functions, it may be necessary to evaluate the integrals numerically.

Now let us suppose that there is an infinite sequence of functions $\phi_1, \ldots, \phi_n, \ldots$, that are continuous and orthonormal on the interval $0 \le x \le 1$. Suppose further that as n increases without bound, the mean square error R_n approaches zero. In this event the infinite series

$$\sum_{i=1}^{\infty} a_i \phi_i(x)$$

is said to converge in the **mean square sense** (or, more simply, in the **mean**) to $f(x)$. Mean convergence is an essentially different type of convergence from the pointwise convergence considered up to now. A series may converge in the mean without converging at each point. This is plausible geometrically because the area between two curves, which behaves in the same way as the mean square error, may be zero even though the functions are not the same at every point. They may differ on any finite set of points, for example, without affecting the mean square error. It is less obvious, but also true, that even if an infinite series converges at every point, it may not converge in the mean. Indeed, the mean square error may even become unbounded. An example of this phenomenon is given in Problem 4.

Now suppose that we wish to know what class of functions, defined on $0 \le x \le 1$, can be represented as an infinite series of the orthonormal set ϕ_i, $i = 1, 2, \ldots$. The answer depends on what kind of convergence we require. We say that the set $\phi_1, \ldots, \phi_n, \ldots$ is **complete** with respect to mean square convergence for a set of functions $\mathcal{F}$, if for each function f in $\mathcal{F}$, the series

$$f(x) = \sum_{i=1}^{\infty} a_i \phi_i(x), \tag{10}$$

with coefficients given by Eq. (9), converges in the mean. There is a similar definition for completeness with respect to pointwise convergence.

Theorems having to do with the convergence of series such as that in Eq. (10) can now be restated in terms of the idea of completeness. For example, Theorem 10.2.4 can be restated as follows: the eigenfunctions of the Sturm–Liouville problem

$$-[p(x)y']' + q(x)y = \lambda r(x)y, \qquad 0 < x < 1, \tag{11}$$

$$\alpha_1 y(0) + \alpha_2 y'(0) = 0, \qquad \beta_1 y(1) + \beta_2 y'(1) = 0 \tag{12}$$

are complete with respect to ordinary pointwise convergence for the set of functions that are continuous on $0 \le x \le 1$ and that have piecewise continuous derivatives there.

If pointwise convergence is replaced by mean convergence, Theorem 10.2.4 can be considerably generalized. Before we state such a companion theorem to Theorem 10.2.4, we first define what is meant by a square integrable function. A function f is said to be **square integrable** on the interval $0 \le x \le 1$ if both f and f^2 are integrable[9] on that interval. The following theorem is similar to Theorem 10.2.4 except that it involves mean convergence.

THEOREM 10.6.1

The eigenfunctions ϕ_i of the Sturm–Liouville problem (11), (12) are complete with respect to mean convergence for the set of functions that are square integrable on $0 \le x \le 1$. In other words, given any square integrable function f, the series (10), whose coefficients are given by Eq. (9), converges to $f(x)$ in the mean square sense.

It is significant that the class of functions specified in Theorem 10.6.1 is very large indeed. The class of square integrable functions contains some functions with many discontinuities, including some kinds of infinite discontinuities, as well as some functions that are not differentiable at any point. All these functions have mean convergent expansions in the eigenfunctions of the boundary value problem (11), (12). However, in many cases these series do not converge pointwise, at least not at every point. Thus mean convergence is more naturally associated with series of orthogonal functions, such as eigenfunctions, than ordinary pointwise convergence.

---

[9] For the Riemann integral used in elementary calculus the hypotheses that f and f^2 are integrable are independent; that is, there are functions such that f is integrable but f^2 is not, and conversely (see Problem 6). A generalized integral, known as the Lebesgue integral, has the property (among others) that if f^2 is integrable, then f is also necessarily integrable. The term *square integrable* came into common use in connection with this type of integration.

The theory of Fourier series discussed in Chapter 9 is just a special case of the general theory of Sturm–Liouville problems. For instance, the functions

$$\phi_n(x) = \sqrt{2}\sin n\pi x \tag{13}$$

are the normalized eigenfunctions of the Sturm–Liouville problem

$$y'' + \lambda y = 0, \qquad y(0) = 0, \quad y(1) = 0. \tag{14}$$

Thus, if f is a given square integrable function on $0 \le x \le 1$, then according to Theorem 10.6.1, the series

$$f(x) = \sum_{m=1}^{\infty} b_m \phi_m(x) = \sqrt{2}\sum_{m=1}^{\infty} b_m \sin m\pi x, \tag{15}$$

where

$$b_m = \int_0^1 f(x)\phi_m(x)\,dx = \sqrt{2}\int_0^1 f(x)\sin m\pi x\,dx, \tag{16}$$

converges in the mean. The series (15) is precisely the Fourier sine series discussed in Section 9.4. If f satisfies the further conditions stated in Theorem 10.2.4, then this series converges pointwise, as well as in the mean. Similarly, a Fourier cosine series is associated with the Sturm–Liouville problem

$$y'' + \lambda y = 0, \qquad y'(0) = 0, \quad y'(1) = 0. \tag{17}$$

EXAMPLE 1

Let $f(x) = 1$ for $0 < x < 1$. Expand $f(x)$ using the eigenfunctions (13) and discuss the pointwise and mean square convergence of the resulting series.

The series has the form (15) and its coefficients b_m are given by Eq. (16). Thus

$$b_m = \sqrt{2}\int_0^1 \sin m\pi x\,dx = \frac{\sqrt{2}}{m\pi}(1 - \cos m\pi), \tag{18}$$

and the nth partial sum of the series is

$$S_n(x) = 2\sum_{m=1}^{n} \frac{1 - \cos m\pi}{m\pi}\sin m\pi x. \tag{19}$$

The mean square error is then

$$R_n = \int_0^1 [f(x) - S_n(x)]^2\,dx. \tag{20}$$

By calculating R_n for several values of n and plotting the results, we obtain Figure 10.6.2. This figure indicates that R_n steadily decreases as n increases. Of course, Theorem 10.6.1 asserts that $R_n \to 0$ as $n \to \infty$. Pointwise, we know that $S_n(x) \to f(x) = 1$ as $n \to \infty$ for $0 < x < 1$; further, $S_n(x)$ has the value zero for $x = 0$ or $x = 1$ for every n. Although the series converges pointwise for each value of x, the least upper bound of the error does not diminish as n increases. For each n there are points close to $x = 0$ and $x = 1$ where the error is arbitrarily close to 1.

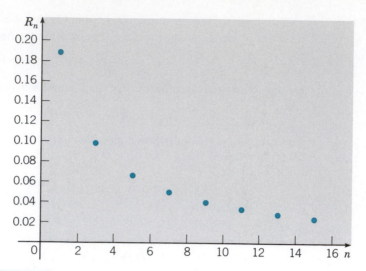

FIGURE 10.6.2 Dependence of the mean square error R_n on n in Example 1.

Theorem 10.6.1 can be extended to cover self-adjoint boundary value problems having periodic boundary conditions, such as the problem

$$y'' + \lambda y = 0, \tag{21}$$

$$y(-L) - y(L) = 0, \qquad y'(-L) - y'(L) = 0 \tag{22}$$

considered in Example 4 of Section 10.2. The eigenfunctions of the problem (21), (22) are $\phi_n(x) = \cos(n\pi x/L)$ for $n = 0, 1, 2, \ldots$ and $\psi_n(x) = \sin(n\pi x/L)$ for $n = 1, 2, \ldots$. If f is a given square integrable function on $-L \leq x \leq L$, then its expansion in terms of the eigenfunctions ϕ_n and ψ_n is of the form

$$f(x) = \frac{a_0}{2} + \sum_{n=1}^{\infty} \left(a_n \cos \frac{n\pi x}{L} + b_n \sin \frac{n\pi x}{L} \right), \tag{23}$$

where

$$a_n = \frac{1}{L} \int_{-L}^{L} f(x) \cos \frac{n\pi x}{L} \, dx, \qquad n = 0, 1, 2, \ldots, \tag{24}$$

$$b_n = \frac{1}{L} \int_{-L}^{L} f(x) \sin \frac{n\pi x}{L} \, dx, \qquad n = 1, 2, \ldots. \tag{25}$$

This expansion is exactly the Fourier series for f discussed in Sections 9.2 and 9.3. According to the generalization of Theorem 10.6.1, the series (23) converges in the mean for any square integrable function f, even though f may not satisfy the conditions of Theorem 9.3.1, which ensure pointwise convergence.

PROBLEMS

CAS 1. Extend the results of Example 1 by finding the smallest value of n for which $R_n < 0.02$, where R_n is given by Eq. (20).

CAS 2. Let $f(x) = x$ for $0 < x < 1$ and let $\phi_m(x) = \sqrt{2} \sin m\pi x$.

(a) Find the coefficients b_m in the expansion of $f(x)$ in terms of $\phi_1(x), \phi_2(x), \ldots$.

(b) Calculate the mean square error R_n for several values of n and plot the results.

(c) Find the smallest value of n for which $R_n < 0.01$.

CAS 3. Follow the instructions for Problem 2 using $f(x) = x(1 - x)$ for $0 < x < 1$.

4. In this problem we show that pointwise convergence of a sequence $S_n(x)$ does not imply mean convergence, and conversely.

(a) Let $S_n(x) = n\sqrt{x}e^{-nx^2/2}$, $0 \leq x \leq 1$. Show that $S_n(x) \to 0$ as $n \to \infty$ for each x in $0 \leq x \leq 1$. Show also that

$$R_n = \int_0^1 [0 - S_n(x)]^2\, dx = \frac{n}{2}(1 - e^{-n})$$

and hence $R_n \to \infty$ as $n \to \infty$. Thus pointwise convergence does not imply mean convergence.

(b) Let $S_n(x) = x^n$ for $0 \leq x \leq 1$ and let $f(x) = 0$ for $0 \leq x \leq 1$. Show that

$$R_n = \int_0^1 [f(x) - S_n(x)]^2\, dx = \frac{1}{2n+1},$$

and hence $S_n(x)$ converges to $f(x)$ in the mean. Also show that $S_n(x)$ does not converge to $f(x)$ pointwise throughout $0 \leq x \leq 1$. Thus mean convergence does not imply pointwise convergence.

5. Suppose that the functions $\phi_1, \ldots, \phi_n$ satisfy the orthonormality relation (1) and that a given function f is to be approximated by $S_n(x) = c_1\phi_1(x) + \cdots + c_n\phi_n(x)$, where the coefficients c_i are not necessarily those of Eq. (9). Show that the mean square error R_n given by Eq. (6) may be written in the form

$$R_n = \int_0^1 r(x)f^2(x)\, dx - \sum_{i=1}^n a_i^2 + \sum_{i=1}^n (c_i - a_i)^2,$$

where the a_i are the Fourier coefficients given by Eq. (9). Show that R_n is minimized if $c_i = a_i$ for each i.

6. In this problem we show by examples that the (Riemann) integrability of f and f^2 is independent.

(a) Let $f(x) = \begin{cases} x^{-1/2}, & 0 < x < 1, \\ 0, & x = 0. \end{cases}$

Show that $\int_0^1 f(x)\, dx$ exists as an improper integral, but $\int_0^1 f^2(x)\, dx$ does not.

(b) Let $f(x) = \begin{cases} 1, & x \text{ rational}, \\ -1, & x \text{ irrational}. \end{cases}$

Show that $\int_0^1 f^2(x)\, dx$ exists, but $\int_0^1 f(x)\, dx$ does not.

7. Suppose that it is desired to construct a set of polynomials $f_0(x)$, $f_1(x)$, $f_2(x)$, $\ldots$, $f_k(x)$, $\ldots$, where $f_k(x)$ is of degree k, that are orthonormal on the interval $0 \leq x \leq 1$. That is, the set of polynomials must satisfy

$$(f_j, f_k) = \int_0^1 f_j(x)f_k(x)\, dx = \delta_{jk}.$$

(a) Find $f_0(x)$ by choosing the polynomial of degree zero such that $(f_0, f_0) = 1$.

(b) Find $f_1(x)$ by determining the polynomial of degree one such that $(f_0, f_1) = 0$ and $(f_1, f_1) = 1$.

(c) Find $f_2(x)$.

(d) The normalization condition $(f_k, f_k) = 1$ is somewhat awkward to apply. Let $g_0(x)$, $g_1(x)$, $\ldots$, $g_k(x)$, $\ldots$ be the sequence of polynomials that are orthogonal on $0 \leq x \leq 1$ and that are normalized by the condition $g_k(1) = 1$. Find $g_0(x)$, $g_1(x)$, and $g_2(x)$ and compare them with $f_0(x)$, $f_1(x)$, and $f_2(x)$.

8. Suppose that it is desired to construct a set of polynomials $P_0(x)$, $P_1(x)$, $\ldots$, $P_k(x)$, $\ldots$, where $P_k(x)$ is of degree k, that are orthogonal on the interval $-1 \leq x \leq 1$; see Problem 7. Suppose further that $P_k(x)$ is normalized by the condition $P_k(1) = 1$. Find $P_0(x)$, $P_1(x)$, $P_2(x)$, and $P_3(x)$. Note that these are the first four Legendre polynomials (see Problem 24 in Section 8.3).

9. This problem develops some further results associated with mean convergence. Let $R_n(a_1, \ldots, a_n)$, $S_n(x)$, and a_i be defined by Eqs. (6), (2), and (9), respectively.

(a) Show that

$$R_n = \int_0^1 r(x)f^2(x)\, dx - \sum_{i=1}^n a_i^2.$$

Hint: Substitute for $S_n(x)$ in Eq. (6) and integrate, using the orthogonality relation (1).

(b) Show that $\sum_{i=1}^n a_i^2 \leq \int_0^1 r(x)f^2(x)\, dx$. This result is known as Bessel's inequality.

(c) Show that $\sum_{i=1}^\infty a_i^2$ converges.

(d) Show that $\lim_{n\to\infty} R_n = \int_0^1 r(x)f^2(x)\, dx - \sum_{i=1}^\infty a_i^2$.

(e) Show that $\sum_{i=1}^{\infty} a_i \phi_i(x)$ converges to $f(x)$ in the mean if and only if

$$\int_0^1 r(x) f^2(x)\, dx = \sum_{i=1}^{\infty} a_i^2.$$

This result is known as Parseval's equation. In Problems 10 through 12 let $\phi_1, \phi_2, \ldots, \phi_n, \ldots$ be the normalized eigenfunctions of the Sturm–Liouville problem (11), (12).

10. Show that if a_n is the nth Fourier coefficient of a square integrable function f, then $\lim_{n\to\infty} a_n = 0$.
Hint: Use Bessel's inequality, Problem 9(b).

11. Show that the series

$$\phi_1(x) + \phi_2(x) + \cdots + \phi_n(x) + \cdots$$

cannot be the eigenfunction series for any square integrable function.
Hint: See Problem 10.

12. Show that the series

$$\phi_1(x) + \frac{\phi_2(x)}{\sqrt{2}} + \cdots + \frac{\phi_n(x)}{\sqrt{n}} + \cdots$$

is not the eigenfunction series for any square integrable function.
Hint: Use Bessel's inequality, Problem 9(b).

13. Show that Parseval's equation in Problem 9(e) is obtained formally by squaring the series (10) corresponding to f, multiplying by the weight function r, and integrating term by term.

CHAPTER SUMMARY

Section 10.1 The Occurrence of Two-Point Boundary Value Problems

Section 10.2 Sturm–Liouville Boundary Value Problems

▶ Many of the partial differential equations that can be solved by the method of separation of variables require the solution of an eigenvalue problem that lies in the class of **regular Sturm–Liouville boundary value problems**:

$$[p(x)y']' - q(x)y + \lambda r(x)y = 0, \qquad \alpha < x < \beta,$$

$$\alpha_1 y(\alpha) + \alpha_2 y'(\alpha) = 0, \qquad \beta_1 y(\beta) + \beta_2 y'(\beta) = 0,$$

where the functions p, p', q, and r are continuous on the interval $\alpha \le x \le \beta$ and where $p(x) > 0$ and $r(x) > 0$ at all points in $\alpha \le x \le \beta$. Regular Sturm–Liouville problems constitute a special case of a general class known as **self-adjoint problems**.

▶ Properties of the eigenvalues and eigenvectors of a regular Sturm–Liouville problem are:

▶ The eigenvalues and eigenfunctions are real.
▶ Eigenfunctions belonging to distinct eigenvalues are orthogonal with respect to the weighted inner product defined by

$$(u, v)_r = \int_\alpha^\beta r(x) u(x) v(x)\, dx = 0.$$

▶ The eigenvalues form an infinite sequence and can be ordered according to increasing magnitude so that

$$\lambda_1 < \lambda_2 < \lambda_3 < \cdots < \lambda_n < \cdots.$$

Moreover, $\lambda_n \to \infty$ as $n \to \infty$.
▶ The eigenvalues are simple, that is, there is only one independent eigenfunction associated with each eigenvalue.
▶ If f and f' are piecewise continuous functions on the interval $[\alpha, \beta]$, and $\phi_1, \phi_2, \ldots$ are the eigenfunctions of a regular Sturm–Liouville problem, then the **generalized**

Fourier series expansion

$$f(x) = \sum_{n=1}^{\infty} c_n \phi_n(x)$$

with coefficients c_n given by

$$c_n = \frac{(f, \phi_n)_r}{(\phi_n, \phi_n)_r} = \frac{\int_\alpha^\beta r(x) f(x) \phi_n(x)\, dx}{\int_\alpha^\beta r(x) \phi_n(x) \phi_n(x)\, dx}, \qquad \text{for } n = 1, 2, \ldots,$$

converges to $[f(x+) + f(x-)]/2$ at each $x \in (\alpha, \beta)$. If f is continuous at x, then $[f(x+) + f(x-)]/2 = f(x)$.

Section 10.3   Nonhomogeneous Boundary Value Problems

▶ If λ_n and ϕ_n are the eigenvalues and eigenvectors, respectively, of the Sturm–Liouville problem

$$L[y] = -[p(x)y']' + q(x)y = \lambda r(x)y,$$

$$\alpha_1 y(0) + \alpha_2 y'(0) = 0, \qquad \beta_1 y(1) + \beta_2 y'(1) = 0,$$

then the associated nonhomogeneous boundary value problem

$$L[y] = -[p(x)y']' + q(x)y = \mu r(x)y + f(x),$$

$$\alpha_1 y(0) + \alpha_2 y'(0) = 0, \qquad \beta_1 y(1) + \beta_2 y'(1) = 0,$$

may be solved by assuming a solution of the form $y = \phi(x) = \sum_{n=1}^{\infty} b_n \phi_n(x)$. Substituting this expansion into the nonhomogeneous equation, using the relation $L[\phi_n] = \lambda_n r \phi_n$, and using the orthogonality conditions $\int_0^1 r(x)\phi_n(x)\phi_m(x)\,dx = 0$ for $n \neq m$ yield the following result:

$$y = \phi(x) = \begin{cases} \sum_{n=1}^{\infty} \dfrac{c_n}{\lambda_n - \mu} \phi_n(x), & \text{if } \mu \notin \{\lambda_1, \lambda_2, \ldots\}, \\[2ex] \sum_{n \neq m} \dfrac{c_n}{\lambda_n - \mu} \phi_n(x) + c\, \phi_m(x), & \text{if } \mu = \lambda_m \text{ and } \int_0^1 f(x)\phi_m(x)\,dx = 0, \end{cases}$$

where $c_n = \int_0^1 f(x)\phi_n(x)\,dx$ and c is an arbitrary constant.

If $\mu = \lambda_m$ and $\int_0^1 f(x)\phi_m(x)\,dx \neq 0$, then the nonhomogeneous problem has no solution.

▶ **Normal Mode Solutions.** The solution of a partial differential equation based on an expansion in terms of an orthogonal system of eigenfunctions (**normal modes**) arising from a Sturm–Liouville problem associated with the partial differential equation and its boundary conditions is often referred to as a **normal mode solution**. The required steps are:

▶ Set up the problem (using a simple transformation, if necessary, to transfer any nonhomogenous boundary conditions to a possibly nonhomogeneous partial differential equation) so that the boundary conditions are homogeneous:

$$r(x)u_t = [p(x)u_x]_x - q(x)u + F(x, t),$$

$$u_x(0, t) - h_1 u(0, t) = 0, \qquad u_x(1, t) + h_2 u(1, t) = 0,$$

$$u(x, 0) = f(x).$$

▶ Find the eigenvalues λ_n and the orthogonal set of eigenfunctions ϕ_n of an associated Sturm–Liouville problem:

$$L[y] = -[p(x)y']' + q(x)y = \lambda r y,$$

$$y'(0) - h_1 y(0) = 0, \qquad y'(1) + h_2 y(1) = 0.$$

▶ Substitute the series expansion $u(x,t) = \sum_{n=1}^{\infty} b_n(t)\phi_n(x)$ into the partial differential equation and combine terms using the relation, $L[\phi_n] = \lambda_n r \phi_n$ for each n, to obtain

$$\sum_{n=1}^{\infty} [b_n' + \lambda_n b_n] r \phi_n = F(x,t).$$

Orthogonality properties of the ϕ_n require that the time-dependent coefficients $b_n' + \lambda_n b_n$ in this expansion satisfy

$$b_n' + \lambda_n b_n = \gamma_n(t) = \frac{\int_0^1 F(x,t)\phi_n(x)\,dx}{\int_0^1 r(x)[\phi_n(x)]^2\,dx}.$$

▶ Initial conditions for $b_n' + \lambda_n b_n = \gamma_n(t)$ are found by substituting $u(x,t) = \sum_{n=1}^{\infty} b_n(t)\phi_n(x)$ into the initial condition $u(x,0) = f(x)$

$$\sum_{n=1}^{\infty} b_n(0)\phi_n(x) = f(x),$$

and using the orthogonality properties of the ϕ_n to obtain

$$b_n(0) = B_n = \frac{\int_0^1 r(x)f(x)\phi_n(x)\,dx}{\int_0^1 r(x)[\phi_n(x)]^2\,dx}.$$

▶ Solve the initial value problem $b_n' + \lambda_n b_n = \gamma_n(t)$, $b_n(0) = B_n$, for each n.

Section 10.4 Singular Sturm–Liouville Problems

The differential operator $L[y] = -[p(x)y']' + q(x)y$ is said to be **formally self-adjoint**. The corresponding differential equation

$$L[y] = -[p(x)y']' + q(x)y = \lambda r(x)y, \qquad 0 < x < 1,$$

is said to be in **self-adjoint form**. If

(i) p, p', q, and r are continuous on the interval $0 \leq x \leq 1$ and

(ii) $p(x) > 0$ and $r(x) > 0$ at all points in $0 \leq x \leq 1$,

except at one or both endpoints of the interval $[0,1]$, then a boundary value problem for the equation $L[y] = \lambda r(x)y$ is said to be a **singular Sturm–Liouville problem** on $(0,1)$.

▶ Conditions for singular Sturm–Liouville problems that commonly occur in applications are:

▶ Left endpoint singular, regular boundary condition at the right endpoint:

$$p(0) = 0, \qquad y'(1) + h_2 y(1) = 0,$$

▶ Right endpoint singular, regular boundary condition at the left endpoint:

$$y'(0) - h_1 y(0) = 0, \qquad p(1) = 0,$$

▶ Both endpoints singular: $p(0) = p(1) = 0$ and no boundary condition specified at $x = 0$ or $x = 1$.

▶ If there is a sufficiently large class of twice-continuously differentiable functions on the interval $(0, 1)$, such that

$$\int_0^1 \{L[u]v - uL[v]\}\, dx = 0$$

is true, in the sense of an improper integral, for all u and v in the class, then the singular Sturm–Liouville problem is **self-adjoint**. The set E of λ for which there are nonzero solutions of $L[\phi] = -[p(x)\phi']' + q(x)\phi = \lambda r(x)\phi$ in this class is called the **spectrum** of L. The set E consists of only real numbers; it may be a discrete set of points (eigenvalues); it may be a subinterval of the real line (a **continuous spectrum**); or it may be a mixture of discrete eigenvalues and a continuous spectrum.

Section 10.5   A Bessel Series Expansion

▶ The method of separation of variables applied to partial differential equations in cylindrical coordinates often leads to a singular Sturm–Liouville problem involving Bessel's equation of order p:

$$(xy')' + \left(\lambda^2 x - \frac{p^2}{x}\right) y = 0, \qquad y(0) \text{ bounded}, \quad y(1) = 0.$$

For each fixed $p \geq 0$, this problem has a discrete set of eigenvalues λ_n^2, where $\lambda_1 < \lambda_2 < \cdots$ are the positive zeros of $J_p(x)$ and corresponding eigenfunctions $\phi_n(x) = J_p(\lambda_n x)$, $n = 1, 2, \ldots$ with the following properties:

▶

$$\int_0^1 x J_p(\lambda_n x) J_p(\lambda_m x)\, dx = \begin{cases} 0, & n \neq m, \\ \frac{1}{2}[J_{p+1}(\lambda_n)]^2, & n = m; \end{cases}$$

▶ If f and f' are piecewise continuous on $[0, 1]$, then the **Bessel series expansion**

$$\sum_{n=1}^{\infty} c_n J_p(\lambda_n x),$$

where

$$c_n = \frac{2}{[J_{p+1}(\lambda_n)]^2} \int_0^1 x f(x) J_p(\lambda_n x)\, dx,$$

converges to $f(x)$ at each point where f is continuous and to $[f(x+) + f(x-)]/2$ at each point where f is discontinuous.

▶ Using these general results, if we assume a normal mode expansion

$$u(r, t) = \sum_{n=1}^{\infty} c_n(t) J_0(\lambda_n r)$$

for the solution of the initial boundary value problem

$$a^2 \left(u_{rr} + \frac{1}{r} u_r \right) = u_{tt}, \qquad 0 \leq r < 1, \quad t > 0,$$

$$u(1, t) = 0, \qquad t \geq 0,$$

$$u(r, 0) = f(r), \qquad u_t(r, 0) = 0, \qquad 0 \leq r \leq 1,$$

we obtain the following initial value problem for the mode amplitude coefficients $c_n(t)$:

$$c_n'' + \lambda_n^2 a^2 c_n = 0, \qquad c_n(0) = \frac{2}{[J_1(\lambda_n)]^2} \int_0^1 r f(r) J_0(\lambda_n r)\, dr, \qquad c_n'(0) = 0.$$

Thus, $c_n(t) = c_n(0) \cos \lambda_n a t$.

Section 10.6   Series of Orthogonal Functions: Mean Convergence

▶ A starting point that is often used to approximate a function $f(x)$ on $0 \le x \le 1$ is to assume an expansion of the form

$$S_n(x) = \sum_{i=1}^{n} a_i \phi_i(x),$$

in terms of a given set of linearly independent base functions $\phi_1(x), \phi_2(x), \dots$. Some commonly used methods for choosing the coefficients $a_1, a_2, \dots$ are:

▶ **Method of Collocation.** Solve the linear algebraic system

$$\sum_{i=1}^{n} a_i \phi_i(x_j) = f(x_j), \qquad j = 1, \dots, n,$$

where $0 \le x_1 < x_2 < \cdots < x_n \le 1$.

▶ Minimizing the **mean square error**

$$R_n(a_1, \dots, a_n) = \int_0^1 r(x)[f(x) - S_n(x)]^2\, dx$$

leads to the linear algebraic system

$$\sum_{i=1}^{n} \int_0^1 r(x)\phi_i(x)\phi_j(x)\, dx\, a_i = \int_0^1 r(x)f(x)\phi_j(x)\, dx, \qquad j = 1, \dots, n.$$

If

$$\int_0^1 r(x)\phi_i(x)\phi_j(x)\, dx = \begin{cases} 0, & i \ne j, \\ 1, & i = j, \end{cases}$$

then

$$a_i = \int_0^1 r(x)f(x)\phi_i(x)\, dx, \qquad i = 1, \dots, n.$$

▶ If

$$\lim_{n \to \infty} R_n(a_1, \dots, a_n) = \lim_{n \to \infty} \int_0^1 r(x)[f(x) - S_n(x)]^2\, dx = 0,$$

then $\sum_{i=1}^{\infty} a_i \phi_i(x)$ **converges in the mean** to $f(x)$.

▶ The eigenfunctions of any regular Sturm–Liouville problem on $(0, 1)$ are **complete** with respect to mean square convergence in the space of square-integrable functions on $(0, 1)$.

PROJECTS

Project 1 Dynamic Behavior of a Hanging Cable

One end of a long cable, or heavy rope, with linear mass density μ and total length L is attached to a fixed point on an overhead beam. The other end is allowed to hang freely (Figure 10.P.1). In this project we derive a wave equation that describes the propagation of small amplitude pulses along the cable, and then solve an initial boundary value problem for the equation using the method of separation of variables.

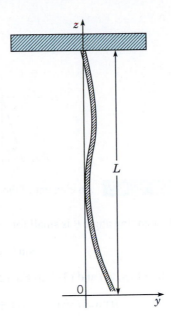

FIGURE 10.P.1   A cable of length L hanging freely from an overhead beam.

Assume a Cartesian coordinate system in which the positive z-axis points vertically upward and the positive y-axis points to the right. If we further assume that the origin coincides with the free end of the cable in its equilibrium state, then the coordinates of the point of attachment to the overhead beam are $(y, z) = (0, L)$. In the limiting case of small horizontal displacements from equilibrium, vertical motions of the material elements of the cable are negligible. Therefore, we denote horizontal displacement of the material elements by $y(z, t)$. Since tension in the cable acts mainly in the vertical direction in the limit of small displacements from equilibrium, tension T in the cable, due to its own weight, is given by

$$T(z) = \mu g z. \tag{1}$$

Thus, tension increases linearly from zero at the free end of the cable to $\mu g L$ at the upper end.

The differential equation governing the horizontal displacement of the cable[10] is derived by considering the forces acting on a small element of the cable of length Δz lying between the points z and $z + \Delta z$ (see Figure 10.P.2). Since the horizontal components of the tension on the cable element of length Δz in the right and left directions are $T(z + \Delta z) \sin(\theta + \Delta\theta)$ and $T(z) \sin\theta$, respectively, Newton's law applied to the element is

$$\mu y_{tt}(z, t)\Delta z = T(z + \Delta z) \sin(\theta + \Delta\theta) - T(z) \sin\theta. \tag{2}$$

[10]See Appendix 9.B for a more detailed derivation of the wave equation.

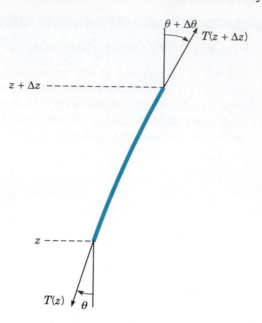

FIGURE 10.P.2 An element of the displaced cable.

Since the bending angle θ is small for small displacements from equilibrium, it follows that

$$\sin \theta \cong \tan \theta = y_z(z, t). \tag{3}$$

Substituting Eqs. (1) and (3) into Eq. (2) gives

$$\mu y_{tt}(z, t)\Delta y = \mu g(z + \Delta z)y_z(z + \Delta z, t) - \mu g z y_z(z, t). \tag{4}$$

Dividing Eq. (4) by $\mu \Delta z$ and passing to the limit as $\Delta z \to 0$ then yields

$$y_{tt} = g(zy_z)_z. \tag{5}$$

Boundary Conditions. Since the upper end of the cable is attached to an overhead beam, no horizontal displacement can occur at $z = L$. Thus, the appropriate boundary condition at the upper end of the cable is

$$y(L, t) = 0, \qquad t \geq 0. \tag{6}$$

On the other hand, the lower end of the freely hanging cable can experience horizontal displacement that, in general, will depend on t. Although we do not have *a priori* knowledge of the time dependence of this displacement, we do know that $|y(0, t)|$ cannot exceed the length L of the cable. Therefore, we impose the condition

$$|y(0, t)| < \infty, \qquad t \geq 0, \tag{7}$$

at $z = 0$.

Initial Conditions. If we note that the differential equation (5) is of second order with respect to t, it is necessary to specify the initial displacement profile of the cable:

$$y(z, 0) = f(z), \qquad 0 \leq z \leq L, \tag{8}$$

and its initial velocity profile:

$$y_t(z, 0) = g(z), \qquad 0 \leq z \leq L, \tag{9}$$

where f and g are prescribed functions. In order for Eqs. (8) and (9) to be consistent with the boundary conditions, and to make physical sense, we require that

$$|f(0)| < \infty, \quad |g(0)| < \infty, \quad f(L) = g(L) = 0. \tag{10}$$

In the following problems, we ask you to use the method of separation of variables to solve the initial boundary value problem consisting of Eqs. (5–10).

Project 1 PROBLEMS

CAS 1. (a) Assume solutions to Eq. (5) of the form $y(z, t) = Z(z)T(t)$. Anticipating that the separation constant is negative by writing it as $-\nu^2$ with $\nu > 0$, show that $Z(z)$ and $T(t)$ must satisfy the differential equations

$$T'' + \nu^2 g T = 0, \tag{i}$$

$$z Z'' + Z' + \nu^2 Z = 0. \tag{ii}$$

(b) Show that the boundary conditions for Eq. (ii) are

$$|Z(0)| < \infty, \quad Z(L) = 0. \tag{iii}$$

(c) Show that if we introduce the independent variable $x = \sqrt{z/L}$, then Eq. (ii) becomes

$$Z''(x) + \frac{1}{x} Z'(x) + \lambda^2 Z(x) = 0, \quad 0 < x < 1, \tag{iv}$$

where $\lambda = 2\nu\sqrt{L}$.

(d) Show that if we make the substitution $\xi = \lambda x$ in Eq. (iv), we obtain

$$Z''(\xi) + \frac{1}{\xi} Z'(\xi) + Z(\xi) = 0, \tag{v}$$

which is Bessel's equation of order 0. The general solution of Eq. (v) for $\xi > 0$ is

$$Z = c_1 J_0(\xi) + c_2 Y_0(\xi),$$

where J_0 and Y_0 are Bessel functions of the first and second kinds, respectively, of order zero (see Section 10.4). Therefore, in terms of z, the general solution of Eq. (ii) is

$$Z = c_1 J_0(\lambda\sqrt{z/L}) + c_2 Y_0(\lambda\sqrt{z/L}).$$

Using the discussion of the vibrating circular membrane in Section 10.5 as a guide, explain why the boundary conditions (iii) imply that the eigenvalues of the singular Sturm–Liouville problem (ii) and (iii) are $\nu_n^2 = \lambda_n^2/(4L)$, $n = 1, 2, 3, \ldots,$ where $0 < \lambda_1 < \lambda_2 < \cdots < \lambda_n < \cdots$ are the positive zeros of J_0, and

the corresponding eigenfunctions are $\phi_n(z) = J_0(\lambda_n\sqrt{z/L})$. Then conclude that $y(z, t)$ can be expressed as an infinite series

$$y(z, t) = \sum_{n=1}^{\infty} [a_n \cos(\nu_n\sqrt{g}\, t)\, J_0(\lambda_n\sqrt{z/L})$$
$$+ b_n \sin(\nu_n\sqrt{g}\, t)\, J_0(\lambda_n\sqrt{z/L})]. \tag{vi}$$

(e) Show that the terms in the series (vi) can be written in the form

$$y(z, t) = \sum_{n=1}^{\infty} A_n \cos(\nu_n\sqrt{g}\, t - \delta_n)\, J_0(\lambda_n\sqrt{x/L}), \tag{vii}$$

where $A_n = \sqrt{a_n^2 + b_n^2}$, $\cos \delta_n = a_n/\sqrt{a_n^2 + b_n^2}$, and $\sin \delta_n = b_n/\sqrt{a_n^2 + b_n^2}$.

(f) Equation (vii) represents the horizontal displacement of the cable as a superposition of **standing waves**. The nth standing wave, $A_n \cos(\nu_n\sqrt{g}\, t - \delta_n)\, J_0(\lambda_n\sqrt{z/L})$, has a shape $J_0(\lambda_n\sqrt{z/L})$ that oscillates with time-varying amplitude $A_n \cos(\nu_n\sqrt{g}\, t - \delta_n)$. Plot the graph of the shape of each of the first four standing waves as a function of z/L for $0 \le z \le L$. If $L = 100$ ft, determine the frequency and period of each of the first four standing waves.

(g) If your computer algebra system contains the appropriate functions, visualize the dynamic behavior of the first four standing waves by creating for each one an animation of the oscillations over a time interval consisting of several periods.

As discussed in Section 10.4, the eigenfunctions $\phi_n(x) = J_0(\lambda_n x)$ of the singular Sturm–Liouville problem

$$(xy')' + \lambda^2 x y, \quad 0 < x < 1,$$

$$y, \ y' \text{ bounded as } x \to 0, \quad y(1) = 0,$$

have the same properties as the eigenfunctions of the regular Sturm–Liouville problems discussed in

THEOREM 10.P.1

Let $\lambda_1 < \lambda_2 < \cdots$ be the positive zeros of $J_0(x)$. Then the set of functions $J_0(\lambda_1 x)$, $J_0(\lambda_2 x)$, ... satisfy the orthogonality conditions

$$\int_0^1 x J_0(\lambda_i x) J_0(\lambda_j x)\, dx = \begin{cases} 0, & i \neq j, \\ \frac{1}{2}[J_1(\lambda_i)]^2, & i = j. \end{cases} \qquad \text{(i)}$$

Each piecewise continuous function f in the interval $[0, 1]$ can be written uniquely in the form

$$f(x) = \sum_{n=1}^{\infty} f_n J_0(\lambda_n x), \qquad \text{(ii)}$$

where the coefficients f_n are given by

$$f_n = \frac{2}{[J_1(\lambda_n)]^2} \int_0^1 x f(x) J_0(\lambda_n x)\, dx \qquad \text{(iii)}$$

and the series (ii) converges in the mean to f. Moreover, if f' is also piecewise continuous on $[0, 1]$, then this series converges pointwise to $f(x)$ at all points where f is continuous, and to

$$\frac{1}{2}[f(x+) + f(x-)]$$

at all points where f is discontinuous. The convergence is uniform on every closed subinterval of $(0, 1)$ that does not contain a point of discontinuity of f.

Section 10.2. Theorem 10.P.1, a special case of a much more general theorem concerning Bessel functions J_p for any $p \geq 0$, summarizes those results that can be used to help justify the series representation in Eq. (vi), and that we will use to complete the solution of the initial boundary value problem specified in Eqs. (5–10).

2. **(a)** Use Theorem 10.P.1 to show that the eigenfunctions $\phi_n(z) = J_0(\lambda_n \sqrt{z/L})$ of the singular Sturm–Liouville problem (ii) and (iii) in Problem 1 satisfy the orthogonality relations

$$\int_0^L \phi_i(z)\phi_j(z)\, dz = \begin{cases} 0, & i \neq j, \\ L[J_1(\lambda_i)]^2, & i = j. \end{cases} \qquad \text{(i)}$$

(b) Assuming that f and g are continuous and that f' and g' are piecewise continuous in $[0, L]$, show that the solution of the initial value problem (5–10) is

$$y(z, t) = \sum_{n=1}^{\infty} [a_n \cos(\nu_n \sqrt{g}\, t)\, J_0(\lambda_n \sqrt{z/L})$$
$$+ b_n \sin(\nu_n \sqrt{g}\, t)\, J_0(\lambda_n \sqrt{z/L})], \qquad \text{(ii)}$$

where

$$a_n = \frac{1}{L[J_1(\lambda_n)]^2} \int_0^L f(z) J_0(\lambda_n \sqrt{z/L})\, dz$$

and

$$b_n = \frac{1}{\nu_n L \sqrt{g}[J_1(\lambda_n)]^2} \int_0^L g(z) J_0(\lambda_n \sqrt{z/L})\, dz.$$

Project 2 Advection Dispersion: A Model for Solute Transport in Saturated Porous Media

Differential equations are used to model transport of solutes (inorganic or organic constituents) through porous materials. A **porous material**, or **porous medium**, consists of solid material (sometimes referred to as a solid matrix) permeated by an interconnected network of pores (voids or channels), filled with liquid or gas (see Figure 10.P.3). Porous materials commonly encountered in hydrogeology are sand, silt, clay, gravel, and rocks.

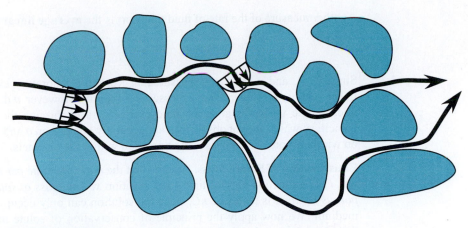

FIGURE 10.P.3 Fluid particles can take different paths and experience changes in velocity as they move through a porous medium.

Certain biological tissues (for example, bones and kidneys), cements, and foams are also treated as porous media.

Porosity. A porous medium consists of solid material and void space. If, in an element of porous material, the total void space is denoted by V_v and the total volume of solid material is denoted by V_s, then the **porosity** n of the material is defined by $n = V_v/V_T$, where $V_T = V_s + V_v$ is the total volume of the element. Porosity, usually expressed as a decimal fraction or a percent, can range from near zero to 0.6. For example, values of n for sand range from 0.25 to 0.5, whereas values for clay range from 0.3 to 0.6. If the void space is filled with fluid in its liquid phase, such as water, we say that the porous medium is **saturated**.

▶ **Derivation of the One-Dimensional Advection-Transport Equation.** Consider a cylindrical column of uniform cross-sectional area A that is filled with a homogeneous, isotropic, porous material with constant porosity n (see Figure 10.P.4). Assume that the medium is saturated with a fluid that is in a state of steady uniform flow, and that a nonreactive solute (such as a tracer) is dissolved in the fluid. If the volumetric flow rate (with physical dimensions length³/time) of solution through the column of porous media is denoted by Q and the cross-sectional area of the column is A, then one measure of the velocity of fluid flow through the porous material is the **specific discharge**[11] v_s through the cylinder, defined by $v_s = Q/A$.

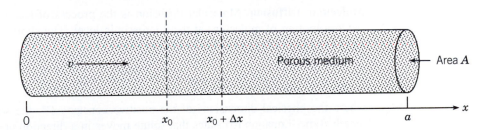

FIGURE 10.P.4 Fluid with constant average linear velocity v flowing through a cylindrical column of porous material.

---

[11] The specific discharge is also called the **Darcy velocity** or **Darcy flux**.

Another measure of the rate of fluid transport is the **average linear velocity** v, defined by

$$v = \frac{Q}{nA} = \frac{v_s}{n}. \tag{1}$$

Since Q, A, and n are easily measured, at least in a laboratory experiment, the average linear velocity v is regarded as a measurable quantity. However, v does not represent actual microscopic velocities at which fluid particles travel through the pore channels. These velocities, which are not observable, are generally larger than v and are strongly dependent on width of pore channels as well as position within the channels.

Conservation of Solute Mass. Assume that the *mass of solute per unit volume of solution* $c(x, t)$ is a function only of time t and position x. The *mass of solute per unit volume of porous medium* is therefore nC, since the solution can only occupy the void space of the medium. We now apply the principle of conservation of solute mass to the representative elementary volume (REV) lying between planar interfaces, orthogonal to the axis of the cylindrical column, at x_0 and $x_0 + \Delta x$. Since the solute is nonreactive, the only way that the solute can appear or disappear in the REV is by entering or leaving through the interfaces at x_0 and $x_0 + \Delta x$. We therefore introduce the flux density $J(x, t)$, which gives the net rate per unit area at which c is crossing x at time t. Note that the physical dimensions of J are mass/(length^2-time). If $J(x, t) > 0$, then net solute movement occurs from left to right and vice versa. The mathematical expression that quantifies the principle of mass conservation is

$$\frac{d}{dt} \int_{x_0}^{x_0 + \Delta x} ncA\, dx = J(x_0, t)A - J(x_0 + \Delta x, t)A, \tag{2}$$

that is, the time rate of change of solute mass within the REV equals the rate at which it crosses the planes at x_0 and $x_0 + \Delta x$. Dividing Eq. (2) by $A\Delta x$ and letting $\Delta x \to 0$ yield

$$n\frac{\partial c}{\partial t} = -\frac{\partial J}{\partial x}. \tag{3}$$

We now consider contributions to J due to three distinct physical processes: advection, diffusion, and mechanical dispersion.

Advection. Advection refers to solute movement due to the bulk motion of the fluid flow. The solute is carried along by the surrounding fluid. If the average linear velocity is v, which we assume to be constant, the flux density due to advection is given by

$$J_a = \frac{Qc}{A} = nvc. \tag{4}$$

Molecular Diffusion. Molecular diffusion is the process of solute transport due to the random motion of molecules at the microscopic level. The simplest expression for flux density due to molecular diffusion is **Fick's law**:

$$J_{\text{diff}} = -D^* \frac{\partial nc}{\partial x} = -nD^* \frac{\partial c}{\partial x}, \tag{5}$$

where the physical dimensions of the positive constant D^*, the **solute diffusivity**, are length^2/time. Equation (5) states that solute moves in a direction opposite to the concentration gradient, that is, from regions of relatively high concentration toward regions of relatively low concentration. Note that we are using solute mass per unit volume nc in Eq. (5). To see that the linear dependence on n in Eq. (5) is necessary, consider what happens if n were reduced by a factor of 1/2. Halving the void space reduces the amount of solute per unit volume by a factor of 1/2. If the number of solute molecules per unit volume is reduced by a factor of 1/2, then the flux density due to molecular diffusion must also be reduced by a factor of 1/2.

Mechanical Dispersion. Mechanical, or hydrodynamic, dispersion refers to the tendency of the solute to spread due to the array of pore channel pathways available to the fluid as it moves through the porous medium. In effect, the pore network mixes the solute as the fluid flows through the medium. At the microscopic level, there are three mechanisms responsible for mechanical dispersion: (1) drag exerted on the fluid by the roughness of the pore channel boundaries causes molecules near those boundaries to travel more slowly than molecules near the center of the channel, (2) bulk fluid velocity within a pore channel depends on the channel diameter and the roughness of the boundary, and (3) mixing due to the tortuosity, branching, and inter-fingering of the pore channels. Spreading of the solute in the direction of the bulk flow is known as *longitudinal dispersion*. A commonly used empirical formula for flux density due to mechanical dispersion is

$$J_{\text{disp}} = -D_x \frac{\partial nc}{\partial x} = -nD_x \frac{\partial c}{\partial x}, \tag{6}$$

where the **dispersion coefficient** $D_x = \alpha_x |v|$ has physical dimensions of length^2/time. The constant of proportionality α_x is called the **longitudinal dispersivity**. Although Eq. (6) has the same form as Fick's law, the underlying physics of molecular diffusion are quite different from the underlying physics of hydrodynamic dispersion. It is reasonable to assume that solute is transported, via the mechanisms described, from regions of relatively high concentration toward regions of relatively low concentration. It is also observed that dispersion increases with the magnitude of average linear velocity. We therefore accept Eq. (6) as a simple, but plausible model for a very complex transport process, with the understanding that more complex formulas for J_{disp} could be considered.[12]

Adding flux densities due to advection, diffusion, and dispersion gives the total flux density

$$J = J_a + J_{\text{diff}} + J_{\text{disp}} = nvc - nD_1 \frac{\partial c}{\partial x}, \tag{7}$$

where $D_1 = D^* + D_x$. The value of D^* is usually negligible compared to $D_x = \alpha_x |v|$ whenever v is nonzero. Substituting the right-hand side of Eq. (7) for J in Eq. (3) then yields the one-dimensional advection-dispersion equation

$$c_t + vc_x = D_1 c_{xx}. \tag{8}$$

The term vc_x is referred to as the **advective term**, while $D_1 c_{xx}$ is referred to as the **dispersion term**. Generally, advection is the more important transport process. In the problems below you are asked to find solutions of this equation in different experimental scenarios.

▶ Advection Dispersion in Two Dimensions. In addition to longitudinal dispersion, mechanical mixing also causes spreading of the solute in directions perpendicular to the flow. However, spreading in directions perpendicular to the flow, called **transverse dispersion**, is normally much smaller than longitudinal dispersion. The advection-diffusion equation must be extended to two or three space dimensions to incorporate the effects of transverse dispersion. In two dimensions, the advection-diffusion equation follows from applying the principle of mass balance to a small rectangular region in the xy-plane (see Problem 3). In this case, the

---

[12]Dispersion, as a process of fluid mixing, is an ongoing topic of research.

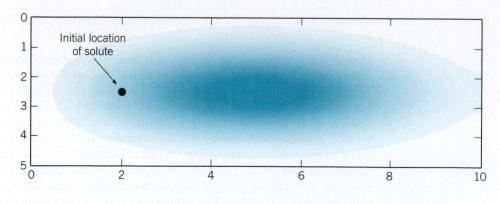

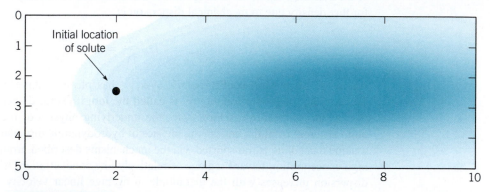

FIGURE 10.P.5 Computer images at two successive times of the solution of the initial boundary value problem described by Eqs. (i–iv) in Problem 4, in which $D_1/D_2 = 10$.

differential equation for the mass of solute per unit volume of solution $c(x, y, t)$ is

$$\frac{\partial c}{\partial t} + v\frac{\partial c}{\partial x} = D_1\frac{\partial^2 c}{\partial x^2} + D_2\frac{\partial^2 c}{\partial y^2}, \tag{9}$$

where $D_1 = D^* + \alpha_x|v|$ and $D_2 = D^* + \alpha_y|v|$. The dispersion coefficient in the direction of flow α_x is much larger than the dispersion coefficient transverse to the flow α_y. Therefore, in Eq. (9), D_1 is much larger than D_2. Consequently, the solute spreads more quickly in the direction of flow than it does in the direction transverse to the flow (see Figure 10.P.5).

Project 2 PROBLEMS

 1. An amount M of nonreactive tracer is uniformly distributed across a thin section of a cylindrical column of a granular medium at a location x_0. The initial boundary value problem for the tracer concentration $c(x, t)$ is

$$c_t + vc_x = D_1c_{xx}, \qquad 0 < x < a, \tag{i}$$

$$c(0, t) = 0, \quad c_x(a, t) = 0, \qquad t > 0, \tag{ii}$$

$$c(x, 0) = \frac{M}{A}\delta(x - x_0), \qquad 0 < x, \ x_0 < a. \tag{iii}$$

(a) Assume that $c(x, t) = \phi(x)T(t)$ and use the method of separation of variables to show that the eigenvalue problem for ϕ can be written as the Sturm–Liouville problem

$$[p(x)\phi']' + \lambda r(x)\phi = 0, \qquad 0 < x < a, \tag{iv}$$

$$\phi(0) = 0, \quad \phi'(a) = 0, \tag{v}$$

where $p(x) = r(x) = \exp(-vx/D_1)$.

(b) Show that the eigenfunctions of the Sturm–Liouville problem (iv–v) are

$$\phi_n(x) = e^{vx/2D_1} \sin\sqrt{\lambda_n - v^2/4D_1^2}\, x,$$

$$n = 1, 2, \ldots,\qquad\qquad\text{(vi)}$$

where the eigenvalues λ_n are solutions of

$$F(\lambda) = \frac{v}{2D_1}\sin\left(\sqrt{\lambda - v^2/4D_1^2}\, a\right)$$

$$+ \sqrt{\lambda - v^2/4D_1^2}\cos\left(\sqrt{\lambda - v^2/4D_1^2}\, a\right) = 0,$$

$$\text{(vii)}$$

or equivalently,

$$\tan\left(\sqrt{\lambda - v^2/4D_1^2}\, a\right) = -\frac{2D_1}{v}\sqrt{\lambda - v^2/4D_1^2}.$$

$$\text{(viii)}$$

(c) Show graphically that Eq. (viii) has a countably infinite number of positive solutions, and that for large values of n,

$$\lambda_n \cong \frac{v^2}{4D_1^2} + \left[\frac{(2n-1)\pi}{2a}\right]^2.\qquad\text{(ix)}$$

(d) Show that

$$\|\phi_n\|^2 = \int_0^a r(x)\phi_n^2(x)\,dx = \frac{a}{2} + \frac{v}{4\lambda_n D_1}.$$

$$\text{(x)}$$

[*Hint:* Use Eq. (viii) to show that $\cos^2\left(\sqrt{\lambda - v^2/4D_1^2}\, a\right) = v^2/\left(4\lambda_n D_1^2\right)$.]

(e) Show that an orthogonal series representation of the solution to Eqs. (i–iii) is

$$c(x, t) = \frac{M}{A}\sum_{n=1}^{\infty}\frac{\phi_n(x_0)}{\|\phi_n\|^2}e^{-D_1\lambda_n t}\phi_n(x).\qquad\text{(xi)}$$

(f) On the same set of coordinate axes, plot graphs of numerical approximations of $c(x, t)$ versus x at the times $t_1 = 0.05$, $t_2 = 0.4$, $t_3 = 1$, and $t_4 = 6$. Use the following parameter values: $M/A = 1$, $v = 1$ m/day, $D_1 = 0.5$ m^2/day, $x_0 = 3$ m, and $a = 10$ m. You will need to numerically determine the solutions of Eq. (vii). You should sum at least 60 terms of the series solution to achieve reasonable accuracy. Experiment to see what happens if too few terms in the orthogonal series are used to approximate the solution.

(g) Compare the approximate solutions in part **(f)** with

$$\hat{c}(x, t) = \frac{M}{A}\frac{1}{\sqrt{4\pi D_1 t}}e^{-(x-x_0-vt)^2/(4D_1 t)},$$

the exact solution to the initial value problem for a cylinder of infinite length:

$$c_t + vc_x = D_1 c_{xx},\qquad c(0, t) = \frac{M}{A}\delta(x - x_0),$$

$$-\infty < x < \infty.$$

2. A nonreactive tracer at concentration c_0 is continuously introduced into a steady-state flow regime at the upstream end of a column packed with a homogeneous granular medium. Assume that the tracer concentration in the column prior to the introduction of the tracer is zero. The boundary value problem that models this experiment is **CAS**

$$c_t + vc_x = D_1 c_{xx},\qquad 0 < x < a,\ t > 0,\qquad\text{(i)}$$

$$c(0, t) = c_0,\qquad c_x(a, t) = 0,\qquad t > 0,\qquad\text{(ii)}$$

$$c(x, 0) = 0,\qquad 0 < x,\ x_0 < a.\qquad\text{(iii)}$$

(a) Set $c(x, t) = c_0 - u(x, t)$ in (i–iii) and find the initial boundary value problem satisfied by $u(x, t)$. Use the eigenfunctions found in Problem 1 to find an orthogonal series representation for $u(x, t)$.

(b) Use a computer software package to plot the graphs of $c(x, t)$ versus x at the times $t_1 = 1$, $t_2 = 2$, $t_3 = 4$, and $t_4 = 6$. Use the following parameter values: $c_0 = 1$, $v = 1$ m/day, $D_1 = 0.5$ m^2/day, and $a = 10$ m.

3. Confining ourselves to two dimensions, denote by $c(x, y, t)$ the mass of solute per unit volume of solution in a porous medium. Assume that the average linear flow velocity is $v\,\mathbf{i} + 0\,\mathbf{j}$, where v is constant, and that the solute transport flux density vector is $\mathbf{J} = (nvc - nD_1 c_x)\,\mathbf{i} - nD_2 c_y\,\mathbf{j}$. Apply the principle of solute mass conservation to the REV in Figure 10.P.6 (assume the REV is of unit thickness in the z-direction) to derive the two-dimensional advection-diffusion equation

$$c_t + vc_x = D_1 c_{xx} + D_2 c_{yy}.$$

4. Consider the following initial boundary value problem for a confined two-dimensional aquifer:

$$c_t + vc_x = D_1 c_{xx} + D_2 c_{yy},$$
$$0 < x < a,\ 0 < y < b,\ t > 0,\qquad\text{(i)}$$

$$c(0, y, t) = 0,\qquad c_x(a, y, t) = 0,$$
$$0 < y < b,\ t > 0\qquad\qquad\text{(ii)}$$

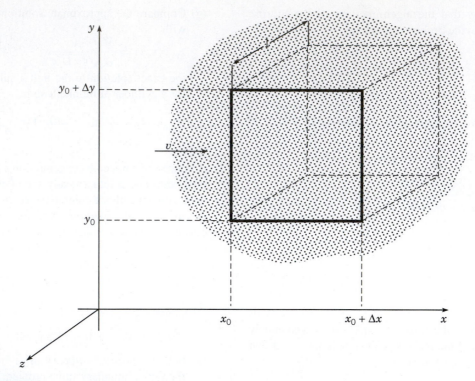

FIGURE 10.P.6   A representative elementary volume used to derive $c_t + vc_x = D_1 c_{xx} + D_2 c_{yy}$.

$$c_y(x, 0, t) = 0, \quad c_y(x, b, t) = 0,$$

$$0 < x < a, \ t > 0 \qquad \text{(iii)}$$

$$c(x, y, 0) = M\delta(x - x_0)\delta(y - y_0),$$

$$0 < x, x_0 < a, \ 0 < y, y_0 < b, \qquad \text{(iv)}$$

where M is mass per unit length of solution in the z-direction.

(a) Show that each term in the biorthogonal series

$$c(x, y, t) = \sum_{n=1}^{\infty}\sum_{m=0}^{\infty} c_{nm} e^{-(\lambda_n D_1 + \nu_m D_2)t} \phi_n(x)\psi_m(y)$$

$$\text{(v)}$$

satisfies Eqs. (i–iii), where λ_n and ϕ_n are the eigenvalues and eigenfunctions found for the

Sturm–Liouville problem (iv–v) of Problem 1, and ν_m and ψ_m are the eigenvalues and eigenfunctions for the problem

$$\psi'' + \nu\psi = 0, \qquad \psi'(0) = \psi'(b) = 0.$$

(b) To satisfy the initial condition (iv) in part **(a)**, show that it is necessary to choose the coefficients c_{nm} such that

$$c_{nm} = M\frac{\phi_n(x_0)\psi_m(y_0)}{||\phi_n||^2 ||\psi_m||_1^2},$$

where $||\phi_n||^2$ is given by Eq. (x) in Problem 1 and

$$||\psi_m||_1^2 = \int_0^b \psi^2(y)\,dy = \begin{cases} b, & m = 1, \\ \frac{1}{2}b, & m \geq 2. \end{cases}$$

Project 3 Fisher's Equation for Population Growth and Dispersion

Recall from Section 2.5 the **logistic equation** for population growth,

$$\frac{du}{dt} = r(1 - u)u, \qquad \text{(1)}$$

where $r > 0$ is called the **intrinsic growth rate**, and we have set the carrying capacity K equal to 1 [see Eq. (7), Section 2.5]. From the general solution of Eq. (1),

$$u = \frac{C}{C + (1 - C)e^{-rt}},\tag{2}$$

or by using geometric phase line analysis, it follows that $u = \hat{u}_0(t) = 0$ is an unstable equilibrium solution, whereas $u = \hat{u}_1(t) = 1$ is an asymptotically stable equilibrium solution. On the other hand, if $r < 0$, $\hat{u}_0$ is asymptotically stable, while $\hat{u}_1$ is unstable.

▶ **Linear Stability Analysis.** An alternative analytical approach to ascertaining stability properties of an equilibrium solution of Eq. (1) is to examine whether any small perturbation v to an equilibrium solution $\hat{u}$ tends to grow or decay in time. Substituting $u = \hat{u} + v$ into Eq. (1) yields

$$\frac{dv}{dt} = r(1 - \hat{u} - v)(\hat{u} + v) = r(1 - \hat{u})\hat{u} + r(1 - 2\hat{u})v - rv^2\tag{3}$$

or

$$\frac{dv}{dt} = r(1 - 2\hat{u})v - rv^2,\tag{4}$$

where we have used the fact that an equilibrium solution $\hat{u}$ satisfies $r(1 - \hat{u})\hat{u} = 0$. Equation (4) is nonlinear due to the presence of the quadratic term rv^2. Note that $v = u - \hat{u}$ is a measure of the departure of u from the equilibrium solution. Since we are assuming that v is a small perturbation, the term rv^2, which contains the product of two small terms, should be negligible in comparison with the other terms in Eq. (4). If we delete the nonlinear term, so only terms that are linear in v or its derivative v_t are retained, we obtain the **linearization** of Eq. (4):

$$\frac{dv}{dt} = r(1 - 2\hat{u})v.\tag{5}$$

The general solution to Eq. (5) in the case that $\hat{u} = \hat{u}_0(t) = 0$ is $v = v_0 e^{rt}$. Thus, if $r < 0$, $v(t) \to 0$ as $t \to \infty$ for any initial condition v_0, and we conclude that $\hat{u}_0$ is asymptotically stable, based on this linear stability test. On the other hand, if $r > 0$ and $v_0 \neq 0$, $|v(t)| \to \infty$ as $t \to \infty$. We conclude that there are initial conditions for which the perturbation v grows as $t \to \infty$. In this case we classify $\hat{u}_0$ as unstable. A similar analysis can be applied to $\hat{u}_1$.

▶ **Fisher's Equation.** A simple extension to Eq. (1) that allows for unidirectional spatial movement of members of the population, in addition to growth and decay, is

$$\frac{\partial u}{\partial t} = D\frac{\partial^2 u}{\partial x^2} + r(1 - u)u,\tag{6}$$

where we have included the diffusion term $D\partial^2 u/\partial x^2$. The dependent variable $u(x, t)$ now depends on spatial position x as well as time t, and is interpreted to be the concentration of the population. Equation (6), known as Fisher's equation in population biology, is also a special case of a reaction-diffusion equation. In the latter case $u(x, t)$ represents the concentration of a chemical species and $r(1 - u)u$ is referred to as the reaction term. It is convenient to think of $u(x, t)$ as the concentration of a population of particles (number of particles per unit length) that are transported by diffusion and can increase or decrease in number as modeled by the reaction term, with the understanding that the particles can be, for example, bacteria or molecules.

Although Eq. (6) may be studied on the real line $-\infty < x < \infty$, where its solutions exhibit interesting behavior such as traveling waves, we will restrict our attention in this project to the interval $0 \le x \le 1$. We choose to impose the boundary conditions

$$u(0, t) = 0, \qquad u(1, t) = 0, \qquad t > 0. \tag{7}$$

Physically, these boundary conditions correspond to particles exiting the domain $(0, 1)$ when they reach either boundary point, $x = 0$ or $x = 1$. In summary, the model represented by Eqs. (6) and (7) describes a population of particles that can grow or decay in number, and experience transport by diffusion, in the domain, $0 < x < 1$. Whenever a particle reaches a boundary point, it is removed from the population.

Equilibrium solutions of Eqs. (6) and (7) must satisfy the nonlinear two-point boundary value problem

$$D\frac{\partial^2 u}{\partial x^2} + r(1 - u)u = 0, \qquad u(0) = 0, \quad u(1) = 0. \tag{8}$$

By inspection, we see that $u = \hat{u}_0(x, r) = 0$ satisfies Eq. (8), and is therefore an equilibrium solution of Eqs. (6) and (7) for all values of r. There may be others that are not as easy to find. In the problems below, we use linear stability analysis to study the stability properties of $\hat{u}_0$ with respect to the intrinsic growth rate r. As a consequence of this analysis, we will infer the existence of additional equilibrium solutions to Eqs. (6) and (7). Numerical methods can then be used to find approximations to these solutions and to infer whether they are asymptotically stable or unstable.

Project 3 PROBLEMS

In Problems 1 and 2 we ask you to apply linear stability analysis to the equilibrium solution $\hat{u}_0$ of Eqs. (6) and (7).

1. Assume a solution to Eqs. (6) and (7) of the form $u = \hat{u}_0(x, r) + v(x, t)$, where $\hat{u}_0(x, r) = 0$ and v is a small perturbation. By substituting $\hat{u}_0 + v$ for u in Eqs. (6) and (7), and retaining only terms that are linear in v and its derivatives, show that the resulting linearized problem for v is

$$\frac{\partial v}{\partial t} = D\frac{\partial^2 v}{\partial x^2} + rv, \qquad 0 < x < 1, \tag{i}$$

$$v(0, t) = 0, \qquad v(1, t) = 0, \qquad t > 0. \tag{ii}$$

2. **(a)** Use the method of separation of variables to show that the solution of Eqs (i) and (ii) in Problem 1, subject to the initial condition $v(x, 0) = f(x)$, is

$$v(x, t) = \sum_{n=1}^{\infty} v_n e^{-(\lambda_n - r)t} \sin n\pi x, \tag{i}$$

where $\lambda_n = n^2\pi^2 D$ and $v_n = 2\int_0^1 f(x)\sin n\pi x\,dx$, for $n = 1, 2, 3, \ldots$.

(b) Explain why the solution (i) in part **(a)** implies that $\hat{u}_0(\cdot, r)$ is asymptotically stable (according to linear stability theory) if $r < \lambda_1$, but unstable

for $r > \lambda_1$. Discuss the dependence of stability properties of $\hat{u}_0(\cdot, r)$ on r as D varies from nearly 0 to ∞.

If an equilibrium solution is unstable and, since disturbances are inevitable, the following question arises. Will a disturbance to an unstable equilibrium state evolve into some other equilibrium state, or into a dynamic state? To answer this question, and to confirm the results of the linear stability test in Problem 2, in Problem 3 we ask you to compute the long time behavior of disturbances to $\hat{u}_0(\cdot, r)$ for values of r both less than λ_1 and greater than λ_1.

3. **(a)** **Method of Lines.** An easy to implement numerical method for approximating solutions of initial boundary value problems associated with Eqs. (6) and (7) is obtained by replacing the partial derivative on the right side of Eq. (6) by a difference quotient,[13]

$$\frac{\partial^2 u}{\partial x^2}(x, t) \cong \frac{u(x + h, t) - 2u(x, t) + u(x - h, t)}{h^2}, \tag{i}$$

to obtain a discrete space, continuous time approximation method. Let $0 = x_0 < x_1 < x_2 < \cdots < x_N < x_{N+1} = 1$ be a grid of $N + 2$ equally spaced points in the interval $[0, 1]$, where $x_n = nh$ for $n = 0, \ldots, N + 1$ with

---

[13] See Problem 1 in Project 3, Chapter 9.

$h = 1/(N + 1)$. Defining $u_n(t) = u(x_n, t)$ for $n = 0, \ldots, N + 1$, show that if the partial derivative on the right side of Eq. (6) is replaced by the difference quotient in Eq. (i), then the components of $\mathbf{u}_N(t) = (u_1(t), u_2(t), \ldots, u_N(t))^T$ satisfy the following system of ordinary differential equations:

$$u_1' = D(-2u_1 + u_2)/h^2 + ru_1(1 - u_1),$$
$$u_n' = D(u_{n-1} - 2u_n + u_{n+1})/$$
$$\qquad h^2 + ru_n(1 - u_n), \quad n = 2, \ldots, N - 1,$$
$$u_N' = D(u_{N-1} - 2u_N)/h^2 + ru_N(1 - u_N),$$
$$\text{(ii)}$$

where we have set $u_0(t) = 0$ and $u_N(t) = 0$ as required by the boundary conditions (7).

(b) For a few representative values of $r < \lambda_1$, use an initial value problem solver to compute solutions to the system (ii) in Problem 3(a) in an interval $0 \le t \le T$, where T is large enough to graphically ascertain whether the components of $\mathbf{u}_N(T)$ are near constant limiting values, for example, $T = 1$ or 2. Assume that $D = 1$ and use $N = 24$. Experiment with different choices of initial conditions for the system (ii), choosing

initial values for components that may be positive, negative, large, or small. Do your computer experiments support the results of the linear stability analysis in Problem 2?

(c) Repeat the computer experiment in Problem 3(b) for a few representative values of $r > \lambda_1$. Consider initial conditions such that initial values for all components of $\mathbf{u}_N$ are nonnegative, with $u_n(0) > 0$ for at least one $n \in \{1, \ldots, N\}$. What happens if initial values for all components of $\mathbf{u}_N$ are zero but with $u_n(0) < 0$ for at least one $n \in \{1, \ldots, N\}$? Do your computer experiments support the results of the linear stability analysis in Problem 2?

(d) For a few representative values of $r > \lambda_1$, plot the graphs of $u_n(T)$ versus n, where T is large enough to graphically ascertain that the components of $\mathbf{u}_N(T)$ are near constant limiting values. Choose initial conditions such that all components of $\mathbf{u}_N$ are nonnegative, with $u_n(0) > 0$ for at least one $n \in \{1, \ldots, N\}$.

Bifurcation. From the computer experiments we infer the existence of an asymptotically stable equilibrium solution $u = \hat{u}_1(x, r) \ne \hat{u}_0(x, r)$ to Eqs. (6) and (7), for values of $r > \lambda_1$, such that $\hat{u}_1(x_n, r) \cong u_n(T, r)$,

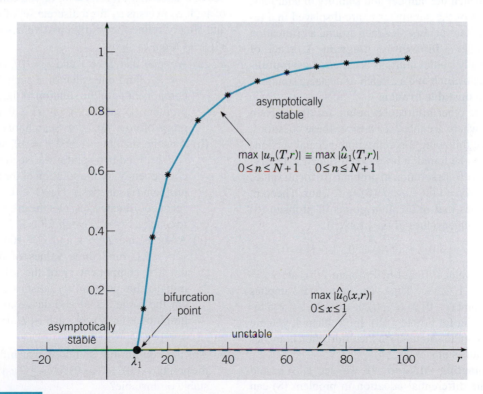

FIGURE 10.P.7 Bifurcation diagram for solutions of the boundary value problem (8).

THEOREM 10.P.2.

Consider the boundary value problem

$$Du'' + rf(u) = 0, \quad 0 < x < 1, \qquad u(0) = 0, \quad u(1) = 0. \tag{i}$$

Assume that $f(0) = 0$ so that $u = 0$ is a solution for all r. Assume also that $f'(u)$ and $f''(u)$ exist and are continuous and that $f'(0) \neq 0$. If $\hat{\lambda} = \hat{r} f'(0)$ is an eigenvalue with eigenvector $\hat{u}$ of the linearized problem

$$Dv'' + rf'(0)v = 0, \quad 0 < x < 1, \qquad v(0) = 0, \quad v(1) = 0,$$

then $(r, u) = (\hat{\lambda}/f'(0), 0)$ is a bifurcation point for the problem (i). Thus, there exists a branch of solutions $\hat{u}(\cdot, r) \neq 0$ to problem (i) for r in a neighborhood of $\hat{r} = \hat{\lambda}/f'(0)$.

$0 \leq n \leq N$, where $u_n(T, r)$ are the components computed in Problem 3(**d**).

The results of Problems 2 and 3 are summarized in Figure 10.P.7 in which we plot the graph of $\max_{0 \leq x \leq 1} |\hat{u}_0(x, r)|$ versus r and the computer-generated graph of $\max_{0 \leq n \leq N} |u_n(T, r)| \cong \max_{0 \leq n \leq N} |\hat{u}_1(x_n, r)|$ versus r.

The point $(\lambda_1, 0)$ is called a **bifurcation point** for the boundary value problem (8). More precisely, the graph in Figure 10.P.7 indicates that, in any ϵ-neighborhood of $(\lambda_1, 0)$, there exist two qualitatively different solutions, or states, $\hat{u}_0(\cdot, r)$ and $\hat{u}_1(\cdot, r)$. The diagram in Figure 10.P.7, in which the number and stability of solutions, as functions of the parameter r, are displayed in a region of the state-parameter space around a bifurcation point, is called a **bifurcation diagram**. A branch of asymptotically stable solutions is called an **asymptotically stable branch** and a branch of unstable solutions is called an **unstable branch**.

Are there other bifurcation points for the problem (8)? If so, where are they? To answer these questions, we state the following theorem, more general versions of which can be found in books on bifurcation theory.

Since $u = 0$ is a solution of problem (8) for all values of r, and $f(u) = u(1 - u)$, $f'(0) = 1$. Thus, Theorem 10.P.2 implies that bifurcation points of problem (8) occur at the eigenvalues of the problem

$$Dv'' + rv = 0, \qquad v(0) = 0, \quad v(1) = 0.$$

These eigenvalues, found in Problem 2(b), are $\lambda_n = n^2\pi^2 D$, $n = 1, 2, \ldots$. Thus, Theorem 10.P.2 implies that each point $(r_n, 0) = (\lambda_n, 0)$, $n = 1, 2, \ldots$, is a bifurcation point for problem (8). In Problem 4, you are asked to compute solutions on the bifurcation branches for some values of $r > \lambda_2$.

The Shooting Method. If we set $y_1 = u$ and $y_2 = u'$, the differential equation in problem (8) can

be associated with the initial value problem

$$y_1' = y_2, \quad y_1(0) = 0,$$
$$y_2' = -ry_1(1 - y_1)/D, \quad y_2(0) = \alpha, \tag{i}$$

where α is fixed but arbitrary. The value of y_1 at $x = 1$ is a function of α, $y_1(1, \alpha)$. In the shooting method, we search for values of $\alpha \neq 0$ such that $y_1(1, \alpha) \cong 0$. For each value of α such that $y_1(1, \alpha) \cong 0$, $y_1(x, \alpha)$ will be an approximation to a nontrivial solution of problem (8).

In Problem 4 you can use trial and error to find values of α such that $y_1(1, \alpha) \cong 0$, or you can plot values of $y_1(1, \alpha_k)$ versus α_k for a discrete set of values of α, and simply refine the grid around values of α, where $y_1(1, \alpha)$ changes sign.

4. **(a)** Assume that $D = 1$ and $r = 20$, so that $\lambda_1 < r < \lambda_2$. Find one value of α such that the first component y_1 of the solution of the initial value problem (i) satisfies $y_1(1, \alpha) \cong 0$ and plot the graph of $y_1(x, \alpha)$. *Hint*: Search for $\alpha \in [1, 2.5]$. **CAS**

   (b) Assume that $D = 1$ and $r = 50$, so that $\lambda_2 < r < \lambda_3$. Find two values of α such that the first component y_1 of the solution of the initial value problem (i) satisfies $y_1(1, \alpha) \cong 0$. On the same set of coordinate axes, plot the graph of $y_1(x, \alpha)$ for each α. *Hint*: Search for $\alpha \in [2, 4.1]$.

   (c) Assume that $D = 1$ and $r = 100$, so that $\lambda_3 < r < \lambda_4$. Find three values of α such that the first component y_1 of the solution of the initial value problem (i) satisfies $y_1(1, \alpha) \cong 0$. On the same set of coordinate axes, plot the graph of $y_1(x, \alpha)$ for each α. *Hint*: Search for $\alpha \in [2.2, 5.78]$.

5. Are the equilibrium solutions that bifurcate from the points $(r_n, 0) = (\lambda_n, 0)$ for $n \geq 2$ asymptotically stable or unstable? **CAS**

Matrices and Linear Algebra

E lementary matrix methods were introduced in Chapter 3 to facilitate the study of first order linear systems of dimension 2. In Appendix A, we present several additional concepts and tools from matrix theory necessary for the study of first order linear systems of dimension $n \geq 2$ taken up in Chapter 6. Readers with an adequate background in matrix methods and linear algebra may wish to proceed directly to Chapter 6, referring back to this Appendix for reference or review if necessary.

A.1 Matrices

The idea of the matrix as a mathematical object is most easily conveyed by considering a type of problem that frequently occurs in science and engineering, the linear algebraic system of equations. For example, consider the system

$$
\begin{aligned}
x_1 + 2x_2 - x_3 + 2x_4 &= 0, \\
2x_1 + x_2 + 2x_3 - 3x_4 &= 0, \\
-x_1 + 3x_2 + x_3 + x_4 &= 0, \\
x_1 + 9x_2 + 3x_3 + x_4 &= 0.
\end{aligned}
\tag{1}
$$

Solving the system (1) means explicitly characterizing the set of all values of x_1, x_2, x_3, and x_4 that simultaneously satisfy all four equations. The solution may simply consist of numerical values for the unknowns, or it may require a mathematical description of a more complex solution set. All information required to solve the problem resides in the coefficients of the unknowns while the names for the unknowns are largely irrelevant. Not only is it particularly convenient to store the coefficients in a rectangular array called a matrix, but efficient algorithms exist for systematically manipulating the entries in the array to arrive at a mathematical characterization of all solutions. The matrix of coefficients for the system (1), displayed in an array identical to the pattern in which they appear is

$$
\begin{pmatrix}
1 & 2 & -1 & 2 \\
2 & 1 & 2 & -3 \\
-1 & 3 & 1 & 1 \\
1 & 9 & 3 & 1
\end{pmatrix}
\tag{?}
$$

Solution algorithms are applied to arrays of this type.

Matrices also play an important role in differential equations. Recall from Chapter 3 that the solution for a linear, constant coefficient system of dimension 2,

$$\mathbf{x}' = \begin{pmatrix} a_{11} & a_{12} \\ a_{21} & a_{22} \end{pmatrix} \mathbf{x} = \mathbf{A}\mathbf{x}, \tag{3}$$

is based on the eigenvalue method, that is, finding the eigenvalues and eigenvectors of the 2×2 matrix $\mathbf{A}$. In Chapter 6, the eigenvalue method is also used to find solutions of linear, constant coefficient systems of dimension n. In this appendix, we develop those matrix concepts and methods pursuant to that goal.

DEFINITION A.1.1	A matrix is an array of mathematical objects, or elements, arranged in m rows and n columns.

The elements may be real or complex numbers such as the matrix (2). The elements may also be functions or contain symbols such as

$$\begin{pmatrix} e^{-t} & e^{2t} \\ e^{-t} & 2e^{2t} \end{pmatrix}$$

and

$$\begin{pmatrix} 3-\lambda & -1 & 2 \\ 0 & -1-\lambda & -4 \\ 1 & -2 & 1-\lambda \end{pmatrix}.$$

We will designate matrices by boldfaced capitals $\mathbf{A}, \mathbf{B}, \mathbf{C}, \ldots$ such as

$$\mathbf{A} = \begin{pmatrix} a_{11} & a_{12} & \cdots & a_{1n} \\ a_{21} & a_{22} & \cdots & a_{2n} \\ \vdots & \vdots & & \vdots \\ a_{m1} & a_{m2} & \cdots & a_{mn} \end{pmatrix}. \tag{4}$$

The rows of $\mathbf{A}$ are the m horizontal n-tuples

$$(a_{11}, a_{12}, \ldots, a_{1n}), \ (a_{21}, a_{22}, \ldots, a_{2n}), \ \ldots, \ (a_{m1}, a_{m2}, \ldots, a_{mn}).$$

The columns of $\mathbf{A}$ are the n vertical m-tuples

$$\begin{pmatrix} a_{11} \\ a_{21} \\ \vdots \\ a_{m1} \end{pmatrix}, \ \begin{pmatrix} a_{12} \\ a_{22} \\ \vdots \\ a_{m2} \end{pmatrix}, \ \ldots, \ \begin{pmatrix} a_{1n} \\ a_{2n} \\ \vdots \\ a_{mn} \end{pmatrix}.$$

The element lying in the ith row and jth column is denoted by a_{ij}, the first subscript identifying its row and the second its column. Frequently the condensed notation $\mathbf{A} = (a_{ij})$ is used, indicating that $\mathbf{A}$ is the matrix whose generic element is a_{ij}. In this case, the **dimensions** of $\mathbf{A}$ (number of rows and columns), if not explicitly stated, are understood by the context in which the matrix is used.

Two matrices of special importance are square matrices and vectors:

1. Matrices which have the same number of rows and columns—that is, $m = n$—are called **square matrices**. The matrix (2) is an example of a square matrix of dimension 4.

2. Matrices which have only one column—that is, the dimensions are $n \times 1$, are called **column vectors** or simply **vectors**. A vector with n entries is called an n-vector. Examples of column vectors are the following 2-vector, 3-vector, and 4-vector, respectively,

$$\begin{pmatrix} 1 \\ 2 \end{pmatrix}, \quad \begin{pmatrix} -1 \\ 0 \\ 3 \end{pmatrix}, \quad \begin{pmatrix} 1 \\ 2 \\ -4 \\ 3 \end{pmatrix}.$$

We denote (column) vectors by boldfaced lower-case letters, $\mathbf{x}, \mathbf{y}, \mathbf{z}, \ldots$. The general form of an n-vector is

$$\mathbf{x} = \begin{pmatrix} x_1 \\ x_2 \\ \vdots \\ x_n \end{pmatrix}.$$

The number x_j is commonly called the jth **component** of the column vector.

▶ **Matrix Algebra.**

DEFINITION A.1.2	**Equality.** Two $m \times n$ matrices $\mathbf{A} = (a_{ij})$ and $\mathbf{B} = (b_{ij})$ are *equal* if all corresponding elements are equal—that is, if $a_{ij} = b_{ij}$ for each i and j.

DEFINITION A.1.3	**Zero Matrix.** The **zero matrix,** denoted by $\mathbf{0}$, is the matrix (or vector) whose elements are zero. Thus $\mathbf{0} = (z_{ij})$ where $z_{ij} = 0$ for each i and j.

Some of the usual arithmetic operations for numbers can be extended to matrices in a natural way.

DEFINITION A.1.4	**Matrix Addition.** The sum of two $m \times n$ matrices $\mathbf{A}$ and $\mathbf{B}$ is defined to be the matrix obtained by adding corresponding elements: $$\mathbf{A} + \mathbf{B} = (a_{ij}) + (b_{ij}) = (a_{ij} + b_{ij}). \tag{5}$$

Thus, if $\mathbf{C} = \mathbf{A} + \mathbf{B}$, the (i, j)th element of $\mathbf{C}$ is the sum of the (i, j)th element of $\mathbf{A}$ and the (i, j)th element of $\mathbf{B}$. For example,

$$\begin{pmatrix} -1 & 3 & 0 \\ 2 & 1 & 4 \end{pmatrix} + \begin{pmatrix} 2 & 5 & -7 \\ -3 & 4 & 1 \end{pmatrix} = \begin{pmatrix} 1 & 8 & -7 \\ -1 & 5 & 5 \end{pmatrix}.$$

If the dimensions of two matrices are not identical, the matrices cannot be added.

Note that $\mathbf{A} + \mathbf{0} = \mathbf{A}$. For this reason, the zero matrix, $\mathbf{0}$, is often called the **additive identity**.

In matrix algebra, it is traditional to use the term **scalar** to refer to a real or complex number. The product of a matrix and a scalar is defined in a natural way.

DEFINITION A.1.5

Product of a Matrix and a Scalar. The product of a matrix $\mathbf{A}$ and a scalar α is defined as follows:

$$\alpha\mathbf{A} = \alpha(a_{ij}) = (\alpha a_{ij}). \tag{6}$$

Therefore the (i, j)th entry of $\alpha\mathbf{A}$ is simply the (i, j)th entry of $\mathbf{A}$ multiplied by α. For example,

$$-3\begin{pmatrix} -2 & 1 & 0 \\ 2 & -1 & 4 \\ 1 & 0 & 5 \end{pmatrix} = \begin{pmatrix} 6 & -3 & 0 \\ -6 & 3 & -12 \\ -3 & 0 & -15 \end{pmatrix}.$$

The negative of a matrix $\mathbf{A} = (a_{ij})$, denoted by $-\mathbf{A}$, is the matrix whose (i, j)th entry is $-a_{ij}$. We then define the **difference** of two matrices, $\mathbf{A} - \mathbf{B}$, by

$$\mathbf{A} - \mathbf{B} = \mathbf{A} + (-\mathbf{B}).$$

Thus

$$\mathbf{A} - \mathbf{B} = (a_{ij}) + (-b_{ij}) = (a_{ij} - b_{ij}).$$

The following theorem gives the algebraic properties of matrices for operations restricted to matrix addition and multiplication of matrices by scalars. The properties follow readily from the properties of real or complex numbers and the above definitions for matrix addition and multiplication of matrices by scalars.

THEOREM A.1.6

If $\mathbf{A}$, $\mathbf{B}$, and $\mathbf{C}$ are $m \times n$ matrices and α and β are scalars, then

a. $\mathbf{A} + \mathbf{B} = \mathbf{B} + \mathbf{A}$

b. $\mathbf{A} + (\mathbf{B} + \mathbf{C}) = (\mathbf{A} + \mathbf{B}) + \mathbf{C}$

c. $\alpha(\beta\mathbf{A}) = (\alpha\beta)\mathbf{A}$

d. $(\alpha + \beta)\mathbf{A} = \alpha\mathbf{A} + \beta\mathbf{A}$

e. $\alpha(\mathbf{A} + \mathbf{B}) = \alpha\mathbf{A} + \alpha\mathbf{B}$

f. $(-1)\mathbf{A} = -\mathbf{A}$

g. $-(-\mathbf{A}) = \mathbf{A}$

h. $0\mathbf{A} = \mathbf{0}$

i. $\alpha\mathbf{0} = \mathbf{0}$.

Note that the properties stated in Theorem A.1.6 simply extend the properties of equivalent operations for real or complex numbers. However, the operation of matrix multiplication does not have the same properties as multiplication of real or complex numbers.

We first define the matrix product of a $1 \times n$ matrix $\mathbf{x}$ with an $n \times 1$ matrix $\mathbf{y}$ by

$$\mathbf{x}\mathbf{y} = (x_1, \ldots, x_n)\begin{pmatrix} y_1 \\ \vdots \\ y_n \end{pmatrix} = x_1 y_1 + \cdots + x_n y_n.$$

This is simply the familiar dot or inner product from calculus extended to vectors with n components. The product of an $m \times n$ matrix and an $n \times p$ matrix is defined next.

DEFINITION A.1.7

Matrix Multiplication. If $\mathbf{A}$ is an $m \times n$ matrix and $\mathbf{B}$ is an $n \times p$ matrix, then $\mathbf{AB}$ is defined to be the $m \times p$ matrix $\mathbf{C} = (c_{ij})$ satisfying

$$c_{ij} = \sum_{k=1}^{n} a_{ik} b_{kj}, \qquad 1 \le i \le m, \quad 1 \le j \le p. \tag{7}$$

Note that the matrix product $\mathbf{AB}$ is defined only if the number of columns in the first factor $\mathbf{A}$ is equal to the number of rows in the second factor $\mathbf{B}$. It is convenient to think of c_{ij} defined in Eq. (7) as the dot product of the ith row of $\mathbf{A}$ with the jth column of $\mathbf{B}$,

$$c_{ij} = (a_{i1}, \ldots, a_{in}) \begin{pmatrix} b_{1j} \\ \vdots \\ b_{nj} \end{pmatrix} = a_{i1} b_{1j} + \cdots + a_{in} b_{nj}.$$

A simple mnemonic device for determining if the matrix product is defined as well as the dimensions of the resultant product, consists of writing the dimensions of each factor adjacent to one another in the same order that the matrices are to be multiplied,

$$\underbrace{\mathbf{A}}_{m \times n} \quad \underbrace{\mathbf{B}}_{n \times p} \quad = \quad \underbrace{\mathbf{C}}_{m \times p}. \tag{8}$$

The product is defined if the interior dimensions n match as indicated in Eq. (8).

EXAMPLE 1

Let

$$\mathbf{A} = \begin{pmatrix} 2 & -1 \\ 3 & 1 \end{pmatrix} \quad \text{and} \quad \mathbf{B} = \begin{pmatrix} 1 & 0 & 2 \\ 2 & 1 & -1 \end{pmatrix}.$$

Since $\mathbf{A}$ is 2×2 and $\mathbf{B}$ is 2×3, $\mathbf{AB}$ is defined and is 2×3. From Eq. (7), we have

$$\mathbf{AB} = \begin{pmatrix} 2-2 & 0-1 & 4+1 \\ 3+2 & 0+1 & 6-1 \end{pmatrix} = \begin{pmatrix} 0 & -1 & 5 \\ 5 & 1 & 5 \end{pmatrix}.$$

The product $\mathbf{BA}$ is not defined because the number of columns of $\mathbf{B}$ (three) is not equal to the number of rows of $\mathbf{A}$ (two).

Example 1 shows that matrix multiplication does not share the property of commutativity possessed by multiplication of numbers since $\mathbf{BA}$ is not defined. Even in the case that both products $\mathbf{AB}$ and $\mathbf{BA}$ exist, the two products are usually not equal.

EXAMPLE 2

Consider the matrices

$$\mathbf{A} = \begin{pmatrix} 1 & -2 & 1 \\ 0 & 2 & -1 \\ 2 & 1 & 1 \end{pmatrix} \quad \text{and} \quad \mathbf{B} = \begin{pmatrix} 2 & 1 & -1 \\ 1 & -1 & 0 \\ 2 & -1 & 1 \end{pmatrix}.$$

Again, from the definition of multiplication given in Eq. (7), we have

$$\mathbf{AB} = \begin{pmatrix} 2-2+2 & 1+2-1 & -1+0+1 \\ 0+2-2 & 0-2+1 & 0+0-1 \\ 4+1+2 & 2-1-1 & -2+0+1 \end{pmatrix}$$

$$= \begin{pmatrix} 2 & 2 & 0 \\ 0 & -1 & -1 \\ 7 & 0 & -1 \end{pmatrix}.$$

Similarly, we find that

$$\mathbf{BA} = \begin{pmatrix} 0 & -3 & 0 \\ 1 & -4 & 2 \\ 4 & -5 & 4 \end{pmatrix}.$$

Clearly, $\mathbf{AB} \neq \mathbf{BA}$.

From these examples we conclude that, in general,

$$\mathbf{AB} \neq \mathbf{BA}.$$

The following theorem states properties of matrix multiplication that are shared in common with multiplication of numbers.

THEOREM A.1.8

Suppose that $\mathbf{A}$, $\mathbf{B}$, and $\mathbf{C}$ are matrices for which the following products are defined and let α be a scalar. Then

a. $(\mathbf{AB})\mathbf{C} = \mathbf{A}(\mathbf{BC})$

b. $\mathbf{A}(\alpha\mathbf{B}) = (\alpha\mathbf{A})\mathbf{B} = \alpha(\mathbf{AB})$

c. $\mathbf{A}(\mathbf{B} + \mathbf{C}) = \mathbf{AB} + \mathbf{AC}$

d. $(\mathbf{A} + \mathbf{B})\mathbf{C} = \mathbf{AC} + \mathbf{BC}$.

▶ **Special Matrices.** **The identity matrix.** Denote the Kronecker delta by δ_{ij},

$$\delta_{ij} = \begin{cases} 1, & \text{if } i = j \\ 0, & \text{if } i \neq j \end{cases}.$$

The $n \times n$ matrix

$$\mathbf{I}_n = (\delta_{ij}) = \begin{pmatrix} 1 & 0 & \cdots & 0 \\ 0 & 1 & \cdots & 0 \\ \vdots & \vdots & \ddots & \vdots \\ 0 & 0 & \cdots & 1 \end{pmatrix} \tag{9}$$

is called the **multiplicative identity**, or simply the **identity matrix**. If $\mathbf{A}$ is an $n \times n$ matrix, then from the definition of matrix multiplication we have

$$\mathbf{I}_n\mathbf{A} = \mathbf{AI}_n = \mathbf{A}.$$

To see that $\mathbf{I}_n\mathbf{A} = \mathbf{A}$, we note that the (i, j)th entry of $\mathbf{I}_n\mathbf{A}$ is

$$\sum_{k=1}^{n} \delta_{ik}a_{kj} = a_{ij}$$

since $\delta_{ik} = 0$ for all $k \neq i$. The equality $\mathbf{AI}_n = \mathbf{A}$ is established in the same way. In particular, for any n-vector $\mathbf{x}$,

$$\mathbf{I}_n\mathbf{x} = \mathbf{x}.$$

Diagonal and Triangular Matrices. If $\mathbf{A}$ is an $n \times n$ matrix, the set of elements $a_{jj}, j = 1, \ldots, n$ is referred to as the **principal diagonal**, or simply, the **diagonal**, of the matrix. If all elements off the diagonal are zero, that is, $a_{ij} = 0$ whenever $i \neq j$, then the matrix is called a **diagonal matrix**. For example,

$$\mathbf{A} = \begin{pmatrix} a_{11} & 0 & 0 & 0 \\ 0 & a_{22} & 0 & 0 \\ 0 & 0 & a_{33} & 0 \\ 0 & 0 & 0 & a_{44} \end{pmatrix}$$

is a diagonal matrix, which we can write in abbreviated form as $\mathbf{A} = \text{diag}(a_{11}, a_{22}, a_{33}, a_{44})$.

If all the elements of a square matrix below (or above) the principal diagonal are zero, then the matrix is called an **upper (lower) triangular matrix**. Examples of upper and lower triangular matrices are

$$\begin{pmatrix} a_{11} & a_{12} & a_{13} \\ 0 & a_{22} & a_{23} \\ 0 & 0 & a_{33} \end{pmatrix} \quad \text{and} \quad \begin{pmatrix} a_{11} & 0 & 0 \\ a_{21} & a_{22} & 0 \\ a_{31} & a_{32} & a_{33} \end{pmatrix},$$

respectively.

The Transpose of a Matrix. If A is an $m \times n$ matrix, then the $n \times m$ matrix formed by interchanging the rows and columns of $\mathbf{A}$ is called the **transpose** of $\mathbf{A}$ and is denoted by $\mathbf{A}^T$. Thus if $\mathbf{A} = (a_{ij})$, $\mathbf{A}^T = (a_{ji})$. For example, if

$$\mathbf{A} = \begin{pmatrix} -1 & 4 & -3 \\ 2 & 5 & -7 \end{pmatrix},$$

then

$$\mathbf{A}^T = \begin{pmatrix} -1 & 2 \\ 4 & 5 \\ -3 & -7 \end{pmatrix}.$$

THEOREM A.1.9

Properties of Transposes.

a. $\mathbf{I}_n^T = \mathbf{I}_n$

b. For any matrix $\mathbf{A}$, $(\mathbf{A}^T)^T = \mathbf{A}$.

c. If $\mathbf{AB}$ is defined, then $(\mathbf{AB})^T = \mathbf{B}^T\mathbf{A}^T$.

Proof

Parts (a) and (b) are obvious. To prove part (c), assume that $\mathbf{A}$ is $m \times n$ and that $\mathbf{B}$ is $n \times p$. Note that the ijth entry of $(\mathbf{AB})^T$ is the jith entry of $\mathbf{AB}$, $\sum_{k=1}^{n} a_{jk}b_{ki}$. On the other hand, the ijth entry of $\mathbf{B}^T\mathbf{A}^T$ is the dot product of the ith row of $\mathbf{B}^T$ with the jth column of $\mathbf{A}^T$, or equivalently, the dot product of the ith column of $\mathbf{B}$ with the jth row of $\mathbf{A}$, $\sum_{k=1}^{n} b_{ki}a_{jk} = \sum_{k=1}^{n} a_{jk}b_{ki}$. Thus, for each i and j, the (i,j)th entry of $(\mathbf{AB})^T$ and $\mathbf{B}^T\mathbf{A}^T$ are equal. ∎

Symmetric Matrices. An $n \times n$ matrix $\mathbf{A}$ is said to be symmetric if $\mathbf{A} = \mathbf{A}^T$. In this case, $a_{ij} = a_{ji}$ for all i and j with the principal diagonal serving as the axis of symmetry. An example of a symmetric matrix is

$$A = \begin{pmatrix} 2 & -3 & 1 & -2 \\ -3 & 0 & 4 & -5 \\ 1 & 4 & -3 & 8 \\ -2 & -5 & 8 & 1 \end{pmatrix}.$$

Matrix Inverse. The square matrix $\mathbf{A}$ is said to be **nonsingular** or **invertible** if there is another matrix $\mathbf{B}$ such that

$$\mathbf{AB} = \mathbf{BA} = \mathbf{I}_n.$$

If there is such a $\mathbf{B}$, there is only one. To see this, suppose that there is also a matrix $\mathbf{C}$ such that

$$\mathbf{AC} = \mathbf{CA} = \mathbf{I}_n.$$

Then

$$\mathbf{B} = \mathbf{BI}_n = \mathbf{B(AC)} = \mathbf{(BA)C} = \mathbf{I}_n\mathbf{C} = \mathbf{C}$$

where the third equality follows from part (a) of Theorem A.1.8. The unique inverse of $\mathbf{A}$, if it exists, is called the **multiplicative inverse**, or simply the **inverse**, of $\mathbf{A}$, and we write $\mathbf{B} = \mathbf{A}^{-1}$. Then

$$\mathbf{AA}^{-1} = \mathbf{A}^{-1}\mathbf{A} = \mathbf{I}_n. \tag{10}$$

Whereas all nonzero scalars possess inverses, there are lots of nonzero matrices that do not have inverses. Matrices that do not have an inverse are called **singular** or **noninvertible**.

EXAMPLE 3

Demonstrate that

$$\mathbf{A} = \begin{pmatrix} 1 & 2 \\ 2 & 2 \end{pmatrix} \quad \text{and} \quad \mathbf{B} = \begin{pmatrix} -1 & 1 \\ 1 & -\frac{1}{2} \end{pmatrix}$$

are both nonsingular by showing that $\mathbf{AB} = \mathbf{BA} = \mathbf{I}_2$.

We have

$$\begin{pmatrix} 1 & 2 \\ 2 & 2 \end{pmatrix} \begin{pmatrix} -1 & 1 \\ 1 & -\frac{1}{2} \end{pmatrix} = \begin{pmatrix} 1 & 0 \\ 0 & 1 \end{pmatrix} = \begin{pmatrix} -1 & 1 \\ 1 & -\frac{1}{2} \end{pmatrix} \begin{pmatrix} 1 & 2 \\ 2 & 2 \end{pmatrix}.$$

Thus both $\mathbf{A}$ and $\mathbf{B}$ are nonsingular, $\mathbf{A}^{-1} = \mathbf{B}$, and $\mathbf{B}^{-1} = \mathbf{A}$.

THEOREM A.1.10

Properties of Matrix Inverses.

a. I_n is nonsingular and $I_n^{-1} = I_n$.

b. If A and B are nonsingular $n \times n$ matrices, then AB is nonsingular and

$$(AB)^{-1} = B^{-1}A^{-1}.$$

c. If A is nonsingular, so is A^{-1}, and $(A^{-1})^{-1} = A$.

d. If A is nonsingular, so is A^T, and $(A^T)^{-1} = (A^{-1})^T$.

e. If A and B are $n \times n$ matrices, either of which in singular, then AB and BA are both singular.

Proof

Part (a) follows from the fact that $I_n I_n = I_n$.

To prove part (b), we use the fact that A^{-1} and B^{-1} exist to write

$$(AB)(B^{-1}A^{-1}) = A(B^{-1}B)A^{-1} = AI_nA^{-1} = AA^{-1} = I_n.$$

Similarly, $(B^{-1}A^{-1})(AB) = I_n$, so $B^{-1}A^{-1}$ is the unique inverse of AB.

Part (c) follows from $AA^{-1} = A^{-1}A = I_n$.

Part (d) follows from the two statements

$$I_n = I_n^T = (AA^{-1})^T = (A^{-1})^T A^T,$$

and

$$I_n = I_n^T = (A^{-1}A)^T = A^T(A^{-1})^T,$$

where we have used part (c) of Theorem A.1.9. Thus the unique inverse of A^T is $(A^{-1})^T$. A proof of (e) will be possible once we have discussed determinants. ∎

PROBLEMS

1. Given the matrices

$$A = \begin{pmatrix} 2 & 2 \\ -1 & 3 \end{pmatrix}, \quad B = \begin{pmatrix} -1 & 1 & -2 \\ 1 & 3 & 4 \end{pmatrix},$$

$$x = \begin{pmatrix} 2 \\ 3 \end{pmatrix}, \quad y = (1, 2, -1),$$

compute each of the following for which the indicated operations are defined.

(a) $4A$ (b) $-2x$ (c) $A + B$

(d) Ax (e) Dx (f) AB

(g) BA (h) $B^T A$ (i) yB^T

(j) $(AB)^T$ (k) $x^T A x$ (l) xBy^T

2. If $A = \begin{pmatrix} 1 & -2 & 0 \\ 3 & -2 & -1 \\ -2 & 1 & 3 \end{pmatrix}$ and if

$B = \begin{pmatrix} 1 & 0 & 2 \\ 0 & 1 & -1 \\ 2 & -1 & 3 \end{pmatrix}$, find

(a) $2A + B$ (b) $A - 4B$

(c) AB (d) BA

(e) $(A + B)^T$ (f) $(B + A)A^T$

(g) $A^2 = AA$ (h) $AB - BA$

(i) $B^3 - 5B^2 + 2B + 2I_3$

3. Demonstrate that

$$\mathbf{A} = \begin{pmatrix} 2 & 2 & 3 \\ 1 & 0 & 1 \\ 1 & 1 & 1 \end{pmatrix} \quad \text{and} \quad \mathbf{B} = \begin{pmatrix} -1 & 1 & 2 \\ 0 & -1 & 1 \\ 1 & 0 & -2 \end{pmatrix}$$

are both nonsingular by showing that $\mathbf{AB} = \mathbf{BA} = \mathbf{I}_3$.

4. Prove each of the following laws of matrix algebra.

(a) $\mathbf{A} + \mathbf{B} = \mathbf{B} + \mathbf{A}$

(b) $\mathbf{A} + (\mathbf{B} + \mathbf{C}) = (\mathbf{A} + \mathbf{B}) + \mathbf{C}$

(c) $\alpha(\mathbf{A} + \mathbf{B}) = \alpha\mathbf{A} + \alpha\mathbf{B}$

(d) $(\alpha + \beta)\mathbf{A} = \alpha\mathbf{A} + \beta\mathbf{A}$

(e) $\mathbf{A}(\mathbf{BC}) = (\mathbf{AB})\mathbf{C}$

(f) $\mathbf{A}(\mathbf{B} + \mathbf{C}) = \mathbf{AB} + \mathbf{BC}$

5. If $\mathbf{A} = \text{diag}(a_{11}, \ldots, a_{nn})$, under what conditions is $\mathbf{A}$ nonsingular? What is $\mathbf{A}^{-1}$ in this case?

6. Prove that sums and products of upper (lower) triangular matrices are upper (lower) triangular.

7. Let $\mathbf{A} = \text{diag}(a_{11}, \ldots, a_{nn})$ be a diagonal matrix. Prove that $\mathbf{A}^k = \text{diag}(a_{11}^k, \ldots, a_{nn}^k)$ where $\mathbf{A}^2 = \mathbf{AA}$, $\mathbf{A}^3 = \mathbf{AAA}, \ldots$.

8. Prove that if $\mathbf{A}$ is symmetric and nonsingular, then $\mathbf{A}^{-1}$ is symmetric.

9. Two square matrices $\mathbf{A}$ and $\mathbf{B}$ are said to commute if $\mathbf{AB} = \mathbf{BA}$. If $\mathbf{A}$ and $\mathbf{B}$ are symmetric, prove that $\mathbf{AB}$ is symmetric if and only if $\mathbf{A}$ and $\mathbf{B}$ commute.

10. If $\mathbf{A}$ is any square matrix, show each of the following.

(a) $\mathbf{AA}^T$ and $\mathbf{A}^T\mathbf{A}$ are both symmetric.

(b) $\mathbf{A} + \mathbf{A}^T$ is symmetric.

A.2 Systems of Linear Algebraic Equations, Linear Independence, and Rank

In this section, we describe an efficient algorithm for finding solutions, if they exist, to a set of m simultaneous linear algebraic equations in n variables,

$$\begin{aligned} a_{11}x_1 &+ a_{12}x_2 + \cdots + a_{1n}x_n = b_1, \\ a_{21}x_1 &+ a_{22}x_2 + \cdots + a_{2n}x_n = b_2, \\ &\vdots \qquad \vdots \qquad\qquad \vdots \qquad \vdots \\ a_{m1}x_1 &+ a_{m2}x_2 + \cdots + a_{mn}x_n = b_m, \end{aligned} \tag{1}$$

or, in matrix notation,

$$\mathbf{Ax} = \mathbf{b}. \tag{2}$$

We begin with an example to help motivate an extremely useful algorithm for solving linear algebraic systems of equations.

EXAMPLE 1

Use the method of elimination to solve the system of equations

$$\begin{aligned} x_1 &- x_2 + 2x_3 + x_4 = 3, \\ 2x_1 &- 3x_2 + 6x_3 + 5x_4 = 4, \\ -2x_1 &+ 4x_2 - 8x_3 - 8x_4 = -2. \end{aligned} \tag{3}$$

If we subtract two times the first row from the second row, we obtain

$$
\begin{array}{rcl}
x_1 - x_2 + 2x_3 + x_4 &=& 3, \\
0x_1 - x_2 + 2x_3 + 3x_4 &=& -2, \\
-2x_1 + 4x_2 - 8x_3 - 8x_4 &=& -2.
\end{array}
$$

Adding two times the first row to the third row gives

$$
\begin{array}{rcl}
x_1 - x_2 + 2x_3 + x_4 &=& 3, \\
0x_1 - x_2 + 2x_3 + 3x_4 &=& -2, \\
0x_1 + 2x_2 - 4x_3 - 6x_4 &=& 4.
\end{array}
$$

Multiplying the second row by -1 yields

$$
\begin{array}{rcl}
x_1 - x_2 + 2x_3 + x_4 &=& 3, \\
0x_1 + x_2 - 2x_3 - 3x_4 &=& 2, \\
0x_1 + 2x_2 - 4x_3 - 6x_4 &=& 4.
\end{array}
$$

Next, subtract two times row two from the third row to obtain

$$
\begin{array}{rcl}
x_1 - x_2 + 2x_3 + x_4 &=& 3, \\
0x_1 + x_2 - 2x_3 - 3x_4 &=& 2, \\
0x_1 + 0x_2 + 0x_3 + 0x_4 &=& 0.
\end{array}
$$

Thus, the nonzero equations satisfied by the unknowns are

$$
\begin{array}{rcl}
x_1 - x_2 + 2x_3 + x_4 &=& 3, \\
x_2 - 2x_3 - 3x_4 &=& 2.
\end{array}
\tag{4}
$$

To solve the system (4), we can choose two of the unknowns arbitrarily and then use the equations to solve for the remaining two unknowns. If we let $x_3 = \alpha$ and $x_4 = \beta$, where α and β are arbitrary, it then follows that

$$
\begin{array}{l}
x_2 = 2\alpha + 3\beta + 2, \\
x_1 = 2\alpha + 3\beta + 2 - 2\alpha - \beta + 3 = 2\beta + 5.
\end{array}
$$

If we write the solution in vector notation, we have

$$
\mathbf{x} = \begin{pmatrix} 2\beta + 5 \\ 2\alpha + 3\beta + 2 \\ \alpha \\ \beta \end{pmatrix} = \alpha \underbrace{\begin{pmatrix} 0 \\ 2 \\ 1 \\ 0 \end{pmatrix}}_{\mathbf{v}_1} + \beta \underbrace{\begin{pmatrix} 2 \\ 3 \\ 0 \\ 1 \end{pmatrix}}_{\mathbf{v}_2} + \underbrace{\begin{pmatrix} 5 \\ 2 \\ 0 \\ 0 \end{pmatrix}}_{\mathbf{v}_p} .
\tag{5}
$$

The solution set of Eq. (3) is therefore

$$
\{\mathbf{x} : \mathbf{x} = \alpha \mathbf{v}_1 + \beta \mathbf{v}_2 + \mathbf{v}_p, \ -\infty < \alpha, \beta < \infty\}
$$

where the vectors $\mathbf{v}_1$, $\mathbf{v}_2$, and $\mathbf{v}_p$ are as indicated in Eq. (5). This example illustrates both a systematic solution algorithm and the general structure of solution sets (provided any solutions actually exist) for linear systems of equations. If we denote the matrix of coefficients in Eq. (3) by

$$\mathbf{A} = \begin{pmatrix} 1 & -1 & 2 & 1 \\ 2 & -3 & 6 & 5 \\ -2 & 4 & -8 & -8 \end{pmatrix}$$

and the vector on the right-hand side of Eq. (3) by

$$\mathbf{b} = \begin{pmatrix} 3 \\ 4 \\ -2 \end{pmatrix},$$

it is easy to verify that $\mathbf{v}_1$ and $\mathbf{v}_2$ are both solutions of the homogeneous equation $\mathbf{Ax} = \mathbf{0}$ while $\mathbf{v}_p$ is a solution of the nonhomogeneous equation $\mathbf{Ax} = \mathbf{b}$. It is common to refer to the solution representation (5) as the **general solution** of Eq. (3). It is a two-parameter family of solutions (the parameters are α and β) representing all possible solutions of $\mathbf{Ax} = \mathbf{b}$.

Motivated by the above example, we now describe, in more detail, a simple but powerful algorithm for solving linear systems called **Gaussian elimination** or **row reduction**. It is a systematic formalization of the operations applied to the system in the example above. We first note that all of the work is performed on the coefficients and the right-hand sides of the equations. The computations are made more efficient by ignoring the names of the unknowns and dealing only with the **augmented matrix** obtained by appending the column m-vector $\mathbf{b}$ of right-hand sides to the $m \times n$ coefficient matrix $\mathbf{A}$. We will denote the augmented matrix by

$$(\mathbf{A}|\mathbf{b}). \tag{6}$$

For example, the augmented matrix for the system (3) is

$$\begin{pmatrix} 1 & -1 & 2 & 1 & 3 \\ 2 & -3 & 6 & 5 & 4 \\ -2 & 4 & -8 & -8 & -2 \end{pmatrix}.$$

In the augmented matrix, a vertical line is normally used to separate the right-hand side of the system from the matrix of coefficients $\mathbf{A}$. The augmented matrix is a rectangular array of numbers, that is, a matrix, to which the elementary row operations described below may be applied. Always keep in mind that $(\mathbf{A}|\mathbf{b})$ is really equivalent to $\mathbf{Ax} = \mathbf{b}$ which is, in turn, matrix notation for the set of equations (1). In the case of a homogeneous system of equations, that is, when $\mathbf{b} = \mathbf{0}$, it is not necessary to augment $\mathbf{A}$ with a column of zeros but sometimes it may be done for the sake of clarity. The permissible row operations used in Gaussian elimination are formally described by the following definition.

DEFINITION A.2.1

The **elementary row operations** used in Gaussian elimination are:

E1. Any row of Eq. (6) may be multiplied by a scalar and the result added to another row of Eq. (6).

E2. Any two rows of Eq. (6) may be interchanged.

E3. Any row of Eq. (6) may be multiplied by a nonzero scalar.

DEFINITION A.2.2

The **solution set** of the system $Ax = b$ is the set of all n-vectors x that satisfy $Ax = b$.

Part of the utility of Gaussian elimination is explained by the following theorem.

THEOREM A.2.3

If an elementary row operation is applied to Eq. (6), the solution set of the resulting system of equations is identical to the solution set of Eq. (6). That is, elementary row operations do not alter the solution set of a linear system of equations.

Proof

Suppose $\hat{x} = (\hat{x}_1, \ldots, \hat{x}_n)^T$ is a solution of $Ax = b$ and α times row i is added to row j so that row j is replaced by

$$(a_{j1} + \alpha a_{i1})x_1 + \cdots + (a_{jn} + \alpha a_{in})x_n = b_j + \alpha b_i. \tag{7}$$

Since

$$a_{j1}\hat{x}_1 + \cdots + a_{jn}\hat{x}_n = b_j \tag{8}$$

and

$$a_{i1}\hat{x}_1 + \cdots + a_{in}\hat{x}_n = b_i, \tag{9}$$

it is obvious that

$$(a_{j1} + \alpha a_{i1})\hat{x}_1 + \cdots + (a_{jn} + \alpha a_{in})\hat{x}_n = b_j + \alpha b_i. \tag{10}$$

On the other hand, if $\hat{x}$ satisfies Eq. (7) and satisfies

$$a_{i1}x_1 + \cdots + a_{in}x_n = b_i,$$

then $\hat{x}$ also satisfies

$$a_{j1}x_1 + \cdots + a_{jn}x_n = b_j$$

so no new solutions are introduced by a row operation of type E1. We leave it to you to show that the operations of interchanging two rows or multiplying a row by a nonzero scalar will also not alter the solution set. ∎

DEFINITION A.2.4

If a matrix B can be obtained from a matrix A by a finite sequence of elementary row operations, the matrices A and B are said to be **row equivalent**.

In view of Theorem A.2.3, we see that if $(A|b)$ and $(B|c)$ are row equivalent, then $Ax = b$ and $Bx = c$ have identical solution sets.

▶ **The Row Echelon Form of a Matrix.** Certain types of structures present in linear systems of equations may be usefully exploited in the solution process. For example, suppose the augmented matrix of a system of equations has the form

$$\begin{pmatrix} 1 & -2 & 3 & \bigm| & 7 \\ 0 & 1 & -1 & \bigm| & -2 \\ 0 & 0 & 1 & \bigm| & 1 \end{pmatrix}. \tag{11}$$

The system of equations represented by the augmented matrix (11) has an upper triangular structure,

$$
\begin{aligned}
x_1 - 2x_2 + 3x_3 &= 7, \\
x_2 - x_3 &= -2, \\
x_3 &= 1.
\end{aligned}
\tag{12}
$$

In solving for the unknowns, it is most sensible to begin with the third equation which gives us $x_3 = 1$ directly. Substituting this value into the second equation we obtain

$$
x_2 = -2 + x_3 = -2 + 1 = -1.
$$

Finally, substituting the values obtained for x_2 and x_3 into the first equation gives

$$
x_1 = 7 + 2x_2 - 3x_3 = 7 - 2 - 3 = 2.
$$

Thus, we obtain the solution of the system (12)

$$
x = \begin{pmatrix} 2 \\ -1 \\ 1 \end{pmatrix}.
$$

The technique used to solve for the unknowns in this example is known as the **method of backsubstitution**. It can be applied to systems of equations that have an upper triangular structure such as Eqs. (12) or, more generally, a row echelon structure that will be described below. This example suggests the following strategy for solving a system of equations that is not necessarily in the desired form:

1. Use elementary row operations, which do not alter the solution set, to reduce the system to upper triangular form or, more generally, row echelon form.
2. Solve the resulting system using backsubstitution.

We illustrate the algorithm on the following system of equations,

$$
\begin{aligned}
x_1 - 2x_2 + 3x_3 &= 7, \\
-x_1 + x_2 - 2x_3 &= -5, \\
2x_1 - x_2 - x_3 &= 4.
\end{aligned}
\tag{13}
$$

The augmented matrix for the system (13) is

$$
\left(\begin{array}{ccc|c}
1 & -2 & 3 & 7 \\
-1 & 1 & -2 & -5 \\
2 & -1 & -1 & 4
\end{array} \right).
\tag{14}
$$

We now perform row operations on the matrix (14) with the goal of introducing zeros in the lower left part of the matrix. Each step is described and the result recorded below.

a. Add the first row to the second row and add (-2) times the first row to the third row.

$$
\left(\begin{array}{ccc|c}
1 & -2 & 3 & 7 \\
0 & -1 & 1 & 2 \\
0 & 3 & -7 & -10
\end{array} \right)
$$

b. Multiply the second row by -1.

$$\left(\begin{array}{ccc|c} 1 & -2 & 3 & 7 \\ 0 & 1 & -1 & -2 \\ 0 & 3 & -7 & -10 \end{array}\right)$$

c. Add (-3) times the second row to the third row.

$$\left(\begin{array}{ccc|c} 1 & -2 & 3 & 7 \\ 0 & 1 & -1 & -2 \\ 0 & 0 & -4 & -4 \end{array}\right)$$

d. Divide the third row by -4.

$$\left(\begin{array}{ccc|c} 1 & -2 & 3 & 7 \\ 0 & 1 & -1 & -2 \\ 0 & 0 & 1 & 1 \end{array}\right) \tag{15}$$

Thus we have arrived at the augmented matrix (11) for the upper triangular system (12) on which we illustrated the method of backsubstitution. This demonstrates the feasibility of the algorithm on a particular example.

We now describe the desired structure of the target matrix of the solution algorithm based on elementary row operations in more detail. Let $\mathbf{A}$ be an $m \times n$ matrix. Whether the matrix is augmented or not is irrelevant. We will say that a row of $\mathbf{A}$ is a **zero row** if all of its elements are zero. If a row has at least one nonzero entry, then that row is a **nonzero row**. The **leading entry** of a nonzero row is the first nonzero entry in the row, reading from left to right. A zero row has no leading (nonzero) entry. As an example, consider the matrix

$$\left(\begin{array}{cccc} 0 & 0 & 0 & 1 \\ 0 & -1 & 0 & 1 \\ 0 & 0 & 0 & 0 \\ 2 & 0 & 0 & 0 \end{array}\right).$$

Rows 1, 2, and 4 are nonzero rows while row 3 is a zero row. The leading entries in rows 1, 2, and 4 are 1, -1, and 2, respectively.

DEFINITION A.2.5	An $m \times n$ matrix $\mathbf{A}$ is said to be in **row echelon form** (or just echelon form) if it has the following properties:

    i. For some integer r, $0 \le r \le m$, the first r rows of $\mathbf{A}$ are nonzero.

    ii. Each leading entry lies to the right of the leading entry of the previous row.

    iii. Below each leading entry is a column of zeros.

    iv. After the first r rows, each row consists entirely of zeros, that is, zero rows are placed at the bottom of the matrix.

For example, the following matrices are all in row echelon form.

$$\begin{pmatrix} 2 & -2 \\ 0 & 3 \end{pmatrix} \qquad \begin{pmatrix} 0 & 3 & 0 & 2 \\ 0 & 0 & 2 & -3 \\ 0 & 0 & 0 & 7 \end{pmatrix} \qquad \begin{pmatrix} 1 & -2 & 4 \\ 0 & 0 & 5 \\ 0 & 0 & 0 \\ 0 & 0 & 0 \end{pmatrix}.$$

However, the matrix

$$\begin{pmatrix} 2 & 3 & -2 \\ 0 & 0 & 0 \\ 1 & 5 & -9 \end{pmatrix}$$

is not in row echelon form, since a nonzero row appears beneath a zero row and the leading entry in the third row does not lie to the right of the leading entry in the first row.

The transformation of a given matrix into a matrix that is in row echelon form by a sequence of elementary row operations is referred to as **row reduction** or **Gaussian elimination**.

EXAMPLE 2

Use Gaussian elimination to find the general solution of the linear system

$$\begin{pmatrix} 1 & -2 & 1 & 0 \\ 2 & 1 & -1 & 3 \\ 1 & 3 & -2 & 3 \end{pmatrix} \mathbf{x} = \begin{pmatrix} 1 \\ -1 \\ -2 \end{pmatrix} \tag{16}$$

by reducing the augmented matrix

$$\left(\begin{array}{cccc|c} 1 & -2 & 1 & 0 & 1 \\ 2 & 1 & -1 & 3 & -1 \\ 1 & 3 & -2 & 3 & -2 \end{array} \right) \tag{17}$$

to row echelon form.

Row echelon form is attained by the following row operations.

a. Subtract 2 times row one from row two.

$$\left(\begin{array}{cccc|c} 1 & -2 & 1 & 0 & 1 \\ 0 & 5 & -3 & 3 & -3 \\ 1 & 3 & -2 & 3 & -2 \end{array} \right)$$

b. Subtract row one from row three.

$$\left(\begin{array}{cccc|c} 1 & -2 & 1 & 0 & 1 \\ 0 & 5 & -3 & 3 & -3 \\ 0 & 5 & -3 & 3 & -3 \end{array} \right)$$

c. Subtract row two from row three.

$$\left(\begin{array}{cccc|c} 1 & -2 & 1 & 0 & 1 \\ 0 & 5 & -3 & 3 & -3 \\ 0 & 0 & 0 & 0 & 0 \end{array} \right) \tag{18}$$

System (18) is in row echelon form and represents the system

$$x_1 - 2x_2 + x_3 \qquad = \quad 1,$$
$$5x_2 - 3x_3 + 3x_4 = -3. \tag{19}$$

Equation (19) consists of 2 equations for 4 unknowns. We are free to choose any two of the unknowns arbitrarily. We set $x_3 = \alpha$ and $x_4 = \beta$ where α and β are arbitrary scalars. The second equation in (19) yields $x_2 = -\frac{3}{5} + \frac{3}{5}\alpha - \frac{3}{5}\beta$. Backsubstituting this expression for x_2 into the first equation in (19) then yields $x_1 = -\frac{1}{5} + \frac{1}{5}\alpha - \frac{6}{5}\beta$. The solution $\mathbf{x}$ may be expressed in the form

$$\mathbf{x} = \alpha \underbrace{\begin{pmatrix} 1/5 \\ 3/5 \\ 1 \\ 0 \end{pmatrix}}_{\mathbf{v}_1} + \beta \underbrace{\begin{pmatrix} -6/5 \\ -3/5 \\ 0 \\ 1 \end{pmatrix}}_{\mathbf{v}_2} + \underbrace{\begin{pmatrix} -1/5 \\ -3/5 \\ 0 \\ 0 \end{pmatrix}}_{\mathbf{v}_p}. \tag{20}$$

If $\mathbf{A}$ is the matrix of coefficients and $\mathbf{b}$ is the vector of right-hand sides in Eq. (16), then $\mathbf{A}\mathbf{v}_1 = \mathbf{0}$, $\mathbf{A}\mathbf{v}_2 = \mathbf{0}$, and $\mathbf{A}\mathbf{v}_p = \mathbf{b}$.

▶ Linearly Dependent and Linearly Independent Sets of Vectors.

DEFINITION A.2.6

A **linear combination** of vectors $\mathbf{v}_1, \ldots, \mathbf{v}_k$ is an expression of the form

$$c_1 \mathbf{v}_1 + \cdots + c_k \mathbf{v}_k$$

where $c_1, \ldots, c_k$ are any scalars.

DEFINITION A.2.7

A set of k vectors $\mathbf{v}_1, \ldots, \mathbf{v}_k$ is said to be **linearly dependent** if there exists a set of (real or complex) numbers $c_1, \ldots, c_k$, at least one of which is nonzero, such that

$$c_1 \mathbf{v}_1 + \cdots + c_k \mathbf{v}_k = \mathbf{0}. \tag{21}$$

On the other hand, if the only set $c_1, \ldots, c_k$ for which Eq. (21) is satisfied is $c_1 = c_2 = \cdots = c_k = 0$, then the set $\mathbf{x}_1, \ldots, \mathbf{x}_k$ is said to be **linearly independent**.

Assuming that each $\mathbf{v}_j, j = 1, \ldots, k$ is an m-vector, using matrix notation, the equation (21) can be written as

$$\begin{pmatrix} v_{11}c_1 & + & \cdots & + & v_{1k}c_k \\ \vdots & & & & \vdots \\ v_{m1}c_1 & + & \cdots & + & v_{mk}c_k \end{pmatrix} = \mathbf{V}\mathbf{c} = \mathbf{0}. \tag{22}$$

Therefore, to test whether the set $\mathbf{v}_1, \ldots, \mathbf{v}_k$ is linearly dependent or linearly independent we find the general solution of $\mathbf{V}\mathbf{c} = \mathbf{0}$. If the only solution is the zero vector, $\mathbf{c} = \mathbf{0}$, then the set is linearly independent. If there are nonzero solutions, then the set is linearly dependent.

EXAMPLE 3

Determine whether the set of vectors

$$\mathbf{v}_1 = \begin{pmatrix} 1 \\ 2 \\ -1 \end{pmatrix}, \quad \mathbf{v}_2 = \begin{pmatrix} 2 \\ 1 \\ 3 \end{pmatrix}, \quad \mathbf{v}_3 = \begin{pmatrix} -4 \\ 1 \\ -11 \end{pmatrix} \tag{23}$$

is linearly independent or linearly dependent.

To determine whether the set is linearly dependent, we seek constants c_1, c_2, c_3 such that $c_1 \mathbf{v}_1 + c_2 \mathbf{v}_2 + c_3 \mathbf{v}_3 = \mathbf{V}\mathbf{c} = \mathbf{0}$. Equivalently, we look for the general solution of

$$\begin{pmatrix} 1 & 2 & -4 \\ 2 & 1 & 1 \\ -1 & 3 & -11 \end{pmatrix} \begin{pmatrix} c_1 \\ c_2 \\ c_3 \end{pmatrix} = \begin{pmatrix} 0 \\ 0 \\ 0 \end{pmatrix}. \tag{24}$$

We use elementary row operations to reduce the corresponding augmented matrix

$$\left(\begin{array}{ccc|c} 1 & 2 & -4 & 0 \\ 2 & 1 & 1 & 0 \\ -1 & 3 & -11 & 0 \end{array} \right)$$

to row echelon form.

a. Add (-2) times the first row to the second row, and add the first row to the third row.

$$\left(\begin{array}{ccc|c} 1 & 2 & -4 & 0 \\ 0 & -3 & 9 & 0 \\ 0 & 5 & -15 & 0 \end{array} \right)$$

b. Divide the second row by (-3), then add (-5) times the second row to the third row

$$\left(\begin{array}{ccc|c} 1 & 2 & -4 & 0 \\ 0 & 1 & -3 & 0 \\ 0 & 0 & 0 & 0 \end{array} \right)$$

The last matrix is equivalent to the system

$$\begin{aligned} c_1 + 2c_2 - 4c_3 &= 0 \\ c_2 - 3c_3 &= 0. \end{aligned} \tag{25}$$

There are two equations and three unknowns so we are able to set $c_3 = \alpha$ where α is an arbitrary number. From the second equation in the system (25), we have $c_2 = 3c_3 = 3\alpha$, and from the first equation, we obtain $c_1 = 4c_3 - 2c_2 = -2\alpha$. The general solution of Eq. (24) is $\mathbf{c} = \alpha(-2, 3, 1)^T$. Choosing α to be any nonzero scalar, say $\alpha = 1$, exhibits a nonzero solution $\mathbf{c} = (-2, 3, 1)^T$ for Eq. (24). Thus $-2\mathbf{v}_1 + 3\mathbf{v}_2 + \mathbf{v}_3 = \mathbf{0}$, so the given set of vectors is linearly dependent.

▶ **The Rank of a Matrix.** We now introduce a property of a matrix that is very useful in the study of linear algebraic systems.

**DEFINITION
A.2.8**

Let $\mathbf{B}$ be any matrix that is row equivalent to $\mathbf{A}$ and is in row echelon form. Then the rank of the matrix $\mathbf{A}$, denoted by rank($\mathbf{A}$), is the number of nonzero rows in the matrix $\mathbf{B}$.

For the rank of an $m \times n$ matrix $\mathbf{A}$ to be well-defined, it is necessary that all matrices row equivalent to $\mathbf{A}$ have the same rank, and therefore rank refers to a property of the set of all matrices row equivalent to $\mathbf{A}$. The property is indeed well-defined although we do not prove it here.

**EXAMPLE
4**

Find the rank of the matrix

$$\mathbf{A} = \begin{pmatrix} 2 & 0 & 1 & -1 \\ 1 & -3 & 4 & 2 \\ 3 & 3 & -2 & -4 \\ 4 & 0 & 2 & -2 \end{pmatrix}.$$

Using elementary row operations, $\mathbf{A}$ is reduced to the following matrix in row echelon form,

$$\begin{pmatrix} 1 & -3 & 4 & 2 \\ 0 & 6 & -7 & -5 \\ 0 & 0 & 0 & 0 \\ 0 & 0 & 0 & 0 \end{pmatrix} \tag{26}$$

and therefore rank($\mathbf{A}$) = 2.

**THEOREM
A.2.9**

If $\mathbf{B}$ is an $n \times n$ matrix in row echelon form, then the columns of $\mathbf{B}$ are linearly independent if and only if rank($\mathbf{B}$) = n.

Proof

Suppose that the columns of $\mathbf{B}$ are linearly independent. Then the only solution of $\mathbf{Bc} = \mathbf{0}$ is $\mathbf{c} = \mathbf{0}$. If rank($\mathbf{B}$) = $r < n$, then $\mathbf{B}$ must have the form

$$\mathbf{B} = \begin{pmatrix} * & * & & & & & * \\ 0 & * & * & & & & * \\ \vdots & & \ddots & & & & * \\ 0 & \cdots & 0 & * & * & \cdots & * \\ 0 & \cdots & 0 & 0 & 0 & \cdots & 0 \\ \vdots & & \vdots & \vdots & \vdots & & \vdots \\ 0 & \cdots & 0 & 0 & 0 & \cdots & 0 \end{pmatrix} \tag{27}$$

with one or more zero rows. Nonzero rows in the matrix (27) are those containing asterisks although not all elements represented by an $*$ are necessarily nonzero. Therefore, it is possible to choose $n - r \geq 1$ components of $\mathbf{c}$ arbitrarily with the remaining r components determined by the r nonzero equations using backsubstitution. Any nonzero choice for these coefficients shows that the columns of $\mathbf{B}$ are linearly dependent, contradicting the original assumption. Thus rank$(\mathbf{B}) = n$.

Conversely, suppose that rank$(\mathbf{B}) = n$. Then the form of $\mathbf{B}$ must be

$$\mathbf{B} = \begin{pmatrix} \times & * & * & \cdots & * \\ 0 & \times & * & \cdots & * \\ \vdots & & & & \vdots \\ 0 & \cdots & & & \times \end{pmatrix}$$

where each entry denoted by $\times$ is nonzero while entries denoted by $*$ may be zero or nonzero. Using backsubstitution, the upper triangular structure of $\mathbf{Bc} = \mathbf{0}$ implies that $\mathbf{c} = \mathbf{0}$ is the only solution, and therefore the columns of $\mathbf{B}$ are linearly independent. ∎

**THEOREM
A.2.10**

Let $\mathbf{A}$ be an $n \times n$ matrix. Then rank$(\mathbf{A}) = n$ if and only if the columns of $\mathbf{A}$ are linearly independent.

Proof

Assume that rank$(\mathbf{A}) = n$ and consider the equation $\mathbf{Ac} = \mathbf{0}$. Let $\mathbf{B}$ be a matrix that is row equivalent to $\mathbf{A}$ and in row echelon form. By Theorem A.2.9 rank$(\mathbf{B}) = n$ so the only solution of $\mathbf{Bc} = \mathbf{0}$ is $\mathbf{c} = \mathbf{0}$. It follows that the only solution of $\mathbf{Ac} = \mathbf{0}$ is $\mathbf{c} = \mathbf{0}$ since the solution sets of $\mathbf{Ac} = \mathbf{0}$ and $\mathbf{Bc} = \mathbf{0}$ are identical.

On the other hand, if the columns of $\mathbf{A}$ are linearly independent, then the only solution of $\mathbf{Ac} = \mathbf{0}$ is $\mathbf{c} = \mathbf{0}$. Therefore the only solution of $\mathbf{Bc} = \mathbf{0}$ is $\mathbf{c} = \mathbf{0}$ so rank$(\mathbf{B}) = n$ by Theorem A.2.9. It follows that rank$(\mathbf{A}) = $ rank$(\mathbf{B}) = n$ by the definition of rank. ∎

▶ Solution Sets of $\mathbf{Ax} = \mathbf{0}$.

**THEOREM
A.2.11**

Assume that $\mathbf{A}$ is an $n \times n$ matrix.

(i) If rank$(\mathbf{A}) = n$, then the unique solution of $\mathbf{Ax} = \mathbf{0}$ is $\mathbf{x} = \mathbf{0}$.

(ii) If rank$(\mathbf{A}) = r < n$ then the general solution of $\mathbf{Ax} = \mathbf{0}$ has the form

$$\mathbf{x} = \alpha_1 \mathbf{v}_1 + \cdots + \alpha_{n-r} \mathbf{v}_{n-r}$$

where $\mathbf{v}_1, \ldots, \mathbf{v}_{n-r}$ is a set of linearly independent solutions of $\mathbf{Ax} = \mathbf{0}$.

Proof

If rank$(\mathbf{A}) = n$, the columns of $\mathbf{A}$ are linearly independent by Theorem A.2.10. Thus $\mathbf{x} = \mathbf{0}$ is the only solution of $\mathbf{Ax} = \mathbf{0}$.

If rank$(\mathbf{A}) = r < n$, any matrix $\mathbf{B}$ that is row equivalent to $\mathbf{A}$ and in row echelon form will have r nonzero rows and $n - r \geq 1$ rows of zeros,

$$
\mathbf{B} = \begin{pmatrix}
* & * & & & & & & * \\
0 & * & * & & & & & * \\
\vdots & & \ddots & & & & & \vdots \\
0 & \cdots & 0 & * & * & \cdots & & * \\
0 & \cdots & 0 & 0 & 0 & \cdots & & 0 \\
\vdots & & & \vdots & \vdots & \vdots & & \vdots \\
0 & \cdots & & 0 & 0 & 0 & \cdots & 0
\end{pmatrix}
$$

where the rows containing the symbols $*$ are nonzero but not all elements represented by $*$ are necessarily nonzero. Since the system $\mathbf{Bx} = \mathbf{0}$ consists of n unknowns and r equations, we are able to choose $n - r$ components of $\mathbf{x}$ to be arbitrary constants, say $\alpha_1, \ldots, \alpha_{n-r}$. The remaining r components of $\mathbf{x}$ are obtained by backsubstitution using the r nonzero equations and each of these components is a linear combination of $\alpha_1, \ldots, \alpha_{n-r}$. Solutions of $\mathbf{Bx} = \mathbf{0}$ may then be represented in the form

$$\mathbf{x} = \alpha_1 \mathbf{v}_1 + \cdots + \alpha_{n-r} \mathbf{v}_{n-r}. \tag{28}$$

Since $\alpha_1, \ldots, \alpha_{n-r}$ are arbitrary, for each $j = 1, \ldots, n - r$, we set $\alpha_j = 1$ and $\alpha_k = 0$ if $k \neq j$ to show that $\mathbf{Av}_j = \mathbf{0}$ for each $j = 1, \ldots, n - r$. Furthermore, the set $\mathbf{v}_1, \ldots, \mathbf{v}_{n-r}$ is linearly independent. If not, then at least one vector in the set, say $\mathbf{v}_{n-r}$, may be expressed as a linear combination of the other vectors,

$$\mathbf{v}_{n-r} = c_1 \mathbf{v}_1 + \cdots + c_{n-r-1} \mathbf{v}_{n-r-1}. \tag{29}$$

Substituting the right-hand side of Eq. (29) for $\mathbf{v}_{n-r}$ in Eq. (28) shows that the general solution of $\mathbf{Bx} = \mathbf{0}$ would then contain only $n - r - 1$ arbitrary parameters

$$\mathbf{x} = (\alpha_1 + \alpha_{n-r} c_1) \mathbf{v}_1 + \cdots + (\alpha_{n-r-1} + \alpha_{n-r} c_{n-r-1}) \mathbf{v}_{n-r-1} = \beta_1 \mathbf{v}_1 + \cdots + \beta_{n-r-1} \mathbf{v}_{n-r-1}$$

implying that rank$(\mathbf{A}) = r + 1$ instead of r as we had originally assumed. Since the solution sets of $\mathbf{Ax} = \mathbf{0}$ and $\mathbf{Bx} = \mathbf{0}$ are identical, the general solution of $\mathbf{Ax} = \mathbf{0}$ is also given by Eq. (28). ∎

EXAMPLE 5

Find the general solution of $\mathbf{Ax} = \mathbf{0}$ where $\mathbf{A}$ is the matrix in Example 4.

The matrix (26) that is row equivalent to $\mathbf{A}$ and in row echelon form represents the set of equations

$$
\begin{aligned}
x_1 - 3x_2 + 4x_3 + 2x_4 &= 0, \\
6x_2 - 7x_3 - 5x_4 &= 0,
\end{aligned}
$$

Since $n = 4$ and $r = \text{rank}(\mathbf{A}) = 2$, we can choose $n - r = 2$ components of $\mathbf{x}$ arbitrarily, say $x_3 = \alpha_1$ and $x_4 = \alpha_2$. It follows from the second equation that $x_2 = (7/6)\alpha_1 + (5/6)\alpha_2$. From the first equation, we find that $x_1 = -(1/2)\alpha_1 + (1/2)\alpha_2$. Thus the general solution of $\mathbf{Ax} = \mathbf{0}$ is

$$\mathbf{x} = \begin{pmatrix} -\alpha_1/2 + \alpha_2/2 \\ 7\alpha_1/6 + 5\alpha_2/6 \\ \alpha_1 \\ \alpha_2 \end{pmatrix} = \alpha_1 \underbrace{\begin{pmatrix} -1/2 \\ 7/6 \\ 1 \\ 0 \end{pmatrix}}_{\mathbf{v}_1} + \alpha_2 \underbrace{\begin{pmatrix} 1/2 \\ 5/6 \\ 0 \\ 1 \end{pmatrix}}_{\mathbf{v}_2}.$$

It is easy to check that $\mathbf{v}_1$ and $\mathbf{v}_2$ are linearly independent and satisfy $\mathbf{Ax} = \mathbf{0}$.

▶ **Solution Sets of $\mathbf{Ax} = \mathbf{b}$.** We introduce some convenient terminology for discussing the problem of solving the nonhomogeneous system $\mathbf{Ax} = \mathbf{b}$.

DEFINITION A.2.12	The **span** of a set of vectors $\{\mathbf{v}_1, \ldots, \mathbf{v}_k\}$ is the set of all possible linear combinations of the given vectors. The span is denoted by span$\{\mathbf{v}_1, \ldots, \mathbf{v}_k\}$ and a vector $\mathbf{b}$ is in this span if $$\mathbf{b} = x_1\mathbf{v}_1 + x_2\mathbf{v}_2 + \cdots + x_k\mathbf{v}_k$$ for some scalars $x_1, \ldots, x_k$.

Let the columns of the $n \times n$ matrix $\mathbf{A}$ be denoted by the set of n-vectors $\mathbf{a}_1, \ldots, \mathbf{a}_n$. We will denote the set of all possible linear combinations of the columns of $\mathbf{A}$ by $S_{\mathrm{col}}(\mathbf{A})$,

$$S_{\mathrm{col}}(\mathbf{A}) = \mathrm{span}\,\{\mathbf{a}_1, \ldots, \mathbf{a}_n\}.$$

In terms of the columns of $\mathbf{A}$, the equation $\mathbf{Ax} = \mathbf{b}$ may be expressed in the form

$$x_1\mathbf{a}_1 + x_2\mathbf{a}_2 + \cdots + x_n\mathbf{a}_n = \mathbf{b}.$$

Thus $\mathbf{Ax} = \mathbf{b}$ has a solution if and only if $\mathbf{b} \in S_{\mathrm{col}}(\mathbf{A})$.

Since the solution set of $\mathbf{Ax} = \mathbf{b}$ is unaltered by elementary row operations, we examine the solvability problem under the assumption that the augmented matrix $(\mathbf{A}|\mathbf{b})$ has been brought into row echelon form $(\mathbf{A}^*|\mathbf{b}^*)$ where $\mathbf{A}^* = (a_{ij}^*)$ and $\mathbf{b}^* = (b_i^*)$. If $(\mathbf{A}^*|\mathbf{b}^*)$ is in row echelon form, it is easy to ascertain whether $\mathbf{b}^* \in S_{\mathrm{col}}(\mathbf{A}^*)$ and therefore whether $\mathbf{Ax} = \mathbf{b}$ has any solutions. A few examples using 3×3 matrices help make this clear.

If $\mathrm{rank}(\mathbf{A}) = 3$ then $(\mathbf{A}^*|\mathbf{b}^*)$ must have the form

$$\begin{pmatrix} a_{11}^* & a_{12}^* & a_{13}^* & b_1^* \\ 0 & a_{22}^* & a_{23}^* & b_2^* \\ 0 & 0 & a_{33}^* & b_3^* \end{pmatrix} \tag{30}$$

where each of the diagonal entries a_{11}^*, a_{22}^*, and a_{33}^* is nonzero. It is clear that, for any 3-vector $\mathbf{b}^*$ the equations represented by the augmented matrix (30),

$$a_{11}^* x_1 + a_{12}^* x_2 + a_{13}^* x_3 = b_1^*,$$
$$a_{22}^* x_2 + a_{23}^* x_3 = b_2^*,$$
$$a_{33}^* x_3 = b_3^*,$$

have a solution $\hat{\mathbf{x}} = (\hat{x}_1, \hat{x}_2, \hat{x}_3)$ that can be found by backsubstitution. Thus for any 3-vector $\mathbf{b}^*$ there exist scalars $\hat{x}_1$, $\hat{x}_2$, and $\hat{x}_3$ such that

$$\mathbf{A}^*\hat{\mathbf{x}} = \hat{x}_1 \begin{pmatrix} a_{11}^* \\ 0 \\ 0 \end{pmatrix} + \hat{x}_2 \begin{pmatrix} a_{12}^* \\ a_{22}^* \\ 0 \end{pmatrix} + \hat{x}_3 \begin{pmatrix} a_{13}^* \\ a_{23}^* \\ a_{33}^* \end{pmatrix} = \begin{pmatrix} b_1^* \\ b_2^* \\ b_3^* \end{pmatrix} = \mathbf{b}^*;$$

equivalently, $\mathbf{b}^* \in S_{\text{col}}(\mathbf{A}^*)$. Since $\mathbf{A}\mathbf{x} = \mathbf{b}$ is row equivalent to $\mathbf{A}^*\mathbf{x} = \mathbf{b}^*$, it follows that $\mathbf{A}\hat{\mathbf{x}} = \mathbf{b}$ and $\mathbf{b} \in S_{\text{col}}(\mathbf{A})$.

Next suppose that $\text{rank}(\mathbf{A}) = 2$, $\text{rank}(\mathbf{A}|\mathbf{b}) = 2$, and $(\mathbf{A}^*|\mathbf{b}^*)$ has the form

$$\begin{pmatrix} a_{11}^* & a_{12}^* & a_{13}^* & b_1^* \\ 0 & a_{22}^* & a_{23}^* & b_2^* \\ 0 & 0 & 0 & 0 \end{pmatrix} \tag{31}$$

where $a_{11}^* \neq 0$ and at least one of a_{22}^* and a_{23}^* is nonzero. The equations represented by the augmented matrix (31) are

$$a_{11}^* x_1 + a_{12}^* x_2 + a_{13}^* x_3 = b_1^*,$$
$$a_{22}^* x_2 + a_{23}^* x_3 = b_2^*.$$

Thus if $\mathbf{b}^* = (b_1^*, b_2^*, 0)^T$, there exist scalars $\hat{x}_1$, $\hat{x}_2$, and $\hat{x}_3$ such that

$$\mathbf{A}^*\hat{\mathbf{x}} = \hat{x}_1 \begin{pmatrix} a_{11}^* \\ 0 \\ 0 \end{pmatrix} + \hat{x}_2 \begin{pmatrix} a_{12}^* \\ a_{22}^* \\ 0 \end{pmatrix} + \hat{x}_3 \begin{pmatrix} a_{13}^* \\ a_{23}^* \\ 0 \end{pmatrix} = \begin{pmatrix} b_1^* \\ b_2^* \\ 0 \end{pmatrix} = \mathbf{b}^*;$$

equivalently, $\mathbf{b}^* \in S_{\text{col}}(\mathbf{A}^*)$. By row equivalence of $\mathbf{A}^*\mathbf{x} = \mathbf{b}^*$ and $\mathbf{A}\mathbf{x} = \mathbf{b}$, we have $\mathbf{A}\hat{\mathbf{x}} = \mathbf{b}$. The same conclusion holds if $\text{rank}(\mathbf{A}) = \text{rank}(\mathbf{A}|\mathbf{b}) = 2$ and $(\mathbf{A}^*|\mathbf{b}^*)$ has the form

$$\begin{pmatrix} 0 & a_{12}^* & a_{13}^* & b_1^* \\ 0 & 0 & a_{23}^* & b_2^* \\ 0 & 0 & 0 & 0 \end{pmatrix}.$$

Now suppose $\text{rank}(\mathbf{A}) = 2$ but $\text{rank}(\mathbf{A}|\mathbf{b}) = 3$ and $(\mathbf{A}^*|\mathbf{b}^*)$ has the form

$$\begin{pmatrix} a_{11}^* & a_{12}^* & a_{13}^* & b_1^* \\ 0 & a_{22}^* & a_{23}^* & b_2^* \\ 0 & 0 & 0 & b_3^* \end{pmatrix},$$

which represents the equations

$$x_1 \begin{pmatrix} a_{11}^* \\ 0 \\ 0 \end{pmatrix} + x_2 \begin{pmatrix} a_{12}^* \\ a_{22}^* \\ 0 \end{pmatrix} + x_3 \begin{pmatrix} a_{13}^* \\ a_{23}^* \\ 0 \end{pmatrix} = \begin{pmatrix} b_1^* \\ b_2^* \\ b_3^* \end{pmatrix}. \tag{32}$$

Since the third component of each column of $\mathbf{A}^*$ is equal to 0 while the third component of $\mathbf{b}^*$ is nonzero, it is clear that there is no choice of x_1, x_2, and x_3 that will make statement (32) true. The same conclusion holds if $(\mathbf{A}^*|\mathbf{b}^*)$ has the form

$$\begin{pmatrix} 0 & a_{12}^* & a_{13}^* & b_1^* \\ 0 & 0 & a_{23}^* & b_2^* \\ 0 & 0 & 0 & b_3^* \end{pmatrix}.$$

Thus, whenever $\text{rank}(\mathbf{A}) = 2$ and $\text{rank}(\mathbf{A}|\mathbf{b}) = 3$, then $\mathbf{b}^* \notin S_{\text{col}}(\mathbf{A}^*)$ and therefore $\mathbf{Ax} = \mathbf{b}$ has no solution. We proceed to the general case.

THEOREM A.2.13

If $\mathbf{A}$ is an $n \times n$ matrix such that $\text{rank}(\mathbf{A}) = n$ and $\mathbf{b}$ is an n-vector, then $\mathbf{Ax} = \mathbf{b}$ has a unique solution.

Proof

If $\text{rank}(\mathbf{A}) = n$, elementary row operations reduce $(\mathbf{A}|\mathbf{b})$ to row echelon form

$$
\begin{pmatrix}
\times & * & * & \cdots & * & \bigg| & * \\
0 & \times & * & \cdots & * & \bigg| & * \\
\vdots & & & & \vdots & \bigg| & \vdots \\
0 & \cdots & & & \times & \bigg| & *
\end{pmatrix}
$$

where each of the entries denoted by $\times$ is nonzero, otherwise $\text{rank}(\mathbf{A}) < n$. It follows that a solution of this system, say $\hat{\mathbf{x}}_1$, can be found by backsubstitution. This solution is also a solution of the row equivalent system $\mathbf{Ax} = \mathbf{b}$. If there were another solution of $\mathbf{Ax} = \mathbf{b}$, say $\hat{\mathbf{x}}_2$, then $\mathbf{A}(\hat{\mathbf{x}}_1 - \hat{\mathbf{x}}_2) = \mathbf{A}\hat{\mathbf{x}}_1 - \mathbf{A}\hat{\mathbf{x}}_2 = \mathbf{b} - \mathbf{b} = \mathbf{0}$. Since $\text{rank}(\mathbf{A}) = n$, the columns of $\mathbf{A}$ are linearly independent so it must be the case that $\hat{\mathbf{x}}_1 - \hat{\mathbf{x}}_2 = \mathbf{0}$. ∎

THEOREM A.2.14

Let $\mathbf{A}$ be an $n \times n$ matrix and let $\mathbf{b}$ be an n-vector.

(i) If $\text{rank}(\mathbf{A}) = \text{rank}(\mathbf{A}|\mathbf{b}) = r < n$, then $\mathbf{Ax} = \mathbf{b}$ has a general solution of the form

$$\mathbf{x} = \alpha_1 \mathbf{v}_1 + \cdots + \alpha_{n-r} \mathbf{v}_{n-r} + \mathbf{v}_p$$

where the n-vectors $\mathbf{v}_1, \ldots, \mathbf{v}_{n-r}$ are linearly independent solutions of $\mathbf{Ax} = \mathbf{0}$, and $\mathbf{v}_p$ satisfies $\mathbf{Av}_p = \mathbf{b}$.

(ii) If $\text{rank}(\mathbf{A}) = r < n$ but $\text{rank}(\mathbf{A}|\mathbf{b}) = r + 1$, then $\mathbf{Ax} = \mathbf{b}$ has no solution.

Proof

Suppose that $\text{rank}(\mathbf{A}) = r < n$. If the augmented matrix $(\mathbf{A}|\mathbf{b})$ is reduced to row echelon form, there are only two possibilities. Either $\text{rank}(\mathbf{A}|\mathbf{b}) = r$,

$$
\begin{pmatrix}
\times & \times & & & & \times & \bigg| & * \\
0 & \times & \times & & & \times & \bigg| & * \\
\vdots & & \ddots & & & \times & \bigg| & \vdots \\
0 & \cdots & 0 & \times & \times & \cdots & \times & \bigg| & * \\
0 & \cdots & 0 & 0 & 0 & \cdots & 0 & \bigg| & 0 \\
\vdots & & \vdots & \vdots & \vdots & & \vdots & \bigg| & \vdots \\
0 & \cdots & 0 & 0 & 0 & \cdots & 0 & \bigg| & 0
\end{pmatrix} ,
$$

or rank$(\mathbf{A}|\mathbf{b}) = r + 1$,

$$
\left(
\begin{array}{cccccccc|c}
\times & \times & & & & & \times & & * \\
0 & \times & \times & & & & \times & & * \\
\vdots & & \ddots & & & & \times & & \vdots \\
0 & \cdots & 0 & \times & \times & \cdots & \times & & * \\
0 & \cdots & 0 & 0 & 0 & \cdots & 0 & & \times \\
\vdots & & \vdots & \vdots & \vdots & & \vdots & & \vdots \\
0 & \cdots & 0 & 0 & 0 & \cdots & 0 & & 0
\end{array}
\right)
\tag{33}
$$

where in each row at least one of the elements denoted by the symbol $\times$ is nonzero. In the former case, it is possible by backsubstitution to find a $\mathbf{v}_p$ such that $\mathbf{A}\mathbf{v}_p = \mathbf{b}$. Now let $\hat{\mathbf{x}}$ be any vector that satisfies $\mathbf{A}\mathbf{x} = \mathbf{b}$. Then $\mathbf{A}(\hat{\mathbf{x}} - \mathbf{v}_p) = \mathbf{A}\hat{\mathbf{x}} - \mathbf{A}\mathbf{v}_p = \mathbf{b} - \mathbf{b} = \mathbf{0}$. From statement (ii) of Theorem A.2.11, it follows that

$$
\hat{\mathbf{x}} - \mathbf{v}_p = \alpha_1 \mathbf{v}_1 + \cdots + \alpha_{n-r}\mathbf{v}_{n-r}
$$

where $\mathbf{v}_1, \ldots, \mathbf{v}_{n-r}$ are linearly independent and $\mathbf{A}\mathbf{v}_j = \mathbf{0}$, $j = 1, \ldots, n - r$. It follows that any solution $\hat{\mathbf{x}}$ of $\mathbf{A}\mathbf{x} = \mathbf{b}$ must be of the form

$$
\hat{\mathbf{x}} = \alpha_1 \mathbf{v}_1 + \cdots + \alpha_{n-r}\mathbf{v}_{n-r} + \mathbf{v}_p
$$

where set $\mathbf{v}_1, \ldots, \mathbf{v}_{n-r}$ are linearly independent solutions of $\mathbf{A}\mathbf{x} = \mathbf{0}$ and $\mathbf{v}_p$ is a particular solution of $\mathbf{A}\mathbf{x} = \mathbf{b}$, that is, $\mathbf{A}\mathbf{v}_p = \mathbf{b}$.

If the row echelon form of $(\mathbf{A}|\mathbf{b})$ is given by the augmented matrix (33), the last column cannot be represented as a linear combination of the columns to the left of the last column. Thus, there exists no solution to $\mathbf{A}\mathbf{x} = \mathbf{b}$. ∎

EXAMPLE 6

Consider the matrix $\mathbf{A}$ and the vectors $\mathbf{b}_1$ and $\mathbf{b}_2$ given by

$$
\mathbf{A} = \begin{pmatrix} 1 & 2 & 1 \\ 0 & 1 & -1 \\ -1 & 1 & -4 \end{pmatrix}, \quad \mathbf{b}_1 = \begin{pmatrix} 4 \\ 0 \\ -4 \end{pmatrix}, \quad \mathbf{b}_2 = \begin{pmatrix} 1 \\ -1 \\ 1 \end{pmatrix}.
$$

Find the general solution of $\mathbf{A}\mathbf{x} = \mathbf{b}_j$, $j = 1, 2$ or else determine that there is no solution. We augment the matrix $\mathbf{A}$ with both $\mathbf{b}_1$ and $\mathbf{b}_2$,

$$
\left(\begin{array}{ccc|cc} 1 & 2 & 1 & 4 & 1 \\ 0 & 1 & -1 & 0 & -1 \\ -1 & 1 & -4 & -4 & 1 \end{array} \right).
$$

Using elementary row operations, we reduce this matrix to one in row echelon form,

$$
\left(\begin{array}{ccc|cc} 1 & 2 & 1 & 4 & 1 \\ 0 & 1 & -1 & 0 & -1 \\ 0 & 0 & 0 & 0 & 5 \end{array} \right).
\tag{34}
$$

From Eq. (34), we see that $\text{rank}(\mathbf{A}|\mathbf{b}_1) = \text{rank}(\mathbf{A}) = 2$ while $\text{rank}(\mathbf{A}|\mathbf{b}_2) = 3 = \text{rank}(\mathbf{A})$ $+ 1$. Thus $\mathbf{b}_2 \notin S_{\text{col}}(\mathbf{A})$ and there is no solution to $\mathbf{Ax} = \mathbf{b}_2$. However, the reduced system of equations row equivalent to $\mathbf{Ax} = \mathbf{b}_1$ consists of $x_1 + 2x_2 + x_3 = 4$ and $x_2 - x_3 = 0$ which has the general solution

$$\mathbf{x} = \alpha \begin{pmatrix} -3 \\ 1 \\ 1 \end{pmatrix} + \begin{pmatrix} 4 \\ 0 \\ 0 \end{pmatrix}.$$

PROBLEMS

1. In each case, reduce $\mathbf{A}$ to row echelon form and determine $\text{rank}(\mathbf{A})$.

   (a) $\mathbf{A} = \begin{pmatrix} 1 & -3 & 4 \\ 2 & -6 & 8 \end{pmatrix}$

   (b) $\mathbf{A} = \begin{pmatrix} 0 & 0 & 1 \\ 0 & 1 & 0 \\ 1 & 0 & 0 \end{pmatrix}$

   (c) $\mathbf{A} = \begin{pmatrix} 1 & 2 & 1 \\ 2 & -1 & 3 \\ 1 & -5 & 3 \end{pmatrix}$

   (d) $\mathbf{A} = \begin{pmatrix} 1 & 0 & -2 & -1 \\ -1 & 1 & 1 & 0 \\ 1 & 1 & -3 & -2 \\ 2 & -3 & -1 & 1 \end{pmatrix}$

In each of Problems 2 through 5, if there exist solutions of the homogeneous system of linear equations other than $\mathbf{x} = \mathbf{0}$, express the general solution as a linear combination of linearly independent column vectors.

2. $\begin{aligned} x_1 \quad\quad - \quad x_3 &= 0 \\ 3x_1 + x_2 + \quad x_3 &= 0 \\ -x_1 + x_2 + 2x_3 &= 0 \end{aligned}$

3. $\begin{aligned} x_1 + 2x_2 - \quad x_3 &= 0 \\ 2x_1 + \quad x_2 + \quad x_3 &= 0 \\ x_1 - \quad x_2 + 2x_3 &= 0 \end{aligned}$

4. $\begin{aligned} 3x_1 \quad\quad - \quad x_3 &= 0 \\ 2x_1 - \quad x_2 + 2x_3 &= 0 \\ x_1 + \quad x_2 - 3x_3 &= 0 \end{aligned}$

5. $\begin{aligned} x_1 - 2x_2 \quad\quad + \quad x_4 &= 0 \\ 2x_1 + \quad x_2 + \quad x_3 - \quad x_4 &= 0 \\ x_1 + 2x_2 + \quad x_3 - 2x_4 &= 0 \\ 3x_1 + 3x_2 + 2x_3 - 3x_4 &= 0 \end{aligned}$

In each of Problems 6 through 9, find the general solution of the given set of equations, or else show that there is no solution.

6. $\begin{aligned} 2x_1 + x_2 + \quad x_3 &= 2 \\ -x_1 \quad\quad + \quad x_3 &= 1 \\ x_1 + x_2 + 2x_3 &= 3 \end{aligned}$

7. $\begin{aligned} 2x_1 + x_2 + \quad x_3 &= 0 \\ -x_1 \quad\quad + \quad x_3 &= -1 \\ x_1 + x_2 + 2x_3 &= 1 \end{aligned}$

8. $\begin{aligned} -2x_1 \quad\quad + \quad x_3 &= 1 \\ 3x_2 - \quad x_3 &= 2 \\ -x_1 + \quad x_2 + 2x_3 &= 3 \end{aligned}$

9. $\begin{aligned} x_1 - \quad x_2 + \quad x_3 + \quad x_4 &= -1 \\ x_2 + \quad x_3 + 3x_4 &= 2 \\ x_1 \quad\quad + 2x_3 + 4x_4 &= 1 \\ x_2 + \quad x_3 + 3x_4 &= 2 \end{aligned}$

In each of Problems 10 through 14, determine whether the members of the given set of vectors are linearly independent. If they are linearly dependent, find a linear relation among them.

10. $\mathbf{v}_1 = (1, 1, 0)^T$, $\mathbf{v}_2 = (0, 1, 1)^T$, $\mathbf{v}_3 = (1, 0, 1)^T$

11. $\mathbf{v}_1 = (2, 1, 0)^T$, $\mathbf{v}_2 = (0, 1, 0)^T$, $\mathbf{v}_3 = (-1, 2, 0)^T$

12. $\mathbf{v}_1 = (1, 2, 2, 3)^T$, $\mathbf{v}_2 = (-1, 0, 3, 1)^T$,
    $\mathbf{v}_3 = (-2, -1, 1, 0)^T$, $\mathbf{v}_4 = (-3, 0, -1, 3)^T$

13. $\mathbf{v}_1 = (1, 2, -1, 0)^T$, $\mathbf{v}_2 = (2, 3, 1, -1)^T$,
    $\mathbf{v}_3 = (-1, 0, 2, 2)^T$, $\mathbf{v}_4 = (3, -1, 1, 3)^T$

14. $\mathbf{v}_1 = (1, 2, -2)^T$, $\mathbf{v}_2 = (3, 1, 0)^T$,
    $\mathbf{v}_3 = (2, -1, 1)^T$, $\mathbf{v}_4 = (4, 3, -2)^T$

In each of Problems 15 through 17, determine whether $\mathbf{b}_j \in S_{\text{col}}(\mathbf{A})$, $j = 1, 2$. If so, express $\mathbf{b}_j$ as a linear combination of the columns of $\mathbf{A}$.

15. $\mathbf{A} = \begin{pmatrix} 2 & 2 & -2 \\ -1 & -3 & -1 \\ 1 & -1 & -2 \end{pmatrix}$, $\mathbf{b}_1 = \begin{pmatrix} 2 \\ 5 \\ 3 \end{pmatrix}$,

    $\mathbf{b}_2 = \begin{pmatrix} 1 \\ 1 \\ 1 \end{pmatrix}$

16. $\mathbf{A} = \begin{pmatrix} 1 & -1 & -3 & 3 \\ 2 & 0 & -4 & 2 \\ -1 & 2 & 4 & -5 \end{pmatrix}$, $\mathbf{b}_1 = \begin{pmatrix} -1 \\ 1 \\ 0 \end{pmatrix}$,

    $\mathbf{b}_2 = \begin{pmatrix} 2 \\ 0 \\ -4 \end{pmatrix}$

17. $\mathbf{A} = \begin{pmatrix} 1 & 2 & -1 & 2 \\ -1 & 1 & 2 & 2 \\ 0 & 1 & 3 & 4 \\ 2 & 0 & -1 & 1 \end{pmatrix}$, $\mathbf{b}_1 = \begin{pmatrix} 4 \\ -5 \\ -3 \\ 7 \end{pmatrix}$,

    $\mathbf{b}_2 = \begin{pmatrix} 0 \\ 1 \\ 1 \\ 2 \end{pmatrix}$

A.3 Determinants and Inverses

With each square matrix $\mathbf{A}$, we associate a number called its **determinant**, denoted by det $\mathbf{A}$. When $\mathbf{A}$ is represented in array form

$$\mathbf{A} = \begin{pmatrix} a_{11} & a_{12} & \cdots & a_{1n} \\ a_{21} & a_{22} & \cdots & a_{2n} \\ \vdots & \vdots & & \vdots \\ a_{n1} & a_{n2} & \cdots & a_{nn} \end{pmatrix},$$

we will denote its determinant by enclosing the array between vertical bars,

$$\det \mathbf{A} = |\mathbf{A}| = \begin{vmatrix} a_{11} & a_{12} & \cdots & a_{1n} \\ a_{21} & a_{22} & \cdots & a_{2n} \\ \vdots & \vdots & & \vdots \\ a_{n1} & a_{n2} & \cdots & a_{nn} \end{vmatrix}.$$

Definitions of determinants of 1×1 and 2×2 matrices are

$$\det(a_{11}) = a_{11}, \qquad \begin{vmatrix} a_{11} & a_{12} \\ a_{21} & a_{22} \end{vmatrix} = a_{11}a_{22} - a_{12}a_{21},$$

respectively. If $\mathbf{A}$ is 3×3, then det $\mathbf{A}$ is defined by

$$|\mathbf{A}| = a_{11}a_{22}a_{33} - a_{11}a_{23}a_{32} - a_{12}a_{21}a_{33}$$
$$+ a_{12}a_{23}a_{31} + a_{13}a_{21}a_{32} - a_{13}a_{22}a_{31}. \tag{1}$$

It is easy to check that the sum on the right-hand side of Eq. (1) is equal to the sum

$$|\mathbf{A}| = (-1)^{(1+1)}a_{11}\underbrace{\begin{vmatrix} a_{22} & a_{23} \\ a_{32} & a_{33} \end{vmatrix}}_{M_{11}} + (-1)^{(1+2)}a_{12}\underbrace{\begin{vmatrix} a_{21} & a_{23} \\ a_{31} & a_{33} \end{vmatrix}}_{M_{12}} + (-1)^{(1+3)}a_{13}\underbrace{\begin{vmatrix} a_{21} & a_{22} \\ a_{31} & a_{32} \end{vmatrix}}_{M_{13}}$$

which can, in turn, be written as

$$|\mathbf{A}| = \sum_{j=1}^{3}(-1)^{1+j}a_{1j}M_{1j} \tag{2}$$

where M_{1j} is the determinant of the 2×2 submatrix obtained by deleting the 1st row and jth column of $\mathbf{A}$, that is, the row and column in which a_{1j} resides. The same value for the determinant is obtained by other expansions similar to the pattern exhibited in Eq. (2). For example, the sum

$$\sum_{i=1}^{3}(-1)^{i+2}a_{i2}M_{i2}$$

$$= (-1)^{(1+2)}a_{12}\underbrace{\begin{vmatrix} a_{21} & a_{23} \\ a_{31} & a_{33} \end{vmatrix}}_{M_{12}} + (-1)^{(2+2)}a_{22}\underbrace{\begin{vmatrix} a_{11} & a_{13} \\ a_{31} & a_{33} \end{vmatrix}}_{M_{22}} + (-1)^{(3+2)}a_{32}\underbrace{\begin{vmatrix} a_{11} & a_{13} \\ a_{21} & a_{23} \end{vmatrix}}_{M_{32}},$$

where M_{i2} is the determinant of the 2×2 submatrix obtained by deleting the ith row and 2nd column of $\mathbf{A}$, is also easily shown to yield the sum on the right-hand side of Eq. (1).

Determinants of square matrices of higher order are defined recursively by following the pattern illustrated above for 3×3 matrices. If $\mathbf{A}$ is $n \times n$, denote by M_{ij} the determinant of the $(n-1) \times (n-1)$ submatrix of $\mathbf{A}$ obtained by deleting the ith row and jth column from $\mathbf{A}$. M_{ij} is called the **minor** of a_{ij}. The **cofactor** of a_{ij} is the quantity $(-1)^{i+j}M_{ij}$. Then the **cofactor expansion** of $|\mathbf{A}|$ along the ith row is defined by

$$|\mathbf{A}| = \sum_{j=1}^{n}(-1)^{i+j}a_{ij}M_{ij}. \tag{3}$$

Similarly, the **cofactor expansion** of $|\mathbf{A}|$ along the jth column is defined by

$$|\mathbf{A}| = \sum_{i=1}^{n}(-1)^{i+j}a_{ij}M_{ij}. \tag{4}$$

The determinant of each $(n-1) \times (n-1)$ minor in Eqs. (3) and (4) is in turn defined by a cofactor expansion along either a row or column. For each minor, this will require a sum of $(n-1)$ minors of dimension $(n-2) \times (n-2)$. This reduction process continues until one gets down to a sum of $n!/2$ determinants of 2×2 matrices. Though we do not prove it here, the value of the determinant obtained by a cofactor expansion is independent of the row or column along which the expansions are performed.

EXAMPLE 1

Evaluate $|\mathbf{A}|$ if

$$\mathbf{A} = \begin{pmatrix} 1 & -1 & 2 & 4 \\ -1 & 3 & -2 & 1 \\ 0 & 2 & 1 & 0 \\ -3 & 1 & 1 & -1 \end{pmatrix}.$$

To evaluate $|\mathbf{A}|$, we can expand by cofactors along any row or column. It is prudent to expand along the third row because this row has two zero elements, thereby reducing the labor. Thus

$$|\mathbf{A}| = \sum_{j=1}^{4} a_{3j} M_{3j}$$

$$= -2 \begin{vmatrix} 1 & 2 & 4 \\ -1 & -2 & 1 \\ -3 & 1 & -1 \end{vmatrix} + 1 \begin{vmatrix} 1 & -1 & 4 \\ -1 & 3 & 1 \\ -3 & 1 & -1 \end{vmatrix}.$$

These 3×3 determinants can be evaluated directly from the definition of a 3×3 determinant, or we can expand each (say along the first row) to get

$$|\mathbf{A}| = -2 \left\{ \begin{vmatrix} -2 & 1 \\ 1 & -1 \end{vmatrix} - 2 \begin{vmatrix} -1 & 1 \\ -3 & -1 \end{vmatrix} + 4 \begin{vmatrix} -1 & -2 \\ -3 & 1 \end{vmatrix} \right\}$$

$$+ 1 \left\{ \begin{vmatrix} 3 & 1 \\ 1 & -1 \end{vmatrix} + \begin{vmatrix} -1 & 1 \\ -3 & -1 \end{vmatrix} + 4 \begin{vmatrix} -1 & 3 \\ -3 & 1 \end{vmatrix} \right\}$$

$$= -2 \{1 - 8 - 28\} + \{-4 + 4 + 32\} = 102.$$

Given a square matrix, most commonly used software packages and many calculators provide functions for evaluating determinants. The following theorem lists a number of important properties of determinants that can be used to simplify the computations required for their evaluation.

THEOREM A.3.1

Properties of Determinants. Let $\mathbf{A}$ and $\mathbf{B}$ be $n \times n$ matrices. Then

1. If $\mathbf{B}$ is obtained from $\mathbf{A}$ by adding a constant multiple of one row (or column) to another row (or column), then $|\mathbf{B}| = |\mathbf{A}|$.
2. If $\mathbf{B}$ is obtained from $\mathbf{A}$ by interchanging two rows or two columns, then $|\mathbf{B}| = -|\mathbf{A}|$.
3. If $\mathbf{B}$ is obtained from $\mathbf{A}$ by multiplying any row or any column by a scalar α, then $|\mathbf{B}| = \alpha|\mathbf{A}|$.
4. If $\mathbf{A}$ has a zero row or zero column, then $|\mathbf{A}| = 0$.
5. If $\mathbf{A}$ has two identical rows (or two identical columns), then $|\mathbf{A}| = 0$.
6. If one row (or column) of $\mathbf{A}$ is a constant multiple of another row (or column), then $|\mathbf{A}| = 0$.

In the previous section, we found that triangular matrices were very important. The determinant of a triangular matrix is also very easy to evaluate.

THEOREM A.3.2	The determinant of a triangular matrix is the product of its diagonal elements.

Proof

Suppose that $\mathbf{A}$ is upper triangular. By repeatedly computing cofactor expansions along the first column we find that

$$
\begin{vmatrix}
a_{11} & a_{12} & \cdots & a_{1n} \\
0 & a_{22} & \cdots & a_{2n} \\
0 & 0 & \cdots & a_{3n} \\
\vdots & \vdots & & \vdots \\
0 & 0 & \cdots & a_{nn}
\end{vmatrix}
= a_{11}
\begin{vmatrix}
a_{22} & \cdots & a_{2n} \\
\vdots & \ddots & \vdots \\
0 & \cdots & a_{nn}
\end{vmatrix}
$$

$$
= a_{11}a_{22}
\begin{vmatrix}
a_{33} & \cdots & a_{3n} \\
\vdots & \ddots & \vdots \\
0 & \cdots & a_{nn}
\end{vmatrix}
$$

$$
= \cdots = a_{11}a_{22}\cdots a_{nn}.
$$

In a similar way, it can be shown that the determinant of a lower triangular matrix is also the product of its diagonal elements. ■

A matrix can often be reduced to upper triangular form by a sequence of elementary row operations of type E1, that is, adding scalar multiples of one row to another row. By Property **1** of Theorem A.3.1, the determinant of the resultant triangular matrix will be equal to the determinant of the original matrix. Thus, using elementary row operations of type E1 combined with Property **1** of Theorem A.3.1, provides a computationally efficient method for computing the determinant of a matrix.

EXAMPLE 2

In the matrix $\mathbf{A}$ of Example 1 if we add row one to row two and then add 3 times row one to row four we get

$$
\det \mathbf{A} =
\begin{vmatrix}
1 & -1 & 2 & 4 \\
-1 & 3 & -2 & 1 \\
0 & 2 & 1 & 0 \\
-3 & 1 & 1 & -1
\end{vmatrix}
=
\begin{vmatrix}
1 & -1 & 2 & 4 \\
0 & 2 & 0 & 5 \\
0 & 2 & 1 & 0 \\
0 & -2 & 7 & 11
\end{vmatrix},
$$

where we have used Property (1) of Theorem A.3.1. Then, subtracting row two from row three followed by adding row two to row four gives us

$$
\det \mathbf{A} =
\begin{vmatrix}
1 & -1 & 2 & 4 \\
0 & 2 & 0 & 5 \\
0 & 0 & 1 & -5 \\
0 & 0 & 7 & 16
\end{vmatrix},
$$

where we have again used Property (1) of Theorem A.3.1. By subtracting 7 times row three from row four in the last result we reduce the problem to evaluating the determinant of an upper triangular matrix,

$$\det \mathbf{A} = \begin{vmatrix} 1 & -1 & 2 & 4 \\ 0 & 2 & 0 & 5 \\ 0 & 0 & 1 & -5 \\ 0 & 0 & 0 & 51 \end{vmatrix} = 1 \cdot 2 \cdot 1 \cdot 51 = 102.$$

THEOREM A.3.3

$|\mathbf{A}^T| = |\mathbf{A}|$.

Proof

The cofactor expansion along the ith column of $\mathbf{A}^T$ is equal to the cofactor expansion along the ith row of $\mathbf{A}$. ∎

We state without proof the following important and frequently used result.

THEOREM A.3.4

$|\mathbf{AB}| = |\mathbf{A}||\mathbf{B}|$.

The next theorem shows that determinants of two row equivalent $n \times n$ matrices are either both zero or both nonzero.

THEOREM A.3.5

Let $\mathbf{A}$ be an $n \times n$ matrix and let $\mathbf{B}$ be row equivalent to $\mathbf{A}$. Then $|\mathbf{A}| = 0$ if and only if $|\mathbf{B}| = 0$. In particular, if $\mathbf{A}_E$ is a matrix in row echelon form that is row equivalent to $\mathbf{A}$, then $|\mathbf{A}| = 0$ if and only if $|\mathbf{A}_E| = 0$.

Proof

We examine the effect of each type of elementary row operation on the value of a determinant.

If $\mathbf{B}$ is obtained from $\mathbf{A}$ by adding a scalar multiple of one row to another row, then $|\mathbf{B}| = |\mathbf{A}|$ by Property (1) of Theorem A.3.1.

If $\mathbf{B}$ is obtained from $\mathbf{A}$ by interchanging two rows of $\mathbf{A}$, then $|\mathbf{B}| = -|\mathbf{A}|$ by Property (2) of Theorem A.3.1.

If $\mathbf{B}$ is obtained from $\mathbf{A}$ by multiplying a row of $\mathbf{A}$ by a nonzero scalar α, then $|\mathbf{B}| = \alpha|\mathbf{A}|$ by Property (3) of Theorem A.3.1.

In general, if $\mathbf{B}$ is obtained from $\mathbf{A}$ by a finite sequence of elementary row operations, then

$$|\mathbf{B}| = \alpha_k \alpha_{k-1} \cdots \alpha_1 |\mathbf{A}|.$$

where each α_j, $j = 1, \ldots, k$ is a nonzero constant. Thus, if $|\mathbf{B}| \neq 0$, then $|\mathbf{A}| \neq 0$. On the other hand, if $|\mathbf{B}| = 0$, then $|\mathbf{A}| = 0$. As a special case, this result holds if $\mathbf{B} = \mathbf{A}_E$. ∎

The determinant can be used to test whether the set of column vectors of a square matrix is linearly independent or linearly dependent.

| THEOREM A.3.6 | The columns of $\mathbf{A}$ are linearly independent if and only if $|\mathbf{A}| \neq 0$. |

Proof

Suppose that the columns of $\mathbf{A}$ are linearly independent. Then the only solution of $\mathbf{Ac} = \mathbf{0}$ is $\mathbf{c} = \mathbf{0}$. Let $\mathbf{B}$ be row equivalent to $\mathbf{A}$ and in row echelon form. Since the solution sets of $\mathbf{Bc} = \mathbf{0}$ and $\mathbf{Ac} = \mathbf{0}$ are identical, the only solution of $\mathbf{Bc} = \mathbf{0}$ is $\mathbf{c} = \mathbf{0}$. This means that the columns of $\mathbf{B}$ are linearly independent. By Theorem A.2.9, rank$(\mathbf{B}) = n$. Since $\mathbf{B}$ is triangular, the diagonal entries must all be nonzero. Consequently, $|\mathbf{B}| \neq 0$ and the row equivalence of $\mathbf{A}$ and $\mathbf{B}$ implies that $|\mathbf{A}| \neq 0$ by Theorem A.3.5.

Conversely, if $|\mathbf{A}| \neq 0$ and $\mathbf{B}$ is row equivalent to $\mathbf{A}$ and in row echelon form, then $|\mathbf{B}| \neq 0$ by Theorem A.3.5. Thus the diagonal entries of $\mathbf{B}$ must all be nonzero which implies that rank$(\mathbf{B}) = n$. By Theorem A.2.9, the columns $\mathbf{B}$ are linearly independent and the only solution of $\mathbf{Bc} = \mathbf{0}$ is $\mathbf{c} = \mathbf{0}$. Thus the only solution of $\mathbf{Ac} = \mathbf{0}$ is $\mathbf{c} = \mathbf{0}$. Consequently, the columns of $\mathbf{A}$ are linearly independent. ∎

EXAMPLE 3

Use the determinant to show that the three vectors in Example 3 of Section A.2,

$$\mathbf{v}_1 = \begin{pmatrix} 1 \\ 2 \\ -1 \end{pmatrix}, \quad \mathbf{v}_2 = \begin{pmatrix} 2 \\ 1 \\ 3 \end{pmatrix}, \quad \mathbf{v}_3 = \begin{pmatrix} -4 \\ 1 \\ 11 \end{pmatrix}$$

are linearly dependent.

We juxtapose the vectors to form the 3×3 matrix $\mathbf{A} = [\mathbf{v}_1, \mathbf{v}_2, \mathbf{v}_3]$. Calculation of the determinant is then simplified by adding appropriate scalar multiples of the first row to the second and third rows followed by a cofactor expansion along the first column,

$$|\mathbf{A}| = \begin{vmatrix} 1 & 2 & -4 \\ 2 & 1 & 1 \\ -1 & 3 & -11 \end{vmatrix} = \begin{vmatrix} 1 & 2 & -4 \\ 0 & -3 & 9 \\ 0 & 5 & -15 \end{vmatrix} = \begin{vmatrix} -3 & 9 \\ 5 & -15 \end{vmatrix} = 45 - 45 = 0.$$

Since $|\mathbf{A}| = 0$, the set $\{\mathbf{v}_1, \mathbf{v}_2, \mathbf{v}_3\}$ is linearly dependent.

▶ **The Determinant Test for Solvability of $\mathbf{Ax} = \mathbf{b}$.**

| THEOREM A.3.7 | Let $\mathbf{A}$ be an $n \times n$ matrix and let $\mathbf{b}$ be an n-vector.

(i) If $|\mathbf{A}| \neq 0$, then rank$(\mathbf{A}) = n$ and $\mathbf{Ax} = \mathbf{b}$ has a unique solution. In particular, the unique solution of $\mathbf{Ax} = \mathbf{0}$ is $\mathbf{x} = \mathbf{0}$.

(ii) If $|\mathbf{A}| = 0$, then $r = $ rank$(\mathbf{A}) < n$. In this case either

    (a) rank$(\mathbf{A}|\mathbf{b}) = r + 1$ and $\mathbf{Ax} = \mathbf{b}$ has no solution, or

    (b) rank$(\mathbf{A}|\mathbf{b}) = r$ and the general solution of $\mathbf{Ax} = \mathbf{b}$ is of the form

$$\mathbf{x} = \alpha_1 \mathbf{v}_1 + \cdots + \alpha_{n-r} \mathbf{v}_{n-r} + \mathbf{v}_p, \qquad (5)$$

where $\mathbf{v}_1, \ldots, \mathbf{v}_{n-r}$ are linearly independent solutions of $\mathbf{Ax} = \mathbf{0}$, and $\mathbf{Av}_p = \mathbf{b}$. |

In particular, if $|\mathbf{A}| = 0$, then the general solution of the homogeneous equation $\mathbf{Ax} = \mathbf{0}$ is of the form

$$\mathbf{x} = \alpha_1\mathbf{v}_1 + \cdots + \alpha_{n-r}\mathbf{v}_{n-r},$$

where $\mathbf{v}_1, \ldots, \mathbf{v}_{n-r}$ are linearly independent solutions of $\mathbf{Ax} = \mathbf{0}$.

Proof

Assume that $|\mathbf{A}| \neq 0$. If $\mathbf{A}_E$ is row equivalent to $\mathbf{A}$ and in row echelon form, then $|\mathbf{A}_E| \neq 0$ by Theorem A.3.5. Since $\mathbf{A}_E$ is upper triangular and $|\mathbf{A}_E| \neq 0$, all of the diagonal entries are nonzero. Thus rank$(\mathbf{A}_E) = $ rank$(\mathbf{A}) = n$. It follows from Theorem A.2.13 that $\mathbf{Ax} = \mathbf{b}$ has a unique solution. If $\mathbf{b} = \mathbf{0}$, that unique solution is obviously $\mathbf{x} = \mathbf{0}$. Thus statement (i) of the theorem is proved.

Now suppose that $|\mathbf{A}| = 0$. Then $|\mathbf{A}_E| = 0$ by Theorem A.3.5. Since $\mathbf{A}_E$ is upper triangular and $|\mathbf{A}_E| = 0$, at least one of the diagonal entries is equal to zero. Consequently, $r = $ rank$(\mathbf{A}) < n$. If rank $(\mathbf{A}|\mathbf{b}) = r + 1$, then $\mathbf{Ax} = \mathbf{b}$ has no solution by part (ii) of Theorem A.2.14. On the other hand, if rank $(\mathbf{A}|\mathbf{b}) = r$, then by part (i) of Theorem A.2.14 the general solution of $\mathbf{Ax} = \mathbf{b}$ has the form

$$\mathbf{x} = \alpha_1\mathbf{v}_1 + \cdots + \alpha_{n-r}\mathbf{v}_{n-r} + \mathbf{v}_p, \tag{6}$$

where $\mathbf{v}_1, \ldots, \mathbf{v}_{n-r}$ are linearly independent solutions of $\mathbf{Ax} = \mathbf{0}$, and $\mathbf{Av}_p = \mathbf{b}$. Thus statement (ii) of the theorem is proved.

In the special case of the homogeneous equation $\mathbf{Ax} = \mathbf{0}$, if $|\mathbf{A}| = 0$, then by part (ii) of Theorem A.2.11 the general solution of $\mathbf{Ax} = \mathbf{0}$ has the form

$$\mathbf{x} = \alpha_1\mathbf{v}_1 + \cdots + \alpha_{n-r}\mathbf{v}_{n-r}, \tag{7}$$

where $\mathbf{v}_1, \ldots, \mathbf{v}_{n-r}$ are linearly independent solutions of $\mathbf{Ax} = \mathbf{0}$. ∎

▶ **Matrix Inverses.**

THEOREM A.3.8

An $n \times n$ matrix is nonsingular if and only if $|\mathbf{A}| \neq 0$.

Proof

Assume that $|\mathbf{A}| \neq 0$. We seek a matrix $\mathbf{B} = [\mathbf{b}_1, \ldots, \mathbf{b}_n]$ such that

$$\mathbf{AB} = [\mathbf{Ab}_1, \ldots, \mathbf{Ab}_n] = \mathbf{I}_n = [\mathbf{e}_1, \ldots, \mathbf{e}_n],$$

where

$$\mathbf{e}_1 = (1, 0, 0, \ldots, 0)^T, \mathbf{e}_2 = (0, 1, 0, \ldots, 0)^T, \ldots, \mathbf{e}_n = (0, \ldots, 0, 1)^T.$$

Since $|\mathbf{A}| \neq 0$, Theorem A.3.7 implies that there is a unique solution $\mathbf{b}_k$ to each of the systems $\mathbf{Ab}_k = \mathbf{e}_k, k = 1, \ldots, n$. Thus, there is a unique matrix $\mathbf{B}$ such that $\mathbf{AB} = \mathbf{I}_n$. Since $|\mathbf{A}^T| = |\mathbf{A}| \neq 0$, the same argument produces a unique matrix $\mathbf{C}^T$ such that $\mathbf{A}^T\mathbf{C}^T = \mathbf{I}_n$, or equivalently, $\mathbf{CA} = \mathbf{I}$. It follows that $\mathbf{C} = \mathbf{CI}_n = \mathbf{C}(\mathbf{AB}) = (\mathbf{CA})\mathbf{B} = \mathbf{I}_n\mathbf{B} = \mathbf{B}$.

Conversely, if $\mathbf{A}$ is nonsingular, there is a matrix $\mathbf{B}$ such that $\mathbf{AB} = \mathbf{I}_n$. By Theorem A.3.4, $|\mathbf{AB}| = |\mathbf{A}||\mathbf{B}| = 1$, so $|\mathbf{A}| \neq 0$. ∎

Thus, whenever $|\mathbf{A}| \neq 0$, the inverse of $\mathbf{A}$ can be found by reducing the augmented matrix $(\mathbf{A}|\mathbf{I}_n)$ to $(\mathbf{I}_n|\mathbf{B})$ using elementary row operations. Since the jth column of $\mathbf{B}$ is the solution of $\mathbf{A}\mathbf{b}_j = \mathbf{e}_j$, $j = 1, \dots, n$, it follows that $\mathbf{A}^{-1} = \mathbf{B}$.

EXAMPLE 4

Find the inverse of

$$\mathbf{A} = \begin{pmatrix} 1 & -1 & -1 \\ 3 & -1 & 2 \\ 2 & 2 & 3 \end{pmatrix}.$$

We begin by forming the augmented matrix $(\mathbf{A}|\mathbf{I}_3)$:

$$(\mathbf{A}|\mathbf{I}_3) = \left(\begin{array}{ccc|ccc} 1 & -1 & -1 & 1 & 0 & 0 \\ 3 & -1 & 2 & 0 & 1 & 0 \\ 2 & 2 & 3 & 0 & 0 & 1 \end{array} \right).$$

The matrix $\mathbf{A}$ can be transformed into $\mathbf{I}_3$ by the following sequence of elementary row operations, and at the same time, $\mathbf{I}_3$ is transformed into $\mathbf{A}^{-1}$.

a. Obtain zeros in the off-diagonal positions in the first column by adding (-3) times the first row to the second row and adding (-2) times the first row to the third row.

$$\left(\begin{array}{ccc|ccc} 1 & -1 & -1 & 1 & 0 & 0 \\ 0 & 2 & 5 & -3 & 1 & 0 \\ 0 & 4 & 5 & -2 & 0 & 1 \end{array} \right)$$

b. Obtain a 1 in the diagonal position in the second column by multiplying the second row by 1/2.

$$\left(\begin{array}{ccc|ccc} 1 & -1 & -1 & 1 & 0 & 0 \\ 0 & 1 & \frac{5}{2} & -\frac{3}{2} & \frac{1}{2} & 0 \\ 0 & 4 & 5 & -2 & 0 & 1 \end{array} \right)$$

c. Obtain zeros in the off-diagonal positions in the second column by adding the second row to the first row and adding (-4) times the second row to the third row.

$$\left(\begin{array}{ccc|ccc} 1 & 0 & \frac{3}{2} & -\frac{1}{2} & \frac{1}{2} & 0 \\ 0 & 1 & \frac{5}{2} & -\frac{3}{2} & \frac{1}{2} & 0 \\ 0 & 0 & -5 & 4 & -2 & 1 \end{array} \right)$$

d. Obtain a 1 in the diagonal position of the third column by multiplying the third row by $-1/5$.

$$\left(\begin{array}{ccc|ccc} 1 & 0 & \frac{3}{2} & -\frac{1}{2} & \frac{1}{2} & 0 \\ 0 & 1 & \frac{5}{2} & -\frac{3}{2} & \frac{1}{2} & 0 \\ 0 & 0 & 1 & -\frac{4}{5} & \frac{2}{5} & -\frac{1}{5} \end{array} \right)$$

e. Obtain zeros in the off-diagonal positions in the third column by adding $(-3/2)$ times the third row to the first row and adding $(-5/2)$ times the third row to the second row.

$$\left(\begin{array}{ccc|ccc} 1 & 0 & 0 & \frac{7}{10} & -\frac{1}{10} & \frac{3}{10} \\ 0 & 1 & 0 & \frac{1}{2} & -\frac{1}{2} & \frac{1}{2} \\ 0 & 0 & 1 & -\frac{4}{5} & \frac{2}{5} & -\frac{1}{5} \end{array} \right)$$

The last of these matrices is $(\mathbf{I}|\mathbf{A}^{-1})$, a fact that can be verified directly by multiplying the matrices $\mathbf{A}^{-1}$ and $\mathbf{A}$. This example was made slightly simpler by the fact that the original matrix $\mathbf{A}$ had a 1 in the upper left corner ($a_{11} = 1$). If this is not the case, then the first step is to produce a 1 there by multiplying the first row by $1/a_{11}$, as long as $a_{11} \neq 0$. If $a_{11} = 0$, then the first row must be interchanged with some other row to bring a nonzero element into the upper left position before proceeding.

We finally note that if $\mathbf{A}$ or $\mathbf{B}$ is singular, then $|\mathbf{AB}| = |\mathbf{A}| \, |\mathbf{B}| = |\mathbf{BA}| = 0$ and therefore Theorem A.3.8 implies that both $\mathbf{AB}$ and $\mathbf{BA}$ are singular. Thus part (e) of Theorem A.1.10 is proved.

PROBLEMS

In each of Problems 1 through 10, use elementary row and column operations to simplify the task of evaluating the determinant by cofactor expansions. If $\mathbf{A}$ is nonsingular, find $\mathbf{A}^{-1}$.

1. $\mathbf{A} = \begin{pmatrix} 1 & 4 \\ -2 & 3 \end{pmatrix}$

2. $\mathbf{A} = \begin{pmatrix} 3 & -1 \\ 6 & 2 \end{pmatrix}$

3. $\mathbf{A} = \begin{pmatrix} 1 & 2 & 3 \\ 2 & 4 & 5 \\ 3 & 5 & 6 \end{pmatrix}$

4. $\mathbf{A} = \begin{pmatrix} 1 & 1 & -1 \\ 2 & -1 & 1 \\ 1 & 1 & 2 \end{pmatrix}$

5. $\mathbf{A} = \begin{pmatrix} 1 & 2 & 1 \\ -2 & 1 & 8 \\ 1 & -2 & -7 \end{pmatrix}$

6. $\mathbf{A} = \begin{pmatrix} 2 & 1 & 0 \\ 0 & 2 & 1 \\ 0 & 0 & 2 \end{pmatrix}$

7. $\mathbf{A} = \begin{pmatrix} 1 & 1 & -1 \\ 2 & 1 & 0 \\ 3 & -2 & 1 \end{pmatrix}$

8. $\mathbf{A} = \begin{pmatrix} 2 & 3 & 1 \\ -1 & 2 & 1 \\ 4 & -1 & -1 \end{pmatrix}$

9. $\mathbf{A} = \begin{pmatrix} 1 & 0 & 0 & -1 \\ 0 & -1 & 1 & 0 \\ -1 & 0 & 1 & 0 \\ 0 & 1 & -1 & 1 \end{pmatrix}$

10. $\mathbf{A} = \begin{pmatrix} 1 & -1 & 2 & 0 \\ -1 & 2 & -4 & 2 \\ 1 & 0 & 1 & 3 \\ -2 & 2 & 0 & -1 \end{pmatrix}$

11. Let

$$\mathbf{A} = \begin{pmatrix} 1 & 3 & 0 \\ -1 & 2 & 4 \\ 1 & -2 & 0 \end{pmatrix} \quad \text{and} \quad \mathbf{B} = \begin{pmatrix} 2 & 0 & 1 \\ -1 & 3 & 2 \\ 4 & 0 & 0 \end{pmatrix}.$$

Verify that $|\mathbf{AB}| = |\mathbf{A}| \, |\mathbf{B}|$.

12. If $\mathbf{A}$ is nonsingular, show that $|\mathbf{A}^{-1}| = 1/|\mathbf{A}|$.

In each of Exercises 13 through 16 find all values of λ such that the given matrix is singular.

13. $\begin{pmatrix} -\lambda & -1 & -3 \\ 2 & 3-\lambda & 3 \\ -2 & 1 & 1-\lambda \end{pmatrix}$

14. $\begin{pmatrix} 4-\lambda & 6 & 6 \\ 1 & 3-\lambda & 2 \\ -1 & -4 & -3-\lambda \end{pmatrix}$

15. $\begin{pmatrix} 1-\lambda & 1 & 0 & -1 \\ 0 & -1-\lambda & 3 & 4 \\ 0 & 0 & -2-\lambda & -3 \\ 0 & 0 & 0 & 1-\lambda \end{pmatrix}$

A.4 The Eigenvalue Problem

Recall that in Chapter 3 we used the eigenvalue method to find solutions of homogeneous linear first order systems of dimension two,

$$\mathbf{x}' = \begin{pmatrix} a_{11} & a_{12} \\ a_{21} & a_{22} \end{pmatrix} \mathbf{x} = \mathbf{A}\mathbf{x}. \tag{1}$$

Nonzero solutions of Eq. (1) of the form $\mathbf{x} = e^{\lambda t}\mathbf{v}$, where $\mathbf{v}$ is a nonzero 2-vector, exist if and only if the scalar λ and the vector $\mathbf{v}$ satisfy the equation

$$\mathbf{A}\mathbf{v} = \lambda\mathbf{v},$$

or equivalently,

$$(\mathbf{A} - \lambda\mathbf{I}_2)\mathbf{v} = \mathbf{0}. \tag{2}$$

The eigenvalue method generalizes, in a natural way, to homogeneous linear constant coefficient systems of dimension n

$$\mathbf{x}' = \begin{pmatrix} a_{11} & \cdots & a_{1n} \\ \vdots & & \vdots \\ a_{n1} & \cdots & a_{nn} \end{pmatrix} \mathbf{x}. \tag{3}$$

As in the case for systems of dimension 2, nonzero solutions of Eq. (3) of the form $\mathbf{x} = e^{\lambda t}\mathbf{v}$, where $\mathbf{v}$ is a nonzero n-vector, exist if and only if λ and $\mathbf{v}$ satisfy

$$(\mathbf{A} - \lambda\mathbf{I}_n)\mathbf{v} = \mathbf{0} \tag{4}$$

where $\mathbf{A}$ is the $n \times n$ matrix of coefficients on the right-hand side of Eq. (3),

$$\mathbf{A} = \begin{pmatrix} a_{11} & \cdots & a_{1n} \\ \vdots & & \vdots \\ a_{n1} & \cdots & a_{nn} \end{pmatrix}.$$

From Theorem A.3.7, nonzero solutions of Eq. (4) exist if and only if

$$\det(\mathbf{A} - \lambda\mathbf{I}_n) = 0. \tag{5}$$

Just as in Chapter 3, values of λ that satisfy Eq. (5) are called **eigenvalues** of the matrix $\mathbf{A}$, and the nonzero solutions of Eq. (4) that are obtained by using such a value of λ are called the **eigenvectors** corresponding to that eigenvalue.

In the general $n \times n$ case, Eq. (5) is a polynomial equation $p(\lambda) = 0$ where

$$p(\lambda) = (-1)^n(\lambda^n + p_{n-1}\lambda^{n-1} + \cdots + p_1\lambda + p_0) \tag{6}$$

is called the **characteristic polynomial** of $\mathbf{A}$. The equation $p(\lambda) = 0$ is called the **characteristic equation** of $\mathbf{A}$. By the Fundamental Theorem of Algebra, we know that $p(\lambda)$ has a factorization over the complex numbers of the form

$$p(\lambda) = (-1)^n(\lambda - \lambda_1)^{m_1} \cdots (\lambda - \lambda_k)^{m_k} \tag{7}$$

where $\lambda_i \neq \lambda_j$ if $i \neq j$. Thus, the eigenvalues of $\mathbf{A}$ are λ_j, $j = 1, \ldots, k$. For each $j = 1, \ldots, k$, the power m_j of the factor $\lambda - \lambda_j$ in Eq. (7) is called the **algebraic multiplicity** of the eigenvalue λ_j. Since $p(\lambda)$ is a polynomial of degree n, it must be the case that

$$m_1 + \cdots + m_k = n.$$

Having found the eigenvalues $\lambda_1, \ldots, \lambda_k$ and their algebraic multiplicities by completely factoring $p(\lambda)$, the next step is to find the maximum number of linearly independent eigenvectors that belong to each eigenvalue. Thus, for each $j = 1, \ldots, k$ we find the general solution of

$$(\mathbf{A} - \lambda_j \mathbf{I}_n)\mathbf{v} = \mathbf{0}. \tag{8}$$

If we denote the rank of $(\mathbf{A} - \lambda_j \mathbf{I}_n)$ by r_j, then from Theorem A.2.11 we know that the general solution of Eq. (8) will have the form

$$\mathbf{v} = \alpha_1 \mathbf{v}_1^{(j)} + \cdots + \alpha_{n-r_j} \mathbf{v}_{n-r_j}^{(j)} \tag{9}$$

where $\mathbf{v}_1^{(j)}, \ldots, \mathbf{v}_{n-r_j}^{(j)}$ are linearly independent solutions of $(\mathbf{A} - \lambda_j \mathbf{I}_n)\mathbf{v} = \mathbf{0}$. The vectors $\mathbf{v}_1^{(j)}, \ldots, \mathbf{v}_{n-r_j}^{(j)}$ are the maximal set of linearly independent eigenvectors that belong to the eigenvalue λ_j. The maximal number of linearly independent eigenvectors, $g_j = n - r_j$, that belong to the eigenvalue λ_j is called the **geometric multiplicity** of the eigenvalue λ_j. If the geometric multiplicity of eigenvalue λ_j is equal to the algebraic multiplicity, that is, $g_j = m_j$, the eigenvalue λ_j is said to be **nondefective**. Otherwise λ_j is said to be **defective**. If $g_j = m_j$ for each $j = 1, \ldots, k$, then $\mathbf{A}$ is said to be a **nondefective matrix**. If $g_j < m_j$ for some j, then the matrix $\mathbf{A}$ is said to be **defective**.

EXAMPLE 1

Find the eigenvalues and eigenvectors of the matrix

$$\mathbf{A} = \begin{pmatrix} 2 & -3 & -1 \\ 0 & -1 & 0 \\ -1 & 1 & 2 \end{pmatrix}. \tag{10}$$

The eigenvalues λ and the eigenvectors $\mathbf{v}$ satisfy the equation $(\mathbf{A} - \lambda \mathbf{I}_3)\mathbf{v} = \mathbf{0}$, or

$$\begin{pmatrix} 2-\lambda & -3 & -1 \\ 0 & -1-\lambda & 0 \\ -1 & 1 & 2-\lambda \end{pmatrix} \begin{pmatrix} v_1 \\ v_2 \\ v_3 \end{pmatrix} = \begin{pmatrix} 0 \\ 0 \\ 0 \end{pmatrix}. \tag{11}$$

The eigenvalues are the roots of the characteristic equation

$$p(\lambda) = \det(\mathbf{A} - \lambda \mathbf{I}_3) = \begin{vmatrix} 2-\lambda & -3 & -1 \\ 0 & -1-\lambda & 0 \\ -1 & 1 & 2-\lambda \end{vmatrix} = -\lambda^3 + 3\lambda^2 + \lambda - 3. \tag{12}$$

The roots of Eq. (12) are $\lambda_1 = -1$, $\lambda_2 = 1$, and $\lambda_3 = 3$. Equivalently, the characteristic polynomial of $\mathbf{A}$, $p(\lambda)$, has the factorization $p(\lambda) = -(\lambda + 1)(\lambda - 1)(\lambda - 3)$. The eigenvalues are said to be **simple** since the algebraic multiplicity of each one is equal to 1. Since each eigenvalue has at least one eigenvector, the geometric multiplicity of each eigenvalue will also be 1. Hence, the matrix $\mathbf{A}$ is nondefective and possesses three linearly independent eigenvectors.

To find the eigenvector corresponding to the eigenvalue λ_1, we substitute $\lambda = -1$ in Eq. (11). This gives the system

$$\begin{pmatrix} 3 & -3 & -1 \\ 0 & 0 & 0 \\ -1 & 1 & 3 \end{pmatrix} \begin{pmatrix} v_1 \\ v_2 \\ v_3 \end{pmatrix} = \begin{pmatrix} 0 \\ 0 \\ 0 \end{pmatrix}.$$

We can reduce this to the equivalent system

$$\begin{pmatrix} 1 & -1 & -3 \\ 0 & 0 & 1 \\ 0 & 0 & 0 \end{pmatrix} \begin{pmatrix} v_1 \\ v_2 \\ v_3 \end{pmatrix} = \begin{pmatrix} 0 \\ 0 \\ 0 \end{pmatrix}$$

by elementary row operations. Thus the components of $\mathbf{v}$ must satisfy the equations $v_1 - v_2 - 3v_3 = 0$ and $v_3 = 0$. The general solution is given by $\mathbf{v} = \alpha(1, 1, 0)^T$ where α is arbitrary. Setting $\alpha = 1$ yields the eigenvector

$$\mathbf{v}_1 = \begin{pmatrix} 1 \\ 1 \\ 0 \end{pmatrix}$$

corresponding to the eigenvalue $\lambda = -1$.

For $\lambda = 1$, Eqs. (11) reduce to the pair of equations $v_1 - v_2 - v_3 = 0$ and $v_2 = 0$ with general solution $\mathbf{v} = \alpha(1, 0, 1)^T$. Choosing $\alpha = 1$ gives the eigenvector

$$\mathbf{v}_2 = \begin{pmatrix} 1 \\ 0 \\ 1 \end{pmatrix}$$

that belongs to the eigenvalue $\lambda = 1$.

Finally, substituting $\lambda = 3$ into Eq. (11) and reducing the system to row echelon form by elementary row operations yields $v_1 - v_2 + v_3 = 0$ and $v_2 = 0$ with general solution $\mathbf{v} = \alpha(1, 0, -1)^T$. Setting $\alpha = 1$ gives the eigenvector

$$\mathbf{v}_3 = \begin{pmatrix} 1 \\ 0 \\ -1 \end{pmatrix}$$

associated with the eigenvalue $\lambda = 3$.

Since $g_j = m_j = 1$, $j = 1, \ldots 3$, all of the eigenvalues of $\mathbf{A}$ are nondefective and therefore $\mathbf{A}$ is a nondefective matrix.

EXAMPLE 2

Find the eigenvalues and eigenvectors of the matrix

$$\mathbf{A} = \begin{pmatrix} 4 & 6 & 6 \\ 1 & 3 & 2 \\ -1 & -5 & -2 \end{pmatrix}. \tag{13}$$

The characteristic polynomial $p(\lambda) = \det(\mathbf{A} - \lambda \mathbf{I}_3)$ of $\mathbf{A}$ is

$$p(\lambda) = -\lambda^3 + 5\lambda^2 - 8\lambda + 4 = -(\lambda - 1)(\lambda - 2)^2$$

so the eigenvalues are $\lambda_1 = 1$ and $\lambda_2 = 2$ with algebraic multiplicities 1 and 2, respectively. The system $(\mathbf{A} - \lambda_1 \mathbf{I}_3)\mathbf{v} = \mathbf{0}$ is reduced by elementary row operations to $v_1 + 2v_2 + 2v_3 = 0$ and $3v_2 + v_3 = 0$ with general solution $\mathbf{v} = \alpha(4, 1, -3)^T$. Setting $\alpha = 1$ gives us the eigenvector

$$\mathbf{v}_1 = \begin{pmatrix} 4 \\ 1 \\ -3 \end{pmatrix}$$

belonging to λ_1.

To find the eigenvector(s) belonging to λ_2, we find the general solution of $(\mathbf{A} - \lambda_2\mathbf{I}_3)\mathbf{v} = \mathbf{0}$. Row reduction leads to the system $v_1 + v_2 + 2v_3 = 0$ and $2v_2 + v_3 = 0$. Since rank $(\mathbf{A} - \lambda_2\mathbf{I}_3) = 2$, we realize that $\lambda = 2$ can have only one eigenvector, that is, λ_2 has a geometric multiplicity equal to 1 and is therefore a defective eigenvalue. The general solution of $(\mathbf{A} - \lambda_2\mathbf{I}_3)\mathbf{v} = \mathbf{0}$ is $\mathbf{v} = \alpha(3, 1, -2)^T$. Setting $\alpha = 1$ yields the eigenvector

$$\mathbf{v}_2 = \begin{pmatrix} 3 \\ 1 \\ -2 \end{pmatrix}$$

for the eigenvalue λ_2. Since $g_2 = 1$ and $m_2 = 2$, the eigenvalue λ_2 is defective. Therefore, the matrix $\mathbf{A}$ in this example is a defective matrix.

▶ **Complex Eigenvalues.** If the entries of the matrix $\mathbf{A}$ are real, then the coefficients of the characteristic polynomial $p(\lambda) = \det(\mathbf{A} - \lambda\mathbf{I}_n)$ of $\mathbf{A}$ are also real. Thus if $\mu + i\nu$ is a complex eigenvalue of $\mathbf{A}$ and $\nu \neq 0$, then $\mu - i\nu$ must also be an eigenvalue of $\mathbf{A}$. To see this, suppose $\lambda = \mu + i\nu$ with $\nu \neq 0$ is an eigenvalue of $\mathbf{A}$ with corresponding eigenvector $\mathbf{v}$ so that

$$\mathbf{A}\mathbf{v} = (\mu + i\nu)\mathbf{v}.$$

If we take the complex conjugate on both sides of this equation we get

$$\mathbf{A}\bar{\mathbf{v}} = (\mu - i\nu)\bar{\mathbf{v}}, \tag{14}$$

where we have used the fact that $\bar{\mathbf{A}} = \mathbf{A}$ since $\mathbf{A}$ is real along with the property that, if z_1 and z_2 are complex numbers, then $\overline{z_1 z_2} = \bar{z}_1 \bar{z}_2$. Eq. (14) shows that not only is $\bar{\lambda} = \mu - i\nu$ an eigenvalue of $\mathbf{A}$, its eigenvector $\bar{\mathbf{v}}$ is the conjugate of the eigenvector belonging to $\lambda = \mu + i\nu$.

EXAMPLE 3

Find the eigenvalues and eigenvectors of the matrix

$$\mathbf{A} = \begin{pmatrix} -\frac{1}{2} & \frac{7}{2} & -3 \\ -\frac{1}{2} & -\frac{1}{2} & -1 \\ \frac{1}{2} & \frac{3}{2} & -2 \end{pmatrix}.$$

The characteristic polynomial of $\mathbf{A}$ is

$$p(\lambda) = -\lambda^3 - 3\lambda^2 - 7\lambda - 5 = -(\lambda + 1)[\lambda - (-1 + 2i)][\lambda - (-1 - 2i)]$$

so the eigenvalues of $\mathbf{A}$ are $\lambda_1 = -1$, $\lambda_2 = -1 + 2i$, and $\lambda_3 = \bar{\lambda}_2 = -1 - 2i$. The eigenvector belonging to λ_1 is found to be

$$\mathbf{v}_1 = \begin{pmatrix} 1 \\ -1 \\ -1 \end{pmatrix}.$$

The equations represented by $(\mathbf{A} - \lambda_2\mathbf{I}_3)\mathbf{v} = \mathbf{0}$ are reduced, via elementary row operations and following the rules of complex arithmetic, to $v_1 - (2 + i)v_3 = 0$ and $v_2 - iv_3 = 0$. The general solution of this system of equations is $\mathbf{v} = \alpha(2 + i, i, 1)^T$. Choosing $\alpha = 1$ yields the complex eigenvector

$$\mathbf{v}_2 = \begin{pmatrix} 2+i \\ i \\ 1 \end{pmatrix}.$$

The eigenvector belonging to λ_3 is therefore

$$\mathbf{v}_3 = \bar{\mathbf{v}}_2 = \begin{pmatrix} 2-i \\ -i \\ 1 \end{pmatrix}.$$

▶ **Symmetric Matrices.** Recall that a symmetric matrix $\mathbf{A}$ satisfies $\mathbf{A}^T = \mathbf{A}$, that is, $a_{ji} = a_{ij}$. The properties of eigenvectors of symmetric matrices require that we introduce the notion of angle between vectors. We define the **inner** or **dot product** of two n-vectors $\mathbf{u}$ and $\mathbf{v}$ by

$$\langle \mathbf{u}, \mathbf{v} \rangle = u_1 v_1 + u_2 v_2 + \cdots + u_n v_n.$$

Note that if $\mathbf{v} \neq \mathbf{0}$, then $\langle \mathbf{v}, \mathbf{v} \rangle = v_1^2 + \cdots + v_n^2 > 0$.

DEFINITION A.4.1	Two n-vectors $\mathbf{u}$ and v are said to be **orthogonal** if $\langle \mathbf{u}, \mathbf{v} \rangle = 0$.

If a set of nonzero n-vectors $\{\mathbf{v}_1, \cdots, \mathbf{v}_k\}$ are pairwise orthogonal, that is, $\langle \mathbf{v}_i, \mathbf{v}_j \rangle = 0$ whenever $i \neq j$, then the set is also linearly independent. To see this, we take the inner product of both sides of the equation

$$c_1 \mathbf{v}_1 + \cdots + c_k \mathbf{v}_k = \mathbf{0}$$

with $\mathbf{v}_j$ for each $j = 1, \ldots, k$,

$$c_1 \langle \mathbf{v}_1, \mathbf{v}_j \rangle + \cdots + c_j \langle \mathbf{v}_j, \mathbf{v}_j \rangle + \cdots + c_k \langle \mathbf{v}_k, \mathbf{v}_j \rangle = \langle \mathbf{0}, \mathbf{v}_j \rangle,$$

or since $\langle \mathbf{v}_i, \mathbf{v}_j \rangle = 0$ for all $i \neq j$, the equation reduces to

$$c_j \langle \mathbf{v}_j, \mathbf{v}_j \rangle = 0.$$

Since $\langle \mathbf{v}_j, \mathbf{v}_j \rangle > 0$, $c_j = 0$ for each $j = 1, \ldots, k$.

Eigenvalues and eigenvectors of symmetric matrices have the following useful properties:

1. All eigenvalues are real.

2. There always exists a full, or complete, set of n linearly independent eigenvectors, regardless of the algebraic multiplicities of the eigenvalues. Thus, symmetric matrices are always nondefective.

3. If $\mathbf{v}_1$ and $\mathbf{v}_2$ are eigenvectors that correspond to different eigenvalues, then $\langle \mathbf{v}_1, \mathbf{v}_2 \rangle = 0$. Thus, if all eigenvalues are simple, then the associated eigenvectors form an orthogonal set of vectors.

4. Corresponding to an eigenvalue of algebraic multiplicity m, it is possible to choose m eigenvectors that are mutually orthogonal. Thus, the complete set of n eigenvectors of a symmetric matrix can always be chosen to be orthogonal.

EXAMPLE 4

Find the eigenvalues and eigenvectors of the matrix

$$\mathbf{A} = \begin{pmatrix} 0 & 1 & 1 \\ 1 & 0 & 1 \\ 1 & 1 & 0 \end{pmatrix}. \tag{15}$$

The characteristic polynomial of $\mathbf{A}$ is

$$p(\lambda) = \det(\mathbf{A} - \lambda \mathbf{I}_3) = -\lambda^3 + 3\lambda + 2 = -(\lambda - 2)(\lambda + 1)^2.$$

Thus the eigenvalues of $\mathbf{A}$ are $\lambda_1 = 2$ and $\lambda_2 = -1$ with algebraic multiplicities 1 and 2, respectively. To find the eigenvector corresponding to λ_1, we use elementary row operations to reduce the system $(\mathbf{A} - \lambda_1 \mathbf{I}_3)\mathbf{v} = \mathbf{0}$ to the system $2v_1 - v_2 - v_3 = 0$ and $v_2 - v_3 = 0$ whose general solution then yields the eigenvector

$$\mathbf{v}_1 = \begin{pmatrix} 1 \\ 1 \\ 1 \end{pmatrix}.$$

Elementary row operations reduce $(\mathbf{A} - \lambda_2 \mathbf{I}_3)\mathbf{v} = \mathbf{0}$ to the single nonzero equation $v_1 + v_2 + v_3 = 0$ which has the general solution

$$\mathbf{v} = \alpha \begin{pmatrix} 1 \\ 0 \\ -1 \end{pmatrix} + \beta \begin{pmatrix} 0 \\ 1 \\ -1 \end{pmatrix}$$

where α and β are arbitrary scalars. First choosing $\alpha = 1$ and $\beta = 0$ and then $\alpha = 0$ and $\beta = 1$ yields a pair of linearly independent eigenvectors

$$\mathbf{v}_2 = \begin{pmatrix} 1 \\ 0 \\ -1 \end{pmatrix}, \qquad \hat{\mathbf{v}}_3 = \begin{pmatrix} 0 \\ 1 \\ -1 \end{pmatrix}$$

associated with λ_2. Since $\mathbf{v}_1$ belongs to a different eigenvalue, $\langle \mathbf{v}_1, \mathbf{v}_2 \rangle = 0$ and $\langle \mathbf{v}_1, \hat{\mathbf{v}}_3 \rangle = 0$ in accordance with property 3 of eigenvectors of symmetric matrices. However, $\mathbf{v}_2$ and $\hat{\mathbf{v}}_3$ are not orthogonal since $\langle \mathbf{v}_2, \hat{\mathbf{v}}_3 \rangle = 1$. For a second eigenvector belonging to λ_2, we could assume that $\mathbf{v}_3$ is the linear combination

$$\mathbf{v}_3 = \alpha \begin{pmatrix} 1 \\ 0 \\ -1 \end{pmatrix} + \beta \begin{pmatrix} 0 \\ 1 \\ -1 \end{pmatrix}$$

instead, and choose α and β to satisfy the orthogonality condition

$$\langle \mathbf{v}_2, \mathbf{v}_3 \rangle = 2\alpha + \beta = 0.$$

Then $\mathbf{v}_2$ and $\mathbf{v}_3$ will be orthogonal eigenvectors of λ_2 provided that $\beta = -2\alpha \neq 0$. For example, choosing $\alpha = 1$ requires that $\beta = -2$ and gives

$$\mathbf{v}_3 = \begin{pmatrix} 1 \\ -2 \\ 1 \end{pmatrix}.$$

Then, in agreement with property 4 above, $\{\mathbf{v}_1, \mathbf{v}_2, \mathbf{v}_3\}$ is a complete orthogonal set of eigenvectors for the matrix $\mathbf{A}$.

▶ **Diagonalizable Matrices.** If an $n \times n$ matrix $\mathbf{A}$ has n linearly independent eigenvectors, $\mathbf{v}_1, \ldots, \mathbf{v}_n$, and we form a matrix $\mathbf{T}$ using these vectors as columns, that is,

$$\mathbf{T} = [\mathbf{v}_1, \ldots, \mathbf{v}_n] = \begin{pmatrix} v_{11} & \cdots & v_{1n} \\ \vdots & & \vdots \\ v_{n1} & \cdots & v_{nn} \end{pmatrix}, \tag{16}$$

then

$$\mathbf{T}^{-1}\mathbf{A}\mathbf{T} = \mathbf{D} = \begin{pmatrix} \lambda_1 & 0 & \cdots & 0 \\ 0 & \lambda_2 & \cdots & 0 \\ \vdots & & & \vdots \\ 0 & 0 & \cdots & \lambda_n \end{pmatrix}. \tag{17}$$

To see why Eq. (17) is true, we first note that $\mathbf{T}^{-1}\mathbf{v}_j = \mathbf{e}_j$ for each $j = 1, \ldots, n$ where $\mathbf{e}_1, \ldots, \mathbf{e}_n$ are the column vectors of $\mathbf{I}_n$. This is evident by comparing, column by column, the right-hand sides of

$$\mathbf{T}^{-1}\mathbf{T} = \mathbf{T}^{-1}[\mathbf{v}_1, \ldots, \mathbf{v}_n] = [\mathbf{T}^{-1}\mathbf{v}_1, \ldots, \mathbf{T}^{-1}\mathbf{v}_n]$$

and

$$\mathbf{T}^{-1}\mathbf{T} = \mathbf{I}_n = [\mathbf{e}_1, \ldots, \mathbf{e}_n].$$

We now write the product $\mathbf{A}\mathbf{T}$ in the form

$$\mathbf{A}\mathbf{T} = \mathbf{A}[\mathbf{v}_1, \ldots, \mathbf{v}_n] = [\mathbf{A}\mathbf{v}_1, \ldots, \mathbf{A}\mathbf{v}_n] = [\lambda_1\mathbf{v}_1, \ldots, \lambda_n\mathbf{v}_n]$$

by using the relations $\mathbf{A}\mathbf{v}_j = \lambda_j\mathbf{v}_j$, $j = 1, \ldots, n$. Premultiplying $\mathbf{A}\mathbf{T}$ by $\mathbf{T}^{-1}$ then gives

$$\mathbf{T}^{-1}\mathbf{A}\mathbf{T} = \mathbf{T}^{-1}[\lambda_1\mathbf{v}_1, \ldots, \lambda_n\mathbf{v}_n] = [\lambda_1\mathbf{T}^{-1}\mathbf{v}_1, \ldots, \lambda_n\mathbf{T}^{-1}\mathbf{v}_n]$$
$$= [\lambda_1\mathbf{e}_1, \ldots, \lambda_n\mathbf{e}_n] = \mathbf{I}_n\mathbf{D} = \mathbf{D}.$$

Thus, if $\mathbf{A}$ has n linearly independent eigenvectors, $\mathbf{A}$ can be transformed into a diagonal matrix by the process shown in Eq. (17). In this case, we say that the matrix $\mathbf{A}$ is **diagonalizable**. In linear algebra, two $n \times n$ matrices $\mathbf{A}$ and $\mathbf{B}$ are said to be **similar** if there exists an invertible $n \times n$ matrix $\mathbf{T}$ such that $\mathbf{T}^{-1}\mathbf{A}\mathbf{T} = \mathbf{B}$, and the transformation is known as a **similarity transformation**. Thus Eq. (17) states that nondefective matrices are similar to diagonal matrices.

▶ **Jordan Forms.** As discussed above, an $n \times n$ matrix $\mathbf{A}$ can be diagonalized if it has a full complement of n linearly independent eigenvectors. If there is a shortage of eigenvectors (because one or more eigenvalues are defective), then $\mathbf{A}$ can always be transformed, via a similarity transformation, into a nearly diagonal matrix called its Jordan form, which has the eigenvalues of $\mathbf{A}$ on the main diagonal, ones in certain positions on the diagonal above the main diagonal, and zeros elsewhere. For example, we consider the matrix

$$\mathbf{A} = \begin{pmatrix} -3 & 0 & -1 \\ 1 & -2 & 1 \\ 1 & 0 & -1 \end{pmatrix} \tag{18}$$

with characteristic polynomial $p(\lambda) = -(\lambda + 2)^3$. Thus $\lambda_1 = -2$ is an eigenvalue with algebraic multiplicity $m_1 = 3$. Applying row reduction to $(\mathbf{A} - \lambda_1\mathbf{I}_3)\mathbf{v} = \mathbf{0}$ leads to the system $v_1 + v_3 = 0$ so the general solution is

$$\mathbf{v} = \alpha \begin{pmatrix} -1 \\ 0 \\ 1 \end{pmatrix} + \beta \begin{pmatrix} 0 \\ 1 \\ 0 \end{pmatrix}. \tag{19}$$

Choosing $\alpha = 1$ and $\beta = 1$ gives the eigenvector

$$\mathbf{v}_1 = \begin{pmatrix} -1 \\ 1 \\ 1 \end{pmatrix}$$

while choosing $\alpha = 1$ and $\beta = 0$ gives a second linearly independent eigenvector

$$\mathbf{v}_2 = \begin{pmatrix} -1 \\ 0 \\ 1 \end{pmatrix}.$$

Thus the geometric multiplicity of $\lambda_1 = -2$ is $g_1 = 2$ and $\mathbf{A}$ is not diagonalizable. However, if

$$\mathbf{w} = \begin{pmatrix} 0 \\ 0 \\ 1 \end{pmatrix}$$

is then selected as a solution of

$$(\mathbf{A} - \lambda_1 \mathbf{I}_3)\mathbf{w} = \mathbf{v}_1 \tag{20}$$

and the matrix

$$\mathbf{T} = [\mathbf{v}_1, \mathbf{w}, \mathbf{v}_2] = \begin{pmatrix} -1 & 0 & -1 \\ 1 & 0 & 0 \\ 1 & 1 & 1 \end{pmatrix}$$

is formed from $\mathbf{v}_1$, $\mathbf{w}$, and $\mathbf{v}_2$, then

$$\mathbf{T}^{-1}\mathbf{A}\mathbf{T} = \mathbf{J} = \begin{pmatrix} -2 & 1 & 0 \\ 0 & -2 & 0 \\ 0 & 0 & -2 \end{pmatrix}.$$

Thus, $\mathbf{A}$ is similar to a matrix with the eigenvalue $\lambda_1 = -2$ along the diagonal, a one in the second column of the first row, and zeros elsewhere. The vector $\mathbf{w}$ satisfying Eq. (20) is called a **generalized eigenvector** belonging to λ_1.

However, instead of pursuing a discussion of generalized eigenvectors here, it is knowledge of the existence and structure of the Jordan form that is most relevant to our approach to finding fundamental sets of solutions of $\mathbf{x}' = \mathbf{A}\mathbf{x}$ for the case of defective $\mathbf{A}$ in Section 6.7. If the distinct eigenvalues of $\mathbf{A}$ are denoted by $\lambda_1, \ldots, \lambda_r$ with corresponding algebraic multiplicities $m_1, \ldots, m_r$ (such that $m_1 + \cdots + m_r = n$), then there is a nonsingular $n \times n$ matrix $\mathbf{T}$ such that

$$\mathbf{T}^{-1}\mathbf{A}\mathbf{T} = \mathbf{J} \tag{21}$$

where $\mathbf{J}$ is block diagonal,

$$\mathbf{J} = \begin{pmatrix} \mathbf{J}_1 & \mathbf{0} & \mathbf{0} & \cdots & \mathbf{0} \\ \mathbf{0} & \mathbf{J}_2 & \mathbf{0} & \cdots & \mathbf{0} \\ \mathbf{0} & \mathbf{0} & \mathbf{J}_3 & \cdots & \mathbf{0} \\ \vdots & \vdots & \vdots & \ddots & \vdots \\ \mathbf{0} & \mathbf{0} & \mathbf{0} & \cdots & \mathbf{J}_r \end{pmatrix}. \tag{22}$$

For each $k = 1, \ldots, r$, the Jordan block $\mathbf{J}_k$ is an $m_k \times m_k$ matrix of the form

$$\mathbf{J}_k = \begin{pmatrix} \lambda_k & * & 0 & 0 & \cdots & 0 \\ 0 & \lambda_k & * & 0 & \cdots & 0 \\ 0 & 0 & \ddots & \ddots & & 0 \\ \vdots & \vdots & & & & \vdots \\ & & & & \lambda_k & * \\ 0 & 0 & & \cdots & & \lambda_k \end{pmatrix} \tag{23}$$

where the asterisks represent ones or zeros. Note that the $\mathbf{0}$'s that appear in Eq. (22) are block matrices of appropriate dimension that consist entirely of zero entries. This structural result underlies the following theorem which is used in Section 6.7 to help construct fundamental sets of solutions for $\mathbf{x}' = \mathbf{A}\mathbf{x}$ in the case that $\mathbf{A}$ is defective.

THEOREM A.4.2

If λ is an eigenvalue of $\mathbf{A}$ and the algebraic multiplicity of λ equals m, then $\operatorname{rank}(\mathbf{A} - \lambda \mathbf{I}_n)^m = n - m$. Thus the general solution of

$$(\mathbf{A} - \lambda \mathbf{I}_n)^m \mathbf{v} = \mathbf{0} \tag{24}$$

can be expressed as

$$\mathbf{v} = \alpha_1 \mathbf{v}_1 + \cdots + \alpha_m \mathbf{v}_m$$

where $\{\mathbf{v}_1, \ldots, \mathbf{v}_m\}$ is a linearly independent set of solutions to Eq. (24).

Rather than prove this theorem, we consider the following example which shows why it is true and how it depends on being able to relate $\mathbf{A}$ to its Jordan form. Suppose that

$$\mathbf{T}^{-1} \mathbf{A} \mathbf{T} = \mathbf{J} \tag{25}$$

where

$$\mathbf{J} = \begin{pmatrix} \mathbf{J}_1 & \mathbf{0} \\ \mathbf{0} & \mathbf{J}_2 \end{pmatrix}, \quad \mathbf{J}_1 = \begin{pmatrix} \lambda_1 & * & 0 \\ 0 & \lambda_1 & * \\ 0 & 0 & \lambda_1 \end{pmatrix}, \quad \text{and} \quad \mathbf{J}_2 = \begin{pmatrix} \lambda_2 & * \\ 0 & \lambda_2 \end{pmatrix},$$

and the asterisks represent either 0's or 1's in the Jordan blocks $\mathbf{J}_1$ and $\mathbf{J}_2$. Thus the algebraic multiplicity of λ_1 is $m_1 = 3$ and the algebraic multiplicity of λ_2 is $m_2 = 2$. From Eq. (25), we may write

$$\mathbf{T}^{-1} (\mathbf{A} - \lambda_1 \mathbf{I}_5) \mathbf{T} = \mathbf{J} - \lambda_1 \mathbf{I}_5$$

and consequently

$$\mathbf{T}^{-1} (\mathbf{A} - \lambda_1 \mathbf{I}_5)^3 \mathbf{T} = (\mathbf{J}\lambda_1 - \mathbf{I}_5)^3.$$

We state, without proof, that, since $\mathbf{T}$ is nonsingular, the rank of $(\mathbf{A} - \lambda_1 \mathbf{I}_5)^3$ is equal to the rank of $(\mathbf{J} - \lambda_1 \mathbf{I}_5)^3$. Thus it is only necessary to examine the rank of $(\mathbf{J} - \lambda_1 \mathbf{I}_5)^3$. The block diagonal structure of $\mathbf{J} - \lambda_1 \mathbf{I}_5$ allows us to write

$$(\mathbf{J} - \lambda_1 \mathbf{I}_5)^3 = \begin{pmatrix} (\mathbf{J}_1 - \lambda_1 \mathbf{I}_3)^3 & \mathbf{0} \\ \mathbf{0} & (\mathbf{J}_2 - \lambda_1 \mathbf{I}_2)^3 \end{pmatrix}.$$

Since

$$\mathbf{J}_1 - \lambda_1 \mathbf{I}_3 = \begin{pmatrix} 0 & * & 0 \\ 0 & 0 & * \\ 0 & 0 & 0 \end{pmatrix},$$

it follows that

$$(\mathbf{J}_1 - \lambda_1 \mathbf{I}_3)^2 = \begin{pmatrix} 0 & 0 & * \\ 0 & 0 & 0 \\ 0 & 0 & 0 \end{pmatrix}$$

and

$$(\mathbf{J}_1 - \lambda_1 \mathbf{I}_3)^3 = \begin{pmatrix} 0 & 0 & 0 \\ 0 & 0 & 0 \\ 0 & 0 & 0 \end{pmatrix}.$$

On the other hand,

$$(\mathbf{J}_2 - \lambda_1 \mathbf{I}_2)^3 = \begin{pmatrix} (\lambda_2 - \lambda_1)^3 & \times \\ 0 & (\lambda_2 - \lambda_1)^3 \end{pmatrix}$$

where the value of the entry represented by the symbol $\times$ is irrelevant. Since λ_1 and λ_2 are distinct, $\lambda_2 - \lambda_1 \neq 0$. Thus $\text{rank}(\mathbf{A} - \lambda_1 \mathbf{I}_5)^3 = \text{rank}(\mathbf{J} - \lambda_1 \mathbf{I}_5)^3 = 5 - 3 = 2$. Note that we are comparing the number of unknowns with the number of independent nonzero equations when we attempt to find the general solution of $(\mathbf{A} - \lambda_1 \mathbf{I}_5)^3 \mathbf{v} = \mathbf{0}$. Since there are only two independent nonzero rows in the reduced form of $(\mathbf{A} - \lambda_1 \mathbf{I}_5)^3 \mathbf{v} = \mathbf{0}$ and there are five unknowns, we are able to choose three components of $\mathbf{v}$ arbitrarily. The nonzero rows in the reduced form of $(\mathbf{A} - \lambda_1 \mathbf{I}_5)^3 \mathbf{v} = \mathbf{0}$ will determine the remaining two unknown components. Thus, the general solution of $(\mathbf{A} - \lambda_1 \mathbf{I}_5)^3 \mathbf{v} = \mathbf{0}$ will be of the form $\mathbf{v} = \alpha_1 \mathbf{v}_1 + \alpha_2 \mathbf{v}_2 + \alpha_3 \mathbf{v}_3$ where $\mathbf{v}_1$, $\mathbf{v}_2$, and $\mathbf{v}_3$ are linearly independent.

PROBLEMS

In each of Problems 1 through 10, find all eigenvalues and eigenvectors of the given matrix.

1. $\begin{pmatrix} 5 & -1 \\ 3 & 1 \end{pmatrix}$

2. $\begin{pmatrix} 3 & -2 \\ 4 & -1 \end{pmatrix}$

3. $\begin{pmatrix} 1 & 0 & 0 \\ 2 & 1 & -2 \\ 3 & 2 & 1 \end{pmatrix}$

4. $\begin{pmatrix} 3 & 2 & 2 \\ 1 & 4 & 1 \\ -2 & -4 & -1 \end{pmatrix}$

5. $\begin{pmatrix} -3 & -7 & -5 \\ 2 & 4 & 3 \\ 1 & 2 & 2 \end{pmatrix}$

6. $\begin{pmatrix} 7 & -2 & -4 \\ 3 & 0 & -2 \\ 6 & -2 & -3 \end{pmatrix}$

7. $\begin{pmatrix} -2 & 2 & -3 \\ 2 & 1 & -6 \\ -1 & -2 & 0 \end{pmatrix}$

8. $\begin{pmatrix} 4 & 2 & 3 \\ 2 & 1 & 0 \\ 1 & -2 & 0 \end{pmatrix}$

9. $\begin{pmatrix} 11 & -4 & -7 \\ 7 & -2 & -5 \\ 10 & -4 & -6 \end{pmatrix}$
10. $\begin{pmatrix} 1 & 0 & 1 \\ 0 & 2 & 0 \\ 1 & 0 & -1 \end{pmatrix}$

17. $\begin{pmatrix} -1 & 0 & -1 & 0 \\ 0 & 1 & 1 & 0 \\ -1 & 1 & 2 & 1 \\ 0 & 0 & 1 & -1 \end{pmatrix}$

In each of Problems 11 through 16, find the eigenvalues and a complete orthogonal set of eigenvectors for the given symmetric matrix.

11. $\begin{pmatrix} 3 & 0 & 0 \\ 0 & 3 & 4 \\ 0 & 4 & 3 \end{pmatrix}$
12. $\begin{pmatrix} 2 & 4 & -6 \\ 4 & 2 & -6 \\ -6 & -6 & -15 \end{pmatrix}$

18. $\begin{pmatrix} -3 & 2 & 1 & -2 \\ 0 & -1 & 0 & 0 \\ -2 & 2 & 0 & -2 \\ 0 & 0 & 0 & -1 \end{pmatrix}$

13. $\begin{pmatrix} 0 & 2 & 2 \\ 2 & 0 & 2 \\ 2 & 2 & 0 \end{pmatrix}$
14. $\begin{pmatrix} 1 & 1 & 1 \\ 1 & 1 & 1 \\ 1 & 1 & 1 \end{pmatrix}$

19. $\begin{pmatrix} 5 & -5 & -3 & 4 \\ 0 & -1 & 0 & 0 \\ 6 & -6 & -4 & 6 \\ 0 & -1 & 0 & 1 \end{pmatrix}$

15. $\begin{pmatrix} 1 & 0 & 0 \\ 0 & 1 & 0 \\ 0 & 0 & 1 \end{pmatrix}$
16. $\begin{pmatrix} 3 & 2 & 4 \\ 2 & 0 & 2 \\ 4 & 2 & 3 \end{pmatrix}$

In each of Problems 17 through 20, use a computer to find the eigenvalues and eigenvectors for the given matrix.

20. $\begin{pmatrix} -2 & \frac{1}{3} & \frac{2}{3} & -\frac{2}{3} \\ 2 & -3 & -2 & 0 \\ -2 & \frac{4}{3} & \frac{2}{3} & -\frac{2}{3} \\ -1 & \frac{1}{3} & \frac{2}{3} & -\frac{5}{3} \end{pmatrix}$

Complex Variables

▶ **Properties of Complex Numbers.** The symbol for a complex number z is $z = x + iy$ where x and y are real numbers and i satisfies $i^2 = -1$. The real number x is called the **real part** of z and is denoted by $x = \mathrm{Re}\, z$. The real number y is called the **imaginary part** of z and is denoted by $y = \mathrm{Im}\, z$. We now state several important properties of complex numbers.

1. Two complex numbers $z_1 = x_1 + iy_1$ and $z_2 = x_2 + iy_2$ are equal if and only if $x_1 = x_2$ and $y_1 = y_2$. In particular, $z = 0$ if and only if $\mathrm{Re}\, z = 0$ and $\mathrm{Im}\, z = 0$.

2. Addition of two complex numbers $z_1 = x_1 + iy_1$ and $z_2 = x_2 + iy_2$ is defined by

$$z_1 + z_2 = x_1 + x_2 + i(y_1 + y_2),$$

and multiplication by a real number c is defined by

$$cz_1 = cx_1 + icy_1.$$

Subtraction is then defined by

$$z_1 - z_2 = z_1 + (-1z_2) = x_1 - x_2 + i(y_1 - y_2).$$

3. The **complex conjugate** of $z = x + iy$ is the number $\bar{z} = x - iy$, that is, the imaginary part of z is multiplied by -1. Thus $\overline{3 - 2i} = 3 + 2i$. Note that z is real if and only if $z = \bar{z}$. Addition and subtraction of a complex number and its complex conjugate can be used to obtain the real and imaginary parts in the following way,

$$x = \mathrm{Re}\, z = \frac{z + \bar{z}}{2},$$

and

$$y = \mathrm{Im}\, z = \frac{z - \bar{z}}{2i}.$$

4. We multiply two complex numbers just as if we were multiplying two binomials,

$$(x_1 + iy_1)(x_2 + iy_2) = x_1 x_2 + ix_1 y_2 + iy_1 x_2 + i^2 y_1 y_2$$
$$= x_1 x_2 - y_1 y_2 + i(x_1 y_2 + y_1 x_2)$$

where we have used $i^2 = -1$. In particular, $z\bar{z} = (x + iy)(x - iy) = x^2 + y^2$. The nonnegative real number $|z| = \sqrt{z\bar{z}} = \sqrt{x^2 + y^2}$ is called the **absolute value** or **modulus** of z. Note that $z = 0$ if and only if $|z| = \sqrt{x^2 + y^2} = 0$.

5. If $z_2 \neq 0$, we define the quotient z_1/z_2 by

$$\frac{z_1}{z_2} = \frac{z_1}{z_2} \frac{\bar{z}_2}{\bar{z}_2} = \frac{z_1 \bar{z}_2}{|z_2|^2}.$$

For example, $1/(2 + 2i) = (2 - 2i)/(4 + 4) = 1/4 - i/4$.

6. The following properties are a direct consequence of the definition of the complex conjugate:

$$\overline{z_1 \pm z_2} = \bar{z}_1 \pm \bar{z}_2,$$

$$\overline{z_1 z_2} = \bar{z}_1 \bar{z}_2,$$

$$\overline{\left(\frac{z_1}{z_2}\right)} = \frac{\bar{z}_1}{\bar{z}_2}.$$

Using $|z| = \sqrt{z\bar{z}}$, the last two properties imply that $|z_1 z_2| = |z_1||z_2|$ and $|z_1/z_2| = |z_1|/|z_2|$.

7. It is convenient to represent a complex number $z = x + iy$ as a vector $\langle x, y \rangle$ in the Euclidean plane by associating it with the directed line segment from the origin O to the point $P(x, y)$ as shown in Figure B.1. The Euclidean plane with the horizontal axis associated with the real part of complex numbers and the vertical axis associated with the imaginary part of complex numbers is called the **complex plane**. The horizontal and vertical axes of the complex plane are referred to as the **real axis** and **imaginary axis**, respectively.

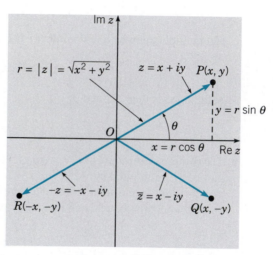

<u>**FIGURE B.1**</u>   Geometric representation of the complex numbers $z = x + iy$, $\bar{z} = x - iy$, and $-z = -x - iy$.

In this geometric representation, the absolute value $r = |x + iy| = \sqrt{x^2 + y^2}$ of $x + iy$ is simply the length of the vector $\overrightarrow{OP} = \langle x, y \rangle$. The angle θ between the positive real axis and the vector $\overrightarrow{OP}$ is called the **argument** of $x + iy$, denoted by $\arg(x + iy)$. It is determined only up to multiples of 2π. To avoid ambiguities, it is at times convenient to select a **principal branch** of the argument, either $0 \le \theta < 2\pi$ or $-\pi < \theta \le \pi$. From Figure B.1, it is clear that

$$x = r \cos\theta, \quad y = r \sin\theta \tag{1}$$

and

$$z = x + iy = r\cos\theta + ir\sin\theta = r(\cos\theta + i\sin\theta). \tag{2}$$

The right side of Eq. (2) is referred to as the **polar coordinate representation** of $z = x + iy$.

▶ Eulers Formula.   We wish to extend the definition of the exponential function e^x, where x is a real variable, to the complex exponential function e^z where $z = x + iy$ is a complex variable. Of course, we want the definition to reduce to the familiar real exponential function when the exponent is real. There are several ways to accomplish this extension of the exponential function. Here, we use a method based on infinite series.

Recall from calculus that the Taylor series for e^t about $t = 0$ is

$$e^t = \sum_{n=0}^{\infty} \frac{t^n}{n!}, \qquad -\infty < t < \infty. \tag{3}$$

If we now assume that we can substitute it for t in Eq. (3), then we have

$$e^{it} = \sum_{n=0}^{\infty} \frac{(it)^n}{n!}$$

$$= \sum_{n=0}^{\infty} \frac{(-1)^n t^{2n}}{(2n)!} + i \sum_{n=1}^{\infty} \frac{(-1)^{n-1} t^{2n-1}}{(2n-1)!}, \tag{4}$$

where we have separated the sum into its real and imaginary parts, making use of the fact that $i^2 = -1, i^3 = -i, i^4 = 1$, and so forth. The first series in Eq. (4) is precisely the Taylor series for $\cos t$ about $t = 0$, and the second is the Taylor series for $\sin t$ about $t = 0$. Thus we have

$$\boxed{e^{it} = \cos t + i \sin t.} \tag{5}$$

Equation (5) is known as **Eulers formula** and is an extremely important mathematical relationship. Although our derivation of Eq. (5) is based on the unverified assumption that the series (3) can be used for complex as well as real values of the independent variable, our intention is to use this derivation only to make Eq. (5) seem plausible. We now put matters on a firm foundation by adopting Eq. (5) as the *definition* of e^{it}. In other words, whenever we write e^{it}, we mean the expression on the right side of Eq. (5).

There are some variations of Eulers formula that are also worth noting. If we replace t by $-t$ in Eq. (5) and recall that $\cos(-t) = \cos t$ and $\sin(-t) = -\sin t$, then we have

$$\boxed{e^{-it} = \cos t - i \sin t.} \tag{6}$$

In terms of complex conjugates, $e^{-it} = \cos t - i \sin t = \overline{\cos t + i \sin t} = \overline{e^{it}}$. Further, if t is replaced by νt in Eq. (5), then we obtain a generalized version of Eulers formula, namely,

$$\boxed{e^{i\nu t} = \cos \nu t + i \sin \nu t.} \tag{7}$$

Thus, we have defined $z = e^{i\nu t}$ to be the complex number which has absolute value equal to one, $|z| = 1$, and has an argument equal to νt, $\arg z = \nu t$. All complex exponentials of the form $e^{i\nu t}$ where νt is real lie on the unit circle in the complex plane, that is, $|e^{i\nu t}| = 1$ (Figure B.2).

Next, we want to extend the definition of the exponential function to arbitrary complex exponents of the form $(\mu + i\nu)t$. Since we want the usual properties of the exponential function to hold for complex exponents, we certainly want $\exp[(\mu + i\nu)t]$ to satisfy

$$e^{(\mu+i\nu)t} = e^{\mu t} e^{i\nu t}. \tag{8}$$

Then, substituting for $e^{i\nu t}$ from Eq. (7), we obtain

$$\boxed{e^{(\mu+i\nu)t} = e^{\mu t}(\cos \nu t + i \sin \nu t) = e^{\mu t} \cos \nu t + i e^{\mu t} \sin \nu t} \tag{9}$$

We now take Eq. (9) as the definition of $\exp[(\mu + i\nu)t]$. The value of the exponential function with a complex exponent is a complex number whose real and imaginary parts are given by the terms on the right side of Eq. (9). The effect of the term μt in the exponent is

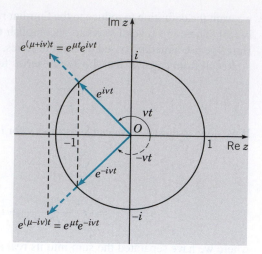

FIGURE B.2   The complex exponential $e^{i\nu t} = \cos \nu t + i \sin \nu t$ and its complex conjugate $e^{-i\nu t} = \cos \nu t - i \sin \nu t$ lie on the unit circle and have arguments νt and $-\nu t$, respectively. The real factor $e^{\mu t}$ merely changes the length of $e^{\pm i\nu t}$.

to multiply the length of $e^{\pm i\nu t}$ by the real scale factor $e^{\mu t}$ as shown in Figure B.2. Note that $|e^{(\mu \pm i\nu)t}| = |e^{\mu t} e^{\pm i\nu t}| = |e^{\mu t}||e^{\pm i\nu t}| = e^{\mu t}$ since $|e^{\pm i\nu t}| = 1$ and $e^{\mu t} > 0$.

With the definitions (5) and (9), it is straightforward to show that the usual laws of exponents are valid for the complex exponential function. It is also easy to verify that the differentiation formula

$$\frac{d}{dt}(e^{\lambda t}) = \lambda e^{\lambda t} \tag{10}$$

holds for complex values of λ.

A N S W E R S

CHAPTER 1

Section 1.1 *page 9*

1. $y \to 3/2$ as $t \to \infty$

2. y diverges from $3/2$ as $t \to \infty$

3. y diverges from $-3/2$ as $t \to \infty$

4. $y \to -1/2$ as $t \to \infty$

5. y diverges from $-1/2$ as $t \to \infty$

6. y diverges from -2 as $t \to \infty$

7. $y' = 3 - y$

8. $y' = 2 - 3y$

9. $y' = y - 2$

10. $y' = 3y - 1$

11. $y = 0$ and $y = 4$ are equilibrium solutions; $y \to 4$ if initial value is positive; y diverges from 0 if initial value is negative.

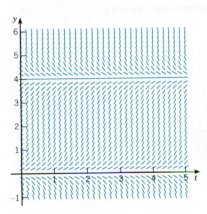

Direction field for Problem 11.

12. $y = 0$ and $y = 5$ are equilibrium solutions; y diverges from 5 if initial value is greater than 5; $y \to 0$ if initial value is less than 5.

13. $y = 0$ is equilibrium solution; $y \to 0$ if initial value is negative; y diverges from 0 if initial value is positive.

14. $y = 0$ and $y = 2$ are equilibrium solutions; y diverges from 0 if initial value is negative; $y \to 2$ if initial value is between 0 and 2; y diverges from 2 if initial value is greater than 2.

15. (j)

16. (c)

17. (g)

18. (b)

19. (h)

20. (e)

21. (a) $dq/dt = 300(10^{-2} - q \; 10^{-6})$; q in g, t in hr   (b) $q \to 10^4$ g; no

22. $dV/dt = -kV^{2/3}$ for some $k > 0$.

23. $du/dt = -0.05(u - 70)$; u in °F, t in min

24. (a) $dq/dt = 500 - 0.4q$; q in mg, t in hr   (b) $q \to 1250$ mg

25. (a) $mv' = mg - kv^2$ (b) $v \to \sqrt{mg/k}$ (c) $k = 0.0002$ kg/m

26. y is asymptotic to $t - 3$ as $t \to \infty$ (see figure on page 828)

27. $y \to 0$ as $t \to \infty$

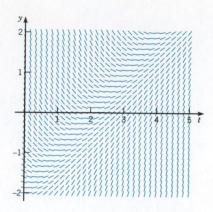

Direction field for Problem 26.

28. $y \to \infty$, 0, or $-\infty$ depending on the initial value of y

29. $y \to \infty$ or $-\infty$ depending on the initial value of y

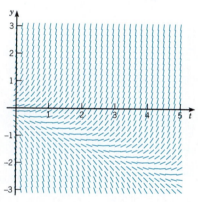

Direction field for Problem 29.

30. $y \to \infty$ or $-\infty$ or y oscillates depending on the initial value of y

31. $y \to -\infty$ or is asymptotic to $\sqrt{2t-1}$ depending on the initial value of y

32. $y \to 0$ and then fails to exist after some $t_f \geq 0$

33. $y \to \infty$ or $-\infty$ depending on the initial value of y

Section 1.2, *page 20*

1. (a) $y = 5 + (y_0 - 5)e^{-t}$ (b) $y = (5/2) + [y_0 - (5/2)]e^{-2t}$
   (c) $y = 5 + (y_0 - 5)e^{-2t}$
   Equilibrium solution is $y = 5$ in (a) and (c), $y = 5/2$ in (b); solution approaches equilibrium faster in (b) and (c) than in (a).

2. (a) $y = 5 + (y_0 - 5)e^{t}$ (b) $y = (5/2) + [y_0 - (5/2)]e^{2t}$
   (c) $y = 5 + (y_0 - 5)e^{2t}$
   Equilibrium solution is $y = 5$ in (a) and (c), $y = 5/2$ in (b); solution diverges from equilibrium faster in (b) and (c) than in (a).

3. (a) $y = ce^{-at} + (b/a)$
   (c) (i) Equilibrium is lower and is approached more rapidly. (ii) Equilibrium is higher. (iii) Equilibrium remains the same and is approached more rapidly.

4. (a) $y_e = b/a$ (b) $Y' = aY$ 5. $y = ce^{-2t} + \frac{1}{2}t - \frac{7}{4}$

6. $y = ce^{3t} - \frac{1}{4}e^{-t}$

7. $y = ce^{-t} + \frac{3}{5}\cos 2t + \frac{6}{5}\sin 2t$

8. $y = ce^{2t} - \frac{2}{5}\cos t - \frac{4}{5}\sin t$

9. $y = ce^{-2t} + t - \frac{1}{2} - \frac{3}{5}\cos t + \frac{6}{5}\sin t$

10. $y = ce^{2t} - 3e^t - \frac{1}{2}t^2 - \frac{1}{2}t - \frac{3}{4}$

11. (a) $T = 2\ln 18 \cong 5.78$ months   (b) $T = 2\ln[900/(900 - p_0)]$ months
    (c) $p_0 = 900(1 - e^{-6}) \cong 897.8$

12. (a) $r = (\ln 2)/30$ day^{-1} (b) $r = (\ln 2)/N$ day^{-1}

13. (a) $T = (\ln 50)/0.28 \cong 13.97$ sec   (b) 366.5 m

14. (a) $dv/dt = 9.8$, $v(0) = 0$ (b) $T = \sqrt{300/4.9} \cong 7.82$ sec   (c) $v \cong 76.68$ m/sec

15. (b) $v = 35\tanh(7t/25)$ m/sec   (e) $x = 125\ln\cosh(7t/25)$ m
    (f) $T = (25/7)\text{arccosh}(e^{12/5}) \cong 11.04$ sec

16. (a) $r \cong 0.02828$ day^{-1} (b) $Q(t) = 100e^{-0.02828t}$ (c) $T \cong 24.5$ days

18. $1620\ln(4/3)/\ln 2 \cong 672.4$ years

19. (a) $u = T + (u_0 - T)e^{-kt}$ (b) $k\tau = \ln 2$

20. 6.69 hr

21. (a) $q(t) = CV(1 - e^{-t/RC})$ (b) $q(t) \to CV = q_L$
    (c) $q(t) = CV\exp[-(t - t_1)/RC]$

22. (a) $Q' = 3(1 - 10^{-4}Q)$, $Q(0) = 0$
    (b) $Q(t) = 10^4(1 - e^{-3t/10^4})$, t in hrs; after 1 year $Q \cong 9277.77$ g
    (c) $Q' = -3Q/10^4$, $Q(0) = 9277.77$
    (d) $Q(t) = 9277.77e^{-3t/10^4}$, t in hrs; after 1 year $Q \cong 670.07$ g
    (e) $T \cong 2.60$ years

23. (a) $q' = -q/300$, $q(0) = 5000$ g   (b) $q(t) = 5000e^{-t/300}$ (c) no
    (d) $T = 300\ln(25/6) \cong 428.13$ min
    (e) $r = 250\ln(25/6) \cong 356.78$ gal/min

Section 1.3, *page 32*

1. (a) 1.2, 1.39, 1.571, 1.7439
   (b) 1.1975, 1.38549, 1.56491, 1.73658
   (c) 1.19631, 1.38335, 1.56200, 1.73308
   (d) 1.19516, 1.38127, 1.55918, 1.72968

2. (a) 1.1, 1.22, 1.364, 1.5368
   (b) 1.105, 1.23205, 1.38578, 1.57179
   (c) 1.10775, 1.23873, 1.39793, 1.59144
   (d) 1.1107, 1.24591, 1.41106, 1.61277

3. (a) 1.25, 1.54, 1.878, 2.2736
   (b) 1.26, 1.5641, 1.92156, 2.34359
   (c) 1.26551, 1.57746, 1.94586, 2.38287
   (d) 1.2714, 1.59182, 1.97212, 2.42554

4. (a) 0.3, 0.538501, 0.724821, 0.866458
   (b) 0.284813, 0.513339, 0.693451, 0.831571
   (c) 0.277920, 0.501813, 0.678949, 0.815302
   (d) 0.271428, 0.490897, 0.665142, 0.799729

5. Converge for $y \geq 0$; undefined for $y < 0$

6. Converge for $y \geq 0$; diverge for $y < 0$ 7. Converge

8. Converge for $|y(0)| < 2.37$ (approximately); diverge otherwise

9. Diverge   10. Diverge

11. (a) 2.30800, 2.49006, 2.60023, 2.66773, 2.70939, 2.73521
    (b) 2.30167, 2.48263, 2.59352, 2.66227, 2.70519, 2.73209
    (c) 2.29864, 2.47903, 2.59024, 2.65958, 2.70310, 2.73053
    (d) 2.29686, 2.47691, 2.58830, 2.65798, 2.70185, 2.72959

12. (a) 1.70308, 3.06605, 2.44030, 1.77204, 1.37348, 1.11925
    (b) 1.79548, 3.06051, 2.43292, 1.77807, 1.37795, 1.12191
    (c) 1.84579, 3.05769, 2.42905, 1.78074, 1.38017, 1.12328
    (d) 1.87734, 3.05607, 2.42672, 1.78224, 1.38150, 1.12411

13. (a) -1.48849, -0.412339, 1.04687, 1.43176, 1.54438, 1.51971
    (b) -1.46909, -0.287883, 1.05351, 1.42003, 1.53000, 1.50549
    (c) -1.45865, -0.217545, 1.05715, 1.41486, 1.52334, 1.49879
    (d) -1.45212, -0.173376, 1.05941, 1.41197, 1.51949, 1.49490

14. (a) 0.950517, 0.687550, 0.369188, 0.145990, 0.0421429, 0.00872877
    (b) 0.938298, 0.672145, 0.362640, 0.147659, 0.0454100, 0.0104931
    (c) 0.932253, 0.664778, 0.359567, 0.148416, 0.0469514, 0.0113722
    (d) 0.928649, 0.660463, 0.357783, 0.148848, 0.0478492, 0.0118978

15. (a) -0.166134, -0.410872, -0.804660, 4.15867
    (b) -0.174652, -0.434238, -0.889140, -3.09810

16. A reasonable estimate for y at $t = 0.8$ is between 5.5 and 6. No reliable estimate is possible at $t = 1$ from the specified data.

17. A reasonable estimate for y at $t = 2.5$ is between 18 and 19. No reliable estimate is possible at $t = 3$ from the specified data.

18. (b) $2.37 < \alpha_0 < 2.38$ 19. (b) $0.67 < \alpha_0 < 0.68$

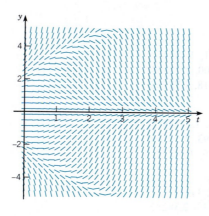

Direction field for Problem 18.

Section 1.4, *page 38*

1. Second order, linear   2. Second order, nonlinear   3. Fourth order, linear

4. First order, nonlinear   5. Second order, nonlinear   6. Third order, linear

15. $r = -2$ 16. $r = \pm 1$ 17. $r = 2, -3$

18. $r = 0, 1, 2$ 19. $r = -1, -2$ 20. $r = 1, 4$

CHAPTER 2

Section 2.1, *page 47*

1. (c) $y = ce^{-3t} + (t/3) - (1/9) + e^{-2t}$; y is asymptotic to $t/3 - 1/9$ as $t \to \infty$

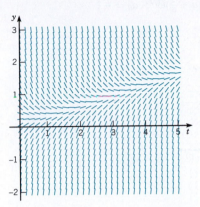

Direction field for Problem 1.

2. (c) $y = ce^{2t} + t^3 e^{2t}/3$; $y \to \infty$ as $t \to \infty$

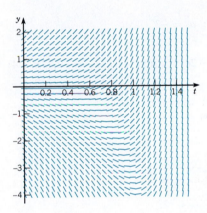

Direction field for Problem 2.

3. (c) $y = ce^{-t} + 1 + t^2 e^{-t}/2$; $y \to 1$ as $t \to \infty$

4. (c) $y = (c/t) + (3\cos 2t)/4t + (3\sin 2t)/2$; y is asymptotic to $(3\sin 2t)/2$ as $t \to \infty$

5. (c) $y = ce^{2t} - 3e^t$; $y \to \infty$ or $-\infty$ as $t \to \infty$

6. (c) $y = (c - t\cos t + \sin t)/t^2$; $y \to 0$ as $t \to \infty$

7. (c) $y = t^2 e^{-t^2} + ce^{-t^2}$; $y \to 0$ as $t \to \infty$

8. (c) $y = (\arctan t + c)/(1 + t^2)^2$; $y \to 0$ as $t \to \infty$

9. (c) $y = ce^{-t/2} + 3t - 6$; y is asymptotic to $3t - 6$ as $t \to \infty$

10. (c) $y = -te^{-t} + ct$; $y \to \infty, 0,$ or $-\infty$ as $t \to \infty$

11. (c) $y = ce^{-t} + \sin 2t - 2\cos 2t$; y is asymptotic to $\sin 2t - 2\cos 2t$ as $t \to \infty$

12. (c) $y = ce^{-t/2} + 3t^2 - 12t + 24$; y is asymptotic to $3t^2 - 12t + 24$ as $t \to \infty$

13. $y = 3e^t + 2(t - 1)e^{2t}$ 14. $y = (t^2 - 1)e^{-2t}/2$

15. $y = (3t^4 - 4t^3 + 6t^2 + 1)/12t^2$ 16. $y = (\sin t)/t^2$

17. $y = (t + 2)e^{2t}$

18. $y = t^{-2}[(\pi^2/4) - 1 - t \cos t + \sin t]$

19. $y = -(1 + t)e^{-t}/t^4, \quad t \neq 0$

20. $y = (t - 1 + 2e^{-t})/t, \quad t \neq 0$

21. (b) $y = -\frac{4}{5} \cos t + \frac{8}{5} \sin t + (a + \frac{4}{5})e^{t/2}; \quad a_0 = -\frac{4}{5}$
    (c) y oscillates for $a = a_0$

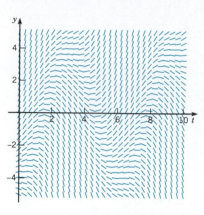

Direction field for Problem 21.

22. (b) $y = -3e^{t/3} + (a + 3)e^{t/2}; \quad a_0 = -3$
    (c) $y \to -\infty$ for $a = a_0$

23. (b) $y = [2 + a(3\pi + 4)e^{2t/3} - 2e^{-\pi t/2}]/(3\pi + 4); \quad a_0 = -2/(3\pi + 4)$
    (c) $y \to 0$ for $a = a_0$

24. (b) $y = te^{-t} + (ea - 1)e^{-t}/t; \quad a_0 = 1/e$
    (c) $y \to 0$ as $t \to 0$ for $a = a_0$

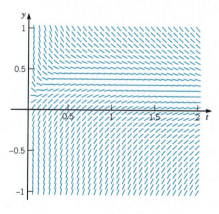

Direction field for Problem 24.

25. (b) $y = -(\cos t)/t^2 + \pi^2 a/4t^2; \quad a_0 = 4/\pi^2$
    (c) $y \to \frac{1}{2}$ as $t \to 0$ for $a = a_0$

26. (b) $y = (e^t - e + a \sin 1)/ \sin t; \quad a_0 = (e - 1)/ \sin 1$
    (c) $y \to 1$ for $a = a_0$

27. $(t, y) = (1.364312, 0.820082)$

28. $y_0 = -1.642876$

29. (a) $y = 12 + \frac{8}{65} \cos 2t + \frac{64}{65} \sin 2t - \frac{788}{65} e^{-t/4}; \quad y$ oscillates about 12 as $t \to \infty$
    (b) $t = 10.065778$

30. $y_0 = -5/2$

31. $y_0 = -16/3$; $y \to -\infty$ as $t \to \infty$ for $y_0 = -16/3$

40. See Problem 2.   41. See Problem 4.

42. See Problem 6.   43. See Problem 12.

Section 2.2, *page 54*

1. $3y^2 - 2x^3 = c$; $\quad y \neq 0$

2. $3y^2 - 2\ln|1 + x^3| = c$; $\quad x \neq -1, y \neq 0$

3. $y^{-1} + \cos x = c$ if $y \neq 0$; $\quad$ also $y = 0$; $\quad$ everywhere

4. $3y + y^2 - x^3 + x = c$; $\quad y \neq -3/2$

5. $2\tan 2y - 2x - \sin 2x = c$ if $\cos 2y \neq 0$; $\quad$ also $y = \pm(2n + 1)\pi/4$ for any integer n; everywhere

6. $y = \sin[\ln|x| + c]$ if $x \neq 0$ and $|y| < 1$; $\quad$ also $y = \pm 1$

7. $y^2 - x^2 + 2(e^y - e^{-x}) = c$; $\quad y + e^y \neq 0$ $\qquad$ 8. $3y + y^3 - x^3 = c$; $\quad$ everywhere

9. (a) $y = 1/(x^2 - x - 6)$ $\qquad\qquad\qquad\qquad$ 10. (a) $y = -\sqrt{2x - 2x^2 + 4}$
   (c) $-2 < x < 3$ $\qquad\qquad\qquad\qquad\qquad\qquad\quad$ (c) $-1 < x < 2$

11. (a) $y = [2(1 - x)e^x - 1]^{1/2}$ $\qquad\qquad\qquad$ 12. (a) $r = 2/(1 - 2\ln\theta)$
    (c) $-1.68 < x < 0.77$ approximately $\qquad\qquad$ (c) $0 < \theta < \sqrt{e}$

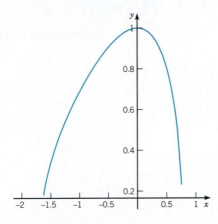

Graph of solution for Problem 11.

13. (a) $y = -[2\ln(1 + x^2) + 4]^{1/2}$ $\qquad\qquad$ 14. (a) $y = [3 - 2\sqrt{1 + x^2}]^{-1/2}$
    (c) $-\infty < x < \infty$ $\qquad\qquad\qquad\qquad\qquad$ (c) $|x| < \frac{1}{2}\sqrt{5}$ (see figure on page 834)

15. (a) $y = -\frac{1}{2} + \frac{1}{2}\sqrt{4x^2 - 15}$ $\qquad\qquad\quad$ 16. (a) $y = -\sqrt{(x^2 + 1)/2}$
    (c) $x > \frac{1}{2}\sqrt{15}$ $\qquad\qquad\qquad\qquad\qquad\qquad$ (c) $-\infty < x < \infty$

17. (a) $y = 5/2 - \sqrt{x^3 - e^x + 13/4}$ $\qquad\quad$ 18. (a) $y = -\frac{3}{4} + \frac{1}{4}\sqrt{65 - 8e^x - 8e^{-x}}$
    (c) $-1.4445 < x < 4.6297$ approximately $\quad$ (c) $|x| < 2.0794$ approximately

19. (a) $y = [\pi - \arcsin(3\cos^2 x)]/3$ $\qquad\qquad$ 20. (a) $y = \left[\frac{3}{2}(\arcsin x)^2\right]^{1/3}$
    (c) $|x - \pi/2| < 0.6155$ $\qquad\qquad\qquad\qquad\quad$ (c) $-1 < x < 1$

21. $y^3 - 3y^2 - x - x^3 + 2 = 0$, $\quad |x| < 1$

22. $y^3 - 4y - x^3 = -1$, $\quad |x^3 - 1| < 16/3\sqrt{3}$ or $-1.28 < x < 1.60$

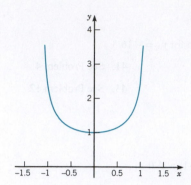

Graph of solution for Problem 14.

23. $y = -1/(x^2/2 + 2x - 1); \quad x = -2$

24. $y = -3/2 + \sqrt{2x - e^x + 13/4}; \quad x = \ln 2$

25. $y = -3/2 + \sqrt{\sin 2x + 1/4}; \quad x = \pi/4$

26. $y = \tan(x^2 + 2x); \quad x = -1$

27. (a) $y \to 4$ if $y_0 > 0; \quad y = 0$ if $y_0 = 0; \quad y \to -\infty$ if $y_0 < 0$
    (b) $T = 3.29527$

28. (a) $y \to 4$ as $t \to \infty$
    (b) $T = 2.84367$
    (c) $3.6622 < y_0 < 4.4042$

29. $x = \dfrac{c}{a}y + \dfrac{ad - bc}{a^2} \ln|ay + b| + k; \quad a \neq 0, \; ay + b \neq 0$

30. (e) $|y + 2x|^3|y - 2x| = c$ 31. (b) $\arctan(y/x) - \ln|x| = c$

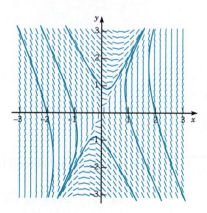

Some solution curves for Problem 30.

32. (b) $x^2 + y^2 - cx^3 = 0$

33. (b) $|y - x| = c|y + 3x|^5; \quad$ also $y = -3x$

34. (b) $|y + x|\,|y + 4x|^2 = c$

35. (b) $2x/(x + y) + \ln|x + y| = c; \quad$ also $y = -x$

36. (b) $x/(x + y) + \ln|x| = c; \quad$ also $y = -x$

37. (b) $|x|^3|x^2 - 5y^2| = c$

38. (b) $c|x|^3 = |y^2 - x^2|$

Section 2.3, *page 65*

1. $t = 100 \ln 100 \, \text{min} \cong 460.5 \, \text{min}$

2. $Q(t) = 120\gamma[1 - \exp(-t/60)]; \quad 120\gamma$

3. $Q = 50e^{-0.2}(1 - e^{-0.2}) \, \text{lb} \cong 7.42 \, \text{lb}$

4. $Q(t) = 200 + t - [100(200)^2/(200 + t)^2] \, \text{lb}, \quad t < 300; \quad c = 121/125 \, \text{lb/gal};$
   $\lim\limits_{t \to \infty} c = 1 \, \text{lb/gal}$

5. (a) $Q(t) = \frac{63{,}150}{2501} e^{-t/50} + 25 - \frac{625}{2501} \cos t + \frac{25}{5002} \sin t$
   (c) level $= 25$; amplitude $= 25\sqrt{2501}/5002 \cong 0.24995$

6. (c) 130.41 sec

7. (a) $(\ln 2)/r$ years   (b) 9.90 years   (c) 8.66%

8. (a) $k(e^{rt} - 1)/r$ (b) $k \cong \$3930$ (c) 9.77%

9. (a) $\$89{,}034.79$ (b) $\$102{,}965.21$

10. (a) $t \cong 135.36$ months
    (b) $\$152{,}698.56$

11. (b) $S_1 = RS_0 - k, \quad S_2 = R^2 S_0 - (1 + R)k, \quad S_3 = R^3 S_0 - (1 + R + R^2)k$
    (d) $k = \$488.26$

12. (a) $0.00012097 \, \text{year}^{-1}$ (b) $Q_0 \exp(-0.00012097t), \quad t$ in years   (c) 13,305 years

13. $P = 201{,}977.31 - 1977.31 e^{(\ln 2)t}, \quad 0 \le t \le t_f \cong 6.6745 \, \text{(weeks)}$

14. (a) $\tau \cong 2.9632$; no
    (b) $\tau = 10 \ln 2 \cong 6.9315$
    (c) $\tau = 6.3805$

15. (b) $y_c \cong 0.83$

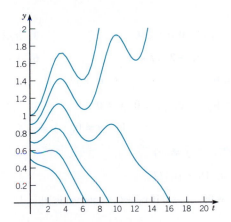

Some solution curves for Problem 15(a).

16. $t = \ln \frac{13}{8} / \ln \frac{13}{12} \, \text{min} \cong 6.07 \, \text{min}$

17. (a) $u(t) = 2000/(1 + 0.048t)^{1/3}$ (c) $\tau \cong 750.77$ sec

18. (a) $u(t) = ce^{-kt} + T_0 + kT_1(k \cos \omega t + \omega \sin \omega t)/(k^2 + \omega^2)$
    (b) $R \cong 9.11°F; \quad \tau \cong 3.51 \, \text{hr}$
    (c) $R = kT_1/\sqrt{k^2 + \omega^2}; \quad \tau = (1/\omega)\arctan(\omega/k)$

19. (a) $c = k + (P/r) + [c_0 - k - (P/r)]e^{-rt/V}$; $\lim_{t \to \infty} c = k + (P/r)$
    (b) $T = (V \ln 2)/r$; $T = (V \ln 10)/r$
    (c) Superior, $T = 431$ years;   Michigan, $T = 71.4$ years;   Erie, $T = 6.05$ years;   Ontario, $T = 17.6$ years

20. (a) 50.408 m   (b) 5.248 sec   21. (a) 45.783 m   (b) 5.129 sec

22. (a) 48.562 m   (b) 5.194 sec

23. (a) 176.7 ft/sec   (b) 1074.5 ft   (c) 15 ft/sec   (d) 256.6 sec

24. (a) $dv/dx = -\mu v$ (b) $\mu = (66/25) \ln 10 \text{ mi}^{-1} \cong 6.0788 \text{ mi}^{-1}$
    (c) $\tau = 900/(11 \ln 10) \text{ sec} \cong 35.533 \text{ sec}$

25. (a) $x_m = -\dfrac{m^2 g}{k^2} \ln\left(1 + \dfrac{kv_0}{mg}\right) + \dfrac{mv_0}{k}$; $t_m = \dfrac{m}{k} \ln\left(1 + \dfrac{kv_0}{mg}\right)$

26. (a) $v = -(mg/k) + [v_0 + (mg/k)] \exp(-kt/m)$ (b) $v = v_0 - gt$;   yes
    (c) $v = 0$ for $t > 0$

27. (a) $v_L = 2a^2 g(\rho - \rho')/9\mu$ (b) $e = 4\pi a^3 g(\rho - \rho')/3E$

28. (a) 11.58 m/sec   (b) 13.45 m   (c) $k \geq 0.2394$ kg/sec

29. (a) $v = R\sqrt{2g/(R + x)}$ (b) 50.6 hr

30. (b) $x = ut \cos A$, $y = -gt^2/2 + ut \sin A + h$
    (d) $-16L^2/(u^2 \cos^2 A) + L \tan A + 3 \geq H$
    (e) $0.63 \text{ rad} \leq A \leq 0.96 \text{ rad}$
    (f) $u = 106.89$ ft/sec, $A = 0.7954$ rad

31. (a) $v = (u \cos A)e^{-rt}$, $w = -g/r + (u \sin A + g/r)e^{-rt}$
    (b) $x = u \cos A(1 - e^{-rt})/r$, $y = -gt/r + (u \sin A + g/r)(1 - e^{-rt})/r + h$
    (d) $u = 145.3$ ft/sec, $A = 0.644$ rad

32. (d) $k = 2.193$

Section 2.4, *page 79*

1. $0 < t < 3$ 2. $0 < t < 4$ 3. $\pi/2 < t < 3\pi/2$

4. $-\infty < t < -2$ 5. $-2 < t < 2$ 6. $1 < t < \pi$

7. $2t + 5y > 0$ or $2t + 5y < 0$ 8. $t^2 + y^2 < 1$

9. $1 - t^2 + y^2 > 0$ or $1 - t^2 + y^2 < 0$, $t \neq 0$, $y \neq 0$

10. Everywhere   11. $y \neq 0$, $y \neq 3$

12. $t \neq n\pi$ for $n = 0, \pm 1, \pm 2, \ldots$; $y \neq -1$

13. $y = \pm\sqrt{y_0^2 - 4t^2}$ if $y_0 \neq 0$; $|t| < |y_0|/2$

14. $y = [(1/y_0) - t^2]^{-1}$ if $y_0 \neq 0$; $y = 0$ if $y_0 = 0$;   interval is $|t| < 1/\sqrt{y_0}$ if $y_0 > 0$;
    $-\infty < t < \infty$ if $y_0 \leq 0$

15. $y = y_0/\sqrt{2ty_0^2 + 1}$ if $y_0 \neq 0$; $y = 0$ if $y_0 = 0$;   interval   is $-1/2y_0^2 < t < \infty$ if
    $y_0 \neq 0$; $-\infty < t < \infty$ if $y_0 = 0$

16. $y = \pm\sqrt{\frac{2}{3} \ln(1 + t^3) + y_0^2}$; $-[1 - \exp(-3y_0^2/2)]^{1/3} < t < \infty$

17. $y \to 3$ if $y_0 > 0$; $y = 0$ if $y_0 = 0$; $y \to -\infty$ if $y_0 < 0$

18. $y \to -\infty$ if $y_0 < 0$; $y \to 0$ if $y_0 \geq 0$ 19. $y \to 0$ if $y_0 \leq 9$; $y \to \infty$ if $y_0 > 9$
    (see figure on page 837)

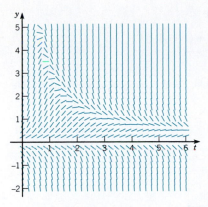

Direction field for Problem 18.

20. $y \to -\infty$ if $y_0 < y_c \approx -0.019$; otherwise y is asymptotic to $\sqrt{t-1}$

21. (a) No   (b) Yes; set $t_0 = 1/2$ in Eq. (19) in text.
(c) $|y| \le (4/3)^{3/2} \cong 1.5396$

22. (a) $y_1(t)$ is a solution for $t \ge 2$; $y_2(t)$ is a solution for all t
(b) f_y is not continuous at $(2, -1)$

26. (a) $y_1(t) = \dfrac{1}{\mu(t)}$; $y_2(t) = \dfrac{1}{\mu(t)} \displaystyle\int_{t_0}^{t} \mu(s)g(s)\,ds$

28. $y = \pm[5t/(2+5ct^5)]^{1/2}$ 29. $y = r/(k + cre^{-rt})$

30. $y = \pm[\epsilon/(\sigma + c\epsilon e^{-2\epsilon t})]^{1/2}$

31. $y = \pm \left\{ \mu(t) \Big/ \left[2 \displaystyle\int_{t_0}^{t} \mu(s)\,ds + c \right] \right\}^{1/2}$, where $\mu(t) = \exp(2\Gamma \sin t + 2Tt)$

32. $y = \frac{1}{2}(1 - e^{-2t})$ for $0 \le t \le 1$; $y = \frac{1}{2}(e^2 - 1)e^{-2t}$ for $t > 1$

33. $y = e^{-2t}$ for $0 \le t \le 1$; $y = e^{-(t+1)}$ for $t > 1$

Section 2.5, *page 91*

1. $y = 0$ is unstable

2. $y = -a/b$ is asymptotically stable, $y = 0$ is unstable

3. $y = 1$ is asymptotically stable, $y = 0$ and $y = 2$ are unstable

4. $y = 0$ is unstable   5. $y = 0$ is asymptotically stable

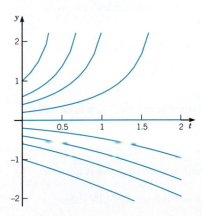

Some solution curves for Problem 4.

6. $y = 0$ is asymptotically stable

7. (c) $y = [y_0 + (1 - y_0)kt]/[1 + (1 - y_0)kt]$

8. $y = 1$ is semistable

9. $y = -1$ is asymptotically stable, $y = 0$ is semistable, $y = 1$ is unstable

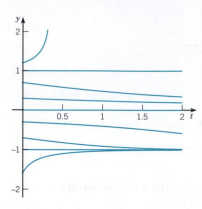

Some solution curves for Problem 9.

10. $y = -1$ and $y = 1$ are asymptotically stable, $y = 0$ is unstable

11. $y = 0$ is asymptotically stable, $y = b^2/a^2$ is unstable

12. $y = 2$ is asymptotically stable, $y = 0$ is semistable, $y = -2$ is unstable

13. $y = 0$ and $y = 1$ are semistable

14. (a) $\tau = (1/r) \ln 4$; 55.452 years
    (b) $T = (1/r) \ln[\beta(1 - \alpha)/(1 - \beta)\alpha]$; 175.78 years

15. (a) $y = 0$ is unstable, $y = K$ is asymptotically stable
    (b) Concave up for $0 < y \le K/e$, concave down for $K/e \le y < K$

16. (a) $y = K \exp\{[\ln(y_0/K)]e^{-rt}\}$ (b) $y(2) \cong 0.7153K \cong 57.6 \times 10^6$ kg
    (c) $\tau \cong 2.215$ years

17. (b) $y = (5 \pm \sqrt{13})/3 \cong 0.4648, 2.8685$
    (c) $y(5) \cong 3.625$ (d) $t = 7.97$

18. (b) $(h/a)\sqrt{k/\alpha\pi}$; yes   19. (b) $k^2/2g(\alpha a)^2$
    (c) $k/\alpha \le \pi a^2$

20. (a) $y = 0$ is unstable, $y = 1$ is asymptotically stable
    (b) $y = y_0/[y_0 + (1 - y_0)e^{-\alpha t}]$

21. (a) $y = y_0 e^{-\beta t}$ (b) $x = x_0 \exp[-\alpha y_0(1 - e^{-\beta t})/\beta]$ (c) $x_0 \exp(-\alpha y_0/\beta)$

22. (b) $z = 1/[\nu + (1 - \nu)e^{\beta t}]$ (c) 0.0927

23. (a,b) $a = 0$: $y = 0$ is semistable.
    $a > 0$: $y = \sqrt{a}$ is asymptotically stable and $y = -\sqrt{a}$ is unstable.

24. (a) $a \le 0$: $y = 0$ is asymptotically stable.
    $a > 0$: $y = 0$ is unstable; $y = \sqrt{a}$ and $y = -\sqrt{a}$ are asymptotically stable.

25. (a) $a < 0$: $y = 0$ is asymptotically stable and $y = a$ is unstable.
    $a = 0$: $y = 0$ is semistable.
    $a > 0$: $y = 0$ is unstable and $y = a$ is asymptotically stable.

26. (a) $\lim\limits_{t\to\infty} x(t) = \min(p, q);$ $x(t) = \dfrac{pq[e^{\alpha(q-p)t} - 1]}{qe^{\alpha(q-p)t} - p}$

(b) $\lim\limits_{t\to\infty} x(t) = p;$ $x(t) = \dfrac{p^2\alpha t}{p\alpha t + 1}$

Section 2.6, *page 102*

1. (a) Exact   (b) $x^2 + 3x + y^2 - 2y = c$ 2. (a) Not exact

3. (a) Exact   (b) $x^3 - x^2y + 2x + 2y^3 + 3y = c$

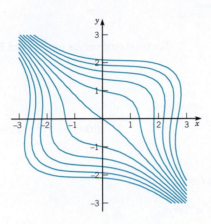

Some integral curves for Problem 3.

4. (a) Exact   (b) $x^2y^2 + 2xy = c$

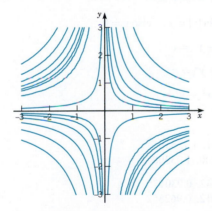

Some integral curves for Problem 4.

5. (a) Exact   (b) $ax^2 + 2bxy + cy^2 = k$ 6. (a) Not exact

7. (a) Exact   (b) $e^x \sin y + 2y \cos x = c$;   also $y = 0$ (see figure on page 840)

8. (a) Not exact   9. (a) Exact   (b) $e^{xy} \cos 2x + x^2 - 3y = c$

10. (a) Exact   (b) $y \ln x + 3x^2 - 2y = c$ 11. (a) Not exact

12. (a) Exact   (b) $x^2 + y^2 = c$

13. $y = [x + \sqrt{28 - 3x^2}]/2,$ $|x| < \sqrt{28/3}$

14. $y = [x - (24x^3 + x^2 - 8x - 16)^{1/2}]/4,$ $x > 0.9846$

15. $b = 3;$ $x^2y^2 + 2x^3y = c$

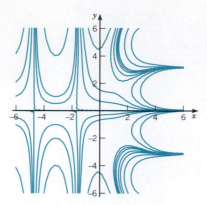

Some integral curves for Problem 7.

16. $b = 1$; $\quad e^{2xy} + x^2 = c$

19. (b) $x^2 + 2\ln|y| - y^{-2} = c$; $\quad$ also $y = 0$

20. (b) $e^x \sin y + 2y \cos x = c$ $\qquad\qquad$ 21. (b) $xy^2 - (y^2 - 2y + 2)e^y = c$

22. (b) $x^2 e^x \sin y = c$

24. $\mu(t) = \exp \int R(t)\, dt$, where $t = xy$

25. (a) $\mu(x) = e^{3x}$; $\quad (3x^2 y + y^3)e^{3x} = c$

26. (a) $\mu(x) = e^{-x}$; $\quad y = ce^x + 1 + e^{2x}$

27. (a) $\mu(y) = y$; $\quad xy + y \cos y - \sin y = c$

28. (a) $\mu(y) = e^{2y}/y$; $\quad xe^{2y} - \ln|y| = c$; $\quad$ also $y = 0$

29. (a) $\mu(y) = \sin y$; $\quad e^x \sin y + y^2 = c$

30. (a) $\mu(y) = y^2$; $\quad x^4 + 3xy + y^4 = c$

31. (a) $\mu(x, y) = xy$; $\quad x^3 y + 3x^2 + y^3 = c$

Section 2.7, *page 109*

1. (a) 1.1975, 1.38549, 1.56491, 1.73658
   (b) 1.19631, 1.38335, 1.56200, 1.73308

2. (a) 1.59980, 1.29288, 1.07242, 0.930175
   (b) 1.61124, 1.31361, 1.10012, 0.962552

3. (a) 1.2025, 1.41603, 1.64289, 1.88590
   (b) 1.20388, 1.41936, 1.64896, 1.89572

4. (a) 1.10244, 1.21426, 1.33484, 1.46399
   (b) 1.10365, 1.21656, 1.33817, 1.46832

5. (a) 0.509239, 0.522187, 0.539023, 0.559936
   (b) 0.509701, 0.523155, 0.540550, 0.562089

6. (a) −0.920498, −0.857538, −0.808030, −0.770038
   (b) −0.922575, −0.860923, −0.812300, −0.774965

7. (a) 2.90330, 7.53999, 19.4292, 50.5614
   (b) 2.93506, 7.70957, 20.1081, 52.9779

8. (a) 0.891830, 1.25225, 2.37818, 4.07257
   (b) 0.908902, 1.26872, 2.39336, 4.08799

9. (a) 3.95713, 5.09853, 6.41548, 7.90174
   (b) 3.95965, 5.10371, 6.42343, 7.91255

10. (a) 1.60729, 2.46830, 3.72167, 5.45963
    (b) 1.60996, 2.47460, 3.73356, 5.47774

11. (a) -1.45865, -0.217545, 1.05715, 1.41487
    (b) -1.45322, -0.180813, 1.05903, 1.41244

12. (a) 0.587987, 0.791589, 1.14743, 1.70973
    (b) 0.589440, 0.795758, 1.15693, 1.72955

14. 1.595, 2.4636

15. $e_{n+1} = [2\phi(\bar{t}_n) - 1]h^2$, $\quad |e_{n+1}| \le \left[1 + 2\max_{0 \le t \le 1} |\phi(t)|\right] h^2$,

    $e_{n+1} = e^{2\bar{t}_n} h^2$, $\quad |e_1| \le 0.012$, $\quad |e_4| \le 0.022$

16. $e_{n+1} = [2\phi(\bar{t}_n) - \bar{t}_n]h^2$, $\quad |e_{n+1}| \le \left[1 + 2\max_{0 \le t \le 1} |\phi(t)|\right] h^2$,

    $e_{n+1} = 2e^{2\bar{t}_n} h^2$, $\quad |e_1| \le 0.024$, $\quad |e_4| \le 0.045$

17. $e_{n+1} = [\bar{t}_n + \bar{t}_n^2 \phi(\bar{t}_n) + \phi^3(\bar{t}_n)]h^2$

18. $e_{n+1} = [19 - 15\bar{t}_n \phi^{-1/2}(\bar{t}_n)]h^2/4$

19. $e_{n+1} = \{1 + [\bar{t}_n + \phi(\bar{t}_n)]^{-1/2}\}h^2/4$

20. $e_{n+1} = \{2 - [\phi(\bar{t}_n) + 2\bar{t}_n^2]\exp[-\bar{t}_n\phi(\bar{t}_n)] - \bar{t}_n \exp[-2\bar{t}_n\phi(\bar{t}_n)]\}h^2/2$

21. (a) $\phi(t) = 1 + (1/5\pi)\sin 5\pi t$ (b) 1.2, 1.0, 1.2
    (c) 1.1, 1.1, 1.0, 1.0 (d) $h < 1/\sqrt{50\pi} \cong 0.08$

22. (a) 1.55, 2.34, 3.46, 5.07
    (b) 1.20, 1.39, 1.57, 1.74
    (c) 1.20, 1.42, 1.65, 1.90

23. (a) 0 (b) 60 (c) -92.16

24. $0.224 \ne 0.225$

Section 2.8, *page 117*

1. (a) 1.19512, 1.38120, 1.55909, 1.72956
   (b) 1.19515, 1.38125, 1.55916, 1.72965
   (c) 1.19516, 1.38126, 1.55918, 1.72967
   (d) 1.19516, 1.38127, 1.55918, 1.72968
   (e) 1.19516, 1.38127, 1.55918, 1.72968

2. (a) 1.62283, 1.33460, 1.12820, 0.995445
   (b) 1.62243, 1.33386, 1.12718, 0.994215
   (c) 1.62234, 1.33368, 1.12693, 0.993921
   (d) 1.62231, 1.33362, 1.12686, 0.993839
   (e) 1.62230, 1.33362, 1.12685, 0.993826

3. (a) 1.20526, 1.42273, 1.65511, 1.90570
   (b) 1.20533, 1.42290, 1.65542, 1.90621
   (c) 1.20534, 1.42294, 1.65550, 1.90634
   (d) 1.20535, 1.42295, 1.65553, 1.90638
   (e) 1.20535, 1.42296, 1.65553, 1.90638

4. (a) 1.10483, 1.21882, 1.34146, 1.47263
   (b) 1.10484, 1.21884, 1.34147, 1.47262
   (c) 1.10484, 1.21884, 1.34147, 1.47262
   (d) 1.10484, 1.21884, 1.34147, 1.47262
   (e) 1.10484, 1.21884, 1.34147, 1.47262

5. (a) 0.510164, 0.524126, 0.542083, 0.564251
   (b) 0.510168, 0.524135, 0.542100, 0.564277
   (c) 0.510169, 0.524137, 0.542104, 0.564284
   (d) 0.510170, 0.524138, 0.542105, 0.564286
   (e) 0.520169, 0.524138, 0.542105, 0.564286

6. (a) −0.924650, −0.864338, −0.816642, −0.780008
   (b) −0.924550, −0.864177, −0.816442, −0.779781
   (c) −0.924525, −0.864138, −0.816393, −0.779725
   (d) −0.924517, −0.864125, −0.816377, −0.779706
   (e) −0.924517, −0.864125, −0.816377, −0.779706

7. (a) 2.96719, 7.88313, 20.8114, 55.5106
   (b) 2.96800, 7.88755, 20.8294, 55.5758
   (c) 2.96825, 7.88889, 20.8349, 55.5957
   (d) 2.96828, 7.88904, 20.8355, 55.5980

8. (a) 0.926139, 1.28558, 2.40898, 4.10386
   (b) 0.925815, 1.28525, 2.40869, 4.10359
   (c) 0.925725, 1.28516, 2.40860, 4.10350
   (d) 0.925711, 1.28515, 2.40860, 4.10350

9. (a) 3.96217, 5.10887, 6.43134, 7.92332
   (b) 3.96218, 5.10889, 6.43138, 7.92337
   (c) 3.96219, 5.10890, 6.43139, 7.92338
   (d) 3.96219, 5.10890, 6.43139, 7.92338

10. (a) 1.61263, 2.48097, 3.74556, 5.49595
    (b) 1.61263, 2.48092, 3.74550, 5.49589
    (c) 1.61262, 2.48091, 3.74548, 5.49587
    (d) 1.61262, 2.48091, 3.74548, 5.49587

11. (a) −1.44768, −0.144478, 1.06004, 1.40960
    (b) −1.44765, −0.143690, 1.06072, 1.40999
    (c) −1.44764, −0.143543, 1.06089, 1.41008
    (d) −1.44764, −0.143427, 1.06095, 1.41011

12. (a) 0.590897, 0.799950, 1.16653, 1.74969
    (b) 0.590906, 0.799988, 1.16663, 1.74992
    (c) 0.590909, 0.800000, 1.166667, 1.75000
    (d) 0.590909, 0.800000, 1.166667, 1.75000

18. $e_{n+1} = (38h^3/3)\exp(4\bar{t}_n)$, $|e_{n+1}| \le 37{,}758.8h^3$ on $0 \le t \le 2$, $|e_1| \le 0.00193389$

19. $e_{n+1} = (2h^3/3)\exp(2\bar{t}_n)$, $|e_{n+1}| \le 4.92604h^3$ on $0 \le t \le 1$, $|e_1| \le 0.000814269$

20. $e_{n+1} = (4h^3/3)\exp(2\bar{t}_n)$, $|e_{n+1}| \le 9.85207h^3$ on $0 \le t \le 1$, $|e_1| \le 0.00162854$

21. $h \cong 0.071$

22. $h \cong 0.023$

23. $h \cong 0.081$

24. $h \cong 0.117$

CHAPTER 3

Section 3.1, *page 138*

1. (a) $x_1 = 2,\quad x_2 = 1$

2. (a) $x_1 = 6,\quad x_2 = -2$

3. (a) $x_1 = 0,\quad x_2 = 0$

4. (a) no solution

5. (a) $x_1 = -1,\quad x_2 = -2$

6. (a) $x_1 = c,\quad x_2 = 3c/2;\quad c$ arbitrary

7. (a) $x_1 = c,\quad x_2 = (2c - 6)/3;\ c$ arbitrary

8. (a) $x_1 = -3/4,\ x_2 = 3$

9. (a) $x_1 = 2,\quad x_2 = 2$

10. (a) $x_1 = -2,\quad x_2 = 1$

11. (a) $x_1 = 0,\quad x_2 = 0$

12. (a) $x_1 = c,\quad x_2 = -2c/5;\quad c$ arbitrary

13. $\lambda_1 = 2,\quad \mathbf{v}_1 = \begin{pmatrix} 2 \\ 1 \end{pmatrix};\quad \lambda_2 = -1,\quad \mathbf{v}_2 = \begin{pmatrix} 1 \\ 2 \end{pmatrix}$

14. $\lambda_1 = 1 + 2i,\quad \mathbf{v}_1 = \begin{pmatrix} 1 + i \\ 2 \end{pmatrix};\quad \lambda_2 = 1 - 2i,\quad \mathbf{v}_2 = \begin{pmatrix} 1 - i \\ 2 \end{pmatrix}$

15. $\lambda_1 = \lambda_2 = 1,\quad \mathbf{v}_1 = \begin{pmatrix} 2 \\ 1 \end{pmatrix}$

16. $\lambda_1 = -1,\quad \mathbf{v}_1 = \begin{pmatrix} 1 \\ 1 \end{pmatrix};\quad \lambda_2 = -2,\quad \mathbf{v}_2 = \begin{pmatrix} 2 \\ 3 \end{pmatrix}$

17. $\lambda_1 = -1 + 2i,\quad \mathbf{v}_1 = \begin{pmatrix} 2 \\ -i \end{pmatrix};\quad \lambda_2 = -1 - 2i,\quad \mathbf{v}_2 = \begin{pmatrix} 2 \\ i \end{pmatrix}$

18. $\lambda_1 = \lambda_2 = 1/2,\quad \mathbf{v}_1 = \begin{pmatrix} -1 \\ 1 \end{pmatrix}$

19. $\lambda_1 = \lambda_2 = -1,\quad \mathbf{v}_1 = \begin{pmatrix} 2 \\ 1 \end{pmatrix}$

20. $\lambda_1 = -1,\quad \mathbf{v}_1 = \begin{pmatrix} 1 \\ 3 \end{pmatrix};\quad \lambda_2 = 1,\quad \mathbf{v}_2 = \begin{pmatrix} 1 \\ 1 \end{pmatrix}$

21. $\lambda_1 = i,\quad \mathbf{v}_1 = \begin{pmatrix} 2 + i \\ 1 \end{pmatrix};\quad \lambda_2 = -i,\quad \mathbf{v}_2 = \begin{pmatrix} 2 - i \\ 1 \end{pmatrix}$

22. $\lambda_1 = 7,\quad \mathbf{v}_1 = \begin{pmatrix} 3 \\ 1 \end{pmatrix};\quad \lambda_2 = 0,\quad \mathbf{v}_2 = \begin{pmatrix} 1 \\ -2 \end{pmatrix}$

23. $\lambda_1 = 2,\quad \mathbf{v}_1 = \begin{pmatrix} 1 \\ 1 \end{pmatrix};\quad \lambda_2 = -3,\quad \mathbf{v}_2 = \begin{pmatrix} 1 \\ -4 \end{pmatrix}$

24. $\lambda_1 = (1 + 3i)/2,\quad \mathbf{v}_1 = \begin{pmatrix} 5 \\ 3 - 3i \end{pmatrix};\quad \lambda_2 = (1 - 3i)/2,\quad \mathbf{v}_2 = \begin{pmatrix} 5 \\ 3 + 3i \end{pmatrix}$

25. $\lambda_1 = \lambda_2 = -1/2,\quad \mathbf{v}_1 = \begin{pmatrix} 1 \\ 1 \end{pmatrix}$

26. $\lambda_1 = -1 + i,\quad \mathbf{v}_1 = \begin{pmatrix} 2 + i \\ 5 \end{pmatrix},\quad \lambda_2 = -1 - i,\quad \mathbf{v}_2 = \begin{pmatrix} 2 - i \\ 5 \end{pmatrix}$

27. $\lambda_1 = -2,\quad \mathbf{v}_1 = \begin{pmatrix} 4 \\ -9 \end{pmatrix};\quad \lambda_2 = 0,\quad \mathbf{v}_2 = \begin{pmatrix} -4 \\ 3 \end{pmatrix}$

28. $\lambda_1 = -1$, $\mathbf{v}_1 = \begin{pmatrix} 1 \\ 1 \end{pmatrix}$; $\lambda_2 = -3$, $\mathbf{v}_2 = \begin{pmatrix} -1 \\ 1 \end{pmatrix}$

29. $\lambda_1 = 3i$, $\mathbf{v}_1 = \begin{pmatrix} 2 \\ -1 + 3i \end{pmatrix}$; $\lambda_2 = -3i$, $\mathbf{v}_2 = \begin{pmatrix} 2 \\ -1 - 3i \end{pmatrix}$

30. $\lambda_1 = \lambda_2 = -2$, $\mathbf{v}_1 = \begin{pmatrix} 1 \\ 2 \end{pmatrix}$

31. $\lambda_1 = 1/2$, $\mathbf{v}_1 = \begin{pmatrix} -1 \\ 1 \end{pmatrix}$; $\lambda_2 = 2$, $\mathbf{v}_2 = \begin{pmatrix} 1 \\ 1 \end{pmatrix}$

32. $\lambda_1 = \lambda_2 = 3/2$, $\mathbf{v}_1 = \begin{pmatrix} 1 \\ -1 \end{pmatrix}$

33. (a) $\lambda = (-1 \pm \sqrt{25 + 4\alpha})/2$

34. (a) $\lambda = (5 \pm \sqrt{1 - 16\alpha})/2$

35. (a) $\lambda = (1 + \alpha \pm \sqrt{\alpha^2 - 2\alpha + 25})/2$

36. (a) $\lambda = 2 \pm \sqrt{1 - 2\alpha^2}$

Section 3.2, *page 146*

1. autonomous, nonhomogeneous

2. nonautonomous, nonhomogeneous

3. nonautonomous, homogeneous

4. autonomous, nonhomogeneous

5. autonomous, homogeneous

6. nonautonomous, homogeneous

7. nonautonomous, nonhomogeneous

8. autonomous, homogeneous

9. (a) (2, 1)   (c) Almost all solutions depart from the critical point.

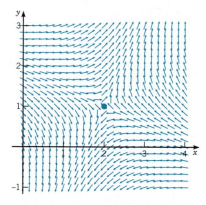

Direction field for Problem 9.

10. (a) (4, −2)   (c) Other solutions spiral toward the critical point. (See figure on page 845)

11. (a) (5, 9)   (c) Other solutions depart from the critical point.

12. (a) (−3, 5)   (c) Almost all solutions depart from the critical point.

13. (a) (2, 1)   (c) Other solutions spiral away from the critical point.

14. (a) (−3, 5)   (c) Other solutions approach the critical point.

15. $x_1' = x_2$, $x_2' = -2x_1 - 0.5x_2$

16. $x_1' = x_2$, $x_2' = -4x_1 - 0.25x_2 + 3\sin 2t$

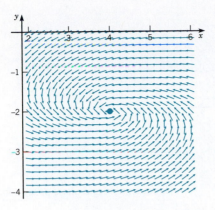

Direction field for Problem 10.

17. $x_1' = x_2, \quad x_2' = -\frac{(t^2 - 0.25)}{t^2} x_1 - \frac{1}{t} x_2$

18. $x_1' = x_2, \quad x_2' = -\frac{5}{t^2} x_1 - \frac{3}{t} x_2 + 1 + \frac{4}{t^2}$

19. $x_1' = x_2, \quad x_2' = -4x_1 - 0.25x_2 + 2\cos 3t; \qquad x_1(0) = 1, \quad x_2(0) = -2$

20. $x_1' = x_2, \quad x_2' = -x_1 - \frac{1}{t} x_2; \qquad x_1(1) = 1, \quad x_2(1) = 0$

24. (a) $Q_1' = 3q_1 - \frac{1}{15}Q_1 + \frac{1}{100}Q_2, \quad Q_1(0) = Q_1^0$

$\qquad Q_2' = q_2 + \frac{1}{30}Q_1 - \frac{3}{100}Q_2, \quad Q_2(0) = Q_2^0$

(b) $Q_1^E = 6(9q_1 + q_2), \quad Q_2^E = 20(3q_1 + 2q_2)$

(c) No

(d) $\frac{10}{9} \leq Q_2^E / Q_1^E \leq \frac{20}{3}$

Section 3.3, *page 162*

1. $\mathbf{x} = c_1 e^{-t} \begin{pmatrix} 1 \\ 2 \end{pmatrix} + c_2 e^{2t} \begin{pmatrix} 2 \\ 1 \end{pmatrix}$

2. $\mathbf{x} = c_1 e^{-t} \begin{pmatrix} 1 \\ 1 \end{pmatrix} + c_2 e^{-2t} \begin{pmatrix} 2 \\ 3 \end{pmatrix}$

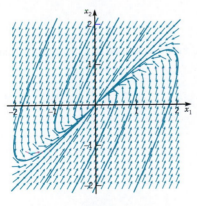

Direction field and phase portrait for Problem 2.

3. $\mathbf{x} = c_1 e^{t} \begin{pmatrix} 1 \\ 1 \end{pmatrix} + c_2 e^{-t} \begin{pmatrix} 1 \\ 3 \end{pmatrix}$

(see figure on page 846)

4. $\mathbf{x} = c_1 e^{-3t} \begin{pmatrix} 1 \\ -4 \end{pmatrix} + c_2 e^{2t} \begin{pmatrix} 1 \\ 1 \end{pmatrix}$

5. $\mathbf{x} = c_1 e^{-2t} \begin{pmatrix} 1 \\ 2 \end{pmatrix} + c_2 \begin{pmatrix} 3 \\ 4 \end{pmatrix}$

6. $\mathbf{x} = c_1 e^{-3t} \begin{pmatrix} 1 \\ -1 \end{pmatrix} + c_2 e^{-t} \begin{pmatrix} 1 \\ 1 \end{pmatrix}$

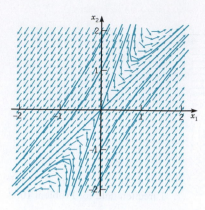

Direction field and phase portrait for Problem 3.

7. $\mathbf{x} = c_1 e^{2t} \begin{pmatrix} 1 \\ 1 \end{pmatrix} + c_2 e^{t/2} \begin{pmatrix} 1 \\ -1 \end{pmatrix}$

8. $\mathbf{x} = c_1 e^{-t/2} \begin{pmatrix} 7 \\ -1 \end{pmatrix} + c_2 e^{t} \begin{pmatrix} 1 \\ -1 \end{pmatrix}$

9. $\mathbf{x} = c_1 e^{t/2} \begin{pmatrix} 1 \\ -1 \end{pmatrix} + c_2 e^{t/4} \begin{pmatrix} 3 \\ -2 \end{pmatrix}$

10. $\mathbf{x} = c_1 e^{2t} \begin{pmatrix} 1 \\ 3 \end{pmatrix} + c_2 e^{4t} \begin{pmatrix} 1 \\ 1 \end{pmatrix}$

11. $\mathbf{x} = c_1 e^{-t} \begin{pmatrix} 1 \\ 1 \end{pmatrix} + c_2 e^{3t} \begin{pmatrix} 1 \\ 5 \end{pmatrix}$

12. $\mathbf{x} = c_1 e^{t} \begin{pmatrix} 3 \\ -1 \end{pmatrix} + c_2 \begin{pmatrix} 2 \\ -1 \end{pmatrix}$

13. $\mathbf{x} = 7 e^{-t} \begin{pmatrix} 1 \\ 1 \end{pmatrix} - 2 e^{-2t} \begin{pmatrix} 2 \\ 3 \end{pmatrix}$

14. $\mathbf{x} = \frac{1}{2} e^{t} \begin{pmatrix} 1 \\ 1 \end{pmatrix} + \frac{3}{2} e^{-t} \begin{pmatrix} 1 \\ 3 \end{pmatrix}$

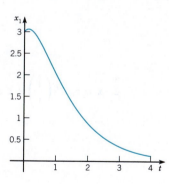

Component plot of x_1 versus t for Problem 13.

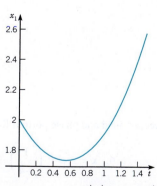

Component plot of x_1 versus t for Problem 14.

15. $\mathbf{x} = -\frac{3}{2} e^{2t} \begin{pmatrix} 1 \\ 3 \end{pmatrix} + \frac{7}{2} e^{4t} \begin{pmatrix} 1 \\ 1 \end{pmatrix}$

16. $\mathbf{x} = \frac{1}{2} e^{-t} \begin{pmatrix} 1 \\ 1 \end{pmatrix} + \frac{1}{2} e^{3t} \begin{pmatrix} 1 \\ 5 \end{pmatrix}$

25. $T \cong 41.57$ min

26. **(a)** $\mathbf{x} = c_1 e^{(-2+\sqrt{2})t/2} \begin{pmatrix} -\sqrt{2} \\ 1 \end{pmatrix} + c_2 e^{(-2-\sqrt{2})t/2} \begin{pmatrix} \sqrt{2} \\ 1 \end{pmatrix}; \quad \lambda_{1,2} = (-2 \pm \sqrt{2})/2; \quad$ node

(b) $\mathbf{x} = c_1 e^{(-1+\sqrt{2})t} \begin{pmatrix} -1 \\ \sqrt{2} \end{pmatrix} + c_2 e^{(-1-\sqrt{2})t} \begin{pmatrix} 1 \\ \sqrt{2} \end{pmatrix}; \quad \lambda_{1,2} = -1 \pm \sqrt{2}; \quad$ saddle point

(c) $\lambda_{1,2} = -1 \pm \sqrt{\alpha}; \quad \alpha = 1$

27. **(a)** $\begin{pmatrix} i \\ v \end{pmatrix} = c_1 e^{-2t} \begin{pmatrix} 1 \\ 3 \end{pmatrix} + c_2 e^{-t} \begin{pmatrix} 1 \\ 1 \end{pmatrix}$

28. **(a)** $\left(\dfrac{1}{CR_2} - \dfrac{R_1}{L} \right)^2 - \dfrac{4}{CL} > 0$

Section 3.4, *page 173*

1. $\mathbf{x} = c_1 e^t \begin{pmatrix} \cos 2t \\ \cos 2t + \sin 2t \end{pmatrix} + c_2 e^t \begin{pmatrix} \sin 2t \\ -\cos 2t + \sin 2t \end{pmatrix}$

2. $\mathbf{x} = c_1 e^{-t} \begin{pmatrix} 2\cos 2t \\ \sin 2t \end{pmatrix} + c_2 e^{-t} \begin{pmatrix} -2\sin 2t \\ \cos 2t \end{pmatrix}$

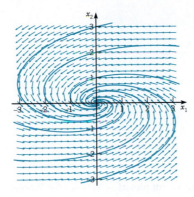

Direction field and phase portrait for Problem 2.

3. $\mathbf{x} = c_1 \begin{pmatrix} 5\cos t \\ 2\cos t + \sin t \end{pmatrix} + c_2 \begin{pmatrix} 5\sin t \\ -\cos t + 2\sin t \end{pmatrix}$

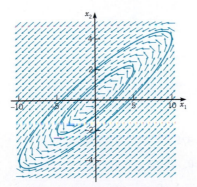

Direction field and phase portrait for Problem 3.

4. $\mathbf{x} = c_1 e^{t/2} \begin{pmatrix} 5\cos\frac{3}{2}t \\ 3(\cos\frac{3}{2}t + \sin\frac{3}{2}t) \end{pmatrix} + c_2 e^{t/2} \begin{pmatrix} 5\sin\frac{3}{2}t \\ 3(-\cos\frac{3}{2}t + \sin\frac{3}{2}t) \end{pmatrix}$

5. $\mathbf{x} = c_1 e^{-t} \begin{pmatrix} \cos t \\ 2\cos t + \sin t \end{pmatrix} + c_2 e^{-t} \begin{pmatrix} \sin t \\ -\cos t + 2\sin t \end{pmatrix}$

6. $\mathbf{x} = c_1 \begin{pmatrix} -2\cos 3t \\ \cos 3t + 3\sin 3t \end{pmatrix} + c_2 \begin{pmatrix} -2\sin 3t \\ \sin 3t - 3\cos 3t \end{pmatrix}$

7. $\mathbf{x} = e^{-t} \begin{pmatrix} 4\cos 2t + 6\sin 2t \\ -3\cos 2t + 2\sin 2t \end{pmatrix}$

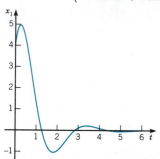

Component plot of x_1 versus t for Problem 7.

8. $\mathbf{x} = \begin{pmatrix} 3\cos t - 4\sin t \\ 2\cos t - \sin t \end{pmatrix}$

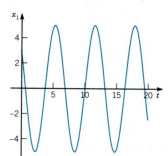

Component plot of x_1 versus t for Problem 8.

9. $\mathbf{x} = e^{-t} \begin{pmatrix} \cos t - 3\sin t \\ \cos t - \sin t \end{pmatrix}$

10. $\mathbf{x} = e^{-2t} \begin{pmatrix} \cos t - 5\sin t \\ -2\cos t - 3\sin t \end{pmatrix}$

11. (a) $\lambda = -\frac{1}{4} \pm i$

12. (a) $\lambda = \frac{1}{5} \pm i$

13. (a) $\lambda = \alpha \pm i$ (b) $\alpha = 0$

14. (a) $\lambda = (\alpha \pm \sqrt{\alpha^2 - 20})/2$ (b) $\alpha = -\sqrt{20},\ 0,\ \sqrt{20}$

15. (a) $\lambda = \pm\sqrt{4 - 5\alpha}$ (b) $\alpha = 4/5$

16. (a) $\lambda = \frac{5}{4} \pm \frac{1}{2}\sqrt{3\alpha}$ (b) $\alpha = 0,\ 25/12$

17. (a) $\lambda = -1 \pm \sqrt{-\alpha}$ (b) $\alpha = -1,\ 0$

18. (a) $\lambda = -\frac{1}{2} \pm \frac{1}{2}\sqrt{49 - 24\alpha}$ (b) $\alpha = 2,\ 49/24$

19. (a) $\lambda = \frac{1}{2}\alpha - 2 \pm \sqrt{\alpha^2 + 8\alpha - 24}$ (b) $\alpha = -4 - 2\sqrt{10},\ -4 + 2\sqrt{10},\ 5/2$

20. (a) $\lambda = -1 \pm \sqrt{25 + 8\alpha}$ (b) $\alpha = -25/8,\ -3$

21. (b) $\begin{pmatrix} i \\ v \end{pmatrix} = c_1 e^{-t/2} \begin{pmatrix} \cos(t/2) \\ 4\sin(t/2) \end{pmatrix} + c_2 e^{-t/2} \begin{pmatrix} \sin(t/2) \\ -4\cos(t/2) \end{pmatrix}$

    (c) Use $c_1 = 2,\quad c_2 = -\frac{3}{4}$ in answer to part (b).

    (d) $\lim_{t\to\infty} i(t) = \lim_{t\to\infty} v(t) = 0$; no

22. (b) $\begin{pmatrix} i \\ v \end{pmatrix} = c_1 e^{-t} \begin{pmatrix} \cos t \\ -\cos t - \sin t \end{pmatrix} + c_2 e^{-t} \begin{pmatrix} \sin t \\ -\sin t + \cos t \end{pmatrix}$

    (c) Use $c_1 = 2$ and $c_2 = 3$ in answer to part (b).

    (d) $\lim_{t\to\infty} i(t) = \lim_{t\to\infty} v(t) = 0$; no

Section 3.5, *page 186*

1. $\mathbf{x} = c_1 e^t \begin{pmatrix} 2 \\ 1 \end{pmatrix} + c_2 \left[t e^t \begin{pmatrix} 2 \\ 1 \end{pmatrix} + e^t \begin{pmatrix} 1 \\ 0 \end{pmatrix} \right]$

2. $\mathbf{x} = c_1 e^{t/2} \begin{pmatrix} 1 \\ -1 \end{pmatrix} + c_2 \left[t e^{t/2} \begin{pmatrix} 1 \\ -1 \end{pmatrix} + e^{t/2} \begin{pmatrix} 0 \\ 4/3 \end{pmatrix} \right]$

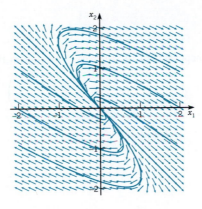

Direction field and phase portrait for Problem 2.

3. $\mathbf{x} = c_1 e^{-t} \begin{pmatrix} 2 \\ 1 \end{pmatrix} + c_2 \left[t e^{-t} \begin{pmatrix} 2 \\ 1 \end{pmatrix} + e^{-t} \begin{pmatrix} 0 \\ 2 \end{pmatrix} \right]$

4. $\mathbf{x} = c_1 e^{-t/2} \begin{pmatrix} 1 \\ 1 \end{pmatrix} + c_2 \left[t e^{-t/2} \begin{pmatrix} 1 \\ 1 \end{pmatrix} + e^{-t/2} \begin{pmatrix} 0 \\ \frac{2}{5} \end{pmatrix} \right]$

    (see figure on page 850)

5. $\mathbf{x} = c_1 e^{-2t} \begin{pmatrix} 1 \\ 2 \end{pmatrix} + c_2 \left[t e^{-2t} \begin{pmatrix} 1 \\ 2 \end{pmatrix} + e^{-2t} \begin{pmatrix} 0 \\ -2 \end{pmatrix} \right]$

6. $\mathbf{x} = c_1 e^{3t/2} \begin{pmatrix} 1 \\ -1 \end{pmatrix} + c_2 \left[t e^{3t/2} \begin{pmatrix} 1 \\ -1 \end{pmatrix} + e^{3t/2} \begin{pmatrix} 0 \\ 2 \end{pmatrix} \right]$

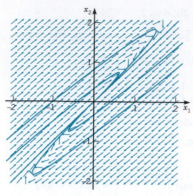

Direction field and phase portrait for Problem 4.

7. $\mathbf{x} = e^{-3t} \begin{pmatrix} 3 + 4t \\ 2 + 4t \end{pmatrix}$

8. $\mathbf{x} = e^{-t} \begin{pmatrix} 3 \\ -1 \end{pmatrix} - 6te^{-t} \begin{pmatrix} 1 \\ 1 \end{pmatrix}$

9. $\mathbf{x} = e^{t/2} \begin{pmatrix} 3 \\ -2 \end{pmatrix} + \frac{3}{2}te^{t/2} \begin{pmatrix} 1 \\ -1 \end{pmatrix}$

10. $\mathbf{x} = e^{t/2} \begin{pmatrix} 3 + \frac{15}{4}t \\ 2 - \frac{15}{4}t \end{pmatrix}$

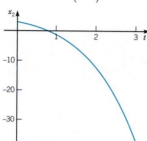

Component plot of x_2 versus t for Problem 10.

11. $\mathbf{x} = e^{-t/2} \begin{pmatrix} 3 - 5t \\ 1 - 5t \end{pmatrix}$

12. $\mathbf{x} = e^{3t/2} \begin{pmatrix} 1 + 2t \\ 3 - 2t \end{pmatrix}$

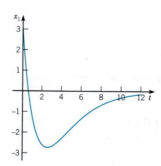

Component plot of x_2 versus t for Problem 11.

14. (b) $\begin{pmatrix} i \\ v \end{pmatrix} = -e^{-t/2} \begin{pmatrix} 1 \\ -2 \end{pmatrix} + \left[te^{-t/2} \begin{pmatrix} 1 \\ -2 \end{pmatrix} + e^{-t/2} \begin{pmatrix} 2 \\ 0 \end{pmatrix} \right]$

Section 3.6, *page 193*

1. (a) $H(x, y) = y/x^2$

2. (a) $H(x, y) = yx^2$

3. (a) $H(x, y) = yx^2$

4. (a) $H(x, y) = y^2 - 4x^2$

5. (a) $H(x, y) = y^2 - 4x^2$

6. (a) $H(x, y) = 4x^2 + y^2$

7. (a) $(0, 0)$ (b) $H(x, y) = x^2 - 4xy + y^2$ (d) $(0, 0)$ is a saddle point.

8. (a) $(0, 0)$ (b) $H(x, y) = x^2 + 2xy - y^2$ (d) $(0, 0)$ is a saddle point.

9. (a) $(0, 0)$ (b) $H(x, y) = x^2 - 2xy + 2y^2$ (d) $(0, 0)$ is a center.

10. (a) $(0, 0)$, $(1, 0)$, $(0.5, 0.25)$ (b) $H(x, y) = -2x^2y - y^2 + 2xy$ (d) $(0, 0)$ and $(1, 0)$ are saddle points; $(0.5, 0.25)$ is a center.

11. (a) $(0, 0)$, $(2, 3)$, $(-2, 3)$ (b) $H(x, y) = x^2y^2 - 3x^2y - 2y^2$ (d) $(0, 0), (2, 3),$ and $(-2, 3)$ are saddle points.

12. (a) $(0, \frac{2}{3})$, $(3, -\frac{2}{3})$ (b) $H(x, y) = x^2y - 3xy + 2x$ (d) $(0, \frac{2}{3})$ and $(3, -\frac{2}{3})$ are saddle points.

13. (a) $(-\frac{1}{2}, 1)$, $(0, 0)$ (c) $(-\frac{1}{2}, 1)$ is a saddle point; $(0, 0)$ is an unstable node.

14. (a) $(\sqrt{2}, 2)$, $(-\sqrt{2}, 2)$ (c) $(\sqrt{2}, 2)$ is a saddle point; $(-\sqrt{2}, 2)$ is an unstable spiral point.

15. (a) $(0, 0)$, $(0, 2)$, $(\frac{1}{2}, \frac{1}{2})$, $(1, 0)$ (c) $(0, 0)$ is an unstable node; $(1, 0)$ and $(0, 2)$ are asymptotically stable nodes; $(\frac{1}{2}, \frac{1}{2})$ is a saddle point.

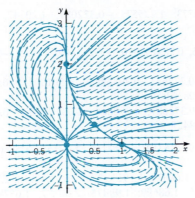

Direction field and phase portrait for Problem 15.

16. (a) $(0, 0)$, $(0, 1)$, $(-2, -2)$, $(3, -2)$ (c) $(0, 0)$ is a saddle point; $(0, 1)$ is an asymptotically stable spiral point; $(-2, -2)$ is an asymptotically stable node; $(3, -2)$ is an unstable node.

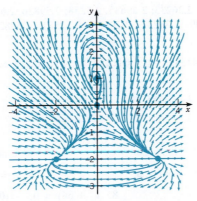

Direction field and phase portrait for Problem 16.

17. (a) $(0, 0)$, $(1 - \sqrt{2}, 1 + \sqrt{2})$, $(1 + \sqrt{2}, 1 - \sqrt{2})$ (c) $(0, 0)$ is an asymptotically stable spiral point; $(1 - \sqrt{2}, 1 + \sqrt{2})$ and $(1 + \sqrt{2}, 1 - \sqrt{2})$ are saddle points.

18. (a) $(0, 0)$, $(2, 2)$, $(-1, -1)$, $(-2, 0)$ (c) $(0, 0)$ and $(-2, 0)$ are saddle points; $(2, 2)$ and $(-1, -1)$ are asymptotically stable spiral points.

19. (a) $(0, 0)$, $(0, 1)$, $(\frac{1}{2}, \frac{1}{2})$, $(-\frac{1}{2}, \frac{1}{2})$ (c) $(0, 0)$ and $(0, 1)$ are saddle points; $(\frac{1}{2}, \frac{1}{2})$ and $(-\frac{1}{2}, \frac{1}{2})$ are centers.

20. (a) $(0, 0)$, $(\sqrt{6}, 0)$, $(-\sqrt{6}, 0)$ (c) $(0, 0)$ is a saddle point; $(\sqrt{6}, 0)$ and $(-\sqrt{6}, 0)$ are asymptotically stable spiral points.

Section 3.7, *page 197*

1. (a) 1.26, 0.76; 1.7714, 1.4824; 2.58991, 2.3703; 3.82374, 3.60413; 5.64246, 5.38885
   (b) 1.32493, 0.758933; 1.93679, 1.57919; 2.93414, 2.66099; 4.48318, 4.22639; 6.84236, 6.56452
   (c) 1.32489, 0.759516; 1.9369, 1.57999; 2.93459, 2.66201; 4.48422, 4.22784; 6.8444, 6.56684

2. (a) 1.451, 1.232; 2.16133, 1.65988; 3.29292, 2.55559; 5.16361, 4.7916; 8.54951, 12.0464
   (b) 1.51844, 1.28089; 2.37684, 1.87711; 3.85039, 3.44859; 6.6956, 9.50309; 15.0987, 64.074
   (c) 1.51855, 1.2809; 2.3773, 1.87729; 3.85247, 3.45126; 6.71282, 9.56846; 15.6384, 70.3792

3. (a) 0.582, 1.18; 0.117969, 1.27344; -0.336912, 1.27382; -0.730007, 1.18572; -1.02134, 1.02371
   (b) 0.568451, 1.15775; 0.109776, 1.22556; -0.32208, 1.20347; -0.681296, 1.10162; -0.937852, 0.937852
   (c) 0.56845, 1.15775; 0.109773, 1.22557; -0.322081, 1.20347; -0.681291, 1.10161; -0.937841, 0.93784

4. (a) -0.198, 0.618; -0.378796, 0.28329; -0.51932, -0.0321025; -0.594324, -0.326801; -0.588278, -0.57545
   (b) -0.196904, 0.630936; -0.372643, 0.298888; -0.501302, -0.0111429; -0.561270, -0.288943; -0.547053, -0.508303
   (c) -0.196935, 0.630939; -0.372687, 0.298866; -0.501345, -0.0112184; -0.561292, -0.28907; -0.547031, -0.508427

5. (a) 2.96225, 1.34538; 2.34119, 1.67121; 1.90236, 1.97158; 1.56602, 2.23895; 1.29768, 2.46732
   (b) 3.06339, 1.34858; 2.44497, 1.68638; 1.9911, 2.00036; 1.63818, 2.27981; 1.3555, 2.5175
   (c) 3.06314, 1.34899; 2.44465, 1.68699; 1.99075, 2.00107; 1.63781, 2.28057; 1.35514, 2.51827

6. (a) 1.42386, 2.18957; 1.82234, 2.36791; 2.21728, 2.53329; 2.61118, 2.68763; 2.9955, 2.83354
   (b) 1.41513, 2.18699; 1.81208, 2.36233; 2.20635, 2.5258; 2.59826, 2.6794; 2.97806, 2.82487
   (c) 1.41513, 2.18699; 1.81209, 2.36233; 2.20635, 2.52581; 2.59826, 2.67941; 2.97806, 2.82488

7. For $h = 0.05$ and 0.025: $x = 1.43383$, $y = 0.642230$. These results agree with the exact solution to six digits.

8. 1.543, 0.0707503; 1.14743, -1.3885

CHAPTER 4

Section 4.1, *page 221*

1. Linear
2. Nonlinear
3. Linear
4. Linear
5. Nonlinear
6. $k = 16$ lb/ft
7. $k = 140$ N/m
8. $y'' + 64y = 0$, $y(0) = 1/4$, $y'(0) = 0$
9. $y'' + 196y = 0$, $y(0) = 0$, $y'(0) = 0.1$
10. $y'' + 128y = 0$, $y(0) = -1/12$, $y'(0) = 2$
11. $Q'' + (4 \times 10^6)Q = 0$, $Q(0) = 10^{-6}$, $Q'(0) = 0$
12. $y'' + 20y' + 196y = 0$, $y(0) = 2$, $y'(0) = 0$
13. $y'' + 4y' + 128y = 0$, $y(0) = 0$, $y'(0) = 1/4$

14. $y'' + 0.3y' + 15y = 0$, $y(0) = 0.05$, $y'(0) = 0.10$

15. $Q'' + 1500Q' + (5 \times 10^5)Q = 0$, $Q(0) = 10^{-6}$, $Q'(0) = 0$

16. $mx'' + \gamma x' + kx = 0$, no gravitational force acts on m

17. (a) $mx'' + kx + \epsilon x^3 = 0$, (b) $mx'' + kx = 0$

18. $mx'' + \gamma x' + (k_1 + k_2)x = 0$ 19. $\rho l u'' + \rho_0 g u = 0$

20. Center with clockwise rotation, stable

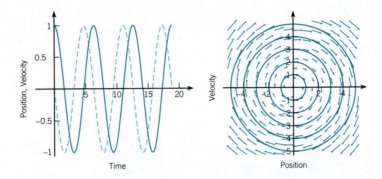

21. Center with clockwise rotation, stable

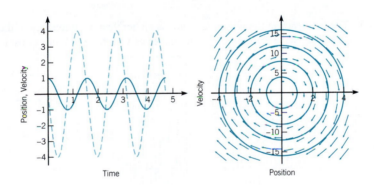

22. Spiral point, asymptotically stable

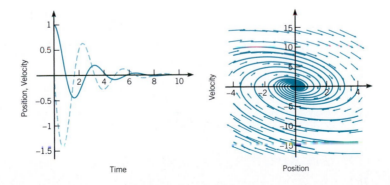

23. Node, asymptotically stable

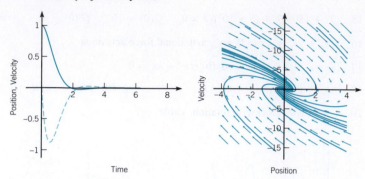

24. Spiral point, unstable

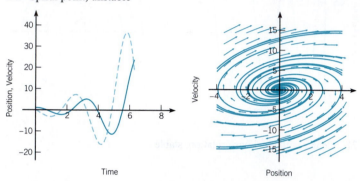

25. The frequency of the linear pendulum is higher than the frequency of the nonlinear pendulum. The difference between frequencies increases with the amplitude of oscillations.

26. (a) $r = \sqrt{k/m}$ (b) (i) $y = \cos t - \sin t$,   (ii) $y = \cos 4t - \dfrac{1}{4}\sin 4t$
(c)

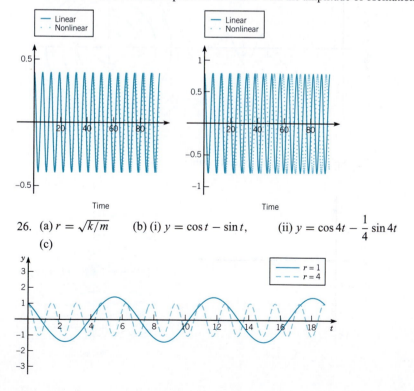

Section 4.2, *page 231*

1. $I = (0, \infty)$
2. $I = (-\infty, 1)$
3. $I = (0, 4)$
4. $I = (0, \infty)$
5. $I = (0, 3)$
6. $I = (2, 3\pi/2)$
7. $-\dfrac{7}{2}e^{t/2}$
8. 1
9. e^{-4t}
10. $x^2 e^x$
11. $-e^{2t}$
12. 0
15. No
16. $3te^{2t} + ce^{2t}$
17. $te^t + ct$
18. $5W(f, g)$
19. $-4(t \cos t - \sin t)$
20. Yes
21. Yes
22. Yes
23. Yes

24. (b) Yes
   (c) $[y_1(t), y_3(t)]$ and $[y_1(t), y_4(t)]$ are fundamental sets of solutions; $[y_2(t), y_3(t)]$ and $[y_4(t), y_5(t)]$ are not

Section 4.3, *page 242*

1. (a) $y = c_1 e^{-3t} + c_2 e^t$

   (b) $\mathbf{x} = c_1 e^{-3t} \begin{pmatrix} 1 \\ -3 \end{pmatrix} + c_2 e^t \begin{pmatrix} 1 \\ 1 \end{pmatrix}$

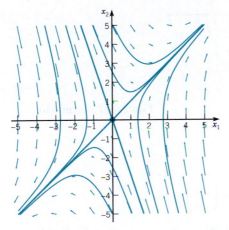

   (c) $(0, 0)$ saddle point, unstable

2. (a) $y = c_1 e^{-2t} + c_2 e^{-t}$

   (b) $\mathbf{x} = c_1 e^{-2t} \begin{pmatrix} 1 \\ -2 \end{pmatrix} + c_2 e^{-t} \begin{pmatrix} 1 \\ -1 \end{pmatrix}$; see figure on page 856

   (c) $(0, 0)$ node, asymptotically stable

3. (a) $y = c_1 e^{t/2} + c_2 e^{-t/3}$

   (b) $\mathbf{x} = c_1 e^{t/2} \begin{pmatrix} 1 \\ 1/2 \end{pmatrix} + c_2 e^{-t/3} \begin{pmatrix} 1 \\ -1/3 \end{pmatrix}$

   (c) $(0, 0)$ saddle point, unstable

4. (a) $y = c_1 e^{t/2} + c_2 e^t$

   (b) $\mathbf{x} = c_1 e^{t/2} \begin{pmatrix} 1 \\ 1/2 \end{pmatrix} + c_2 e^t \begin{pmatrix} 1 \\ 1 \end{pmatrix}$; see figure on page 856

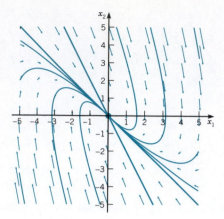

Figure for Problem 2(b).

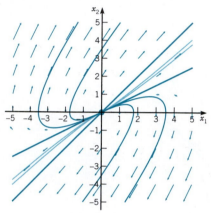

Figure for Problem 4(b).

(c) $(0, 0)$ node, unstable

5. (a) $y = c_1 + c_2 e^{-5t}$

(b) $\mathbf{x} = c_1 \begin{pmatrix} 1 \\ 0 \end{pmatrix} + c_2 e^{-5t} \begin{pmatrix} 1 \\ -5 \end{pmatrix}$

(c) $(x_1, 0)$ nonisolated, stable

6. (a) $y = c_1 e^{-3t/2} + c_2 e^{3t/2}$

(b) $\mathbf{x} = c_1 e^{-3t/2} \begin{pmatrix} 1 \\ -3/2 \end{pmatrix} + c_2 e^{3t/2} \begin{pmatrix} 1 \\ 3/2 \end{pmatrix}$

(c) $(0, 0)$ saddle point, unstable

7. (a) $\lambda_1 = (9 + 3\sqrt{5})/2, \lambda_2 = (9 - 3\sqrt{5})/2;$

$y = c_1 e^{\lambda_1 t} + c_2 e^{\lambda_2 t}$ (b) $\mathbf{x} = c_1 e^{\lambda_1 t} \begin{pmatrix} 1 \\ \lambda_1 \end{pmatrix} + c_2 e^{\lambda_2 t} \begin{pmatrix} 1 \\ \lambda_2 \end{pmatrix}$

(c) $(0, 0)$ node, unstable

8. (a) $\lambda_1 = (1 + \sqrt{3}), \lambda_2 = (1 - \sqrt{3}); \quad y = c_1 e^{\lambda_1 t} + c_2 e^{\lambda_2 t}$

(b) $\mathbf{x} = c_1 e^{\lambda_1 t} \begin{pmatrix} 1 \\ \lambda_1 \end{pmatrix} + c_2 e^{\lambda_2 t} \begin{pmatrix} 1 \\ \lambda_2 \end{pmatrix}$

(c) $(0, 0)$ saddle point, unstable

9. (a) $y = c_1 e^t + c_2 t e^t$

   (b) $\mathbf{x} = c_1 e^t \begin{pmatrix} 1 \\ 1 \end{pmatrix} + c_2 e^t \begin{pmatrix} t \\ 1+t \end{pmatrix}$

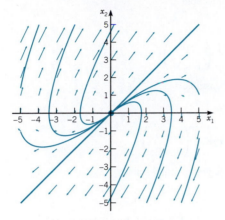

   (c) $(0, 0)$ improper node, unstable

10. (a) $y = c_1 e^{-t/3} + c_2 t e^{-t/3}$

    (b) $\mathbf{x} = c_1 e^{-t/3} \begin{pmatrix} 1 \\ -1/3 \end{pmatrix} + c_2 e^{-t/3} \begin{pmatrix} t \\ 1 - t/3 \end{pmatrix}$

    (c) $(0, 0)$ improper node, asymptotically stable

11. (a) $y = c_1 e^{t/2} + c_2 t e^{t/2}$

    (b) $\mathbf{x} = c_1 e^{t/2} \begin{pmatrix} 1 \\ 1/2 \end{pmatrix} + c_2 e^{t/2} \begin{pmatrix} t \\ 1 + t/2 \end{pmatrix}$

    (c) $(0, 0)$ improper node, unstable

12. (a) $y = c_1 e^{-3t/2} + c_2 t e^{-3t/2}$

    (b) $\mathbf{x} = c_1 e^{-3t/2} \begin{pmatrix} 1 \\ -3/2 \end{pmatrix} + c_2 e^{-3t/2} \begin{pmatrix} t \\ 1 - 3t/2 \end{pmatrix}$

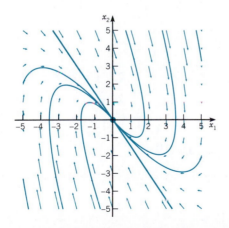

    (c) $(0, 0)$ improper node, asymptotically stable

13. (a) $y = c_1 e^{2t/5} + c_2 t e^{2t/5}$

(b) $\mathbf{x} = c_1 e^{2t/5} \begin{pmatrix} 1 \\ 2/5 \end{pmatrix} + c_2 e^{2t/5} \begin{pmatrix} t \\ 1 + 2t/5 \end{pmatrix}$

(c) $(0, 0)$ improper node, unstable

14. (a) $y = c_1 e^{3t} + c_2 t e^{3t}$

(b) $\mathbf{x} = c_1 e^{3t} \begin{pmatrix} 1 \\ 3 \end{pmatrix} + c_2 e^{3t} \begin{pmatrix} t \\ 1 + 3t \end{pmatrix}$

(c) $(0, 0)$ improper node, unstable

15. (a) $y = c_1 e^{-2t} + c_2 t e^{-2t}$

(b) $\mathbf{x} = c_1 e^{-2t} \begin{pmatrix} 1 \\ -2 \end{pmatrix} + c_2 e^{-2t} \begin{pmatrix} t \\ 1 - 2t \end{pmatrix}$

(c) $(0, 0)$ improper node, asymptotically stable

16. (a) $y = c_1 e^{4t/3} + c_2 t e^{4t/3}$

(b) $\mathbf{x} = c_1 e^{4t/3} \begin{pmatrix} 1 \\ 4/3 \end{pmatrix} + c_2 e^{4t/3} \begin{pmatrix} t \\ 1 + 4t/3 \end{pmatrix}$

(c) $(0, 0)$ improper node, unstable

19. $y = e^t$

20. $y = 2e^{2t/3} - (7/3)t e^{2t/3}$

21. $y = -(1/2)e^{-3t} + (5/2)e^{-t}$

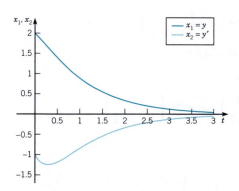

22. $y = -8e^{t/2} + 12e^{t/3}$ 23. $y = 2t e^{3t}$

24. $y = -1 - e^{-3t}$ 25. $y = 7e^{-2(t+1)} + 5t e^{-2(t+1)}$

26. $y = \left(\frac{1}{2} + \frac{5}{26} \sqrt{13} \right) e^{\frac{1}{2}(-5+\sqrt{13})t} + \left(-\frac{5}{26} \sqrt{13} + \frac{1}{2} \right) e^{-\frac{1}{2}(5+\sqrt{13})t}$

27. $y = \frac{2}{33} \sqrt{33} e^{\frac{1}{4}(-1+\sqrt{33})t} - \frac{2}{33} \sqrt{33} e^{-\frac{1}{4}(1+\sqrt{33})t}$

28. $y = \frac{1}{10} e^{9-9t} + \frac{9}{10} e^{-1+t}$

29. $y = -\frac{1}{2} e^{1+\frac{1}{2}t} + \frac{3}{2} e^{-1-\frac{1}{2}t}$

30. $y'' + y' - 6y = 0$

31. $y'' + 4y' + 4 = 0$

32. $y \to 0$ for $\alpha < 0$; y becomes unbounded for $\alpha > 1$

33. $y \to 0$ for $\alpha < 1$; there is no α for which all nonzero solutions become unbounded

35. (a) $y = d/c$ (b) $aY'' + bY' + cY = 0$

36. (a) $b > 0$ and $0 < c < b^2/4a$ (b) $c < 0$ (c) $b < 0$ and $0 < c < b^2/4a$

37. $y_2 = t^3$ 38. $y_2 = t^{-2}$ 39. $y_2 = t^{-1} \ln t$

40. $y_2 = te^t$ 41. $y_2 = \cos x^2$ 42. $y_2 = x$

43. $y_2 = x^{1/4} e^{-2\sqrt{x}}$ 44. $y_2 = x^{-1/2} \cos x$ 46. $y = c_1 e^{-\delta x^2/2} \int_0^x e^{\delta s^2/2}\, ds$
                                                                              $+ c_2 e^{-\delta x^2/2}$

Section 4.4, *page 250*

1. (b) $z_1 z_2 = (x_1 + iy_1)(x_2 + iy_2) = (x_1 x_2 - y_1 y_2) + i(x_1 y_2 + x_2 y_1) \Rightarrow \overline{z_1 z_2} = (x_1 x_2 - y_1 y_2)$
     $- i(x_1 y_2 - x_2 y_1)$ while $\bar{z}_1 \bar{z}_2 = (x_1 - iy_1)(x_2 - iy_2) = (x_1 x_2 - y_1 y_2) - i(x_1 y_2 - x_2 y_1)$
     (f) Using 1(b) and 1(c), $\left|\dfrac{z_1}{z_2}\right|^2 = \dfrac{z_1 \bar{z}_1}{z_2 \bar{z}_2} = \dfrac{|z_1|^2}{|z_2|^2} \Rightarrow \left|\dfrac{z_1}{z_2}\right| = \dfrac{|z_1|}{|z_2|}$.

4. $e \cos 2 + ie \sin 2 \cong -1.1312 + 2.4717i$

5. $e^2 \cos 3 - ie^2 \sin 3 \cong -7.3151 - 1.0427i$

6. -1 7. $-ie^2 \cong -7.3891i$

8. $2\cos(\ln 2) - 2i \sin(\ln 2) \cong 1.5385 - 1.2779i$

9. $\pi^{-1} \cos(2 \ln \pi) + i\pi^{-1} \sin(2 \ln 2) \cong -0.20957 + 0.23959i$

11. (a) $y = c_1 e^t \sin t + c_2 e^t \cos t$

     (b) $\mathbf{x} = c_1 e^t \begin{pmatrix} \sin t \\ \sin t + \cos t \end{pmatrix} + c_2 e^t \begin{pmatrix} \cos t \\ \cos t - \sin t \end{pmatrix}$

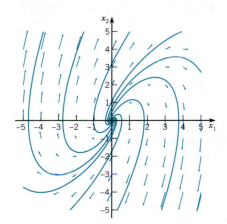

     (c) $(0, 0)$ spiral point, unstable

12. (a) $y = c_1 e^t \sin \sqrt{5}t + c_2 e^t \cos \sqrt{5}t$

     (b) $\mathbf{x} = c_1 e^t \begin{pmatrix} \sin \sqrt{5}t \\ \sin \sqrt{5}t + \sqrt{5}\cos \sqrt{5}t \end{pmatrix} + c_2 e^t \begin{pmatrix} \cos \sqrt{5}t \\ \cos \sqrt{5}t - \sqrt{5}\sin \sqrt{5}t \end{pmatrix}$

     (c) $(0, 0)$ spiral point, unstable

13. (a) $y = c_1 e^{-4t} + c_2 e^{2t}$

(b) $\mathbf{x} = c_1 e^{-4t} \begin{pmatrix} 1 \\ -4 \end{pmatrix} + c_2 e^{2t} \begin{pmatrix} 1 \\ 2 \end{pmatrix}$

(c) $(0, 0)$ saddle point, unstable

14. (a) $y = c_1 e^{-t} \sin t + c_2 e^{-t} \cos t$

(b) $\mathbf{x} = c_1 e^{-t} \begin{pmatrix} \sin t \\ \cos t - \sin t \end{pmatrix} - c_2 e^{-t} \begin{pmatrix} \cos t \\ \cos t + \sin t \end{pmatrix}$

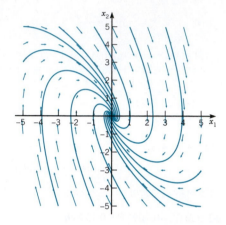

(c) $(0, 0)$ spiral point, asymptotically stable

15. (a) $y = c_1 e^{-3t} \sin 2t + c_2 e^{-3t} \cos 2t$

(b) $\mathbf{x} = c_1 e^{-3t} \begin{pmatrix} \sin 2t \\ 2 \cos 2t - 3 \sin 2t \end{pmatrix} - c_2 e^{-3t} \begin{pmatrix} \cos 2t \\ 3 \cos 2t + 2 \sin 2t \end{pmatrix}$

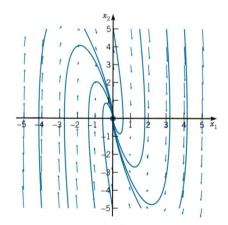

(c) $(0, 0)$ spiral point, asymptotically stable

16. (a) $y = c_1 e^{-t} \sin(t/2) + c_2 e^{-t} \cos(t/2)$

(b) $\mathbf{x} = c_1 e^{-t} \begin{pmatrix} \sin(t/2) \\ \frac{1}{2} \cos(t/2) - \sin(t/2) \end{pmatrix} - c_2 e^{-t} \begin{pmatrix} \cos(t/2) \\ \cos(t/2) + \frac{1}{2} \sin(t/2) \end{pmatrix}$

(c) $(0, 0)$ spiral point, asymptotically stable

17. (a) $y = c_1 \sin \frac{3}{2}t + c_2 \cos \frac{3}{2}t$

(b) $\mathbf{x} = c_1 \begin{pmatrix} \sin \frac{3}{2}t \\ \frac{3}{2} \cos \frac{3}{2}t \end{pmatrix} + c_2 \begin{pmatrix} \cos \frac{3}{2}t \\ -\frac{3}{2} \sin \frac{3}{2}t \end{pmatrix}$

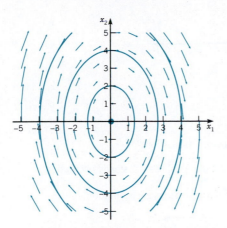

(c) $(0, 0)$ center, stable

18. (a) $y = c_1 e^{-4t/3} + c_2 e^{t/3}$

(b) $\mathbf{x} = c_1 e^{-4t/3} \begin{pmatrix} 1 \\ -4/3 \end{pmatrix} + c_2 e^{t/3} \begin{pmatrix} 1 \\ 1/3 \end{pmatrix}$

(c) $(0, 0)$ saddle point, unstable

19. (a) $y = c_1 e^{-t/2} \sin t + c_2 e^{-t/2} \cos t$

(b) $\mathbf{x} = c_1 e^{-t/2} \begin{pmatrix} \sin t \\ \cos t - \frac{1}{2} \sin t \end{pmatrix} - c_2 e^{-t/2} \begin{pmatrix} \cos t \\ \sin t + \frac{1}{2} \cos t \end{pmatrix}$

(c) $(0, 0)$ spiral point, asymptotically stable

20. (a) $y = c_1 e^{-2t} \sin \frac{3}{2}t + c_2 e^{-2t} \cos \frac{3}{2}t$

(b) $\mathbf{x} = c_1 e^{-2t} \begin{pmatrix} \sin \frac{3}{2}t \\ \frac{3}{2} \cos \frac{3}{2}t - 2 \sin \frac{3}{2}t \end{pmatrix} - c_2 e^{-2t} \begin{pmatrix} \cos \frac{3}{2}t \\ 2 \cos \frac{3}{2}t + \frac{3}{2} \sin \frac{3}{2}t \end{pmatrix}$

(c) $(0, 0)$ spiral point, asymptotically stable

21. $y = \frac{1}{2} \sin(2t)$; steady oscillation

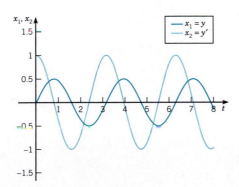

22. $y(t) = 2e^{-2t} \sin t + e^{-2t} \cos t$;   decaying oscillation

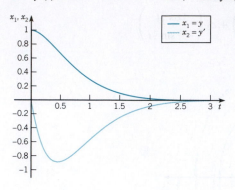

23. $y = -e^{t-\pi/2} \sin 2t$;   growing oscillation

24. $y = \left(-2 + \sqrt{3}\right) \sin t + \left(2\sqrt{3} + 1\right) \cos t$;   steady oscillation

25. $y = \frac{5}{2} e^{-t/2} \sin t + 3 e^{-t/2} \cos t$;   decaying oscillation

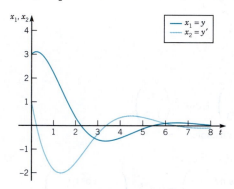

26. $y = \sqrt{2}e^{-(t-\pi/4)} \cos t + \sqrt{2}e^{-(t-\pi/4)} \sin t$;   decaying oscillation

34. $y = c_1 \sin(\ln x) + c_2 \cos(\ln x)$

35. $y = c_1/x^2 + c_2/x$

36. $y = c_1 x^{-1} \sin\left(\frac{1}{2} \ln x\right) + c_2 x^{-1} \cos\left(\frac{1}{2} \ln x\right)$

37. $y = c_1 x^{-1} + c_2 x^6$

38. $y = c_1 x^{-1} + c_2 x^2$

39. $y = c_1 x^2 + c_2 x^2 \ln x$

40. $y = c_1 x^{-1/2} \sin\left(\frac{1}{2} \sqrt{15} \ln x\right) + c_2 x^{-1/2} \cos\left(\frac{1}{2} \sqrt{15} \ln x\right)$

41. $y = c_1 x^{3/2} \sin\left(\frac{\sqrt{3}}{2} \ln(x)\right) + c_2 x^{3/2} \cos\left(\frac{\sqrt{3}}{2} \ln(x)\right)$

42. $y = \frac{1}{5}x^{-1} + \frac{4}{5}x^{3/2};$ $\lim_{x\to 0} y(x) = \infty$

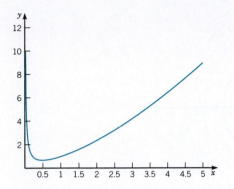

43. $y = -x^{-1/2}\sin(2\ln x) + 2x^{-1/2}\cos(2\ln x);$ $\lim_{x\to 0} y(x) = \infty$

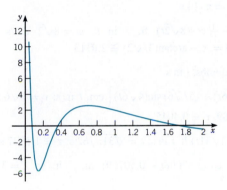

44. $y = 2x^2 + 7x^2\ln(-x)$

45. $y = x^{-1}\cos(2\ln x)$

Section 4.5, *page 260*

1. $y = 5\cos(2t - \delta),$ $\delta = \arctan(4/3) \cong 0.9273$

2. $y = 2\cos(t - 2\pi/3)$

3. $y = 2\sqrt{5}\cos(3t - \delta),$ $\delta = -\arctan(1/2) \cong -0.4636$

4. $y = \sqrt{13}\cos(\pi t - \delta),$ $\delta = \pi + \arctan(3/2) \cong 4.1244$

5. (a) $y = \frac{1}{4}\cos(8t)$ ft, t in s, $\omega = 8$ rad/s, $T = \pi/4$ s, $R = 1/4$ ft

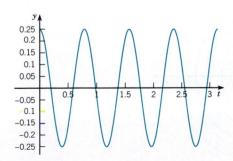

(b)

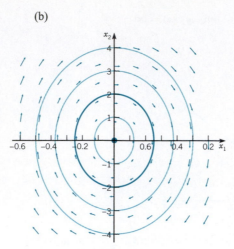

6. (a) $y = \frac{5}{7}\sin(14t)$ cm, $t = \pi/14$ s

7. $y = (\sqrt{2}/8)\sin(8\sqrt{2}t) - \frac{1}{12}\cos(8\sqrt{2}t)$ ft, t in s, $\omega = 8\sqrt{2}$ rad/s, $T = \sqrt{2}\pi/8$ s, $R = \sqrt{11/288} \cong 0.1954$ ft, $\delta = \pi - \arctan(3/\sqrt{2}) \cong 2.0113$

8. $Q = 10^{-6}\cos 2000t$ coulombs, t in s

9. (a) $y = e^{-10t}\left[2\cos(4\sqrt{6}t) + (5/\sqrt{6})\sin(4\sqrt{6}t)\right]$ cm, t in s; $\nu = 4\sqrt{6}$ rad/s, $T_d = \sqrt{6}\pi/12$ s, $T_d/T = 7/2\sqrt{6} \cong 1.4289$, $\tau \cong 0.4045$

10. $y = (\sqrt{31}/248)e^{-2t}\sin(2\sqrt{31}t)$ ft, t in s; $t = \sqrt{31}\pi/62$ s, $\tau \cong 1.5927$ s

11. (a) $y \cong 0.057198e^{-0.15t}\cos(3.87008t - 0.50709)$ m, t in s; $\nu = 3.87008$ rad/s, $\nu/\omega_0 = 3.87008/\sqrt{15} \cong 0.99925$

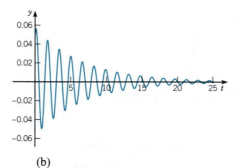

(b)

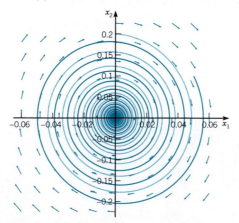

12. $Q = 10^{-6}(2e^{-500t} - e^{-1000t})$ coulombs, t in s

13. $\gamma = \sqrt{20/9} \cong 1.4907$

16. $r = \sqrt{A^2 + B^2}, \ r\cos\theta = B, \ r\sin\theta = -A; \ R = r; \ \delta = \theta + (4n+1)\pi/2, n = 0, 1, 2, \ldots$

17. $\gamma = 8$ lb-s/ft   18. $R = 10^3$ ohms   20. $v_0 < -\gamma y_0/2m$

22. $2\pi/\sqrt{31}$ 23. $\gamma = 5$ lb-s/ft   24. $k = 6, v = \pm 2\sqrt{5}$

25. (a) $\tau \cong 41.715$ (d) $\gamma_0 \cong 1.73$, min $\tau \cong 4.87$ (e) $\tau = (2/\gamma)\ln(400/\sqrt{4-\gamma^2})$

26. (a) $y(t) = e^{-\gamma t/2m}\left[y_0\sqrt{4km-\gamma^2}\cos vt + (2mv_0 + \gamma y_0)\sin vt\right]/\sqrt{4km-\gamma^2}$
    (b) $R^2 = 4m(ky_0^2 + \gamma y_0 v_0 + mv_0^2)/(4km - \gamma^2)$

27. $\rho l u'' + \rho_0 g u = 0, T = 2\pi\sqrt{\rho l/\rho_0 g}$

28. (a) $y = \sqrt{2}\sin\sqrt{2}t$ (c) clockwise

29. (a) $y = (16/\sqrt{127})e^{-t/8}\sin(\sqrt{127}t/8)$ (c) clockwise

30. (b) $y = a\cos(\sqrt{k/m}t) + b\sqrt{m/k}\sin(\sqrt{k/m}t)$

32. (b) $y = \sin t, A = 1, T = 2\pi$ (c)$A = 0.98, T = 6.07$
    (d) $\epsilon = 0.2, A = 0.96, T = 5.90; \quad \epsilon = 0.3, A = 0.94, T = 5.74$
    (f) $\epsilon = -0.1, A = 1.03, T = 6.55; \quad \epsilon = -0.2, A = 1.06, T = 6.90;$
    $\epsilon = -0.3, A = 1.11, T = 7.41$

Section 4.6, *page 271*

1. $y = c_1 e^{3t} + c_2 e^{-t} - e^{2t}$

2. $y = c_1 e^{-t}\cos 2t + c_2 e^{-t}\sin 2t + \frac{3}{17}\sin 2t - \frac{12}{17}\cos 2t$

3. $y = c_1 e^{3t} + c_2 e^{-t} + \frac{3}{16}te^{-t} + \frac{3}{8}t^2 e^{-t}$

4. $y = c_1 + c_2 e^{-2t} + \frac{3}{2}t - \frac{1}{2}\sin 2t - \frac{1}{2}\cos 2t$

5. $y = c_1\cos 3t + c_2\sin 3t + \frac{1}{162}(9t^2 - 6t + 1)e^{3t} + \frac{2}{3}$

6. $y = c_1 e^{-t} + c_2 te^{-t} + t^2 e^{-t}$

7. $y = c_1 e^{-t} + c_2 e^{-t/2} + t^2 - 6t + 14 - \frac{3}{10}\sin t - \frac{9}{10}\cos t$

8. $y = c_1\cos t + c_2\sin t - \frac{1}{2}t\cos 2t - \frac{5}{9}\sin 2t$

9. $u = c_1\cos\omega_0 t + c_2\sin\omega_0 t + (\omega_0^2 - \omega^2)^{-1}\cos\omega t$

10. $u = c_1\cos\omega_0 t + c_2\sin\omega_0 t + (1/2\omega_0)t\sin\omega_0 t$

11. $y = c_1 e^{-t/2}\cos(\sqrt{15}t/2) + c_2 c_1 e^{-t/2}\sin(\sqrt{15}t/2) + \frac{1}{6}e^t - \frac{1}{4}e^{-t}$

12. $y = c_1 e^{-t} + c_2 e^{2t} + \frac{1}{6}te^{2t} + \frac{1}{8}e^{-2t}$

13. $y = c_1 e^t - \frac{1}{2}e^{-2t} - t - \frac{1}{2}$

14. $y = \frac{7}{10}\sin 2t - \frac{19}{40}\cos 2t + \frac{1}{4}t^2 - 1/8 + \frac{3}{5}e^t$

15. $y = -3e^t + 4te^t + \frac{1}{6}t^3 e^t + 4$

16. $v = \frac{2}{3}e^{-t} + e^{3t} - \frac{1}{3}(2+3t)e^{2t}$

17. $y = -\frac{1}{8}\sin 2t + 2\cos 2t - \frac{3}{4}t\cos 2t$

18. $y = \frac{1}{2}e^{-t}\sin 2t + e^{-t}\cos 2t + te^{-t}\sin 2t$

19. (a) $Y(t) = t(A_0 t^4 + A_1 t^3 + A_2 t^2 + A_3 t + A_4) + t(B_0 t^2 + B_1 t + B_2)e^{-3t} + D \sin 3t$
    $+ E \cos 3t$
    (b) $A_0 = 2/15$, $A_1 = -2/9$, $A_2 = 8/27$, $A_3 = -8/27$, $A_4 = 16/81$, $B_0 = -1/9$, $B_1 = -1/9$,
    $B_2 = -2/27$, $D = -1/18$, $E = -1/18$

20. (a) $Y(t) = A_0 t + A_1 + t(B_0 t + B_1) \sin t + t(D_0 t + D_1) \cos t$
    (b) $A_0 = 1$, $A_1 = 0$, $B_0 = 0$, $B_1 = 1/4$, $D_0 = -1/4$, $D_1 = 0$

21. (a) $Y(t) = e^t(A \cos 2t + B \sin 2t) + (D_0 t + D_1)e^{2t} \sin t + (E_0 t + E_1)e^{2t} \cos t$
    (b) $A = -1/20$, $B = -3/20$, $D_0 = -3/2$, $D_1 = -5$, $E_0 = 3/2$, $E_1 = 1/2$

22. (a) $Y(t) = Ae^{-t} + t(B_0 t^2 + B_1 t + B_2)e^{-t} \cos t + t(D_0 t^2 + D_1 t + D_2)e^{-t} \sin t$
    (b) $A = 3$, $B_0 = -2/3$, $B_1 = 0$, $B_2 = 1$, $D_0 = 0$, $D_1 = 1$, $D_2 = 1$

23. (a) $Y(t) = A_0 t^2 + A_1 t + A_2 + t^2(B_0 t + B_1)e^{2t} + (D_0 t + D_1) \sin 2t + (E_0 t + E_1) \cos 2t$
    (b) $A_0 = 1/2$, $A_1 = 1$, $A_2 = 3/4$, $B_0 = 2/3$, $B_1 = 0$, $D_0 = 0$, $D_1 = -1/16$, $E_0 = 1/8$,
    $E_1 = 1/16$

24. (a) $Y(t) = t(A_0 t^2 + A_1 t + A_2) \sin 2t + t(B_0 t^2 + B_1 t + B_2) \cos 2t$
    (b) $A_0 = 0$, $A_1 = 13/16$, $A_2 = 7/4$, $B_0 = -1/12$, $B_1 = 0$, $B_2 = 13/32$

25. (a) $Y(t) = (A_0 t^2 + A_1 t + A_2)e^t \sin 2t + (B_0 t^2 + B_1 t + B_2)e^t \cos 2t + e^{-t}(D \cos t + E \sin t)$
    $+ Fe^t$
    (b) $A_0 = 1/52$, $A_1 = 10/169$, $A_2 = -1233/35152$, $B_0 = -5/52$, $B_1 = 73/676$,
    $B_2 = -4105/35152$, $D = -3/2$, $E = 3/2$, $F = 2/3$

26. (a) $Y(t) = t(A_0 t + A_1)e^{-t} \cos 2t + t(B_0 t + B_1) \sin 2t + (D_0 t + D_1)e^{-2t} \cos t + (E_0 t$
    $+ E_1)e^{-2t} \sin t$
    (b) $A_0 = 0$, $A_1 = 3/16$, $B_0 = 3/8$, $B_1 = 0$, $D_0 = -2/5$, $D_1 = -7/25$, $E_0 = 1/5$, $E_1 = 1/25$

27. (b) $w = -\frac{2}{5} + c_1 e^{5t}$

28. $y = c_1 x^2 + c_2 x^2 \ln x + \frac{1}{4} \ln x + \frac{1}{4}$

29. $y = c_1 x^{-1} + c_2 x^{-5} + \frac{1}{12} x$

30. $y = c_1 x + c_2 x^2 + 3x^3 \ln x + \ln x + \frac{3}{2}$

31. $y = c_1 \cos(2 \ln x) + c_2 \sin(2 \ln x) + \frac{1}{3} \sin(\ln x)$

32. $y = c_1 \cos \lambda t + c_2 \sin \lambda t + \sum_{m=1}^{N} [a_m/(\lambda^2 - m^2 \pi^2)] \sin m\pi t$

33. $y = \begin{cases} t, & 0 \le t \le \pi \\ -(1 + \pi/2) \sin t - (\pi/2) \cos t + (\pi/2)e^{\pi-t}, & t > \pi \end{cases}$

34. $y = \begin{cases} \frac{1}{5} - \frac{1}{10}e^{-t} \sin 2t - \frac{1}{5}e^{-t} \cos 2t, & 0 \le t \le \pi/2 \\ -\frac{1}{5}(1 + e^{\pi/2})e^{-t} \cos 2t - \frac{1}{10}(1 + e^{\pi/2})e^{-t} \sin 2t, & t > \pi/2 \end{cases}$

Section 4.7, *page 281*

1. $-2 \sin 8t \sin t$

2. $2 \sin(t/2) \cos(13t/2)$

3. $2 \cos(3\pi t/2) \cos(\pi t/2)$

4. $2 \sin(7t/2) \cos(t/2)$

5. $y'' + 256y = 16 \cos 3t$, $y(0) = \frac{1}{6}$, $y'(0) = 0$ y in ft, t in s

6. $y'' + 10y' + 98y = 2 \sin(t/2)$, $y(0) = 0$, $y'(0) = 0.03$, y in m, t in s

7. (a) $y = \frac{183}{988} \cos 16t + \frac{16}{247} \cos 3t$

(b)

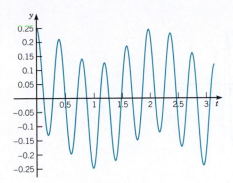

(c) $G(i\omega) = \frac{1}{256-\omega^2}$, $|G(i\omega)| = \frac{1}{|256-\omega^2|}$, $\phi(\omega) = \begin{cases} 0 & \omega < 16 \\ \pi/2 & \omega = 16 \\ \pi & \omega > 16 \end{cases}$, $\omega_{max} = 16$

8. (a) $y = \frac{383443}{1118951300} e^{-5t} \sin\left(\sqrt{73}t\right)\sqrt{73} + \frac{160}{153281} e^{-5t} \cos\left(\sqrt{73}t\right) + \frac{3128}{153281} \sin(t/2)$
$- \frac{160}{153281} \cos(t/2)$

(b) The first two terms are transient.

(c)

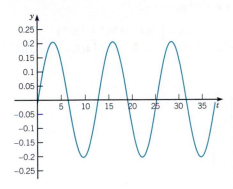

(d) $G(i\omega) = \frac{98-\omega^2-10\omega i}{(98-\omega^2)^2+100\omega^2}$, $|G(i\omega)| = \frac{1}{\sqrt{(98-\omega^2)^2+100\omega^2}}$, $\phi(\omega) = \cos^{-1}\left[\frac{98-\omega^2}{\sqrt{(98-\omega^2)^2+100\omega^2}}\right]$,

$\omega_{max} = 4\sqrt{3}$

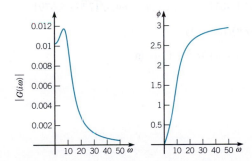

9. $y = \frac{64}{45}(\cos 7t - \cos 8t) = \frac{128}{45}\sin(t/2)\sin(15t/2)$ ft, t in s

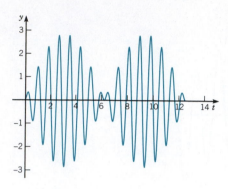

10. $y = (\cos 8t + \sin 8t - 8t\cos 8t)/4$ ft, t in s; $1/8, \pi/8, \pi/4, 3\pi/8$ s

11. (a) $\frac{8}{901}(30\cos 2t + \sin 2t)$ ft, t in s   (b) $m = 4$ slug

12. $y = (\sqrt{2}/6)\cos(3t - 3\pi/4)$ m, t in s

14. $y = \begin{cases} A(t - \sin t), & 0 \le t \le \pi \\ A[(2\pi - t) - 3\sin t], & \pi < t \le 2\pi \\ -4A\sin t, & 2\pi < t < \infty \end{cases}$

15. $Q(t) = 10^{-6}(e^{-4000t} - 4e^{-1000t} + 3)$ coulombs, t in s, $Q(0.001) \cong 1.5468 \times 10^{-6}$; $Q(0.01) \cong 2.9998 \times 10^{-6}$; $Q(t) \to 3 \times 10^{-6}$ as $t \to \infty$

16. (a) $y = [32(2 - \omega^2)\cos\omega t + 8\omega\sin\omega t]/(64 - 63\omega^2 + 16\omega^4)$
    (b) $|G(i\omega)| = 4/\sqrt{64 - 63\omega^2 + 16\omega^4}$, $\phi(\omega) = \cos^{-1}\left[4(2 - \omega^2)/\sqrt{64 - 63\omega^2 + 16\omega^4}\right]$
    (c)

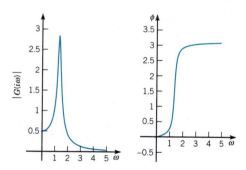

    (d) $\omega_{\max} = 3\sqrt{14}/8 \cong 1.4031$, $|G(i\omega_{\max})| = 64/\sqrt{127} \cong 5.6791$

17. (a) $y = 3(\cos t - \cos\omega t)/(\omega^2 - 1)$

18. (a) $[(\omega^2 + 2)\cos t - 3\cos\omega t]/(\omega^2 - 1) + \sin t$

20. (a)

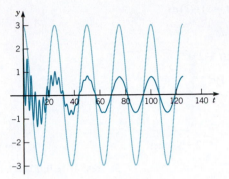

Section 4.8, *page 289*

2. $\mathbf{x} = \begin{pmatrix} \frac{1}{2}t \\ \frac{3}{2}t - \frac{1}{2} \end{pmatrix}$

3. $\mathbf{x} = \begin{pmatrix} 2t + te^{-t} \\ -(1+t) \end{pmatrix}$

4. $\mathbf{x} = \begin{pmatrix} -1 \\ t \end{pmatrix}$

5. $\mathbf{x} = \begin{pmatrix} t\cos t \\ -t\sin t \end{pmatrix}$

6. $\mathbf{x} = \begin{pmatrix} \frac{1}{2}t + \frac{5}{4}e^{-4t} + \frac{3}{4}e^{-2t} \\ -\frac{1}{2} - \frac{5}{4}e^{-4t} + \frac{3}{4}e^{-2t} + \frac{3}{2}t \end{pmatrix}$

7. $\mathbf{x} = \begin{pmatrix} 2t - 2e^t + e^{-t}t + e^{-t} \\ -t - 1 + 2e^t \end{pmatrix}$

8. $\mathbf{x} = \begin{pmatrix} -1 + 2e^{-t} \\ -2e^{-t}t + t \end{pmatrix}$

9. $\mathbf{x} = \begin{pmatrix} \sin t + \cos t + t\cos t \\ \cos t - \sin t - t\sin t \end{pmatrix}$

10. $y = e^t$

11. $y = -\frac{2}{3}te^{-t}$

12. $y = \frac{3}{2}t^2 e^{-t}$

13. $y = 2t^2 e^{t/2}$

14. $y = c_1 \cos t + c_2 \sin t - (\cos t)\ln(\tan t + \sec t)$

15. $y = c_1 \cos 3t + c_2 \sin 3t + (\sin 3t)\ln(\tan 3t + \sec 3t) - 1$

16. $y = c_1 e^{-2t} + c_2 te^{-2t} - e^{-2t}\ln t$

17. $y = c_1 \cos 2t + c_2 \sin 2t + \frac{3}{4}(\sin 2t)\ln(\sin 2t) - \frac{3}{2}t\cos 2t$

18. $y = c_1 \cos(t/2) + c_2 \sin(t/2) + t\sin(t/2) + 2[\ln\cos(t/2)]\cos(t/2)$

19. $y = c_1 e^t + c_2 te^t - \frac{1}{2}e^t \ln(1+t^2) + te^t \arctan t$

20. $y = c_1 e^{2t} + c_2 e^{3t} + \int [e^{3(t-s)} - e^{2(t-s)}]g(s)\,ds$

21. $y = c_1 \cos 2t + c_2 \sin 2t + \frac{1}{2}\int [\sin 2(t-s)]g(s)\,ds$

22. $Y(t) = -2t^2$

23. $Y(t) = \frac{1}{2}(t-1)e^{2t}$

24. $Y(t) = -\frac{1}{2}(2t-1)e^{-t}$

25. $Y(x) = -\frac{3}{2}x^{1/2}\cos x$

26. $Y(x) = \int \dfrac{xe^t - te^x}{(1-t)^2 e^t}\,g(t)\,dt$

27. $Y(x) = x^{-1/2}\int t^{-3/2}\sin(x-t)\,g(t)\,dt$

28. $y = c_1 t^2 + c_2 t^{-1} + t^2 \ln t + 1/2$

29. $y = c_1 x^2 + c_2 x^2 \ln x + \frac{1}{6}x^2(\ln x)^3$

30. $y = c_1 t + c_2 t^2 + 4t^2(\ln t - 1)$

31. $y = c_1 t^{-1} + c_2 t^{-5} + \frac{1}{2}t$

34. (b) $y = y_0 \cos t + y_1 \sin t + \int_{t_0}^t \sin(t-s)g(s)\,ds$

35. $y = (b-a)^{-1} \int_{t_0}^{t} [e^{b(t-s)} - e^{a(t-s)}] g(s)\, ds$

36. $y = \mu^{-1} \int_{t_0}^{t} e^{\lambda(t-s)} \sin \mu(t-s)\, g(s)\, ds$

37. $y = \int_{t_0}^{t} (t-s) e^{a(t-s)}\, g(s)\, ds$

40. $y = c_1(1+t) + c_2 e^t + \frac{1}{2}(t-1)e^{2t}$

41. $y = c_1 e^t + c_2 t - \frac{1}{2}(2t-1)e^{-t}$

CHAPTER 5

Section 5.1, *page 314*

1. Piecewise continuous   2. Neither   3. Continuous

4. Piecewise continuous

5. Yes; $M = 0, K = 3, a = 5$

6. Yes; $M = 0, K = 1, a = -3$

7. Yes; $M = 0, K = 1, a = 2$

8. Yes; $M = 1, K = 1, a = 10$

9. No

10. Yes; $M = 0, K = 1, a = 0$

11. Yes; $M = 0, K = 1, a = 0$

12. No

13. (a) $1/s$ (b) $2/s^2$ (c) $n!/s^{n+1}$

14. $\dfrac{1}{s^2} - \dfrac{e^{-s}}{s^2}$

15. $\dfrac{e^{-s}}{s} - \dfrac{e^{-2s}}{s}$

16. $\dfrac{e^{-(s+1)}}{s+1}$

17. $\dfrac{2}{s^3} + \left(\dfrac{1}{s} - \dfrac{3}{s^2} - \dfrac{2}{s^3}\right)e^{-s} + \dfrac{e^{-2s}}{s^2}$

18. $\dfrac{s}{s^2 - b^2}, \; s > |b|$

19. $\dfrac{b}{s^2 - b^2}, \; s > |b|$

20. $\dfrac{s-a}{(s-a)^2 - b^2}, \; s - a > |b|$

21. $\dfrac{b}{(s-a)^2 - b^2}, \; s - a > |b|$

22. $\dfrac{s}{s^2 + b^2}, \; s > 0$

23. $\dfrac{b}{(s-a)^2 + b^2}, \; s > a$

24. $\dfrac{s-a}{(s-a)^2 + b^2}, \; s > a$

25. $\dfrac{1}{(s-a)^2}, \; s > a$

26. $\dfrac{2as}{(s^2 + a^2)^2}, \; s > 0$

27. $\dfrac{s^2 + a^2}{(s-a)^2(s+a)^2}, \; s > |a|$

28. $\dfrac{n!}{(s-a)^{n+1}}, \; s > a$

29. $\dfrac{2a(3s^2 - a^2)}{(s^2 + a^2)^3}, \; s > 0$

30. $\dfrac{2a(3s^2 + a^2)}{(s^2 - a^2)^3}, \; s > |a|$

31. Converges

32. Converges

33. Diverges

34. Converges

36. (d) $\Gamma(3/2) = \sqrt{\pi}/2;$ $\Gamma(11/2) = 945\sqrt{\pi}/32$

Section 5.2, *page 321*

1. $\dfrac{4}{(s+2)^2 + 16}$

2. $\dfrac{s-3}{(s-3)^2 + 4}$

3. $\dfrac{6}{s^4} - \dfrac{8}{s^3} + \dfrac{5}{s}$

4. $\dfrac{s^2 - 1}{(s^2 + 1)^2}$

5. $\dfrac{24}{(s+4)^5} + \dfrac{4}{(s+4)^3} + \dfrac{1}{s+4}$

6. $\dfrac{5!}{(s-2)^6}$

7. $2\,\dfrac{3s^2 b - b^3}{(s^2 + b^2)^3}$

8. $\dfrac{n!}{(s-a)^{n+1}}$

9. $2\dfrac{b\,(s-a)}{\left[(s-a)^2+b^2\right]^2}$

10. $\dfrac{(s-a)^2-b^2}{\left[(s-a)^2+b^2\right]^2}$

11. (b) Let $g_k(t)=\int_0^t\int_0^{t_k}\cdots\int_0^{t_2} f(t_1)\,dt_1\cdots dt_k$ and use mathematical induction.

12. $\dfrac{s+3}{s^2+2s-2}$

13. $\dfrac{18\,s+15}{9\,s^2+12\,s+4}$

14. $\dfrac{2\,s+7}{s^2+4\,s+3}$

15. $\dfrac{24\,s+20}{6\,s^2+5\,s+1}$

16. $\dfrac{s^5-4\,s^4+10\,s^3-16\,s^2+15\,s-4}{(s-1)^3\,s\,(s^2-2\,s+2)}$

17. $\dfrac{s^4-2\,s^3+4\,s^2+2}{s^3\,(s^2-2\,s-3)}$

18. $\dfrac{2\,s^3+7\,s^2+12\,s-2}{(s^2+4\,s+8)\,(s^2+4)}$

19. $\dfrac{s^5+8\,s^3+16\,s+2\,s^4+17\,s^2+28}{(s^2+4)^2\,(s^2+2\,s+5)}$

20. $\dfrac{s^2+s-1}{s^3+s^2+s+1}$

21. $\dfrac{1}{s^4-1}$

23. $\dfrac{s^2+1}{s(s^2+4)}-\dfrac{e^{-\pi s}}{s(s^2+4)}$

24. $\dfrac{1}{s^2(s^2+1)}-\dfrac{(1+s)e^{-s}}{s^2(s^2+1)}$

25. $\dfrac{1}{s^2(s^2+4)}-\dfrac{e^{-s}}{s^2(s^2+4)}$

26. $\dfrac{1}{s\left(s+\frac{1}{50}\right)}-\dfrac{e^{-10s}}{s\left(s+\frac{1}{50}\right)}$

27. $F_0\left[\dfrac{1}{s^2(ms^2+\gamma s+k)}-\dfrac{e^{-sT}(1+sT)}{s^2(ms^2+\gamma s+k)}\right]$

29. (a) $Y'+s^2Y=s$ (b) $s^2Y''+2sY'-[s^2+\alpha(\alpha+1)]Y=-1$

Section 5.3, *page 331*

1. $a=4,\ b=-3$

2. $a_1=3,\ a_2=-2$

3. $a=2,\ b=-5,\ c=3$

4. $a_1=1,\ a_2=2,\ a_3=2$

5. $a=2,\ b=1,\ c=-2$

6. $a_1=0,\ a_2=2\ b=2,\ c=-4$

7. $a_1=-1,\ a_2=10,\ a_3=-24,\ b_1=0,\ b_2=0,\ b_3=0$

8. $a_1=1,\ a_2=-4,\ b_1=0,\ b_2=0$

9. $\frac{3}{2}\sin 2t$

10. $2t^2e^t$

11. $-\frac{2}{5}e^{-4t}+\frac{2}{5}e^t$

12. $\frac{9}{5}e^{3t}+\frac{6}{5}e^{-2t}$

13. $2e^{-t}\cos 2t$

14. $2\cosh 2t-\frac{3}{2}\sinh 2t$

15. $2e^t\cos t+3e^t\sin t$

16. $5\cos 2t-2\sin 2t+3$

17. $-2e^{-2t}\cos t+5e^{-2t}\sin t$

18. $2e^{-t}\cos 3t-\frac{5}{3}e^{-t}\sin 3t$

19. $2e^{2t}-3e^{-2t}+e^{-t}$

20. $-4t+e^{2t}+e^{-2t}$

21. $2e^{2t}\sin 2t+4e^{-3t}$

22. $2\cos t+3\sin 2t$

23. $-3t+e^{-t}\cos t$

24. $\left(\frac{5}{2}-t\right)e^{-t}\sin t-\frac{3}{2}te^{-t}\cos t$

25. $(1-\frac{7}{3}t)e^{-t}\cos 3t-(\frac{8}{9}+\frac{4}{3}t)e^{-t}\sin 3t$

26. $-\left(\frac{1016}{3125}+\frac{549}{625}t+\frac{118}{125}t^2+\frac{9}{25}t^3+\frac{1}{20}t^4\right)+\frac{1016}{3125}e^{2t}\cos t+\frac{713}{3125}e^{2t}\sin t$

27. $\left(\frac{1}{250}+\frac{13}{25}t-\frac{1}{40}t^2\right)e^t\cos t+\left(-\frac{261}{500}+\frac{3}{200}t+\frac{1}{5}t^2\right)e^t\sin t-\left(\frac{1}{250}+\frac{1}{100}t\right)e^{-2t}$

28. $-\left(\frac{59}{4000}-\frac{21}{400}t-\frac{3}{160}t^2\right)e^{3t}+\left(\frac{59}{4000}+\frac{1}{80}t\right)e^{-t}\cos 2t-\left(\frac{3}{1000}+\frac{7}{320}t\right)e^{-t}\sin 2t$

Section 5.4, *page 339*

1. $\frac{4}{5}e^{-2t} + \frac{1}{5}e^{3t}$

2. $\frac{1}{2}t - \frac{3}{4} - \frac{5}{4}e^{-2t} + 3e^{-t}$

3. $e^t \sin t$

4. $e^{2t} - e^{2t}t$

5. $2e^t \cos\sqrt{3}t - \frac{2}{3}\sqrt{3}e^t \sin\sqrt{3}t$

6. $\frac{5}{8}e^{-t}\sin 2t + 2e^{-t}\cos 2t - \frac{1}{4}te^{-t}\cos 2t$

7. $(\omega^2 - 4)^{-1}\left[(\omega^2 - 5)\cos\omega t + \cos 2t\right]$

8. $-\frac{2}{5}e^t \sin t + \frac{4}{5}e^t \cos t - \frac{2}{5}\sin t + \frac{1}{5}\cos t$

9. $\frac{7}{5}e^t \sin t - \frac{1}{5}e^t \cos t + \frac{1}{5}e^{-t}$

10. $2e^{-t} + te^{-t} + 2t^2 e^{-t}$

11. $e^t t - t^2 e^t + \frac{2}{3}t^3 e^t$

12. $\frac{1}{2}e^t + \frac{1}{2}e^{-t}$

13. $\cos\sqrt{2}t$

14. $\mathbf{y} = \begin{pmatrix} 2e^{-t} - e^t \\ e^{-t} - e^t \end{pmatrix}$

15. $\mathbf{y} = \begin{pmatrix} 4e^{2t} - 3e^t \\ 6e^{2t} - 6e^t \end{pmatrix}$

16. $\mathbf{y} = \begin{pmatrix} \cos 2t + 2\sin 2t \\ \frac{5}{2}\sin 2t \end{pmatrix}$

17. $\mathbf{y} = \begin{pmatrix} \cos t + \sin t \\ \cos t - \sin t \end{pmatrix}$

18. $\mathbf{y} = \begin{pmatrix} (1-t)e^{-3t} \\ te^{-3t} \end{pmatrix}$

19. $\mathbf{y} = \begin{pmatrix} -7te^{-2t} \\ (1-7t)e^{-2t} \end{pmatrix}$

20. $\mathbf{y} = \begin{pmatrix} \left(\frac{2005}{2704} - \frac{151}{52}t\right)e^{-3t} + \frac{3}{16}e^t + \frac{12}{169}\cos(2t) - \frac{5}{169}\sin(2t) \\ -\frac{38}{169}\cos(2t) + \left(\frac{5847}{2704} + \frac{151}{52}t\right)e^{-3t} + \frac{44}{169}\sin(2t) + \frac{1}{16}e^t \end{pmatrix}$

21. $\mathbf{y} = \begin{pmatrix} -\frac{8}{25}e^{-t} + \left(-\frac{74}{15}t - \frac{578}{225}\right)e^{4t} - \frac{1}{9}e^t \\ \frac{2}{25}e^{-t} + \left(\frac{532}{225} - \frac{74}{15}t\right)e^{4t} - \frac{4}{9}e^t \end{pmatrix}$

22. $\mathbf{y} = \begin{pmatrix} \frac{11}{2}e^t \cos t - \frac{7}{2}e^t \sin t - \frac{9}{2} + 5(2\cos t - 4\sin t)e^{t/2}\sinh\frac{1}{2}t \\ -\frac{3}{2}e^t \cos t + \frac{15}{2}e^t \sin t + \frac{3}{2} - \cos t - 3\sin t \end{pmatrix}$

23. $\mathbf{y} = \begin{pmatrix} -\frac{1}{20}e^{-3t} + \frac{1}{4}e^{-t} - \frac{1}{5}\cos t + \frac{1}{10}\sin t \\ -\frac{3}{10}\cos t + \frac{2}{5}\sin t + \frac{1}{20}e^{-3t} + \frac{1}{4}e^{-t} \end{pmatrix}$

24. $\mathbf{y} = \begin{pmatrix} (-1+t)e^{-t} + 2\cosh t - 3\sinh t \\ 1 + \left(-t + \frac{3}{2}\right)e^{-t} + \frac{3}{2}e^t - 2\cosh t + 3\sinh t + 2e^{t/2}\sinh\frac{1}{2}t \\ 2e^{t/2}\sinh\frac{1}{2}t + 1 + \cosh t + 2\sinh t + e^t \end{pmatrix}$

25. $\begin{pmatrix} x \\ y \end{pmatrix} = \begin{pmatrix} -\frac{1}{5}\sqrt{10}\sin\left(\frac{1}{2}\sqrt{10}t\right) + \sin t \\ \frac{1}{5}\sqrt{10}\sin\left(\frac{1}{2}\sqrt{10}t\right) \end{pmatrix}$

26. $\mathbf{m} = \begin{pmatrix} m_0 e^{-k_1 t} \\ \dfrac{k_1 m_0 \left(e^{-k_2 t} - e^{-k_1 t}\right)}{k_1 - k_2} \\ m_0 \left(\dfrac{k_2 e^{-k_1 t} - k_1 e^{-k_2 t}}{k_1 - k_2} + 1\right) \end{pmatrix}$, $k_1 \neq k_2$; $\mathbf{m} = \begin{pmatrix} m_0 e^{-k_1 t} \\ m_0 k_1 t e^{-k_1 t} \\ m_0 \pm m_0(1 + k_1 t)e^{-k_1 t} \end{pmatrix}$, $k_1 = k_2$

Section 5.5, *page 348*

1.

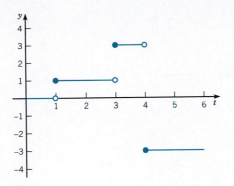

2.

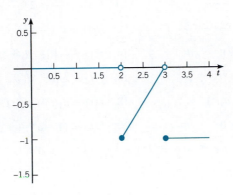

4.

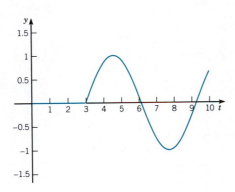

6.

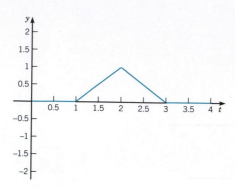

7. $F(s) = \frac{2e^{-2s}}{s^3}$

8. $F(s) = \frac{e^{-s}}{s} + 2\frac{e^{-s}}{s^3}$

9. $F(s) = \frac{e^{-s\pi}}{s^2} - \frac{\pi e^{-2s\pi}}{s} - \frac{e^{-2s\pi}}{s^2}$

10. $F(s) = \frac{e^{-s}}{s} + 2\frac{e^{-3s}}{s} - 6\frac{e^{-4s}}{s}$

11. $F(s) = -\frac{e^{-2s}}{s} + \frac{e^{-2s}}{s^2} - \frac{e^{-3s}}{s} - \frac{e^{-3s}}{s^2}$

12. $F(s) = s^{-2} - \frac{e^{-s}}{s^2}$

13. $f(t) = (t-1)^3 e^{2(t-1)} u_1(t)$

14. $f(t) = -\frac{1}{3}\left(e^{-2(t-2)} - e^{t-2}\right) u_2(t)$

15. $f(t) = 2e^{t-2}\cos(t-2)u_2(t)$

16. $f(t) = \frac{1}{2}\left(e^{2(t-2)} - e^{-2(t-2)}\right) u_2(t)$

17. $f(t) = \frac{1}{2}\left(e^{3(t-1)} + e^{t-1}\right) u_1(t)$

18. $f(t) = u_1(t) + u_2(t) - u_3(t) - u_4(t)$

19. $F(s) = (1 - e^{-s})/s, \; s > 0$

20. $F(s) = (1 - e^{-s} + e^{-2s} - e^{-3s})/s, \; s > 0$

21. $F(s) = \frac{1}{s}\left[1 - e^{-s} + \cdots + e^{-2ns} - e^{-(2n+1)s}\right] = \frac{1 - e^{-(2n+2)s}}{s(1+e^{-s})}, \; s > 0$

22. $\mathcal{L}\{f(t)\} = \frac{1 - e^{-s}}{s(1+e^{-s})}, \; s > 0$

23. $\mathcal{L}\{f(t)\} = \frac{1 - (1+s)e^{-s}}{s^2(1-e^{-s})}, \; s > 0$

24. $\mathcal{L}\{f(t)\} = \frac{1 + e^{-\pi s}}{(1+s^2)(1-e^{-\pi s})}, \; s > 0$

25. (a) $\mathcal{L}\{f(t)\} = s^{-1}(1 - e^{-s}), \; s > 0$
    (b) $\mathcal{L}\{g(t)\} = s^{-2}(1 - e^{-s}), \; s > 0$
    (c) $\mathcal{L}\{h(t)\} = s^{-2}(1 - e^{-s})^2, \; s > 0$

26. (b) $\mathcal{L}\{p(t)\} = \frac{1 - e^{-s}}{s^2(1+e^{-s})}, \; s > 0$

Section 5.6, *page 353*

1. $y = 1 - \cos t + \sin t - u_{\pi/2}(1 - \sin t)$

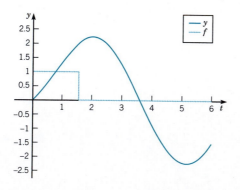

2. $y = e^{-t} \sin t + \frac{1}{2}u_\pi(t)[1 + e^{-(t-\pi)} \cos t + e^{-(t-\pi)} \sin t]$
$\quad - \frac{1}{2}u_{2\pi}(t)[1 - e^{-(t-2\pi)} \cos t - e^{-(t-2\pi)} \sin t]$

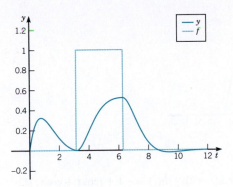

3. $y = \frac{1}{6}[1 - u_{2\pi}](2 \sin t - \sin 2t)$

4. $y = \frac{1}{6}(2 \sin t - \sin 2t) - \frac{1}{6}u_\pi(t)(2 \sin t + \sin 2t)$

5. $y = \frac{1}{2} + \frac{1}{2}e^{-2t} - e^{-t} - u_{10}(t)[\frac{1}{2} + \frac{1}{2}e^{-2(t-10)} - e^{-(t-10)}]$

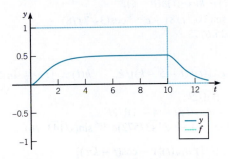

6. $y = e^{-t} - e^{-2t} + u_2(t)\left(-e^{-t+2} + \frac{1}{2}e^{-2t+4} + \frac{1}{2}\right)$

7. $y = \cos t + u_{3\pi}(t)[1 - \cos(t - 3\pi)]$

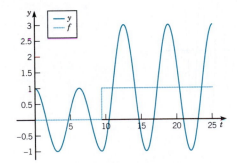

8. $y = h(t) - u_{\pi/2}(t)h(t - \pi/2), \quad h(t) = \frac{4}{25}(-4 + 5t + 4e^{-t/2} \cos t - 3e^{-t/2} \sin t)$

9. $y = \frac{1}{2} \sin t + \frac{1}{2}t - \frac{1}{2}u_6(t)[t - 6 - \sin(t - 6)]$

10. $y = h(t) + u_\pi(t)h(t - \pi), \quad h(t) = \frac{4}{17}[-4 \cos t + \sin t + 4e^{-t/2} \cos t + e^{-t/2} \sin t]$

11. $y = u_\pi(t)[\frac{1}{4} - \frac{1}{4}\cos(2t - 2\pi)] - u_{3\pi}(t)[\frac{1}{4} - \frac{1}{4}\cos(2t - 6\pi)]$

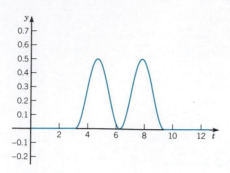

12. $y = u_1(t)h(t - 1) - u_2(t)h(t - 2), \quad h(t) = -1 + (\cos t + \cosh t)/2$

13. $y = h(t) - u_\pi(t)h(t - \pi), \quad h(t) = (3 - 4\cos t + \cos 2t)/12$

14. $f(t) = [u_{t_0}(t)(t - t_0) - u_{t_0+k}(t)(t - t_0 - k)](h/k)$

15. $g(t) = [u_{t_0}(t)(t - t_0) - 2u_{t_0+k}(t)(t - t_0 - k) + u_{t_0+2k}(t - t_0 - 2k)](h/k)$

16. (b) $u(t) = ku_{3/2}(t)h(t - \frac{3}{2}) - ku_{5/2}(t)h(t - \frac{5}{2})$,
$h(t) = 1 - (\sqrt{7}/21)e^{-t/8}\sin(3\sqrt{7}t/8) - e^{-t/8}\cos(3\sqrt{7}t/8)$
(d) $k = 2.51$ (e) $\tau = 25.6773$

17. (a) $k = 5$
(b) $y = [u_5(t)h(t - 5) - u_{5+k}(t)h(t - 5 - k)]/k, \quad h(t) = \frac{1}{4}t - \frac{1}{8}\sin 2t$

18. (b) $f_k(t) = [u_{4-k}(t) - u_{4+k}(t)]/2k$;
$y = [u_{4-k}(t)h(t - 4 + k) - u_{4+k}(t)(t - 4 - k)]/2k$
$h(t) = \frac{1}{4} - \frac{1}{4}e^{-t/6}\cos(\sqrt{143}t/6) - (\sqrt{143}/572)e^{-t/6}\sin(\sqrt{143}t/6)$

19. (a) $y = 1 - \cos t + 2\sum_{k=1}^{n}(-1)^k u_{k\pi}(t)[1 - \cos(t - k\pi)]$
(b)

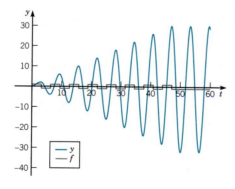

20. (a) $y = h(t) + 2\sum_{k=1}^{n}(-1)^k u_{k\pi}(t)h(t - k\pi)$,
$h(t) = 1 - e^{-t/20}\cos\frac{\sqrt{399}}{20}t - \frac{1}{\sqrt{399}}e^{-t/20}\sin\frac{\sqrt{399}}{20}t$

(b)

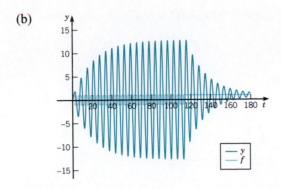

21. (b) $y = 1 - \cos t + \sum_{k=1}^{n}(-1)^k u_{k\pi}(t)[1 - \cos(t - k\pi)]$

23. (a) $y = 1 - \cos t + 2\sum_{k=1}^{n}(-1)^k u_{11k/4}(t)[1 - \cos(t - \frac{11}{4}k)]$

(b)

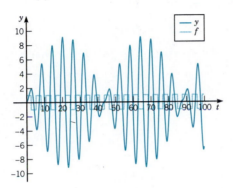

Section 5.7, *page 361*

1. $y = e^{-t}\cos t + e^{-t}\sin t + u_\pi(t)e^{-(t-\pi)}\sin(t - \pi)$

2. $y = \frac{1}{2}u_\pi(t)\sin 2(t - \pi) - \frac{1}{2}u_{2\pi}(t)\sin 2(t - 2\pi)$

3. $y = -\frac{1}{2}e^{-2t} + \frac{1}{2}e^{-t} + u_5(t)[-e^{-2(t-5)} + e^{-(t-5)}] + u_{10}(t)[\frac{1}{2} + \frac{1}{2}e^{-2(t-10)} - e^{-(t-10)}]$

4. $y = \cosh t - 20u_3(t)\sinh(t - 3)$

5. $y = \frac{1}{4}\sin t - \frac{1}{4}\cos t + \frac{1}{4}e^{-t}\cos\sqrt{2}t + (1/\sqrt{2})u_{3\pi}(t)e^{-(t-3\pi)}\sin\sqrt{2}(t - 3\pi)$

6. $y = \frac{1}{2}\cos 2t + \frac{1}{2}u_{4\pi}(t)\sin 2(t - 4\pi)$

7. $y = \sin t + u_{2\pi}\sin(t - 2\pi)$

8. $y = u_{\pi/4}(t)\sin 2(t - \pi/4)$

9. $y = u_{\pi/2}(t)[1 - \cos(t - \pi/2)] + 3u_{3\pi/2}(t)\sin(t - 3\pi/2) - u_{2\pi}(t)[1 - \cos(t - 2\pi)]$

10. $y = (1/\sqrt{31})u_{\pi/6}(t)\exp[-\frac{1}{4}(t - \pi/6)]\sin(\sqrt{31}/4)(t - \pi/6)$

11. $y = \frac{1}{5}\cos t + \frac{2}{5}\sin t - \frac{1}{5}e^{-t}\cos t - \frac{3}{5}e^{-t}\sin t + u_{\pi/2}(t)e^{-(t-\pi/2)}\sin(t - \pi/2)$

12. $y = u_1(t)[\sinh(t - 1) - \sin(t - 1)]/2$

13. (a) $-e^{-T/4}\delta(t - 5 - T)$, $T = 8\pi/\sqrt{15}$

14. (a) $y = (4/\sqrt{15})u_1(t)e^{-(t-1)/4}\sin(\sqrt{15}/4)(t-1)$

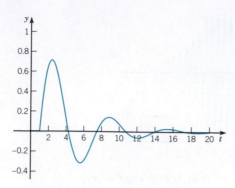

(b) $t_1 \cong 2.3613$, $y_1 \cong 0.71153$
(c) $y = (8\sqrt{7}/21)u_1(t)e^{-(t-1)/8}\sin(3\sqrt{7}/8)(t-1)$; $t_1 \cong 2.4569$, $y_1 \cong 0.83351$
(d) $t_1 = 1 + \pi/2 \cong 2.5708$, $y_1 = 1$

15. (a) $k_1 \cong 2.8108$,   (b) $k_1 \cong 2.3995$ (c) $k_1 = 2$

16. (a) $\phi(t, k) = [u_{4-k}(t)h(t-4+k) - u_{4+k}(t)h(t-4-k)]/2k$, $h(t) = 1 - \cos t$
   (b) $u_4(t)\sin(t-4)$ (c) Yes

17. (b) $y = \sum_{k=1}^{20} u_{k\pi}(t)\sin(t - k\pi)$

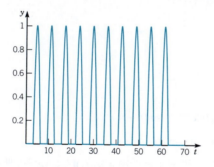

18. (b) $y = \sum_{k=1}^{20} (-1)^{k+1} u_{k\pi}(t)\sin(t - k\pi)$

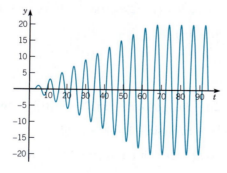

19. (b) $y = \sum_{k=1}^{20} u_{k\pi/2}(t)\sin(t - k\pi/2)$

20. (b) $y = \sum_{k=1}^{20} (-1)^{k+1} u_{k\pi/2}(t)\sin(t - k\pi/2)$

21. (b) $y = \sum_{k=1}^{15} u_{(2k-1)\pi}(t) \sin[t - (2k-1)\pi]$

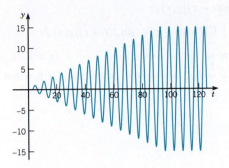

22. (b) $y = \sum_{k=1}^{40} (-1)^{k+1} u_{11k/4}(t) \sin(t - \frac{11}{4}k)$

23. (b) $y = \frac{20}{\sqrt{399}} \sum_{k=1}^{20} (-1)^{k+1} u_{k\pi}(t) e^{-(t-k\pi)/20} \sin[\sqrt{399}(t - k\pi)/20]$

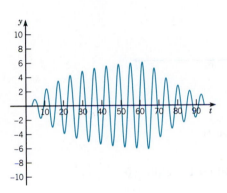

24. (b) $y = \frac{20}{\sqrt{399}} \sum_{k=1}^{15} u_{(2k-1)\pi}(t) e^{-[t-(2k-1)\pi]/20} \sin[\sqrt{399}[t - (2k-1)\pi]/20]$

Section 5.8, *page 370*

2. $\sin t * \sin t = \frac{1}{2}(\sin t - t\cos t)$ is negative when $t = 2\pi$, for example.

3. $F(s) = 2/s^2(s^2 + 4)$

4. $F(s) = 1/(s+1)(s^2 + 1)$

5. $F(s) = 1/s^2(s - 1)$

6. $F(s) = s/(s^2 + 1)^2$

7. $f(t) = \frac{1}{6} \int_0^t (t - \tau)^3 \sin \tau \, d\tau$

8. $f(t) = \int_0^t e^{-(t-\tau)} \cos 2\tau \, d\tau$

9. $f(t) = \frac{1}{2} \int_0^t (t - \tau) e^{-(t-\tau)} \sin 2\tau \, d\tau$

10. $f(t) = \int_0^t \sin(t - \tau) g(\tau) \, d\tau$

11. $f(t) = \int_0^t \sin(t - \tau) \sin(\tau) \, d\tau$

12. $f(t) = \int_0^t \sin(t - \tau) \cos(\tau) \, d\tau$

13. (c) $\int_0^1 u^m (1 - u)^n \, du = \dfrac{\Gamma(m+1)\Gamma(n+1)}{\Gamma(m+n+2)}$

14. $y = \dfrac{1}{\omega} \sin \omega t + \dfrac{1}{\omega} \int_0^t \sin \omega(t - \tau) g(\tau) \, d\tau$

15. $y = \int_0^t e^{-(t-\tau)} \sin(t - \tau) \sin \alpha \tau \, d\tau$

16. $y = \frac{1}{8} \int_0^t e^{(t-\tau)/2} \sin 2(t - \tau) g(\tau) \, d\tau$

17. $y = e^{-t/2} \cos t - \frac{1}{2} e^{-t/2} \sin t + \int_0^t e^{-(t-\tau)/2} \sin(t - \tau)[1 - u_\pi(\tau)] \, d\tau$

18. $y = 2e^{-2t} + te^{-2t} + \int_0^t (t - \tau) e^{-2(t-\tau)} g(\tau) \, d\tau$

19. $y = 2e^{-t} - e^{-2t} + \int_0^t [e^{-(t-\tau)} - e^{-2(t-\tau)}] \cos \alpha\tau \, d\tau$

20. $y = \frac{1}{2} \int_0^t [\sinh(t-\tau) - \sin(t-\tau)] g(\tau) \, d\tau$

21. $y = \frac{4}{3} \cos t - \frac{1}{3} \cos 2t + \frac{1}{6} \int_0^t [2 \sin(t-\tau) - \sin 2(t-\tau)] g(\tau) \, d\tau$

22. (a) The Laplace transform of the solution of $ay'' + by' + cy = u(t)$, $y(0) = 0$, $y'(0) = 0$ is
$$Y(s) = \frac{1}{s(as^2 + bs + c)} = \frac{1}{s} H(s)$$
where $H(s)$ is the transfer function for the system. Thus $y(t) = \int_0^t h(t) \, dt$.

(b)

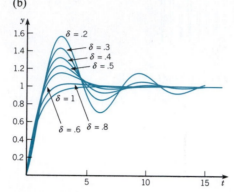

23. $\Phi(s) = \dfrac{F(s)}{1 + K(s)}$

24. (a) $\phi(t) = \frac{1}{3}(4 \sin 2t - 2 \sin t)$

25. (a) $\phi(t) = \cos t$
(b) $\phi''(t) + \phi(t) = 0$, $\quad \phi(0) = 1$, $\quad \phi'(0) = 0$

26. (a) $\phi(t) = \cosh t$
(b) $\phi''(t) - \phi(t) = 0$, $\quad \phi(0) = 1$, $\quad \phi'(0) = 0$

27. (a) $\phi(t) = (1 - 2t + t^2)e^{-t}$
(b) $\phi''(t) + 2\phi'(t) + \phi(t) = 2e^{-t}$, $\quad \phi(0) = 1$, $\quad \phi'(0) = -3$

28. (a) $\phi(t) = \frac{1}{3}e^{-t} - \frac{1}{3}e^{t/2} \cos(\sqrt{3}t/2) + \frac{1}{\sqrt{3}}e^{t/2} \sin(\sqrt{3}t/2)$
(b) $\phi'''(t) + \phi(t) = 0$, $\quad \phi(0) = 0$, $\quad \phi'(0) = 0$, $\quad \phi''(0) = 1$

29. (a) $\phi(t) = \cos t$
(b) $\phi^{(4)}(t) - \phi(t) = 0$, $\quad \phi(0) = 1$, $\quad \phi'(0) = 0$, $\quad \phi''(0) = -1$, $\quad \phi'''(0) = 0$

30. (a) $\phi(t) = 1 - \frac{2}{\sqrt{3}}e^{-t/2} \sin(\sqrt{3}t/2)$
(b) $\phi'''(t) + \phi''(t) + \phi'(t) = 0$, $\quad \phi(0) = 1$, $\quad \phi'(0) = -1$, $\quad \phi''(0) = 1$

Section 5.9, page 379

1. $H = \dfrac{H_1 + G_2}{1 + H_1 G_1}$

2. $H = \dfrac{H_1 H_2 H_3}{1 + G_1 H_2 H_3 + G_2 H_1 H_2}$

3. Use mathematical induction and the fact that $|h(t)| \le t^k e^{\alpha t}$ for all three functions.

4. Products of linear factors $s + \lambda_i$ and quadratic factors $s^2 + 2\alpha_j s + \alpha_j^2 + \beta_j^2$ are polynomials in s with positive coefficients, assuming that each $\lambda_i > 0$ and $\alpha_j > 0$.

5. $a_0, a_1, a_2 > 0$ and $a_1a_2 > a_0$ 6. 2

7. 0   8. 1   9. 2

10. 0   11. 2   12. $K > 15/2$

13. $-4 < K < 20$ 14. $-8 < K < 22$ 15. $-16 < K < 29$

CHAPTER 6

Section 6.1, *page 401*

1. (a) $\begin{pmatrix} 7e^t & 5e^{-t} & 10e^{2t} \\ -e^t & 7e^{-t} & 2e^{2t} \\ 8e^t & 0 & -e^{2t} \end{pmatrix}$

(b) $\begin{pmatrix} 2e^{2t} - 2 + 3e^{3t} & 1 + 4e^{-2t} - e^t & 3e^{3t} + 2e^t - e^{4t} \\ 4e^{2t} - 1 - 3e^{3t} & 2 + 2e^{-2t} + e^t & 6e^{3t} + e^t + e^{4t} \\ -2e^{2t} - 3 + 6e^{3t} & -1 + 6e^{-2t} - 2e^t & -3e^{3t} + 3e^t - 2e^{4t} \end{pmatrix}$

(c) $\begin{pmatrix} e^t & -2e^{-t} & 2e^{2t} \\ 2e^t & -e^{-t} & -2e^{2t} \\ -e^t & -3e^{-t} & 4e^{2t} \end{pmatrix}$ (d) $(e-1)\begin{pmatrix} 1 & 2e^{-1} & \frac{1}{2}(e+1) \\ 2 & e^{-1} & -\frac{1}{2}(e+1) \\ -1 & 3e^{-1} & e+1 \end{pmatrix}$

4. $\mathbf{x}' = \begin{pmatrix} 0 & 1 & 0 & 0 \\ 0 & 0 & 1 & 0 \\ 0 & 0 & 0 & 1 \\ -3 & 0 & 0 & -4 \end{pmatrix}\mathbf{x} + \begin{pmatrix} 0 \\ 0 \\ 0 \\ t \end{pmatrix}$

5. $\mathbf{x}' = \begin{pmatrix} 0 & 1 & 0 \\ 0 & 0 & 1 \\ -3/t & 0 & -\sin t/t \end{pmatrix}\mathbf{x} + \begin{pmatrix} 0 \\ 0 \\ \cos t/t \end{pmatrix}$

6. $\mathbf{x}' = \begin{pmatrix} 0 & 1 & 0 & 0 \\ 0 & 0 & 1 & 0 \\ 0 & 0 & 0 & 1 \\ -4t/(t-1) & 0 & -e^t/(t(t-1)) & 0 \end{pmatrix}\mathbf{x}$

7. $\mathbf{x}' = \begin{pmatrix} 0 & 1 & 0 \\ 0 & 0 & 1 \\ -t^2 & -t^2 & -t \end{pmatrix}\mathbf{x} + \begin{pmatrix} 0 \\ 0 \\ \ln t \end{pmatrix}$

8. $\mathbf{x}' = \begin{pmatrix} 0 & 1 & 0 & 0 \\ 0 & 0 & 1 & 0 \\ 0 & 0 & 0 & 1 \\ -\tan x/(x-1) & 0 & -(x+1)/(x-1) & 0 \end{pmatrix}\mathbf{x}$

9. $\mathbf{x}' = \begin{pmatrix} 0 & 1 & 0 & 0 & 0 & 0 \\ 0 & 0 & 1 & 0 & 0 & 0 \\ 0 & 0 & 0 & 1 & 0 & 0 \\ 0 & 0 & 0 & 0 & 1 & 0 \\ 0 & 0 & 0 & 0 & 0 & 1 \\ -9/(x^2-4) & 0 & -x^2/(x^2-4) & 0 & 0 & 0 \end{pmatrix}\mathbf{x}$

11. $\mathbf{x}' = \begin{pmatrix} -k_{01} - k_{21} - k_{31} & k_{12} & k_{13} \\ k_{21} & -k_{02} - k_{12} & 0 \\ k_{31} & 0 & -k_{13} \end{pmatrix} \mathbf{x} + \begin{pmatrix} L \\ 0 \\ 0 \end{pmatrix}$

12. $\dfrac{d}{dt} \begin{pmatrix} v_1 \\ v_2 \\ i_1 \\ i_2 \end{pmatrix} = \begin{pmatrix} 0 & 0 & 1/C_1 & -1/C_1 \\ 0 & 0 & 0 & 1/C_2 \\ -1/L_1 & 0 & 0 & 0 \\ 1/L_2 & -1/L_2 & 0 & -R/L_2 \end{pmatrix} \begin{pmatrix} v_1 \\ v_2 \\ i_1 \\ i_2 \end{pmatrix} + \begin{pmatrix} 0 \\ 0 \\ e(t) \\ 0 \end{pmatrix}$

13. (b) $\mathbf{x} = \dfrac{\alpha}{L_{12} + L_{21}} \begin{pmatrix} L_{12} + L_{21} \exp[-(L_{12} + L_{21})t] \\ L_{21} - L_{21} \exp[-(L_{12} + L_{21})t] \end{pmatrix}$

(c) $\bar{x}_1 = \alpha L_{12}/(L_{12} + L_{21})$ $\bar{x}_2 = \alpha L_{21}/(L_{12} + L_{21})$; Increasing $L_{12} + L_{21}$ increases the rate of approach to equilibrium.

(d)

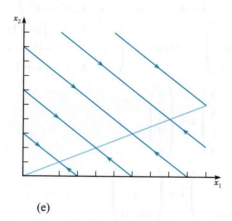

(e)

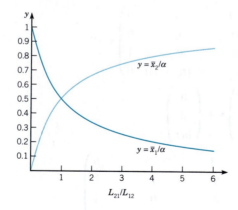

The ratio of fractional turnover rates determines the proportions of the initial amount of tracer in each compartment at equilibrium.

14. (a) $\mathbf{K} = \begin{pmatrix} -(k_1 + k_2)/m_1 & k_2/m_1 \\ k_2/m_2 & -(k_2 + k_3)/m_2 \end{pmatrix}$ $\mathbf{f}(t) = \begin{pmatrix} F_1(t)/m_1 \\ F_2(t)/m_2 \end{pmatrix}$

(b) $m_1 = m_2$

15. $C = (0 \quad 1 \quad 0 \quad 0)$

Section 6.2, *page 411*

1. $-\infty < t < \infty$ 2. $t > 0$ or $t < 0$

3. $t > 1$, or $0 < t < 1$, or $t < 0$ 4. $t > 0$

5. $\ldots, -3\pi/2 < x < -\pi/2, \quad -\pi/2 < x < 1, \quad 1 < x < \pi/2, \quad \pi/2 < x < 3\pi/2, \ldots$

6. $-\infty < x < -2, \quad -2 < x < 2, \quad 2 < x < \infty$

8. No 9. Yes

10. (a) $W(t) = ce^{\int \text{tr}(\mathbf{P}(t))\,dt}$ (b) $W' = -p_1(t)W$

11. $W = 1$ 12. $W = 1$ 13. $W = -6e^{-2t}$

14. $W = e^{-2t}$ 15. $W = 6x$ 16. $W = 6x^{-1}$

Section 6.3, *page 421*

1. $\mathbf{x} = c_1 e^{-3t} \begin{pmatrix} 1 \\ 1 \\ 1 \end{pmatrix} + c_2 e^{-4t} \begin{pmatrix} 1 \\ 0 \\ -1 \end{pmatrix} + c_3 e^{-6t} \begin{pmatrix} 1 \\ -2 \\ 1 \end{pmatrix}$

2. $\mathbf{x} = c_1 e^{t} \begin{pmatrix} 1 \\ 0 \\ 0 \end{pmatrix} + c_2 e^{-t} \begin{pmatrix} 0 \\ 1 \\ -1 \end{pmatrix} + c_3 e^{5t} \begin{pmatrix} 2 \\ 1 \\ 1 \end{pmatrix}$

3. $\mathbf{x} = c_1 e^{-2t} \begin{pmatrix} 1 \\ 1 \\ 0 \end{pmatrix} + c_2 e^{-2t} \begin{pmatrix} 0 \\ 1 \\ 2 \end{pmatrix} + c_3 e^{7t} \begin{pmatrix} 2 \\ -2 \\ 1 \end{pmatrix}$

4. $\mathbf{x} = c_1 e^{-t} \begin{pmatrix} 1 \\ 1 \\ 1 \end{pmatrix} + c_2 e^{-t} \begin{pmatrix} 1 \\ 0 \\ -1 \end{pmatrix} + c_3 \begin{pmatrix} 1 \\ 2 \\ 2 \end{pmatrix}$

5. $\mathbf{x} = c_1 e^{t} \begin{pmatrix} 1 \\ -2 \\ 1 \end{pmatrix} + c_2 e^{-t} \begin{pmatrix} 1 \\ 0 \\ -1 \end{pmatrix} + c_3 e^{4t} \begin{pmatrix} 1 \\ 1 \\ 1 \end{pmatrix}$

6. $\mathbf{x} = c_1 e^{-t} \begin{pmatrix} 1 \\ 0 \\ -1 \end{pmatrix} + c_2 e^{-t} \begin{pmatrix} 1 \\ -2 \\ 0 \end{pmatrix} + c_3 e^{8t} \begin{pmatrix} 2 \\ 1 \\ 2 \end{pmatrix}$

7. $\mathbf{x} = c_1 e^{-t} \begin{pmatrix} 3 \\ -4 \\ -2 \end{pmatrix} + c_2 e^{-2t} \begin{pmatrix} 4 \\ -5 \\ -7 \end{pmatrix} + c_3 e^{2t} \begin{pmatrix} 0 \\ 1 \\ -1 \end{pmatrix}$

8. $\mathbf{x} = c_1 e^{t} \begin{pmatrix} 1 \\ -4 \\ -1 \end{pmatrix} + c_2 e^{-2t} \begin{pmatrix} 1 \\ -1 \\ -1 \end{pmatrix} + c_3 e^{3t} \begin{pmatrix} 1 \\ 2 \\ 1 \end{pmatrix}$

9. $\mathbf{x} = 2e^{2t} \begin{pmatrix} 1 \\ 1 \\ 0 \end{pmatrix} + e^{t} \begin{pmatrix} 0 \\ -2 \\ 1 \end{pmatrix}$

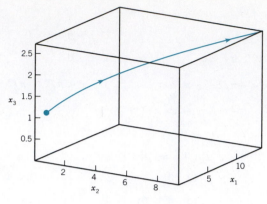

10. $\mathbf{x} = 6e^t \begin{pmatrix} 1 \\ 2 \\ -1 \end{pmatrix} + 3e^{-t} \begin{pmatrix} 1 \\ -2 \\ 1 \end{pmatrix} + e^{4t} \begin{pmatrix} -2 \\ -1 \\ 8 \end{pmatrix}$

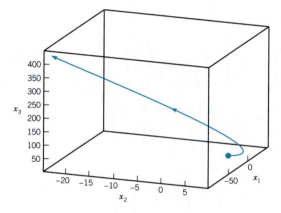

11. $\mathbf{x} = e^{-2t} \begin{pmatrix} 0 \\ -1 \\ 0 \end{pmatrix} + 2e^{-4t} \begin{pmatrix} 1 \\ 0 \\ -1 \end{pmatrix}$

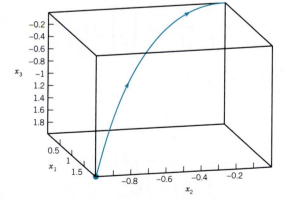

12. $\mathbf{x} = e^{-t/2} \begin{pmatrix} 1 \\ 1 \\ 0 \end{pmatrix} + e^{-t} \begin{pmatrix} 1 \\ 0 \\ 1 \end{pmatrix}$

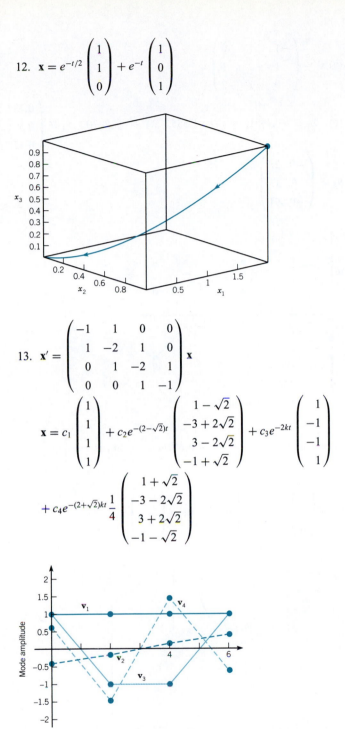

13. $\mathbf{x}' = \begin{pmatrix} -1 & 1 & 0 & 0 \\ 1 & -2 & 1 & 0 \\ 0 & 1 & -2 & 1 \\ 0 & 0 & 1 & -1 \end{pmatrix} \mathbf{x}$

$\mathbf{x} = c_1 \begin{pmatrix} 1 \\ 1 \\ 1 \\ 1 \end{pmatrix} + c_2 e^{-(2-\sqrt{2})t} \begin{pmatrix} 1-\sqrt{2} \\ -3+2\sqrt{2} \\ 3-2\sqrt{2} \\ -1+\sqrt{2} \end{pmatrix} + c_3 e^{-2kt} \begin{pmatrix} 1 \\ -1 \\ -1 \\ 1 \end{pmatrix}$

$+ c_4 e^{-(2+\sqrt{2})kt} \dfrac{1}{4} \begin{pmatrix} 1+\sqrt{2} \\ -3-2\sqrt{2} \\ 3+2\sqrt{2} \\ -1-\sqrt{2} \end{pmatrix}$

Long term decay rate is controlled by the negative eigenvalue nearest zero, $-2+\sqrt{2}$.

14. (a) $\mathbf{x}' = \begin{pmatrix} -1 & 1 & 0 \\ 1 & -2 & 1 \\ 0 & 1 & -2 \end{pmatrix} \mathbf{x}$

(c) $\lambda_1 \cong -0.1981, \quad \lambda_2 \cong -1.555, \quad \lambda_3 \cong -3.2470, \quad \mathbf{v}_1 \cong \begin{pmatrix} 0.737 \\ 0.591 \\ 0.328 \end{pmatrix}, \quad \mathbf{v}_2 \cong \begin{pmatrix} -0.591 \\ 0.328 \\ 0.737 \end{pmatrix},$

$\mathbf{v}_3 \cong \begin{pmatrix} -0.328 \\ 0.737 \\ -0.591 \end{pmatrix}$

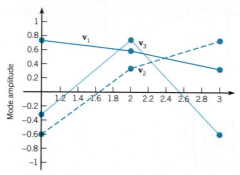

(d) $\mathbf{x} = c_1 e^{k\lambda_1 t} \mathbf{v}_1 + c_2 e^{k\lambda_2 t} \mathbf{v}_2 + c_3 e^{k\lambda_3 t} \mathbf{v}_3, \qquad \lim_{t \to \infty} \mathbf{x}(t) = \begin{pmatrix} 0 \\ 0 \\ 0 \end{pmatrix}$

15. $\mathbf{u}_1 = \begin{pmatrix} 1 \\ 1 \\ 0 \end{pmatrix}, \quad \mathbf{u}_2 = \begin{pmatrix} 1 \\ 0 \\ 1 \end{pmatrix}, \quad \mathbf{u}_3 = \begin{pmatrix} -1 \\ -1 \\ 1 \end{pmatrix}$

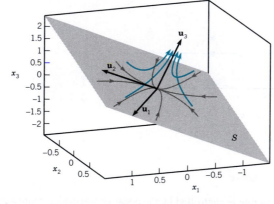

$\mathbf{x} = c_1 e^{-2t} \mathbf{u}_1 + c_2 e^{-t} \mathbf{u}_2 + c_3 e^t \mathbf{u}_3$. If $\mathbf{x}_0 \notin S$, $\mathbf{x}(t)$ is asymptotic to line determined by $\mathbf{u}_3$ as $t \to \infty$.

16. $\mathbf{u}_1 = \begin{pmatrix} 1 \\ 0 \\ -1 \\ 0 \end{pmatrix}$, $\mathbf{u}_2 = \begin{pmatrix} 0 \\ 1 \\ 0 \\ 1 \end{pmatrix}$, $\mathbf{u}_3 = \begin{pmatrix} 1 \\ -1 \\ 1 \\ 0 \end{pmatrix}$

17.

$$\mathbf{m} = m_0 \begin{pmatrix} 0 \\ 0 \\ 1 \end{pmatrix} + m_0 e^{-k_1 t} \begin{pmatrix} 1 \\ k_1/(k_2 - k_1) \\ -k_2/(k_2 - k_1) \end{pmatrix} + m_0 e^{-k_2 t} \begin{pmatrix} 0 \\ -k_1/(k_2 - k_1) \\ k_1/(k_2 - k_1) \end{pmatrix}, \quad k_2 \neq k_1$$

$$\mathbf{m} = m_0 \begin{pmatrix} 0 \\ 0 \\ 1 \end{pmatrix} + m_0 e^{-k_1 t} \begin{pmatrix} 1 \\ k_1 t \\ -1 - k_1 t \end{pmatrix}, \quad k_2 = k_1$$

18. $e^{-4t} \begin{pmatrix} 1 \\ 0 \\ 0 \\ -1 \end{pmatrix}$, $e^{-3t} \begin{pmatrix} 0 \\ 1 \\ 0 \\ 1 \end{pmatrix}$, $e^{-2t} \begin{pmatrix} 1 \\ 0 \\ -1 \\ 1 \end{pmatrix}$, $e^{-t} \begin{pmatrix} 1 \\ 0 \\ -1 \\ 0 \end{pmatrix}$

19. $e^{-3t} \begin{pmatrix} 1 \\ -2 \\ 0 \\ 1 \end{pmatrix}$, $e^{3t} \begin{pmatrix} 0 \\ 0 \\ 1 \\ 0 \end{pmatrix}$, $e^{3t} \begin{pmatrix} 1 \\ 0 \\ 0 \\ -1 \end{pmatrix}$, $e^{3t} \begin{pmatrix} 1 \\ 1 \\ 0 \\ 1 \end{pmatrix}$

20. $e^{-6t} \begin{pmatrix} -1 \\ -2 \\ 4 \\ 1 \end{pmatrix}$, $\begin{pmatrix} 1 \\ -2 \\ 3 \\ 0 \end{pmatrix}$, $\begin{pmatrix} 1 \\ 1 \\ 0 \\ -3 \end{pmatrix}$, $e^{9t} \begin{pmatrix} 1 \\ 2 \\ -1 \\ -1 \end{pmatrix}$

21. $e^{-4t} \begin{pmatrix} 1 \\ 0 \\ 0 \\ 1 \end{pmatrix}$, $e^{-2t} \begin{pmatrix} 1 \\ 0 \\ -1 \\ 0 \end{pmatrix}$, $e^{2t} \begin{pmatrix} 1 \\ 1 \\ -1 \\ -1 \end{pmatrix}$, $e^{4t} \begin{pmatrix} 0 \\ 1 \\ 0 \\ 1 \end{pmatrix}$

22. $e^{-4t} \begin{pmatrix} 1 \\ 0 \\ 0 \\ -1 \\ 1 \end{pmatrix}$, $e^{-3t} \begin{pmatrix} 0 \\ 1 \\ 0 \\ 1 \\ 0 \end{pmatrix}$, $e^{-2t} \begin{pmatrix} 1 \\ 0 \\ -1 \\ 1 \\ 0 \end{pmatrix}$, $e^{-2t} \begin{pmatrix} 1 \\ 0 \\ 0 \\ 0 \\ 1 \end{pmatrix}$, $e^{-t} \begin{pmatrix} -1 \\ 0 \\ 1 \\ 0 \\ -1 \end{pmatrix}$

23. $e^{-2t} \begin{pmatrix} 1 \\ -2 \\ 2 \\ -2 \\ 1 \end{pmatrix}$, $e^{-2t} \begin{pmatrix} 0 \\ 1 \\ 0 \\ 1 \\ 0 \end{pmatrix}$, $e^{-t} \begin{pmatrix} -1 \\ 0 \\ 0 \\ 1 \\ -1 \end{pmatrix}$, $e^{t} \begin{pmatrix} 2 \\ 0 \\ 0 \\ 0 \\ 1 \end{pmatrix}$, $e^{2t} \begin{pmatrix} 1 \\ 1 \\ -1 \\ 1 \\ 0 \end{pmatrix}$

Section 6.4, *page 431*

1. $\mathbf{x} = c_1 e^{-t} \begin{pmatrix} 1 \\ 0 \\ 1 \end{pmatrix} + c_2 e^{-t} \begin{pmatrix} \cos t + \sin t \\ 2\cos t \\ -2\cos t \end{pmatrix} + c_3 e^{-t} \begin{pmatrix} \sin t - \cos t \\ 2\sin t \\ -2\sin t \end{pmatrix}$

2. $\mathbf{x} = c_1 e^{-t} \begin{pmatrix} 1 \\ 1 \\ -1 \end{pmatrix} + c_2 e^{t} \begin{pmatrix} -\sin 4t \\ \cos 4t \\ -\sin 4t \end{pmatrix} + c_3 e^{t} \begin{pmatrix} \cos 4t \\ \sin 4t \\ \cos 4t \end{pmatrix}$

3. $\mathbf{x} = c_1 e^{-t} \begin{pmatrix} 1 \\ 1 \\ -1 \end{pmatrix} + c_2 e^{-t} \begin{pmatrix} -\sin 2t \\ \cos 2t \\ -\sin 2t \end{pmatrix} + c_3 e^{-t} \begin{pmatrix} \cos 2t \\ \sin 2t \\ \cos 2t \end{pmatrix}$

4. $\mathbf{x} = c_1 e^{-2t} \begin{pmatrix} 1 \\ 0 \\ -2 \end{pmatrix} + c_2 e^{-2t} \begin{pmatrix} \cos 4t + \sin 4t \\ 3\cos 4t \\ 2\sin 4t \end{pmatrix} + c_3 e^{-2t} \begin{pmatrix} \sin 4t - \cos 4t \\ 3\sin 4t \\ -2\cos 4t \end{pmatrix}$

5. $\mathbf{x} = c_1 e^{-t} \begin{pmatrix} 1 \\ 0 \\ -1 \end{pmatrix} + c_2 e^{-t} \begin{pmatrix} 2\cos 3t + 2\sin 3t \\ 3\cos 3t \\ -\sin 3t \end{pmatrix} + c_3 e^{-t} \begin{pmatrix} 2\sin 3t - 2\cos 3t \\ 3\sin 3t \\ \cos 3t \end{pmatrix}$

6. $\mathbf{x} = c_1 e^{-t} \begin{pmatrix} 1 \\ 1 \\ 1 \end{pmatrix} + c_2 \begin{pmatrix} \cos 2t - \sin 2t \\ -2\cos 2t \\ \cos 2t + \sin 2t \end{pmatrix} + c_3 \begin{pmatrix} \cos 2t + \sin 2t \\ -2\sin 2t \\ \sin 2t - \cos 2t \end{pmatrix}$

7. $\mathbf{x} = c_1 e^{-2t} \begin{pmatrix} 4 \\ -5 \\ -7 \end{pmatrix} + c_2 e^{-t} \begin{pmatrix} 3 \\ -4 \\ -2 \end{pmatrix} + c_3 e^{2t} \begin{pmatrix} 0 \\ 1 \\ -1 \end{pmatrix}$

8. $\mathbf{x} = c_1 e^{-2t} \begin{pmatrix} 1 \\ -1 \\ -1 \end{pmatrix} + c_2 e^{t} \begin{pmatrix} 1 \\ -4 \\ -1 \end{pmatrix} + c_3 e^{3t} \begin{pmatrix} 1 \\ 2 \\ 1 \end{pmatrix}$

9. (a) $\mathbf{a} = \begin{pmatrix} 5 \\ 1 \\ 2 \end{pmatrix}, \quad \mathbf{b} = \begin{pmatrix} 0 \\ -2 \\ 1 \end{pmatrix}$

(b) and (c)

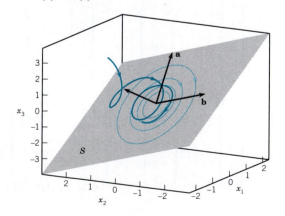

$$\mathbf{x} = c_1 e^{-t} \begin{pmatrix} 1 \\ -1 \\ -1 \end{pmatrix} + c_2 \begin{pmatrix} 5\cos 4t \\ \cos 4t - 2\sin 4t \\ 2\cos 4t + \sin 4t \end{pmatrix} + c_3 \begin{pmatrix} 5\sin 4t \\ \sin 4t + 2\cos 4t \\ 2\sin 4t - \cos 4t \end{pmatrix}$$

If $\mathbf{x}_0 \notin S$, the solution trajectory approaches a closed curve in S.

10. (a) $\mathbf{a} = \begin{pmatrix} 0 \\ 1 \\ 1 \\ 0 \end{pmatrix}$, $\mathbf{b} = \begin{pmatrix} -1 \\ 0 \\ 0 \\ -1 \end{pmatrix}$ (b) $\mathbf{x} = c_1 \begin{pmatrix} \sin 3t \\ \cos 3t \\ \cos 3t \\ \sin 3t \end{pmatrix} + c_2 \begin{pmatrix} -\cos 3t \\ \sin 3t \\ \sin 3t \\ -\cos 3t \end{pmatrix}$

$$+ c_3 e^{-t} \begin{pmatrix} -3\cos 2t + \sin 2t \\ -\cos 2t - 3\sin 2t \\ 2\cos 2t + \sin 2t \\ -5\cos 2t \end{pmatrix} + c_4 e^{-t} \begin{pmatrix} -3\sin 2t - \cos 2t \\ -\sin 2t + 3\cos 2t \\ 2\sin 2t - \cos 2t \\ -5\sin 2t \end{pmatrix}$$

so all solutions approach closed curves in S as $t \to \infty$.

12. (c) $\omega_1^2 = -1$, $\mathbf{u}_1 = \begin{pmatrix} 3 \\ 2 \end{pmatrix}$; $\omega_2^2 = -4$, $\mathbf{u}_2 = \begin{pmatrix} 3 \\ -4 \end{pmatrix}$

(d) $y_1 = 3c_1\cos t + 3c_2\sin t + 3c_3\cos 2t + 3c_4\sin 2t$
$y_2 = 2c_1\cos t + 2c_2\sin t - 4c_3\cos 2t - 4c_4\sin 2t$
(e) $y_1' = -3c_1\sin t + 3c_2\cos t - 6c_3\sin 2t + 6c_4\cos 2t$
$y_2' = -2c_1\sin t + 2c_2\cos t + 8c_3\sin 2t - 8c_4\cos 2t$

13. (a) $\lambda_1 = i$, $\mathbf{v}_1 = \begin{pmatrix} 1 \\ 1 \\ i \\ i \end{pmatrix}$; $\lambda_2 = -i$, $\mathbf{v}_2 = \begin{pmatrix} 1 \\ 1 \\ -i \\ -i \end{pmatrix}$; $\lambda_3 = \sqrt{3}i$, $\mathbf{v}_3 = \begin{pmatrix} 1 \\ -1 \\ \sqrt{3}i \\ -\sqrt{3}i \end{pmatrix}$;

$$\lambda_4 = -\sqrt{3}i, \mathbf{v}_4 = \begin{pmatrix} 1 \\ -1 \\ -\sqrt{3}i \\ \sqrt{3}i \end{pmatrix}$$

(b) $\mathbf{x} = c_1 \begin{pmatrix} \cos t \\ \cos t \\ -\sin t \\ -\sin t \end{pmatrix} + c_2 \begin{pmatrix} \sin t \\ \sin t \\ \cos t \\ \cos t \end{pmatrix} + c_3 \begin{pmatrix} \cos\sqrt{3}t \\ -\cos\sqrt{3}t \\ -\sqrt{3}\sin\sqrt{3}t \\ \sqrt{3}\sin\sqrt{3}t \end{pmatrix} + c_4 \begin{pmatrix} \sin\sqrt{3}t \\ -\sin\sqrt{3}t \\ \sqrt{3}\cos\sqrt{3}t \\ -\sqrt{3}\cos\sqrt{3}t \end{pmatrix}$

(c)

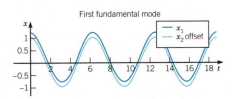

First fundamental mode

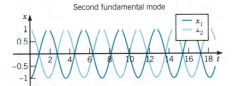

Second fundamental mode

(d)

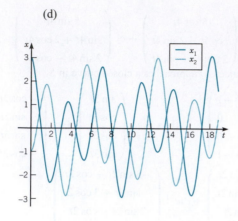

14. $\mathbf{x} = c_1 \begin{pmatrix} \cos \omega t \\ \cos \omega t \\ -\omega \sin \omega t \\ -\omega \sin \omega t \end{pmatrix} + c_2 \begin{pmatrix} \sin \omega t \\ \sin \omega t \\ \omega \cos \omega t \\ \omega \cos \omega t \end{pmatrix} + c_3 \begin{pmatrix} \cos \sqrt{3}\omega t \\ -\cos \sqrt{3}\omega t \\ -\sqrt{3}\omega \sin \sqrt{3}\omega t \\ \sqrt{3}\omega \sin \sqrt{3}\omega t \end{pmatrix} + $

$c_4 \begin{pmatrix} \sin \sqrt{3}\omega t \\ -\sin \sqrt{3}\omega t \\ \sqrt{3}\omega \cos \sqrt{3}\omega t \\ -\sqrt{3}\omega \cos \sqrt{3}\omega t \end{pmatrix}$

First mode: $c_3 = c_4 = 0$, frequency $= \omega$;   Second mode: $c_1 = c_2 = 0$, frequency $= \sqrt{3}\omega$

15. $e^{-t} \begin{pmatrix} 1 \\ 0 \\ 0 \\ -1 \end{pmatrix}$, $e^{t} \begin{pmatrix} 1 \\ 1 \\ 0 \\ 1 \end{pmatrix}$, $e^{-t} \begin{pmatrix} \cos 4t \\ -\sin 4t \\ \cos 4t \\ -\sin 4t \end{pmatrix}$, $e^{-t} \begin{pmatrix} \sin 4t \\ \cos 4t \\ \sin 4t \\ \cos 4t \end{pmatrix}$

16. $e^{-2t} \begin{pmatrix} \cos t - \sin t \\ -2\sin t \\ -2\sin t \\ 0 \end{pmatrix}$, $e^{-2t} \begin{pmatrix} \sin t + \cos t \\ 2\cos t \\ 2\cos t \\ 0 \end{pmatrix}$, $e^{-t} \begin{pmatrix} \cos 2t - \sin 2t \\ \cos 2t + \sin 2t \\ -\cos 2t \\ \cos 2t \end{pmatrix}$,

$e^{-t} \begin{pmatrix} \sin 2t + \cos 2t \\ \sin 2t - \cos 2t \\ -\sin 2t \\ \sin 2t \end{pmatrix}$

17. $\begin{pmatrix} \cos 2t - \sin 2t \\ \cos 2t + \sin 2t \\ -\cos 2t \\ \cos 2t \end{pmatrix}$, $\begin{pmatrix} \sin 2t + \cos 2t \\ \sin 2t - \cos 2t \\ -\sin 2t \\ \sin 2t \end{pmatrix}$, $\begin{pmatrix} \cos 3t - \sin 3t \\ -2\sin 3t \\ -2\sin 3t \\ 0 \end{pmatrix}$, $\begin{pmatrix} \sin 3t + \cos 3t \\ 2\cos 3t \\ 2\cos 3t \\ 0 \end{pmatrix}$

18. $e^{-2t} \begin{pmatrix} \cos 3t \\ -\sin 3t \\ -\sin 3t \\ \cos 3t \end{pmatrix}$, $e^{-2t} \begin{pmatrix} \sin 3t \\ \cos 3t \\ \cos 3t \\ \sin 3t \end{pmatrix}$, $e^{-2t} \begin{pmatrix} 1 \\ 0 \\ -1 \\ 1 \end{pmatrix}$, $e^{-t} \begin{pmatrix} 1 \\ -1 \\ 0 \\ -1 \end{pmatrix}$

Section 6.5, *page 443*

1. $e^{At} = \begin{pmatrix} -\frac{1}{3}e^{-t} + \frac{4}{3}e^{2t} & \frac{2}{3}e^{-t} - \frac{2}{3}e^{2t} \\ -\frac{2}{3}e^{-t} + \frac{2}{3}e^{2t} & \frac{4}{3}e^{-t} - \frac{1}{3}e^{2t} \end{pmatrix}$

2. $e^{At} = \begin{pmatrix} \frac{1}{2}e^{-t/2} + \frac{1}{2}e^{-t} & e^{-t/2} - e^{-t} \\ \frac{1}{4}e^{-t/2} - \frac{1}{4}e^{-t} & \frac{1}{2}e^{-t/2} + \frac{1}{2}e^{-t} \end{pmatrix}$

3. $e^{At} = \begin{pmatrix} (1+2t)e^t & -4te^t \\ te^t & (-2t+1)e^t \end{pmatrix}$

4. $e^{At} = \begin{pmatrix} 4t+1 & -2t \\ 8t & -4t+1 \end{pmatrix}$

5. $e^{At} = \begin{pmatrix} \cos t + 2\sin t & -5\sin t \\ \sin t & \cos t - 2\sin t \end{pmatrix}$

6. $e^{At} = \begin{pmatrix} e^{-t}\cos 2t & -2e^{-t}\sin 2t \\ 1/2\,e^{-t}\sin 2t & e^{-t}\cos 2t \end{pmatrix}$

7. $e^{At} = \begin{pmatrix} e^{3t}(\cosh t + 2\sinh t) & -\sinh t\,e^{3t} \\ 3\sinh t\,e^{3t} & e^{3t}(\cosh t - 2\sinh t) \end{pmatrix}$

8. $e^{At} = \begin{pmatrix} e^{-t}\cos t + 2e^{-t}\sin t & -e^{-t}\sin t \\ 5e^{-t}\sin t & e^{-t}\cos t - 2e^{-t}\sin t \end{pmatrix}$

9. $e^{At} = \begin{pmatrix} \cosh t + 2\sinh t & -\sinh t \\ 3\sinh t & \cosh t - 2\sinh t \end{pmatrix}$

10. $e^{At} = \begin{pmatrix} e^{-1/2t}\left(\cosh\frac{5}{2}t + 3/5\sinh\frac{5}{2}t\right) & 2/5\,e^{-1/2t}\sinh\frac{5}{2}t \\ 8/5\sinh\frac{5}{2}t\,e^{-1/2t} & e^{-1/2t}\left(\cosh\frac{5}{2}t - 3/5\sinh\frac{5}{2}t\right) \end{pmatrix}$

11. $e^{At} = \begin{pmatrix} \left(1 - \frac{1}{2}t\right)e^{-t} & te^{-t} \\ -\frac{1}{4}te^{-t} & \left(\frac{1}{2}t+1\right)e^{-t} \end{pmatrix}$

12. $e^{At} = \begin{pmatrix} \left(1 - \frac{5}{2}t\right)e^{t/2} & \frac{5}{2}te^{t/2} \\ -\frac{5}{2}te^{t/2} & \left(\frac{5}{2}t+1\right)e^{t/2} \end{pmatrix}$

13. $e^{At} = \begin{pmatrix} -2e^{-2t} + 3e^{-t} & 2e^{-3t/2}\sinh(1/2\,t) & 2e^{-3t/2}\sinh\left(\frac{1}{2}t\right) \\ \frac{3}{2}e^{2t} + \frac{5}{2}e^{-2t} - 4e^{-t} & \frac{13}{12}e^{2t} + \frac{5}{4}e^{-2t} - \frac{4}{3}e^{-t} & \frac{5}{4}e^{-2t} - \frac{4}{3}e^{-t} + \frac{1}{12}e^{2t} \\ -\frac{3}{2}e^{2t} + \frac{7}{2}e^{-2t} - 2e^{-t} & \frac{7}{4}e^{-2t} - \frac{2}{3}e^{-t} - \frac{13}{12}e^{2t} & -\frac{1}{12}e^{2t} + \frac{7}{4}e^{-2t} - \frac{2}{3}e^{-t} \end{pmatrix}$

14. $e^{At} = \begin{pmatrix} \frac{1}{6}e^t + \frac{1}{3}e^{-2t} + \frac{1}{2}e^{3t} & -\frac{2}{3}e^{-2/t}\sinh\left(\frac{3}{2}t\right) & -e^{-2t} + \frac{1}{2}e^t + \frac{1}{2}e^{3t} \\ e^{3t} - \frac{1}{3}e^{-2t} - \frac{2}{3}e^t & e^{-2/t}\left(\cosh\left(\frac{3}{2}t\right) + \frac{5}{3}\sinh\left(\frac{3}{2}t\right)\right) & -2e^t + e^{-2t} + e^{3t} \\ -\frac{1}{6}e^t + \frac{1}{6}e^{3t} - \frac{1}{3}e^{-2t} & \frac{2}{3}e^{-2/t}\sinh\left(\frac{3}{2}t\right) & \frac{1}{2}e^{3t} - \frac{1}{2}e^t + e^{-2t} \end{pmatrix}$

15. $\mathbf{x} = \begin{pmatrix} 3e^{-t}\cos 2t - 2e^{-t}\sin 2t \\ \frac{3}{2}e^{-t}\sin 2t + e^{-t}\cos 2t \end{pmatrix}$

16. $\mathbf{x} = \begin{pmatrix} 2\cosh t + 5\sinh t \\ 8\sinh t - \cosh t \end{pmatrix}$

17. $e^{At} = \begin{pmatrix} (-t+1)e^{-3t} & -te^{-3t} \\ te^{-3t} & (t+1)e^{-3t} \end{pmatrix}$

18. $e^{At} = \begin{pmatrix} (1+t)e^{4t} & -te^{4t} \\ te^{4t} & (1-t)e^{4t} \end{pmatrix}$

19. $e^{At} = \begin{pmatrix} e^t\cos t - 2e^t\sin t & -5e^t\sin t \\ e^t\sin t & e^t\cos t + 2e^t\sin t \end{pmatrix}$

20. $e^{At} = \begin{pmatrix} 1 & 2e^{-t/2}\sinh\left(\frac{1}{2}t\right) & -2e^{-t/2}\sinh\left(\frac{1}{2}t\right) \\ 2e^{t/2}\sinh\left(\frac{1}{2}t\right) & -1+2\cosh(t) & 2e^{-1/2t}\sinh\left(\frac{1}{2}t\right) \\ 2e^{t/2}\sinh\left(\frac{1}{2}t\right) & 2e^{t/2}\sinh\left(\frac{1}{2}t\right) & 1 \end{pmatrix}$

21. $\mathbf{x} = c_1 = \begin{pmatrix} u_0 \\ v_0 \end{pmatrix}\cos\omega t + \begin{pmatrix} v_0 \\ -\omega^2 u_0 \end{pmatrix}\dfrac{\sin\omega t}{\omega}$

22. $e^{At} = \begin{pmatrix} e^{-k_1 t} & 0 & 0 \\ \dfrac{k_1\left(e^{-k_2 t}-e^{-k_1 t}\right)}{k_1-k_2} & e^{-k_2 t} & 0 \\ \dfrac{e^{-k_2 t}k_1 - e^{-k_1 t}k_2}{k_1-k_2} - 1 & 2e^{-1/2 k_1 t}\sinh\left(1/2\,k_2 t\right) & 1 \end{pmatrix}$,

$\mathbf{m} = m_0 \begin{pmatrix} e^{-k_1 t} \\ \dfrac{k_1\left(e^{-k_2 t}-e^{-k_1 t}\right)}{k_1-k_2} \\ \dfrac{e^{-k_2 t}k_1 - e^{-k_1 t}k_2}{k_1-k_2} - 1 \end{pmatrix}$ if $k_1 \neq k_2$

$e^{At} = \begin{pmatrix} e^{-k_1 t} & 0 & 0 \\ k_1 te^{-k_1 t} & e^{-k_1 t} & 0 \\ 1-(1+k_1 t)e^{-k_1 t} & 2e^{-1/2 k_1 t}\sinh\left(1/2\,k_1 t\right) & 1 \end{pmatrix}$,

$\mathbf{m} = m_0 \begin{pmatrix} e^{-k_1 t} \\ k_1 te^{-k_1 t} \\ 1-(1+k_1 t)e^{-k_1 t} \end{pmatrix}$ if $k_2 = k_1$

23. (a) $\mathbf{A} = \mathbf{TDT}^{-1}$, $\mathbf{A}^2 = (\mathbf{TDT}^{-1})(\mathbf{TDT}^{-1}) = \mathbf{TD}(\mathbf{T}^{-1}\mathbf{T})\mathbf{DT}^{-1} = \mathbf{TDI_n DT}^{-1} = \mathbf{TD}^2\mathbf{T}^{-1}$, etc.

   (b) $e^{At} = \sum_{k=0}^{\infty} \dfrac{\mathbf{A}^k t^k}{k!} = \sum_{k=0}^{\infty} \dfrac{\mathbf{TD}^k\mathbf{T}^{-1} t^k}{k!} = \mathbf{T}\left(\sum_{k=0}^{\infty}\dfrac{\mathbf{D}^k t^k}{k!}\right)\mathbf{T}^{-1} = \mathbf{T}e^{\mathbf{D}t}\mathbf{T}^{-1}$

Section 6.6, *page 449*

2. $\mathbf{x} = c_1 e^t \begin{pmatrix} 1 \\ 1 \end{pmatrix} + c_2 e^{-t} \begin{pmatrix} 1 \\ 3 \end{pmatrix} + \begin{pmatrix} t+\frac{3}{2}te^t - \frac{1}{4}e^t \\ 2t-1-\frac{3}{4}e^t + \frac{3}{2}te^t \end{pmatrix}$

3. $\mathbf{x} = c_1 e^{2t} \begin{pmatrix} \sqrt{3} \\ 1 \end{pmatrix} + c_2 e^{-2t} \begin{pmatrix} 1 \\ -\sqrt{3} \end{pmatrix} + \begin{pmatrix} -\frac{2}{3}e^t - e^{-t} \\ -\frac{1}{\sqrt{3}}e^t + \frac{2}{\sqrt{3}}e^{-t} \end{pmatrix}$

4. $\mathbf{x} = c_1 \begin{pmatrix} \cos t \\ \frac{2}{5}\cos t + \frac{1}{5}\sin t \end{pmatrix} + c_2 \begin{pmatrix} \sin t \\ \frac{2}{5}\sin t - \frac{1}{5}\cos t \end{pmatrix} + \begin{pmatrix} 2t\cos t - t\sin t - \cos t \\ t\cos t - \cos t \end{pmatrix}$

5. $\mathbf{x} = c_1 e^{2t} \begin{pmatrix} 1 \\ 1 \end{pmatrix} + c_2 e^{-3t} \begin{pmatrix} 1 \\ -4 \end{pmatrix} + \begin{pmatrix} \frac{1}{2}e^t \\ -e^{-2t} \end{pmatrix}$

6. $\mathbf{x} = c_1 e^{2t} \begin{pmatrix} 1 \\ 0 \\ 2 \end{pmatrix} + c_2 e^{t} \begin{pmatrix} 1 \\ 0 \\ 1 \end{pmatrix} + c_3 e^{2t} \begin{pmatrix} 0 \\ 1 \\ 1 \end{pmatrix} + \begin{pmatrix} \frac{1}{6}e^{-t} - \frac{1}{2}t - \frac{9}{4} \\ -\frac{1}{2}t - \frac{1}{4} \\ -\frac{1}{2}t - \frac{7}{4} - \frac{1}{6}e^{-t} \end{pmatrix}$

7. $\mathbf{x} = c_1 e^{-t} \begin{pmatrix} 0 \\ 1 \\ 1 \end{pmatrix} + c_2 e^{-2t} \begin{pmatrix} -1 \\ 2 \\ -1 \end{pmatrix} + c_3 e^{-t} \begin{pmatrix} 1 \\ 0 \\ 1 \end{pmatrix} + \begin{pmatrix} \frac{7}{8} + \frac{1}{4}t - \frac{1}{4}e^{-3t} \\ -\frac{3}{4} + \frac{1}{2}e^{-3t} + \frac{1}{2}t \\ -\frac{1}{8} - \frac{3}{4}e^{-3t} + \frac{1}{4}t \end{pmatrix}$

8. $\mathbf{x} = c_1 e^{-2t} \begin{pmatrix} 3 \\ 0 \\ 2 \end{pmatrix} + c_2 e^{-t} \begin{pmatrix} 1 \\ 0 \\ 1 \end{pmatrix} + c_3 e^{-2t} \begin{pmatrix} 0 \\ 3 \\ -1 \end{pmatrix} + \begin{pmatrix} \frac{5}{4} - \frac{1}{2}t + \frac{3}{10}\cos t + \frac{9}{10}\sin t \\ 0 \\ \frac{11}{10}\sin t + \frac{7}{10}\cos t + \frac{3}{2} - t \end{pmatrix}$

9. $\mathbf{x} = c_1 e^{t} \begin{pmatrix} 1 \\ 1 \\ 1 \end{pmatrix} + c_2 e^{-t} \begin{pmatrix} 1 \\ 0 \\ -1 \end{pmatrix} + c_3 e^{-2t} \begin{pmatrix} 1 \\ -2 \\ 1 \end{pmatrix} + \begin{pmatrix} -\frac{1}{10}\cos t - \frac{3}{10}\sin t \\ \frac{1}{10}\sin t - \frac{3}{10}\cos t \\ -\frac{1}{10}\cos t - \frac{3}{10}\sin t \end{pmatrix}$

10. (a)

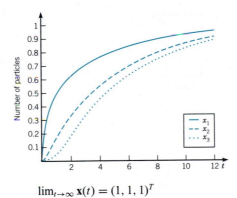

$\lim_{t\to\infty} \mathbf{x}(t) = (1, 1, 1)^T$

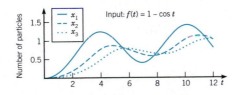

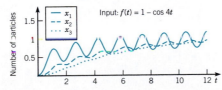

11. (a)

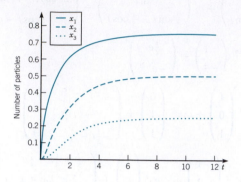

$$\lim_{t \to \infty} \mathbf{x}(t) = (3/4, 1/2, 1/4)^T$$

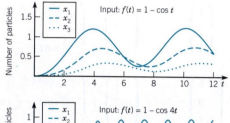

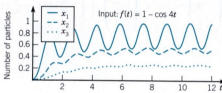

13.

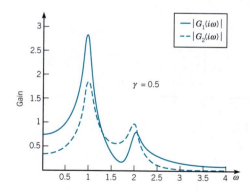

14. $\mathbf{a} = \begin{pmatrix} 1 \\ 2 \end{pmatrix}$, $\mathbf{b} = \begin{pmatrix} -4/3 \\ -5/3 \end{pmatrix}$

15. $\mathbf{a} = \begin{pmatrix} -3/10 \\ -1/5 \end{pmatrix}$, $\mathbf{b} = \begin{pmatrix} 2/5 \\ 1/10 \end{pmatrix}$

16. $\mathbf{a} = \begin{pmatrix} 1 \\ 1 \end{pmatrix}$, $\mathbf{b} = \begin{pmatrix} 1 \\ 0 \end{pmatrix}$

Section 6.7, *page 457*

1. $e^{At} = \begin{pmatrix} (2t+1)e^t & -4te^t \\ te^t & (-2t+1)e^t \end{pmatrix}$

2. $e^{At} = \begin{pmatrix} 1+4t & -2t \\ 8t & -4t+1 \end{pmatrix}$

3. $e^{At} = \begin{pmatrix} (-t+1)e^{2t} & te^{2t} & te^{2t} \\ -\frac{1}{2}\left(-4t+t^2\right)e^{2t} & \left(\frac{1}{2}t^2-t+1\right)e^{2t} & \frac{1}{2}\left(-2t+t^2\right)e^{2t} \\ \frac{1}{2}\left(t^2-6t\right)e^{2t} & -\frac{1}{2}\left(-4t+t^2\right)e^{2t} & \left(1+2t-\frac{1}{2}t^2\right)e^{2t} \end{pmatrix}$

4. $e^{At} = \begin{pmatrix} (4t+1)e^t & -3te^t & -2te^t \\ 8te^t & (-6t+1)e^t & -4te^t \\ -4te^t & 3te^t & (1+2t)e^t \end{pmatrix}$

5. $e^{At} = \begin{pmatrix} \frac{7}{3}e^{-t}+(-2t-4/3)e^{2t} & -2e^{-t}+(3t+2)e^{2t} & e^{-t}+(-1-3t)e^{2t} \\ \frac{14}{9}e^{-t}+\left(-\frac{10}{3}t-\frac{14}{9}\right)e^{2t} & -\frac{4}{3}e^{-t}+\left(\frac{7}{3}+5t\right)e^{2t} & \frac{2}{3}e^{-t}+\left(-\frac{2}{3}-5t\right)e^{2t} \\ -2te^{2t} & 3te^{2t} & (1-3t)e^{2t} \end{pmatrix}$

6. $e^{At} = \begin{pmatrix} (1+4t)e^t & 6te^t & 2te^t \\ -2te^t & (-3t+1)e^t & -te^t \\ -2te^t & -3te^t & (-t+1)e^t \end{pmatrix}$

7. $e^{At} =$

$\begin{pmatrix} -5e^{-t}\cos t - e^{-t}\sin t + 6e^{-t} & -9e^{-t}\cos t - 7e^{-t}\sin t + 9e^{-t} \\ (-4t+3)e^{-t} - 3e^{-t}\cos t + e^{-t}\sin t & (-6t+8)e^{-t} - e^{-t}\sin t - 7e^{-t}\cos t \\ (-1+4t)e^{-t} + e^{-t}\cos t - 3e^{-t}\sin t & (-5+6t)e^{-t} - 5e^{-t}\sin t + 5e^{-t}\cos t \\ 4e^{-t}\cos t + 2e^{-t}\sin t - 4e^{-t} & 6e^{-t}\cos t + 8e^{-t}\sin t - 6e^{-t} \end{pmatrix}$

$\begin{pmatrix} -9e^{-t}\cos t - 7e^{-t}\sin t + 9e^{-t} & 12e^{-t} - 12e^{-t}\cos t - 5e^{-t}\sin t \\ (-6t+7)e^{-t} - e^{-t}\sin t - 7e^{-t}\cos t & (-8t+8)e^{-t} + e^{-t}\sin t - 8e^{-t}\cos t \\ (-4+6t)e^{-t} - 5e^{-t}\sin t + 5e^{-t}\cos t & (-4+8t)e^{-t} - 7e^{-t}\sin t + 4e^{-t}\cos t \\ 6e^{-t}\cos t + 8e^{-t}\sin t - 6e^{-t} & -8e^{-t} + 9e^{-t}\cos t + 7e^{-t}\sin t \end{pmatrix}$

8. $e^{At} = \begin{pmatrix} (-t^2+1+2t)e^{-t} & (-t+t^2)e^{-t} & 2(-t+t^2)e^{-t} & -(t^2-3t)e^{-t} \\ -\frac{1}{2}\left(-4t+3t^2\right)e^{-t} & \left(\frac{3}{2}t^2+1-\frac{1}{2}t\right)e^{-t} & (3t^2-t)e^{-t} & -\frac{1}{2}(3t^2-7t)e^{-t} \\ \frac{1}{2}\left(-2t+t^2\right)e^{-t} & -\frac{1}{2}(-t+t^2)e^{-t} & (t-t^2+1)e^{-t} & \frac{1}{2}(t^2-3t)e^{-t} \\ \frac{1}{2}\left(t^2-4t\right)e^{-t} & -\frac{1}{2}(t^2-3t)e^{-t} & (t^2-3t)e^{-t} & \left(\frac{5}{2}t+1+\frac{1}{2}t^2\right)e^{-t} \end{pmatrix}$

9. $\mathbf{x} = e^{-3t}\begin{pmatrix} 3+4t \\ 2+4t \end{pmatrix}$

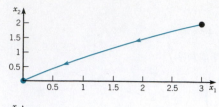

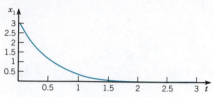

10. $\mathbf{x} = e^t \begin{pmatrix} -2 - 20t \\ 4 - 10t \end{pmatrix}$

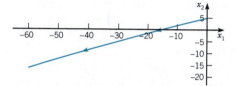

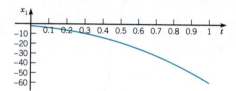

11. $\mathbf{x} = e^t \begin{pmatrix} 1 + 5t \\ -10 \\ 4 - 5t \end{pmatrix} + e^{2t} \begin{pmatrix} 0 \\ 9 \\ -3 \end{pmatrix}$

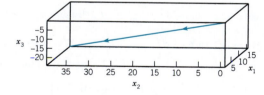

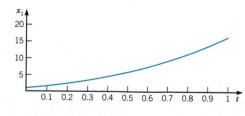

12. $\mathbf{x} = e^{-2t} \begin{pmatrix} 5 + 22t - \frac{29}{2}t^2 \\ 7 - 95t + \frac{87}{2}t^2 \\ -4 + 10t - \frac{145}{2}t^2 \end{pmatrix}$

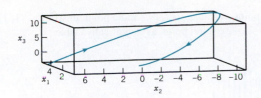

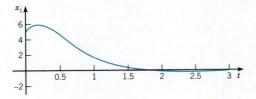

CHAPTER 7

Section 7.1, *page 480*

1. (a, c) $(1, -\frac{1}{2})$, saddle point, unstable; $(0, 0)$, center, stable

2. (a, c) $(-\sqrt{3}/3, -\frac{1}{2})$, saddle point, unstable; $(\sqrt{3}/3, -\frac{1}{2})$, center, stable

3. (a, c) $(0, 0)$, node, unstable; $(0, \frac{3}{2})$, saddle point, unstable;
   $(2, 0)$, node, asymptotically stable; $(-1, 3)$, node, asymptotically stable

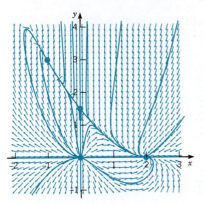

Direction field and phase portrait for Problem 3.

4. (a, c) $(0, 0)$, spiral point, asymptotically stable;
   $(0, 1)$, saddle point, unstable; $(-2, -2)$, saddle point, unstable; $(3, -2)$, saddle point, unstable

5. (a, c) $(0, 0)$, node, unstable;
   $(1, 1)$, saddle point, unstable; $(3, -1)$, spiral point, asymptotically stable

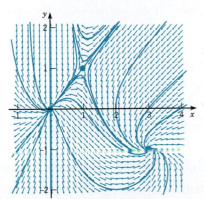

Direction field and phase portrait for Problem 5.

6. (a, c) (0, 0), saddle point, unstable;   (2, 0), saddle point, unstable;   (1, 1), spiral point, asymptotically stable;   (−2, −2), spiral point, asymptotically stable

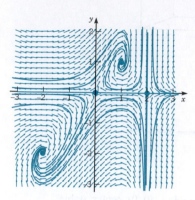

Direction field and phase portrait for Problem 6.

7. (a, c) (0, 0), spiral point, unstable;   (−1, −1), saddle point, unstable;   (−1, 2), saddle point, unstable;   (−2, 2), spiral point, asymptotically stable

8. (a, c) (0, 1), saddle point, unstable;   (1, 1), node, asymptotically stable;   (−2, 4), spiral point, unstable

9. (a, c) (0, 0), spiral point, unstable;   (−2, −2), saddle point, unstable;   (3, 1), saddle point, unstable

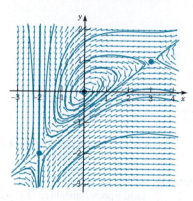

Direction field and phase portrait for Problem 9.

10. (a, c) (1, 0), saddle point, unstable;   (2, −1), node, asymptotically stable;   (−1, −1), node, unstable

11. (a, c) (0, 0), saddle point, unstable;   (1, 0), node, asymptotically stable;   (2, 0), saddle point, unstable

12. (a, c) (−2, 1), node, unstable;   (−2, 1/2), saddle point, unstable;   (1/3, −2/3), spiral point, asymptotically stable;   (2, 1), saddle point, unstable

13. (a, c) (0, 0), node, unstable;   (−2, 4/3), saddle point, unstable;   (2, 0), saddle point, unstable;   (3, −1/3), spiral point, asymptotically stable

14. (a, c) (−1, 0), saddle point, unstable;   (1, 1), spiral point, unstable;   (2, −1), spiral point, asymptotically stable

15. (a, c) (0, 1), center, stable;   (1, 2), saddle point, unstable;   (−2, 2), saddle point, unstable

16. (a, c) (−3, 3), spiral point, unstable;   (1, 3), saddle point, unstable;   (1, −3), saddle point, unstable;   (3, −3), spiral point, asymptotically stable

17. (a, c) (0, 0), saddle point, unstable; $(\sqrt{6}, 0)$, center, stable; $(-\sqrt{6}, 0)$, center, stable

18. (a, c) (0, 0), saddle point, unstable;   (1, 0), saddle point, unstable;   (1/2, 1/4), center, stable

20. (a) $R = A$, $T \cong 3.17$ (b) $R = A$, $T \cong 3.20$,   3.35,   3.63,   4.17
    (c) $T \to \pi$ as $A \to 0$ (d) $A = \pi$

21. (b) $v_c \cong 4.90$

Section 7.2, *page 491*

1. (a, b, c) (0, 0); $u' = -2u + w$, $w' = -w$; $\lambda = -1, -2$;   node, asymptotically stable
   (2, 4); $u' = -2u + w$, $w' = 4u - w$; $\lambda = (-3 \pm \sqrt{17})/2$;   saddle point, unstable

2. (a, b, c) (2, 2); $u' = u - w$, $w' = 3u$; $\lambda = (1 \pm \sqrt{11}i)/2$;   spiral point, unstable
   (-1, -1); $u' = u - w$, $w' = -w$; $\lambda = 1, -1$;   saddle point, unstable

3. (a, b, c) (0, 0); $u' = u$, $w' = u + 2w$; $\lambda = 1, 2$;   node, unstable
   (-4, 2); $u' = u + 4w$, $w' = u + 2w$; $\lambda = (3 \pm \sqrt{33})/2$;   saddle point, unstable

4. (a, b, c) (0, 0); $u' = u$, $w' = u - 2w$; $\lambda = 1, -2$;   saddle point, unstable
   (1, 1); $u' = u - 2w$, $w' = 3u - 2w$; $\lambda = (-1 \pm \sqrt{15}i)/2$;   spiral point, asymptotically stable

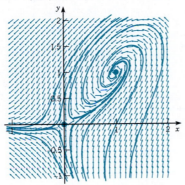

Direction field and phase portrait for Problem 4.

5. (a, b, c) (0, 0); $u' = -2u + 2w$, $w' = 4u + 4w$; $\lambda = 1 \pm \sqrt{17}$;   saddle point, unstable
   (-2, 2); $u' = 4u$, $w' = 6u + 6w$; $\lambda = 4, 6$; node, unstable
   (4, 4); $u' = -6u + 6w$, $w' = -8u$; $\lambda = -3 \pm \sqrt{39}\,i$;   spiral point, asymptotically stable

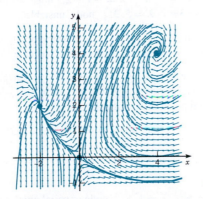

Direction field and phase portrait for Problem 5.

6. (a, b, c) (0, 0); $u' = u$, $w' = 3w$; $\lambda = 1, 3$;   node, unstable
   (1, 0); $u' = -u - w$, $w' = 2w$; $\lambda = -1, 2$;   saddle point, unstable
   $(0, \frac{3}{2})$; $u' = -\frac{1}{2}u$, $w' = -\frac{3}{2}u - 3w$; $\lambda = -\frac{1}{2}, -3$;   node, asymptotically stable
   (-1, 2); $u' = u + w$, $w' = -2u - 4w$; $\lambda = (-3 \pm \sqrt{17})/2$;   saddle point, unstable
   see figure on page 900

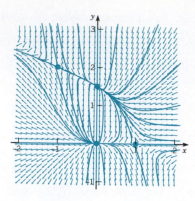

Direction field and phase portrait for Problem 6.

7. (a, b, c) $(1, 1)$; $u' = -w$, $w' = 2u - 2w$; $\lambda = -1 \pm i$; spiral point, asymptotically stable
$(-1, 1)$; $u' = -w$, $w' = -2u - 2w$; $\lambda = -1 \pm \sqrt{3}$; saddle point, unstable

8. (a, b, c) $(0, 0)$; $u' = u$, $w' = -w$; $\lambda = 1, -1$; saddle point, unstable
$(1, 0)$; $u' = -u - 2w$, $w' = -2w$; $\lambda = -1, -2$; node, asymptotically stable
$(-1, 1)$; $u' = u + 2w$, $w' = -u$; $\lambda = (1 \pm \sqrt{3}\,i)/2$; spiral point, unstable

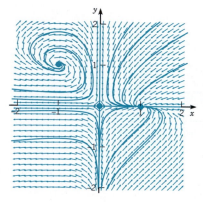

Direction field and phase portrait for Problem 8.

9. (a, b, c) $(0, 0)$; $u' = -u + w$, $w' = 2u$; $\lambda = -2, 1$; saddle point, unstable
$(0, 1)$; $u' = -u - w$, $w' = 3u$; $\lambda = (-1 \pm \sqrt{11}\,i)/2$; spiral point, asymptotically stable
$(-2, -2)$; $u' = -5u + 5w$, $w' = -2w$; $\lambda = -5, -2$; node, asymptotically stable
$(3, -2)$; $u' = 5u + 5w$, $w' = 3w$; $\lambda = 5, 3$; node, unstable

10. (a, b, c) $(0, 0)$; $u' = u$, $w' = w$; $\lambda = 1, 1$; node or spiral point, unstable
$(-1, 0)$; $u' = -u$, $w' = 2w$; $\lambda = -1, 2$; saddle point, unstable

11. (a, b, c) $(0, 0)$; $u' = 2u + w$, $w' = u - 2w$; $\lambda = \pm\sqrt{5}$; saddle point, unstable
$(-1.1935, -1.4797)$; $u' = -1.2399u - 6.8393w$, $w' = 2.4797u - 0.80655w$;
$\lambda = -1.0232 \pm 4.1125i$; spiral point, asymptotically stable

12. (a, b, c) $(0, \pm 2n\pi)$, $n = 0, 1, 2, \ldots$; $u' = w$, $w' = -u$; $\lambda = \pm i$; center or spiral point, indeterminate
$[2, \pm(2n - 1)\pi]$, $n = 1, 2, 3, \ldots$; $u' = -3w$, $w' = -u$; $\lambda = \pm\sqrt{3}$;
saddle point, unstable

13. (a, b, c) $(0, 0)$; $u' = u$, $w' = w$; $\lambda = 1, 1$; node or spiral point, unstable
$(1, 1)$; $u' = u - 2w$, $w' = -2u + w$; $\lambda = 3, -1$; saddle point, unstable

14. (a, b, c) $(1, 1)$; $u' = -u - w$, $w' = u - 3w$; $\lambda = -2, -2$; node or spiral point, asymptotically stable
$(-1, -1)$; $u' = u + w$, $w' = u - 3w$; $\lambda = -1 \pm \sqrt{5}$; saddle point, unstable

15. (a, b, c) $(0, 0)$; $\;u' = -2u - w$, $\;w' = u - w$; $\;\lambda = (-3 \pm \sqrt{3}\,i)/2$; spiral point, asymptotically stable $(-0.33076,\ 1.0924)$ and $(0.33076,\ -1.0924)$; $\;u' = -3.5216u - 0.27735w$, $w' = 0.27735u + 2.6895w$; $\;\lambda = -3.5092,\ 2.6771$; saddle point, unstable

16. (a, b, c) $(0, 0)$; $\;u' = u + w$, $\;w' = -u + w$; $\;\lambda = 1 \pm i$; spiral point, unstable

17. (a, b, c) $(0, 0)$; $\;u' = -u + 2w$, $\;w' = u + 2w$; $\;\lambda = 2, -1$; saddle point, unstable
$(2, 1)$; $\;u' = -\frac{3}{2}u + 2w$, $\;w' = -2u$; $\;\lambda = (-6 \pm \sqrt{55}\,i)/4$; spiral point, asymptotically stable
$(2, -2)$; $\;u' = -3w$, $\;w' = u$; $\;\lambda = \pm\sqrt{3}\,i$ center or spiral point, indeterminate
$(4, -2)$; $\;u' = -4w$, $\;w' = -u - 2w$; $\;\lambda = -1 \pm \sqrt{5}$; saddle point, unstable

18. (a, b, c) $(2, 2)$; $\;u' = -4w$, $\;w' = -\frac{7}{2}u + \frac{7}{2}w$; $\;\lambda = (7 \pm \sqrt{273})/4$; saddle point, unstable
$(-2, -2)$; $\;u' = 4w$, $\;w' = \frac{1}{2}u - \frac{1}{2}w$; $\;\lambda = (-1 \pm \sqrt{33})/4$; saddle point, unstable
$(-\frac{3}{2}, 2)$; $\;u' = -4w$, $\;w' = \frac{7}{2}u$; $\;\lambda = \pm\sqrt{14}\,i$; center or spiral point, indeterminate
$(-\frac{3}{2}, -2)$; $\;u' = 4w$, $\;w' = -\frac{1}{2}u$; $\;\lambda = \pm\sqrt{2}\,i$; center or spiral point, indeterminate

19. (a, b, c) $(0, 0)$; $\;u' = 2u - w$, $\;w' = 2u - 4w$; $\;\lambda = -1 \pm \sqrt{7}$; saddle point, unstable
$(2, 1)$; $\;u' = -3w$, $\;w' = 4u - 8w$; $\;\lambda = -2, -6$; node, asymptotically stable
$(-2, 1)$; $\;u' = 5w$, $\;w' = -4u$; $\;\lambda = \pm 2\sqrt{5}\,i$; center or spiral point, indeterminate
$(-2, -4)$; $\;u' = 10u - 5w$, $\;w' = 6u$; $\;\lambda = 5 \pm \sqrt{5}\,i$; spiral point, unstable

20. (a, b, c) $(0, 0)$; $\;u' = -4w$, $\;w' = 0$; $\;\lambda = 0, 0$; no conclusion, indeterminate
$(2, 3)$; $\;u' = 12u + 4w$, $\;w' = -12w$; $\;\lambda = 12, -12$; saddle point, unstable
$(-2, 3)$; $\;u' = -12u + 4w$, $\;w' = 12w$; $\;\lambda = 12, -12$; saddle point, unstable

21. (b) $x' = y$, $\;y' = x$; $\;dy/dx = x/y$; $\;y^2 - x^2 = c$
(c) $dy/dx = (x + 2x^3)/y$; $\;y^2 - x^2 - x^4 = c$

22. (b) $x' = x$, $\;y' = -2y$; $\;dy/dx = -2y/x$; $\;x^2 y = c$
(c) $dy/dx = (-2y + x^3)/x$; $\;x^2 \left[y - (x^3/5) \right] = c$

23. (b) $v_c \cong 5.41$

27. (a) $dx/dt = y$, $\;dy/dt = -g(x) - c(x)y$
(b) The linear system is $dx/dt = y$, $\;dy/dt = -g'(0)x - c(0)y$.
(c) The eigenvalues satisfy $\lambda^2 + c(0)\lambda + g'(0) = 0$.

Section 7.3, *page 502*

1. (b, c) $(0, 0)$; $\;u' = \frac{3}{2}u$, $\;w' = 2w$; $\;\lambda = \frac{3}{2}, 2$; node, unstable
$(0, 2)$; $\;u' = \frac{1}{2}u$, $\;w' = -\frac{3}{2}u - 2w$; $\;\lambda = \frac{1}{2}, -2$; saddle point, unstable
$(\frac{3}{2}, 0)$; $\;u' = -\frac{3}{2}u - \frac{3}{4}w$, $\;w' = \frac{7}{8}w$; $\;\lambda = -\frac{3}{2}, \frac{7}{8}$; saddle point, unstable
$(\frac{4}{5}, \frac{7}{5})$; $\;u' = -\frac{4}{5}u - \frac{2}{5}w$, $\;w' = -\frac{21}{20}u - \frac{7}{5}w$; $\;\lambda = (-22 \pm \sqrt{204})/20$; node, asymptotically stable

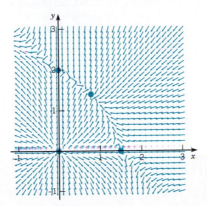

Direction field for Problem 1.

2. (b, c) $(0, 0)$; $u' = \frac{3}{2}u$, $w' = 2w$; $\lambda = \frac{3}{2}, 2$; node, unstable

$(0, 4)$; $u' = -\frac{1}{2}u$, $w' = -6u - 2w$; $\lambda = -\frac{1}{2}, -2$; node, asymptotically stable

$(\frac{3}{2}, 0)$; $u' = -\frac{3}{2}u - \frac{3}{4}w$, $w' = -\frac{1}{4}w$; $\lambda = -\frac{1}{4}, -\frac{3}{2}$; node, asymptotically stable

$(1, 1)$; $u' = -u - \frac{1}{2}w$, $w' = -\frac{3}{2}u - \frac{1}{2}w$; $\lambda = (-3 \pm \sqrt{13})/4$; saddle point, unstable

3. (b, c) $(0, 0)$; $u' = \frac{3}{2}u$, $w' = 2w$; $\lambda = \frac{3}{2}, 2$; node, unstable

$(0, 2)$; $u' = -\frac{1}{2}u$, $w' = -\frac{9}{4}u - 2w$; $\lambda = -\frac{1}{2}, -2$; node, asymptotically stable

$(3, 0)$; $u' = -\frac{3}{2}u - 3w$, $w' = -\frac{11}{8}w$; $\lambda = -\frac{3}{2}, -\frac{11}{8}$; node, asymptotically stable

$(\frac{4}{5}, \frac{11}{10})$; $u' = -\frac{2}{5}u - \frac{4}{5}w$, $w' = -\frac{99}{80}u - \frac{11}{10}w$; $\lambda = -1.80475, 0.30475$; saddle point, unstable

4. (b, c) $(0, 0)$; $u' = \frac{3}{2}u$, $w' = \frac{3}{4}w$; $\lambda = \frac{3}{2}, \frac{3}{4}$; node, unstable

$(0, \frac{3}{4})$; $u' = \frac{3}{4}u$, $w' = -\frac{3}{4}w$; $\lambda = \pm\frac{3}{4}$; saddle point, unstable

$(3, 0)$; $u' = -\frac{3}{2}u - 3w$, $w' = \frac{3}{8}w$; $\lambda = -\frac{3}{2}, \frac{3}{8}$; saddle point, unstable

$(2, \frac{1}{2})$; $u' = -u - 2w$, $w' = -\frac{1}{16}u - \frac{1}{2}w$; $\lambda = -1.18301, -0.31699$; node, asymptotically stable

5. (b, c) $(0, 0)$; $u' = u$, $w' = \frac{3}{2}w$; $\lambda = 1, \frac{3}{2}$; node, unstable

$(0, \frac{3}{2})$; $u' = -\frac{1}{2}u$, $w' = -\frac{3}{2}u - \frac{3}{2}w$; $\lambda = -\frac{1}{2}, -\frac{3}{2}$; node, asymptotically stable

$(1, 0)$; $u' = -u - w$, $w' = \frac{1}{2}w$; $\lambda = -1, \frac{1}{2}$; saddle point, unstable

6. (b, c) $(0, 0)$; $u' = u$, $w' = \frac{5}{2}w$; $\lambda = 1, \frac{5}{2}$; node, unstable

$(0, \frac{5}{3})$; $u' = \frac{11}{6}u$, $w' = \frac{5}{12}u - \frac{5}{2}w$; $\lambda = \frac{11}{6}, -\frac{5}{2}$; saddle point, unstable

$(1, 0)$; $u' = -u + \frac{1}{2}w$, $w' = \frac{11}{4}w$; $\lambda = -1, \frac{11}{4}$; saddle point, unstable

$(2, 2)$; $u' = -2u + w$, $w' = \frac{1}{4}u - 3w$; $\lambda = (-5 \pm \sqrt{3})/2$; node, asymptotically stable

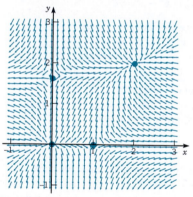

Direction field for Problem 6.

8. (a) $\sigma_1\epsilon_2 - \alpha_2\epsilon_1 \neq 0$: $(0, 0), (0, \epsilon_2/\sigma_2), (\epsilon_1/\sigma_1, 0)$

$\sigma_1\epsilon_2 - \alpha_2\epsilon_1 = 0$: $(0, 0)$, and all points on the line $\sigma_1 x + \alpha_1 y = \epsilon_1$

(b) $\sigma_1\epsilon_2 - \alpha_2\epsilon_1 > 0$: $(0, 0)$ is unstable node; $(\epsilon_1/\sigma_1, 0)$ is saddle point; $(0, \epsilon_2/\sigma_2)$ is asymptotically stable node.

$\sigma_1\epsilon_2 - \alpha_2\epsilon_1 < 0$: $(0, 0)$ is unstable node; $(0, \epsilon_2/\sigma_2)$ is saddle point; $(\epsilon_1/\sigma_1, 0)$ is asymptotically stable node.

(c) $(0, 0)$ is unstable node; points on the line $\sigma_1 x + \alpha_1 y = \epsilon_1$ are stable, nonisolated critical points.

10. (a) $(0,0)$, saddle point; $(0.15, 0)$, spiral point if $\gamma^2 < 1.11$, node if $\gamma^2 > 1.11$; $(2, 0)$, saddle point (c) $\gamma \cong 1.20$

11. (b) $(2 - \sqrt{4 - \frac{3}{2}\alpha}, \frac{3}{2}\alpha)$, $(2 + \sqrt{4 - \frac{3}{2}\alpha}, \frac{3}{2}\alpha)$

(c) $(1, 3)$ is an asymptotically stable node; $(3, 3)$ is a saddle point

(d) $\alpha_0 = 8/3$; critical point is $(2, 4)$; $\lambda = 0, -1$

12. (b) $(2 - \sqrt{4 - \frac{3}{2}\alpha}, \frac{3}{2}\alpha)$, $(2 + \sqrt{4 - \frac{3}{2}\alpha}, \frac{3}{2}\alpha)$

(c) 1, 3) is a saddle point; (3, 3) is an unstable spiral point
(d) $\alpha_0 = 8/3$; critical point is $(2, 4)$; $\lambda = 0, 1$

13. (b) $([3 - \sqrt{9 - 4\alpha}]/2, \ [3 + 2\alpha - \sqrt{9 - 4\alpha}]/2)$,
    $([3 + \sqrt{9 - 4\alpha}]/2, \ [3 + 2\alpha + \sqrt{9 - 4\alpha}]/2)$
    (c) $(1, 3)$ is a saddle point; $(2, 4)$ is an unstable spiral point
    (d) $\alpha_0 = 9/4$; critical point is $(3/2, 15/4)$; $\lambda = 0, 0$

14. (b) $([3 - \sqrt{9 - 4\alpha}]/2, \ [3 + 2\alpha - \sqrt{9 - 4\alpha}]/2)$,
    $([3 + \sqrt{9 - 4\alpha}]/2, \ [3 + 2\alpha + \sqrt{9 - 4\alpha}]/2)$
    (c) $(1, 3)$ is a center of the linear approximation and also of the nonlinear system; $(2, 4)$ is a
    saddle point
    (d) $\alpha_0 = 9/4$; critical point is $(3/2, 15/4)$; $\lambda = 0, 0$

15. (a) $P_1(2, 1)$, $P_2(-2, 1)$, $P_3(-2, 1/\alpha)$, $P_4[(\alpha - 1)/(\alpha + 1), -2/(\alpha + 1)]$
    (b) P_2 and P_3 coincide when $\alpha_0 = 1$.
    (c) For $\alpha < 1$, P_2 is a saddle point and P_3 is an unstable node; for $\alpha > 1$, P_2 is an unstable node
    and P_3 is a saddle point.

16. (a) $P_1(2, 0)$, $P_2[2, (4 - \alpha)/3]$, $P_3[(\alpha - 9)/2, -3]$
    (b) P_2 coincides with P_1 for $\alpha_0 = 4$; P_2 coincides with P_3 for $\alpha_0 = 13$.
    (c) Near $\alpha = 4$: P_1 is a saddle point for $\alpha < 4$ and an unstable spiral point for $\alpha > 4$; P_2 is an
    asymptotically stable spiral point for $\alpha < 4$ and a saddle point for $\alpha > 4$.
    Near $\alpha = 13$: P_2 is a saddle point for $\alpha < 13$ and an unstable node for $\alpha > 13$;
    P_3 is an unstable node for $\alpha < 13$ and a saddle point for $\alpha > 13$.

17. (a) $(0, 0)$, $(0, 2 + 2\alpha)$, $(4, 0)$, $(2, 2)$
    (b) $\alpha = 0.75$, asymptotically stable node; $\alpha = 1.25$, (unstable) saddle point
    (c) $u' = -2u - 2w$, $w' = -2\alpha u - 2w$
    (d) $\lambda = -2 \pm 2\sqrt{\alpha}$; $\alpha_0 = 1$

Section 7.4, *page 510*

1. (b, c) $(0, 0)$; $u' = \frac{3}{2}u$, $w' = -\frac{1}{2}w$; $\lambda = \frac{3}{2}, -\frac{1}{2}$; saddle point, unstable
   $(\frac{1}{2}, 3)$; $u' = -\frac{1}{4}w$, $w' = 3u$; $\lambda = \pm\sqrt{3}\,i/2$; center or spiral point, indeterminate

2. (b, c) $(0, 0)$; $u' = u$, $w' = -\frac{1}{4}w$; $\lambda = 1, -\frac{1}{4}$; saddle point, unstable
   $(\frac{1}{2}, 2)$; $u' = -\frac{1}{4}w$, $w' = u$; $\lambda = \pm\frac{1}{2}i$; center or spiral point, indeterminate

3. (b, c) $(0, 0)$; $u' = u$, $w' = -\frac{1}{4}w$; $\lambda = 1, -\frac{1}{4}$; saddle point, unstable
   $(2, 0)$; $u' = -u - w$, $w' = \frac{3}{4}w$; $\lambda = -1, \frac{3}{4}$; saddle point, unstable
   $(\frac{1}{2}, \frac{3}{2})$; $u' = -\frac{1}{4}u - \frac{1}{4}w$, $w' = \frac{3}{4}u$; $\lambda = (-1 \pm \sqrt{11}\,i)/8$; spiral point, asymptotically
   stable

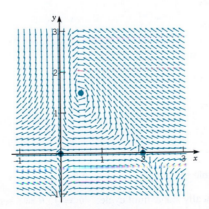

Direction field for Problem 3.

4. (b, c) $(0, 0)$; $u' = \frac{9}{8}u$, $w' = -w$; $\lambda = \frac{9}{8}, -1$; saddle point, unstable
   $(\frac{9}{8}, 0)$; $u' = -\frac{9}{8}u - \frac{9}{16}w$, $w' = \frac{1}{8}w$; $\lambda = -\frac{9}{8}, \frac{1}{8}$; saddle point, unstable
   $(1, \frac{1}{4})$; $u' = -u - \frac{1}{2}w$, $w' = \frac{1}{4}u$; $\lambda = (-1 \pm \sqrt{0.5})/2$; node, asymptotically stable

5. (b, c) $(0, 0)$; $u' = -u$, $w' = -\frac{3}{2}w$; $\lambda = -1, -\frac{3}{2}$; node, asymptotically stable
   $(\frac{1}{2}, 0)$; $u' = \frac{3}{4}u - \frac{3}{20}w$, $w' = -w$; $\lambda = -1, \frac{3}{4}$; saddle point, unstable
   $(2, 0)$; $u' = -3u - \frac{3}{5}w$, $w' = \frac{1}{2}w$; $\lambda = -3, \frac{1}{2}$; saddle point, unstable
   $(\frac{3}{2}, \frac{5}{3})$; $u' = -\frac{3}{4}u - \frac{9}{20}w$, $w' = \frac{5}{3}u$; $\lambda = (-3 \pm \sqrt{39}\,i)/8$; spiral point, asymptotically stable

6.
   $t = 0, T, 2T, \ldots :$ H is a max., dP/dt is a max.
   $t = T/4, 5T/4, \ldots :$ dH/dt is a min., P is a max.
   $t = T/2, 3T/2, \ldots :$ H is a min., dP/dt is a min.
   $t = 3T/4, 7T/4, \ldots :$ dH/dt is a max., P is a min.

7. (a) $\sqrt{c}\,\alpha/\sqrt{a}\,\gamma$ (b) $\sqrt{3}$
   (d) The ratio of prey amplitude to predator amplitude increases very slowly as the initial point moves away from the equilibrium point.

8. (a) $4\pi/\sqrt{3} \cong 7.2552$
   (c) The period increases slowly as the initial point moves away from the equilibrium point.

9. (a) $T \cong 6.5$ (b) $T \cong 3.7, T \cong 11.5$ (c) $T \cong 3.8, T \cong 11.1$

11. (a) $P_1(0, 0)$, $P_2(1/\sigma, 0)$, $P_3(3, 2 - 6\sigma)$; P_2 moves to the left and P_3 moves down; they coincide at $(3, 0)$ when $\sigma = 1/3$.
    (b) P_1 is a saddle point. P_2 is a saddle point for $\sigma < 1/3$ and an asymptotically stable node for $\sigma > 1/3$. P_3 is an asymptotically stable spiral point for $\sigma < \sigma_1 = (\sqrt{7/3} - 1)/2 \cong 0.2638$, an asymptotically stable node for $\sigma_1 < \sigma < 1/3$, and a saddle point for $\sigma > 1/3$.

12. (a) $P_1(0, 0)$, $P_2(a/\sigma, 0)$, $P_3[c/\gamma, (a/\alpha) - (c\sigma/\alpha\gamma)]$; P_2 moves to the left and P_3 moves down; they coincide at $(c/\gamma, 0)$ when $\sigma = a\gamma/c$.
    (b) P_1 is a saddle point. P_2 is a saddle point for $\sigma < a\gamma/c$ and an asymptotically stable node for $\sigma > a\gamma/c$. P_3 is an asymptotically stable spiral point for sufficiently small values of σ and becomes an asymptotically stable node at a certain value $\sigma_1 < a\gamma/c$. P_3 is a saddle point for $\sigma > a\gamma/c$.

13. (a, b) $P_1(0, 0)$ is a saddle point; $P_2(5, 0)$ is a saddle point; $P_3(2, 2.4)$ is an asymptotically stable spiral point.

14. (b) same prey, fewer predators
    (c) more prey, same predators
    (d) more prey, fewer predators

15. (a) same prey, fewer predators
    (b) more prey, fewer predators
    (c) more prey, even fewer predators

16. (b) same prey, fewer predators
    (c) more prey, same predators
    (d) more prey, fewer predators

Section 7.5, *page 521*

1. $r = 1$, $\theta = t + t_0$, asymptotically stable limit cycle

2. $r = 1$, $\theta = -t + t_0$, semistable limit cycle

3. $r = 1$, $\theta = t + t_0$, asymptotically stable limit cycle; $r = 3$, $\theta = t + t_0$, unstable periodic solution

4. $r = 1$, $\theta = -t + t_0$, unstable periodic solution; $r = 2$, $\theta = -t + t_0$, asymptotically stable limit cycle

5. $r = 2n - 1$, $\theta = t + t_0$, $n = 1, 2, 3, \ldots$, asymptotically stable limit cycle;
   $r = 2n$, $\theta = t + t_0$, $n = 1, 2, 3, \ldots$, unstable periodic solution

6. $r = 2$, $\theta = -t + t_0$, semistable limit cycle;
   $r = 3$, $\theta = -t + t_0$, unstable periodic solution

8. (a) Counterclockwise
   (b) $r = 1$, $\theta = t + t_0$, asymptotically stable limit cycle; $r = 2$, $\theta = t + t_0$, semistable limit cycle; $r = 3$, $\theta = t + t_0$, unstable periodic solution

9. $r = \sqrt{2}$, $\theta = -t + t_0$, unstable periodic solution

14. (a) $\mu = 0.2$, $T \cong 6.29$; $\mu = 1$, $T \cong 6.66$; $\mu = 5$, $T \cong 11.60$

15. (a) $x' = y$, $y' = -x + \mu y - \mu y^3/3$
    (b) $0 < \mu < 2$, unstable spiral point; $\mu \geq 2$, unstable node
    (c) $A \cong 2.16$, $T \cong 6.65$
    (d) $\mu = 0.2$, $A \cong 1.99$, $T \cong 6.31$; $\mu = 0.5$, $A \cong 2.03$, $T \cong 6.39$;
    $\mu = 2$, $A \cong 2.60$, $T \cong 7.65$; $\mu = 5$, $A \cong 4.36$, $T \cong 11.60$

16. (b) $x' = \mu x + y$, $y' = -x + \mu y$; $\lambda = \mu \pm i$; the origin is an asymptotically stable spiral point for $\mu < 0$ and an unstable spiral point for $\mu > 0$.
    (c) $r' = r(\mu - r^2)$, $\theta' = -1$

17. (a) The origin ia an asymptotically stable node for $\mu < -2$, an asymptotically stable spiral point for $-2 < \mu < 0$, an unstable spiral point for $0 < \mu < 2$, and an unstable node for $\mu > 2$.

18. (a, b) $(0, 0)$ is a saddle point; $(12, 0)$ is a saddle point; $(2, 8)$ is an unstable spiral point.

19. (a) $(0, 0)$, $(5a, 0)$, $(2, 4a - 1.6)$
    (b) $\lambda = -0.25 + 0.125a \pm 0.25\sqrt{220 - 400a + 25a^2}$; $a_0 = 2$

20. (b) $\lambda = \left[-(5/4 - b) \pm \sqrt{(5/4 - b)^2 - 1} \right]/2$
    (c) $0 < b < 1/4$: asymptotically stable node; $1/4 < b < 5/4$: asymptotically stable spiral point; $5/4 < b < 9/4$: unstable spiral point; $9/4 < b$; unstable node
    (d) $b_0 = 5/4$

21. (b) $k = 0$, $(1.1994, -0.62426)$; $k = 0.5$, $(0.80485, -0.13106)$
    (c) $k_0 \cong 0.3465$, $(0.95450, -0.31813)$
    (d) $k = 0.4$, $T \cong 11.23$; $k = 0.5$, $T \cong 10.37$; $k = 0.6$, $T \cong 9.93$
    (e) $k_1 \cong 1.4035$

Section 7.6, *page 530*

1. (b) $\lambda = \lambda_1$, $\mathbf{v}_1 = (0, 0, 1)^T$; $\lambda = \lambda_2$, $\mathbf{v}_2 = (20, 9 - \sqrt{81 + 40r}, 0)^T$;
   $\lambda = \lambda_3$, $\mathbf{v}_3 = (20, 9 + \sqrt{81 + 40r}, 0)^T$

   (c) $\lambda_1 \cong -2.6667$, $\mathbf{v}_1 = (0, 0, 1)^T$; $\lambda_2 \cong -22.8277$, $\mathbf{v}_2 \cong (20, -25.6554, 0)^T$;
   $\lambda_3 \cong 11.8277$, $\mathbf{v}_3 \cong (20, 43.6554, 0)^T$

2. (c) $\lambda_1 \cong -13.8546$; $\lambda_2, \lambda_3 \cong 0.0939556 \pm 10.1945i$

4. (a) $dV/dt = -2\sigma[rx^2 + y^2 + b(z - r)^2 - br^2]$

CHAPTER 8   Section 8.1, *page 546*

1. $\rho = 1$ 2. $\rho = 2$ 3. $\rho = \infty$

4. $\rho = \dfrac{1}{2}$ 5. $\rho = \dfrac{1}{2}$ 6. $\rho = 1$

7. $\rho = 3$

8. $\rho = e$

9. $\sum_{n=0}^{\infty} \frac{(-1)^n x^{2n+1}}{(2n+1)!}, \quad \rho = \infty$

10. $\sum_{n=0}^{\infty} \frac{x^n}{n!}, \quad \rho = \infty$

11. $1 + (x - 1), \quad \rho = \infty$

12. $1 - 2(x+1) + (x+1)^2, \quad \rho = \infty$

13. $\sum_{n=1}^{\infty} (-1)^{n+1} \frac{(x-1)^n}{n}, \quad \rho = 1$

14. $\sum_{n=0}^{\infty} (-1)^n x^n, \quad \rho = 1$

15. $\sum_{n=0}^{\infty} \frac{(-1)^n}{3^{2n+2}} x^{4n+1}, \quad \rho = \sqrt{3}$

16. $\sum_{n=0}^{\infty} (-1)^{n+1} (x-2)^n, \quad \rho = 1$

17. $y' = 1 + 2^2 x + 3^2 x^2 + 4^2 x^3 + \cdots + (n+1)^2 x^n + \cdots$

   $y'' = 2^2 + 3^2 \cdot 2x + 4^2 \cdot 3x^2 + 5^2 \cdot 4x^3 + \cdots + (n+2)^2 (n+1) x^n + \cdots$

18. $y' = a_1 + 2a_2 x + 3a_3 x^2 + 4a_4 x^3 + \cdots + (n+1)a_{n+1} x^n + \cdots$

   $= \sum_{n=1}^{\infty} n a_n x^{n-1} = \sum_{n=0}^{\infty} (n+1) a_{n+1} x^n$

   $y'' = 2a_2 + 6a_3 x + 12a_4 x^2 + 20a_5 x^3 + \cdots + (n+2)(n+1) a_{n+2} x^n + \cdots$

   $= \sum_{n=2}^{\infty} n(n-1) a_n x^{n-2} = \sum_{n=0}^{\infty} (n+2)(n+1) a_{n+2} x^n$

21. $\sum_{n=0}^{\infty} (n+2)(n+1) a_{n+2} x^n$

22. $\sum_{n=2}^{\infty} a_{n-2} x^n$

23. $\sum_{n=0}^{\infty} (n+1) a_n x^n$

24. $\sum_{n=0}^{\infty} [(n+2)(n+1) a_{n+2} - n(n-1) a_n] x^n$

25. $\sum_{n=0}^{\infty} [(n+2)(n+1) a_{n+2} + n a_n] x^n$

26. $a_1 + \sum_{n=1}^{\infty} [(n+1) a_{n+1} + a_{n-1}] x^n$

27. $\sum_{n=0}^{\infty} [(n+1) n a_{n+1} + a_n] x^n$

28. $a_n = (-2)^n a_0 / n!, \quad n = 1, 2, \ldots; \quad a_0 e^{-2x}$

Section 8.2, *page 557*

1. (a) $a_{n+2} = a_n / (n+2)(n+1)$

   (b,d) $y_1(x) = 1 + \frac{x^2}{2!} + \frac{x^4}{4!} + \frac{x^6}{6!} + \cdots = \sum_{n=0}^{\infty} \frac{x^{2n}}{(2n)!} = \cosh x$

   $y_2(x) = x + \frac{x^3}{3!} + \frac{x^5}{5!} + \frac{x^7}{7!} + \cdots = \sum_{n=0}^{\infty} \frac{x^{2n+1}}{(2n+1)!} = \sinh x$

2. (a) $a_{n+2} = a_n / (n+2)$

   (b,d) $y_1(x) = 1 + \frac{x^2}{2} + \frac{x^4}{2 \cdot 4} + \frac{x^6}{2 \cdot 4 \cdot 6} + \cdots = \sum_{n=0}^{\infty} \frac{x^{2n}}{2^n n!}$

   $y_2(x) = x + \frac{x^3}{3} + \frac{x^5}{3 \cdot 5} + \frac{x^7}{3 \cdot 5 \cdot 7} + \cdots = \sum_{n=0}^{\infty} \frac{2^n n! x^{2n+1}}{(2n+1)!}$

3. (a) $(n+2) a_{n+2} - a_{n+1} - a_n = 0$

(b) $y_1(x) = 1 + \dfrac{1}{2}(x-1)^2 + \dfrac{1}{6}(x-1)^3 + \dfrac{1}{6}(x-1)^4 + \cdots$

$y_2(x) = (x-1) + \dfrac{1}{2}(x-1)^2 + \dfrac{1}{2}(x-1)^3 + \dfrac{1}{4}(x-1)^4 + \cdots$

4. (a) $a_{n+4} = -k^2 a_n/(n+4)(n+3); \quad a_2 = a_3 = 0$

(b,d) $y_1(x) = 1 - \dfrac{k^2 x^4}{3 \cdot 4} + \dfrac{k^4 x^8}{3 \cdot 4 \cdot 7 \cdot 8} - \dfrac{k^6 x^{12}}{3 \cdot 4 \cdot 7 \cdot 8 \cdot 11 \cdot 12} + \cdots$

$= 1 + \displaystyle\sum_{m=0}^{\infty} \dfrac{(-1)^{m+1}(k^2 x^4)^{m+1}}{3 \cdot 4 \cdot 7 \cdot 8 \cdots (4m+3)(4m+4)}$

$y_2(x) = x - \dfrac{k^2 x^5}{4 \cdot 5} + \dfrac{k^4 x^9}{4 \cdot 5 \cdot 8 \cdot 9} - \dfrac{k^6 x^{13}}{4 \cdot 5 \cdot 8 \cdot 9 \cdot 12 \cdot 13} + \cdots$

$= x \left[1 + \displaystyle\sum_{m=0}^{\infty} \dfrac{(-1)^{m+1}(k^2 x^4)^{m+1}}{4 \cdot 5 \cdot 8 \cdot 9 \cdots (4m+4)(4m+5)} \right]$

Hint: Let $n = 4m$ in the recurrence relation, $m = 1, 2, 3, \ldots$.

5. (a) $(n+2)(n+1)a_{n+2} - n(n+1)a_{n+1} + a_n = 0, \quad n \geq 1; \quad a_2 = -\dfrac{1}{2}a_0$

(b) $y_1(x) = 1 - \dfrac{1}{2}x^2 - \dfrac{1}{6}x^3 - \dfrac{1}{24}x^4 + \cdots, \quad y_2(x) = x - \dfrac{1}{6}x^3 - \dfrac{1}{12}x^4 - \dfrac{1}{24}x^5 + \cdots$

6. (a) $a_{n+2} = -(n^2 - 2n + 4)a_n/[2(n+1)(n+2)], \quad n \geq 2; \quad a_2 = -a_0, \quad a_3 = -\dfrac{1}{4}a_1$

(b) $y_1(x) = 1 - x^2 + \dfrac{1}{6}x^4 - \dfrac{1}{30}x^6 + \cdots,$

$y_2(x) = x - \dfrac{1}{4}x^3 + \dfrac{7}{160}x^5 - \dfrac{19}{1920}x^7 + \cdots$

7. (a) $a_{n+2} = -a_n/(n+1), \quad n = 0, 1, 2, \ldots$

(b,d) $y_1(x) = 1 - \dfrac{x^2}{1} + \dfrac{x^4}{1 \cdot 3} - \dfrac{x^6}{1 \cdot 3 \cdot 5} + \cdots = 1 + \displaystyle\sum_{n=1}^{\infty} \dfrac{(-1)^n x^{2n}}{1 \cdot 3 \cdot 5 \cdots (2n-1)}$

$y_2(x) = x - \dfrac{x^3}{2} + \dfrac{x^5}{2 \cdot 4} - \dfrac{x^7}{2 \cdot 4 \cdot 6} + \cdots = x + \displaystyle\sum_{n=1}^{\infty} \dfrac{(-1)^n x^{2n+1}}{2 \cdot 4 \cdot 6 \cdots (2n)}$

8. (a) $a_{n+2} = -[(n+1)^2 a_{n+1} + a_n + a_{n-1}]/(n+1)(n+2), \quad n = 1, 2, \ldots$

$a_2 = -(a_0 + a_1)/2$

(b) $y_1(x) = 1 - \dfrac{1}{2}(x-1)^2 + \dfrac{1}{6}(x-1)^3 - \dfrac{1}{12}(x-1)^4 + \cdots$

$y_2(x) = (x-1) - \dfrac{1}{2}(x-1)^2 + \dfrac{1}{6}(x-1)^3 - \dfrac{1}{6}(x-1)^4 + \cdots$

9. (a) $(n+2)(n+1)a_{n+2} + (n-2)(n-3)a_n = 0; \quad n = 0, 1, 2, \ldots$

(b) $y_1(x) = 1 - 3x^2, \quad y_2(x) = x - x^3/3$

10. (a) $4(n+2)a_{n+2} - (n-2)a_n = 0; \quad n = 0, 1, 2, \ldots$

(b,d) $y_1(x) = 1 - \dfrac{x^2}{4}, \quad y_2(x) = x - \dfrac{x^3}{12} - \dfrac{x^5}{240} - \dfrac{x^7}{2240} - \cdots - \dfrac{x^{2n+1}}{4^n(2n-1)(2n+1)} - \cdots$

11. (a) $3(n+2)a_{n+2} - (n+1)a_n = 0; \quad n = 0, 1, 2, \ldots$

(b,d) $y_1(x) = 1 + \dfrac{x^2}{6} + \dfrac{x^4}{24} + \dfrac{5}{432}x^6 + \cdots + \dfrac{3 \cdot 5 \cdots (2n-1)}{3^n \cdot 2 \cdot 4 \cdots (2n)}x^{2n} + \cdots$

$y_2(x) = x + \dfrac{2}{9}x^3 + \dfrac{8}{135}x^5 + \dfrac{16}{945}x^7 + \cdots + \dfrac{2 \cdot 4 \cdots (2n)}{3^n \cdot 3 \cdot 5 \cdots (2n+1)}x^{2n+1} + \cdots$

12. (a) $(n+2)(n+1)a_{n+2} - (n+1)n a_{n+1} + (n-1)a_n = 0; \quad n = 0, 1, 2, \ldots$

(b,d) $y_1(x) = 1 + \dfrac{x^2}{2} + \dfrac{x^3}{6} + \dfrac{x^4}{24} + \cdots + \dfrac{x^n}{n!} + \cdots, \quad y_2(x) = x$

13. (a) $2(n+2)(n+1)a_{n+2} + (n+3)a_n = 0; \ n = 0, 1, 2, \ldots$

(b,d) $y_1(x) = 1 - \dfrac{3}{4}x^2 + \dfrac{5}{32}x^4 - \dfrac{7}{384}x^6 + \cdots + (-1)^n \dfrac{3 \cdot 5 \cdots (2n+1)}{2^n(2n)!}x^{2n} + \cdots$

$y_2(x) = x - \dfrac{x^3}{3} + \dfrac{x^5}{20} - \dfrac{x^7}{210} + \cdots + (-1)^n \dfrac{4 \cdot 6 \cdots (2n+2)}{2^n(2n+1)!}x^{2n+1} + \cdots$

14. (a) $2(n+2)(n+1)a_{n+2} + 3(n+1)a_{n+1} + (n+3)a_n = 0; \ n = 0, 1, 2, \ldots$

(b) $y_1(x) = 1 - \dfrac{3}{4}(x-2)^2 + \dfrac{3}{8}(x-2)^3 + \dfrac{1}{64}(x-2)^4 + \cdots$

$y_2(x) = (x-2) - \dfrac{3}{4}(x-2)^2 + \dfrac{1}{24}(x-2)^3 + \dfrac{9}{64}(x-2)^4 + \cdots$

15. (a) $y = 2 + x + x^2 + \dfrac{1}{3}x^3 + \dfrac{1}{4}x^4 + \cdots$ (c) about $|x| < 0.7$

16. (a) $y = -1 + 3x + x^2 - \dfrac{3}{4}x^3 - \dfrac{1}{6}x^4 + \cdots$ (c) about $|x| < 0.7$

17. (a) $y = 4 - x - 4x^2 + \dfrac{1}{2}x^3 + \dfrac{4}{3}x^4 + \cdots$ (c) about $|x| < 0.5$

18. (a) $y = -3 + 2x - \dfrac{3}{2}x^2 - \dfrac{1}{2}x^3 - \dfrac{1}{8}x^4 + \cdots$ (c) about $|x| < 0.9$

19. (a) $y_1(x) = 1 - \dfrac{1}{3}(x-1)^3 - \dfrac{1}{12}(x-1)^4 + \dfrac{1}{18}(x-1)^6 + \cdots$

$y_2(x) = (x-1) - \dfrac{1}{4}(x-1)^4 - \dfrac{1}{20}(x-1)^5 + \dfrac{1}{28}(x-1)^7 + \cdots$

21. (a) $y_1(x) = 1 - \dfrac{\lambda}{2!}x^2 + \dfrac{\lambda(\lambda-4)}{4!}x^4 - \dfrac{\lambda(\lambda-4)(\lambda-8)}{6!}x^6 + \cdots$

$y_2(x) = x - \dfrac{\lambda-2}{3!}x^3 + \dfrac{(\lambda-2)(\lambda-6)}{5!}x^5 - \dfrac{(\lambda-2)(\lambda-6)(\lambda-10)}{7!}x^7 + \cdots$

(b) $1, \ x, \ 1 - 2x^2, \ x - \dfrac{2}{3}x^3, \ 1 - 4x^2 + \dfrac{4}{3}x^4, \ x - \dfrac{4}{3}x^3 + \dfrac{4}{15}x^5$

(c) $1, \ 2x, \ 4x^2 - 2, \ 8x^3 - 12x, \ 16x^4 - 48x^2 + 12, \ 32x^5 - 160x^3 + 120x$

22. (b) $y = x - x^3/6 + \cdots$

29. (a) Let $\hat{x} = |x| - \lfloor|x|/(2\pi)\rfloor \cdot 2\pi$ where $\lfloor|x|/(2\pi)\rfloor =$ the greatest integer less than equal to $|x|/(2\pi)$, so $0 \le \hat{x} < 2\pi$. Let $f(\hat{x})$ be the extension of $\sin \hat{x}$ from the interval $[0, \pi/2]$ to $[0, 2\pi)$ using the sine function in such a way that its argument always lies in the interval $[0, \pi/2]$,

$$f(\hat{x}) = \begin{cases} \sin \hat{x}, & 0 \le \hat{x} \le \pi/2, \\ \sin(\pi - \hat{x}), & \pi/2 < \hat{x} \le \pi, \\ -\sin(\hat{x} - \pi), & \pi < \hat{x} \le 3\pi/2, \\ -\sin(2\pi - \hat{x}), & 3\pi/2 < \hat{x} < 2\pi. \end{cases}$$

Then

$$\sin x = \begin{cases} -f(\hat{x}), & x < 0, \\ f(\hat{x}), & x \ge 0. \end{cases}$$

Use $\cos x = \sin(x + \pi/2)$ to get values for the cosine function from the sine function. Finally, $\tan x$, $\cot x$, $\sec x$, and $\csc x$ can be obtained from the sine and cosine functions.

(b) $|S(x) - S_n(x)| \le |x|^{(2n+3)}/(2n+3)!$. With $n = 7$, $\max_{0 \le x \le \pi/2}|x|^{17}/17! \le (\pi/2)^{17}/17! = 6.0669 \times 10^{-12}$.

(c) If $h(n+1) = (2n+3)\ln 100 - \displaystyle\sum_{k=1}^{2n+3}\ln k - 11\ln 10$, then $h(145) = 1.74$ while $h(146) = -0.4066$.

Section 8.3, *page 562*

1. $\phi''(0) = -1$, $\quad \phi'''(0) = 0$, $\qquad \phi^{(4)}(0) = 3$

2. $\phi''(0) = 0$, $\quad \phi'''(0) = -2$, $\qquad \phi^{(4)}(0) = 0$

3. $\phi''(1) = 0$, $\quad \phi'''(1) = -6$, $\qquad \phi^{(4)}(1) = 42$

4. $\phi''(0) = 0$, $\quad \phi'''(0) = -a_0$, $\qquad \phi^{(4)}(0) = -4a_1$

5. $\rho = \infty$, $\rho = \infty$ $\qquad\qquad\qquad$ 6. $\rho = 1$, $\rho = 3$, $\rho = 1$

7. $\rho = 1$, $\rho = \sqrt{3}$ $\qquad\qquad\qquad$ 8. $\rho = 1$

9. (a) $\rho = \infty$ $\quad$ (b) $\rho = \infty$ $\quad$ (c) $\rho = \infty$ $\quad$ (d) $\rho = \infty$ $\quad$ (e) $\rho = 1$
   (f) $\rho = \sqrt{2}$ $\quad$ (g) $\rho = \infty$ $\quad$ (h) $\rho = 1$ $\quad$ (i) $\rho = 1$ $\qquad$ (j) $\rho = 2$
   (k) $\rho = \sqrt{3}$ $\quad$ (l) $\rho = 1$ $\quad\;\,$ (m) $\rho = \infty$ $\quad$ (n) $\rho = \infty$

10. (a) $y_1(x) = 1 - \dfrac{\alpha^2}{2!}x^2 - \dfrac{(2^2 - \alpha^2)\alpha^2}{4!}x^4 - \dfrac{(4^2 - \alpha^2)(2^2 - \alpha^2)\alpha^2}{6!}x^6 - \cdots$

$$- \frac{[(2m-2)^2 - \alpha^2] \cdots (2^2 - \alpha^2)\alpha^2}{(2m)!}x^{2m} - \cdots$$

$$y_2(x) = x + \frac{1 - \alpha^2}{3!}x^3 + \frac{(3^2 - \alpha^2)(1 - \alpha^2)}{5!}x^5 + \cdots$$

$$+ \frac{[(2m-1)^2 - \alpha^2] \cdots (1 - \alpha^2)}{(2m+1)!}x^{2m+1} + \cdots$$

(b) $y_1(x)$ or $y_2(x)$ terminates with x^n as $\alpha = n$ is even or odd

(c) $n = 0, y = 1$; $\quad n = 1, y = x$; $\quad n = 2, y = 1 - 2x^2$; $\quad n = 3, y = x - \dfrac{4}{3}x^3$

11. $y_1(x) = 1 - \dfrac{1}{6}x^3 + \dfrac{1}{120}x^5 + \dfrac{1}{180}x^6 + \cdots$, $\quad y_2(x) = x - \dfrac{1}{12}x^4 + \dfrac{1}{180}x^6 + \dfrac{1}{504}x^7 + \cdots$,
    $\rho = \infty$

12. $y_1(x) = 1 - \dfrac{1}{6}x^3 + \dfrac{1}{12}x^4 - \dfrac{1}{40}x^5 + \cdots$, $\quad y_2(x) = x - \dfrac{1}{12}x^4 + \dfrac{1}{20}x^5 - \dfrac{1}{60}x^6 + \cdots$, $\quad \rho = \infty$

13. $y_1(x) = 1 + x^2 + \dfrac{1}{12}x^4 + \dfrac{1}{120}x^6 + \cdots$, $\quad y_2(x) = x + \dfrac{1}{6}x^3 + \dfrac{1}{60}x^5 + \dfrac{1}{560}x^7 + \cdots, \rho = \pi/2$

14. $y_1(x) = 1 + \dfrac{1}{6}x^3 + \dfrac{1}{12}x^4 - \dfrac{1}{120}x^6 + \cdots$, $\quad y_2(x) = x - \dfrac{1}{6}x^3 + \dfrac{1}{24}x^4 + \dfrac{7}{120}x^5 + \cdots, \rho = 1$

15. Cannot specify arbitrary initial conditions at $x = 0$; hence $x = 0$ is a singular point.

16. $y = 1 + x + \dfrac{x^2}{2!} + \cdots + \dfrac{x^n}{n!} + \cdots = e^x$

17. $y = 1 + \dfrac{x^2}{2} + \dfrac{x^4}{2 \cdot 4} + \dfrac{x^6}{2 \cdot 4 \cdot 6} + \cdots + \dfrac{x^{2n}}{2^n \cdot n!} + \cdots$

18. $y = 1 + x + \dfrac{1}{2}x^2 + \dfrac{1}{2}x^3 + \cdots$

19. $y = 1 + x + x^2 + \cdots + x^n + \cdots = \dfrac{1}{1 - x}$

20. $y = a_0 \left(1 + x + \dfrac{x^2}{2!} + \cdots + \dfrac{x^n}{n!} + \cdots\right) + 2\left(\dfrac{x^3}{3!} + \dfrac{x^4}{4!} + \cdots + \dfrac{x^n}{n!} + \cdots\right)$

$$= a_0 e^x + 2\left(e^x - 1 - x - \dfrac{x^2}{2}\right) = ce^x - 2 - 2x - x^2$$

21. $y = a_0 \left(1 - \dfrac{x^2}{2} + \dfrac{x^4}{2^2 2!} - \dfrac{x^6}{2^3 3!} + \cdots + \dfrac{(-1)^n x^{2n}}{2^n n!} + \cdots \right)$

$\qquad + \left(x + \dfrac{x^2}{2} - \dfrac{x^3}{3} - \dfrac{x^4}{2 \cdot 4} + \dfrac{x^5}{3 \cdot 5} + \cdots \right)$

$\qquad = a_0 e^{-x^2/2} + \left(x + \dfrac{x^2}{2} - \dfrac{x^3}{3} - \dfrac{x^4}{2 \cdot 4} + \dfrac{x^5}{3 \cdot 5} + \cdots \right)$

23. $1, 1 - 3x^2, 1 - 10x^2 + \dfrac{35}{3}x^4;\quad x, x - \dfrac{5}{3}x^3, x - \dfrac{14}{3}x^3 + \dfrac{21}{5}x^5$

24. (a) $1, x, (3x^2 - 1)/2, (5x^3 - 3x)/2, (35x^4 - 30x^2 + 3)/8, (63x^5 - 70x^3 + 15x)/8$
    (c) $P_1, 0; P_2, \pm 0.57735; P_3, 0, \pm 0.77460; P_4, \pm 0.33998, \pm 0.86114; P_5, 0, \pm 0.53847, \pm 0.90618$

Section 8.4, *page 569*

1. $y = c_1 x^{-1} + c_2 x^{-2}$

2. $y = c_1 |x + 1|^{-1/2} + c_2 |x + 1|^{-3/2}$

3. $y = c_1 x^2 + c_2 x^2 \ln|x|$

4. $y = c_1 x^{-1} \cos(2 \ln|x|) + c_2 x^{-1} \sin(2 \ln|x|)$

5. $y = c_1 x + c_2 x \ln|x|$

6. $y = c_1 (x - 1)^{-3} + c_2 (x - 1)^{-4}$

7. $y = c_1 |x|^{(-5 + \sqrt{29})/2} + c_2 |x|^{(-5 - \sqrt{29})/2}$

8. $y = c_1 |x|^{3/2} \cos\left(\dfrac{1}{2} \sqrt{3} \ln|x| \right) + c_2 |x|^{3/2} \sin\left(\dfrac{1}{2} \sqrt{3} \ln|x| \right)$

9. $y = c_1 x^3 + c_2 x^3 \ln|x|$

10. $y = c_1 (x - 2)^{-2} \cos(2 \ln|x - 2|) + c_2 (x - 2)^{-2} \sin(2 \ln|x - 2|)$

11. $y = c_1 |x|^{-1/2} \cos\left(\dfrac{1}{2} \sqrt{15} \ln|x| \right) + c_2 |x|^{-1/2} \sin\left(\dfrac{1}{2} \sqrt{15} \ln|x| \right)$

12. $y = c_1 x + c_2 x^4$

13. $y = 2x^{3/2} - x^{-1}$

14. $y = 2x^{-1/2} \cos(2 \ln x) - x^{-1/2} \sin(2 \ln x)$

15. $y = 2x^2 - 7x^2 \ln|x|$

16. $y = x^{-1} \cos(2 \ln x)$

17. $x = 0$, regular

18. $x = 0$, regular; $x = 1$, irregular

19. $x = 0$, irregular; $x = 1$, regular

20. $x = 0$, irregular; $x = \pm 1$, regular

21. $x = 1$, regular; $x = -1$, irregular

22. $x = 0$, regular

23. $x = -3$, regular

24. $x = 0, -1$, regular; $x = 1$, irregular

25. $x = 1$, regular; $x = -2$, irregular

26. $x = 0, 3$, regular

27. $x = 1, -2$, regular

28. $x = 0$, regular

29. $x = 0$, irregular

30. $x = 0$, regular

31. $x = 0$, regular

32. $x = 0, \pm n\pi$, regular

33. $x = 0, \pm n\pi$, regular

34. $x = 0$, irregular; $x = \pm n\pi$, regular

35. $\alpha < 1$

36. $\beta > 0$

37. $\gamma = 2$

38. $\alpha > 1$

39. (a) $\alpha < 1$ and $\beta > 0$
    (b) $\alpha < 1$ and $\beta \geq 0$, or $\alpha = 1$ and $\beta > 0$
    (c) $\alpha > 1$ and $\beta > 0$
    (d) $\alpha > 1$ and $\beta \geq 0$, or $\alpha = 1$ and $\beta > 0$
    (e) $\alpha = 1$ and $\beta > 0$

41. $y = a_0 \left(1 - \dfrac{x^2}{2 \cdot 5} + \dfrac{x^4}{2 \cdot 4 \cdot 5 \cdot 9} - \cdots \right)$

44. Irregular singular point   45. Regular singular point

46. Regular singular point   47. Irregular singular point

48. Irregular singular point   49. Irregular singular point

Section 8.5, *page 575*

1. (b) $r(2r - 1) = 0$; $a_n = -\dfrac{a_{n-2}}{(n+r)[2(n+r)-1]}$; $r_1 = \dfrac{1}{2}$, $r_2 = 0$

   (c) $y_1(x) = x^{1/2} \left[1 - \dfrac{x^2}{2 \cdot 5} + \dfrac{x^4}{2 \cdot 4 \cdot 5 \cdot 9} - \dfrac{x^6}{2 \cdot 4 \cdot 6 \cdot 5 \cdot 9 \cdot 13} + \cdots \right.$
   $$+ \frac{(-1)^n x^{2n}}{2^n n! \, 5 \cdot 9 \cdot 13 \cdots (4n+1)} + \cdots \Bigg]$$

   (d) $y_2(x) = 1 - \dfrac{x^2}{2 \cdot 3} + \dfrac{x^4}{2 \cdot 4 \cdot 3 \cdot 7} - \dfrac{x^6}{2 \cdot 4 \cdot 6 \cdot 3 \cdot 7 \cdot 11} + \cdots$
   $$+ \frac{(-1)^n x^{2n}}{2^n n! \, 3 \cdot 7 \cdot 11 \cdots (4n-1)} + \cdots$$

2. (b) $r^2 - \dfrac{1}{9} = 0$; $a_n = -\dfrac{a_{n-2}}{(n+r)^2 - \dfrac{1}{9}}$; $r_1 = \dfrac{1}{3}$, $r_2 = -\dfrac{1}{3}$

   (c) $y_1(x) = x^{1/3} \left[1 - \dfrac{1}{1!\left(1+\dfrac{1}{3}\right)}\left(\dfrac{x}{2}\right)^2 + \dfrac{1}{2!\left(1+\dfrac{1}{3}\right)\left(2+\dfrac{1}{3}\right)}\left(\dfrac{x}{2}\right)^4 + \cdots \right.$
   $$\left. + \frac{(-1)^m}{m!\left(1+\dfrac{1}{3}\right)\left(2+\dfrac{1}{3}\right)\cdots\left(m+\dfrac{1}{3}\right)}\left(\frac{x}{2}\right)^{2m} + \cdots \right]$$

   (d) $y_2(x) = x^{-1/3} \left[1 - \dfrac{1}{1!\left(1-\dfrac{1}{3}\right)}\left(\dfrac{x}{2}\right)^2 + \dfrac{1}{2!\left(1-\dfrac{1}{3}\right)\left(2-\dfrac{1}{3}\right)}\left(\dfrac{x}{2}\right)^4 + \cdots \right.$
   $$\left. + \frac{(-1)^m}{m!\left(1-\dfrac{1}{3}\right)\left(2-\dfrac{1}{3}\right)\cdots\left(m-\dfrac{1}{3}\right)}\left(\frac{x}{2}\right)^{2m} + \cdots \right]$$

   Hint: Let $n = 2m$ in the recurrence relation, $m = 1, 2, 3, \ldots$.

3. (b) $r(r - 1) = 0$; $a_n = -\dfrac{a_{n-1}}{(n+r)(n+r-1)}$; $r_1 = 1$, $r_2 = 0$

   (c) $y_1(x) = x \left[1 - \dfrac{x}{1!2!} + \dfrac{x^2}{2!3!} + \cdots + \dfrac{(-1)^n}{n!(n+1)!} x^n + \cdots \right]$

4. (b) $r^2 = 0$; $\quad a_n = \dfrac{a_{n-1}}{(n+r)^2}$; $\quad r_1 = r_2 = 0$

(c) $y_1(x) = 1 + \dfrac{x}{(1!)^2} + \dfrac{x^2}{(2!)^2} + \cdots + \dfrac{x^n}{(n!)^2} + \cdots$

5. (b) $r(3r - 1) = 0$; $\quad a_n = -\dfrac{a_{n-2}}{(n+r)[3(n+r)-1]}$; $\quad r_1 = \dfrac{1}{3}, r_2 = 0$

(c) $y_1(x) = x^{1/3}\left[1 - \dfrac{1}{1!7}\left(\dfrac{x^2}{2}\right) + \dfrac{1}{2!7 \cdot 13}\left(\dfrac{x^2}{2}\right)^2 + \cdots\right.$

$\left. + \dfrac{(-1)^m}{m!7 \cdot 13 \cdots (6m+1)}\left(\dfrac{x^2}{2}\right)^m + \cdots\right]$

(d) $y_2(x) = 1 - \dfrac{1}{1!5}\left(\dfrac{x^2}{2}\right) + \dfrac{1}{2!5 \cdot 11}\left(\dfrac{x^2}{2}\right)^2 + \cdots + \dfrac{(-1)^m}{m!5 \cdot 11 \cdots (6m-1)}\left(\dfrac{x^2}{2}\right)^m + \cdots$

Hint: Let $n = 2m$ in the recurrence relation, $m = 1, 2, 3, \ldots$.

6. (b) $r^2 - 2 = 0$; $\quad a_n = -\dfrac{a_{n-1}}{(n+r)^2 - 2}$; $\quad r_1 = \sqrt{2}, r_2 = -\sqrt{2}$

(c) $y_1(x) = x^{\sqrt{2}}\left[1 - \dfrac{x}{1(1+2\sqrt{2})} + \dfrac{x^2}{2!(1+2\sqrt{2})(2+2\sqrt{2})} + \cdots\right.$

$\left. + \dfrac{(-1)^n}{n!(1+2\sqrt{2})(2+2\sqrt{2})\cdots(n+2\sqrt{2})}x^n + \cdots\right]$

(d) $y_2(x) = x^{-\sqrt{2}}\left[1 - \dfrac{x}{1(1-2\sqrt{2})} + \dfrac{x^2}{2!(1-2\sqrt{2})(2-2\sqrt{2})} + \cdots\right.$

$\left. + \dfrac{(-1)^n}{n!(1-2\sqrt{2})(2-2\sqrt{2})\cdots(n-2\sqrt{2})}x^n + \cdots\right]$

7. (b) $r^2 = 0$; $\quad (n+r)a_n = a_{n-1}$; $\quad r_1 = r_2 = 0$

(c) $y_1(x) = 1 + x + \dfrac{x^2}{2!} + \dfrac{x^3}{3!} + \cdots + \dfrac{x^n}{n!} + \cdots = e^x$

8. (b) $2r^2 + r - 1 = 0$; $\quad (2n + 2r - 1)(n + r + 1)a_n + 2a_{n-2} = 0$; $r_1 = \dfrac{1}{2}, r_2 = -1$

(c) $y_1(x) = x^{1/2}\left(1 - \dfrac{x^2}{7} + \dfrac{x^4}{2!7 \cdot 11} - \cdots + \dfrac{(-1)^m x^{2m}}{m!7 \cdot 11 \cdots (4m+3)} + \cdots\right)$

(d) $y_2(x) = x^{-1}\left(1 - x^2 + \dfrac{x^4}{2!5} - \cdots + \dfrac{(-1)^m x^{2m}}{m!5 \cdot 9 \cdots (4m-3)} + \cdots\right)$

9. (b) $r^2 - 4r + 3 = 0$; $\quad (n+r-3)(n+r-1)a_n - (n+r-2)a_{n-1} = 0$; $r_1 = 3, r_2 = 1$

(c) $y_1(x) = x^3\left(1 + \dfrac{2}{3}x + \dfrac{x^2}{4} + \cdots + \dfrac{2x^n}{n!(n+2)} + \cdots\right)$

10. (b) $r^2 - r + \dfrac{1}{4} = 0$; $\quad \left(n + r - \dfrac{1}{2}\right)^2 a_n + a_{n-2} = 0$; $\quad r_1 = r_2 = 1/2$

(c) $y_1(x) = x^{1/2}\left(1 - \dfrac{x^2}{2^2} + \dfrac{x^4}{2^2 4^2} - \cdots + \dfrac{(-1)^m x^{2m}}{2^{2m}(m!)^2} + \cdots\right)$

11. (a) $r^2 = 0$; $\quad r_1 = 0, r_2 = 0$

(b) $y_1(x) = 1 + \dfrac{\alpha(\alpha+1)}{2 \cdot 1^2}(x-1) - \dfrac{\alpha(\alpha+1)[1 \cdot 2 - \alpha(\alpha+1)]}{(2 \cdot 1^2)(2 \cdot 2^2)}(x-1)^2 + \cdots$

$+ (-1)^{n+1}\dfrac{\alpha(\alpha+1)[1 \cdot 2 - \alpha(\alpha+1)] \cdots [n(n-1) - \alpha(\alpha+1)]}{2^n (n!)^2}(x-1)^n + \cdots$

12. (a) $r_1 = \dfrac{1}{2}$, $r_2 = 0$ at both $x = \pm 1$

(b) $y_1(x) = |x - 1|^{1/2}$

$$\times \left[1 + \sum_{n=1}^{\infty} \frac{(-1)^n(1 + 2\alpha)\cdots(2n - 1 + 2\alpha)(1 - 2\alpha)\cdots(2n - 1 - 2\alpha)}{2^n(2n + 1)!}(x - 1)^n \right]$$

$$y_2(x) = 1 + \sum_{n=1}^{\infty} \frac{(-1)^n\alpha(1 + \alpha)\cdots(n - 1 + \alpha)(-\alpha)(1 - \alpha)\cdots(n - 1 - \alpha)}{n!\, 1 \cdot 3 \cdot 5 \cdots (2n - 1)}(x - 1)^n$$

13. (b) $r^2 = 0$; $r_1 = 0$, $r_2 = 0$; $a_n = \dfrac{(n - 1 - \lambda)a_{n-1}}{n^2}$

(c) $y_1(x) = 1 + \dfrac{-\lambda}{(1!)^2}x + \dfrac{(-\lambda)(1 - \lambda)}{(2!)^2}x^2 + \cdots + \dfrac{(-\lambda)(1 - \lambda)\cdots(n - 1 - \lambda)}{(n!)^2}x^n + \cdots$

For $\lambda = n$, the coefficients of all terms past x^n are zero.

16. (e) $[(n - 1)^2 - 1]b_n = -b_{n-2}$, and it is impossible to determine b_2.

Section 8.6, *page 583*

1. (a) $x = 0$;　(b) $r(r - 1) = 0$;　$r_1 = 1$, $r_2 = 0$

2. (a) $x = 0$;　(b) $r^2 - 3r + 2 = 0$;　$r_1 = 2$, $r_2 = 1$

3. (a) $x = 0$;　(b) $r(r - 1) = 0$;　$r_1 = 1$, $r_2 = 0$
　　(a) $x = 1$;　(b) $r(r + 5) = 0$;　$r_1 = 0$, $r_2 = -5$

4. None

5. (a) $x = 0$;
　　(b) $r^2 + 2r - 2 = 0$;　$r_1 = -1 + \sqrt{3} \cong 0.732$, $r_2 = -1 - \sqrt{3} \cong -2.73$

6. (a) $x = 0$;　(b) $r\left(r - \dfrac{3}{4}\right) = 0$;　$r_1 = \dfrac{3}{4}$, $r_2 = 0$

　　(a) $x = -2$;　(b) $r\left(r - \dfrac{5}{4}\right) = 0$;　$r_1 = \dfrac{5}{4}$, $r_2 = 0$

7. (a) $x = 0$;　(b) $r^2 + 1 = 0$;　$r_1 = i$, $r_2 = -i$

8. (a) $x = -1$;
　　(b) $r^2 - 7r + 3 = 0$;　$r_1 = (7 + \sqrt{37})/2 \cong 6.54$, $r_2 = (7 - \sqrt{37})/2 \cong 0.459$

9. (a) $x = 1$;　(b) $r^2 + r = 0$;　$r_1 = 0$, $r_2 = -1$

10. (a) $x = -2$;　(b) $r^2 - (5/4)r = 0$;　$r_1 = 5/4$, $r_2 = 0$

11. (a) $x = 2$;　(b) $r^2 - 2r = 0$;　$r_1 = 2$, $r_2 = 0$
　　(a) $x = -2$;　(b) $r^2 - 2r = 0$;　$r_1 = 2$, $r_2 = 0$

12. (a) $x = 0$;　(b) $r^2 - (5/3)r = 0$;　$r_1 = 5/3$, $r_2 = 0$
　　(a) $x = -3$;　(b) $r^2 - (r/3) - 1 = 0$;　$r_1 = (1 + \sqrt{37})/6 \cong 1.18$,
　　$r_2 = (1 - \sqrt{37})/6 \cong -0.847$

13. (b) $r_1 = 0$, $r_2 = 0$
　　(c) $y_1(x) = 1 + x + \dfrac{1}{4}x^2 + \dfrac{1}{36}x^3 + \cdots$

　　　$y_2(x) = y_1(x)\ln x - 2x - \dfrac{3}{4}x^2 - \dfrac{11}{108}x^3 + \cdots$

14. (b) $r_1 = 1$, $r_2 = 0$
　　(c) $y_1(x) = x - 4x^2 + \dfrac{17}{3}x^3 - \dfrac{47}{12}x^4 + \cdots$

　　　$y_2(x) = -6y_1(x)\ln x + 1 - 33x^2 + \dfrac{449}{6}x^3 + \cdots$

15. (b) $r_1 = 1$, $r_2 = 0$

   (c) $y_1(x) = x + \dfrac{3}{2}x^2 + \dfrac{9}{4}x^3 + \dfrac{51}{16}x^4 + \cdots$

   $y_2(x) = 3y_1(x)\ln x + 1 - \dfrac{21}{4}x^2 - \dfrac{19}{4}x^3 + \cdots$

16. (b) $r_1 = 1$, $r_2 = 0$

   (c) $y_1(x) = x - \dfrac{1}{2}x^2 + \dfrac{1}{12}x^3 - \dfrac{1}{144}x^4 + \cdots$

   $y_2(x) = -y_1(x)\ln x + 1 - \dfrac{3}{4}x^2 + \dfrac{7}{36}x^3 - \dfrac{35}{1728}x^4 + \cdots$

17. (b) $r_1 = 1$, $r_2 = -1$

   (c) $y_1(x) = x - \dfrac{1}{24}x^3 + \dfrac{1}{720}x^5 + \cdots$

   $y_2(x) = -\dfrac{1}{3}y_1(x)\ln x + x^{-1} - \dfrac{1}{90}x^3 + \cdots$

18. (b) $r_1 = \dfrac{1}{2}$, $r_2 = 0$

   (c) $y_1(x) = (x-1)^{1/2}\left[1 - \dfrac{3}{4}(x-1) + \dfrac{53}{480}(x-1)^2 + \cdots\right]$,   (d) $\rho = 1$

19. (c) *Hint:* $(n-1)(n-2) + (1+\alpha+\beta)(n-1) + \alpha\beta = (n-1+\alpha)(n-1+\beta)$
   (d) *Hint:* $(n-\gamma)(n-1-\gamma) + (1+\alpha+\beta)(n-\gamma) + \alpha\beta = (n-\gamma+\alpha)(n-\gamma+\beta)$

Section 8.7, *page 594*

1. $y_1(x) = \displaystyle\sum_{n=0}^{\infty} \dfrac{(-1)^n x^n}{n!(n+1)!}$

   $y_2(x) = -y_1(x)\ln x + \dfrac{1}{x}\left[1 - \displaystyle\sum_{n=1}^{\infty} \dfrac{H_n + H_{n-1}}{n!(n-1)!}(-1)^n x^n\right]$

2. $y_1(x) = \dfrac{1}{x}\displaystyle\sum_{n=0}^{\infty} \dfrac{(-1)^n x^n}{(n!)^2}$, $y_2(x) = y_1(x)\ln x - \dfrac{2}{x}\displaystyle\sum_{n=1}^{\infty} \dfrac{(-1)^n H_n}{(n!)^2}x^n$

3. $y_1(x) = \displaystyle\sum_{n=0}^{\infty} \dfrac{(-1)^n 2^n}{(n!)^2}x^n$, $y_2(x) = y_1(x)\ln x - 2\displaystyle\sum_{n=1}^{\infty} \dfrac{(-1)^n 2^n H_n}{(n!)^2}x^n$

4. $y_1(x) = \dfrac{1}{x}\displaystyle\sum_{n=0}^{\infty} \dfrac{(-1)^n}{n!(n+1)!}x^n$

   $y_2(x) = -y_1(x)\ln x + \dfrac{1}{x^2}\left[1 - \displaystyle\sum_{n=1}^{\infty} \dfrac{H_n + H_{n-1}}{n!(n-1)!}(-1)^n x^n\right]$

5. $y_1(x) = x^{3/2}\left[1 + \displaystyle\sum_{m=1}^{\infty} \dfrac{(-1)^m}{m!\left(1+\dfrac{3}{2}\right)\left(2+\dfrac{3}{2}\right)\cdots\left(m+\dfrac{3}{2}\right)}\left(\dfrac{x}{2}\right)^{2m}\right]$

   $y_2(x) = x^{-3/2}\left[1 + \displaystyle\sum_{m=1}^{\infty} \dfrac{(-1)^m}{m!\left(1-\dfrac{3}{2}\right)\left(2-\dfrac{3}{2}\right)\cdots\left(m-\dfrac{3}{2}\right)}\left(\dfrac{x}{2}\right)^{2m}\right]$

   Hint: Let $n = 2m$ in the recurrence relation, $m = 1, 2, 3, \ldots$.
   For $r = -\dfrac{3}{2}$, $a_1 = 0$ and a_3 is arbitrary.

CHAPTER 9

Section 9.1, *page 618*

1. $y = -\sin x$

2. $y = (\cot\sqrt{2}\pi\cos\sqrt{2}x + \sin\sqrt{2}x)/\sqrt{2}$

3. $y = 0$ for all L; $y = c_2\sin x$ if $\sin L = 0$

4. $y = -\tan L\cos x + \sin x$ if $\cos L \neq 0$; no solution if $\cos L = 0$

5. No solution

6. $y = (-\pi\sin\sqrt{2}x + x\sin\sqrt{2}\pi)/2\sin\sqrt{2}\pi$

7. No solution

8. $y = c_2\sin 2x + \frac{1}{3}\sin x$

9. $y = c_1\cos 2x + \frac{1}{3}\cos x$

10. $y = \frac{1}{2}\cos x$

11. $y = -\frac{5}{2}x + \frac{3}{2}x^2$

12. $y = -\frac{1}{9}x^{-1} + \frac{1}{9}(1 - e^3)x^{-1}\ln x + \frac{1}{9}x^2$

13. No solution

14. $\lambda_n = [(2n - 1)/2]^2$, $y_n(x) = \sin[(2n - 1)x/2]$; $n = 1, 2, 3, \ldots$

15. $\lambda_n = [(2n - 1)/2]^2$, $y_n(x) = \cos[(2n - 1)x/2]$; $n = 1, 2, 3, \ldots$

16. $\lambda_0 = 0$, $y_0(x) = 1$; $\lambda_n = n^2$, $y_n(x) = \cos nx$; $n = 1, 2, 3, \ldots$

17. $\lambda_n = [(2n - 1)\pi/2L]^2$, $y_n(x) = \cos[(2n - 1)\pi x/2L]$; $n = 1, 2, 3, \ldots$

18. $\lambda_0 = 0$, $y_0(x) = 1$; $\lambda_n = (n\pi/L)^2$, $y_n(x) = \cos(n\pi x/L)$; $n = 1, 2, 3, \ldots$

19. $\lambda_n = -[(2n - 1)\pi/2L]^2$, $y_n(x) = \sin[(2n - 1)\pi x/2L]$; $n = 1, 2, 3, \ldots$

20. $\lambda_n = 1 + (n\pi/\ln L)^2$, $y_n(x) = x\sin(n\pi\ln x/\ln L)$; $n = 1, 2, 3, \ldots$

21. (a) $w(r) = G(R^2 - r^2)/4\mu$ (c) Q is reduced to 0.3164 of its original value.

22. (a) $y = k(x^4 - 2Lx^3 + L^3x)/24$
    (b) $y = k(x^4 - 2Lx^3 + L^2x^2)/24$
    (c) $y = k(x^4 - 4Lx^3 + 6L^2x^2)/24$

Section 9.2, *page 627*

1. $T = 2\pi/5$

2. $T = 1$

3. Not periodic

4. $T = 2L$

5. $T = 1$

6. Not periodic

7. $T = 2$

8. $T = 4$

9. $f(x) = 2L - x$ in $L < x < 2L$; $f(x) = -2L - x$ in $-3L < x < -2L$

10. $f(x) = x - 1$ in $1 < x < 2$; $f(x) = x - 8$ in $8 < x < 9$

11. $f(x) = -L - x$ in $-L < x < 0$

13. (b) $f(x) = \dfrac{2L}{\pi}\sum_{n=1}^{\infty}\dfrac{(-1)^n}{n}\sin\dfrac{n\pi x}{L}$

14. (b) $f(x) = \dfrac{1}{2} - \dfrac{2}{\pi}\sum_{n=1}^{\infty}\dfrac{\sin[(2n - 1)\pi x/L]}{2n - 1}$

15. (b) $f(x) = -\dfrac{\pi}{4} + \sum_{n=1}^{\infty}\left[\dfrac{2\cos(2n - 1)x}{\pi(2n - 1)^2} + \dfrac{(-1)^{n+1}\sin nx}{n}\right]$

16. (b) $f(x) = \dfrac{1}{2} + \dfrac{4}{\pi^2}\sum_{n=1}^{\infty}\dfrac{\cos(2n - 1)\pi x}{(2n - 1)^2}$

17. (b) $f(x) = \dfrac{3L}{4} + \sum_{n=1}^{\infty}\left[\dfrac{2L\cos[(2n - 1)\pi x/L]}{(2n - 1)^2\pi^2} + \dfrac{(-1)^{n+1}L\sin(n\pi x/L)}{n\pi}\right]$

18. (b) $f(x) = \sum_{n=1}^{\infty} \left[-\frac{2}{n\pi}\cos\frac{n\pi}{2} + \left(\frac{2}{n\pi}\right)^2 \sin\frac{n\pi}{2} \right] \sin\frac{n\pi x}{2}$

19. (b) $f(x) = \frac{4}{\pi}\sum_{n=1}^{\infty}\frac{\sin[(2n-1)\pi x/2]}{2n-1}$ 20. (b) $f(x) = \frac{2}{\pi}\sum_{n=1}^{\infty}\frac{(-1)^{n+1}}{n}\sin n\pi x$

21. (b) $f(x) = \frac{2}{3} + \frac{8}{\pi^2}\sum_{n=1}^{\infty}\frac{(-1)^n}{n^2}\cos\frac{n\pi x}{2}$

22. (b) $f(x) = \frac{1}{2} + \frac{12}{\pi^2}\sum_{n=1}^{\infty}\frac{\cos[(2n-1)\pi x/2]}{(2n-1)^2} + \frac{2}{\pi}\sum_{n=1}^{\infty}\frac{(-1)^n}{n}\sin\frac{n\pi x}{2}$

23. (b) $f(x) = \frac{11}{12} + \frac{1}{\pi^2}\sum_{n=1}^{\infty}\frac{(-1)^n - 5}{n^2}\cos\frac{n\pi x}{2} + \sum_{n=1}^{\infty}\left[\frac{4[1-(-1)^n]}{n^3\pi^3} - \frac{(-1)^n}{n\pi}\right]\sin\frac{n\pi x}{2}$

24. (b) $f(x) = \frac{9}{8} + \sum_{n=1}^{\infty}\left[\frac{162[(-1)^n - 1]}{n^4\pi^4} - \frac{27(-1)^n}{n^2\pi^2}\right]\cos\frac{n\pi x}{3} - \sum_{n=1}^{\infty}\frac{108(-1)^n + 54}{n^3\pi^3}\sin\frac{n\pi x}{3}$

25. (b) $m = 81$ 26. (b) $m = 27$

28. $\int_0^x f(t)\,dt$ may not be periodic; for example, let $f(t) = 1 + \cos t$.

Section 9.3, *page 633*

1. (a) $f(x) = \frac{4}{\pi}\sum_{n=1}^{\infty}\frac{\sin(2n-1)\pi x}{2n-1}$

2. (a) $f(x) = \frac{\pi}{4} - \sum_{n=1}^{\infty}\left[\frac{2}{(2n-1)^2\pi}\cos(2n-1)x + \frac{(-1)^n}{n}\sin nx\right]$

3. (a) $f(x) = \frac{L}{2} + \frac{4L}{\pi^2}\sum_{n=1}^{\infty}\frac{\cos[(2n-1)\pi x/L]}{(2n-1)^2}$

4. (a) $f(x) = \frac{2}{3} + \frac{4}{\pi^2}\sum_{n=1}^{\infty}\frac{(-1)^{n+1}}{n^2}\cos n\pi x$

5. (a) $f(x) = \frac{1}{2} + \frac{2}{\pi}\sum_{n=1}^{\infty}\frac{(-1)^{n-1}}{2n-1}\cos(2n-1)x$

6. (a) $f(x) = \frac{a_0}{2} + \sum_{n=1}^{\infty}(a_n\cos n\pi x + b_n\sin n\pi x)$;

   $a_0 = \frac{1}{3},\qquad a_n = \frac{2(-1)^n}{n^2\pi^2},\qquad b_n = \begin{cases} -1/n\pi, & n\text{ even} \\ 1/n\pi - 4/n^3\pi^3, & n\text{ odd} \end{cases}$

7. (a) $f(x) = -\frac{\pi}{4} + \sum_{n=1}^{\infty}\left[\frac{1-\cos n\pi}{\pi n^2}\cos nx - \frac{(-1)^n}{n}\sin nx\right]$

   (b) $n = 10$;   max$|e| = 1.6025$ at $x = \pm\pi$
       $n = 20$;   max$|e| = 1.5867$ at $x = \pm\pi$
       $n = 40$;   max$|e| = 1.5788$ at $x = \pm\pi$
   (c) Not possible

8. (a) $f(x) = \frac{1}{2} + \frac{2}{\pi^2}\sum_{n=1}^{\infty}\frac{1-\cos n\pi}{n^2}\cos n\pi x$

   (b) $n = 10$;   max$|e| = 0.02020$ at $x = 0, \pm1$
       $n = 20$;   max$|e| = 0.01012$ at $x = 0, \pm1$
       $n = 40$;   max$|e| = 0.005065$ at $x = 0, \pm1$
   (c) $n = 21$

9. (a) $f(x) = \dfrac{2}{\pi} \displaystyle\sum_{n=1}^{\infty} \dfrac{(-1)^{n+1}}{n} \sin n\pi x$

   (b) $n = 10, 20, 40$; $\max|e| = 1$ at $x \pm 1$

   (c) Not possible

10. (a) $f(x) = \dfrac{1}{2} + \displaystyle\sum_{n=1}^{\infty} \left[\dfrac{6(1 - \cos n\pi)}{n^2 \pi^2} \cos \dfrac{n\pi x}{2} + \dfrac{2 \cos n\pi}{n\pi} \sin \dfrac{n\pi x}{2} \right]$

   (b) $n = 10$; $\text{lub}|e| = 1.0606$ as $x \to 2$
   $n = 20$; $\text{lub}|e| = 1.0304$ as $x \to 2$
   $n = 40$; $\text{lub}|e| = 1.0152$ as $x \to 2$

   (c) Not possible

11. (a) $f(x) = \dfrac{1}{6} + \displaystyle\sum_{n=1}^{\infty} \left[\dfrac{2 \cos n\pi}{n^2 \pi^2} \cos n\pi x - \dfrac{2 - 2 \cos n\pi + n^2 \pi^2 \cos n\pi}{n^3 \pi^3} \sin n\pi x \right]$

   (b) $n = 10$; $\text{lub}|e| = 0.5193$ as $x \to 1$
   $n = 20$; $\text{lub}|e| = 0.5099$ as $x \to 1$
   $n = 40$; $\text{lub}|e| = 0.5050$ as $x \to 1$

   (c) Not possible

12. (a) $f(x) = -\dfrac{12}{\pi^3} \displaystyle\sum_{n=1}^{\infty} \dfrac{(-1)^n}{n^3} \sin n\pi x$

   (b) $n = 10$; $\max|e| = 0.001345$ at $x = \pm 0.9735$
   $n = 20$; $\max|e| = 0.0003534$ at $x = \pm 0.9864$
   $n = 40$; $\max|e| = 0.00009058$ at $x = \pm 0.9931$

   (c) $n = 4$

13. $y = (\omega \sin nt - n \sin \omega t)/\omega(\omega^2 - n^2)$, $\omega^2 \neq n^2$
   $y = (\sin nt - nt \cos nt)/2n^2$, $\omega^2 = n^2$

14. $y = \displaystyle\sum_{n=1}^{\infty} b_n(\omega \sin nt - n \sin \omega t)/\omega(\omega^2 - n^2)$, $\omega \neq 1, 2, 3, \ldots$
   $y = \displaystyle\sum_{\substack{n=1 \\ n \neq m}}^{\infty} b_n(m \sin nt - n \sin mt)/m(m^2 - n^2) + b_m(\sin mt - mt \cos mt)/2m^2$, $\omega = m$

15. $y = \dfrac{4}{\pi} \displaystyle\sum_{n=1}^{\infty} \dfrac{1}{\omega^2 - (2n-1)^2} \left[\dfrac{1}{2n-1} \sin(2n-1)t - \dfrac{1}{\omega} \sin \omega t \right]$

16. $y = \cos \omega t + \dfrac{1}{2\omega^2}(1 - \cos \omega t) + \dfrac{4}{\pi^2} \displaystyle\sum_{n=1}^{\infty} \dfrac{\cos(2n-1)\pi t - \cos \omega t}{(2n-1)^2[\omega^2 - (2n-1)^2 \pi^2]}$

Section 9.4, *page 640*

1. Odd   2. Neither   3. Odd

4. Even   5. Even   6. Neither

14. $f(x) = \dfrac{1}{4} + \dfrac{4}{\pi^2} \displaystyle\sum_{n=1}^{\infty} \dfrac{1 - \cos(n\pi/2)}{n^2} \cos \dfrac{n\pi x}{2}$

   $f(x) = \dfrac{4}{\pi^2} \displaystyle\sum_{n=1}^{\infty} \dfrac{(n\pi/2) - \sin(n\pi/2)}{n^2} \sin \dfrac{n\pi x}{2}$

15. (a) $f(x) = \dfrac{1}{2} + \dfrac{2}{\pi} \displaystyle\sum_{n=1}^{\infty} \dfrac{(-1)^{n-1}}{2n-1} \cos \dfrac{(2n-1)\pi x}{2}$

16. (a) $f(x) = \displaystyle\sum_{n=1}^{\infty} \dfrac{2}{n\pi} \left(-\cos n\pi + \dfrac{2}{n\pi} \sin \dfrac{n\pi}{2} \right) \sin \dfrac{n\pi x}{2}$

17. (a) $f(x) = 1$

18. (a) $f(x) = \dfrac{4}{\pi} \displaystyle\sum_{n=1}^{\infty} \dfrac{\sin(2n-1)x}{2n-1}$

19. (a) $f(x) = \displaystyle\sum_{n=1}^{\infty} \dfrac{2}{n\pi}\left(\cos\dfrac{n\pi}{3} + \cos\dfrac{2n\pi}{3} - 2\cos n\pi\right)\sin\dfrac{nx}{3}$

20. (a) $f(x) = \dfrac{1}{2} - \dfrac{1}{\pi}\displaystyle\sum_{n=1}^{\infty}\dfrac{\sin 2n\pi x}{n}$

21. (a) $f(x) = \dfrac{L}{2} + \dfrac{4L}{\pi^2}\displaystyle\sum_{n=1}^{\infty}\dfrac{\cos[(2n-1)\pi x/L]}{(2n-1)^2}$

22. (a) $f(x) = \dfrac{2L}{\pi}\displaystyle\sum_{n=1}^{\infty}\dfrac{\sin(n\pi x/L)}{n}$

23. (a) $f(x) = \dfrac{\pi}{4} + \dfrac{1}{\pi}\displaystyle\sum_{n=1}^{\infty}\left[\dfrac{2\pi}{n}\sin\dfrac{n\pi}{2} + \dfrac{4}{n^2}\left(\cos\dfrac{n\pi}{2} - 1\right)\right]\cos\dfrac{nx}{2}$

24. (a) $f(x) = 2\displaystyle\sum_{n=1}^{\infty}\dfrac{(-1)^n}{n}\sin nx$

25. (a) $f(x) = \displaystyle\sum_{n=1}^{\infty}\left[\dfrac{4n^2\pi^2(1+\cos n\pi)}{n^3\pi^3} + \dfrac{16(1-\cos n\pi)}{n^3\pi^3}\right]\sin\dfrac{n\pi x}{2}$

26. (a) $f(x) = \dfrac{4}{3} + \dfrac{16}{\pi^2}\displaystyle\sum_{n=1}^{\infty}\dfrac{1+3\cos n\pi}{n^2}\cos\dfrac{n\pi x}{4}$

27. (b) $g(x) = \dfrac{3}{2} + \dfrac{6}{\pi^2}\displaystyle\sum_{n=1}^{\infty}\dfrac{1-\cos n\pi}{n^2}\cos\dfrac{n\pi x}{3}$

   $h(x) = \dfrac{6}{\pi}\displaystyle\sum_{n=1}^{\infty}\dfrac{1}{n}\sin\dfrac{n\pi x}{3}$

28. (b) $g(x) = \dfrac{1}{4} + \displaystyle\sum_{n=1}^{\infty}\dfrac{4\cos(n\pi/2)+2n\pi\sin(n\pi/2)-4}{n^2\pi^2}\cos\dfrac{n\pi x}{2}$

   $h(x) = \displaystyle\sum_{n=1}^{\infty}\dfrac{4\sin(n\pi/2)-2n\pi\cos(n\pi/2)}{n^2\pi^2}\sin\dfrac{n\pi x}{2}$

29. (b) $g(x) = -\dfrac{5}{12} + \displaystyle\sum_{n=1}^{\infty}\dfrac{12\cos n\pi + 4}{n^2\pi^2}\cos\dfrac{n\pi x}{2}$

   $h(x) = -\dfrac{1}{2}\displaystyle\sum_{n=1}^{\infty}\dfrac{n^2\pi^2(3+5\cos n\pi)+32(1-\cos n\pi)}{n^3\pi^3}\sin\dfrac{n\pi x}{2}$

30. (b) $g(x) = \dfrac{1}{4} + \displaystyle\sum_{n=1}^{\infty}\dfrac{6n^2\pi^2(2\cos n\pi - 5)+324(1-\cos n\pi)}{n^4\pi^4}\cos\dfrac{n\pi x}{3}$

   $h(x) = \displaystyle\sum_{n=1}^{\infty}\left[\dfrac{4\cos n\pi + 2}{n\pi} + \dfrac{144\cos n\pi + 180}{n^3\pi^3}\right]\sin\dfrac{n\pi x}{3}$

40. (a) Extend $f(x)$ antisymmetrically into $(L, 2L)$; that is, so that $f(2L - x) = -f(x)$ for $0 \le x < L$. Then extend this function as an even function into $(-2L, 0)$.

Section 9.5, *page 649*

1. $xX'' - \lambda X = 0$, $T' + \lambda T = 0$

2. $X'' - \lambda x X = 0$, $T' + \lambda t T = 0$

3. $X'' - \lambda(X' + X) = 0$, $T' + \lambda T = 0$

4. $[p(x)X']' + \lambda r(x)X = 0$, $T'' + \lambda T = 0$

5. Not separable

6. $X'' + (x + \lambda)X = 0$, $Y'' - \lambda Y = 0$

7. $u(x, t) = e^{-400\pi^2 t}\sin 2\pi x - e^{-2500\pi^2 t}\sin 5\pi x$

8. $u(x, t) = 2e^{-\pi^2 t/16}\sin(\pi x/2) - e^{-\pi^2 t/4}\sin \pi x + 4e^{-\pi^2 t}\sin 2\pi x$

9. $u(x,t) = \dfrac{100}{\pi} \displaystyle\sum_{n=1}^{\infty} \dfrac{1-\cos n\pi}{n} e^{-n^2\pi^2 t/1600} \sin\dfrac{n\pi x}{40}$

10. $u(x,t) = \dfrac{160}{\pi^2} \displaystyle\sum_{n=1}^{\infty} \dfrac{\sin(n\pi/2)}{n^2} e^{-n^2\pi^2 t/1600} \sin\dfrac{n\pi x}{40}$

11. $u(x,t) = \dfrac{100}{\pi} \displaystyle\sum_{n=1}^{\infty} \dfrac{\cos(n\pi/4)-\cos(3n\pi/4)}{n} e^{-n^2\pi^2 t/1600} \sin\dfrac{n\pi x}{40}$

12. $u(x,t) = \dfrac{80}{\pi} \displaystyle\sum_{n=1}^{\infty} \dfrac{(-1)^{n+1}}{n} e^{-n^2\pi^2 t/1600} \sin\dfrac{n\pi x}{40}$

13. $t=5, n=16; t=20, n=8; t=80, n=4$

14. (d) $t=673.35$ 15. (d) $t=451.60$

16. (d) $t=617.17$

17. (b) $t=5, x=33.20; \quad t=10, x=31.13; \quad t=20, x=28.62; \quad t=40, x=25.73;$
    $t=100, x=21.95; \quad t=200, x=20.31$
    (e) $t=524.81$

18. $u(x,t) = \dfrac{200}{\pi} \displaystyle\sum_{n=1}^{\infty} \dfrac{1-\cos n\pi}{n} e^{-n^2\pi^2\alpha^2 t/400} \sin\dfrac{n\pi x}{20}$
    (a) $35.91°C$ (b) $67.23°C$ (c) $99.96°C$

19. (a) 76.73 s   (b) 152.56 s   (c) 1093.36 s

21. (a) $aw_{xx} - bw_t + (c-b\delta)w = 0$ (b) $\delta = c/b$ if $b \neq 0$

22. $X'' + \mu^2 X = 0, \quad Y'' + (\lambda^2 - \mu^2)Y = 0, \quad T' + \alpha^2\lambda^2 T = 0$

23. $r^2 R'' + rR' + (\lambda^2 r^2 - \mu^2)R = 0, \quad \Theta'' + \mu^2\Theta = 0, \quad T' + \alpha^2\lambda^2 T = 0$

Section 9.6, *page 658*

1. $u = 10 + \frac{3}{5}x$ 2. $u = 30 - \frac{5}{4}x$

3. $u = 0$ 4. $u = T$

5. $u = 0$ 6. $u = T$

7. $u = T(1+x)/(1+L)$ 8. $u = T(1+L-x)/(1+L)$

9. (a) $u(x,t) = 3x + \displaystyle\sum_{n=1}^{\infty} \dfrac{70\cos n\pi + 50}{n\pi} e^{-0.86 n^2\pi^2 t/400} \sin\dfrac{n\pi x}{20}$
   (d) 160.29 s

10. (a) $f(x) = 2x, \ 0 \leq x \leq 50; \quad f(x) = 200 - 2x, \ 50 < x \leq 100$
    (b) $u(x,t) = 20 - \dfrac{x}{5} + \displaystyle\sum_{n=1}^{\infty} c_n e^{-1.14 n^2\pi^2 t/(100)^2} \sin\dfrac{n\pi x}{100},$
    $c_n = \dfrac{800}{n^2\pi^2}\sin\dfrac{n\pi}{2} - \dfrac{40}{n\pi}$
    (d) $u(50,t) \to 10$ as $t \to \infty$; 3754 s

11. (a) $u(x,t) = 30 - x + \displaystyle\sum_{n=1}^{\infty} c_n e^{-n^2\pi^2 t/900} \sin\dfrac{n\pi x}{30},$
    $c_n = \dfrac{60}{n^3\pi^3}[2(1-\cos n\pi) - n^2\pi^2(1+\cos n\pi)]$

12. (a) $u(x, t) = \dfrac{2}{\pi} + \displaystyle\sum_{n=1}^{\infty} c_n e^{-n^2\pi^2\alpha^2 t/L^2} \cos\dfrac{n\pi x}{L}$, $c_n = \begin{cases} 0, & n \text{ odd;} \\ -4/(n^2-1)\pi, & n \text{ even} \end{cases}$

(b) $\lim_{t\to\infty} u(x, t) = 2/\pi$

13. (a) $u(x, t) = \dfrac{200}{9} + \displaystyle\sum_{n=1}^{\infty} c_n e^{-n^2\pi^2 t/6400} \cos\dfrac{n\pi x}{40}$, $c_n = -\dfrac{160}{3n^2\pi^2}(3 + \cos n\pi)$

(c) 200/9

(d) 1543 s

14. (a) $u(x, t) = \dfrac{25}{6} + \displaystyle\sum_{n=1}^{\infty} c_n e^{-n^2\pi^2 t/900} \cos\dfrac{n\pi x}{30}$, $c_n = \dfrac{50}{n\pi}\left(\sin\dfrac{n\pi}{3} - \sin\dfrac{n\pi}{6}\right)$

15. (b) $u(x, t) = \displaystyle\sum_{n=1}^{\infty} c_n e^{-(2n-1)^2\pi^2\alpha^2 t/4L^2} \sin\dfrac{(2n-1)\pi x}{2L}$, $c_n = \dfrac{2}{L}\displaystyle\int_0^L f(x)\sin\dfrac{(2n-1)\pi x}{2L}\,dx$

16. (a) $u(x, t) = \displaystyle\sum_{n=1}^{\infty} c_n e^{-(2n-1)^2\pi^2 t/3600} \sin\dfrac{(2n-1)\pi x}{60}$, $c_n = \dfrac{120}{(2n-1)^2\pi^2}[2\cos n\pi + (2n-1)\pi]$

(c) x_m increases from $x = 0$ and reaches $x = 30$ when $t = 104.4$.

17. (a) $u(x, t) = 40 + \displaystyle\sum_{n=1}^{\infty} c_n e^{-(2n-1)^2\pi^2 t/3600} \sin\dfrac{(2n-1)\pi x}{60}$,

$c_n = \dfrac{40}{(2n-1)^2\pi^2}[6\cos n\pi - (2n-1)\pi]$

19. $u(x) = \begin{cases} T\dfrac{x}{a}\left[\dfrac{\xi}{\xi + (L/a) - 1}\right], & 0 \le x \le a, \\[2mm] T\left[1 - \dfrac{L-x}{a}\dfrac{1}{\xi + (L/a) - 1}\right], & a \le x \le L, \end{cases}$ where $\xi = \kappa_2 A_2/\kappa_1 A_1$

20. (e) $u_n(x, t) = e^{-\mu_n^2\alpha^2 t} \sin\mu_n x$

21. $\alpha^2 v'' + s(x) = 0$; $v(0) = T_1$, $v(L) = T_2$
    $w_t = \alpha^2 w_{xx}$; $w(0, t) = 0$, $w(L, t) = 0$, $w(x, 0) = f(x) - v(x)$

22. (a) $v(x) = T_1 + (T_2 - T_1)(x/L) + kLx/2 - kx^2/2$

(b) $w(x, t) = \displaystyle\sum_{n=1}^{\infty} c_n e^{-n^2\pi^2 t/400} \sin\dfrac{n\pi x}{20}$, $c_n = \dfrac{160(\cos n\pi - 1)}{n^3\pi^3}$

23. (a) $v(x) = T_1 + (T_2 - T_1)x/L + kLx/6 - kx^3/6L$

(b) $w(x, t) = \displaystyle\sum_{n=1}^{\infty} c_n e^{-n^2\pi^2 t/400} \sin\dfrac{n\pi x}{20}$, $c_n = \dfrac{20}{3}\left[\dfrac{3m^3\pi^3(3\cos m\pi - 1) + 60\cos m\pi}{m^4\pi^4}\right]$

Section 9.7, *page 670*

1. (a) $u(x, t) = \dfrac{8}{\pi^2}\displaystyle\sum_{n=1}^{\infty}\dfrac{1}{n^2}\sin\dfrac{n\pi}{2}\sin\dfrac{n\pi x}{L}\cos\dfrac{n\pi a t}{L}$

2. (a) $u(x, t) = \dfrac{8}{\pi^2}\displaystyle\sum_{n=1}^{\infty}\dfrac{1}{n^2}\left(\sin\dfrac{n\pi}{4} + \sin\dfrac{3n\pi}{4}\right)\sin\dfrac{n\pi x}{L}\cos\dfrac{n\pi a t}{L}$

3. (a) $u(x, t) = \dfrac{32}{\pi^3}\displaystyle\sum_{n=1}^{\infty}\dfrac{2 + \cos n\pi}{n^3}\sin\dfrac{n\pi x}{L}\cos\dfrac{n\pi a t}{L}$

4. (a) $u(x, t) = \dfrac{4}{\pi}\displaystyle\sum_{n=1}^{\infty}\dfrac{\sin(n\pi/2)\sin(n\pi/L)}{n}\sin\dfrac{n\pi x}{L}\cos\dfrac{n\pi a t}{L}$

5. (a) $u(x, t) = \dfrac{8L}{a\pi^3}\displaystyle\sum_{n=1}^{\infty}\dfrac{1}{n^3}\sin\dfrac{n\pi}{2}\sin\dfrac{n\pi x}{L}\sin\dfrac{n\pi a t}{L}$

6. (a) $u(x,t) = \dfrac{8L}{a\pi^3} \displaystyle\sum_{n=1}^{\infty} \dfrac{\sin(n\pi/4) + \sin(3n\pi/4)}{n^3} \sin\dfrac{n\pi x}{L} \sin\dfrac{n\pi at}{L}$

7. (a) $u(x,t) = \dfrac{32L}{a\pi^4} \displaystyle\sum_{n=1}^{\infty} \dfrac{\cos n\pi + 2}{n^4} \sin\dfrac{n\pi x}{L} \sin\dfrac{n\pi at}{L}$

8. (a) $u(x,t) = \dfrac{4L}{a\pi^2} \displaystyle\sum_{n=1}^{\infty} \dfrac{\sin(n\pi/2)\sin(n\pi/L)}{n^2} \sin\dfrac{n\pi x}{L} \sin\dfrac{n\pi at}{L}$

9. $u(x,t) = \displaystyle\sum_{n=1}^{\infty} c_n \sin\dfrac{(2n-1)\pi x}{2L} \cos\dfrac{(2n-1)\pi at}{2L}$, $\quad c_n = \dfrac{2}{L}\displaystyle\int_0^L f(x)\sin\dfrac{(2n-1)\pi x}{2L}\,dx$

10. (a) $u(x,t) = \dfrac{8}{\pi} \displaystyle\sum_{n=1}^{\infty} \dfrac{1}{2n-1} \sin\dfrac{(2n-1)\pi}{4} \sin\dfrac{(2n-1)\pi}{2L} \sin\dfrac{(2n-1)\pi x}{2L} \cos\dfrac{(2n-1)\pi at}{2L}$

11. (a) $u(x,t) = \dfrac{512}{\pi^4} \displaystyle\sum_{n=1}^{\infty} \dfrac{(2n-1)\pi + 3\cos n\pi}{(2n-1)^4} \sin\dfrac{(2n-1)\pi x}{2L} \cos\dfrac{(2n-1)\pi at}{2L}$

14. (b) $\phi(x+at)$ represents a wave moving in the negative x direction with speed $a > 0$.

15. (a) 248 ft/s (b) $49.6\pi n$ rad/s (c) Frequencies increase; modes are unchanged.

21. $r^2 R'' + rR' + (\lambda^2 r^2 - \mu^2)R = 0$, $\quad \Theta'' + \mu^2\Theta = 0$, $\quad T'' + \lambda^2 a^2 T = 0$

23. (b) $a_n = a\sqrt{1 + (\gamma^2 L^2/n^2\pi^2)}$ (c) $\gamma = 0$

24. (a) $c_n = \dfrac{20}{n^2\pi^2}\left(2\sin\dfrac{n\pi}{2} - \sin\dfrac{2n\pi}{5} - \sin\dfrac{3n\pi}{5}\right)$

Section 9.8, *page 681*

1. (a) $u(x,y) = \displaystyle\sum_{n=1}^{\infty} c_n \sin\dfrac{n\pi x}{a} \sinh\dfrac{n\pi y}{a}$, $\quad c_n = \dfrac{2/a}{\sinh(n\pi b/a)}\displaystyle\int_0^a g(x)\sin\dfrac{n\pi x}{a}\,dx$

   (b) $u(x,y) = \dfrac{4a}{\pi^2}\displaystyle\sum_{n=1}^{\infty}\dfrac{1}{n^2}\dfrac{\sin(n\pi/2)}{\sinh(n\pi b/a)}\sin\dfrac{n\pi x}{a}\sinh\dfrac{n\pi y}{a}$

2. $u(x,y) = \displaystyle\sum_{n=1}^{\infty} c_n \sin\dfrac{n\pi x}{a} \sinh\dfrac{n\pi(b-y)}{a}$, $\quad c_n = \dfrac{2/a}{\sinh(n\pi b/a)}\displaystyle\int_0^a h(x)\sin\dfrac{n\pi x}{a}\,dx$

3. (a) $u(x,y) = \displaystyle\sum_{n=1}^{\infty} c_n^{(1)} \sinh\dfrac{n\pi x}{b} \sin\dfrac{n\pi y}{b} + \displaystyle\sum_{n=1}^{\infty} c_n^{(2)} \sin\dfrac{n\pi x}{a} \sinh\dfrac{n\pi(b-y)}{a}$,

   $c_n^{(1)} = \dfrac{2/b}{\sinh(n\pi a/b)}\displaystyle\int_0^b f(y)\sin\dfrac{n\pi y}{b}\,dy$, $\; c_n^{(2)} = \dfrac{2/a}{\sinh(n\pi b/a)}\displaystyle\int_0^a h(x)\sin\dfrac{n\pi x}{a}\,dx$

   (b) $c_n^{(1)} = \dfrac{2}{n\pi\sinh(n\pi a/b)}$, $\; c_n^{(2)} = -\dfrac{2}{n^3\pi^3}\dfrac{(n^2\pi^2-2)\cos n\pi + 2}{\sinh(n\pi b/a)}$

5. $u(r,\theta) = \dfrac{c_0}{2} + \displaystyle\sum_{n=1}^{\infty} r^{-n}(c_n\cos n\theta + k_n\sin n\theta)$;

   $c_n = \dfrac{a^n}{\pi}\displaystyle\int_0^{2\pi} f(\theta)\cos n\theta\,d\theta$, $\; k_n = \dfrac{a^n}{\pi}\displaystyle\int_0^{2\pi} f(\theta)\sin n\theta\,d\theta$

6. (a) $u(r,\theta) = \displaystyle\sum_{n=1}^{\infty} c_n r^n \sin n\theta$, $\; c_n = \dfrac{2}{\pi a^n}\displaystyle\int_0^{\pi} f(\theta)\sin n\theta\,d\theta$

   (b) $c_n = \dfrac{4}{\pi a^n}\dfrac{1-\cos n\pi}{n^3}$

7. $u(r,\theta) = \displaystyle\sum_{n=1}^{\infty} c_n r^{n\pi/\alpha} \sin\dfrac{n\pi\theta}{\alpha}$, $\; c_n = (2/\alpha)a^{-n\pi/\alpha}\displaystyle\int_0^{\alpha} f(\theta)\sin\dfrac{n\pi\theta}{\alpha}\,d\theta$

8. (a) $u(x,y) = \displaystyle\sum_{n=1}^{\infty} c_n e^{-n\pi y/a} \sin\dfrac{n\pi x}{a}$, $\; c_n = \dfrac{2}{a}\displaystyle\int_0^a f(x)\sin\dfrac{n\pi x}{a}\,dx$

(b) $c_n = \dfrac{4a^2}{n^3\pi^3}(1 - \cos n\pi)$ (c) $y_0 \cong 6.6315$

10. (b) $u(x, y) = c_0 + \displaystyle\sum_{n=1}^{\infty} c_n\cosh\dfrac{n\pi x}{b}\cos\dfrac{n\pi y}{b}$, $c_n = \dfrac{2/n\pi}{\sinh(n\pi a/b)}\displaystyle\int_0^b f(y)\cos\dfrac{n\pi y}{b}\,dy$

11. $u(r, \theta) = c_0 + \displaystyle\sum_{n=1}^{\infty} r^n(c_n\cos n\theta + k_n\sin n\theta), c_n = \dfrac{1}{n\pi a^{n-1}}\displaystyle\int_0^{2\pi} g(\theta)\cos n\theta\,d\theta$,

$k_n = \dfrac{1}{n\pi a^{n-1}}\displaystyle\int_0^{2\pi} g(\theta)\sin n\theta\,d\theta$; necessary condition is $\displaystyle\int_0^{2\pi} g(\theta)\,d\theta = 0$.

12. (a) $u(x, y) = \displaystyle\sum_{n=1}^{\infty} c_n\sin\dfrac{n\pi x}{a}\cosh\dfrac{n\pi y}{a}$, $c_n = \dfrac{2/a}{\cosh(n\pi b/a)}\displaystyle\int_0^a g(x)\sin\dfrac{n\pi x}{a}\,dx$

(b) $c_n = \dfrac{4a\sin(n\pi/2)}{n^2\pi^2\cosh(n\pi b/a)}$

13. (a) $u(x, y) = \displaystyle\sum_{n=1}^{\infty} c_n\sinh\dfrac{(2n-1)\pi x}{2b}\sin\dfrac{(2n-1)\pi y}{2b}$,

$c_n = \dfrac{2/b}{\sinh[(2n-1)\pi a/2b]}\displaystyle\int_0^b f(y)\sin\dfrac{(2n-1)\pi y}{2b}\,dy$

(b) $c_n = \dfrac{32b^2}{(2n-1)^3\pi^3\sinh[(2n-1)\pi a/2b]}$

14. (a) $u(x, y) = \dfrac{c_0 y}{2} + \displaystyle\sum_{n=1}^{\infty} c_n\cos\dfrac{n\pi x}{a}\sinh\dfrac{n\pi y}{a}$,

$c_0 = \dfrac{2}{ab}\displaystyle\int_0^a g(x)\,dx$, $c_n = \dfrac{2/a}{\sinh(n\pi b/a)}\displaystyle\int_0^a g(x)\cos\dfrac{n\pi x}{a}\,dx$

(b) $c_0 = \dfrac{2}{b}\left(1 + \dfrac{a^4}{30}\right)$, $c_n = -\dfrac{24a^4(1 + \cos n\pi)}{n^4\pi^4\sinh(n\pi b/a)}$

16. (a) $u(x, z) = b + \dfrac{\alpha a}{2} - \dfrac{4\alpha a}{\pi^2}\displaystyle\sum_{n=1}^{\infty}\dfrac{\cos[(2n-1)\pi x/a]\cosh[(2n-1)\pi z/a]}{(2n-1)^2\cosh[(2n-1)\pi b/a]}$

CHAPTER 10

Section 10.1, *page 711*

1. Homogeneous

2. Nonhomogeneous

3. Nonhomogeneous

4. Homogeneous

5. Nonhomogeneous

6. Homogeneous

7. (a) $\phi_n(x) = \sin\sqrt{\lambda_n}\,x$, where $\sqrt{\lambda_n}$ satisfies $\sqrt{\lambda} = -\tan\sqrt{\lambda}\,\pi$; (b) No
   (c) $\lambda_1 \cong 0.6204$, $\lambda_2 \cong 2.7943$
   (d) $\lambda_n \cong (2n-1)^2/4$ for large n

8. (a) $\phi_n(x) = \cos\sqrt{\lambda_n}\,x$, where $\sqrt{\lambda_n}$ satisfies $\sqrt{\lambda} = \cot\sqrt{\lambda}$; (b) No
   (c) $\lambda_1 \cong 0.7402$, $\lambda_2 \cong 11.7349$
   (d) $\lambda_n \cong (n-1)^2\pi^2$ for large n

9. (a) $\phi_n(x) = \sin\sqrt{\lambda_n}\,x + \sqrt{\lambda_n}\cos\sqrt{\lambda_n}\,x$, where $\sqrt{\lambda_n}$ satisfies
   $(\lambda - 1)\sin\sqrt{\lambda} - 2\sqrt{\lambda}\cos\sqrt{\lambda} = 0$; (b) No
   (c) $\lambda_1 \cong 1.7071$, $\lambda_2 \cong 13.4924$
   (d) $\lambda_n \cong (n-1)^2\pi^2$ for large n

10. (a) For $n = 1, 2, 3, \ldots$, $\phi_n(x) = \sin\mu_n x - \mu_n\cos\mu_n x$ and $\lambda_n = -\mu_n^2$, where μ_n satisfies $\mu = \tan\mu$.
    (b) Yes; $\lambda_0 = 0$, $\phi_0(x) = 1 - x$
    (c) $\lambda_1 \cong -20.1907$, $\lambda_2 \cong -59.6795$ (d) $\lambda_n \cong -(2n+1)^2\pi^2/4$ for large n

12. $\mu(x) = e^{-x^2}$

13. $\mu(x) = 1/x$

14. $\mu(x) = e^{-x}$

15. $\mu(x) = (1 - x^2)^{-1/2}$

16. $X'' + \lambda X = 0, \quad T'' + cT' + (k + \lambda a^2)T = 0$

17. (a) $s(x) = e^x$ (b) $\lambda_n = n^2\pi^2, \quad \phi_n(x) = e^x \sin n\pi x; \quad n = 1, 2, 3, \ldots$

18. Positive eigenvalues $\lambda = \lambda_n$, where $\sqrt{\lambda_n}$ satisfies $\sqrt{\lambda} = \frac{2}{3}\tan 3\sqrt{\lambda}L$; corresponding eigenfunctions are $\phi_n(x) = e^{-2x} \sin 3\sqrt{\lambda_n}\, x$. If $L = \frac{1}{2}, \lambda_0 = 0$ is eigenvalue, $\phi_0(x) = xe^{-2x}$ is eigenfunction; if $L \neq \frac{1}{2}, \lambda = 0$ is not eigenvalue. If $L \leq \frac{1}{2}$, there are no negative eigenvalues; if $L > \frac{1}{2}$, there is one negative eigenvalue $\lambda = -\mu^2$, where μ is a root of $\mu = \frac{2}{3}\tanh 3\mu L$; corresponding eigenfunction is $\phi_{-1}(x) = e^{-2x}\sinh 3\mu x$.

19. No real eigenvalues.

20. Only eigenvalue is $\lambda = 0$; eigenfunction is $\phi(x) = x - 1$.

21. (a) $2\sin\sqrt{\lambda} - \sqrt{\lambda}\cos\sqrt{\lambda} = 0$
    (c) $\lambda_1 \cong 18.2738, \quad \lambda_2 \cong 57.7075$
    (d) $2\sinh\sqrt{\mu} - \sqrt{\mu}\cosh\sqrt{\mu} = 0, \quad \mu = -\lambda$
    (e) $\lambda_{-1} \cong -3.6673$

24. (a) $\lambda_n = \mu_n^4$, where μ_n is a root of $\sin\mu L\sinh\mu L = 0$, hence $\lambda_n = (n\pi/L)^4$;
    $\lambda_1 \cong 97.409/L^4, \quad \lambda_2 \cong 1558.5/L^4, \quad \phi_n(x) = \sin(n\pi x/L)$
    (b) $\lambda_n = \mu_n^4$, where μ_n is a root of $\sin\mu L\cosh\mu L - \cos\mu L\sinh\mu L = 0$;
    $\lambda_1 \cong 237.72/L^4, \lambda_2 \cong 2496.5/L^4, \phi_n = \dfrac{\sin\mu_n x\sinh\mu_n L - \sin\mu_n L\sinh\mu_n x}{\sinh\mu_n L}$
    (c) $\lambda_n = \mu_n^4$, where μ_n is a root of $1 + \cosh\mu L\cos\mu L = 0; \quad \lambda_1 \cong 12.362/L^4$,
    $\lambda_2 \cong 485.52/L^4$
    $$\phi_n(x) = \frac{[(\sin\mu_n x - \sinh\mu_n x)(\cos\mu_n L + \cosh\mu_n L) + (\sin\mu_n L + \sinh\mu_n L)(\cosh\mu_n x - \cos\mu_n x)]}{\cos\mu_n L + \cosh\mu_n L}$$

25. (c) $\phi_n(x) = \sin\sqrt{\lambda_n}\, x$, where λ_n satisfies $\cos\sqrt{\lambda_n}\, L - \gamma\sqrt{\lambda_n}\, L\sin\sqrt{\lambda_n}\, L = 0$
    (d) $\lambda_1 \cong 1.1597/L^2, \quad \lambda_2 \cong 13.276/L^2$

Section 10.2, *page 723*

1. $\phi_n(x) = \sqrt{2}\sin(n - \frac{1}{2})\pi x; \quad n = 1, 2, \ldots$ 2. $\phi_n(x) = \sqrt{2}\cos(n - \frac{1}{2})\pi x; \quad n = 1, 2, \ldots$

3. $\phi_0(x) = 1, \quad \phi_n(x) = \sqrt{2}\cos n\pi x; \quad n = 1, 2, \ldots$

4. $\phi_n(x) = \dfrac{\sqrt{2}\cos\sqrt{\lambda_n}\, x}{(1 + \sin^2\sqrt{\lambda_n})^{1/2}}$, where λ_n satisfies $\cos\sqrt{\lambda_n} - \sqrt{\lambda_n}\sin\sqrt{\lambda_n} = 0$

5. $\phi_n(x) = \sqrt{2}\, e^x \sin n\pi x; \quad n = 1, 2, \ldots$ 6. $a_n = \dfrac{2\sqrt{2}}{(2n - 1)\pi}; \quad n = 1, 2, \ldots$

7. $a_n = \dfrac{4\sqrt{2}(-1)^{n-1}}{(2n - 1)^2\pi^2}; \quad n = 1, 2, \ldots$

8. $a_n = \dfrac{2\sqrt{2}}{(2n - 1)\pi}\{1 - \cos[(2n - 1)\pi/4]\}; \quad n = 1, 2, \ldots$

9. $a_n = \dfrac{2\sqrt{2}\sin(n - \frac{1}{2})(\pi/2)}{(n - \frac{1}{2})^2\pi^2}; \quad n = 1, 2, \ldots$

In Problems 10 through 13, $\alpha_n = (1 + \sin^2\sqrt{\lambda_n})^{1/2}$ and $\cos\sqrt{\lambda_n} - \sqrt{\lambda_n}\sin\sqrt{\lambda_n} = 0$.

10. $a_n = \dfrac{\sqrt{2}\sin\sqrt{\lambda_n}}{\sqrt{\lambda_n}\alpha_n}; \quad n = 1, 2, \ldots$ 11. $a_n = \dfrac{\sqrt{2}(2\cos\sqrt{\lambda_n} - 1)}{\lambda_n\alpha_n}; \quad n = 1, 2, \ldots$

12. $a_n = \dfrac{\sqrt{2}(1 - \cos\sqrt{\lambda_n})}{\lambda_n\alpha_n}$; $n = 1, 2, \ldots$ 13. $a_n = \dfrac{\sqrt{2}\sin(\sqrt{\lambda_n}/2)}{\sqrt{\lambda_n}\alpha_n}$; $n = 1, 2, \ldots$

14. Not self-adjoint   15. Self-adjoint

16. Not self-adjoint   17. Self-adjoint

18. Self-adjoint

21. (a) If $a_2 = 0$ or $b_2 = 0$, then the corresponding boundary term is missing.

25. (a) $\lambda_1 = \pi^2/L^2$; $\phi_1(x) = \sin(\pi x/L)$

     (b) $\lambda_1 \cong (4.4934)^2/L^2$; $\phi_1(x) = \sin\sqrt{\lambda_1}\,x - \sqrt{\lambda_1}\,x\cos\sqrt{\lambda_1}\,L$

     (c) $\lambda_1 = (2\pi)^2/L^2$; $\phi_1(x) = 1 - \cos(2\pi x/L)$

26. $\lambda_1 = \pi^2/4L^2$; $\phi_1(x) = 1 - \cos(\pi x/2L)$

27. (a) $X'' - (v/D)X' + \lambda X = 0$, $X(0) = 0$, $X'(L) = 0$; $T' + \lambda DT = 0$

     (e) $c(x, t) = \displaystyle\sum_{n=1}^{\infty} a_n e^{-\lambda_n Dt} e^{vx/2D} \sin\mu_n x$, where $\lambda_n = \mu_n^2 + (v^2/4D^2)$;

$$a_n = \dfrac{4D\mu_n^2 \displaystyle\int_0^L e^{-vx/2D} f(x)\sin\mu_n x\,dx}{(2LD\mu_n^2 + v\sin^2\mu_n L)}$$

28. (a) $u_t + vu_x = Du_{xx}$, $u(0, t) = 0$, $u_x(L, t) = 0$, $u(x, 0) = -c_0$

     (b) $u(x, t) = \displaystyle\sum_{n=1}^{\infty} b_n e^{-\lambda_n Dt} e^{vx/2D} \sin\mu_n x$, where $\lambda_n = \mu_n^2 + (v^2/4D^2)$;

$$b_n = \dfrac{8c_0 D^2\mu^2(2D\mu_n e^{-vL/2D}\cos\mu_n L + ve^{-vL/2D}\sin\mu_n L - 2D\mu_n)}{(v^2 + 4D^2\mu_n^2)(2LD\mu_n^2 + v\sin^2\mu_n L)}$$

Section 10.3, *page 735*

1. $y = 2\displaystyle\sum_{n=1}^{\infty} \dfrac{(-1)^{n+1}\sin n\pi x}{(n^2\pi^2 - 2)n\pi}$ 2. $y = 2\displaystyle\sum_{n=1}^{\infty} \dfrac{(-1)^{n+1}\sin(n - \frac{1}{2})\pi x}{[(n - \frac{1}{2})^2\pi^2 - 2](n - \frac{1}{2})^2\pi^2}$

3. $y = -\dfrac{1}{4} - 4\displaystyle\sum_{n=1}^{\infty} \dfrac{\cos(2n - 1)\pi x}{[(2n - 1)^2\pi^2 - 2](2n - 1)^2\pi^2}$

4. $y = 2\displaystyle\sum_{n=1}^{\infty} \dfrac{(2\cos\sqrt{\lambda_n} - 1)\cos\sqrt{\lambda_n}\,x}{\lambda_n(\lambda_n - 2)(1 + \sin^2\sqrt{\lambda_n})}$ 5. $y = 8\displaystyle\sum_{n=1}^{\infty} \dfrac{\sin(n\pi/2)\sin n\pi x}{(n^2\pi^2 - 2)n^2\pi^2}$

6–9. For each problem the solution is

$$y = \sum_{n=1}^{\infty} \dfrac{c_n}{\lambda_n - \mu}\phi_n(x), \quad c_n = \int_0^1 f(x)\phi_n(x)\,dx, \quad \mu \neq \lambda_n,$$

where $\phi_n(x)$ is given in Problems 1–4, respectively, in Section 11.2, and λ_n is the corresponding eigenvalue. In Problem 8 summation starts at $n = 0$.

10. $a = -\dfrac{1}{2}$, $y = \dfrac{1}{2\pi^2}\cos\pi x + \dfrac{1}{\pi^2}\left(x - \dfrac{1}{2}\right) + c\sin\pi x$

11. No solution   12. a is arbitrary, $y = c\cos\pi x + a/\pi^2$

13. $a = 0$, $y = c\sin\pi x - (x/2\pi)\sin\pi x$ 17. $v(x) = a + (b - a)x$

18. $v(x) = 1 - \frac{3}{2}x$

19. $u(x,t) = \sqrt{2}\left[-\dfrac{4c_1}{\pi^2} + \left(\dfrac{4c_1}{\pi^2} + \dfrac{1}{\sqrt{2}}\right)e^{-\pi^2 t/4}\right]\sin\dfrac{\pi x}{2}$

$\qquad -\sqrt{2}\displaystyle\sum_{n=2}^{\infty}\dfrac{4c_n}{(2n-1)^2\pi^2}[1 - e^{-(n-1/2)^2\pi^2 t}]\sin(n-\tfrac{1}{2})\pi x,$

$\qquad c_n = \dfrac{4\sqrt{2}(-1)^{n+1}}{(2n-1)^2\pi^2},\quad n = 1,2,\ldots$

20. $u(x,t) = \sqrt{2}\displaystyle\sum_{n=1}^{\infty}\left[\dfrac{c_n}{\lambda_n - 1}(e^{-t} - e^{-\lambda_n t}) + \alpha_n e^{-\lambda_n t}\right]\dfrac{\cos\sqrt{\lambda_n}\,x}{(1 + \sin^2\sqrt{\lambda_n})^{1/2}},$

$\qquad c_n = \dfrac{\sqrt{2}\sin\sqrt{\lambda_n}}{\sqrt{\lambda_n}(1 + \sin^2\sqrt{\lambda_n})^{1/2}},\quad \alpha_n = \dfrac{\sqrt{2}(1 - \cos\sqrt{\lambda_n})}{\lambda_n(1 + \sin^2\sqrt{\lambda_n})^{1/2}},$

$\qquad$ and λ_n satisfies $\cos\sqrt{\lambda_n} - \sqrt{\lambda_n}\sin\sqrt{\lambda_n} = 0.$

21. $u(x,t) = 8\displaystyle\sum_{n=1}^{\infty}\dfrac{\sin(n\pi/2)}{n^4\pi^4}(1 - e^{-n^2\pi^2 t})\sin n\pi x$

22. $u(x,t) = \sqrt{2}\displaystyle\sum_{n=1}^{\infty}\dfrac{c_n(e^{-t} - e^{-(n-1/2)^2\pi^2 t})\sin(n-\tfrac{1}{2})\pi x}{(n-\tfrac{1}{2})^2\pi^2 - 1},$

$\qquad c_n = \dfrac{2\sqrt{2}(2n-1)\pi + 4\sqrt{2}(-1)^n}{(2n-1)^2\pi^2}$

23. (a) $r(x)w_t = [p(x)w_x]_x - q(x)w,\quad w(0,t) = 0,\quad w(1,t) = 0,\quad w(x,0) = f(x) - v(x)$

24. $u(x,t) = x^2 - 2x + 1 + \dfrac{4}{\pi}\displaystyle\sum_{n=1}^{\infty}\dfrac{e^{-(2n-1)^2\pi^2 t}\sin(2n-1)\pi x}{2n-1}$

25. $u(x,t) = -\cos\pi x + e^{-9\pi^2 t/4}\cos(3\pi x/2)$

31–34. In all cases solution is $y = \displaystyle\int_0^1 G(x,s)f(s)\,ds$, where $G(x,s)$ is given below.

31. $G(x,s) = \begin{cases} 1-x, & 0 \le s \le x \\ 1-s, & x \le s \le 1 \end{cases}$

32. $G(x,s) = \begin{cases} s(2-x)/2, & 0 \le s \le x \\ x(2-s)/2, & x \le s \le 1 \end{cases}$

33. $G(x,s) = \begin{cases} \cos s\sin(1-x)/\cos 1, & 0 \le s \le x \\ \sin(1-s)\cos x/\cos 1, & x \le s \le 1 \end{cases}$

34. $G(x,s) = \begin{cases} s, & 0 \le s \le x \\ x, & x \le s \le 1 \end{cases}$

Section 10.4, *page 744*

1. $y = \displaystyle\sum_{n=1}^{\infty}\dfrac{c_n}{\lambda_n - \mu}J_0(\sqrt{\lambda_n}\,x),\quad c_n = \int_0^1 f(x)J_0(\sqrt{\lambda_n}\,x)\,dx\Big/\int_0^1 xJ_0^2(\sqrt{\lambda_n}\,x)\,dx,$
$\qquad \sqrt{\lambda_n}$ satisfies $J_0(\sqrt{\lambda}) = 0.$

2. (c) $y = -\dfrac{c_0}{\mu} + \displaystyle\sum_{n=1}^{\infty}\dfrac{c_n}{\lambda_n - \mu}J_0(\sqrt{\lambda_n}\,x);$
$\qquad c_0 = 2\int_0^1 f(x)\,dx;\ c_n = \int_0^1 f(x)J_0(\sqrt{\lambda_n}\,x)\,dx\Big/\int_0^1 xJ_0^2(\sqrt{\lambda_n}\,x)\,dx,\ n = 1,2,\ldots;$
$\qquad \sqrt{\lambda_n}$ satisfies $J_0'(\sqrt{\lambda}) = 0.$

3. (d) $a_n = \displaystyle\int_0^1 x J_k(\sqrt{\lambda_n}\, x) f(x)\, dx \Big/ \int_0^1 x J_k^2(\sqrt{\lambda_n}\, x)\, dx$

(e) $y = \displaystyle\sum_{n=1}^{\infty} \frac{c_n}{\lambda_n - \mu} J_k(\sqrt{\lambda_n}\, x), \quad c_n = \int_0^1 f(x) J_k(\sqrt{\lambda_n}\, x)\, dx \Big/ \int_0^1 x J_k^2(\sqrt{\lambda_n}\, x)\, dx$

4. (b) $y = \displaystyle\sum_{n=1}^{\infty} \frac{c_n}{\lambda_n - \mu} P_{2n-1}(x), \quad c_n = \int_0^1 f(x) P_{2n-1}(x)\, dx \Big/ \int_0^1 P_{2n-1}^2(x)\, dx$

Section 10.5, *page 749*

1. (b) $u(\xi, 2) = f(\xi + 1), \quad u(\xi, 0) = 0, \quad 0 \le \xi \le 2$
$u(0, \eta) = u(2, \eta) = 0, \quad 0 \le \eta \le 2$

2. $u(r, t) = \displaystyle\sum_{n=1}^{\infty} k_n J_0(\lambda_n r) \sin \lambda_n a t, \quad k_n = \frac{1}{\lambda_n a} \int_0^1 r J_0(\lambda_n r) g(r)\, dr \Big/ \int_0^1 r J_0^2(\lambda_n r)\, dr$

3. Superpose the solution of Problem 2 and the solution [Eq. (21)] of the example in the text.

6. $u(r, z) = \displaystyle\sum_{n=1}^{\infty} c_n e^{-\lambda_n z} J_0(\lambda_n r), \quad c_n = \int_0^1 r J_0(\lambda_n r) f(r)\, dr \Big/ \int_0^1 r J_0^2(\lambda_n r)\, dr,$
and λ_n satisfies $J_0(\lambda) = 0$.

7. (b) $v(r, \theta) = \frac{1}{2} c_0 J_0(kr) + \displaystyle\sum_{m=1}^{\infty} J_m(kr)(b_m \sin m\theta + c_m \cos m\theta),$

$b_m = \dfrac{1}{\pi J_m(kc)} \displaystyle\int_0^{2\pi} f(\theta) \sin m\theta\, d\theta; \quad m = 1, 2, \ldots$

$c_m = \dfrac{1}{\pi J_m(kc)} \displaystyle\int_0^{2\pi} f(\theta) \cos m\theta\, d\theta; \quad m = 0, 1, 2, \ldots$

8. $c_n = \displaystyle\int_0^1 r f(r) J_0(\lambda_n r)\, dr \Big/ \int_0^1 r J_0^2(\lambda_n r)\, dr$

10. $u(\rho, s) = \displaystyle\sum_{n=0}^{\infty} c_n \rho^n P_n(s), \quad \text{where } c_n = \int_{-1}^1 f(\arccos s) P_n(s)\, ds \Big/ \int_{-1}^1 P_n^2(s)\, ds;$
P_n is the nth Legendre polynomial and $s = \cos \phi$.

Section 10.6, *page 756*

1. $n = 21$ 2. (a) $b_m = (-1)^{m+1} \sqrt{2}/m\pi$ (c) $n = 20$

3. (a) $b_m = 2\sqrt{2}(1 - \cos m\pi)/m^3\pi^3$ (c) $n = 1$

7. (a) $f_0(x) = 1$ (b) $f_1(x) = \sqrt{3}(1 - 2x)$ (c) $f_2(x) = \sqrt{5}(-1 + 6x - 6x^2)$
(d) $g_0(x) = 1, \quad g_1(x) = 2x - 1, \quad g_2(x) = 6x^2 - 6x + 1$

8. $P_0(x) = 1, \quad P_1(x) = x, \quad P_2(x) = (3x^2 - 1)/2, \quad P_3(x) = (5x^3 - 3x)/2$

APPENDIX A.1, *page 785*

1. (a) $\begin{pmatrix} 8 & 8 \\ -4 & 12 \end{pmatrix}$ (b) $\begin{pmatrix} -4 \\ -6 \end{pmatrix}$ (c) not defined

(d) $\begin{pmatrix} 10 \\ 7 \end{pmatrix}$ (e) not defined   (f) $\begin{pmatrix} 0 & 8 & 4 \\ 4 & 8 & 14 \end{pmatrix}$

(g) not defined

(h) $\begin{pmatrix} -3 & 1 \\ -1 & 11 \\ -8 & 8 \end{pmatrix}$

(i) $\begin{pmatrix} 3 & 3 \end{pmatrix}$

(j) $\begin{pmatrix} 0 & 4 \\ 8 & 8 \\ 4 & 14 \end{pmatrix}$

(k) 41

(l) not defined

2. (a) $\begin{pmatrix} 3 & -4 & 2 \\ 6 & -3 & -3 \\ -2 & 1 & 9 \end{pmatrix}$

(b) $\begin{pmatrix} -3 & -2 & -8 \\ 3 & -6 & 3 \\ -10 & 5 & -9 \end{pmatrix}$

(c) $\begin{pmatrix} 1 & -2 & 4 \\ 1 & -1 & 5 \\ 4 & -2 & 4 \end{pmatrix}$

(d) $\begin{pmatrix} -3 & 0 & 6 \\ 5 & -3 & -4 \\ -7 & 1 & 10 \end{pmatrix}$

(e) $\begin{pmatrix} 2 & 3 & 0 \\ -2 & -1 & 0 \\ 2 & -2 & 6 \end{pmatrix}$

(f) $\begin{pmatrix} 6 & 8 & 0 \\ 5 & 13 & -13 \\ 0 & -6 & 18 \end{pmatrix}$

(g) $\begin{pmatrix} -5 & 2 & 2 \\ -1 & -3 & -1 \\ -5 & 5 & 8 \end{pmatrix}$

(h) $\begin{pmatrix} 4 & -2 & -2 \\ -4 & 2 & 9 \\ 11 & -3 & -6 \end{pmatrix}$

(i) $\begin{pmatrix} 0 & 0 & 0 \\ 0 & 0 & 0 \\ 0 & 0 & 0 \end{pmatrix}$

5. $\mathbf{A}^{-1} = \mathrm{diag}(a_{11}^{-1}, \ldots, a_{nn}^{-1})$ if $a_{kk} \neq 0$ for $k = 1, \ldots n$

6. If $\mathbf{A}$ and $\mathbf{B}$ are upper triangular, $a_{ij} = b_{ij} = 0$ for $i > j$ implies $c_{ij} = a_{ij} + b_{ij} = 0$ for $i > j$. If $i > j$, then $c_{ij} = \sum_{k=1}^{n} a_{ik}b_{kj} = 0$ since $a_{ik} = 0$ for $k < i$ and $b_{kj} = 0$ if $k > j$.

7. Let $\delta_{ij} = 0$ if $i \neq j$ and $\delta_{ij} = 1$ if $i = j$. Then diagonal $\mathbf{A}$ and $\mathbf{B}$ can be written as $\mathbf{A} = (a_{ii}\delta_{ij})$ and $\mathbf{B} = (\delta_{ij}b_{jj})$. If $\mathbf{C} = \mathbf{AB}$, then $c_{ij} = \sum_{k=1}^{n} a_{ii}\delta_{ik}\delta_{kj}b_{jj}$ is zero if $i \neq j$ and $c_{ii} = a_{ii}b_{ii}$. Use induction with $\mathbf{B} = \mathbf{A}^{n-1} = (a_{ii}^{n-1}\delta_{ij})$.

8. Take the transpose of $\mathbf{AA}^{-1} = \mathbf{I}_n$ to get $(\mathbf{A}^{-1})^T\mathbf{A}^T = (\mathbf{A}^{-1})^T\mathbf{A} = \mathbf{I}_n$. Similarly, take the transpose of $\mathbf{A}^{-1}\mathbf{A} = \mathbf{I}_n$ to get $\mathbf{A}^T(\mathbf{A}^{-1})^T = \mathbf{A}(\mathbf{A}^{-1})^T = \mathbf{I}_n$. It follows that $(\mathbf{A}^{-1})^T = \mathbf{A}^{-1}$.

9. If $\mathbf{A}$, $\mathbf{B}$, and $\mathbf{AB}$ are symmetric, then $\mathbf{BA} = \mathbf{B}^T\mathbf{A}^T = (\mathbf{AB})^T = \mathbf{AB}$. Conversely, if $\mathbf{A}$ and $\mathbf{B}$ are symmetric and $\mathbf{A}$ and $\mathbf{B}$ commute, then $\mathbf{AB} = \mathbf{BA} = \mathbf{B}^T\mathbf{A}^T = (\mathbf{AB})^T$.

10. (a) $(\mathbf{AA}^T)^T = (\mathbf{A}^T)^T\mathbf{A}^T = \mathbf{AA}^T$ and $(\mathbf{A}^T\mathbf{A})^T = \mathbf{A}^T(\mathbf{A}^T)^T = \mathbf{A}^T\mathbf{A}$.
(b) $(\mathbf{A} + \mathbf{A}^T)^T = \mathbf{A}^T + ((\mathbf{A})^T)^T = \mathbf{A}^T + \mathbf{A}$.

APPENDIX A.2, *page 802*

1. (a) $\begin{pmatrix} 1 & -3 & 4 \\ 0 & 0 & 0 \end{pmatrix}$, rank($\mathbf{A}$) = 1

(b) $\begin{pmatrix} 1 & 0 & 0 \\ 0 & 1 & 0 \\ 0 & 0 & 1 \end{pmatrix}$, rank($\mathbf{A}$) = 3

(c) $\begin{pmatrix} 1 & 0 & 0 \\ 0 & 1 & 0 \\ 0 & 0 & 1 \end{pmatrix}$, rank($\mathbf{A}$) = 3

(d) $\begin{pmatrix} 1 & 0 & -2 & -1 \\ 0 & 1 & -1 & -1 \\ 0 & 0 & 0 & 0 \\ 0 & 0 & 0 & 0 \end{pmatrix}$, rank($\mathbf{A}$) = 2

2. $\mathbf{x} = \mathbf{0}$

3. $\mathbf{x} = \alpha \begin{pmatrix} 1 \\ -1 \\ -1 \end{pmatrix}$

4. $\mathbf{x} = \alpha \begin{pmatrix} 1 \\ 8 \\ 3 \end{pmatrix}$

5. $\mathbf{x} = \alpha \begin{pmatrix} -1 \\ 0 \\ 3 \\ 1 \end{pmatrix}$

6. $\mathbf{x} = c_1 \begin{pmatrix} 1 \\ -3 \\ 1 \end{pmatrix} + \begin{pmatrix} -1 \\ 4 \\ 0 \end{pmatrix}$

7. no solution

8. $\mathbf{x} = \begin{pmatrix} 0 \\ 1 \\ 1 \end{pmatrix}$

9. $\mathbf{x} = c_1 \begin{pmatrix} -2 \\ -1 \\ 1 \\ 0 \end{pmatrix} + c_2 \begin{pmatrix} -4 \\ -3 \\ 0 \\ 1 \end{pmatrix} + \begin{pmatrix} 1 \\ 2 \\ 0 \\ 0 \end{pmatrix}$

10. linearly independent

11. $\mathbf{v}_1 - 5\mathbf{v}_2 + 2\mathbf{v}_3 = \mathbf{0}$

12. $-2\mathbf{v}_1 + 3\mathbf{v}_2 - 4\mathbf{v}_3 + \mathbf{v}_4 = \mathbf{0}$

13. linearly independent

14. $\mathbf{v}_1 + \mathbf{v}_2 - \mathbf{v}_4 = \mathbf{0}$

15. $\mathbf{A} \begin{pmatrix} -4 \\ 1 \\ -4 \end{pmatrix} = \mathbf{b}_1; \qquad \mathbf{A} \begin{pmatrix} -3/4 \\ 1/4 \\ -1 \end{pmatrix} = \mathbf{b}_2$

16. $\mathbf{b}_1 \notin S_{\text{col}}(\mathbf{A})$, general solution of $\mathbf{Ax} = \mathbf{b}_2$ is $\mathbf{x} = c_1 \begin{pmatrix} 2 \\ -1 \\ 1 \\ 0 \end{pmatrix} + c_2 \begin{pmatrix} -1 \\ 2 \\ 0 \\ 1 \end{pmatrix} + \begin{pmatrix} 0 \\ -2 \\ 0 \\ 0 \end{pmatrix}$

17. $\mathbf{b}_2 \notin S_{\text{col}}(\mathbf{A})$, general solution of $\mathbf{Ax} = \mathbf{b}_1$ is $\mathbf{x} = c_1 \begin{pmatrix} 1 \\ 1 \\ 1 \\ -1 \end{pmatrix} + \begin{pmatrix} 3 \\ 0 \\ -1 \\ 0 \end{pmatrix}$

APPENDIX A.3, *page 811*

1. $|\mathbf{A}| = 11$, $\qquad \mathbf{A}^{-1} = \dfrac{1}{11} \begin{pmatrix} 3 & -4 \\ 2 & 1 \end{pmatrix}$

2. $|\mathbf{A}| = 12$, $\qquad \mathbf{A}^{-1} = \dfrac{1}{12} \begin{pmatrix} 2 & 1 \\ -6 & 3 \end{pmatrix}$

3. $|\mathbf{A}| = -1$, $\qquad \mathbf{A}^{-1} = \begin{pmatrix} 1 & -3 & 2 \\ -3 & 3 & -1 \\ 2 & -1 & 0 \end{pmatrix}$

4. $|\mathbf{A}| = -9$, $\qquad \mathbf{A}^{-1} = \begin{pmatrix} 1/3 & 1/3 & 0 \\ 1/3 & -1/3 & 1/3 \\ -1/3 & 0 & 1/3 \end{pmatrix}$

5. $|\mathbf{A}| = 0$

6. $|\mathbf{A}| = 8$, $\mathbf{A}^{-1} = \begin{pmatrix} 1/2 & -1/4 & 1/8 \\ 0 & 1/2 & -1/4 \\ 0 & 0 & 1/2 \end{pmatrix}$

7. $|\mathbf{A}| = 6$, $\mathbf{A}^{-1} = \begin{pmatrix} 1/6 & 1/6 & 1/6 \\ -1/3 & 2/3 & -1/3 \\ -7/6 & 5/6 & -1/6 \end{pmatrix}$

8. $|\mathbf{A}| = 0$

9. $|\mathbf{A}| = -1$, $\mathbf{A}^{-1} = \begin{pmatrix} 1 & 1 & 0 & 1 \\ 1 & 0 & 1 & 1 \\ 1 & 1 & 1 & 1 \\ 0 & 1 & 0 & 1 \end{pmatrix}$

10. $|\mathbf{A}| = -5$, $\mathbf{A}^{-1} = \begin{pmatrix} 6 & 13/5 & -8/5 & 2/5 \\ 5 & 11/5 & -6/5 & 4/5 \\ 0 & -1/5 & 1/5 & 1/5 \\ -2 & -4/5 & 4/5 & -1/5 \end{pmatrix}$

12. $1 = |\mathbf{I}_n| = |\mathbf{A}\mathbf{A}^{-1}| = |\mathbf{A}||\mathbf{A}^{-1}|$ so $|\mathbf{A}^{-1}| = 1/|\mathbf{A}|$

13. $\lambda = -2, 2, 4$ 14. $\lambda = -1, 1, 4$ 15. $\lambda = -2, -1, 1$

APPENDIX A.4, *page 821*

1. $\lambda_1 = 2, \mathbf{v}_1 = \begin{pmatrix} 1 \\ 3 \end{pmatrix}$; $\lambda_1 = 4, \mathbf{v}_2 = \begin{pmatrix} 1 \\ 1 \end{pmatrix}$

2. $\lambda_1 = 1 + 2i, \mathbf{v}_1 = \begin{pmatrix} 1+i \\ 2 \end{pmatrix}$; $\lambda_2 = 1 - 2i, \mathbf{v}_2 = \begin{pmatrix} 1-i \\ 2 \end{pmatrix}$

3. $\lambda_1 = 1 + 2i, \mathbf{v}_1 = \begin{pmatrix} 0 \\ i \\ 1 \end{pmatrix}$; $\lambda_2 = 1 - 2i, \mathbf{v}_2 = \begin{pmatrix} 0 \\ -i \\ 1 \end{pmatrix}$; $\lambda_3 = 1, \mathbf{v}_3 = \begin{pmatrix} 2 \\ -3 \\ 2 \end{pmatrix}$

4. $\lambda_1 = 2, \mathbf{v}_1 = \begin{pmatrix} -2 \\ 1 \\ 0 \end{pmatrix}$; $\lambda_2 = 3, \mathbf{v}_2 = \begin{pmatrix} 0 \\ -1 \\ 1 \end{pmatrix}$; $\lambda_3 = 1, \mathbf{v}_3 = \begin{pmatrix} -1 \\ 0 \\ 1 \end{pmatrix}$

5. $\lambda_1 = 1, \mathbf{v}_1 = \begin{pmatrix} -3 \\ 1 \\ 1 \end{pmatrix}$

6. $\lambda_1 = 1, \mathbf{v}_1 = \begin{pmatrix} 1 \\ 3 \\ 0 \end{pmatrix}$; $\quad \lambda_2 = 1, \mathbf{v}_2 = \begin{pmatrix} 2 \\ 0 \\ 3 \end{pmatrix}$; $\quad \lambda_3 = 2, \mathbf{v}_3 = \begin{pmatrix} 2 \\ 1 \\ 2 \end{pmatrix}$

7. $\lambda_1 = -3, \mathbf{v}_1 = \begin{pmatrix} 3 \\ 0 \\ 1 \end{pmatrix}$; $\quad \lambda_2 = -3, \mathbf{v}_2 = \begin{pmatrix} -2 \\ 1 \\ 0 \end{pmatrix}$; $\quad \lambda_3 = 5, \mathbf{v}_3 = \begin{pmatrix} -1 \\ -2 \\ 1 \end{pmatrix}$

8. $\lambda_1 = \sqrt{3}, \mathbf{v}_1 = \begin{pmatrix} 3\sqrt{3} - 3 \\ 6 \\ 3 - 5\sqrt{3} \end{pmatrix}$; $\quad \lambda_2 = -\sqrt{3},$

$\mathbf{v}_2 = \begin{pmatrix} -3\sqrt{3} - 3 \\ 6 \\ 3 + 5\sqrt{3} \end{pmatrix}$; $\quad \lambda_3 = 5, \mathbf{v}_3 = \begin{pmatrix} 2 \\ 1 \\ 0 \end{pmatrix}$

9. $\lambda_1 = 1, \mathbf{v}_1 = \begin{pmatrix} 1 \\ -1 \\ 2 \end{pmatrix}$; $\quad \lambda_2 = 0, \mathbf{v}_2 = \begin{pmatrix} 1 \\ 1 \\ 1 \end{pmatrix}$; $\quad \lambda_3 = 2, \mathbf{v}_3 = \begin{pmatrix} 2 \\ 1 \\ 2 \end{pmatrix}$

10. $\lambda_1 = 2, \mathbf{v}_1 = \begin{pmatrix} 0 \\ 1 \\ 0 \end{pmatrix}$; $\quad \lambda_2 = \sqrt{2}, \mathbf{v}_2 = \begin{pmatrix} 1 + \sqrt{2} \\ 0 \\ 1 \end{pmatrix}$; $\quad \lambda_3 = -\sqrt{2},$

$\mathbf{v}_3 = \begin{pmatrix} 1 + \sqrt{2} \\ 0 \\ 1 \end{pmatrix}$

11. $\lambda_1 = 7, \mathbf{v}_1 = \begin{pmatrix} 0 \\ 1 \\ 1 \end{pmatrix}$; $\quad \lambda_2 = 3, \mathbf{v}_2 = \begin{pmatrix} 1 \\ 0 \\ 0 \end{pmatrix}$; $\quad \lambda_3 = -1, \mathbf{v}_3 = \begin{pmatrix} 0 \\ -1 \\ 1 \end{pmatrix}$

12. $\lambda_1 = 9, \mathbf{v}_1 = \begin{pmatrix} 2 \\ 2 \\ -1 \end{pmatrix}$; $\quad \lambda_2 = -2, \mathbf{v}_2 = \begin{pmatrix} 1 \\ -1 \\ 0 \end{pmatrix}$; $\quad \lambda_3 = -18, \mathbf{v}_3 = \begin{pmatrix} 1 \\ 1 \\ 4 \end{pmatrix}$

13. $\lambda_1 = 4, \mathbf{v}_1 = \begin{pmatrix} 1 \\ 1 \\ 1 \end{pmatrix}$; $\quad \lambda_2 = -2, \mathbf{v}_2 = \begin{pmatrix} -1 \\ 0 \\ 1 \end{pmatrix}$; $\quad \lambda_3 = -2, \mathbf{v}_3 = \begin{pmatrix} 1 \\ -2 \\ 1 \end{pmatrix}$

14. $\lambda_1 = 0, \mathbf{v}_1 = \begin{pmatrix} -1 \\ 0 \\ 1 \end{pmatrix}$; $\quad \lambda_2 = 0, \mathbf{v}_2 = \begin{pmatrix} -1 \\ 1 \\ 0 \end{pmatrix}$; $\quad \lambda_3 = 3, \mathbf{v}_3 = \begin{pmatrix} 1 \\ 1 \\ 1 \end{pmatrix}$

15. $\lambda_1 = 1, \mathbf{v}_1 = \begin{pmatrix} 1 \\ 0 \\ 0 \end{pmatrix}$; $\quad \lambda_2 = 1, \mathbf{v}_2 = \begin{pmatrix} 0 \\ 1 \\ 0 \end{pmatrix}$; $\quad \lambda_3 = 1, \mathbf{v}_3 = \begin{pmatrix} 0 \\ 0 \\ 1 \end{pmatrix}$

16. $\lambda_1 = -1, \mathbf{v}_1 = \begin{pmatrix} 4 \\ 2 \\ -5 \end{pmatrix}$; $\quad \lambda_2 = -1, \mathbf{v}_2 = \begin{pmatrix} -1 \\ 2 \\ 0 \end{pmatrix}$; $\quad \lambda_3 = 8, \mathbf{v}_3 = \begin{pmatrix} 2 \\ 1 \\ 2 \end{pmatrix}$

17. $\lambda_1 = 3, \mathbf{v}_1 = \begin{pmatrix} -1 \\ 2 \\ 4 \\ 1 \end{pmatrix}$; $\quad \lambda_2 = (-1 + \sqrt{5})/2, \mathbf{v}_2 = \begin{pmatrix} -1 \\ 1 \\ \dfrac{2 - \sqrt{5}}{2} \\ \dfrac{2}{-1 + \sqrt{5}} \\ 1 \end{pmatrix}$;

$\lambda_3 = (-1 - \sqrt{5})/2, \mathbf{v}_3 = \begin{pmatrix} -1 \\ 1 \\ \dfrac{2 + \sqrt{5}}{2} \\ \dfrac{2}{-1 - \sqrt{5}} \\ 1 \end{pmatrix}$;

$\lambda_4 = -1, \mathbf{v}_4 = \begin{pmatrix} 1 \\ 0 \\ 0 \\ 1 \end{pmatrix}$

18. $\lambda_1 = -1, \mathbf{v}_1 = \begin{pmatrix} 1 \\ 0 \\ 2 \\ 0 \end{pmatrix}$; $\quad \lambda_2 = -1, \mathbf{v}_2 = \begin{pmatrix} 1 \\ 1 \\ 0 \\ 0 \end{pmatrix}$; $\quad \lambda_3 = -1, \mathbf{v}_3 = \begin{pmatrix} -1 \\ 0 \\ 0 \\ 1 \end{pmatrix}$;

$\lambda_4 = -2, \mathbf{v}_4 = \begin{pmatrix} 1 \\ 0 \\ 1 \\ 0 \end{pmatrix}$

19. $\lambda_1 = 1, \mathbf{v}_1 = \begin{pmatrix} -1 \\ 0 \\ 0 \\ 1 \end{pmatrix}$; $\quad \lambda_2 = 2, \mathbf{v}_2 = \begin{pmatrix} 1 \\ 0 \\ 1 \\ 0 \end{pmatrix}$; $\quad \lambda_3 = -1, \mathbf{v}_3 = \begin{pmatrix} 1 \\ 2 \\ 0 \\ 1 \end{pmatrix}$;

$\lambda_4 = -1, \mathbf{v}_4 = \begin{pmatrix} 1 \\ 0 \\ 2 \\ 0 \end{pmatrix}$

20. $\lambda_1 = -1, \mathbf{v}_1 = \begin{pmatrix} 1 \\ -1 \\ 2 \\ 0 \end{pmatrix}$; $\quad \lambda_2 = -1, \mathbf{v}_2 = \begin{pmatrix} -1 \\ -1 \\ 0 \\ 1 \end{pmatrix}$; $\quad \lambda_3 = -2, \mathbf{v}_3 = \begin{pmatrix} 1 \\ 2 \\ 0 \\ 1 \end{pmatrix}$;

$\lambda_4 = -2, \mathbf{v}_4 = \begin{pmatrix} 0 \\ -2 \\ 1 \\ 0 \end{pmatrix}$

REFERENCES

Differential Equations and Dynamical Systems

Arnol'd, Vladimer I., *Ordinary Differential Equations* (New York/Berlin: Springer-Verlag, 1992). Translation by Roger Cooke of the third Russian edition.

Birkhoff, G., and Rota, G.-C., *Ordinary Differential Equations* (4th ed.) (New York: Wiley, 1989).

Brauer, Fred and Nohel, John A., *The Qualitative Theory of Ordinary Differential Equations: An Introduction* (New York: W. A. Benjamin, Inc., 1969; New York: Dover, 1989).

Coddington, E. A., *An Introduction to Ordinary Differential Equations* (Englewood Cliffs, NJ: Prentice-Hall, 1961; New York: Dover, 1989).

Coddington, E. A., and Levinson, N., *Theory of Ordinary Differential Equations* (New York: McGraw-Hill, 1955).

Cole, R. H., *Theory of Ordinary Differential Equations* (New York: Irvington, 1968).

Guckenheimer, John and Holmes, Philip, *Nonlinear Oscillations, Dynamical Systems, and Bifurcations of Vector Fields* (New York/Berlin: Springer-Verlag, 1983).

Hirsch, Morris W., Smale, Stephen, and Devaney, Robert L., *Differential Equations, Dynamical Systems, and an Introduction to Chaos* (2nd ed.) (San Diego, CA/London: Elsevier Academic Press, 2004).

Hochstadt, H., *Differential Equations: A Modern Approach* (New York: Holt, 1964; New York: Dover, 1975).

Miller, R. K., and Michel, A. N., *Ordinary Differential Equations* (New York: Academic Press, 1982).

Rainville, E. D., *Intermediate Differential Equations* (2nd ed.) (New York: Macmillan, 1964).

Sparrow, Colin, *The Lorenz Equations: Bifurcations, Chaos, and Strange Attractors* (New York/Berlin: Springer-Verlag, 1982).

Strogatz, Steven H., *Nonlinear Dyamics and Chaos* (Reading, MA: Addison-Wesley, 1994).

Yosida, K., *Lectures on Differential and Integral Equations* (New York: Wiley-Interscience, 1960; New York: Dover, 1991).

Perturbation Methods, Dimensional Analysis, and Scaling

Holmes, M. H., *Introduction to Perturbation Methods* (New York: Springer-Verlag, 1995).

Lin, C. C. and Segel, L. A., *Mathematics Applied to Deterministic Problems in the Natural Sciences* (Philadelphia: SIAM, 1989).

Handbooks, Tables, and Special Functions

Abramowitz, M., and Stegun, I. A. (eds.), *Handbook of Mathematical Functions* (New York: Dover, 1965); originally published by the National Bureau of Standards, Washington, DC, 1964.

Copson, E. T., *An Introduction to the Theory of a Complex Variable* (Oxford: Oxford University, 1935).

Hochstadt, H., *Special Functions of Mathematical Physics* (New York: Holt, 1961).

Jahnke, E., and Emde, F., *Tables of Functions with Formulae and Curves* (Leipzig: Teubner, 1938; New York: Dover, 1945).

Advanced Calculus

Buck, R. C., and Buck, E. F., *Advanced Calculus* (3rd ed.) (New York: McGraw-Hill, 1978).

Courant, R., and John, F., *Introduction to Calculus and Analysis* (New York: Wiley-Interscience, 1965; reprinted by Springer-Verlag, New York, 1989).

Kaplan, W., *Advanced Calculus* (5th ed.) (Reading, MA: Addison-Wesley, 2003).

Fourier Series

Carslaw, H. S., *Introduction to the Theory of Fourier's Series and Integrals* (3rd ed.) (Cambridge: Cambridge University Press, 1930; New York: Dover, 1952).

Churchill, R. V., and Brown, J. W., *Fourier Series and Boundary Value Problems* (6th ed.) (New York: McGraw-Hill, 2000).

Partial Differential Equations

Haberman, R., *Elementary Applied Partial Differential Equations* (3rd ed.) (Englewood Cliffs, NJ: Prentice-Hall, 1998).

Pinsky, M. A., *Partial Differential Equations and Boundary Value Problems with Applications* (3rd ed.) (Boston: WCB/McGraw-Hill, 1998).

Powers, D. L., *Boundary Value Problems* (4th ed.) (San Diego: Academic Press, 1999).

Sagan, H., *Boundary and Eigenvalue Problems in Mathematical Physics* (New York: Wiley, 1961; New York: Dover, 1989).

Strauss, W. A., *Partial Differential Equations, an Introduction* (New York: Wiley, 1992).

Weinberger, H. F., *A First Course in Partial Differential Equations* (New York: Wiley, 1965; New York: Dover, 1995).

Numerical Analysis

Ascher, Uri M., and Petzold, Linda R., *Computer Methods for Ordinary Differential Equations and Differential-Algebraic Equations* (Philadelphia: Society for Industrial and Applied Mathematics, 1998).

Henrici, Peter, *Discrete Variable Methods in Ordinary Differential Equations* (New York: Wiley, 1962).

Shampine, Lawrence F., *Numerical Solution of Ordinary Differential Equations* (New York: Chapman and Hall, 1994).

Mathematical Modeling

Bailey, N. T. J., *The Mathematical Theory of Infectious Diseases and Its Applications* (2nd ed.) (New York: Hafner Press, 1975).

Brauer, Fred and Castillo-Chávez, Carlos, *Mathematical Models in Population Biology and Epidemiology* (New York/Berlin: Springer-Verlag, 2001).

Clark, Colin W., *Mathematical Bioeconomics* (2nd ed.) (New York: Wiley-Interscience, 1990).

Odum, E. P., *Fundamentals of Ecology* (3rd ed.) (Philadelphia: Saunders, 1971).

PHOTO CREDITS

INDEX